Encyclopedia of Geomorphology

Encyclopedia of Geomorphology

Volume 1

A–I

Edited by A.S. Goudie

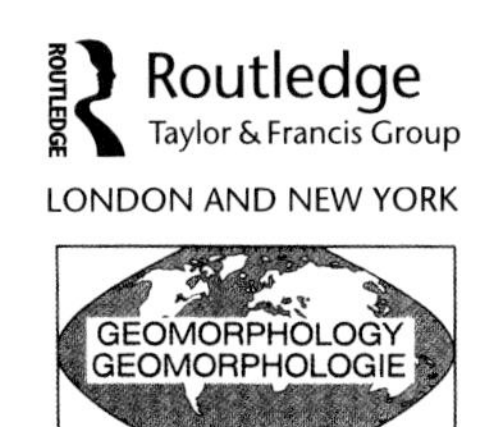

International Association of Geomorphologists

To the Founders of the International Association of Geomorphologists and to the IAG Senior Fellows:

1989: Harley J. Walker (USA)
1993: Hanna Bremer (Germany), Ross Mackay (Canada), Anders Rapp (Sweden)
1997: Denys Brunsden (UK), Richard Chorley (UK), Luna Leopold (USA)
2001: Stanley A. Schumm (USA), Torao Yoshikawa (Japan)

First published 2004
by Routledge
11 New Fetter Lane, London EC4P 4EE
Simultaneously published in the USA and Canada
by Routledge
29 West 35th Street, New York, NY 10001

Routledge is an imprint of the Taylor & Francis Group

Typeset in Times by Newgen Imaging Systems (P) Ltd, Chennai, India
Printed and bound in Great Britain by
TJ International Ltd, Padstow, Cornwall

British Library Cataloguing in Publication Data
A catalogue record for this book is available from the British Library

Library of Congress Cataloging in Publication Data
Encyclopedia of geomorphology / edited by A.S. Goudie.
p. cm.
Includes bibliographical references and index.
1. Geomorphology–Encyclopedias. I. Goudie, Andrew.

GB400.3.E53 2003
551.41′03–dc21 2003046892

ISBN 0–415–27298–X (set)
ISBN 0–415–32737–7 (volume one)
ISBN 0–415–32738–5 (volume two)

Contents

Editorial team

Editor-in-Chief

A.S. Goudie
University of Oxford, UK

Assistant editor
Jan Burke

Consultant editors

Richard Dikau
University of Bonn, Germany

Basil Gomez
Indiana State University, USA

Piotr Migoń
University of Wrocław, Poland

Olav Slaymaker
University of British Columbia, Canada

Alan Trenhaile
University of Windsor, Canada

Paul W. Williams
University of Auckland, New Zealand

Contributors

Adeeba E. Al-Hurban
Kuwait University

J.R.L. Allen
University of Reading, UK

John T. Andrews
University of Colorado at Boulder, USA

Peter Ashmore
University of Western Ontario, Canada

Philip J. Ashworth
University of Brighton, UK

Andres Aslan
Mesa State College, USA

Paul Augustinus
University of Auckland, New Zealand

Richard Bailey
University of Oxford, UK

Victor R. Baker
University of Arizona, USA

Colin K. Ballantyne
University of St Andrews, UK

Mark D. Bateman
University of Sheffield, UK

James C. Bathurst
University of Newcastle upon Tyne, UK

Bernard O. Bauer
University of Southern California, USA

Mohamed Tahar Benazzouz
University of Constantine, Algeria

Sean J. Bennett
US Department of Agriculture

Jerry M. Bernard
US Department of Agriculture, Natural Resources Conservation Service

Ivar Berthling
Norwegian Water Resources and Energy Directorate

Jim Best
University of Leeds, UK

Eric C.F. Bird
Geostudies, Australia

Paul Bishop
University of Glasgow, UK

Helgi Bjornsson
University of Iceland

Ryszard K. Borówka
Szczecin University, Poland

Stuart C. Boucher
Monash University, Australia

Mary C. Bourke
University of Oxford, UK

Michael J. Bovis
University of British Columbia, Canada

D.Q. Bowen
Cardiff University, UK

Louise Bracken (née Bull)
University of Durham, UK

Robert W. Brander
University of New South Wales, Australia

Hanna Bremer
Geographisches Institut der Universität zu Köln, Germany

Tracy A. Brennand
Simon Fraser University, Canada

John S. Bridge
Binghamton University, USA

Gary Brierley
Macquarie University, Australia

Denys Brunsden
King's College, University of London, UK

Rorke Bryan
University of Toronto, Canada

Kenneth L. Buchan
Geological Survey of Canada

Thomas Buffin-Bélanger
University of Western Ontario, Canada

C.R. Burn
Carleton University, Canada

Tim Burt
Durham University, UK

David R. Butler
Southwest Texas State University, USA

Doriano Castaldini
Università di Modena e Reggio Emilia, Italy

John A. Catt
University College London, UK

John Chappell
Australian National University

Anne Chin
Texas A&M University, USA

Michael Church
University of British Columbia, Canada

John P. Coakley
National Water Research Institute, Canada

Eric A. Colhoun
University of Newcastle, Australia

Arthur Conacher
University of Western Australia

Michael J. Crozier
Victoria University, New Zealand

Joanna C. Curran
Southwest Texas State University, USA

Donald A. Davidson
University of Stirling, UK

Alastair G. Dawson
Coventry University, UK

Michael J. Day
University of Wisconsin–Milwaukee, USA

Tommaso De Pippo
Università di Napoli, Italy

John Dearing
University of Liverpool, UK

Richard Dikau
University of Bonn, Germany

Jean Claude Dionne
Université Laval, Canada

John C. Dixon
University of Arkansas, USA

Stefan H. Doerr
University of Wales, Swansea, UK

Ronald I. Dorn
Arizona State University, USA

Ian Douglas
University of Manchester, UK

Terry Douglas
University of Northumbria, UK

Julian A. Dowdeswell
University of Cambridge, UK

G.A.T. Duller
University of Wales, Aberystwyth, UK

Alan P. Dykes
University of Huddersfield, UK

Frank Eckardt
University of Botswana

Judy Ehlen
USA Engineer Research and Development Center

Nabil S. Embabi
Ain Shams University, Egypt

Christine Embleton-Hamann
Universität Wien, Austria

Richard E. Ernst
Geological Survey of Canada,

Bernd Etzelmüller
University of Oslo, Norway

David J.A. Evans
University of Glasgow, UK

Ian S. Evans
University of Durham, UK

Martin G. Evans
University of Manchester, UK

David Favis-Mortlock
Queen's University Belfast, Northern Ireland

Helen Fay
Staffordshire University, UK

Rob Ferguson
University of Sheffield, UK

Timothy G. Fisher
University of Toledo, USA

Blair Fitzharris
University of Otago, New Zealand

Derek C. Ford
McMaster University, Canada

Paolo Forti
University of Bologna, Italy

Ian D.L. Foster
Coventry University, UK

J.R. French
University College London, UK

Lynne Frostick
University of Hull, UK

Kirstie Fryirs
Macquarie University, Australia

Michael A. Fullen
University of Wolverhampton, UK

Jérôme Gaillardet
Institut de Physique du Globe de Paris, France

José M. García-Ruiz
Instituto Pirenaico de Ecología, Spain

Martin Gibling
Dalhousie University, Canada

Daniel A. Gilewitch
Arizona State University, USA

Helen S. Goldie
University of Durham, UK

Douglas Goldsack
Laurentian University, Canada

Basil Gomez
Indiana State University, USA

Craig N. Goodwin
Utah State University, USA

Steven J. Gordon
Arizona State University, USA

A.S. Goudie
University of Oxford, UK

Gerard Govers
Katholieke Universiteit Leuven, Belgium

Stefan Grab
University of the Witwatersrand, South Africa

William L. Graf
University of South Carolina, USA

Brian Greenwood
University of Toronto at Scarborough, Canada

Kenneth J. Gregory
University of Southampton, UK

James S. Griffiths
University of Plymouth, UK

John Gunn
University of Huddersfield, UK

Angela Gurnell
King's College London, UK

Stephen D. Gurney
University of Reading, UK

M. Gutierrez-Elorza
Universitad de Zaragoza, Spain

Wilfried Haeberli
University of Zurich, Switzerland

Jon Ove Hagen
University of Oslo, Norway

Darren Ham
University of British Columbia, Canada

Jim Hansom
University of Glasgow, UK

Jon Harbor
Purdue University, USA

Carol Harden
University of Tennessee, USA

Charles Harris
Cardiff University, UK

Jane K. Hart
University of Southampton, UK

Adrian Harvey
University of Liverpool, UK

Nick Harvey
University of Adelaide, Australia

Louise Heathwaite
University of Sheffield, UK

Patrick Hesp
Louisiana State University, USA

Paul Hesse
Macquarie University, Sydney, Australia

Edward J. Hickin
Simon Fraser University, Canada

D. Murray Hicks
National Institute of Water and Atmospheric Research, New Zealand

David Higgitt
University of Durham, UK

Carl H. Hobbs III
College of William and Mary, USA

Trevor B. Hoey
University of Glasgow, UK

Diane Horn
Birkbeck College, University of London, UK

He Qing Huang
University of Oxford, UK

David Huddart
Liverpool John Moores University, UK

Richard Huggett
University of Manchester, UK

Oldrich Hungr
University of British Columbia, Canada

Dorina Camelia Ilies
University of Oradea, Romania

Richard M. Iverson
US Geological Survey

N.L. Jackson
New Jersey Institute of Technology, USA

Matthias Jakob
Kerr Wood Leidal Associates Ltd, Canada

Jacek Jania
University of Silesia, Poland

Vibhash C. Jha
Visva-Bharati University, India

Vincent Jomelli
Laboratoire Géographie Physique, France

David K.C. Jones
London School of Economics, UK

J. Anthony A. Jones
University of Wales, Aberystwyth,UK

Barbara A. Kennedy
University of Oxford, UK

James W. Kirchner
University of California at Berkeley, USA

Alistair D. Kirkbride
Université de Montréal, Canada

Mike Kirkby
University of Leeds, UK

Andrew Klein
Texas A&M University, USA

Brian Klinkenberg
University of British Columbia, Canada

Peter G. Knight
Keele University, UK

Gary Kocurek
University of Texas, USA

Oliver Korup
Victoria University of Wellington, New Zealand

Michel Lacroix
Université Pierre et Marie Curie, France

Julie E. Laity
California State University – Northridge, USA

Nick Lancaster
University of Nevada System, USA

Stuart Lane
University of Leeds, UK

Andreas Lang
University of Liverpool, UK

Damian Lawler
University of Birmingham, UK

Wendy Lawson
University of Canterbury, New Zealand

Marcel Leach
Laurentian University, Canada

Karna Lidmar-Bergström
Stockholm University, Sweden

Thomas E. Lisle
US Forest Service, USA

Ian Livingstone
University College Northampton, UK

Dénes Lóczy
University of Pécs, Hungary

Brian Luckman
University of Western Ontario, Canada

Elvidio Lupia-Palmieri
Università di Roma, La Sapienza, Italy

Brian G. McAdoo
Vassar College, USA

Danny McCarroll
University of Wales, Swansea, UK

T. S. McCarthy
University of Witwatersrand, South Africa

Sue McLaren
University of Leicester, UK

Roger F. McLean
Australian Defence Force Academy – UNSW, Australia

Chris MacLeod
Cardiff University, UK

Jon J. Major
US Geological Survey

Mauro Marchetti
Università di Modena e Reggio Emilia, Italy

W. Andrew Marcus
University of Oregon, USA

Susan B. Marriott
University of the West of England, Bristol, UK

Richard A. Marston
Oklahoma State University, USA

Yvonne Martin
University of Calgary, Canada

Gerhard Masselink
Loughborough University, UK

Anne E. Mather
University of Plymouth, UK

Norikazu Matsuoka
University of Tsukuba, Japan

John Menzies
Brock University, Canada

Katerina Michaelides
King's College London, UK

Nicholas Middleton
University of Oxford, UK

Piotr Migoń
University of Wrocław, Poland

David R. Montgomery
University of Washington, USA

Gerald C. Nanson
University of Wollongong, Australia

David J. Nash
University of Brighton, UK

Larissa Naylor
Komex, UK

Vincent E. Neall
Massey University, New Zealand

Atle Nesje
University of Bergen, Norway

Scott Nichol
University of Auckland, New Zealand

Dawn T. Nicholson
Manchester Metropolitan University, UK

Karl F. Nordstrom
Rutgers University, USA

Patrick D. Nunn
University of the South Pacific, Fiji

Colm Ó Cofaigh
University of Cambridge, UK

Cliff Ollier
University of Western Australia, Australia

Carolyn G. Olson
US Department of Agriculture

Clive Oppenheimer
University of Cambridge, UK

Julian Orford
Queen's University, Northern Ireland, UK

W.R. Osterkamp
US Geological Survey

Ian Owens
University of Canterbury, New Zealand

André Ozer
Université de Liège, Belgium

Ken Page
Charles Sturt University, Australia

Mario Panizza
Università di Modena e Reggio Emilia, Italy

Chris Paola
University of Minnesota, USA

Gary Parker
University of Minnesota, USA

Kevin Parnell
James Cook University, Australia

A.J. Parsons
University of Leicester, UK

Frank J. Pazzaglia
Lehigh University, USA

Marcus E. Pearson
Oklahoma State University, USA

Ellen L. Petticrew
University of Northern British Columbia, Canada

Geoffrey Petts
University of Birmingham, UK

William M. Phillips
University of Edinburgh, UK

Richard J. Pike
US Geological Survey

Jan A. Piotrowski
University of Aarhus, Denmark

P.A. Pirazzoli
CNRS-Laboratoire Géographie Physique, France

Albert Pissart
Université de Liège, Belgium

Ross D. Powell
Northern Illinois University, USA

Nick Preston
Landcare Research, New Zealand

Angélique Prick
The University Centre on Svalbard, Norway

Roland E. Randall
University of Cambridge, UK

Stefan Rasemann
University of Bonn, Germany

Ian Reid
Loughborough University, UK

Chris S. Renschler
The State University of New York at Buffalo, USA

Emmanuel Reynard
University of Lausanne, Switzerland

Stephen Rice
Loughborough University, UK

Neil Roberts
University of Plymouth, UK

Robert J. Rogerson
The University of Lethbridge, Canada

Charles L. Rosenfeld
Oregon State University, USA

Jürgen Runge
Johann Wolfgang Goethe-Universität Frankfurt am Main, Germany

Maria Sala
University of Barcelona, Spain

Gregory H. Sambrook Smith
University of Birmingham, UK

Erik Schiefer
University of British Columbia, Canada

Jochen Schmidt
Landcare Research, New Zealand

Karl-Heinz Schmidt
Universität Halle, Germany

Jacques Schroeder
Université du Quebec à Montréal, Canada

Stanley A. Schumm
Mussetter Engineering, Inc., USA

Matti Seppälä
University of Finland

Jamshid Shahabpour
Shahid Bahonar University of Kerman, Iran

Richard A. Shakesby
University of Wales, Swansea, UK

Andrew D. Short
University of Sydney, Australia

Michael Slattery
Texas Christian University, USA

Olav Slaymaker
University of British Columbia, Canada

Rudy Slingerland
Penn State University, USA

Ian Smalley
University of Leicester, UK

Mauro Soldati
Università di Modena e Reggio Emilia, Italy

Catherine Souch
Indiana University – Purdue, USA

James A. Spotila
Virginia Polytechnic Institute and State University, USA

Iain S. Stewart
University of Glasgow, UK

Chris R. Stokes
University of Reading, UK

Esther Stouthamer
Utrecht University, The Netherlands

Arjen P. Stroeven
Stockholm University, Sweden

Mike Summerfield
University of Edinburgh, UK

Takasuke Suzuki
Chuo University, Japan

D.T.A. Symons
University of Windsor, Canada

James Syvitski
University of Colorado at Boulder, USA

David G. Tarboton
Utah State University, USA

Graham Taylor
University of Canberra, Australia

Mark Patrick Taylor
Macquarie University, Australia

Vatche P. Tchakerian
Texas A&M University, USA

David S.G. Thomas
University of Sheffield, UK

Michael F. Thomas
University of Stirling, UK

Douglas M. Thompson
Connecticut College, USA

Colin E. Thorn
University of Illinois, USA

Keith J. Tinkler
Brock University, Canada

Michael Tooley
University of Durham, UK

Stephen Tooth
University of Wales, Aberystwyth, UK

Alan Trenhaile
University of Windsor, Canada

Steve Trudgill
University of Cambridge, UK

Greg Tucker
University of Oxford, UK

Alice Turkington
University of Kentucky, USA

C.R. Twidale
University of Adelaide, Australia

Christian Valentin
IRD, France

Juan Ramon Vidal-Romani
Universidade da Coruna, Spain

Heather A. Viles
University of Oxford, UK

John Wainwright
King's College London, UK

John Walden
University of St Andrews, UK

H. Jesse Walker
Louisiana State University, USA

Tony Waltham
Nottingham Trent University, UK

Jeff Warburton
Durham University, UK

Steve Ward
University of Oxford, UK

Andrew Warren
University College London, UK

Charles Warren
University of St Andrews, UK

Neil A. Wells
Kent State University, USA

Brian Whalley
Queen's University, Northern Ireland, UK

Geraldene Wharton
Queen Mary, University of London, UK

Kelin X. Whipple
Massachusetts Institute of Technology, USA

Kevin White
University of Reading, UK

Mike Widdowson
Open University, UK

Giles F.S. Wiggs
University of Sheffield, UK

Paul W. Williams
University of Auckland, New Zealand

Peter J. Williams
Carleton University, Canada

Colin J.N. Wilson
Institute of Geological and Nuclear Sciences, New Zealand

Vanessa Winchester
University of Oxford, UK

Pia Windland
University of Oxford, UK

Ellen E. Wohl
Colorado State University, USA

Colin Woodroffe
University of Wollongong, Australia

Peter Worsley
University of Oxford, UK

Robert Wray
University of Wollongong, Australia

L.D. Wright
College of William and Mary, USA

R. W. Young
University of Wollongong, Australia

Witold Zuchiewicz
Jagiellonian University, Poland

Zbigniew Zwolinski
Adam Mickiewicz University, Poland

Illustrations

Figures

Plates

Tables

Foreword

As president of the International Association of Geomorphologists (IAG), a body that seeks to provide a forum for the promotion of geomorphology internationally, I am delighted that, in association with Routledge, it has been possible to produce this great encyclopedia. It is written by contributors from some thirty countries, all of whom have generously agreed that their royalties should go to the IAG. This will add substantially to the financial resources of the Association. The IAG is grateful to the editorial team, and in particular to Andrew Goudie, for the work they have done to bring it to fruition and I am sure that it will be an invaluable resource for the international geomorphological community.

Mario Panizza
Modena, Italy
October 2002

Preface

The term 'geomorphology' arose in the Geological Survey in the USA in the 1880s and was possibly coined by those two great pioneers, J.W. Powell and WJ McGee.

In 1891 McGee wrote: 'The phenomena of degradation form the subject of geomorphology, the novel branch of geology.' He plainly regarded geomorphology as being that part of geology which enabled the practitioner to reconstruct Earth history by looking at the evidence for past erosion, writing:

> A new period in the development of geologic science has dawned within a decade. In at least two American centres and one abroad it has come to be recognised that the later history of world growth may be read from the configuration of the hills as well as from the sediments and fossils of ancient oceans… The field of science is thereby broadened by the addition of a coordinate province – by the birth of a new geology which is destined to rank with the old. This is geomorphic geology, or geomorphology.

Of course, many scientists had studied the development of erosional landforms (see the magisterial history of Chorley *et al.* 1964) before the term was thus defined and since that time its meaning has become broader. Many geomorphologists believe that the purpose of geomorphology goes beyond reconstructing Earth history and that the core of the subject is the comprehension of the form of the ground surface and the processes which mould it. In recent years there has been a tendency for geomorphologists to become more deeply involved with understanding the processes of erosion, weathering, transport and deposition, with measuring the rates at which such processes operate, and with quantitative analysis of the forms of the ground surface (morphometry) and of the materials of which they are composed. Geomorphology now has many component branches and involves the study of a huge range of phenomena.

In 1968 Rhodes W. Fairbridge edited a large and invaluable encyclopedia of geomorphology that explored this diversity. However, geomorphology has changed greatly since that time, not least because of the plate tectonics paradigm, the revolution in our knowledge of the Quaternary Era brought about by new dating and environmental reconstruction techniques, the development of modelling and systems thinking, appreciation of the importance of organisms, application of geomorphology to the study of engineering problems and global change, a greater appreciation of the nature of geomorphological processes, and availability of a whole range of new technologies for analysis of data and materials, the development of satellite-borne remote sensing, and the exploration of space.

Over that time, due to the inspiration of the people to whom this book is dedicated, Geomorphology has for the first time organized itself internationally so that the geomorphological traditions that have grown up in different countries (see Walker and Grabau 1993) can interact as never before. It was therefore felt at the International Geomorphological Congress in Tokyo in August 2002 that the International Association of Geomorphologists (itself officially founded in 1989) should seek to publish a new and truly international Encyclopedia of Geomorphology that could survey the nature of the discipline at the turn of a new millennium. I am indebted to my Consultant Editors and the contributors from some thirty or so countries, who have made this endeavour possible.

Andrew S. Goudie
Oxford

References

Chorley, R.J., Dunn, A.J. and Beckinsale, R.P. (1964) *The History of the Study of Landforms*, Vol. 1, London: Methuen.

Fairbridge, R.W. (ed.) (1968) *Encyclopedia of Geomorphology*, New York: Reinhold.

McGee, W.J. (1891) The Pleistocene history of northeastern Iowa, *Eleventh Annual Report of the US Geological Survey*, 189–577.

Walker, H.J. and Grabau, W.E. (1993) *The Evolution of Geomorphology. A Nation-by-Nation Summary of Development*, Chichester: Wiley.

Thematic entry list

Aeolian

Adhesion
Aeolation
Aeolian geomorphology
Aeolian processes
Aeolianite
Aligned drainage
Barchan
Beach-dune interaction
Bedform
Bounding surface
Deflation
Desert geomorphology
Draa
Dune, aeolian
Dune, coastal
Dune mobility
Dune, snow
Dust storm
Glaciaeolian
Interdune
Loess
Lunette
Nebkha
Niveo-aeolian activity
Pan
Parna
Ripple
Saltation
Sand ramp
Sand sea and dunefield
Sandsheet
Sastrugi
Singing sand
Stone pavement
Ventifact
Wind erosion
Wind tunnels in geomorphology
Yardang

Biogeomorphology

Beach rock
Biogeomorphology
Biokarst
Boring organism
Brousse tigrée
Coral reef
Corniche
Crusting of soil
Desert varnish
Forest geomorphology
Large woody debris
Mangrove swamp
Microatoll
Mima mound
Mire
Mud flat and muddy coast
Nebkha
Organic weathering
Oyster reef
Peat erosion
Reef
Riparian geomorphology
Saltmarsh
Serpulid reef
Stromatolite (stromatolith)
Termites and termitaria
Tree fall
Turf exfoliation
Vermetid reef and boiler
Zoogeomorphology

Coastal and marine

Atoll
Bar, coastal
Barrier and barrier island
Base level
Beach
Beach cusp
Beach nourishment
Beach ridge
Beach rock
Beach–dune interaction
Beach sediment transport
Blowhole
Blue hole
Boring organism
Bruun rule
Calanque
Cay
Chenier ridge
Cliff, coastal
Coastal classification
Coastal geomorphology
Continental shelf
Coral reef
Corniche
Current
Cuspate foreland
Dune, coastal
Equilibrium shoreline
Estuary
Eustasy
Fjord
Fringing reef
Glacimarine
Groyne
Guyot
Integrated coastal management

Concept

Fluvial

Glacial

Hazards and environmental geomorphology

Karst

Lacustrine

Palaeogeomorphology

Inheritance
Inverted relief
Lichenometry
Neoglaciation
Neotectonics
Palaeochannel
Palaeoclimate
Palaeoflood
Palaeohydrology
Palaeokarst and relict karst
Palaeosol
Peneplain
Planation surface
Pluvial lake
Postglacial transgression
Raised beach
Rejuvenation
Relief generation
Sea level
(Uranium-Thorium)/Helium analysis

Periglacial

Alas
Asymmetric valley
Avalanche boulder tongue
Avalanche, snow
Blockfield and blockstream
Cambering and valley bulging
Coulee
Coversand
Cryoplanation
Cryostatic pressure
Dune, snow
Freeze-thaw cycle
Frost and frost weathering
Frost heave
Geocryology
Grèze litée
Hummock
Hydro-laccolith
Ice wedge and related structures
Icing
Lithalsa
Needle-ice
Nivation
Niveo-aeolian activity
Oriented lake
Palsa
Patterned ground
Periglacial geomorphology
Permafrost
Pingo
Ploughing block and boulder
Protalus rampart
Rock glacier
Slushflow
Solifluction
Thermokarst

Slopes and mass movements

Aspect and geomorphology
Asymmetric valley
Butte
Cambering and valley bulging
Caprock
Colluvium
Debris flow
Debris torrent
Decollement
Deep-seated gravitational slope deformation
Equilibrium slope
Factor of safety
Failure
Fall line
Flat iron
Fluidization
Grèze litée
Ground water
Hamada
Hillslope-channel coupling
Hillslope, form
Hillslope hollow
Hillslope, process
Lahar
Landslide
Landslide dam
Liquefaction
Mass movement
Method of slices
Overland flow
Pediment
Peneplain
Pore-water pressure
Quickclay
Quicksand
Raindrop impact, splash and wash
Repose, angle of
Residual slope
Richter denudation slope
Riedel shear
Rockfall
Scree
Sensitive clay
Shear and shear surface
Sheet erosion, sheet flow, sheet wash
Slickenside
Slope, evolution
Slope stability
Slopewash
Soil creep
Soil erosion
Solifluction (solifluxion)
Sturzstrom
Submarine landslide geomorphology
Talus
Terracette
Toreva block
Unequal slopes, law of
Uniclinal shifting

Soils and materials

Aeolianite
Alluvium
Beach rock
Calcrete
Caliche (sodium nitrate)
Caprock
Case hardening
Clay-with-flint
Colluvium
Crusting of soil
Desert varnish
Desiccation cracks and polygons
Diamictite
Drape, silt and mud
Duricrust
Effective stress
Eluvium and eluviation
Erosivity
Expansive soil
Fabric analysis
Fech-fech
Ferricrete
Fragipan
Gilgai
Gypcrete
Hydrophobic soil (water repellency)

Structural

Techniques

Tectonic and volcanic

Weathering

ABRASION

The mechanical wearing down, scraping, or grinding away of a rock surface by friction, ensuing from collision between particles during their transport in wind, ice, running water, waves or gravity. The effectiveness of abrasion depends upon the concentration, hardness and kinetic energy of the impacting particles, alongside the resistance of the bedrock surface. Abrasion may scour, polish, scratch or smooth existing rock faces. Abrasion ramps are seaward sloping platforms (typically 1° gradient) formed at the base of cliffs in intertidal environments due to continued wave abrasion.

Further reading

Hamblin, W.K. and Christiansen, E.H. (2001) *Earth's Dynamic Systems*, 9th edition, Upper Saddle River, NJ: Prentice Hall.

STEVE WARD

ACCRETION

The gradual enlargement of an area of land through the natural accumulation of sediment, washed up from a river, lake or sea. Sediment accretion is the basic process of wetland formation, as continuous flooding and subsequent receding river flows emplace sediment which then provides the soil base for wetlands.

Accretion also refers to the theory that continents have increased their surface area during geological history by the addition of marine sediments at their boundaries via tectonic collision with other oceanic or continental plates.

Further reading

Pye, K. (1994) *Sediment Transport and Depositional Processes*, Oxford: Blackwell Scientific.

STEVE WARD

ACTIVE AND CAPABLE FAULT

Currently no universally accepted definition has been agreed upon for 'active fault', nor have the principles and criteria for the identification of active faults and their ranking been worked out. As a result, the various definitions of fault activity terms are the source of some confusion and discussion, both in literature and in practice (see FAULT AND FAULT SCARP).

An important review on this topic was presented by Slemmons and McKinney (1977) who, after examining numerous papers, suggested the following definitions. An 'active fault' is a fault that has slipped during the present seismotectonic regime and is therefore likely to show renewed displacement in the future. Fault activity may be indicated by historical, geological, seismological, geodetic or other geophysical evidence. The definitions for 'capable faults', which were specified for siting nuclear reactors, restrict this term to faults that have been displaced once during the past 35,000 years, or movements of a recurring nature within the past 500,000 years or faults which have been active during the Late Quaternary.

The term active fault in Japan was defined as a fault which has moved repeatedly in recent geological times and could resume activity in the future. Subsequently, this term has been used for faults which have moved during the Quaternary. The analysis of topographic features has provided the most important clues in the work of recognizing active faults (RGAFJ 1980).

A particular definition was given by Panizza and Castaldini (1987) who distinguish two categories: (1) active fault: proven displacement of rocks and/or significant forms; (2) fault held to be active: on the basis of supporting geomorphological or other evidence, but showing no visible displacement of rock or other significant forms. Rocks and/or 'significant' landforms are those included in the neotectonic period considered. The distinction between 'active' and 'held to be active' faults is finalized to constrain in a more precise and less subjective way the concept of fault activity.

The 'World Map of Major Active Faults' shows five fault age categories (historical to <1.6 Ma). Slip rate, which is used as a proxy for fault activity, is classified in four categories ranging from $<0.2\,\text{mm}\,\text{year}^{-1}$ to $>5\,\text{mm}\,\text{year}^{-1}$. The maps are accompanied by a database which describes evidence for Quaternary faulting, geomorphic expression and paleoseismic parameters (Trifonov and Machette 1993).

In some glossaries, an active fault is defined as 'A fault along which there is recurrent movement, which is usually indicated by small, periodic displacements or seismic activity' (Bates and Jackson 1987), or as 'A fault likely to move at the present day' (Ollier 1988).

A paper on the most commonly used terms associated with seismogenetic faults in the United States was published by Machette (2000). The author notes that the three following terms are used in a variety of ways and for different reasons or applications:

- Active fault: one demonstrating current movement or action (what is meant by 'current'? Contemporary, historical, Holocene or Quaternary?).
- Capable fault: one having the capability for movements.
- A potentially active fault: one capable of being or becoming active (this definition is very similar to that of capable fault).

On the Internet various definitions pinpointing the indeterminateness of the term can be found, such as:

1 The definition of active fault is not straightforward. In some cases, the maximum age that can be determined by means of Carbon 14 analyses (35,000–50,000 years) is used as a time span for such measurements: if a fault can be shown not to have been active within this time span then it is not active (http://www.geol.binghamton.edu/class/geo205/html/faults.html).
2 Active faults are structures along which displacements are expected to occur. By definition, since a shallow earthquake is a process that produces displacement across a fault, all shallow earthquakes occur on active faults (http:// www.eas.slu.edu/People/CJAmmon/HTML/Classes/introQuakes/Notes/faults.html).
3 'A fault that is likely to undergo displacement by another earthquake sometime in the future.' Faults are commonly considered to be active if they have moved one or more times in the last 10,000 years (http://earthquake.usgs.gov/image_glossary).
4 An active fault is one that has moved at least once in geologically recent times. In the Californian definition it means a movement occurring within the last 11,000 years, rather than the longer period of 125,000 years used on New Zealand maps (http://www.gsnz.org.nz/gsprfa.htm).

In short, on the concepts of 'active fault' and 'capable fault' the following remarks can be made:

1 the terms are used to indicate faults which have been subject to movement in recent geological time or which might move at present or in the future;
2 their age limits vary depending on the authors;
3 active faults are often associated with strong earthquakes.

Identification of active and capable faults can be based on direct and/or indirect criteria: historical, geological, geomorphological, geomorphic, seismological, geodetic, geochemical, geophysical and volcanic.

Finally, apart from the terminological aspects, some of the major active faults around the world include: the North Anatolian fault in Turkey, the Dead Sea Valley between Israel and Jordan, the Philippine fault, the San Andreas fault in California, the Red River fault in China and the South Island alpine fault in New Zealand.

References

Bates, R.L. and Jackson, J.A. (1987) *Glossary of Geology*, American Geological Institute, Alexandria, VA.

Machette, M.N. (2000) Active, capable and potentially active faults: a paleoseismologic prospective, *Journal of Geodynamics* 29, 387–392.
Ollier, C.D. (1988) *Glossary of Morphotectonics*, 3rd edition, Dept of Geography and Planning, University of New England, Armidale, Australia.
Panizza, M. and Castaldini, D. (1987) Neotectonic research in applied geomorphological studies, *Zeitschrift für Geomorphologie Supplementband* 63, 173–211.
RGAFJ (The Research Group for Active Faults of Japan) (1980) Active faults in and around Japan: distribution and degree of activity, *Journal of Natural Disaster Science* 2(2), 61–99.
Slemmons, D.B. and McKinney, R. (1977) Definition of 'active fault', *US Army Engineer Waterways Experiment Station, Soils and Pavements Laboratory, miscellaneous paper S*, 77–8, Vicksburg.
Trifonov, V.G. and Machette, M.N. (1993) The world map of major active faults, *Annali di Geofisica* 36(3–4), 225–236.

DORIANO CASTALDINI AND
DORINA CAMELIA ILIES

ACTIVE LAYER

Ground above PERMAFROST which thaws in summer and freezes again in winter. In the northern hemisphere, it reaches its full depth each year in late August or September. The active layer is critical to the ecology of permafrost terrain, as it provides a rooting zone for plants and is a seasonal aquifer. An ice-rich zone, commonly below the base of the active layer, is responsible for the sensitivity of permafrost terrain to disturbance. Deepening of the active layer and melting of the ground ice leads to subsidence in flat terrain, and landslides with accelerated erosion on slopes (Mackay 1970).

The active layer thaws once air temperature is above 0 °C and the snow cover has melted. The total depth depends on the length and surface temperature of the thawing season, the ice content of the ground, the thermal conductivity of soil materials, and the temperature of near-surface permafrost. The active layer is thickest in bedrock, where there is little ice to melt. In unconsolidated sediments, the thickness is greatest in dry, sandy soils or gravel, where the depth may be enhanced by heat advected in groundwater, and thinnest in peat. Local variation in soil materials may be reflected in active-layer depth, as in hummocky terrain, where the base of the active layer forms a mirror image of the ground surface, with depth greatest beneath the mineral-soil centres and least beneath the organic-rich circumference of the hummocks.

At the end of the thaw season, freezing of the active layer usually begins from the bottom upwards. Upfreezing commonly accounts for up to 10 per cent of the thickness. During upfreezing, moisture is drawn downward into permafrost from the base of the active layer, leading to development of the ice-rich zone. Simultaneously, soil water is drawn upwards from the rest of the active layer, to freeze near the ground surface. As a result, the centre of the active layer tends to be dry when frozen. Stones and structures embedded in the active layer may be pulled upwards as the ground freezes. Characteristically these objects are supported from below during thawing the following summer, leading to their progressive jacking out of the ground.

Freezing and thawing of the active layer modifies the annual propagation of surface temperature into permafrost. Cooling of permafrost in autumn is delayed by freezing, which may take several months, depending primarily on the water content and snow cover. Mean annual temperature decreases with depth in the active layer, due to the seasonal difference in soil thermal properties wrought by freezing and thawing. The difference in mean annual temperature, or thermal offset, between the ground surface and the top of permafrost may be over 2 °C, increasing with water content and depth of active layer (Romanovsky and Osterkamp 1995).

References

Mackay, J.R. (1970) Disturbances to the tundra and forest tundra environment of the western Arctic, *Canadian Geotechnical Journal* 7, 420–432.
Romanovsky, V.E. and Osterkamp, T.E. (1995) Interannual variations of the thermal regime of the active layer and near-surface permafrost in northern Alaska, *Permafrost and Periglacial Processes* 6, 313–335.

C.R. BURN

ACTIVE MARGIN

In plate tectonic theory ocean crust is created by SEAFLOOR SPREADING, and old crust is consumed at subduction sites. A continental margin where subduction occurs is called an active continental margin. Active margins occupy essentially the borders of the Pacific: the west Pacific borders are

ISLAND ARC type; the western margins of the Americas are the other type, to be described here.

The spreading of the Atlantic causes America to move west, where it overrides the seafloor, which is subducted. Plate collision is thought to fold and uplift the continental edge to form mountains and their internal structures, and also create a deep trench offshore where sediments are deposited. These may be scraped off the subducted slab to form an accretionary prism, or subducted where they may produce granites, and andesitic magma which erupts as volcanoes.

The Pacific border of the Americas falls into three main units with different MORPHOTECTONICS: South America, Central America and North America.

The Andes run along the entire western side of South America, divided for most of their length into Eastern and Western Cordilleras, with a graben between called the Inter-Andean Depression. Bedrock is folded and faulted Palaeozoic and Mesozoic rocks, with granite intrusions. The region was largely eroded to a plain before ignimbrites spread over large areas, and planation was complete in the Neogene. The area was uplifted as linear fault blocks in the Plio-Pleistocene, or earlier in some places. The large strato-volcanoes are of Quaternary age, erupted onto the planation surface. Major thrust faults diverge from the centre of the Andes in a symmetrical way, hard to explain by one-sided subduction.

Offshore a deep trench extends as far north as Mexico. Trenches have many graben and normal faults indicating extension. Sediments are usually horizontal, and some trenches are almost empty. Mesozoic plutons constitute the world's greatest granite batholith which runs the length of the Andes, covering 15 per cent of the Andes surface. The alignment parallel to the coast suggests some control on the location and possibly the origin of the Andes, but the plutons took over 70 million years to rise and intrusion ceased about 30 million years ago, long before the uplift of the Andes (Gansser 1973).

The many great volcanoes found along the Andes (with some gaps, and some double lines) are Quaternary. They are on the top of horsts, usually close to the Inter-Andean Depression.

Central America can be regarded as the Middle America arc. The trench has no accretionary prism, and sediments are horizontal.

A basement of metamorphic rocks and granites is exposed in northern Honduras. This is block faulted, and split by the Honduras Depression consisting of north–south graben that opened in the early Pliocene. The chain of volcanoes close to the south coast consists of five straight-line segments. These young cones are built on a basement of older volcanics. The same basement forms the Nicaraguan volcanic upland, separated from the young volcanoes by a major fault scarp. To the north these volcanics overlap the Honduras Massif.

The Isthmian link to Panama is not the young volcanic chain or even rocks of the Nicaraguan volcanic upland, but consists of even older volcanics. Block faulting is common.

Western North America is largely a collage of exotic terranes (Howell 1989). There are abundant strike-slip faults (such as the San Andreas fault) with movement of hundreds of kilometres. There is no offshore trench, but an offshore topography of basins and swells, possibly related to strike-slip fault blocks. These differences perhaps occur because the mid-ocean ridge runs aground near the Mexico/USA border. To the north the transform faults associated with seafloor spreading affect the continental margin as they run nearly parallel to it.

The Pacific border region of the USA consists of two main ranges: in the west are the Coast Ranges, in the east to the north are the Cascades and to the south the Sierra Nevada. The Coast Range seems to have formed as a large but rather simple arch. The Cascade Range is mainly a huge pile of volcanic rocks, with many famous strato-volcanoes such as Mount Shasta and Mount St Helens. The Sierra Nevada is a huge tilt block, mainly uplifted in the Quaternary. The Coast Ranges of Canada are a continuation of the Cascades of the United States, and also consist of a simple arch.

Planation surfaces are common on the North American cordillera (Ollier and Pain 2000). The mountains of North America were uplifted in the Neogene, mostly within the past 5 million years, though subduction has presumably been going on for the life of the Pacific, at least 200 Ma.

Plate tectonic theory has been applied not only to coastal ranges, but to mountains 1,500 km inland (Miller and Gans 1997). The Rocky Mountains consist of elongated blocks aligned in all directions, including east–west (Uinta Mountains).The blocks have Precambrian cores, and divergent thrust faults on both sides. They are too far inland to be explained by subduction, separated from the Pacific by the extensional

Basin and Range Province, and the uplift occurred in the last few million years.

As Gansser (1973) explained, plate tectonic theories that use the Andes as a model adopt simplified assumptions which neglect the fact that only the recent morphogenic uplift made the apparently uniform Andes, masking a very complicated geological history. The same seems true of North and Central America.

References

Gansser, A. (1973) Facts and theories on the Andes, *Journal of the Geological Society of London* 129, 93–131.
Howell, D.G. (1989) *Tectonics of Suspect Terranes: Mountain Building and Continental Growth*, London: Chapman and Hall.
Miller, E.L. and Gans, P.B. (1997) The North American Cordillera, in B.A. van der Pluijm and S. Marshak (eds) *Earth Structure: An Introduction to Structural Geology and Tectonics*, 424–429, New York: WCB/McGraw-Hill.
Ollier, C.D. and Pain, C.F. (2000) *The Origin of Mountains*, London: Routledge.

Further reading

Leggett, J.K. (1982) *Trench-Forearc Geology: Sedimentation and Tectonics in Modern and Ancient Active Plate Margins*, Oxford: Blackwell.
Nairn, A.E.M., Stehli, F.G. and Uyeda, S. (1985) *The Ocean Basins and Margins. Vol. 7A The Pacific Ocean*, New York: Plenum.

SEE ALSO: mountain geomorphology; plate tectonics

CLIFF OLLIER

ACTUALISM

Actualism is a concept based on the premise that present causes of environmental change are sufficient to explain events of the past. Causes of changes in the past differ not in kind, but often in energy, from those now in operation. The French term *actualisme* and the German terms *aktualismus* or *aktualitatsprinsip* are commonly used in Europe in opposition to catastrophism. Hooykaas (1970) makes a distinction between actualistic methodology and actualistic historical description. Tidal variation over geological time provides an instructive example. Actualist methodology leads to the conclusion that the Moon and Earth were very much closer and that gravitational attraction was therefore greater before 3.5 billion years BP. Huge tidal ranges require a catastrophist historical description.

Reference

Hooykaas, R. (1970) *Catastrophism in Geology: Its Scientific Character in Relation to Actualism and Uniformitarianism*, Amsterdam: North-Holland Publishing Co.

SEE ALSO: catastrophism; uniformitarianism

OLAV SLAYMAKER

ADHESION

Adhesion refers to the adhering of wind-blown sand to a wet or damp surface. Adhesion is most common in damp or wet INTERDUNES between active dunes, but also occurs on SANDSHEETS, beaches, riverbanks and damp portions of dunes. Adhesion ripples and plane bed are the most common surface features that result from adhesion, and each forms a distinctive sedimentary structure with deposition. A related feature is formed by adhesion of sediment to salt during periods of high humidity.

Further reading

Kocurek, G. and Fielder, G. (1982) Adhesion structures, *Journal of Sedimentary Petrology* 52, 1,229–1,241.

GARY KOCUREK

AEOLATION

The moulding of desert landscapes by the erosional action of wind (see WIND EROSION OF SOIL). At the start of the twentieth century there was a phase of what has been termed 'extravagant aeolation' (Cooke and Warren 1973). This had its roots in the work undertaken in Africa by French and German geomorphologists, such as Walther and Passarge, but was put forward in its most exuberant form in the USA by Keyes (1912), who believed that material weakened by thermoclasty (INSOLATION WEATHERING) would be evacuated by wind and deposited as dust sheets on desert margins. He argued that the end result of such activity would be the formation of great plains, mountain ranges without foothills, and towering eminences. As he remarked (p. 551):

> Under conditions of aridity plain meets mountain sharply. The bevelled rock-floor of many intermont plains throughout the dry regions is explicable on no known activity of water action in such situations. Existence of isolated plateau

plains rising abruptly out of the general plains surface far from any sight of running water is an anomaly met with only in the desert.

Not all American geomorphologists took such a firm view as Keyes, and Tolman (1909), for instance, in his study of the Arizona BOLSON region, considered the role of both fluvial and aeolian processes to be important and recognized that STONE PAVEMENTs 'fortified' large tracts of the arid region of the south-west of the USA against wind attack.

Aeolianist views declined in popularity so that from about 1920 onwards the belief that entire landscapes were shaped by wind became less acceptable. The reasons for the decline of aeolianist views were many.

First, the great PEDIMENT landscapes of the American deserts were seen, following the work of McGee (1897) and others, as being attributable to planation by sheetflood activity. The second reason for the decline of aeolianist views was that many desert landscapes were thought to have been moulded by fluvial processes that had been more powerful and widespread during the pluvial phases that were held to be a feature of the Pleistocene. Third, doubt was expressed about the power of thermoclasty as a process capable of preparing desert surfaces for subsequent aeolian attack. Such doubt largely arose because of laboratory simulations. Fourth, it was widely held that lag gravels (stone pavements) and salt and clay crusts would limit the extent to which aeolian processes could cause excavation of surfaces below the water table. Fifth, it became apparent that many of the world's great LOESS deposits, in North America, China and the erstwhile USSR, were the product of deflation from glacial areas rather than from deserts. Glacial grinding was thought to be the most efficient way of producing silt-sized quartz particles. Sixth, it was recognized that not all deserts had either adequate supplies of abrasive sand or of frequent high-velocity winds for wind erosion to be achieved with any degree of facility. Finally, features that were conceded to have an aeolian origin (e.g. YARDANGS, VENTIFACTs and pedestal rocks) were thought to be but minor, bizarre embellishments of otherwise fluvial environments, whilst other possibly aeolian features (notably stone pavements and closed depressions) were also explicable by other means. STONE PAVEMENTs, for example, could be the product of the removal of fine sediments by sheetflood activity or they could result from vertical sorting processes associated with wetting and drying, dust inputs, salt hydration or freezing and thawing. Deflational removal of fines to leave a lag was just one possible formation mechanism. In the same way, closed depressions could be attributed to wind excavation, but might also be explained by tectonic, solutional or zoogenic processes.

Nevertheless, the power of wind erosion cannot be dismissed. Closed depressions (PANs) and wind moulded landforms (yardangs) are important landforms in some arid areas and wind erosion plays a significant role in their development.

References

Cooke, R.U. and Warren, A. (1973) *Geomorphology in Deserts*, London: Batsford.

Keyes, C.R. (1912) Deflative scheme of the geographic cycle in an arid climate, *Geological Society of America Bulletin* 23, 537–562.

McGee, W.J. (1897) Sheetflood erosion, *Geological Society of America Bulletin* 8, 87–112.

Tolman, C.F. (1909) Erosion and deposition in the southern Arizona bolson region, *Journal of Geology* 17, 126–163.

A.S. GOUDIE

AEOLIAN GEOMORPHOLOGY

Aeolian geomorphology is the study of the effect of the wind on Earth surface processes and landforms. It encompasses studies of the fundamental physical mechanisms and movement of materials at the scale of a single grain, studies of the development of landforms such as dunes (see DUNE, AEOLIAN) and YARDANGs, and studies of the wider effect of the wind at the regional scale of SAND SEA AND DUNEFIELD, SANDSHEETs and LOESS deposition. It is also concerned with applied aspects of aeolian activity. In recent years this has led to a particular focus on the erosion by wind of soils in agricultural lands, especially in the semi-arid lands (see WIND EROSION OF SOIL and DEFLATION). Aeolian geomorphology also includes the study of the palaeoenvironmental significance of aeolian features, for landscapes can be just as sensitive to changes in the activity of the wind as they are to water and ice. As with other elements of geomorphology, technological changes in recent years have led to considerable advances in the understanding of aeolian geomorphology.

At the scale of the movement of individual sand and dust particles, the benchmark work was undertaken by Bagnold and summarized in his

Physics of Blown Sand and Desert Dunes (1941). Since Bagnold's work very considerable progress has been made in the use of wind tunnels and field instruments (see WIND TUNNELS IN GEOMORPHOLOGY). Although the fundamental physics of aeolian sand transport is much as Bagnold described it, very considerable detail has been added in recent years (see also AEOLIAN PROCESSES and SALTATION).

At the scale of individual landforms considerable progress has also been made. As a result of the improvement of technologies for data capture there has been a spate of studies of wind flow and sand flux on single dunes (e.g. Tsoar 1983; Walker 1999) reviewed by Wiggs (2001). Often these are coupled with improving surveying techniques that enable accurate measurement of change (e.g. Stokes *et al.* 1999). In addition, ground penetrating radar (GPR) is now being routinely used to ascertain the internal sedimentary structure of dunes as an important indication of the evolutionary history of dunes (e.g. Bristow *et al.* 2000). Studies have also investigated erosional features such as YARDANGs and VENTIFACTs (e.g. Laity 1994).

At the regional scale the development of remote sensing has enabled a better grasp of the relationships between landforms. The advance of remote sensing investigations in dryland areas has been of particular importance because desert areas are often difficult to access. The pioneering work of McKee and co-workers (McKee 1979) has been followed by numerous applications of remote sensing in aeolian studies. Imagery has been used to map dune patterns (e.g. Al-Dabi *et al.* 1997), detect small changes in dune morphology using high resolution synthetic aperture radar (SAR) imagery (e.g. Blumberg 1998), map and detect dust emissions from dryland pan systems (e.g. Eckardt *et al.* 2001) and detect mineral assemblages (e.g. White *et al.* 1997).

Aeolian features also hold considerable palaeoenvironmental information because aeolian activity is sensitive to changes in environmental controls such as wind energy and moisture availability. The extent of dunefields at the last glacial maximum was used as a surrogate indicator of global aridity by Sarnthein (1978) but a basic on/off classification of aeolian activity is now seen as too simplistic (Livingstone and Thomas 1993). Kocurek and Lancaster (1999), for instance, have sought to incorporate variability of sediment availability along with wind energy in discussions of past aeolian activity in the Mojave Desert.

A profound impact on aeolian studies has been the development of luminescence dating techniques (see DATING METHODS). Many aeolian deposits lack organic matter and so have not been susceptible to radiocarbon dating. Since the early 1980s luminescence dating primarily of quartz grains has enabled dating of aeolian deposits such as dunes and loess, and luminescence dates in aeolian studies are now commonplace (e.g. Stokes *et al.* 1997).

Improvement of dating techniques has led to considerable interest in the palaeoenvironmental information stored in LOESS (terrestrial deposits of aeolian dust). The best documented of these are the deposits of the Chinese loess plateau. Here mineral magnetism has been used as a proxy for weathering of PALAEOSOLs, patterns of magnetic reversals have been used to date deposits covering the past 2.5 million years and loess particle size has been used as an indicator of palaeo wind speeds. Techniques developed on the Chinese deposits have been extended to loess deposits elsewhere and knowledge of the extent and nature of world loess deposits has steadily increased (e.g. Derbyshire 2001).

Aeolian geomorphology has moved on considerably since the claims of Keyes in the early part of the twentieth century (see AEOLATION). The task that faces aeolian geomorphologists is to move from studies of individual landforms formed predominantly by aeolian activity to consider the wider role of the wind alongside water and ice in forming landscapes (e.g. Bullard and Livingstone 2002).

References

Al-Dabi, H., Koch, M., Al-Sarawi, M. and El-Baz, F. (1997) Evolution of sand dune patterns in space and time in north-western Kuwait using Landsat images, *Journal of Arid Environments* 36, 15–24.

Bagnold, R.A. (1941) *The Physics of Blown Sand and Desert Dunes*, London: Methuen.

Blumberg, D.G. (1998) Remote sensing of desert dune forms by polarimetric synthetic aperture radar (SAR), *Remote Sensing of the Environment* 65, 204–216.

Bristow, C.S., Bailey, S.D. and Lancaster, N. (2000) The sedimentary structure of linear sand dunes, *Nature* 406, 56–59.

Bullard, J.E. and Livingstone, I. (2002) Interactions between aeolian and fluvial systems in dryland environments, *Area* 34, 8–16.

Derbyshire, E. (2001) Recent research on loess and palaeosols, pure and applied: a preface, *Earth-Science Reviews* 54, 1–4.

Eckardt, F., Drake, N., Goudie, A.S., White, K. and Viles, H. (2001) The role of playas in pedogenic gypsum crust formation in the central Namib desert: a theoretical model, *Earth Surface Processes and Landforms* 26, 1,177–1,193.
Kocurek, G. and Lancaster, N. (1999) Aeolian system sediment states: theory and Mojave Desert Kelso Dune field example, *Sedimentology* 46, 505–515.
Laity, J.E. (1994) Landforms of aeolian erosion, in A.D. Abrahams and A.J. Parsons (eds) *Geomorphology of Desert Environments*, 506–535, London: Chapman and Hall.
Livingstone, I. and Thomas, D.S.G. (1993) Modes of linear dune activity and their palaeoenvironmental significance: an evaluation with reference to southern African examples, in K. Pye (ed.) *The Dynamics and Environmental Context of Aeolian Sedimentary Systems*, Geological Society Special Publication 72, 91–101, London: Geological Society.
McKee, E.D. (ed.) (1979) A study of global sand seas, *United States Geological Survey Professional Paper* 1052.
Sarnthein, M. (1978) Sand deserts during glacial maximum and climatic optimum, *Nature* 272, 43–46.
Stokes, S., Thomas, D.S.G. and Washington, R. (1997) Multiple episodes of aridity in southern Africa since the last interglacial period, *Nature* 388, 154–158.
Stokes, S., Goudie, A.S., Ballard, J., Gifford, C., Samieh, S., Embabi, N. and El-Rashidi, O.A. (1999) Accurate dune displacement and morphometric data using kinematic GPS, *Zeitschrift für Geomorphologie Supplementband* 116, 195–214.
Tsoar, H. (1983) Dynamic processes acting on a longitudinal (seif) dune, *Sedimentology* 30, 567–578.
Walker, I.J. (1999) Secondary airflow and sediment transport in the lee of a reversing dune, *Earth Surface Processes and Landforms* 24, 437–448.
White, K., Walden, J., Drake, N., Eckardt, F. and Settle, J. (1997) Mapping the iron oxide content of dune sands, Namib Sand Sea, Namibia, using Landsat Thematic Mapper data, *Remote Sensing of the Environment* 62, 30–39.
Wiggs, G.F.S. (2001) Desert dune processes and dynamics, *Progress in Physical Geography* 25, 53–79.

Further reading

Goudie, A.S., Livingstone, I. and Stokes, S. (eds) (1999) *Aeolian Environments, Sediments and Landforms*, Chichester: Wiley.
Livingstone, I. and Warren, A. (1996) *Aeolian Geomorphology*, London: Longman.

IAN LIVINGSTONE AND GILES F.S. WIGGS

AEOLIAN PROCESSES

Wind is the movement of the mixture of gases that constitute the air. It is a fluid like water and obeys the same fundamental physical mechanisms as water. However, there are clear distinctions between the effects of water and wind at the Earth's surface. Air is 100 times less dense than water and consequently is only able to carry small clastic material. However, wind is not constrained by channels in the way that much water action is and consequently its influence can be much wider spread. Paradoxically, it is this wide spread of activity that means that aeolian activity sometimes goes unrecognized. A few millimetres of erosion or deposition of material over a large area is much less obvious than the erosion of rills and gullies or deposition of bars in channels even though the total amount of material moved may be similar.

Controls on aeolian processes

While aeolian activity is often associated with hot deserts, it is not restricted to these areas, although these are among the regions with the most favourable conditions for aeolian activity. Primary requirements for aeolian activity are: sufficient wind energy; material of a size that can be transported; and surface conditions that make that material available to the wind. Aeolian activity is therefore controlled by transport capacity, sediment supply and sediment availability (Kocurek and Lancaster 1999).

- *Transport capacity* Most places on the Earth's surface experience sufficient wind energy for aeolian processes to operate, so wind energy is rarely a limiting factor. The high levels of aeolian activity in low-latitude deserts do not occur because they are windier than other places. In fact the windiest places on Earth are close to the poles and around coastlines.
- *Sediment supply* Because wind is not as dense or viscous as water it is much more selective about the size of material that it can carry. The size of material most readily entrained by the wind is fine sand (see below). Wind rarely carries material above sand size, although transport of gravel-sized particles (>2 mm) has been reported from the dry valleys in Antarctica where wind speeds are very high and the extremely cold air is dense. The size-selectivity of wind means that surface materials must usually be sand- or dust-sized to be entrained. Often this requires that the materials are pre-sorted by other fluvial, glacial or marine processes. Some of the best sources of aeolian material are alluvial fans, glacial outwash plains and beaches (Bullard and Livingstone 2002).

- *Sediment availability* Provided with sufficient wind energy and material of the right size, the remaining control on aeolian processes is the surface conditions. Deserts have high levels of aeolian activity because soil, vegetation or moisture do not seal the surface. Conversely, aeolian activity is more rare in mid-latitudes, not because of lack of wind or material of the right size, but because surface conditions prevent the wind from entraining material.

Processes of wind erosion

Erosion of materials at the Earth's surface by the wind occurs as a result of two processes: deflation and abrasion (both of these mechanisms also occur in flowing water although the equivalent term 'fluid stressing' is used instead of deflation in fluvial geomorphology). DEFLATION is very simply the entrainment of material by the wind. Surfaces on which dust- or sand-sized material are exposed are particularly susceptible and in some places agricultural land with sandy or dusty soils where farmers expose the soil by ploughing is subject to considerable deflation by the wind (see WIND EROSION OF SOIL). Abrasion is caused by bombardment by particles being transported by the wind, most usually by SALTATION (see below). The impact of these transported grains can cause considerable sculpting of natural and built features. YARDANGS and VENTIFACTS are the geomorphological features most affected by abrasion.

Sediment entrainment

Aeolian sediment entrainment on a stable non-eroding surface occurs when the shear stress of the wind (a function of wind speed, turbulent energy and surface roughness) overcomes forces of particle cohesion, packing and weight. The principal erosive forces include lift, form drag and surface drag. The first two of these forces both result from air pressure differences around an individual particle. Higher velocity winds are associated with lower air pressure, so where wind flow is accelerated over a particle lying on the surface there is also a decrease in pressure above the particle resulting in a lift force. Similarly, form drag results from the high wind pressure on the upwind side of the particle contrasting with the decreased pressure in the downwind region. These two pressure forces combine with the surface drag resulting directly from the shearing stress of the wind to shake the particle loose before spinning it up into the airstream.

The relationship between erosivity and entrainment can effectively be simplified to two parameters, critical wind shear (u_{*ct}) and particle diameter (d) (Bagnold 1941):

$$u_{*ct}=A\sqrt{\frac{(\sigma-\rho)}{\rho}g\cdot d}$$

where: σ = particle density, g = acceleration due to gravity, A = constant dependent upon the grain Reynolds number (≈ 0.1).

Generally, larger grains require a greater wind shear to dislodge them. However, as shown in Figure 1, this relationship is reversed for particles smaller than about 0.06 mm (dust-sized) where increased electrostatic and molecular cohesion require larger erosive forces for entrainment. Figure 1 also demonstrates that the grain sizes most susceptible to entrainment have diameters between 0.06 and 0.40 mm, sand-sized particles. It is this susceptibility of sand to entrainment that allows the accumulation of extensive dunefields (see SAND SEA AND DUNEFIELD) and SANDSHEETs in dryland regions.

A further process important in the entrainment of sand grains is the bombardment of the surface by grains that are already in transport. Once a few grains have been entrained by the wind they may be transported by the process of saltation,

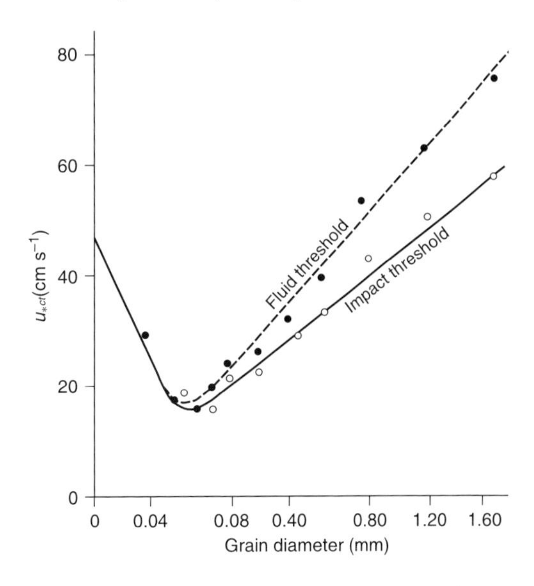

Figure 1 The relationship between particle size and the threshold shear velocity required for entrainment (after Chepil 1945)

Source: Thomas, D.S.G. (1997) *Arid Zone Geomorphology*, 2nd edition. © John Wiley & Sons Limited. Reproduced with premission

bouncing along the surface with ballistic trajectories (see below). Each saltating grain gathers momentum from the wind and then imparts it to the sand surface on impact. This impact can 'splash-out' up to ten other grains that may also become entrained by the wind. A few saltating grains can quickly induce mass transport of sediment in a cascading system (Nickling 1988). Two thresholds of entrainment may therefore be identified: one (the fluid threshold) relates only to the drag and lift forces of the wind, the second (the impact threshold) is lower and combines wind forces with additional forces provided by impacting grains already in transport (see Figure 1). Once a sediment surface has begun to be eroded by wind forces at the fluid threshold, sediment transport is maintained at the lower impact threshold because energy is also available from the saltating grains. Wind shear stress may therefore reduce once entrainment has begun, but sediment transport will continue until the wind drops below the new impact threshold.

Transport mechanisms

The grain size of entrained particles also determines the mode of transport undertaken. Although dust-sized particles are not the easiest to entrain, they have very low settling velocities in comparison to potential wind lift and turbulent velocities, so can be transported in *suspension*. Particles suspended in the atmosphere may be held aloft for several days and hence travel long distances. An example of this is the deposition of Saharan dust in the south-eastern USA (Prospero 1999). Often this transport in suspension is in barely visible dust haze, but sometimes there are more concentrated dust plumes which are clearly seen on spectacular satellite images, and still less frequently suspended aeolian material is concentrated as dust storms which can lead to 'blackout' conditions.

Particles up to about 1.0 mm are commonly transported in SALTATION. Figure 2 shows the typical trajectory of saltating particles with a progressively increasing forward velocity from entrainment to impact as the particle draws momentum from the wind.

The actual trajectory of a particle depends on the height of its bounce. Wind velocity increases at a logarithmic rate away from the surface and so a particle that bounces higher into the wind will be able to draw greater momentum from it and so travel further and faster. The length of jump is thought to be about 12–15 times the height of bounce, or further if the particle spins

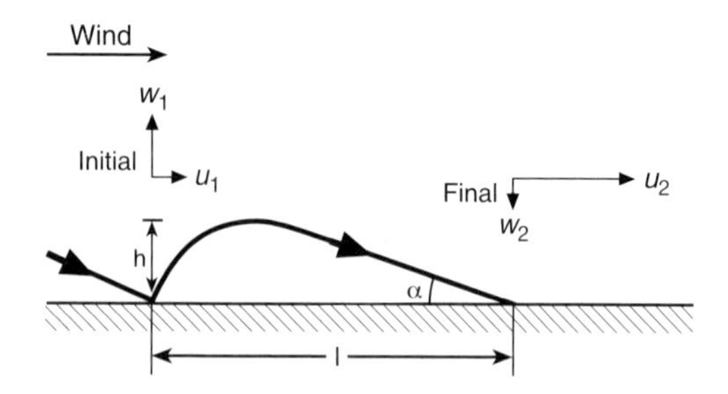

Figure 2 The ballistic trajectory of a saltating sand grain. w and u represent vertical and horizontal velocities, respectively (after Bagnold 1941)

Source: Thomas, D.S.G. (1997) *Arid Zone Geomorphology*, 2nd edition. © John Wiley & Sons Limited. Reproduced with premission

and induces an additional lift force (called the Magnus effect, White and Schultz 1977). The height of the saltation layer is dependent both on the wind velocity and also on the hardness of the surface over which the particles are saltating. Sand that is saltating over a rock or pebbly surface loses much less of its momentum on impact and so tends to bounce higher, reaching up to 3.0 m. An average saltation height, however, is about 0.2 m. The amount of sand in transport declines exponentially with height and so up to 80 per cent of all saltation activity takes place within 0.02 m of the surface (Butterfield 1991).

Grains which are ejected into the airflow as a result of the impact of a saltating grain may also enter the saltation system. However, some of these ejected grains may not have sufficient velocity fully to enter saltation and hence take only a single jump in a downwind direction. This process is termed *reptation* and, whilst much further research is required into its operation, it may be very significant in near-surface aeolian transport (Anderson *et al.* 1991).

The final mode of sand transport is *creep* and this describes the downwind rolling of larger sand particles (usually >0.5 mm). Such a process results both from the drag of wind on the surface of the particles and also the high velocity impact of saltating grains. It is thought that the process of creep may account for up to one-quarter of the bedload (saltation plus creep) transport rate.

Sand flux

The mass flux of sand (q) transported during an erosion event is often calculated as a cubic function of wind shear velocity (u_*). Most relationships are derived from theoretical analyses or wind tunnel experiments and a popular expression is that of Lettau and Lettau (1978):

$$q = C\left(\frac{d}{D}\right)^{0.5}(u_{*} - u_{*ct})u_{*}\frac{2\rho}{g}$$

where: C = constant (4.2), d = grain diameter, D = standard grain diameter (0.25 mm), u_{*ct} = shear velocity threshold of grain entrainment, ρ = air density, g = acceleration due to gravity.

There has been little empirical testing of relationships like the one above and that which has been accomplished shows considerable variation between observed and predicted rates. Such variation is to be expected when the complex nature of the saltation system is considered and the fact that the predictive expressions available rarely account for variations in terrain, vegetation, surface moisture or wind turbulence. Furthermore, the accurate measurement of sand flux in the natural environment is very difficult, with the published efficiencies of sand traps varying between 20 and 70 per cent (Jones and Willetts 1979).

Deposition

Just as material is entrained when shear stress overcomes inhibiting forces, so material is deposited when shear stress is no longer greater than these forces. This manifests itself both at the large scale where, if regional wind patterns lead to a decrease of wind speed, SANDSHEETs or dunefields (see SAND SEA AND DUNEFIELD) are formed, but also at the much smaller scale where surface irregularities may be responsible for the deposition of sand patches. Finer-grained material carried in suspension is often deposited as LOESS.

Although a lack of vegetation is usually important in the aeolian entrainment of material, paradoxically its presence can be important in trapping dust and sand in depositional features. Dust is only deposited as loess where it is prevented from re-entrainment, often by vegetation, and vegetation can also be important in stabilizing coastal dune ridges.

References

Anderson, R.S., Sørensen, M. and Willetts, B.B. (1991) A review of recent progress in our understanding of aeolian sediment transport, *Acta Mechanica Supplement* 1, 1–19.

Bagnold, R.A. (1941) *The Physics of Blown Sand and Desert Dunes*, London: Methuen.

Bullard, J.E. and Livingstone, I. (2002) Interactions between aeolian and fluvial systems in dryland environments, *Area* 34, 8–16.

Butterfield, G.R. (1991) Grain transport rates in steady and unsteady turbulent airflows, *Acta Mechanica Supplement* 1, 97–122.

Chepil, W.S. (1945) Dynamics of wind erosion: 1. Nature of movement of soil by wind, *Soil Science* 60, 305–320.

Jones, J.R. and Willetts, B.B. (1979) Errors in measuring uniform aeolian sandflow by means of an adjustable trap, *Sedimentology* 26, 463–468.

Kocurek, G. and Lancaster, N. (1999) Aeolian system sediment states: theory and Mojave Desert Kelso Dune Field example, *Sedimentology* 46, 505–515.

Lettau, K. and Lettau, H.H. (1978) Experimental and micrometeorological field studies on dune migration, in H.H. Lettau and K. Lettau (eds) *Exploring the World's Driest Climates*, University of Wisconsin-Madison, Institute for Environmental Studies, Report 101, 110–147.

Nickling, W.G. (1988) The initiation of particle movement by wind, *Sedimentology* 35, 499–511.

Prospero, J.M. (1999) Long-term measurements of the transport of African mineral dust to the south-eastern United States: implications for regional air quality, *Journal of Geophysical Research* 104, 15,917–15,927.

White, B.R. and Schultz, J.C. (1977) Magnus effect in saltation, *Journal of Fluid Mechanics* 81, 497–512.

Further reading

Livingstone, I. and Warren, A. (1996) *Aeolian Geomorphology*, London: Longman.

Nickling, W.G. (1994) Aeolian sediment transport and deposition, in K. Pye (ed.) *Sediment Transport and Depositional Processes*, 293–350, Oxford: Blackwell Scientific.

Wiggs, G.F.S. (1997) Sediment mobilisation by the wind, in D.S.G. Thomas (ed.) *Arid Zone Geomorphology*, 2nd edition, 351–372, London: Wiley.

GILES F.S. WIGGS AND IAN LIVINGSTONE

AEOLIANITE

Aeolianite is a cemented sandstone that has formed as a result of the processes of entrainment, transportation and deposition by wind. This rock type has various names including eolianite (US), miliolite (India and the Middle East), dunerock (South Africa), kurkar (Israel) and grès dunnaire (Mediterranean). Aeolianites of Quaternary age are most commonly found between 20° and 40° either side of the equator although examples have been found as far north as about 60°. Most aeolianites in the geological record are Quaternary in age and these tend to range between about 0.5 m to about 100 m in thickness.

Most aeolianites are rich in carbonate although silica-dominated forms also exist. Carbonate-rich aeolianites are largely associated with coastal sources of sediment and are thus close to modern or palaeo-shorelines. Semi-arid to sub-humid tropical shorelines are the most suitable locations for aeolianites as the oceans are productive in the formation of shelly biogenic grains or ooliths; strong onshore currents breakdown and move the sediment onshore; and climatic conditions are conducive for subsequent DIAGENESIS. In arid environments where there are strong onshore winds and a high sediment supply, dunes may be transported several hundred kilometres inland.

Along coastlines aeolianites often form elongate shore parallel or oblique bodies deposited as transverse ridges. The dunes are often stacked up against one another and may coalesce. The sediment size of aeolianites is typically coarse silt to sand-sized. Clays are rare because those that are entrained by the wind tend to be removed by suspension. Insufficient wind energy to transport grain sizes coarser than 2 mm generally limits the upper size of the clasts.

Aeolianites have distinctive bedding structures such as cross-bedding and laminations, which represent the progradation and growth of the dunes. Steeply dipping units up to 30°–34° reflect the former dune slipface. As a result of erosion the palaeodune bedforms are commonly lost and the dune type and direction of sand movement has to be deduced largely from the internal structures.

Lithification under freshwater vadose, mixed and/or phreatic conditions may occur. The main diagenetic processes result in alteration of unstable aragonite and high-Mg calcite clasts to low-Mg calcite clasts and cement. The balance between dissolution of carbonate grains by leaching and the production of cement is the prime control on the degree of dune induration. The main sources for the cement come from biogenic skeletal remains (e.g. molluscs, foraminifera, echinoderms, algae and coral fragments), ooliths, biota, sea spray, dust, bedrock and ground water.

Alteration of aeolianites by diagenetic processes in the vadose environment is the most common, occurring in three ways: (1) by loss of Mg^{2+} from the crystal lattices of high-Mg calcite; (2) by dissolution of aragonite, the loss of some strontium and reprecipitation as low-Mg calcite; and (3) by calcitization of aragonite *in situ*. A wide range of controls results in significant variability in aeolianite diagenesis in terms of both causal factors and diagenetic product (Gardner and McLaren 1994). Such controls include climate, sea level and time at the macro-scale; sea spray, plants and texture at the meso-scale and at the micro-scale the amount, rate of movement and chemistry of pore waters.

Major unconformities in aeolianites are often marked by PALAEOSOLS that develop as a result of solution and weathering. Commonly these soils are *terra rossa* and red latosols that have developed *in situ* but may contain inputs from wind-blown dust. Weathering may subsequently result in a solution of carbonate products and karstification. In semi-arid environments surface crusts and thin laminar CALCRETES often form as a result of solution and rapid reprecipitation.

Radiometric dating of aeolianites is notoriously difficult. The effects of diagenesis mean that there are chances of contamination from secondary calcite. Occasionally unaltered shells are found which have allowed radiocarbon dating (e.g. McLaren and Gardner 2000). In addition, uranium series dating, amino acid racemization, luminescence and electron spin resonance dating have been used with varying success (see Brooke 2001).

Lithification increases the aeolianites' resistance to erosion and enhances their preservation potential in the geological record. The cement types (such as meniscus, rim, pore filling and needle fibre), amount and distribution, along with geochemistry, can aid interpretations concerning palaeoenvironments such as identifying palaeo-water tables, palaeo-erosion surfaces or degree of exposure to marine environments.

References

Brooke, B. (2001) The distribution of carbonate eolianite, *Earth-Science Reviews* 55, 135–164.

Gardner, R.A.M. and McLaren, S. (1994) Variability in early vadose carbonate diagenesis in sandstones, *Earth-Science Reviews* 36, 27–45.

McLaren, S. and Gardner, R.A.M. (2000) New radiocarbon dates from a Holocene aeolianite, Isla Cancun, Quintana Roo, Mexico, *Holocene* 10, 757–761.

SEE ALSO: karst

SUE McLAREN

AGGRADATION

The long-term accumulation of sediment in a channel and the readjustment of the stream profile where there is a vertical growth of the land surface in response. Some possible agents of this process are running water, waves, glaciers and wind. Aggradation can occur at a variety of spatial scales and temporal scales (gradual or PUNCTUATED AGGRADATION), and may take place under constrained or unconstrained conditions. As aggradation is a long-term process, short-term fluctuations in sediment transport have no relevance.

Further reading

Easterbrook, D.J. (1999) *Surface Processes and Landforms*, 2nd edition, Upper Saddle River, NJ: Prentice Hall.

STEVE WARD

ALAS

The periglacial landform term alas, which is of Yakutian origin, was first introduced by the Russian worker P.A. Soloviev in the 1960s. It refers to the substantial circular and oval depressions with steep sides and flat floors, sometimes occupied with lakes, which characterize the geomorphology of the higher river terraces in central Yakutia (65°N, 125°E). Alases typically have diameters from 0.1 km up to 15 km and depths in the 3–40 m range. In morphological expression a tract of alas depressions on a river terrace surface is not dissimilar to a suite of KETTLE AND KETTLE HOLE forming a pitted glacial outwash plain. Genetically, an alas is a type of THERMOKARST feature, i.e. a subsidence landform arising from the degradation and settlement of ice-rich frozen ground. An essential prerequisite for alas development is terrain with a high ground ice content. The natural vegetation cover is taiga (coniferous forest) although the alas floors are often grass covered.

Coalescence processes amongst individual alas depressions can lead to the development of alas valleys. These valleys are characterized by a variable width with an alternation of narrow sections marking the former location of watersheds and wide sections together with branches with no outlets. The longitudinal profiles are not necessarily graded, reflecting the fact that they are thermokarst landforms rather than normal river cut features. In the Yakutian lowlands drained by the Lena River the spread of alas-related depressions has affected some 40 per cent of the higher river terrace surfaces. Surprisingly, the alas valleys form pockets of cultivated land where hardy strains of some grains along with some root crops are produced during the relatively warm summers.

Some prominence has been given in periglacial texts to a hypothetical reconstructed sequence of alas development (Soloviev 1973) although it needs to be emphasized that its applicability outside Yakutia has yet to be established. An important factor is that Yakutia is unglaciated yet nevertheless sustained permafrost throughout much of the Quaternary. Accordingly its ground ice history is complex with, for example, massive syngenetic ice wedges attaining sizes well in excess of those formed epigenetically.

The initial stage in alas development is a disturbance to the ground surface thermal regime's equilibrium state, such as can arise from the destruction of the natural vegetation as the result of climatic change, a forest fire or human activities. The upset thermal balance invariably leads to the degradation of the ice wedge tops beneath their surface polygonal troughs and the resultant growth of an enhanced hummocky surface morphology. Once ponded water accumulates between the mounds, further ice wedge decay is inevitable as in summer the water quickly warms and heat is transferred to the ice beneath. Thaw settlement in conjunction with progressive amalgamation of the ponds accelerates the melting process and leads to the creation of a thermokarst lake at the bottom of a major flat-bottomed depression. With time stability may be attained and lakes occupying old alas depressions may disappear through either sedimentation or drainage. Either process causes the sub-lake taliks to shrink leading to the growth of one or more PINGOs beneath the now dry hollow floor.

Reference

Soloviev, P.A. (1973) Thermokarst phenomena and landforms due to frost heaving in central Yakutia, *Builetyn perglacjalny* 23, 135–155.

SEE ALSO: ice wedge and related structures; permafrost; pingo; thermokarst

PETER WORSLEY

ALIGNED DRAINAGE

A parallelism of drainage lines. In some cases parallel or aligned drainage (rather than more normal dendritic patterns) covers great areas. As W.L. Russell (1929: 249) wrote:

> One of the most remarkable features of the northwestern Great Plains is the well-defined northwest–southeast alignment of the valleys

and ridges. The prevailing direction of the valleys and ridges is nearly identical over such great areas that it is evident that the causes or forces which produced the parallelism must have operated on a grand scale.

Using available maps, Russell showed that the alignment was developed in parts of western South Dakota, western Nebraska, western North Dakota, western Montana and eastern Wyoming. He suggested that this alignment was not caused by any structural control (though in other areas this is a perfectly valid hypothesis) but was associated with the former presence of sand dunes and associated interdunal channelling of erosion, particularly in the susceptible Pierre Shale. The aeolian hypothesis was endorsed by Flint (1955) who pointed out that alignment could be produced either by the former existence of linear dunes or by deflation of susceptible materials such as the Pierre Shale. Aligned drainage also occurs on the High Plains of Texas.

Similarly large expanses of aligned drainage occur in parts of Africa and are associated with the former greater extents of the Kalahari and Sahara deserts. In southern Angola, for example, even in areas where the current mean annual precipitation is as high as 1,200 mm, aligned stream channels (many of which are tens of km in length) run from east to west, as do the old dunes of the Mega-Kalahari (Thomas and Shaw 1991). In west Africa aligned drainage, related to the Pleistocene expansion of the Sahara, occurs as far south as southern Nigeria and Cameroon (Nichol 1998).

References

Flint, R.F. (1955) Pleistocene geology of eastern South Dakota, *US Geological Survey Professional Paper* 262.

Nichol, J.E. (1998) Quaternary climate and landscape development in west Africa: evidence from satellite images, *Zeitschrift für Geomorphologie NF* 42(3), 329–347.

Russell, W.L. (1929) Drainage alignment in the western Great Plains, *Journal of Geology* 37, 240–255.

Thomas, D.S.G. and Shaw, P.A. (1991) *The Kalahari Environment*, Cambridge: Cambridge University Press.

A.S. GOUDIE

ALLOMETRY

Allometry is the measurement of proportional changes in parts of an organism and correlated with variation in size of the total organism (Gould 1966). Church and Mark (1980) have provided the most comprehensive discussion of the geomorphological applications of allometry.

Allometric relations are usually described by power laws, such as $y = ax^b$ where x is an index of system scale, y is an attribute of the system, a is a constant and the exponent b is the ratio of x and y.

Four distinctions need to be made:

1 *Allometry and isometry* If x and y have the same dimensions, isometry obtains when $b = 1$. Under this condition, there is no change in the relative proportions of x and y with increasing scale and the system is described as self-similar. When $b \neq 1$, the relation is allometric, implying a scale-related distortion of geometry. For example, in Bull's (1964) analyses of the areas of alluvial fans compared with their contributing drainage areas, $b = 0.9$ and allometry obtains. This is an indication that larger drainage basins have a relatively greater tendency to store sediment than smaller basins.
2 *Negative and positive allometry* If $b > 1$, the relation is positively allometric; if $b < 1$, the relation is negatively allometric. However, care must be taken to check that the dimensions of x and y are the same. If, for example, y is a length (L) and x is an area (L^2), then a value of b of 0.5 would indicate isometry and values of $b >$ or < 0.5 would indicate positive and negative allometry respectively.
3 *Dynamic and static allometry* In biology it is relatively easy to compare organisms at various stages of growth (dynamic allometry). Typically, landforms are compared at one moment in time with little control over their absolute ages (static allometry). There are serious limitations to static allometry, not least of which is the spatial heterogeneity of geological materials and the difficulty of defining drainage basins with similar growth histories.
4 *Simple and compound allometry* If the b value of an allometric relation changes as system scale changes compound allometry obtains. There is increasing evidence that compound allometry results from dominant process change between slope-dominated and channel-dominated basins.

References

Bull, W.B. (1964) Geomorphology of segmented alluvial fans in western Fresno County, California, *US Geological Survey Professional Paper* 352-E, 89–129.

Church, M.A. and Mark, D.M. (1980) On size and scale in geomorphology, *Progress in Physical Geography* 4, 342–390.
Gould, S.J. (1966) Allometry and size in ontogeny and phylogeny, *Cambridge Philosophical Society Biological Reviews* 41, 587–640.

SEE ALSO: fractals

OLAV SLAYMAKER

ALLUVIAL FAN

Alluvial fans are depositional landforms created where steep high-power channels enter a zone of reduced STREAM POWER. Typically they range in scale from axial lengths of tens of metres to tens of kilometres. They are usually cone-shaped forms with surface slopes radiating away from an apex, located at the point where the feeder channel enters the fan. This form can be modified by the presence of confining neighbouring fans or valley walls. In addition, the burial of the fan apex area can cause backfilling into the mountain catchment. Alluvial fans are subaerial features, however if they extend into water they are known as fan deltas.

Many of the classic studies of alluvial fans, which established the basic properties, were carried out in the basin-and-range terrain of the deserts of the American south-west (Blackwelder 1928; Blissenbach 1954; Hooke 1967), culminating in Bull's (1977) review paper. Since then there have been many studies of other (mostly) dry-region fans, with emphasis on relations between sedimentary processes and morphology (Wells and Harvey 1987; Blair and McPherson 1994) and on fan dynamics (Harvey 1997), in addition to studies of fans in humid regions (see Rachocki and Church 1990).

Fan occurrence

Alluvial fans occur in two characteristic situations: at mountain fronts and at tributary junctions. In both cases, high sediment loads encounter zones of reduced stream power, with accommodation space for deposition. These conditions are controlled by long-term landform evolution, including the tectonic setting and erosional history. Mountain fronts may be fault-controlled or erosional, in which case the fans may bury an older PEDIMENT surface. Tributary-junction settings are controlled by the long-term dissectional history. A common fan setting occurs in glaciated mountain terrain, where steep tributary valleys join wide formerly glaciated valleys.

Much of the literature emphasizes the importance of alluvial fans in desert mountain areas. In such areas, FLASH FLOODS transport abundant coarse sediment, and the depositional setting created by regional tectonics may be enhanced by the tendency for desert floods to lose power downstream. However, neither active tectonics nor aridity are prerequisites for fan formation. Fans can occur in mountain areas in all climatic settings and in tectonically stable areas, provided there is juxtaposition of high coarse-sediment transport and a sudden downstream loss in transporting power. But, as outlined below, as fan morphology tends to respond to climatically controlled water and sediment supply, climatic change can induce a change in fan processes and morphology.

Fan processes

Processes on fans include four groups. Primary processes deliver sediment to the fan, principally by DEBRIS FLOWS, or fluvial processes (by channelized and/or sheet flows). These processes are expressed by the sediments comprising the fan and by the surface morphology. Debris flows are massive, usually matrix-supported coarse sediments with clasts up to boulder size. Depositional features may include lobate and levee forms. Fluvial sediments in fan environments are usually moderately sorted gravels and cobbles, stratified or lensed in channelized or sheet bodies. Depositional features may include a range of channel forms or shallow bar and swale topography.

Fans have been classified on the basis of the primary processes into debris-flow and fluvially dominant fans. These processes are catchment controlled and depend on the water:sediment mix fed to the fan during flood events. Debris flows operate as sediment-rich flows, but under conditions of greater dilution become transitional or HYPERCONCENTRATED FLOWS then fluvial flows. Debris flows are most common where sediment concentrations are high, e.g. from small, steep catchments (Kostaschuk *et al.* 1986). Fluvial processes are more common from large, less steep catchments. The old-fashioned distinction between 'dry' and 'wet' fans, interpreted on the basis of climate, is outmoded – the primary processes are controlled mainly by catchment characteristics.

Secondary processes rework the sediment on the fan by fluvial, or in arid areas by AEOLIAN

PROCESSES. Third, stabilization processes involve surface modification by soil formation and vegetation colonization. Such processes may influence the hydrology of the fan surface, but are important in fan studies as they allow the relative ages of fan segments to be assessed (see McFadden *et al.* 1989). In arid and semi-arid areas such processes include surface modification by desert

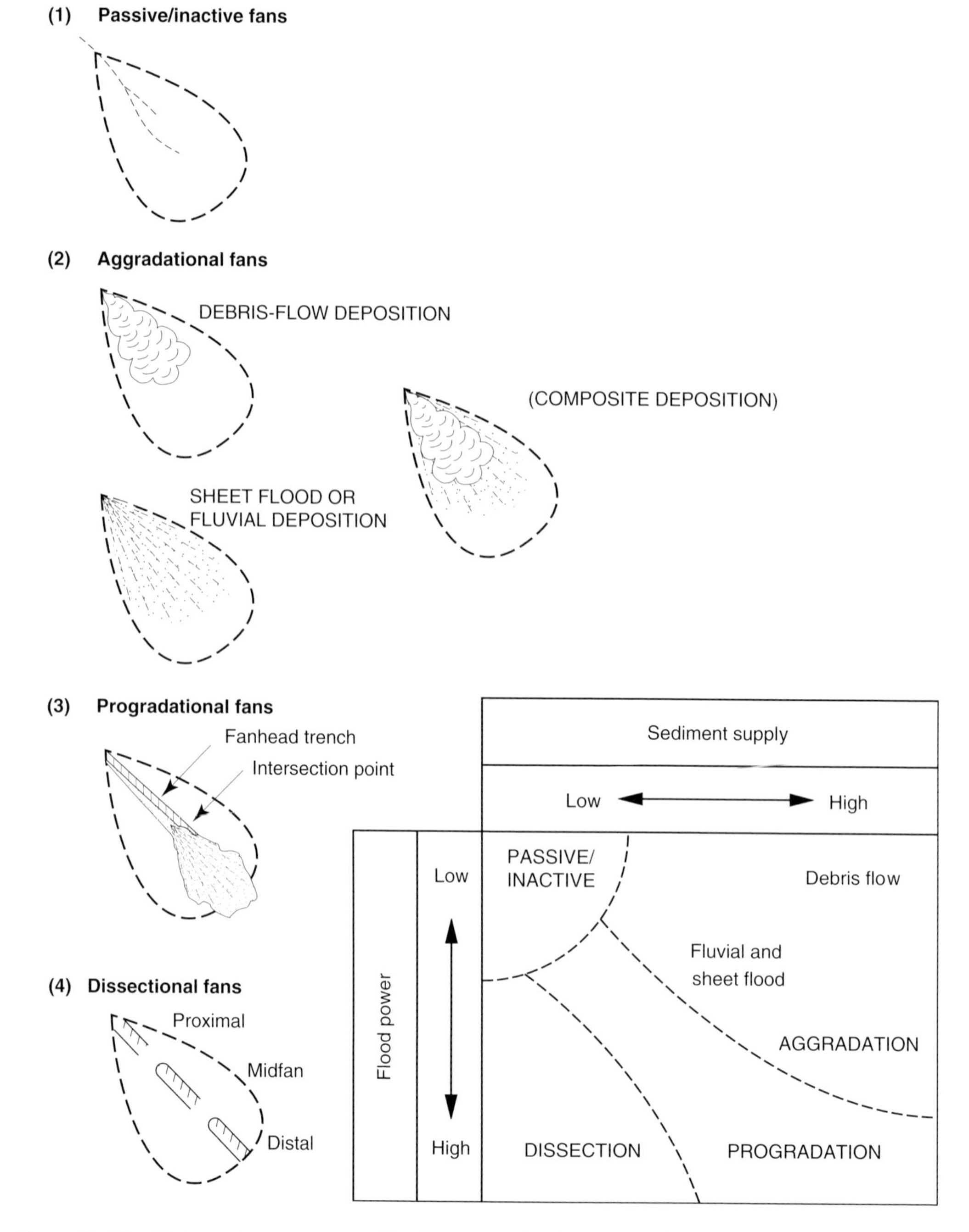

Figure 3 Alluvial fan styles: response to flood power and sediment supply (modified from Harvey 2002c)

pavement (see STONE PAVEMENT) formation and DESERT VARNISH development, and pedogenic processes leading for example to carbonate accumulation and CALCRETE formation. In humid areas lichen colonization and colonization by higher plants may be important as well as soil formation. Finally, dissection processes may erode the fan surface. Dissection may simply increase with fan surface age, or be accelerated by climatic or base-level change.

Fan morphology

Within the context of the topographic setting, fan morphology reflects fan processes and evolution. The relationships between erosion and deposition on the fan can be described as fan style (Figure 3), which in turn depends on the relationship between flood power and sediment supply. Under conditions of low power and little sediment supply the fan may be inactive. Under conditions of excess sediment supply the fan will aggrade, by debris-flow or fluvial processes dependent on the water:sediment mix fed to the fan. Such AGGRADATION will occur from the fan apex downfan. Commonly, both power and sediment supply are moderate. The feeder channel incises into the fan surface to form a fanhead trench, which emerges onto the fan surface at a midfan intersection point (Plate 1), beyond which deposition occurs. Such fans are described as 'telescopic' and may extend by progradation. A zone of coalescent deposition from adjacent prograding fans is known as a BAJADA. If power is excessive, either through high runoff or sediment starvation, erosion may dominate. Erosion may be concentrated within the fanhead area, in midfan, or in the case of base-level induced erosion, at the fan toe.

Plate 1 Characteristic alluvial fan morphology: Death Valley, California (photo: A.M. Harvey)

Notes: f = fanhead trench; i = intersection point; o = older fan surfaces; y = younger fan surfaces; a = active depositional segment

On many fans, BASE LEVELS are stable, at least over moderate timescales, and fan processes are primarily proximally controlled – by water and sediment supply from the catchment. A climatic (or other environmental change, e.g. related to human activity) causing changes in water and sediment supply may result in a change in fan style towards greater erosion or deposition.

Two aspects of fan morphometry have been demonstrated to reflect fan context, processes and evolution. General relationships of fan area and fan gradient to drainage areas have the forms:

$$A_f = p\,A_d^{\,q} \qquad (1)$$
$$G_f = a\,A_d^{\,-b} \qquad (2)$$

(where A is area, G is gradient, f of the fan, d of the drainage basin, pqab are constants). For the fan-area relationship, exponent q generally ranges between 0.7 and 1.1, and the value of the constant p reflects fan age, degree of confinement, basin area, geology and climate. For the fan-gradient relationship exponent b generally ranges between −0.15 and −0.35, and the value of constant a primarily reflects sedimentary processes (Harvey 1997). Debris-flow fans are steeper than fluvial fans. Fan-surface and fan-channel profile relationships (Figure 4) reflect erosion and deposition histories, and the interaction between proximal climate- and sediment-led controls and distal base-level controls.

Fan dynamics

Three sets of factors affect the geomorphology of alluvial fans: (1) context and locational factors, particularly tectonics and geomorphic history; (2) water and sediment delivery to the fan, controlled in the context of catchment geology size and relief, largely by climatic factors; (3) factors affecting the fan environment itself, especially base level.

Interactions between tectonic, climatic and base-level factors form a major thrust of alluvial fan research. Tectonics and gross geomorphology may control the fan setting, but the consensus is that, at least for Quaternary fans, climate appears to have the primary role in causing changes in fan

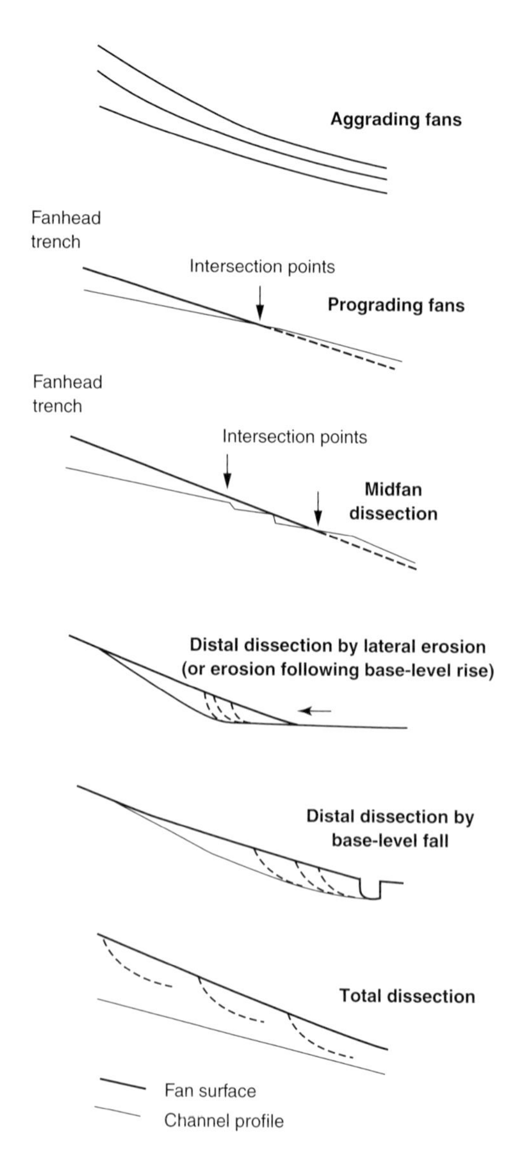

Figure 4 Fan surface and fan channel relationships

behaviour and dynamics (Frostick and Reid 1989; Ritter *et al.* 1995), modified by base-level conditions (Harvey 2002a).

Significant changes in fan processes in response to Quaternary climatic changes have been identified in many areas, including dry regions (e.g. Wells *et al.* 1987; Bull 1991), and humid temperate regions, especially in a PARAGLACIAL context in mountain areas glaciated during the Pleistocene (Ryder 1971).

Alluvial fans are important features within mountain fluvial systems. They act as sediment stores, modifying the transmission of coarse sediments through the fluvial system. They have a profound effect on the buffering/coupling relationships of fluvial systems (Harvey 1997, 2002b). Similarly they preserve a sensitive sedimentary record of environmental change within the mountain source areas.

References

Blackwelder, E. (1928) Mudflow as a geologic agent in semi-arid mountains, *Geological Society of America Bulletin* 39, 465–484.

Blair, T.C. and McPherson, J.G. (1994) Alluvial fan processes and forms, in A.D. Abrahams and A.J. Parsons (eds) *Geomorphology of Desert Environments*, 354–402, London: Chapman and Hall.

Blissenbach, E. (1954) Geology of alluvial fans in semi-arid regions, *Geological Society of America Bulletin* 65, 175–190.

Bull, W.B. (1977) The alluvial fan environment, *Progress in Physical Geography* 1, 222–270.

——(1991) *Geomorphic Responses to Climatic Change*, Oxford: Oxford University Press.

Frostick, L.E. and Reid, I. (1989) Climatic versus tectonic controls of fan sequences: lessons from the Dead Sea, Israel, *Journal of the Geological Society, London* 146, 527–538.

Harvey, A.M. (1997) The role of alluvial fans in arid zone fluvial systems, in D.S.G. Thomas (ed.) *Arid Zone Geomorphology: Process, Form and Change in Drylands*, 2nd edition, 231–259, Chichester: Wiley.

——(2002a) The role of base-level change in the dissection of alluvial fans: case studies from southeast Spain and Nevada, *Geomorphology* 45, 67–87.

——(2002b) Effective timescales of coupling within fluvial systems, *Geomorphology* 44, 175–201.

——(2002c) Factors influencing the geomorphology of dry-region alluvial fans: a review, in A. Perez-Gonzalez, J. Vegas and M.J. Machado (eds) *Aportaciones a la Geomorfologia de Espana en el Inicio del Tercer Mileno*, 59–75, Madrid: Instituto Geologico y Minero de Espana.

Hooke, R. le B. (1967) Processes on arid region alluvial fans, *Journal of Geology* 75, 438–460.

Kostaschuk, R.A., MacDonald, G.M. and Putnam, P.E. (1986) Depositional processes and alluvial fan – drainage basin morphometric relationships near Banff, Alberta, Canada, *Earth Surface Processes and Landforms* 11, 471–484.

McFadden, L.D., Ritter, J.B. and Wells, S.G. (1989) Use of multiparameter relative-age methods for age estimation and correlation of alluvial fan surfaces on a desert piedmont, eastern Mojave Desert, California, *Quaternary Research* 32, 276–290.

Rachocki, A.H. and Church, M. (eds) (1990) *Alluvial Fans: A Field Approach*, Chichester: Wiley.

Ritter, J.B., Miller, J.R., Enzel, Y. and Wells, S.G. (1995) Reconciling the roles of tectonism and climate in Quaternary alluvial fan evolution, *Geology* 23, 245–248.

Ryder, J.N. (1971) The stratigraphy and morphology of paraglacial alluvial fans in south central British

Columbia, *Canadian Journal of Earth Sciences* 8, 279–298.
Wells, S.G. and Harvey, A.M. (1987) Sedimentologic and geomorphic variations in storm generated alluvial fans, Howgill Fells, northwest England, *Geological Society of America Bulletin* 98, 182–198.
Wells, S.G., McFadden, L.D. and Dohrenwend, J.C. (1987) Influence of late Quaternary climatic change on geomorphic and pedogenic processes on a desert piedmont, eastern Mojave Desert, California, *Quaternary Research* 27, 130–146.

ADRIAN HARVEY

ALLUVIUM

Alluvium (neuter of the Latin adjective *alluvius*, meaning washed against) is the term used for the sediments that are deposited by flowing water in river valleys and deltas. Alluvial sediment originates ultimately from the breakdown (weathering) of pre-existing rocks on land. This sediment is then transported downslope by mass wasting, overland water flow, river flow and floodplain flow, and deposited in areas of the river valley where the water flow decelerated. Alluvium is deposited in distinctive landforms (e.g. channel bars, channel fills, levees, crevasse splays, flood basins, fans and deltas). Alluvial sediments are normally stratified gravels, sands, silts and clays, and the texture and stratification of the sediments are determined by the associated landform and the mode of deposition and subsequent erosion. Alluvium has been deposited on the Earth's surface for as long as rivers have existed. The sedimentary characteristics of modern alluvium are used to interpret the origin of ancient alluvium. Alluvium of modern rivers and floodplains commonly is fertile agriculturally, and a source for ground water, sand and gravel. Ancient alluvium commonly contains economically important resources such as water, gas, oil, coal and placer minerals. Reviews of the origin and nature of alluvial deposits are given by Bridge (2002), Carling and Dawson (1996) and Miall (1996).

Nature of transport and deposition of alluvium

Water flow in alluvial river channels and floodplains is turbulent. Turbulent water flow results in relatively coarse sediment (sand and gravel) being transported near the sediment bed as bedload, and finer-grained sediment (sand, silt and clay) being transported within the flow as suspended load. Depending on the sediment transport rate, the bed is normally moulded into various types of bedform, such as ripples, dunes and antidunes. Near-plane beds also occur in sands when the sediment transport rate is relatively high and in gravels at low sediment transport rates. The geometry of ripples is dependent on bed-sediment size. The geometry of dunes and antidunes is related to flow depth. Bars are larger bedforms that occur in channels, and their geometry is controlled mainly by channel width. Ripples, dunes and antidunes may be superimposed on bars.

Deposition of alluvium occurs mainly due to spatial (but also temporal) decrease in water flow velocity (actually bed shear stress) and sediment transport rate. This deposition occurs over a large range of spatial scales in areas of flow expansion and deceleration such as the lee sides of bedforms, at the edge of channels as water moves onto the floodplain, abandoned channels, zones of tectonic subsidence, and where water flows into lakes and the sea (forming deltas). Most deposition occurs during floods when flow velocity, bed shear stress and sediment transport rate are large. If sediment transport rate is large, spatial decrease in sediment transport rate will cause relatively high deposition rate. However, much deposited sediment is subsequently re-eroded during the same or subsequent floods. The coarsest sediments are deposited from bedload in places where bed shear stress is large (typically channels), whereas the finest grained sediments are deposited from suspended load (typically in abandoned channels, flood basins and lakes). Intermediate sediment sizes are deposited from bedload and suspended load. It is common for the size of channel-bed sediment to decrease down valley, primarily due to downstream decrease in channel slope and bed shear stress. It is also common for bed-sediment size to decrease laterally from the channel to the distal edge of the floodplain, also due to decreasing bed shear stress.

Alluvial landforms

Alluvial river channels contain various types of bars, the geometry and evolution of which determine the plan form of the channel. Simple (unit) bars occur in all alluvial channels. In meandering channels, the unit bars combine to give compound point bars on the inside of channel bends. In braided channels, the unit bars combine to give mid-channel compound bars (braid bars) in addition to point bars. As the supply of water and/or sediment increases, alluvial channels change from meandering to braided. Straight channels are rare and occur when the stream is not powerful enough to erode its banks. Channels change position by bank erosion and bar deposition, or by channel diversions. Channels can be diverted within their channel belts by cutoff, and channel belts can be diverted to different positions within their floodplains (avulsion). Channels abandoned by cutoff or avulsion become blocked with bars and eventually become elongate lakes.

Floodplains are the areas adjacent to channels that are inundated with water during seasonal floods. LEVEES are wedge-shaped accumulations of sediment that form floodplain ridges adjacent to channels. Crevasse channels cut levees in places and pass downstream into lobate sediment accumulations called crevasse splays. Some levees are composed of laterally adjacent crevasse splays. Crevasse splays are fan shaped in plan and contain a system of distributive and or anastomosing channels. The active and abandoned channels, levees and crevasse splays constitute the alluvial ridge that stands above the adjacent flood basin. The alluvial ridge exists because deposition rate is greatest in and around the main channel. The flood basin contains floodplain channels, both ephemeral and permanent lakes, and abandoned channel belts (alluvial ridges).

Alluvial fans and deltas are areas of alluvial deposition that are distinctive because of their plan shapes and distributive and/or anastomosing channels bordered by floodplains. Deltas build into standing bodies of water. If fans build into standing bodies of water, they are referred to as fan deltas. The term terminal fan has been used for fans in arid areas where water flow percolates into the ground before reaching beyond the fan margins. Alluvial fans occur in all climates where a confined channel passes from an area of high slope to an unconfined area of lower slope. The abrupt change of slope results in a downstream decrease in bed shear stress and sediment transport rate, which leads to deposition. ALLUVIAL FANS commonly occur adjacent to fault scarps, and the preservation of fan deposits is enhanced by the subsidence of the hanging wall. Usually, one channel is active on a fan surface at any time, but avulsion is a common process and many wholly or partially abandoned channels occur on fan surfaces. Where fan surfaces are steep (and relatively coarse grained), sediment gravity flows are common depositional processes in addition to water flows.

A RIVER DELTA is a mound of sediment deposited where a river channel enters a body of water (such as a lake or sea) and supplies more sediment than can be carried away by currents in the water body. At the river mouth, the previously confined flow expands and decelerates, depositing its sediment load. The coarse bedload is deposited close to the mouth (as a mouth bar), whereas the finer sediment in suspension is carried further into the water body before being deposited. Currents in the body of water (perhaps associated with tides, wind waves, geostrophic flows or turbidity currents) may subsequently rework and move the deposited sediment. The morphology and sediments of deltas reflect the balance between these different stages of delta formation.

River terraces (see TERRACE, RIVER) are remnants of floodplains, fans or delta plains that have become elevated relative to the modern river and floodplain, as a result of widespread channel incision. Different episodes of incision and deposition can result in a series of terraces of different height, and valley fills with a complicated internal structure.

Alluvial deposits

Depending on the availability of different sediment sizes, channel deposits are usually mainly gravels and sands. Floodplain deposits are mainly sands, silts and clays. Different scales of stratification in alluvial deposits depend on the scale of topographic feature associated with the deposit: ripples form small-scale cross strata (set thickness < 30 mm); dunes form medium-scale cross strata (set thickness 30 mm to metres); unit bars form simple sets of large-scale inclined strata (set thickness normally decimetres to metres); compound bars form compound sets of large-scale inclined strata (set thickness metres to tens of metres). Channel fills are composed of bar deposits overlain by lacustrine silts and clays. Channel belts are composed of superimposed bars and channel fills, and are commonly metres to tens of metres thick and hundreds to thousands of metres wide. Levees, crevasse splays and lacustrine deltas may be metres thick and hundreds to thousands of metres long and wide, composed mainly of sands and silts. Floodplain-channel fills typically are up to metres deep and tens to hundreds of metres across. Silty and clayey deposits of flood basins and lakes commonly occur in metre-thick sequences. Floodplain deposits are normally subjected to pedogenesis, and soil horizons are ubiquitous in alluvium.

The nature and degree of soil development varies in time and space as a function of floodplain deposition rate, parent materials, groundwater composition, climate and vegetation.

The proportion of channel-belt deposits (coarse sediments) relative to floodplain deposits (fine sediments) in a valley fill depends on factors such as the frequency of channel-belt diversions (avulsions), the width of the channel belt relative to the floodplain width, the overall deposition rate, and tectonic subsidence or uplift within the valley. High proportions of channel deposits in valley fills typically occur on the upstream parts of alluvial fans where deposition rate and avulsion frequency are locally high, in parts of valleys that are narrow relative to the channel-belt width (e.g. incised valleys), and in areas of the valley where tectonic subsidence has attracted avulsing channel belts. Low proportions of channel deposits in valley fills occur typically where floodplain (valley) width is large relative to channel-belt width (e.g. on delta plains), and in tectonically uplifted parts of floodplains.

As avulsion frequency, relative widths of channel belts and floodplains, overall deposition rate, and subsidence or uplift rate are controlled by climate, eustatic sea-level change, and tectonism, the nature of the valley fill will also be controlled by these factors. Furthermore, spatial and temporal variations in the effects of climate, eustatic sea-level change and tectonism on deposition and erosion of alluvium result in spatial variations in its texture and internal structure. These spatial variations are commonly cyclic.

References

Bridge, J.S. (2002) *Rivers and Floodplains*, Oxford: Blackwells.

Carling, P.A. and Dawson, M.R. (eds) (1996) *Advances in Fluvial Dynamics and Stratigraphy*, Chichester: Wiley.

Miall, A.D. (1996) *The Geology of Fluvial Deposits*, New York: Springer-Verlag.

SEE ALSO: aggradation; anabranching and anastomosing river; antidune; avulsion; bank erosion; bar, river; bedform; bedload; braided river; channel, alluvial; current, downstream fining; erosion; dune, fluvial; flood; flood plain; point bar; suspended load

JOHN S. BRIDGE

AMPHITHEATRE

Some early studies of curved valley heads of non-glacial origin attributed them to erosion under arid climates, but they are also widespread in humid areas. They occur mainly where a valley extends headward through gently inclined sedimentary rocks, or through dissected volcanic domes, such as those of Hawaii. Unless angular morphology is maintained by strong rectangular fracturing, curved planimetry develops as a strong CAPROCK is undercut either by seepage or by mass failure.

An amphitheatre can be likened to an arch lying on its side, because lateral stresses hold blocks in place on the curved rock face. This is especially so where the dominant stresses in a rock mass are essentially horizontal, and keep the rock face in compression. Amphitheatres thus tend to be more stable than straight cliff lines. The development of the curvature seems to be linked to the three-dimensional distribution of stresses on the rock face. Experimental studies for open-cut mining show that slopes are most stable where the radius of curvature approximates the height on the back wall, but that stability decreases markedly as the radius of curvature increases to about four times the height. Similar relationships occur in many natural amphitheatres. South of Sydney, Australia, 90 per cent of amphitheatres have a radius-to-height ratio below 5:1, with approximately 20 per cent of them below 2:1. The dimensions of amphitheatres are far from random, and are indicative of an equilibrium between form and stress distribution.

Further reading

Laity, J. and Malin, M.C. (1985) Sapping processes and the development of theater-headed valley networks on the Colorado Plateau, *Geological Society of America Bulletin* 96, 203–217.

Young, R. and Young, A. (1992) *Sandstone Landforms*, Berlin: Springer.

R.W. YOUNG

ANABRANCHING AND ANASTOMOSING RIVER

An *anabranching* alluvial river is a system of multiple channels characterized by vegetated or otherwise stable alluvial islands that divide flows at discharges up to bankfull (Plate 2). The islands may be developed from within-channel deposition, excised by channel AVULSION from extant floodplain, or formed by prograding distributary-channel accretion on splays or deltas. A specific

Plate 2 An aerial view of muddy anabranching channels at South Galway on Cooper Creek in western Queensland, Australia. Standing water is pale grey and the recently wetted channel boundary is darker and about 20–30 m wide. The islands were not quite over-topped by this bankfull flow

subset of distinctive low-energy anabranching systems associated with mostly fine-grained or organic sedimentation are defined as *anastomosing* rivers (Smith and Smith 1980; Knighton and Nanson 1993; Makaske 2001). Neither of these terms now applies to BRAIDED RIVERS where divided flow is strongly stage dependent around bars that are unconsolidated, ephemeral, poorly vegetated and overtopped at less than bankfull. However, some confusion remains because an individual low-flow channel in a braided system is sometimes referred to as an anabranch. The islands in an anabranching river are about the same elevation as the adjacent floodplain, persist for decades to centuries, have relatively resistant banks, and support mature vegetation.

Anabranching bedrock rivers can occur where the individual channels follow joint and fracture patterns. However, bankfull flow is unclearly defined making such rivers difficult to compare to their alluvial counterparts. Channels are often sediment free with pools, cataracts and waterfalls. Van Niekerk *et al.* (1999) found that bedrock anabranching channels on the Sabie River in South Africa have a significantly greater potential to transport sediment than do all the other channel types along that river. At present, relatively little is known about bedrock anabranching systems.

Anabranching is not a mutually exclusive category for it occurs in association with other patterns whereby individual anabranches braid, meander (see MEANDERING) or are straight, and it occupies a wide range of environments, from low to high energy, and in arctic, alpine, temperate, humid tropical and arid climatic settings. Anabranching rivers are more common than has been recognized previously; a total of more than 90 per cent by length of the alluvial reaches of the world's five largest rivers anabranch and it is a particularly widespread river pattern in inland Australia for both large and small rivers. In Europe many rivers used to anabranch but most of these have now been modified to provide more convenient single-thread systems in densely populated and heavily utilized valleys.

Determining the fundamental cause of anabranching remains ellusive but it is understood that in some cases, the advantage of anabranching over a single wide channel is that islands concentrate stream flow and maximize bed-sediment transport per unit of stream power, thereby maintaining equilibrium conditions. This occurs particularly where there is little or no opportunity to increase channel gradient (Nanson and Huang 1999) or where vegetation increases channel roughness (Tooth and Nanson 2000). In other words, some anabranching rivers appear to exhibit MAXIMUM FLOW EFFICIENCY and LEAST ACTION PRINCIPLE (Huang and Nanson 2000). However, there are also cases where anabranching is associated with non-equilibrium sediment transport and inefficient flow, exhibiting extensive overbank flooding, the dispersal of sediment over extensive floodplains (Plate 2), and rapid vertical accretion (Makaske 2001; Abbado *et al.* 2003). As with meandering and braiding rivers, it is apparent that anabranching systems can exhibit equilibrium or non-equilibrium behaviour.

Classification

Six types of anabranching river have been recognized by Nanson and Knighton (1996) on the basis of stream energy, sediment size and morphological characteristics: Types 1–3 are lower energy and Types 4–6 are higher energy systems. Figure 5 illustrates the planform expressions for various types of anabranching river. Type 1 consists of *cohesive sediment* rivers (commonly termed anastomosing rivers) with low w/d ratio channels that exhibit little or no lateral migration. Type 2 consists of *sand-dominated island forming* rivers and Type 3 consists of *mixed load laterally active*

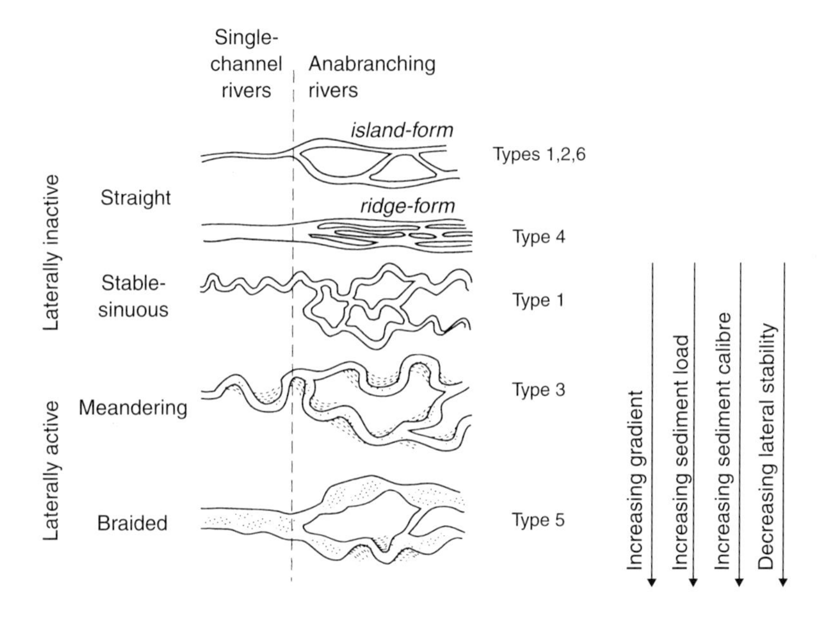

Figure 5 A classification of river channel patterns including single channel and anabranching planforms. Laterally inactive channels consist of straight and sinuous forms whereas laterally active channels consist of meandering and braided forms. The anabranching types are described in the text (after Nanson and Knighton 1996)

meandering rivers. Type 4 consists of *sand-dominated ridge form* rivers characterized by long, parallel channel-dividing ridges. Type 5 consists of *gravel-dominated laterally active* rivers that interface between meandering and braiding in mountainous regions. These have been described as wandering gravel-bed rivers (Church 1983). Type 6 consists of *gravel-dominated stable* rivers that occur as non-migrating channels in small, relatively steep basins.

Anastomosing rivers

Anastomosing rivers are an economically important subgroup of anabranching rivers and consequently have been studied in detail by sedimentologists because of their fine-grained nature and tendency to accumulate a substantial organic (coal) stratigraphy. Anastomosing commonly occurs in the lower fine-textured reaches of rivers, or in depositional basins, where vertical accretion can be rapid and hence their preservation potential is high. Crevasse splays and thick natural levees may be common. Makaske (2001) describes them as forming by avulsion and the islands as having flood basins, but these characteristics are not so apparent in some arid environments (Knighton and Nanson 1993). Modern examples were first described in detail in the alpine and humid environment of the Rocky Mountains of western Canada (e.g. Smith 1973; Smith and Smith 1980) but have subsequently been described in a wide variety of settings including arid environments (e.g. Knighton and Nanson 1993; Gibling *et al.* 1998; Makaske 2001).

Anastomosing river stratigraphy

In rapidly accreting humid settings, peats can accumulate in floodplain lakes and swamps to form coal, and sandy palaeochannels may act as reservoirs for hydrocarbons. However, not all anabranching rivers are rapidly vertically accreting and in arid environments they do not accumulate organics. Makaske (2001) found no standard sedimentary succession for anastomosing rivers, although he described them in three different settings and showed some common characteristics. The Columbia River is an example of the style of stratigraphy in a rapidly vertically accreting humid montane setting with organic-clastic accumulation (Figure 6). Such anastomosing rivers (and delta distributary

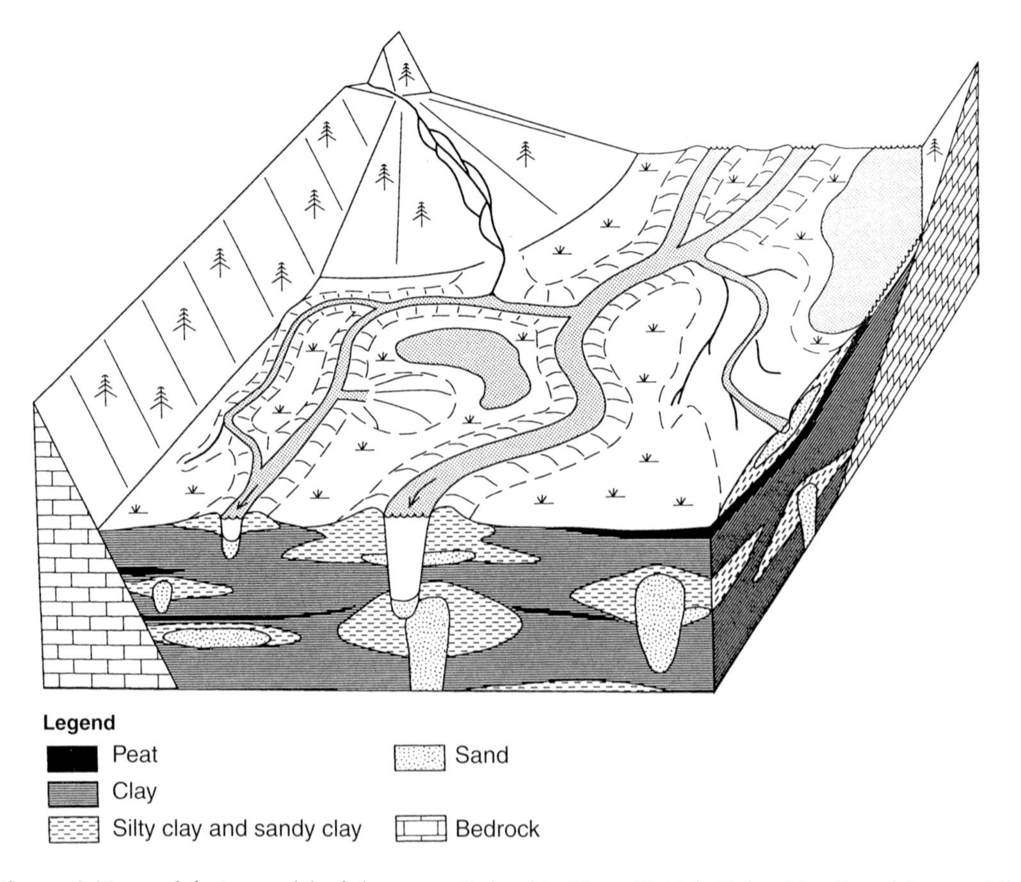

Figure 6 Textural facies model of the upper Columbia River (British Columbia, Canada), a rapidly aggrading anastomosing system in a temperate humid montane setting. Scale is approximately 2 km in width and alluvial thickness ~10 m (after Makaske 2001)

channels) tend to have fixed channels that aggrade with only limited lateral migration, thus generating ribbons or narrow sheets. Many of these deposits lie in rapidly subsiding settings, especially foreland and extensional basins characterized by large sediment flux and low gradients. The avulsion of a major channel into wetlands may generate a splay complex with suites of small, transient anastomosing channels, with the eventual establishment of a stable, single-channel course. In arid environments, alluvial and aeolian deposits can be juxtaposed, whereas in vertically accreting humid environments channel fills are flanked by silty levee deposits, lacustrine clay and coal. However, because it is difficult to show that the palaeochannels formed a synchronous anastomosing network at a single point in time (Makaske 2001), assessing the truly anabranching origin of such stratigraphies may be sometimes an educated guess.

Vegetation

Vegetation plays a major role in the development and maintenance of anabranching rivers. Indeed, it is very likely that truly anabranching rivers did not exist prior to the Devonian Period when the evolution of land plants and their associated role in the weathering of clays and the stabilization of the land surface became important. The establishment and maintenance of channels and islands with stable, often near vertical, banks means that the channels, instead of widening as a simple function of shear stress and limited alluvial strength, maintain narrow, deep and flow-efficient channels. Smith (1976) demonstrated the enormous increase in erosional resistance that plant roots can offer riverbanks. In some dryland rivers anabranching has been shown to increase in intensity below tributary junctions due to irrigation of the often dry channel floor and the greater flow and sediment-transport resistance offered by trees growing on the channel bed (Tooth and Nanson 1999). Such anabranching, resulting from the progressive evolution of within channel bars to ridges, organizes the flow into well-defined multiple channels, narrower in total than the adjacent single-thread reaches (Wende and Nanson 1999; Tooth and Nanson 2000). In certain dryland environments where bankline trees are less dense, then cohesive mud plays an important role in producing stable multiple-channel systems (Gibling *et al.* 1998).

Conclusion

Anabranching characterizes a disparate group of alluvial systems from low-energy organic or fine sediment-textured, to high-energy gravel transporting rivers, and even occurs in bedrock systems. It is a widely represented – even the dominant – style along the world's largest alluvial rivers. Alluvial anabranching rivers can be equilibrium systems that maintain their sediment flux by confining bankfull flows, or non-equilibrium systems that very effectively distribute and deposit excess sediment over extensive depositional surfaces. Anabranching is commonly associated with flood-dominated flow regimes and well-vegetated, erosion-resistant banks. As such they sometimes exhibit mechanisms to block or constrict channels and induce channel avulsion. Some develop as erosional systems that scour channels into floodplains or jointed bedrock, while others build long-lived, stable islands or ridges within existing channels. On deltas they can build floodplains vertically around initially subaqueous channels. Anabranching rivers are commonly laterally stable but individual channels can meander, braid or be straight, and as such they represent a diverse river style.

References

Abbado, D., Slingerland, R. and Smith, N.D. (2003) The origin of anastomosis on the Columbia River, British Columbia, Canada, in M.D. Blum, S.B. Marriott and S.M. Leclair (eds) *Fluvial Sedimentology VII, Proceedings of the 7th International Conference on Fluvial Sedimentology*, Special Publication of the International Association of Sedimentologists, 35.

Church, M. (1983) Anastomosed fluvial deposits: modern examples from Western Canada, in J. Collinson and J. Lewin (eds) *Modern and Ancient Fluvial Systems*, 155–168, Special Publication of the International Association of Sedimentologists, 6, Oxford: Blackwell.

Gibling, M.R., Nanson, G.C. and Maroulis, J.C. (1998) Anastomosing river sedimentation in the Channel Country of central Australia, *Sedimentology* 45, 595–619.

Huang, H.Q. and Nanson G.C. (2000) Hydraulic geometry and maximum flow efficiency as products of the principle of least action, *Earth Surface Processes and Landforms* 25, 1–16.

Knighton, A.D. and Nanson, G.C. (1993) Anastomosis and the continuum of channel pattern, *Earth Surface Processes and Landforms* 18, 613–625.

Makaske, B. (2001) Anastomosing rivers: a review of their classification, origin and sedimentary products, *Earth-Science Reviews* 53, 149–196.

Nanson, G.C. and Huang, H.Q. (1999) Anabranching rivers: divided efficiency leading to fluvial diversity, in A.J. Miller and A. Gupta (eds) *Varieties of Fluvial Form*, 219–248, Chichester: Wiley.

Nanson, G.C. and Knighton, A.D. (1996) Anabranching rivers: their cause, character and classification, *Earth Surface Processes and Landforms* 21, 217–239.

Smith, D.G. (1973) Aggradation of the Alexandria–North Saskatchewan River, Banff Park, Alberta, in M. Morisawa (ed.) *Fluvial Geomorphology*, 201–219, Binghamton, NY: Publications in Geomorphology, New York State University.

——(1976) Effect of vegetation on lateral migration of anastomosed channels of a glacial meltwater river, *Geological Society of America Bulletin* 86, 857–860.

Smith, D.G. and Smith, N.D. (1980) Sedimentation in anastomosed river systems: examples from alluvial valleys near Banff, Alberta, *Journal of Sedimentary Petrology* 50, 157–164.

Tooth, S.J. and Nanson, G.C. (1999) Anabranching rivers on the Northern Plains of arid central Australia, *Geomorphology* 29, 211–233.

——(2000) The role of vegetation in the formation of anabranching channels in an ephemeral river, Northern plain, arid central Australia, *Hydrological Processes* 14, 3,099–3,117.

Van Niekerk, A.W., Heritage, G.L., Broadhurst, L.J. and Moon, B.P. (1999) Bedrock anastomosing channel systems: morphology and dynamics in the Sabie River, Mpumalanga Province, South Africa, in A.J. Miller and A. Gupta (eds) *Varieties of Fluvial Form*, 33–51, Chichester: Wiley.

Wende, R. and Nanson, G.C. (1999) Anabranching rivers: ridge-form alluvial channels in tropical northern Australia, *Geomorphology* 22, 205–224.

SEE ALSO: avulsion; bedrock channel; braided river; floodplain; meandering

GERALD C. NANSON AND MARTIN GIBLING

ANTHROPOGEOMORPHOLOGY

Anthropogeomorphology is the study of the human role in creating landforms and modifying the operation of geomorphological processes such as weathering, erosion, transport and deposition (see, for example, Brown 1970; Nir 1983;

Goudie 1993). Some landforms are produced by direct anthropogenic processes. These tend to be relatively obvious in form and are frequently created deliberately and knowingly. They include landforms produced by: construction (e.g. spoil tips, bunds, embankments), excavation (e.g. road cuttings, and open-cast and strip mines, etc.), hydrological interference (e.g. reservoirs, ditches, channelized river reaches and canals) and farming (e.g. terraces; see Plate 3).

Landforms produced by indirect anthropogenic processes are often less easy to recognize, not least because they tend to involve not the operation of new processes, but the acceleration of natural processes. They are the result of environmental changes brought about inadvertently by human actions. By removing or modifying land cover – through cutting, bulldozing, burning and grazing – humans have accelerated rates of erosion and sedimentation (see SOIL EROSION). Sometimes the results will be spectacular, for example when major gully systems rapidly develop (see ARROYO, DONGA). By other indirect means humans may create subsidence features (Johnson 1991), cause lake desiccation (Gill 1996), trigger mass movements like landslides, and influence the operation of phenomena like earthquakes through the impoundment of large reservoirs (Meade 1991). Rates of rock weathering may be modified because of the acidification of precipitation caused by accelerated sulphate emissions (see SULPHATION) or because of accelerated salinization in areas of irrigation (Goudie and Viles 1998).

Plate 3 Strip lynchets in Dorset, southern England, are a manifestation of the impact that agricultural activities can have on the geomorphology of slopes. Many of the lynchets are the result of ploughing in medieval times

There are situations where, through a lack of understanding of the operation of geomorphological systems, humans may deliberately and directly alter landforms and processes and thereby set in train a series of events which were not anticipated or desired. There are, for example, many records of attempts to reduce coast erosion by important and expensive hard engineering solutions, which, far from solving erosion problems, only exacerbated them (Bird 1979).

As so often with environmental change, it is seldom easy to disentangle changes that are anthropogenic from those that are natural (Brookfield 1999). There has, for example, been a long-continued debate about the origin of deeply incised gullies, called ARROYOS, which developed in the southwestern United States over a relatively short period in the late nineteenth century. Some workers have championed human actions (e.g. overgrazing) as the cause of this erosion spasm, while others have championed the importance of natural environmental changes, noting that arroyo cutting had occurred repeatedly before the arrival of Europeans in the area. Among the natural changes that could promote the phenomenon are a trend towards aridity (which depletes the cover of protective vegetation) or increased frequencies of high-intensity storms (which generates erosive runoff).

Another example of the complexity of causation is posed by a consideration of the potential causes of loss of land to the sea in coastal Louisiana (Walker *et al.* 1987), something that appears to be proceeding at a rapid rate at the present time. Among the factors that need to be considered are the natural ones of sea-level change, subsidence, progressive compaction of sediments, changes in the locations of deltaic depocentres, hurricane attack and degradation by marsh fauna. Equally, however, one has to consider a range of human actions, including the role that dams and levees have played in reducing the amount and texture of sediment reaching the coast, the role of canal and highway construction and subsidence caused by fluid withdrawals.

In many cases, however, as with the USA Dust Bowl in the 1930s, it is a conjunction in time of human actions (the busting of the sod) with a climatic perturbation (a great drought) that produces change.

The possibility that the build-up of greenhouse gases in the atmosphere may cause enhanced global warming in coming decades has many implications for anthropogeomorphology (see

GLOBAL GEOMORPHOLOGY). Increased sea-surface temperatures may change the geographical spread, frequency and wind speeds of hurricanes – highly important geomorphological agents, particularly in terms of river channels and mass movements. Warmer temperatures will cause sea ice to melt and may lead to the retreat of alpine glaciers and the melting of permafrost. Vegetation belts will change latitudinally and altitudinally and this will also influence the operation of geomorphological processes. Changes in temperature, precipitation amounts, and the timing and form of precipitation (e.g. whether it is rain or snow) will have a whole suite of important hydrological consequences. Some parts of the world may become moister (e.g. high latitudes and some parts of the tropics) while other parts (e.g. some of the world's drylands) may become drier. The latter would suffer from declines in river flow, lake desiccation, reactivation of sand dunes and increasing dust storm frequencies.

However, among the most important potential future anthropogeomorphological changes are those associated with sea-level change caused by the steric effect and by the melting of land ice. Low-lying coastal areas (e.g. saltmarshes, mangrove swamps, sabkhas, deltas, atolls) would tend to be particularly susceptible. Moreover, rising sea levels could promote beach erosion, as is suggested by the BRUUN RULE.

Some landscapes – 'geomorphological hot spots' (Goudie 1996) – will be especially sensitive because they are located in areas where it is forecast that climate will change to an above average degree. This is the case, for instance, in the high latitudes of Canada or Russia, where the degree of warming may be 3–4 times greater than the global average. It may also be the case with respect to some areas where particularly substantial changes in precipitation may occur. For example, various scenarios portray the High Plains of the USA as becoming markedly more arid. Other landscapes will be especially sensitive because certain landscape-forming processes are closely controlled by climatic conditions. If such landscapes are close to particular climatic thresholds then quite modest amounts of climatic change can switch them from one state to another.

References

Bird, E.C.F. (1979) Coastal processes, in K.J. Gregory and D.G. Walling (eds) *Man and Environmental Processes*, 82–101, Folkestone: Dawson.

Brookfield, H. (1999) Environmental damage: distinguishing human from geophysical causes, *Environmental Hazards* 1, 3–11.

Brown, E.H. (1970) Man shapes the Earth, *Geographical Journal* 136, 74–85.

Gill, T.E. (1996) Eolian sediments generated by anthropogenic disturbance of playas: human impacts on the geomorphic system and geomorphic impacts on the human system, *Geomorphology* 17, 207–228.

Goudie, A.S. (1993) Human influence in geomorphology, *Geomorphology* 7, 37–59.

——(1996) Geomorphological 'hotspots' and global warming, *Interdisciplinary Science Reviews* 21, 253–259.

Goudie, A.S. and Viles, H.A. (1998) *Salt Weathering Hazards*, Chichester: Wiley.

Johnson, A.I. (ed.) (1991) Land subsidence, *Publication, International Association of Hydrological Sciences*, No. 200.

Meade, R.B. (1991) Reservoirs and earthquakes, *Engineering Geology* 30, 245–262.

Nir, D. (1983) *Man, A Geomorphological Agent, An Introduction to Anthropic Geomorphology*, Jerusalem: Keter.

Walker, H.J., Coleman, J.M., Roberts, H.H. and Tye, R.S. (1987) Wetland loss in Louisiana, *Geografiska Annaler* 69A, 189–200.

A.S. GOUDIE

ANTIDUNE

A symmetrical fluvial BEDFORM produced by near-critical flows, forming in broad shallow channels, and comparable to a sand dune. However, antidunes are more temporary and are less common than dunes. Antidune formation requires a Froude Number (quantifying the relationship between the bedform and flow regime) greater than 0.8, with development often dependent upon channel depth and bed material. Antidunes migrate upstream as sediment is lost from their downstream side more rapidly than it is deposited, though they can also move downstream or remain stationary. Antidunes form directly in phase with standing waves on the water's surface, and are characterized by shallow foresets which dip upstream at an angle of about 10°. They show low resistance to flow and are rare in the rock record, probably due to reworking. Where antidunes are observed in ancient sediments they are characterized by fine, poorly developed laminae.

Further reading

Barwis, J.H. and Hayes, M.O. (1983) Genesis and preservation of antidune stratification in modern and ancient washover deposits, *Association of American Petroleum Geologists' Bulletin* 7(3), 419–420.

Mehrotra, S.C. (1983) Antidune movement, *Journal of Hydraulic Engineering – ASCE* 109, 302–304.

STEVE WARD

APPLIED GEOMORPHOLOGY

The application of geomorphology to the solution of miscellaneous problems, especially to the development of resources and the diminution of hazards (Goudie 2001), to planning, conservation and specific engineering or environmental issues (Brunsden 2002). It incorporates what is sometimes called 'ENGINEERING GEOMORPHOLOGY' (Coates 1971).

In the last three decades applied geomorphology has become a much more central and accepted part of the discipline and a variety of texts have now appeared that review its nature (e.g. Hails 1977; Cooke and Doornkamp 1974; Thorne *et al.* 1997).

The reasons for what Jones (1980: 49) calls this 'significant transformation' are various. He cites three main reasons:

1 An increasing awareness on the part of environmental decision-makers as to the complexity of environmental conditions and the significance of geomorphological hazards (e.g. landslides and floods).
2 Demand from engineers for more information on ground conditions for construction purposes and for engineering geomorphological maps.
3 A decreasing level of insularity among geomorphologists and their feeling that they needed to justify their existence in a society that increasingly measured value in terms of practical achievement.

Other reasons have been the change of emphasis in geomorphology towards the study of contemporary processes and changes at the Earth's surface. The second has been the development of more precise techniques for mapping, monitoring and analysis. A third has been growing awareness of the finite limits to some resources and the importance of a seemingly growing number of environmental crises and catastrophes. Growing urban centres face many serious geomorphological hazards. A fourth stimulus has been the development in various countries of the need for environmental impact assessments.

A major new stimulus to applied geomorphological research has been a concern with global environmental change and the potential consequences of global warming. Matters such as the stability of the Antarctic ice sheets, the susceptibility of permafrost to thermokarst development, the sensitivity of coastal wetlands to sea-level rise, and the possible reactivation of sand seas are some of the major issues that have been investigated and for which land management solutions may be required. At a more local scale, human activities are modifying the rate and extent of particular geomorphological processes including soil erosion, salt weathering and river channel form.

Geomorphologists have a variety of applicable skills which, while they may not individually be unique, as a package are distinctive.

1 'An eye for country' and the ability to interpret landscapes and identify landforms.
2 The ability to interpret and produce maps, for these uniquely effective means of imparting spatial information are central to applied geomorphology. GEOMORPHOLOGICAL MAPPING, based on field surveys and the use of topographic base maps and remote sensing techniques, have for long been used by applied geomorphologists. Cartographic skills have been revolutionized in recent years through the use of new technologies, including differential GPS, GIS, DIGITAL ELEVATION MODELS and LIDAR. Maps are especially important for land use planning and zoning.
3 Competence in the use of techniques to measure the operation of geomorphological processes.
4 Appreciation of the relationships between environmental phenomena. This enables applied geomorphologists to see a site in its broader context and to appreciate that change in one place will have ramifications elsewhere. Thus an engineering scheme (e.g. the construction of an erosion control device on a coastline) can have a range of unintended impacts on slope stability or on downdrift beach nourishment.
5 Recognition of the importance of spatial scale. Geomorphologists appreciate, for example, that rates of sediment yield vary according to the area studied and that small erosion plots may give different orders of magnitude of rates than whole catchment studies in a large basin.
6 Recognition that all places are different and that a practice which may be appropriate in one place may not be appropriate in another.

Thus some areas may be peculiarly aggressive while others may be especially sensitive to change. In a permafrost area, for example, there may be profound local differences in permafrost stability because of local soil or microclimatic conditions.

7 Recognition that the landscape is subject to change at all temporal scales and that not only is the present always changing but also that the present is a poor guide to either past or future conditions. An example of such a skill is the recognition of the need to reconstruct long-term discharge records for rivers using a range of dating and sedimentological techniques.

8 Recognition of the importance of human activities and human attitudes. A natural science/social science mix is a unique attribute that will be of particular importance in the field of environmental management (Jones 1980: 70).

Plate 5 The demolition of a railway line in Swaziland, southern Africa, caused by floods associated with a large tropical cyclone. One role of the applied geomorphologist is to undertake post-event surveys in order to ascertain past discharges

The roles of the applied geomorphologist

The various roles of the applied geomorphologist are shown in Table 1.

1 A very basic, but highly important role is to map geomorphological phenomena as a basis for TERRAIN EVALUATION. Landforms, especially depositional ones, may be important sources of useful materials for construction, while maps of slope categories may help in the planning of land use and maps of hazardous ground may facilitate the optimal location of engineering structures.

Plate 4 The main railway line from Swakopmund to Walvis Bay in Namibia illustrates the hazard posed by sand and dune movement. One role of the applied geomorphologist is to identify optimum route corridors and to advise on their management

2 Use landforms as the basis for mapping other aspects of the environment, the distribution of which is related to their position on different landforms. This is important because landforms are relatively easily recognized on air photographs and other types of remote sensing imagery. An important example of the use of landforms as surrogates for other phenomena is the use of landform mapping to provide the basis of a soil map through the use of the CATENA concept and soil toposequences.

3 Recognize and measure the speed at which geomorphological change is taking place. Such changes (Table 2) may be hazardous to humans (e.g. coastal retreat, movement of river bluffs, surges of glaciers). By using sequential maps and remote sensing images, archival information or by monitoring processes with appropriate instrumentation, areas at potential risk can be identified, and predictions can be made as to the amount and direction of change.

4 Assess the causes of observed and measured changes and hazards, for without a knowledge of cause, attempts at amelioration and management may have limited success. There is an increasing need to assess the role that humans are playing in modifying rates

Table 1 The roles of the applied geomorphologist

1	Mapping of landforms, resources and hazards
2	Use of maps of landforms as surrogates for other phenomena (e.g. soils)
3	Establishment of rates of geomorphological change by direct monitoring, use of sequential maps, archives, etc.
4	Establishing causes of change
5	Assessment of management options
6	Post-construction assessment of engineering schemes
7	Post-event evaluations (e.g. palaeodischarges)
8	Prediction of future events and changes

Table 2 Examples of geomorphological hazards

Arid zones	*Coastal*
Dune encroachment	Sea-level change
Soil deflation	Dune blowouts
Arroyo formation	Cliff retreat
Dust storms	Saltmarsh siltation
Fan entrenchment	Coastal progradation
Flash floods	Spit growth and breaching
Salt weathering	
Ground subsidence	
Tundra areas	*General*
Thermokarst formation	Mass movement
Frost leave	Karstic collapse
Thaw floods and ice jams	River floods
Glacier surges and glacier dams	Shifting river courses
Avalanches	Lake sedimentation
Jökulhaups	Soil erosion
	Riverbank erosion
	Neotectonic activity

of geomorphological processes, particularly as a result of land-cover changes.

5 Having decided on the speed, location and causes of change, appropriate management solutions need to be adopted. Although the management solution to a particular geomorphological problem may involve the building of an engineering structure (e.g. a sand fence, a sea wall, a check dam, a shelterbelt), these structures may themselves create problems and their relative effectiveness needs to be assessed. The applied geomorphologist may make certain recommendations as to the likely consequences of building, for example, GROYNES to reduce coastal erosion. Examples of engineering solutions having unforeseen environmental consequences, sometimes to the extent that the original problem is heightened and intensified rather than reduced, are all too common, especially in coastal situations (Viles and Spencer 1995). Management issues involve a consideration of ecological issues, such as when one decides on the most appropriate form for a river channelization scheme, and are likely to become increasingly important as decisions have to be made about how to manage the landscape in the face of global climate change. More and more alternatives are being sought to ecologically injurious 'hard engineering' solutions.

6 Related to environmental management and the use of engineering solutions, is the field of assessment of the success of particular schemes. An audit of performance is required as the basis for formulating best practice.

7 Undertake 'after-the-event' surveys. It is important to put on record the magnitude and consequences of extreme events as a basis for improving engineering designs and land-zoning policies. For instance, establishing the Holocene flood histories of rivers by surveying and dating slack water deposits give an important tool for predicting possible future flood peaks, especially in ungauged catchments.

8 This brings us to the final role of the applied geomorphologist, which is to look forward and to predict. When is a particular glacier likely to surge, how long will it take for an irrigation canal to be blocked by a wandering barchan, when is this slope likely to fail, how quickly will this reservoir be rendered useless through sedimentation, will the surface of a delta be built up by fluvial sediment inputs more quickly than sea-level rises? These are examples of where geomorphologists can help to answer questions about the future. Their answers can be based on studies of the past rate of operation of geomorphological processes or by developing their modelling capability.

References

Brunsden, D. (2002) Geomorphological roulette for engineers and planners: some insights into an old

game, *Quarterly Journal of Engineering Geology and Hydrogeology* 35, 101–142.
Coates, D.R. (ed.) (1971) *Environmental Geomorphology*, Binghamton: State University of New York.
Cooke, R.U. and Doornkamp, J.C. (1974) *Geomorphology in Environmental Management*, Oxford: Oxford University Press.
Goudie, A.S. (2001) Applied geomorphology: an introduction, *Zeitschrift für Geomorphologie Supplementband* 124, 101–110.
Hails, J.R. (ed.) (1977) *Applied Geomorphology; A Perspective of the Contribution of Geomorphology to Interdisciplinary Studies and Environmental Management*, Amsterdam: Elsevier Science Publishers.
Jones, D.K.C. (1980) British applied geomorphology: an appraisal, *Zeitschrift für Geomorphologie Supplementband* 36, 48–73.
Thorne, C.R., Hey, R.D. and Newson, M. (eds) (1997) *Applied Fluvial Geomorphology for River Engineering and Management*, Chichester: Wiley.
Viles, H.A. and Spencer, T. (1995) *Coastal Problems*, London: Arnold.

A.S. GOUDIE

ARCH, NATURAL

Natural arches are formed when weathering, together with mass collapse, and in arid areas with wind erosion, creates a tunnel through a slab of rock. They can thus be distinguished from NATURAL BRIDGES which are formed by fluvial or marine erosion. They most commonly occur in sandstone, which has sufficient permeability to provide the seepage that promotes weathering, yet which has the necessary cohesion for an arch to develop. Arches are most numerous where long and closely spaced joints have been eroded to form narrow fins of rock that are readily pierced by weathering. These characteristics, and thus a very high concentration of arches, occur in the Entrada and Cedar Mesa Sandstones of the Colorado Plateau.

In strongly bedded rock, widening of the initial tunnel may result in the development of a long slab or lintel. The load of the undercut rock creates tensional stress on the lower face of the slab. If the space continues to grow, the stress may exceed the tensile strength of the rock, causing the slab to collapse. An upward curving form, rather than a slab, will develop where there are curved joints in the rock, or, more commonly, where concave stress patterns in the undercut rock result in minor failures or surface spalling. The curved form of the true arch is much more stable than a slab, because the load is transmitted to the abutments, and virtually the entire structure is in compression. This is so even when the arch is split by joints, for compressive stress on each of the joint-bounded blocks keeps them in place.

Natural arches may take various forms, but will remain stable provided the load is transmitted into the abutments. This condition is met so long as the thrust line of the load remains within the arch. Arches are therefore very stable features. However, continued erosion may result in an unstable form, and the arch may then collapse by folding in on itself at several hinge points. Erosional weakening of the abutments into which the load is transmitted can also cause an arch to collapse. Conversely, rock pinnacles transmitting a vertical load down into the abutments, or natural buttresses supporting them laterally, increase the stability of the arch.

Especially where joints are widely spaced, the hollow developed in a cliff may not penetrate through the rock mass. Instead of a true arch, an alcove or apse develops. These are much more widespread than arches, but they form in essentially the same manner, being particularly well developed where seepage issues from massive sandstone.

Further reading

Robinson, E.R. (1970) Mechanical disintegration of the Navajo sandstone in Zion Canyon, Utah, *Geological Society of America Bulletin* 81, 2,799–2,806.
Young, R. and Young, A. (1992) *Sandstone Landforms*, Berlin: Springer.

R.W. YOUNG

ARÊTE

A landform composed of a fretted, steep-sided ridge that separates valley or cirque GLACIERS. Arêtes are the result of glacial undercutting – basal sapping – of rock slopes. They are common whenever mountains and peaks rise above glaciers, as in the case with NUNATAKS.

A.S. GOUDIE

ARMOURED MUD BALL

Roughly spherical lumps of cohesive sediment which generally have a diameter of a few centimetres (Bell 1940). They are also called mud balls, mud pebbles, pudding balls, till balls and clay balls. Many examples are lumps of clay or cohesive mud that have been eroded by vigorous currents from stream beds or banks. They often occur in areas of badland topography and along ephemeral streams, but can also be found on beaches (Kale and Awasthi 1993), in tidal channels, and as ice-transported debris dumped on the seafloor (Goldschmidt 1994).

References

Bell, H.S. (1940) Armoured mud balls: their origin, properties and role in sedimentation, *Journal of Geology* 48, 1–31.
Goldschmidt, P.M. (1994) Armoured and unarmoured till balls from the Greenland sea-floor, *Marine Geology* 121, 121–128.
Kale, V.S. and Awasthi, A. (1993) Morphology and formation of armoured mud balls on Revadanda Beach, Western India, *Journal of Sedimentary Petrology* 63, 809–813.

A.S. GOUDIE

ARMOURING

'Armouring', the process whereby a clastic deposit develops a surface layer that is coarser than the substrate, is most commonly associated with warm deserts and gravel-bed rivers (see BOULDER PAVEMENT, STONE PAVEMENT, FLUVIAL ARMOUR). DEFLATION, the removal of fine-grained material by wind; the winnowing of fines by surface wash (see SHEET EROSION, SHEET FLOW, SHEET WASH); the upward migration of coarse particles, as a result of alternate wetting and drying and the associated swelling and shrinking of fine debris, or FREEZE–THAW CYCLE activity at high altitudes; the upward displacement of gravel clasts as fines accumulate; and the preferential weathering and breakdown of coarse debris at depth have all been proposed as mechanisms that produce concentrations of coarse particles at the ground surface in desert environments (Cooke 1970; Dan *et al.* 1982; McFadden *et al.* 1987). The process(es) operating in any given location depend on climate, geomorphic setting, and the nature of the clastic particles and local soils. The gravel clasts involved may be produced by mechanical weathering of the local bedrock, or be of fluvial origin. In rivers, armouring may involve the concentration of coarse clasts at the base of the active layer or the preferential winnowing of finer sediment from the surface during degradation; and vertical winnowing during active BEDLOAD transport which compensates for the disparity in mobility between coarse and fine particles (Andrews and Parker 1987; Parker and Sutherland 1990). Size segregation which produces concentrations of large particles at the surface, also occurs in gravity-driven, granular mass flows, including DEBRIS FLOWS and PYROCLASTIC FLOW DEPOSITS. Segregation mechanisms include size percolation, in which fine grains infiltrate beneath coarser particles; size exclusion, where coarse grains are excluded from narrow, convective downwellings; and cascading segregation, in which larger particles roll more rapidly downslope than small ones (Shinbrot and Muzzio 2000; Vallance and Savage 2000).

References

Andrews, E.D. and Parker, G. (1987) Formation of a coarse surface layer as the response to gravel mobility, in C.R. Thorne, J.C. Bathurst and R.D. Hey (eds) *Sediment Transport in Gravel-Bed Rivers*, 269–300, Chichester: Wiley.
Cooke, R.U. (1970) Stone pavements in deserts, *Annals of the Association of American Geographers* 60, 560–577.
Dan, J., Yaalon, D.H., Moshe, R. and Nissim, S. (1982) Evolution of reg soils in southern Israel and Sinai, *Geoderma* 28, 173–202.
McFadden, L.D., Wells, S.G. and Jercinovich, M.J. (1987) Influences of eolian and pedogenic processes on the origin and evolution of desert pavements, *Geology* 15, 504–508.
Parker, G. and Sutherland, A.J. (1990) Fluvial armor, *Journal of Hydraulics Research* 28, 529–544.
Shinbrot, T. and Muzzio, F.J. (2000) Nonequilibrium patterns in granular mixing and segregation, *Physics Today* 53, 25–30.
Vallance, J.W. and Savage, S.B. (2000) Particle segregation in granular flows down chutes, in A. Rosato and D. Blackmore (eds) *International Union of Theoretical and Applied Mechanics Symposium on Segregation in Granular Flows*, Dordrecht, The Netherlands, 31–51.

BASIL GOMEZ

ARROYO

An incised valley bottom, particularly in the western USA (Figure 7), where many broad valleys and plains became deeply incised with valley-bottom gullies (arroyos) over a short period between 1865 and 1915, with the 1880s being especially important (Cooke and Reeves 1976). The arroyos can be cut as deeply as 20 m, be over 50 m wide and tens or even hundreds of kilometres long. There has been a long history of debate as to the causes of incision (Elliott *et al.* 1999) and an increasing

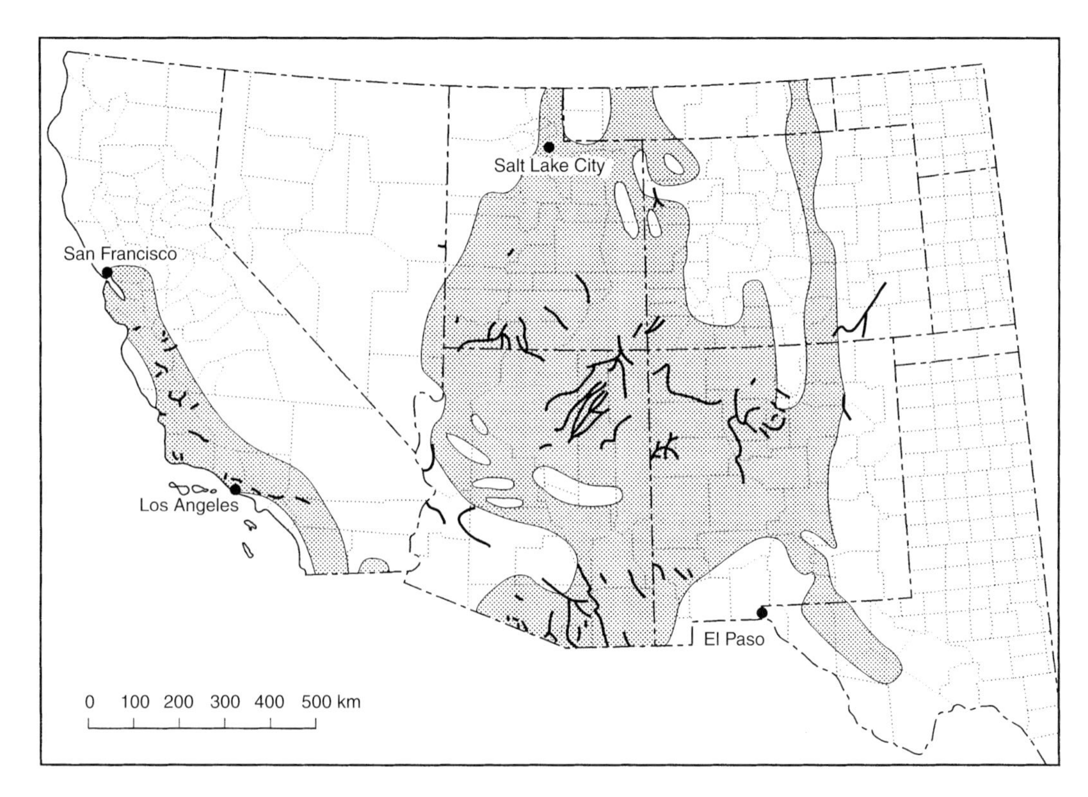

Figure 7 The distribution of arroyos in the southwestern USA (shaded), showing the course of some large examples (dark lines)

appreciation of the scale and frequency of climatic changes in the Holocene (McFadden and McAuliffe 1997), which could have led to changes in channel and slope behaviour. For example, Waters and Haynes (2001) have argued that arroyos first appeared in the American south-west after *c*.8,000 years ago, and that a dramatic increase in cutting and filling episodes occurred after *c*.4,000 years ago. They believe that this intensification could be related to a change in the frequency and strength of El Niño events.

Many students of this phenomenon have believed that human actions caused the entrenchment, and the apparent coincidence of white settlement and arroyo development in the late nineteenth century tended to give credence to this viewpoint. The range of actions that could have been culpable is large: timber-felling, overgrazing, cutting grass for hay in valley bottoms, compaction along well-travelled routes, channelling of runoff from trails and railways, disruption of valley-bottom sods by animals' feet, and the invasion of grasslands by scrub.

On the other hand, study of the long-term history of the valley fills shows that there have been repeated phases of aggradation and incision and that some of these took place before the influence of humans could have been a significant factor. Elliott *et al.* (1999) recognize various Holocene phases of channel incision at 700–1,200 BP, 1,700–2,300 BP and 6,500–7,400 BP.

A climatic interpretation was advanced by Leopold (1951), which involved a change in rainfall intensity. He indicated that a reduced frequency of low-intensity rains would weaken the vegetation cover, while an increased frequency of heavy rains at the same time would increase the incidence of erosion. Balling and Wells (1990), working in New Mexico, attributed early twentieth-century arroyo trenching to a run of years with intense and erosive rainfall that succeeded a phase of drought conditions in which the protective ability of the vegetation had declined. Large floods have also been important causative agents (Hereford 1986). Erosion and entrenchment result from a larger flood regime, with streams having a large sediment transport capacity. With lower flood regimes, a reduction in channel width and sediment storage occur, but if there are no floods, no alluviation of floodplains is possible. It is also possible, as Schumm *et al.* (1984) have pointed out, that arroyo incision could result from neither climatic change nor human influence. It could be the result of some intrinsic natural geomorphological threshold (see THRESHOLD, GEOMORPHIC) (such as stream gradient) being crossed. Under this argument, conditions of valley-floor stability decrease slowly over time until some triggering event initiates incision of the previously 'stable' reach.

It is possible that arroyo incision and alluviation result from a whole range of causes (Gonzalez 2001), that the timing of events will have varied from area to area and that individual arroyos will have had unique histories.

References

Balling, R.C. and Wells, S.G. (1990) Historical rainfall patterns and rainfall activity within the Zuni River drainage basin, New Mexico, *Annals of the Association of American Geographers* 80, 603–617.

Cooke, R.U. and Reeves, R.W. (1976) *Arroyos and Environmental Change in the American South-west*, Oxford: Clarendon Press.

Elliott, J.G., Gillis, A.C. and Aby, S.B. (1999) Evolution of arroyos: incised channels of the southwestern United States, in S.E. Darby and A. Simon (eds) *Incised River Channels*, 153–185, Chichester: Wiley.

Gonzalez, M.A. (2001) Recent formation of arroyos in the Little Missouri Badlands of southwestern Dakota, *Geomorphology* 38, 63–84.

Hereford, R. (1986) Modern alluvial history of the Paria River drainage basin, southern Utah, *Quaternary Research* 25, 293–311.

Leopold, L.B. (1951) Rainfall frequency: an aspect of climate variation, *Transactions of the American Geophysical Union* 32, 347–357.

McFadden, L.D. and McAuliffe, J.R. (1997) Lithologically influenced geomorphic responses to Holocene climatic changes in the southern Colorado Plateau, Arizona: a soil-geomorphic and ecologic perspective, *Geomorphology* 19, 303–332.

Schumm, S.A., Harvey, M.D. and Watson, C.C. (1984) *Incised Channels: Morphology, Dynamics and Control*, Littleton, CO: Water Resources Publications.

Waters, M.R. and Haynes, C.V. (2001) Late Quaternary arroyo formation and climate change in the American southwest, *Geology* 29, 399–402.

A.S. GOUDIE

ASPECT AND GEOMORPHOLOGY

As the sun moves across the sky, through the course of each day and through the seasons, the intensity of short wave radiation at a point on the hillside changes. At night, there is little radiation. In the daytime, radiation is greatest when the sun is un-obscured, and not reduced by cloud cover or where the hillside is shaded by surrounding hills. Because, north of the equator, the sun is highest in the sky towards the south, sunny south-facing slopes receive more short wave radiation than north-facing slopes, while east- and west-facing slopes receive intermediate amounts, east-facing slopes receiving more in the mornings and west-facing in the evenings. In the southern hemisphere, relationships are exchanged, north-facing slopes receiving most radiation, although east-facing slopes still face the morning sun.

Some solar radiation is lost in passing through the atmosphere, partly through the scattering which gives blue sky light, and much more if there are clouds in front of the sun. The radiation from both a cloudy sky and a clear blue sky is diffuse, and comes from all directions, although some light is lost by shading in deep valleys. The direct beam of the un-obscured sun is strongly directional, and its intensity on the surface is directly proportional to the cosine of the angle between the sun's rays and a perpendicular to the slope surface. Thus solar radiation is highest where rays fall squarely on the surface, and is greatly reduced when the rays graze the surface.

The sun's path through the sky changes in a regular way through the year, so that the amount of radiation on a hillside can be computed trigonometrically, from the latitude, the slope gradient and the direction in which the slope faces.

The sun's azimuth Φ (bearing to the sun's position in the sky) and elevation θ (angle above the horizon) can be calculated with reasonable accuracy as:

$$\sin\theta = \sin\lambda \sin\beta - \cos\lambda \cos\beta \cos\gamma$$

$$\tan\Phi = \frac{\cos\beta \sin\gamma}{\sin\lambda \cos\beta \cos\gamma + \cos\lambda \sin\beta}$$

where λ is the latitude in degrees North, β is the sun's declination ~ $-23.5 \cos J$ on Julian day J (0–360) and γ is $15\,h$ at hour h (0–24 hr local sun time). Even under a clear sun, some light is diffused (about 15 per cent under unpolluted skies) to provide blue sky light. Corrections must also be made for cloudiness and shading by any hills which form the local horizon. Making these calculations, Figure 8 shows that the difference in radiation received from clear skies on north- and south-facing slopes is greatest at about 60° latitude, but because cloudiness also increases with latitude on the continents from 30°, particularly in summer, the actual difference in radiation received is greatest at latitudes of 30°–40°.

Aspect affects geomorphology through the contrasts in radiation, most strongly between north- and south-facing slopes, which leads to differences in hydrology and sediment transport rates. Table 3 summarizes the main differences for the northern hemisphere, and north and south should be consistently exchanged for the southern hemisphere.

The effect of aspect differences is generally to create differences in the intensity of geomorphic processes between the two opposing hillsides. For example the greater radiation on south-facing semi-arid slopes increases evapotranspiration rates, so that water stress occurs in vegetation more quickly after rain. As a result, the vegetation cover is sparser and the species more drought adapted. Sparse vegetation encourages greater crusting of the soil surface, more overland flow runoff and more erosion by wash erosion. On north-facing slopes, soil moisture is maintained after rain for a longer period, so that humid vegetation can grow, usually providing greater ground cover, and better conditions for soil accumulation. Although these conditions improve infiltration rates and reduce overland flow and wash erosion, they can also provide better conditions for mass movements due to the greater depth of soil and higher moisture content.

In the short term, an increase in erosion may lead to steepening of the slope profile, but the longer term implications, as slope profiles approach some form of equilibrium, are less clear, although process differences due to aspect are commonly associated with ASYMMETRIC VALLEYS. Where there is pronounced slope asymmetry, short steep slopes on one side of the valley are matched by longer and gentler slopes on the opposite side. Two factors influence the form of the asymmetric valley cross-section. First, sediment transport depends on both slope length and slope gradient, so that the steeper slope does not necessarily deliver the more sediment. Second, at equilibrium, the valley form may not only be cutting vertically downwards, but may also be migrating laterally. For both these reasons, the hillside with the more intense process activity, due to aspect differences, may not become the gentler slope to compensate for its more intense geomorphological activity. Observations of semi-arid slopes generally suggest that radiation differences tend to maintain steep bedrock slopes on south-facing aspects, and gentler slopes mantled with soil and vegetation on north-facing aspects,

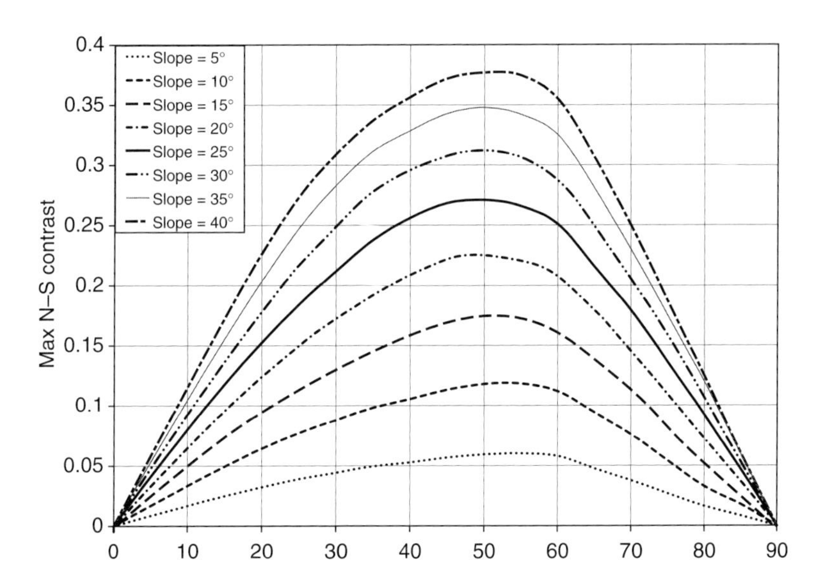

Figure 8 Difference in total annual radiation between north and south-facing slopes for clear sky conditions

Table 3 Summary of effects of aspect differences

Climatic regime	North-facing	South-facing	Geomorphic impact
Very cold (arctic or high altitude)	Permanently frozen	Some freeze-thaw	Greater solifluction and other activity on S-facing slopes
Moderately cold	Some freeze-thaw	Mainly unfrozen	Greater disturbance of vegetation and solifluction on N-facing slopes
Moist temperate	Cooler and moister	Warmer and dryer	Where water is not limiting, differences due to aspect are weak
Warm semi-arid	Cooler and moister	Warmer and dryer	S-facing slopes have sparser and more xeric vegetation, and greater runoff and erosion

although the strength of asymmetry is affected by a number of other factors, particularly geological structure and the meandering activity of rivers.

Further reading

Kirkby, M.J., Atkinson, K. and Lockwood, J.G. (1990) Aspect, vegetation cover and erosion on semi-arid hillslopes, in J.B. Thornes (ed.) *Vegetation and Erosion*, 25–39, Chichester: Wiley.

Robinson, N. (1966) *Solar Radiation*, Amsterdam: Elsevier.

MIKE KIRKBY

ASTROBLEME

The term *astrobleme* (literally 'star-wound') was introduced by Robert S. Dietz (1960) in reference to ancient erosional scars, usually circular in outline, that form on the Earth's surface through the impact of a cosmic body. This origin was recognized because of the presence of highly disturbed rocks that display evidence of intense shock (Dietz 1961). In the early debates over the origins of these features, it was not clear whether the intense pressures responsible for the disturbed rocks resulted from a bolide (an exploding meteor or comet) or from a volcanic explosion. Structures formed in the latter manner were termed *cryptovolcanic* by Branco and Frass (1901). However, the nongenetic term, *cryptoexplosion structure* (Dietz 1959), is preferred when the origin is uncertain. Nevertheless, modern research methods can nearly always confirm or reject an origin by meteor or comet impact.

The sites of relatively recent impacts on Earth, Quaternary to late Tertiary in age, will generally preserve the morphology of impact craters that is so commonly observed on the surfaces of other rocky planetary bodies in the solar system. The lack of long-term preservation of distinctive crater morphologies on the Earth's surface is the result of long-acting and relatively rapid erosional and depositional processes, when compared to circumstances on the other planetary objects. The ancient eroded impact structures of Earth (Plate 6) include circular features that are much larger than the better preserved, young impact craters. Debates over the cryptovolcanic versus impact origin of these large features raged until about the 1960s, when mineralogical studies confirmed that one of these structures, the Ries Kessel in Germany, was clearly the result of a large impact.

During the impact cratering process, immense pressures are imparted on target rocks by the high-velocity projectile. The highest pressures vaporize and melt rocks upon their release. Indeed, some large astroblemes, like Ries Kessel, are associated with huge amounts of impact melt, which early workers found difficult to distinguish from igneous rocks. Somewhat lower pressures are responsible for the metamorphic alteration of quartz to coesite and stishovite, minerals which do not form in the tectonic and volcanic processes of Earth's interior. Even lower pressures produce distinctive planar features in crystals, shocked quartz, and a distinctive cone-in-cone fracture pattern in target rocks, called *shatter cones*. The study of such features, along with their structural and geological settings, has led to the discovery of

Plate 6 Central uplift of the deeply eroded Gosses Bluff impact structure, an astrobleme in central Australia. The bluff comprises a ring of resistant sandstone, about 5 km in diameter, that was uplifted in the centre of a much larger transient crater created during the early Cretaceous (Milton *et al.* 1972). The larger structure has a diameter of about 22 km, but it is has been eroded to a nearly level plane. An ancient, higher planation surface bevels the crests of the sandstone ridges that mark the central uplift

well over a hundred terrestrial astroblemes over the last several decades.

Perhaps the most famous astrobleme is the Chicxulub structure, which is buried beneath cover rocks at the northern end of the Yucatan Peninsula, Mexico (Hildebrand *et al.* 1991). The recognition of this feature and its significance illustrates the highly interdisciplinary character of planetary science studies in application to the Earth. The story begins with the discovery in the late 1970s of an enrichment in the element iridium in a 3-cm thick layer of clay at the Cretaceous–Tertiary boundary in a thick section of marine sediments at Gubbio, Italy (Alvarez *et al.* 1980). This geochemical anomaly led the discoverers to propose that a 10-km diameter comet or asteroid collided with Earth 65 million years ago, ending the Cretaceous era and causing one of the most extensive mass extinctions of organisms in geological history, including the demise of the dinosaurs. This was indeed a provocative hypothesis, of immense potential importance to our understanding of Earth history. How could it be verified?

The iridium anomaly was subsequently identified at numerous other Cretaceous–Tertiary boundary sites around the world. Associated with the iridium were other, somewhat exotic elements in concentrations typical for chrondritic meteorites, as would be expected from the composition of the impactor. Also found were shocked quartz grains, stishovite, coesite and small glass spherules. The latter are interpreted as microtektites. Long considered a geological curiosity, relatively large, pebble-sized tektites have been found over extensive surfaces in local regions. They are clearly melted silicates, but their streamlined shapes showed that they had fallen through the atmosphere. Modern understanding of impact cratering mechanics shows that tektites are droplets of impact melt that achieve widespread ballistic dispersal from very large impact events.

The geochemical evidence all pointed to an object that would have produced a crater about 200 km in diameter, which was considerably larger than any astrobleme that had yet been identified on Earth. By following various indicators of proximity to the impact source, including tsunami deposits, tektite sizes and other features of world Cretaceous–Tertiary boundary deposits, the assembled evidence all pointed to the Caribbean and Gulf of Mexico as the likely target area. Interest then moved toward a previously obscure circular structural anomaly in northern Yucatan. The Chicxulub feature is about 180 km in diameter. Though buried, it has surface expression in a ring of cenotes (karstic sinkholes), and it is well displayed in geophysical surveys of the subsurface structure.

The discovery of astroblemes is accelerating. New features are being found on the ocean floor, aided by the extensive exploration for hydrocarbon resources. The techniques for identifying these anomalous forms make use of classical geomorphological reasoning. Moreover, it is now clear that impact cratering, the most prevalent geomorphological process on the rocky planetary bodies of the solar system, is not so rare on Earth as was once believed. It is just that the immense timescales involved for the larger impacts means that their landform consequences mostly appear as eroded, buried, and/or exhumed features that are intimately associated with the Earth's long-term geological record.

References

Alvarez, L.W., Alvarez, W., Asaro, W. and Michel, H.V. (1980) Extraterrestrial cause for the Cretaceous–Tertiary extinction, *Science* 208, 1,095–1,108.

Branco, W. and Frass, E. (1901) *Das vulcanische ries bei Nördlingen in seiner Bedeutung für Fragen der Allgemeinen Geologie*, Berlin: Akademie der Wissenschaften.

Dietz, R.S. (1959) Shatter cones in crytoexplosion structures (meteorite impact?), *Journal of Geology* 67, 496–505.
——(1960) Meteorite impact suggested by shatter cones in rock, *Science* 131, 1,781–1,784.
——(1961) Astroblemes, *Scientific American* 205, 50–58.
Hildebrand, A.R., Penfield G.T., Kring, D.A., Pilkington, M., Camargo, Z.A., Jacobsen, S.B. and Boynton, W.W. (1991) Chicxulub Crater: a possible Cretaceous/Tertiary boundary impact crater on the Yucatan Peninsula, Mexico, *Geology* 19, 867–871.
Milton, D.J., Barlow, B.C., Brett, R., Brown, A.Y., Glikson, A.Y., Manwaring, E.A. *et al.* (1972) Gosses Bluff impact structure, Australia, *Science* 175, 1,199–1,207.

SEE ALSO: crater; cryptovolcano; extraterrestrial geomorphology

VICTOR R. BAKER

ASYMMETRIC VALLEY

In very few valleys are the profiles of the opposite sides exact mirror images about the axis of the thalweg; the geomorphological definition of valley asymmetry, however, requires substantial differences in the shape and/or steepness of the two hillsides. This asymmetry may be localized, e.g. where a meander creates a river cliff opposite a slip-off slope, or valley-wide, e.g. in the case of the UNICLINAL SHIFTING characteristic of scarp and vale scenery. The ultimate in asymmetry is the case of 'one-sided' valleys, such as those of glaciated regions where the missing side was once provided by an ice sheet.

Asymmetry can, then, be the product of a whole range of circumstances relating to the orientation of valley axes and hillsides with respect to both the underlying geology and the past and present subaerial processes. Kennedy (1976) lists eight factors which have been considered to produce valley asymmetry: Coriolis force; differences in insolation and precipitation receipts; differences in slope dimensions; variable lithology; geologic structure; warping; evolution of the drainage net; and glaciation. Of these the role of geologic structure and of aspect-induced variations (see ASPECT AND GEOMORPHOLOGY) in microclimate are the two most commonly attributed causes of asymmetry.

To deal with geologic structure: faulting is evidently capable of producing dramatic asymmetry, either by opposing a fault scarp to a lower-angle hillside, or by creating hillsides with contrasting lithologies. More generally, it is accepted that the low-angle dips of domes such as the English Weald can lead to preferential down-dip migration of rivers, in the process of uniclinal shifting, resulting in broad and broadly asymmetric valleys. Whilst this is a widely observed geologic control, the question of any more general influence exerted by the dip of beds on the movement of stream channels has never been fully explored. M.J. Selby's ROCK MASS STRENGTH classification includes the dip of joints (and bedding planes), but his concept of strength-equilibrium slopes excludes those undercut by streams (1993: 104).

Far more attention has been directed towards the role of microclimatic variability and the asymmetry of slope processes which results. This was explicitly tested by A.N. Strahler (1950) in his quantitative investigation of the Davisian explanatory trio of 'structure, process and stage'. Working in the Verdugo Hills, California, Strahler found that marked vegetation contrasts between north- and south-facing hillsides were not reflected in significant angular differences. This study was extended and refined by M.A. Melton (1960) who revealed statistically significant asymmetry associated both with profile orientation and with the location of stream channels in the Laramie Mountains, Wyoming; the steepening of undercut profiles was shown to be additively linked to that associated with slope aspect (north-facing steeper).

Kennedy (1976) summarizes evidence for the presence or absence of localized and valley-wide asymmetry in seven areas of North America, ranging from 69°N to 31°N. There is no simple pattern, with the exception of the greater prevalence of valley-wide asymmetry in basins whose axes trend east-west, rather than north-south. This suggests strongly that it is the radiation balance, rather than differential precipitation inputs – at least in these cases – which is crucial to the development of process asymmetry. What is of particular interest, however, is the finding (Kennedy and Melton 1972) that an area of modern permafrost (the Caribou Hills, Northwest Territory) shows distinct, topographically determined cases where either north-facing or south-facing slopes are steeper. This must cast some doubt on the persistent attempts (cf. French 1996) to identify distinctive 'periglacial' asymmetry in terms of the orientation of steeper slopes. Kennedy found steeper north-facing slopes as far south as Kentucky (38°N), where it would seem improbable that they represent any legacy of periglaciation.

If there is any generalization to be made about the role of aspect in inducing valley-wide asymmetry, it is probably that it will develop in cases where the overall moisture balance is in some sense marginal: where this is the case, small topographic differences (or – cf. Schumm 1956 – lithologic ones) may create relatively dramatic variations in infiltration, runoff and mass movements and, ultimately, angular differences. That said, one must largely agree with Selby's assessment: 'few [studies] are based on critical examination of all slope units . . . and even in those that are, it has proved impossible to relate hillslope asymmetry to processes, because hillslopes develop over long periods' (Selby 1993: 289–290).

References

French, H.M. (1996) *The Periglacial Environment*, 2nd edition, Harlow Addison-Wesley.
Kennedy, B.A. (1976) Valley-side slopes and climate, in E. Derbyshire (ed.) *Geomorphology and Climate*, 171–201, Chichester: Wiley.
Kennedy, B.A. and Melton, M.A. (1972) Valley asymmetry and slope forms of a permafrost area in the Northwest Territories, Canada, *Special Publication, Institute of British Geographers* 4, 107–121.
Melton, M.A. (1960) Intravalley variation in slope angles related to microclimate and erosional environment, *Geological Society of America Bulletin* 71, 133–144.
Schumm, S.A. (1956) The role of creep and rainwash on the retreat of badland slopes, *American Journal of Science* 254, 693–706.
Selby, M.J. (1993) *Hillslope Materials and Processes*, 2nd edition, Oxford: Oxford University Press.
Strahler, A.N. (1950) Equilibrium theory of erosional slopes, approached by frequency distribution analysis, *American Journal of Science* 248, 673–696.

Further reading

Bloom, A.L. (1998) *Geomorphology: A Systematic Analysis of Late Cenozoic Landforms*, 3rd edition, Upper Saddle River, NJ: Prentice Hall.
Parsons, A.J. (1988) *Hillslope Form*, London: Routledge.
Tricart, J. (1963) *Géomorphologie des régions froides*, Paris: Presses Universitaires de France.

SEE ALSO: aspect and geomorphology; cross-profile, valley; rock mass strength

BARBARA A. KENNEDY

ATOLL

Atolls are generally sub-circular rings of CORAL REEF surrounding a lagoon with no dry land other than occasional islands (called *motu*) made from sand and gravel-sized detritus thrown up on the reef during storms (Nunn 1994). The word 'atoll' should strictly be applied only to the reef and lagoon (as it is here) but is sometimes used more loosely to refer to *motu*.

On first encounter, it may come as a surprise how ancient many atolls are. In the Pacific, where some of the world's oldest atolls exist, many have reef foundations dating from at least the Oligocene. It may be even more of a surprise to hear how apparently strong such organic structures are, remaining intact despite the continuous buffering of storm waves, earthquakes and even nuclear weapons tests. Yet as we learn more about the structural history of such islands, so it is becoming clear that atolls do, even without such stresses, occasionally experience catastrophic collapses. Thus Johnston Atoll in the central Pacific, where the US chemical weapon stocks are being destroyed, lost its southern flank in a series of huge landslides predating its discovery by humans. On the other hand, part of Moruroa Atoll in French Polynesia, where 98 subterranean tests of nuclear bombs were carried out between 1981 and 1991, has subsided as a direct consequence and growing concern exists about the stability of the remainder – and the possibility of radioactive residues leaking out into the ocean from the test chambers (Keating 1998).

Atoll origins

The classic exposition of atoll origins was that given by Charles Darwin who, having reached Tahiti in 1835 on the *Beagle*, climbed the slopes above Papeete and, looking across at neighbouring Mo'orea Island surrounded by its barrier reef, realized that if the island disappeared, only the REEF would remain. Thus an atoll would be formed. So, even before he had seen an atoll, Darwin set out his Theory of Atoll Development which involved the upward growth of a coral reef in response to the subsidence of its foundations (Darwin 1842). Darwin figured, with considerable prescience, that it was the tendency of ancient volcanic islands in the world's oceans to subside but that their coral fringe could stay alive only if it was able to grow upwards at the same rate. Thus modern atoll reefs (and FRINGING REEFS and CORAL REEFs) had only veneers of living coral growing atop a coral framework composed largely of the skeletal remains of dead hermatypic (reef-building) corals.

For Darwin, atoll lagoons were places where the volcanic foundations of atolls were buried by reef detritus washed over the reef during storms. Organic and mechanical forces combined to make these lagoon sediments finer over time.

What Darwin was unaware of was that sea level had oscillated with amplitudes of 100 m or more during much of the past few million years and that, although this fact did not invalidate his basic model (which is still regarded as essentially correct), sea-level change needs to be incorporated into models of atoll formation. During every period of low sea level, atolls would be converted into high limestone islands analogous to modern Niue Island in the central Pacific and others. The surfaces of these limestone islands would be reduced by KARST erosion and, when sea level rose once again at the end of the low sea-level stand, the reef would begin growing on the reduced surface (Purdy 1974).

It is worth taking a closer look at what would happen when the reef began growing once again on these surfaces during postglacial periods, taking the last as an example. In places where oceanographic and other conditions were most favourable, the upward growth of coral reef was able to 'keep up' with sea-level rise. Yet in most places, it seems that coral could not grow upwards as fast as sea level rose and that only later did the reef surface 'catch up' with sea level. In some cases, reef upgrowth was altogether too slow to keep pace with sea-level rise and the reef 'gave up' resulting in the formation of a drowned atoll (Neumann and MacIntyre 1985). The presence of drowned atolls in many parts of the Pacific and Indian oceans may be a result of Holocene reefs 'giving up' an unequal race as well as their latitude, a proxy for calcium carbonate production (Grigg 1982) and other conditions (Flood 2001).

Atoll forms

It seems most likely that the aerial form of any atoll reflects the subsurface form of the island from which it grew most recently (Purdy 1974). But within that general principle, there is considerable variation of atoll form which cannot be so readily explained.

Like barrier reefs, atoll reefs tend to be broader and more biotically diverse along their windward sides. These are also the places where atoll *motu* are generally more abundant. Thus atolls in the central Pacific easterly wind and swell belt tend to have broader reefs and more (extensive and continuous) *motu* along their eastern sides than their western sides. In contrast, Diego Garcia Atoll in the central Indian Ocean, which experiences a reversal in swell direction every six months, has a symmetrical reef with a continuous *motu*.

On some atolls, particularly those with completely enclosed lagoons, *motu* are extending lagoonwards and beginning to fill in lagoons. Some islands in Tuvalu, such as Nanumaga, which have only a few small lakes and depressions in their central parts are thought to have formed in this manner.

Humans and atolls

It is clear that most atolls only became habitable in the late Holocene because a fall of sea level exposed the tops of atoll reefs which then became foci for the accumulation of sediment dredged up from reef-talus slopes by large waves to form *motu*. Thus the existence of atoll *motu* is clear evidence for the occurrence of a higher-than-present sea level during the middle Holocene, about 4,000 cal. yr BP (Nunn 1994).

Humans have occupied many atolls continuously since that time but stress comes today from many sources. Not only has atoll life become more difficult as populations have increased and demands on naturally resource-poor environments have become more complicated, but today the fabric of atoll islands is threatened by sea-level rise. The rise of sea level during the twentieth century has caused erosion along many atoll shorelines, although often a direct link has been difficult to demonstrate because of a lack of understanding of lagoon sediment dynamics and because of the construction of artificial structures like causeways to link *motu*. In this regard, the effects of creating or enlarging reef passes to enable larger vessels to enter atoll lagoons has created problems for some atoll communities (Nunn 1994).

For the future, there is a widespread perception that the projected rise in sea level in the twenty-first century will result both in the comprehensive destruction of many atoll islands and in many atoll dwellers becoming environmental refugees in consequence. There is certainly cause for concern; one of the best geomorphological studies of recent years (Dickinson 1999) showed that the sea was currently attacking the lithified foundation of Funafuti Atoll in Tuvalu but that soon it was likely that this level would be overtopped and the sea would find itself eroding only the unconsolidated cover of the *motu* resulting in their rapid removal.

References

Darwin, C.R. (1842) *Structure and Distribution of Coral Reefs*, London: Smith, Elder.
Dickinson, W.R. (1999) Holocene sea-level record on Funafuti and potential impact of global warming on central Pacific atolls, *Quaternary Research* 51, 124–132.
Flood, P.G. (2001) The 'Darwin Point' of Pacific Ocean atolls and guyots: a reappraisal, *Palaeogeography, Palaeoclimatology, Palaeoecology* 175, 147–152.
Grigg, R.W. (1982) Darwin Point: a threshold for atoll formation, *Coral Reefs* 1, 29–34.
Keating, B.H. (1998) Nuclear testing in the Pacific from a geologic perspective, in J.P. Terry (ed.) *Climate and Environmental Change in the Pacific*, 113–144, Suva: SSED, The University of the South Pacific.
Neumann, A.C. and MacIntyre, I. (1985) Reef response to sea-level rise: keep-up, catch-up or give-up, in *Proceedings of the 5th International Coral Reef Congress* 3, 105–110.
Nunn, P.D. (1994) *Oceanic Islands*, Oxford: Blackwell.
Purdy, E.G. (1974) Reef configuration: cause and effect, *Society of Economic Paleontologists and Mineralogists, Special Publication* 18, 9–76.

PATRICK D. NUNN

AVALANCHE BOULDER TONGUE

Avalanche boulder tongues are distinctive accumulations of coarse debris resulting from the long continued, downslope transport of debris by snow avalanches (see AVALANCHE, SNOW). Two basic forms were identified (Rapp 1959). Fan tongues are thin veneers of angular debris in the avalanche runout zone. Many larger boulders and vegetation have a scattered surface cover of smaller 'perched' boulders that have been let down in precarious positions from an ablating snow cover. These fans may extend for several hundred metres across slopes of as little as 8°. Similar low-angled fans may also result from the activity of SLUSHFLOW in subarctic environments.

Where avalanches run across accumulations of loose debris (e.g. SCREE slopes below major couloirs) they erode loose debris from the upper surface of these slopes and, redistributing it downslope, produce a raised tongue of debris extending from the base of the original deposit. These 'roadbank' tongues have a pronounced basal concavity and are often asymmetric in cross section with a smoothed bevelled top, flanked by steep side slopes and a lobate front.

Reference

Rapp, A. (1959) Avalanche boulder tongues in Lappland, *Geografiska Annaler* 41, 34–48.

Further reading

Luckman, B.H. (1978) Geomorphic work of snow avalanches in the Canadian Rockies, *Arctic and Alpine Research* 10, 261–276.

SEE ALSO: hillslope, form; hillslope, process; mass movement; slushflow

BRIAN LUCKMAN

AVALANCHE, SNOW

Controls and characteristics of snow avalanches

In mountainous areas above the snowline, topographical controls result in snow avalanches recurring in locations referred to as avalanche paths. The paths are conventionally divided into the starting zone, track and runout zone. Starting zones occur at elevations above the winter snowline, have slopes in the range of 30° to 45° and are generally lee to the main storm wind directions. Below local treeline elevation, presence of forest cover also influences avalanche location while some secondary topographic factors such as slope form and roughness also play a role. Downslope convexities and transitions from anchoring points to smooth slopes are zones of tension often reflected in avalanche starting points. Across slopes, concavities (bowls and gullies) provide local snow accumulation areas. Tracks and runout zones are at lower angles. The extent of avalanche influence can be determined by their effect on vegetation and contribution to fan-shaped landforms (see Plate 7).

In addition to highest frequencies in winter and spring, avalanche timing is often related to storm events though strong melt may also be significant. Avalanches may be classified in relation to storms as either direct action or climax. The

Plate 7 An avalanche path in the Canadian Rocky Mountains showing a broad bowl-type starting zone, a track devoid of vegetation and a fan-type runout zone

former are initiated by storms (particularly heavy snowfall) and involve only the new snow while the latter owe more to instabilities that develop within the snowpack over periods of at least several days. Climax avalanches may be triggered by new snowfalls but they involve old as well as new snow and are common in spring when the snowpack may contain significant quantities of liquid water.

Snow avalanches are conventionally described in terms of a number of criteria including type of snow, form of motion, snow wetness and depth of failure. The type of snow effects the form of release, such that loose snow avalanches form in cohesionless dry or wet snow, beginning from a point and broadening downslope, while slab avalanches resulting from the existence of a strong layer of snow over a weakly resistant layer, propagate first in a line across the slope. The form of movement after initiation depends on slope steepness, roughness and the nature of the snow. For smooth, relatively gentle slopes and/or wet snow, movement is by flowing in contact with the surface. If the snow is dry and slopes steep and rough, then airborne flow (a powder avalanche) is likely, though most powder avalanches also contain a surface flowing component. Although dry avalanches will generally travel more quickly than wet snow avalanches, wet snow avalanches are capable of transporting considerable quantities of debris and are more likely to be full-depth avalanches. The extreme case of wet avalanches are slush avalanches or SLUSHFLOWS.

Snow avalanches and landforms

Studies of the accumulation of debris transported by avalanches in Scandinavia (Rapp 1960) and Canada (Luckman 1977, 1988; Gardner 1983) show that they may contribute up to several tens of $\mathrm{mm\,a^{-1}}$ of accretion on debris slopes with higher rates near the apex. Erosion also occurs

Table 4 Geomorphic effects of snow avalanches

Direct effects				
	Erosional		Depositional	
Bedrock abrasion, transport	Avalanche impact at breaks of slope		Surface sediment redistribution	
Chutes	*Pits and pools*	*Mounds and ramparts* *Debris tails*	*Boulder tongues* *Road-bank*	*Fan*
Indirect effects				
Sediments	Soils	Hydrology, glacier nourishment, snow melt floods	Nivation	

Note: Landforms in italics

though the effects are much more variable and difficult to investigate. While they often occur in association with other processes typical of mountainous areas such as rockfall and DEBRIS FLOWS, in some areas, for example the Himalayas, snow avalanches may be the dominant process in specific elevation zones (Hewitt 1989).

The geomorphic effects of snow avalanches may be either direct or indirect (Table 4). The former may involve both erosional and depositional processes and forms and in the case of some landforms, elements of both. The latter relate to aspects of the environment that influence other geomorphic processes.

The presence of downslope aligned alternating ribs and furrows in rock slopes affected by snow avalanches has long been known (Allix 1924). Abrasion of bedrock by rocks in avalanches is thought to play a role in chute formation but it is probably secondary to the transport of material loosened by other processes (Luckman 1977).

Where snow avalanches occur in locations with significant concave breaks of slope (often where formerly glaciated valleys have been subsequently filled with alluvium), they may generate great impact pressures on the landscape resulting in features collectively referred to as avalanche impact landforms (Luckman *et al.* 1994). The erosional parts of these may form circular or elongated pits but more commonly the pits intersect the local water table or the landforms may occur in water bodies such as lakes or rivers to form pools. They have been described for many areas of the world but seem particularly characteristic in areas of North America, Norway and New Zealand where resistant bedrock has resulted in the preservation of steep-sided formerly glaciated valleys.

Mounds and ramparts are made up of material scooped out by avalanche impacts and often form arcuate ridges at the distal edge of pits or pools. However, there may also be some contribution to their formation by accumulation of debris from frequent small 'dirty' avalanches once the pit or pool is full of previously avalanched snow.

In mountainous areas above the treeline and the winter snowline, snow avalanches redistribute material on debris slopes in association with other processes such as DEBRIS FLOWS and rockfall. Where avalanche transport of debris onto slopes of about 20° to 30° dominates, landforms referred to as avalanche boulder tongues occur. In the pioneering study of these features, Rapp (1959) identified two types – road-bank and fan. Road-banks are smooth flat-topped accumulations of debris often with an asymmetrical profile while fans are fan-shaped tongues of debris extending on relatively low angles towards valley floors. Rapp suggested that fan tongues result from larger avalanches and that the asymmetry of road-bank tongues resulted from preferential deposition of wind-drifted snow on the down valley side of the deposit. However, subsequent studies have suggested that other factors may lead to differences in form with road-bank tongues tending to be favoured where there is a plentiful debris supply and a confined track (Luckman 1977). Boulder tongues are characteristically 100 to 1,000 m long, up to 200 m wide and 10 to 30 m thick. They generally have a strongly concave long profile by which they can be distinguished from debris slopes formed by other processes.

Debris tails are small-scale forms which often occur on boulder tongues where there is a large range in debris sizes. They take the form of streamlined deposits of small to medium-sized particles usually downslope and more rarely upslope of large boulders. As indicated in Table 4, both erosion and deposition may play a role in their formation.

The most significant of the indirect effects of snow avalanches in relation to geomorphology is their influence on the characteristics of sediments and soils, as these may often be used to infer which processes have been responsible for building debris slopes under present or past conditions. Blikra and Nemec (1998) showed that while snow avalanches may transfer surface debris in a similar manner to debris flows, there are significant differences in the resulting sedimentary deposits, including the existence of precariously perched melt-out debris. In addition, snow avalanche deposits are often patchy, ranging from lobes of unsegregated debris to areas with better sorted sands or granules arising from water deposition following snow flows. Snow avalanche deposits may be distinguished from those of rockfalls which are characterized by openwork structure, weaker fabric strength (see FABRIC ANALYSIS) and existence of pronounced downslope increase in sediment size, as shown by Blikra and Nemec (1998) in Norway and Jomelli and Francou (2000) in the French Alps.

References

Allix, A. (1924) Avalanches, *Geographical Review* 14, 519–560.

Blikra, L.H. and Nemec, W. (1998) Post-glacial colluvium in western Norway: depositional processes, facies and paleoclimatic record, *Sedimentology* 45, 909–959.

Gardner, J.S. (1983) Accretion rates on some debris slopes in the Mt. Rae area, Canadian Rocky Mountains, *Earth Surface Processes and Landforms* 6, 347–355.

Hewitt, K. (1989) The altitudinal organisation of Karakoram geomorphic processes and depositional environments, *Zeitschrift für Geomorphologie, NF, Supplementband* 76, 9–32.

Jomelli, V. and Francou, B. (2000) Comparing the characteristics of rockfall talus and snow avalanche landforms in an alpine environment using a new methodological approach; Massif des Ecrins, French Alps, *Geomorphology* 35, 181–192.

Luckman, B.H. (1977) The geomorphic activity of snow avalanches, *Geografiska Annaler* 59A, 31–48.

——(1988) Debris accumulation patterns on talus slopes in Surprise Valley, Alberta, *Géographie physique et Quaternaire* 42, 247–278.

Luckman, B.H., Matthews, J.A., Smith, D.J., McCarroll, D. and McCarthy, D.P. (1994) Snow-avalanche impact landforms – a brief discussion of terminology, *Arctic and Alpine Research* 26, 128–129.

Rapp, A. (1959) Avalanche boulder tongues in Lappland: a description of little-known landforms of periglacial debris accumulation, *Geografiska Annaler* 41, 34–48.

——(1960) Recent development of mountain slopes in Karkevagge and surroundings, northern Scandinavia, *Geografiska Annaler* 42, 73–200.

Further reading

McClung, D. and Schaerer, P. (1993) *The Avalanche Handbook*, Seattle: Mountaineers.

IAN OWENS

AVULSION

Shift of a part or the whole of a river channel to another location on the FLOODPLAIN. It seems to be caused by local superelevation of part of the channel (see CHANNEL, ALLUVIAL) or channel system above the floodplain, as a result of river AGGRADATION, creating a local gradient advantage. Most avulsions occur when a triggering event forces a river across the stability threshold (see THRESHOLD, GEOMORPHIC). The closer the river is to the threshold, the smaller the trigger needed to initiate an avulsion (Jones and Schumm 1999). Local short-term processes triggering avulsions include: tectonic movements, variations in discharge and SEDIMENT LOAD AND YIELD, mass failure, aeolian processes (e.g. the formation of dunes) and log or ice jams. Regional long-term factors controlling avulsion include: BASE LEVEL change, climatic change, tectonic movements and discharge variation (Stouthamer and Berendsen 2000).

References

Jones, L.S. and Schumm, S.A. (1999) Causes of avulsion: an overview, in N.D. Smith and J. Rogers (eds) *Fluvial Sedimentology VI*, Special Publication of the International Association of Sedimentologists 28, 171–178.

Stouthamer, E. and Berendsen, H.J.A. (2000) Factors controlling the Holocene avulsion history of the Rhine–Meuse delta (The Netherlands), *Journal of Sedimentary Research* 70(5), 1,051–1,064.

SEE ALSO: anabranching and anastomosing river; palaeochannel; river delta

ESTHER STOUTHAMER

B

BADLAND

Badlands are deeply dissected erosional landscapes (Plate 8), formed in softrock terrain, commonly but not exclusively in semi-arid regions. Badland processes are dominated by overland-flow erosion. Badlands usually have a high DRAINAGE DENSITY of rill and gully systems, and at most support sparse vegetation. Badlands may comprise zones of coalesced hillslope gullies (see GULLY) within which little of the pre-gullying terrain remains. Badlands are common in areas with at least seasonal drought, in semi-arid and arid areas, Mediterranean and dry-season tropical areas. However, they also occur in humid regions, for example on eroding coastal and river cliffs. Badlands may result from natural processes, but their extent may be accentuated by human activity. Some badlands may be the result of human-induced soil erosion. Two prerequisites for badland development are erodible rock, typically marl, clay, or shale, and available relief. Badlands are common in uplifted and dissected softrock terrain (Plate 8).

Plate 8 Extensive badland development, Tabernas, south-east Spain. These badlands are cut into older pediment surfaces (o), and owe their origin to tectonically induced dissection of an uplifted Neogene sedimentary basin, under semi-arid climatic conditions. The badland slopes are dominated by surface processes, but note the strong aspect-related contrasts. Note also the differences between micropediment-based badland slopes (p) and gully-based badland slopes (g)

Badland processes and morphology

Processes on badland slopes are dominated by surface erosion by Hortonian (see HORTON'S LAWS) OVERLAND FLOW (Horton 1945), created by rainfall intensity exceeding infiltration capacity. Away from the drainage divide, runoff increases in depth, at first as non-erosive laminar flow, then as turbulent sheet flow. Then, when shear stresses exceed the resistance of the surface, erosion is possible. Initially erosion is by surface winnowing as sheet erosion (see SHEET EROSION, SHEET FLOW, SHEET WASH) or linearly as RILL erosion. As runoff increases further downslope, shear stresses exceed the strength of the underlying material, and channel incision is possible. At that point, defined by a minimum drainage area (Schumm's (1956a) constant of channel maintenance), sheet flow and rill flow give way to open channel flow, and sheet and rill erosion to gully erosion and stream channel processes. The requirements for Hortonian processes are intense storm rainfall, little vegetation cover, low infiltration capacity, easily erodible material and relatively steep slopes.

On slopes dominated by Hortonian processes, smooth rounded divides (Horton's (1945) belt of no erosion) give way downslope to straight rilled slopes, which in turn feed v-shaped gullies at the slope base. Rill and gully networks generally accord with the 'laws' of drainage composition (Schumm 1956a; Strahler 1957) (see HORTON'S LAWS). Local variations in the drainage density of the rill networks or other microtopographic features, such as erosion pinnacles (see HOODOO), reflect local lithological variations in infiltration capacity or erosional resistance.

On many badlands, other processes also operate. Repeated WETTING AND DRYING WEATHERING, and in some areas freezing and thawing, weather the surface materials. These processes have two main effects. Desiccation cracking or frost heave may greatly increase infiltration capacity, so that rainfall-excess overland flow becomes unlikely. In that case surface water infiltrates, and reaches rill systems by lateral flow through the weathered surface layers, reducing interill, but not rill erosion. These processes may be exacerbated by the geochemistry of the material, particularly the presence of exchangeable sodium salts. On wetting, such materials may be prone to slaking, greatly enhancing the potential for mudslides (Benito *et al.* 1993).

Weathering processes, especially on materials rich in swelling clays (see EXPANSIVE SOIL), create their own microtopography of pinnacles (Finlayson *et al.* 1987), crack patterns and so-called 'popcorn textures' (Hodges and Bryan 1982). At a larger scale, the relation between weathering and removal rates (i.e. WEATHERING-LIMITED AND TRANSPORT-LIMITED conditions) may be important. Under transport-limited conditions a weathered mantle may accumulate, ultimately to fail as a shallow mass movement, whereas on an equivalent weathering-limited slope Hortonian processes may dominate. Aspect (see ASPECT AND GEOMORPHOLOGY) often controls these processes (Plate 8), either directly or through its influence on vegetation.

A major process in some badland areas is subsurface erosion by piping (Bryan and Yair 1982; Harvey 1982) (see PIPE AND PIPING; TUNNEL EROSION). Pipes may be induced mechanically by the channelling of overland flow below the surface along animal burrows, vegetation rootways and, particularly in dissected terrain, down tension cracks. Piping is enhanced by the geochemical properties mentioned above (Gutierrez *et al.* 1988; Faulkner *et al.* 2000).

On piped badlands surfaces may have lower rill network densities, and pipe inlets and outlets may add to the morphological diversity. There may be modifications to channel alignments, when the major gullies result from pipe collapse rather than from Hortonian processes.

On some badlands, processes are dominated by single (Hortonian) processes, but on many badlands interactions between several processes take place (Schumm 1956b; Faulkner 1987; Harvey and Calvo-Cases 1991). Spatially, process interactions include on-slope interactions between weathering, Hortonian runoff, mass movements and piping, and also include HILLSLOPE-CHANNEL COUPLING relationships involving interactions between the slope processes and basal stream or gully activity (Harvey 2002). This may involve the build-up of material derived from slope erosion, and its periodic flushing by stream processes, maintaining erosional activity at the slope base (Harvey 1992).

Temporal characteristics of process interactions result from discrepancies between effectiveness, rates and frequency of the various processes. They may relate to individual storm events and recovery periods. However, a common timescale is one of seasonal cyclicity, often related to a seasonal process regime, generating for example seasonal rill development cycles (Schumm 1956a; Harvey 1992). Another type of seasonality or longer-term cyclicity may relate to flushing, when there are different frequencies of sediment generation and removal rates (Harvey 1992; Faulkner 1994). Cyclicity may also be due to discrepancies between weathering and removal rates (Harvey and Calvo-Cases 1991). Over an even longer term there may be progressive changes related to longer-term morphological development (Harvey 1992).

In a wider context, badlands show relationships to GULLY systems. Gully systems may develop as valley-floor (ARROYOS) or hillslope gullies (Campbell 1997). Badlands result from the coalescence of both basally- and midslope-induced hillslope gullies. Of fundamental importance for badland morphology and development is the local base level. An incising or laterally migrating gully channel maintains an active base level, influencing all badland processes, surface processes, slope stability, sediment removal, and even subsurface processes through its influence on hydraulic gradients. Basally-induced gullying and badland development are more likely to have effective base-level control, but even there slope retreat may transform gully/channel-based badlands by micropediment-based badlands (Plate 8).

Badland dynamics

Badlands have two major roles within the context of the wider geomorphic system: (1) as major sediment sources to the fluvial system, and (2) as a major influence on slope evolution. Badlands, especially when coupled with the stream network, represent a zone of drainage network expansion, an increase in drainage density and an increase in stream order (Strahler 1957). This has hydrological implications, but above all increases the sediment supply to the fluvial system to the extent that a zone of badlands may dominate sediment dynamics of a drainage basin (Campbell 1997). Badlands may act as a major influence on slope evolution, especially in semi-arid areas, producing extensive pediment areas at the base of retreating escarpments.

Badlands are erosional not equilibrium forms. In addition to the results of process interactions, badland morphology progressively changes as the badlands develop. This, in turn, modifies the processes. Ultimately badland development depends on the relative rates of extension and stabilization. Harvey (1992) has demonstrated that in a humid environment, once a gully system is decoupled from a basal stream, stabilization by vegetation colonization operates faster than gully extension. Those gullies do not develop into badlands. However, in many semi-arid areas, although auto-stabilization mechanisms have been recognized (Alexander *et al.* 1994), stabilization processes are slower than gully extension so that badlands develop and persist. Under conditions of incising base levels, basal v-shaped gully systems would maintain characteristic badland processes and morphology on the slopes. Under stable base-level conditions the badland slopes progressively retreat, forming pediment-based badlands, which if they do not self-stabilize, would ultimately produce a landscape of extensive pediments and small badland residuals.

One factor of fundamental importance is the interaction between vegetation and geomorphic processes which affects both the generation of overland flow and the stabilization of eroded slopes (Alexander *et al.* 1994; Gallert *et al.* 2002). However, of the main factors influencing badland geomorphology, it is the interaction of base level with the surface processes that has the greatest influence on badland evolution.

References

Alexander, R.W., Harvey, A.M., Calvo A., James, P.A. and Cerda, A. (1994) Natural stabilisation mechanisms on Badland slopes, Tabernas, Almeria, Spain, in A.C. Millington and K. Pye (eds) *Environmental Change in Drylands: Biogeographical and Geomorphological Perspectives*, 85–111, Chichester: Wiley.

Benito, G., Gutierrez, M. and Sancho, C. (1993) The influence of physico-chemical properties on erosion processes in badland areas, Ebro basin, NE Spain, *Zeitschrift für Geomorphologie* 37, 199–214.

Bryan, R. and Yair, A. (eds) (1982) *Badland Geomorphology and Piping*, Norwich: Geobooks.

Campbell, I.A. (1997) Badlands and badland gullies, in D.S.G. Thomas (ed.) *Arid-Zone Geomorphology*, 2nd edition, 261–291, Chichester: Wiley.

Faulkner, H. (1987) Gully evolution in response to both snowmelt and flash flood erosion, western Colorado, in V. Gardiner (ed.) *International Geomorphology*, vol. 1, 947–969, Chichester: Wiley.

——(1994) Spatial and temporal variations of sediment processes in the alpine semi-arid basin of Alkali Creek, Colorado, USA, *Catena* 9, 203–222.

Faulkner, H., Spivey, D. and Alexander, R.W. (2000) The role of some site geochemical processes in the development and stabilisation of three badland sites in Almeria, *Geomorphology* 35, 87–99.

Finlayson, B.L., Gerits, J.J.P. and van Wesermael, B. (1987) Crusted microtopography on badland slopes in southeast Spain, *Catena* 14, 131–144.

Gallert, F., Sole, A., Puigdefabregas, J. and Lazaro, R. (2002) Badland systems in the Mediterranean, in L.J. Bull and M.J. Kirkby (eds) *Dryland Rivers: Hydrology and Geomorphology of Semi-arid Channels*, 299–326, Chichester: Wiley.

Gutierrez, M., Benito, G. and Rodriguez, J. (1988) Piping in badland areas of the middle Ebro basin, Spain, *Catena Supplement* 13, 49–60.

Harvey, A.M. (1982) The role of piping in the development of badlands and gully systems in south east Spain, in R. Bryan and A. Yair (eds) *Badland Geomorphology and Piping*, 317–335, Norwich: Geobooks.

——(1992) Process interactions, temporal scales and the development of hillslope gully systems: Howgill Fells, northwest England, *Geomorphology* 5, 323–344.

——(2002) Effective timescales of coupling within fluvial systems, *Geomorphology* 44, 175–201.

Harvey, A.M. and Calvo-Cases, A. (1991) Process interactions and rill development on badland and gully slopes, *Zeitschrift für Geomorphologie, Supplementband* 83, 175–194.

Hodges, W.K. and Bryan, R.B. (1982) The influence of material behaviour on runoff initiation in the Dinosaur Badlands, Canada, in R. Bryan and A. Yair (eds) *Badland Geomorphology and Piping*, 13–46, Norwich: Geobooks.

Horton, R.E. (1945) Erosional development of streams and their drainage basins, Hydrophysical approach to quantitative morphology, *Geological Society of America Bulletin* 56, 275–370.

Schumm, S.A. (1956a) Evolution of drainage systems and slopes in badlands at Perth Amboy, New Jersey, *Geological Society of America Bulletin* 67, 597–646.
——(1956b) The role of creep and rainwash on the retreat of Badland slopes, *American Journal of Science* 254, 693–700.
Strahler, A.N. (1957) Quantitative analysis of watershed geomorphology, *American Geophysical Union, Transactions* 38, 913–920.

ADRIAN HARVEY

BAJADA

The broad zone of coalesced or compound ALLUVIAL FANS that form a more or less continuous piedmont alluvial apron lying between the mountain front and the basin floor in areas like the semi-arid south western United States. They are in contrast to rock-cut PEDIMENTS. The term was introduced by Tolman (1909).

Reference

Tolman, C.F. (1909) Erosion and deposition in the southern Arizona bolson region, *Journal of Geology* 17, 136–163.

A.S. GOUDIE

BANK EROSION

Bank erosion is the detachment and entrainment of bank material as grains, aggregates or blocks by fluvial, subaerial or geotechnical processes. Riverbank erosion processes are important for the evolution of MEANDERING and BRAIDED RIVER systems and FLOODPLAINS, catchment sediment output and biodiversity on floodplains. Bank erosion can also lead to loss of agricultural land and riparian structures, exacerbated sedimentation problems and riverine boundary disputes, sometimes necessitating bank stabilization works.

Bank erosion measurement

The many methods of bank retreat measurement can be classified into long-timescale, medium-timescale, and short-timescale techniques (Lawler 1993a). Long-timescale methods employ sedimentological, botanical or cartographic evidence to reveal channel change over decades or thousands of years. For example, sequences of channel movements can be preserved in datable fluvial deposits, or quantified through dendrochronological analysis (see DENDROCHRONOLOGY) of trees colonizing bar (see BAR, RIVER) surfaces (e.g. Hickin and Nanson 1984). River course changes can be quantified by superimposing early maps, aerial photographs and satellite imagery (e.g. Hooke and Redmond 1989; Lewin 1987), often using analytical photogrammetry and GIS (e.g. Lane *et al.* 1993). Medium-timescale techniques include the periodic field resurvey of bank lines with theodolites, EDMs (Electronic Distance Measurers), Total Stations or GPS (Global Positioning Systems) (Lawler 1993a). Cross-section resurvey, however, using levelling or Total Station techniques, can be more sensitive to subtle changes. Airborne laser altimetry and side-scan sonar methods have also been applied.

The following short-timescale techniques are more useful for process studies, because the geomorphological change can be related to forcing hydrological and meteorological events. Erosion pins can be inserted into banks: erosion then exposes more pin, the length of which is recorded periodically (Lawler 1993a). Terrestrial photogrammetric monitoring involves the repeated capture of ground photograms using stereometric cameras, from which the three-dimensional bank form (DIGITAL ELEVATION MODEL (DEM)) is derived. 'Subtracting' successive DEMs reveals the intervening bank erosion rate (Lawler 1993a; Lane *et al.* 1993). However, all the methods above reveal little about the *timing* of bank erosion *events*, knowledge which is crucial to process inference. Lawler (1992), therefore, developed the *automatic* Photo-Electronic Erosion Pin (PEEP) system. When erosion occurs, the PEEP signal increases; if deposition occurs, voltages are decreased. The system thus allows the magnitude, timing and frequency of erosional and depositional activity to be monitored more precisely, including for TIDAL CREEK (Lawler *et al.* 2001), and is now used by twenty research groups worldwide. The example in Figure 9 shows how the PEEP approach, for the first time, fixes the time of an erosion event to forty-three hours *after* the hydrograph peak; this suggests the operation of mass failure processes rather than fluid entrainment.

Bank erosion rates

Bank erosion rates range from 0–1,000 m a^{-1} and tend to increase with boundary shear stress, STREAM POWER, FREEZE–THAW CYCLE activity and for silty or

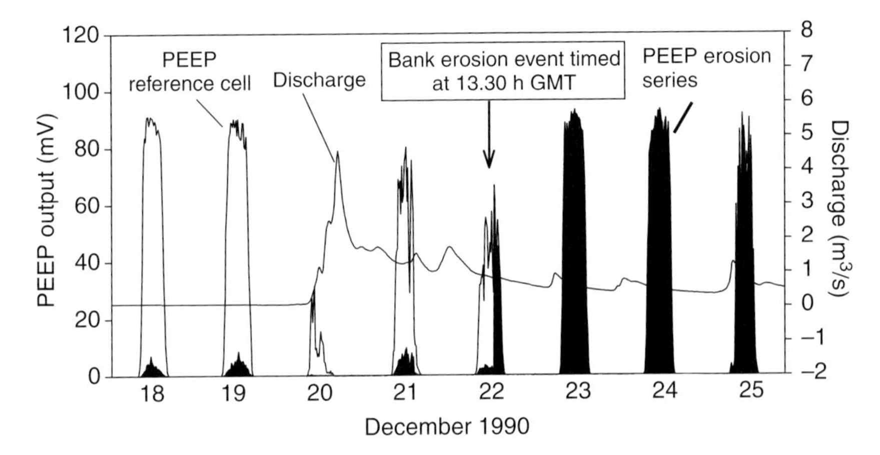

Figure 9 Timing of a bank erosion event at 13.30 h GMT on 22 December 1990 detected by the automatic Photo-Electronic Erosion Pin (PEEP) monitoring system on the Upper River Severn, UK. The bank erosion event lags the flow peak by forty-three hours, suggesting mass failure processes are responsible (from Lawler *et al.* 1997)

saturated bank materials of low cohesion (Lawler *et al.* 1997). Within some basins bank erosion rates peak in the *middle* reaches (e.g. Lewin 1987; Lawler *et al.* 1999), possibly related to stream power increases (Lawler 1992; Abernethy and Rutherfurd, 1998; Knighton, 1999).

Riverbank erosion processes

The many bank erosion mechanisms identified can be grouped into fluid entrainment, preparation or mass failure processes (Lawler 1992; Lawler *et al.* 1997; Prosser *et al.* 2000).

FLUID ENTRAINMENT PROCESSES

Entrainment occurs when the motivating forces due to the flow (lift and drag, often indexed by boundary shear stress) and particle mass exceed the friction and cohesion forces holding the particle in place (Lawler *et al.* 1997). On non-cohesive (e.g. sandy) banks, material is usually entrained grain by grain, while on cohesive, fine-grained banks, material is eroded as aggregates or crumbs bound by cohesive forces. Cohesion results from a combination of physico-chemical, inter-granular, bonding forces, driven by the mineralogy, dispersivity, moisture content and particle size distribution of the bank material, and the temperature, pH and electrical conductivities of the pore fluid and river water (Osman and Thorne 1988). Cohesive banks are normally much more resistant to fluid entrainment than non-cohesive ones.

PREPARATION PROCESSES

The ERODIBILITY of cohesive bank materials, however, can change because of preparation or weakening processes. Crucially, then, the critical shear stress value for entrainment varies with antecedent material preparation (Lawler 1992; Prosser *et al.* 2000). For example, desiccated banks may crack as moisture is thermally driven off and clay minerals shrink (Plate 9). For instance, in summer, east-facing banks of the river Arrow, Warwickshire, UK reach early-morning warming rates of 7 °C h^{-1}, peak daily temperatures above 30 °C and diurnal temperature ranges of 20 °C (Lawler 1992). Flowing waters then exploit cracks to enhance erosion (e.g. Lawler 1992). Freeze–thaw activity takes many forms (e.g. Lawler 1993b; Prosser *et al.* 2000). For example, NEEDLE-ICE can lift or incorporate material, and transport it downslope on ablation. Much disturbed sediment remains, though, to be readily removed by subsequent flow rises (Lawler 1993b).

MASS FAILURE PROCESSES

Mass failure occurs when blocks of material collapse or slide towards the bank toe (Plate 9). Banks are vulnerable to mass failure if steep, high, fine-grained, of high bulk unit weight and subject to high or fluctuating PORE-WATER PRESSURES – indeed any variable which increases the mass of material above a potential failure

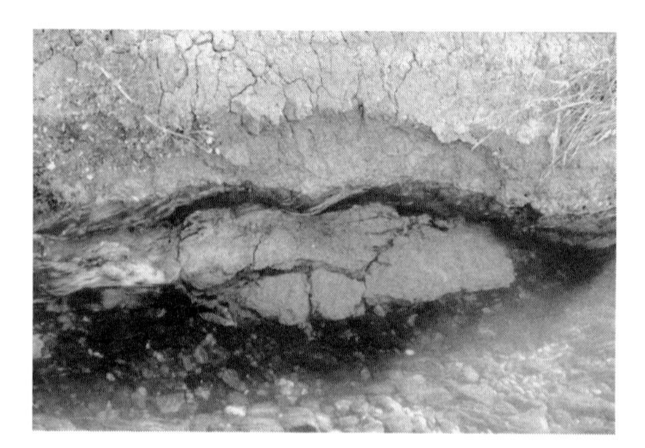

Plate 9 Bank erosion on the lowland river Arrow, Warwickshire, UK. A failed block of bank material (length ~2 m) lies under water at the bank toe, with the scar of the failure surface visible above. Desiccation cracking is above the scar. Flow is from right to left

surface. Hence, bed scour can induce bank failure by increasing bank height and angle. Also, failure can follow increases in block mass due to moisture uptake, often after submergence. Bank failures should thus occur on hydrograph recession limbs, following saturation. This is confirmed by the PEEP automatic erosion monitoring system (Figure 9). Figure 10 shows bank failure characteristics. For the most common type, the failure surface is almost planar (Plate 9 and Figure 10). Cantilever failure occurs on composite riverbanks if, because of faster erosion of the lower more erodible layers, an overhang develops then collapses (Lawler *et al.* 1997) (Figure 10g and h).

Mass failures can be analysed using geotechnical slope stability theory (e.g. Osman and Thorne 1988; Darby and Thorne 1996; the CONCEPTS model of the United States Department of Agriculture (USDA)). One example is the Culmann formula for planar failure (Figure 10c), which predicts the critical height for a bank, H_{crit}:

$$H_{crit} = \frac{4c}{\gamma} \cdot \frac{\sin\alpha \cdot \cos\phi}{\{1 - \cos(\alpha - \phi)\}}$$

where c = material cohesion (k Pa), γ = *in situ* unit weight of material ($kN\,m^{-3}$), α = slope angle (°), ϕ = friction angle (°). Many of the data required can be collected using the Stream Reconnaissance Record Sheets of Thorne *et al.* (1996).

Bank processes may change in a longitudinal direction (Lawler 1992). This idea, developed into the DOCPROBE model (DOwnstream Change in the PROcesses of Bank Erosion), suggests that, in upstream reaches, preparation processes are most effective, because stream power and bank heights are too low for significant fluid entrainment and mass failure respectively. In middle reaches, where stream power is high, fluid entrainment dominates. Further downstream, bank heights and material properties exceed critical values and mass failure processes prevail. Evidence has now emerged to support this model (e.g. Lawler 1992; Knighton 1999; Abernethy and Rutherfurd 1998). Vegetation can considerably reduce fluid erosion, partly through root reinforcement or foliage protection (Thorne 1990; Abernethy and Rutherfurd 1998; Simon and Collinson 2002). However, forest canopy shade may suppress shorter riparian vegetation, increasing erosion.

A much richer mix of bank erosion processes is now recognized. Though flow processes are important, research has shifted to temporal change in bank erodibility, vegetation, riparian hydrology and the dynamics of bank erosion events, often using novel automated monitoring techniques.

References

Abernethy, B. and Rutherfurd, I.D. (1998) Where along a river's length will vegetation most effectively stabilize stream banks? *Geomorphology* 23, 55–75.

Darby, S.E. and Thorne, C.R. (1996) Development and testing of riverbank-stability analysis, *Proceedings of the American Society of Civil Engineers, Journal of Hydraulic Engineering* 122, 443–454.

Hey, R.D., Heritage, G.L., Tovey, N.K., Boar, R.R., Grant, N. and Turner, R.K. (1991) Streambank protection in England and Wales, Research and Development Note 22, London: National Rivers Authority.

Hickin, E.J. and Nanson, G.C. (1984) Lateral migration of river bends, *Proceedings of the American Society of Civil Engineers, Journal of Hydraulic Engineering* 110, 1,557–1,567.

Hooke, J.M. and Redmond, C.E. (1989) River-channel changes in England and Wales, *Journal of the Institute of Water Engineers and Managers* 3, 328–335.

Knighton, A.D. (1999) Downstream variation in stream power, *Geomorphology* 29, 293–306.

Lane, S.N., Richards, K.S. and Chandler, J.H. (1993) Developments in photogrammetry; the geomorphological potential, *Progress in Physical Geography* 17, 306–328.

Lawler, D.M. (1992) Process dominance in bank erosion systems, in P. Carling and G.E. Petts (eds)

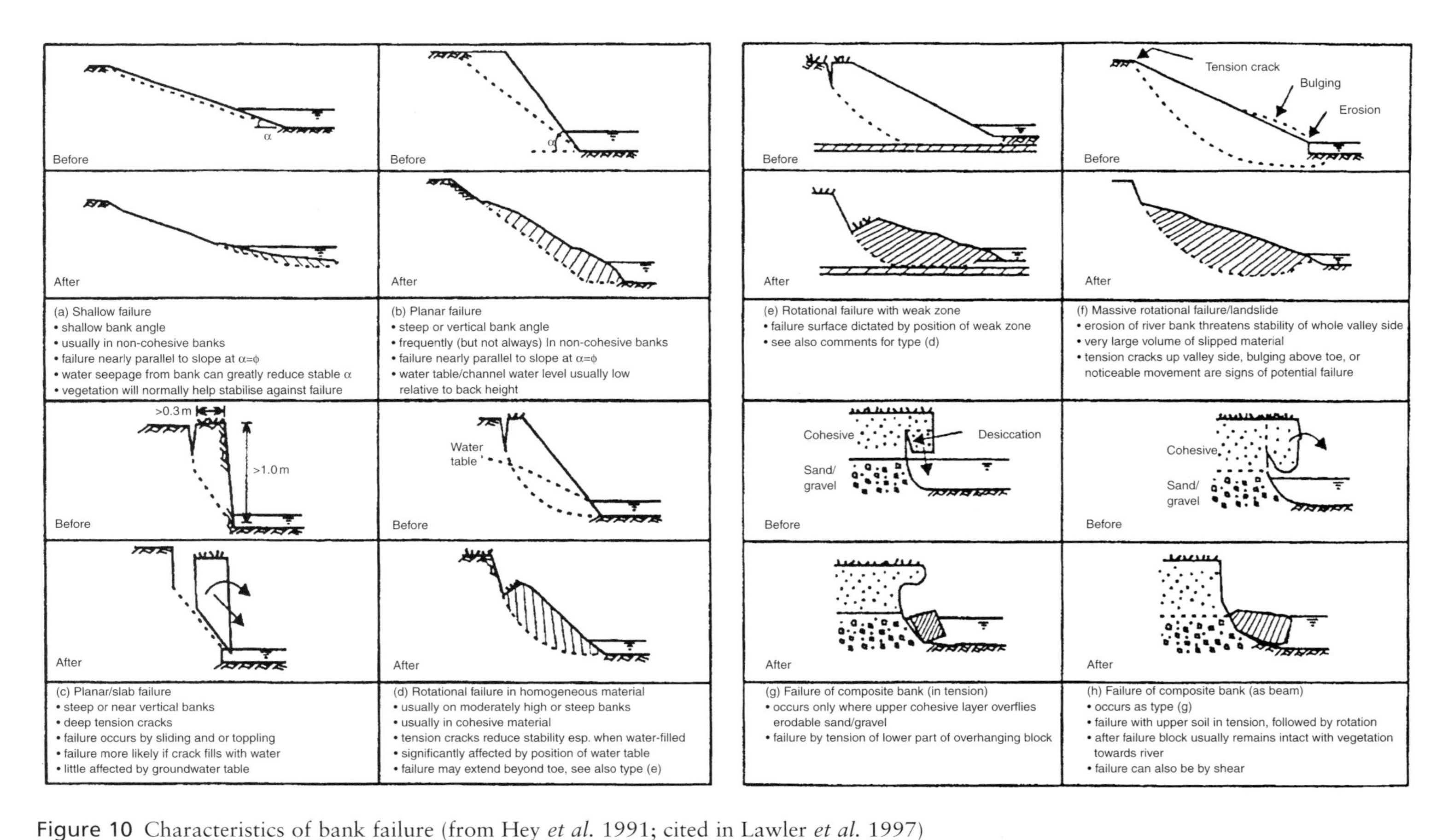

Figure 10 Characteristics of bank failure (from Hey *et al.* 1991; cited in Lawler *et al.* 1997)

Lowland Floodplain Rivers: Geomorphological Perspectives, 117–143, Chichester: Wiley.
Lawler, D.M. (1993a) The measurement of river bank erosion and lateral channel change: a review, *Earth Surface Processes and Landforms* 18, 777–821.
——(1993b) Needle-ice processes and sediment mobilization on river banks: the River Ilston, West Glamorgan, UK, *Journal of Hydrology* 150, 81–114.
Lawler, D.M., Grove, J., Couperthwaite, J.S. and Leeks, G.J.L. (1999) Downstream change in river bank erosion rates in the Swale–Ouse system, northern England, *Hydrological Processes* 13, 977–992.
Lawler, D.M., Thorne, C.R. and Hooke, J.M. (1997) Bank erosion and instability, in C.R. Thorne, R.D. Hey and M.D. Newson (eds) *Applied Fluvial Geomorphology for River Engineering and Management*, 137–172, Chichester: Wiley.
Lawler, D.M., West, J.R., Couperthwaite, J.S. and Mitchell, S.B. (2001) Application of a novel automatic erosion and deposition monitoring system at a channel bank site on the tidal River Trent, UK, *Estuarine, Coastal and Shelf Science* 53, 237–247.
Lewin, J. (1987) Historical river channel changes, in K.J. Gregory, J. Lewin and J.B. Thornes (eds) *Palaeohydrology in Practice*, 161–175, Chichester: Wiley.
Osman, A.M. and Thorne, C.R. (1988) Riverbank stability analysis. I: Theory, *Proceedings of the American Society of Civil Engineers Journal of Hydraulic Engineering* 114, 134–150.
Prosser, I.P., Hughes, A.O. and Rutherfurd, I.D. (2000) Bank erosion of an incised upland channel by subaerial processes: Tasmania, Australia, *Earth Surface Processes and Landforms* 25, 1,085–1,101.
Simon, A. and Collinson, A.J.C. (2002) Quantifying the mechanical and hydrologic effects of riparian vegetation on streambank stability, *Earth Surface Processes and Landforms* 27, 527–546.
Thorne, C.R. (1990) Effects of vegetation on riverbank erosion and stability, in J.B. Thornes (ed.) *Vegetation and Erosion*, 125–144, Chichester: Wiley.
Thorne, C.R., Allen, R.G. and Simon, A. (1996) Geomorphological river channel reconnaissance for river analysis, engineering and management, *Transactions of the Institute of British Geographers* 21, 469–483.
USDA CONCEPTS model at: http://msa.ars.usda.gov/ms/oxford/nsl/agnps/concepts/concepts_dl.html

SEE ALSO: fluvial geomorphology; hydraulic geometry

DAMIAN LAWLER

BANKFULL DISCHARGE

Bankfull discharge, a hydrologic term, is the flow rate when the stage (height) of a stream is coincident with the uppermost level of the banks – the water level at channel capacity, or bankfull stage. Bankfull stage is a fluvial-geomorphic term (see FLUVIAL GEOMORPHOLOGY) requiring an interpretation of site-specific landforms. In this context, bank typically refers to a sloping margin of a natural, stream-formed, alluvial channel (see CHANNEL, ALLUVIAL) that confines discharge during non-flood flow. Although the term bankfull stage can refer to various channel-bank levels, it generally applies to alluvial-stream channels (1) having sizes and shapes adjusted to recent fluxes of water and sediment, (2) that are principal conduits for discharges moving through a length of alluvial bottomland, and (3) that are bounded by FLOODPLAINS upon which water and sediment spill when the flow rate exceeds that of bankfull discharge. Thus, the concept of bankfull discharge, which often approximates the mean annual flood for perennial streams, includes the floodplain as a unique, identifiable geomorphic surface, all higher surfaces of alluvial bottomlands being terraces (see TERRACE, RIVER) (generally former floodplain surfaces), and acknowledgement that bankfull discharge occurs only when stream stage is at floodplain level.

Previous studies

Numerous discussions, only a few of which are cited here, have addressed the bankfull concept. A review by Williams (1978) documented a variety of published definitions for bankfull discharge and listed a wide range of flood frequencies related to the bankfull stage; almost without doubt the range resulted from observer misidentification of surfaces higher and lower than the floodplain as that of bankfull. Radecki-Pawlik (2002), among many others, showed that field determinations of bankfull stage and therefore floodplain level are interpretive, and thus the related bankfull discharge may differ greatly from that of the mean annual flood.

Papers by Woodyer (1968) and Osterkamp and Hupp (1984) recognized a variety of bottomland surfaces, including the floodplain, and related an approximate flow characteristic to each. Petit and Pauquet (1997) and Castro and Jackson (2001), respectively, determined return intervals for bankfull discharge of generally 1.2 to 3.3 years at sites of gravel-bed streams in France, and return intervals of 1.0 to 3.11 years for bankfull discharge at seventy-five stream sites of the north-western United States.

Significance

Bankfull discharge is significant owing to the hydraulic and related physical changes that occur for most alluvial stream channels when flow increases from in-channel to overbank conditions. Resistance to flow 'decreases with increasing water depth to reach a minimum at bankfull stage, so that the channel operates most efficiently with regard to water conveyance when the flow is at bankfull level' (Petts and Foster 1985: 150). Thus, the change in hydraulics as flow depth increases exerts a basic control on the geomorphic processes that are related to floodplain formation, regardless of the return period that may be associated empirically with the discharge at bankfull stage. The approximate height above the channel bed at which overbank flow begins is the level to which riverine bars develop (see BAR, RIVER), a process that can be described in terms of flow field, channel bathymetry and characteristics of sediment size and transport (Nelson and Smith 1989). Significant also is the observation that numerous data collected from perennial streams suggest that floodplain level is roughly equivalent to the stage of the mean annual flood, about 2.3 years (Wolman and Leopold 1957).

Floodplain formation

Floodplains typically form through POINT BAR deposition and, generally to a lesser extent, by deposition of sediment during overbank flows. Overbank deposits underlying floodplain and alluvial-terrace surfaces are typically poorly sorted and generally exhibit thinly bedded and alternating layers of silt, sand and possibly gravel. When a succession of floods causes overbank deposition, each flood elevating the surface higher above the channel, the deposits tend to grade from relatively coarse particles at the bottom upward to finer sizes. Because the thickness of overbank sediment deposited by large floods is generally small, averaging about 20 mm (Wolman and Leopold 1957), numerous episodes of overbank deposition are ordinarily needed to result in the accumulation of sediment on a gravel bar to a flood-plain level. AGGRADATION above flood-plain level is minimal owing to the infrequency of overtopping discharges, scour of flood-plain sediment by large floods and EROSION of accumulated deposits by lateral channel migration (Wolman and Leopold 1957). Nanson (1986), among others, suggested that processes resulting in the development of floodplains are influenced by prevailing conditions of energy (largely channel gradient) and the availability of sediment for entrainment. It follows, therefore, that many high-gradient streams have little potential for lateral ACCRETION (point-bar formation) and that flood-plain development occurs principally by vertical accretion.

Considerations and problems

The legitimate application of bankfull stage, bankfull discharge and floodplain to hydrologic and geomorphic studies requires accurate field determination of bankfull (flood-plain) level, which may be difficult if streamflow data are unavailable. Bankfull level is recognized easily along channels with point-bar deposits, especially if recent overbank deposits overlie point-bar sediment. For channels lacking point-bar features, interpretation of bankfull stage must rely on observations of channel morphology and gradient, bed and bank sediment, vegetation, root exposure and indications of flood processes.

The bankfull concept has been valuable as a means of describing clearly and effectively the processes and landforms of perennial-stream channels in humid areas. In recent decades, however, the bankfull concept has been so prevalent that it often has been overextended by Earth scientists who have confused the floodplain with other prominent alluvial landforms and corresponding flow frequencies. Thus, 'bankfull' should be used cautiously and consistently within the constraints by which it was described.

Common difficulties of the bankfull concept include its misapplication to non-alluvial conditions, such as streams incising till or debris-flow deposits, where fluvial adjustment is incomplete and bank height may be poorly related to flood frequency. Alluvial surfaces adjacent to high-energy (especially alpine and subalpine) streams, which typically correspond to the approximate stage of mean discharge (Osterkamp and Hupp 1984; Hupp 1986), commonly are misidentified as floodplain. As noted previously, the approximate correlation of bankfull stage (flood-plain level) and discharge with the mean annual flood pertains principally to perennial streams of moist areas; in drier regions with intermittent to highly ephemeral streams, the floodplain may correspond to floods with return periods of 100 years or more.

Flows smaller than bankfull discharge, those that occur more frequently than that of bankfull stage, typically cause in-channel features unrelated to bankfull discharge (such as BEDFORMS and bars). Processes and channel features resulting from these common events should not be confused with processes that correspond to bankfull discharge, and it should be recognized that all flows transport and sort sediment, thereby modifying the stream-channel morphology. Inappropriate emphasis on the bankfull concept has given rise to terms such as 'dominant discharge' and 'channel-forming discharge'. Such terminology, which focuses on a single flow rate, fails to recognize that all flows contribute to channel shape. As demonstrated by Wolman and Miller (1960), bankfull discharge, if related to geomorphic work accomplished, may indeed be dominant when applied to well-adjusted channels of perennial streams, but the dominance of a bankfull discharge becomes increasingly questionable as rates of precipitation and runoff decrease. Because all flows alter the shape of alluvial-stream channels, the term 'channel-forming discharge' may be inappropriate.

References

Castro, J.N. and Jackson, P.L. (2001) Bankfull discharge recurrence intervals and regional hydraulic geometry relationships: patterns in the Pacific Northwest, USA, *Journal of the American Water Resources Association* 37, 1,249–1,262.

Hupp, C.R. (1986) The headward extent of fluvial landforms and associated vegetation on Massanutten Mountain, Virginia, *Earth Surface Processes and Landforms* 11, 545–555.

Nanson, G.C. (1986) Episodes of vertical accretion and catastrophic stripping: a model of disequilibrium flood-plain development, *Geological Society of America Bulletin* 97, 1,467–1,475.

Nelson, J.M. and Smith, J.D. (1989) Evolution of erodible channel beds, in S. Ikeda and G. Parker (eds) *River Meandering*, 321–77, Washington, DC: AGU Water Resources Monograph 12.

Osterkamp, W.R. and Hupp, C.R. (1984) Geomorphic and vegetative characteristics along three northern Virginia streams, *Geological Society of America Bulletin* 95, 1,093–1,101.

Petit, F. and Pauqet, A. (1997) Bankfull discharge recurrence interval in gravel-bed rivers, *Earth Surface Processes and Landforms* 22, 685–693.

Petts, Geoff and Foster, Ian (1985) *Rivers and Landscape*, London: Edward Arnold.

Radecki-Pawlik, Artur (2002) Bankfull discharge in mountain streams: theory and practice, *Earth Surface Processes and Landforms* 27, 115–123.

Williams, G.P. (1978) Bank-full discharge of rivers, *Water Resources Research* 14, 1,141–1,154.

Wolman, M.G. and Leopold, L.B. (1957) *River Flood Plains: Some Observations on their Formation*, Washington, DC: US Geological Survey Professional Paper 282-C.

Wolman, M.G. and Miller, J.P. (1960) Magnitude and frequency of forces in geomorphic processes, *Journal of Geology* 68, 54–74.

Woodyer, K.D. (1968) Bankfull frequency in rivers, *Journal of Hydrology* 6, 114–142.

W.R. OSTERKAMP

BAR, COASTAL

Coastal bars can broadly be defined as aggradational ridges of sediments whose formation, morphology and behaviour are determined by interactions between WAVES, CURRENTS, tides, local slope and grain size. Some confusion exists regarding usage of the terms *bar* and *ridge*, however, since features such as BEACH RIDGES and CHENIER RIDGES are not normally considered bars. Furthermore, bars are found in BEACH, RIVER DELTA, ESTUARY and CONTINENTAL SHELF environments with a wide range of sizes, types and orientation. However, comparatively more COASTAL GEOMORPHOLOGY research studies have focused on bars which exist in the nearshore zone of sandy wave-dominated beaches.

Early studies (e.g. Shepard 1950) identified a seasonal cycle of beach morphology with winter storms promoting offshore sediment transport and bar formation and calmer conditions in summer favouring landward migration of the bar and eventual welding to the beach face. However, the existence of such 'winter' and 'summer' profiles is not universal as both barred and non-barred profiles occur at all times in some areas, while in others only one type may persist throughout the year. Furthermore, cyclic beach response at timescales much shorter than seasons can result in barred/non-barred profiles (Short 1979).

Cross-shore barred profiles are most commonly asymmetrical, having a distinct crest and a steeper landward slope than seaward slope (Figure 11a). Types of bars are often distinguished based on their alongshore planform shape and orientation relative to the shoreline. They may be linear (also referred to as longshore or shore-parallel; see Figure 11b), sinuous or crescentic (often termed

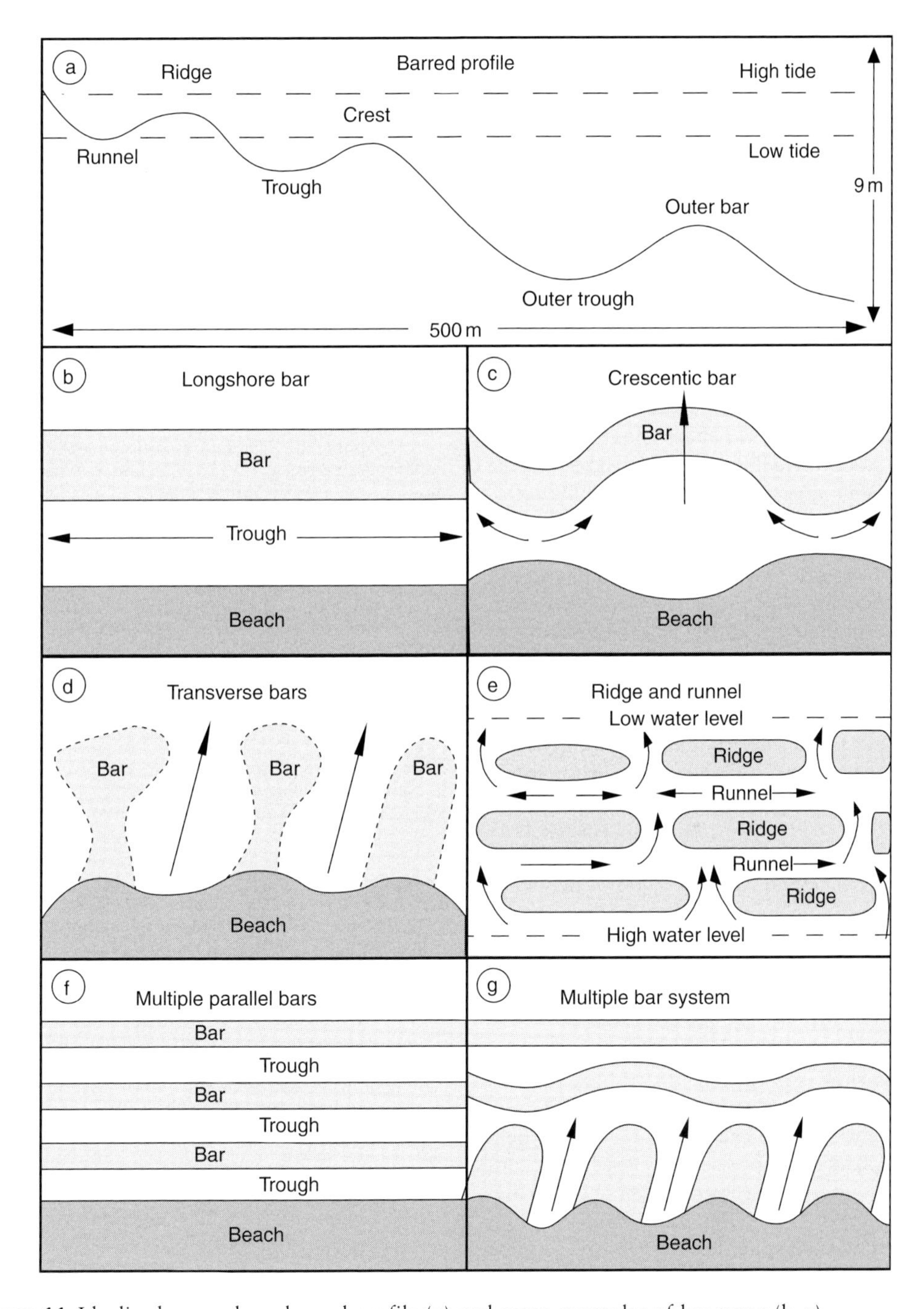

Figure 11 Idealized cross-shore barred profile (a) and some examples of bar types (b–g)

rhythmic) with a trough separating them from the shoreline (Figure 11c), or they may consist of alternating transverse bars, which are welded to the shoreline and are separated by channels occupied by RIP CURRENTs (Figure 11d). Bar type appears to be strongly related to wave energy level with linear bars developing under high-energy conditions, crescentic bars during intermediate energy, and transverse bars during lower wave energy levels. Under very low-energy conditions, a bar may become fully welded to the beach and appear as a flat terrace at low tide. These types of bar configurations are common on microtidal beaches and may grade into each other as energy levels vary. A number of classifications exist describing both bar types and the continuum of bar evolution (e.g. Greenwood and Davidson-Arnott 1979; Short and Aagaard 1993; Wijnberg and Kroon 2002).

Many beaches are characterized by the existence of multiple offshore bars (Figure 11f, g), ranging in number from two to over a dozen in some cases. Although high-energy swell wave environments can exhibit two or three bars, multi-barred profiles are most commonly developed in storm wave dominated sea and lacustrine environments where the outer bars are only active during short intense storms, remaining stationary during longer periods of low-wave energy. The number of bars seems to be related to the overall nearshore slope with lower gradients characterized by more bars. Both the spacing and size of bars has been observed to increase offshore. Bars are commonly absent on steep beaches.

Sandy beaches characterized by significant tidal ranges and low-energy conditions typically have gentle gradients and although some of the previously described bar types may be present, the presence of RIDGE AND RUNNEL TOPOGRAPHY (Figure 11e) in the intertidal zone is more common (Masselink and Anthony 2001). These exist as a series of low amplitude bars (ridges), which are usually stable in form and position, separated by subdued channels (runnels) associated with tidal drainage, and should not be confused with the multi-barred profiles described above.

As reviewed by Komar (1998) and Aagaard and Masselink (1999), some uncertainty remains regarding the formation of nearshore bars despite considerable theoretical, laboratory and field research. Bars develop as a result of sediment convergence and most mechanisms for their formation attempt to explain this. An early theory, that vortices under plunging breakers move sediment seaward forming a bar just seaward of the breakpoint, has largely been discounted since the breakpoint location on natural beaches with irregular waves varies considerably. It is more likely that sediment convergence results from onshore sediment transport outside the surf zone due to wave asymmetry and offshore transport in the surf zone due to bed return flow, with the bar forming somewhere near the breakpoint. Single and multiple bar formation has also been attributed to net sediment transport patterns associated with standing infragravity waves. According to this theory, if the sediment transport is predominantly bedload the bars will form at nodal positions, whereas if suspension dominates, they will form at antinodal positions (Bowen 1980). This theory has also proven useful in explaining rhythmic bar morphology (Holman and Bowen 1982) although nearshore cell circulation and rip currents are also important factors.

References

Aagaard, T. and Masselink, G. (1999) The surf zone, in A.D. Short (ed.) *Handbook of Beach and Shoreface Morphodynamics*, 72–118, Chichester: Wiley.

Bowen, A.J. (1980) Simple models of nearshore sedimentation; beach profiles and longshore bars, in S.B. McCann (ed.) *The Coastline of Canada*, 1–11, Geological Survey of Canada, Paper 80–10.

Greenwood, B. and Davidson-Arnott, R.G.D. (1979) Sedimentation and equilibrium in wave-formed bars: a review and case study, *Canadian Journal of Earth Sciences* 18, 424–433.

Holman, R.A. and Bowen, A.J. (1982) Bars, bumps and holes: models for the generation of complex beach topography, *Journal of Geophysical Research* 87, 457–468.

Komar, P.D. (1998) *Beach Processes and Sedimentation*, New Jersey: Prentice Hall.

Masselink, G. and Anthony, E.J. (2001) Location and height of intertidal bars on macrotidal ridge and runnel beaches, *Earth Surface Processes and Landforms* 26, 759–774.

Shepard, F.P. (1950) *Beach Cycles in Southern California*, US Army Corps of Engineers, Beach Erosion Board, Technical Memo No. 15.

Short, A.D. (1979) Three dimensional beach-stage model, *Journal of Geology* 87, 553–571.

Short, A.D. and Aagaard, T. (1993) Single and multi-bar beach change models, *Journal of Coastal Research*, Special Issue 15, 141–157.

Wijnberg, K.M. and Kroon, A. (2002) Barred beaches, *Geomorphology* 48, 103–120.

SEE ALSO: beach; beach sediment transport; current; ridge and runnel topography; wave

ROBERT W. BRANDER

BAR, RIVER

The nature and distribution of alluvial instream geomorphic units is fashioned by the interaction between unit stream power along a river reach and sediment calibre and availability. If a reach has excess energy relative to available sediment of sufficient size, flushing is likely to occur.

Alternatively, with excess sediment availability or insufficient flow energy, continuous instream sedimentation is likely to occur, commonly in the form of near-homogenous sheets. In most cases, rivers fall somewhere between these extremes, with transient sediment stores of differing calibre and bed material organization in differing landforms along the channel bed.

The most common instream geomorphic units are accumulations of deposits referred to as bars. These areas of net sedimentation of comparable size to the channels in which they occur are key indicators of within-channel processes. Interpretation of bar type is often critical in elucidation of river character and behaviour. There are two main components in bar form. The basal feature, or platform, is made up of coarse material and is overlain by supraplatform deposits of varying forms which is subject to removal and replacement during floods. Bars are readily reworked as channels shift position over the valley floor. Bank-attached features are much less likely to be reworked than mid-channel forms. The long-term preservation of bars is conditioned by factors such as the aggradational regime and the manner of channel movement.

Bars adopt many varied morphologies, ranging from simple unit bars (Smith 1970) to complex compound features (Brierley 1991, 1996). Bar character is controlled primarily by local-scale flow and grain size characteristics. Unit bars are simple features composed of one depositional style. The sediments of a unit bar (whether they be sand or gravel in composition) tend to fine in a downstream direction. As unit bars are found at characteristic locations along long profiles under particular sets of flow energy (stream power) and bed material texture relationships, a 'typical' down-valley transition in forms can be discerned (Church and Jones 1982). Bed material character, and the competence of flow to transport it, determine formation of longitudinal bars as flow divides around a tear-drop shaped feature. When flow is oriented obliquely to the long axis of the bar, a diagonal feature is produced. This is commonly associated with a dissected riffle. In highly sediment-charged sandy conditions, flow divergence results in transverse or linguoid bars, which extend across rather than down the channel (Collinson 1970; Cant and Walker 1978). Alternatively, the entire channel bed may comprise a homogenous sandsheet.

Instances in which patterns of sedimentation are dominated by within-channel bars reflect situations in which the material on the channel bed is either too coarse to be transported or the volume of material is too great to be transported. These scenarios are generally associated with gravel and sand bed systems respectively, such that competence and capacity limits are exceeded and flow divides around sediment stored in the channel zone.

In contrast to various mid-channel sedimentation features, rivers that are more readily able to accommodate their sediment load or have lower available energy are commonly characterized by bank-attached bars. Dependent on channel/flow alignment, lateral and POINT BARS are found at channel margins under both sand and gravel conditions. These features record sediment accretion on the convex slopes of river bends. Lateral bars tend to occur along straighter river reaches, while point bars are formed on bends. Scroll bars on the inside of bends may form a distinct element in themselves, while former positions of the channel may be recorded by a series of accretionary ridges and intervening swales (Nanson 1980). A range of bar forms have also been characterized for laterally constrained sinuous channels, such as point dunes (Hickin 1969), gravel counterpoint bars (Smith 1987) and convex bar deposits (Goodwin and Steidtmann 1981).

Most river bars are not simple unit features, but are complex, compound features made up of a mosaic of depositional forms such as bar platforms, ridges, chute channels, etc. Compound bars can be differentiated into mid-channel and bank-attached forms. On mid-channel compound bars, chute channels may dissect the bar surface into a chaotic pattern of remnant units. Variants of within-channel compound bar features in sand-bed channels include linguoid bars (Collinson 1970), macroforms (Crowley 1983), sand flats (Cant and Walker 1978), sand waves (Coleman 1969) and sandsheets (Smith 1970). Vegetated mid-channel compound bars are referred to as islands. The array of smaller-scale geomorphic units that make up an island provides key insights into its formation and reworking. On bank-attached compound bars a range of erosional and depositional features such as chute channels, ridges and ramps can be formed under varying flow conditions. Chute channels short-circuit the main body of flow in a river. Enlargement of the chute channel and plugging of the old channel proceed gradually, resulting in a chute cut-off. Because of the small angular difference between the old channel and the chute channel, the stream continues to flow through the old channel for some time, depositing bedload sediment at the upstream and downstream ends and on the floor and sides until terminal closure of the cut-off is complete. Ramp units are depositional forms that result from deposition of coarse gravels within a partially-filled chute channel. These features have a steep upstream facing surface that effectively plugs the chute channel, disconnecting it from the downstream outlet. These chute channel fills are notably straighter in outline than either meander cut-offs or swales.

In quite different environmental settings, bedrock accretionary forms occur on low slopes. These bedrock core bars are characterized by bedrock ridges atop which alluvial materials have been deposited during the waning stages of floods. A positive feedback mechanism is induced when vegetation colonizes these surfaces inducing further deposition and the vertical building of a bar feature. These features are common along bedrock-anastomosing rivers (e.g. Van Niekerk *et al.* 1999).

References

Brierley, G.J. (1991) Bar sedimentology of the Squamish River, British Columbia: definition and application of morphostratigraphic units, *Journal of Sedimentary Petrology* 61, 211–225.

——(1996) Channel morphology and element assemblages: A constructivist approach to facies modelling, in P. Carling, and M. Dawson (eds) *Advances in Fluvial Dynamics and Stratigraphy*, 263–298, Chichester: Wiley Interscience.

Cant, D.J. and Walker, R.G. (1978) Fluvial processes and facies sequences in the sandy braided South Saskatchewan River, Canada, *Sedimentology* 25, 625–648.

Church, M. and Jones, D. (1982) Channel bars in gravel-bed rivers, in R.D. Hey, J.C. Bathurst and C.R. Thorne (eds) *Gravel-bed Rivers: Fluvial Processes, Engineering and Management*, 291–338, Chichester: Wiley.

Coleman, J.D. (1969) Brahmaputra River: channel processes and sedimentation, *Sedimentary Geology* 3, 129–239.

Collinson, J.D. (1970) Bedforms of the Tana River: Norway, *Geografiska Annaler* 52A, 31–55.

Crowley, K.D. (1983) Large-scale bed configurations (macroforms), Platte River Basin, Colorado and Nebraska: primary structures and formative processes, *Geological Society of America Bulletin* 94, 117–133.

Goodwin, C.G. and Steidtmann, J.R. (1981) The convex bar: member of the alluvial channel side-bar continuum, *Journal of Sedimentary Petrology* 51, 129–136.

Hickin, E.J. (1969) A newly identified process of point bar formation in natural streams, *American Journal of Science* 267, 999–1,010.

Nanson, G.C. (1980) Point bar and floodplain formation of the meandering Beatton River, northeastern British Columbia, Canada, *Sedimentology* 27, 3–30.

Smith, N.D. (1970) The braided stream depositional environment: comparison of the Platte River with some Silurian clastic rocks, north central Appalachians, *Geological Society of America Bulletin* 81, 2,993–3,014.

Smith, S.A. (1987) Gravel counterpoint bars: examples from the River Tywi, South Wales, in F.G. Ethridge, R.M. Flores and M.D. Harvey (eds) *Recent Developments in Fluvial Sedimentology*, Society of Economic Paleontologists and Mineralogists Special Publication Number 39, 75–81.

Van Niekerk, A.W., Heritage, G.L., Broadhurst, L.J. and Moon, B.P. (1999) Bedrock anastomosing channel systems: morphology and dynamics in the Sabie River, Mpumalanga Province, South Africa, in A.J. Miller, and A. Gupta (eds) *Varieties of Fluvial Form*, 33–51, Chichester: Wiley.

SEE ALSO: channel, alluvial; fluvial geomorphology, point bar

KIRSTIE FRYIRS AND GARY BRIERLEY

BARCHAN

Barchan is an active crescentic-shaped dune (see DUNE, AEOLIAN), developing in areas of unidirectional winds and limited sand supply. Most barchans are arranged in belts, in which they follow each other. Such belts contain dunes of different sizes: small dunes, which do not exceed several tens of metres in length or width and a few metres in height, develop in a short time; and meso-dunes, which rise up to 40 m and attain several hundred metres in length or width. In various deserts such as in the Western Desert of Egypt (Embabi 1982), Qatar, Peru and California, dimensions of simple barchans show strong linear allometric relationships, such as between length of windward side and dune height, and dune length and width of horns (Plate 10). Mega-barchans that attain heights up to 120 m and lengths of 2–4 km are less common than small and meso-barchans, and are recorded in areas such as the northern parts of Rub' al-Khali in Arabia, and Taklamakan. Sand supply, wind environment, atmospheric motion and age are the main controlling factors of barchan size.

Slope form is concave–convex on the windward side of simple barchans, with angles varying between 1° and 10°. As the barchan grows in size, concave segments occupy a higher percentage of the total length of the windward side. The form of the leeward side changes from

Plate 10 Barchans are crescentic dunes, the horns of which point downwind. These examples are in the Western Desert of Egypt in the Kharga Depression

convex–concave to convex–straight to mostly straight when it acquires the angle of repose.

The internal structure of barchans reflects the dynamics of sand removal and deposition on both dune sides. The dominant structure is composed of thin steeply dipping cross-strata resulting from grainflow and grainfall deposited on the slip face, and is preserved as the dune migrates downwind. A secondary horizontal to low dipping structure develops due to deposition on the top of the dune. The sets of cross-strata are separated by horizontal to steeply dipping bounding surfaces.

Barchan moves in the downwind direction due to the dynamics of sand removal and deposition on dune sides. Sand is removed from the lower part of the windward side, and is deposited on the dune crest or on the leeward side. Accumulation of sand on the dune crest leads to periodic sand avalanches on the surface of the slip face. By time, sand removed from the windward side is deposited on the leeward side/slip face, resulting in dune advancement in the downwind direction. Average annual net migration of barchans varies between a few metres to 100 m. Wind energy, dune size and surface relief are the most significant controlling factors in dune advancement. As barchans move forward, they encroach on highways, railways, fields and buildings, representing a permanent hazard to all sorts of human activities, unless checked.

References

Embabi, N.S. (1982) Barchans of the Kharga Depression, in F. El-Baz and T.A. Maxwell (eds) *Desert Landforms of SW Egypt: A Basis for Comparison with Mars*, 141–155, Washington, DC: NASA.

Embabi, N.S. and Ashour, M.M. (1993) Barchan dunes in Qatar, *Journal of Arid Environment* 25, 49–69.

Lancaster, N. (1995) *Geomorphology of Desert Dunes*, London: Routledge.

SEE ALSO: dune, aeolian

NABIL S. EMBABI

BARRIER AND BARRIER ISLAND

Coastal barriers and SPITS are often regarded as similar coastal forms in terms of beach deposition projecting across coastal re-entrants/bays. While barriers tend to bridge the re-entrant by joining the mainland at each end, spits are only attached at one end. This distinction is not rigid, as many barriers show cross-barrier breaks or breaches through which the sea may enter on a permanent or intermittent basis, thus forming barrier segments or barrier islands. The degree of segmentation required to allow the 'island' nomenclature is not defined. Studies on the US eastern seaboard (Johnson 1925; Hoyt 1967; Kraft 1971; Leatherman 1979) have concentrated on barrier formation and reworking, though studies on barrier stability and defending barriers in the face of rising sea levels have come to the fore, pressured by the extensive and expensive real estate located on them (Titus 1990).

Coastal barriers are complex constructional morphology involving deposition by waves, wave-generated currents, tidal currents and wind activity (Hayes 1979). By morphological definition a barrier must exhibit two morpho-dynamic environments/units: a seaward beach face and a landward facing back-barrier slope (Plate 11). These two units are exposed most clearly when the barrier is gravel-dominated (Orford *et al.* 1996). As sand becomes the dominant sediment, a third environment comprising aeolian dunes can appear at the top of the beach face (barrier crest) and spread onto the backslope. Current flows may have been responsible for the initial submarine platform under the barrier, but over time and with sea-level rise, wave action forcing barrier onshore-migration generates much of the later barrier's basal-stratigraphy in combination with fine sedimentation characteristic of the low-energy back-barrier bay. Tidal currents become influential once barrier islands appear.

Plate 11 Sand-dominated coastal barrier, Long Island, New York State. Photo by permission of Whittles Publishing, Caithness

Some barriers along the US coast are thought to have evolved out of spits (Fisher 1967). However, other barrier origin mechanisms have been proposed that involve onshore rather than longshore sediment supply. Offshore shoals and longshore bars have been proposed as accretional cores to barriers, but as bars are now recognized as being beach face dependent, it is unlikely that bars can appear before the barrier defines the beach face. Roy *et al.* (1994) suggested that some Australian barriers emerged through shoreface accretion when sea level achieved stationarity after major early-Holocene fluctuations and the shelf area was reworked with excess sediment moving onshore. This latter position sees barriers accreting throughout their length, their plan-view position being set by wave refraction of constructional swell waves. Regardless of origin per se, barriers should be seen as multi-phase morphology relating to changes in controlling variables. Barriers should also be seen in terms of a palimpsest imposed by the structure and roughness of the terrestrial platform that they are superimposed on.

Sediment type and supply are major controls on barrier development with a behavioural distinction to be drawn between sand-dominated barriers (SDB) and gravel-dominated barriers (GDB). This distinction has a spatial basis with GDB being more prevalent in mid-upper latitudes compared to SDB, a dominance reflecting the greater potential of coarse material in high latitudes as a residue of late Quaternary glacigenic processes. SDB were often associated with low angle coastal plains, but barriers are potentially viable wherever sand sinks can be found, e.g. Holocene restructuring of deltas.

Barriers are generally seen as having a Holocene timescale, though this may be truncated to periods since the last major eustatic sea-level variation occurred. In particular many barriers around the North Atlantic have a history commensurate with the mid-Holocene decline in the rate of relative sea-level (RSL) rise. This emphasizes RSL as a datum control on wave activity and barrier development, and the rate of RSL change as a control on the tempo of barrier migration. Early work on the US east coast barriers identified their development with a Holocene TRANSGRESSION that swept up available sediment and concentrated it into the barriers. This perspective has been challenged in that, although it sounds intuitively correct, the actual initial mechanism for onshore concentration of sediment in the surf zone with rising RSL has not been verified, hence the switch to spit elongation as a more coherent model of barrier building in the face of rising RSL. An alternative perspective is to consider aggraded barriers as a consequence of falling RSL. This suggestion has been made for some Florida and Texas barriers, identifying a higher than present sea level during the mid to late Holocene, as a consequence of which barriers developed and aggraded during the subsequent regression. The lack of an obvious mechanism for barrier build-up during a transgression should not be confused with the more understandable behaviour of an existing barrier during subsequent transgressive phases. Jennings *et al.* (1998) suggest that the longshore coherency of GDB relates to the rate of RSL rise: slower rates ($<2\,mm\,a^{-1}$) mean reduced longshore sediment supply and the cannibalization of existing barrier segments to the point of barrier breaching. Barrier migration rates generally relate to RSL rise rate, though severely reduced GDB may be overwhelmed by the ambient RSL rise and flattened, to be rebuilt further onshore (i.e. overstepped; Forbes *et al.* 1991).

Wave climate is a major constraint on barriers. Many barriers are prominent in areas that are dominated by oceanic swell. These swells are likely to be transformed as constructional breakers, refracting parallel to existing shorelines and minimizing offshore sediment losses. This does not mean that barriers cannot emerge in storm-wave dominated areas, indeed such areas often show GDB, which tend to move onshore during storms, e.g. Atlantic Canada (Orford *et al.* 1996) and Patagonia (Isla and Bujalesky 2000). Storms are of crucial importance to barrier development. Most storms are expected to work at moving sediment offshore (gravel barriers aside), however as the severity of the storm goes upscale, then the emphasis of sediment transport switches from offshore to onshore. This threshold is reached when run-up physically reaches beyond the barrier crest and transports beach face material over the crest on to the back slope. This process is overwashing and its product is known as washover sedimentation. The latter is most obvious where overwash generates distinct flow channels (throats) through the crest and down the back slope ending in fan splay deposition beyond the previous back-barrier shoreline; such splays are the basis for barrier retreat. The position of throats and fans are partially dictated by the longshore gradients of the breaking wave and morphology of the barrier's seaward face. Beach faces showing cuspate morphology can preferentially set a rhythmic spacing to overwash, which in turn forces a consistent barrier retreat. As storm severity increases then the volume and depth of surface flows over the crest cause lateral extension of throats to the point of coalescence and the overt channels are lost in the face of generalized mobilization (sluicing overwash) of the barrier top into the back-barrier bay area. As sediment is washed into the back-barrier bay, it helps to build up a sub-aqueous sedimentation base for later marsh sedimentation (pads). These pads clearly help to fill in an accommodation space over which the barrier will migrate, such that the shallower the back-barrier area becomes the faster the potential for the barrier to migrate. Some controversy has been generated by the perceived influence of dunes on this migration (and hence survival) process, as dunes will block overwash, or spatially defuse the longshore consistency of overwash and hence reduce migration rate. This led to a short-lived management policy on the US barriers of advocating bulldozing the dune so as to promote overwash – scarcely a recipe for short-term barrier stability. The reverse is now in favour, that of promoting dune sedimentation as a 'sustainable' coastal defence.

Severe storms are also important for the development of cross-barrier breaches. US east coast barriers are vulnerable to hurricanes generating storm surge flow whose elevations are higher than the barrier crest. Hurricane overwash can back-up in the back-barrier bays, impounded by onshore surges, and flow laterally to escape; (1) through old breaches in the barrier; (2) along old overwash channels and (3) at any topographic low point on the barrier that erodes to form a new cross-barrier breach. These breaches can be quickly sealed up by post-hurricane beach face longshore sediment transport, or breaches may persist over decades. Sediment can be transported either way through breaches and surge deposits on the bay side can act as shallow water platforms for future barrier retreat. It is these overwash events that are responsible for most coarser sediment (i.e. high-energy) found in low-energy back-barrier environments.

Tidal range is considered as an indirect control of barrier development. As the tidal range expands, then so does the effectiveness of the tidal current flow regime. Increased tidal prisms maintain the storm-generated breaches. This tidal regime prevents post-storm breach healing and diverts longshore sediment into stores formed as flood or ebb deltas offset from the barrier. The larger the tidal range, the more breaches can be maintained in terms of hydraulic efficiency. Texas has one of the longest barriers in the USA – Padre Island is over 100 km in length and though subjected to hurricane attacks and suffering some breaches (mostly sealed) does not have a sufficient tidal prism (micro-tidal) to maintain the hurricane openings. New Jersey also has storm breaches with a low tidal range, but low longshore sediment availability reinforced by human interference is holding open breaches that would be closed elsewhere. South Carolina experiences a meso-tidal regime that maintains a greater longshore density of breaches or tidal passes, sufficient to define barrier islands. As a concomitant of inlet development and maintenance, there are substantial flood and ebb deltas linking the barrier islands into a hydraulically efficient network conditioned by tidal prism. If the prism alters due to back-barrier reclamation then inlet dimensions will alter, e.g. the Friesian Islands, German Wadden Sea (FitzGerald *et al.* 1984). The ebb/flow characteristics may be conditioned by saltmarsh growth within the bay, acting as a retardant to balanced flood/ebb tidal flow asymmetry. It is rare to find barriers in macro-tidal ranges, but when they do occur (north Norfolk, England) the forcing of the barrier is more to do with longshore sediment supply than tidal inlet forcing, however subsequent evolution in the face of diminishing sediment may mean that the potential for segmentation is great and the apparent morphology of barrier islands may be superimposed.

References

Fisher, J.J. (1967) Origin of barrier island chain shorelines, Middle Atlantic states (abs.) *Geological Society of America*, Special Paper 115, 66–67.

FitzGerald, D.M., Penland, S. and Nummedal, D. (1984) Changes in tidal inlet geometry due to back-barrier filling: East Friesian islands, West Germany, *Shore and Beach*, 52, 3–8.

Forbes, D.L., Taylor, R.B., Orford, J.D., Carter, R.W.G. and Shaw, J. (1991) Gravel barrier migration and overstepping, *Marine Geology* 97, 305–313.

Hayes, M.O. (1979) Barrier island morphology as a function of tidal and wave regime, in S.P. Leatherman (ed.) *Barrier Islands*, New York: Academic Press.

Hoyt, J.H. (1967) Barrier island formation, *Geological Society of America Bulletin* 78, 1,125–1,136.

Isla, F.I. and Bujalesky, G.G. (2000) Cannibalisation of Holocene gravel beach-ridge plains, northern Tierra del Fuego, Argentina, *Marine Geology* 170, 105–122.

Jennings, S.C., Orford, J.D., Canti, M., Devoy, R.J.N. and Straker, V. (1998) The role of relative sea-level rise and changing sediment supply on Holocene gravel barrier development; the example of Porlock, Somerset, UK, *Holocene* 8, 165–181.

Johnson, D.W. (1925) *The New England Arcadian Shoreline*, New York: Wiley.

Kraft, J.C. (1971) Sedimentary facies pattern and geologic history of Holocene marine transgression, *Geological Society of America Bulletin* 82, 2,131–2,158.

Leatherman, S.P. (ed) (1979) *Barrier Islands*, New York: Academic Press.

Orford, J.D., Carter, R.W.G. and Jennings, S.C. (1996) Control domains and morphological phases in gravel-dominated coastal barriers, *Journal of Coastal Research* 12, 589–605.

Roy, P.S., Cowell, M.A., Ferland, M.A. and Thom, B.G. (1994) Wave dominated coasts, in R.W.G. Carter and C.D. Woodroffe (eds) *Coastal Evolution: Late Quaternary Shoreline Morphodynamics*, 121–186, Cambridge: Cambridge University Press.

Titus, J. (1990) Greenhouse effect, sea-level rise and barrier island: case study of Long Beach Island New Jersey, *Coastal Management* 18, 65–90.

JULIAN ORFORD

BASE LEVEL

The concept that there is an effective lower limit to erosional processes. Powell (1875) first named the concept: 'we may consider the level of the sea to be a grand base level, below which the dry lands cannot be eroded; but we may also have, for local and temporary purposes, other base levels of erosion, which are the levels of the beds of the principal streams which carry away the products of erosion.' Chorley and Beckinsale (1968) recognized four main interpretations of the term.

1. Grand base level or 'ultimate base level' which is the plane surface forming the extension of the sea under the lands.
2. Temporary or structural base level, whereby there is a limit to downward erosion of an ephemeral character imposed headwards of a resistant outcrop.
3. Base-levelled surface, which is an ultimate or penultimate topographic surface.
4. Local base level, as for example in areas of interior drainage under an arid cycle.

The first of these usages occupied a fundamental place in the cycle of erosion concept of W.M. Davis (1902). Base-level changes are also crucial in the study of fluvial terraces, deltas and other depositional systems (Koss *et al.* 1994). They can result from tectonic activity, sea-level change and river capture (Mather 2000).

References

Chorley, R.J. and Beckinsale, R.P. (1968) Base-level, in R.W. Fairbridge (ed.) *The Encyclopedia of Geomorphology*, 58–60, New York:Reinhold.

Davis, W.M. (1902) Base-level, grade and peneplain, *Journal of Geology* 10, 77–111.

Koss, J.E., Ethridge, F.G. and Schumm, S.A. (1994) An experimental study of the effect of base-level change on fluvial, coastal plain and shelf systems, *Journal of Sedimentary Research* B, 64, 90–98.

Mather, A.E. (2000) Adjustment of a drainage network to capture induced base-level change. An example from the Sorbas Basin, S.E. Spain, *Geomorphology* 34, 271–289.

Powell, J.W. (1875) *Exploration of the Colorado River of the West*, New York.

A.S. GOUDIE

BEACH

A beach is a wave deposited accumulation of sediment located between modal wave base and the upper swash limit. Beaches may be composed of fine sand through boulders and may range from low energy, narrow strips of sand lapped by low wind waves, to high energy systems exposed to persistent 2–3 m high swell which breaks across 500 m wide surf zones. Beach systems are also located in all tide ranges, in all latitudes, in all climates and on all manner of coast, from the low coastal plains fronted by long beaches to small pockets of sand wedged in at the base of massive cliffs. Therefore beaches exist in a wide spectrum of wave, tide and sediment combinations and geological settings.

In two dimensions, however, all beaches will contain three dynamic zones – of wave shoaling, wave breaking and swash–backwash. The wave shoaling zone extends from the modal wave base where average waves can entrain and move sediment shoreward, to the outer breakpoint. The wave shoaling zone is dominated by asymmetrical wave orbital motions which produce a concave upward profile sloping less than 1°, dominated by wave ripples which become increasingly three dimensional close to the breaker zone. The depth and width of this zone is dependent on wave height and sediment size. On high energy coasts it extends out to depths of 30 m or more which may lie 2–3 km offshore, while on low energy systems it may only extend to low tide a few metres from the shore. Sediment is often graded across the system and fines seaward.

The surf zone is located between the breakpoint and shoreline. The surf zone has the greatest potential for complex dynamic processes and resulting topography and bedforms. The dominant processes start with wave breaking. Onshore currents are generated by wave bores and orbital currents associated with reformed waves. Longshore currents result from oblique waves and bi-directional rip feeder currents. Offshore flows are driven by wave reflection, discrete rip currents and bed return flow. The width of the surf zone is dependent on surf zone gradient, a function of sand size and wave height. It will be as narrow as a few metres on a steep reflective beach, typically 50–100 m wide on a single bar intermediate beach and up to several hundred metres on a high energy dissipative beach. The presence and number of bars will increase with decreasing wave period and decreasing gradient. Longshore variations in form and processes are driven by longshore changes in wave conditions, and by three-dimensional bar and rip topography.

Surf zone topographic features include shore parallel bars and troughs with waves breaking over the bars and reforming in the troughs. Swell coasts rarely have more than two bars (see BAR, COASTAL), while energetic sea coasts may have several bars. On intermediate beaches with cellular rip circulation, alternating shore transverse bars and rips, also known as crescentic bars, can occur. When present they dominate the inner bar and can lead to a rearranging of the shoreline to produce rhythmic topography. Surf zone BEDFORMs reflect the changing velocity and direction of currents and depth of water, and range from flat bed over the shallow bars, to wave orbital and shore perpendicular current ripples in the troughs, to shore parallel seaward migrating ripples in the rip channels.

The swash zone extends up the beach from the shoreline, from where the wave breaks or bore collapses, to the limit of swash. The swash zone is always an upward sloping zone of wave uprush (swash) and backwash. Its slope is directly related to grain size and inversely related to wave height and may range from 1° to 10°. It is usually featureless, with ripples only produced by strong backwash in the lower swash zone. The swash may, however, be superimposed on high tide beach cusps or a berm, and mesoscale megacusps. Beyond the limit of normal swash is the backshore, a zone of either wind-blown aeolian deposits and/or storm wave overwash.

In three dimensions beaches respond to a greater range of variables and become increasingly complex. First is the alignment of the shoreline to the dominant wave crest, which produces swash aligned beaches. As waves refract around headlands and nearshore topography, the wave crests bend to parallel the contours. At the shoreline the beach also moves to parallel the wave crests so as to minimize longshore transport, and thereby produce a more stable shoreline in equilibrium with the wave crest. Where waves arrive at a persistent oblique angle to the shore, particularly on long beaches, then sediment is moved downdrift by the longshore surf zone currents generated by the waves, producing a drift aligned shore.

Beach type

Beach type refers to the morphodynamic character of a beach system, which is a product of the interaction of waves, tide and sediment. Beaches may be of three types: wave dominated, tide modified and tide dominated. Wave-dominated beaches occur where waves are high relative to the tide range. This can be defined quantitatively by the relative tide range

$$RTR = TR/H_b \quad (1)$$

where TR is the spring tide range and H_b the average breaker wave height. When $RTR < 3$ beaches are tide dominated, when $3 < RTR < 15$ they are tide modified and when the $RTR > 15$ they become tide dominated.

Within each of these beach types a range of wave and sediment combinations can occur which will influence the actual state of the beach. The dimensionless fall velocity

$$\Omega = H_b/T\,W_s \quad (2)$$

where T is wave period (s) and Ws sediment fall velocity ($m\,s^{-1}$) can be used to quantify beach state. They range from the lower energy reflective ($\Omega < 1$) favoured by low waves, longer periods and coarser sediments, to dissipative ($\Omega > 6$) with high waves, shorter periods and fine sand. In between are the more rip dominated intermediate beaches ($\Omega = 2$–5) produced by moderate to high wave conditions.

WAVE-DOMINATED BEACHES

Wave-dominated beaches consist of three types: reflective, intermediate and dissipative.

Reflective beaches

Reflective beaches are produced by combinations of lower waves, longer periods and particularly coarser sands. They occur on sandy open swell coasts when waves average less than 0.5 m, and on all coasts when beach sediments are composed of coarse sand or coarser, including all gravel and boulder beaches, even under higher waves. However, they are all characterized by a concave upward nearshore zone of wave shoaling that extends to the shoreline. Waves then break by plunging and/or surging across the base of the beach face. The ensuing strong swash rushes up the beach, combining with the coarse sediment to build a steep beach face (4°–10°), commonly capped by well-developed beach cusps and/or a *berm* (Plate 12), possibly backed by a runnel where the high tide swash may temporarily accumulate. When the sediment consists of a range of grain sizes, the coarser grains accumulate as a coarser steep *step* below the zone of wave breaking, at the base of the beach face.

The cusps are a product of cellular circulation on the high tide beach resulting from sub-harmonic edge waves produced from the interaction of the incoming and reflected backwash. The high degree of incident wave reflection off the beach face is responsible for the naming of this beach type, i.e. reflective. Apart from the cosmetic beach cusps and swash circulation these are essentially two-dimensional beaches with no longshore variation in either processes or morphology. On sand beaches they also represent the lower energy end of the beach spectrum and as such are relatively stable systems only responding to an increase in wave height. Such an increase induces a growth in the swash energy and erosion of the swash zone.

Plate 12 A lower energy reflective beach with wave surging up the moderately steep beach, Horseshoe Bay, South Australia (Andrew D. Short)

Intermediate beaches

Intermediate beaches are called such as they represent a suite of beach types between the lower energy reflective and higher energy dissipative. They are the beaches that form under moderate to high waves, on swell and sea coasts, in fine to medium sand. The two most distinguishing characteristics of intermediate beaches are (1) a surf zone, and (2) cellular rip circulation (see RIP CURRENT) commonly associated with rhythmic bar and beach topography (Plate 13). Since intermediate beaches can occur across a wide range of wave conditions, they consist of four beach states ranging from the lower energy low tide terrace to the rip dominated transverse bar and rip and rhythmic bar and beach, and the high energy straighter longshore bar and trough.

Intermediate beaches are controlled by processes related to wave dissipation across the surf zone which transfers energy from incident

Plate 13 Well-developed transverse bar and rips, Lighthouse Beach, New South Wales, Australia (Andrew D. Short)

waves with periods of 2 to 20 s, to longer infragravity waves with periods >30 s. As a consequence, incoming long waves associated with wave groupiness, increase in energy and amplitude across the surf zone and are manifest at the shoreline as wave set up (crest) and set down (trough). The long wave then reflects off the beach leading an interaction between the incoming and outgoing waves to produce a standing wave across the surf zone. It is believed that standing edge waves trapped in the surf zone are responsible for the cellular circulation that develops into rip current circulation and associated transverse bars and rips. These are in turn responsible for the high degree of spatial and temporal variability in intermediate beach morphodynamics.

The low tide terrace (or ridge and runnel) beaches are characterized by a continuous attached bar or terrace located at low tide. They form under lower waves (0.5–1 m) and usually undergo temporal variation between low tide when the waves break and dissipate across the bar, while at high tide they may remain unbroken and surge up the reflective high tide beach face. Weak rips may occur at mid to low tide.

Transverse bar and rip beaches form under moderate waves (1–1.5 m) on swell coasts and consist of well-developed rip channels, which are separated by shallow bars, the bars attached and perpendicular or transverse to the beach (Plate 13). Variable wave breaking and refraction across the shallow bars and deeper rip channels lead to a longshore variation in swash height and approach, which reworks the beach to form prominent megacusp horns in lee of bars, and embayments in lee of channels. Water tends to flow shoreward over the bars, then into the rip feeder channels. The flow moves close to the shoreline and converges laterally in the rip embayment. It then moves seaward in the rip channel as a relatively narrow (few metres), strong flow (0.5–1 m s^{-1}), called a *rip current*. This beach state has extreme spatial-longshore variation in wave breaking, surf zone and swash circulation and beach and surf zone topography, leading to a highly unstable and variable beach system.

The rhythmic bar and beach state forms during periods of moderate to high waves (1.5–2 m) on swell coasts. The high waves lead to greater surf zone discharges that require deeper and wider rip feeder and rip channels to accommodate the flows. Rips flow in well-developed rip channels, separated by transverse bars, however the bars are detached from the beach by the wider feeder channels.

The longshore bar and trough systems are a product of periods of higher waves which excavate a continuous longshore trough between the bar and the beach. Waves break heavily on the outer bar, reform in the trough and then break again at the shoreline, often producing a steep reflective beach face (coarser sand) or low tide terrace (finer sand). Surf zone circulation consists of both cellular rip flows as well as increasingly shore normal bed return flows (see below).

Dissipative beaches

Dissipative beaches represent the high-energy end of the beach spectrum. They occur in areas of high waves, prefer short wave periods, and must have fine sand. They are relatively common in exposed sea environments where occasional periods of high, short storm waves produce multi-barred dissipative beach systems, as in the North and Baltic seas. They also occur on high-energy mid-latitude swell and storm wave coasts as in northwest USA, southern Africa, southern Australia and New Zealand. On swell coasts waves must exceed 2–3 m for weeks to generate fully dissipative beaches. They are characterized by a wide long gradient beach face and surf zone, with two and more shore parallel bars forming across the surf zone (Plate 14). The

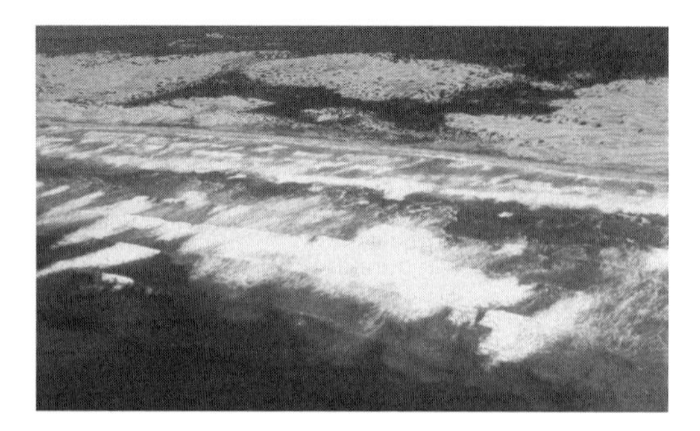

Plate 14 High energy dissipative beach containing an inner bar, trough and wide outer bar, Dog Fence Beach, South Australia (Andrew D. Short)

low gradient is a product of both the fine sand, as well as the dominance of lower frequency infragravity swash and surf zone circulation, which act to plane down the beach. The name comes from the fact that waves dissipate their energy across the many bars and wide surf zone.

The dissipation of the incident wave energy leads to a growth in the longer period infragravity energy, which becomes manifest as a strong set up and set down at the shoreline. The standing wave generated by the interaction of the incoming and outgoing waves may have two and more nodes across the surf zone. It is believed that the bar crests form under the standing wave nodes and troughs under the antinodes. Surf zone circulation is vertically segregated. Wave bores move water shoreward toward the surface of the water column. This water builds up against the shoreline as wave set up. As it sets down the return flows tends to concentrate toward the bed, which propels a current across the bed of the surf zone (below the wave bores) called *bed return flow*.

Like the reflective end of the beach spectrum dissipative beaches are remarkably stable systems. They are designed to accommodate high waves, and can accommodate still higher waves by simply widening their surf zone and increasing the amplitude of the standing waves, while periods of lower waves are often too short to permit substantial onshore sediment migration.

TIDE-MODIFIED BEACHES

Most of the world's beaches are affected by tide. On most open coasts where tides are low (<2 m) waves dominate and tidal impacts are minimized. However, as tide range increases and/or wave height decreases tidal influences become increasingly important. To accommodate these influences beaches, still by definition wave-formed, can be divided into tide-modified and tide-dominated types, as defined by equation 1 (see p. 64).

The major impact of increasing tide range is to shift the location of the shoreline between high and low tide, which – depending on the shoreline gradient – will be tens to hundreds of metres. This shift not only moves the shoreline and accompanying swash zone, but also the surf zone, if present, and the nearshore zone. While wave-dominated beaches have a relatively 'fixed' swash–surf–nearshore zone, on tide-modified beaches they are more mobile and transient. The net result is a smearing of the three dominant wave processes of shoaling (nearshore zone), breaking (surf zone) and swash (swash zone). A section of intertidal beach can be exposed to all three processes at different states of the tide. Second, because all three zones are mobile, except for a brief period at high and low tide, there is a reduction in the time any one process can fully imprint its dynamics on a particular part of the beach. As a consequence there is a tendency for swash processes only to dominate the spring high tide beach, for surf zone processes only to dominate the beach morphology around low tide, during the turn of the low tide, while shoaling wave processes become increasingly dominant overall, producing a lower gradient, featureless, concave beach cross section.

Tide-modified beaches can contain three states. When waves are lower ($\Omega < 1–2$) they consist of a steep reflective high tide beach face fronted by a wide low gradient low tide terrace, often with a sharp break in slope between the two. At low tide waves dissipate across the terrace, while at high tide they pass unbroken across the now submerged terrace to surge up the steep reflective high tide beach. In areas of moderate waves ($\Omega = 2–5$) and tide range (RTR = 3–7) the tide-modified beaches consist of a high tide reflective beach, a usually wider intertidal zone, and a low tide zone dominated by surf zone morphology, which may include transverse and rhythmic bars and rips. In moderate energy sea environments a series of shore parallel ridges and runnels may develop (Plate 15). Higher energy tide-modified beaches ($\Omega > 5$) composed of fine sand are characterized by a wide, low gradient concave upward, flat and featureless, beach and intertidal system, called an ultradissipative beach (Plate 16).

Plate 15 Reflective high tide beach (foreground) fronted by three ridges and runnels, Omaha Beach, France (Andrew D. Short)

Plate 16 Rhossili Beach, Wales, a high energy ultradissipative tide-modified beach, shown here at low tide (Andrew D. Short)

TIDE-DOMINATED BEACHES

Tide-dominated beaches occur when the RTR > 15, that is the tide range is more than 15 times the wave height. As the maximum global tide range is about 12 m, and usually much less, this means that most tide-dominated beaches also receive low (< 1 m) to very low waves, and are commonly dominated by locally generated wind waves. Tide-dominated beaches are characterized by a usually steep, narrow high tide beach, and a wide, low gradient (< 1°) sandy intertidal zone, which in temperate to tropical locations is usually bordered by subtidal seagrass meadows. They consist with decreasing wave energy of three states: (1) a beach and sand ridges, containing multiple, low amplitude, shore parallel sand ridges across the intertidal; (2) a reflective beach fronted by a usually wide, flat, featureless intertidal sand flat; (3) a tidal sand flat, which is a beach in so far as it has a high tide beach. However, the fronting often wide tidal flats are dominated by the tides and not waves, and may contain intertidal biota and tidal drainage features. It is part of the transition between the beaches and the often muddy, tidal flats.

Further reading

Carter, R.W.G. (1988) *Coastal Environments*, London: Academic Press.
Hardisty, J. (1990) *Beaches: Form and Process*, London: Unwin and Hyman.
Horikawa, K. (ed.) (1988) *Nearshore Dynamics and Coastal Processes*, Tokyo: University of Tokyo Press.
Komar, P.D. (1998) *Beach Processes and Sedimentation*, 2nd edition, Upper Saddle River, NJ: Prentice Hall.
Short, A.D. (ed.) (1999) *Handbook of Beach and Shoreface Morphodynamics*, Chichester: Wiley.

SEE ALSO: bar, coastal; bedform; rip current

ANDREW D. SHORT

BEACH CUSP

Beach cusps are crescentic concave-seaward regularly spaced features occurring along the shoreline. The term has been used for features with spacing ranging from 10 cm to many hundreds of metres, although larger examples tend to be called rhythmic beach features with the term swash cusp being used for features with a spacing less than tens of metres (Hughes and Turner 1999). Beach cusps (swash cusps) are most commonly associated with medium to coarse sands, shingle or mixed sand-shingle sediments, on steep beaches demonstrating significant wave reflection. Their amplitude ranges from almost zero to over 1 m. On beaches with high tidal range, multiple sets of cusps may be present at different levels. Beach cusps consist of embayments or swales separated by triangular horns which are normally comprised of coarser sediments.

A number of different swash flow patterns in and around cusps have been reported in the literature. Under low energy conditions, oscillatory flows (with swash largely unaffected by cusp morphology), horn divergent flows (uprush flow separation at the horn with water returning from the embayment), and horn convergent flows (uprush entering the cusp in the embayment and returning along the sides of the horn, converging at its apex) have all been reported. Under high energy conditions sweeping flow (alongshore directed water movement) and swash-jet flows (where incoming swash is held back by the backwash until it develops sufficient head to break through as a jet in the centre of the embayment) can also occur (Masselink and Pattiaratchi 1998).

Numerous conflicting ideas have been proposed for the mode of formation of beach cusps, including processes of accretion, processes of erosion or a combination of both, and theories based on instabilities on wave breaking, alongshore sediment transport, and intersecting wave trains. Debate continues on the formation of beach cusps, with two theories based on fundamentally different mechanisms dominating.

Cusp development caused by standing edge WAVES at either twice the period (subharmonic) of, or synchronous with the incident waves has been proposed. This hypothesis can explain regular spacing (cusp spacing equal to edge wave length for synchronous edge waves, or half edge wave length for sub-harmonic edge waves), but can also explain complex quasi-regular spacing if more than one mode of edge wave is present. Werner and Fink (1993), however, proposed a self-organization model for beach cusp formation, with topographic depressions resulting in feedback mechanisms between swash and morphology resulting in self-emergence of beach cusps. Cusp spacing is proposed as being proportional to swash excursion length. Predications of similar beach cusp spacing based on the quite distinct self-organization and edge wave theories of cusp formation have meant that field studies have been unable to discriminate between the two models. It is also possible that edge waves may initiate cusp development with self-organization then allowing for the growth of the features. Many field studies have reinforced the importance of feedback processes between swash and morphology (Masselink *et al.* 1997).

References

Hughes, M. and Turner, I. (1999) The beachface, in A.D. Short (ed.) *Handbook of Beach and Shoreface Morphodynamics*, 119–144, Chichester: Wiley.

Masselink, G. and Pattiaratchi, C.B. (1998) Morphological evolution of beach cusps and associated swash circulation patterns, *Marine Geology* 98, 93–113.

Masselink, G., Hegge, B.J. and Pattiaratchi, C.B. (1997) Beach cusp morphodynamics, *Earth Surface Processes and Landforms* 22, 1,139–1,155.

Werner, B.T and Fink, T.M. (1993) Beach cusps as self-organized patterns, *Science* 260, 968–971.

Further reading

Komar, P.D. (1998) *Beach Processes and Sedimentation*, 2nd edition, Upper Saddle River, NJ: Prentice Hall.

SEE ALSO: beach; wave

KEVIN PARNELL

BEACH–DUNE INTERACTION

The original generation of the wave–beach–dune model of beach and dune interactions was formulated by Hesp (1982). It was followed by the publication of a robust micro-tidal beach model with reasonably high predictability (Wright and Short 1984). The beach model enabled scientists to classify micro-tidal beaches into six states with characteristic morphologies, mobility, and modes of erosion and accretion. Subsequent research has extended the original model to meso- and macro-tidal beaches (see BEACH). An understanding of beach and backshore morphology for different surfzone-beach types allowed Hesp (1982) to develop actual and theoretical links between backshore morphology, potential aeolian transport, foredune state and morphology, and dune-field type and development.

Surfzone-beach state

The micro-tidal beach models classified beaches into six states, with the dissipative state at the high wave energy (>2.5 m) extreme and reflective state at the low wave energy (<1 m) extreme. Four intermediate beach states occur between these states. Dissipative beaches are characteristically high wave energy beaches and have the highest potential onshore sediment supply (Hesp 1988). Note, however, that beaches may also be dissipative because of the presence of very fine sand (hence low gradient) or abundant sand, so some dissipative beaches may, in fact, be low wave energy beaches. They are typically wide, display flat to concave morphologies (no berms), low gradients and minimal backshore mobility. The latter refers to the coefficient of variation of mean shoreline position, and in reality refers to the amount of volumetric and profile change the beach and backshore experiences over time and through erosion to accretion phases. Reflective beaches are characteristically low wave energy beaches with low potential onshore wave driven sediment transport. Note that they may also be moderate to high wave energy where sediments are coarse sand to boulders. They are relatively steep, narrow, linear to terraced (i.e. display a berm form) morphologies, with low backshore mobility. Intermediate beaches range from wide, relatively flat beaches with low mobility at the higher energy end of the spectrum, through moderate width beaches with pronounced berms and high mobility to narrow beaches and moderate to low mobility berms at the reflective end of the range. Rips dominate surfzone processes in the intermediate range.

Beach-backshore width and morphology, fetch and potential aeolian transport

Beach width is important in determining fetch which is critical for determining the volume of sand delivered across the backshore and to dunes (Davidson-Arnott 1988). Beach morphology is important because the greater the morphological variability, the more likely that wind velocity decelerations and variations take place across the backshore. Hesp (1982, 1999) showed that the wind flow across a wide, low gradient, dissipative beach displayed minimal flow variation and gradually accelerated across the backshore, thus maximizing potential aeolian transport. The wind flow over the berm crest of an intermediate beach was accelerated but decelerated leeward of the berm crest. High narrow berms typical of some reflective beaches display significant flow disturbance and deceleration leeward of the berm crest (Short and Hesp 1982). Sherman and Lyons (1994) modelled wind flow and potential sediment transport on a flat beach, low berm and high berm profiles, and found that sand transport off the dissipative beach was 20 per cent higher than off the reflective beach if just slope and grain size were taken into account. When moisture content was added, transport rates were nearly two orders of magnitude higher off the dissipative beach compared to the reflective beach. Note, however, that each beach had the same width (100 m wide), whereas actual reflective beaches and many intermediate beaches are considerably narrower.

Beach mobility is important because the greater the beach mobility, the greater the morphological variability. The latter affects the fetch such that the beach width is at times quite narrow, at times quite wide, particularly for intermediate beaches. It is also important because alternating episodes of cut-and-fill result in varying beach morphologies which then affect airflow and sediment transport as indicated above.

Thus, the link between surfzone beach state, aeolian sediment transport and landward dunes is that modal dissipative beaches have maximum potential aeolian sediment transport, reflective beaches minimal potential aeolian sediment transport, and intermediate beaches range from relatively high potential at the dissipative end to low potential at the reflective end. Note that a minimal sediment supply ('minimal' is currently undetermined) is required.

Aeolian sediment transport and foredune morphology

An examination of foredune heights and volumes on dissipative to reflective beaches allows one to examine the validity of the links above. Since established foredunes occupy a foremost backshore position, they are a medium-term indicator of beach and backshore processes. Hesp (1988) measured incipient and established foredune volumetric changes over several years at Myall Lakes National Park in New South Wales, Australia to find that a modal reflective beach with the same wind exposure as a modal dissipative beach received 60 per cent less sand than the dissipative beach over the same survey period. Intermediate beach volumes ranged from relatively high to relatively low between the dissipative and reflective beaches.

Surveys of established foredunes, which have been present for potentially several hundred years, provide further evidence that there is a strong link between surfzone-beach type and foredune height and volume. Hesp (1982, 1988) demonstrated that in the Myall Lakes National Park the smallest established foredunes, with lowest sediment volumes were found on reflective beaches, while the highest and largest foredunes occurred on dissipative beaches. Similar results are reported by Davidson-Arnott and Law (1990). Intermediate beaches followed a trend from low to high volumes on lowest to highest energy intermediate beaches respectively (see reviews in Sherman and Bauer 1993 and Bauer and Sherman 1999).

Foredune ecology

The vegetation cover, species richness and zonation of foredunes is determined by several factors, but sediment supply and sand deposition rate, and salt spray aerosol levels are two important factors (Hesp 1991). Simultaneous studies carried out on adjoining reflective, intermediate and dissipative beaches show that salt spray aerosol levels are related to surfzone-beach type. Dissipative beaches have the widest surfzones, the greatest number of breaking waves, and highest wave heights and the highest salt aerosol levels. Reflective beaches often have only one breaking wave, narrow to very narrow surfzones, and low wave heights and the lowest salt aerosol levels. All other factors being equal, foredune species richness and zonation tends to be greatest and narrowest respectively on reflective beaches (low sediment supply and salt aerosols), and lowest and widest on dissipative beaches (highest sediment supply, high salt aerosol levels) (Hesp 1988).

Foredune stability and type, erosion processes and dunefield development

Foredunes bear a morphological imprint dictated, in part, by modal surfzone-beach erosion and accretion modes, and the wind often accentuates this morphological imprint. Dissipative beaches are typically eroded by swash bores and undertow commonly associated with elevated water levels and storm surge. Beach erosion and dune scarping is laterally continuous alongshore, and at times catastrophic. Hesp (1988) and Short and Hesp (1982) theorized that such laterally continuous alongshore, large-scale foredune scarping would on occasions lead to large-scale foredune destabilization. Transgressive dunefields would most likely result from the breakdown of the large established foredune. In fact, transgressive dunefields are most commonly found on high energy dissipative surfzone-beach systems (e.g. Australian and South African coasts below the tropics; west coast USA; west coast North Island, New Zealand).

Intermediate beaches are characterized by localized, arcuate rip embayment erosion during storms. Such arcuate erosion extends well into the foredune during extreme events resulting in large-scale but localized foredune scarping. Topographic funnelling of the wind may result in the evolution of blowouts and eventually parabolic dunes at these locations. On average, many higher energy intermediate beaches display parabolic dune complexes (Hesp 1982, 1988; Short and Hesp 1982).

On south-east Australian beach systems where overwash events are minor to absent, where sediment supply is generally not limited, and where an aggressive pioneer grass (*Spinifex* sp.) exists, relict foredune plains are common, particularly on the moderate energy intermediate beaches. Here established foredune stability is maintained to various degrees, and progradation over the last 6,000–7,000 years has led to the development of foredune plains.

Reflective beaches are characterized by accentuated swash during storms and laterally continuous alongshore beach erosion. Recovery is fairly rapid. Foredunes remain relatively stable over time, and because they are typically small, with limited sediment supply, little dune transgression results. Thus reflective beaches are characterized by a single foredune, or a few relict foredunes.

Role of sediment supply and other factors

There is no doubt that sediment supply, wind energy, sea-level state (transgressive, stable, regressive), return interval and magnitude of extreme storm events, and Pleistocene inheritance factors will all, at times, and in some places, be a controlling variable in beach–dune interactions. If sediment supply is limited, sea level is rising, and coastal erosion is the general rule, the model as outlined above may not work in part or perhaps at all (Psuty 1988).

References

Bauer, B.O. and Sherman, D.J. (1999) Coastal dune dynamics: problems and prospects, in A.S. Goudie, I. Livingstone and S. Stokes (eds) *Aeolian Environments, Sediments and Landforms*, 71–104, Chichester: Wiley.

Davidson-Arnott, R.G.D. (1988) Temporal and spatial controls on beach/dune interaction, Long Point, Lake Erie, in N.P. Psuty (ed.) *Dune/Beach Interaction, Journal of Coastal Research Special Issue* 3, 131–136.

Davidson-Arnott, R.G.D. and Law, M.N. (1990) Seasonal patterns and controls on sediment supply to coastal foredunes, Long Point, Lake Erie, in K.F. Nordstrom, N.P. Psuty and R.W.G. Carter (eds)

Coastal Dunes: Form and Process, 177–200, Chichester: Wiley.
Hesp, P.A. (1982) *Morphology and Dynamics of Foredunes in S.E. Australia*, unpublished Ph.D. Thesis, Dept. Geography, University of Sydney.
——(1988) Surfzone, beach and foredune interactions on the Australian south east coast, *Journal of Coastal Research Special Issue* 3, 15–25.
——(1991) Ecological processes and plant adaptations on coastal dunes, *Journal of Arid Environments* 21, 165–191.
——(1999) The beach backshore and beyond, in A.D. Short (ed.) *Handbook of Beach and Shoreface Morphodynamics*, 145–170, Chichester: Wiley.
Psuty, N.P. (1988) Sediment budget and beach/dune interaction, in N.P. Psuty (ed.) Dune/Beach Interaction, *Journal of Coastal Research Special Issue* 3, 1–4.
Sherman, D.J. and Bauer, B.O. (1993) Dynamics of beach–dune interaction, *Progress in Physical Geography* 17, 413–447.
Sherman, D.J. and Lyons, W. (1994) Beach-state controls on aeolian sand delivery to coastal dunes, *Physical Geography* 15, 381–395.
Short, A.D. and Hesp, P.A. (1982) Wave, beach and dune interactions in south eastern Australia, *Marine Geology* 48, 259–284.
Wright, L.D. and Short, A.D. (1984) Morphodynamic variability of beaches and surfzones: A synthesis, *Marine Geology* 56, 92–118.

PATRICK HESP

BEACH NOURISHMENT

Beach nourishment is the act of placing sediment (termed fill) on a beach by artificial means using sources outside the nourished area. Nourishment is primarily conducted to overcome a sediment deficit and create a beach of sufficient width to protect existing buildings and infrastructure from wave attack, but it also can enhance the value of urban locations for tourism or create new natural environments.

Debates have occurred over the cost effectiveness of nourishment and whether nourished beaches erode more rapidly than predicted in design studies (Houston 1991; Pilkey 1992), but projects are increasingly implemented, and nourishment is now the principal option for shore protection in some countries. Nourishment is conducted at all levels of management, from the national level to private homeowner groups. Borrow areas for fill materials include offshore, inlets, backbays, rivers and glaciated uplands. Opportunistic sources from dredging of harbours, marinas, lagoons and inland construction sites are also used (Nordstrom 2000).

Nourishment occurs on the upper beach, the nearshore, offshore on stable berms (designed to alter wave conditions) and active berms (that change shape or migrate onshore), and on existing dunes or on the backbeach to create new dunes. Large-scale nourishment operations on the upper beach commonly use a pipeline to transport a sand/water slurry. Small-scale operations transport sediment in dump trucks. The nourished beach is then often reshaped by bulldozers. The result of upper beach nourishment is a high, wide beach with an unstable shape, but the fill is easy to place, provides good initial protection against wave overwash, and creates a wide recreation platform. Conspicuous losses may subsequently occur on the upper beach as the fill adjusts to a more natural equilibrium shape. Fill sediments on the foreshore are reworked by waves and often become similar to pre-nourished sediments in size and sorting, but fill sediments on the backbeach above the zone of wave reworking retain characteristics that differ from native sediments.

Nearshore nourishment occurs by spraying sediment as a sand/water slurry or dumping it from shallow-draught barges. By placing sediments directly in the dynamic surf zones, losses through time are not visible and aesthetics are not spoiled by different sediments on the backbeach, but a beach nourished this way evolves slowly. Offshore berms are often implemented as disposal areas for sediment dredged from navigation channels, and more study is required to evaluate their use as protection structures.

Sediment bypassing (artificially transporting sediment to the downdrift side of obstacles to littoral drift) and backpassing (artificially transporting sediment from downdrift deposits to updrift eroding zones) may also be considered nourishment projects. Bypassing is gradually becoming more common at inlets where jetties or dredging of navigation channels interrupt longshore sediment transport. Backpassing is now most frequently conducted in small-scale trucking operations, but it may become more significant in the future as ready supplies of external sediment for nourishment projects become exhausted.

Nourishment of the upper beach can alter aeolian transport by (1) increasing the source width for entrainment of sediments; (2) adding fine sediment that is more readily transported; (3) changing moisture-retention characteristics; (4) changing the shape of the beach or dune profile; and (5) changing the likelihood of marine erosion of the incipient foredune (van der Wal 1998). Rapid dune growth can occur on nourished beaches, especially when sand fences and vegetation plantings are used to trap sand.

Dunes may be created directly by mechanically depositing sediment. Most dune nourishment operations place the new fill in front of the existing dune to create a sacrificial structure or on top of it to increase the level of protection against wave overwash; more rarely, a new dune may be built behind an existing foredune (Nordstrom 2000). Dunes built and used as protection structures can evolve into a condition that functions naturally or appears natural in terms of surface vegetation (Nordstrom *et al.* 2002).

Nourished beaches benefit threatened species by providing habitat that would otherwise be unavailable, but detrimental ecological impacts can occur due to (1) mechanical removal of habitat in borrow areas; (2) burial of habitat in nourished areas; (3) increased turbidity and sedimentation; (4) disruptions to foraging, nesting, nursing and breeding; (5) change in sediment characteristics, wave action and beach state; and (6) change in community structure and evolutionary trajectories, including enhancement of undesirable species (Nelson 1993; National Research Council 1995). Detrimental effects are often considered temporary, but little is known about long-term, cumulative impacts and critical thresholds.

Human activities, such as driving on the beach or raking the beach to remove litter, can eliminate incipient topography and vegetation and prevent formation of natural landforms, so true restoration of landforms and habitats may not occur in the absence of controls on subsequent human activities (Nordstrom 2000). The great importance of nourishment as a form of shore protection and as a sediment resource that can evolve into naturally functioning landforms makes this option an important area of future geomorphological research. To be effective, the nourished beach must be considered as a landform in its own right and as a source of sediment for evolution of other landforms landward and downdrift of it, rather than merely as an engineering structure or recreation platform.

References

Houston, J.R. (1991) Beachfill performance, *Shore and Beach* 59(3), 15–24.

National Research Council (1995) *Beach Nourishment and Protection*, Washington, DC: National Academy Press.

Nelson, W.G. (1993) Beach restoration in the southeastern US: environmental effects and biological monitoring, *Ocean and Coastal Management* 19, 157–182.

Nordstrom, K.F. (2000) *Beaches and Dunes of Developed Coasts*, Cambridge: Cambridge University Press.

Nordstrom, K.F., Jackson, N.L., Bruno, M.S. and de Butts, H.A. (2002) Municipal initiatives for managing dunes in coastal residential areas: a case study of Avalon, New Jersey, USA, *Geomorphology*, 48, 147–162.

Pilkey, O.H. (1992) Another view of beachfill performance, *Shore and Beach* 60(2), 20–25.

van der Wal, D. (1998) The impact of the grain-size distribution of nourishment sand on aeolian sand transport, *Journal of Coastal Research* 14, 620–631.

KARL F. NORDSTROM

BEACH RIDGE

Beach ridges are azonal accumulation forms created on the shores of seas and lakes. They are usually subparallel ridges of sand, gravel or pebble, as well as detritus of shell, situated in the foreshore zone, which is the boundary of low and high water's range. Older, subfossil complexes of beach ridges may actually appear in the backshore zone, which lies above the high water's range. Beach ridges forming at the present day are roughly parallel to the coast. The height of their crests is usually a little bit higher than mean high tide or storm, and the bottom of the adjacent troughs or swales have elevations not far from mean low water (Stapor 1975).

In Carter's (1986) opinion two types of beach ridges may develop on a progradational sea coast. The first type is a result of gradual accretion and coalescing of swash bars during a deposit's transport by wave action. This type of beach ridge is constructed by seaward dipping laminae of sand or gravel (swash lamination). The second type is connected with longshore bar emergence during low wave energy conditions and simultaneous fall of sea level. The morphology of these ridges is more complicated. They are constructed mainly by landward dipping laminae. However, tabular planar cross-lamination connected with landward migration of the lee slope of the emerge feature are also situated here. On the seaward slope of this type of beach ridge a thin layer of swash lamination can be present.

Beach ridges are also partially created by the processes of aeolian deflation and accumulation. There is often an accumulated cover of aeolian deposits on earlier formed ridges, stabilized by vegetation. As a result, on the beach ridges irregular hummock dunes or parallel foredune ridges can be situated. In this case, the sediments of beach ridges are usually separated from aeolian covers by fossil erosional surfaces with a shell or gravel pavement (Carter and Wilson 1990).

The formation of beach ridges is very dependent on conditions of beach supply by littoral deposits. Beach ridges are created only when wave action, and connected with it sea currents, provide more deposits to the beach than the waves can remove (Johnson 1919). Important factors during the creation of beach ridges are the bathymetry of the inner shelf, abundant sediment supply in the nearshore zone and also the wave energy regime and fluctuations of sea level. The average size of a beach's material is also a very important factor. On sandy beaches, the beach ridges accumulate during the low wave energy events, but on gravelly beaches the formation of ridges is usually connected with high wave energy events.

Beach ridges may appear as a single form, as well as a complex of forms, creating often expansive plains of beach ridges. These plains are especially characteristic of progradational coasts. The relief of the individual beach ridges is different. Their height may reach values from a few dozens of centimetres to about a few metres. The distance between two different beach ridges also varies. It is usually thought that the smaller, closely spaced ridges are formed during rapid beach progradation, and that the ridges of larger dimensions and greater spacing are connected with a relatively slower rate of growth (Taylor and Stone 1996).

Beach ridges are good palaeogeography indicators of past wave regimes, sediment supply, sediment source, climatic conditions, sea-level change and also isostatic emergence or submergence of land. If we can measure the absolute age of beach ridges, e.g. using radiocarbon or archaeological methods, we will be able to reconstruct the ancient shorelines' position as well as speed of coast progradation. Beach ridges can also be used to understand past relative sea-level changes and the history of deposit availability and abundance within the inner shelf.

References

Carter, R.W.G. (1986) The morphodynamics of beach ridge formation; Magilligan, Northern Ireland, *Marine Geology* 73, 191–214.

Carter, R.W.G. and Wilson, P. (1990) The geomorphological, ecological and pedological development of coastal foredunes at Magilligan Point, Northern Ireland, in K.F. Nordstrom, N.P. Psuty and R.W.G. Carter (eds) *Coastal Dunes: Form and Process*, 129–157, Chichester: Wiley.

Johnson, D.W. (1919) *Shore Processes and Shoreline Development*, New York: Wiley.

Stapor, F.W. (1975) Holocene beach ridge plain development, Northwest Florida, *Zeitschrift für Geomorphologie, Supplementband* 22, 116–141.

Taylor, M. and Stone, G.W. (1996) Beach-ridges: a review, *Journal of Coastal Research* 12(3), 612–621.

SEE ALSO: beach; beach–dune interaction; beach sediment transport; chenier ridge

RYSZARD K. BORÓWKA

BEACH ROCK

'A consolidated deposit that results from lithification by calcium carbonate of sediment in the intertidal and spray zones of mainly tropical coasts' (Scoffin and Stoddart 1983: 401). Some authors have also used the term 'cay sandstone' to distinguish between rocks that are formed in the supratidal zone and beach rock or beach sandstone formed in the intertidal zone of coral reefs (see, for example, Gischler and Lomando 1997). Although the latitudinal limits of most contemporary beach rocks are approximately 35°N to 35°S, from time to time they do occur in higher latitudes, for example in north-west Scotland (Kneale and Viles 2000).

Beach rock is also widespread on beaches around the Mediterranean Sea (Plate 17) but is perhaps best known for its association with the calcium carbonate beaches of coral reef islands. Beach rock is geomorphologically important in that it preserves coastal landforms, provides a record of former sea levels, creates distinctive pavements and forms very rapidly. It also displays suites of characteristic erosional landforms that include micro-scarps, ridges and runnels and various weathering forms produced by biological processes, chemical erosion and salt attack.

The origin of beach rock has perplexed investigators ever since it was described in the early nineteenth century by travellers like Admiral Beaufort and Charles Darwin. Proposed mechanisms of formation include both physico-chemical and biological models. The former involve cement precipitation resulting from evaporation, CO_2 degassing owing to wave agitation and increasing temperature, and mixing of alkaline fresh water with sea water. Such models tend to have dominated the literature. However, the role of micro-organisms is now being seriously considered as a result of both

Plate 17 Beach rock developed on the south-east coast of Turkey at Arsuz near Iskenderun

field (Webb *et al.* 1999) and laboratory evidence (Neumeier 1999). High Mg calcite (often micritic) and aragonite are the commonest cement types, although low Mg calcite cements are common from temperate zone beach rocks. The cements occur most commonly as clean isopachous fringes of acicular crystals around grains, but meniscus and gravitational cements are also known (Scoffin and Stoddart 1983).

References

Gischler, E. and Lomando, A.J. (1997) Holocene cemented beach deposits in Belize, *Sedimentary Geology* 110, 277–297.

Kneale, D. and Viles, H.A. (2000) Beach cement: incipient $CaCO_3$ – cemented beach rock development in the upper intertidal zone, North Uist, Scotland, *Sedimentary Geology* 132, 165–170.

Neumeier, U. (1999) Experimental modelling of beach rock cementation under microbial influence, *Sedimentary Geology* 126, 35–46.

Scoffin, T.P. and Stoddart, D.R. (1983) Beach rock and intertidal cements, in A.S. Goudie and K. Pye (eds) *Chemical Sediments and Geomorphology*, 401–425, London: Academic Press.

Webb, G.E., Jell, J.S. and Baker, J.C. (1999) Cryptic intertidal microbialites in beach rock, Heron Island, Great Barrier Reef: implications for the origin of microcrystalline beach rock cement, *Sedimentary Geology* 126, 317–334.

A.S. GOUDIE

BEACH SEDIMENT TRANSPORT

Beach sediment transport occurs over the whole area where wave-induced currents are capable of moving sediment: in the shoaling zone, surf zone, breakers and swash zone, and also as aeolian sediment transport on beaches. Spatial gradients in the sediment transport rate determine positions of erosion and deposition and thus three-dimensional changes in the beach shape. Research on beach sediment transport concentrates on predicting the mechanics of transport, direction of transport, transport rate, transport volume, and changes in beach morphology.

The essential difference between sediment transport under a steady unidirectional flow and under waves is that oscillatory BOUNDARY LAYERS show a temporal variation which is not present under steady flow, growing and decaying twice every wave cycle as flows accelerate, decelerate and change direction. Boundary layers are not able to develop fully under oscillatory flow, and consequently are always thinner than under an equivalent unidirectional flow. This means that for a given free-stream velocity and bed roughness, the bed shear stress in oscillatory flow is always larger than beneath steady flow.

Beach sediment transport can be in the form of SUSPENDED LOAD, BEDLOAD and sheet flow (see SHEET EROSION, SHEET FLOW, SHEET WASH). Bedload transport is often modelled as a function of the shear stress acting on the sediment grains, while suspended transport is generally calculated as the product of velocity and concentration profiles. Suspended sediment transport generally receives more attention than bedload transport; however, this is mainly because instruments such as the optical backscatter sensor and acoustic backscatter sensor have been developed which are capable of making high-frequency measurements of suspended sediment concentrations simultaneously with velocity measurements. Because only suspended sediment transport can be measured in this way, many researchers conclude that the most reasonable approach at present is to assume that suspended sediment transport dominates when strong wave motion is present. However, this assumption will remain essentially untested until instruments are developed which are capable of high-frequency measurement of bedload transport.

A number of mechanisms contribute to beach sediment transport, including turbulence, mean currents, currents generated by oscillatory waves at incident and infragravity frequencies, and wave–current interaction. Wave-induced currents include both unidirectional currents (such as longshore currents, RIP CURRENTS and undertow) and rapidly reversing asymmetrical cross-shore currents which flow onshore under the crest of the wave and offshore under the trough. In combined flows, oscillatory wave motion is generally assumed to entrain sediment which is then moved by a steady current. Beach sediment transport is also affected by local bed slope and the presence of RIPPLEs and other bedforms, and can be modified by human activity such as BEACH NOURISHMENT and coastal engineering structures.

The easiest approach to beach sediment transport is to consider cross-shore and longshore transport separately. Longshore transport is responsible for changes in the beach plan shape, and is usually considered to be unidirectional in the direction of the longshore current. In the simplest formulation, the longshore sediment transport rate is proportional to the longshore wave energy flux at the breakpoint. (See LONGSHORE (LITTORAL) DRIFT.)

Cross-shore sediment transport is responsible for changes in the beach profile. This includes features on the subaerial profile such as beach face slope and berms, and submerged features such as nearshore bars. Net cross-shore sediment transport is difficult to calculate because it occurs as an accumulation of small differences between the large values of onshore and offshore transport, which must be evaluated separately and correctly. Most field data and model predictions indicate that offshore sediment transport dominates under breaking waves due to the seaward-directed undertow. During non-breaking wave conditions, transport is generally onshore-directed due to the effects of incident wave asymmetry.

Governing equations based on fundamental physics have not yet been established, and no unified theory of sediment transport presently exists that is valid for all water depths and fluid motions in the nearshore. Instead, there are many sediment transport models, ranging from quasi-steady formulas such as the energetics approach described below to complex numerical models involving higher-order turbulence closure schemes that attempt to resolve the flow field at small scales. Models can be classified by direction (cross-shore or longshore), driving force (e.g. bottom fluid velocity, bed shear stress, wave energy dissipation, stream power, etc.), or mode of transport (bedload, suspended load, total load). Reviews of sediment transport models are given by Schoones and Theron (1995), Bayram *et al.* (2001) and Davies *et al.* (2002), and measurement of coastal sediment transport is reviewed by White (1998).

One of the preferred approaches to modelling both longshore and cross-shore wave and current-induced sediment transport is based on the energetics approach of Bagnold (1963), formulated for a time-varying flow field (e.g. Bailard 1981). The energetics approach assumes that sediment is mobilized by the oscillatory flow under waves and is related to some power of the instantaneous velocity. Once mobilized, sediment can be transported by a number of different mechanisms: time-averaged flows (undertow or longshore currents), asymmetric orbital velocities and gravity in the downslope direction. The fluid forces which drive the energetics model are based on the calculation of various moments of the fluid velocity, which give the direction and magnitude of both oscillatory and mean flows. Use of the energetics model requires knowledge of the moments of the instantaneous flow field, often decomposed into mean, gravity and infragravity band components and then time-averaged. Net transport is calculated from the integral of the instantaneous rate through a particular time interval.

The energetics model is regarded as one of the best theoretical models available at the moment for time-dependent nearshore sediment transport because of its capacity to represent a wide variety of transport conditions in a computationally efficient manner. However, it does not include a number of factors which are believed to be important in beach sediment transport, such as turbulence, fluid accelerations, threshold of motion, and transport over bedforms. In particular, the energetics model may not be appropriate for use in the swash zone, where beach accretion is most likely to occur. Additional processes are likely to be important in swash sediment transport, such as infiltration/exfiltration, bore-generated turbulence, water depth, inertial forces on coarse grains, and sediment advection and convection.

References

Bagnold, R.A. (1963) Mechanics of marine sedimentation, in M.V. Hill (ed.) *The Sea*, Volume 3, 507–528, New York: Wiley.

Bailard, J.A. (1981) An energetics total load sediment transport model for a plane sloping beach, *Journal of Geophysical Research* 86, 10,938–10,954.

Bayram, A., Larson, M., Miller, H.C. and Kraus, N.C. (2001) Cross-shore distribution of longshore sediment transport: comparison between predictive formulas and field measurements, *Coastal Engineering* 44, 79–99.

Davies, A.G., van Rijn, L.C., Damgaard, J.S., van de Graaff, J. and Ribberink, J. (2002) Intercomparison

of research and practical sand transport models, *Coastal Engineering* 46, 1–23.
Schoones, J.S. and Theron, A.K. (1995) Evaluation of 10 cross-shore sediment transport/morphological models, *Coastal Engineering* 25, 1–41.
White, T.E. (1998) Status of measurement techniques for coastal sediment transport, *Coastal Engineering* 35, 17–45.

Further reading

Komar, P.D. (1998) *Beach Processes and Sedimentation*, 2nd edition, Englewood Cliffs, NJ: Prentice Hall.
Pethick, J.S. (1984) *An Introduction to Coastal Geomorphology*, London: Edward Arnold.
Short, A. (1999) *Handbook of Beach and Shoreface Morphodynamics*, Chichester: Wiley.

SEE ALSO: bar, coastal; beach; beach–dune interaction; beach nourishment; bedload; current; longshore (littoral) drift; sediment budget; sediment transport; sheet erosion, sheet flow, sheet wash; suspended load; wave

DIANE HORN

BEDFORM

The transport of sand or gravel as BEDLOAD (and see SEDIMENT LOAD AND YIELD) creates on the bed a variety of features the size, form and relative orientation of which depend through complex interactions on the density, shape and coarseness of the sediment particles, and on the strength, uniformity and steadiness of the current. These features are bedforms. They participate in the sediment transport and are normally very small or small in height compared to flow thickness. Although the main kinds are, generally speaking, agreed upon, there is little uniformity as to their nomenclature. Bedforms are known from rivers, tidal systems (especially estuaries) and the deep sea, and are most widely familiar from deserts (see DUNE, AEOLIAN). They contribute significantly to contemporary landscapes and seascapes and, where preserved in Quaternary sediments, are valuable indicators of environment and sediment transport strength and direction. Bedforms and their internal structures can be used to establish sediment-transport directions not only on a regional scale but also locally, where changes in strength and direction have occurred on small geographical and stratigraphical scales. For the river and irrigation engineer, bedforms are among the most important determinants of channel hydraulic ROUGHNESS and resistance to flow.

Extensive field and laboratory studies show that river bedforms and their distinctive patterns of internal stratification can be placed in a number of fields defined more or less closely by grain size and flow strength (Figure 12). At flow strengths below the entrainment threshold (see INITIATION OF MOTION) there is neither sediment transport nor bedforms. As flow strength rises, the bedforms first to appear are current ripples (medium- and finer-grained quartz-density sands) and lower-stage plane beds (coarser sands, granules, gravel). At equilibrium, current ripples are ridges of linguoid plan about 0.02 m high and 0.1–0.2 m in wavelength, which increases with sediment size (see RIPPLE). They move very slowly beneath the current as grains are eroded from the long upstream slope and then deposited by settling and avalanching on the steep leeward face (see REPOSE, ANGLE OF) overlain by a turbulent, recirculating vortex. An internal pattern of cross-lamination records the successive positions of the migrating downstream face. Lower-stage plane beds are underlain by subhorizontal parallel laminae on a millimetre to coarser scale. In sediments coarser than medium-grained sand, but at increasingly large flow strengths for progressively finer grades, lower-stage plane beds and current ripples are replaced by dunes. Like current ripples, these forms are transverse ridges which migrate by the erosion of particles from the upstream side followed by their deposition on the downstream face after settling through and avalanching beneath a recirculating, leeward vortex (see REPOSE, ANGLE OF). Patterns of cross-bedding occur internally, the foresets dipping in the sediment-transport direction. Unlike current ripples, dunes scale on flow depth, varying in height from about one-twentieth to about one-fifth of the depth. In large rivers, such as the Mississippi and Brahmaputra, they are several metres high and one to two hundred metres in wavelength. Extensive fields of even larger dunes, composed of cobble and boulder gravel, have been created by some major catastrophic floods (see DAM; ICE DAM, GLACIER DAM; OUTBURST FLOOD). At large enough flow strengths, current ripples and dunes become increasingly round-crested and flat, and are replaced by upper-stage plane beds over which sediment transport, in the form of very low bed waves, is intense. Internally, forms of subhorizontal laminae and bedding occur within such beds. In the case of sands, the surfaces of the laminae carry faint flow-parallel ridges, called parting or primary current

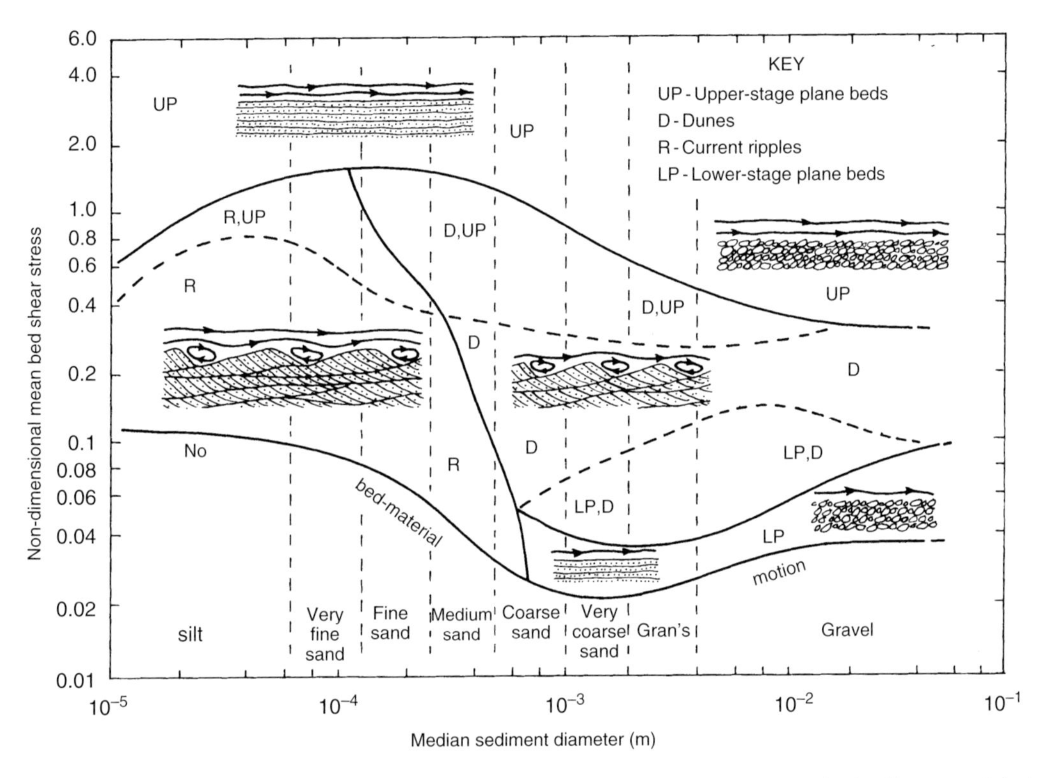

Figure 12 Existence fields, defined by flow strength and median grain diameter, for bedforms and their internal sedimentary structure as shaped by unidirectional water currents in quartz-density sediments. The diagram is based on many hundreds of individual observations. Flow strength is given in a non-dimensional form, defined as the quotient of the mean bed shear stress and the product of the relative particle density, acceleration due to gravity and median particle diameter

lineations, which may be related to flow patterns within the grain-dense lower part of the turbulent BOUNDARY LAYER.

These bedforms are restricted to subcritical flows (see ANTIDUNE), marked by comparatively smooth water surfaces and low values of flow velocity compared to flow depth. Supercritical flows over loose beds consist of unstable, transverse, surface waves more or less in phase with similar waves on the bed, but of a lower height. These are antidune bedforms. At high levels of supercriticality, various very flat bed waves arise, which include rhomboid ripples and dunes related to surface shocks. As supercriticality is promoted by shallow depths, supercritical bedforms can arise at almost any flow strength capable of sediment transport.

Tidal currents, reversing and rotating on a semi-diurnal or diurnal scale, are as non-uniform as river flows, but hugely more unsteady, and this factor complicates the shape, internal structure and relationship to flow conditions of the bedforms encountered in estuaries (see ESTUARY) and shallow seas. Additional kinds of bedform are recorded from these environments, as well as those familiar from rivers. Current ripples, sand and gravel dunes, upper-stage plane beds and antidunes are chiefly restricted to the shallower channels and intertidal shoals of estuaries. As an expression of the unsteady conditions, drapes of mud deposited when the water is slack may accumulate in the troughs of the ripples and dunes, and later become preserved within the cross-stratified interiors of the forms. Large bedforms – sand ribbons and sand waves – are found below tide level in the deeper channels and on tide-swept floors of confined seas such as the English Channel, the Southern Bight of the North Sea and Cook Inlet, Alaska. Subtidal sand waves were discovered in the 1920s and 1930s as the result of detailed hydrographic surveys and the appearance of practical echo-sounders. A few decades elapsed before the development of side-scan sonar allowed sand ribbons to be recognized.

Sand ribbons are long, flow-parallel belts of ripple- or dune-covered sand or fine gravel of low relief with a spacing across the current of a few to several times the flow depth. They express bedload transport under conditions of restricted sediment supply. Sand waves are series of long-crested ridges of sediment arranged transversely to the stronger of the tidal currents. The largest, found in the deepest waters, measure 5 m or more in height and a few hundred metres in spacing. They assume a roughly symmetrical, trochoidal profile where flood and ebb tidal streams are comparably strong, the small- to medium-sized dunes on their backs reversing in direction of travel with each change in tidal phase. Sand waves become increasingly asymmetrical in profile as the ebb and flood tidal streams become more unequal in their ability to transport sediment, and the dunes they carry may migrate only in the direction of the stronger flow, although being slightly rounded by the weaker stream. The internal structure of sand waves is not well known but is certainly complex, reflecting the presence of superimposed dunes which may change and reverse their direction of movement as the tidal streams reverse and rotate. Internally, the more symmetrical waves seem to consist of comparatively thin cross-bedded units recording sediment transport in many different but largely opposed directions. The more asymmetrical ones reveal internally a 'master-bedding' that dips gently in the direction of the stronger tidal stream and between which appears to lie cross-bedding facing chiefly in that same direction.

Currents strong enough to transport sand-sized particles in places affect large areas of the ocean floor and the deeper parts of open continental shelves. These variable but essentially unidirectional flows are not of tidal origin but depend on various thermohaline effects. Sand ribbons and transverse structures which have been called sand waves (very large dunes) have been described from many of the places where these currents operate, such as the long and intricate narrows between the Baltic Sea and the North Sea, the ocean floor immediately west of Gibraltar, swept by the Mediterranean Undercurrent, the continental shelf of south-east Africa, affected by the Agulhas Current, and the level tops of several oceanic guyots. The sediments involved are of diverse origins. They range from terrigenous sands, in some cases reworked after being introduced from shallower depths by TURBIDITY CURRENTS, to bioclastic debris (chiefly shells or foraminifera) eroded from adjacent parts of the ocean floor. Other than their location, general form and link with strong currents, little is known or understood about these deep-sea bedforms.

Further reading

Alexander, J., Bridge, J.S., Cheel, R.J. and LeClair, S.F. (2001) Bedforms and associated sedimentary structures formed under supercritical water flows over aggrading sand beds, *Sedimentology* 48, 133–152.

Allen, J.R.L. (1982) *Sedimentary Structures*, Amsterdam: Elsevier.

Allen, J.R.L., Friend, P.F., Lloyd, A. and Wells, H. (1994) Morphodynamics of intertidal dunes: a year-long study at Lifeboat Station Bank, Wells-next-the-Sea, Eastern England, *Philosophical Transactions of the Royal Society of London* A347, 291–345.

Ashley, G.M. (1990) Classification of large-scale subaqueous bedforms: a new look at an old problem, *Journal of Sedimentary Petrology* 60, 160–172.

Berné, S., Castaing, P., Le Drezen, E. and Lericolais, G. (1993) Morphology, internal structure and reversal of asymmetry of large subtidal dunes in the entrance to the Gironde Estuary (France), *Journal of Sedimentary Petrology* 63, 780–793.

Bridge, J.S. and Best, J.L. (1988) Flow, sediment transport and bedform dynamics over the transition from dunes to upper-stage plane beds: implications for the formation of planar laminae, *Sedimentology* 35, 753–763.

Bridge, J.S. and Best, J. (1997) Preservation of planar lamination due to migration of low-relief bed waves over aggrading upper-stage plane beds: comparison of experimental data with theory, *Sedimentology* 44, 253–262.

Carling, P.A. (1999) Subaqueous gravel dunes, *Journal of Sedimentary Research* 69, 534–545.

Carling, P.A., Gölz, E., Orr, H.G. and Redecki-Pawlik, A. (2000) The morphodynamics of fluvial sand dunes in the River Rhine, near Mainz, Germany. I. Sedimentology and morphology, *Sedimentology* 47, 227–252.

Dalrymple, R.W. and Noble, R.N. (1995) Estuarine dunes and bars, in G.M.E. Perillo (ed.) *Geomorphology and Sedimentology of Estuaries*, 359–422, Amsterdam: Elsevier.

Flemming, B.W. (1980) Sand transport and bedform patterns on the continental shelf between Durban and Port Elizabeth (southern African continental margin), *Sedimentary Geology* 26, 179–205.

Kenyon, N.H. and Belderson, R.H. (1973) Bedforms in the Mediterranean undercurrent observed with side-scan sonar, *Sedimentary Geology* 9, 77–99.

Simons, D.B. and Richardson, E.V. (1966) Resistance to flow in alluvial channels, *United States Geological Survey Professional Paper* 422-J, 1–61.

Stride, A.H. (ed.) (1982) *Offshore Tidal Sands*, London: Chapman and Hall.

Weedman, S.D. and Slingerland, R. (1985) Experimental study of sand streaks formed in turbulent boundary layers, *Sedimentology* 32, 133–145.

J.R.L. ALLEN

BEDLOAD

By mode of transport, the sediment load is divided into SUSPENDED LOAD and bedload. The bedload typically consists of coarse particles derived from the bed material. The immersed weight of these particles is supported by a combination of fluid forces and solid reactive forces exerted at intermittent or continuous contacts with the bed (Abbott and Francis 1977). Bedload is dispersed in a zone immediately above the bed surface and is transported in the rolling/sliding or saltating modes. Particles comprising the bedload continually move in and out of storage on the bed. The pattern of particle motion can be characterized as a series of relatively short steps of random length, each of which is followed by a rest period of random duration, and each particle spends a negligible time in motion compared to the time spent at rest. Consequently, the virtual velocity of bedload is much lower than the flow velocity; in water, for example, it is only of the order of metres per hour, compared to the flow velocity which may be of the order of metres per second (Haschenburger and Church 1998).

In wind, most sand particles move by SALTATION. The saltating particles interact with the bed surface and disturb stationary grains (Anderson and Haff 1991). This not only reduces the threshold of particle motion, it also promotes reptation (the movement of particles impacted by saltating grains over short distances) and surface creep. Within the saltation layer most grains move within 1 to 2 cm of the sand surface. The size of the mobile grains decreases and there is an exponential decline in sediment and mass flux concentration with height (Anderson and Hallet 1986). Accurate data on sediment fluxes are difficult to obtain (Greeley *et al.* 1996; Iversen and Rasmussen 1999), and sand transport equations are frequently used to predict aeolian sand transport rates (Sarre 1989). Above the threshold for sand movement, sand transport rates are commonly assumed to be proportional to the wind (shear) velocity. Aeolian sand transport in bulk is associated with the formation of BEDFORMS of varying size, that develop as regularly repeating patterns (Anderson 1987; Lancaster 1988), and field observation suggests there is close agreement between measured and simulated patterns of erosion and deposition on dunes and wind velocity and direction (Howard *et al.* 1978).

In water, the maximum size of sediment that can be moved by a given flow condition defines flow competence, but the size and amount of sediment moved as bedload is constrained by a river's transport capacity. Transport rates may also be limited by sediment availability. Continuity of bedload transport typically is not maintained along a river, because transport capacity usually does not match the sediment supply. This may promote scour or fill of the river bed and other adjustments to channel geometry. For this reason, although bedload typically constitutes only a few per cent of the total sediment load of most rivers, bedload transport is a very significant process as it governs virtually all aspects of morphological change in river channels. Downstream through a drainage basin, the bed material generally becomes finer through the action of sorting and abrasion; in consequence, the suspended load increasingly dominates over the bedload.

At the lower limit of active transport, where rolling is the dominant transport mode, bed pocket geometry determines which particle sizes are mobile (Andrews 1994). When conditions are below the threshold for general bed motion or the sediment supply is limited the bedload transport rate is moderated by the interaction of coarse and fine size fractions in the bed material, as well as by the available shear stress (Gomez 1995). ARMOURING compensates for the intrinsically lower mobility of coarse particles relative to that of fine grains and renders all particle sizes on the bed surface equally mobile (Parker and Klingeman 1982). Equal mobility arises as a consequence of the shielding of small grains from the flow and the preferential exposure of large particles, coupled with their relative abundance on the bed surface. The adjustments combine to counteract the absolute size effect of particle weight by making coarse particles more available to the flow, and enhancing their probability for entrainment. There are two facets to equal mobility. Equal entrainment mobility is defined as the case when all particle sizes comprising the bed material begin to move at the same flow strength. Equal transport mobility refers to the situation where all particle sizes are transported according to their relative proportions in the bed material, so that the bedload and bed material grain-size distributions are identical. Departures from these conditions give rise to differences in the transport rate of individual size fractions (Wilcock and McArdell 1993), and to hydraulic sorting. Hydraulic sorting is known to occur during the entrainment, transport and deposition of heterogeneous bedload; it is important because of its links to channel armouring and downstream fining (Paola *et al.* 1992).

In most rivers, bedload transport is highly variable in time and space. Temporal variability in bedload transport rates, which is independent of variations in flow conditions, arises from three main sources (Gomez *et al.* 1989). First, variations may result from long- to intermediate-term changes in the rate at which sediment is supplied to or distributed within a channel or reach. Second, short-term, often quasi-cyclic, variations in bedload-transport rates may occur in response to the temporary exhaustion of the supply of transportable material, to the migration of BEDFORMS or groups of particles, or to processes such as ARMOURING. Third, instantaneous fluctuations in bedload transport rates result from the inherently stochastic nature of the physical processes that govern the entrainment and transport of bedload. Spatial variability in bedload transport rates results from downstream and cross-channel changes in the transport field, that occur primarily in response to differences in the shear stress and to changes in the local relation between boundary shear stress and sediment transport (Dietrich and Whiting 1989).

Commonly utilized approaches for gaining knowledge of the bedload transfer through a river reach involve field sampling or measurement, and the application of a formula. Sampling involves the collection of discrete quantities of bedload at various points across a channel, over limited time intervals. Measurement involves the continuous or time-integrated monitoring of bedload over the entire cross-section or reach. The presence of a sampling device on the river bed necessarily alters the pattern of the flow and sediment transport in its vicinity. Thus, bedload samplers must be calibrated to determine their efficiency under different hydraulic and sediment transport conditions. Determining the hydraulic efficiency of a bedload sampler has proved to be a relatively simple task, but determining the sampling efficiency is considerably more complex. Consequently, the sampler calibration process remains incomplete because none of the tests performed to date on any sampling device has provided definitive results (Thomas and Lewis 1993). Since bedload transport rates vary across channel and with time, appropriate temporal and spatial sampling strategies also are required to minimize sampling errors, which decrease as the number of samples collected increases and the number of traverses of the channel over which the samples are collected increases (Gomez and Troutman 1997).

Measurements are usually regarded as exact, and most commonly are obtained using a pit trap. Traps also have a distinct advantage over samplers in as much as, if the trap spans the entire width of the river, it is not only possible to catch all the bedload that passes through the measuring section in a given period of time but also to continuously measure the rate at which sediment accumulates. The simplest traps consist of lined pits or slots in the streambed in which the bedload collects over one or more events (Church *et al.* 1991). More sophisticated traps continuously weigh the mass of sediment (Reid *et al.* 1980).

Bedload formulae equate the rate at which bedload is transported with a specific set of hydraulic and sedimentological variables, and predict bedload transport capacity under given flow conditions. The underlying physics appear quite straightforward (Bagnold 1966), but the conditions governing fluvial bedload transport are complex and there is little consensus about the fundamental hydraulic and sedimentological quantities involved. Consequently, there are numerous bedload transport formulae (Gomez and Church 1989), and none has been universally accepted or recognized as being especially appropriate for practical application.

References

Abbott, J.E. and Francis, J.R.D. (1977) Saltation and suspension trajectories of solid grains in a water stream, *Philosophical Transactions of the Royal Society of London* A284, 225–254.

Anderson, R.S. (1987) A theoretical model for aeolian impact ripples, *Earth Science Reviews* 10, 263–342.

Anderson, R.S. and Haff, P.K. (1991) Wind modification and bed response during saltation of sand in air, *Acta Mechanica Supplement* 1, 21–51.

Anderson, R.S. and Hallet, B. (1986) Sediment transport by winds: toward a general model, *Geological Society of America Bulletin* 97, 523–535.

Andrews, E.D. (1994) Marginal bed load transport in a gravel bed stream, Sagehen Creek, California, *Water Resources Research* 30, 2,241–2,250.
Bagnold, R.A. (1966) An approach to the sediment transport problem from general physics, US *Geological Survey Professional Paper* 422-I.
Church, M.A., Wolcott, J.F. and Fletcher, W.K. (1991) A test of equal mobility in fluvial sediment transport: behaviour of the sand fraction, *Water Resources Research* 27, 2,941–2,951.
Dietrich, W.E. and Whiting, P.J. (1989) Boundary shear stress and sediment transport in river meanders of sand and gravel, in S. Ikeda and G. Parker (eds) *River Meandering*, 1–50, Washington, DC: American Geophysical Union.
Gomez, B. (1995) Bedload transport and changing grain size distributions, in A.M. Gurnell and G. Petts (eds) *Changing River Channels*, 177–199, Chichester: Wiley.
Gomez, B. and Church, M. (1989) An assessment of bedload sediment transport formulae for gravel-bed rivers, *Water Resources Research* 25, 1,161–1,186.
Gomez, B. and Troutman, B.M. (1997) Evaluation of process errors in bed load sampling using a dune model, *Water Resources Research* 33, 2,387–2,398.
Gomez, B., Naff, R.L. and Hubbell, D.W. (1989) Temporal variations in bedload transport rates associated with the migration of bedforms, *Earth Surface Processes and Landforms* 14, 135–156.
Greeley, R., Blumberg, D.G. and Williams, S.H. (1996) Field measurement of the flux and speed of wind-blown sand, *Sedimentology* 43, 41–52.
Haschenburger, J.K. and Church, M.A. (1998) Bed material transport estimated from the virtual velocity of sediment, *Earth Surface Processes and Landforms* 23, 791–808.
Howard, A.D., Morton, J.B., Gad-el-Hak, M. and Pierce, D.B. (1978) Sand transport model of barchan dune equilibrium, *Sedimentology* 25, 307–338.
Iversen, J.D. and Rasmussen, K.R. (1999) The effect of wind speed and bed slope on sand transport, *Sedimentology* 46, 723–731.
Lancaster, N. (1988) Controls on aeolian dune size and spacing, *Geology* 16, 972–975.
Parker, G. and Klingeman, P.C. (1982) On why gravel-bed streams are paved, *Water Resources Research* 18, 1,409–1,423.
Paola, C., Parker, G., Seal, R., Sinha, S.K., Southard, J.B. and Wilcock, P.R. (1992) Downstream fining by selective deposition in a laboratory flume, *Science* 258, 1,757–1,760.
Reid, I., Layman, J.T. and Frostick, L.E. (1980) The continuous measurement of bedload discharge, *Journal of Hydraulics Research* 18, 243–249.
Sarre, R.D. (1989) Aeolian sand transport, *Progress in Physical Geography* 11, 157–182.
Thomas, R.B. and Lewis, J. (1993) A new model for bed-load sampler calibration to replace the probability-matching method, *Water Resources Research* 29, 583–597.
Wilcock, P.R. and McArdell, B.W. (1993) Surface-based fractional transport rates: mobilization thresholds and partial transport of a sand-gravel sediment, *Water Resources Research* 29, 1,297–1,312.

BASIL GOMEZ

BEDROCK CHANNEL

Bedrock channels are those with frequent exposures of bedrock in their bed and banks. More precisely, these channels lack a coherent cover of alluvial sediments, even at low flow, although a thin and patchy alluvial cover may be present. However, short-term pulses of rapid sediment delivery may produce temporary sediment fills (see SEDIMENT ROUTING; SEDIMENT WAVE). Bedrock channels exist only where transport capacity exceeds sediment supply over the long term. Contrary to classical definitions, bedrock channels are self-formed. Bed and banks are not composed of transportable sediment, but are erodible. Flow, sediment flux and base-level conditions (see BASE LEVEL) dictate self-adjusted combinations of channel gradient, width and bed morphology.

Bedrock channels are important because (1) they set much of the RELIEF structure of unglaciated mountain ranges, and (2) the controls on their incision rates largely dictate the relationships among climate, lithology, tectonics and topography. Moreover, because river incision rate sets the boundary condition for hillslope EROSION, regional DENUDATION rates and patterns are dictated by the bedrock river network (see HILLSLOPE-CHANNEL COUPLING). Finally, bedrock rivers transmit signals of base level (tectonic/eustatic (see EUSTASY)) and climate change through landscape, and therefore set the timescale of response to perturbation.

Similar to alluvial channels, the longitudinal profiles of bedrock channels are typically smoothly concave-up (see LONG PROFILE, RIVER). These profiles are well described by Flint's law relating local channel gradient (S) to upstream drainage area (A):

$$S = k_s A^{-\theta} \quad (1)$$

where k_s is the steepness index and θ the concavity index. Steepness index is known to be a function of rock uplift rate, lithology and climate (see GRADE, CONCEPT OF). The concavity index is typically in the range 0.4–0.6, and is apparently insensitive to differences in uplift rate, lithology and climate where these are uniform within a DRAINAGE BASIN. However, θ does vary beyond this typical range, usually where downstream fining is particularly strong, or where lithology or uplift rate vary systematically downstream.

Bedrock channel width also varies with drainage area in a manner similar to that observed in alluvial channels (see HYDRAULIC GEOMETRY):

$$W \propto A^{0.3-0.4} \quad (2)$$

Bed morphology also appears to be dynamically adjusted to hydraulic and sediment-flux conditions, and in bedrock-dominated reaches includes STEP-POOL SYSTEMS, plane bed and incised inner channel forms. Discrete KNICKPOINTS and erosional forms such as flutes, POT-HOLES, longitudinal grooves and undulating canyon walls are common.

Processes of erosion in bedrock channels include plucking, macro-abrasion, wear, chemical and physical WEATHERING, and possibly CAVITATION (see CORROSION; FROST AND FROST WEATHERING). These processes all include critical thresholds (see THRESHOLD, GEOMORPHIC) and most work is probably done by large storms (see INITIATION OF MOTION; MAGNITUDE–FREQUENCY CONCEPT). The relative roles of extraction of joint blocks (plucking plus macro-abrasion) and incremental wear are debated, but appear to depend on properties of the substrate lithology and flow conditions (see ROCK MASS STRENGTH). The relative contributions of BEDLOAD and SUSPENDED LOAD to ABRASION (macro- and wear) are also debated, but most agree sediment flux plays a dual role: providing tools for abrasion, but protecting the bed when overly abundant. The exact nature of the dependence of incision rate on sediment flux and grain size, and the different mechanics of plucking, macro-abrasion and wear all have far-reaching consequences for the relations among climate, tectonics and topography. Both DEBRIS FLOWS and FLOODS likely contribute to bedrock channel erosion in mountainous areas. Their relative importance is not well known, but apparently depends on position in the landscape and setting (tectonics, lithology and climate).

Rates of incision of bedrock rivers are highly variable (from mMa^{-1} to cmyr^{-1}), and depend primarily on tectonic setting and other controls on base-level fall. Where they have both been measured, long-term bedrock river incision rates match the highest rock uplift rates. Burbank *et al.* (1996) estimated incision rates up to $12\,\mathrm{mmyr}^{-1}$ on the basis of cosmogenic exposure ages of strath terraces along the Indus River in north-west Pakistan (see COSMOGENIC DATING; TERRACE, RIVER). Short-term incision rates up to $10\,\mathrm{cmyr}^{-1}$ have been measured under extreme circumstances.

Incision rates are positively correlated with channel gradient and drainage area, and are often modelled as a function of bed shear stress. The best known, semi-successful, bedrock river incision model is the shear stress or unit STREAM POWER model:

$$E = KA^mS^n \quad (3)$$

where E denotes vertical incision rate (L/T), A upstream drainage area (L^2), S channel gradient, K is a dimensional coefficient of erosion (L^{1-2m}/T) (see EROSIVITY), and m and n are positive constants that depend on erosion process, channel hydraulics and basin hydrology. Although this simple model has been useful for exploration of interactions among erosion, topography, climate and tectonics, much uncertainty remains regarding the physical controls on the model parameters K, m and n. In addition, equation (3) neglects an incision threshold and therefore misses an important aspect of the physics. Further field and laboratory studies are needed to resolve important outstanding issues.

Reference

Burbank, D.W., Leland, J., Fielding, E., Anderson, R.S., Brozovic, N., Reid, M.R. and Duncan, C. (1996) Bedrock incision, rock uplift and threshold hillslopes in the northwestern Himalayas, *Nature* 379, 505–510.

Further reading

Howard, A.D., Seidl, M.A. and Dietrich, W.E. (1994) Modeling fluvial erosion on regional to continental scales, *Journal of Geophysical Research* 99, 13,971–13,986.

Sklar, L.S. and Dietrich, W.E. (2001) Sediment and rock strength controls on river incision into bedrock, *Geology* 29, 1,087–1,090.

Stock, J.D. and Montgomery, D.R. (1999) Geologic constraints on bedrock river incision using the stream power law, *Journal of Geophysical Research* 104, 4,983–4,993.

Tinkler, K. and Wohl, E.E. (eds) (1998) *Rivers over Rock: Fluvial Processes in Bedrock Channels*, Washington, DC: AGU Press.

Whipple, K.X., Hancock, G.S. and Anderson, R.S. (2000) River incision into bedrock: mechanics and relative efficacy of plucking, abrasion, and cavitation, *Geological Society of America Bulletin* 112, 490–503.

Whipple, K.X. and Tucker, G.E. (1999) Dynamics of the stream-power river incision model: implications for height limits of mountain ranges, landscape response timescales, and research needs, *Journal of Geophysical Research* 104, 17,661–17,674.

SEE ALSO: channel, alluvial; dynamic equilibrium; palaeoflood; tectonic geomorphology; valley

KELIN X. WHIPPLE

BEHEADED VALLEY

Fluvial valleys running across an active strike-slip fault react to lateral movement along the fault in the way that their lower reaches, located downstream from the fault, become horizontally displaced in relation to the upper reaches, situated upstream from the fault. In this way the continuity of the valley is lost and the lower reach becomes beheaded. Streams may deflect at the fault and follow the fault zone until they turn into the displaced lower reach, or abandon the original valley and continue without deflection. In the latter case the beheaded section of a valley becomes dry. It is usually only small, narrow valleys occupied by minor streams which become beheaded. For larger rivers, floodplains are normally sufficiently wide to retain spatial continuity.

If a beheaded valley can be clearly defined in the field and contains an alluvial suite which can be dated, then slip rate along the fault can be determined.

Further reading

Keller, E.A., Bonkowski, M.S., Korsch, R.J., Shlemon, R.J. (1982) Tectonic geomorphology of the San Andreas fault zone in the southern Indio Hills, Coachella Valley, California, *Geological Society of America Bulletin* 93, 46–56.

SEE ALSO: seismotectonic geomorphology; tectonic geomorphology

PIOTR MIGOŃ

BERGSCHRUND

Deep, transverse or extensional crevasses that occur at the heads of valley or cirque glaciers (see CIRQUE, GLACIAL) are called bergschrunds. They differ from randklufts in that they occur in glacier ice rather than between the ice and the bedrock headwall. The randkluft of a glacier exists due to a combination of preferential ablation adjacent to warm rock surfaces and ice movement away from the rock wall. Both types of crevasse form formidable barriers for climbers in glacierized mountainous terrain and are particularly dangerous when covered in snow. Although numerous studies have suggested that the bergschrund of a glacier separates immobile, cold based ice at the head of a glacier from the active, sliding ice lower down, Mair and Kuhn (1994) have demonstrated that ice was sliding both above and below the position of a bergschrund in a glacier in the Austrian Alps. Early work on bergschrunds suggested that they were the location of intense FREEZE–THAW CYCLE activity and were, therefore, crucial to the excavation of cirques. Several problems with this hypothesis became apparent once bergschrunds were visited more frequently, most notably by W.R.B. Battle. Essentially, the base of bergschrunds, where bedrock is only occasionally encountered, do not experience appreciable freeze–thaw cycles. It is now accepted that the most effective conditions for freeze–thaw activity lie in the randkluft rather than in the bergschrund of a glacier (Gardner 1987).

References

Gardner, J.S. (1987) Evidence for headwall weathering zones, Boundary Glacier, Canadian Rocky Mountains, *Journal of Glaciology* 33, 60–67.
Mair, R. and Kuhn, M. (1994) Temperature and movement measurements at a bergschrund, *Journal of Glaciology* 40, 561–565.

Further reading

Battle, W.R.B. and Lewis, W.V. (1951) Temperature observations in bergschrunds and their relationship to cirque erosion, *Journal of Geology* 59, 537–545.
Embleton, C. and King, C.A.M. (1975) *Glacial Geomorphology*, London: Edward Arnold.
Lewis, W.V. (ed.) (1960) *Norwegian Cirque Glaciers*, Royal Geographical Society Research Series No. 4.
Thompson, H.R. and Bonnlander, B.H. (1956) Temperature measurements at a cirque bergschrund in Baffin Island: some results of W.R.B. Battle's work in 1953, *Journal of Glaciology* 2, 762–769.

DAVID J.A. EVANS

BIOGEOMORPHOLOGY

Biogeomorphology is sometimes used as an umbrella term to describe studies which focus on the linkages between ecology and geomorphology. Because biogeomorphology deals with the interface between two disciplines it is necessarily diverse, interdisciplinary and hard to define in detail. Biogeomorphological studies have a long history, with several nineteenth-century workers focusing on the interrelationships between communities and landscapes at a range of scales, although the term itself was only coined in the late twentieth century (Viles 1990). Amongst nineteenth-century pioneers,

Charles Lyell in 1835 noted the importance of the agency of organic beings in causing superficial modifications on the Earth's surface, and Charles Darwin undertook a classic piece of work on the role of earthworms in influencing soils (Darwin 1881). In recent years several volumes of collected papers on biogeomorphology have been published which provide varying pictures of the scope and nature of biogeomorphological research (see, for example, Viles 1988; Thornes 1990; Hupp *et al.* 1995; Viles and Naylor 2002). Papers within these volumes cover a whole range of organism: geomorphology interactions in riparian, hillslope and coastal settings in environments ranging from arid to humid tropical.

Similar terms have also been used in the literature, including zoogeomorphology or the interrelationship between animals and geomorphology (Butler 1995) and phytogeomorphology (Howard and Mitchell 1985) which investigates the influence of topography on plant communities. Furthermore, geoecology is a commonly used term, especially in the European literature, which also encompasses work addressing interactions between ecology and geomorphology (often at a large scale). Dendrogeomorphology, or the use of tree ring and allied evidence to study geomorphic processes, makes use of the influence of geomorphic processes on plant growth to throw light on the nature and timing of those processes – a neat way of linking ecology and geomorphology from a rather different perspective. Biogeomorphology and these other, similar, umbrella terms reflect a recent trend within the Earth and environmental sciences to investigate links between biotic and abiotic processes (as shown by the flowering of biogeochemistry as a study area, and the growth of interest in Gaia).

Three common themes within present-day biogeomorphological research are the effects of organisms on geomorphic processes, the contribution made by organic processes to the development of landforms, and the impact of geomorphological processes on ecological community development. Many studies have been made in recent years within these themes. For example, in terms of the impact of organisms on geomorphic processes studies have been made of the role of isopods and other fauna in sediment movement in the Negev desert by Yair (1995); the role of *Sabellaria alveolata* reefs in storing coastal sand on the Welsh coast by Naylor and Viles (2000), and the role of plants in influencing splash erosion in Mediterranean matorral environments by Bochet *et al.* (2002). Examples of studies of the role of organic processes in landform development include the study of Fiol *et al.* (1996) which investigates the role of biological weathering in the creation of solutional rillenkarren, and the work of Whitford and Kay (1999) on the role of mammal bioturbation in the production of long-lived mound structures (often called mima mounds). Investigations of the influence of geomorphic processes on ecosystems have been carried out by many ecologists and geomorphologists, such as Scatena and Lugo (1995) in subtropical forests and Hayden *et al.* (1995) on coastal barrier islands. Overall, research into these three major biogeomorphological themes is characterized to date by being largely empirical, field based and focused on a limited range of interactions. There are clear links between the three themes, as for example mammal burrowing produces mima mounds which then influence subsequent vegetation patterning.

Geomorphology and ecology are linked in detail in a range of different ways and understanding and measuring these links has provided much work for biogeomorphologists. Looking at the impact of ecology on geomorphology, organisms can have passive and/or active impacts on geomorphological processes. For example, a micro-organic biofilm can produce chemical weathering of the underlying rock (an active link) whilst also retarding the action of other weathering processes (passively). Biological impacts of geomorphological processes are often referred to by specific terms such as bioerosion, bioweathering, bioturbation, bioconstruction and bioprotection (see Naylor *et al.* 2002 for further details). Considerable research effort has gone into defining these terms and developing ways of studying and quantifying these processes. For example, bioerosion of coastal rocks by a suite of sessile and motile organisms has necessitated measurement of burrow dimensions and calculation of ages of the organisms creating them, as well as quantification of grazing trails through measurement of faecal contents. On the other side of the equation, geomorphology can exert an active and/or passive control on ecosystems. For example, topography influences microclimate which in turn affects plant communities (a passive geomorphological impact) whilst geomorphic processes such as mudflows and rockfalls provide an active control on vegetation development. A whole host of different techniques have been developed to study such influences, often involving mapping and correlation.

All exogenetic geomorphological processes have the potential to be influenced by biological activity; even in some quite hostile environments, as work on subglacial bacterial involvement in chemical weathering has demonstrated (Sharp *et al.* 1999). Indeed, there have been some suggestions that the harsher the environment the more closely interlinked biotic and geomorphic processes are, as organisms extract nutrients, shelter and water from sediments and rocks (Viles 1995). The whole spectrum of biological life forms is involved in biogeomorphological interactions, with animals, plants and a host of micro-organisms all recorded as influencing geomorphic processes. Bacteria have been found to contribute to the precipitation of sinter and travertine in hot spring environments, for example, and tree roots commonly enhance riverbank stability, whilst beaver dams have been recorded as having major impacts on some river networks. As a general rule, micro-organic and plant impacts are more widespread and important to geomorphology than those of animals, which tend to be spatially and temporally patchy in occurrence. Biogeomorphic interactions range greatly in scale and complexity: from the impact of one single organism on rock weathering at the microscale to the involvement of dynamic forest communities in whole catchments. One of the biggest challenges awaiting biogeomorphology in the future is to develop further studies of large-scale ecosystem: geomorphological system interactions over hundreds to thousands of year timescales.

Biogeomorphology is not simply an esoteric scientific backwater dealing with a few bizarre processes (although there are some notably weird examples of biogeomorphic studies such as the work of Splettstoesser in 1985 which discusses the role of rockhopper penguin (*Eudyptes crestatus*) feet in sandstone weathering); it has many applications. Identification of current biological:geomorphological linkages can help geologists interpret unusual sedimentary structures. Recognition of distinctive signatures of biogenic contributions to geomorphic processes on Earth can help scientists search for evidence of former life on other planets such as Mars. More practically, environmental engineering can harness the protective role of organisms in many environments to retard the action of some geomorphological processes. For example, stabilization of coastal dunes through revegetation is an essentially biogeomorphological project. At the smaller scale, understanding links between biofilms and rock surface weathering can aid the conservation of cultural heritage through reducing the threat of biodeterioration.

The future development of biogeomorphology depends both upon its capacity to answer fundamental questions and its ability to provide practical solutions to environmental problems. In some areas, such as the riparian environment, biogeomorphological studies are blossoming and providing much practical information on the mechanical role of roots, the influence of fluvial processes on seed banks and the biochemical role of riparian vegetation. In other areas, biogeomorphological studies remain more narrowly focused on unusual links between single organisms and one geomorphic process. In order to prosper further biogeomorphological studies need to establish novel research methodologies and techniques to investigate the varied links between the biotic and geomorphic worlds, many of which have proved quite hard to quantify and monitor. Furthermore, biogeomorphic studies need to move away from simple empirical, short-term studies to looking at longer term and larger scale situations. For this, numerical modelling may provide a way forward. Also, biogeomorphic studies must try and encompass the evolving two-way interplay between geomorphic and ecological processes, rather than simply focus on the impact of organisms on geomorphic processes or the influence of geomorphology on ecosystem development. Finally, biogeomorphological studies need to continue and expand their essential bridging role – by considering the links between a whole host of organic and inorganic processes in a wide range of environments within a broadly defined Earth surface systems science. The term biogeomorphology is far less important than the scientific terrain it describes – one part of the fertile, dynamic, boundary between the inorganic and organic worlds.

References

Bochet, E., Poesen, J. and Rubio, J.L. (2002) Influence of plant morphology on splash erosion in a Mediterranean matorral, *Zeitschrift für Geomorphologie* 46, 223–243.

Butler, D.R. (1995) *Zoogeomorphology: Animals as Geomorphic Agents*, Cambridge: Cambridge University Press.

Darwin, C. (1881) *Vegetable Mould and Earthworms*, London: John Murray.

Fiol, L., Fornos, J.-J. and Gines, A. (1996) Effects of biokarstic processes on the evolution of solutional

rillenkarren in limestone rocks, *Earth Surface Processes and Landforms* 21, 447–452.
Hayden, B.P., Santos, M.C.F.V., Shao, G. and Kochel, R.C. (1995) Geomorphological controls on coastal vegetation at the Virginia Coast Reserve, *Geomorphology* 13, 283–300.
Howard, J.A. and Mitchell, C.W. (1985) *Phytogeomorphology*, New York: Wiley.
Hupp, C.R., Osterkamp, W.R. and Howard, A.D. (eds) (1995) Biogeomorphology, terrestrial and freshwater systems, *Geomorphology Special Issue* 13, 1–347.
Lyell, C. (1835) *Principles of Geology*, 4th edition, London: John Murray.
Naylor, L.A. and Viles, H.A. (2000) A temperate reef builder: an evaluation of the growth, morphology and composition of *Sabellaria alveolata* (L.) colonies on carbonate platforms in South Wales, in E. Insalaco, P.W. Skelton and T.J. Lamer (eds) *Carbonate Platform Systems: Components and Interactions*, Geological Society Special Publication No. 178, 9–19.
Naylor, L.A., Viles, H.A. and Carter, N.E.A. (2002) Biogeomorphology revisited: looking towards the future, *Geomorphology* 47, 3–14.
Scatena, F.N. and Lugo, A.E. (1995) Geomorphology, disturbance, and the soil and vegetation of two subtropical steepland watersheds of Puerto Rico, *Geomorphology* 13, 199–213.
Sharp, M., Parkes, J., Cragg, B., Fairchild, I.J., Lamb, H. and Tranter, M. (1999) Widespread bacterial populations at glacier beds and their relationship to rock weathering and carbon cycling, *Geology* 27, 107–110.
Splettstoesser, J.F. (1985) Note on rock striations caused by penguin feet, Falkland Islands, *Arctic and Alpine Research* 17, 107–111.
Thornes, J.B. (ed.) (1990) *Vegetation and Erosion: Processes and Environments*, Chichester: Wiley.
Viles, H.A. (ed.) (1988) *Biogeomorphology*, Oxford: Blackwell.
——(1990) 'The agency of organic beings': A selective review of recent work in biogeomorphology, in J.B. Thornes (ed.) *Vegetation and Erosion*, 5–24, Chichester: Wiley.
——(1995) Ecological perspectives on rock surface weathering: towards a conceptual model, *Geomorphology* 13, 21–35.
Viles, H.A. and Naylor, L.A. (2002) Biogeomorphology Special Issue. *Geomorphology* 47, 1, 1–94.
Whitford, W.G. and Kay, F.R. (1999) Biopedturbation by mammals in deserts: a review, *Journal of Arid Environments* 41, 203–230.
Yair, A. (1995) Short- and long-term effects of bioturbation on soil erosion, water resources and soil development in an arid environment, *Geomorphology* 13, 87–100.

HEATHER A. VILES

BIOKARST

Biokarst refers to karst landforms created, or influenced to a significant degree, by biological processes. In turn, the processes involved in the formation of such landforms are often called biokarstic. Biokarst features can be erosional or depositional, or involve a combination of the two processes, and are commonly found on exposed limestone surfaces in a range of environmental settings. An early paper by Jones (1965) described many of the erosional features found on limestone pavements as being at least partly biokarstic in origin. Some TUFAS AND TRAVERTINES are largely influenced by biological processes and thus can be seen to be biokarstic, as can some organically influenced cave deposits. Most landforms recognized as biokarst are quite small (maximum of tens of metres in dimensions), but there is an indirect biokarstic element to most karst landscapes as organisms play a key influence on soil acidity and CO_2 levels which in turn are a vital control of karst development.

Similar terms in the literature include phytokarst, which is more narrowly defined as karst landforms produced by the action of plants, and zookarst, which refers to features produced by animal action. Both phytokarst and zookarst are subsumed within biokarst which can be produced by animal, plant or micro-organism action (and commonly involves a combination of organisms). The classic phytokarst landscape is that described by Folk *et al.* (1973) at Hell, Grand Cayman Island. Here, a series of limestone pinnacles in a low-lying swampy environment have been blackened and dissected in a random spongework pattern which Folk *et al.* ascribe to the action of cyanobacteria (blue-green algae). Another commonly identified type of phytokarst are the light-oriented erosional pinnacles found in the lit zone of many cave entrances (as reported by Bull and Laverty in 1982 in Mulu, Borneo, for example, and sometimes given the alternative name of photokarren). Other phytokarst features are the root holes produced in many limestone surfaces. Zookarstic features are rather rare and localized, but include small-scale erosional relief produced by rock wallaby urine in parts of Australia, and grooves produced by the giant tortoise (*Geochelone gigantea*) on Aldabra Atoll, Indian Ocean. By far the most important group of organisms contributing to biokarstic processes are micro-organisms and lower plants, which in mixed biofilm communities coat most subaerial limestone surfaces in a wide range of environments. Such biofilms play a range of active and passive roles in geochemical transformations, aiding both solution and re-precipitation of calcite.

Biokarst has been recorded from most karst areas, with many studies emanating from the great Chinese karst landscapes. Spectacular biologically influenced erosional relief is also found on many coastal limestone platforms, where bioerosion by a range of organisms produces a complex coastal biokarst. Although karst scientists have often been quick to note biokarst features, it has proved difficult to provide convincing process-form links in order to identify the exact nature and importance of biological influences. The major reason for this difficulty is the multi-factorial nature of karst development, which makes it impossible to untangle the interaction of interlinked processes and emerging forms. Some progress has been made with experimental studies, for example the work of Fiol *et al.* (1996) on the influence of rock surface micro-organism communities in rillenkarren development and the work of Moses and Smith (1993) on the role of physical weathering by lichens on kamenitza evolution.

The small-scale nature of many biokarstic features and the difficulties of positively ascribing their genesis to specific biological processes has made several karst scientists doubt their importance either to karst landscape development or as diagnostic landforms. The most important goal for future work is to provide a more general explanation of the role of a whole range of organisms and biological processes in karst landscape development as a whole, rather than worry whether any individual landform can be defined as biokarstic.

References

Bull, P.A and Laverty, M. (1982) Observations on phytokarst, *Zeitschrift für Geomorphologie* 26, 437–457.

Fiol, L., Fornos, J.J. and Gines, A. (1996) Effects of biokarstic processes on the development of solutional rillenkarren in limestone rocks, *Earth Surface Processes and Landforms* 21, 447–452.

Folk, R.L., Roberts, H.H. and Moore, C.H. (1973) Black phytokarst from Hell, Cayman Islands, British West Indies, *Bulletin, Geological Society of America* 84, 2,351–2,360.

Jones, R.J. (1965) Aspects of the biological weathering of limestone pavements, *Proceedings, Geologists' Association* 76, 421–433.

Moses, C.A. and Smith, B.J. (1993) A note on the role of *Collema auriforma* in solution basin development on a Carboniferous limestone substrate, *Earth Surface Processes and Landforms* 18, 363–368.

Further reading

Viles, H.A. (1984) Biokarst: review and prospect, *Progress in Physical Geography* 8, 523–542.

Viles, H.A. (1988) Organisms and karst geomorphology, in H.A. Viles (ed.) *Biogeomorphology*, 319–350, Oxford: Blackwell.

HEATHER A. VILES

BLIND VALLEY

This is a valley formed by fluvial processes that terminates downstream against a steep and sometimes precipitous slope at the foot of which the stream that carved the valley disappears underground into a cave system. The headwaters are usually on relatively impervious rocks such as sandstones or granites and the surface stream disappears underground when it crosses a lithological contact onto a KARST rock such as limestone. The larger the stream, the further it penetrates into the karst before sinking underground, and hence the longer the associated blind valley. In the early stages of development of blind valleys, the downstream wall is not very steep or high, so if the capacity of the stream-sink (swallow hole, ponore) is exceeded during flood the excess water will overflow downstream along its former course, which is usually dry and abandoned. Such cases are referred to as semi-blind valleys. The incision of the blind valley is controlled by the rate of lowering of the cave system into which it drains. This can proceed in stages as the cave stream breaks through to lower levels. Incision is propagated upstream into the blind valley and results in stream terraces. Thus terraces are often found in blind valleys that grade to the position of a former stream-sink in the terminal face high above the modern swallow hole. Over 10^4 to 10^5 years blind valley incision can attain tens to hundreds of metres.

PAUL W. WILLIAMS

BLOCKFIELD AND BLOCKSTREAM

The term blockfield (or block field) is used to describe an extensive cover of coarse rubble on flat or gently sloping terrain, with an absence of fine material at the ground surface. The German term *felsenmeer* ('stone sea') is sometimes used to describe the same phenomenon. Three types of blockfield are recognized: autochthonous blockfields, formed *in situ* by WEATHERING of the underlying bedrock; para-autochthonous blockfields, in which boulders produced by weathering of bedrock have undergone downslope mass movement over low gradients; and allochthonous blockfields derived from GLACIAL DEPOSITION by upfreezing of boulders and washing out of fine sediments. Blockstreams (or block streams) are covers of coarse debris that have accumulated by mass movement on valley floors.

Most blockfields and blockstreams occur in areas of present or former periglacial conditions (see PERIGLACIAL GEOMORPHOLOGY), particularly in arctic environments and on mid-latitude mountains that lay in the periglacial zone outside the limits of the last Pleistocene ice sheets. Blockfields are particularly widespread on mid- and high-latitude plateaux such as those of Scandinavia and Scotland.

Blockfields and blockstreams occur on a wide range of rock types, but are particularly common on well-jointed igneous and metamorphic rocks that have weathered to produce abundant boulders but only limited amounts of fine sediment. Most blockfields comprise boulders less than 1–2 m in length. In autochthonous blockfields the largest boulders usually occur at the surface and boulder size diminishes with depth. Below the openwork surface layer, blockfields and blockstreams usually contain an infill or matrix of fine sediment (sand, silt and clay), and interstitial organic material has also been recorded. Plateau blockfields tend to be 0.5–4.0 m deep, but blockstreams consisting of accumulated valley-floor boulder deposits reach depths of 10 m or more.

Surface boulders in blockfields may be angular or, more commonly, edge-rounded by GRANULAR DISINTEGRATION. Where downslope mass movement has occurred, elongate boulders often exhibit preferred downslope orientation and upslope imbrication. PATTERNED GROUND may be present in the form of large sorted circles on level ground and sorted stripes on slopes, and blockstreams sometimes support lobate structures indicative of movement by SOLIFLUCTION of underlying fine sediments.

In a perceptive early (1906) account of blockfields and blockstreams on the Falkland Islands, J.G. Anderrson attributed their formation to frost weathering (see FROST AND FROST WEATHERING) of the underlying bedrock, slow downslope movement of the weathered debris by solifluction, and immobilization by eluviation (see ELUVIUM AND ELUVIATION) of fine sediment from the upper layers. Upheaving of boulders and frost-sorting also appear necessary to produce downward fining of the openwork boulder layer and the formation of sorted patterned ground. Although this general model is widely accepted, some researchers have suggested that autochthonous and para-autochthonous blockfields and blockstreams are of polygenetic origin. In particular, it has been proposed that some plateau blockfields evolved from chemically-weathered (see CHEMICAL WEATHERING) REGOLITH mantles, of interglacial or Tertiary age, that were subsequently modified by frost action (e.g. Nesje 1989; Rea *et al.* 1996; Dredge 2000). This view is based on the location of blockfields on Tertiary erosion surfaces, and the presence in the subsurface fine fraction of clay minerals indicative of prolonged chemical weathering. On certain lithologies, however, blockfields have developed on glacially eroded bedrock since the last glacial maximum (Ballantyne 1998) implying formation under periglacial conditions alone within the last 20,000 years. Some blockfields also show evidence for modification by glacier ice or glacial meltwater (Dredge 2000).

Although there is evidence for blockfield formation during the Holocene in arctic permafrost environments (Dredge 1992), mid-latitude blockfields and blockstreams are manifestly relict. Exposed boulder surfaces have been edge-rounded by prolonged granular disintegration and many support a cover of mosses and lichens. The relationship between such relict blockfields and former ICE SHEETs has been vigorously debated. In some areas, such as western Norway and north-west Scotland, the lower limits of autochthonous blockfields descend regularly along former glacier flow-lines and have been interpreted as trimlines (see TRIMLINE, GLACIAL) marking the maximum vertical extent of the last ice sheets in these areas (Nesje 1989; Nesje and Dahl 1990; Ballantyne *et al.* 1998). Elsewhere, however, there is convincing evidence that blockfields survived the last glacial maximum under a cover of cold-based glacier ice that was frozen to the underlying substrate and hence accomplished little or no erosion (Kleman and Borgström 1990; Dredge 2000). Thus not only does the age and evolution of blockfields and blockstreams vary from area to area, but also their significance in relation to the dimensions of former ice sheets is dependent on the thermal regime of these ice masses.

References

Anderrson, J.G. (1906) Soliflution, a component of subaerial denudation, *Journal of Geology* 14, 91–112.
Ballantyne, C.K. (1998) Age and significance of mountain-top detritus, *Permafrost and Periglacial Processes* 9, 327–345.
Ballantyne, C.K., McCarroll, D., Nesje, A., Dahl, S.O. and Stone, J.O. (1998) The last ice sheet in North-West Scotland: reconstruction and implications, *Quaternary Science Reviews* 17, 1,149–1,184.
Dredge, L.A. (1992) Breakup of limestone bedrock by frost shattering and chemical weathering, eastern Canadian arctic, *Arctic and Alpine Research* 24, 314–323.
——(2000) Age and origin of upland blockfields on the Melville Peninsula, Eastern Canadian Arctic, *Geografiska Annaler* 82A, 443–454.
Kleman, J. and Borgström, I. (1990) The boulderfields of Mt Fulfjället, west-central Sweden, *Geografiska Annaler* 72A, 63–78.
Nesje, A. (1989) The geographical and altitudinal distribution of block fields in southern Norway and its significance to the Pleistocene ice sheets, *Zeitschrift für Geomorphologie, Supplementband* 72, 41–53.
Nesje, A. and Dahl, S.O. (1990) Autochthonous block fields in southern Norway: implications for the geometry, thickness and isostatic loading of the Late Weichselian Scandinavian ice sheet, *Journal of Quaternary Science* 5, 225–234.
Rea, B.R., Whalley, W.B., Rainey, M.M. and Gordon, J.E. (1996) Blockfields, old or new? Evidence and implications from some plateaus in northern Norway, *Geomorphology* 15, 109–121.

Further reading

Ballantyne, C.K. and Harris, C. (1994) *The Periglaciation of Great Britain*, Cambridge: Cambridge University Press.

SEE ALSO: frost and frost weathering; frost heave; mechanical weathering; periglacial geomorphology

COLIN K. BALLANTYNE

BLOWHOLE

Fountains of spray are emitted through blowholes during storms and high tidal periods when large breakers surge into tunnel-like caves connected to the surface. Many blowholes develop along joint (see JOINTING) or fault-controlled shafts, but particularly spectacular examples result from marine invasion of KARST tunnels and sinkholes in limestone regions, and lava tubes or tunnels in volcanic areas. Blowholes are also common on CORAL REEFs, where encrusting coralline algae can enclose spur and groove systems and surge channels running through algal ridges.

Further reading

Trenhaile, A.S. (1987) *The Geomorphology of Rock Coasts*, Oxford: Oxford University Press.

ALAN TRENHAILE

BLUE HOLE

Likened to sapphires set in turquoise, they are submarine, circular, steep-sided holes which occur in coral reefs.

The classic examples come from the Bahamas (Dill 1977), but other instances are known from Belize and the Great Barrier Reef of Australia (Backshall *et al.* 1979). Although volcanicity and meteorite impact have both been proposed as mechanisms of formation, the most favoured view is that they are the product of karstic processes (i.e. they are a DOLINE or CENOTE) which acted at times of low glacial sea levels when the reefs were exposed to subaerial processes. Subsequently they were submerged by the Flandrian Transgression of the Holocene.

References

Backshall, D.G., Barnett, J. and Davies, P.J. (1979) Drowned dolines – the blue holes of the Pompey Reefs, Great Barrier Reef, BMR *Journal of Australian Geology and Geophysics* 4, 99–109.
Dill, R.F. (1977) The blue holes – geologically significant sink holes and caves off British Honduras and Andros, Bahama Islands, *Proceedings of the 3rd International Coral Reef Symposium*, Miami, 2, 238–242.

A.S. GOUDIE

BOLSON

Derived from the Spanish word for 'purse', bolsons are depressions with centripetal drainage that are surrounded by hills and mountains (Tight 1905). At their centre there is normally a saline playa or PAN, but if the low-lying area is drained by an ephemeral stream the basin may then be termed a 'semi-bolson' (Tolman 1909). Bolsons are a feature of semi-arid basin-and-range terrain and may contain such landform types as PEDIMENTs, ALLUVIAL FANs and BAJADAs.

References

Tight, W.G. (1905) Bolson plains of the southwest, *American Geologist* 36, 271–284.

Tolman, C.F. (1909) Erosion and deposition in the Southern Arizona bolson region, *Journal of Geology* 17, 136–163.

A.S. GOUDIE

BORING ORGANISM

Several life forms have evolved a means of penetrating a variety of substances for security and protection, wood and softer rocks being common subjects of such actions. The geomophological interest is largely focused on the rock-boring organisms because their activity acts as a direct erosional agent and can also weaken the rock, making it more susceptible to erosion by other means. There exist terrestrial boring organisms, mainly algae and the fungal component of lichens, and particular interest has been shown in the marine borers, especially around the intertidal zone where they can lead to the formation of an undercut notch on rocky coasts. Some erosive mechanisms appear to be mechanical but many also appear to be chemical, attacking the more soluble rocks. Thus much of the interest in boring organisms lies in the field of the production of surface textures and smaller scale landforms in terrestrially exposed limestones (Trudgill 1985, Ch. 2, 3, 4, 8; Viles 1988) and in coastal limestone geomorphology where significant features, such as undercut notches of up to a few metres in dimension, can be formed (Trudgill 1985, Ch. 9, 10).

In studies involving environmental reconstruction, the occurrence of fossil intertidal boring organisms either above or below present sea level can provide evidence of former sea levels. This is particularly the case when undercut notches are found on dry land, considerably above present sea level. In some situations, these could have been formed by river action but where there are fossil perforations made by boring organisms, this confirms a marine origin for the undercut as assemblages of boring organisms in hard rock are unusual in fresh-water situations. This is especially the case if fossil boring bivalve shells are still present and the species can be identified and confirmed as marine organisms. In some cases, the shell material can be extracted and used for dating purposes and thus if there is a sequence of raised shorelines, palaeoenvironmental reconstruction is greatly assisted. Rowland and Hopkins (1971) noted that the boring bivalve mollusc *Hiatella arctica* can be found widely in the Arctic and Atlantic oceans and also in the Pacific Ocean from Alaska to Mexico. They noted the potential for the use of its fossil shells in paleoclimatic reconstruction.

Boring algae

Boring algae may be found in the first few millimetres of very many rock surfaces, and indeed the darker colour of rock surfaces which have been exposed for any length of time is often ascribable to this algal layer. The algae are frequently found to be blue-green algae (cyanobacteria). The algae associated with rock surfaces can be described as epiliths which live on the surface or as endoliths which live below it, with a further distinction being made for the chasmoliths which exist in the interstitial spaces between the rock grains – thus only endoliths which penetrate grains, or perforants, can be regarded as borers. The presence of endolithic perforant algae in limestone leads to the formation of very fretted surfaces known as phytokarst (Viles 1988). This can give a ferociously sharp, intricate and spongy rock surface. In cave entrances, the phytokarst is directional and angled to the light source and the borings are a product of erosion by phototrophic algae.

The benefit to the algae is to have access to moisture within the rock; however they still have to photosynthesize so they are found in thin layers just below and parallel to the surface at depths where moisture is present and where light can still penetrate. This optimum depth is termed the light compensation depth or LCD. The access to moisture is especially important in harsh environments and so while endolithic algae are found very widely over the rock surfaces of the Earth, they can occur in extreme environments including Antarctica (Friedmann and Ocampo 1976) where they can be important primary producers.

In the marine environment they commonly dominate in the mid- to upper shore, beyond which (inland) it is too dry and below which (towards the sea) it is wet enough for other organisms also to occur. They provide food for a wide range of rasping molluscs which contribute to rock erosion by ingesting rock with the algae. The algae then penetrate further into the rock to achieve an optimum LCD.

Boring endolithic algae are most usually found in carbonate rocks and the mechanism by which they bore is thought to be one where the algal filaments, about 10 μm wide, release extracellular chelating or acid fluids from the terminal cell. Using a high-powered electron microscope it can be established that up to 50 per cent of a rock surface bored by algae can be void space. Such boring can give rise to an extremely fretted dissected surface.

Boring fungi and lichens

The fungal portions of lichens can penetrate into rocks, again mainly in the carbonate rocks, in both intertidal and terrestrial environments. In both cases they exploit the weakness of calcite crystal interfaces and can also make larger pits.

Boring sponges

Boring sponges, commonly of the species *Cliona* are less able to withstand desiccation than boring algae and thus their distribution is from the mid- to lower intertidal. Extensions of the sponge tissue, termed etching amoebocytes, which, using acid secretions, are able to penetrate calcite in semicircular cuttings about 60–80 μm wide. On the surface, small 'keyhole' slots of 0.5–1 mm long are visible to the naked eye.

Boring bivalves and boring barnacles

Species of bivalve molluscs produce tubular borings which may penetrate into carbonate rock by several centimetres; they may also bore into live coral, sandstone, clay, peat and wood. There is evidence for acid secretion in carbonate substrates but they can also excavate the substrate by mechanical means, combining a rocking or rotational movement, which acts to grind the substrate, and a pumping motion using muscular contractions. In tropical areas the commonest boring bivalve is *Lithophaga* and the commonest boring barnacle is *Lithotrya*. In temperate regions the boring bivalve *Hiatella arctica* is a frequent borer of limestone in low intertidal and subtidal locations (Trudgill and Crabtree 1987). Additionally there also exist boring sipunculid worms and polychaete worms which generally make much thinner borings than the boring bivalve molluscs or boring barnacles.

Boring echinoderms

In the lower intertidal, subtidal and in rock pools several species of boring echinoderms exist, in temperate regions commonly *Paracentrotus lividus* (Trudgill *et al.* 1987) and in tropical regions *Echinometra lucunter* is common. The former make semicircular pits a few centimetres in diameter and the latter effect grooves in the rock surface. It is evident that echinoderms bore for protection, such as on exposed coasts of Carboniferous Limestone in Co. Clare, Eire, *Paracentrotus lividus* bores at rates between 0.25–1.5 cm a^{-1} whereas on sheltered coasts they may exist on unbored surfaces or in shallow depressions with lower excavation rates (Trudgill *et al.* 1987).

Rates of boring

The rates of erosion by boring organisms can be measured in a linear fashion (mm a^{-1}) where there is surface retreat or a single boring, or in a cubic fashion (cm^3 a^{-1}) where there are more diffuse excavations. Published rates of the erosion of limestones by boring organisms (Spencer 1988) include the following:

Echinoderms	0.25–14.0 cm^3 a^{-1}
Sponges	1.0–1.4 cm^3 a^{-1}
Hiatella	5–10 mm a^{-1}
Lithophaga	9–15 mm a^{-1}
Lithotrya	8–9 mm a^{-1}

Given that overall surface retreat ranges from around 0.5 to 4 mm a^{-1} and commonly is 1.0 mm a^{-1}, it can be seen that boring organisms are highly significant erosive organisms. Indeed, where boring organisms are present, this leads to the formation of a horizontal undercut or notch in the coastline, not only through the direct action of the organisms themselves but also through the removal of the mechanically weakened rock by wave action. The mechanical boring of hard rock by the larger shelled organisms produces significant quantities of fine carbonate sediment.

Zonation of boring organisms

The zonation of intertidal organisms is of interest to the geomorphologist if it is proposed that there is a cause and effect relationship between biological zonation and morphological zonation (Trudgill 1987). Originally it was thought that biological zonation was a response solely to the ability of different species to withstand emersion and desiccation. However, theories of intertidal distribution have become less environmentally deterministic and now involve concepts of interspecific competition and predation. Thus species distribution can vary markedly in any one tidal zone according to the presence or absence of predators and other species. This suggests that while there may be a general zonation of boring organisms and hence morphological types produced by them, the distribution of individual boring organisms is liable to vary at different locations in relation to predation and competition rather than just to tidal zonation.

In addition, the variation of the landform itself may provide different micro-habitats which afford protection or sites which are too exposed, leading to local variability and that feedback effects can occur. In particular, on flatter, near-horizontal surfaces boring can produce low-lying areas which facilitate water retention and hence survival, meaning that they then become even deeper as boring activity and number of boring organisms can intensify.

Geomorphological significance

Schneider (1976) suggests that in the Adriatic, moisture conditions provide the limiting factor on boring activities. He sees three stages:

1. The primary depressions in rock surfaces are first colonized since they retain moisture longer than their higher surroundings. Conditions for boring and grazing prevail longest here. As a result each depression becomes the site of more intense biological erosion.
2. The pools enlarge laterally and small depressions coalesce, wet areas are preferentially bioeroded and relief is thus intensified.
3. This is the stage of maximal relief. It shows maximal contrast in ecological conditions and thus in the destructive processes. The water in the depressions is changed at most high tides and thus brings fresh sea-water to organisms. Each pool may enlarge and break into the next.

The sequence may be restarted if a deeper bedding plane or other weakness is reached, thus draining the pool; alternatively the rim of the pool may be breached.

Tidal range itself and the degree of exposure are also important considerations. In areas with limited tidal range, as, say, in the Mediterranean, there is commonly a deep undercut notch formed by boring organisms limited to 15–20 cm in height in the mid-intertidal which in itself may only have an amplitude of some 30 cm. The same ratio of notch to range applies as tidal ranges expand. Where coasts are exposed to larger waves and storms, boring organisms may be present but contribute quantitatively far less to the overall erosion of the coast, mechanical erosion tends to dominate, and, correspondingly, the undercut notch is either weakly developed or absent. Here the coast has the appearance of a sloping ramp rather than of a vertical cliff with recess or undercut notch as tends to be the case in sheltered locations.

References

Friedmann, E.I. and Ocampo, R. (1976) Endolithic blue-green algae in the dry valleys: primary producers in the Antarctic desert ecosystem, *Science* 193, 1,247–1,249.

Rowland, R.W. and Hopkins, D.M. (1971) Comments on the use of *Hiatella arctica* for determining Cenozoic sea temperatures, *Paleogeography, Paleoclimatology, Paleoecology* 19, 59–64.

Schneider, J. (1976) Biological and inorganic factors in the destruction of limestone coasts, *Contributions to Sedimentology* 6, 1–112.

Spencer, T. (1988) Coastal biogeomorphology, in H. Viles (ed.) *Biogeomorphology* 255–318, Oxford: Blackwell.

Trudgill, S.T. (1985) *Limestone Geomorphology*, Harlow: Longman.

Trudgill, S.T., Smart, P.L, Friederich, H. and Crabtree, R.W. (1987) Bioerosion of intertidal limestone, Co. Clare, Eire. 1: *Paracentrotus lividus*, *Marine Geology* 74, 85–98.

Trudgill, S.T. and Crabtree, R.W. (1987) Bioerosion of intertidal limestone, Co. Clare, Eire. 2: *Hiatella arctica*, *Marine Geology* 74, 99–109.

Trudgill, S.T. (1987) Bioerosion of intertidal limestone, Co. Clare, Eire. 3: zonation, process and form, *Marine Geology* 74, 111–121.

Viles, H.A. (1988) Organisms and karst geomorphology, in H. Viles (ed.) *Biogeomorphology*, 319–350 Oxford: Blackwell.

STEVE TRUDGILL

BORNHARDT

Bornhardts are dome-shaped, steep-sided hills, usually built of massive igneous rocks such as granite or ryolite, with bare convex slopes covered with very little talus and flattened summit surface. Bornhardts form due to differential weathering and erosion, in the course of which the surrounding less massive rock is eroded away leaving massive, sparsely jointed compartments. Many bornhardts have probably formed through selective DEEP WEATHERING followed by stripping of the SAPROLITE, but they can also emerge through gradual lowering of the surrounding terrain in the absence of deep weathering. Bornhardts are structure-controlled landforms and occur in every climatic zone; the existence of massive rock compartments is the necessary factor.

Characteristic features of bornhardts are slope-parallel joints called sheeting joints. They tend to be considered as resultant from unloading and would develop at shallow depth and after exposure, although it is also argued that sheeting develops in deeper parts of the crust, in response to compressional stress (Vidal Romani and Twidale 1999). Gradual opening of joints promotes slope instability, therefore rock slides and falls involving large masses of rock are common on bornhardts. Consequently, whereas upper slopes are bare and talus-free, footslopes may be covered by big blocks derived from upslope.

One of the persistent problems in the literature is the distinction between a bornhardt and an INSELBERG. The term 'bornhardt' was used by B. Willis in the 1930s to honour a German explorer from the turn of the nineteenth century, W. Bornhardt, who had introduced the name 'inselberg', but primarily to emphasize a special category of massive, dome-shaped inselbergs. The term subsequently evolved to describe monolithic domes regardless of their degree of isolation in the landscape. Therefore, although the terms are occasionally used as synonyms, these two categories of hills should not be confused. Nor is it justified to restrict bornhardts to granite lithology. Whereas 'inselberg' emphasizes isolation in space, bornhardts need to have distinctive domed shapes. Hence there is only partial overlap between the two and there exist bornhardts which are not inselbergs, and vice versa.

Classic examples of bornhardts include domes in Rio de Janeiro, Half Dome in Yosemite Valley and Ayers Rock in Australia. They are also abundant within African shields.

Reference

Vidal Romani, J.R. and Twidale, C.R. (1999) Sheet fractures, other stress forms and some engineering implications, *Geomorphology* 31, 13–27.

Further reading

Selby, M.J. (1982) Form and origin of some bornhardts of the Namib Desert, *Zeitschrift für Geomorphologie N.F.* 26, 1–15.

Thomas, M.F. (1965) Some aspects of the geomorphology of domes and tors in Nigeria, *Zeitschrift für Geomorphologie N.F.* 9, 63–82.

Twidale, C.R. and Bourne, J.A. (1978) Bornhardts, *Zeitschrift für Geomorphologie N.F. Supplementband* 31, 111–137.

PIOTR MIGOŃ

BOULDER PAVEMENT

Striated boulder pavements form on intertidal surfaces affected by floating ice (Hansom 1983; Forbes and Taylor 1994), in ice-affected fluvial environments (Mackay and Mackay 1977) and at the base of glaciers or grounded ice sheets (Eyles 1988). Pavements deposited subglacially are the result of accretion of boulders around an obstacle and carry striations that are largely unidirectional, similar to fluvially derived striations produced by debris-charged floating ice. Striations on intertidal pavements are controlled by the direction of grounding of floating ice together with rotational striations imparted on stranding. Intertidal boulder pavements are composed of smoothed and highly polished boulders, often up to 1 m in diameter, tightly packed together as an undulating mosaic. They are often interrupted by bedrock outcrops together with furrows and polygonal depressions up to 5 m across. The main process seems to be the bulldozing and packing of loose boulders in the intertidal zone of a low-gradient boulder-strewn shore and the abrasion and striation of boulder surfaces by rock-shod floating ice. Prerequisites for their development appear to be a boulder source, frequent onshore movement of floating ice and a low-gradient intertidal zone. The degree of development is controlled by the frequency of onshore ice movement, well-formed pavements occurring in environments subject to high frequencies of freely moving ice.

References

Eyles, C.H. (1988) A model for striated boulder pavement formation on glaciated shallow-marine shelves: an example from the Yakataga Formation, Alaska, *Journal of Sedimentary Petrology* 58, 62–71.

Forbes, D.L. and Taylor, R.B. (1994) Ice in the shore zone and the geomorphology of cold coasts, *Progress in Physical Geography* 18, 59–89.

Hansom, J.D. (1983) Ice-formed intertidal boulder pavements in the sub-Antarctic, *Journal of Sedimentary Petrology* 53, 1035–1045.

Mackay, J.R. and Mackay, D.R. (1977) The stability of ice-push features, Mackenzie River, Canada, *Canadian Journal of Science* 14, 2,213–2,225.

JIM HANSOM

BOUNDARY LAYER

Emergence of boundary layer theory

The German engineer Ludwig Prandtl (1875–1953) presented a seminal paper to the 1904 Mathematical Congress in Heidelberg entitled 'Fluid Motion with very Small Shear' (Schlichting 1968). Prandtl showed that, with the aid of theoretical considerations and simple experiments, fluid flow over or around a solid body such as a sphere, cylinder or flat plate could be divided into two distinct regions. One region is relatively close to the body (or boundary), is relatively thin, and is characterized by large velocity gradients and viscous shear stresses. That is, fluid friction plays an important role in determining the physical characteristics of the layer. The second region is relatively far away from the boundary, and it is characterized by small velocity gradients and viscous shear stresses. That is, fluid friction may be neglected. This conceptualization termed boundary layer theory, as presented by Prandtl and further expanded by Geoffrey I. Taylor (1886–1975) and Theodor von Kármán (1881–1963), proved to become the foundation for modern fluid mechanics (Schlichting 1968).

A boundary layer can be defined as that part of the flow markedly affected by the presence of the boundary (Middleton and Wilcock 1994). Here, a flow refers to the motion of almost any kind of Newtonian fluid. Most real flows of geomorphic interest, such as flowing water in a river or blowing wind over a sand dune, are considered boundary layers because much of the flow is strongly affected by the boundary.

Laminar and turbulent flow

Osborne Reynolds (1842–1912) was the first to distinguish between two types of flow regime: laminar and turbulent (Schlichting 1968; Tritton 1988). In the laminar regime, the entire flow region appears to be divided into a series of fluid layers, each layer bounded by stream surfaces conforming to the boundary. In the case of two-dimensional flows, the traces of these surfaces on the flow plane are called streamlines (Figure 13). The rate of flow between two adjacent streamlines remains constant, although their spacing and orientation may vary, and velocity at-a-point does not vary or fluctuate in time. The transfer of fluid momentum, which results from the acceleration of slower fluid layers by faster moving layers, occurs at the molecular scale (Schlichting 1968).

In a turbulent flow, fluid particle paths are sinuous, intertwining and disordered. Fluid mixing occurs at both molecular and macroscopic scales. At this larger scale, fluid mixing commonly involves three-dimensional flow structures called eddies or vortices (turbulence), which are hairpin or horseshoe shaped rotating parcels of fluid moving away from or toward a boundary (Smith 1996). Because these vortices are relatively large and energetic, the time and length scales of turbulence are large and hence the turbulent transfer of fluid momentum is

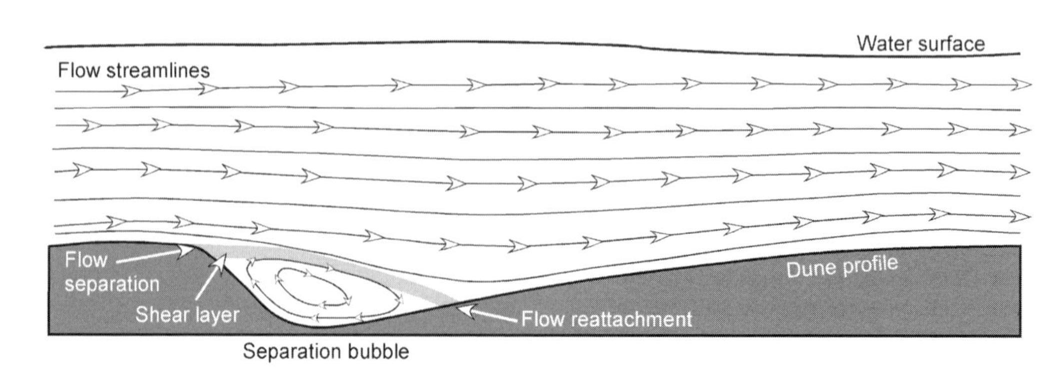

Figure 13 Time-averaged flow over a dune bedform (from Bennett and Best 1995). Flow is from left to right

large compared to that of molecular diffusion. Within a turbulent flow, velocity at-a-point can fluctuate greatly as a result of passing vortices. Turbulence intensity typically is a measure of the magnitude of the velocity fluctuations compared to the time-average value at-a-point.

The boundary Reynolds number

Reynolds found both experimentally and through dimensional analysis that the transition between laminar flow and turbulent flow occurs when the ratio of the inertia fluid forces is significantly larger than the viscous or frictional fluid forces (Tritton 1988). The inertia forces can be defined as the product ρud where ρ is fluid density, u is the mean flow velocity and d is the mean flow depth. The frictional forces can be characterized by the molecular viscosity of the flow, μ. This dimensionless ratio is called the boundary Reynolds number, Re. Flows are considered laminar when $Re < 500$, turbulent when $Re > 2000$, and transitional when $500 < Re < 2000$ (Tritton 1988). When a flow is laminar, any small disturbance to the flow, such as a protruding particle or a small change in bed topography, will not cause a change to the flow path lines or velocity and the disturbance will be damped by viscous forces. When a flow is turbulent, all disturbances to the flow will produce an effect throughout the boundary layer. When a flow is transitional, only select disturbances will affect the flow.

Few natural flows are laminar because most are deep and fast enough for the boundary Reynolds number to be very large. For example, the Mississippi River near its mouth has a Reynolds number near 10^7.

Flow separation and reattachment

As a flow moves along a curved boundary or over an obstacle, the boundary layer may separate and move away from the wall. Separation occurs when the pressure gradient in the downstream direction is adverse or unfavourable (Tritton 1988). A pressure gradient is considered adverse when a flow is expanding, diverging and decelerating in the longitudinal direction, and the pressure acting on the boundary is increasing, such as on the rear part of a streamlined pier. Both laminar and turbulent boundary layers can separate. Laminar flows usually require only a relatively short region of adverse pressure gradient to produce separation, whereas turbulent flows separate less readily.

Reattachment is the opposite of separation. There is a tendency for separation to be followed by reattachment unless the adverse pressure gradient continues long enough to prevent it (Tritton 1988). This separation–reattachment phenomenon is associated with a separation bubble or a region of flow recirculation. A typical example of flow separation can be found downstream of a ripple or a dune (Figure 13; Bennett and Best 1995). A recirculation bubble extends from the ripple or dune crest point to a distance downstream of about 5 to 7 times the bedform height, where flow reattaches to the boundary.

Downstream of the line of separation, there is a region of intense shear between the faster-moving outer part and the slower-moving or counter-rotating inner part of the boundary layer. Consequently, the flow along this shear layer is unstable, and turbulence is produced (Figure 13; Tritton 1988; Middleton and Wilcock 1994). Turbulent wakes, or regions of high turbulence, are typical of flow past any kind of obstruction at a high Reynolds number. The mixing layer present above a ripple or dune just downstream of the bedform brink is dominated by shear layer turbulence.

Structure of turbulent boundary layers

A turbulent boundary layer can be subdivided into three distinct zones: an inner layer, an outer layer and a wake region (Schlichting 1968). The inner region of a turbulent boundary layer is composed of a viscous sublayer (up to $y^+ = yu*/\nu = 10$, where y is distance from the boundary, $u*$ is shear velocity, $u* = (\tau/\rho)^{0.5}$, τ is bed shear stress, and ν is the kinematic viscosity of the flow) and a buffer layer (from $10 < y^+ < 40$). In the viscous sublayer, viscous forces dominate, yet very weak turbulent motions occur. These motions, called viscous sublayer streaks, are longitudinally oriented rotating tubes of alternating high- and low-speed fluid (Smith 1996). In some flows, sand streaks can be observed on the bed surface and these demarcate the location of low-speed streaks that tend to accumulate sand. Such sand streaks have been observed in the rock record and are called parting or current lineations.

The buffer layer is where the turbulent bursting process takes place. Low-speed streaks, in the general shape of a hairpin vortex, are lifted from the boundary and into the buffer region.

This lifted low-speed streak creates a thin shear layer that becomes unstable, oscillates, and is energetically ejected into the outer region of the flow (called a burst or an ejection event). Immediately following an ejection event, high-speed fluid from the outer region rushes in to replace the ejected fluid, impinging the bed (called a sweep event; see Smith 1996). This two-stage phenomenon is called the bursting process and can account for 70 per cent or more of all turbulence production within a boundary layer. Turbulent bursts or ejections are energetic enough to suspend sediment from the bed in river and airflows. Sweep events, with their high instantaneous drag forces, can entrain sediment particles resting on a bed surface.

In the outer region, representing the lower 10 to 20 per cent of the flow depth, there is a region where the velocity distribution varies logarithmically with distance from the bed, thus termed the logarithmic zone. Prandtl first conceptualized this velocity distribution in his 1925 mixing length theory (Schlichting 1968). Prandtl visualized a simple mechanism of fluid motion where parcels of fluid would move upwards or downwards, accelerating or decelerating the surrounding fluid. The distance over which the fluid is mixed is called the mixing length. Von Kármán expanded this theory by assuming that the mixing length varies as a simple function of distance from the bed multiplied by a dimensionless, universal coefficient (von Kármán's coefficient, $\kappa \sim 0.41$; Schlichting 1968). The final result is the Kármán–Prandtl law of the wall, $u/u_* = 1/\kappa \ln(y/y_0)$, where u is the velocity at a distance y from the wall and y_0 is the roughness height where velocity goes to zero. This velocity distribution has been shown applicable to a wide variety of flows such as pipes, rivers, near-shore environments, aeolian environments and in the near-bed region of gravity currents. Common uses of the law of the wall are the determinations of bed shear stress, roughness height and the turbulent mixing characteristics of the flow.

Finally, the velocity distribution in the outer 80 per cent of the turbulent boundary layer deviates from the logarithmic law. Here the velocity distribution is similar in shape to the velocity-defect profile in wakes (law of the wake; Coles 1956). For many straight rivers with relatively flat beds, the law of the wall is applicable over the entire flow depth (Middleton and Wilcock 1994). However, the presence of bedforms will significantly increase roughness length scales and velocity distributions.

Turbulent boundary layers are further qualified based on the roughness of the bed surface. This roughness parameter is called a grain Reynolds number Re_G, and it is defined as $Re_G = u_* k_s/\nu$, where k_s is the equivalent sand roughness height, which is approximately equal to the bed grain size (Schlichting 1968; Bridge and Bennett 1992). If the bed sediment is relatively small or absent, such that the grains are completely immersed in the viscous sublayer, then $Re_G < 11$, the sediment particles are subjected to viscous fluid forces only, and the boundary is considered hydraulically smooth. If the bed sediment is relatively large, such that the grains are larger than the viscous sublayer, then $Re_G > 70$, the sediment particles are subjected to turbulent fluid forces, and the boundary is considered hydraulically rough. Turbulent boundary layers are considered hydraulically transitional if some but not all grains are immersed in the viscous sublayer and $11 < Re_G < 70$. There are slightly different versions of the law of the wall and the determination of the equivalent sand roughness height depending on the roughness of the turbulent boundary layer. In general, beds that are composed of larger grains have relatively higher turbulent intensities and greater flow resistance.

The shape of the oft-used Shields curve for the dimensionless threshold of particle entrainment reflects this effect of grain Reynolds number on boundary layer characteristics (Bridge and Bennett 1992). Very small grains ($Re_G < 10$; hydraulically smooth flows) immersed in the viscous sublayer require higher dimensionless shear stresses for particle entrainment than larger grains that protrude higher in the turbulent boundary layer ($Re_G > 100$; hydraulically rough flows).

References

Bennett, S.J. and Best, J.L. (1995) Mean flow and turbulence structure over fixed, two-dimensional dunes: implications for sediment transport and bedform stability, *Sedimentology* 42, 491–513.

Bridge, J.S. and Bennett, S.J. (1992) A model for the entrainment and transport of sediment grains of

mixed sizes, shapes and densities, *Water Resources Research* 28, 337–363.
Coles, D. (1956) The law of the wake in the turbulent boundary layer, *Journal of Fluid Mechanics* 1, 191–226.
Middleton, G.V. and Wilcock, P.R. (1994) *Mechanics in the Earth and Environmental Sciences*, Cambridge: Cambridge University Press.
Schlichting, H. (1968) *Boundary-Layer Theory*, 6th edition, New York: McGraw-Hill.
Smith, C.R. (1996) Coherent flow structures in smooth-wall turbulent boundary layers: facts, mechanisms and speculation, in P.J. Ashworth, S.J. Bennett, J.L. Best and S.J. McLelland (eds) *Coherent Flow Structures in Open Channels*, 1–39, Chichester: Wiley.
Tritton, D.J. (1988) *Physical Fluid Dynamics*, 2nd edition, Oxford: Oxford Science Publications.

SEAN J. BENNETT

BOUNDING SURFACE

Bounding surfaces represent discontinuities in sedimentation. Surfaces occur within all environments, and form an integral part of the geomorphic landscape and the rock record.

For example, migrating aeolian dunes produce three bedform-scale bounding surfaces. Reactivation surfaces form when the lee faces of dunes are eroded, such as when the dune reverses. Where associated with seasonal winds, these surfaces define cycles within the dune cross-strata. Superposition surfaces occur where smaller dunes migrate over the lee face of the main bedform. The surface is produced by scour associated with the passage of the INTERDUNE troughs of the superimposed dunes. Interdune surfaces begin with deflation of the stoss (windward) slopes of migrating dunes and culminate at the interdune floor.

At the dunefield scale, sequence or super surfaces form when accumulation in the field ceases. Examples include stabilization of the field by vegetation, and deflation to a planar surface defined by the water table.

Further reading

Kocurek, G. (1996) Desert aeolian systems, in H.G. Reading (ed.) *Sedimentary Environments: Processes, Facies and Stratigraphy*, 125–153, Oxford: Blackwell.
Rubin, D.M. (1987) *Cross-bedding, Bedforms, and Paleocurrents*, Tulsa: Society of Economic Paleontologists and Mineralogists.

SEE ALSO: interdune

GARY KOCUREK

BOWEN'S REACTION SERIES

In the early twentieth century, N.L. Bowen (1928) developed an idealized model, now called Bowen's Reaction Series, to describe the evolution or differentiation of igneous rocks. Recognizing that the types of minerals that form, and the sequence in which they crystallize, depend on the chemical composition of the magma and the temperature and pressure range over which the magma crystallizes, Bowen described two separate reaction sequences at high temperatures that eventually merge into a single series at cooler temperatures (see Figure 14).

The discontinuous series (left-hand side), involves the formation of chemically unique minerals at discrete temperature intervals from iron and magnesium-rich mafic magma. The first rocks to form are composed primarily of the mineral olivine. Continued temperature decreases, and fractionation of the magma (the early formed minerals are removed from the liquid by gravity), change the dominant minerals which form from pyroxene, to amphibole, and then to biotite.

The continuous series (right-hand side), involves the mineral plagioclase feldspar. At high temperatures, these minerals are dominated with calcium. With continued cooling, calcium and aluminum are exchanged for sodium and silicon. The convergence of both series occurs with a continued drop in magma temperature. Crystallizing rocks become richer in potassium and silica. The last mineral to crystallize in the Bowen's Reaction Series is quartz.

Examples of complete igneous sequences from basalt to granite are rare and other mechanisms are now known to produce differentiation sequences. Bowen himself acknowledged that the series was a simplification of very complex reactions and could be misleading if taken at face value. The reaction series also is used to explain susceptibility of minerals to weathering (see GOLDICH WEATHERING SERIES).

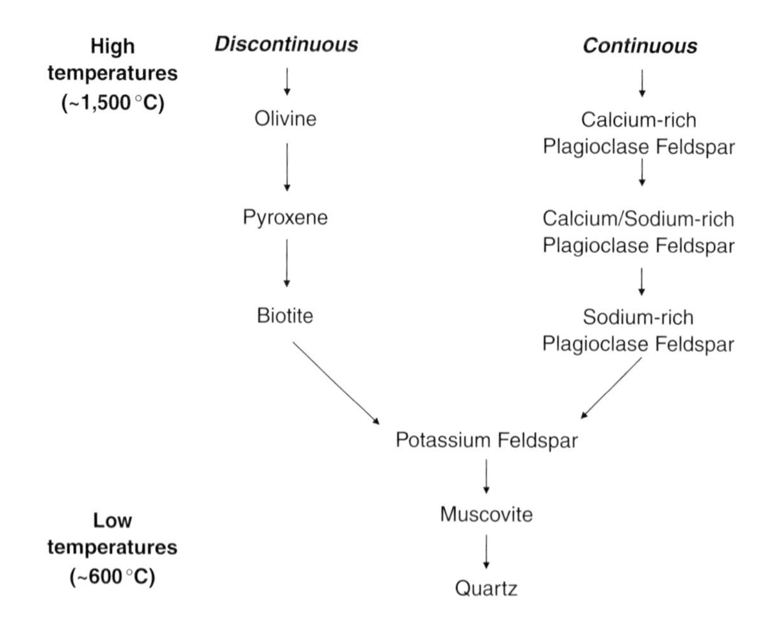

Figure 14 The Bowen's Reaction Series

Reference

Bowen, N.L. (1928) *The Evolution of the Igneous Rocks*, Princeton, NJ: Princeton University Press.

SEE ALSO: Goldich weathering series; chemical weathering

CATHERINE SOUCH

BOX VALLEY

A box valley has a broad flat floor, bounded by steep slopes which form a sharp piedmont angle. They are common in periglacial areas (see PERIGLACIAL GEOMORPHOLOGY), where they are formed by rapid lateral migration of braided channels, and by MECHANICAL WEATHERING by an 'ice rind' beneath the floodplain. Box valleys in mid-latitudes have been interpreted as relics of former cold climates. However, they also occur in tropical and arid lands, where they are formed by intense weathering or sheet wash.

Further reading

Büdel, J. (1982) *Climatic Geomorphology*, Princeton: Princeton University Press.
Young, R. (1987) Sandstone landforms of the tropical East Kimberley region, northwestern Australia, *Journal of Geology* 95, 205–218.

R.W. YOUNG

BRAIDED RIVER

Phenomenology of braiding

The hallmark of braided rivers is the presence of multiple active channels that divide and rejoin to form a pattern of gently curved channel segments separated by exposed bars (Plate 18). Braided rivers are marked equally by temporal dynamism: gradients in sediment flux associated with the complex spatial topography change local slopes, leading the flow to continually adjust its path as

Plate 18 The braided Rakaia River, New Zealand

it picks its way through the network. Even when external conditions are constant, the braided pattern is continually changing, yet statistically consistent: a true dynamic equilibrium.

Braided rivers are known from around the world, but they are most common today at high latitudes. Often, braided rivers are GRAVEL-BED RIVERS, but prominent exceptions include some of the largest braided rivers in the world, such as braided sections of the Huang He and Ganges–Brahmaputra rivers. It has been suggested that braiding was the dominant river pattern on Earth before the first appearance of land plants in late Silurian time (Schumm 1968). Braided patterns have been observed on Mars, and vegetal patterns that resemble braiding are known from some bogs. However, although many morphological features of rivers are reproduced under oceans and lakes by the action of density currents, braiding appears to be rare or absent in the subaqueous realm.

It is worth distinguishing braided rivers from anastomosing channels, the other main type of anabranching channel (see ANABRANCHING AND ANASTOMOSING RIVER). In anastomosing channel networks, the typical width of the channels is much smaller than that of the bars, whereas in braided rivers these two length scales are comparable. Thus, the braided channel pattern is more space-filling than the anastomosed pattern. It is also worth distinguishing two somewhat different ways in which braided stream patterns can develop. In one case, 'confined braiding', there is a well-defined channelway that fills with water during floods and develops a pattern of submerged bars. As stage decreases, the bar tops emerge, producing a braided channel pattern. In 'free braiding', the braiding develops on an effectively unconfined plain. As discharge increases, an increasing number of channels is occupied, but the braid plain is never completely submerged. The relation between these two types of braiding is still not clear.

The dynamic character of braided networks owes much to interplay of their three basic elements (Ashmore 1991b): channel segments (anabranches), confluences and in-channel bars. Generally, these morphological elements are associated with locally parallel, converging and diverging bank geometries respectively. Channel segments may be straight or gently sinuous. Curved channel segments increase their SINUOSITY through erosion of their outer bank much as MEANDERING channels do, though braid anabranches often widen as they do so. Confluences are associated with channel narrowing, elevated velocities and local scour (Best 1988; Roy and Bergeron 1990). Bars are associated with channel expansion and widening, deposition (mainly on the bar periphery) and eventual splitting of the flow via scour along the bar sides. The dynamics of stream braiding largely results from the strongly nonlinear relation between flow strength and sediment flux. Confluences become scour sites because narrowing and acceleration of the flow increase its capacity for sediment transport. The scour further accentuates the narrowing and acceleration – an example of positive feedback. The converse is true in divergences. This tendency of the bed to accentuate local variability in the flow means the system never develops a static, steady-state configuration, even if water and sediment are supplied at a constant rate. This 'dynamic equilibrium' applies also to the flow of sediment through the braided network. As bars grow and are then incised by channels, sediment is impounded and released, producing highly variable sediment flow even if external conditions are steady (Ashmore 1991a).

Why do rivers braid?

Conditions commonly associated with the occurrence of braided rivers in nature include steep slopes, variable water discharge, coarse grain size and high rates of sediment supply. Empirically, we can identify sets of variables that discriminate braided from straight or meandering rivers. The most common of these discriminant plots is slope versus discharge, in which braided rivers appear at higher slopes for the same discharge than meandering rivers do. Empirical relations such as these provide hints as to the important variables, but little physical insight into the actual cause of braiding.

Historically, a major step in analysis of the causes of river patterns like braiding came with the application of stability analysis to the problem (Fredsoe 1978; Parker 1976). In stability analysis, one asks mathematically how a system responds to small perturbations. In analysing river planform the starting system is a straight channel, referred to as the 'base state'. Then one adds perturbations to the bed (and in some cases the banks), generally represented as one- or two-dimensional sine waves of infinitesimally small amplitude, and investigates how the perturbations change the flow and sediment-transport fields. If any of these perturbations changes the system in such a way as to produce its own growth, we have positive feedback and the system is unstable. This approach is based on the idea that natural systems are constantly being 'probed' by random disturbances – a tree falls in the river, for instance – that include a wide spectrum of wavelengths. A system that could not recover from such a disturbance would not last long in the real world.

In the case of rivers, the main control on planform stability turns out to be the channel aspect (width:depth) ratio. Channels narrower than about 20 times the depth tend to remain straight; those with widths roughly between 15 and 150 depths develop alternate bars, presumed to lead to meandering; and channels wider than about 150 depths develop multiple bars that are interpreted as leading to braiding.

Stability analysis was a major advance in that it provided a mechanistic foundation for understanding the origin of braiding and meandering. It also raised a number of new questions. For most rivers, the channel aspect ratio is not imposed from outside but is set by the dynamics of the channel itself. Unfortunately, the dynamics of channel width remains one of the fundamental unsolved problems of fluvial geomorphology. But it does seem clear that one of the strongest controls on width is the total effective sediment discharge (i.e. excluding the washload, suitably defined). Thus, high effective sediment loads are critical to braiding in two ways. First, high effective loads directly increase the width, directly increasing the aspect (width:depth) ratio. Second, for a given water supply, increasing the ratio of sediment to water discharge increases the slope, leading to smaller depths and thus further increasing the aspect ratio. This analysis helps explain why plots using slope and water discharge can discriminate a braiding 'regime' but suggests that neither variable is the fundamental control per se.

Chaos, complexity and braiding

The core idea of *chaos* in the scientific sense is that a fully deterministic system nonetheless can be effectively unpredictable. Surprisingly little has been done to analyse braided rivers formally as chaotic systems. It is clear that they are governed by a set of reasonably well-known deterministic equations, and they certainly appear to be unpredictable to any level of detail on timescales much longer than that required for migration of an anabranch or bar a significant fraction of its width. One especially fruitful line of analysis (Foufoula-Georgiou and Sapozhnikov 1998; Sapozhnikov and Foufoula-Georgiou 1996) has shown that braided-river plan patterns are fractal (specifically, *self-affine* fractals). The self-similarity or self-affinity that defines a pattern as fractal can occur either within one river (a small part of the river looks like a larger part), or between two different rivers (a small river looks like a larger river). An easily seen manifestation of similarity between large and small rivers is that the braided patterns one might see around town or on the beach share many basic dynamical characteristics with full-scale braided rivers. The similarity of large and small braided rivers makes braided rivers accessible to experimental study (Ashmore 1982).

Braided rivers also show a time–space scaling according to which the time evolution of a small part of a braided channel system is statistically indistinguishable from that of a larger part of the system, provided time is scaled (imagine speeding up or slowing down a film) according to a power of the ratio of the two areas being compared (Sapozhnikov and Foufoula-Georgiou 1997). These scaling results are not chaos per se, but power-law scaling of this kind is a common byproduct of chaotic dynamics. The time–space scaling also implies that braided rivers may be self-organized critical systems.

Chaos theory arose from the study of atmospheric convection, and turbulent fluid flow remains one of the archetypes of chaotic behaviour. Braiding as a phenomenon seems analogous to turbulence in some respects (Paola 1996). In effect, a braided channel pattern is to a straight channel as turbulent flow is to laminar flow. Increasing the Reynolds number of laminar shear flow increases the momentum flux, which produces unstable high velocity gradients. The instability leads to a new, chaotic state that is more efficient at transferring momentum than the original laminar flow. In a straight river channel, increasing the sediment flux increases the width and (indirectly) decreases the depth, leading to unstable high channel aspect ratios. This instability leads to a new, chaotic state (braiding) that is more efficient at transferring sediment than a straight channel. (The nonlinearity of sediment flux as a function of flow velocity means that a flow system with high-speed and low-speed regions transports more sediment on average than a uniform stream with the same mean velocity.) The main stability parameter in braiding, and hence the equivalent of the Reynolds number, is the width:depth (aspect) ratio.

These observations help us understand aspects of the phenomenon of braiding not well captured in either empirical analyses or stability theory. The appearance of alternate bars in stability analyses is generally interpreted as implying a meandering plan pattern. But experimentally, meandering in channels without cohesive sediment is a transient phenomenon. Left to its own devices, a channel with alternate bars and noncohesive banks eventually evolves into a braided pattern with a low braid index. Channelized flow over noncohesive sediment cannot produce fully developed meandering, regardless of the channel aspect ratio. Evidently, just as pipe flow has two fundamental states (laminar and turbulent), channel flow in noncohesive sediment has two fundamental states: straight and braided. A fully realized meandering state requires that channels with alternate bars be stabilized, for example by cohesive sediment or vegetation.

We seem to be on the threshold of major advances in the theoretical modelling of braided rivers (see MODELS). The new field of complexity theory, which seeks unifying theoretical ideas and common behavioural patterns across a range of nonlinear systems, may help us to develop better theories of stream braiding. It is clear at this point that some aspects of the phenomenon of braiding can be captured in models that greatly simplify the detailed mechanics of flow and sediment transport (Murray and Paola 1994), while other aspects cannot. An insightful synthetic approach based on abstracting the detailed mechanics, perhaps ordered in the kind of hierarchical structure that has been used to study other complex systems, may be the most effective way of modelling braided rivers. It also may be that the best approach will simply be to develop numerical tools to solve the governing flow and sediment-flux equations on a sufficiently fine and adaptable mesh to allow for detailed simulation of the complex physics of braiding. Either way, newly emerging 'synoptic' field and laboratory data sets that capture the co-evolution of flow and topography over a whole river reach rather than a small area will be the standard against which new theoretical ideas will be tested. Of the main river types, braided rivers have proved to be the most challenging to analyse formally. The next edition of this encyclopedia will no doubt show dramatic results from some of the simulation efforts now under way.

References

Ashmore, P.E. (1982) Laboratory modelling of gravel braided stream morphology, *Earth Surface Processes and Landforms* 7, 201–225.

——(1991a) Channel morphology and bed load pulses in braided, gravel-bed streams, *Geografiska Annaler* 73A, 37–52.

——(1991b) How do gravel-bed rivers braid? *Canadian Journal of Sciences* 28, 326–341.

Best, J.L. (1988) Sediment transport and bed morphology at river channel confluences, *Sedimentology* 35, 481–498.

Foufoula-Georgiou, E. and Sapozhnikov, V.B. (1998) Anisotropic scaling in braided rivers: an integrated theoretical framework and results from application to an experimental river, *Water Resources Research* 34(4), 863–867.

Fredsoe, J. (1978) Meandering and braiding of rivers, *Journal of Fluid Mechanics* 84, 609–624.

Murray, A.B. and Paola, C. (1994) A cellular model of braided rivers, *Nature* 371, 54–57.

Paola, C. (1996) Incoherent structure: turbulence as a metaphor for stream braiding, in P.J. Ashworth, S.J. Bennett, J.L. Best and S.J. McLelland (eds) *Coherent Flow Structures in Open Channels* 705–723, Chichester: Wiley.

Parker, G. (1976) On the cause and characteristic scales of meandering and braiding in rivers, *Journal of Fluid Mechanics* 76, 457–480.

Roy, A.G. and Bergeron, N. (1990) Flow and particle paths at a natural river confluence with coarse bed material, *Geomorphology* 3, 99–112.

Sapozhnikov, V.B. and Foufoula-Georgiou, E. (1996) Self-affinity in braided rivers, *Water Resources Research* 32(5), 1,429–1,439.

Sapozhnikov, V.B. and Foufoula-Georgiou, E. (1997) Experimental evidence of dynamic scaling and indications of self-organized criticality in braided rivers, *Water Resources Research* 33, 1,983–1,991.

Schumm, S.A. (1968) Speculations concerning paleohydrologic control of terrestial sedimentation, *Geological Society of America Bulletin* 79, 1,573–1,588.

CHRIS PAOLA

BROUSSE TIGRÉE

One of the most striking forms of PATTERNED GROUND is the brousse tigrée as identified from aerial photographs in West Africa (Clos-Arceduc 1956). This pattern is composed of alternating bands of vegetation and bare grounds aligned at the contour. From the air, these bands or arcs form a distinctive pattern similar to the pelt of a tiger.

Similar patterns have been recognized from aerial photos from many parts of the world. They were called *mulga groves* in Australia and *mogote* in Mexico. Ground truth may differ since banded vegetation can consist either of grass (Mauritania, Somalia, Sudan), shrubs (Australia, Mexico), or trees (Australia, Mali, Niger). They occur only where the co-occurrence of several critical conditions is met: low annual rainfall (75–650 mm), gentle and uniform slope (0.2–2 per cent) and crusting soils. These factors favour water runoff sufficient to produce sheet OVERLAND FLOW over a distance of a few tens of metres but insufficient to trigger the concentration of runoff into RILLS. In flatter landscapes, the vegetation is no longer banded but spotted because of the nondirectional runoff pattern. Slope also controls the wavelength (band plus interband width) of the pattern even at a local scale. The wavelength decreases exponentially with increasing slope gradient. Differences observed in the soils of bands and associated interbands are a consequence rather than a cause of banded ground (Bromley *et al.* 1997).

For a given slope, the mean annual rainfall determines the ratio between the width of the vegetation bands of arcs and the width of the bare bands. The bands accumulate runoff water and function as if they were in a higher rainfall climatic regime. The optimal rainfall for band development increases with increasing percentage of high rainfall event and decreasing duration of the rainy season. This optimal annual rainfall increases from 250 mm in central Australia to 550 mm in south-west Niger.

These banded patterns are natural examples demonstrating the principles of water, soil and nutrients conservation in space and time. Although the role of wind cannot be overlooked in certain circumstances, surface hydrological processes are critical to the ongoing functioning of banded landscapes. Three main processes are involved: differential infiltration, obstruction to overland flow, and efficient nutrient cycling. Soil crusts dominate in the interbands, resulting in low infiltration, whereas vegetation, litter and bioturbation effects facilitate high infiltration rates in the bands and arcs. The banded patterns act as a natural water harvesting system, the overland flow produced from the bare and impermeable interbands running onto the bands. Vegetation bands tend to obstruct or regulate sheet flow so that sediments and organic matter are continually being deposited and conserved within the bands, forming a natural bench structure that limits soil erosion. Due to the rainwater redistribution, the bands receive from two (in south-eastern Australia) to four times (in southwest Niger) the rainfall at the site. The centre of the bands has abundant biopores enabling effective water capture from the interband. The soils in the bands also concentrate more soil nutrients and organic matter than the adjacent interbands. This resource concentration enables the formation of a forest system, the productivity of which equals and can even double that of adjacent non-banded landscapes.

These systems can persist in the face of severe drought by adjusting the proportion of runoff and runon areas. They can also resist the stress and disturbance caused by moderate land use. The earliest indicator of deterioration is the decline in the contrast between the two mosaic phases. The late stage in degradation is characterized by disruption of the band pattern. Overgrazing is considered to be the prime cause of deterioration of banded landscapes in Australia. Firewood and timber harvesting threaten the brousse tigrée in West Africa.

Models have demonstrated that these patterns may result either from landscape degradation or rehabilitation, but the natural initiation of banded landscapes has never been observed. The slow upslope migration of the bands is also a debated topic. It is linked to the runoff/runon theory that underpins the basic functioning of banded vegetation. The obstruction of overland flow by the bands would favour the upslope germination of pioneer plants in this upslope edge and the decline of vegetation due to resource shortage at the downslope edge. This notion of upslope band migration is strongly supported by an array of arguments such as the seedling concentration on the upslope edge of the band, the decaying vegetation in the downslope edge, the sequence of soil crust types across the interbands, and the marked gradient in soil organic matter. The migration 'velocity' of bands has been assessed using a variety of methods including field monitoring with benchmarks, digitized aerial photographs, age distribution of trees with dendrochronology, and TRACERS (residual ^{137}Caesium) distribution in the soil, under a wide range of climatic and topographic conditions. The fastest observed migration was 1.5 m yr^{-1} for grass bands and 0.8 m yr^{-1} for trees and shrubs. Because of some stationary systems, the migration of vegetation bands cannot be regarded as an invariable property of the banded systems.

In the arid and semi-arid environment, the banded patterns are clear examples of heterogeneous landscapes that are more sustainable than homogeneous systems. The lessons drawn from them lead to the recognition of the ecological value of water harvesting and runoff farming.

References

Bromley, J., Brouwer, J., Barker, T., Gaze, S. and Valentin, C. (1997) The role of surface water redistribution in an area of patterned vegetation in South West Niger, *Journal of Hydrology* 198, 1–29.

Clos-Arceduc, M. (1956) Étude sur photographies aériennes d'une formation végétale sahélienne: la brousse tigrée, *Bulletin de l'IFAN*, série A 7(3), 677–684.

Further reading

Tongway, D.J. and Seghieri, J. (eds) (1999) *Acta Oecologica* 20(3), entire issue.

Tongway, D.J., Valentin, C. and Seghieri, J. (2001) *Banded Vegetation Patterning in Arid and Semiarid Environments. Ecological Processes and Consequences for Management*. Ecological Studies 149. New York: Springer.

Valentin, C. and Poesen, J. (eds) (1999) The significance of soil, water and landscape processes in banded vegetation patterning, *Catena* 37(1–2), entire issue.

SEE ALSO: crusting of soil

CHRISTIAN VALENTIN

BRUUN RULE

The prospect of accelerated sea-level rise as a consequence of global warming has renewed interest in models that link sea-level rise and coastal change, such as the shoreline translation model of Cowell and Thom (1994). But to date, no model is better known or more widely accepted than that of Per Bruun.

In 1962 Bruun (1962) proposed that with a rise in sea level, the profile of a beach and its nearshore zone would move landward and upward, and that the quantity of sediment eroded from the upper part of the profile would be transported seaward to build up the adjacent seafloor by an amount equivalent to the sea-level rise (Figure 15). In this model the retreat of the beach (R) is given as:

$$R = XS/Y,$$

where R is the difference in distance between the initial sea level–profile intercept and the intercept after sea-level rise, X is the horizontal length from shore to the limiting depth, S is the sea-level rise, and Y is the vertical dimension of the profile, which is the sum of the limiting depth below sea level and the top of the foredune above sea level.

Early testing of the model in a small-scale laboratory wave-table experiment, as well as sequential field measurements of shore profiles around Cape Cod, Chesapeake Bay and Lake Michigan during and after episodes of rising water level, tended to support the basic tenets of the model, such that in the 1970s it became known as the Bruun Rule, after temporarily being declared a 'theory' by Schwartz (1967). However, such status has not been without its critics, for rarely are the model's basic assumptions satisfied in the real world of multidimensional coastal morphodynamics (Healy 1991).

The Bruun Rule is a two-dimensional cross-shore model applicable to long straight sandy shorelines that adjust to a rise in sea level over decadal to centennial timescales. The original model assumes that the initial and final profiles are in equilibrium, that a constant profile shape is preserved over the period considered, that the total quantity of sediment in the cross-section is conserved, and that a constant water depth is maintained in the offshore zone as sea level rises. Clearly, these assumptions are unrealistic. For instance, given the fact that beach erosion is already widespread, it is unlikely that it would be possible to determine an EQUILIBRIUM SHORELINE, or that the shape of the profile would not change through time. It is also unrealistic to expect a beach profile to conserve sediment and

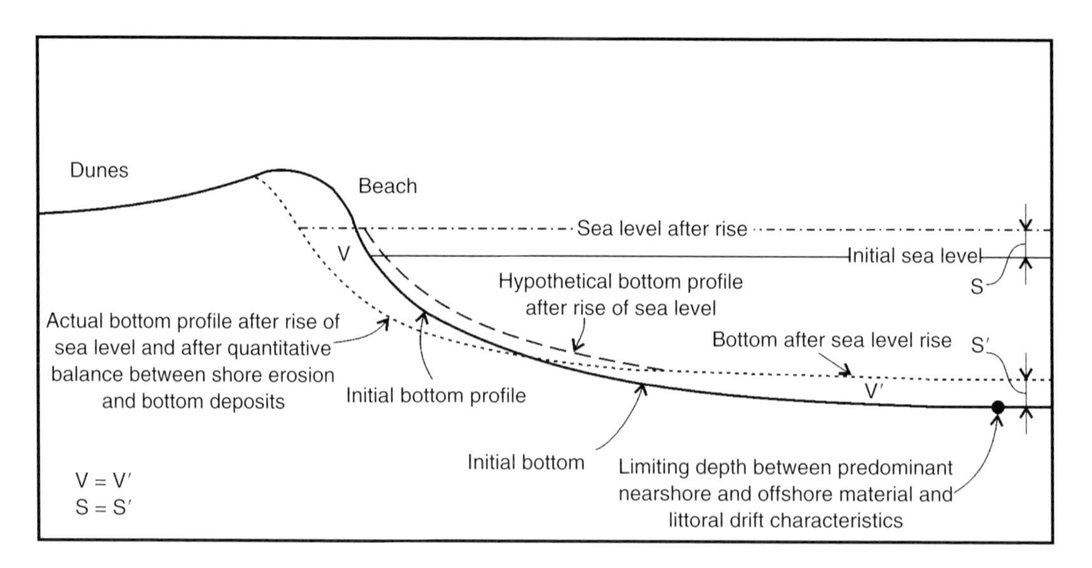

Figure 15 The Bruun Rule implies that the sediment volume removed from the beach and nearshore (V) must equal the sediment volume on the lower shoreface (V′) and that the lower shoreface aggrades in direct proportion (S′) to the rise in sea level (S). (After Bruun 1962 and Dubois 2002)

not to have sediment leakage, either as gains or losses resulting from LONGSHORE (LITTORAL) DRIFT, which is such a common process on sandy beaches. Similarly, wind erosion is not included in the Bruun Rule even though it can result in significant profile change through deflation of the beach and accumulation on a foredune at the landward end of a profile. Determination of the seaward end of a profile (closure depth) is equally problematical and maintenance of a constant water depth as implied in the Bruun Rule would result in a bathymetric or sediment discontinuity that in reality is difficult to define.

In spite of such difficulties, the Bruun Rule remains particularly attractive for several reasons. First, it is simple in concept and intuitively attractive given the fact that over the past hundred years or so global sea level has been rising at rates of around 1–2 mm per year, and that during that time approximately 70 per cent of the world's sandy shorelines have been eroding. Second, it can give quantitative results such that shore retreat will be 50 to 100 times the rise in sea level. For instance, a rise in mean sea level of 50 cm would result in beach recession of 25 to 50 m. And, third, the Bruun Rule has proven flexible enough to spawn a number of derivative models. Some refinements were proposed by Bruun (1983, 1988) himself, others by Dean and Maurmeyer (1983) who upscaled the concept to account for the landward and upward migration of an entire barrier island system. But the most persistent alternative model builder has been Dubois (1992, 2002).

Because the Bruun concept is not dependent on the shape of the shore profile, more complex topographies than in the original figure (Figure 15) can be incorporated (Dubois 1992). In Figure 16 zones of erosion and deposition are identified associated with onshore bar migration after shoreward displacement of the whole profile resulting from sea-level rise. Moreover, in this alternative model the eroded beach material is not only displaced seaward (as in the original Bruun model) but is also moved in a landward direction and washed or deflated on to the dune face or into a backing swale or lagoon. While overall conservation of sediment should be maintained within the whole profile, fine suspended sediment can be carried seaward and deposited in deeper water such that the lower shoreface and ramp do not accrete following the rise in sea level, but are simply abandoned by wave action (Figure 16).

In a comprehensive review of the response of beaches to sea-level changes, Working Group 89 of the Scientific Committee for Ocean Research (SCOR 1991) concluded that the quantitative predictions of shore change based on the Bruun

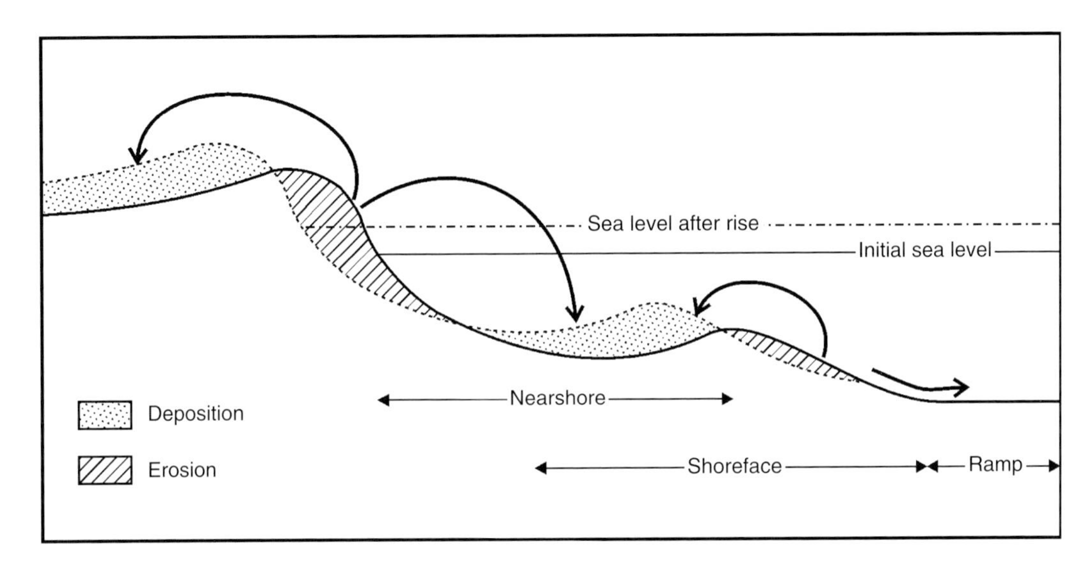

Figure 16 A two-dimensional model of a shore profile responding to a rise in sea level. Arrows show potential directions of sediment transport. Note that Bruun's Rule is embedded in this model, although it contributes only a small amount to the total shore erosion caused by rising sea level. (After Dubois 1992)

model are dependent on a number of parameters that are difficult to define, that there may be a significant time lag of beach response to sea-level rise, though the principal hindrance in achieving acceptable predictions is that the model does not include other sediment budget components that can result in either coastal accretion or erosion. Nevertheless, the SCOR group did suggest that the Bruun Rule could be used, though only for order-of-magnitude estimates of potential shore recession rates in appropriate coastal settings.

There is little doubt that globally sea level is rising and that sandy shores around the world are continuing to erode, including barrier beaches and barrier islands. The Bruun Rule gives an insight into how sea-level rise and coastal erosion are coupled, though we also know there are a host of other factors that contribute to coastal erosion independent of sea level.

References

Bruun, Per (1962) Sea-level rise as a cause of shore erosion, *Journal of the Waterways and Harbours Division*, Proceedings American Society of Civil Engineers 88, WW1, 117–130.

——(1983) A review of conditions for the use of the Bruun rule of erosion, *Coastal Engineering* 7, 77–89.

Bruun, Per (1988) The Bruun Rule of erosion by sea-level rise: a discussion of large-scale two- and three-dimensional usages, *Journal of Coastal Research* 4, 627–648.

Cowell, P.J. and Thom, B.G. (1994) Morphodynamics of coastal evolution, in R.W.G. Carter and C.D. Woodroffe (eds) *Coastal Evolution: Late Quaternary Shoreline Morphodynamics*, 33–86, Cambridge: Cambridge University Press.

Dean, R.G. and Maurmeyer, E.M. (1983) Models for beach profile response, in P.D. Komar (ed.) *Handbook of Coastal Processes and Erosion*, 151–166, Boca Raton, Florida: CRC Press.

Dubois, R.N. (1992) A re-evaluation of Bruun's rule and supporting evidence, *Journal of Coastal Research* 8, 618–628.

——(2002) How does a barrier shoreface respond to a sea-level rise? *Journal of Coastal Research* 18(2), iii–v.

Healy, T. (1991) Coastal erosion and sea level rise, *Zeitchrift für Geomorphologie* 81, 15–29.

Schwartz, M.L. (1967) The Bruun theory of sea-level rise as a cause of shore erosion, *Journal of Geology* 75, 76–92.

SCOR Working Group 89 (1991) The response of beaches to sea-level changes: a review of predictive models, *Journal of Coastal Research* 7, 895–921.

Further reading

Bird, Eric (2000) *Coastal Geomorphology: An Introduction*, Chichester: Wiley.

Van Rijn, Leo C. (1998) *Principles of Coastal Morphology*, Amsterdam: Aqua Publications.

SEE ALSO: barrier and barrier island; beach–dune interaction; equilibrium shoreline; longshore (littoral) drift

ROGER F. McLEAN

BUBNOFF UNIT

Unit providing a useful means to quantify the rate of operation of diverse geomorphological processes as a rate of ground loss (perpendicular to the surface) or slope retreat. A unit equals 1 mm per 1,000 years, equivalent to $1\,m^3\,km^{-2}\,a^{-1}$ (Fischer 1969).

Reference

Fischer, A.G. (1969) Geological time–distance rates: the Bubnoff unit, *Geological Society of America Bulletin* 80, 594–652.

A.S. GOUDIE

BURIED VALLEY

A buried valley is the bedrock expression of a valley buried by more recent deposits. These features are surprisingly common but are not well known as they have no surface expression. They are usually identified following borehole information or other sub-surface investigations employing geophysical techniques. The identification of these features is often an important element in the reconstruction of the geomorphological history of an area.

A number of types of buried valley can be identified. First, there are those buried valleys which are the result of glaciation. These can be sub-aerially eroded valleys buried by deposits of glacial origin. A good example of this type is afforded in the English Midlands by the Proto-Soar valley. This broad sub-drift valley has its head between Stratford-on-Avon and Warwick and heads northeastwards towards Leicester where it underlies the contemporary river Soar, a major tributary of the Trent. Before the glacial event which buried this surface, the main watershed of England between Avon/Severn drainage to the west and Soar/Trent drainage to the east lay some 30 km at least to the west of its present position which is on top of a thick plug of glacial deposits. Elsewhere, buried valleys have been described which themselves have been created by subglacial processes and then subsequently been buried. In interpreting the buried valleys identified widely in East Anglia, Woodland (1970) drew attention to the 'tunnel valleys' of Denmark and northern Germany. In East Anglia, many of the buried valleys appear to be quite narrow, up to 500 m wide and often 100 m deep, whereas the features in Denmark are broader and shallower. Woodland, however, attributes a similar origin to these features – subglacial erosion beneath an ice sheet. In the case of the East Anglian examples, they have been infilled by a wide range of often complex sediments which mask the former topography. Other authors have preferred to explain the excavation of these tunnel valleys through glacial modification of existing valleys (West and Whiteman 1986).

Second, there are buried valleys resulting from changes in SEA LEVEL following the last glacial event. These are widespread around many coastlines where marine transgressions, estuarine deposition and alluvial fill have buried a landscape graded to a lower sea level. Thus under many existing rivers can be found buried valleys that represent the valley of the former river draining into a sea that might have been 100 m or so lower. Boreholes can reveal that the essentially level surface of the contemporary alluvium conceals an irregular surface comprising valley forms which often, but not always, parallel the existing drainage.

Finally, mention must be made of the numerous buried valleys, usually in urban areas, where human activity has been responsible for modification of the valley topography. In some cities, rubble has been used to infill minor valleys to create flat land for urban land uses.

References

West, R.G. and Whiteman, C.A. (1986) *The Nar Valley and North Norfolk*, Cambridge: Quaternary Research Association.

Woodland, A.W. (1970) The Buried Tunnel-Valleys of East Anglia, *Proceedings of the Yorkshire Geological Society* 37, 521–578.

TERRY DOUGLAS

BUTTE

Butte is a small steep-sided and flat-topped hill, built of flat-lying soft rocks capped by a more resistant layer of sedimentary rock, lava flow or duricrust, surrounded by a plain. Butte is smaller than MESA and may be considered as a more advanced stage of mesa degradation, although there are no formal criteria to distinguish between the two. Together with mesas, buttes are outliers, indicative of long-term scarp retreat. They occur in front of CUESTAs and plateau margins, their morphology being best pronounced in arid and semi-arid regions.

PIOTR MIGOŃ

C

CALANQUE

Coastal inlets (such as those to the east of Marseilles) which tend to be of a gorge-like form. They are widespread around the Mediterranean Sea and may be karstic dry valleys which have been partially drowned, as a result of the Flandrian transgression of the Holocene. Their positions may be fault controlled. In Mallorca they are called *calas*.

Further reading

Nicod, J. (1951) Le problème de la classification des 'calanques' parmi les formes de côtes de submersion, *Revue de Géomorphologie Dynamique* 2, 120–127.

Paskoff, R. and Sanlaville, P. (1978) Observations géomorphologiques sur les côtes de l'archipel Maltais, *Zeitschrift für Geomorphologie NF* 22, 310–328.

A.S. GOUDIE

CALCRETE

A term, proposed by Lamplugh (1902), to describe a terrestrial near-surface accumulation of predominantly calcium carbonate ($CaCO_3$) which occurs in a variety of forms ranging from powdery to nodular to highly indurated. It results from low temperature physico-chemical processes operating within the zone of WEATHERING which lead to the displacive and/or replacive introduction of $CaCO_3$ into a soil profile, sediment, rock or weathered material. Calcretes develop as a result of carbonates in solution moving laterally and vertically through vadose and shallow phreatic groundwater systems until they become, over time, saturated with respect to $CaCO_3$ and precipitate as calcite crystals (Wright and Tucker 1991). Calcretes often occur within soil profiles, where they may form single or multiple horizons, but they are not a type of soil. The term is synonymous with CALICHE (SODIUM NITRATE) and kunkur but distinct from other $CaCO_3$ cemented materials such as cave SPELEOTHEMS, lacustrine algal STROMATOLITES (STROMATOLITHS), TUFA AND TRAVERTINE, BEACH ROCK or AEOLIANITE.

Calcretes are estimated to underlie 13 per cent of the Earth's land surface and are most widespread in semi-arid regions. They form an important component of many contemporary dryland landscapes, and, where well indurated, may act as a threshold (see THRESHOLD, GEOMORPHIC) to erosion. Important areas of occurrence include the High Plains of the USA (e.g. Gile *et al.* 1966; Machette 1985), Africa north of the Sahara (e.g. Goudie 1973), the Kalahari of southern Africa (e.g. Watts 1980; Netterberg 1980), central and western Australia (e.g. Mann and Horwitz 1979; Milnes and Hutton 1983), and parts of southern Europe (e.g. Nash and Smith 1998). The close association between calcrete distribution and present-day dryland regions has led to the widespread use of calcretes in the geological record as indicators of past aridity. However, it is critical that the mode of origin of any calcrete is identified before it can be interpreted in this way. Whilst carbonate accumulation within soil may require a semi-arid climate, calcretes developed by other mechanisms may form under much wetter conditions. Non-pedogenic Holocene calcretes have, for example, been found in temperate locations such as the UK (Strong *et al.* 1992). Furthermore, calcrete accumulation is closely controlled by carbonate supply.

Calcretes are highly variable in appearance and range from thin rock coatings to massive horizons. Thickness varies with the mode of origin and stage of development, with laminar calcretes rarely exceeding 0.25 m whilst multiple pedogenic and groundwater profiles may reach tens of metres thickness. Most calcretes are white, cream or grey in colour, though mottling and banding is common. Calcretes are predominantly cemented by calcite with some $CaMg(CO_3)_2$ (dolomite) often present. The size and shape of calcite crystals is dependent upon the composition of the host material, the duration of wetting and the influence of biological mechanisms. Cements are typically dominated by microcrystalline carbonate (or micrite) although larger crystals of sparry calcite may be present. If significant biological fixation of carbonate occurred during development, the calcrete is likely to exhibit a complex beta fabric dominated by organic structures when viewed in microscopic thin-section as opposed to simpler alpha fabrics developed by inorganic mechanisms (Wright and Tucker 1991). The mean global chemical composition of calcrete is *c.*78 per cent $CaCO_3$, 12 per cent SiO_2, 3 per cent MgO, 2 per cent Fe_2O_3 and 2 per cent Al_2O_3 (Goudie 1973), although variations occur dependent upon the host material chemistry, cement type, presence of authigenic silica and silicates, mode of origin, and stage of development.

There are a range of classification schemes for calcrete, of which the most widely employed use morphological criteria. Netterberg (1980), for example, recognized a range of forms including calcareous and calcified soils, powder, nodular, honeycomb, hardpan, laminar and boulder calcretes, a sequence which also reflects the phases of development of many calcretes. Gile *et al.* (1966) and Machette (1985) have proposed a scheme to assist the identification of calcretes at different stages of development, with stage I–III calcretes consisting of morphologically simple carbonate accumulations within soils progressing to more mature horizons by stages IV–VI. Calcretes have also been classified on the basis of their hydrological setting, with vadose, capillary fringe and

Table 5 A genetic classification of calcrete types

Environment of formation	Calcrete type	Incorporated calcrete types	Mode of formation
Pedogenic	Pedogenic calcrete	Caliche; kunkar; nari; petrocalcic horizons	Developed by vertical redistribution of calcium carbonate within a soil profile
Non-pedogenic	Non-pedogenic superficial calcrete	Laminar crusts; case hardening; gully bed cementation	Formed by surficial transport of calcium carbonate
Non-pedogenic	Non-pedogenic gravitational zone calcrete	Gravitational zone calcrete	Formed by downward accumulation of calcium carbonate in irregular permeability channels
Non-pedogenic	Non-pedogenic groundwater calcrete	Valley calcrete; channel calcrete; deltaic calcrete; lake margin calcrete; alluvial fan calcrete; fault trace and other groundwater calcretes	Formed by lateral transport of calcium carbonate
Non-pedogenic	Detrital and reconstituted calcrete	Recemented transported calcrete; calcretes which are brecciated and recemented *in situ*	Formed by recementation of existing fragmented or brecciated calcrete

Source: After Carlisle (1983)

phreatic types identified. Other schemes have subdivided calcretes by their dolomite content (Netterberg 1980) or by the relative abundance of alpha and beta cements (Wright and Tucker 1991). However, none of these classifications completely distinguishes between calcretes formed by different mechanisms. As such, the most helpful scheme is Carlisle's (1983) genetic classification (Table 5) which subdivides calcretes into pedogenic and non-pedogenic forms using geomorphological, chemical, macro- and micromorphological criteria.

Pedogenic calcretes develop near the land surface, usually in areas of low slope angle, through the mobilization, redistribution and relative accumulation of $CaCO_3$ within a soil profile. Formation may also involve some absolute accumulation of $CaCO_3$ if there are additional carbonate inputs to the profile. Such calcretes commonly show enrichment in $CaCO_3$ up-profile and consist of a powdery or nodular basal section overlain by a more massive hardpan which may, in turn, be capped by a laminar crust. Cements are usually dominated by micrite and, because of the mechanisms by which they develop, exhibit a complex micromorphology. Non-pedogenic calcretes encompass a wide variety of types, ranging from laminar crusts developed on rock or other calcrete surfaces by evaporation and/or biological fixing of $CaCO_3$, to detrital and reconstituted calcretes formed by the cementation of pre-existing fragmented crusts. By far the largest group are the groundwater calcretes. These are calcretes developed in channel, valley, alluvial fan, delta and lake marginal sediments, usually in the absence of soil-forming processes and sometimes at depths of tens of metres beneath the land surface. They can be distinguished from pedogenic calcretes by their lack of profile development, normally simple micromorphology and the presence of more crystalline calcite cements, especially where formation occurred at or below the water table (Nash and Smith 1998).

Despite the wide range of mechanisms by which they can develop, all calcretes result from the solution, movement and subsequent precipitation of $CaCO_3$, described by the following chemical reaction:

$$CO_2 + H_2O + CaCO_3 \leftrightarrow Ca^{2+} + 2HCO_3^-$$

Calcretes require a carbonate source, usually released as a result of $CaCO_3$ dissolution (Goudie 1983). $CaCO_3$ solubility is closely linked to environmental pH, with solubility rapidly increasing below pH 9.0. Mechanisms which lower pH and drive the reaction to the right, such as the introduction of weak carbonic acid ($CO_2 + H_2O$ in the equation) or an increase in soil CO_2 partial pressure, will trigger dissolution. Sources of carbonate can be distant or local to the site of formation, and include weathered bedrock, volcanic (and other) dust and organic remains. Once carbonate is in solution, it may be moved laterally and/or vertically to the site of formation. Lateral transfer mechanisms may include transport in solution via ephemeral or perennial rivers as well as in shallow or deep groundwater systems. Vertical transfers include percolation of surface water or capillary rise from the water table. Carbonate precipitation (where the above reaction proceeds to the left) may be triggered by a variety of factors which lead to the concentration of carbonate-rich solutions and/or cause environmental pH to increase. Foremost amongst these are evapotranspiration, biological processes, decreases in CO_2 partial pressure, CO_2 degassing, and the common ion effect (Goudie 1983; Salomons and Mook 1986).

References

Carlisle, D. (1983) Concentration of uranium and vanadium in calcretes and gypcretes, in R.C.L. Wilson (ed.) *Residual Deposits: Surface Related Weathering Processes and Materials*, 185–195, London: Geological Society of London.

Gile, L.H., Peterson, F.F. and Grossman, R.B. (1966) Morphological and genetic sequences of carbonate accumulation in desert soils, *Soil Science* 101, 347–360.

Goudie, A.S. (1973) *Duricrusts in Tropical and Subtropical Landscapes*, Oxford: Clarendon Press.

——(1983) Calcrete, in A.S. Goudie and K. Pye (eds) *Chemical Sediments and Geomorphology*, 93–131, London: Academic Press.

Lamplugh, G.W. (1902) Calcrete, *Geological Magazine* 9, 575.

Machette, M.N. (1985) Calcic soils in the southwestern United States, in D.L. Weide (ed.) *Soils and Quaternary Geology of the Southwestern United States*, Geological Society of America Special Paper 203, 1–21, Boulder, CO: Geological Society of America.

Mann, A.W. and Horwitz, R.C. (1979) Groundwater calcrete deposits in Australia: some observations from Western Australia, *Journal of the Geological Society of Australia* 26, 293–303.

Milnes, A.R. and Hutton, J.T. (1983) Calcretes in Australia, in *Soils: An Australian Viewpoint*, 119–162, Melbourne: CSIRO/Academic Press.

Nash, D.J. and Smith, R.F. (1998) Multiple calcrete profiles in the Tabernas Basin, southeast Spain: their

origins and geomorphic implications, *Earth Surface Processes and Landforms* 23, 1,009–1,029.
Netterberg, F. (1980) Geology of southern African calcretes: terminology, description, macrofeatures and classification, *Transactions of the Geological Society of South Africa* 83, 255–283.
Salomons, W. and Mook, W.G. (1986) Isotope geochemistry of carbonates in the weathering zone, in P. Fritz and J.Ch. Fontes (eds) *Handbook of Environmental Isotope Geochemistry*, Volume 2, 239–269, Amsterdam: Elsevier.
Strong, G.E., Giles, J.R.A. and Wright, V.P. (1992) A Holocene calcrete from North Yorkshire, England: implications for interpreting palaeoclimates using calcretes, *Sedimentology* 39, 333–347.
Watts, N.L. (1980) Quaternary pedogenic calcretes from the Kalahari (southern Africa): mineralogy, genesis and diagenesis, *Sedimentology* 27, 661–686.
Wright, V.P. and Tucker, M.E. (1991) Calcretes: an introduction, in V.P. Wright and M.E. Tucker (eds) *Calcretes*, 1–21, Oxford: Blackwell Scientific.

SEE ALSO: duricrust; silcrete

DAVID J. NASH

CALDERA

Calderas are large circular or elliptical volcanic depressions whose diameter (typically several or several tens of kilometres) greatly exceeds those of any included vents. They are formed by the evacuation of a magma chamber within the crust, and subsidence of the overlying rocks. Calderas can form on volcanoes of different magma composition, from low silica (mafic) to high silica (silicic), though the mechanisms vary. Though there are no hard and fast distinctions, the term crater tends to be reserved for smaller features created as a result of excavation of rock during explosive eruptions or smaller-scale collapse (collapse pits). The highest magnitude explosive eruptions on Earth, sometimes called super-eruptions, have generated the largest calderas. Volcanoes that have experienced more than one super-eruption are colloquially known as super-volcanoes. The geometries and structures of the resulting nested and overlapping calderas can be complex, and obscured by post-collapse uplift, volcanism and erosion.

Origins and development

Super-eruptions involve magma volumes of several thousand km^3 (masses up to 10^{16} kg). The rapid removal of such an amount of material from a crustal magma chamber invariably induces failure of the overlying rocks. There is a rough correspondence between the volume of magma erupted and that of the hole left in the ground by the caldera collapse. The largest known Quaternary eruption occurred about 74,000 years ago, and expelled an estimated 7×10^{15} kg (2,800 km^3) of silicic magma, making a significant contribution to the 100 km × 30 km caldera complex occupied today by Lake Toba in northern Sumatra (Oppenheimer 2002). Toba can certainly be classed as a super-volcano, since at least two similar events occurred around 840,000 and 500,000 years ago, as can Yellowstone (USA), whose last super-eruption took place about 600,000 years ago.

Important insights into caldera evolution associated with explosive eruptions have been gained from detailed investigations of a number of much smaller historic and prehistoric calderas, for example at Crater Lake (Oregon, USA; Bacon 1983), and Santorini (Greece (Plate 19); Druitt *et al.* 1999), and also ancient examples such as Scafell caldera of the English Lake District, Ordovician in age but revealing much of its structure thanks to erosion (Branney and Kokelaar 1994). Several subsidence processes have been recognized (Lipman 2000). Larger calderas tend to involve piston-like (plate) collapse, where the floor remains largely undeformed. The collapse occurs along steep ring faults, with vertical displacements of about 1 km. In contrast, downsag subsidence does not preserve the coherence of the developing caldera floor, which is instead tilted and flexed. Intermediate between these two is trapdoor subsidence, which occurs when the caldera floor remains hinged along part of its length but elsewhere has subsided in plate fashion. Geometrically complex systems of arcuate faults and subsiding blocks reflect, in some cases, the breakup of the floor during eruption but prior to ring-fault subsidence, and are referred to as piecemeal calderas (Branney and Kokelaar 1994). Relatively small-scale collapses resulting from modest explosive eruptions from a central vent, such as that of Pinatubo (Philippines) in 1991, are sometimes referred to as funnel calderas. These lack a bounding ring fault or coherent subsided plate.

Calderas associated with low silica (mafic) volcanoes such as those found in Hawai'i have somewhat different origins to their silicic counterparts. Some interpret their development as a late stage in the growth of Hawaiian shields; others see them as a recurrent process. Clearly, those on

Plate 19 Santorini volcano, Greece. The last major caldera-forming eruption occurred in the mid-seventeenth century BC but is only the most recent in a series of eruptions exceeding 1,014 kg in magnitude (Druitt *et al.* 1999). The caldera rim is partly submerged such that the caldera is open to the sea

Mauna Loa and Kīlauea are very young features suggesting that they are rejuvenated at intervals of a few centuries, rather than the many millennia that can separate subsidence events on silicic volcanoes. Again, unlike silicic systems, the volumes of caldera subsidence on mafic volcanoes do not bear any obvious relationship with erupted volumes; the volumes of the young Hawaiian calderas are larger than any known Hawaiian eruptions. While Mauna Loa and Kīlauea appear superficially to be well-defined piston subsidence structures, Walker (1988) has suggested on the basis of mapping of the 2 Myr old Koolau volcano on Oahu, where erosion has exposed its 1 km depth roots, that downsagging may prevail in the central parts of funnel-shaped subsidence structures underlying the calderas. He suggests that, rather than simple piston subsidence into an evacuated magma chamber, the Hawaiian calderas develop as the dense intrusive rocks associated with the magma chambers sink into the warm lithosphere beneath.

Geomorphology

Calderas are enclosed by a topographic rim that is simply the head of the escarpment bounding the caldera. The inner walls can form cliffs in young calderas but retreat through time as a result of landslides. In map view, most large calderas reveal bites in the rim due to larger slope failures. The rock redistributed by landslides may form a collapse collar on the caldera floor. The arcuate bounding faults (ring faults) are sometimes exposed in deeply eroded calderas, and where observed, generally dip near-vertically or steeply inwards.

The largest calderas often enclose a central elevated massif. The vertical extent of the upheaval in these resurgent calderas can be 1 km or more. Toba is a good example; a lake occupies much of the caldera but an island – Samosir Island – occupies much of the lake. The island is composed substantially of the pyroclastic intracaldera fill from the Younger Toba Tuff eruption. Lacustrine sediments can be found several hundred metres above the present lake level and testify to substantial post-caldera uplift. Yellowstone is another example of a resurgent caldera. The mechanisms of resurgence, however, are not well understood. Refilling of the magma chamber is one possibility, along with the exsolution of remaining volatiles in the caldera eruption chamber and bubble formation (vesiculation) causing an increase in volume, expressed at the surface in the form of uplift (Marsh 1984).

Erosion rates of the pyroclastic deposits of caldera-forming eruptions can be rapid, especially for non-welded portions. Where developed, columnar jointing, akin to that observed in mafic lava plateaux (see LAVA LANDFORMS), can strongly influence the drainage fabric. In arid environments, the outflow sheets of ash flow deposits (i.e. those outside the caldera) may experience rapid aeolian erosion. This is evident in the wind-sculpted morphology of the Central Andean ignimbrites, which are adorned with YARDANGs and deflationary hollows. Wigwams or tent rocks are another common feature of the pyroclastic deposits of caldera-forming eruptions. These may develop by the intersection of drainage channels, or more commonly by the action of a resistant block of lava within the deposit, protecting the underlying material from erosion.

Hazards and climate change

Along with bolide impacts, large caldera-forming eruptions are the most catastrophic geologic events that affect the Earth's surface. Compared in terms of energy release, they are more frequent than bolides, however. A super-eruption today would have devastating impacts on regional populations and the global economy. The massive quantities of sulphur gases that can be released in super-eruptions strongly perturb atmospheric chemistry and radiation, with potentially global-scale climatic consequences (Rampino and Self 1993). Currently, little is known about the precursory events that might lead up to a super-eruption but numerous calderas worldwide have shown signs of unrest in the historic period (Newhall and Dzurisin 1988).

References

Bacon, C.R. (1983) Eruptive history of Mount Mazama and Crater Lake caldera, Cascade Range, U.S.A., *Journal of Volcanology and Geothermal Research* 18, 57–115.

Branney, M.J. and Kokelaar, P. (1994) Volcanotectonic faulting, soft-state deformation and rheomorphism of tuffs during development of a piecemeal caldera: English Lake District, *Geological Society of America Bulletin* 106, 507–530.

Druitt, T.H., Edwards, L., Mellors, R.M., Pyle, D.M., Sparks, R.S.J., Lanphere, M., Davies, M. and Barriero, B. (1999) *Santorini Volcano*, London: Geological Society.

Lipman, P.W. (2000) Calderas, in H. Sigurdsson, B.F. Houghton, S.R. McNutt, H. Rymer and J. Stix (eds) *Encyclopedia of Volcanoes*, 643–662, San Diego: Academic Press.

Marsh, B.D. (1984) On the mechanics of caldera resurgence, *Journal of Geophysical Research* 89, 8,245–8,251.

Newhall, C.G. and Dzurisin, D. (1988) *Historical Unrest at Large Calderas of the World*, US Geological Survey Bulletin 1,855, 2 volumes.

Oppenheimer, C. (2002) Limited global change due to largest known Quaternary eruption, Toba ≈74 kyr BP? *Quaternary Science Review* 21, 1,593–1,609.

Rampino, M.R. and Self, S. (1993) Climate-volcanism feedback and the Toba eruption of ~74,000 years ago, *Quaternary Research* 40, 269–280.

Walker, G.P.L. (1988) Three Hawaiian calderas: an origin through loading by shallow intrusions? *Journal of Geophysical Research* 93, 14,773–14,784.

Further reading

Francis, P. (1993) *Volcanoes: A Planetary Perspective*, Oxford: Oxford University Press.

Friedrich, W.L. (2000) *Fire in the Sea*, Cambridge: Cambridge University Press.

Long Valley Observatory http://lvo.wr.usgs.gov/

Williams, H. (1941) Calderas and their origin, *Bulletin of the Department of Geological Sciences, University of California* 25, 239–346.

Yellowstone Volcano Observatory http://volcanoes. usgs.gov/yvo/

SEE ALSO: lava landform; volcano

CLIVE OPPENHEIMER

CALICHE (SODIUM NITRATE)

The term has been used for both CALCRETE and for sodium nitrate deposits. The most famous and important deposits of the latter in the world occur in the Atacama Desert (Ericksen 1981), though others are known in California and Antarctica. Chile possesses a band of nitrate-containing terrain up to 30 km wide and 700 km long. Much of the Coastal Range is mantled with nitrate-bearing saline-cemented regolith, commonly ranging from a few tens of centimetres to a few metres in thickness. The petrography of the deposits, which cause extreme bedrock disintegration, is described by Searl and Rankin (1993). Most of the deposits lie at altitudes below 2,000 m, though some occur as high as 4,000 m. Low-grade deposits are also known in the coastal desert of Peru several hundred kilometres north of the Chilean border. Many of the deposits were extensively worked in the late nineteenth and early twentieth centuries. The landscape is now scarred with the pits from which nitrate was dug and is dotted with waste heaps.

The reason for the localization of the nitrate deposits appears to be the extreme aridity of the area, for sodium nitrate is more soluble in water than most common crust materials. The Atacama is among the driest and oldest of the world's deserts, and the average annual rainfall is less than 1 mm in the areas where the nitrate deposits are most prevalent. In any given part of the desert measurable rainfall (1 mm or more) may be as infrequent as once every 5–20 years. Heavy rainfall of a few centimetres or more may occur only a few times each century.

The Chilean nitrate fields occur typically in areas of low relief characterized by rounded hills and ridges and by broad, shallow debris-filled valleys. Significantly for their origin, they occur in all topographic positions from tops of hills and ridges to the centres of the broad valleys, though the richest deposits that have been worked most extensively tend to be on the lower slopes of hills.

Such a catholicity of geomorphological siting tends to imply that the nitrates have been derived as atmospheric inputs from the sea or from volcanic emissions, a mechanism supported by the fact that the nitrates occur on all rock and sediment types (Ericksen 1981). The model proposed by Ericksen (1981: 32) is that there has been long-term accumulation of atmospherically derived saline material for perhaps 10–15 million years (i.e. since the Mid-Miocene, under conditions of general extreme aridity). The sources of the material would include sea spray, volcanic emanations, photochemical reactions and dust from *salars* (salt lakes). The ore-grade nitrate deposits are formed by accumulation of saline materials on very old, flat to gently inclined or undulating landsurfaces, where rainwater dissolved the more soluble component and redeposited them at deeper soil levels. Ericksen's model of nitrate bedformation as a result of long-term atmospheric deposition has been confirmed by recent stable isotope studies (Böhlke *et al.* 1997).

References

Böhlke, J.K., Ericksen, G.E. and Revesz, K. (1997) Stable isotopes evidence for an atmospheric origin of desert nitrate deposits in northern Chile and southern California, USA, *Chemical Geology* 136, 135–152.

Ericksen, G.E. (1981) Geology and origin of the Chilean nitrate deposits, *United States Geological Survey Professional Paper* 1,188.

Searl, A. and Rankin, S. (1993) A preliminary petrographic study of the Chilean nitrates, *Geological Magazine* 130, 319–333.

A.S. GOUDIE

CALVING GLACIER

Calving GLACIERS terminate in water and lose mass by calving, the process whereby masses of ice break off to form ICEBERGS. Since they may be temperate or polar (see glaciers), grounded or floating, and may flow into the sea or into lakes, many types exist. They are widely distributed, but while lake-calving glaciers may exist in any glacierized mountain range, tidewater glaciers are currently confined to latitudes higher than 45°. Typically calving glaciers are fast flowing and characterized by extensional (stretching) flow near their termini, resulting in profuse crevassing. They terminate at near-vertical ice cliffs up to 80 m high. Calving activity above the waterline comprises a continuum from small fragments of ice to pillars the full height of the cliff. Below the waterline much ice may be lost through melting, but in deep water buoyancy causes infrequent but high magnitude calving events. In lakes, thermal erosion (melting) at the waterline can cause calving by undercutting the cliff. Calving permits much larger volumes of ice to be lost over a given time than melting.

Glaciers calve faster in deeper water. This correlation between calving rate (u_c in metres per annum) and water depth (h_w) is linear, and can be simply expressed as $u_c = ch_w$. The value of the coefficient c varies greatly in different settings, being highest for temperate glaciers and lowest for polar glaciers. Also, for any given water depth, calving is an order of magnitude faster in FJORDS than in lakes. The established correlation between calving rate and water depth may or may not imply that faster calving is *caused* by deeper water.

Calving glaciers are significant for three main reasons:

1 *Glacier dynamics* Calving glaciers comprise the most dynamic elements of many of the world's ice masses, and calving is the major means of ice loss from the two continental ICE SHEETS of Antarctica and Greenland. During the waning stages of Quaternary glacials (see ICE AGES (INTERGLACIALS, INTERSTADIALS AND STADIALS)), calving was the dominant process of mass loss around the mid-latitude ice sheets; the efficiency of calving helps to explain the catastrophic rates of ice sheet disintegration. Armadas of icebergs discharged during ice sheet collapses are believed to have caused global climate change by altering oceanic circulation.
2 *Non-climatic behaviour* Calving glacier fluctuations are highly sensitive to topographic controls (see PINNING POINT). Some tidewater glaciers fluctuate cyclically, in ways unrelated to climate, over distances of tens of kilometres and over timescales of centuries to millennia. Therefore neither the contemporary behaviour of calving glaciers nor the geomorphological records from past fluctuations (see MORAINE) are reliable indicators of climatic change.
3 *Socio-economic impacts* Calving glaciers constitute significant resources and GEOMORPHOLOGICAL HAZARDS for society. Resources include tourism and the potential for

harnessing Antarctic tabular icebergs as a source of freshwater, while hazards include icebergs and OUTBURST FLOODS.

Further reading

Van der Ween, C.J. (2002) Calving glaciers, *Progress in Physical Geography* 26, 96–122.
Warren, C.R. (1992) Iceberg calving and the glacio-climatic record, *Progress in Physical Geography* 16, 253–282.

CHARLES WARREN

CAMBERING AND VALLEY BULGING

Cambering occurs where large-scale valley incision has exposed gently dipping bedrock in which competent strata overlie less competent strata. Cambered strata consist of an attenuating drape of competent caprock extending down the valley sides. This drape of CAPROCK shows evidence of extension, accommodated by deep fractures termed *gulls* (Figure 17). Gulls run parallel to the contours and separate intact caprock blocks. These blocks tend to tilt forward, increasing apparent dip and producing the widely reported 'dip-and-fault structure' (Figure 17). Cambering is often associated with the development of anticlinal deformation within the less competent strata beneath the valley axis, the resulting structure being termed *valley bulging*. Cambering and valley bulge structures are thought to have formed during the Quaternary Period. They reflect ice segregation processes within the less competent strata (usually clays) during PERMAFROST aggradation at depth, and subsequent thaw consolidation processes caused by thawing at the base of the permafrost (underthaw) during permafrost degradation in the transition from cold glacial to warm interglacial stages. Hutchinson (1991) has provided a detailed review of processes and structures involved in cambering and valley bulge development.

Classic descriptions of cambering and valley bulging come from the Jurassic Limestones and underlying Upper Lias clay in England (Chandler *et al.* 1976; Hollingworth *et al.* 1945; Horswill and Horton 1976). At the Empingham Reservoir, the cambered limestone strata are draped across much of the valley side. Gulls separate cambered blocks, which display classic dip-and-fault structure. Valley bulging is highlighted by disturbance of marker horizons within the Upper Lias clays. Disturbed brecciated clay extends to a depth of 25 to 30 m, the base lying parallel to the present ground surface, suggesting that the phase of brecciation occurred after the main valley relief was established. The disturbed Lias is underlain by a sheared plane of decollement at a depth of approximately 60 m under the valley crest and 30 m under the valley bottom. It is likely that the brecciated clay fabrics reflect the combined effects of ice segregation, creep and shear.

Hutchinson (1991: Figure 5.5) provides a model for cambering and valley bulging based on estimated displacements in the Lias at

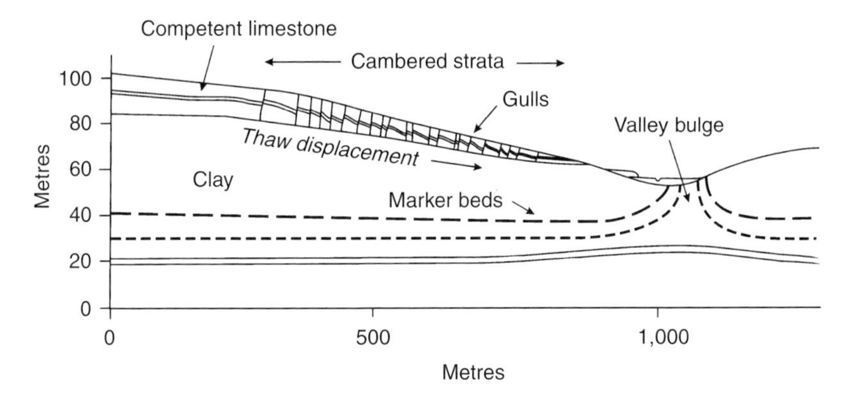

Figure 17 Cambering and valley bulging, based on the Gwash Valley, Lincolnshire, England (Horswill and Horton 1976)

Empingham (Vaughan 1976). The following stages are envisaged:

1 Initial incision of the river leading to unloading and valley rebound.
2 Permafrost develops, with ice segregation processes increasing ice contents of the frozen clay.
3 Valley-ward creep of the frozen clay occurs in response to lateral stresses. This causes extension and initial cambering of caprocks.
4 Permafrost degradation due to large-scale climate warming leads to thaw from the surface downwards, enhanced surface solifluction, and displacement of caprock down the valley sides over the thaw-softened clays beneath.
5 Thawing of the permafrost base due to geothermal heat flux is associated with thaw consolidation and high pore pressures within an effectively confined thawed stratum.
6 Lateral extrusion of the thawed clay at the permafrost base towards the valley bottom leads to compression along the valley axis and pronounced increase in the valley bulge structure.

Cambering and valley bulging represent some of the largest structures attributed to permafrost. Their presence as Quaternary relict features may be of considerable significance to engineering works such as the design of foundations for buildings which are particularly affected by the presence of gulls, and dam construction, where voids may increase water seepage, and deep-seated shearing may affect foundations.

References

Chandler, R.J., Kellaway, G.A., Skempton, A.W. and Wyatt, R.J. (1976) Valley slope sections in Jurassic strata near Bath, Somerset, *Philosophical Transactions of the Royal Society* A283, 527–556.

Hollingworth, S.E., Taylor, J.H. and Kellaway, G.A. (1945) Large-scale superficial structures in the Northamptonshire Ironstone Field, *Quarterly Journal of the Geological Society* 100, 1–35.

Horswill, P. and Horton, A. (1976) Cambering and valley bulging in the Gwash Valley at Empingham, Rutland *Philosophical Transactions of the Royal Society* A283, 451–461.

Hutchinson, J.N. (1991) Periglacial slope processes, in A. Forster, M.G. Culshaw, J.C. Cripps, J.A. Little and C.F. Moon (eds) *Quaternary Engineering Geology*, Geological Society Engineering Geology Special Publication 7, 283–331.

Vaughan, P.R. (1976) The deformations of the Empingham Valley slope, Appendix to P. Horswill and A. Horton (1976) Cambering and valley bulging in the Gwash Valley at Empingham, Rutland, *Philosophical Transactions of the Royal Society* A283, 451–461.

CHARLES HARRIS

CANYON

A canyon is a long, deep, relatively narrow, steep-sided valley, often cut through bedrock which forms precipitous cliffs along the valley walls. The word comes from the Spanish cañon. Canyons are formed by running water. The term is typically used for such features in arid and semi-arid regions, such as the western United States (e.g. the Grand Canyon in Arizona, USA). Canyons are similar to gorges (see GORGE AND RAVINE), but the side-walls are usually not as steep, and canyons are typically larger than gorges (e.g. the Grand Canyon contains an 'Inner Gorge' through which the Colorado River runs). Canyons are typical of mountainous regions, but are also found cutting high-elevation plateau (e.g. the Black Canyon of the Gunnison on the Colorado Plateau in Colorado, USA). They occur where stream erosion significantly outpaces weathering. Streams in canyons frequently flow through BEDROCK CHANNELS.

JUDY EHLEN

CAPROCK

Geomorphologically resistant lithological units, which protect underlying less resistant rocks from erosion and denudation, are called caprocks. Tablelands, CUESTAS, MESAS, buttes and hogbacks are examples of landforms which are composed of a backslope (dipslope), supported by a competent caprock, and a bipartite scarp slope. The scarp slope consists of an upper slope in the caprock and a moderately inclined lower slope in the less resistant rock below. It originates from fluvial downcutting or fault scarp development. In composite landforms like canyons and stepped cuesta landscapes whole sequences of caprocks and soft rocks can be found as in the Grand Canyon, in the Giant Staircase of southern Utah and northern Arizona or in the scarplands of southwestern Germany.

There is no specific method for defining caprock resistance on a metric scale, but there have been attempts to describe the resistance on an ordinal scale (Schmidt 1991). A scarp-forming rock must be relatively more resistant than the underlying soft rock. Caprock resistance can be connected with lithological attributes such as: mechanical hardness protecting the rock against the direct effects of weathering and erosion; porosity and JOINTING resulting in greater water permeability. The attributes of the more resistant scarp-forming rock are most effective when the less resistant rock possesses contrasting characteristics such as mechanical weakness, easy disintegration and low permeability (Ahnert 1998: 239). The effects of different resistance are most visible in dry climates with selective weathering and erosion, where even minor lithological variations are reflected in slope geometry.

In most cases, especially in climates with greater surface water availability, permeability is more important for determining caprock resistance than mechanical strength. Due to their perviousness caprock outcrops are generally characterized by a lack of surface water and low values of drainage density. Water infiltrates and percolates through the caprock body until it reaches the impermeable lower slope rock. At the caprock/soft rock interface it reappears in springs and seepage zones, sometimes connected with sapping processes and slope undercutting. The missing to low activity of surface water erosion on the caprock-protected backslopes reduces mechanisms of denudational downwearing.

Most caprocks are sedimentary rocks like sandstones, conglomerates and limestones. The properties of the cementing materials (carbonates, iron oxides, clay minerals) control the erosion and weathering susceptibility of the sandstones and conglomerates. Karst processes are effective in carbonate caprocks. Joints and fissures are widened by solution resulting in increased permeability. In the scarplands the softer rocks below the caprocks are often clays, marls and fine, densely layered sandstones.

Volcanic rocks can also act as caprocks. This especially occurs in the case of lava flows which moved down former valley floors and covered older sedimentary rocks. The valley slopes in sedimentary material at the sides of the lava flows were subsequently removed by erosion and denudation, and the lava flows, due to their resistance, survived as caprocks of residual hills. This process is called relief inversion (see INVERTED RELIEF). Examples of this geomorphological process combination can be found in Tertiary volcanics in the Ore Mountains in Saxony and in Cenozoic volcanics at the margins of the Colorado Plateau in Utah and Arizona.

Scarp formation is also possible with DURICRUSTs as caprocks. The resistant crusts develop in homogeneous lithological units by weathering and soil-forming processes. Scarps of this kind are called homolithic scarps and are most frequently found in semi-arid areas (calcretes as caprocks) and in the tropics (silcretes and ferricretes). Some rock types can only have the function of caprocks under specific climatic conditions. Gypsum, for instance, can be a caprock in arid regions (e.g. southern Morocco), but in more humid climates rapid sulphate dissolution makes it an incompetent lithological unit.

The geomorphology of the upper scarp slope and the backslope is controlled by the attributes of the caprock. The upper scarp slope has a steep to cliff-like morphology. Its upper end forms a sharp crest (Trauf), especially in horizontally bedded caprocks with vertical joints. Caprock slopes, which are controlled by mass movement activity, also have a sharp crest at their top. In poorly cemented sandstones (e.g. Navajo Sandstone on the Colorado Plateau) rounded upper slopes are developed, which can also be found in more humid and colder areas, where sheet wash or solifluction processes are active on the upper slope.

The backslope begins at the scarp crest and follows the direction of caprock dip. In dry regions with highly selective weathering and erosion the inclination of the backslope is the same as the dip angle. Here the backslopes are stripped surfaces. The overlying less resistant rocks have been removed by erosion and denudation; there is a close conformity of topographic and structural relief. In the scarplands of more humid climates (e.g. Central Europe) the inclination of the backslope is generally less than the dip. The overlying strata are truncated at a low angle in the distal parts of the backslope. It resembles an inclined planation surface, but it is a caprock-controlled structural relief element of the cuesta landscape.

Especially in dry climates there is no mechanism of denudational downwearing of the caprocks. Their erosion is accomplished by aquatic and gravitational processes, which work on the scarp slope and result in lateral backwearing by parallel scarp recession. The area of the caprock outcrop is consumed by scarp retreat, at the margins of residual outliers by circumdenudation. The rate of retreat is controlled by caprock lithology and thickness (Schmidt 1989). It is not surprising that the backslope length is also controlled by these variables, and additionally by structural dip and the lithology of the overlying rocks (Schmidt 1991).

References

Ahnert, F. (1998) *Introduction to Geomorphology*, London: Arnold.
Schmidt, K.-H. (1989) The significance of scarp retreat for Cenozoic landform evolution on the Colorado Plateau, USA, *Earth Surface Processes and Landforms* 14, 93–105.
—— (1991) Lithological differentiation of structural landforms on the Colorado Plateau, USA, *Zeitschrift für Geomorphologie Supplementband* 82, 153–161.

Further reading

Blume, H. (1971) *Probleme der Schichtstufenlandschaft*, Darmstadt: Wissenschaftliche Buchgesellschaft.
Schmidt, K.-H. (1988) Die Reliefentwicklung des Colorado Plateaus, *Berliner Geograpische Abhandlungen* 49, 1–183.

SEE ALSO: butte; cuesta; hogback; mesa; structural landform

KARL-HEINZ SCHMIDT

CASE HARDENING

Case hardening describes rocks with outer shells more resistant to erosion than interior material. This hardening is sometimes called induration. Although case hardening is occasionally used as a synonym for DURICRUST, case hardening most frequently refers to differential weathering of the same rock type – often associated with intricate weathering features such as tafoni (Campbell 1999).

Two general types of processes create the appearance of case hardening: core softening of the interior; and case hardening of the exterior. James Conca proposed a lithologically based explanation. Crystalline rocks such as granite tend to core soften, whereas clastic rocks such as sandstone tend to case harden. The dichotomy has to do with the way the minerals bond together. Since sandstone grains are held together by cementing agents, a greater accumulation of cements at the surface causes case hardening. In contrast, the greatest change in hardness in a crystalline rock takes place when bonds are broken by CHEMICAL WEATHERING. Core-softened boulders are seen in locales of most intense chemical weathering.

The early literature advocates the view that hardening occurs by solutions that are mobilized from the rock's WEATHERING rind, drawn out by evaporative stresses, and then reprecipitated in the rock's outer shell. A growing body of evidence indicates that external agents also penetrate into the outer shell of the host rock, hardening the surface. A variety of different hardening agents have been found within host rocks lacking these agents. Amorphous silica, calcite, calcium borate, clay minerals such as kaolinite, iron hydroxides, oxalate minerals, rock varnish and other internally and externally derived agents penetrate about a millimetre to harden the very surface of the rock.

Case hardening, by definition, is not a ROCK COATING. However, a wide variety of rock coatings can act as case-hardening agents. Glazes of mostly silica and aluminum with some iron, only 20–30 μm thick, impede erosion of greenschist in southern England (Mottershead and Pye 1994). The role of silica glaze can be striking for temperate sandstones: '[o]ne of the most important characteristics of many porous sandstones is their tendency to case-harden owing to the development of a surface crust or rind' (Robinson and Williams 1994: 382). In Antarctica, iron-stained silica glaze reduces permeability and channels moisture towards uncoated rock surfaces. Thus, rock weathering is concentrated away from the rock coatings (Conca and Astor 1987). Lichen-generated oxalates protect sandstone surfaces in the Roman Theatre of Petra. Dark coatings of silica, oxides of iron/manganese, and charcoal case hardened rock faces at Yarwondutta Rock, Australia (Twidale 1982). While lichens are usually erosional agents, these epilithic (rock surface) organisms sometimes protect the underlying rock from erosion.

Although case hardening is most commonly noted in warm deserts where little soil covers rock surfaces, case-hardened rocks occur in all terrestrial weathering environments. In the wet tropics, for example, case hardening is frequently seen on bedrock along rivers at stages only reached by wet-season floods. In alpine settings, case hardening helps preserve glacial polish. Silica glaze is an important case-hardening agent in temperate (Mottershead and Pye 1994; Robinson and Williams 1994) and Antarctic areas (Conca and Astor 1987). Iron films can be seen splitting apart and also holding together weathering rinds in northern Scandinavia (Dixon *et al.* 2002).

Case hardening (Plate 20) often has subsurface origins in JOINTING. Mottershead and Pye (1994), for example, discerned a three-stage process. First, the host rock hardens along joint faces within the subsurface. Silica, aluminum and some iron comprise the bulk of the case-hardening agent. Second, DENUDATION of the land surface exposes joint faces at the surface. Third, erosion of rock underneath the case-hardened surface creates cavities called TAFONI, that highlight the case hardening. Rock engravings (petroglyphs) also emphasize planar JOINTING surfaces that were case hardened while in the subsurface. Road cuts of granitic rocks that weather to GRUS often reveal case-hardened subsurface joints. Geothermal and other DIASTROPHISM processes often leave behind case-hardened joints.

Considerable disagreement exists over how long it takes case hardening to form, with assertions in the literature ranging from months to thousands of years. James Conca studied rates of hardening in the Mono Basin of eastern California, finding that changes take place on the timescale of thousands to tens of thousands of years. Rates of hardening, however, vary with climate and the particular hardening process.

Plate 20 Case hardening on a *c.*140,000-year-old moraine boulder of the Sierra Nevada, California, where a combination of processes produce the differential weathering seen in the top image. Core softening of the host granodiorite boulder is the most important process. The electron microscope image and corresponding map shows a close-up of the area around the tip of the rock hammer. Some softening comes from chemical weathering and some hardening takes place as a result of the penetration of desert varnish into the weathering rind

References

Campbell, S.W. (1999) Chemical weathering associated with tafoni at Papago Park, Central Arizona, *Earth Surface Processes and Landforms* 24, 271–278.

Conca, J.L. and Astor, A.M. (1987) Capillary moisture flow and the origin of cavernous weathering in dolerites of Bull Pass Antarctica, *Geology* 15, 151–154.

Dixon, J.C., Thorn, C.E., Darmody, R.G. and Campbell, S.W. (2002) Weathering rinds and rock coatings from an Arctic alpine environment, northern Scandinavia, *Geological Society of America Bulletin* 114, 226–238.

Mottershead, D.N. and Pye, K. (1994) Tafoni on coastal slopes, South Devon, U.K., *Earth Surface Processes and Landforms* 19, 543–563.

Robinson, D.A. and Williams, R.B.G. (1994) Sandstone weathering and landforms in Britain and Europe, in D.A. Robinson and R.B.G. Williams (eds) *Rock Weathering and Landform Evolution*, 371–391, Chichester: Wiley.

Twidale, C.R. (1982) *Granite Landforms*, Amsterdam: Elsevier.

Further reading

Conca, J.L. and Rossman, G.R. (1982) Case-hardening of sandstone, *Geology* 10, 520–523.

Paradise, T.R. (1995) Sandstone weathering thresholds in Petra, Jordan, *Physical Geography* 16, 205–222.

SEE ALSO: chemical weathering; denudation; grus; jointing; rock coating; tafoni; weathering

RONALD I. DORN

CATACLASIS

A natural process whereby a faulted rock is deformed as a result of mechanical forces in the crust, such as fracturing, shearing, grain boundary sliding, and granulation. Cataclasis transforms a simple fault into a zone of fracturing and deformation without chemical alteration, and causes a decrease in the porosity of the rock alongside rock volume. Cataclasis takes place at low temperature–low pressure conditions, and high strain rates. The product of cataclasis in sediment is a cataclasite, a metamorphic rock composed of angular fragments (e.g. tectonic breccia) and a structureless rock powder fabric. When considered on a regional scale, cataclasis has also been interpreted as a flow mechanism.

STEVE WARD

CATACLINAL

A dip stream or a valley that runs in the same direction as the dip of the surrounding rock strata. Cataclinal slopes can be further classified into overdip slopes, underdip slopes, and dip slopes (steeper than, shallower than, and following the dip of surrounding strata, respectively). They may follow an individual rock layer from the base of a mountain to its peak (e.g. Mount Rundle, Canadian Rockies) (Cruden 2000). In contrast, anaclinal slopes dip in the opposite direction to the surrounding strata.

Reference

Cruden, D.M. (2000) Some forms of mountain peaks in the Canadian Rockies controlled by their rock structure, *Quaternary International* 68–71, 59–65.

STEVE WARD

CATASTROPHISM

Catastrophes in common parlance are unexpected major events with negative outcomes affecting large numbers of people. Large earthquakes, floods, hurricanes and wars are good examples of catastrophes, not all of which have geomorphic relevance. In the slightly more technical language of natural hazards, catastrophes are high magnitude, low frequency geophysical events which impact the socio-economic environment negatively. At a third level of technicality, catastrophes can be defined in the context of a mathematical approach to the analysis of inherently unstable Earth surface systems. This mathematical approach is called bifurcation theory and a special branch of bifurcation theory has been called catastrophe theory (Thom 1975). The essential concept is that many Earth systems are inherently unstable and are called dissipative structures. Such systems are characterized by a series of thresholds, below which the solutions to equations governing system dynamics are unique; beyond the threshold (or bifurcation point) the system loses its structural stability and undergoes a sudden or catastrophic change to a new form. Examples in the geomorphic literature include stream junctions in the Henry Mountains, Utah (Graf 1979), sediment transport processes in rivers (Thornes 1983) and accretion–degradation processes on a beach (Chappell 1978).

Catastrophism is a mode of thought which was common in the eighteenth and early nineteenth centuries and had as its basic premise that Earth history consisted of a series of high magnitude events separated by periods of quiescence. This mode of thought contrasts with gradualism in which small-scale, commonly acting processes are thought to be the dominant mode of geomorphic evolution.

The origins of catastrophism have often been traced to Baron Georges Cuvier (1769–1832) but, as we shall show below, his was only one brand of catastrophism and there were many catastrophist predecessors. Cuvier was the father of comparative anatomy and proposed that the Earth had suffered many catastrophes in the form of global earthquakes, each of which had changed the global landscape and annihilated almost all the flora and fauna. After each catastrophe, a new set of flora and fauna appeared: 'Thus we shall seek in vain among the various forces which still operate on the surface of our Earth, for causes competent to the production of those revolutions and catastrophes of which its external crust exhibits so many traces' (Cuvier 1817: 36–37). Huggett (1990) summarizes the issue by saying that there is either abrupt and violent change (catastrophism) or gradual and gentle change (gradualism). But the issue is more complex. There are different styles of catastrophism, and the style is determined by the premises of the author with respect to the following dichotomies: actualism v. non-actualism; directionalism v. steady state and, in dealing with organic change, internalism v. externalism. Although dichotomies tend to oversimplify the situation and in reality theorists of the Earth occupy positions along a spectrum of ideas, some simplification is necessary within the word constraints of this encyclopedia entry. The issue between actualists and non-actualists is whether past processes have differed in kind from those now in operation; non-actualists took the view that past processes could differ in kind from those presently in operation. The issue between directionalists and steady statists has been eloquently discussed by Gould (1987) in terms of time's arrow versus time's cycle; directionalists took the view that monotonic change, whether progress or regress, could be detected in Earth history. The issue between internalists and externalists, which arose only in the context of organic change, was whether the motor of organic change was an inner drive or external environmental factors; externalists argued that the environment forced change. Using these dichotomies as a basis for classification of styles of catastrophism, Huggett (1990) came up with eight categories of catastrophism which could be distinguished in the early history of environmental science. Six of these categories are reproduced here:

1 *Actualistic directional catastrophism* The Wernerian system of Earth history, following Abraham Werner (1749–1817), is a classic example of this kind of catastrophism. He envisaged five periods of Earth history, punctuated by intermittent and catastrophic ocean subsidence and precipitation of crustal rocks. The five periods were consecutive and demonstrated directional change.
2 *Non-actualistic directional catastrophism* René Descartes (1596–1650) described the origin of the Earth as an incandescent ball, followed by collapse of the Earth's outer crust and the release of massive volumes of water. His system involved directional evolution from original chaos created out of nothing by God through to an ordered universe which evolved according to natural laws invested in the original particles by God. The so-called Scriptural geologists, such as William Buckland (for most of his active career), Adam Sedgwick (1785–1873), William Conybeare (1787–1857) and Robert Murchison (1792–1871) all fell into this category in the sense that they believed in God's special intervention in the regular course of Nature through geological catastrophes and the sudden rise of species.
3 *Non-actualistic steady-state catastrophism* Baron Georges Cuvier (1769–1832) was the leading protagonist of this school of thought. He saw in the fossil and stratigraphic record evidence of catastrophic changes too great to be explained by the ordinary, slow-acting processes on the Earth's surface. Each catastrophe had changed the global landscape and a new set of plants and animals had appeared, with no particular connection with the previous flora and fauna.
4 *Inner-driven directional catastrophism* Louis Agassiz (1807–1873) espoused this position for most of his career. The underlying premise is that organisms have an immanent quality leading to progressive but discontinuous change. The progression of life was the unfolding of God's plan through catastrophes in the inorganic world and punctuations in the organic world, but with no causal connection between the two.
5 *Environmentally driven directional catastrophism* The mature William Buckland was a proponent of this position, that the Earth had suffered a series of catastrophes and that a new set of species was created after each mass extinction. Each new creation was an improvement on the previous one and the improvements placed the organisms in better harmony with the changed environment.
6 *Environmentally driven steady-state catastrophism* Baron Georges Cuvier was the leading advocate of this position. He could not accept a progression of organisms. He recognized four chief branches of animals, the members in each of which were fixed and designed to meet all environmental conditions. His catastrophes were essentially sudden environmental changes in the distribution of land and sea. The motor of biotic change is sudden environmental change.

The other two of Huggett's categories, the actualistic and inner-driven steady-state catastrophisms, were rarely expressed positions in the early nineteenth century but became more popular in the late twentieth century with the rise of neocatastrophism.

Since the 1960s, one of the key assumptions of catastrophism has been making a strong comeback in the form of neocatastrophism. This key assumption is that high magnitude, low frequency events are cumulatively more important in Earth history than low magnitude, high frequency events. Some reasons for this changed perspective are:

(1) Improved precision in geochronology has demonstrated unexpectedly rapid past changes.
(2) The exploration of mass extinctions in the past has intensified.
(3) Some geomorphological features, such as the channelled scablands of eastern Washington, are more amenable to explanation by low frequency, high magnitude events than by gradual, semi-continuous processes.
(4) Space exploration has generated a strong interest in galactic-scale events.
(5) Interest in global environmental change has provided evidence of rapid past changes, such as found in the polar ice caps and the oceanic deep sediments.
(6) The rise of non-linear dynamics and chaos theory is beginning to provide ways of synthesizing gradualism and catastrophism.

It seems self-evident that Earth history contains a combination of catastrophic and gradual events; accepting the occurrence of catastrophes is not to deny the effectiveness of gradual processes. The main reason for concerns about catastrophism expressed by Earth scientists since the mid-nineteenth century has been a fear of the reintroduction of religious beliefs into the canon of modern science because of the long-held views about the Noachian flood in western thinking. Largely for this reason, the geological establishment of the day refused to countenance the work of Harlen Bretz (1923) in his account of the origins of the channelled scablands of eastern Washington. He suggested that these SCABLANDS (massively and regionally gullied lands) could be explained best by the action of a single large flood over a period of only a few days. The fact that they have been shown subsequently to be caused by a succession of pulses associated with the draining of glacial Lake Missoula (Baker 1973) was complete vindication for Bretz. It is important to recognize that in no sense did the catastrophists violate the principle of uniformity of law (i.e. that natural laws are invariant in time and space). On the other hand, the non-actualists did violate the principle of simplicity, a principle which states that no unknown causes should be invoked if available processes are adequate. This guideline is known as the uniformity of process.

References

Baker, V.R. (1973) Palaeohydrology and sedimentology of Lake Missoula flooding in eastern Washington, *Geological Society of America Special Paper* 144.

Bretz, J.H. (1923) The channelled scablands of the Columbia Plateau, *Journal of Geology* 31, 617–649.

Chappell, J. (1978) On process-landform models from Papua New Guinea and elsewhere, in J.L. Davies and M.A.J. Williams (eds) *Landform Evolution in Australasia*, 348–361, Canberra: Australian National University Press.

Cuvier, G. (1817) *Essay on the Theory of the Earth*. With mineralogical notes an account of Cuvier's geological discoveries by Professor Jameson. Edinburgh: William Blackwood.

Gould, S.J. (1987) *Time's Arrow, Time's Cycle. Myth and Metaphor in the Discovery of Geological Time*, Cambridge: Harvard University Press.

Graf, W.L. (1979) Catastrophe theory as a model for changes in fluvial systems, in D.D. Rhodes and G.P. Williams (eds) *Adjustments of the Fluvial System*, 13–32, Dubuque: Kendall Hunt.

Huggett, R. (1990) *Catastrophism: Systems of Earth History*, London: Edward Arnold.

Thom, R. (1975) *Structural Stability and Morphogenesis*, New York: Benjamin.

Thornes, J.B. (1983) Evolutionary geomorphology, *Geography* 68, 225–235.

SEE ALSO: actualism; neocatastrophism; uniformitarianism

OLAV SLAYMAKER

CATENA

The concept of a catena is one of linkage – catena being the Latin for chain. So, by analogy individual elements are linked in some way and have something in common. In the case of a soil catena, the soils are linked by having the same parent material and age but are differentiated by their position on a slope which alone gives them different characteristics, especially in relation to drainage. Thus a formal definition is: 'A sequence of soils of about the same age, derived from similar parent material, and occurring under similar climatic conditions, but having different characteristics because of variations in relief and in drainage.'

A more generalized term which can be used is the toposequence, which refers to any sequence of soils which varies with topography. A more extensive, related concept is soil association, which is any repeated pattern in the landscape which may not necessarily be repeated with topography, as with a catena, but could be linked to geology, geomorphology and indicated by vegetation in any geomorphic element of the landscape.

The term soil catena was first proposed by Milne (1935a,b; 1936) for topographically linked soils in East Africa. On the sides of large valleys different soils were found on the upslope crest, the lower downslope and the footslope. Milne felt that the profiles of the soil types changed character downslope according to both drainage and the past history of the land surface. Here, there is an essentially uniform lithology throughout the slope and the differences derive from the shedding of moisture upslope and wetter conditions downslope as well as the movement of solutes in that water and the physical movement of eroded particles and their accumulation downslope.

Milne felt that the upper soils might also be older and more remnant and that the downslope soils might be younger, with a fresher accumulation of deposits, and also that the rocks might actually differ downslope. Ruhe (1960) further expanded on this by differentiating between the 'classic' catena on similar parent materials and a sequence which could be formed on two or more geological formations where the lithology actually varied. Here the downslope series is still linked by drainage and transport of material and shares a common physiographic history and geomorphic evolution despite being on different parent materials.

The catena concept is useful in soil mapping because it implies a regularly occurring relationship of the soils with topography, giving an expectation or prediction of what is liable to be present. An example of a general schema is in Table 6. With mechanical movement of particles, the upslope soils might be more coarse grained, having lost fines downslope, while the downslope soils could be more alluvial, with the accumulation of fines and/or with much coarser particles accumulating, according to the amount of mechanical action on the slope. Generally the midslope soils tend to be prone to erosion and are much thinner than the slope foot, where accumulation occurs.

The more complex and nutrient-rich montmorillonite clays may be found where nutrients accumulate at the base of the slope. In warmer areas, montmorillonite clays also survive on the soil crest and plateau but in wetter, cooler areas, the simpler kaolinite clays are found in these positions due to greater leaching. In hot, humid or semi-humid regions, the upslope leached soils are often red due to OXIDATION and at the slope foot where drainage is impeded the colour changes downslope through yellow to grey. The effect of topography can thus be expressed in the tropics with high rainfall where red soils with kaolonitic clays are found in the better drained upslope areas with montmorillonite and black, organic soils in the less well-drained depressions. On a larger scale, which includes tropical mountains, in areas of highland surrounded by lower arid land, the sequence with decreasing height is one of upland podzols and brown earths to chernozems to desert and semi-desert soils; for uplands surrounded by lower land with high rainfall the downslope sequence is one of podzols and brown earths to yellow, red and black soils of the humid lowlands.

The significance of a catenary sequence for agriculture and land-use planning is again predictive in that more drought resistant crops can be grown on the upper slopes and moisture-tolerant crops on the lower slope where the aerated zone is nearer the surface.

Hillslope hydrologists have also used the catena concept when devising predictive models for hillslope runoff generation. Here hydrological processes, such as infiltration, overland flow and throughflow, can be predicted in relation to soil profile properties which can be predicted to vary systematically downslope (McCaig 1985). Anderson (1985) also uses the concept of catena in predicting the mechanical, load-bearing properties of soils. Regression modelling has been attempted to quantify the relationship between slope position and soil properties with varying degrees of success (Furley 1971).

References

Anderson, M.G. (1985) Forecasting the trafficability of soils, in K.S. Richards, R.R. Arnett and S. Ellis (eds) *Geomorphology and Soils*, 396–416, London: George Allen and Unwin.

Furley, P.A. (1971) Relationships between slope form and soil properties developed over chalk parent materials, in D. Brunsden (ed.) *Slopes: Form and Process*, Institute of British Geographers, 141–163.

McCaig, M. (1985) Soil properties and subsurface hydrology, in K.S. Richards, R.R. Arnett and S. Ellis (eds) *Geomorphology and Soils*, 121–140, London: George Allen and Unwin.

Milne, G. (1935a) Some suggested units of classification and mapping, particularly for East African soils, *Soil Research* Berlin, 4, 183–198.

——(1935b) Complex units for the mapping of complex soil associations, *Transactions Third International Congress of Soil Science* 1, 345–1,347.

Table 6 The relationship of soils and topography

	Wetter, cooler climates	Wet climates	Drier climates
Upslope/ slope crest	Peat, podzolization	Podzolization	Leached soil
Midslope	Brown earth	Brown earth	Non-calcareous
Downslope	Peat, gley	Gley	Calcareous soil

Milne, G. (1936) A soil reconnaissance journey through parts of Tanganyika territory. *Memoirs of the Agricultural Research Station, Amani*. Reprinted 1947. *Journal of Ecology* 35, 192–265.
Ruhe, R.V. (1960) Elements of the soil landscape, *7th International Congress of Soil Science* 4, 165–170.

STEVE TRUDGILL

CAVE

The standard definition of a cave is 'an underground opening large enough for human entry' (Oxford English Dictionary and others). As such, natural caves occur in most consolidated rocks or compacted sediments and in most geomorphic settings, created by a variety of processes. This entry focuses upon KARST caves, which are the most important in terms of their magnitude, frequency, diversity of form, and role in general geomorphology. They are formed where dissolution is either the quantitatively predominant or essential trigger process of rock removal. Karst caves thus are restricted to the comparatively soluble rocks. In descending order of solubility, these are salt, gypsum and anhydrite, limestone and dolostone and (to a much lesser extent) quartzites, calcareous and siliceous sandstones. Limestone caves display the greatest range in size and form. Although 'enterable by humans' is the criterion, initial solution caves are much smaller (~1 cm diameter is a reasonable minimum) and the rules of genesis do not change significantly as they are enlarged to enterable size.

Caves created with little or no dissolution are PSEUDOKARSTic. CAVERNOUS WEATHERING voids can be considered transitional. Many pseudokarst caves are created by mechanical processes, including piping (see PIPE AND PIPING), THERMOKARST collapse, frost riving, wave action and CAVITATION along coasts. Such caves are rarely longer than a few tens of metres; most do not pass out of the range of daylight. Stream melting and sublimation in GLACIER ice and firn creates greater caves that are nearly identical in form and scale to the vadose shafts, canyons and simple phreatic passages found in many dissolutional caves. The lengthiest pseudokarst caves are lava tubes (see LAVA LANDFORM), formed by channelled discharge of still-molten lava within consolidating flows; single tubes, dendritic networks and anastomosing mazes are known, some extending for 10 km or more (Gillieson 1996).

Karst caves

Modern classification (Klimchouk *et al.* 2000) recognizes three principal genetic settings: (1) coastal caves in young carbonate rocks; (2) hypogene caves, formed by waters ascending out of artesian traps ('confined groundwaters') in any soluble rock; (3) unconfined meteoric water caves in soluble rocks, the most abundant and significant class. Figure 18 shows a basic range of plan patterns in these caves, relating them to type of recharge and the most effective (transmissive) porosity existing at the onset of dissolution. In most known hypogene and unconfined caves intergranular ('matrix') porosity is low, (<5 per cent), solvent water being transmitted via penetrable bedding planes, joints and faults which control the loci of the solutional passages. The relevant chemistry and kinetics of aqueous dissolution are discussed in Ford and Williams (1989: 42–126); Klimchouk *et al.* (2000: 124–223).

EOGENETIC COASTAL CAVES

'Eogenetic' describes very young limestone and dolostone accumulations where consolidation by compaction and interstitial cementation (i.e. diagenesis) is still limited, with the consequence that intergranular porosity offers principal or, at least, significant routes for solvent water flow. Such rocks are found in tropical/subtropical coastal settings today, e.g. Florida, Yucatan, Bahamas, many Pacific atolls, etc., and are chiefly Pleistocene in age. The matrix porosity yields cave patterns similar to those of cavernous weathering in non-karst strata.

'Syngenetic' caves (Jennings 1985) form in calcareous sand dunes when surficial sand becomes case hardened by cementation (i.e. earliest diagenesis), following which storm waves or surface streams breach the casing and wash out the non-cemented sand behind it, creating cavities sometimes many metres in length or height. This is one end of the spectrum of karst caves, where mechanical washout (piping) is quantitatively predominant but dissolution and cementation must precede it.

Much more widespread are caves formed where fresh and salt waters mix along the water table at the coast itself (flank margin caves) or at the halocline beneath the freshwater lens further inland (Klimchouk *et al.* 2000: 226–233). Flank margin caves display large entrance chambers dividing and tapering to blind endings a few metres or tens

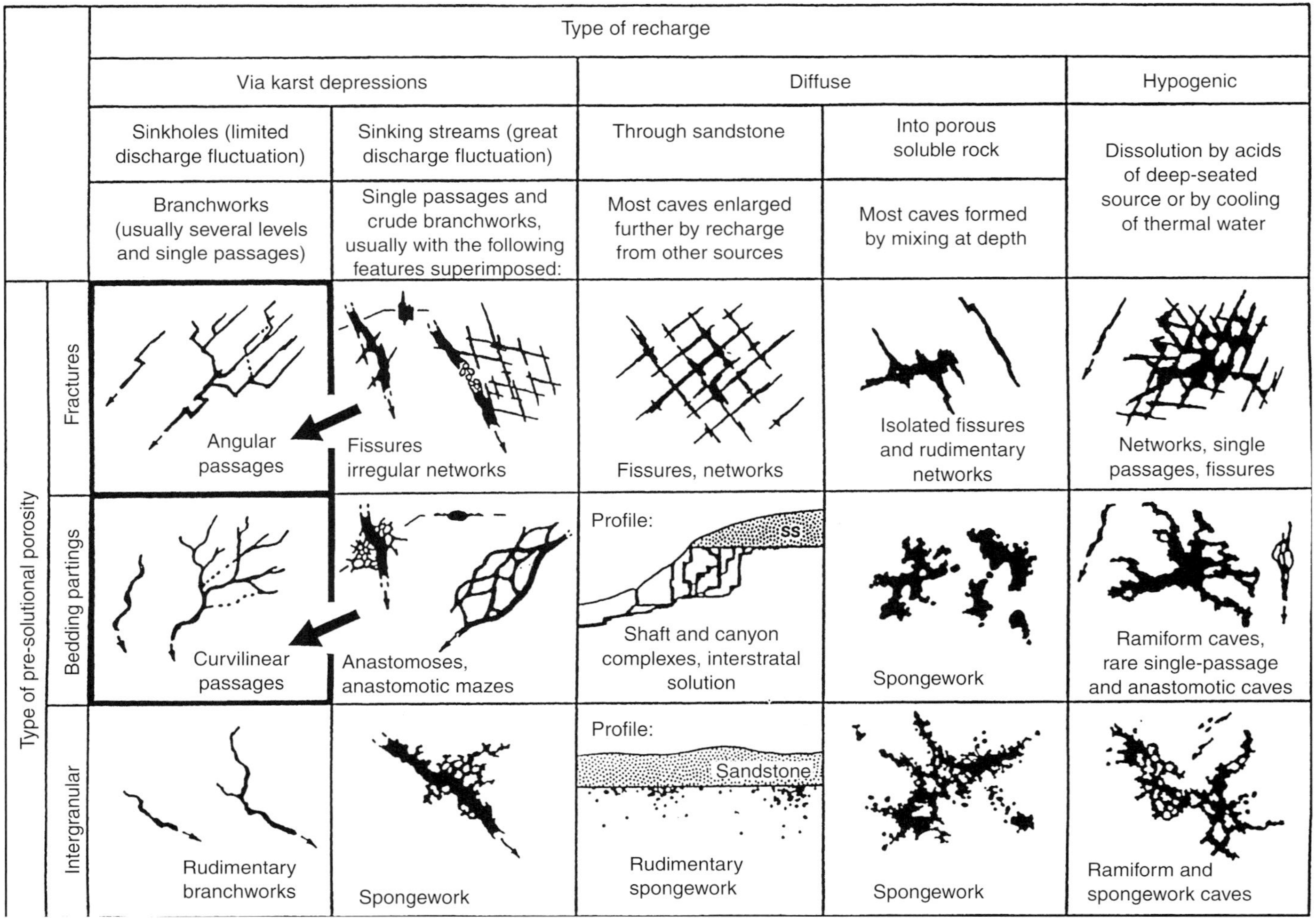

Type of pre-solutional porosity	Type of recharge				
	Via karst depressions		Diffuse		Hypogenic
	Sinkholes (limited discharge fluctuation)	Sinking streams (great discharge fluctuation)	Through sandstone	Into porous soluble rock	Dissolution by acids of deep-seated source or by cooling of thermal water
	Branchworks (usually several levels and single passages)	Single passages and crude branchworks, usually with the following features superimposed:	Most caves enlarged further by recharge from other sources	Most caves formed by mixing at depth	
Fractures	Angular passages	Fissures irregular networks	Fissures, networks	Isolated fissures and rudimentary networks	Networks, single passages, fissures
Bedding partings	Curvilinear passages	Anastomoses, anastomotic mazes	Profile: SS Shaft and canyon complexes, interstratal solution	Spongework	Ramiform caves, rare single-passage and anastomotic caves
Intergranular	Rudimentary branchworks	Spongework	Profile: Sandstone Rudimentary spongework	Spongework	Ramiform and spongework caves

Figure 18 Basic plan patterns of caves shown in relation to types of pre-solutional porosity and conditions of recharge; slightly modified from Palmer (1991)

of metres inland. Halocline caves have more complex spongework patterns, in part due to the shifting of salt/fresh mixing zones as Quaternary sea levels moved up and down; aggregate passage lengths as great as 1 km are rare.

Cave systems many tens of km in length and extending 5 km or more inland are being explored along the Caribbean coast of the Yucatan Peninsula; although in young limestones, they are of unconfined meteoric origin, modified in form by the salt waters that now inundate them.

HYPOGENE CAVES

In hypogene caves the waters may have circulated deeply within karst strata alone (due to synclinal or graben-type structural traps), or be ascending into the karst rocks from underlying, non-karst aquifers (interstratal flow). They may be thermal (>4 °C warmer than mean temperatures in the rock to be dissolved) or ambient.

A majority of hypogene caves are excavated under phreatic conditions, i.e. beneath any water table. The most simple form is a vertical/near-vertical chimney on a fracture, up which the water flows to discharge at a surface spring or into overlying, more porous, strata. Active instances include thermal springs in Mexico from shafts 40+ m in diameter and plumbed to −300 m. Deeper examples are known, but drained by uplift and erosion; some contain economic minerals precipitated on the walls, e.g Tyuya Muyun, Kazakhstan. More complex in form are arborescent chimney caves, branching upwards from basal reservoir chambers; Satorkopuszta Cave, Hungary, is a spectacular instance with later spheroidal rooms of condensation corrosion origin branching from an original phreatic shaft (Klimchouk *et al.* 2000: 292–303).

Fracture-guided network caves (Figure 18) are common. In western Ukraine local meteoric waters passing up through a ~14 m stratum of gypsum from an underlying sand aquifer created joint-guided mazes with intersections every 2–5 m; 212 km of contiguous passages are mapped in Opitimists' Cave. More complex are multistorey rectilinear mazes in the Black Hills, South Dakota, where thermal waters converged on Carboniferous palaeokarst (see PALAEOKARST AND RELICT KARST) preserved under clastic strata. The waters discharge through weaknesses in the clastics today, enlarging their routes and lowering the water table in Wind Cave 14 km distant at 40 cm k yr^{-1}. Jewel Cave, ~40 km distant, is fully drained; its 200 km of mapped passages are crusted with 10–20 cm of calcite spar deposited as the waters declined. Large-scale groundwater invasion and dissolution of limestones, gypsum and salt such as this, but deeply buried under later rocks, is associated with formation of solution breccias hosting oil and gas, lead/zinc and other mineral deposits, or creating breccia pipes that can stope upwards through 1,000+ m of overlying rocks. The greatest reported hypogene cavity is in Archean-Proterozoic marbles of the Rhodope Mountains, Bulgaria. It has an estimated volume of 237 million m^3 and a roof-to-floor depth believed to exceed 1,340 m (Klimchouk *et al.* 2000: 304–306).

A very distinctive type is the cave formed by sulphuric acid from H_2S. In most known instances the gas migrated from adjoining coal or oil basins where it was produced by bacterial reduction of gypsum. Reaching carbonate rocks it oxidized to H_2SO_4 at and just below the water table. Small H_2S caves tend to be linear outlets to springs. Large caves ramify about the gas inlets (Figure 18). Big chambers are created by lateral corrosion at the water table plus condensation corrosion above it that may convert limestone walls to gypsum to depths as great as one metre. Shift of inlet points and lowering of springs leads to multilevel development of spectacular systems such as Lechuguilla Cave, New Mexico (172 km, ~480 m in depth; Widmer 1998).

Unconfined caves

These are the principal type known to explorers and geomorphologists. The caves extend from surface water input points such as KARRENfields, DOLINEs, river sinks or POLJEs at the karst margins, to springs that are lower in elevation. Although it is rare for cavers to be able to follow the water all the way, dye tracing and other analysis invariably confirms that dissolutional conduits are continuous between sink and spring and regulate the flow. The most simple caves are single dissolutional pipes between sinks and springs. There are many instances in underground meander cut-offs or river short-cuts across narrow horsts or anticlines. However, the majority of caves have multiple inlets that, in plan view, link up to form crudely dendritic patterns. These are angular where joints are dominant controls and sinuous in bedding planes; many caves exhibit mixtures of the two. Joint mazes and bedding plane anastomoses

(Figure 18) are usually subsidiary components in the dendritic plans, formed where there is rapid flooding at stream sinkpoints or underground blockages. The published plans of many cave systems appear more complicated than these simple combinations because the systems are multiphase; relict passages from higher levels cross those that are still active Modelling of plan pattern genesis is quite advanced, (see Ford and Williams 1989: 249–261; Klimchouk *et al.* 2000: 175–223).

Cave morphology in long section (i.e. length × depth) is closely related to geologic structure and groundwater zonation. In young mountain terrains if karst aquifers are thick and rapid uplift opens vertical fractures widely, caves are sequences of shafts down the steepest fractures, linked together by short, sinuous, stream canyons. Gravitational control of flow is predominant. A majority of the world's caves deeper than 1 km gain most of their depth in this upper vadose zone. Voronja Cave, Caucasus – currently the explored depth record holder at −1,710 m – is a fine example.

Where the karst formations are relatively thin and/or were little stressed during uplift, such conditions may never have existed. Instead, the uppermost zone is waterfilled initially but drains progressively as caves propagate through it and become enlarged – the 'drawdown vadose zone'. The caves display initial phreatic features such as elliptical cross sections in bedding planes, but with subsequent gravitational entrenchments beneath them. This combination can be found locally in young mountains also where passages are perched on shale bands or other obstructions. Entire cave systems, from sinkpoints to springs, can develop wholly within these two vadose settings, especially where karst strata are perched on insoluble rocks above regional base levels. There can be deep gravitational entrenchment into the insolubles; e.g. some 'contact caves' in Greenbrier County, West Virginia, have 90+ per cent of their volume in erodible shales beneath limestones that hosted the initial passages.

Most extensive cave systems have substantial water table or phreatic (sub-watertable) sections, however. Their length is usually greater than that of the vadose parts, and the geometry more complex and varied. There are four basic possibilities ('four state model'; Ford and Williams 1989: 261–274). If the density of penetrable, interconnected fissures is very low, geologic structure may compel the conduits to follow courses below the elevation of the springs or water table ('phreatic loops'). A State 1 system passes from the vadose zone to the spring in one loop. State 2 is a sequence of loops whose crests fix local elevations of the water table. Where fissure frequency and interconnection are greater, caves display mixtures of loops with gently graded passages at the water table (State 3). Very high frequency permits continuous development along the water table (State 4), similar to flank caves in young, porous limestones. There is probably phreatic looping to depths of 1,000+ m in some State 1 caves. Individual loops greater than 250 m deep are common in State 2. In regions subject to large magnitude, abrupt flooding such as alpine mountains, overflow ('epiphreatic') passages develop above the low stage conduits (Audra 1995).

Multi-level (multi-phase) caves

Most extensive caves have two or more 'levels' that developed to drain to successively lower springs – 'level' denoting the historical succession but not implying that the galleries must be near-horizontal; often, there is State 2 or 3 geometry. In vadose caves the lower levels may be simple entrenchments beneath the older passages. Where there is water table or phreatic development, the new springs are usually offset laterally tens to hundreds of metres or more and may have distributaries. The new springs steepen the hydraulic gradient in the downstream section of the old cave, which then adjusts to its new 'level' in a sequence of breakthrough undercaptures (French – *soutirages*) that move the hydraulic steepening progressively upstream like a river knickpoint. Portions of individual capture links are incorporated into the final dendritic pattern of the new level but others are left redundant, becoming drained relicts or silted backwaters.

Superimposition of successive levels, redundant links and invasion vadose caves from new sinkpoints, make maps of great systems such as Mammoth Cave, Kentucky (556 km – the longest mapped) more complex in appearance than almost any other geomorphic or hydrogeologic phenomena.

References

Audra, P. (1995) Karst alpins; Genèse des grands reseaux souterraines, *Karstologia Memoires*, 5.

Ford, D.C. and Williams, P.W. (1989) *Karst Geomorphology and Hydrology*, London: Chapman and Hall.

Gillieson, D. (1996) *Caves: Processes, Development, Management*, Oxford: Blackwell.
Jennings, J.N. (1985) *Karst Geomorphology*, Oxford: Blackwell.
Klimchouk, A.V., Ford, D.C., Palmer, A.N. and Dreybrodt, W. (eds) (2000) *Speleogenesis; Evolution of Karst Aquifers*, Huntsville, AL: National Speleological Society of America.
Palmer, A.N. (1991) Origin and morphology of limestone caves, *Geological Society of America Bulletin*, 43.
Widmer, U. (ed.) (1998) *Lechuguilla, Jewel of the Underground*, Basel: Caving Publications International.

DEREK C. FORD

CAVERNOUS WEATHERING

Cavernous weathering is a process which causes the hollowing-out of rock outcrops and boulders on vertical and near-vertical faces. The hollows, or caverns, may take one of two forms. The first are known as 'tafoni' (singular: tafone), a term derived from the Sicilian word meaning 'windows'. TAFONI typically conform to a spherical or elliptical shape, have arched-shaped entrances, concave inner walls, overhanging visors and gently sloping debris-covered floors. Tafoni range in size from several centimetres to several metres in diameter and depth, may coalesce or interconnect and second order tafoni may develop on the back-walls of larger forms. The second form caverns may take is commonly called 'alveoli' (singular: alveole). Alveoli are formed by similar processes, termed HONEYCOMB WEATHERING, which involves the progressive development of closely spaced cavities in rock faces. These small-scale caverns are separated by narrow, intricate walls, creating a surface reminiscent of honeycomb. Alveoli are usually several centimetres in diameter and rarely are larger than one metre wide. The relationship between alveoli and tafoni has not been clearly defined and the distinction is therefore one of size and shape. Although cavernous and honeycomb weathering frequently occur together, their independent existence and differences in form have led some geomorphologists to contend that they are not genetically related, but rather derive from different modes of origin (Mustoe 1982). Cavernous weathering cannot be defined on the basis of geographical, lithological or climatological occurrence, as caverns may be found in many environments and on most rock types.

Caverns cannot be categorized on the basis of their occurrence on specific rock types or in specific climate regimes, as they occur under cold, temperate, hot, humid and arid environments, and are found on a variety of rock types. Cavernous weathering was previously considered to be a diagnostic feature of arid environments (Blackwelder 1929). Tafoni are most prolific in salt-rich environments, and have been documented most often in deserts (McGreevy 1985) and coastal zones (Mottershead and Pye 1994). Common factors in this disparate range of environments are high salt concentrations and frequent or occasional desiccating conditions. The occurrence of tafoni on sandstone surfaces (Young and Young 1992), and on granite surfaces (Dragovich 1969) has frequently been documented. Aside from these siliceous rock types, tafoni have also been recognized on tonalite, dolerite, lacustrine silts and conglomerates. A range of weathering processes may be responsible for the occurrence of tafoni, and there is clearly convergence of form. Considerable literature has accumulated on the nature of both tafoni and alveoli, but as more information has been presented, their possible origins, rather than being clarified, seem to have become more confused.

There are two types of tafoni: 'basal' and 'sidewall' tafoni. Basal tafoni are often found, as the name suggests, on outcrops and boulders at ground level. Sidewall tafoni are present on vertical and near-vertical outcrop surfaces where strong rock discontinuities are not present. Tafoni which develop along discontinuities, above ground level, may also be considered to be basal tafoni, and are characterized by a higher rate of back weathering than upwards progression.

Early studies of cavernous weathering suggested that caverns were created by the action of the wind, which was also responsible for the removal of weathering products. It is now widely accepted, however, that aeolian deflation is not responsible for the hollowing out of boulders or pitting of rock faces. Disintegration of cavern walls generally proceeds by flaking and granular disintegration, and numerous weathering processes have been invoked, including insolation weathering, WETTING AND DRYING WEATHERING, frost weathering (see FROST AND FROST WEATHERING), solution and chemical alteration of rock minerals, and SALT WEATHERING.

CHEMICAL WEATHERING processes are considered to be important in the development of tafoni in some circumstances. Tafoni in sandstone appear to be partly the result of the reaction of water and organic acids with iron and silica. Caverns in limestone are produced by a solution of calcium and magnesium carbonates; chemical weathering of dolerite and tonalite has been identified in caverns. Other chemical weathering processes which may have contributed to cavernous weathering include solution, HYDROLYSIS and hydration reactions, induced by microclimatological differences between caverns and exposed rock surfaces.

Of all the processes which may create caverns, salt weathering is the most commonly invoked. The importance of salt weathering is indicated by the presence of salt crystals on walls of coastal and desert tafoni and alveoli. Salts are evident in seepage, in granular debris and flakes being detached from cavern walls, in floor sediments and in crevices within caverns in many locations, testifying to the role of salt crystal growth in cavern development. Salt crystallization, hydration and thermal expansion may contribute to disaggregation, but given the wide geographical, climatological and lithological range of tafoni and alveoli, it is more likely that several weathering processes are involved in cavernous weathering.

A common feature of cavernously weathered surfaces is a case-hardened layer on exposed rock surfaces, penetrated by tafoni. Some researchers conclude that the presence of a hardened crust and weakened interior is a fundamental reason for tafoni occurrence (Conca and Rossman 1982). Conca and Rossman (1985) postulated that the presence of a case-hardening cement is of secondary importance compared to core softening, a result of differential weathering rates between the rock interior and exterior. Many other examples of tafoni in the absence of case hardening have been presented, however, so it appears that a hardened outer crust is not a prerequisite for tafoni formation.

A problem that has beset studies of tafoni weathering has been the tendency to relate it to one single formative process, whereas in many cases cavern development can only be satisfactorily explained by invoking the operation of a range of weathering mechanisms. There may be processes which are active in all cases, but it is likely that the relative importance of physical and chemical weathering processes will vary with different environmental conditions, which may operate under different catalytic conditions, and may act synergistically.

The relative significance of lithological controls is unclear in the context of tafoni origin. Caverns may be initiated along pre-existing joints or bedding planes, or may be distributed randomly across rock surfaces. This random (or pseudo-random) distribution may reflect points of mineralogical weakness on the rock surface, but this idea cannot be tested as the initial surface has been lost in the creation of the hollow.

Once initiated, the sheltering afforded by caverns may provide temperature ranges which are less extreme than on rock surfaces exposed to direct insolation, and higher relative humidities, a significant factor influencing the deposition, absorption and evaporation of moisture on rock surfaces. Microclimatological differences created in shadow zones of tafoni may accelerate the effects of weathering processes. Conversely, the exposed surfaces may shed moisture and solutes more rapidly, creating a negative, or self-regulating, feedback in the weathering system. Cavernous weathering may therefore represent the response of the weathering system to dynamical instabilities, where the positive feedback produced by material loss encourages accelerated weathering within caverns and the system responds by divergent evolution of the rock surface into hollows and exposed stable surfaces.

References

Blackwelder, E. (1929) Cavernous rock surfaces of the desert, *American Journal of Science* 17(101), 394–399.

Conca, J.L. and Rossman, G.R. (1982) Case hardening of sandstone, *Geology* 10, 520–523.

Conca, J.L. and Rossman, G.R. (1985) Core softening in cavernously weathered tonalite, *Journal of Geology* 93, 59–73.

Dragovich, D. (1969) The origin of cavernous surfaces (tafoni) in granitic rocks of southern South Australia, *Zeitschrift für Geomorphologie NF* 13, 163–181.

McGreevy, J.P. (1985) A preliminary Scanning Electron Microscope study of honeycomb weathering of sandstone in a coastal environment, *Earth Surface Processes and Landforms* 10, 509–518.

Mottershead, D.N. and Pye, K. (1994) Tafoni on coastal slopes, south Devon, U.K., *Earth Surface Processes and Landforms* 19, 543–563.

Mustoe, G.E. (1982) The origin of honeycomb weathering, *Geological Society of America Bulletin* 93, 108–115.

Young, R.W. and Young, A.R.M. (1992) *Sandstone Landforms*, Berlin: Springer-Verlag.

Further reading

Goudie, A.S. (1997) Weathering processes, in D.S.G. Thomas (ed.) *Arid Zone Geomorphology* (2nd edn), 25–39, Chichester: Wiley.

Goudie, A.S. and Viles, H. (1997) *Salt Weathering Hazards*, Chichester: Wiley.
Trenhaile, A.S. (1987) *The Geomorphology of Rock Coasts*, Oxford: Clarendon Press.

SEE ALSO: case hardening; weathering

ALICE TURKINGTON

CAVITATION

A form of erosion that can occur in rapidly moving water. Local areas of low pressure may be created in the water and as Drewry (1986: 68) has explained:

> If the pressure falls as low as the vapour pressure of the water at bulk temperature, macroscopic bubbles of vapour (cavities) will form. The cavitation bubbles grow and are moved along in the fluid flow until they reach a region of slightly higher local pressure where they will suddenly collapse. If cavity collapse is adjacent to the channel wall localized but very high impact forces are produced against the rock. This action may give rise to mechanical failure of the channel.

The destructive action of cavitation is probably due to the shock waves created when the bubbles collapse. Cavitation is of great significance in the malfunction of hydraulic machinery (e.g. turbine blades, ships' propellors, etc.) but its geomorphological effects may also be substantial (Barnes 1956). Sufficiently high velocities for its operation can occur in such situations as waterfalls, rapids, bedrock channels, beneath glaciers and on TSUNAMI scoured surfaces (Aalto *et al.* 1999). Mean flow velocities necessary to initiate the process are generally higher than about $10\,m\,s^{-1}$ for flow depths greater than about 4 m. Cavitation may contribute to the fluting and potholing of massive, unjointed rocks in bedrock channels (Whipple *et al.* 2000).

The term cavitation has a second and unrelated meaning in geomorphology, namely the formation of cavity at the bed or a sliding glacier. Their formation can enhance basal sliding (Lliboutry 1968).

References

Aalto, K.R., Aalto, R., Garrison-Laney, C.E. and Abramson, H.F. (1999) Tsumani(?) Sculpturing of the pebble beach wave-cut platform, Crescent City area, California, *Journal of Geology* 107, 607–622.
Barnes, H.L. (1956) Cavitation as a geological agent, *American Journal of Science* 254, 493–505.
Drewry, D. (1986) *Glacial Geologic Processes*, London: Arnold.
Lliboutry, L. (1968) General theory of subglacial cavitation and sliding of temperate glaciers, *Journal of Glaciology* 7, 21–58.
Whipple, K.X., Hancock, G.S. and Anderson, R.S. (2000) River incision into bedrock: mechanics and relative efficacy of plucking, abrasion and cavitation, *Bulletin of the Geological Society of America* 112, 480–503.

A.S. GOUDIE

CAY

Cays are the general terms for islands which develop on CORAL REEFs but can be divided for geomorphological purposes into true cays and *motu* (Nunn 1994). Both types are dominated by clastic materials scooped off the front of a coral reef, particularly off reef-talus slopes by large-amplitude waves, and dumped on reef surfaces. Such deposits have been observed to migrate across reef surfaces away from the oceans until they reach a point where they accumulate.

True cays are more transient, sometimes in existence for less than one year, compared to *motu*, some of which are 3,000–4,000 years old. In general, true cays are confined to narrow reef flats, commonly in either high-energy wave environments and/or in places affected annually by storm surges. *Motu* tend to be confined to broader reef surfaces, typically those outside the hurricane (tropical-cyclone) belt, where Holocene sea level exceeded its present level around 4,000 cal. yr BP.

True cays

Being ephemeral and transient reef islands, true cays are generally distinguished by being bare of vegetation and regularly overtopped by waves. They are also characterized by the absence of cemented deposits which renders them vulnerable to obliteration by large waves.

Although we know that true cays usually form as a result of storm surge (or tsunami) deposition of reef-front sediments, we do not clearly understand why this happens only in some instances while in others cays can be removed. It is likely that cay formation occurs when waves are coming from a direction where, in running up the submarine island slope, they can (and do) pick up a lot of material which is carried on to the reef surface. Cay removal may occur at higher velocities but may also be when the wave picks up little material during run-up so that the main outcome of its passing across a reef surface is erosion rather than deposition. But there are other factors involved, particularly to do with reef morphology, sediment character and wave aspect which may lend a cyclical dimension to cay formation and removal. Studies of cays on Ontong Java Atoll in the western Pacific were an important step towards understanding this process (Bayliss-Smith 1988).

One characteristic of cays (and to a lesser extent of *motu*) is their mobility across reef flats. Historical data show that cays change shape regularly and even migrate across reef flats with erosion along the windward side commonly being compensated by progradation along the leeward side. A good example is that of Sand Island off St Croix in the Caribbean (Gerhard 1981). Such movements are the bane of cay-based tourist resorts.

Many cays endure for more than a few decades because they grow large enough to become vegetated and are in appropriate locations to develop beachrock. Such cays are better referred to as *motu*.

Motu

The main way in which *motu* can be distinguished from true cays is by the inclusion of shingle ridges within their fabric (Steers and Stoddart 1977). Such shingle ridges tend to be the residuals of rubble banks thrown up on the outer (ocean) sides of reefs by storm surges. As these ridges migrate landwards or lagoonwards, so the finer material is removed by wave wash so only the coarser fractions remain. Since they are so difficult to shift, particularly when located on the least vulnerable parts of a reef, these ridges often form the core of an atoll *motu*. The migration of the rubble bank thrown up on the Funafuti (Tuvalu) reef during Tropical Cyclone Bebe in 1972 was monitored by Baines and McLean (1976). Later work on the other atolls of Tuvalu demonstrated that such coarse shingle banks were integral parts of *motu*, particularly along windward reefs (McLean and Hosking 1991).

Motu also persist longer than true cays because they develop various forms of physical protection against wave erosion. These include emerged reef, BEACH ROCK, conglomerate platforms (*pakakota*) and phosphate rock, all of which greatly increase the resistance of *motu* against wave and/or precipitation attack, particularly during storms.

Those CORAL REEFS which were able to 'keep up' with Holocene sea-level rise grew to levels of 1–1.5 m above present levels in most parts of the tropics (except apparently the Caribbean) around 4,000 cal. yr BP (Nunn 2001). When the sea level fell by this amount in the later Holocene, the surface of these reefs was exposed and died. Subaerial erosion reduced them and wave erosion trimmed them, but many remained to act as foci for the accumulation of reef detritus. These fossil-reef cores underlie many *motu* today in the central Pacific, for example, and explain their persistence and suitability for human habitation.

Beach rock forms in a variety of ways beneath the surface of sandy beaches within the regularly inundated zone. For beach rock to form also usually requires a critical mass of sediment (equated with minimum *motu* size) so that ground water can flow through the beach sand.

Conglomerate platforms or breccia ramparts are cemented features of many *motu* thought to have formed at present sea level. Although there is clearly some unexplored genetic diversity amongst these features, most are believed to have originated as rubble banks which were subsequently cemented and planed down (McLean and Hosking 1991).

Phosphate rock also forms through the lithification of unconsolidated sediments, but on those *motu* where large numbers of seabirds roost (Stoddart and Scoffin 1983).

The future of cays

Many cays (including *motu*) have experienced apparently unprecedented morphological changes during the twentieth century, many of which can be attributed to the sea-level rise of ~15 cm. It is projected that twenty-first century sea-level rise may be 3–4 times as much, which has resulted in many gloomy prognoses for the future of cays (Roy and Connell 1989).

Should projections of sea-level rise prove correct, then it is likely that the numbers of cays worldwide will decrease hugely. First they may decrease because sea-level rise will, through the Bruun Effect, cause the erosion of sandy shorelines. It may become more common to see lines of beach rock exposed across reef flats marking places where cays once existed. Also, because of the rise of mean sea level and the likely inability of most oceanic reefs to respond immediately (despite some optimistic forecasts), it is probable that sediment of every grade presently lying on reef surface will become more mobile.

For many people occupying cays, particularly in independent countries like the Maldives (Indian Ocean) and Kiribati, Marshall Islands, Tokelau and Tuvalu (Pacific Ocean), it is unlikely that they will be able to continue living in such environments and will become 'environmental refugees'. Questions about national sovereignty and whether or not the Exclusive Economic Zones (EEZs) of these countries will be redrawn as a consequence are exercising the minds of many decision-makers.

References

Baines, G.B.K. and McLean, R.F. (1976) Sequential studies of hurricane deposit evolution at Funafuti atoll, *Marine Geology* 21, M1–M8.

Bayliss-Smith, T. (1988) The role of hurricanes in the development of reef islands, Ontong Java atoll, Solomon Islands, *Geographical Journal* 154, 377–391.

Gerhard, L.C. (1981) Origin and evolution of the Candlelight reef-sand cay system, St. Croix, *Atoll Research Bulletin* 242.

McLean, R.F. and Hosking, P.L. (1991) Geomorphology of reef islands and atoll motu in Tuvalu', *South Pacific Journal of Natural Science* 11, 167–189.

Nunn, P.D. (1994) *Oceanic Islands*, Oxford: Blackwell.

——(2001) Sea-level change in the Pacific, in J. Noye and M. Grzechnik (eds) *Sea-Level Changes and their Effects*, 1–23, Singapore: World Scientific Publishing.

Roy, P.S. and Connell, J. (1989) The Greenhouse Effect: where have all the islands gone? *Pacific Islands Monthly* 59, 16–21.

Steers, J.A. and Stoddart, D.R. (1977) The origin of fringing reefs, barrier reefs and atolls, in O.A. Jones and R. Endean (eds) *Biology and Geology of Coral Reefs*, Volume 4, 21–57, New York: Academic Press.

Stoddart, D.R. and Scoffin, T.P. (1983) Phosphate rock on coral reef islands, in A.S. Goudie and K. Pye (eds) *Chemical Sediments and Geomorphology*, 369–400, London: Academic Press.

SEE ALSO: coral reef

PATRICK D. NUNN

CENOTE

A cenote is a distinctive type of DOLINE or sinkhole, formed by the dissolution of limestone or other soluble rocks in subdued KARST plains (Plate 21). The type example occurs in the northern Yucatan Peninsula of Mexico, where the Mayan word dzonot, from which cenote is derived, means 'water cave'. Cenotes also occur in south-eastern Australia, in Africa, in Papua New Guinea, in Florida and in north-western Canada (Marker 1976).

The typical Yucatan cenote is a near-circular, water-filled shaft with vertical or overhanging walls extending up to 100 m downward from the ground surface (Pearse *et al.* 1936; Corbel 1959; Gerstenhauer 1968; Doering and Butler 1974). Some Yucatan cenotes resemble cylindrical shafts, but others are flooded bell-shaped chambers with bulbous bases, relatively small surface openings and thin roofs (Reddell 1977). Some have horizontal cave passages leading off from the walls, although these are often blocked by fallen rock (breakdown). The upper portions of cenote walls generally are pitted by dissolution, but the lower walls are blocky and overhanging, suggesting collapse (Whitaker 1998).

The development of cenotes is incompletely understood. Earlier hypotheses suggested that they developed through the local focusing of downward surface dissolution, but it now appears more likely that they have evolved through localized upward dissolution along fractures intersecting groundwater-filled caves or by stoping of cave ceilings, ultimately leading to surface collapse. Global sea-level oscillations have probably played a significant role too, since most cenotes are developed in Tertiary or younger reef limestones in low-lying coastal areas (Marker 1976). Sea-level lowering would have encouraged collapse by reducing the buoyant support of cenote rock walls and ceilings. Cenote-like flooded shafts, known as BLUE HOLES, occur in

Plate 21 A steep-sided cenote formed in dolomitic limestones at Otjikoto near Tsumeb, in northern Namibia

offshore reefs in the Caribbean and Australia (Mylroie *et al.* 1995). These may be drowned cenotes, although some of them have other origins (Ford and Williams 1989).

The distribution of cenotes may be unrelated to other karst landforms, but they are often distributed in a linear pattern that may reflect underground fracture patterns or the paths of major cave passages. It has been suggested that the arcuate pattern of the Yucatan cenotes represents fracturing around the perimeter of the Chicxulub impact crater, which formed at the end of the Cretaceous period, some 65 million years ago (Hildebrand *et al.* 1995).

References

Corbel, J. (1959) Karst du Yucatan et de la Floride, *Bulletin de l'Association Géographique de France* 282/283, 2–14.

Doering, D.O. and Butler, J.H. (1974) Hydrogeologic constraints on Yucatan's development, *Science* 186(4,164), 591–595.

Ford, D.C. and Williams, P.W. (1989) *Karst Geomorphology and Hydrology*, Boston: Unwin Hyman.

Gerstenhauer, A. (1968) Ein Karstmorphologischer Vergleich zwishen Florida und Yucatan, *Deutscher Geographisher*, 322–344, Bad Godesburg: Wissen Abhandlungen.

Hildebrand, A.R., Pilkington, M., Connors, M., Ortiz-Aleman, C. and Chavez, R.E. (1995) Size and structure of the Chicxulub crater revealed by horizontal gravity gradients and cenotes, *Nature* 376(6,539), 415–418.

Marker, M.E. (1976) Cenotes: a class of enclosed karst hollows, *Zeitschrift für Geomorpholgie N.F. Supplementband* 26, 104–123.

Mylroie, J.E., Carew, J.L. and Moore, A.I. (1995) Blue holes: definition and genesis, *Carbonates and Evaporites* 10, 225–233.

Pearse, A.S., Creaser, E.P. and Hall, F.G. (1936) *The Cenotes of Yucatan*, Carnegie Institute Publication 457. Washington, DC: Carnegie Institute.

Reddell, J.R. (1977) A preliminary survey of the caves of the Yucatan Peninsula, *Association for Mexican Cave Studies Bulletin* 6, 219–296.

Whitaker, F.F. (1998) The blue holes of the Bahamas: an overview and introduction to the Andros Project, *Cave and Karst Science* 25(2), 53–56.

MICHAEL J. DAY

CHANNEL, ALLUVIAL

Unconsolidated sediment deposited by rivers in subaerial settings is called ALLUVIUM and river channels formed of alluvium usually have a mobile boundary and are self-adjusting in response to changing conditions. An alluvial channel is commonly parabolic or trapezoid in cross section with adjacent roughly horizontal FLOODPLAINs that are inundated when the channel exceeds *bankfull capacity* (see BANKFULL DISCHARGE). Due largely to tributary contributions, channels generally increase in size and discharge downstream.

Alluvial channel morphology is the product of complex fluid mechanical processes. It was not until the late nineteenth and early twentieth centuries that river channels received widespread and detailed investigation when research was undertaken into partially self-adjusting canals built by the colonial British on the Indian subcontinent. The most important subsequent advances in understanding natural river-channel form and process originated from research in the USA by L.B. Leopold, M.G. Wolman, J.M. Miller, S.A. Schumm and their associates in the 1950s and 1960s.

Channel gradient and knickpoints

Channel gradients are the result of two broad controls. An *independent* gradient is imposed on the stream by antecedent VALLEY forms, products of geological and hydrological history. However, an adjustable and therefore *dependent* component develops from the interaction of channel discharge, width, depth, velocity, sediment size, sediment load, boundary roughness and path sinuosity. Mackin (1948) stated that a *graded* (or equilibrium) stream is one in which, over a period of years, the slope is delicately adjusted to provide, with available discharge and prevailing channel characteristics, just the velocity required for the transportation of the material supplied from upstream. Due to bedrock constraints (see BEDROCK CHANNEL), confined upland channels are generally not at equilibrium gradient, whereas in the middle or lower reaches the valley is wider and a channel can more readily adjust gradient by altering sinuosity and hence path length. Following this original emphasis on gradient, later work has shown that slope provides adjustments in concert with a variety of other morphological and hydraulic parameters (Leopold *et al.* 1964).

For three reasons the long profiles of natural rivers show a strong tendency for upwards concavity. First, downstream discharge increases as a cubic function whereas the resisting channel boundary increases only as a squared function, so if gradient did not decline the growing imbalance between impelling and resisting forces downstream would cause flow to accelerate rapidly. Second, because grain size commonly declines downstream, the equilibrium gradient required for sediment transport must also decline. Third, antecedent relief and potential energy conditions along a river from headwaters to mouth cause random-walk models to develop concavity as the *most probable* profile-shorter streams tend to have less concave profiles and streams with greater relief exhibit greater concavity.

Marked downstream increases in channel gradient are termed KNICKPOINTs and may reflect changes in bedrock erosional resistance, changes in sediment load from tributaries, tectonic activity, meander cutoffs, removal of LARGE WOODY DEBRIS (LWD), or base-level changes in the past. In confined valleys, knickpoints as concentrated zones of erosion can migrate considerable distances upstream. Unconfined channels, however, can adjust more readily by increasing sinuosity and thereby locally reducing knickpoint gradients.

Channel equilibrium and threshold conditions

Because alluvial channels are open systems with mobile and deformable boundaries, they have the ability to self-regulate to the imposed flow and sediment load. This reflects DYNAMIC EQUILIBRIUM, a condition first described for rivers in the nineteenth century by G.K. Gilbert. If one variable is altered, the others adjust in a way that minimizes the effect of the change and the system will return to something like its original condition (*homeostasis*).

While rivers in dynamic equilibrium generally resist change, Schumm (1973) has shown that an *extrinsic* THRESHOLD can be reached when a progressive change in an external variable triggers a sudden change in the system's response. At an imposed critical change of slope or sediment load, a meandering channel can change abruptly to a braided channel (Schumm and Kahn 1972). Similarly, a gradual and progressive increase in the flow velocity may suddenly achieve the threshold for sediment entrainment, after which the whole channel bed becomes mobile. Changes can be initiated intrinsically when, with no external change, one of the variables reaches a critical condition (an *intrinsic threshold*). A meander cutoff is an example where gradual, ongoing adjustments to equilibrium conditions prevail in a channel until an intrinsic threshold is reached.

Biota, soils and channel form

Prior to the middle Palaeozoic, subaerial erosion was dominantly physical and produced abundant coarse material forming mostly braided river channels. In the late Silurian and Devonian the evolution of terrestrial plant communities and associated soils greatly enhanced chemical weathering and the production of clays. The development of cohesive banks of muddy sediment and stabilizing root systems must have changed river channels dramatically. There is a growing appreciation of the importance and complexity of river–vegetation interactions. Particular attention has been given to the influence of within-channel vegetation and large woody debris (LWD) on flow resistance, and of bankline vegetation on bank strength and channel morphology.

The evolution of animals, including dinosaurs in the Mesozoic, has undoubtedly played a part in channel formation. Large mammals (e.g. American buffalo, African hippopotamus and domestic cattle) as well as smaller mammals (such as beavers) have been documented trampling sediments, creating paths down banks and damming channels, leading to channel avulsion and initiation.

Hydraulic geometry, regime theory and dominant discharge

Acceleration due to gravity acts to move water and sediment downslope while flow resistance opposes such motion. The interaction of these two forces ultimately determines the ability of flow to erode and transport sediment and to shape the boundary of an alluvial channel.

Flow velocity is usually fastest at or just below the surface near the centre of a straight channel and declines towards the bed and banks, the flow field deforming through river bends (Figure 19). A narrow deep channel usually exhibits a relatively gentle velocity gradient towards a fine-grained erodible bed, and directs relatively steep gradients to banks that are often cohesive, well vegetated and therefore erosion resistant. Wide shallow channels tend to exhibit erodible banks and coarse and/or abundant bedload that requires high shear stress for transport, braided rivers being a classic type.

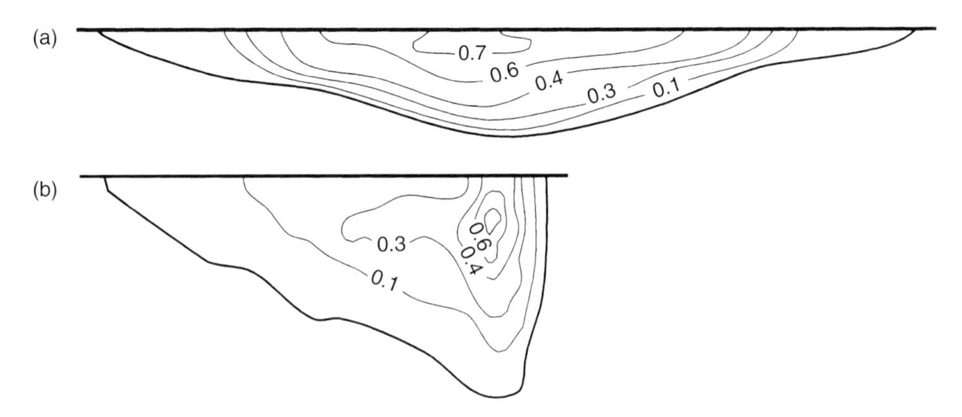

Figure 19 Velocity fields in cross sections of: (a) a wide shallow channel (note the steep velocity gradient to the bed); and (b) a narrow deep channel bend viewed downstream and curving to the left (note the steep velocity gradient against the outer cutbank)

Channel geometry is the cross-sectional form of a stream channel (width, depth, cross-sectional area) fashioned over a period of time in response to formative discharges and sediment characteristics. Because the above three geometric parameters and the additional four flow parameters (velocity, water-surface slope, flow resistance and sediment concentration) vary with discharge, the term HYDRAULIC GEOMETRY or *regime theory* is used to describe the relationships of all seven parameters to discharge as the independent variable (Figure 20). Consequently, stable alluvial rivers exhibiting consistent and predictable hydraulic geometries are said to be *in equilibrium* or *in regime*.

Discharge changes can be measured increasing at-a-station as the channel fills during a flood, or at bankfull in the downstream direction. There are significantly different relationships for *at-a-station* and *downstream* hydraulic geometry (Figure 20). Holding discharge constant, at-a-station hydraulic geometry is controlled mostly by variations in bank strength and available sediment load. Channels with low sediment loads and cohesive or well-vegetated banks tend to be relatively narrow and deep whereas those with abundant loads and weak banks tend to be wide and shallow. However, because bank strength has only a moderate range but river discharges vary by many orders of magnitude, hydraulic geometry is remarkably consistent across the full range of river discharges (Figure 20). Furthermore, because channel depth is greatly restricted by the limited strength of alluvial banks, rivers increase in width relative to their depth as their size and discharge increases – a prominent downstream tendency (Church 1992).

Hydraulic geometry shows that river channel dimensions are closely adjusted to water discharge. However, discharge varies from perhaps no flow in droughts through to catastrophic flood events, so which discharge(s) define a channel's characteristics? Leopold *et al.* (1964) showed that, in the USA, bankfull flows occur with the surprising regularity of about once every 1–2 years across a diverse range of rivers, something that would be an extraordinary coincidence if bankfull flows did not in themselves play a large part in determining channel dimensions. It has also been shown that with increasing at-a-station discharge, flow velocity tends to increase until near bankfull flow conditions and then stabilizes at higher discharges because of a marked increase in roughness near the bank crests and over the floodplains. In other words, most flows beyond bankfull are not notably more effective in altering the channel and transporting sediment than is bankfull flow. Furthermore, while in some cases exceptional floods may undertake significant work in the form of sediment transport and channel reconstruction, they are sufficiently rare that on an average annual basis, they usually achieve far less than do smaller but more frequent events of about bankfull stage (Figure 21) (Wolman and Miller 1960).

In some environments, particularly in confined alluvial settings, high-velocity events can cause considerable channel enlargement, followed by

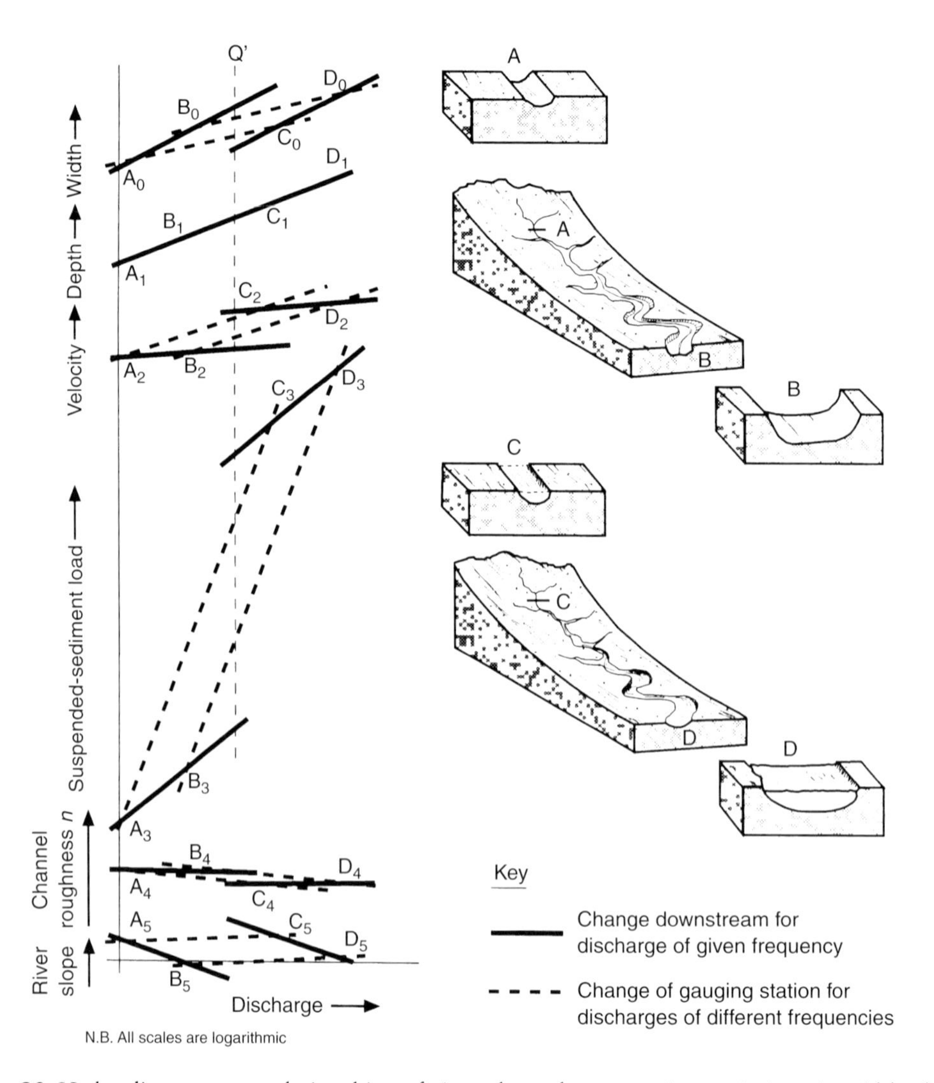

Figure 20 Hydraulic geometry relationships of river channels, comparing variations in width, depth, velocity, suspended load, toughness and slope to variations in discharge, both at-a-station and downstream (after Leopold *et al.* 1964)

a long period of 'recovery' from smaller flows. Thus, channel dimensions at a given time in such an environment may reflect considerable 'memory' of the last extreme event.

While empirical research into stochastic relationships has shown that, as flows vary, rivers construct highly predictable channel forms and sedimentary structures, a truly rational or deterministic explanation for such consistency has not been obtained. A lack of mathematical closure results from there being four flow variables (width, depth, velocity and slope) but only three determining equations (continuity, resistance and sediment transport). As a consequence, solutions have been sought by adopting *extremal hypotheses*, such as maximum sediment transport rate or minimum stream power. In a recent reassessment of some of these approaches, Huang and Nanson (2000) have demonstrated mathematically that straight reaches of alluvial rivers appear to operate at MAXIMUM FLOW EFFICIENCY (MFE) and illustrate the basic physical LEAST ACTION PRINCIPLE. However, although research into this principle is ongoing, such theoretical proposals remain contentious and are not uniformly accepted.

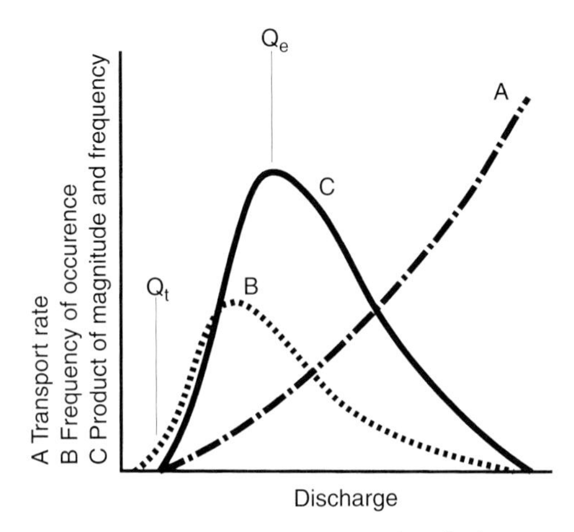

Figure 21 Dominant discharge. Curve A is the transport rate of suspended load rising with discharge. Curve B is the frequency of the full range of possible discharges. Curve C is the product of curves A and B and shows that the most effective discharges for transporting a river's load are moderate floods, generally occurring about once every 1 to 5 years (after Wolman and Miller 1960)

Channel patterns

River channels respond to imposed discharges and sediment loads by adjusting pattern or planform in conjunction with their hydraulic geometry. Because channel patterns are so easily recognized on air photos and maps they have become a primary basis for river classification from which it is possible to generalize less obvious channel characteristics such as lateral stability, sediment load, sediment size, bed/suspended load ratio and width/depth ratio.

Leopold *et al.* (1964) proposed the first widely adopted geomorphological classification with their concept of a continuum of river channel patterns from *straight* to MEANDERING to BRAIDED RIVERS, although a significant problem is that these are not mutually exclusive. For example, meanders sometimes also braid. Both meandering and braiding patterns appear to reflect a need to consume excess energy where the valley slope is greater than that required for an equilibrium channel slope (Bettess and White 1983; Schumm and Kahn 1972).

The term *anabranching* describes rivers that flow in multiple channels separated by stable, vegetated, alluvial islands that divide flows up to bankfull, regardless of their energy or sediment size, with *anastomosing* rivers simply a low-energy fine-grained type (see ANABRANCHING AND ANASTOMOSING RIVERS) (Nanson and Knighton 1996). Importantly, neither of these terms is now used as a synonym for braided rivers in which the flow is divided by unstable braid bars overtopped below bankfull.

Straight rivers consist of a single channel with a sinuosity of <1.1 (sinuosity is the ratio of channel length to valley length), a condition rarely persisting in an alluvial reach for distances of more than 7–10 channel widths. Consequently, straight reaches are classed as those without significant bends for more than this distance. Flume experiments suggest that straight channels form at very low gradients (Schumm and Kahn 1972). Where naturally sinuous channels in readily erodible material have been artificially straightened, alternate bars usually form rapidly and subsequent bank erosion leads to the development of a meandering pattern.

Braided rivers are relatively high-energy systems with large width–depth ratios and at low stage have multiple channels that divide and rejoin around alluvial bars. They tend to occur in settings with steep gradients, weakly cohesive banks, abundant coarse sediment (usually gravel and sand), and variable discharge (Leopold *et al.* 1964; Knighton 1998). Leopold *et al.* (1964) argued that braiding is an equilibrium adjustment to erodible banks and excessive load whereas Bettess and White (1983) see it as a pattern consuming energy in excessively steep valleys. Both explanations probably apply under different circumstances.

Meandering rivers consist of a single channel of moderate to low gradient with a sinuosity >1.3 and moderate width–depth ratios. Point bars commonly develop on the convex bank of a bend whereas the concave bank is typically erosional and adjacent to a pool. The locus of the lowest point in the channel (the thalweg) regularly oscillates laterally, switching channel sides at the riffle (crossover), a shallow zone in the long profile between each bend (pool). While there is no general agreement as to exactly how or why streams meander, they are self-similar over a wide range of scales. Width and wavelength in particular can be related to channel discharge (see Knighton 1998, Table 5.9). Using 'probability theory', Langbein and Leopold (1966) proposed that meanders reduce stream gradients to an equilibrium slope for the transport of an imposed sediment load (see also Bettess and White 1983), producing a longer path length with minimum variance and minimum total work. Meandering rivers tend to have cohesive and/or well-vegetated banks, mixed loads of sand (sometimes gravel) and mud, and commonly perennial flow. Channel lateral migration rates are most rapid where bends have a radius of curvature to channel width ratios of about 2 to 3 (Hickin and Nanson 1975).

Anabranching rivers are a system of multiple channels characterized by vegetated, stable alluvial islands which are either excised by avulsion from a previously continuous floodplain, or formed by the accretion of sediment in a previously wide channel. The islands divide flow up to bankfull (Nanson and Knighton 1996). Anabranching rivers include a wide range of subarctic, alpine, temperate, wet tropical and semi-arid settings and individual channels can be straight, meandering or braided. *Anastomosing* channels are low gradient, laterally stable, straight (most common) to highly sinuous variants, with low width–depth ratio and well-vegetated or highly cohesive banks. Anabranching rivers can confine flow and maintain an equilibrium bedload flux over low gradients, however, they can also distribute sediment over wide floodplains in disequilibrium, vertically accreting locations.

Church (1992) noted the problem of including rivers from mountains to basins within one classification scheme. He divided the full range of alluvial and non-alluvial channels into small, intermediate and large categories based not on channel dimensions but on the relationship between grain diameter (D) and depth (d). This approach offers opportunities for better classifying aquatic habitats but is less visually appealing.

Sediment transport and channel sedimentation

River channels transport their sediment load in essentially four ways; *bedload (traction load), saltating load, suspended load and dissolved load*. BEDLOAD is the coarsest fraction and moves short distances during relatively infrequent, high magnitude flows. It is commonly the smallest proportion of transported sediment (often <5 per cent of the total load), yet is of great geomorphic importance. It is largely bedload that controls channel configuration because its transport is a function of shear stress acting on the channel bed, and this is controlled by channel gradient (adjustable with sinuosity) and channel geometry. The capacity of alluvial rivers with unconstrained mobile boundaries to transport bedload is usually hydraulically defined, but few flows reach their capacity for transporting suspended load that is determined largely by the rate of supply. In other words, the character of a river is first determined by its imposed bedload, with suspended load and vegetation influencing bank cohesion and form. Because sediment character has a profound influence on river-channel morphology, Schumm (1960) developed a highly influential classification of rivers based on bedload, mixed load and suspended load systems, with width–depth ratios of >40, 10–40 and <10, respectively.

Alluvium results from fluvial sedimentation. This takes place both inbank and overbank as velocity wanes locally. The coarsest fractions are deposited first and as a result, sediment sizes are sorted vertically and laterally within the channel and floodplain. In laterally migrating meandering channels, upward fining successions within point bar and floodplain deposits result from flow velocities that decline from near the deepest part of channel (the thalweg) and adjacent point bar (depositing gravel or coarse sand), to the upper point bar and floodplain surface (depositing fine sand and mud). In braided rivers, coarse braid bars characterize the lowermost deposits while braid-channel and braid-bar migration or abandonment, overbank fines and channel fills characterize the uppermost deposits. Adjacent to laterally stable channels, or on floodplains away from the zone of active channels, floodplain strata broadly fine upward as each successive stratum makes the surface higher and less accessible to channel flows.

Secondary currents play a major role in producing the broad spatial variations of sediments in channel bends and bars, as well as numerous smaller flow structures. In gravel streams, prolonged flows near critical entrainment conditions can winnow fines and armour the surface, thereby lifting the threshold of bed motion during the next flood.

Conclusion

Alluvial channels represent continuum of forms that are classifiable on the basis of their cross-sectional shape, planform and associated processes. Whereas early research focused on stochastic relationships between channel form and process, there is a growing appreciation of rational explanations based on mechanics and accepted physical theory. Research into the operation and maintenance of alluvial channels remains a major area of pure and applied research within fluvial geomorphology.

References

Bettess, R. and White, W.R. (1983) Meandering and braiding of alluvial channels, *Proceedings of the Institution of Civil Engineers* 75, 525–538.

Church, M. (1992) Channel morphology and typology, in P. Calow and G.E. Petts (eds) *The River's Handbook: Hydrological and Ecological Principles*, 126–143, Washington, DC: Blackwell.

Hickin, E.J. and Nanson, G.C. (1975) The character of channel migration on the Beatton River, north-east British Columbia, Canada, *Geological Society of America Bulletin* 86, 487–494.

Huang, H.Q. and Nanson, G.C. (2000) Hydraulic geometry and maximum flow efficiency as products of the principle of least action, *Earth Surface Processes and Landforms* 25, 1–16.

Knighton, D. (1998) *Fluvial Form and Processes: A New Perspective*, London: Arnold.

Langbein, W.B. and Leopold, L.B. (1966) River meanders: theory of minimum variance, *United States Geological Survey Professional Paper*, 422H.

Leopold, L.B., Wolman, M.G. and Miller, J.P. (1964) *Fluvial Processes in Geomorphology*, San Francisco: Freeman.

Mackin, J.H. (1948) Concept of the graded river, *Geological Society of America Bulletin* 59, 463–512.

Nanson, G.C. and Knighton, A.D. (1996) Anabranching rivers: their cause, character and classification, *Earth Surface Processes and Landforms* 21, 217–239.

Schumm, S.A. (1960) The shape of alluvial channels in relation to sediment type, *United States Geological Survey Professional Paper* 352B, 17–30.

——(1973) Geomorphic thresholds and complex response of drainage systems, in M. Morisawa (ed.) *Fluvial Geomorphology*, 299–309, Binghamton, New York State University, Publications in Geomorphology.

Schumm, S.A. and Khan, H.R (1972) Experimental study of channel patterns, *Geological Society of American Bulletin* 83, 1,755–1,770.

Wolman, M.G. and Miller, J.P (1960) Magnitude and frequency of forces in geomorphic processes, *Journal of Geology* 68, 54–74.

SEE ALSO: armouring; bank erosion; channelization; confluence, channel and river junction; gravel-bed river; levee; long profile, river; models overflow channel; riparian geomorphology; river continuum; roughness; sediment load and yield; suspended load

GERALD C. NANSON AND MARTIN GIBLING

CHANNELIZATION

Channelization is the term used to describe the modification of river channels (usually alluvial channels, see CHANNEL, ALLUVIAL) by engineering. The aim is to provide flood control, improve land drainage, reduce erosion of channel banks and river beds, improve and maintain river navigation and to relocate channels in situations such as highway construction (see Brookes 1988 for a detailed text on channelized rivers). Some river channels have also been altered to float logs out from forests. River engineering can also create impounded rivers through the construction of DAMS (Petts 1984). Whilst the term channelization is extensively used, there are some equivalent terms used for the same group of engineering methods. These include 'kanalisation' in Germany, 'chenalisation' in France and 'canalization' in the UK.

River channelization has a long history and a large geographical coverage. Its origins can be traced to Mesopotamia and Egypt where there is evidence of river channelization for flood control and water supply as early as the sixth millennium BC. Indeed, most early civilizations constructed flood embankments. By 600 BC, reaches of the Huanghe (Yellow River) in China were embanked, and in Britain, the Romans constructed embankments to provide flood protection in the Fens and Somerset Levels.

Not surprisingly, the highest density of channelization occurs in developed countries associated with industrialization, urbanization and the intensification of agriculture. In the USA, 65 per cent of all channel alteration work is concentrated in Illinois, Indiana, North Dakota, Ohio and Kansas, with 51 per cent of all levee (embankment) work in California, Illinois and Florida (Brookes 1988: 10). In England and Wales, Brookes *et al.* (1983) estimated that for the period 1930–1980, 8,500 km of main rivers underwent major structural river engineering and a further 35,500 km were regularly maintained by dredging and weed cutting. And in Denmark, it is estimated (Brookes 1987) that 97.8 per cent of all streams were straightened by 1987. This is equivalent to a density of modified watercourses of 0.9 km km^{-2} and compares with a density of channelized rivers in England and Wales of 0.06 km km^{-2} (Brookes *et al.* 1983) and 0.003 km km^{-2} for the USA (Leopold 1977). Thus, Denmark has a density of channelized river fifteen times greater than

England and Wales and 300 times greater than the USA, reflecting its intensity of land use.

Engineering techniques for flood control aim to prevent flood discharges overtopping the channel banks and spilling out onto the surrounding floodplain. Channels are designed and engineered to carry a design flood, which has a particular magnitude and frequency. If the 100-year event is selected as the design flood, the river channel will be engineered to contain and transmit a peak flow that will occur on average once every 100 years.

A range of 'structural' or 'hard' river engineering techniques are employed in channelization (Wharton 2000: 24–34) and many projects are comprehensive or composite in nature in that they employ more than one of the following engineering techniques.

Resectioning increases the cross-sectional area of the channel through widening and/or deepening. This allows flood flows that would have previously spread onto the floodplain to be contained and to flow through the channel at a lower and safer level. By combining with a process known as regrading (smoothing out the river bed by removing features such as depositional bars and pool-riffle sequences) the flow velocities are increased and flood levels are further reduced in the engineered reach.

Embankments, also known as levees, floodbanks and stopbanks, are structures built alongside a river to increase the bank height and prevent flooding onto the floodplain. They are normally constructed from material excavated from the channel or from a borrow pit in the floodplain, although imported materials are sometimes used. Detailed design specifications exist for embankments but a major consideration is the height, determined by the design flood.

Lining of channels in artificial materials is undertaken for both flood control and channel stability. Lined channels are common in urban areas and are normally rectangular in cross section with a straightened planform.

Realignment or straightening aims to improve the ease with which water flows through a river reach. The techniques range from removing deposited sediment by dredging (regrading), for example 'rock raking' carried out on gravel-bed rivers in New Zealand, to the removal of meander bends through cutoff programmes, for example the Middle Yangtze and Huanghe rivers in China and the Lower Mississippi river in the USA. River straightening also improves river navigation.

Diversion channels are relief channels constructed to divert flood flows away from an area requiring protection. The Jubilee River (completed in 2002) is a diversion or bypass channel providing flood relief for part of the River Thames catchment, UK. The newly engineered Jubilee River has a maximum capacity of $215\,m^3\,s^{-1}$ and the main channel of the River Thames and existing right bank flood channels can carry up to $300\,m^3\,s^{-1}$. It is predicted that the overall system capacity of $515\,m^3\,s^{-1}$ will protect up to the 1 in 65-year return period flood. For environmental reasons the Jubilee River maintains a flow of $10\,m^3\,s^{-1}$ at all times.

Culverts are structures that encase watercourses to provide flood protection. They may be masonry arches or large-diameter concrete or metal pipes. In many towns and cities, culverted streams flow beneath the streets, for example the rivers Fleet, Westbourne and Tyburn in London (UK).

Bank protection methods and river training works are engineering techniques for controlling river channel adjustments that could threaten settlements and agricultural land and have an impact on river navigation. Deposits from eroded riverbanks can also impede the river flow and increase the risk of flooding. Riverbanks have traditionally been protected by riprap (quarried stone), gabions (rock-filled wire baskets) and revetments (coverings of resistant materials such as concrete, steel or plastic sheeting). Although riprap is usually the preferred option, gabions do have an advantage in that the wire mesh allows the rocks to change position (caused by unstable ground or scouring of the riverbank) without failure. River training works are structures built to extend from the channel banks into the river and provide bank protection by deflecting erosive river flows away from vulnerable areas along the channel banks. The most common structures are groynes (also known as deflectors or dikes). Flows can be deflected onto channel deposits that pose problems for navigation or flood control to promote their removal through the natural process of scouring. Groynes have been used in this way on the Mississippi River to maintain a navigation channel. River training works can also be used to promote sediment trapping and deposition in areas that have suffered erosion. For example, a series of permeable groynes will allow water to pass through the structures but induce deposition of fine suspended sediment between the groynes, whereas impermeable groynes will deflect river flows and promote the trapping of larger bed material.

Dredging, weed cutting, clearing and snagging (collectively known as channel maintenance activities) are routinely undertaken on many rivers to improve the efficiency of water flow through the channel and reduce the flood risk. The removal of 'obstructions' to flow, reduces channel roughness, increases river flow velocity and lowers the flood height for a given discharge. Dredging may simply involve breaking up and loosening material for the river to transport downstream. In contrast, sediment may be removed by mechanical diggers, pumped onto the floodplain or be discharged into river barges before being dumped at selected locations. Weed cutting is practised in many streams, especially nutrient-rich chalk streams, to control the annual growth of submerged and emergent aquatic plants. In addition to physically reducing the capacity of the channel, plants also increase flood risk by increasing the resistance to flow and reducing water velocities. This further promotes the accumulation of sediment within and around the plants. Aquatic vegetation is traditionally controlled by mechanical cutting, but herbicides and grazing fish (such as carp) have also been used. Clearing and snagging refers to the removal of fallen trees and debris dams from the river and the harvesting of timber from the channel banks and floodplains, respectively.

A number of concerns surround river channelization. First, channelization, is unable to provide complete protection against flooding and its associated channel form adjustments. It is simply not possible or economically viable to control the very rare, high-magnitude flood events. To achieve this, all rivers would need to be channelized and all flood defences would need to be designed and constructed to convey a correctly estimated maximum possible flood. Second, there is evidence from developed countries with a long history of channelization that the financial costs of floods are continuing to rise despite ever-increasing expenditure on structural flood defences. This has been attributed, at least in part, to the false sense of security created by flood defences that encourages further floodplain development. And third, river channelization has resulted in changes to the river, many of which were not anticipated at the design stage. These changes can have a damaging impact on the river environment and also necessitate costly maintenance activities to keep the structures operating at their design specifications. Brookes (1988: 81–185) provides a comprehensive review of the main impacts of river channelization. Included in this third set of concerns are fears that river engineering may have worsened flooding on some rivers. Whilst channelization can reduce flood risk in the engineered reach, the reverse may be true downstream.

Brookes (1988) describes the primary impact of channelization as the physical alteration to the river (i.e. its width, depth, slope and planform) by the engineering procedure. These changes then result in secondary effects that encompass changes to the river channel morphology, hydrology, water quality and ecology. Importantly, these impacts are transmitted beyond the channelized river section to downstream and upstream reaches and even along tributary streams. Post-engineering adjustments demonstrate the need for long-term and often costly maintenance operations and also have implications for structures built adjacent to, or across, river channels. For example, bridges may have to be reinforced or even replaced if bank erosion causes the river to migrate and enlarge.

The reporting of channelization impacts and the appraisal of channelization schemes will lead to improved understanding of the various changes that river engineering may cause. Greater recognition of the undesirable consequences of channelization has led to calls for a 'reverence for rivers' (Leopold 1977) with attempts to design with nature (after McHarg 1969) and develop 'geomorphic engineering' (Coates 1976). This has translated into a variety of revised construction and maintenance procedures (see Brookes 1988: 189–209) and the development of more environmentally sensitive flood alleviation schemes, such as the flexible two-stage channel constructed on the River Roding, UK (Raven 1986). In this design, the additional capacity is created by excavating outside the original channel thus leaving it to transport the normal range of flows and remain as natural as possible. Growing concern over the impacts of channelization has also prompted efforts to enhance, rehabilitate and restore river systems (see RIVER RESTORATION).

References

Brookes, A. (1987) The distribution and management of channelized streams in Denmark, *Regulated Rivers* 1, 3–16.

Brookes, A (1988) *Channelized Rivers: Perspectives for Environmental Management*, Chichester: Wiley.
Brookes, A., Gregory, K.J. and Dawson, F.H. (1983) An assessment of river channelization in England and Wales, *Science of the Total Environment* 27, 97–112.
Coates, D.R. (ed.) (1976) *Geomorphology and Engineering*, London: George Allen and Unwin.
Leopold, L.B. (1977) A reverence for rivers, *Geology* 5, 429–430.
McHarg, I.L. (1969) *Design with Nature*, New York: Doubleday.
Petts, G.E. (1984) *Impounded Rivers: Perspectives for Ecological Management*, Chichester: Wiley.
Raven, P.J. (1986) Changes of in-channel vegetation following two-stage channel construction on a small rural clay river, *Journal of Applied Ecology* 23, 333–345.
Wharton, G. (2000) *Managing River Environments*, Cambridge: Cambridge University Press.

SEE ALSO: anthropogeomorphology; bankfull discharge

GERALDENE WHARTON

CHAOS THEORY

Chaos theory has been claimed by some enthusiasts as being one of the great ideas of twentieth century science which, as with relativity and quantum mechanics, has the power to transform our view of the world. As the popular book on chaos by James Gleick (1987) illustrates, chaos theory developed in a series of often unrelated spheres of science as developments in computing power permitted the increasingly sophisticated study of non-linear systems. Non-linear systems are those in which a change in one variable produces a non-linear response in another, and thus have to be represented by non-linear equations. Chaos is a property sometimes exhibited by such non-linear systems where even under simple conditions the system can tend to complex, pseudo-random behaviour. A classic paper by Edward Lorenz in 1963 illustrates the potential for chaos in relatively simple systems. Lorenz developed a simple climatic model of the atmosphere heated from below to produce convection, involving three non-linear equations. The three equations describe the change in x, y and z over time respectively, where x describes the intensity of convective motion, y the horizontal temperature variation and z the vertical temperature variation. Despite its simplicity the modelled system exhibited chaotic behaviour, indicating the unpredictable behaviour of this sort of climatic system.

Chaotic behaviour can be identified in systems through using phase diagrams, which plot the state of the system over time in terms of the system variables. In the case of the Lorenz model above, for example, the phase diagram would plot each point in time of the evolution of the system on x, y and z co-ordinates. A stable system would have a phase diagram which converged on a point, an oscillating or periodic system would have one which resembled a ring. Such shapes on a phase diagram are called attractors. Phase diagrams for chaotic systems are characterized by what are called 'strange attractors' – bifurcating, complex patterns illustrating the many different possible states of the system as it evolves over time. Lorenz's model, for example, has a strange attractor which looks like an owl mask. Strange attractors are fractals (see FRACTAL).

According to Malanson *et al.* (1990) chaos theory has three central tenets. First, that many simple deterministic systems are rarely predictable. Second, that some systems show great sensitivity to initial conditions. Tweaking an input to one equation of a system very slightly at the beginning can thus produce highly divergent outcomes. Third, that the conjunction of the first two tenets produces a seeming randomness which may be quite ordered (as illustrated by the presence of strange attractors in their phase diagrams).

Geomorphologists have been keen to investigate the utility of chaos theory ideas and methods for the study of geomorphic systems, many of which can be shown to be non-linear in nature. For example, in a series of papers Jonathan Phillips has investigated the presence of chaos in surface runoff, hillslope evolution, coastal wetlands and soil systems as reviewed in his book on Earth surface systems (Phillips 1999). Mass movement systems often behave chaotically (Qin *et al.* 2002). Increasingly, geomorphologists suspect that chaotic and self-organized behaviour (see SELF-ORGANIZED CRITICALITY) may be common within Earth surface systems, and that stable states may be relatively uncommon. However, chaotic behaviour may also be scale-dependent, and at other scales ordered behaviour may emerge. For example, at the microscale turbulence (a classic manifestation of chaotic behaviour) characterizes many aeolian-sediment interactions within dunefields, whereas at the larger scale ordered dune systems result. As Phillips (1999: 71) puts it 'Order is an emergent property of the unstable, chaotic system'.

Although chaos theory has undoubtedly stimulated much interesting and useful research and discussion in geomorphology, its application to geomorphic systems is not problem-free. Three key issues are discussed by Baas (2002). First, the presence of random noise within many geomorphic systems can often mask chaotic behaviour and make it almost impossible to analyse what is going on. Second, analysing chaos requires good datasets, which are not necessarily forthcoming in many areas of geomorphology, although the advent of good quality DIGITAL ELEVATION MODELS (DEMs) at a range of scales has started to help enormously in this regard. Finally, there are a range of different interpretations of chaos theory in the scientific literature, and many different methods available to analyse chaotic systems – all of which can be rather confusing to geomorphologists wishing to understand and utilize chaos theory.

References

Baas, A.C.W. (2002) Chaos, fractals and self-organization in coastal geomorphology: simulating dune landscapes in vegetated environments, *Geomorphology* 48, 309–328.
Gleick, J. (1987) *Chaos*, Harmondsworth: Penguin.
Lorenz, E.N. (1963) Deterministic non-periodic flows, *Journal of Atmospheric Sciences* 20, 130–141.
Malanson, G.P., Butler, D.R. and Walsh, S.J. (1990) Chaos theory in physical geography, *Physical Geography* 11, 293–304.
Phillips, J.D. (1999) *Earth Surface Systems: Complexity, Order and Scale*, Oxford: Blackwell.
Qin, S., Jico, J.J. and Wang, S. (2002) A nonlinear dynamical model of landslide evolution, *Geomorphology*, 43, 77–86.

Further reading

Malanson, G.P., Butler, D.R. and Geograkakos, K.P. (1992) Nonequilibrium geomorphic processes and deterministic chaos, *Geomorphology* 5, 311–322.
Sivakumar, B. (2000) Chaos theory in hydrology: important issues and interpretations, *Journal of Hydrology* 227, 1–20.
Turcotte, D.L. (1992) *Fractals and Chaos in Geology and Geophysics*, Cambridge: Cambridge University Press.

HEATHER A. VILES

CHELATION AND CHELUVIATION

Organic compounds, derived through the partial decomposition of organic matter, are important agents in weathering. Some act because they are acid and simply etch into minerals but for others, ions from the mineral actually become incorporated into the chemical structure of the organic compound and it is these compounds which are called chelates. The word is derived from the Latin *chela* and Greek *khele* which means a claw – and it can be readily imagined how the claw of, say, a crab can hold an object in the tips of its pincers and this is analogous to the way in which the chemical compound holds an atom derived from a mineral. The definition of a chelate can now be appreciated: 'a compound containing a ligand (typically organic) bonded to a central metal atom at two or more points'; where a ligand is: 'an ion or molecule attached to a metal atom by co-ordinate bonding'.

In the context of weathering, the metal ions of interest are commonly iron but can be zinc, copper, manganese, calcium or magnesium. Chelation weathering is then the process of the incorporation of these metal atoms into an organic compound derived from the decay of organic matter. The significance of this process is that many minerals are subject to weathering by chelates to a much greater degree than they are in water, even acidified water (Huang and Keller 1972; Huang and Kiang 1972).

Cheluviation is a compound word derived from chelation and eluviation (see ELUVIUM AND ELUVIATION). Since eluviation is the down-washing of material through the soil in mobile soil water, cheluviation involves the down-washing of chelates, with their associated metal cations, from the upper horizons of the soil to the lower horizons. It is in this way that iron can be moved from the upper horizons of a podzolic soil, rendering it a pale colour with an absence of reddish oxidized iron, ferric iron or Iron III. The process involves simultaneous chelation and REDUCTION of the iron to the mobile ferrous (Iron II) form. The iron then may accumulate lower down in the soil as a reddish or, because of the presence of organic matter, blackish layer. Here the reddish oxidized Iron III forms as a result of OXIDATION through a rise in pH which occurs in the lower parts of the soil profile which are less acid than the upper parts which are acidified by organic acids. The redeposition of the iron can be in the form of a hard iron pan, termed a BFe horizon, or a more diffuse reddish horizon. The latter is termed a Bs horizon as it contains sesquioxides which are defined as compounds such as Fe_2O_3 which have a ratio of metal to oxygen of $1{:}1\frac{1}{2}$.

References

Huang, W.H. and Keller, W.D. (1972) Organic acids as agents of chemical weathering of silicate minerals, *Nature (Physical Science)* 239, 149–151.

Huang, W.H. and Kiang, W.C. (1972) Laboratory dissolution of plagioclase feldspars in water and organic acids at room temperature, *American Mineralogist* 57, 1,849–1,859.

STEVE TRUDGILL

CHEMICAL DENUDATION

Central to any understanding of landform change through time is an understanding of DENUDATION rates (the volume of rock removed from a given area in a specified period of time). Knowledge of denudation rates is relevant also to geochemical and sediment mass balance studies, with important implications for global carbon budgets and global climate change. Denudation results from the removal of solid particles (mechanical denudation; Meybeck 1987) and dissolved material (chemical denudation). Overall, chemical denudation has received less attention than mechanical denudation and estimates of its local, regional and global significance often are subject to much uncertainty. The processes of CHEMICAL WEATHERING through which atmospheric, hydrologic and biologic agents act upon and alter mineral constituents of rocks by chemical reactions, thereby releasing dissolved material to be removed, are considered elsewhere. Here, an overview of the methods for studying chemical denudation and the variability of rates from different environments is provided. The role of relief, lithology and climate as controlling factors are discussed.

Methods for studying chemical denudation

SOLUTE YIELDS

Most frequently chemical denudation is calculated from the solute loads of rivers draining large catchments. An estimate of chemical denudation can be achieved simply by multiplying the mean solute concentration from samples of river water by mean discharge. More accurate estimates, however, take into account solute concentration relationships with discharge, particularly through floods using solute rating curves (see SOLUTE LOAD AND RATING CURVE) constructed from equations which best fit the relationship between solute concentrations and discharge. For greater accuracy these rating curves can be used for the rising and falling limbs of flood hydrographs. Solute transport rates can then be calculated by relating the rating curve to either continuous stream-flow data or flow-duration curves based on hourly, daily or even monthly data.

Given the complexity of measuring separately each dissolved constituent in stream water, electrical conductivity (specific conductance) of the water, which is more easily measured, is often used to provide an estimate of solute concentration. Although there is a strong correlation between the concentration of ionic species in solution and electrical conductivity, the exact relationship varies depending on concentrations present of particular dissolved constituents. Moreover, SiO_2, which may be a significant component of many tropical lowland rivers, is not recorded by this technique.

The most significant problem in estimating the contribution of solute transport to denudation is the separation of denudational and non-denudational contributions. Chemical weathering is not the only process affecting solute yields (Figure 22). Dissolved constituents introduced into a catchment from atmospheric wet and dry deposition must be accounted for. These atmospheric deposition fluxes can be quantified by direct measurement, though results often are highly variable with complex spatial patterns in regions with different vegetation types (Drever and Clow 1995). Global estimates of non-denudational atmospheric inputs (details in Summerfield 1991) from precipitation (oceanic salts) and atmospheric CO_2 (incorporated during weathering reactions), average approximately 40 per cent of catchment solute yields. The fraction is highest for the ions of Na, Cl and HCO_3 (50 to 70 per cent), although it is important to caution that these values vary greatly.

Further complications, depending on the timescale of study, relate to changes in the exchange pool of cations and anions in the soil and biomass (Figure 22). In the short term, changes in the soil occur as a result of precipitation events, evapotranspiration and the growth cycles of plants. As plants grow, they extract inorganic nutrients from the soil solution and incorporate them into plant tissue. When plants die and decompose, the process is reversed and the elements are returned to the soil. If an ecosystem

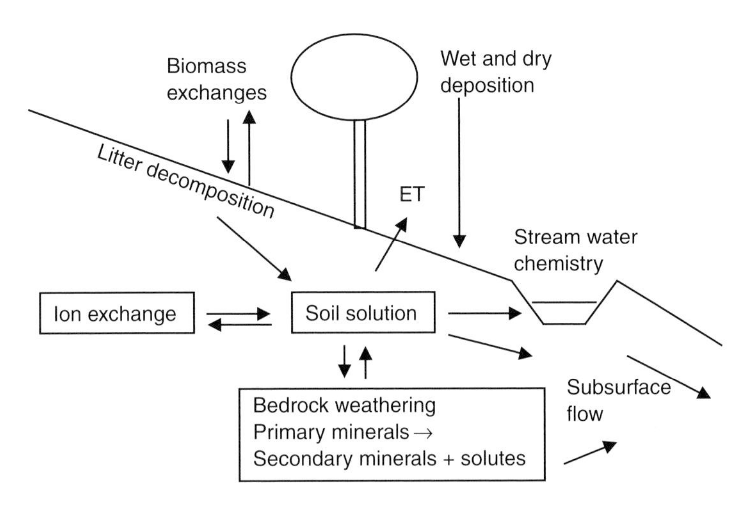

Figure 22 Schematic representation of key factors influencing solute fluxes in a catchment (adapted from Drever and Clow 1995)

is in steady state, new growth is exactly balanced by the death and decay of old vegetation and the biomass is neither a net source or sink. However, in forested catchments the biomass is rarely in steady state. For example, data of Likens *et al.* (1977) in Hubbard Brook indicated that the uptake of Ca by biomass was 45 per cent of the amount released by weathering. For K the value was 86 per cent.

The solute loads of rivers increasingly are being impacted by human inputs, especially where industrial and agricultural activities are concentrated. Elevated acid inputs increase the rate of chemical weathering in watersheds underlain by reactive rock types and cause acidification of water and soil in catchments underlain by non-reactive rocks types. Significant changes in the soil and biomass exchange pools can result, enhancing rates of solute input into stream waters. Such anthropogenic influence further complicates interpretation of solute concentrations in terms of 'natural' chemical denudation rates.

While corrections can be made to solute yields to account for atmospheric, biogenic and anthropogenic processes, the actual volume change (denudation) in the catchment cannot be determined unless the alterations in bulk density that accompany the weathering reactions releasing the solutes are known. While some chemical weathering dissolves bedrock minerals completely with all the products as dissolved species (common with limestone and quartzite) (congruent reactions), many weathering reactions produce both dissolved species and new solids (often clay minerals) with a similar volume but decreased bulk density. Moreover, even the congruent reactions and associated chemical losses may result in a substantial decrease in density as silicate rocks are altered to SAPROLITE (a weathered residuum retaining the structure and layering of the bedrock from which it forms) (density of bedrock 2.5–2.7; saprolite 1.3–1.7; soil 0.8–1.3). Thus there may be no discernible effect on the configuration of the landscape and direct conversion of dissolved loss into surface lowering is unrealistic.

SOIL PROFILE DEPLETION AND MASS BALANCE MODELLING

Soil mineralogy represents the residual product of chemical reactions which integrate the weathering rate over the entire period of soil development. Thus an alternative approach to determine long-term rates of chemical weathering and denudation is to quantify element and mineral losses in a soil profile relative to the initial or parent material. The most common approach is to define the mass ratio (enrichment) of a conservative component whose absolute mass does not change during weathering. As relatively soluble minerals in soils dissolve away, the more immobile elements in soils become increasingly enriched relative to their concentrations in the unweathered parent material. Measurement of enrichment of immobile elements, such as Zr, Ti, of rare Earth elements such as Nb, can reveal the degree of soil weathering and thus can be used to quantify the total dissolution loss from a soil (see examples in White 1995). However, there is considerable disagreement in the literature about the relative mobility of elements in different weathering regimes, and minor elements, such as Zr and Ti, are often concentrated in the small size, heavy fraction which may be subjected to significant fractionation during sediment transport and deposition. Assuming these are not major issues, the average weathering rate can be estimated by dividing the dissolution loss by the soil age. However, because non-eroding soils of known age are rare, this mass balance approach cannot be used in many environments. Riebe *et al.* (2001) show how the soil mass balance approach can be extended to measure long-term weathering rates in eroding landscapes. Physical erosion rates can be inferred from cosmogenic nuclides (see COSMOGENIC DATING), with dissolution losses inferred from rock-to-soil enrichment of insoluble elements.

Regional and global patterns

The distribution of studies of solute loads and chemical denudation is uneven. Most recent work on the rates and significance of chemical weathering in small watersheds has been driven by interest in the effects of acid deposition. Thus numerous studies have been conducted in North America and Europe, yet few data collected at the watershed-scale exist from other parts of the world. Thus estimates at continent-wide or global-scales must be extrapolated. This is usually based on empirical relationships observed between solute transport rates and the factors thought to control these rates, notably rock-type, climate and relief (see further discussion below). Moreover, records of dissolved loads tend to be short and results variable through time. Thus there are also problems extrapolating such data to the longer time periods over which ecosystems, soils, landscapes and climates evolve.

Some of the earliest regional estimates of chemical denudation were attempted by Dole and Stabler (1909) for the United States. Data compiled by Summerfield (1991) are used here to provide a range of estimates of solute load transport and equivalent rates of chemical denudation (see regional summary in Table 7). These estimates are subject to all the errors described above.

The data yield a global average for chemical denudation of 3,700 $Mt\,a^{-1}$. Reducing this value by 40 per cent, to account for non-denudational component of solute loads (see discussion above), the estimate is 2,200 $Mt\,a^{-1}$ for denudational solute load. Globally, this is approximately 15 per cent of natural mechanical denudation. Chemical denudation rates although less variable than mechanical denudation rates, do vary significantly (Table 7). Reported values range from 1 $mm\,ka^{-1}$ in drainage basins such as the Nile, Niger and Rio Grande to 27 $mm\,ka^{-1}$ in the Chiang Jiang basin (Summerfield and Hulton 1994).

Some of the measured variability is related to lithology. Maximum yields of 6,000 $t\,km^{-2}\,a^{-1}$ occur in rare instances (for example in areas underlain by halite). More usual maxima are 1,000 $t\,km^{-2}\,a^{-1}$ in limestone regions. Although few studies have attempted to reconcile laboratory-based experimental studies of weathering rates and catchment scale estimates, the real-world weathering rates of different lithologies do correspond qualitatively to rates measured in the laboratory (Drever and Clow 1995). Many studies have documented a consistent positive correlation between solute load and annual runoff. This results from more water available for chemical reactions in the regolith and solute release, and greater runoff to transport these solutes. The relationship with temperature tends to be very weak (overwhelmed by other variables, especially precipitation and local relief). Relief influences a number of factors which impact the rate of surface runoff, rate of subsurface drainage and therefore rate of leaching of soluble constituents, and rate of erosion of weathered products and thereby rate of exposure of fresh mineral surfaces. In the Amazon basin, a relationship between relief and chemical weathering exists: ~86 per cent of the solutes delivered by the Amazon to the Atlantic come from the Andes mountains (~12 per cent of the area) (Gibbs 1967). The problem, however, is that for the Amazon relief and lithology are highly correlated; outcrops of limestone and evaporates are common in the Andes, whereas most of the remainder of the basin is underlain by silicate rocks. Based on data for externally draining basins exceeding $5 \times 10^5\,km^2$ in area, Summerfield and Hulton (1994) conclude that chemical denudation rates are more strongly associated with relief than climatic factors. This supports the idea that the efficient removal of bedrock in the weathering zone is the critical determinant of the rate of chemical weathering.

Table 7 Solute denudational loads of major rivers in relation to climate and relief

Climate and relief zone	Denudational solute load ($t\,km^{-2}\,a^{-1}$)	Total denudation ($mm\,ka^{-1}$)	Typical solute load as % of total
Mountainous			
High precipitation	70–350	95–740	10
Low precipitation	10–60	45–370	10
Moderate relief			
Temperate or Tropical climate	25–60	30–110	35
Low relief			
Dry climate	3–10	5–35	10
Temperate climate	12–50	15–30	65
Subarctic climate	5–35	5–15	80
Tropical climate	2–15	1.5–10	50

Sources: Adapted from Summerfield (1991) based on Meybeck (1976)

Overall, high rates of chemical denudation are found in humid mountainous regions, where high relief is coupled with high runoff (Table 7). Minimum rates tend to be recorded in semi-arid regions where runoff is very low (although concentrations of dissolved load may be high), and in high latitude lowland terrains where runoff and solute concentrations are low. In some basins, especially those in a predominantly humid lowland environment, chemical denudation exceeds mechanical denudation. The other extreme are basins where extremely high sediment yields mean that chemical denudation represents less than 5 per cent of total denudation. Proportionally chemical denudation tends to become lower in drainage basins experiencing higher total denudation rates (Table 7).

Chemical denudation and global climate change

Given chemical denudation is an important control on the biogeochemistry of ecosystems, its study has implications not only for landform development but also global environmental change, notably issues of water quality, watershed acidification, nutrient cycling, and the greenhouse effect. As described above, chemical denudation is influenced by climate. Over geological time periods, however, chemical weathering also has a significant influence on global climate. During the weathering of carbonates and silicates, atmospheric CO_2 is taken up and converted to dissolved HCO_3^-. The HCO_3^- after delivery to the oceans by rivers can be stored in the form of carbonate minerals or organic matter in sediments. Either way there is a net loss of CO_2 from the atmosphere. Given CO_2 is a greenhouse gas, any changes in its concentration affects radiative exchanges in the Earth's atmosphere (Berner *et al.* 1987). By way of example, increased rates of chemical weathering associated with the Himalayan–Tibetan uplift have been suggested as a primary cause of the late Cenozoic ice ages (Raymo and Ruddiman 1992).

References

Berner, Robert A., Lasaga, Antonio G. and Garrels, Robert M. (1987) The Carbonate-Silicate geochemical cycle and its effect on atmospheric carbon dioxide over the past 100 million years, *American Journal of Science* 283, 641–683.

Drever, J.I. and Clow, D.W. (1995) Weathering rates in catchments, in A.F. White and S.L. Brantley (eds) *Chemical Weathering Rates of Silicate Minerals*, Mineralogical Society of America, *Reviews in Mineralogy* 1, 407–461.

Dole, R.B. and Stabler, H. (1909) *Denudation*, US Geological Survey Water Supply Paper 234, 78–93.

Gibbs, R.J. (1967) The geochemistry of the Amazon system I. The factors that control the salinity and the composition and concentration of the suspended solids, *Geological Society of America Bulletin* 78, 1,203–1,232.

Likens, G.E., Bormann, F.H., Pierce, R.S., Eaton, J.S. and Johnson, J.M. (1977) *Biogeochemistry of a Forested Ecosystem*, New York: Springer-Verlag.

Meybeck, M. (1976) Total dissolved transport by world major rivers, *Hydrological Sciences Bulletin* 21, 265–284.

Meybeck, M. (1987) Global chemical weathering of surficial rocks estimated from river dissolved loads, *American Journal of Science* 287, 401–428.
Raymo, M.E. and Ruddiman, W.F. (1992) Tectonic forcing of late Cenozoic climate, *Nature* 359, 117–122.
Riebe, C.S., Kirchner, J.W., Granger, D.E. and Finkel, R.C. (2001) Strong tectonic and weak climatic control on long-term chemical weathering rates, *Geology* 29, 511–514.
Summerfield, M.A. (1991) *Global Geomorphology*, Harlow, England: Longman.
Summerfield, M.A. and Hulton, N.J. (1994) Natural controls of fluvial denudation rates in major world drainage basins, *Journal of Geophysical Research – Solid Earth* 99, 13,871–13,883.
White, A.F. (1995) Chemical weathering rates of silicate minerals in soils, in A.F. White and S.L. Brantley (eds) *Chemical Weathering Rates of Silicate Minerals*, Mineralogical Society of America, *Reviews in Mineralogy* 1, 407–461.

SEE ALSO: weathering; weathering and climate change

CATHERINE SOUCH

CHEMICAL WEATHERING

The biogeochemical alteration of the Earth's surface and associated processes are called WEATHERING. These processes are usually separated into chemical, physical and biologic weathering for discussion. In reality, these processes are not mutually exclusive. Chemical weathering is the process by which chemical reactions such as hydrolysis, hydration, oxidation-reduction, ion exchange, solution and organic reactions transform rocks and minerals into new chemical combinations that are stable under conditions at or near the Earth's surface. Chemical weathering begins as thermodynamically unstable minerals adjust to the surrounding environment. Rocks and minerals that are not in equilibrium with near-surface conditions of temperature, pressure and water begin to alter to new products that are chemically more stable in the near surface.

Chemical weathering processes are many. The ability to measure these processes has progressed over the years as newer technologies and interdisciplinary research have led to discoveries at all scales from the molecular to the macroscale. Although chemical weathering occurs at many different temperatures and pressures this discussion will focus on a few basic concepts common to weathering under near-surface conditions.

The resistance of rocks and minerals to chemical breakdown influences the stability of individual mineral species in the environment. This stability is related to several mineral properties including cleavage and fracture patterns, particle size and specific surface, solubility and the relative stability of the surrounding environment. Structurally, the resistance to weathering increases as the complexity of silicate linkage increases, particularly the number of shared oxygens, from independent tetrahedral structures (e.g. olivine) to single chain silicates (e.g. enstatite, a pyroxene) to sheet silicates (e.g. talc) to continuous framework silicates (e.g. quartz). A weathering sequence that illustrates this concept for common rock-forming silicate minerals is illustrated in Figure 23. Stability of minerals increases from top to bottom. Additional guides to mineral stability are discussed in Ritter (1986).

Other factors being equal, minerals formed in environments resembling those in which weathering takes place will be the most resistant. This concept is based on thermodynamic principles. For example, olivines and calcium plagioclase feldspars form at higher temperatures and pressures and weather more rapidly than muscovite and quartz-rich minerals which form at lower temperatures. These latter conditions are more similar to near-surface weathering conditions.

Chemical weathering processes

Solution occurs when a mineral dissolves to form ions or dispersed colloidal molecular units. It is one of the simplest of weathering processes. Bicarbonate ($2HCO_3^-$) is derived from the dissociation of carbonic acid (H_2CO_3) that in turn formed from the dissolution of carbonate rock and atmospheric CO_2 dissolved in water:

$$H_2O + CO_2 + CaCO_3 \leftrightarrow Ca^{++} + 2HCO_3^-$$

Bicarbonate is one of the most abundant anions in weathering systems. Its weathering effects have been studied in detail relative to limestone KARST systems. Bicarbonate ions can also form from dissolution of CO_2 in plant and microbial respiration processes:

$$H_2O + CO_2 \leftrightarrow H_2CO_3 \leftrightarrow H^+ + HCO_3$$

HYDROLYSIS is the reaction of compounds with water to produce a weak acid or weak base. Water molecules are attracted to surfaces of

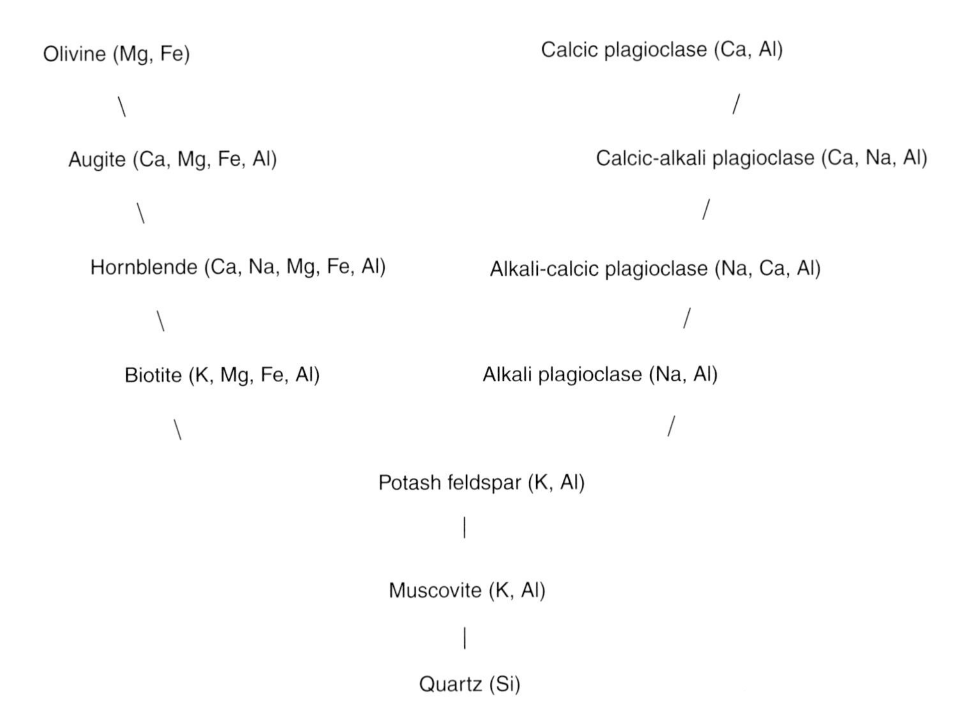

Figure 23 Weathering of common rock-forming silicate minerals
Source: Data from Goldich, S.S. (1938) A study in rock weathering, *Journal of Geology*, 46, 17–58

minerals due to the attraction of polar water molecules to the polar surfaces of many minerals. Here, forsterite hydrolysis produces silicic acid:

$$Mg_2SiO_4 \text{ (forsterite)} + 4H_2O \leftrightarrow 2Mg^{++} + 4OH^- + H_4SiO_4 \text{ (silicic acid)}$$

Natural waters usually contain dissolved CO_2 so reactions often contain carbonic acid as well. A more complete way to write the above reaction in a natural system would be:

$$Mg_2SiO_4 \text{ (forsterite)} + 4H_2CO_3 \leftrightarrow 2Mg^{++} + 4HCO_3^- + H_4SiO_4$$

H^+ (from water) can replace other ions such as K^+, Ca^{++} and Na^+ in mineral structures. The H^+ disrupts the structural bonds. If the H^+ is smaller than the cation it replaces, physical strain occurs in the mineral which in turn accelerates weathering. For example, microcline feldspar reacts with water and loses a potassium ion:

$$KAlSi_3O_8 \text{ (microcline feldspar)} + H_2O \rightarrow HAlSi_3O_8 + K^+ + (OH)^-$$

Lowering the pH increases hydrolysis because the number of H^+ ions in solution increases. For example, organic matter decomposition adds H^+ and speeds hydrolysis as do many other biologic processes such as nutrient uptake, nitrification and sulphur oxidation. Warm temperatures have an effect similar to lowered pH. Higher temperatures increase the dissociation of water molecules and provide additional H^+, potentially increasing hydrolysis in a system. Thus the microcline feldspar in the above example should weather more quickly in a warmer rather than a cold climate and in an acid rather than a more neutral environment.

HYDRATION adds water molecules to mineral structures but the water does not dissociate as in hydrolysis. Gypsum is a hydrated form of anhydrite. The reverse reaction is dehydration:

$$CaSO_4 \text{ (anhydrite)} + 2H_2O \leftrightarrow CaSO_4 \cdot 2H_2O \text{ (gypsum)}$$

Although quartz is a resistant mineral, under specific conditions it can dissolve by hydration:

$$SiO_2 \text{ (quartz)} + 2H_2O \leftrightarrow H_4SiO_4$$

Some minerals may expand during hydration. Commonly smectite hydrates and dehydrates when water molecules enter or leave interlayers, respectively. In an expanded condition, minerals are more porous and become more susceptible to additional weathering.

Ion-exchange reactions are important and are usually related directly to clay mineral weathering and other secondary minerals because these minerals have a high capacity for exchange within the interlayer and with surface ions. During exchange, the basic structure of the mineral is unchanged, but interlayer spacing varies with each cation absorbed into the interlayer. This mechanism has a unique outcome for clay minerals in that the alteration of one clay mineral may produce another. For example, under certain circumstances smectite may form from illite with the loss of interlayer K^+. Ion exchange is an important factor in biogeochemical reactions of rocks and sediments with organic matter and colloids. Ion exchange can also occur in the initial weathering of primary minerals such as silicates.

Ion mobility is key to primary mineral weathering. Hudson (1995) discusses an update of Polynov's 1937 ion mobility series that ranks major elements from very mobile (I) to relatively immobile (V):

$$Cl > SO_4 > Na > Ca > Mg > K > Si > Fe > Al$$

where mobility phase I is Cl and SO_4, II is Na, III is Ca, Mg and K, IV is Si and V is Fe and Al. Mobility depends on charge and charge density.

In a strongly leaching environment only phase V elements would remain. As an environment became drier, phase IV through I elements would become increasingly abundant. For example, gibbsite ($Al[OH]_3$) is a common aluminium hydroxide in sediments and soils assumed to be in the latter stages of weathering where leaching conditions and free drainage occur. This would be equivalent to phase V ion mobility. At this point, silica has been so thoroughly removed from the system that phyllosilicates can no longer form. Aluminium hydroxide-rich sediments are associated with tropical environments today and in the weathering profiles of bauxite deposits of ancient silica-depleted rock systems.

OXIDATION and REDUCTION equilibria, also known as redox reactions, take place when an atom or element gains or loses net charge; oxidation, the loss of electrons and reduction, the gaining of electrons. The availability or absence of one electron acceptor leads to the reduction of another element. Elements must have at least two viable oxidation states to be involved in redox reactions. Only about six elements, oxygen, iron, manganese, sulphur, nitrogen and carbon, are abundant enough in the natural environment to take part in common redox reactions of the near-surface environment. Oxygen plays a role in most oxidation processes. In the reaction below, ferrous iron derived from the hydrolysis of an iron-bearing silicate, is oxidized from +2 to +3 oxidation state to form haematite. Oxygen is reduced from 0 to −2:

$$2Fe^{++} + 4HCO_3^- + \tfrac{1}{2}O_2 \leftrightarrow Fe_2O_3 \text{ (haematite)} + 4CO_2 + 2H_2O$$

Haematite is stable in many environments but goethite, also a ferric iron component and primary constituent of limonite, may occur with the addition of more moisture:

$$Fe_2O_3 \text{ (haematite)} + H_2O \leftrightarrow 2HFeO_3 \text{ (goethite)}$$

Oxidation of pyrite, FeS_2 to iron hydroxides or sulphates and sulphuric acid on exposure to water and oxygen has detrimental consequences. This reaction, often occurring in materials adjacent to mine sites, is a common cause for the sterile biologic conditions in sediments drained by acid waters. The term acid mine drainage is applied to these waters in which the pH can drop below 2.

Oxidation of organic carbon is often due to micro-organisms that play a major role in expediting redox reactions. An example of an organic oxidation reaction in which carbon dioxide is formed is:

$$C_6H_{12}O_6 + 6O_2 \leftrightarrow 6CO_2 + 6H_2O$$

The carbon dioxide formed is available for solution and hydrolysis reactions.

Chelation (see CHELATION AND CHELUVIATION), a form of metal complexation, is the reaction between a metallic ion and a complexing agent, usually organic, resulting in the formation of a ring structure that encompasses the metallic ion effectively removing it from the system. Hydrogen is often released during the process and becomes available for hydrolysis reactions. Chelating agents in contact with rocks or minerals can cause significant weathering (Berthelin 1988). For example, lichens and mosses remove cations from silicate minerals and may produce dissolved or amorphous silica. Some breakdown of minerals occurs from reactions with organic acids produced at the root tips of plants or produced by bacteria acting on decaying material.

Chemical weathering products

Chemical weathering results in either congruent dissolution, in which the material goes completely into solution, or incongruent dissolution, in which at least some weathering products may form new minerals (neoformation or synthesis) or leave a residue or precipitate. If limestone dissolves completely and releases Ca^{++} and HCO_3^- ions into aqueous solution it is a congruent dissolution. However, most limestones are not pure $CaCO_3$ and leave a residue. During chemical changes, particle size decreases, surface area increases and constituents continue to dissolve into aqueous weathering solutions. Water is often the transferring agent and its activity is important.

Berner (1971) and Berner and Berner (1996) emphasize the importance of water flow as a control factor on the intensity of weathering. Berners' example suggests that at moderate flow rates albite alters to kaolinite but at higher flow rates silicic acid is removed so quickly that gibbsite rather than kaolinite may form. When flow rates were very slow, material was removed slowly and if magnesium was available, the product was montmorillonite. This suggests that climate and relief control weathering products. The mineralogy of the rock weathering and the chemical composition of weathering solutions are two additional determining factors. Chadwick *et al.* (2003) present a biogeochemical model for an arid to humid climosequence on Kohala Mt., Hawaii. They found that where mean annual precipitation is high and total sediment pore space is annually full, leaching of soluble base cations and silica is nearly complete. At lower precipitation inputs, leaching losses are progressively lower. Secondary mineral weathering was controlled by metastable non-crystalline weathering products rather than soil solution composition.

Weathering products may be grouped into four categories: (1) soluble constituents; (2) residual primary minerals unaffected by weathering reactions; (3) new stable minerals produced by weathering reactions; (4) organic compounds. Soluble constituents are those that remain in solution at near-surface conditions. Three primary groups of residual minerals remain in weathered soils: (a) phyllosilicate clay minerals; (b) very resistant end products such as sesquioxides of Fe and Al; (c) very resistant primary minerals such as quartz, zircon and rutile. Each group contributes less as weathering progresses. In highly weathered soils and sediments of the humid tropics or subtropics, Al and Fe oxides and low-activity clay minerals with low Si/Al ratios may be all that remains of the original primary minerals. Feldspars, mica, amphiboles and pyroxene minerals alter to clay minerals through hydrolysis, hydration and oxidation. For example, biotite mica weathers as Fe^{++} oxidizes, K^+ leaves the structure to maintain neutrality, the structure begins to weaken and soluble cations in solution such as Ca^{++}, Mg^{++} or Na^+ replace the remaining K^+. A new phyllosilicate such as vermiculite or montmorillonite forms.

Phyllosilicates are commonly stable mineral products of weathering. They are specific clay minerals occurring primarily in the clay-size fraction of a material (see Moore and Reynolds 1997). Phyllosilicates strongly influence the chemical as well as physical properties of sediments, in part due to their unusually small particle size and resulting high surface area but also, as described earlier, due to cation exchange characteristics uniquely related to their crystal structures (see Dixon and Weed 1989; Moore and Reynolds 1997). Linus Pauling (1929, 1930), Kelley (1948) and Grim (1962) were some of the first individuals to recognize the unique chemical properties of phyllosilicate clay minerals.

Chemical weathering and landscapes

Measurement of the total amount of chemical weathering is important to a geologist or geomorphologist because it can provide some estimate of landscape evolution. Although both physical and chemical denudation affects landscapes on a catchment or global scale this discussion centres on chemical weathering. Chemical denudation can be calculated from dissolved stream loads and corrected for atmospheric input because most ions in water come from weathering reactions (Berner and Berner 1987). Annual load is multiplied by annual discharge and divided by basin area. Berner and Berner (1987) have calculated a world average. Garrels and MacKenzie (1971) ranked chemical denudation by continent: Europe $>$ North America $=$ Asia $>$ South America $>$ Africa $\gg$ Australia. Degree of weathering is often calculated on the basis of total chemical analyses comparing fresh parent rock with saprolite or soil derived *in situ* from it. Birkeland (1999) presents a good summary of this approach.

During physical weathering in an open system, the landscape is generally lowered volumetrically because solids are removed but during chemical weathering the landscape may increase volumetrically. Ions removed from a weathering rock mass might be reflected in a bulk density change such that a geomorphic surface is unchanged or is even raised. For example, when a soil forms from rock, the bulk density will decrease, sometimes by 0.5 $g\,cm^{-3}$ or more. This results in an overall volumetric expansion (Birkeland 1999). Brimhall and others (1991) developed a method for assessing chemical change during weathering that gives values for volume change as well as for losses or gains in mass. In some cases this expansion is the catalyst for increased physical weathering. Several researchers have suggested that the formation of grus from granite follows these steps (e.g. Wahrhaftig 1965; Nettleton *et al.* 1970).

References

Berner, E.K. and Berner, R.A. (1987) *The Global Water Cycle; Geochemistry and Environment*, Englewood Cliffs, NJ: Prentice Hall.

Berner, E.K. and Berner, R.A. (1996) *Global Environment: Water, Air, and Geochemical Cycles*, Englewood Cliffs, NJ: Prentice Hall.

Berner, R.A. (1971) *Principles of Chemical Sedimentology*, New York: McGraw-Hill.

Berthelin, J. (1988) Microbial weathering processes in natural environments, in A. Lerman and M. Meybeck (eds) *Physical and Chemical Weathering in Geochemical Cycles*, 33–59, The Netherlands: Kluwer.

Birkeland, P.W. (1999) *Soils and Geomorphology*, New York: Oxford University Press.

Brimhall, G.H., Chadwick, O.A., Lewis, C.J., Compston, W., Williams, I.S., Danti, K.J., Dietrich, W.E., Power, M.E., Hendricks, D. and Bratt, J. (1991) Deformational mass transport and invasive processes in soil evolution, *Science* 255, 695–702.

Chadwick, O.A., Gavenda, R.T., Kelly, E.F., Ziegler, K., Olson, C.G., Elliot, W.C. and Hendricks, D.M. (2003) The impact of climate on the biogeochemical functioning of volcanic soils, *Chemical Geology*, in press.

Dixon, J.B. and Weed, S.B. (eds) (1989) *Minerals in Soil Environments*, 2nd edition, Soil Science Society of America Book Series 1, Madison: Soil Science Society of America.

Garrels, R.M. and MacKenzie, F.T. (1971) *Evolution of Sedimentary Rocks*, New York: W.W. Norton.

Goldich, S.S. (1938) A study in rock weathering, *Journal of Geology* 46.

Grim, R.E. (1962) *Applied Clay Mineralogy*, New York: McGraw-Hill.

Hudson, B.D. (1995) Reassessment of Polynov's ion mobility series, *Soil Science Society of America Journal* 59, 1,101–1,103.

Kelley, W.P. (1948) *Cation Exchange in Soils*, America Chemical Society Monograph, New York: Reinhold.

Moore, D.M. and Reynolds, R.C. Jr. (1997) *X-ray Diffraction and the Identification and Analyses of Clay Minerals*, New York: Oxford University Press.

Nettleton, W.D., Flach, K.W. and Nelson, R.E. (1970) Pedogenic weathering of tonalite in southern California, *Geoderma* 4, 387–402.

Pauling, L. (1929) The principles determining the structure of complex ionic crystals, *Journal of the American Chemical Society* 51, 1,010–1,026.

——(1930) The structure of micas and related minerals, *Proceedings National Academy of Science USA* 16, 123–129.

Ritter, D.F. (1986) *Process Geomorphology*, Dubuque, IA: William Brown.

Wahrhaftig, C. (1965) Stepped topography of the southern Sierra Nevada, California, *Geological Society of America Bulletin* 76, 1,165–1,190.

Further reading

Bartlett, R.J. and James, B.R. (1993) Redox chemistry of soils, *Advances in Agronomy* 50, 151–208.

Brindley, G.W. and Brown, G. (eds) (1980) *Crystal Structures of Clay Minerals and their X-ray Identification*, Monograph No. 5, London: Mineralogical Society.

Brownlow, A.H. (1979) *Geochemistry*, Englewood Cliffs, NJ: Prentice Hall.

Clayton, J.L., Megahan, W.F. and Hampton, D. (1979) *Soil and Bedrock Properties: Weathering and Alteration Products and Processes in the Idaho Batholith*, USDA Forest Service Research Paper INT-237.

Colman, S.M. and Dethier, D.P. (eds) (1986) *Rates of Chemical Weathering of Rocks and Minerals*, Orlando, FL: Academic Press.

Drever, J.I. (1982) *The Geochemistry of Natural Waters*, Englewood Cliffs, NJ: Prentice Hall.

——(ed) (1985) *The Chemistry of Weathering*, NATO-ASI, The Netherlands: Reidel Publishing.

——(1997) Weathering processes, in O.M. Saether and P. de Caritat (eds) *Geochemical Processes, Weathering and Groundwater Recharge in Catchments*, 3–19, Rotterdam: A.A. Balkema.

Garrels, R.M. and Christ, C.M. (1965) *Solutions, Minerals and Equilibria*, New York: Freeman and Cooper.

Greenland, D.J. and Hayes, M.H.B. (1983) *The Chemistry of Soil Constitutents*, New York: John Wiley.

Helling, C.S., Chester, G. and Corey, R.B. (1964) Contribution of organic matter and clay to soil cation-exchange capacity as affected by the pH of the saturation solution, *Soil Science Society of America Proceedings* 28, 517–520.

James, B.R. and Bartlett, R.J. (2000) Redox phenomena, in M.E. Sumner (ed.) *Handbook of Soil Science*, B169–B194, Boca Raton, FL: CRC Press.

Krauskopf, K.B. (1979) *Introduction to Geochemistry*, 2nd edition, New York: McGraw-Hill.

Newman, A.C.D. (ed.) (1987) *Chemistry of Clays and Clay Minerals*, Mineralogical Society of England, Monograph No. 6, Harlow, Essex: Longman.

Sparks, D.L. (1995) *Environmental Soil Chemistry*, New York: Academic Press.
Thurman, E.M. (1985) *Organic Geochemistry of Natural Waters*, Hingham, MA: Kluwer.
White, A.F. and Brantley, S.L. (eds) (1995) Chemical weathering rates of silicate minerals, *Reviews in Mineralogy* 31, 583.
Yuan, T.L., Gammon, N. Jr. and Leighty, R.G. (1967) Relative contribution of organic and clay fractions to cation-exchange capacity of sandy soils from several groups, *Soil Science* 104, 123–128.

SEE ALSO: dissolution; leaching; solubility; weathering

CAROLYN G. OLSON

CHENIER RIDGE

Chenier ridges (cheniers) are sandy or shelly elongate BEACH RIDGES, differentiated from other sand or shell beach ridges by the fact that they are perched on and separated laterally from other cheniers on a chenier plain, by fine-grained, muddy (or sometimes marshy) sediments. Other types of barrier beach plains can be mistaken for cheniers if the presence of underlying and interspersing muddy sediments is not adequately determined (normally by coring). Chenier ridges frequently bend landward at the downdrift end, and branch in a fan-like fashion. The name derives from the French word chêne, meaning oak, which grows on the Louisiana USA chenier ridges. Cheniers can be up to 6 m high, tens of kilometres in length, and hundreds of metres wide. Chenier plains can be tens of kilometres wide. Cheniers are found on generally low wave energy, low gradient, muddy shorelines, in areas where there is an abundant sediment supply. They are frequently associated with river deltas and bayhead situations. Although reported at high latitudes, most examples occur in tropical or subtropical locations. Augustinus (1989) provides an overview of examples and presumed examples of cheniers. Among the most reported examples are: the west Louisiana and Texas coast; Suriname, Guyana and French Guiana; the Gulf of California; New Zealand; northern Australia; east China.

Local variations in sediment supply (such as periods of different river discharge) have been suggested as the likely cause of alternate mudflat progradation and chenier ridge deposition (Otvos and Price 1979), although synchronous development of mudflat and chenier ridges has also been reported (Woodroffe *et al.* 1983; Woodroffe and Grime 1999). Periods of higher wave energy, however, are generally regarded as providing the means by which coarser sediments (including shells) are winnowed out for accumulation in the chenier ridge, with these sediments then moved landward by wave action and OVERWASHING. Some authors argue that 'true cheniers' must result from transgressive processes; however, Otvos (2000) believes the term is appropriate for both stranded (regressive) and transgressive cheniers (see TRANSGRESSION). Otvos (2000) also argues that beach ridges fronting chenier plains must become isolated from the sea and inactive by the deposition of mudflats on their seaward side before they can be considered as cheniers. Chenier plains can provide a sensitive record of changes in sediment supply, sea level and environmental processes.

References

Augustinus, P.G.E.F. (1989) Cheniers and chenier plains: a general introduction, *Marine Geology* 90, 219–229.
Otvos, E.G. (2000) Beach ridges – definitions and significance, *Geomorphology* 32, 83–108.
Otvos, E.G. and Price, W.A. (1979) Problems of chenier genesis and terminology – an overview, *Marine Geology* 31, 251–263.
Woodroffe, C.D. and Grime, D. (1999) Storm impact and evolution of a mangrove-fringed chenier plain, Shoal Bay, Darwin, Australia, *Marine Geology* 159, 303–321.
Woodroffe, C.D, Curtis, R.J. and McLean, R.F. (1983) Development of a chenier plain, Firth of Thanes, New Zealand, *Marine Geology* 53, 1–22.

SEE ALSO: beach ridge; overwashing; raised beach; transgression

KEVIN PARNELL

CHRONOSEQUENCE

The term is used to describe a series of soils that reflect the importance of time for soil formation. *Inter alia*, young soils will differ from mature soils in the degree of weathering of the soil parent material, the development of the soil horizons and the abundance of secondary minerals.

Because the time spans involved in soil development are beyond the time frame of direct observation, usually the development of soils of different age are compared. A chronosequence is thus a sequence of related soils that differ from one another in certain properties primarily as a result of the time available for soil formation. Classical examples are the soils developed on the different members of a flight of terraces, where – except for time – all soil-forming factors (as parent material, landform, climate, etc.) should be rather similar.

Different types of chronosequences can be distinguished: (1) post-incisive, (2) pre-incisive, and (3) time-transgressive. The most frequently studied is case (1) – the example mentioned above – where soils evolve on a sequence of surfaces of different age. In (2) soils that began to develop on a particular surface at the same time, but that were subsequently buried at different times at different places, form a chronosequence. Case (3) relates to a vertical stacking of sediments and PALAEOSOLS, i.e. soils that formed on the same place, but that have been buried after differing periods of development.

Chronosequences have been used to establish quantitative descriptions of soil changes with time, called chronofunctions, and to use the degree of soil development for estimating soil age. To allow for quantitative estimates, soils on dated surfaces (see DATING METHODS) are investigated and numerical indices, such as eluvial-illuvial coefficients and soil development indices, have been developed. There are limits to the range over which chronofunctions can be applied. The rates of development of most soil properties decrease with time; once this degree of development has been achieved, further inferences regarding time cannot be made. In addition, many more complex functions with clear thresholds are involved in soil development. Establishment of chronosequences is difficult in many cases because with the passage of time other soil-forming factors usually also change – as is most clear in the case of climate (see PALAEOCLIMATE). It is often also difficult to rule out the influence of soil disturbances and soil erosion.

Chronosequences have played an important role in establishing relative soil chronologies, which in turn have been used to establish stratigraphic relationships for different geomorphic surfaces. This is especially important where due to the lack of suitable methods or materials modern chronometric dating techniques cannot be applied.

Further reading

Birkeland, W.P. (1999) *Soils and Geomorphology*, 3rd edition, New York: Oxford University Press.

SEE ALSO: catena; soil geomorphology; weathering

ANDREAS LANG

CIRQUE, GLACIAL

Definition and Form

Cirques, also known as corries, coves, combes or cwms, are hollows formed at glacier sources in mountains and partly enclosed by steep, arcuate slopes (headwalls) (Plate 22). Cirque formation requires deepening of the floor by glacial plucking and abrasion, plus glacial removal of plucked or fallen rock encouraging continued headwall retreat. These are aided by basal slip and rotational flow of steep glaciers.

A well-developed 'armchair cirque' has a gently sloping floor and a steep headwall (giving profile closure). At least some of the floor should be gentler than 20°. The headwall curves around the floor, giving plan closure. Ideally, the floor ends in a distinct threshold beyond which the slope steepens, but this may be absent in a trough-head cirque. The headwall should exceed the angle of talus (about 31°–35°) at least in part. We can draw the boundary between headwall and floor at an angle of some 27° (a 2 mm spacing of 10 m contours on a 1 : 10,000 map). A similar gradient can be used to define the cirque crest at the top of the headwall if there is a gentler slope above. It is useful to define a 'cirque focus' in the middle of the threshold. A line from there to the top of the headwall, dividing the cirque into two halves equal in map area, to left and to right, is the median axis: this is used to measure length and overall aspect (Evans and Cox 1995).

Plate 22 East-facing cirques in Ordovician volcanic tuffs on the ridge south of Helvellyn, English Lake District; from left to right, Cock and Ruthwaite Coves, Hard Tarn (a smaller hollow) and Nethermost Cove. The cirques hang above Grisedale trough

CLAY-WITH-FLINT

The chalklands of southern Britain (and northern France) are mantled over extensive areas by a group of deposits called clay-with-flint (*argile à silex*) (Laignel *et al.* 2002).

They are highly variable in composition, ranging 'from heavy reddish brown clays with large unworn flint nodules to almost stoneless yellow or white sands, yellowish to reddish brown silt loams, brightly mottled (red, lilac, green and white) stoneless clays, and beds of rounded flint pebbles' (Catt 1986: 151). Early English geologists tended to regard them as an insoluble residue, left after a long period of dissolution and weathering of the chalk. However, although some of the constituent material of clay-with-flint may have been derived from this source, it is not an adequate explanation of the variability of the material nor of the presence of miscellaneous types of clay, sand and flint shape. Much of it is probably derived and reworked from Palaeogene beds and other Cenozoic deposits, as Jukes-Browne (1906) so astutely recognized.

References

Catt, J.A. (1986) The nature, origin and geomorphological significance of clay-with-flints, in G. de G. Sieveking and M.B. Hart (eds) *The Scientific Study of Flint and Chert*, 151–159, Cambridge: Cambridge University Press.

Jukes-Brown, A.J. (1906) The clay-with-flints: its origin and distribution, *Quarterly Journal of the Geological Society of London* 62, 132–164.

Laignel, B., Quesnel, F. and Meyer, R. (2002) Classification and origin of the clay-with-flints of the western Paris Basin (France), *Zeitschrift für Geomorphologie* 46, 69–91.

A.S. GOUDIE

CLIFF, COASTAL

A cliff is a steep slope (usually >40°, often vertical and sometimes overhanging), exposing rock formations (Plate 24). Most coastal cliffs have been produced by wave ABRASION at the cliff base, but some have been formed by faulting or earlier fluvial or glacial erosion.

Cliffs cut in unconsolidated formations are known as Earth cliffs (May 1972), and those at

Plate 24 Chalk cliffs at Seven Sisters, Sussex, England, retreat as the result of wave abrasion, but are also influenced by solution, bioerosion and rock falls due to freeze–thaw effects and groundwater discharge

the seaward ends of glaciers ice cliffs. Hard rock cliffs, which change very slowly, have been relatively neglected in coastal research. In humid regions soil and vegetation may cover coastal slopes, except on actively receding cliff faces. Vegetated bluffs are not necessarily stable: on the Oregon coast they are cut back as cliffs during occasional severe storms or tsunamis, and then revegetate.

Cliffs rising 100–500 m above sea level are termed high cliffs, and those >500 m (as in Peru and western Ireland) megacliffs (Guilcher 1966). Cliffs less than a metre high are termed microcliffs.

Coastal cliffs recede as the result of basal marine erosion accompanied by subaerial erosion of the cliff face. A sharp angle or notch (see NOTCH, COASTAL) at the cliff base generally indicates active marine erosion. Some cliff profiles are of uniform gradient, others concave or convex, or a combination of these. Concave profiles occur where subaerial erosion exceeds marine erosion and convex profiles where marine erosion has been dominant (Emery and Kuhn 1982), but cliff profiles are also related to the position and inclination of resistant strata. A resistant caprock forms bold cliffs, hard outcrops in the cliff face produce ledges, and a resistant formation at the cliff base slows marine erosion (Figure 25A). A seaward dip facilitates landslides, horizontal strata may form stepped profiles and a landward dip produces an escarpment cliff (Figure 25B). Slope-over-wall profiles may be related to weak above resistant formations or an undercut seaward dipslope (Figure 25C: 1, 2). Joints, bedding planes, faults and intrusions influence cliff morphology, and lateral changes in lithology result in changes in cliff profiles, as on Triassic sandstones and clays in south-east Devon, England. On limestone coasts marine erosion exposes caves and cauldrons produced by earlier karstic dissection.

Cliff outlines in plan are also related to geological structure, with headlands where resistant formations outcrop at the cliff base and bays where weaker formations are excavated by marine erosion; headlands often coincide with ridges and bays with valleys. The Dorset coast, east of Weymouth in southern England illustrates these relationships (Bird 1995).

Cliff-base erosion is achieved by wave quarrying, which dislodges and removes rock material, and abrasion where waves throw sand or gravel against the cliff base. Cliff outcrops may disintegrate as the result of WETTING AND DRYING WEATHERING of surfaces subject to spray, splash and rainwash, or SALT WEATHERING where salt crystallizes from sea splash, notably on arid coasts. Solution by runoff, seepage, spray and sea water contributes to cliff-base erosion, particularly on limestone coasts where distinctive flat-floored solution notches may form, in contrast with sloping ramps where wave abrasion is dominant. Bioerosion (by plants and animals that live on the cliff and shore) also contributes.

Cliff faces may be indurated by calcareous or ferruginous compounds precipitated from groundwater seepage, forming crusts that eventually crack and exfoliate, exposing uncemented rock. Cliff faces are also indurated by carbonates precipitated from sea splash, particularly on headlands. Downwashed sediment may adhere to a cliff face as stalactitic structures, notably on limestone and AEOLIANITE (Hills 1971). By contrast, fine-grained sediment winnowed from a cliff face by onshore winds has been deposited as a cliff-top levee on the Port Campbell coast in south-eastern Australia (Baker 1943).

As a cliff is undercut it may collapse, producing a debris fan below a fresh rock scar. Sediment yield from cliffs depends on the rate of recession and the effects of weathering and erosion. Accumulation of sediment at the cliff base slows recession, but usually the debris fan is dispersed by erosion, and when it has been removed basal undercutting resumes.

MASS MOVEMENTs occur on cliffs where the groundwater load becomes excessive, where stresses develop as the result of freeze and thaw, where a massive caprock exerts pressure on underlying weaker formations, or where there is expansion or base exchange, weakening clay minerals. Breakaways develop at the cliff crest where masses of rock topple down the cliff, and slumping produces irregular topography as rock outcrops disintegrate and material slides, flows or creeps down the slope towards a basal receding cliff. Such cascading systems, with instability transmitted upward to the cliff crest, occur on the Dorset coast (Brunsden and Jones 1980). In Oregon coastal landslides in weathered rock are commoner in winter, when stronger wave action attacks formations saturated by heavy rain, but may also be triggered by tectonic movements or tsunamis generated along the nearby plate edge.

Some cliffs descend to SHORE PLATFORMs cut by marine erosion and weathering processes as the

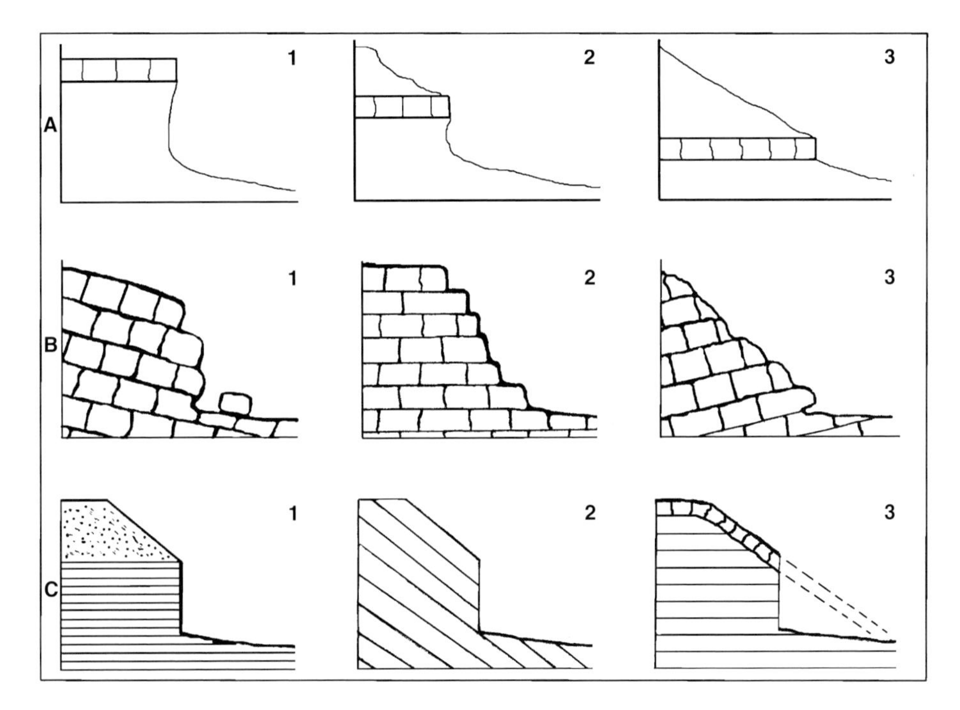

Figure 25 A, the effects of a resistant formation on cliff profiles; B, variations related to the dip of strata; C, slope-over wall cliffs: 1, related to lithology; 2, related to structure; 3, retaining a slope formed by periglacial solifluction

cliff recedes; others are fronted by irregular rocky shores, particularly where the geological formations are of intricate structure with resistant elements; others (plunging cliffs) continue below sea level, either because of partial marine submergence (where they descend to submerged coastlines) or because they formed by faulting, glaciation or vulcanicity.

Some cliffs are actively receding; others are inactive behind persisting basal talus or a prograding beach, or because of lowering of sea level. Inactive cliffs may decline into subaerially shaped slopes which become vegetated. Cliffs stranded by land uplift or sea-level lowering become bluffs behind emerged beaches and shore platforms.

Rates of cliff recession vary with cliff height, rock resistance, structure, weathering and exposure to wave attack. They are usually reported as annual averages, but are generally episodic, related to occasional storms or mass movements. Rapid cliff recession ($>1\,\mathrm{m\,yr^{-1}}$) occurs on soft rock formations, and rates of $>100\,\mathrm{m\,yr^{-1}}$ have been reported on cliffs in volcanic ash and arctic tundra deposits (humates with melting ice), but some hard rock cliffs have shown little or no recession in the period (up to 6,000 years) that the sea has stood at its present level.

Where cliff recession has been slow, features inherited from earlier environments may persist. Examples of this are the slope-over-wall profiles on the Atlantic coasts of Britain, where the slope (which may be convex, a straight bevel or concave) is mantled by earthy gravel (termed Head) formed by periglacial SOLIFLUCTION in cold phases of the Pleistocene, and the wall is a receding undercliff (Figure 25C: 3): the proportion of slope to wall diminishes as exposure to wave attack increases. Active periglaciation forms steep slopes of angular debris on arctic coasts, as on Baffin Island in Canada. In northern Britain and Scandinavia the periglacial slope gives place to slopes formed by glacial erosion or deposition, also undercut by Holocene marine erosion. In the humid tropics slope-over-wall profiles occur where a coastal slope on deeply weathered rock has been undercut by marine erosion, and in arid regions the undercut coastal slope may have been a pediment.

Cliff recession is likely to accelerate (and coastal landslides become more frequent) during a rising relative sea level, and when storminess increases in coastal waters: protective beaches diminish, and wave attack on the cliff base becomes stronger and more sustained.

Human impacts on cliffs include stabilization by the building of basal sea walls or boulder ramparts to halt coastline retreat, the grading, vegetating or concreting of cliff faces, and the introduction of drains to hasten groundwater discharge. By contrast, cliffs become more unstable as the result of the reduction of beaches (when beach sand or gravel are extracted), an increase in groundwater load and levels (when previously dry cliff-top terrain is irrigated) and cliff-top loading by buildings and other structures.

References

Baker, G. (1943) Features of a Victorian limestone coastline, *Journal of Geology* 51, 359–386.

Bird, E.C.F. (1995) *Geology and Scenery of Dorset*, Bradford-on-Avon: Ex-Libris Press.

Brunsden, D. and Jones, D.K.C. (1980) Relative timescales in formative events in coastal landslide systems, *Zeitschrift für Geomorphologie, Supplementband* 34, 1–19.

Emery, K.O. and Kuhn, G.G. (1982) Sea cliffs: their processes, profiles and classification, *Geological Society of America Bulletin* 93, 644–654.

Guilcher, A. (1966) Les grandes falaises et mégafalaises des côtes sud-ouest et ouest de l'Irlande, *Annales de Géographie* 75, 26–38.

Hills, E.S. (1971) A study of cliffy coastal profiles based on examples in Victoria, Australia, *Zeitschrift für Geomorphologie* 15, 137–180.

May, V.J. (1972) Earth cliffs, in R.S.K. Barnes (ed.) *The Coastline*, 215–235, New York: Wiley.

Further reading

Sunamura, T. (1992) *Geomorphology of Rocky Coasts*, Chichester: Wiley.

Trenhaile, A.S. (1987) *The Geomorphology of Rock Coasts*, Oxford: Clarendon.

SEE ALSO: slope, evolution

ERIC C.F. BIRD

CLIMATIC GEOMORPHOLOGY

The part of the discipline that seeks to explain the form and distribution of landforms in terms of climate. It developed during the period of European colonial expansion and exploration at the end of the nineteenth century, when unusual and often spectacular landforms were encountered in deserts, polar regions and the humid tropics. In addition, it was a time when regionalization and classification were major endeavours in geography and cognate subjects. Attempts at climatic, soil and vegetation classifications were being made by scientists like Köppen, Dokuchayev and Schimper. They sought to understand the regional patterns of the phenomena they were classifying, and climate was seen as a major control at their scale of investigation.

In the USA, W.M. Davis recognized 'accidents', whereby non-temperate and non-humid climatic regions were seen as deviants from his normal cycle of erosion and he introduced, for example, his arid cycle (Davis 1905). Some (see Derbyshire 1973) regard Davis as one of the founders of climatic geomorphology, although the leading French climatic geomorphologists Tricart and Cailleux (1972) criticized Davis for his neglect of the climatic factor in landform development. Much important work was undertaken on dividing the world into climatic zones (morphoclimatic regions) with distinctive landform assemblages, in France (e.g. Birot 1968), in Germany (e.g. Büdel 1982) and in New Zealand (Cotton 1942). This version of geomorphology was seen as essentially geographical (Holzner and Weaver 1965).

In the later years of the twentieth century the popularity of climatic geomorphology became less as certain limitations became apparent (see Stoddart 1969).

(1) Much climatic geomorphology was based on inadequate knowledge of rates of processes and on inadequate measurement of process and form. Assumptions were made that, for example, rates of chemical weathering were high in the humid tropics and low in cold regions, whereas subsequent empirical studies have shown that this is far from inevitable.
(2) Some of the climatic parameters used for morphoclimatic regionalization were meaningless or crude from a process viewpoint (e.g. mean annual air temperature).
(3) Macroscale regionalization was seen as having little inherent merit and ceased to be a major goal of geographers, who eschewed 'placing lines that do not exist around areas that do not matter'.
(4) Conversely, and paradoxically, climatic geomorphology had a tendency to concentrate

Table 8 Büdel's morphogenetic zones of the world

Zone	Present climate	Past climate	Active processes (fossil ones in brackets)	Landforms
(1) Of glaciers	Glacial	Glacial	Glaciation	Glacial
(2) Of pronounced valley formation	Polar, tundra	Glacial, polar, tundra	Frost, mechanical weathering, stream erosion (glaciation)	Box valleys, patterned ground, etc.
(3) Of extra-tropical valley formation	Continental, cool temperate	Polar, tundra continental	Stream erosion (frost processes, glaciation)	Valley
(4) Of subtropical pediment and valleys formation	Subtropical (warm; wet or dry)	Continental, subtropical	Pediment formation (stream erosion)	Planation surfaces and valleys
(5) Of tropical plantation surface formation	Tropical (hot; wet or wet–dry)	Subtropical, tropical	Planation, chemical weathering	Planation surfaces and laterites

on bizarre forms found in some 'extreme' environments rather than on the overall features of such areas.

(5) Many landforms that were supposedly diagnostic of climate (e.g. pediments in arid regions or inselbergs in the tropics) are either very ancient relict features that are the product of a range of past climates or they have a form that gives an ambiguous guide to origin.

(6) The impact of the large, frequent and rapid climatic changes of the Quaternary and of the very different climates of the Tertiary has disguised any simple climate–landform relationship. For this reason, Büdel (1982) attempted to explain landforms in terms of fossil as well as present-day climatic influences (Table 8). He recognized that landscape was composed of various 'relief generations' and saw the task of what he termed 'climato-genetic geomorphology' as being to recognize, order and distinguish these relief generations, so as to analyse today's highly complex relief.

Although these tendencies have tended to reduce the relative importance of traditional climatic geomorphology, notable studies still appear that look at the nature of landforms and processes in different climatic settings (e.g. M. Thomas 1994 on the humid tropics; D. Thomas 1998 on arid lands; and French 1999 on periglacial regions). In addition, a concern with GLOBAL WARMING and its geomorphological impact leads to a renewed concern with climate-landform links.

References

Birot, P. (1968) *The Cycle of Erosion in Different Climates*, London: Batsford.

Büdel, J. (1982) *Climatic Geomorphology*, Princeton, NJ: Princeton University Press.

Cotton, C.A. (1942) *Climatic Accidents in Landscape-making*, Christchurch: Whitcombe and Tombs.

Davis, W.M. (1905) The geological cycle in an arid climate, *Journal of Geology* 13, 381–407.

Derbyshire, E. (ed.) (1973) *Climatic Geomorphology*, London: Macmillan.

French, H.M. (1999) *The Periglacial Environment*, 2nd edition, London: Longman.

Holzner, L. and Weaver, G.D. (1965) Geographical evaluation of climatic and climato-genetic geomorphology, *Annals of the Association of American Geographers* 55, 592–602.

Stoddart, D.R. (1969) Climatic geomorphology, in R.J. Chorley (ed.) *Water, Earth and Man*, 473–485, London: Methuen.

Thomas, D.S.G. (ed.) (1998) *Arid Zone Geomorphology*, 2nd edition, Chichester: Wiley.

Thomas, M.F. (1994) *Geomorphology in the Tropics: A Study of Weathering and Denudation in Low Latitudes*, Chichester: Wiley.

Tricart, J. and Cailleux, A. (1972) *Introduction of Climatic Geomorphology*, London: Longman.

Further reading

Elorza, M.G. (2001) *Geomorfología climática*, Barcelona: Ediciones Omega.

A.S. GOUDIE

CLIMATO-GENETIC GEOMORPHOLOGY

Climato-genetic geomorphology is the systematic field investigation of landforms in a certain area according to their evolution. Many different methods may be applied, but the basis for it is the observation of an assemblage of rested relief elements. There are two roots to climato-genetic geomorphology. First, the more palaeoforms were acknowledged, it became clear that their systematic investigation was necessary, not only to explain the relief but also to estimate their influence on recent processes. Second, the system of 'klimatische Geomorphologie' was developed. It is the basis for relief forming processes or process fabric, which is applied to the different RELIEF GENERATIONS, the constituents of climato-genetic geomorphology.

The terminology is not very clear for originally the term 'klimatische Geomorphologie' was introduced to differentiate it from tectonic or structural geomorphology. However, it was misleading. Dynamic geomorphology would have been a much better term, as it is the study of processes mainly at the medium scale. 'Climatic geomorphology' is the literal translation, but this has the very different aim of relating landforms to climatic or hydrological data. 'Klimatische Geomorphologie' investigates the relief forming processes in a certain MORPHOGENETIC REGION. Climatic geomorphology looks more for single forms or processes. More or less similar to 'klimatische Geomorphologie' are 'research in a morphoclimatic zone', 'climato-geomorphology', 'forms of morphoclimates' or 'DYNAMIC GEOMORPHOLOGY'. For the evolution of landforms there are the words 'klimatische Morphogenese' (literally climatic morphogenesis) and 'klimagenetische Geomorphologie' (literally climato-genetic morphology). Thus the position of the words geomorphology, climate and genetic might change without any generally agreed special connotations. There is a distinction though between processes and evolution in the terms. It seems that there is a difference in the English and German use of the word 'genetic' in geomorphology, that is, development and historical outline of natural phenomena.

Processes were deduced rather early on in the search for an explanation of landforms, and their relation to exogene (i.e. climate controlled force) was acknowledged. In Europe the work of glaciers was studied on recent examples and similar landforms and deposits were classified accordingly (ACTUALISM). In the west of the USA early research detected the specific processes of the arid zone. Palaeoforms have been increasingly acknowledged since the early twentieth century. The systematic approach to climatic geomorphology dates from 1948. After the Second World War the overseas research of palaeoclimatology (see PALAEOCLIMATE), deduced from morphological features like moraines and solifluction forms, increased. All these research efforts were the basis for the concept of the development of RELIEF GENERATIONS. There are several possibilities for applying this concept besides the explanation of relief evolution. It may serve to control erosion rates, especially their extrapolation and the distinction of human accelerated rates. On the other hand, the extension and preservation of the different relief generations shows the intensity and specific location of recent land forming activity. This is a good basis for applied questions like soil erosion. In connection with ecological studies, relief generations are a basis for the spatial extent of investigated features, e.g. the distribution of soil types.

The recent process fabric is either observed directly or deduced from fresh landform scars after catastrophic events. Similar forms of different size, and sequences, are extrapolated to get an idea about intensity, recurrence and the forming power of special processes and their interrelation. The relation between denudation and linear erosion is investigated as well as between erosion and deposition. This is counterchecked by the known facts of climatic change and tectonic movements, which give an estimate of the change of the processes. As the process fabric is a systematic combination of single geomorphological activities one can ask for completeness of processes as well as forms. A simple example may illustrate this: in the northern foreland of the Alps the rivers now carry sand and show a rather low activity. The slopes are more or less undisturbed. Thus the younger process fabric is not strong and not widespread. There is a large amount which remains unexplained, which from the analysis of the forms is easily classified as moraines. If a soil is developed on them, they are inactive. Moraines are known from the surroundings of recent glaciers in their form and sedimentary structure. Connected landforms are outwash plains in front of them and overdeepening to the rear of the moraines. With this form assemblage the older process fabric can be extended beyond the moraines. Gravel terraces in front of them are of fluvioglacial origin, while lakes to the rear are a sign of glacial scour. There is a feedback in the analysis of forms and processes. Many more details and several stages of the advancing and retreating glaciers have been classified and mapped, e.g. for the Inn Chiemsee glacier by Carl Troll.

With the advancement of knowledge about relief generations it became clear that older forms are widely distributed. There are a few landscapes, like young volcanoes or badlands, which consist only of one relief generation.

Therefore fieldwork should start with the oldest forms and look for the nested younger form assemblage. By interpolation, the younger process fabric is derived. There are the same feedback mechanisms as those named above. This method has two advantages: the existence of remaining unexplained phenomena can be avoided, which one is always inclined to keep as low as possible. Second 'Mehrzeitformen' (i.e. forms shaped in different climates) are more easily detected. It is self-evident that there is a slight change to all the forms of the older relief generation (e.g. the removal of the soil cover of an old plain), but there are a few forms which were noticeably changed by younger processes, like blockfields in the mid-latitudes (cf. RELIEF GENERATIONS).

Climato-genetic geomorphology works not only by analysing landforms, but also uses a wide range of other methods. As a genetic science the connection to the well-developed soil science is especially close. For the tropical zone soil analysis in the field and in the laboratory can solve problems of allochthonous or autochthonous weathering, of relative age, and especially of palaeofeatures. In the humid mid-latitudes, relics of tropical weathering, the periglacial cover of solifluction and loess are counterchecks for the distinction of relief generations. All direct and indirect dating methods are helpful.

Further reading

Bremer, H. (2002) Tropical weathering, landforms and geomorphological processes: fieldwork and laboratory analysis, *Zeitschrift für Geomorphologie* 46, 273–291.

Büdel, J. (1977) *Klima-Geomorphologie*, Borntraeger: Berlin. Translated by L. Fischer and D. Busche (1982) *Climatic Geomorphology*, Princeton: Princeton University Press.

HANNA BREMER

COASTAL CLASSIFICATION

Coastal classification is the grouping of similar coastal features in categories that distinguish them from dissimilar features. The aim is to elucidate the relationships between coastal landforms and processes and to understand coastal evolution. Simple classifications are implicit in the topics identified in chapter headings in coastal textbooks, and when coastal features are categorized and shown on maps of coastal morphology.

Some attempts to classify coastal landforms (including shores and shoreline features) have been genetic, based on the origin of the landforms, rather than descriptive (e.g. cliffed coasts, delta coasts, mangrove coasts). The difficulty is that genetic classifications can only be applied when the mode of origin of coastal landforms is known, and as only a small proportion of the world's coastline has been investigated in sufficient detail to determine evolution such classifications remain somewhat speculative. Certainly the assumption that particular types or associations of landforms can be used as indicators of particular modes of origin can be misleading, for some coastal landforms (e.g. barrier islands, beach ridges, cuspate forelands, shore platforms) may evolve in more than one way: a phenomenon termed multicausality (Schwartz 1971). Various kinds of coastal classification are now described, with references.

Atlantic and Pacific type coasts

Suess (1906) distinguished Atlantic coasts, which run across the general trend of geological structures, from Pacific coasts, which run parallel to structural trends. The former are characteristic of the Atlantic shores of Britain and Europe; the latter of the Pacific coasts of North and South America.

Cliffed coastlines that transgress geological structures are termed discordant, whereas those that follow the strike of a particular geological formation are termed concordant.

Classification and plate tectonics

Inman and Nordstrom (1971) devised a geophysical classification based on PLATE TECTONICS, recognizing that the Earth's crust is a pattern of plates separated by zones of spreading and zones of convergence, with plate margins moving at rates of up to $15\,\mathrm{cm\,yr^{-1}}$. They contrasted subduction coasts, where one plate is passing beneath another, with trailing-edge coasts on a diverging plate margin and marginal sea coasts on the lee side of island arcs, and described features characteristic of each of these. It was a broad-scale classification, dealing with first-order (continental) features (*c.*1,000 km long × 100 km wide × 10 km high).

Coasts of submergence and emergence

Gulliver (1899) distinguished coasts formed by submergence from coasts formed by emergence. This was developed into a genetic classification by Johnson (1919), who described coastlines (he used the American term shorelines) of submergence, coastlines of emergence, neutral coastlines (with forms due neither to submergence nor emergence, but to deposition, e.g. delta coastlines, alluvial plain coastlines, glacial outwash coastlines and volcanic coastlines) and compound coastlines (with an origin combining two or more of the preceding categories). Most coasts fall into the compound category, because they show evidence of both emergence, following high sea levels in interglacial phases of the Pleistocene, and submergence, due to the Late Quaternary (Flandrian) marine transgression.

Classification based on climate

Aufrère (1936) proposed a coastal classification based on climate, which distinguished coasts with a permanent ice cover (no marine processes), coasts with a seasonal ice cover (seasonal marine processes and abundant sediment from glacial sources), temperate humid coasts (as in Europe), tropical humid coasts (with abundant fluvial sediment in deltas and coastal plains), arid coasts (without rivers; marine sediments dominant) and semi-arid coasts (some river features; SABKHAS). The global distribution of coastal climates shows sector variations related to latitude and wind regime with coastwise transitions that are generally gradual, although there are rapid transitions from humid tropical to arid within comparatively short distances in Ecuador and Colombia, in west Africa and northern Madagascar.

Classification based on coastal processes

Variations in coastal processes effective around the world's coastline were discussed by Davies (1980), who defined and mapped swell and storm wave environments, coasts subject to trade winds, monsoons and tropical cyclones, the distribution of high, moderate and low wave energy coasts, tidal types (semi-diurnal, mixed and diurnal) and mean maximum tide ranges divided into microtidal (<2 m), mesotidal (2–4 m) and macrotidal (>4 m), to which may be added megatidal (>6 m).

Initial and subsequent coasts

A distinction can be made between initial forms, which existed when the present relative levels of land and sea were established and marine processes began work (on most coasts about 6,000 years ago) and sequential forms, those that have since developed as the result of marine action. Shepard (1976) devised a classification on this basis, making a distinction between primary coasts shaped largely by non-marine agencies and secondary coasts that owe their present form to marine action. It was essentially a genetic classification, with descriptive detail inserted to clarify the subdivisions, and it recognized that, because of the worldwide Late Quaternary marine transgression, the sea has not long been at its present level relative to the land, so that many coasts have been little modified by marine processes.

Shepard's aim was to devise a classification that would prove useful in diagnosing the origin and history of coastlines from a study of charts and air photographs, but it is dangerous to assume that the origin and history of a coast can be deduced from such evidence without field investigation. A straight coast may be produced by deposition, faulting, emergence of a featureless seafloor or submergence of a coastal plain; an indented coast by submergence of an undulating or dissected land margin, emergence of an irregular seafloor, differential marine erosion of hard and soft outcrops along the coast or transverse tectonic deformation (folding and faulting) of the land margin. It is doubtful whether configuration can be taken as a reliable indicator of coastal evolution.

Leontyev *et al.* (1975) also considered initial and sequential forms (using the cycle of youth, maturity and old age) in a classification based on coasts not changed by the sea, coasts formed by abrasion or accumulation, and a combination of the two.

Stable and mobile coasts

Cotton (1952) made a distinction between coasts of stable and mobile regions, stable regions being those that escaped the Quaternary tectonic movements that have affected mobile regions, especially around the Pacific rim, where they still continue. On the coasts of stable regions he separated those dominated by features produced by Late Quaternary marine submergence from those dominated by inherited (mainly Pleistocene) features preserved by earlier emergence. On the coasts of mobile regions he separated those where the effects of Late Quaternary marine submergence have not been counteracted by recent uplift of the land from those where recent uplift of the land has caused emergence.

Morphological classification

De Martonne (1909) used a morphological distinction between steep and flat coasts as a basis for classification, suggesting a number of subtypes, some descriptive (estuary coasts, skerry coasts), others genetic (fault coasts, glacially sculptured coasts). Ottmann (1965) followed a similar approach, recognizing three categories of cliffed coast (cliffs plunging to oceanic depths, cliffs with shore platforms and cliffs plunging to submerged platforms), partially submerged uncliffed coasts, and low depositional coasts behind gently shelving seafloors.

Zenkovich (1967) classified depositional coastal features into five categories: attached forms (including beaches and cuspate forelands), free forms (including spits), barriers, looped forms (including tombolos) and detached forms (including barrier islands).

Geology in coastal classification

Russell (1967) advocated classification of rocky coasts on the basis of geology and structure, noting the striking similarity of features developed on crystalline rocks, irrespective of their climatic and ecological environments: granites that outcrop on parts of the coasts of Scandinavia, southwest Australia, South Africa and Brazil all show similar domed surfaces related to large-scale spalling and conspicuous joint-control. Limestones (including chalk and coral), basalts and sandstones also show distinctive kinds of coastal landforms. Bedrock coasts are commoner in cold, arid and temperate regions than in the humid tropics, where there has been deep weathering and depositional aprons are extensive.

Advancing and receding coasts

A coastline may advance because of coastal emergence and/or progradation by deposition, or retreat because of coastal submergence and/or retrogradation by erosion. Valentin (1952) used this analysis as the basis for a system of coastal classification that could be shown on a world map. Coasts that had advanced were divided into those produced by emergence, by organic deposition (mangroves, coral) and by inorganic deposition (marine and fluvial), while coasts that had retreated were divided into those produced by submergence of glaciated landforms and fluvially eroded landforms and those shaped by marine erosion. Bloom (1965) elaborated Valentin's scheme by considering historical evolution where the response to emergence, submergence, erosion and deposition has varied through time. Thus on the Connecticut coast, where radiocarbon dates from buried peat horizons have yielded a chronology of relative changes of land and sea level in Holocene times, there is evidence that at some stages the sea gained on the land during submergence, even though deposition continued, while at other stages deposition was sufficiently rapid to prograde the land during continuing submergence: at present there is widespread erosion on the seaward margins of saltmarshes, possibly because of resumed submergence.

The advantage of such non-cyclic classifications is that they pose problems and stimulate further research instead of trying to fit observed features into presupposed evolutionary sequences.

Composite classifications

McGill (1958) produced a map of the world's coastline which showed the major landforms of the coastal fringe, 8–16 km wide. This was a composite classification in which major coastal landforms were classified in terms of lowland or upland hinterlands, with additional information on selected features (constructional or destructional) in the backshore, foreshore and offshore zones, categorized by the agent responsible: sea, wind, coral or vegetation.

Artificial coastlines

Little attention has been given in coastal classifications to the fact that long sectors of coastline have become artificial during recent decades, partly as the result of engineering works designed to combat erosion and partly as a consequence of embanking or infilling to extend coastal land. On developed coasts the proliferation and extension of anti-erosion works, notably sea walls and boulder ramparts, has resulted in large proportions of artificial coastline: 85 per cent in Belgium, 51 per cent in Japan, 38 per cent in England. Coastal land has been artificially extended on a large scale in Singapore, Hong Kong, Tokyo Bay in Japan, western Malaysia and the Netherlands. The category of artificial coastlines is increasing rapidly, and much more of the world's coastline will become artificial as attempts are made to halt submergence and erosion.

References

Aufrère, L. (1936) Le rôle du climat dans l'activité morphologique littorale, *Proceedings, 14th International Geographical Congress*, Warsaw, 2, 189–195.

Bloom, A.L. (1965) The explanatory description of coasts, *Zeitschrift für Geomorphologie*, 9, 422–436.

Cotton, C.A. (1952) Criteria for the classification of coasts, *Proceedings 17th Conference, International Geographical Union*, Washington, 315–319.

Davies, J.L. (1980) *Geographical Variation in Coastal Development*, London: Longman.

De Martonne, E. (1909) *Traité de Géographie Physique*, Paris: Colin.

Gulliver, F.P. (1899) Shoreline topography, *Proceedings, American Academy of Arts and Sciences*, 34, 151–258.

Inman, D.L. and Nordstrom, C.E. (1971) On the tectonic and morphologic classification of coasts, *Journal of Geology* 79, 1–21.

Johnson, D.W. (1919) *Shore Processes and Shoreline Development*, New York: Wiley.

Leontyev, O.K., Nikiforov, L.G. and Safyanov, G.A. (1975) *The Geomorphology of the Sea Coasts* (in Russian), Moscow: Moscow University.

McGill, J.T. (1958) Map of coastal landforms of the world, *Geographical Review* 48, 402–405.

Ottmann, F. (1965) *Introduction a la Géologie Marine et Littorale*, Paris: Masson.

Russell, R.J. (1967) *River Plains and Sea Coasts*, Berkeley, CA: University of California Press.

Schwartz, M.L. (1971) The multiple causality of barrier islands, *Journal of Geology* 79, 91–93.

Shepard, F.P. (1976) Coastal classification and changing coastlines, *Geoscience and Man* 14, 53–64.

Suess, E. (1906) *The Face of the Earth*, Oxford: Clarendon Press.

Valentin, H. (1952) *Die Küsten der Erde*, Petermanns Geographische Mitteilungen, 246.

Zenkovich, V.P. (1967) *Processes of Coastal Development*, Edinburgh: Oliver and Boyd.

Further reading

Bird, E.C.F. (2000) *Coastal Geomorphology: An Introduction*, Chichester: Wiley.

Schwartz, M.L. (ed.) (1982) *The Encyclopedia of Beaches and Coastal Environments*, Stroudsburg, PA: Hutchinson Ross.

SEE ALSO: coastal geomorphology; global geomorphology

ERIC C.F. BIRD

COASTAL GEOMORPHOLOGY

The industrial, recreational, agricultural and transportational activities of growing human populations are exerting enormous pressures on coastal resources. To manage these activities in the least detrimental way, we need to have a better understanding of the dynamic nature of coastal landforms and the operation and interaction of marine and terrestrial processes. Differences in climate, changes in relative SEA LEVEL, wave environments, tides, winds, the morphology, structure and lithology of the hinterland, terrestrial and marine sediment sources, human activity and numerous other factors provide almost infinite variety to coastal scenery around the world. Coastal regions consist of a mosaic of diverse elements, some of which are contemporary, whereas others are ancient vestiges of periods when climate and sea level may have been similar or different from today's. Small-scale elements of depositional coasts, which can experience rapid changes in morphology, may attain a rough state of balance with their environmental conditions, but other features – particularly on hard rock coasts – require long periods to adjust to changing conditions. Furthermore, even if environmental conditions remain constant, individual coastal landforms still have to adjust to slow changes in the morphology of the coast itself. For example, whereas the profiles of sandy BEACHes respond fairly quickly to changing wave conditions, they may also have to adjust slowly to long-term changes in coastal configuration, sediment budgets, offshore gradients, climate, sea level and increasingly the effects of human interference.

Coastal classification

There have been many attempts to classify coasts, although none are entirely satisfactory. Most COASTAL CLASSIFICATIONS use at least two of three basic variables: the shape of the coast; changes in relative sea level; and the effect of marine processes. Some classifications are genetic, others are descriptive and others combine the two approaches. Genetic classifications are hindered by a lack of relevant data, however, and descriptive classifications, which have to accommodate an enormous variety of coastal types, tend to be cumbersome. Two classifications, which consider the nature of coastal environments and the effect of PLATE TECTONICS on coastal development, are particularly useful.

Davies (1972) proposed that coastal processes are strongly influenced by morphogenic factors that vary in a fairly systematic way around the world. Davies's morphogenic classification was based upon four major wave climates, although differences in coastal characteristics also reflect variations in tidal range, climate and many other factors. The highest WAVEs are usually generated in the storm belts of temperate latitudes. Beaches in storm wave environments tend to have dissipative or gently sloping and barred profiles, and the major constructional features are often composed of coarse clastic material. Constructional features are oriented more by local fetch than by the variable direction of the deep water waves, and mechanical wave erosion is important in the formation of cliffs (see CLIFF, COASTAL) and SHORE PLATFORMS. Long, low constructional waves dominate swell environments between the northern and southern storm wave belts. The beaches have berms, and they tend to be towards the steeper, reflective, non-barred end of the spectrum. The direction of longshore currents is more constant than in storm wave environments, and large, sandy constructional features are oriented toward the approaching swell. Mechanical wave erosion of cliffs and platforms is probably slower than in storm wave environments, and this, combined with warmer climates, makes CHEMICAL WEATHERING and biological WEATHERING more important in swell wave environments. Sheltered, enclosed seas and ice-infested waters are low energy environments. Waves are flat and constructional, and beaches have prominent berms. The orientation of sandy constructional features, which are common in partially enclosed seas, is largely determined by local fetch.

Plate tectonics provide a partial explanation for the distribution of a variety of coastal elements, although the degree of explanation decreases with the decreasing size of the feature. Inman and Nordstrom (1971) proposed that the morphology of the largest, or first-order, coastal elements can be attributed to their position on moving tectonic plates. Three main geotectonic classes were identified: continental and ISLAND ARC collision coasts form along the edges of converging plates; plate-imbedded or trailing edge coasts face spreading centres; and marginal sea coasts develop where island arcs separate and protect continental coasts from the open ocean. The structural grain of collision coasts is parallel to the shore and they are therefore fairly straight and regular. Tectonically mobile collision coasts have narrow continental shelves and high, steep hinterlands, often with flights of raised terraces. The high relief provides an abundant supply of sediment to the coast. Plate-imbedded or trailing edge coasts usually have hilly, plateau, or low hinterlands, and wide continental shelves. The structural grain may be at high angles to the coast, which can therefore be very indented. Marginal seacoasts range from low-lying to hilly, with wide to narrow shelves, and they are often modified by large rivers and RIVER DELTAS.

Coastal modelling

Models provide one of the best ways of investigating the poorly understood components of a coastal system. They provide insights into the interrelationships between and among variables, and they are indispensable in enhancing our efforts to monitor, manage, control and develop the coastal system and its associated resources.

Physical models are simplified and scaled representations of the real world. They can be used to control and isolate variables, to provide insights into phenomena not yet described or understood, to provide measurements to test theoretical results and to measure complicated phenomena that cannot be theoretically analysed. Coastal engineers have constructed a wide variety of fixed-bed hydraulic scale models to study the action of waves, tides and currents, and to assist in the design of coastal structures. Geologists and geomorphologists have used movable bed models to examine sediment transport and the dynamics and formation of bars (see BAR, COASTAL), barriers (see BARRIER AND BARRIER ISLAND) and beaches. Unlike natural oceanic waves, however, the shallow water waves generated in most wave tanks have no orbital kinetic energy and are nearly pure solitons. Physical models therefore have not been able to describe accurately the hydrodynamics and sedimentary processes operating in coastal systems, and the results obtained from them always have to be verified or corroborated with other evidence.

Because of their generality, versatility and flexibility, mathematical models are the most common type used by coastal workers. Unfortunately, however, our lack of knowledge of coastal processes and the frequent reliance on laboratory data to determine the value of coefficients, casts doubt on the applicability of many mathematical models to the real world. There are several types of mathematical model. Deterministic models, which are based on the principles of fluid mechanics, seem to work best in conjunction with laboratory experiments that allow parameters to be held constant while one is varied at a time. Simulation models involve the manipulation of process–response equations on computers, compressing years of coastal development in the prototype into minutes. This allows the behaviour of a system to be determined under a variety of situations and conditions, and to test the sensitivity of the system to changing input parameters. Statistical models can be used to study the relationships between a set of variables, and to verify possible relationships identified by theoretical models. To use equations derived from one area for predictive purposes in another, however, often requires the determination of a different set of coefficients.

Coastal inheritance

There is growing evidence that because interglacial sea levels were similar to today, contemporary coastal features often formed close to, or were superimposed on top of, their ancient counterparts. Although evidence of past sea levels and climates is generally easily obliterated in unconsolidated coastal deposits, many sandy coasts retain sedimentary and morphological elements of former environmental conditions. Coastal deposits from the last interglacial stage are being cannibalized in some areas to provide sediment for the construction and maintenance of modern coastal features, and barrier systems have sometimes developed on top of older Pleistocene barriers, or are located somewhat seaward of them. Most barrier islands on the German North Sea coast and in places on the Atlantic coast of the USA, for example, consist of a core of Pleistocene deposits, mantled by Holocene sediments. In south-eastern Australia, a distinct inner barrier of the last interglacial age is separated from an outer Holocene barrier by a lagoon and swamp tract. Pleistocene dunefields are adjacent, and probably under Holocene coastal dunes (see DUNE, COASTAL) in some places, especially in Australia and the Mediterranean, although they are generally absent in northern Europe, where most dunes were built at different stages during the Holocene. The presence of near-surface discontinuities shows that Holocene limestones, ranging from a few metres up to about 30 m in thickness, also form veneers over foundations of older reef-rock. The concept of INHERITANCE is particularly important on resistant rock coasts which have probably evolved very slowly during successive periods of high interglacial sea level. It has been demonstrated that some cliffs, sea caves, ramps (see RAMP, COASTAL) and shore platforms are at least last interglacial in age, and modelling suggests that many platforms have developed during interglacial stages during the middle and late Pleistocene.

Coastal management

Despite the problems associated with flooding, erosion, pollution and other hazards, and the increasing aesthetic and practical impetus for sustainable coastal management, rising populations and growing economic pressures are accelerating the pace of human interference and degradation on the world's coasts (Plate 25). We lack reliable models, however, that can be usefully employed by managers, planners and decision-makers for INTEGRATED COASTAL MANAGEMENT and to predict the effects of sea-level changes, human activities and other factors on the coast. The available field data on coastal changes are often of questionable reliability and usually too short-term to analyse the interaction of a large number of variables. Coastal changes are also frequently complex and non-linear (see NON-LINEAR DYNAMICS), and may reflect the interaction and exchange of sediment between the coast and the CONTINENTAL SHELF, and between the coast and the land, a relationship that is increasingly influenced by anthropological activities.

Long stretches of coastlines are now essentially artificial, with GROYNEs, breakwalls and other engineering structures (Plate 26). These structures

Plate 25 Crowded beach on the Costa del Sol, southern Spain

Plate 26 Groynes on Gold Beach, Normandy, France

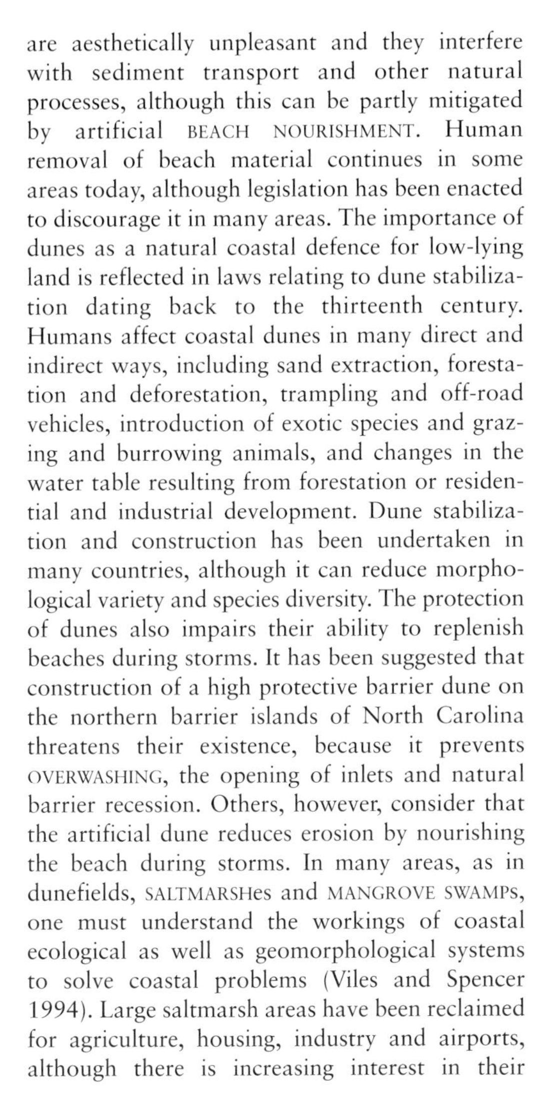

are aesthetically unpleasant and they interfere with sediment transport and other natural processes, although this can be partly mitigated by artificial BEACH NOURISHMENT. Human removal of beach material continues in some areas today, although legislation has been enacted to discourage it in many areas. The importance of dunes as a natural coastal defence for low-lying land is reflected in laws relating to dune stabilization dating back to the thirteenth century. Humans affect coastal dunes in many direct and indirect ways, including sand extraction, forestation and deforestation, trampling and off-road vehicles, introduction of exotic species and grazing and burrowing animals, and changes in the water table resulting from forestation or residential and industrial development. Dune stabilization and construction has been undertaken in many countries, although it can reduce morphological variety and species diversity. The protection of dunes also impairs their ability to replenish beaches during storms. It has been suggested that construction of a high protective barrier dune on the northern barrier islands of North Carolina threatens their existence, because it prevents OVERWASHING, the opening of inlets and natural barrier recession. Others, however, consider that the artificial dune reduces erosion by nourishing the beach during storms. In many areas, as in dunefields, SALTMARSHes and MANGROVE SWAMPs, one must understand the workings of coastal ecological as well as geomorphological systems to solve coastal problems (Viles and Spencer 1994). Large saltmarsh areas have been reclaimed for agriculture, housing, industry and airports, although there is increasing interest in their preservation with the recognition that they are important and productive ecosystems. Human activities, including deforestation for rice paddies, fuel, construction materials and industrial uses, are continuing to cause irreversible damage to coastal mangroves in tropical regions, however, where there is often little appreciation of their value to native populations. Estuarine dynamics and siltation patterns are being affected by deforestation, mining and quarrying, urbanization, DAM construction, sewage discharge, dredging, dock and marina construction, the reclamation of TIDAL DELTAs, flats and marshes and the diversion of water from one watershed into another. Although much human activity is deleterious to deltas, deforestation, agricultural intensification and extensive soil erosion have sometimes been responsible for their formation or growth. Many deltas are receiving less water and sediment as rivers are dammed for irrigation, flood control and power generation. Much of the loss of wetlands in the Mississippi Delta has natural causes, but it is being exacerbated by deforestation of the drainage basin and dam construction, the building of LEVEEs and other attempts to confine and control the Mississippi River for navigation and flood control. Human activity has been modifying the Nile Delta since predynastic time, but, with construction of the Aswan Dams, almost no fluvial sediment now reaches the delta, and this has resulted in accelerated coastal erosion and marine encroachment. Coral communities and reefs are also threatened by a variety of human activities, including dredging, mining, land clearance, effluents from desalination, sewage discharge, the use of chlorine bleach and explosives for fishing, nuclear weapon testing, oil, chemical and sewerage pollution, thermal pollution from electrical generating stations, careless anchoring, boat grounding and the collection of precious corals and other marine organisms. The Great Barrier Reef Marine Park in Australia was created to manage reefs comprehensively, but economic pressures are more severe in developing areas, and conservation policies more difficult to enforce.

Global warming

One of the greatest challenges facing coastal populations will be to plan for, and manage, the effects of rising sea level resulting from global warming. There is continuing debate over the rate and magnitude of the changes that are to be expected, however, although there has been a trend towards progressively more conservative predictions of sea-level rise in this century. The 2001 third assessment report of a working group for the intergovernmental panel on climatic change (IPCC) has concluded that sea level will rise by between 0.09 and 0.88 m between 1990 and 2100.

Global warming and rising sea level will cause tidal flooding and the intrusion of salt water into rivers, estuaries (see ESTUARY) and groundwater, and it will affect tidal range, oceanic currents, upwelling patterns, salinity levels, biological processes, runoff and landmass erosion patterns. Increasing rates of erosion will make cliffs more susceptible to falls, landslides and other MASS MOVEMENTS, exacerbating problems where loose or weak materials are already experiencing rapid recession. Nevertheless, the effect of rising sea level will vary around the world according to the characteristics of the coast, including its slope, wave climate, tidal regime and susceptibility to erosion.

It has been estimated that about half the world's population lives in vulnerable coastal lowlands, subsiding RIVER DELTAS and river floodplains. The effects of climatic change will be particularly acute in these densely populated regions. It is often the rate of sea-level change rather than the absolute amount that determines whether natural systems, such as coastal marshes and CORAL REEFS, can successfully adapt to changing conditions. Human and natural systems can adjust to slowly changing mean climatic conditions, but it is more difficult to accommodate changes in the occurrence of extreme events. It is not yet known, however, whether higher sea temperatures will increase the frequency and intensity of tropical storms and spread their influence further polewards, or whether higher temperature gradients between land and sea will increase the intensity of monsoons and affect their timing.

Human responses to the rise in sea level will depend upon available resources and the value of the land being threatened. High waterfront values will justify economic expenditure to combat rising sea level in cities, but less attention is likely to be paid to the deleterious effects on saltmarshes, mangroves, coral reefs, lagoons and ice-infested Arctic coasts. The decision-making process associated with coastal erosion and flooding is complex, because of constraints imposed by financial considerations and a myriad of physical, social, economic, legal, political and aesthetic factors. There is public and political pressure on coastal planners and managers to be seen to be doing something about the problem, and this can result in engineering projects that provide only short-term benefits, or which may even exacerbate the original problem. Several managerial options are available, however, ranging from the 'do nothing' approach, to the construction of a completely artificial coast.

References

Davies, J.L. (1972) *Geographical Variation in Coastal Development*, Edinburgh: Oliver and Boyd.
Inman, D.L. and Nordstrom, C.E. (1971) On the tectonic and morphologic classification of coasts, *Journal of Geology* 79, 1–21.
Viles, H. and Spencer, T. (1994) *Coastal Problems*, Oxford: Edward Arnold.

Further reading

Carter, R.W.G. (1988) *Coastal Environments*, London: Academic Press.
Carter, R.W.G. and Woodroffe, C.D. (1994) *Coastal Evolution*, Cambridge: Cambridge University Press.
Lakhan, V.C. and Trenhaile, A.S. (eds) (1989) *Applications in Coastal Modelling*, Amsterdam: Elsevier.
Trenhaile, A.S. (1997) *Coastal Dynamics and Landforms*, Oxford: Oxford University Press.

ALAN TRENHAILE

COHESION

The force by which particles are able to stick together. Cohesion is important in soil mechanics, as it is one of two parameters (alongside the angle of internal friction) that characterize a soil's resistance to an applied stress (though the two parameters are not always independent of each other). Soils with high levels of cohesion (termed cohesive soils) commonly contain a significant amount of clay, which are able to cement the soil internally (yet these typically have low frictional strength). Conversely, dry sand is termed non-cohesive (as particles are easily moved in isolation), with the only resistance to shear coming from the internal friction of sand particles. When sand is moist (though unsaturated) the surface tension of the water menisci between the grains provides an apparent cohesiveness to the sand. This is removed when the sand either dries or becomes saturated. Rocks are commonly high in both parameters. Cohesion becomes proportionately stronger as grain size decreases, allowing fine grain sediments (muds and silts, etc.) to remain stable on high-angle slopes.

Reference

Bullock, M.S., Kemper, W.D. and Nelson, S.D. (1988) Soil cohesion as affected by freezing, water content, time and tillage. *Soil Science Society of America Journal* 52(3), 770–776.

SEE ALSO: adhesion

STEVE WARD

COLLUVIUM

Sedimentary material that has been transported across and deposited on slopes as a result of mass movement processes and soil wash. It is frequently derived from the erosion of weathered bedrock (eluvium) and its deposition on low-angle surfaces, and can be differentiated from material which is deposited primarily by fluvial agency (alluvium). Colluvium can be many metres thick and can infill bedrock depressions (Crozier *et al.* 1990). It often contains palaeosols, which represent halts in deposition, crude bedding downslope, and a large range of grain sizes and fabrics (Bertram *et al.* 1997). Cut-and-fill structures may represent phases when stream incision has been more important than colluvial deposition (Price-Williams *et al.* 1982).

Colluvium may provide a rich record of long-term climatic change (see, for example, Nemec and Kazanci 1999), preserve archaeological materials, indicate phases of accelerated anthropogenic soil erosion during the Holocene and act as a medium into which gullies may be incised (see DONGA).

Colluvial deposits are known from almost all climatic zones from former glacial (Blikra and Nemec 1998) and periglacial environments (Mason and Knox 1997) through to the tropics (Thomas 1994).

References

Bertram, P., Hetu, B., Texier, J.P. and van Steijn, H. (1997) Fabric characteristics of subaerial slope deposits, *Sedimentology* 44, 1–16.

Blikra, L.H. and Nemec, W. (1998) Postglacial colluvium in western Norway: depositional processes, facies and palaeoclimatic record, *Sedimentology* 45, 909–959.

Crozier, M.J., Vaughn, C.E. and Tippett, J.M. (1990) Relative instability of colluvium-filled bedrock depressions, *Earth Surface Processes and Landforms* 15, 329–339.

Mason, J.A. and Knox, J.C. (1997) Age of colluvium indicates accelerated late Wisconsinian hillslope erosion in the Upper Mississippi Valley, *Geology* 25, 267–270.

Nemec, W. and Kazanci, N. (1999) Quaternary colluvium in west-central Anatolia: sedimentary facies and palaeoclimatic significance, *Sedimentology* 46, 139–170.

Price-Williams, D., Watson, A. and Goudie, A. (1982) Quaternary colluvial stratigraphy, archaeological sequences and palaeoenvironment in Swaziland, Southern Africa, *Geological Journal* 148, 50–67.

Thomas, M.F. (1994) *Geomorphology in the Tropics: A Study of Weathering and Denudation in Low Latitudes*, Chichester: Wiley.

A.S. GOUDIE

COMMINUTION

Refers to the reduction of rock debris to fine powder or to small pieces. In nature, comminution is usually as a result of ABRASION and attrition, and is often linked with problems of coastal erosion due to reduction of shingle.

Further reading

Kabo, M., Goldsmith, W. and Sackman, J.L. (1977) Impact and comminution processes in soft and hard rock, *Rock Mechanics*, Supplementum 9(4), 213–243.

STEVE WARD

COMPACTION OF SOIL

The term compaction refers to a progressive decrease in the volume of a soil element over time, resulting in an increase in density. Recently deposited sediments tend to exhibit a progressive increase in density over time, as consolidation occurs due to self weight and loads imposed by overlying sediment. A commonly used measure of the relative degree of compaction of a soil within engineering soil mechanics is the overconsolidation ratio: $OCR = \sigma'_{max} / \sigma'_{pres}$. Here σ'_{max} refers to the maximum normal EFFECTIVE STRESS which the soil material has experienced over geologic time, while σ'_{pres} is the present-day normal effective stress. Effective stress is defined as total stress minus ambient PORE-WATER PRESSURE (Barnes 2000). Normally consolidated (NC) soils have $\sigma'_{pres} \approx \sigma'_{max}$, and include most postglacial fluvial and colluvial sediments. Overconsolidated (OC) soils have $\sigma'_{max} \gg \sigma'_{pres}$, and include basal tills and geological strata such as clays and shales which have experienced normal stress reduction caused by erosion of superjacent materials. There are large ranges of OCR from approximately 1.0 to several 100, depending on the history of load changes that the soil has experienced. A transient condition, known as underconsolidation, refers to effective stress below the NC value. This is possible where part or all of the total overburden pressure is borne by the pore fluid, and thus positive excess pore pressures prevail shortly after deposition. It is common where fine-grained, saturated materials are deposited rapidly as QUICKCLAY earthflows or muddy DEBRIS FLOWS. Underconsolidation may also occur where formerly submerged muds become abruptly subaerial, due to either rapid tectonic uplift or lake drainage.

Although the rate of consolidation is controlled strongly by the normal stresses imposed by external loads, soil compaction also varies according to the compressibility of the soil particles themselves, the water content, and the hydraulic conductivity (Barnes 2000). In unsaturated soils, having a high air content, rate of consolidation is controlled primarily by the compressibility of the soil matrix, which is a function of particle shape, sorting, and mineralogy. In saturated soils, rate of consolidation is regulated by soil hydraulic conductivity, since expulsion of virtually incompressible pore fluid is a prerequisite for consolidation. Conductivity varies by several orders of magnitude, depending on particle size and *in situ* density.

Within the normally consolidated class of soils, which comprise many soils worldwide, significant variations in ambient *in situ* density occur as a result of both geomorphic and sedimentological factors. Mixed, poorly sorted materials, such as LANDSLIDE deposits, often possess a relatively high *in situ* density since a wide range of particle sizes ensures that voids between large clasts are filled with finer material (Bement and Selby 1997). It is possible that natural, vibration-induced compaction of rapidly emplaced landslide materials further enhances densification. By contrast, very well-sorted aeolian materials, such as LOESS and DUNE sand, exhibit a much lower *in situ* density, especially if fairly equant grains are dominant in the deposit. Such soils are inherently very compressible.

In the near-surface zone, the effects of geological consolidation are periodically offset by MECHANICAL WEATHERING processes, which lead to a volume increase, and hence a density decrease, relative to that of the unweathered material below. By contrast, the amount of net volume increase brought about by CHEMICAL WEATHERING appears to be slight (Birkeland 1984). In cold regions, FREEZE–THAW CYCLE processes cause seasonal and shorter term cycles of heave and settlement. Thaw and consolidation of the ACTIVE LAYER during spring and summer may produce transient excess pore pressures if the water generated by ice lens melting is slow to escape. This may be due to either a low material conductivity or the existence of an impermeable PERMAFROST table (Williams and Smith 1989). Thaw-consolidation has been credited with the development of very low effective stresses within a thawing active layer, allowing SOLIFLUCTION lobes to move on slope angles as low as $\frac{1}{4}\ \phi'_r$, where ϕ'_r is the residual angle of shearing resistance. Cycles of HYDRATION and dehydration also produce appreciable cyclical volume changes, especially in soils containing montmorillonite clays. However, the magnitudes of the resultant cyclical volume changes are generally far lower than the values attained within seasonally ice-rich sediments.

Rainfall impact, together with infiltration seepage, is also a well-documented soil compacting process, especially in semi-arid environments where it leads to the development of a surface crust of reduced infiltrability. The widespread conversion of grassland and forest soils to arable use has caused significant rainfall compaction of soil, causing reduced infiltrability, and hence accelerated runoff (see RUNOFF GENERATION) and EROSION (Morgan *et al.* 1998). In arable areas, such compaction may be rectified by ploughing and harrowing. In time, uncultivated near-surface soil becomes naturally loosened again by the combined effects of freeze–thaw cycles, bioturbation from soil micro- and macrofauna, in addition to root growth and decay, and downward mixing of low density organic material.

Several problem soils have been identified within engineering soil mechanics based on their poor performance under surcharge stresses or cyclical shear loads. Normally consolidated clays are prone to significant consolidation under structural loads, and may require the placement of fill materials to effect soil consolidation prior to construction (Barnes 2000). NC soils are also more prone to landsliding than are OC materials, since the lesser degree of compaction in the former is generally associated with lower shear strength. A common problem in loess soils is HYDROCOMPACTION (Derbyshire 2001), which involves a localized collapse of soil structure in response to vertical seepage forces. It is a widespread problem where loess is subjected to flood irrigation.

References

Barnes, G.E. (2000) *Soil Mechanics*, 2nd edition, London: Macmillan.

Bement, R.A.P. and Selby, A.R. (1997) Compaction of granular soils by uniform vibration equivalent to vibrodriving of piles, *Geotechnical and Geological Engineering* 15, 121–143.

Birkeland, P.W. (1984) *Soils and Geomorphology*, Oxford: Oxford University Press.

Derbyshire, E. (2001) Geological hazards in loess terrain, with particular reference to the loess regions of China, *Earth Science Reviews* 54, 231–260.

Morgan, R.P.C., Quinton, J.N. *et al.* (1998) The European soil erosion model (EUROSEM): a dynamic approach for predicting sediment transport from fields and catchments, *Earth Surface Processes and Landforms* 23, 527–544.

Williams, P.J. and Smith, M.W. (1989) *The Frozen Earth*, Cambridge: Cambridge University Press.

MICHAEL J. BOVIS

COMPLEX RESPONSE

Landforms respond to the controlling variables of tectonics, sea level, climate and biotic activity over time. They also respond to the changes, rhythms and thresholds of Earth. Available data suggest that over timescales of 10^2 years increases of geomorphological rates of activity change with a frequency of *c.*2,000 years. Over 10^3 years rates and process balances change with a frequency of 30,000–50,000 years and over 10^{4-5} years full system control changes occur every 100,000–150,000 years. Flux in sediment yield and landform adjustment should be regarded as the norm. Regularity of landform may then be the product of polygenetic landform origins. A central proposition of geomorphology, therefore, is that landform change (response) takes place as states of equilibrium, stability or tranquillity are upset by complex episodic changes to the environmental controls. This may be called 'complex cause' (see LANDSCAPE SENSITIVITY).

The *response* to the hierarchy of controls and events also varies on all timescales and are variably distributed in space. Complex response (Schumm 1973, 1975, 1977, 1979, 1981; Schumm and Parker 1973) describes the way in which the internal structure of a system controls the reaction and relaxation of the system after an impulse of change. There are many aspects to be considered: the effect of internal thresholds (see THRESHOLD, GEOMORPHIC) that control sudden change; the fluctuation between cut-and-fill as the capacity of the system dictates temporary storage of eroded sediment; the effect of area as an impulse moves from a point application (e.g. a river mouth base level change), along a sensitive linear pathway (e.g. a channel, a joint) to diffuse over a catchment as a wave of erosional aggression moving inland (e.g. from a sea cliff or an incising river). Such changes occur after every effective event and the direction of change follows every structural instability.

Landform 'evolution' is a never-ending set of adjustments to impulses of change on all temporal and spatial scales. It is complex.

References

Schumm, S.A. (1973) Geomorphic thresholds and complex responses of drainage systems, in M. Morisawa (ed.) *Fluvial Geomorphology*, Binghamton, Publications in Geomorphology 3, 299–310.

——(1975) Episodic erosion: a modification of the geomorphic cycle, in W.N. Melhorn and R.C. Flemal (eds) *Theories of Landform Development*, 69–86, London: George, Allen and Unwin.

——(1977) *The Fluvial System*, Chichester: Wiley.

Schumm, S.A (1979) Geomorphic thresholds: the concept and its applications, *Transactions Institute of British Geographers*, NS 4, 485–515.
——(1981) Evolution and response of the fluvial system, sedimentological implications, *SEPM, Special Publication*, 31, 19–29.
Schumm, S.A. and Parker, R.S. (1973) Implications of complex response of drainage systems for Quaternary alluvial stratigraphy, *Nature* 243, 99–100.

DENYS BRUNSDEN

COMPLEXITY IN GEOMORPHOLOGY

Complexity is a way of describing complicated, irregular patterns that appear random. It is something tangible that is observable in geomorphic systems, such as in turbulent flow in streams. Much chaotic complexity in geomorphology underlies a larger scale geomorphic order, and overlies smaller scale, more orderly and understandable components. Chaotic turbulent flow is part of a larger scale order seen in the predictable rate and direction of mean streamflow; it is also the result of a huge number of well-understood individual particle trajectories describable by the basic laws of physics. Complexity in geomorphic systems is thus often part of a hierarchy of interrelated structures and processes. Similarly, simple geomorphic patterns, such as beach cusps, commonly arise from complex underlying dynamics; at the same time, they are but a part of broader scale complex patterns. Beach cusps result from complex non-linear interactions between beaches and waves or the complicated formation of edge waves (waves trapped at the shoreline by refraction); at the same time, they are a part of irregular coastline geometry.

One line of explanation for complexity rests in non-linear dynamical systems theory, which has revolutionized many branches of science (see Stewart 1997). To understand the general reasoning involved, it may help to define a few terms first. An unstable system is susceptible of small perturbations and is potentially chaotic. A chaotic system behaves in a complex and pseudo-random manner purely because of the way the system components are interrelated, and not because of forcing by external disturbances, or at least independently of those external factors. The equations describing the system generate the chaos, which is deterministic; chance-like (stochastic) events do not. Systems displaying chaotic behaviour through time usually display spatial chaos, too. Therefore, a landscape that starts with a few small perturbations here and there, if subject to chaotic evolution, displays increasing spatial variability as the perturbations grow. This happens when rivers dissect a landscape and relief increases. Self-organization is the tendency of, for example, flat or irregular beds of sand on streambeds or in deserts to organize themselves into regular spaced forms – ripples and dunes – that are rather similar in size and shape. Self-organization also occurs in patterned ground, beach cusps and river channel networks. Self-destruction (non-self-organization) is the tendency of some systems to consume themselves, as when relief is reduced to a plain. An attractor is a system state that controls system changes and into which other system states are drawn.

Many geomorphic systems are complex, but not all are. Some non-linear geomorphic systems are unstable, chaotic and self-organizing, but some are not. Nevertheless, plentiful evidence suggests that complexity is common in geomorphic systems and begs an explanation. The truly puzzling fact is that most geomorphic systems display order and complexity concurrently. Are the complexities (irregularities) merely deviations from an orderly norm, or are they informative in their own right? A growing body of evidence from field studies, laboratory studies, and real-world datasets suggests that in some geomorphic systems complexity is significant in its own right. Signs of complex behaviour in systems include deterministic chaos, instability, increasing variability over time, self-organization, divergence from similar initial conditions and sensitivity to initial conditions (Phillips 1999: 39–57). Evidence exists for all these indicators of complexity.

Several hydrological records, tree rings series and topographic images reveal chaotic patterns. In other cases, field investigations have confirmed chaotic behaviour predicted in models, as in the genesis of Ultisols in eastern North Carolina.

Field examples of dynamical systems' instability and sensitivity to small perturbations abound, including river meander initiation generated by the unstable growth of small flow perturbations.

Some studies demonstrate patterns of spatial variability that become increasingly complex (less uniform) over time: there is a spatial differentiation of the landscape. Desertification appears to involve an increasingly more complex pattern of vegetation and soil-nutrient resources through time.

In some geomorphic systems, orderly self-organizing patterns seem to emerge from complex non-linear dynamics. Field and laboratory work confirms theoretical work showing that sorted nets in non-periglacial environments may develop spontaneously on any piece of unobstructed land with little or no slope, proving it carries a loose and discontinuous cover of pebbles, each of which may move in small steps with equal probability in all directions (Ahnert 1994).

Much field evidence strongly suggests that some geomorphic systems evolve by diverging from the same, or very similar, initial conditions. In the Norfolk marshes, England, vegetated marsh traps more sediment than bare marsh, so reducing the chances of inundation and lowering (or stabilizing) salinity. The bare marsh becomes lower land that traps more water and the salinity rises, inhibiting vegetation colonization and growth.

Several studies indicate that small variations in initial conditions amplify as a geomorphic system evolves. In podzolized soils in Canada, microtopographic variations produce favoured sites for infiltration and 'funnel' effects that eventually create large variations in the thickness of A and B soil horizons (Price 1994).

Related to complexity are the ideas of fractals and self-organized criticality. Fractal landscapes display self-similar patterns repeated across a range of scales. A small section of coastline may be self-similar to a much larger piece of coastline, of which it is part. Drainage networks, sedimentary layers and joint systems in rocks possess fractal patterns. Self-organized criticality is a theory that systems composed of myriad elements will evolve to a critical state, and that once in this state, tiny perturbations may lead to chain reactions that may affect the entire system. The classic example is a pile of sand. Adding grains one by one to a sandbox causes a pile to start growing, the sides of which become increasingly steep. In time, the slope angle becomes critical: one more grain added to the pile triggers an avalanche that fills up empty areas in the sandbox. After adding sufficient grains, the sandbox overflows. When, on average, the number of sand grains entering the pile equals the number of grains leaving the pile, the sand pile has self-organized into a critical state. Landslides, drainage networks, and the magnitude and frequency relations of earthquakes display self-organized criticality.

Phillips (1999) identifies eleven 'principles of Earth surface systems' that follow from theoretical and empirical work on order and complexity in geomorphic systems. Some of these principles appear to conflict, but that is the nature of complexity. In summary, and applied specifically to geomorphic systems, the principles are (see Huggett 2002: 339–41):

1. Geomorphic systems are inherently unstable, chaotic and self-organizing. Many, but definitely not all, geomorphic systems display a tendency to diverge or to become more differentiated through time in some places and at some times, as when an initially uniform mass of weathered rock or sediment develops distinct horizons.
2. Geomorphic systems are inherently orderly. Deterministic chaos in a geomorphic system is governed by an >attractor = that constrains the possible states of the system. Such a geomorphic system displays dynamic instability but does not behave randomly. The dynamic instability has bounds. Beyond these bounds, orderly patterns emerge that include the chaotic patterns inside them. Thus, even a chaotic system must exhibit order at certain scales or under certain circumstances. For example, at local scales, soil formation is sometimes chaotic, with giant spatial variations in soil properties; as the scale is increased, regular soil–landscape relationships emerge.
3. Order and complexity are emergent properties of geomorphic systems. This principle means that, as the spatial or temporal scale is altered, orderly, regular, stable, and non-chaotic patterns and behaviours and irregular, unstable, and chaotic patterns and behaviours appear and disappear. In debris flows, deterministic chaos governs collisions between particles where the flow is highly sheared and the collisions are sensitive to initial conditions and unpredictable. However, the bulk behaviour of granular flows is orderly and predictable from a relationship between kinetic energy (drop height) and travel length. Therefore, the behaviour of a couple of particles is perfectly predictable from basic physical principles; a collection of particles interacting with each other is chaotic; and the aggregate behaviour of the flow at a still broader scale is again predictable.
4. Geomorphic systems have both self-organizing and non-self-organizing modes. This principle

follows from the first three principles. Some geomorphic systems may operate in self-organizing and non-self-organizing modes at the same time. The evolution of topography, for example, may be self-organizing where relief increases, and self-destructing where relief decreases. Mass wasting denudation is a self-destructing process that homogenizes landscapes by decreasing relief and causing elevations to converge. Dissection is a self-organizing process that increases relief and causes elevations to diverge.

5. Both unstable–chaotic and stable–non-chaotic features may coexist in the same landscape at the same time. Because a geomorphic system may operate in either mode, different locations in the system may display different modes simultaneously. This is the idea of >complex response =, in which different parts of a system respond differently at a given time to the same stimulus. An example is channel incision in headwater tributaries occurring concurrently with valley aggradation in trunk streams.
6. Simultaneous order and disorder, observed in real landscapes, may be explained by a view of Earth surface systems as complex non-linear dynamical systems. They may also arise from stochastic forcings and environmental processes.
7. The tendency of small perturbations to persist and grow over finite times and spaces is an inevitable outcome of geomorphic system dynamics. In other words, small changes are sometimes self-reinforcing and lead to big changes. Examples are the growth of nivation hollows and dolines. An understanding of non-linear dynamics helps to determine the circumstances under which some small changes grow and others do not.
8. Geomorphic systems do not necessarily evolve towards increasing complexity. This principle arises from the previous principles and particularly from Principle 4. Geomorphic systems may become more complex or simpler at any given scale, and may do either at a given time.
9. Neither stable, self-destructing nor unstable, self-organizing evolutionary pathways can continue indefinitely in geomorphic systems. No geomorphic system changes ad infinitum. Stable development implies convergence that eventually leads to a lack of differentiation in space or time, as when different elevations in a landscape converge to form a plain. Disturbances disrupt such stable states by reconfiguring the system and resetting the geomorphic clock. Divergent evolution is also self-limiting. For example, base levels ultimately limit landscape dissection.
10. Environmental processes and controls operating at distinctly different spatial and temporal scales are independent. For example, processes of wind transport are effectively independent of tectonic processes, although there are surely remote links between them.
11. Scale independence is a function of the relative rates, frequencies and durations of geomorphic phenomena.

References

Ahnert, F. (1994) Modelling the development of non-periglacial sorted nets, *Catena* 23, 43–63.

Huggett, R.J. (2002) *Fundamentals of Geomorphology*, London: Routledge.

Phillips, J.D. (1999) *Earth Surface Systems: Complexity, Order, and Scale*, Oxford: Blackwell.

Price, A.G. (1994) Measurement and variability of physical properties and soil water distribution in a forest podzol, *Journal of Hydrology* 161, 347–364.

Stewart, I. (1997) *Does God Play Dice? The New Mathematics of Chaos*, new edn, Harmondsworth: Penguin Books.

Further reading

Culling, W.E.H. (1988) A new view of the landscape, *Transactions of the Institute of British Geographers*, New Series 13, 345–360.

Hergarten, S. and Neugebauer, H.J. (2001) Self-organized critical drainage, *Physical Review Letters* 86, 2,689–2,692.

Phillips, J.D. (1999) Divergence, convergence, and self-organization in landscapes, *Annals of the Association of American Geographers* 89, 466–488.

——(2000) Signatures of divergence and self-organization in soils and weathering profiles, *Journal of Geology* 108, 91–102.

Richards, A.E. (2002a) Complexity in physical geography, *Geography* 87, 99–107.

Richards, A.E., Phipps, P. and Lucas, N. (2000) Possible evidence for underlying non-linear dynamics in steep-faced glaciodeltaic progradational successions, *Earth Surface Processes and Landforms* 25, 1,181–1,200.

Xu, T., Moore, I.D., and Gallant, J.C. (1993) Fractals, fractal dimensions and landscapes – a review, *Geomorphology* 8, 245–262.

RICHARD HUGGETT

COMPUTATIONAL FLUID DYNAMICS (CFD)

Fluid motions play a central role in sculpting a great variety of landforms, both terrestrial and submarine. Examples range from river channels to aeolian dunes to barrier islands. Naturally, investigations into landform origins often involve applications of fluid dynamics. Fluid dynamics is that branch of mechanics that concerns the physics of fluid motion. The motion of a non-turbulent, Newtonian fluid is described by the Navier–Stokes equations. These equations express continuity of mass and momentum in three dimensions in a continuum fluid subject to gravitational, inertial, viscous and pressure forces. The equations take on different forms depending on whether the fluid is compressible (e.g. air) or incompressible (e.g. water to a close approximation). Normally the Navier–Stokes equations are combined with models of turbulence (for application to turbulent flows) and with models of boundary friction (for any flows involving contact with a surface, such as a channel bed). Except in special cases, the Navier–Stokes equations cannot be analytically solved. Their solution can, however, be approximated using numerical methods (see, e.g. Cheney and Kincaid 1999; Press *et al.* 1993) combined with a set of specified initial and boundary conditions. Such methods involve dividing up space and time into discrete elements, within which the variables of interest – such as velocity and pressure – are either interpolated or held constant. The development of numerical solution methods for different types of equations, including fluid flow equations, is a major area of research in the fields of mathematics and computing science. Numerical solutions of equations for fluid motion can be quite computationally intensive, and the computer models that implement these solutions are referred to as Computational Fluid Dynamics (CFD) models. Depending on the methods used and the degree of approximation involved, the computer codes can be quite complex, and there are many commercially available packages as well as research codes developed within universities. Applications of CFD are increasingly widespread in geomorphology. CFD models have been of great benefit, for example in understanding interactions between fluid flow, bed morphology and sediment transport in river channels (e.g. Hankin *et al.* 2002; Lane *et al.* 2002; Ma *et al.* 2002). CFD models of airflow dynamics have been used to understand the interactions between airflow and dune morphology (e.g. Walmsley-John and Howard 1985). Other applications have been wide-ranging; examples include water flow in karst conduits (e.g. Hauns *et al.* 2001), circulation and sediment movement in ancient epeiric seas (e.g. Slingerland *et al.* 1996), coastal morphology (e.g. Deigaard and Fredsoe 2001) and paleoflood hydrology (e.g. House and Baker 2001).

References

Cheney, W. and Kincaid, D. (1999) *Numerical Mathematics and Computing*, 4th edition, Pacific Grove, CA: Brooks/Cole.

Deigaard, R. and Fredsoe, J. (2001) The use of numerical models in coastal hydrodynamics and morphology, in G. Seminara, and P. Blondeaux (eds) *River, Coastal and Estuarine Morphodynamics*, 61–92, Berlin: Springer-Verlag.

Hankin, B.G., Holland, M.J., Beven, K.J. and Carling, P. (2002) Computational fluid dynamics modelling of flow and energy fluxes for a natural fluvial dead zone, *Journal of Hydraulic Research* 40(4), 389–402.

Hauns, M., Jeannin, P-Y. and Atteia, O. (2001) Dispersion, retardation, and scale effect in tracer breakthrough curves in karst conduits, *Journal of Hydrology* 241(3–4), 177–193.

House, P.K. and Baker, V.R. (2001) Paleohydrology of flash floods in small desert watersheds in western Arizona, *Water Resources Research* 37(6), 1,825–1,839.

Lane, S.N., Hardy, R.J., Elliot, L. and Ingham, D.B. (2002) High-resolution numerical modelling of three-dimensional flows over complex river bed topography, *Hydrological Processes* 16(11), 2,261–2,272.

Ma, L., Ashworth, P.J., Best, J.L., Elliot, L., Ingham, D.B. and Whitcombe, L.J. (2002) Computational fluid dynamics and the physical modelling of an upland urban river, *Geomorphology* 44(3–4), 375–391.

Press, W.H., Flannery, B.P., Teukolsky, S.A. and Vetterling, W.T. (1993) *Numerical Recipes in C*, 2nd edition, Cambridge: Cambridge University Press.

Slingerland, R., Kump, L.R., Arthur, M., Fawcett, P., Sageman, B. and Barron, E. (1996) Estuarine circulation in the Turonian Western Interior Seaway of North America, *Geological Society of America Bulletin* 108, 941–952.

Walmsley-John, L. and Howard, Alan D. (1985) Application of a boundary-layer model to flow over an eolian dune, *Journal of Geophysical Research, D, Atmospheres* 90(6), 10,631–10,640.

Further reading

Feynman, R.P., Leighton, R.B. and Sands, M.L. (1989) *The Feynman Lectures on Physics: Commemorative Issue* (3 volume set), Redwood City, CA: Addison-Wesley.

White, F.M. (1998) *Fluid Mechanics*, 4th edition, Boston, MA: McGraw-Hill.

GREG TUCKER

CONCHOIDAL FRACTURE

A smoothly curved fracture, marked by concentric rings and resembling a bi-valve shell in shape. Conchoidal fractures are the most common type of fracture, and are also known as clamshell fractures. They occur when bonds between atoms are approximately the same in all directions within a mineral, and result in breakage along smooth, curved surfaces. Conchoidal fractures occur particularly in amorphous materials (i.e. those showing no definite crystalline structure) such as obsidian, and are also common in quartz, chert and glass.

Further reading

Atkinson, B.K. (1987) *Fracture Mechanics of Rock*, London: Academic Geology Series, Academic Press.

STEVE WARD

CONFLUENCE, CHANNEL AND RIVER JUNCTION

River channel confluences, the sites at which two open channels combine, are ubiquitous features of all river networks and channel patterns. These sites mark nodes of significant change in hydraulic geometry (Richards 1980), flow and sediment discharge, and are characterized by a complex three-dimensional flow field and variable bed geometry (Mosley 1976; Best 1988; Bradbrook *et al.* 2000; Rhoads and Sukhodolov 2001). River channel junctions are often points of significant bed scour (e.g. Best and Ashworth 1997), and are critical in considerations of sediment/pollutant dispersal and mixing in channel networks (Plate 27). Study of these complex fluvial sites has progressed through field, physical and numerical modelling and has identified five principal controls on flow, sediment transport and bed morphology at channel confluences: (1) the angle of convergence between the confluent channels; (2) the ratio of discharge, or flow momentum, between the incoming channels; (3) the planform shapes of the junction (for instance, 'Y' or '⊥' shaped) and upstream channels (i.e. curved, straight; single, multiple); (4) the presence of any depth differential between the incoming channels; and (5) the relative roughness of the confluence (ratio of flow depth to grain size), with the hydrodynamic influence of the particles beginning to dominate at larger grain sizes (e.g. Roy *et al.* 1988).

Plate 27 Junction of the Paraná and Paraguay Rivers, Argentina. The Paraguay River enters from the right and is picked out by its higher suspended sediment concentration. The shear layer between the two rivers displays a series of large vortices and the mixing layer remains distinct for many tens of kilometres downstream. Width of Paraguay River inflow at confluence ~1 km

Fluid dynamics

Channel confluences are zones of complex, three-dimensional, turbulent flow where significant local flow acceleration and deceleration may occur due to both the increased combined fluid discharge and the specific fluid dynamics of the confluence region. Experimental, field and numerical studies have shown confluences to be dominated by seven fluid dynamic zones (Figure 26).

1. A zone of flow stagnation near the upstream junction corner; this fluid deceleration is caused by turning and hence centrifugal forcing of the flows as they approach the junction, together with the influence of a pressure gradient within the junction that is generated by water surface superelevation in the junction centre.

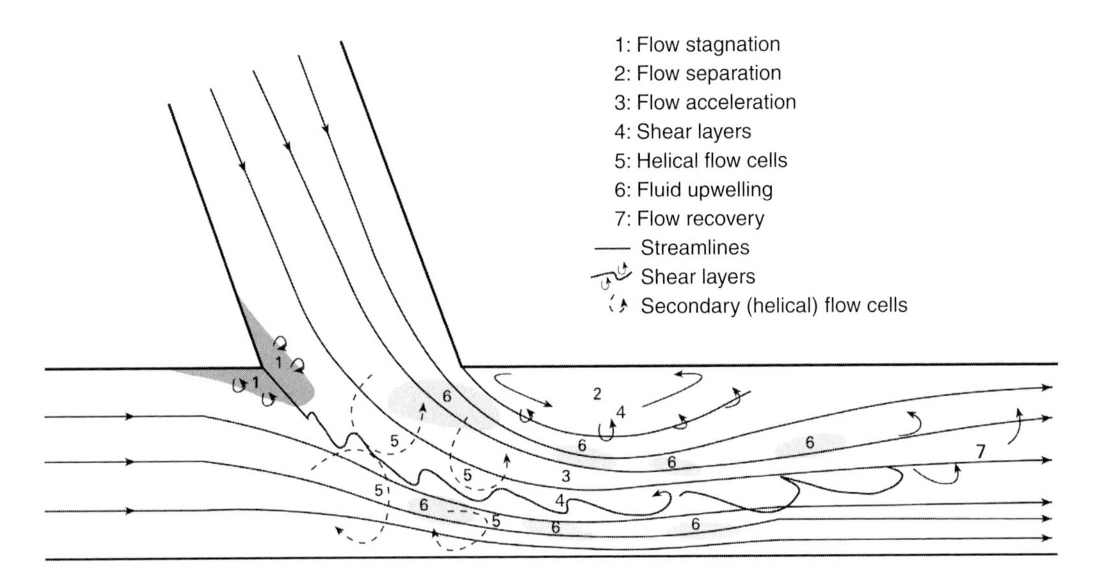

Figure 26 Schematic diagram of the seven principal fluid dynamic zones that may be present at channel confluences

2 A region of flow separation may occur downstream from the downstream junction corner(s); flow cannot remain attached to the boundary at sudden changes in geometry, and an adverse hydrostatic pressure gradient here causes the flow to separate from the wall and form a region of slow, recirculating flow. In symmetrical confluences, the downstream separation zone may form on both sides of the junction. In an asymmetric confluence, the downstream separation zone may only form on the angled (i.e. tributary) side of the junction. The size of the downstream separation zone(s) increase(s) with junction angle and tributary discharge (Best 1988; Bradbrook *et al.* 2000), but may be modified/absent at natural junctions where the angle of bank divergence at the downstream junction corner(s) may be modified by formation of a point bar through sediment deposition (e.g. Rhoads and Sukhodolov 2001) and/or bank erosion.

3 A region of flow acceleration forms at the centre of the confluence that is generated by both the increased fluid discharge passing through the junction (see streamline convergence, Figure 26) and also the constricting influence of any region of flow separation.

4 Distinct shear layers are generated along regions where velocity gradients are severe. Thus, shear layers can be present on either side of the flow stagnation region, along the mixing interface between the two joining flows, bounding any regions of flow separation and also arising from any steep changes in bed topography (such as the avalanche faces that may dip into the central scour – see below). Large, turbulent and 3D flow structures arising along these shear layers, termed Kelvin–Helmholtz instabilities, may give rise to high turbulent shear stresses that are influential in fluid mixing and sediment transport.

5 Helical flow may develop within the junction due to the presence of streamline curvature (Figure 26; streamlines are lines drawn in the fluid of which the tangent at any point is the direction of velocity at that point) and water surface superelevation within the confluence. In ideal cases, where the tributaries are near symmetrical and have equal flow momentum, these secondary flows may be expressed as surface convergent, bed divergent flows much as in placing two meanders back to back, although the duration of streamline curvature through the bend means that it is unlikely that an entire helix is ever completed. The presence of both flow separation at the junction corner, changing pressure gradients or flow separation associated with bed

topography ('topographic forcing' of the flow) or a depth differential between the two incoming tributaries, may lessen the effects or destroy such large-scale secondary flows. Additionally, the time-averaged picture of a series of individual turbulent events, such as fluid upwelling in the confluence, may be manifested as apparent secondary circulation (Lane *et al.* 2000).

6 Regions of distinct fluid upwelling may be generated by both distortion of the shear layer and flow associated with bed topography, such as where the beds of the tributaries are discordant in their height at the junction. This may encourage upwelling of one stream into the other, thus greatly increasing the rate of mixing at the junction (Gaudet and Roy 1995).
7 Finally, a region of flow recovery downstream of the confluence has been observed. This is where the effects of the junction lessen and flow returns to a more uniform cross-stream distribution. However, the flows may remain unmixed for many channel widths downstream if the velocity differential across the shear layer is minimal and the local turbulence at the junction does not mix the flows (see Plate 27).

Bed morphology

The topography of river channel confluences is often characterized by four distinct elements. First, a central scour hole is often present whose orientation approximately bisects the junction angle. The depth of scour increases at both higher junction angles and momentum ratios, and some of the largest alluvial scours are found at these sites. Scours at junctions may reach between two and ten times the depth of the upstream confluent channels: for instance, scour depth at the confluence between the Ganges and Jamuna (Brahmaputra) Rivers in Bangladesh has been recorded as up to 30 m below the upstream bed level in the confluent channels (Best and Ashworth 1997). The position and cause of the confluence scour have been related to (a) flow acceleration in the centre of the confluence (e.g. Roy *et al.* 1988); (b) the influence of turbulence along the shear layer between the flows; (c) downwelling created by secondary flows that may cause higher momentum fluid to be transferred towards the bed at the centre of the confluence; and (d) the differential routing of sediment around the scour.

Second, tributary mouth bars have been observed that terminate at the junction. These bars may possess a steep slipface that dips into the central scour, although the angle of this surface can range from only a few degrees to angle-of-repose for the sediment (~20°–35°). The position of these faces is controlled by the momentum ratio between the confluent channels, with tributary mouth bars migrating further into the junction as the discharge from that channel becomes a greater fraction of the combined confluence flow.

Third, bars may form within regions of flow separation downstream of the junction corners. Flow separation provides a low velocity region into which sediment can accumulate and these bars may show appreciable fining of sediment since only the finer grained sediment can be entrained into this area. Accumulation of sediment in this region will alter the velocity and pressure gradients within this zone and may lead to a lessening of the extent and influence of flow separation.

Finally, mid-channel bars may form in the region of flow deceleration downstream of the junction scour, especially in 'Y'-shaped junctions, or where sediment delivery is high, and they may mark regions of deposition of sediment eroded at the junction scour. Ferguson (1993) has identified the confluence–diffluence unit as a fundamental braided river building block, in which confluence scour creates the sediment that, as the channel widens to cope with the increased discharge, encourages mid-channel bar development and diffluence formation. However, little study has been conducted on sediment transport through confluences, although experimental work suggests that the bed scour may be a zone of reduced transport rates and that sediment may be routed around and not through the scour. This also reflects the streamline pattern within the junction (Figure 26) and the influence of both shear layers and secondary flows within the confluence.

References

Best, J.L. (1988) Sediment transport and bed morphology at river channel confluences, *Sedimentology* 35, 481–498.

Best, J.L. and Ashworth, P.J. (1997) Scour in large braided rivers and the recognition of sequence stratigraphic boundaries, *Nature* 387, 275–277.

Bradbrook, K.F., Lane, S.N. and Richards, K.S. (2000) Numerical simulation of three-dimensional, time-averaged flow structure at river channel confluences, *Water Resources Research* 36, 2,731–2,746.
Ferguson, R.I. (1993) Understanding braiding processes in gravel-bed rivers: progress and unresolved problems, in J.L. Best and C.S. Bristow (eds) *Braided Rivers*, Geological Society of London Special Publication 75, 73–87.
Gaudet, J.M. and Roy, A.G. (1995) Effect of bed morphology on flow mixing length at river confluences, *Nature* 373, 138–139.
Lane, S.N., Bradbrook, K.F., Richards, K.S., Biron, P.M. and Roy, A.G. (2000) Secondary circulation cells in river channel confluences: measurement artefacts or coherent flow structures?, *Hydrological Processes* 14, 2,047–2,071.
Mosley, M.P. (1976) An experimental study of river channel confluences, *Journal of Geology* 84, 535–561.
Rhoads, B.L. and Sukhodolov, A.N. (2001) Field investigation of three-dimensional flow structure at stream confluences: 1: Thermal mixing and time-averaged velocities, *Water Resources Research* 37, 2,393–2,410.
Richards, K.S. (1980) A note on change in geometry at tributary junctions, *Water Resources Research* 16, 241–244.
Roy, A.G., Roy, R. and Bergeron, N. (1988) Hydraulic geometry and changes in flow velocity at a river confluence with coarse bed material, *Earth Surface Processes and Landforms* 13, 583–598.

Further reading

Best, J.L. and Roy, A.G. (1991) Mixing layer distortion at the confluence of channels of different depth, *Nature* 350, 411–413.
Biron, P., Roy, A.G. and Best, J.L. (1996) Turbulent flow structure at concordant and discordant open-channel confluences, *Experiments in Fluids* 21, 437–446.

JIM BEST AND STUART LANE

CONTINENTAL SHELF

The continental shelf generally is defined as the zone adjacent to a continent or around an island that is between the shoreline and a noticeable break in slope, the shelf break, to the steeper continental slope or, where there is no break in slope, to a depth of about 200 m. Along with the coastal plain, continental slope and continental rise, the shelf is considered part of the continental margin (Gary *et al.* 1972: 153) and usually is synonymous with the term continental platform (Baker *et al.* 1966: 38). The division of the shelf into the inner, mid (occasionally) and outer continental shelf is arbitrary and may be based on logical or scientific criteria, such as the depth to which waves agitate the seafloor, or by legal criteria, such as the geographic limit of jurisdiction by a government. In many regions, the continental shelf is physically continuous with the coastal plain; the separation between the two being the location of the shoreline. The dynamic nature of the continental shelf is symbolized by the active character of the shoreline which moves laterally and vertically in spatial and temporal scales that vary by orders of magnitude.

The shelf is important for several reasons. It is a zone of many physical and biological transitions from oceanic to terrestrial conditions and processes. As everything that moves from land to the ocean must pass across the continental shelf, the suite of processes acting on the shelf is vitally important. The shelves are regions of abundant biological activity as there are substantial supplies of nutrients from both upwelling and upland runoff and there generally is good light penetration. Finally, the continental shelves are sites of major economic interest ranging from the commercial and recreational fisheries of the shelf waters to the sands, gravels and other hard minerals of the surficial sediments to oil and gas that have formed from included biotic sediments and accumulated in any of several types of traps within the body of the shelf.

At the smallest scales, the shoreline and, hence, the boundary between the coastal plain and continental shelf shifts within seconds and hours in response to waves and tides. While, probably more importantly, the multi-millennial, glacial-eustatic SEA-LEVEL changes during the Quaternary have moved the shoreline several tens of kilometres laterally and a hundred or so metres vertically. Furthermore, consequences of local or regional tectonic activity, which usually is spasmodic, are additive to eustatic trends. The presence, or absence, and cause of the tectonics contribute to the overall form of the shelf. The proximity of the continental shelf to the edge of a crustal plate (see PLATE TECTONICS) and the type of inter-plate dynamic play crucial roles in the form and function of the shelf.

Perhaps the least geologically mature shelves are those along convergent plate boundaries and other ACTIVE MARGINs as commonly occur around much of the Pacific Ocean and along the northern shore of the Mediterranean Sea. Although this situation presents a potentially complex and geologically interesting continental margin, the rate of tectonic activity tends to limit the length of time during which marine processes are able to act on a specific body of sediment or location on the continental shelf. However, the same processes that restrict the geographic domain of the shelf result in a rapid, gravity driven flux of material between the often steep and high, near-coastal continental areas and the deep ocean. Milliman and Syvitski (1992) indicate that the small drainage, high relief river systems of active margins contribute a vast quantity of sediment to the ocean basins. Residence time of sediment on the shelf is short and the movement of the sediment across the narrow continental shelf mostly is controlled by oceanographic processes that respond to shelf morphology among other factors. As an example, the zone in which WAVES shoal and resuspend bottom sediments is relatively narrow. This narrowness, in turn, results in sharp gradients in the intensity of wave transformation and related processes.

The contrasting situation is a continental shelf on a PASSIVE MARGIN well removed from a spreading centre, as is the situation along much of the Atlantic coasts of North and South America, Europe and Africa. Such broad, gently sloping continental margins can be significant sites of sediment accumulation over an extensive time. Studies along the east coast of North America indicate a kilometres-thick depositional sequence that began with the filling of early Mesozoic rift valleys (see RIFT VALLEY AND RIFTING) or basins and continues through the present.

Large-scale – many tens of metres – changes in sea level play a major role in the development of the continental shelf. Wright (1995), studying the mid-Atlantic shelf of North America, considers the cumulative time during which any portion of the seafloor potentially is subject to wave energy of sufficient magnitude to agitate the bottom sediments. This zone extends from the shoreline/surf zone offshore to a depth determined wave dynamics and assumptions about the likelihood of specific waves occurring within the area. The width of this zone of bottom agitation primarily is a function of the slope of the shelf surface. The rate of movement of the zone across the shelf is a function of both the rate of sea-level change and the bottom slope. In regions such as that studied by Wright (1995), where sea-level history mainly is governed by eustasy, the determination of the duration of potential bottom activity is comparatively straightforward whereas in areas with a complex history, with major tectonic or glacio-isostatic sea-level component (Kelley *et al.* 1992), the process history is more complex.

In addition to growth by upward or outward sedimentation, other factors can influence the trapping of sediments and the subsequent form of the shelf. Lengthy barrier reefs, shore parallel lines of DIAPIRS, fault blocks or folds can form dams to cross-shelf sediment transport with consequent ponding of sediments. In the situation where the offshore shelf dam has substantial relief and catches a significant quantity of sediment, the mass of the accumulated sediments can trigger isostatic subsidence which results in a deepening of the depositional basin and further trapping of sediment. This seems to have been the case with the growth of the up to 15-km thick Baltimore Canyon Trough which appears to have formed both as the fill in Mesozoic grabens or rift basins and, in places, behind a Jurassic/Cretaceous barrier reef (Schlee 1980).

Several factors including shelf width and slope, rate of change of relative sea level, the availability of sediment and the characteristics of that sediment, and the intensity of physical oceanographic processes determine whether a continental shelf builds laterally or vertically or does not accrete while serving as a conduit for sediment moving from the continent to the deep sea. Similarly, the interaction of the rate and locus of sediment deposition on the shelf with the rate of sea-level rise influences whether an area experiences marine transgression or regression. An understanding of the occurrence and forms of RIVER DELTAS may serve as a surrogate for a similar knowledge of continental shelf growth especially as most deltas grow on or across the shelf. These same factors in combination with others, such as climate, determine the character of sediments that are resident on and within the shelf. Hayes (1967) observed that mud is a major constituent of inner continental shelf sediments offshore of areas with high temperature and high rainfall (strong CHEMICAL WEATHERING), coral is most common in areas with high temperatures, gravel is most common offshore of areas with low temperatures (where MECHANICAL WEATHERING dominates and there is substantial ice transport of large particles), and that rock is abundant in cold areas (perhaps due to scouring of sediments by ice) but is strongly correlated with the slope of the inner shelf.

References

Baker, B.B., Jr, Deebel, W.R. and Geisenderfer, R.D. (eds) (1966) *Glossary of Oceanographic Terms*, Washington, DC: US Naval Oceanographic Office.
Gary, M., McAffee, R., Jr and Wolf, C.L. (eds) (1972) *Glossary of Geology*, Washington, DC: American Geological Institute.
Hayes, M.O. (1967) Relationship between climate and bottom sediment type on the inner continental shelf, *Marine Geology* 5, 111–132.
Kelley, J.T., Dickson, S.M., Belknap, D.F. and Stuckenrath, R., Jr (1992) Sea-level change and Late Quaternary sediment accumulation on the Southern Maine inner continental shelf, in C.H. Fletcher, III and J.F. Wehmiller (eds) *Quaternary Coasts of the United States: Marine and Lacustrine Systems*, Tulsa, OK: SEPM (Society of Sedimentary Geology).
Milliman, J.D. and Syvitski, J.P.M. (1992) Geomorphic/tectonic control of sediment discharge to the ocean: the importance of small mountainous rivers', *Journal of Geology* 100, 525–544.
Schlee, J.S. (1980) Seismic stratigraphy of the Baltimore Canyon Trough, *US Geological Survey Open File Report* 80–1,079.
Wright, L.D. (1995) *Morphodynamics of Inner Continental Shelves*, Boca Raton, FL: CRC Press.

CARL H. HOBBS, III

CONTRIBUTING AREA

In hydrological terms a contributing area is the part of a DRAINAGE BASIN that provides stormwater RUNOFF GENERATION. The link between precipitation input and catchment outflow is largely determined by variability in soil moisture storage and the spatial distribution of contributing areas for surface runoff. Almost all stormwater runoff is generated by surface or near-surface flow processes. Therefore runoff-contributing areas within drainage basins are mainly dominated by subsurface stormflow and OVERLAND FLOW. Two processes can generate overland flow. Infiltration-excess overland flow, occurs when precipitation intensity exceeds the rate of water infiltration into the soil. This process tends to occur in catchments in semi-arid regions where natural vegetation is sparse or where there has been disturbance of the land (e.g. extensive agriculture). The second process is saturation-excess overland flow, which occurs when precipitation falls on a saturated soil surface. During a storm, when antecedent soil-moisture conditions in a catchment are high, the water table may temporarily intersect with the ground surface producing saturation-excess overland flow.

The spatial extent and pattern of runoff-contributing areas is affected by climate, soil and topography. Contributing areas of infiltration-excess overland flow are determined by the interaction of rainfall intensity and soil permeability. The least permeable soils in a basin are the most likely to contribute infiltration-excess overland flow. As rainfall intensity increases, areas with more moderate permeability also may contribute overland flow.

However, at the start of rainfall soil moisture is not evenly distributed but is concentrated in the areas adjacent to perennial water courses and in topographic hollows. Overland flow may be generated by return flow when seepage is concentrated and surface soils become fully saturated. Under these circumstances the water table is high and ground water is in close proximity to the surface. These areas preferentially generate storm runoff so the storm hydrograph peak is generated from a relatively small part of the catchment – the partial contributing area (Betson 1964). This runoff-producing area will expand during the course of a storm.

Figure 27 shows the extent of saturation in a small drainage basin at three stages: pre-storm, mid-storm and late storm. Prior to a storm the area of saturation is preferentially concentrated in hollows and in soils adjacent to stream channels. As the storm progresses the saturated area expands into the hillslope hollows at the channel heads creating saturated overland flow from return flow. This coalesces into stream flow resulting in extension of the channel network. By late storm, channel heads are fully saturated and small perennial streams are flowing. The question as to where channels begin has been addressed in a model by Montgomery and Dietrich (1988) who predict the contributing area required for channel initiation in channel heads generated in landslide hollows. It follows that the areas contributing to runoff in a drainage basin are fairly restricted, occurring mainly at the base of slopes or channel heads where subsurface runoff is at its maximum and groundwater tables are very shallow; where subsurface flow converges in the soil in hillslope hollows; and areas of reduced soil moisture storage.

The importance of the contributing area idea is underpinned by several important hydrological concepts. Betson (1964) developed the concept of partial area storm runoff. This was based on a series of simple mathematical models that used Hortonian infiltration theory to predict the areas contributing to runoff during a storm. The

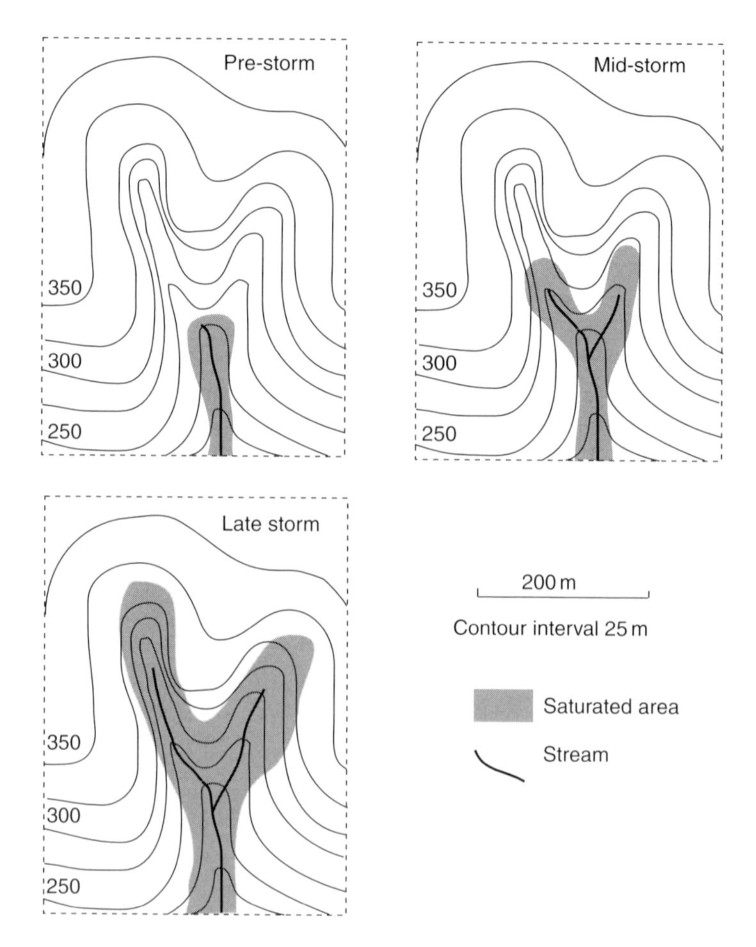

Figure 27 Sequence of expansion of the saturated area of a first-order stream catchment in response to a storm event

equations developed, which can be thought of as functions of apparent watershed infiltration capacity, demonstrate that runoff originates from a small, but relatively consistent, part of the catchment. Using the basic hydrological variables of storm precipitation, storm duration and runoff volume, Betson defined the contributing area as peak stream runoff divided by peak rainfall intensity. This ratio defines the effective runoff-producing area of a drainage basin, which can be expressed as a percentage or proportion of the total catchment area. This is calculated over a storm as total storm runoff divided by total storm rainfall. Typical values are less than 10 per cent for small well-vegetated catchments. The observation that storm runoff frequently occurs from only a small part of the catchment and the size of the runoff-contributing area does not vary greatly within a drainage basin is the basis of the partial area storm runoff concept. However, because this idea is based on Hortonian infiltration-excess runoff theory this is not generally applicable to all catchments.

Working in the humid forests, workers began to recognize the importance of subsurface stormflow generation by throughflow and saturated overland flow at rainfall intensities far less than infiltration excess overland flow (Troendle 1985). This led to the concept of the variable source area, whereby storm runoff was generated in only certain parts of the catchment (Hewlett 1961). The extent of the runoff-generating areas varied from storm to storm and from season to season. On lower slope where groundwater levels are nearest the surface and soil water seepage results in elevated soil moisture storage during a storm, subsurface flow may resurge towards the base as saturated overland flow. Hewlett (1961) and Hewlett and Hibbert (1967), working in forested catchments of North Carolina, demonstrated the importance of this runoff mechanism as opposed to infiltration-excess overland flow so widely popularized by Horton. Other work, particularly by Dunne and Black (1970) working in Vermont, USA clearly established saturation overland flow could be the dominant source of stormwater runoff in a stream.

These ideas are manifest today in the partial contributing area concept which is implicit in the dynamic contributing area concept in recognition of the fact that the area contributing runoff is not fixed but expands during a storm as the saturated areas at the foot of slopes and channel heads extend. When precipitation stops and slopes begin to drain the contributing areas contract. Given that contributing areas are defined by the spatial pattern of surface storm runoff, including overland flow, topography is fundamental in determining the extent. For example, hillslope hollows and swales tend to concentrate saturated overland flow.

Contributing areas of saturation-excess overland flow are determined by the interaction of topography and soil-moisture conditions (Anderson and Burt 1978). The degree of concentration is determined by the area drained per unit contour length (*a*) and the local slope gradient (*s*). The a/s index (Kirkby 1978) defines areas of flow convergence and divergence that dictate local drainage conditions for both saturated overland flow and seepage. This topographic control on saturation-excess overland flow can be quantified for the drainage basin as a whole using the topographic wetness index (TWI) (Wolock and McCabe 1995). The TWI is calculated as $\ln(a/s)$ for all points in a catchment. The areas of a catchment with the highest TWI values are the most likely to contribute saturation-excess overland flow. During dry periods when soil-moisture storage is low, only areas with the very highest TWI values are likely to be saturated and contribute overland flow runoff. Under saturated conditions areas with lower TWI values will contribute to runoff.

Land use strongly affects the nature of runoff within a catchment, both in terms of physical processes and solute dynamics. Factors such as surface vegetation, soil permeability and land management practices determine the relative importance of runoff from different types of land use. Furthermore, the pathway of flow through the soil is likely to alter the solute balance of stormwater runoff (e.g. the take-up of nitrates from agricultural fertilizer). In this respect the land use not only influences runoff pathways but will also be important in controlling the sources, types and amounts of contaminants that enter runoff. Furthermore, the dominance of surface or near-surface flow processes in generating storm runoff is an important consideration in the stability of slopes. Many soil mechanics problems can only be addressed by having a good knowledge of hillslope hydrology.

The concept of contributing areas within drainage basins has provided much better understanding of stormwater runoff mechanisms. This has led to better hydrological predictions and the development of distributed runoff models. These models can now be coupled with sediment transport and erosion models to provide realistic simulations of drainage basin development (Willgoose *et al.* 1991).

References

Anderson, M.G. and Burt, T.P. (1978) The role of topography in controlling throughflow generation, *Earth Surface Processes and Landforms* 3, 331–344.

Betson, R.P. (1964) What is watershed runoff? *Journal of Geophysical Research* 69, 1,541–1,551.

Dunne, T. and Black, R.D. (1970) Partial area contributions to storm runoff in a small New England watershed, *Water Resources Research* 6, 1,296–1,311.

Hewlett, J.D. (1961) Some ideas about storm runoff and baseflow, *United States Department of Agriculture, Forest Service Southeast Forest and Range Experiment Station, Annual Report*, 62–6.

Hewlett, J.D. and Hibbert, A.R. (1967) Factors affecting the response of small watersheds to precipitation in humid areas, in *Proceedings of the International Symposium on Forest Hydrology*, 275–290, Oxford: Pergamon.

Kirkby, M.J. (1978) *Hillslope Hydrology*, Chichester: Wiley.

Montgomery, D.R. and Dietrich, W.E. (1988) Where do channels begin? *Nature* 336, 232–234.

Troendle, C.A. (1985) Variable source are models, in M.G. Anderson and T.P. Burt (eds) *Hydrological Forecasting*, 347–403, Chichester: Wiley.

Willgoose G.R., Bras R.L. and Rodriguez-Iturbe, I. (1991) Results from a new model of river basin evolution, *Earth Surface Processes and Landforms* 16, 237–254.

Wolock, D.M. and McCabe, G.J. (1995) Comparison of single and multiple flow-direction algorithms for computing topographic parameters in TOPMODEL, *Water Resources Research* 31, 1,315–1,324.

SEE ALSO: drainage basin; models; overland flow; runoff generation

JEFF WARBURTON

CORAL REEF

Coral reefs are natural structures of calcium carbonate made largely from the skeletons of hard corals and coralline algae. Some modern reefs have been forming for millions of years and can stretch for hundreds of kilometres off tropical coasts.

Distribution in time and space

Coral reefs are found mainly between 25°N and 25°S. The reef-building (hermatypic) corals and associated organisms live best in sea-surface temperatures between 25 °C and 29 °C. Hermatypic corals mostly live only within the upper few metres of water, the 'photic' zone into which sufficient light can penetrate for their symbiotic algae (zooxanthellae) to be able to photosynthesize.

In a general sense, the distribution of fossil coral reefs suggest that sea-surface temperatures have constrained their spread since their appearance in the early Triassic (Birkeland 1997). At a sub-regional scale, other factors were important. For example, the presence of terranes in the central tropical Pacific aided the dispersal of corals across the wider-than-present Pacific during the Palaeozoic and much of the Mesozoic (Grigg and Hey 1992). West–east ocean currents helped the development of coral reefs in the easternmost Pacific during the Cretaceous, when the gap between the Americas was open but species exchange gradually became less during the Tertiary as the Panama Isthmus rose. In the Hawaii group, coral reefs became established only during the Oligocene following the intensifying of the North Pacific ocean-surface gyral circulation (Grigg 1988). Subsequent changes in species composition may be an effect of episodes of extinction and recolonization associated with Quaternary climate changes.

As temperatures and sea levels oscillated during the Quaternary, coral reefs were alternately exposed and drowned. During glacial periods, when sea levels were low, the distribution of coral reefs was much less and in marginal areas of the modern coral seas (like the Hawaiian Islands; Grigg 1988) reefs died out entirely. Owing to cooler temperatures, coral reefs grew at slower rates, and many were comparatively ephemeral. As temperatures increased and sea levels rose at the end of the glacial periods, reefs gradually became reestablished across wider areas of the coral seas. Depending on oceanographic factors, upward-growing coral reefs were either able to 'keep-up' with rising postglacial sea level, or they later were able to 'catch-up', or they had to 'give-up' and thereby forming a drowned reef (Neumann and MacIntyre 1985).

Drowned reefs occur in many parts of the Pacific and Indian Oceans in particular. Most are thought to have failed to keep up with rising sea level during a period of sea-level rise, for reasons associated with climate and sea-level history, palaeolatitude, seawater temperature, and light (Flood 2001). Other 'drowned' coral reefs, particularly those on the flanks of the Hawaiian Ridge, have slipped hundreds of metres downslope.

In many parts of the world, but especially near convergent plate boundaries, coral reefs are found raised above their modern counterparts and, as such, often provide important insights into reef structure and history (Plate 28). Emerged reef staircases on islands like Sumba in

Plate 28 The Talava Arches on Niue Island, central South Pacific. Niue is a fine example of an uplifted atoll, with a well-preserved atoll reef (now 70 m above the modern reef) and lagoon floor. Around the fringes of the emerged atoll reef are a series of emerged fringing reefs. The emerged reef shown here dates from the Last Interglacial period. The modern reef here is rising and is consequently narrow except in embayments as shown here (Photo by Patrick D. Nunn)

Indonesia have been studied in detail (Pirazzoli *et al.* 1991).

In the Pacific and Indian Oceans during the Holocene, keep-up coral reefs grew above their present levels around 4,000 cal. yr BP and have since been exposed as sea level fell. These fossil reefs form the cores of many reef islands (see CAY) in the central Pacific and have been critical factors in their habitability and persistence.

On the basis of their form, coral reefs can generally be either FRINGING REEFs, barrier REEFs or ATOLL reefs. Fringing reefs are juvenile, sometimes ephemeral, and grow outwards from a coast. Barrier and atoll reefs are older, often being composed of reefs of many different ages; reef upgrowth during postglacial periods has been followed by subaerial exposure and erosion during the following glacial period, followed by renewed upgrowth. A good example is that of Midway Island in Hawaii where reef dating back to the mid-Tertiary has been cored (Lincoln and Schlanger 1987). Barrier reefs are separated from a nearby coast by a lagoon while atoll (or ring) reefs enclose a lagoon. The three types were first linked by Darwin (1842) in his Theory of Atoll Development. In this he envisaged that a young volcanic island would develop a fringing reef. As the island subsided so the fringing reef would become transformed into a barrier reef and finally, as the last vestiges of the island were submerged, an atoll reef. Deep drilling of atolls demonstrated the essential correctness of Darwin's model.

Coral reefs in geomorphological research

Since corals are temperature-sensitive organisms, we can learn a lot about palaeoclimates from studying their fossil distributions (see above). We can also use coral reefs as palaeosea-level indicators, both for the Last Interglacial when it is of interest to know whether or not sea level reached 6 m above its present level, as suggested by studies of the emerged reef series on the Huon Peninsula of Papua New Guinea (Chappell 1983). In the Pacific, studies of Holocene emerged reefs have given us much information about the sea-level maximum about 4,500 cal. yr BP (see Nunn and Peltier 2001, for example) and another about 650 cal. yr BP which marked the start of the 'AD 1300 Event' (Nunn 2000). There have also been successful studies of stable isotopes in long-living corals to generate climate data prior to the start of the instrumental record in key regions such as the South Pacific (Quinn *et al.* 1993).

Much research has focused on modelling the relationship between coral reefs and the shorelines which they commonly adjoin, particularly in terms of sediment production, lagoonal dynamics, beach nourishment and shoreline erosion; good studies are those of Munoz-Perez *et al.* (1999) and Hearn and Atkinson (2001). It is clear, for example, that along many tropical coasts, coral reefs are the main producers of the fine-grained sediments which supply nearby beaches and that, should those reefs become degraded, then these beaches can become starved of sediment and destabilized.

Human impacts on coral reefs

It has been realized only comparatively recently how fragile coral-reef ecosystems are, and how much they have been affected by and/or are vulnerable to a variety of human impacts, direct and indirect (Bryant *et al.* 1998). Recently evidence has been presented suggesting that the first human colonizers of some remote Pacific Island groups ~3,000 years ago inadvertently brought with them alien organisms which occupied coral reefs causing reef-surface growth to cease for several hundred years (Nunn 2001). Modern human impacts are more familiar and better understood. These include direct impacts, ranging from the overexploitation of reef organisms (including corals) for sustenance or sale to the dynamiting of reef waters to maximize fish catches, which commonly cause structural reef damage. Indirect impacts are from pollution, including excessive sediment inputs from logging into nearshore areas and chemical pollutants from mineral processing or domestic waste disposal, for example.

Many coral reefs have become degraded as a result of such impacts, manifested as a loss of corals and associated reef organisms, and a reduction in species diversity. Certain more hardy organisms such as sea grasses and various algae (especially *Halimeda*) often cover such degraded reefs. Sometimes reef degradation allows reef predators like the crown-of-thorns starfish (*Acanthaster*) sufficient access to result in an infestation which then exacerbates the process of degradation.

Tourism along tropical coasts often focuses on coral reefs; around 80 per cent of the visitors to the Maldives in 2001 wanted to dive on their reefs. While reef-associated tourism can be sustainable, in many cases it is not because the effects of constructing tourist infrastructure and the effluents which must be disposed of when a large hotel or resort exists in a particular place all reduce the health of the reef ecosystem.

Coral-reef conservation initiatives, including the establishment of marine-protected areas, are often well intentioned but ineffective. Good examples are found in parts of the Caribbean and tropical Pacific Islands where the idea of marine reserves is anathema to people who have been accustomed to free access to reef areas for susbistence purposes (Birkeland 1997).

The future of coral reefs and the implications for geomorphology

Many coral-reef ecosystems have become significantly degraded as a result of human impacts (see above). Many reefs are now being pushed to the brink of extinction because of the additional stress associated with rising sea-surface temperatures (Hoegh-Guldberg 1999). High levels of stress often cause corals to become bleached, the loss of colouration being associated with the ejection of the symbiotic algae that live within coral polyps. Whole reefs can die as a result of bleaching episodes, and there are no instances where a formerly bleached reef has been able to recover its former state. As sea-surface warming continues over the next few decades, so the instances of bleaching resulting from prolonged periods of high temperatures (often associated with El Niño) are likely to increase. The Great Barrier Reef is likely to be experiencing annual bleaching events by 2030.

The implications of regular bleaching for the world's coral reefs are extremely serious, and will have huge implications for many subsistence coastal dwellers in the tropics, who depend daily on reefs for sustenance, and for those countries which depend heavily on revenue generated from reef-associated tourism. The effects for coastal landscapes will involve drastic reductions in the amounts of fine calcareous sediment being generated in reef-lagoon areas, perhaps with many beaches disappearing as a result. This may in turn increase the vulnerability of sandy shorelines to erosion, also an effect of larger waves crossing reefs which are unable to grow upwards in response to projected sea-level rise (Birkeland 1997).

References

Birkeland, C.E. (ed.) (1997) *Life and Death of Coral Reefs*, New York: Chapman and Hall.

Bryant, D., Burke, L. and McManus, J. (1998) *Reefs at Risk: A Map-based Indicator of Threats to the World's Coral Reefs*, Washington, DC: World Resources Institute.

Chappell, J. (1983) A revised sea-level record for the last 300,000 years from Papua New Guinea, *Search* 14, 99–101.

Darwin, C.R. (1842) *Structure and Distribution of Coral Reefs*, London: Smith, Elder.

Flood, P.G. (2001) The 'Darwin Point' of Pacific Ocean atolls and guyots: a reappraisal, *Palaeogeography, Palaeoclimatology, Palaeoecology* 175, 147–152.

Grigg, R.W. (1988) Paleoceanography of coral reefs in the Hawaiian–Emperor chain, *Science* 240, 1,737–1,743.

Grigg, R.W. and Hey, R. (1992) Paleoceanography of the tropical eastern Pacific Ocean, *Science* 255, 172–178.

Hearn, C.J. and Atkinson, M.J. (2001) Effects of sea-level rise on the hydrodynamics of a coral reef lagoon: Kaneohe Bay, Hawaii, in J. Noye and M. Grzechnik (eds) *Sea-Level Changes and their Effects*, 25–47, Singapore: World Scientific Publishing.

Hoegh-Guldberg, O. (1999) Coral bleaching, climate change and the future of the world's coral reefs, *Review of Marine and Freshwater Research* 50, 839–866.

Lincoln, J.M. and Schlanger, S.O. (1987) Miocene sea-level falls related to the geologic history of Midway atoll, *Geology* 15, 454–457.

Munoz-Perez, J.J., Tejedor, L. and Medina, R. (1999) Equilibrium beach profile model for reef-protected beaches, *Journal of Coastal Research* 15, 950–957.

Neumann, A.C. and MacIntyre, I. (1985) Reef response to sea-level rise: keep-up, catch-up or give-up', in *Proceedings of the 5th International Coral Reef Congress* 3, 105–110.

Nunn, P.D. (2000) Illuminating sea-level fall around AD 1220–1510 (730–440 cal. yr BP) in the Pacific Islands: implications for environmental change and cultural transformation, *New Zealand Geographer* 56, 46–54.

——(2001) Ecological crises or marginal disruptions: the effects of the first humans on Pacific Islands, *New Zealand Geographer* 57, 11–20.

Nunn, P.D. and Peltier, W.R. (2001) Far-field test of the ICE-4G (VM2) model of global isostatic response to deglaciation: empirical and theoretical Holocene sea-level reconstructions for the Fiji Islands, Southwest Pacific, *Quaternary Research* 55, 203–214.

Pirazzoli, P.A., Radtke, U., Hantoro, W.S., Jouannic, C., Hoang, C.T., Causse, C. and Best, M.B. (1991) Quaternary raised coral-reef terraces on Sumba Island, Indonesia, *Science* 252, 1,834–1,836.

Quinn, T.M., Taylor, F.W. and Crowley, T.J. (1993) A 173 stable isotope record from a tropical South Pacific coral, *Quaternary Science Reviews* 12, 407–418.

SEE ALSO: atoll; fringing reef; reef

PATRICK D. NUNN

CORNICHE

Corniches are narrow organic ledges, 0.5 to 2 m in width, growing on steep rock surfaces at about mean sea level. The best examples are on limestones where notches (see NOTCH, COASTAL) develop in the spray zone. Corniches in the northwestern Mediterranean consist of algae, particularly the calcareous alga *Tenarea tortuosa*, although Serpulid (see SERPULID REEF) worms or Vermetid (see VERMETID REEF AND BOILER) gastropod tubes can play a similar role. Although the interiors are generally quite hard, corniches cannot resist very strong waves and they are best developed in inlets on exposed coasts.

Reference

Trenhaile, A.S. (1987) *The Geomorphology of Rock Coasts*, Oxford: Oxford University Press.

ALAN TRENHAILE

CORROSION

Corrosion is synonymous with solutional erosion, the erosion of material by chemical activity. The majority of studies of corrosion have been undertaken on carbonates and these are the primary focus of this entry. However, similar considerations apply to evaporites and the estimation of gypsum corrosion rates is particularly problematic because of the more rapid solution and the consequently greater spatial and temporal variability of dissolution.

Corrosion rates are commonly expressed as mm/1000a, implying that all corrosion contributes to surface lowering and that environmental conditions have remained broadly the same for millennia. The former is incorrect, particularly in karst, while the latter is also highly questionable. The preferred unit is $m^3\ km^{-2}\ a^{-1}$ and $1\ mm\ ka^{-1}$ is equivalent to $1\ m^3\ km^{-2}\ a^{-1}$. Where surface lowering is measured directly then units of $\mu m\ a^{-1}$ are appropriate.

Limestone corrosion rates may be estimated from knowledge of dissolution kinetics, runoff, carbon dioxide and temperature, but there remains a need for field measurements to provide actual values of regional denudation; to compare rates in contrasting environments and by different processes; to understand landform evolution; and to understand how processes operate in a complex natural environment as opposed to the laboratory. When evaluating results from past studies it is important to understand what was actually measured and how the corrosion rates were calculated. Most field measurements of corrosion in carbonate karst were based on spot samples, with denudation being estimated from the Corbel formula. This suffers several problems, the three most important being: the carbonate concentration is frequently the average of a few spot measurements, with the implicit assumption of a linear relationship between carbonate hardness and discharge; carbonates present in solution are assumed to only come from karst denudation; and measurements are usually made at only one point, commonly the output of a drainage basin, with the implicit assumption that this is representative of conditions upstream.

Where water samples have been collected over a range of flow conditions it is apparent that the relationship between dissolved load and discharge is usually non-linear and particularly in small drainage basins may be complicated by hysteresis effects (usually higher concentrations per unit discharge on the rising limb). It is virtually impossible to correct for hysteresis, but by collecting samples over a range of discharges it is usually possible to construct a reliable discharge-concentration or discharge-load rating curve. This can be applied to the discharge curve and the results summed to obtain the total annual solute load. Greater accuracy may be obtained using a logging conductivity meter, developing a conductivity-concentration rating curve, and using this to predict the concentration at each measured discharge.

Having computed the total solute load (TSL) at a point it is important to realize that this is made up of corrosion of karst rocks by both autogenic (CKAu) and allogenic waters (CKAl), less any deposition of previously dissolved material (D), together with corrosion of non-karst rocks by allogenic waters (CNK), solute accessions in rainfall and snowfall (AC), and any anthropogenic inputs such as fertilizers (AN). The gross karst solution is then (CKAu + CKAl) whereas the net karst solution is (CKAu + CKAl − D). Where precipitation of previously dissolved carbonates is minimal then gross and net solution will be similar, but elsewhere failure to account for deposition may result in a significant underestimate of gross denudation, which is the real measure of relief transformation. In contrast, failure to take into account the solution of non-karst rocks and solute accessions in precipitation will result in an overestimate of karst corrosion. Error in estimating corrosion rates can arise from many sources and even in a careful study using hydrochemical budgeting and taking into account non-denudational components potential errors of around 25 per cent are possible.

Corrosion rates for whole drainage basins derived by sampling of water at the basin outlet are unlikely to be representative of any specific location within the basin. This information may best be obtained by an extension of the hydrochemical budgeting method discussed above. Water samples are collected from the full range of sites – bare limestone surfaces, the soil zone, the subcutaneous zone, the main body of bedrock, and cave streams in both vadose and phreatic zones. These, together with estimates of the proportion of water following the various pathways through the system, permit the breaking down of the overall corrosion budget. Those few studies that have been made show that a high proportion of corrosion (50–85 per cent) occurs within several metres of the surface in the soil (if present) and subcutaneous zone (uppermost bedrock). Caves account for very little of the erosion when averaged over the whole basin.

The principal drawback of the hydrochemical approach is that it requires frequent, ideally continuous, measurement of discharge and sufficient samples to establish the pattern and extent of variations in solute concentrations. As this is not always possible alternative methods that integrate erosion over a longer time period have been derived. The two most commonly used are the micro-erosion meter and rock tablets. In contrast to the hydrochemical method these are highly site-specific and may only be used to assess corrosion rates on bare limestone surfaces, in the soil zone, at the soil–bedrock interface, and in cave streams. Tablets have been found to give estimates two orders of magnitude less than those calculated using hydrochemical data. The most likely explanation is that natural rock surfaces come into contact with larger volumes of water than do isolated rock tablets, simply because of their greater lateral flow component. Thus, the two methods measure fundamentally different phenomena and the hydrochemical method provides the only reliable means of estimating corrosion rates on limestone surfaces. Different problems arise if tablets are placed in cave streams as they will project above the natural surface and as a consequence are likely to erode more rapidly. They are also likely to suffer from abrasion as well as corrosion, although this can be exploited by placing the tablets in nylon cages with differing mesh sizes and comparing the erosional losses suffered.

Further reading

Dreybrodt, W. (1988) *Processes in Karst Systems: Physics, Chemistry and Geology*, Berlin and New York: Springer.

Ford, D.C. and Williams, P.W. (1989) *Karst Geomorphology and Hydrology*, London: Unwin Hyman.

White, W.B. (1984) Rate processes: chemical kinetics and karst landform development, in R.G. LaFleur (ed.) *Groundwater as a Geomorphic Agent*, 227–248, London: Allen and Unwin.

SEE ALSO: dissolution

JOHN GUNN

COSMOGENIC DATING

Cosmogenic dating is a group of related techniques for estimating landform ages and erosion rates. It is based upon the generation of rare isotopes within minerals by cosmic rays. Primary cosmic rays composed largely of highly energetic protons interact with gases in the Earth's atmosphere to produce showers of secondary subatomic particles, mostly neutrons and muons. These secondary cosmic rays induce nuclear reactions within the Earth's terrestrial surface, producing cosmogenic nuclides. The length of surface exposure, or alternatively, the rate of surface erosion, is computed from the concentration of cosmogenic nuclides in a landform.

Six cosmogenic nuclides have found widespread application in geomorphology (Table 9). These are stable isotopes of the noble gases helium and neon (^{3}He and ^{21}Ne), and radioactive isotopes of beryllium, carbon, aluminum, and chlorine (^{10}Be, ^{14}C, ^{26}Al , and ^{36}Cl). The nuclides ^{10}Be, ^{14}C and ^{36}Cl are also produced within the atmosphere by cosmic rays. The best known example of atmospheric production is ^{14}C which forms the basis for radiocarbon dating (see DATING METHODS). To avoid confusion, nuclides generated within mineral lattices in the Earth's solid surface are termed *in situ*-produced terrestrial cosmogenic nuclides (TCN).

Most TCN production is from neutron spallation (Lal 1991). TCN spallation occurs when a

Table 9 Properties of *in situ*-produced terrestrial cosmogenic nuclides (TCN)

Nuclide	Mean lifetime (yrs)	Host mineral
^{3}He	Stable	Olivine, clinopyroxene
^{21}Ne	Stable	Olivine, clinopyroxene, quartz
^{10}Be	2.2 Myr	Quartz
^{14}C	0.82 kyr	Quartz, calcite
^{26}Al	1.0 Myr	Quartz
^{36}Cl	430 kyr	Calcite, dolomite, whole rocks

secondary neutron with energy > 10 MeV collides with a target nucleus in a mineral lattice, breaking protons, neutrons or clusters of these particles from the nucleus. Spallation products always consist of an isotope of lower atomic number than the target. As neutrons do not penetrate deeply in rocks, most neutron spallation occurs within about one metre of the surface. Thermal neutrons (energy ~0.025 eV) are absorbed by some target nuclei, causing radioactive decay and production of a cosmogenic isotope. Thermal neutron production is important for cosmogenic ^{36}Cl. Muons also create cosmogenic nuclides but at rates much lower than neutron spallation. Muons penetrate far more deeply than neutrons, creating measurable quantities of cosmogenic nuclides at depths of over 20 metres (Granger and Muzikar 2001).

The production rate of cosmogenic nuclides by all reaction mechanisms is low, ranging (at sea level and latitudes $> 60°$) from about 5 to 6 atoms g^{-1} a^{-1} for ^{10}Be to about 120 atoms g^{-1} a^{-1} for ^{3}He. Cosmic rays are attenuated by the atmosphere and the geomagnetic field; consequently production rates vary significantly with altitude and latitude. For this reason, TCN production rates are always quoted for sea level and high latitude, and scaled to the altitude and latitude of study sites using empirical functions (Lal 1991; Stone 2000; Dunai 2000). Production rates must be precisely known for reliable TCN results. This is a difficult task because both atmospheric shielding and geomagnetic field intensity vary with time. Calibration sites are used to determine production rates. At a calibration site, TCN concentrations are measured in independently dated geomorphic surfaces with near-zero erosion rates such as lava flows, glacially eroded bedrock or large landslides.

Applications of TCN fall into two main categories: surface exposure dating and measurement of erosion rates. Both applications yield model results with accuracy highly dependent on the validity of simplifying assumptions. In exposure dating, the first requirement is that the geomorphic surface must have formed over a short time period. Examples of such surfaces include fault scarps (see FAULT AND FAULT SCARP), LAVA LANDFORMS, LANDSLIDES and ERRATIC boulders. Surfaces forming incrementally over long periods of time have cosmogenic nuclide concentrations best interpreted in terms of erosion rates. The second requirement is that the geomorphic surface be free of TCN at the time of surface formation. Remnant TCN from past periods of surface exposure is termed nuclide inheritance. Lava flows and large glacial erratics generally have little or no nuclide inheritance. The final requirement for accurate exposure dating is that the primary geomorphic surface form must be preserved over the period of exposure. Erosion rates must either be known, or be assumed to be zero. Surface exposure dating therefore requires careful analysis of landscapes and sampling of surfaces experiencing very low rates of erosion. The requirement of near-zero erosion limits the age range of TCN exposure dating. In most geomorphic environments, reliable exposure ages generally range from about 5,000 years to less than 100,000 years. Younger ages are limited by detection limits for measuring TCN while older surfaces are generally destroyed by erosion or buried by sediments. The polar deserts of east Antarctica are a major exception, with exposure ages of over 5 million years. The precision of exposure ages and erosion rates, as estimated by analytical errors, varies with isotope and application but generally ranges between ± 3 per cent to 15 per cent.

In TCN erosion rate studies, an assumption of equilibrium between TCN production and loss by erosion and radioactive decay is made (Bierman and Steig 1996; Granger *et al.* 1996). Under these circumstances, exposure time is not important and TCN concentrations vary inversely with erosion rates. For example, steep hillslopes with high erosion rates have low TCN concentrations because of the short residence time of target minerals within the zone of production. Averaging time is the time necessary to achieve equilibrium conditions. The lower the erosion rate, the longer the averaging time. TCN averaging times for erosion rates typical of temperate climates range from about 100,000 years to about 5,000 years. Averaging erosion over such timescales makes the TCN method very useful for investigating links between climate and tectonics, and for establishing baseline erosion rates unrelated to human activities. Two types of sampling are applied in TCN erosion rate studies. Bedrock samples give information about minimum rates of landscape lowering and the influence of lithology on erosion rates. Alluvial samples average erosion rates for the contributing catchment and therefore are easiest to compare with traditional methods of measuring short-term erosion rates such as suspended sediment studies.

TCN vary greatly in terms of ease and cost of measurement, sample preparation and host minerals. ^{3}He and ^{21}Ne are measured in olivine and clinopyroxene using noble gas mass spectrometry techniques similar to those employed for ^{40}Ar/^{39}Ar studies. They are primarily used for dating mafic volcanic rocks, and for studies of long-term landscape evolution in Antarctica where extremely low rates of erosion require the use of a stable nuclide (Summerfield *et al.* 1999). The most used TCN are ^{10}Be and ^{26}Al. These nuclides are popular because the host mineral (quartz) is present in the majority of geologic settings, production reactions are relatively simple and well understood, and both nuclides can be measured in the same sample. Since the mean lives of ^{10}Be and ^{26}Al differ significantly (Table 9), measurement of the nuclides in the same sample can constrain both erosion rate and exposure time as well as indicate periods of burial (Lal 1991; Bierman *et al.* 1999). It is also possible to use this nuclide pair for dating the burial of sediments (Granger and Muzikar 2001). Measurement of ^{10}Be and ^{26}Al is by accelerator mass spectrometry (AMS). Sample preparation requires preparation of high purity quartz separates and removal of atmospheric ^{10}Be with hydrofluoric acid etching. ^{36}Cl is also widely applied to geomorphic problems, particularly in carbonate and volcanic landscapes where ^{10}Be cannot be used. Production rates for ^{36}Cl are less well established than for other TCN because of more complex production reactions. ^{36}Cl is produced by neutron spallation on K and Ca, and by thermal neutron capture on ^{35}Cl. Production rates vary with rock composition, and major and trace element data are needed to compute rates. AMS is also used to measure ^{36}Cl concentrations. *In situ*-produced ^{14}C has not been widely applied in geomorphology because of problems separating atmospheric contamination. However, applications with this nuclide are likely to increase. The mean life of ^{14}C is much shorter than any other TCN, therefore, when measured in conjunction with ^{10}Be and ^{26}Al, it can be used to establish production rates, estimate erosion corrections, and detect periods of burial by sediment or ice.

References

Bierman, P.R. and Steig, E.J. (1996) Estimating rates of denudation using cosmogenic isotope abundances in sediment, *Earth Surface Processes and Landforms* 21, 125–139.
Bierman, P.R., Marsella, K.A., Paterson, C., Davis, P.T. and Caffee, M. (1999) Mid-Pleistocene cosmogenic minimum age limits for pre-Wisconsin glacial surfaces in southwestern Minnesota and southern Baffin Island: a multiple nuclide approach, *Geomorphology* 27, 25–40.
Dunai, T.J. (2000) Scaling factors for production rates of *in situ* produced cosmogenic nuclides: a critical reevaluation, *Earth and Planetary Science Letters* 176, 157–169.
Granger, D.E. and Muzikar, P.F. (2001) Dating sediment burial with *in situ*-produced cosmogenic nuclides: theory, techniques, and limitations, *Earth and Planetary Science Letters* 188, 269–281.
Granger, D.E., Kirchner, J.W. and Finkel, R. (1996) Spatially averaged long-term erosion notes measured from *in-situ*-produced cosmogenic nuclides in alluvial sediments, *Journal of Geology* 104(3), 249–257.
Lal, D. (1991) Cosmic ray labeling of erosion surfaces: *in situ* nuclide production rates and erosion rates, *Earth and Planetary Science Letters* 104, 424–439.
Stone, J.O. (2000) Air pressure and cosmogenic isotope production, *Journal of Geophysical Research* 105, 22,753–23,759.
Summerfield, M.A., Stuart, F.M., Cockburn, H.A.P., Sugden, D.E., Denton, G.H., Dunai, T. and Marchant, D.R. (1999) Long-term rates of denudation in the Dry Valleys, Transantarctic Mountains, southern Victoria Land, Antarctica, based on *in-situ* produced cosmogenic ^{21}Ne, *Geomorphology* 27, 113–130.

Further reading

Bierman, P.R. (1994) Using *in situ* produced cosmogenic isotopes to estimate rates of landscape evolution: a review from the geomorphic perspective, *Journal of Geophysical Research* 99, 13,885–13,896.
Cerling, T.E. and Craig, H. (1994) Geomorphology and *in-situ* cosmogenic isotopes, *Annual Reviews of Earth and Planetary Sciences* 22, 273–317.
Gosse, J.C. and Phillips, F.M. (2001) Terrestrial *in situ* cosmogenic nuclides: theory and application, *Quaternary Science Reviews* 20, 1,475–1,560.

WILLIAM M. PHILLIPS

COULEE

In western North America, coulee (French *couler*: to flow) is a common term used to describe a dry valley, canyon, gulch or wash. Most coulees were formed rapidly in late glacial times by large discharges of melt water, particularly with the emptying of proglacial lakes (Bretz 1969). Selby (1985: 458) adopts this as a specific origin. Coulees may have ponded water bodies, intermittent or underfit streams. Parallel sets of coulees in southern Alberta, Canada, may have been aligned by regional joint patterns (Babcock 1974), or possibly formed in postglacial time through some imperfectly understood process controlled by prevailing wind direction (Beaty 1975). Less commonly, the term coulee is used to describe a short lobe of viscous lava on the flanks of a volcano and a lobe of debris moved by gelifluction.

References

Babcock, E.A. (1974) Photolineaments and regional joints: lineament density and terrain parameters, south-central Alberta, *Bulletin of Canadian Petroleum Geology* 22, 89–105.

Beaty, C.B. (1975) Coulee alignment and the wind in southern Alberta, Canada, *Geological Society of America Bulletin* 86, 119–128.

Bretz, J.H. (1969) The Lake Missoula floods and the channelled scablands, *Journal of Geology* 77, 505–543.

Selby M.J. (1985) *Earth's Changing Surface*, Oxford: Clarendon Press.

ROBERT J. ROGERSON

COVERSAND

Originally a Dutch term applied to aeolian SANDSHEET deposits overlying older sediments. Its generic nature has led it to be applied to a range of deposits. However, the common denominator has been its application to sandsheet deposits of cold-climate (see PERIGLACIAL GEOMORPHOLOGY) aeolian origin. The latter is proven by the occurrence in coversands of frost cracks, involutions and ice wedge casts (see ICE WEDGE AND RELATED STRUCTURES), as well as from pollen and beetle evidence obtained from intercalated organic deposits. The aeolian origin of coversands has been determined on the basis of their concordance with dune forms (see DUNE, AEOLIAN), occurrence with VENTIFACTs and/or on the basis of particle characteristics (mineralogy, sorting, rounding, surface matting and textures). Whilst predominantly aeolian derived, coversands can incorporate components of sand derived from other processes, e.g. niveo- and/or fluvio-aeolian (see NIVEO-AEOLIAN ACTIVITY).

Coversands in northern Europe (Schwan 1988) and mid-continental north America although relict, are widespread, extending over 10,000s of km^2 as nearly spatially continuous sheets with flat to undulating relief (less than 5 m) and a notable paucity of dunes (Koster 1988). This differentiates coversand from more recent sand deposits which tend to have been formed into dunes, e.g. Drift sands in The Netherlands (Koster *et al.* 1993). The coversand is typically of a uniform thickness of up to several metres; only in valleys, depressions or against topographical barriers is it thicker. The coversands also tend to be (sub)horizontally stratified, composed of thin beds, setting them apart from the high angle bedding of coastal dune (see DUNE, COASTAL) sands and the cross bedding, troughs and ripples of riverine sands.

Detailed examination of coversand stratification in Europe has led to classification of coversands into two types which are in turn subdivided into two: Older coversands I and II and Younger coversands I and II. The Older coversand is characterized by an alternation of well-sorted parallel-laminated beds of greyish loam/fine sand and yellowish fine/medium sand. The Older coversand I has evidence of more cryogenic deformation and frost wedge casts, especially in its upper layers, than Older coversand II and the two facies are commonly separated by a disconformity, e.g. the Beuningen pebble. The Younger coversand is typically a unimodal, well-sorted, parallel-laminated medium sand with a large sand component derived from local sources. The sand is rarely buried or cross-bedded, has a low relief and has no evidence of ice wedge formation in it. The primary differentiation between the two Younger coversand facies is on sedimentary structures which indicate that the Younger coversand II was deposited under drier conditions.

Fragmentary evidence indicates coversands have been deposited during several Pleistocene glacials and are not unique just to the last glacial cycle. The northerly limit of the relict but extensive European coversands found in Britain, The Netherlands, Germany, Denmark, Poland and the Baltic states is broadly coincident with the maximal position of the Late-Weichselian (Devensian) ice sheet. In general, the last era of north-west European coversand activity started after the last interglacial, increasing in intensity throughout the Weichselian period. Two main phases of coversand deposition have been reported: one around 18,000–15,000 years ago (Older Coversand II) and another more intense period between 14,000–11,000 years ago (Younger Coversand) (Koster 1988; Bateman 1998; Singhvi *et al.* 2001). Older coversand I appears to have been dominantly deposited separate to, but contemporary with, the widespread LOESS deposits of north-western and eastern Europe which were mostly deposited just prior to the last glacial maximum and appear to have stabilized by approximately 13,000 years ago (Singhvi *et al.* 2001). However, evidence also suggests localized environmental conditions blurred these discrete aeolian phases with Older coversand type facies still being deposited in places during the so-called Younger coversand phase (Kolstrup *et al.* 1990; Kasse 1997).

Formation and preservation of the Late-Weichselian coversands is thought to have been aided by enhanced sand sources as a result of glaciation, sparse vegetation, low relief and low sand supply due to periodically wet, frozen or cemented depositional surfaces (Kasse 1997). Use of the orientation of dune morphology, bedding inclination and unit thickness has enabled the reconstruction of palaeo-wind directions. The Older coversands type was deposited by predominantly north-westerly to westerly winds and the Younger coversands deposited by more westerly to south-westerly winds. Such information has been used to inform palaeoclimatic models for north-western and central Europe (e.g. Isarin *et al.* 1998).

References

Bateman, M.D. (1998) The origin and age of coversand in north Lincolnshire, UK, *Permafrost and Periglacial Processes* 9, 313–325.

Isarin, R.F.B., Renssen, H. and Vandenberghe, J. (1998) The impact of the North Atlantic Ocean on the Younger Dryas climate in northwestern and central Europe, *Journal of Quaternary Science* 13, 447–453.

Kasse, C. (1997) Cold-Climate aeolian sand-sheet formation in North-Western Europe (*c.*14–12.4 ka); a response to permafrost degradation and increased aridity, *Permafrost and Periglacial Processes* 8, 295–311.

Kolstrup, E., Grun, R., Mejdahl, V., Packman, S.C. and Wintle, A.G. (1990) Stratigraphy and thermoluminescence dating of Late Glacial cover sand in Denmark, *Journal of Quaternary Science* 5, 207–224.

Koster, E.A. (1988) Ancient and Modern cold-climate aeolian sand deposition: a review, *Journal of Quaternary Science* 3, 69–83.

Koster, E.A., Castel, I.I.Y. and Nap, R.L. (1993) Genesis and sedimentary structures of Late Holocene aeolian drift sands in northwest Europe, in K. Pye (ed.) *The Dynamics and Environmental Context of Aeolian Sedimentary Systems*, Geological Society Special Publication 72, 247–267.

Schwan, J. (1988) The structure and genesis of Weichselian to early Holocene aeolian sand sheets in W. Europe, *Sedimentary Geology* 55, 197–232.

Singhvi, A.K., Bluszcz, A., Bateman, M.D. and Someshwar Rao, M. (2001) Luminescence dating of loess-palaeosol sequences and coversands: methodological aspects and palaeoclimatic implications, *Earth-Science Reviews* 54, 193–211.

MARK D. BATEMAN

CRATER

Craters are bowl-shaped, approximately circular depressions that typically form by high-energy impact or explosive activity. There is a fundamental geomorphological problem in distinguishing crater origins by volcanic versus impact processes. The latter involve collision with a planetary surface by meteors, comets and asteroids. It is also possible that a crater can form by the explosion, just above the ground, of a meteor or comet, known in this context as a *bolide*. Volcanic craters generally form at the summits of volcanic cones and result from explosive eruptions or the accumulation of *pyroclastic* material in a rim around a volcanic vent. Of course, human activity can produce explosion craters, perhaps the most spectacular of which resulted from nuclear testing. Interestingly, it was the well-funded study of physics for the latter that ultimately led to considerable advancement in understanding the natural process of impact cratering (Roddy *et al.* 1977).

One of the great controversies in planetary geomorphology concerned the origin of craters on the moon. G.K. Gilbert (1893) used geomorphological reasoning to argue that the moon's craters had an impact origin. Until the 1930s, however, most astronomers thought that the circularity of the moon's craters required an origin by volcanic processes. Objects striking the moon, it was thought, would include many oblique impacts, and these would not be circular. Only later in the twentieth century did the physics of the cratering process come to be well understood enough to show that most oblique impacts produced circular, rather than elliptical, craters. Nevertheless, some astronomers continued to argue for the volcanic origin until the Apollo missions of the 1970s returned incontrovertible proof of the impact origin for nearly all lunar craters.

Volcanic craters

Craters can be a variety of depressions associated with volcanic or pseudovolcanic activity, including mud volcanoes, mound springs, hot springs, and even pingos. The geomorphology of truly volcanic craters was reviewed by Fairbridge (1968), who considered the large complex collapse and explosion structures known as *calderas* separately from other volcanic craters. Magmas rich in silica tend to produce highly explosive activity in which the volcanic materials become fragmented into *pyroclastic* rock. Domes of silica-rich volcanic rock, including obsidian, commonly fills preexisting pyroclastic craters. Explosive activity for basaltic magmas produces spatter cones with craters over rift zones, and a variety of pit craters. One of the most famous of these is Halemaumau, a pit crater on the floor of Kilauea Caldera, on Earth's most continuously active volcano at the southern end of the island of Hawaii. There are also many volcanic craters on other planets, including spectacular calderas on the volcanoes of Venus, Mars and Io (the highly volcanically active satellite of Jupiter).

A special type of crater, known as a *maar*, derives its name from the Rhineland dialect of German. The term was originally applied to volcanic explosion craters near Eifel, Germany. Maar craters may be associated with *diatremes*, which are breccia-filled volcanic pipes that form by gas explosions. They also occur within fields of monogenetic volcanic cones, which develop during single eruptive phases. Maar craters usually have a ring of erupted pyroclastic material, and lakes often occur on their floors. They generally form by the interaction of rising lavas with near-surface ground water.

Another interesting crater form is known as a pseudocrater, or rootless cone. These were first recognized in Iceland, where basaltic lava flows advanced over substrates that were rich in water or ice. The interaction of the lava and water produced pyroclastic explosions that formed the craters. The advancing flow may then separate the crater or cone from its source zone. Such features range from a few tens of metres to hundreds of metres in diameter.

Impact crater morphology

It is one of the major discoveries of recent planetary exploration that the surfaces of rocky objects in the solar system are almost all marked by numerous impact craters. These occur over an immense range of size scales. The smallest are microcraters or pits, which form from the impact of micrometeorites or high-velocity cosmic dust grains on exposed rocks. These only form on bodies that lack atmospheres, which would induce the very small projectiles to burn up before impact. Simple craters are larger, bowl-shaped depressions that form on land surfaces. They range up to several kilometres across, and typically have diameters across their rims that are about five times their depths from rim top to crater floor. Simple craters are familiar to many geomorphologists because they were much in evidence during the Apollo landings of humans on Earth's moon. On Earth one of the most famous simple craters is the 1-km diameter Barringer Crater in northern Arizona, also known as Meteor Crater. It is interesting that a major controversy occurred in regard to its origin, with Gilbert (1896) eventually concluding that it had a volcanic origin, despite making a strong argument for impact as well.

Most of the larger craters visible on planetary and satellite surfaces are complex craters. These have rims marked by terraces along their inner margins. Their floors are broad and flat, and there is often a central peak. Such craters are generally from a few tens to a few hundred kilometres in diameter, and they are well known from observations of the moon (Figure 28). Because of their flat floors and very high ratios of width to depth, these features are usually described as impact basins, rather than craters. Much larger impact structures are also known, and many of these are multi-ring basins. They have multiple concentric rings, each consisting of rugged hilly terrain. The floors of these exceptionally large craters are commonly flooded by lava. They can have diameters of up to two thousand kilometres or more.

Recent work has shown that many of the projectiles generating impact craters arrive in groups, rather than as single projectiles. One of the most spectacular examples of this phenomenon was the comet Shoemaker-Levy 9, which broke into fragments as it collided with Jupiter in July 1994. Asteroids also may break up when they interact with a planet's atmosphere. Among the 150 or so

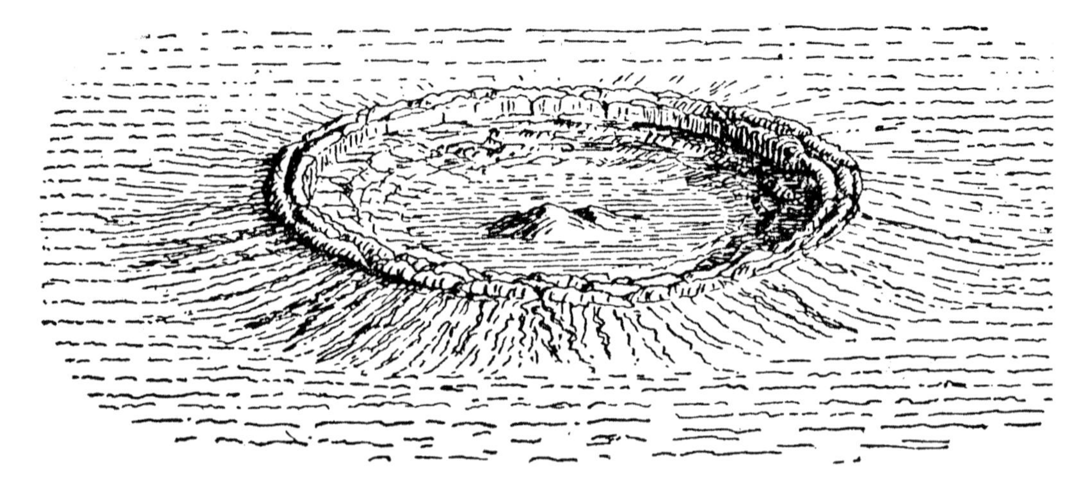

Figure 28 Sketch of a complex lunar crater made by Grove Karl Gilbert (1893: 243)

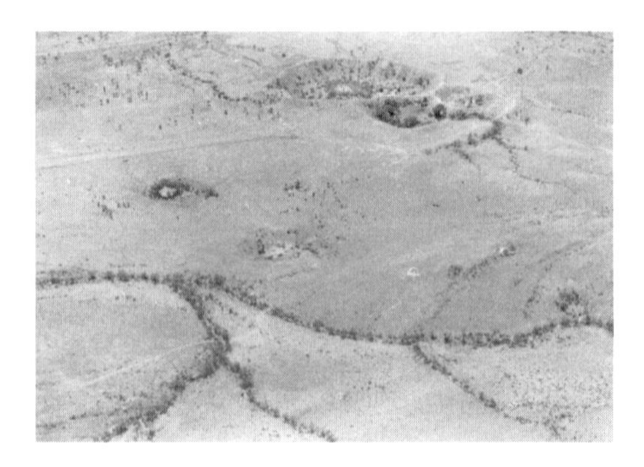

Plate 29 Henbury impact craters in central Australia. These structures formed when a group of meteors struck a pediment surface less than about 5,000 years ago (Milton 1968). The largest of the craters, part of a tight group of four, is about 150 m in diameter and about 10 to 15 m deep. Note the capture of drainage by the craters

Earth impact sites are many that include multiple craters (Plate 29). The Kaali impacts, which struck Estonia about 2,400 to 2,800 years ago, consist of nine craters, the largest of which is 110 m wide and 20 m deep.

Impact crater processes

Meteors and comets arrive at velocities of many metres per second, causing an immense transfer of energy in an exceedingly short period of time as they strike the surface of a planet. The actual cratering process is surprisingly orderly, and very well known from both theoretical and experimental work (Melosh 1989). The initial phase is contact and compression, which lasts only a few times longer than the time it takes for the projectile to traverse its own diameter. This produces prominent very high-velocity jets of highly shocked material that shoot upward from the margins of the deforming projectile. A zone of phenomenally high pressure is produced at the front of the projectile, as it is deformed by contact with the target material. In the inner solar system the target is usually rock, but in the outer solar system the satellites of Jupiter, Saturn, Uranus and Neptune are commonly icy. The ices are so cold that they generally behave like rock.

Contact and compression is followed by an ejection or excavation phase. The projectile is melted or evaporated by a shock wave propagating into it, while another shock wave propagates into the target. The shock wave is followed by rarefaction waves that decompress the material and generate excavation flows that open up a transient crater. This excavation process may last several minutes, depending on the energy level of the original impact. Material ejected from the crater will then comprise an outwardly expanding ejecta curtain, which has the form of an inverted cone, centred on the impact site. Material deposited from this curtain will comprise an ejecta blanket that covers the terrain out to about two crater radii from the rim. Additional large ejecta blocks may create additional impacts, or secondary craters. These have distinctive morphologies because of the slower projectile velocities, highly oblique paths and radial structure in relation to their source craters.

At the end of the excavation stage the transient crater will often experience collapse and modification. For the larger complex craters this produces terraces and central peaks. The terraces develop by slumping of the crater rim after all material has been excavated. The central peaks represent uplift of the floor material beneath the transient crater cavity. A peak ring may form as the central peak grows and collapses.

Cratered landscapes

Cratered landscapes dominate on the surfaces of rocky objects in the solar system. This is mainly because most of those surfaces are very old. In general, the density of impact craters on a surface corresponds approximately to the age of that surface. However, this relationship holds on very long timescales. Moreover, it is not linear. During the early part of solar system history, the impacting rate was extremely high. From the final accretion of planets and many satellites, about 4.5 billion years ago, until about 3.9 billion years ago for the moon, and perhaps a few hundred million years later for Mars, there was a period of intense heavy bombardment. This produced overlapping craters with sizes up to the scale of the multi-ring basins. The scaling is very regular with many more craters of smaller sizes than of larger. After the heavy bombardment, which was caused by many objects left over from solar system formation, the impacting rates declined by more than an order of magnitude. On the moon these timescales of cratering have been confirmed by radiometric dates on rocks returned to Earth by the Apollo missions. The lunar highlands correspond to the heavy bombardment phase, and much lower crater densities mark the younger volcanic plains of the lunar mare, which occur on the floors of very large impact basins. On Mars there is a similar dichotomy between old, heavily cratered highlands and younger, lightly cratered plains. Unlike the moon, however, many Martian craters are highly degraded by erosion, including the action of fluvial, periglacial and glacial processes.

References

Fairbridge, R.W. (1968) Crater, in R.W. Fairbridge (ed.) *Encyclopedia of Geomorphology* 207–218, New York: Reinhold.

Gilbert, G.K. (1893) The Moon's face: a study of the origin of its features, *Philosophical Society of Washington Bulletin* 12, 241–292.

——(1896) The origins of hypotheses illustrated by the discussion of a topographic problem, *Science* n.s. 3, 1–13.

Melosh, H.J. (1989) *Impact Cratering: A Geologic Process*, Oxford: Oxford University Press.

Milton, D.J. (1968) Structural geology of the Henbury meteorite craters, Northern Territory, Australia, *US Geological Survey Professional Paper* 499-C, 1–17.

Roddy, R.J., Pepin, R.O. and Merrill, R.B. (eds) (1977) *Impact and Explosion Cratering*, New York: Pergamon.

SEE ALSO: astrobleme; caldera; extraterrestrial geomorphology; volcano

VICTOR R. BAKER

CRATON

The central core of extensive, stable continental crust in present-day continents that has achieved tectonic stability. All cratons are older than 570 million years, dating from the Precambrian period. Cratons have essentially rigid foundations, composed of predominantly granite and metamorphosed rocks that have been deposited on pre-existing older basement rocks. They are generally low-lying with little relief, as a result of erosion. Cratons have only been affected by EPEIROGENY and are devoid of orogenic features and recent volcanic activity.

The term craton is derived from the Greek word 'kraton' meaning shield and should only be applied to continents and not to oceans, according to the theory of plate tectonics. Cratons are added to by the process of cratonization, an important mechanism for continental growth. Sediments accumulate within thick linear troughs on the cratonic margins. Here, the material is eventually deformed and partially melted onto the existing craton. Early Achaean cratons were smaller and greater in number, yet through the process of cratonization throughout Phanerozoic time cratons became larger and fewer in number as they were fused together.

The area of a craton that becomes exposed is termed a SHIELD. Shields are composed of ancient crystalline basement rocks, and represent the core of the craton. The Canadian Shield is an example;

it is composed of granite and metamorphic rocks (e.g. gneisses), alongside heavily deformed metamorphosed sedimentary (e.g. quartzites) and other volcanic rocks. The term shield is also sometimes employed as a synonym for craton.

The shield is unconformably overlapped at its margins by thin sedimentary units, termed platforms. Platforms are typically *c.*1 km thick and derived from the Palaeozoic and Mesozoic periods, predominantly composed of shallow marine sandstones, limestones and shales.

Since cratons are tectonically stable, sediments tend to spread out widely into any areas of relatively low-lying ground, such as the intra-cratonic basins. These are typically shallow (though can range up to 3,000 m), bowl-shaped, and are characterized by very slow subsidence. Basin sediments thicken regularly towards the centre, yet their fill is discontinuous. As such, the stratigraphy reflects major transgressions across the entire craton, punctuated by periods of stability. Many of them develop as shallow 'sag' lakes, such as Lake Chad in North Africa. The cause of intra-cratonic basins remains contentious.

Further reading

Condie, K.C. (1997) *Plate Tectonics and Crustal Evolution*, 4th edition, Oxford: Butterworth Heinmann.

STEVE WARD

CROSS PROFILE, VALLEY

In most introductory physical geography and geology textbooks a distinction is made between V-shaped valley cross profiles, described as characteristic of a system dominated by active fluvial erosion, and U-shaped valley cross profiles, described as characteristic of a system dominated by GLACIAL EROSION. This process-oriented distinction gained wide popularity as a component of classic Davisian landscape classification, particularly in the first half of the twentieth century, and continues to be used in more modern landscape interpretation and analysis. Morphometric analyses have been used to show that glaciated valley cross-section profiles can be approximated by the mathematical equivalent of the letter U, a parabolic equation, whereas fluvial valley side slopes are more nearly linear. In addition, the amount of rock removal required to convert a V-shaped cross-profile geometry to a U-shape has been used as a measure of the glacial erosion component of valley development, and the extent of valley development towards a particular form has been used as a measure of the degree of valley modification by fluvial or glacial processes. However, valley cross profiles include a much wider variety of forms than the two-fold division into V- and U-shaped suggests, and cross-profile forms are not only controlled by glacial and fluvial erosion, but also by patterns of hillslope erosion and deposition, and by patterns of rock resistance to erosion.

Typical explanations for the development of V-shaped valleys in areas with active fluvial erosion include several components: (1) that river erosion is dominantly vertical; (2) that the river is capable of transporting all of the material supplied to it by hillslope processes; and (3) that valley side slopes are steepened to a critical angle for hillslope transport or failure. This ideal set of conditions results in uniform valley side slope angles either side of a central river that is eroding vertically into the landscape with little or no floodplain, i.e. a V-shaped cross profile. However, if the river is not capable of transporting all the material supplied to it by hillslope processes, if there are significant lithological variations along the slope profile, or if different hillslope processes dominate in separate parts of the slope profile, then more complex hillslopes and valley bottoms will develop than the simple linear form required for a V-shaped cross profile.

Typical explanations for the development of U-shaped valleys as a result of glacial erosion rely on the argument either that glacial 'valleys' are actually glacial channels, and that steep side walls and a relatively flat bottom is a characteristic form for fluid flow in channels, or that the cross-sectional pattern of erosion under a glacier includes a wide central maximum leading to steep side walls and a low gradient profile section in the channel centre. Numerical modelling linking ice dynamics, subglacial erosion patterns and cross-profile form development has demonstrated that U-shaped cross sections can result solely from glacial erosion in homogeneous bedrock. However, when spatial variations in rock resistance to erosion are introduced to the model, a wide variety of cross-section shapes can develop, including V-shaped forms. In addition, many glaciated valleys used to illustrate U-shaped valleys or included in morphometric analyses of valley form include substantial depositional components; the U-shaped form arises from the combination of a low gradient valley floor (fluvioglacial deposition), and a concave talus slope (postglacial and ice marginal slope processes) below steep bedrock walls (glacially modified).

Although the distinction between idealized U-shaped and V-shaped valleys for areas dominated by glacial and fluvial erosion is useful, there is in fact a wide variety of valley cross-profile forms. Other and more complex cross profiles result from temporal and spatial variations in processes across the profile, including both erosion and deposition, and from patterns of surface material resistance to erosion.

Further reading

Augustinus, P.C. (1995) Glacial valley cross-profile development: the influence of *in situ* rock stress and rock mass strength, with examples from Southern Alps, New Zealand, *Geomorphology* 11, 87–97.

Carson, M.A. and Kirkby, M.J. (1972) *Hillslope Form and Process*, Cambridge: Cambridge University Press.

Harbor, J. (1995) Development of glacial-valley cross sections under conditions of spatially variable resistance to erosion, *Geomorphology* 14, 99–107.

Hirano, M. and Aniya, M. (1988) A rational explanation of cross-profile morphology for glacial valleys and of glacial valley development, *Earth Surface Processes and Landforms* 13, 707–716.

SEE ALSO: hillslope, form; hillslope, process; valley

JON HARBOR

CRUSTAL DEFORMATION

Motions of the lithosphere disrupt and modify rocks and the topographic surface. As a manifestation of PLATE TECTONICS, these deformations maintain continental forms that protrude above sea level. Crustal deformations, such as fault offsets and folds, produce diverse constructional landforms dependent on local material properties and surface processes. ACTIVE MARGINS are shaped by competition between deformation and erosion. Deformation occurs at the timescale of plate motions (centimetres per year) but can be slower along individual structures. Recent technologies have revolutionized crustal deformation studies, such as space-based geodesy (e.g. GPS) and seismology that constrain short-term behaviour, dating techniques (e.g. COSMOGENIC DATING) that constrain chronologies of offset geologic markers, and DIGITAL ELEVATION MODELS that permit topographic assessment of large areas.

Crustal deformation leads to OROGENESIS and basin formation over the long term, producing wholesale surface uplift, DENUDATION, and SUBSIDENCE (see TECTONIC GEOMORPHOLOGY). Fluvial systems respond to perturbations in BASE LEVEL where the crust has risen or fallen (Burbank and Anderson 2001). Long-profiles (see LONG PROFILE, RIVER) of stream channels adjust to uplift via KNICKPOINT migration and incision, often leaving behind suites of terraces. Drainage networks may also be modified, as streams can be deflected by zones of uplift or forced to migrate by tilting (see ASYMMETRIC VALLEYS). Sediment loading and gradient changes further influence fluvial form, such as the occurrence of meandering versus BRAIDED RIVER channels. Adjustment to base level in turn affects hillslope processes, leading to increased RELIEF, hillslope length and sediment production. Glacial and coastal erosion similarly respond to uplift and subsidence. Displaced geomorphic features, such as river terraces (see TERRACE, RIVER), shorelines, ALLUVIAL FANs, strata, MORAINEs and PLANATION SURFACEs, serve as markers that are valuable constraints on relative uplift.

Crustal deformation is most commonly associated with faults (see FAULT AND FAULT SCARP). Dislocations occur along lengths of faults during rupture events, producing earthquakes as a side effect. Ruptures that break the surface are

Plate 30 Crustal deformation in alluvium produced locally along the Emerson fault during the 28 June 1992 Landers earthquake in California (M = 7.3). The scarp faces to the south-west and is approximately 1 m high. Its height is locally accentuated by lateral offset of the hilly topography. Dextral offset of ~5 m is evident in the displaced stream course. This photograph was taken several days after the earthquake by Kerry Sieh (California Institute of Technology, USA)

typically tens of kilometres long and involve metres of slip. They are quantified in terms of seismic moment: $M_0 = \mu AD$, where μ is rigidity, A is rupture area, and D is average displacement. Earthquake size is thus partly dependent on rupture length, which is controlled by fault zone geometry and segmentation (Plate 30). Coseismic displacement also scales with rupture length, such as the tendency for slip to be $\sim 10^{-4} - 10^{-5}$ of the length of strike-slip fault ruptures. This scaling is related to the elastic strain the crust adjacent to faults sustains during interseismic periods. The release of accumulated strain provides for moderately regular rupture recurrence. Short-lived faulting events are thus the building blocks by which plate motion translates into long-term deformation (Yeats *et al.* 1997). Over the long term, fault displacements tend to scale as several per cent of the total fault length.

Each of the three main types of plate boundaries consists of faults characterized by certain landforms. Strike-slip faults produced by simple shear involve mainly horizontal displacement and create a minimal degree of topographic disruption. Linear troughs are common along such faults, where weakened fault rocks (see CATACLASIS) are easily eroded by deflected stream courses (e.g. the San Andreas fault). Landforms produced by transpression and transtension at restraining and releasing fault bends include pressure ridges, pull-apart basins (see PULL-APART AND PIGGY-BACK BASIN), and variably faced scarps (scissoring). Strike-slip faults also disrupt geomorphic features horizontally, creating shutter ridges (topographic steps) and deflected or BEHEADED VALLEYs and streams (Sieh and Jahns 1984).

Dip-slip faults involve primarily vertical motion. Normal faults are produced by horizontal extension, where maximum compressive stress is oriented vertically. Resulting fault planes typically dip steeply (~60°). Normal faults juxtapose tilted basement blocks and alluvial valleys in the characteristic basin and range terrain. Vertical separation tends to be asymmetric, with valley subsidence exceeding uplift of basement blocks. Edges of uplifted blocks may preserve FLAT IRONs (triangular facets) related to the fault surface. Mountain fronts typically consist of linear segments interrupted by complex transfer zones, such as the Wasatch front (Machette *et al.* 1992). Parallel normal faults produce down-dropped rift valleys (grabens) (see RIFT VALLEY AND RIFTING) and upthrown blocks (HORSTs).

Reverse or thrust faults are produced by horizontal compression, where the least principal stress is oriented vertically. Thrust fault planes dip shallowly (~30°) and produce irregular mountain fronts that involve wide belts of deformation (Philip and Meghraoui 1983). The degree to which such piedmonts are dominated by erosion, deposition and deformation is represented by numerous geomorphic characteristics, including sinuosity, fan entrenchment and valley geometry. Thrust belts typically involve overlapping arcuate fault segments in parallel series that are connected by secondary structure. These may also involve folding, as typical of foreland fold and thrust belts such as along the Nepal Himalaya (Schelling and Arita 1991). Megathrusts of subduction zones create unique cycles of elastic uplift and subsidence in both hanging wall and footwall, leading to rhythmic perturbation of coastal geomorphology.

Deformation along faults during rupture events can be complex. Fault traces tend to be irregular, such as the characteristic en echelon, anastomosing arrangement of faults within wide (~50 m) ruptures of strike-slip faults (Yeats *et al.* 1997). These shear zones can involve pervasive shearing, although slip tends to concentrate along principal displacement zones. A variety of microgeomorphic features are produced during surface ruptures (see SEISMOTECTONIC GEOMORPHOLOGY). Fault scarps record the vertical separation along faults and portray characteristics linked with fault orientation. Scarp degradation through time occurs predictably by incision and diffusive hillslope creep, such that scarp form is related to scarp age (Avouac and Peltzer 1993). These distinctive landforms record deformation history that can be unravelled using palaeoseismology.

Tectonic strain is also accommodated by FOLDing of rock and sediment, particularly in deep basins. Folding of near surface involves permanent brittle deformation in the form of penetrative intergranular shear or flexural slip between strata. Folds are often associated with blind thrust faults and evolve as faults propagate towards the surface. Fold geometry is closely linked with fault bend and tip geometry. Ongoing deposition around folds can result in piggy-back basins and growth strata that itself becomes folded. Erosion and deposition can also mask the topographic expression of folding in unconsolidated sediment. Processes of diagenetic and pedogenic lithification are thus important for fold preservation. Because strata vary in composition and resistance to erosion, ancient folds can be exhumed by erosion, such as palaeo-folds of the Appalachian Valley and Ridge.

References

Avouac, J.-P. and Peltzer, G. (1993) Active tectonics in southern Xinjiang, China: analysis of terrace riser and normal fault scarp degradation along the Hotan-Qira fault system, *Journal of Geophysical Research* 98, 21,773–21,807.

Burbank, D.W. and Anderson, R.S. (2001) *Tectonic Geomorphology*, Massachusetts: Blackwell Science.

Machette, M.N., Personius, S.F. and Nelson, A.R. (1992) The Wasatch Fault Zone, U.S.A., *Annales Tectonicae* 6, 5–39.

Philip, H. and Meghraoui, M. (1983) Structural analysis and interpretation of the surface deformations of the El Asnam earthquake of October 10, 1980, *Tectonics* 2, 17–49.

Schelling, D. and Arita, K. (1991) Thrust tectonics, crustal shortening, and the structure of the far-eastern Nepal Himalaya, *Tectonics* 10, 851–862.

Sieh, K.E. and Jahns, R.H. (1984) Holocene activity of the San Andreas fault at Wallace Creek, California, *Geological Society of America Bulletin* 95, 883–896.

Yeats, R.S., Sieh, K. and Allen, C.R. (1997) *The Geology of Earthquakes*, Oxford: Oxford University Press.

JAMES A. SPOTILA

CRUSTING OF SOIL

Crusts are thin layers, different in character from the soil beneath, that develop at the interface between the soil and the atmosphere.

One class of crust is often termed inorganic or rain-beat crusts. Large amounts of energy and high transient forces are imparted to the surface of the soil by the impact of raindrops. These break down soil aggregates, compress the surface and dislodge particles. This physical disruption, which may be especially effective where vegetation cover is limited, is aided by certain chemical processes, which include dispersion, which cause further breakdown of soil aggregates. The resulting dense surface layer forms a surface seal, and when this seal dries, a crust forms. Such a crust can have profound effects on runoff and on erosion by wind and water (Poesen and Nearing 1993).

Recently it has been recognized that organic (also called microphytic, microbiotic, cryptogamic or biological) crusts in and on the surfaces of soils play important hydrological and geomorphological roles (Eldridge and Rosentreter 1999). Organic compounds, including plant waxes, can produce hydrophobic (water repellent) substances, as can a range of fungi and soil micro-organisms. Although water repellent soils occur in more humid environments, there are many examples of them that have been reported from semi-arid areas (Doerr *et al.* 2000). These hydrophobic surfaces tend to be zones of reduced soil infiltration capacity and thus of increased overland flow. Following from this is the likelihood that enhanced soil erosion also occurs. Removal of the crusts has been shown to have a dramatic effect on infiltration rates (Eldridge *et al.* 2000).

Likewise biological soil crusts have an influence on aeolian processes. A cover of cyanobacteria, green algae, lichens and mosses is important in stabilizing soils in drylands and thus protects them from wind erosion. They play a role in dune stabilization (Kidron *et al.* 2000). Unlike vascular plants, the cover of organic crusts is not reduced in drought years and they are present the whole year round. However, they are very susceptible to anthropogenic disturbance (Belnap and Gillette 1997). Filamentous cyanobacteria mats are especially effective against wind attack (McKenna-Neuman *et al.* 1996). The filaments and extracellular secretions of cyanobacteria also form water stable aggregates that help soils to resist water erosion and raindrop impact effects (Issa *et al.* 2001). It also needs to be appreciated that not all organic crusts are hydrophobic and that by eliminating the effect of raindrop impact, they prevent the rapid development of a sealed rain-beat crust conducive to runoff generation (Kidron and Yair 1997).

References

Belnap, J. and Gillette, D.A. (1997) Disturbance of biological soil crusts: impacts on potential wind erodibility of sandy desert soils in southeastern Utah, *Land Degradation and Development* 8, 355–362.

Doerr, S.H., Shakesby, R.A. and Walsh, R.P.D. (2000) Soil water repellancy: its causes, characteristics and hydro-geomorphological significance, *Earth-Science Reviews* 51, 33–65.

Eldridge, D.J. and Rosentreter, R. (1999) Morphological groups: a framework for monitoring microphytic crusts in arid landscapes, *Journal of Arid Environments* 41, 11–25.

Eldridge, D.J., Zaady, G. and Shachack, M. (2000) Infiltration through three contrasting biological soil crusts in patterned landscapes in the Negev, Israel, *Catena* 40, 323–336.

Issa, O.M., Le Bissonnais, Y., Défrage, C. and Trichet, J. (2001) Role of cyanobacterial cover on structural stability of sandy soils in the Sahelian part of Western Niger, *Geoderma* 101, 15–30.

Kidron, G.J. and Yair, A. (1997) Rainfall-runoff relationship over encrusted dune surfaces, Nizzama, western Negev, Israel, *Earth Surface Processes and Landforms* 22, 1,169–1,184.

Kidron, G.J., Barzilay, E. and Sachs, E. (2000) Microclimate control upon sand microbiotic crusts, western Negev Desert, Israel, *Geomorphology* 36, 1–18.

McKenna-Neuman, C., Maxwell, C.D. and Bouton, J.W. (1996) Wind transport of sand surface crusted with photoautotrophic micro organisms, *Catena* 27, 229–247.

Poesen, J.W.A. and Nearing, M.A. (eds) (1993) Soil surface sealing and crusting, *Catena Supplement* 24.

A.S. GOUDIE

CRYOPLANATION

Cryoplanation (Bryan 1946) is a morphogenetic term introduced to explain and describe low-angled slope surfaces occurring on higher valley-side and summit positions (cryoplanation terraces, or benches), or in valley-side foot positions (cryopediments) in periglacial regions. Cryoplanation and altiplanation are synonymous, and both are forms of equiplanation. Cryoplanation terrace has subsumed several other periglacial terrace terms, including: goletz, altiplanation, NIVATION and equiplanation.

Alleged cryoplanation terraces have flats or treads of 1° to 12° with a sharp inflection in slope at the upslope limit (sometimes called the knickpoint) where risers are often 25° to 35°. Terrace width is often only a few metres, but claims in excess of one kilometre exist (Demek 1969). Sets of cryoplanation terraces produce a staircase effect on a hillside, and their convergence on a summit from two or more sides may produce a summit flat. Both terrace size and frequency appear to increase with time since deglaciation, but terraces may also occur in unglaciated regions. Terrace relationship to permafrost and rock structure is extremely uncertain, but adjustment to rock type is reported. Transport of debris across entire sets of cryoplanation terraces seems essential in some circumstances. This appears to be problematic as lower terraces would have to export all debris from upslope terraces unless it was shed laterally which seems unlikely.

Cryopediments are subject to the same uncertainties associated with tropical pedimentation. A cryopediment is viewed as expanding by headward incision by freeze–thaw weathering, or nivation more broadly. The flat is viewed as a bedrock surface veneered by debris experiencing common periglacial mass wasting processes, e.g. SOLIFLUCTION.

As a process cryoplanation has no unique elements but appears to be synonymous with nivation (Thorn and Hall, 2003) which itself merits more precise articulation. While emphasis has been placed on nivation during the early stage of cryoplanation specifically (Demek 1969), no other specific process (while implied) has ever been invoked for the mature stages. If large perennial snowpatches are protective rather than erosive, as well may be the case, largeness and/or increasingly cold climate may not favour headward expansion. The presence of patterned ground on the tread or transport surface is often, but not always, invoked as an indicator of inactivity.

While the landforms designated cryoplanation terraces or cryopediments are clearly found in periglacial environments, the general absence of sound process research (but see Hall 1997) renders their origins unknown. This problem is exacerbated by the apparent present inactivity of many such features. Nelson (1989) has suggested that cryoplanation terraces may be a periglacial analogue of cirque glaciation reflecting a precipitation/temperature regime unable to sustain full glaciation. Hall (1998) and Thorn and Hall (2003) have suggested that the distinction between cryoplanation and nivation forms and processes needs careful re-examination as it is presently far from clear.

References

Bryan, K. (1946) Cryopedology – the study of frozen ground and intensive frost-action with suggestions on nomenclature, *American Journal of Science* 244, 622–642.

Demek, J. (1969) Cryoplanation terraces, their geographical distribution, genesis and development, *Československé Akademie Věd Rozpravy, Řada Matematických a Přírodních Věd*, 79(4).

Hall, K. (1997) Observations on 'cryoplanation' benches in Antarctica, *Antarctic Science* 9, 181–187.

——(1998) Nivation or cryoplanation: different terms, same features? *Polar Geography* 22, 1–16.

Nelson, F.E. (1989) Cryoplanation terraces: periglacial cirque analogs, *Geografiska Annaler* 71A, 31–41.

Thorn, C.E. and Hall, K. (2003) Nivation and cryoplanation: the case for scrutiny and Integration, *Progress in Physical Geography* 26, 553–560.

Further reading

Priesnitz, K. (1988) Cryoplanation, in M.J. Clark (ed.) *Advances in Periglacial Geomorphology*, 49–67, Chichester: Wiley.

SEE ALSO: nivation

COLIN E. THORN

CRYOSTATIC PRESSURE

The elevated water potential in saturated, coarse-grained sediments caused by freezing in a closed system. Where the volumetric expansion of water by 9 per cent on becoming ice cannot be accommodated in freezing sediment, pore water is expelled into proximal unfrozen ground, raising the water pressure. Cryostatic pressure is responsible for the uplift of closed-system pingos, beneath which pressures of up to 0.4 MPa have been measured (Mackay 1977). In fine-grained soils, cryostatic pressure may develop at the beginning of laboratory freezing tests, but it has not been measured under field conditions.

Reference

Mackay, J.R. (1977) Pulsating pingos, Tuktoyaktuk Peninsula, N.W.T., *Canadian Journal of Earth Sciences* 14, 209–222.

C.R. BURN

CRYPTOKARST

Cryptokarst is a form of karstification limited to the EPIKARSTIC zone. It is always developed under a cover of superficial formations resulting from deposition (loess, etc.) or weathering (alterite). The quality and the thickness of the superficial formation have a direct influence on cryptokarst activity (Nicod 1994).

The main source of acid in ground water is the surface, with the percolation of humic acid (biological activity) and the gaseous exchanges between atmosphere and rainwater. Thus the epikarstic zone is submitted to intense dissolution (Klimchouk 1995) due to the proximity of the surface. The formation of the cryptokarst is enabled by the layer of superficial formation that distributes the water in a diffuse manner, avoiding the concentration of water with high dissolution potential. The chemical equilibration between water and terranes (Stumm and Morgan 1981) implies that the dissolution capacity will decrease proportionally to the residence time of water in the epikarst zone. To stay in the cryptokarst phase, the karstification has to be aborted before reaching the underlying rocks. It means that there will not be any transmission of aggressive water below the epikarst zone, i.e. no water at all either because there are no fast paths (diaclases) or water is non-aggressive because it has already reached equilibrium with the rocks. Concentration of clayey particles that originate from the weathering of the superficial formations can also lead to the clogging of the bedrock interface. In some circumstances, the superficial formations may be drawn down with the vertical progression of the cryptokarstic front and this may induce surface depressions like dolines. On the other hand, the layer of superficial material protects the cryptokarst from surface mechanical erosion.

The geological and topographical conditions for cryptokarst are found in the chalky Cretaceous formations of the Paris Basin (Rodet 1992), England and Denmark. The chalky basement is slightly tectonized (with a resultant low density of fracturation) and the relief is composed of plateaus separated by DRY VALLEYS. The carbonate components of the chalk are easily dissolved but the argillaceous part remains in place. The argillaceous particles are hardly removed by the horizontal water movement. This causes a reduction of the permeability and of the capacity of the basement to be eroded (Lacroix *et al.* 2002).

References

Klimchouk, A.B. (1995) Karst morphogenesis in the epikarstic zone, *Cave and Karst Science* 21, 45–50.
Lacroix, M., Rodet, J., Wang, H.Q., Laignel, B. and Dupont, J.P. (2002) Microgranulometric approach to a chalk karst, western Paris Basin, France, *Geomorphology* 44, 1–17.
Nicod, J. (1994) Plateaux karstiques sous couverture en France, *Annales de Géographie* 576, 170–194.
Rodet, J. (1992) *La craie et ses karsts*, Caen: Ed. CNEK-Groupe Seine.
Stumm, W. and Morgan, J.J. (1981) *Aquatic Chemistry*, New York: Wiley.

SEE ALSO: chemical weathering; epikarst; ground water; karst; palaeokarst and relict karst

MICHEL LACROIX

CRYPTOVOLCANO

A roughly circular area of greatly disturbed rocks and sediments that is morphologically suggestive of being the result of volcanic activity but does not contain any true volcanic materials. Very often the origin of the features has been a matter of controversy. For example, the Pretoria Salt Pan crater in South Africa and the great Vredefort Dome have in the past sometimes been interpreted as volcanic features, but there is now an accumulation of evidence that they both result from meteorite impact (Reinold *et al.* 1992; Reinold and Coney 2001). Conversely, Upheaval Dome in the Canyonlands National Park, Utah, USA, has variously been attributed to meteoric impact, fluid escape, cryptovolcanic explosion and salt doming, with the last explanation now being favoured (Jackson *et al.* 1998). Some features, including a group of eight structures running in a 200-km straight line across the USA, may be the result of comet or asteroid impact (Rampino and Volk 1996) (see ASTROBLEMES, CRATERS). The cryptovolcanic features discussed above show a great range in size. The Pretoria Salt Pan crater is 1.13 km in diameter, whereas the Vredefort structure is 250–300 km across. The aligned structures in the USA are *c*.3–17 km in diameter.

The problem of establishing the origin of closed depressions and circular structures is made evident when one considers the range of hypotheses that have been put forward to explain the Carolina Bays in the eastern USA (Ross 1987):

1 Spring basins
2 Sand bar dams or drowned valleys
3 Depressions dammed by giant sand ripples
4 Craters of meteor swarm
5 Submarine scour by eddies, currents or undertow
6 Segmentation of lagoons and formation of crescentic keys; original hollows at the foot of marine terraces and between dunes
7 Lakes in sand elongated in direction of maximum wind velocity
8 Solution depression, with wind-drift sand forming the rims
9 Solution depressions, with magnetic highs near bays due to redeposition of iron compounds leached from the basins
10 Basins scoured out by confined gyroscopic eddies
11 Solution basins of artesian springs with lee dunes
12 Fish nests made by giant schools of fish waving their fins in unison over submarine artesian springs
13 Aeolian blowouts
14 Bays are sinks over limestone solution areas streamlined by ground water
15 Oriented lakes of stabilized grassland interridge swales of former beach plains and longitudinal dunefields with some formed from basins in Pleistocene lagoons
16 Black hole striking in Canada (Houston Bay) throwing ice onto coastal plain
17 Cometary fragments exploding above surface, their shock waves creating depressions
18 Drought with subsequent fire in peat bogs followed by aeolian activity.

References

Jackson, M.P.A, Schultz-Ela, D.D., Hudec, M.R., Watson, I.A. and Porter, M.L. (1998) Structure and evolution of Upheaval Dome: a pinched-off salt diapir, *Geological Society of America Bulletin* 110, 1,547–1,573.

Rampino, M.R. and Volt, T. (1996) Multiple impact event in the Paleozoic: collision with a string of comets or asteroids? *Geophysical Research Letters* 23, 49–52.

Reinold, W.V. and Coney, L. (2001) *The Vredefort Impact Structure and Directly Related Subjects: An Updated Bibliography*, Economic Geology Research Institute, University of the Witwatersrand, Johannesburg, Information Circular No. 353.

Reinold, W.V., Koeberl, C., Partridge, T.C. and Kerr, S.J. (1992) Pretoria Saltpan crater: impact origin confirmed, *Geology* 20, 1,079–1,082.

Ross, T.E. (1987) A comprehensive bibliography of the Carolina Bays Literature, *Journal of the Elisha Mitchell Scientific Society* 103, 28–42.

A.S. GOUDIE

CUESTA

An asymmetric ridge built of dipping sedimentary rocks of alternating resistance against weathering and erosion, elongated along the strike of strata, is called a cuesta. The steep front slope is opposite to the dip, whereas the gently sloping backslope is more or less parallel to the dip. The top part of the cuesta face and the backslope are built of more resistant strata; less resistant ones are exposed in the lower part of the front scarp.

Because of contrasting slope and lithology, each side of a cuesta is shaped by different sets of processes. Rapid mass movement and gully erosion predominate on the steeper slope, and fluvial incision and slow mass movement operate on the backslope. Hence in the long term a cuesta both retreats and is worn down. There is a number of theories how cuesta ridges form but most emphasize differential fluvial erosion within a monocline, which leaves outcrops of more resistant strata as divides and initial cuestas, which then begin to retreat. Bevelled ridge tops indicate that cuesta have developed from a former plain through river incision.

Cuesta is an example of a structure-controlled and climate-independent landform. Classic cuesta landscapes include the Colorado Plateau in North America, the Paris Basin in France and Southern German Uplands.

Further reading

Ahnert, F. (1996) *Einführung in die Geomorphologie*, Stuttgart: Ulmer.

Schmidt, K.-H. (1994) The groundplan of cuesta scarps in dry regions as controlled by lithology and structure, in D.A. Robinson and R.B.G. Williams (eds) *Rock Weathering and Landform Evolution*, 355–368, Chichester: Wiley.

SEE ALSO: caprock; escarpment; mesa; sandstone geomorphology; structural landform

PIOTR MIGOŃ

CURRENT

The hydrodynamics responsible for sediment entrainment and transport, erosion and accretion, and morphological change within the coastal zone, consist of *oscillatory motions* associated with WAVES of various frequencies and forms and *quasi-steady, unidirectional currents*. The currents are forced by: (1) a secondary effect of the waves themselves, i.e. *wave drift* or *wave streaming*; (2) *tides*; (3) *wind* stress; (4) *pressure* and *density* gradients; and (5) a variety of motions resulting from the dissipation of wave energy at, and landward of, depth-controlled breaking (*surf zone*). Here the kinetic energy of the waves is transformed into: (a) increased macro and micro turbulence; (b) *drift currents* associated with secondary progressive or standing waves, generally of lower frequency than the incident waves (e.g. edge waves, leaky waves, etc.); (c) *longshore currents*; (d) *rip currents*; (e) *undertow*; (f) *swash*.

Wave streaming (wave drift)

Stokes in 1847 first recognized that WAVE orbital motions were not closed in the case of small amplitude waves in a perfect non-viscous fluid, even in deep water. The fluid particles have a second-order, wave-averaged, mean *Lagrangian* velocity and thus there is a finite *mass flux* of water. Since horizontal velocities increase slightly with distance above the bed, so that particle motion under the crest is slightly larger than under the trough, conservation of mass causes a stratification of flow (Figure 29a). In shallow water, with greater bed friction, the wave orbits become elliptical and drift velocities increase (~0.1 m s^{-1}). A *Eulerian* measure of mass flux can also be obtained by integrating the horizontal velocities beneath the crest and trough over space and time; the same mass flux is obtained although the vertical distribution is different. For real viscous fluids, and waves in finite depth, Longuet-Higgins (1953) showed that there is a time-averaged, net downward transfer of momentum into the boundary layer at the bed, producing a *Eulerian* streaming in addition to the Stokes drift. Again by conserving mass, a stratified profile of the mean current is obtained (Figure 29b); flow is in the direction of propagation at the bed and a reversal occurs at mid-depth; Klopman (1994) has confirmed this pattern through laboratory experiments. In strongly asymmetric flows over steep slopes, shear stresses within the boundary layer may cause a reversal (upwave) mean current at the bed.

Surf zone currents

Currents in the surf zone interact with the instantaneous wave orbital motions (over the full range of short and long period waves) producing a rather complex time-dependent three-dimensional pattern (Svendsen and Lorenz 1989; Figure 30). This is usually disaggregated into a number of distinct components:

Longshore currents are generated when waves break at an oblique angle to the shoreline, and the alongshore component of the onshore directed

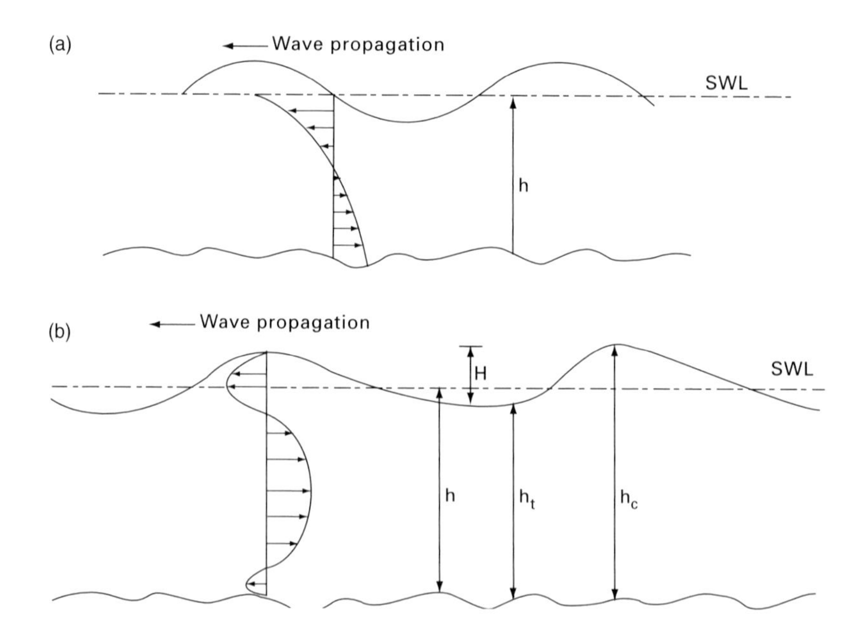

Figure 29 Wave-induced quasi-steady currents: (a) Stokes drift; (b) Longuet-Higgins mass transport

radiation stress (see WAVE) forces a shore-parallel or *longshore current*. Pressure gradients due to water level *set-up* differentials along shore, as well as the shore-parallel component of the onshore *wind stress*, can enhance (or reduce) this forcing. Laboratory and field measurements indicate that the longshore current increases landward from the breakpoint, reaches a maximum around the mid-surf zone and decreases to near zero at the shoreline. Since the radiation stress gradient is a maximum at the breakpoint in the ideal theoretical solution (Longuet-Higgins 1970a,b), a 'smoothing' of the momentum flux across the surf zone, called *lateral mixing*, causes the maximum current to be displaced landward. This mixing also causes *longshore currents* to flow outside the zone of breaking, even though the radiation stress gradients approximate zero. Komar (1998) gives a detailed review of the origins and the spatial and temporal patterns of longshore currents.

Undertow or *near-bed return flow* is a pressure gradient driven, time-averaged, mean current directed seawards near to the bed, everywhere along the shoreline. It is caused by cross-shore differences in the mean water elevation due to wave *set-up* at the shoreline and *set-down* under the breaker zone. *Set-up* and *set-down* result from differences in the local *onshore flux of momentum* by waves (*radiation stress*), which is largest at the breakpoint (where waves are largest) and smallest at the final point of wave dissipation at the shoreline. This gradient forces a displacement of the water from beneath the largest waves towards the shoreline and will be complemented by the onshore *mass flux* of water by the waves, as well as any water moved by *wind stress* acting towards the shoreline. Where nearshore sand bars are present, multiple set-ups and set-downs and associated undertows may be formed by the multiple breaker lines (Greenwood and Osborne 1990). Typically velocities are small, but recordings have been made of undertows up to $0.80\,\mathrm{m\,s^{-1}}$.

Rip currents are discrete, narrow, high velocity *jets* of offshore-directed flow across the surf zone, often forming part of a regular horizontal cellular circulation, with associated shore-parallel to oblique *feeder currents*, and an area of flow expansion, the *rip head*, seaward of wave breaking (Figure 31). *Rips* are often associated with cross-shore oriented depressions (*rip channels*) or breaks in a quasi-shore parallel bar, but are found also on uniformly sloping beaches. Rip currents are not generally *steady state* phenomena, but vary both spatially and temporally.

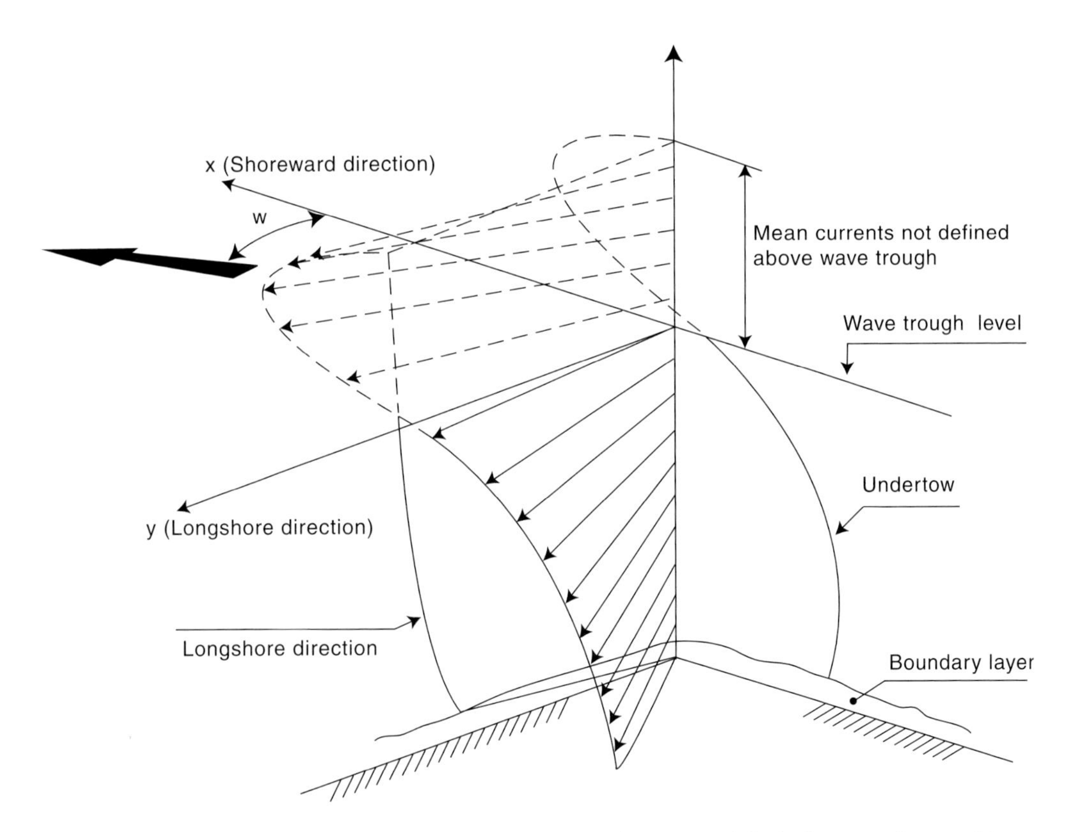

Figure 30 Time-averaged mean velocity vectors in the surf zone (modified after Svendsen and Lorenz 1989)

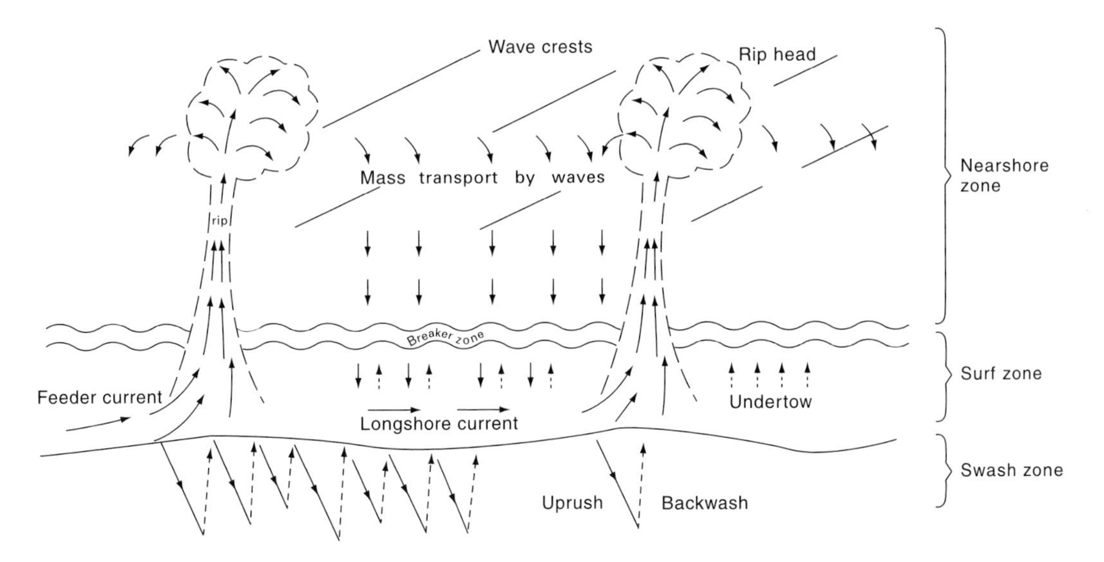

Figure 31 Horizontal cellular circulations in the surf zone

They may be spatially periodic alongshore, with spacing ranging from a few metres (*micro-rips* associated with the embayments of *beach cusps*), to order of 10^2 m (storm cusps), to *mega rips* (Short 1985), which are often single, large-scale, topographically controlled *jets* induced by cellular circulation within the confines of headlands/structures. *Mega rips* may flow offshore for more than one kilometre and can reach speeds $> 3\,m\,s^{-1}$. Theoretically, rip currents result from a periodic modulation of wave heights and thus *wave set-up* alongshore. This modulation has been related to: (1) rhythmic variations in topography (Sonu 1972; Komar 1971); (2) edge waves (Bowen 1969; Bowen and Inman 1969); (3) interference between two incident wave fields (Dalrymple 1975). Rip currents typically pulse at infragravity wave frequencies (< 0.04 Hz), most likely as a result of the increased *radiation stresses* at the *wave group* frequency. Aagaard *et al.* (1997) measured very long period fluctuations of 5–10 minutes. Current speeds may increase with falling tidal levels, especially where topography increasingly confines the current. Rip currents are capable of transporting large volumes of both coarse bedload (especially in migrating megaripples, e.g. Gruszczynski *et al.* 1993) and suspended load from within the surf zone to seaward. Mega rips may well transport nearshore sediments onto the continental shelf below the level of average *wave base* (depth limit of surface waves).

Swash is the *uprush* and *backwash* currents on the beach face and reflects the ultimate dissipation of incident wave energy. Both currents are turbulent, with the former assuming a flow direction coincident with the angle of approach of the breaking wave; the backwash results from gravity acting on the water on the beach and thus flows down the maximum beach slope. This often results in a 'zig-zag' motion of water and sediment. Swash currents depend on the nature of the incoming waves, the beach face slope and the state of the beach water table. If the beach is not saturated there will be a tendency for infiltration of the uprush into the permeable beach face, and thus a reduction in the amount of water in the backwash. Because water can drain from the beach face water table, the backwash tends to last longer and is usually thinner and flows may be supercritical, with hydraulic jumps common. Van Rijn (1998) and Butt and Russell (2000) provide reviews of *swash* hydrodynamics and sediment transport.

Tidal currents or tidal streams

Currents that result from the tidal wave forced by the gravitational tractive stresses generated by the moon and sun are called *tidal currents* or *tidal streams* and reverse their direction either semi-diurnally or diurnally. They may reach speeds up to $6\,m\,s^{-1}$ in coastal waters if they are constrained topographically, but are generally much smaller ($\sim 0.05\,m\,s^{-1}$) and are dominated on open coasts by gravity WAVE oscillations. Tidal currents vary in magnitude in response to the local tidal range and will have a variable phase relationship with tidal elevation depending upon whether the tidal wave is *standing* (maximum flows at mid-tide) or *progressive* (maximum flows at high and low tide). The current direction will be constrained by Coriolis and thus currents generated by the rising (flood) tide will follow a different path than those of the falling (ebb) tide, giving an ellipsoidal pattern of currents. In estuaries, for example, distinct *flood* and *ebb channels* may exist. This creates *residual currents* that may be significant in terms of sediment transport and deposition. In all cases tidal currents vary with depth as a result of bottom friction and a logarithmic velocity profile develops. Tidal currents are most significant in estuaries, inlets and straits and in some sections of continental shelf. Davis and Hayes (1984) discuss the relative role of tides and waves in the development of coastal morphologies.

Other currents in the coastal ocean

Wind-induced currents are formed when wind shear on the surface is transferred into the water column. At the shoreline this will induce a mass transport in the direction of the wind and can be resolved into a shore-normal and shore-parallel component. The former results in an elevation of the mean sea level (*wind set-up*) at the shoreline, which is an addition to the *wave set-up* which causes *undertows*; the latter will enhance the radiation stress driven *longshore current*. However, such currents are often short-lived, as local winds are subject to frequent change in speed and direction. Gradient in *wind set-up* can also generate currents at the scale of the complete coastal ocean boundary layer during large storm events (e.g. hurricanes, intense mid-latitude cyclones). This results in large offshore directed pressure-gradient flows, whose speed and direction are constrained by frictional forces and the Coriolis effect (Swift 1976). In deep water and where wind systems are of long duration (e.g. the Trade Winds, Equatorial Winds, other zonal winds, etc.) they cause large coastal circulation systems (e.g. upwelling and downwelling systems, the Equatorial currents, etc.). The rotational force of the moving Earth (Coriolis) also influences such large-scale flows. Currents tend to move at 45 degrees to the wind at the surface and rotate clockwise (or anticlockwise depending on the hemisphere) with depth, to flow in the opposite direction to the surface wind; this is the *Ekman Spiral*.

Density-induced currents result from density differences due to differences in temperature, salinity or sediment mass concentration. Such gradients force both horizontal and vertical currents, which are also affected by the rotational effect of the Earth. The component of flow driven by the slope of an internal density surface is called the *baroclinic* component; the component driven by the slope of the sea surface is the *barotrophic* component.

Inertial currents are residual currents in large bodies of water, which continue to flow under their own momentum long after the original forcing has ceased.

References

Aagaard, T., Greenwood, B. and Nielsen, J. (1997) Mean currents and sediment transport in a rip channel, *Marine Geology* 140, 25–25.

Bowen, A.J. (1969) The generation of longshore currents on a plane beach, *Journal of Marine Research* 27, 206–214.

Bowen, A.J. and Inman, D.L. (1969) Rip currents, II Laboratory and field observations, *Journal of Geophysical Research* 74, 5,479–5,490.

Butt, T. and Russell, P. (2000) Hydrodynamics and cross-shore sediment transport in the swash zone of natural beaches: a review, *Journal of Coastal Research* 16, 255–268.

Dalrymple, R.A. (1975) A mechanism for rip current generation on an open coast, *Journal of Geophysical Research* 80, 3,485–3,487.

Davis, R.A. Jr. and Hayes, M.O. (1984) What is a wave-dominated coast, *Marine Geology* 60, 313–329.

Greenwood, B. and Osborne, P.D. (1990) Vertical and horizontal structure in cross-shore flows: an example of undertow and wave set-up on a barred beach, *Coastal Engineering* 14, 543–580.

Gruszczynski, M., Rudowski, S., Semil, J., Slominski, J. and Zrobek, J. (1993) Rip currents as a geological tool, *Sedimentology* 40, 217–236.

Klopman, G. (1994) Vertical structure of flow due to waves and currents, *Report H 840.30*, Delft Hydraulics, Delft, The Netherlands.

Komar, P.D. (1971) Nearshore circulation and the formation of giant cusps, *Geological Society of America Bulletin* 82, 2,643–2,650.

——(1998) *Beach Processes and Sedimentation*, Upper Saddle River, NJ: Prentice Hall.

Longuet-Higgins, M.S. (1953) Mass transport in water waves, *Philosophical Transactions of the Royal Society of London, Series* A 245, 535–581.

——(1970a) Longshore currents generated by obliquely incident waves, 1, *Journal of Geophysical Research* 75, 6,778–6,789.

——(1970b) Longshore currents generated by obliquely incident waves, 2, *Journal of Geophysical Research* 75, 6,790–6,801.

Short, A.D. (1985) Rip current type, spacing and persistence, Narrabeen Beach, Australia, *Marine Geology* 65, 47–71.

Sonu, C.J. (1972) Field observations of nearshore circulation and meandering currents, *Journal of Geophysical Research* 77, 3,232–3,247.

Svendsen, I.A. and Lorenz, R.S. (1989) Velocities in combined undertow and longshore currents, *Coastal Engineering* 13, 55–79.

Swift, D.J.P. (1976) Coastal sedimentation, in D.J. Stanley and D.J.P. Swift (eds) *Marine Sediment Transport and Environmental Management*, 255–310, New York: Wiley.

Van Rijn, L.C. (1998) *Principles of Coastal Morphology*, Amsterdam: Aqua Publications.

BRIAN GREENWOOD

CUSPATE FORELAND

Cuspate forelands are large-scale, tooth-shaped coastal promontories. Although erosion plays a part in their evolution and their form, they are basically landforms of accretion, composed of sorted beach sand deposited from littoral transport (see LONGSHORE (LITTORAL) DRIFT). They often enclose a lagoon (see LAGOON, COASTAL) or marsh. There are two basic types.

Recurved cuspate forelands

At sites where the coastline changes direction abruptly landward, littoral transport slows and deposition occurs, creating over time, a broad elongated shoal. This feature, termed 'spit platform' by Meistrell (1966), is the foundation on which the emergent cuspate foreland grows. The original elongated form is sometimes referred to as a 'flying spit', 'spit with recurves' or 'fleche', in that its growth is in a direction continuous with that of the updrift coast, 'flying' offshore into deeper water. On the leeward side, sand deposits washed over during storms or transported around the tip are subjected to wave action from the opposite direction. The result is a series of concave-seaward recurves, or secondary spits, extending at an acute angle from the tip to the downstream coast. Because of the effect of this bi-directional wave climate, the foreland may range in form from symmetrically cuspate, when the wave effect is fairly balanced on both sides, to asymmetrical and elongated, if wave effect on one side predominates. Examples of this type are Cape Canaveral in Florida, Pointe de la Coubre near the Gironde estuary in western France, and the Toronto Islands of Lake Ontario, Canada.

Dungeness-type cuspate forelands

This is the term originally given by Gulliver (1895) and Johnson (1919) and elaborated by Zenkovitch (1967), to refer to symmetric, accretionary forelands that grow at high angles to the shore. Coakley (1976) demonstrated that they usually form at the site of a pre-existing morphological feature that is transverse to the coastline, e.g. a recessional moraine, or low bedrock ridge. This disruption of the coastal orientation causes the accumulation of the spit platform. These forelands are dynamic and are influenced by periodic reversals in littoral drift direction due to changes in the wind/wave climate. Thus, the foreland may be nourished, and be eroded, from both sides. This results in the classic pointed cuspate form with a well-developed complex of beach ridges. The evolution of the foreland may be studied through the pattern of the preserved beach ridges. Good examples are Dungeness on the south-eastern English coast and Point Pelee, Lake Erie, Canada.

References

Coakley, J.P. (1976) The formation and evolution of Point Pelee, western Lake Erie, *Canadian Journal of Earth Science* 13(1), 136–144.

Gulliver, F.P. (1895) Cuspate forelands, *Geological Society of America Bulletin* 7, 399–422.

Johnson, D.W. (1919) *Shore Processes and Shoreline Development*, New York: Wiley.

Meistrell, F.J. (1966) The spit-platform concept: laboratory observation of spit development, in M.L. Schwartz (ed.) *Spits and Bars*, 225–284, Stroudsburg, PA: Dowden, Hutchinson and Ross.

Zenkovitch, V.P. (1967) *Processes of Coastal Development*, New York: Interscience Publishers.

SEE ALSO: beach cusp; tombolo

JOHN P. COAKLEY

CUT-AND-FILL

Cut- (or scour) and-fill is the local cyclic erosion and deposition of sediment in a river channel, usually over short time periods (hours to years). It occurs as part of the process of sediment transport and development of channel morphology and consequently is associated with spatial and/or temporal changes in flow conditions, such as the passage of a flood wave along a channel. Cut-and-fill is distinct from progressive changes in channel elevation over longer time spans and greater distances that are usually referred to as degradation (erosion) and aggradation (deposition) and may produce substantial accumulation of ALLUVIUM and formation of terraces (see TERRACE, RIVER).

Cut-and-fill occurs in alluvial stream channels whenever bed sediment is moved. It is the result of variation in channel topography related to the normal processes of channel development, sediment transport and the response to, and recovery from, events such as large floods. Consequently cut-and-fill occurs for a variety of reasons including changes in flow hydraulics and sediment transport rate along the stream or during a flood event, development and migration of BEDFORMS, and changes in channel pattern, position or overall morphology. Cut-and-fill related to changes in channel morphology or channel migration is well known from studies of braided streams (see BRAIDED RIVERS) and is associated with the formation and migration of scour pools and bars (see BAR, RIVER) and with channel migration and AVULSION.

A cycle of cut-and-fill may occur at a single cross section during a flood, sometimes associated with rising and falling stages of the hydrograph. For example, in a POOL AND RIFFLE channel, cutting followed by filling may occur in pools while the reverse occurs in riffles because of changes in velocity and bed shear stress as discharge rises and falls. In other cases there are distinct areas of the channel in which only cut or fill occurs during a flood event or over longer time periods. Generally, there is compensating cut-and-fill within a channel reach so that the quantity of erosion at one location is matched by deposition nearby, sediment may be transferred from one to the other and this mass conservation means that there is no overall change in channel elevation (Colby 1964; Ashmore and Church 1998; Eaton and Lapointe 2001).

Observations in large SAND-BED RIVERS have shown cut, and subsequent fill, of the order of two or three metres at particular river cross sections during a single flow event (Colby 1964). In small, GRAVEL-BED RIVERS and streams, measurements indicate that the average depth of cut-and-fill during sediment transport events is of the order of about twice the maximum grain size, but local depths may be much greater than this (Hassan 1990; Haschenburger 1999). The average and maximum depth of cut or fill in a particular stream tends to be greater at higher discharges (greater bed shear stress) as does the area of the channel experiencing cut or fill. Where cut-and-fill is related to the development and migration of bedforms and scour pools the depth of activity is determined by the vertical amplitude of the channel topography.

Common methods for measurement of cut-and-fill are depth sounding, survey of topographic changes over an area of channel, and deployment of scour chains or tracers. Sounding provides very high temporal resolution but may be limited in spatial coverage while repeated surveying provides detailed information on the spatial pattern but may underestimate cut-and-fill amounts and rates if there is both erosion and deposition at a given location between surveys. Scour chains can provide both spatial patterns and also some information about the alternation of cut-and-fill at a point during a flow event. Scour chains are inserted vertically into the stream bed so that the increase in length of chain exposed at the bed after a flow event indicates the depth of cutting, while the depth of fill can be inferred from the depth of burial of the vertical section of chain.

Cut-and-fill is fundamentally and practically significant. Fundamentally, it is the result of the direct connection between channel morphology and sediment transport – spatial and temporal variation in transport rate leads to cut-and-fill and therefore change in channel morphology. Furthermore, the rate of transport of bed sediment during a transport event can be defined as the average depth of cut or fill multiplied by the average velocity of the sediment particles (distance moved divided by the duration of the transport event), which is one method for estimating bed sediment transport rate (Ashmore and Church 1998; Haschenburger 1999). Because cut-and-fill is a significant aspect of stream channel dynamics it is also important in a number of other contexts such as sedimentological interpretation of alluvial deposits (Best and Ashworth 1997), engineering design of river structures, and anticipation of the effects of direct or indirect modification of river channels on channel dynamics and stream habitat.

References

Ashmore, P. and Church, M. (1998) Sediment transport and river morphology: a paradigm for study, in P.C. Klingeman, R.L. Beschta, P.D. Komar and J.B. Bradley (eds) *Gravel-Bed Rivers in the Environment*, 115–148, Highlands Ranch, CO: Water Resources Publications.

Best, J.L. and Ashworth, P.J. (1997) Scour in large braided rivers and the recognition of sequence stratigraphic boundaries, *Nature* 387, 275–277.

Colby, B.R. (1964) Scour and fill in sand-bed streams, *United States Geological Survey Professional Paper* 462-D.

Eaton, B.C. and Lapointe, M.F. (2001) Effects of large floods on sediment transport and reach morphology in the cobble-bed Sainte Marguerite River, *Geomorphology* 40, 291–309.

Haschenburger, J.K. (1999) A probability model of scour and fill depths in gravel-bed channels, *Water Resources Research* 35, 2,857–2,869.

Hassan, M.A. (1990) Scour, fill and burial depth of coarse material in gravel-bed streams, *Earth Surface Processes and Landforms* 15, 341–356.

Further reading

Knighton, D. (1998) *Fluvial Forms and Processes*, London: Arnold.

PETER ASHMORE

CYCLE OF EROSION

The Cycle of Erosion or 'The Geographical Cycle' was formulated in the latter years of the nineteenth century by W.M. Davis (e.g. Davis 1899). It was the first widely accepted modern theory of landscape evolution (see SLOPE, EVOLUTION). Davis regarded landscapes as evolving through a progressive sequence of stages, each of which exhibited similar landforms. In the Davisian model it was assumed that uplift takes place quickly. The land is then gradually worn down by the operation of geomorphological processes, without further complications being produced by tectonic movements. It was believed that slopes declined in steepness through time until an extensive flat region was produced close to BASE LEVEL, though locally hills called *monadnocks* might rise above it. This erosion surface was termed a *peneplain*. The reduction in the landscape creates a time sequence of landforms progressing through three stages: youth, maturity and old age.

Initially the Davisian model was postulated in the context of development under humid temperate ('normal') conditions, but it was then extended to other landscapes including arid (Davis 1905), glacial (Davis 1900), coastal (Johnson 1919), karst (Cvijić 1918) and periglacial landscapes (Peltier 1950).

Davis's model was immensely influential and dominated much of thinking in Anglo-Saxon geomorphology in the first half of the twentieth century, contributing to the development of DENUDATION CHRONOLOGY. Davis was a veritable 'Everest' among geomorphologists (Chorley *et al.* 1973). The model was largely deductive and theoretical and suffered from a rather vague understanding of surface processes, from a paucity of data on rates of operation of processes, from a neglect of climate change, and from assumptions he made about the rates and occurrence of uplift. However, it was elegant, simple and tied in with broad, evolutionary concerns in science at the time. Nonetheless, by the mid-1960s the concept was under attack (Chorley 1965).

The Davisian model was never universally accepted in Europe, where the views of W. Penck were more widely adopted. Penck's model involves more complex tectonic changes than that of Davis, and regards slopes as evolving in a different manner (slope replacement rather than slope decline) through time (Penck 1953). An alternative model of slope development by parallel retreat leading to *pediplanation* was put forward by L.C. King (e.g. King 1957). Thorn (1988) provides a comparative analysis of the models of Davis, Penck and King. Another evolutionary model of landscape evolution was produced by Büdel (1982), who developed the concept of ETCHING, ETCHPLAIN AND ETCHPLANATION.

References

Büdel, J. (1982) *Climatic Geomorphology*, Princeton: Princeton University Press.
Chorley, R.J. (1965) A re-evaluation of the Geomorphic System of W.M. Davis, in R.J. Chorley and P. Haggett (eds) *Frontiers in Geographical Teaching*, 21–38, London: Methuen.
Chorley, R.J., Beckinsale, R.P. and Dunn, A.J. (1973) *The History of the Study of Landforms or the Development of Geomorphology. Volume 2. The Life and Work of William Morris Davis*, London: Methuen.
Cvijić, J. (1918) Hydrographie souterraine et evolution morphologique du karst, *Recueil des Travaux de L'Institut Géographie Alpine* (Grenoble), 6(4).
Davis, W.M. (1899) The Geographical Cycle, *Geographical Journal* 14, 481–504.
——(1900) Glacial erosion in France, Switzerland and Norway, *Proceedings of the Boston Society of Natural History* 29, 273–322.
——(1905) The Geographical Cycle in an arid climate, *Journal of Geology* 13, 381–407.
Johnson, D.W. (1919) *Shore Processes and Shoreline Development*, New York: Prentice Hall.
King, L.C. (1957) The uniformitarian nature of hillslopes, *Transactions of the Edinburgh Geological Society* 17, 81–102.
Peltier, L. (1950) The geographic cycle in periglacial regions as it is related to climatic geomorphology, *Annals of the Association of American Geographers* 40, 214–236.
Penck, W. (1953) *Morphological Analysis of Landforms*, London: Macmillan.
Thorn, C.E. (1988) *Introduction to Theoretical Geomorphology*, Boston: Unwin Hyman.

A.S. GOUDIE

CYCLIC TIME

A cycle is a period of time in which events happen in an orderly way. The order repeats itself in time so that there is a recurring series of changes. The term 'cyclic time' is unnecessary and illustrates the confusion that exists in the use of cyclic concepts.

Unfortunately a geomorphological cycle is often regarded as a sequence of *changes* from an initial state, through a series of stages to an ultimate state. In such models it is assumed that the changes taking place are such that the system has a different configuration when observed at different times, in other words, landforms have an observable history.

This emphasis meant that attention was paid to the sequence of changes and the 'stages of evolution' rather than the temporal lengths, frequencies and durations of the cycle and its events. This led to an emphasis on DENUDATION CHRONOLOGY (establishing and dating the stages of change) rather than a real understanding of geomorphological processes, rates of change and event statistics.

The interchangeable use of 'time' as the time during which changes take place, as the sequence of changes or the stage reached in the 'cycle' began with W.M.Davis (1899, 1905). In some passages he uses the correct dictionary sense. After describing the way a river advances through its long life and reduces an uplifted landmass to a PENEPLAIN he states, 'This lapse of time will be called a cycle in the life of a river.'

Unfortunately, he also described the geographical cycle as 'a complete sequence of landforms' but then qualified this as taking place from the uplift (an event) that produced the initial form through a sequence of form changes (responses to process events) to an ultimate form – a plain of low relief. In another passage Davis said that a geographical cycle 'may be divided into parts of unequal duration, each part of which will be characterized by the degree and variety of relief, and by the rate of change that has been accomplished since the start of the cycle'. Davis described how the successive forms of the cycle were dependent on three variable quantities: structure, process and time. He therefore makes it clear that 'time' is the amount of change from the initial form or its stage of development. In other passages the amount of change is 'a function of time' and time is again used as one of the trio of controls.

The period of time involved in a cycle has been poorly thought out. Davis (1899, 1905) estimated that the block mountains of Utah would be peneplained in 20–200 Ma. Wooldridge (personal, communication 1960) estimated up to 100 Ma but stated that the Mio-Pliocene peneplain had been produced in less than 20 Ma (Wooldridge and Linton 1955). Schumm and Lichty (1965) thought in terms of 10^6 years and Schumm (1963) pointed out that the time period to base levelling would be greatly extended by isostatic, erosional rebound. The general conclusion is that denudation cycles involve time spans of geological duration for their completion and recurrence by further uplift.

It is now known that the controls of Earth systems, such as structure, climate and base level, do not remain stable for such long periods of time and that it is preferable to establish the time periods for the frequency of landform creation events, the relaxation times and the landform survival times for the relevant system specifications. Some geomorphologists would argue (Schumm and Lichty 1965) that time can be divided into cyclic, graded and steady time periods. A more recent view (Graf 1977; Brunsden and Thornes 1979; Brunsden 1990) would suggest that the period 'cycle' be dropped in favour of system-based terms.

The name of a cycle (cyclic time?) is taken from the subject matter of the changes involved. General examples are a geographical cycle, a geomorphic cycle, an erosion cycle, a cycle of topographic development, a cycle of denudation, a cycle of life (Davis 1899). More specific uses were the normal cycle (landscapes developed under humid temperate conditions), shoreline development, sedimentation, karst, slope evolution, underground drainage, hydrologic, climatic and cycles of all geomorphological processes regimes.

References

Brunsden, D. (1990) Tablets of Stone: toward the ten commandments of geomorphology, *Zeitschrift für Geomorphologie N.F. Supplementband* 79, 1–37.

Brunsden, D. and Thornes, J.B. (1979) Landscape sensitivity and change, *Transaction Institute British Geographers* NS4, 463–484.

Davis, W.M. (1899) The Geographical Cycle, *Geographical Journal* 14, 481–504.

——(1905) The Geographical Cycle in an arid climate, *Journal of Geology* 13, 381–407.

Graf, W.L. (1977) The rate law in fluvial geomorphology, *American Journal of Science* 277, 178–191.

Schumm, S.A. (1963) Disparity between present rates of denudation and orogeny, *US Geological Survey Professional Paper* 454, 13.

Schumm, S.A. and Lichty, R.W. (1965) Time, space and causality in geomorphology, *American Journal of Science* 263, 110–119.

Wooldridge, S.W. and Linton, D.L. (1955) *Structure, Surface and Drainage in South-East England*, London: George Philip.

DENYS BRUNSDEN

CYMATOGENY

A term introduced by L.C. King (1959) to describe crustal movements intermediate between EPEIROGENY and OROGENESIS. They involve a warping of the Earth's crusts over horizontal distances that range from tens to hundreds of kilometres, and with vertical movements up to thousands of metres. They involve, however, minimal rock deformation. It is thought that the uplift is caused by processes active within the Earth's mantle.

Reference

King, L.C. (1959) Denudation and tectonic relief in southeastern Australia, *Transactions of the Geological Society of South Africa* 62, 113–138.

A.S. GOUDIE

D

DAM

Dams have been used to secure water supplies, to control floods and to generate power for more than a thousand years. The earliest civilizations developed along rivers in arid and semi-arid areas, such as along the Nile, and it is here that the oldest dams were built about 5,000 years ago. Flows in the wet season were stored in reservoirs to supply water for large-scale irrigation agriculture during the dry season. Water security and food security were closely linked and maintained the social, economic and political stability of the developing civilizations.

Today, the flows on most rivers are controlled to some degree by dams (WCD 2000). There are more than 45,000 dams over 15 m high and the largest dams stand more than 200 m high! The first big dam was the 221 m high Hoover Dam on the Colorado River, constructed in 1935. Kariba dam on the Zambezi, closed in 1958, was the first large dam to be constructed in the tropics. Water stored in reservoirs exceeds that stored in natural lakes by more than three times. Major rivers such as the Colorado and Columbia in USA, the Volga in Europe, the Nile in Africa, the Parana in Latin America, and the Murray–Darling in Australia have been intensively developed. Hydro-electric power is a major driver of dam building. Only about 3 per cent of the world's total energy consumption is supplied by water power and some 75 per cent of the hydroelectric power potential of the world's rivers is still to be exploited.

The geomorphological significance of large dams includes reservoir-induced earthquakes that have occurred in a small proportion of cases but have dramatic impacts. Large dams and reservoirs can both increase the frequency of earthquakes in areas prone to seismic activity and cause earthquakes in areas thought to be geologically stable. The mechanism involves the extra water pressure created by the dam and reservoir within faults in underlying rocks. Gupta (1992) records seventy examples of reservoir-induced seismicity. In many cases, the strongest shocks, often exceeding 4 and occasionally 6 on the Richter scale, occurred shortly after the initial filling of the reservoir.

Much more common are the impacts of dams on the fundamental fluvial processes, the flow and sediment transport regimes. These process changes induce adjustments of the size and shape of the river channel, and the form of the floodplain. These changes of the flooding and sedimentation regimes together with the changes to the morphology of the river corridor impact upon plants and wildlife by changing the habitats available for biota.

All dams are designed to capture floodwaters (see FLOOD) and represent perhaps the greatest point-source of hydrological impact. On some rivers reduced flood magnitudes have been experienced for more than 1,000 km below the dam and below the Aswan Dam on the River Nile, the reduction in freshwater flows is seen in the increased salinity of waters offshore of the delta in the south-east Mediterranean Sea. The Colorado River, USA, is dammed along its length, as are its major tributaries, and less than 1 per cent of the virgin flow reaches the river mouth. On the Murray–Darling system in Australia, which is regulated by nine principal storage reservoirs, the natural flow pattern was reversed, with high flows being released from the dams to supply irrigation demands downstream.

The basic concept of flood storage is 'empty space', keeping a reservoir as empty as possible to store floodwaters when they arrive. Water-supply reservoirs need to be kept full to provide water for domestic, industrial or irrigation supplies during dry seasons and dry years. But even when a reservoir is full and spilling over the dam, the flood peak downstream will not be as high as that for the inflow because of temporary storage in the lake as levels rise above the crest level of the overflow weir. Commonly, the size of the mean annual flood below dams has been reduced by between 25 and 50 per cent.

Dams and reservoirs also trap the sediments transported by a river – in many cases permanently storing the entire sediment load supplied by the upstream drainage basin. As the relatively high-velocity, and turbulent, water of a river feeding the reservoir is transferred into the slow-flowing water within the lake the sediment is deposited. Part is deposited in the reservoir itself and part in the channel and valley-bottom upstream, as a result of the backwater effects from the reservoir reducing velocities of river and floodplain flows. The coarser sediments settle out to form a delta. The finer particles, especially the clays, are distributed further out into the lake. Average annual rates of reservoir storage loss are usually less than 0.5 per cent per year but exceptional rates of more than 2 per cent per year have been reported from regions with high SEDIMENT LOAD AND YIELDS. One extreme case is the Heosonghi Reservoir on the Huang Ho, China that lost nearly 20 per cent of its storage capacity within three years of completion.

Flows released from dams or passing the spillway during floods are known as 'clearwater' releases because they are more or less sediment free. However, sometimes the water can appear turbid, not because of suspended sediments, but because of high concentrations of plankton when water is released from the lake surface during summer. This is caused by phytoplankton – algae and diatoms – which can reach high concentrations in relatively warm, surface layers of reservoirs having long retention times. Turbid releases may also be caused by the discharge of deep water during the autumn when stratified lakes mix – the 'overturn'. Such discharges can contain high concentrations of iron, manganese and hydrogen sulphide, giving a bad egg smell. However, in both cases, the quality of the water discharged from a reservoir can be controlled by the selective release of water from different depths within the lake. Occasionally, sediments are deliberately flushed from reservoirs by opening deep valves in the dam, to reduce the rate of storage loss. An example of this operation is the management of the Verbois, Chancy–Pougny and Genissiat reservoirs on the River Rhone in France. During these rare events, suspended sediment concentrations can exceed $1\,g\,l^{-1}$ but such sudden surges of sediment-laden water can cause problems for water quality downstream.

Clearwater releases and the regulated flow regime below dams induce changes of channel morphology. The size and shape of natural river channels are in regime with the flows and sediment loads. Below dams two general types of change in regime can occur, although in detail there are many variations on these (Brandt 2000). The first type of channel change occurs where the dominant change of fluvial process is the reduction in sediment load. The clearwater releases of sediment-free water from reservoirs into channels with alluvial bed and banks can cause rapid erosion, or degradation, that may extend for many kilometres downstream. Typically bed degradation deepens the channel and the banks may also be undermined and sand and gravel bars eroded. An increase in the size of the sediments on the channel bed, which becomes armoured by the selective removal of the finer particles, and the reduction in channel slope as a result of bed incision may limit the amount of bed erosion. The result is a channel of increased cross-sectional area. Reports of degradation rates of more than 100 mm per year over channel lengths of more than 100 km are not uncommon. Rates decline over time until a new 'regime' condition is reached.

The second type of channel response is to the regulated flows, especially the lower flood levels. This induces a reduction of channel capacity most commonly observed as a reduction of channel width. Flow regulation reduces the capacity of a river to transport sediments supplied by sources downstream from the dam. These sources include tributary catchments and any degrading reaches and the dam and reservoir site during construction. Coarse sediments will be deposited on the channel bed but sediments will also accumulate as bars and benches along the channel margin, sometimes creating a new floodplain. The former floodplain is then converted into a river terrace (see TERRACE, RIVER).

The rate of channel narrowing is highly variable but can be particularly rapid in two situations. First, channel change is often rapid along regulated rivers in semi-arid areas where wide, braided rivers are converted into single channels. In these cases, the growth of vegetation such as willows and poplars, sometimes accelerated by the maintenance of higher baseflows than in the natural river, can result in dramatic reductions of channel width (Merritt and Cooper 2000). The second situation is downstream from tributaries that produce high sediment delivery to the regulated channel. Sometimes, the reduced flood levels within the regulated river can accelerate erosion within the tributary increasing sediment supplies until the tributary has reached regime (Germanovski and Ritter 1988).

Each river comprises a sequence of channel reaches, each having a different channel form reflecting the history of the reach over Quaternary, historical and recent timescales. Channel change involves the movement – erosion, transport and deposition – of large volumes of sediments into and through this series of channel reaches over periods of time ranging from years to centuries. Volumes of up to 1 million cubic metres in a one-kilometre reach are not uncommon. In many cases individual reaches of river channel will show a COMPLEX RESPONSE to impoundment (Sherrard and Erskine 1991; Church 1995). This involves alternating phases of degradation and aggradation as the river network, the main channel and its tributaries, continue to adjust to the regulated flow regime by moving sediment through the sequence of reaches until a new 'regime' channel form is established.

Along rivers that have low sediment loads and stable, cohesive bank materials, adjustments of channel form may be very slow. In extreme cases the timescale for channel change to establish a new 'regime' channel may extend to hundreds of years. In these cases, the existing channel form will accommodate the regulated flows and evidence of upstream impoundment may be limited to local sediment accumulation in pools and backwaters, the growth of moss on large stones, and the marginal growth of emergent aquatic plants. An extreme flood may be required to initiate major channel changes in these reaches.

Geomorphology provides a physical template for river, riparian and floodplain ecology (Petts 2000) (see PHYSICAL INTEGRITY OF RIVERS). Variable river flows and sediment loads, and dynamic channels that change position by the processes of deposition and erosion, creating new floodplain patches and eroding others, sustain a diverse and highly productive riverine ecosystem. The channel pattern (see CHANNEL, ALLUVIAL) determines the range of habitat types found along any river but the frequency of erosion and deposition determine the level of disturbance that rejuvenates ecological successions.

Dams reduce the physical dynamism of the downstream riverine ecosystem, simplifying the physical habitat, and reducing both biological diversity and productivity (Ward and Stanford 1995). Advances in the application of geomorphological knowledge to the operational management of regulated rivers through the development of instream flow models (Petts and Maddock 1994) seek to sustain the ecological integrity of rivers below dams. Such models determine three levels of flows that need to be sustained along a regulated river to maintain the physical and, therefore, ecological dynamism of the river corridor. These flows are the floodplain maintenance flow, the channel maintenance flow (usually the BANKFULL DISCHARGE), and flushing flows to prevent the siltation of the channel bed and to prevent vegetation encroachment into the channel.

References

Brandt, S.A. (2000) Classification of geomorphological effects downstream of dams, *Catena* 40, 375–401.

Church, M. (1995) Geomorphic response to river flow regulation: case studies and time scales, *Regulated Rivers* 11, 3–22.

Germanovski, D. and Ritter, D.F. (1988) Tributary response to local base level lowering below a dam, *Regulated Rivers* 2, 11–24.

Gupta, H. (1992) *Reservoir-Induced Earthquakes*, Amsterdam: Elsevier.

Merritt, D.M. and Cooper, D.J. (2000) Riparian vegetation and channel change in response to river regulation: a comparative study of regulated and unregulated streams in the Green River Basin, USA, *Regulated Rivers* 16, 543–564.

Petts, G.E. (2000) A perspective on the abiotic processes sustaining the ecological integrity of running waters, *Hydrobiologia* 422/423, 15–27.

Petts, G.E. and Maddock, I. (1994) Flow allocation for in-river needs, in P. Calow and G.E. Petts (eds) *The Rivers Handbook*, 2, 289–307.

Sherrard, J.J. and Erskine, W.D. (1991) Complex response of a sand-bed stream to upstream impoundment, *Regulated Rivers* 6, 53–70.

Ward, J.V. and Stanford, J.A. (1995) The Serial Discontinuity Concept: extending the model to floodplain rivers, *Regulated Rivers* 10, 159–168.

WCD (2000) *Dams and Development*, London: World Commission on Dams (WCD) and Earthscan.

Further reading

Beyer, P.J. (ed.) (2004) Dams and geomorphology, Binghamton Symposium Special Issue, *Geomorphology*, in press.
Petts, G.E. (1984) *Impounded Rivers*, Chichester: Wiley.
——(1994) Large-scale river regulation, in C.N. Roberts (ed.) *The Changing Global Environment*, 262–283, Oxford: Blackwell.
Williams, G.P. and Wolman, M.G. (1984) Downstream effects of dams on alluvial rivers, *US. Geological Survey Professional Paper* 1,286.
The journal *River Research and Applications* (until 2002 *Regulated Rivers*) has a focus on rivers below dams.

GEOFFREY PETTS

Plate 31 A broad, flat-floored, grassy dambo in west central Zambia

DAMBO

A headwater valley in areas of low relief, particularly in the seasonal tropics, that is channelless and in humid areas may contain swamps. Dambos are also known as *vleis* in southern Africa, *matoro* in Zimbabwe, *baixas* in Amazonia, *bolis* in Sierra Leone, *mbuga* in East Africa and *fadama* in northern Nigeria. German geomorphologists (e.g. Büdel 1982) have called them 'Spülmulden' or *wash depressions*. True dambos tend to be restricted to climates with present-day rainfalls between 600 and 1,500 mm, but the *bolis* of Sierra Leone are found where annual rainfall approaches 2,500 mm. They are also probably best developed on ancient planation surfaces. They occur on a wide range of rock types from unconsolidated Kalahari Sand through to shales, quartzites, schists, gneisses and granites (Thomas and Goudie 1985, Plate 31).

Their hydrology has been described by Bullock (1992), and they are a major source of water supply in rural areas in countries like Zimbabwe. Many of them are now being exploited for agricultural reasons and are suffering degradation, including gullying, as a consequence. Indeed, dambo is a Bantu word meaning 'meadow grazing,' for they are often grass covered and have no true woodland vegetation (Mäckel 1974).

Dambos tend to have low gradients (usually less than 2°). They receive their water either from direct precipitation onto the dambo or by subsurface flow from the surrounding high ground. With regard to the processes that lead to their formation, two main schools of thought exist (Boast 1990). The fluvial school envisages dambos as the simple extensions of the channelled drainage network. Rivers erode their head valleys which may subsequently be infilled by slope colluviation and by channel alluviation. Sheet-wash processes under seasonal rainfall regimes may be especially important. The other school of thought advocates differential chemical and biochemical corrosion or sapping rather than mechanical erosion as the main process. It sees dambo morphology as breaking 'too many fluvial rules' to be explicable in simple fluvial terms. That fluvial processes have operated in some dambos is made clear by the stratigraphy of their floors, which can reveal old alluvial fills. It is evident in many parts of central Africa that the balance between colluviation and alluviation has varied repeatedly in response to climatic changes. However, the two schools of thought are not necessarily mutually exclusive and Thomas (1994: 279) believes that 'Opposition between sapping (or etching) processes and sedimentation in dambos is misplaced.'

References

Boast, R. (1990) Dambos: a review, *Progress in Physical Geography* 14, 153–177.
Büdel, J. (1982) *Climatic Geomorphology*, Princeton, NJ: Princeton University Press.
Bullock, A. (1992) Dambo hydrology in southern Africa – review and assessment, *Journal of Hydrology* 134, 373–396.
Mäckel, R. (1974) Dambos: a study of morphodynamic activity on plateau regions of Zambia, *Catena* 1, 327–365.
Thomas, M.F. (1994) *Geomorphology in the Tropics*, Chichester: Wiley.
Thomas, M.F. and Goudie, A.S. (eds) (1985) Dambos: small channelless valleys in the tropics, *Zeitschrift für Geomorphologie, Supplementband* 52.

A.S. GOUDIE

DATING METHODS

Stratigraphic relationships between landforms or within depositional sequences provide the most common, and simplest, means of deducing the age. Other than under exceptional circumstances, younger landscape features or sediments overlie older ones. However, this approach does not enable rates of processes to be deduced, nor give any idea of the relative or absolute timing of events. A number of dating methods exist based on chemical and biological changes that occur through time. The formation of CHEMICAL WEATHERING rinds and DESERT VARNISH on exposed rock are examples of the former, while amino-acid racemization and LICHENOMETRY are examples of the latter. These are all relative dating methods (i.e. indicating that one landform is approximately twice as old, or three times as old, as another). Another class of dating methods is that based on the correlation of events. For example, periodically the Earth's magnetic field is reversed, with the positions of the north and south magnetic poles switching. The last time that such a reversal occurred was 780,000 years ago. This event is recorded in a number of sedimentary and volcanic records and provides a synchronous marker across the globe, thus allowing one site to be correlated to another. In order to define the numerical age of this event (i.e. the age expressed as the number of years before present) a different class of dating methods are required – absolute age methods.

The discovery of radioactivity at the end of the nineteenth century provided the foundation for a suite of absolute dating techniques collectively known as radioisotopic methods. These all rely upon the fact that the rate at which a radioactive isotope of an element undergoes decay to produce another isotope (known as the daughter product) is constant, unaffected by any external controls such as temperature or pressure.

Radiocarbon dating was the first radioisotopic dating method to be widely applied starting in the 1950s. Carbon occurs as three isotopes, ^{12}C, ^{13}C and ^{14}C. The first two are stable isotopes, while the latter is radioactive, but all react chemically in identical ways. Radiocarbon (^{14}C) is generated in the upper atmosphere by the interaction of high energy cosmic rays with nitrogen atoms. The ^{14}C generated in this way is rapidly oxidized to form carbon dioxide which enters the carbon cycle. Radiocarbon has a half-life (the time taken for half of the atoms of ^{14}C within a sample to undergo radioactive decay) of $5,730 \pm 40$ years, and the concentration of ^{14}C in the atmosphere is a balance between the rate of production and the rate of decay. All living things exchange carbon with some part of the carbon cycle, and thus contain ^{14}C. After death this exchange ceases. The ^{14}C continues to decay according to its half-life, but it is no longer replaced by exchange with any part of the carbon cycle. Measurement of the ^{14}C remaining in a sample allows calculation of the period of time since death. Radiocarbon dating is most appropriate for organic materials, but can also be applied to some carbonates. The method assumes that the concentration of ^{14}C in the various reservoirs of the carbon cycle has remained constant through time. Measurement of the ^{14}C activity of tree rings of known age for the last 11,000 years shows this not to be the case, but these results allow ^{14}C ages to be calibrated to calendar years (Aitken 1990: 98). Between 11,000 years and ~40,000 years, the limit of the method, the ^{14}C calibration is less well known and the uncertainties on the ages larger.

In addition to ^{14}C, a wide variety of other isotopes (^{10}Be, ^{26}Al, ^{36}Cl) are generated both in the atmosphere and at the surface of the Earth by the interaction of cosmic rays. A suite of dating methods based on these cosmogenic isotopes have recently been developed (see COSMOGENIC DATING).

Other radioisotopic methods rely upon the very long half-lives of certain isotopes. Uranium occurs naturally as several isotopes (^{234}U, ^{235}U, ^{238}U). ^{238}U has a half-life of 4.47×10^9 years, comparable with the age of the Earth, and thus a significant quantity persists in the natural environment. Unlike ^{14}C, whose daughter product (^{14}N) is stable, the decay of ^{238}U produces ^{234}Th, which is itself radioactive. This in turn decays to produce ^{234}Pa, which decays to produce ^{234}U, ^{230}Th, ^{226}Ra and so on, producing a decay series until a stable isotope, ^{206}Pb, is produced (Figure 32). Over time the concentration of the different isotopes within the decay chain will alter until a state is reached where the number of decays per unit time from each isotope is identical – this state is termed secular equilibrium. The different chemical characteristics of the elements within the decay series provide a number of radioisotopic dating methods. For instance, when calcite is deposited in KARST environments, trace quantities of uranium are also deposited.

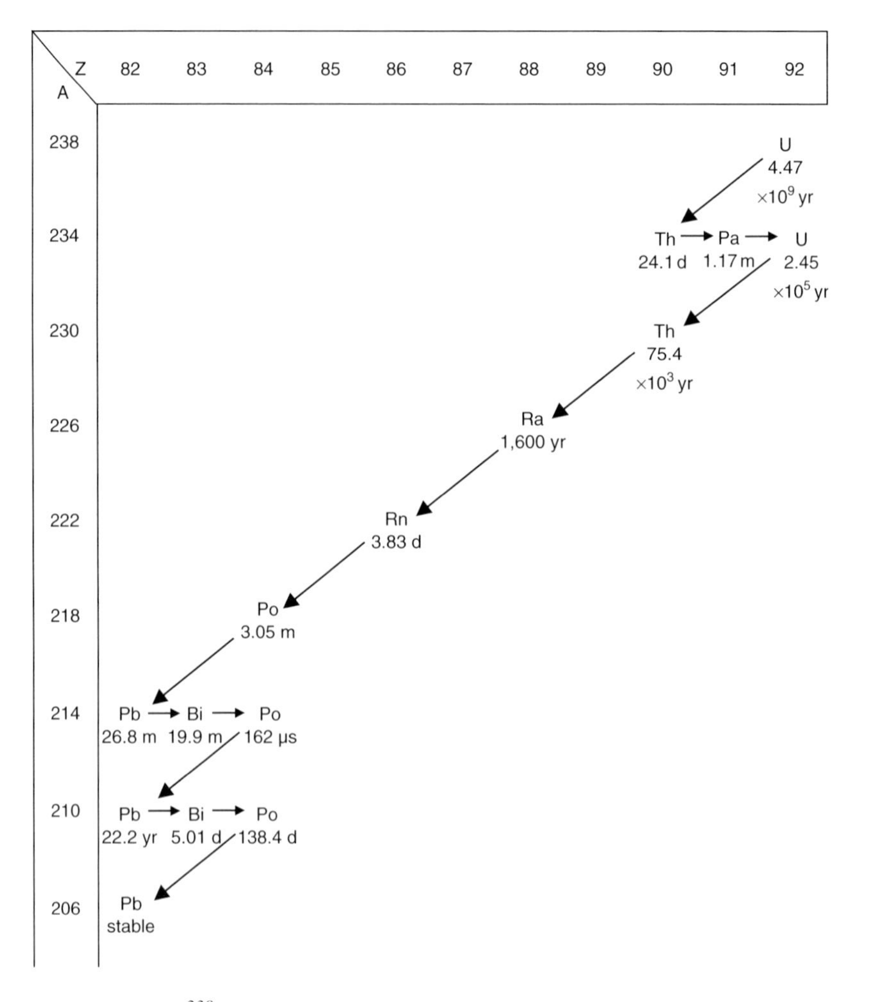

Figure 32 Decay series for ^{238}U. The half-life of each isotope is shown under the isotope. A is the atomic mass, while Z is atomic number

However, little or no thorium is deposited because it is relatively insoluble. Thus within the calcite, uranium will occur but without its thorium daughter products – it is said to be daughter-deficient. Over time, as the uranium undergoes decay, the concentration of thorium will increase. At the time of deposition, the $^{230}Th/^{234}U$ ratio will be zero, and will increase in a predictable manner allowing the age of formation of the calcite to be determined. This process can be used to date the precipitation of calcites over the last 350,000 years. As well as calcite in karst environments, another excellent target for Th/U dating is coral (Muhs 2002). Another part of the uranium decay series ^{210}Pb has a much shorter half-life (22 years) and can be used to provide ages over the last 100 years. In this case, the method is most commonly applied to lake sediments and nearshore marine basins (Appleby and Oldfield 1992). The method relies on the fact that one of the isotopes within the ^{238}U decay series, ^{222}Rn (radon), is a gas. This escapes to the atmosphere where it will undergo decay via a series of short-lived daughter products to produce ^{210}Pb. This falls from the atmosphere, producing a near constant supply to the surface of lakes and the nearshore. This ^{210}Pb is incorporated into the sediment accumulating under the water body, but none of its parent isotopes are present – thus there is a daughter excess.

Like uranium, potassium has an isotope with a long half-life (1.25×10^9 years). A small, but significant, proportion (0.01167 per cent) of all potassium is the radioactive isotope ^{40}K. This forms the basis for the techniques of potassium–argon (K–Ar) and argon–argon (Ar–Ar) dating of volcanic rocks (see VOLCANO). ^{40}K undergoes radioactive decay to produce either ^{40}Ca or ^{40}Ar. Argon is an inert gas, and while magma is molten any ^{40}Ar produced will be driven off, eventually making its way into the atmosphere (where it constitutes ~1 per cent by volume). Once crystallization occurs at the time of eruption, argon is unable to escape and begins to accumulate within the minerals crystal structure. Thus the ratio of the parent isotope (^{40}K) to the daughter product (^{40}Ar) provides a means of dating the volcanic eruption – this is the K–Ar method. The ratio of the parent and daughter isotopes can be measured more precisely by irradiating the sample of volcanic tephra or lava in a neutron beam in a nuclear reactor. This causes a proportion of the potassium to transform to ^{39}Ar, an isotope not found in nature. The age of the sample can then be found by measuring the ratio of two argon isotopes, ^{39}Ar (which is now a measure of the potassium concentration) and ^{40}Ar. Measuring this isotopic ratio is a more precise analytical process than measuring potassium and argon separately. Equally importantly, both argon isotopes are measured on the same subsample, thus allowing samples as small as single tephra crystals to be dated. Using the Ar–Ar method, ages as recent as a few thousands of years can be obtained (e.g. Renne *et al.* 1997, Figure 33).

An alternative approach to dating is not to measure the concentration of radioactive isotopes directly, but instead to look at the effect that the radioactivity has on materials in the natural environment – these are radiogenic methods. One such method is fission track dating. The most common way in which uranium decays is by the emission of an alpha particle (consisting of two neutrons and two protons). However, ^{238}U may also undergo fission, whereby the nucleus (consisting of 92 protons and 146 neutrons) splits into two new nuclei of almost equal masses. A significant amount of energy is released at the same time, and the two nuclei (the fission fragments) recoil away from each other. This leads to ionization of the crystal along these tracks – this damage can be made visible by etching the crystal surface using acids, and the number of fission tracks counted. The method is most commonly applied to volcanic rocks, including far-travelled tephra, and dates the formation of the crystals. Zircons have the advantage of high uranium concentrations (typically between 10 and 1,000 ppm) meaning that the number of tracks produced in a given time will be high. Glass has a much lower uranium concentration (~1 ppm) but is the most abundant component of tephra, and it too can be used for fission track dating providing that a method such as Isothermal Plateau Fission Track Dating (ITPFT) is used which compensates for the ability of glass to naturally anneal fission tracks (Westgate 1989).

Luminescence techniques are also based on the effects of radioactive decay. Alpha, beta and gamma radiation, resulting from the decay of various radioactive elements in the Earth's crust, is ubiquitous. When this radiation is absorbed by commonly occurring minerals such as quartz and feldspar, the energy from the radiation may be used to trap electrons at excited sites within the crystal. In effect, the mineral grains act as dosemeters, integrating the total amount of radioactivity that they are exposed to. In the laboratory, these mineral grains can be stimulated, allowing the trapped electrons to release their stored energy. The energy is released as light emitted from the quartz or feldspar grains – it is this light that is called luminescence. If the mineral grains are stimulated by heating (typically up to 500 °C) then this is termed thermoluminescence (TL). For geological materials it is normally more appropriate to stimulate them using light of a fixed wavelength (e.g. 532 nm from a $NdYVO_4$ laser) in which case optically stimulated luminescence (OSL) is observed. The luminescence signal is light sensitive, and exposure to natural daylight reduces the luminescence signal to a low level. Many subaerial transport processes will entail exposure of mineral grains to daylight (e.g. AEOLIAN PROCESSES) and the sediments deposited by these processes (e.g. DUNE, AEOLIAN; LOESS) are ideally suited to luminescence dating (Stokes 1999). Upon burial the continued exposure to radiation from the natural environment causes the trapped electron population to increase with time. The OSL signal is reset by exposure to daylight more completely than the TL signal, and hence the use of OSL has allowed more precise ages to be obtained and has allowed younger samples to be dated. In environments where the exposure to daylight at deposition can be assumed, events as recent as the last 50–100 years can be routinely dated.

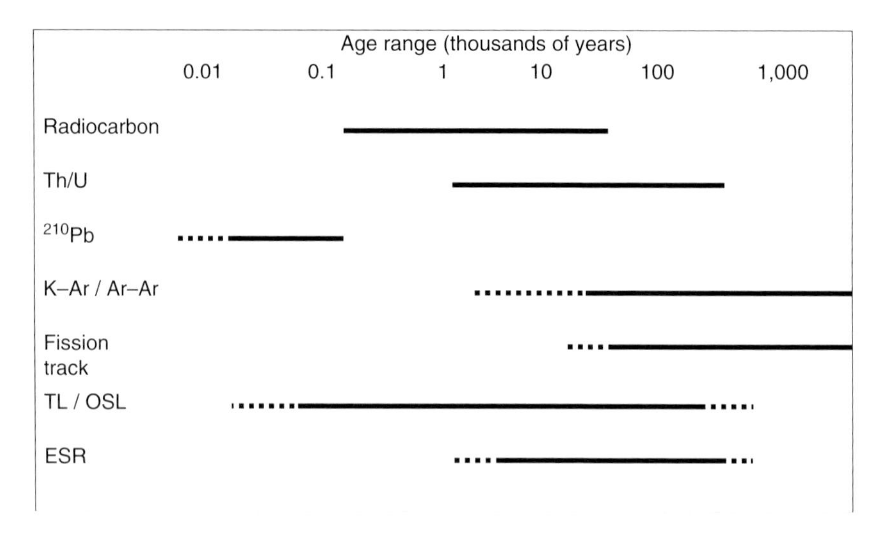

Figure 33 Age ranges over which various radioisotopic and radiogenic dating methods can be applied. The exact limits are often determined by the nature of the material being dated, and the dashed lines reflect this variation from one application to another

Electron spin resonance (ESR) dating is another technique based on measurement of the charge trapped in materials due to radiation from the environment. While TL and OSL are applicable to quartz and feldspar in sediments, ESR can be applied to stalagmites, tooth enamel, corals and sometimes bones.

References

Aitken, M.J. (1990) *Science-based Dating in Archaeology*, Harlow: Longman.

Appleby, P.G. and Oldfield, F. (1992) Application of Lead – 210 to sedimentation studies, in M. Ivanovich and R.S. Harmon (eds) *Uranium-series Disequilibrium, Applications to Earth, Marine and Environmental Sciences*, 2nd edition, 731–778, Oxford: Clarendon Press.

Muhs, D.R. (2002) Evidence for timing and duration of the last interglacial period from high-precision uranium-series ages of corals on tectonically stable coastlines, *Quaternary Research* 58, 36–40.

Renne, P.R., Sharp, W.D., Deino, A.L., Orsi, G. and Civetta, L. (1997) Ar-40/Ar-39 dating into the historical realm: calibration against Pliny the Younger, *Science* 277, 1,279–1,280.

Stokes, S. (1999) Luminescence dating applications in geomorphological research, *Geomorphology* 29, 153–171.

Westgate, J.A. (1989) Isothermal plateau fission-track ages of hydrated glass shards from silicic tephra beds, *Earth and Planetary Science Letters* 95, 226–234.

Further reading

Ivanovich, M. and Harmon, R.S. (eds) (1992) *Uranium-series Disequilibrium, Applications to Earth, Marine and Environmental Sciences*, 2nd edition, Oxford: Clarendon Press.

Noller, J.S., Sowers, J.M. and Lettis, W.R. (eds) (2000) *Quaternary Geochronology: Methods and Applications*, Washington, DC: American Geophysical Union.

Taylor, R.E. and Aitken, M.J. (1997) *Chronometric Dating in Archaeology*, New York: Plenum Press.

Wagner, G.A. (1998) *Age Determination of Young Rocks and Artifacts*, Berlin; New York: Springer.

Wintle, A.G. (1996) Archaeologically-relevant dating techniques for the next century, *Journal of Archaeological Science* 23, 123–138.

SEE ALSO: cosmogenic dating; dendrochronology; lichenometry

G.A.T. DULLER

DAYA

Small, silt-filled, closed solutional depressions found on limestone surfaces in some arid areas of the Middle East and North Africa. They are a type of PAN.

Reference

Mitchell, C.W. and Willmott, S.G. (1974) Dayas of the Moroccan Sahara and other arid regions, *Geographical Journal* 140, 441–453.

A.S. GOUDIE

DEBRIS FLOW

Debris flows are MASS MOVEMENT phenomena transitional between LANDSLIDEs and sediment-laden water floods. They occur commonly in tectonically active regions subject to rapid uplift and erosion. Typically debris flows consist of churning, water-saturated mixtures of poorly sorted sediment and miscellaneous detritus, which rush down slopes and funnel into channels when they reach valley floors. Debris flows generally form abrupt surge fronts, attain peak speeds greater than 10 metres per second, and include up to 70 per cent solid particles by volume. As a consequence, debris flows can denude slopes, damage structures, drastically alter stream channels and endanger human life. Notable debris-flow disasters include those in Armero, Colombia, 1985, and Vargas state, Venezuela, 1999, each of which resulted in more than 20,000 fatalities.

Debris flows have some alternative names. For example, LAHAR is a commonly used Indonesian term for a debris flow that originates on a volcano, and mudflow describes a debris flow that consists predominantly of silt and clay. Such fine-grained flows are rare in SUBAERIAL settings but more common in submarine (see SUBMARINE LANDSLIDE GEOMORPHOLOGY) environments.

Most subaerial debris flows commence as rapid landslides triggered by intense rainfall or rapid snowmelt. A flow may originate from a single, discrete landslide source or from numerous, distributed sources from which debris issues and coalesces. Source areas generally slope more steeply than 25 degrees, but debris flows commonly scour bed and bank sediment from channels that slope as gently as about 8 degrees. On flatter slopes debris flows typically decelerate and form lateral LEVEEs and lobate deposits that are very poorly sorted and readily distinguished from fluvial deposits. Many ALLUVIAL FANs in tectonically active regions are composed largely of debris-flow deposits.

Debris flows have a remarkable ability to flow quite fluidly, despite having grain concentrations comparable to those of static soil. The fluidity of debris flows results principally from a phenomenon called LIQUEFACTION, which occurs when pressure in the intergranular pore water rises to levels sufficient to support the weight of the overlying debris, thereby reducing friction at grain contacts. The reduced friction allows grains to move smoothly past one another, facilitating downslope flow. Liquefaction commences when debris flows begin to mobilize during landsliding of loosely packed soil or sediment, which contracts during shear deformation and transfers pressure to the intergranular pore water. Liquefaction persists in debris-flow bodies because silt and clay-sized sediment impedes pore-pressure dissipation, even if the fine sediment comprises just a few per cent of the debris-flow mass.

Effects of liquefaction are reduced or absent at the heads and lateral margins of debris-flow surges, where high concentrations of coarse debris accumulate. Debris-flow deposition occurs because coarse-grained marginal debris lacks high pore-water pressures and exerts strong frictional resistance to motion.

Further reading

Iverson, R.M. (1997) The physics of debris flows, *Reviews of Geophysics* 35, 245–296.

Iverson, R.M., Reid, M.E. and LaHusen, R.G. (1997) Debris-flow mobilization from landslides, *Annual Review of Earth and Planetary Sciences* 25, 85–138.

Johnson, A.M. (1984) Debris flow, in D. Brunsden and D.B. Prior (eds) *Slope Instability*, 257–361, Chichester: Wiley.

Takahashi, T. (1991) *Debris Flow*, Rotterdam: Balkema.

RICHARD M. IVERSON

DEBRIS TORRENT

Debris torrents are a regional phenomenon, extensively documented in the coastal Pacific north-west of the United States, British Columbia and south-east Alaska. A debris torrent is defined as 'a mass movement event that involves water-charged, predominantly coarse grained inorganic and organic material flowing rapidly down a steep, confined, pre-existing channel' (Van Dine 1985; Slaymaker 1988). This is a North American usage which contrasts with the European usage of the term torrent (*torrent* in French; *torrente* in Italian and *wildbach* in German). In Europe, torrent is descriptive of mountain stream morphology and not of a debris discharge event (Aulitzky 1980). Descroix and Gautier (2002) describe the appearance and disappearance of torrents (in the sense of a distinctive morphology) in the southern French Alps as a function of climate and land use changes.

Swanston (1974) and Hungr *et al.* (1984) have argued that the term 'debris torrent' is highly descriptive and well suited to the particular character of coarse-grained, channelized mass movement events of the Pacific maritime mountains. Slaymaker (1988) has argued that the case for debris torrents as a separate category is that they are a form of channelized debris flow which lack a fine-grained fraction, particularly clay, and have a relatively large organic debris content.

Debris torrents tend to occur in small drainage basins, from 0.1–10 km^2 (Mizuyama 1982); have steep channels, with an initiation zone greater than 25°, an erosion/transport zone (10–25°) and a depositional zone (5–12°); occur in high runoff intensity zones and require substantial amounts of organic and inorganic debris available for mobilization. Triggering mechanisms include storm and/or snowmelt runoff, water release from subglacial or lake storage, log jam bursts, rockfall, debris or snow avalanches from upslope or seismic shaking. The history of sediment accumulation in the channel is also critical (Bovis and Dagg 1987). Little cohesive material is present in debris torrents, a high proportion is gravel and boulders and wood and organic mulch is prominent. A frontal and lateral 'macrostructure' consists of framework supported boulders which are pushed forward by a turbulent slurry. The slurry is extruded through the macrostructure, effectively producing a two-phase flow.

References

Aulitzky, H. (1980) Preliminary two-fold classification of torrent, *Symposium Interpraevent* 4, 285–309.

Bovis, M.J. and Dagg, B. (1987) A model for debris accumulation and mobilization in steep mountain streams, *Hydrological Sciences Journal* 33, 589–605.

Descroix, L. and Gautier, E. (2002) Water erosion in the southern French Alps: climatic and human mechanisms, *Catena* 50, 53–85.

Hungr, O., Morgan, G.C. and Kellerhals, R. (1984) Quantitative analysis of debris torrent hazards, *Canadian Geotechnical Journal* 21, 663–677.

Mizuyama, T. (1982) Analysis of sediment yield and transport data for erosion control works, *International Association of Hydrological Sciences Publication* 137, 177–182.

Slaymaker, O. (1988) The distinctive attributes of debris torrents, *Hydrological Sciences Journal* 33, 567–573.

Swanston, D.N. (1974) Slope and stability problems associated with timber harvesting, *USDA Forest Service, Pacific Northwest, General Technical Report* PNW–21.

Van Dine, D.F. (1985) Debris flows and debris torrents in the southern Canadian Cordillera, *Canadian Geotechnical Journal* 22, 44–68.

SEE ALSO: debris flow; mass movement

OLAV SLAYMAKER

DECOLLEMENT

A fault surface marking where crustal deformation occurs in a parallel fashion, usually between an upper mechanically weak horizon, layer, or boundary, and a lower undeformed boundary. Decollements or decollement surfaces are formed by the upper rock series sliding over the lower during folding, and so is associated with overthrusting. They are typical between crystalline basement rock overlying sedimentary rock, often in thrust faulted regions such as the Alps, the Jura Mountains and the Zagros Mountains of Iran.

Further reading

Ramsay, J.G. and Huber, M.I. (1987) *The Techniques of Modern Structural Geology – Volume 2: Folds and Fractures*, London: Academic Press.

SEE ALSO: crustal deformation

STEVE WARD

DEEP-SEATED GRAVITATIONAL SLOPE DEFORMATION

Deep-seated gravitational slope deformations (DGSDs) are gravity-induced processes which evolve over a very long time interval and usually affect entire slopes, displacing rock volumes up to hundreds of millions of cubic metres over areas of several square kilometres with thicknesses of several tens of metres. The main feature of these processes is the probable absence of a continuous surface of rupture and the presence, at depth, of a zone where displacement takes place mostly through microfracturing of the rock mass (Radbruch-Hall 1978). Before the definition in literature of DGSDs, Terzaghi (1950: 84) contributed to this subject significantly by clarifying the difference between a 'creep' and 'LANDSLIDE' with a statement that is applicable also to deep-seated phenomena:

> A landslide is an event which takes place within a short period of time as soon as the stress conditions for the failure of the ground located beneath the slope are satisfied. By contrast, creep is more or less a continuous process. A landslide represents the movement of a relatively small body of material with well-defined boundaries, whereas creep may involve the ground located beneath all the slopes in a whole region and no sharp boundary exists between stationary and moving material.

Thus the deformation phase may be naturally followed by a sliding phase within which shear planes are recognizable, though the evolution time of these processes is hard to predict and generally extremely long.

DGSDs, thus defined by Malgot (1977), have been documented almost everywhere in the world since the end of the 1960s and described by different authors with different terms, such as sackung, gravity faulting, depth creep of slopes, deep-reaching gravitational deformations, deep-seated creep deformations, gravitational block-type movements, gravitational spreading and gravitational creep (see MASS MOVEMENT). In spite of the variety of terms used, at present the terms most frequently used to identify the main DGSD types are *sackung* and *lateral spreading*.

Sackung

SACKUNG can be described as a sagging of a slope due to visco-plastic deformations taking place at depths which affect high and steep slopes made up of homogeneous, jointed or stratified rock masses showing brittle behaviour (Zischinsky 1969; Bisci *et al.* 1996). Typical morphological features are twin ridges, trenches, gulls and uphill facing scarps in the upper part of the slopes whereas the middle and lower parts of the slopes tend to assume a convex shape because of bulging and cambering. At the foot of the slope sub-horizontal joints can be found. The displacement mechanism, though, has not been well defined. It is thought that the rock mass behaviour at depth is different from that at the surface, owing to the high confining pressure acting all over the material. Two main displacement models have been defined. Most researchers (e.g. Mahr 1977) assume that at depth, in correspondence with the central portion of the slope, a high confining pressure does not allow the formation of well-defined surfaces of rupture, permitting only viscous deformations (non-shearing model). On the contrary, at the top and toe of the slope, where these pressures are lower, such surfaces might develop. Savage and Varnes (1987) assume instead that the zone subject to ductile deformation is indeed interrupted along a shear surface located at the base of the unstable rock mass (plastic failure model).

Lateral spreading

Lateral spreading consists of lateral expansions of rock masses occurring along shear or tensile fractures. Two main types of rock spreading, occurring in different geological situations, can be distinguished (Pasuto and Soldati 1996):

1. *Lateral spreading affecting brittle formations overlying ductile units*, generally due to the deformation of the underlying material. They are characterized by prevalently horizontal movements along tensile fractures or subvertical tectonic discontinuities. Trenches, gulls, grabens, karst-like depressions in the competent rocks and bulges in the clayey material are common features in this type of deformation. The overburden of the rock slabs is generally assumed as the cause of long-term displacements affecting the underlying formations which result in the squeezing out of the weaker rock types and rock block spreading due to tensile stresses. The process may be accelerated by water percolation through the fissures and consequent softening of the clay shales. Downcutting of valleys may then induce rotational slides and rock falls, together with block tilting and rotation which may prepare the way for block slides. The process may continue and cause progressive spreading and dismembering of the rock slab. The spreading may extend for several kilometres back from the edges of plateau.
2. *Lateral spreading in homogeneous rock masses* (usually brittle) without a recognized or well-defined basal shear surface or zone of visco-plastic flow. Typical morphological evidence is given by double ridges, uphill-facing scarps, ridge-top depressions and infilled troughs. This phenomenon has been recognized as prevalent in high mountain areas. The pre-existence of cracks in the rock mass and a high relief energy are considered as favouring factors but the mechanics of the deformation have not yet been well defined.

The evolution scenarios of sackung and lateral spread are different. The former may be considered as an initial stage of rotational– translational slides, with the tendency to evolve into rock or debris avalanches, i.e. processes which may induce high geomorphological risk situations. On the other hand, the latter may correspond to an early phase in the development of block slide-type phenomena, which are usually subject to a slow evolution of displacements.

References

Bisci, C., Dramis, F. and Sorriso-Valvo, M. (1996) Rock flow (sackung), in R. Dikau, D. Brunsden, L. Schrott and M.-L. Ibsen (eds) *Landslide Recognition: Identification, Movement and Causes*, 150–160, Chichester: Wiley.
Mahr, T. (1977) Deep-reaching gravitational deformations of high mountain slopes, *Bulletin International Association of Engineering Geologists* 16, 121–127.
Malgot, J. (1977) Deep-seated gravitational slope deformations in neovolcanic mountain ranges of Slovakia, *Bulletin International Association of Engineering Geologists* 16, 106–109.
Pasuto, A. and Soldati, M. (1996) Rock spreading, in R. Dikau, D. Brunsden, L. Schrott and M.-L. Ibsen (eds) *Landslide Recognition: Identification, Movement and Causes*, 122–136, Chichester: Wiley.
Radbruch-Hall, D.H. (1978) Gravitational creep on rock masses on slopes, in B. Voight (ed.) *Rockslides and Avalanches*, 607–675, Amsterdam: Elsevier.
Savage, W.Z. and Varnes, D.J. (1987) Mechanics of gravitational spreading of steep-sided ridges ('Sackung'), *Bulletin International Association of Engineering Geologists* 35, 31–36.
Terzaghi, K. (1950) Mechanism of landslides, in S. Paige (ed.) *Application of Geology to Engineering Practice*, 83–123, Washington, DC: Geological Society of America.
Zischinsky, Ü. (1969) Über Sackungen, *Rock Mechanics* 1(1), 30–52.

Further reading

Cruden, D.M. and Varnes, D.J. (1996) Landslides types and processes, in A.K. Turner and R.L. Schuster (eds) *Landslides: Investigation and Mitigation*, 36–75, Washington, DC: Transportation Research Board, National Academy of Sciences, Special Report 247.

MAURO SOLDATI

DEEP WEATHERING

Weathering studies have enjoyed a precarious role in geomorphology, at once central and yet often neglected. Rock decay due to chemical and biochemical processes mediates the rate of erosion and destruction of relief in almost all climates, and the dominance of quartz sand in clastic sediments demonstrates its effectiveness. Soil clays are products of these processes and are universally recognized without special comment. However, weathered materials frequently extend well below the classic soil profile to depths of tens of metres, and not infrequently to more than 100 m. The transition from surface soil to fresh rock is described as the REGOLITH or *weathering profile*. While there is no formal definition of what constitutes *deep weathering*, some authors use the term to describe 'exceptional' depths of rock decay (Taylor and Eggleton 2001), but this reflects experience from outside the humid tropics where weathering depths exceeding 30 m are common (Plate 32). A different approach refers to denudation being 'weathering limited' where altered materials are removed more or less instantaneously following breakdown (by chemical and mechanical processes), or 'erosion limited' where stores of non-cohesive, weathered material underlie the landsurface. In the latter case the products of weathering have remained *in situ* for an unspecified period, and it implies that during this time rates of weathering have exceeded rates of erosion. It is these circumstances that lead to the formation of deep weathering profiles, often over periods of 10^6–10^7 y. In the upper zones of many deep weathering profiles the rock has been largely reduced to a mixture of clays, Al and Fe oxides and quartz sand, through which traces of the rock structure can still be seen. This material is termed SAPROLITE.

Plate 32 Deep weathering profile (>50 m) in granite with corestones in east Brazil

It is often stated that deep weathering is mainly associated with ancient landsurfaces of low relief and is the product of a humid tropical or subtropical environment. This reasoning is commonly applied to occurrences of deep weathering found in high latitudes, which are explained as relicts of a formerly extensive mantle of weathered rock formed at the end of the Mesozoic or in the early Cenozoic, when warm moist conditions prevailed to perhaps 60°N. In support of this view, extensive deep weathering of the Scandinavian shield rocks is found below Cretaceous sediments in South Sweden (Lidmar Bergström 1989), and 5–10 m of advanced alteration is found between Palaeogene lava flows in Northern Ireland (Smith and McAlister 1995). Deep saprolites are widely encountered throughout Western Australia, to depths of 100 m in places, and oxygen isotope and other methods have indicated ages from Permean to Miocene, when the Australian plate was far south of its present position and never in tropical latitudes (Bird and Chivas 1988). This led Taylor *et al.* (1992) to argue that time rather than climate might be the main determinant of advanced rock decay to great depths. However, deep saprolites exhibiting advanced weathering are found in Neogene terrain in the humid tropics, and have been cited from Borneo and New Guinea (Thomas 1994; Löffler 1977). Extensive planation is not recorded in these areas, so the profiles indicate high rates of weathering combined with low rates of erosion in a landscape of moderate relief in an equatorial climate under rainforest. In contrast, many deep weathering occurrences in high latitudes present features indicative of incipient rather than advanced decay. These materials are sandy, with a low clay content (typically 2–7 per cent), and are described as *arènes* (French) or GRUS (German). Occurrences of grus are found worldwide in temperate climates, and a similar material may be found at depth beneath clayey saprolites in the tropics. Grus depths are usually <15 m and commonly 3–6 m, but are not confined to landscapes of low relief (Migoń and Lidmar Bergström 2001). When all types and degrees of rock alteration are grouped together, deep weathering is found to be very widespread. It is comparatively rare in hot and cold deserts, and in areas of recent or active tectonics. Most of the regolith mantle has also been removed where there has been severe Pleistocene glacial scour. But deep profiles have been found in north-east Scotland (Hall 1985) and northern Scandinavia, where ice sheets were either cold based and non-erosive or had extended on to low ground.

The formation of deep weathering profiles poses difficult problems. For example, weathering processes are advanced by renewal of ground water and removal of minerals in solution, and will be inhibited by rising concentrations of solutes. Very deep profiles beneath ancient plateaux must, by this reasoning, require very long periods to form and need some means to export minerals in solution. Low solute concentrations in tropical rivers draining weathered landscapes are often cited in support of low weathering rates in these landscapes. The formation of a thick layer of saprolite is, therefore, considered by many to be a self-limiting system experiencing negative feedback. However, we know little about either the deep circulation of water or the potential for long distance migration of ions by diffusion processes. Arguments have been advanced in favour of hydrothermal processes being responsible for much deep rock decay, especially in granites. But many analyses have adduced oxygen isotope evidence for low temperature alteration (70 °C) as at St Austell, south-west England (Sheppard 1977), and hydrothermal mineralization is usually very restricted in extent (Ollier 1983). It is necessary to recognize the importance of interactions between meteoric water penetrating from the Earth's surface and juvenile waters generated by magmatic processes. Both are part of the global water cycle, and rock decay is ultimately a process of adjustment of mineral species to atmospheric conditions at the Earth's surface.

The existence of an extensive mantle of residual weathering products has great significance for engineers, as well as for geomorphologists and pedologists. But the nature of the weathered material is equally important. Grus behaves very differently from a clay-rich saprolite, for example. The transition from fresh rock, upward through the weathered rock towards the surface soil can be complex, but models have been developed to describe the *weathering profile*, as distinct from descriptions of soil profiles (Figures 34, 35). At the base of the profile is the WEATHERING FRONT, often described as the *basal weathering surface* because of the commonly observed, abrupt transition from sound rock to a disaggregated and altered 'saprock'. Very little chemical change is required to cause expansion of rock minerals by hydration and partial hydrolysis, leading to a disruption of the rock fabric. The most commonly described weathering profiles (Figure 34) are based on examples in jointed granites, and similar features are found in basaltic lavas and in feldspathic sandstones. But in banded and foliated metamorphic rock, such as schists, profile subdivisions may be indistinct.

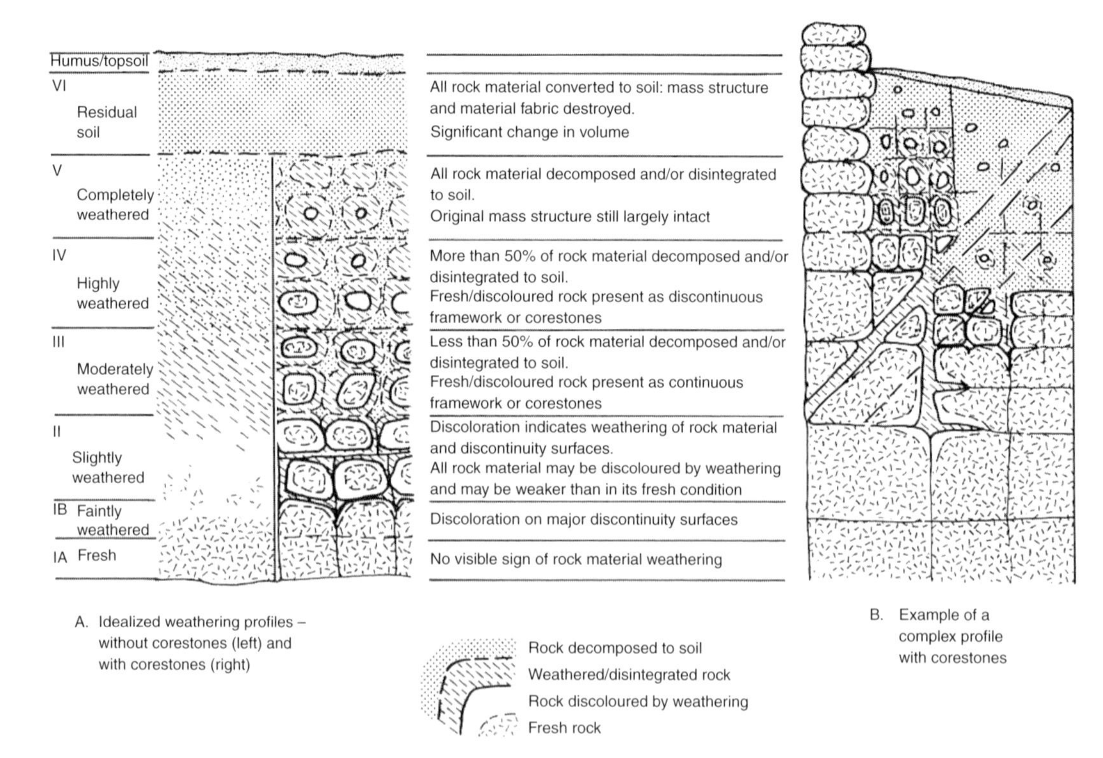

Figure 34 Characteristic weathering profiles with commonly used weathering grades shown in far-left column. Compiled by the author for Fookes (1997)

The 'granite model', first formally described from Hong Kong (Ruxton and Berry 1957), has been refined for the use of engineers (Fookes 1997); other models have been developed to describe mineralogic changes or the occurrence of specific weathering zones, including *laterite* (Figure 35). Chemical and mineralogical changes down profile are important in mineral prospecting, and the nature of the clays predicts engineering behaviour. The understanding of complete regolith profiles can be difficult due to problems of sampling, complexities of rock structure, and the mineral transformations caused by changing hydrologic conditions over long periods. But the issue is important if partly eroded (truncated) profiles, often found in the field, are to be correctly described and understood. The properties of soils in areas of deeply weathered rock, are strongly influenced by the degree of pre-weathering, which limits the availability of cations for plant growth. In many parts of the tropics, several generations of soils may have been formed, lost by erosion and re-formed within deeply weathered parent materials (Ollier 1959).

In TROPICAL GEOMORPHOLOGY, the role of the weathered mantle in determining landscape forms has been widely discussed (Thomas 1994). The balance between the rate of weathering and the rate of erosion is central to questions about the degree of alteration of near-surface weathering products on the one hand and the exposure of fresh rock forms on the other. Estimated rates of weathering on silicate rocks range from 2–50 m Ma^{-1} (mm ka^{-1}). Although surface erosion rates may exceed the highest value by two orders of magnitude, many forested slopes of moderate inclination in the tropics erode at rates less than 5 mm ka^{-1}. But data are sometimes contradictory and it is difficult to generalize. Circumstantial evidence for low rates of erosion in undulating, forested terrain comes from the partial conformity of weathering zones with present-day relief, which often exhibits multiconvex weathered compartments (*demi-oranges*,

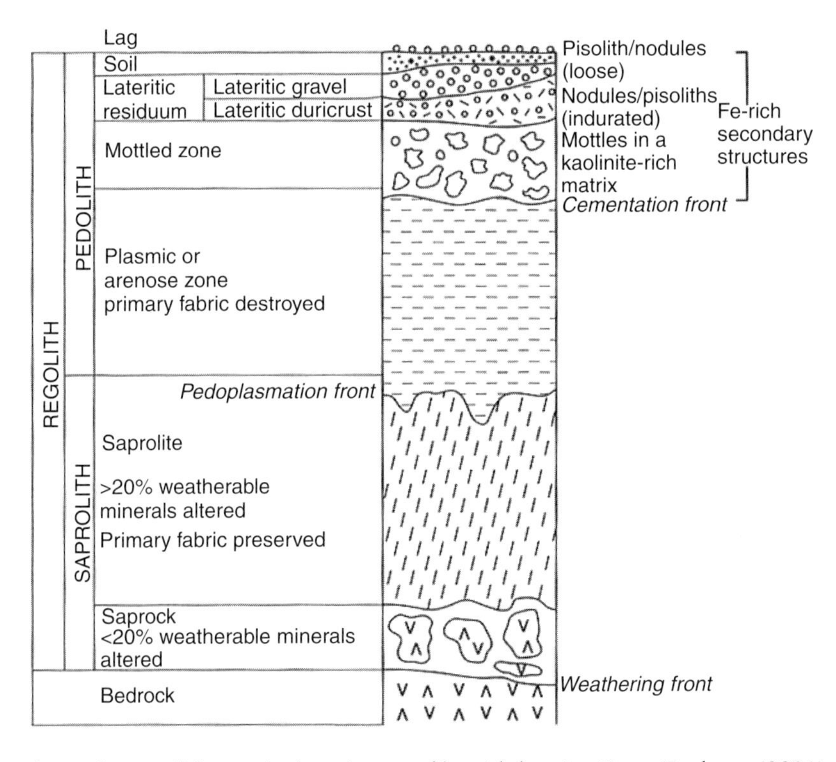

Figure 35 Scheme for regolith terminology in a profile with laterite. From Eggleton (2001)

French; meias laranjas, Portuguese). In the more arid areas of Africa and Australia, weathering profiles have been widely truncated, leaving mesa-shaped tabular hills capped by FERRICRETE or SILCRETE DURICRUSTS or landscapes with shallow, sandy regoliths and frequent outcrops (boulders, TORS, INSELBERGS). In central Australia, reference is made to 'the weathered landsurface' (Mabbutt 1965), and to the varied results of partial stripping of the regolith. Such landscapes have been described as varieties of 'etched plain' (see ETCHING, ETCHPLAIN AND ETCHPLANATION). A wider interpretation views these characteristics as the result of a 'cratonic regime' (Fairbridge and Finkl 1980) involving landscape stability and advanced weathering lasting perhaps 10^7–10^8 y, alternating with periods of erosion and regolith with a duration of 10^5–10^6 y.

In detail, the patterns of deep weathering can be shown to follow petrographic variations and structural weaknesses within rock masses. Ferromagnesian minerals and plagioclase feldspars decay more rapidly than orthoclase and mica in granites, and adjacent plutons containing different mineral suites often show contrasts in weathering. Potassium-rich intrusive rocks and silicified metamorphic gneisses, in particular, resist chemical attack. Intersecting joint patterns often outline basins of deeper rock decomposition. Deep erosion into ancient granite plutons exposes massive compartments under compressive stress that resist weathering although subject to spalling, while younger, higher level intrusives are usually subdivided along many joint directions. Where geology is uniform, patterns of weathering often respond to the relief, deeper weathering being found beneath convex summits in forested environments. While this can result from dissection into an extensive, deep saprolite mantle, the better drained conditions beneath upper slopes contribute to more rapid decay. Most perennial rivers flow in bedrock channels, but some channels in plateau landscapes alternate between anastomosing reaches containing rapids marking exposed fresh rock and meandering channels where the river flows above saprolite.

Deeply weathered landscapes were formally very extensive in Europe (and elsewhere). So-called 'lateritic' weathering covers were partially stripped from the Hercynian massifs of Europe during the Cenozoic, and are found today in the deposits of the Aquitaine, Paris and many other sedimentary basins. These were described by Millot (1970) as 'siderolithic facies', and elsewhere as 'laterite derived facies' (Goldberry 1979) and 'red beds'. During the Neogene a renewal of the regolith cover occurred patchily in the broken relief, resulting from Alpine tectonics. But the short duration of this period and the cooler climates of higher latitudes resulted in thinner, poorly differentiated sandy 'grus'. In the tropics, the breakup of Gondwanaland and drifting apart of the continents during the last 100 Ma also led to deep dissection and the infilling of downwarped and faulted sedimentary basins with the detritus of Mesozoic weathering. These are known as the Continental Terminal in west Africa and the Barreiras Formation in South America. Climatic vicissitudes have involved aridification of many tropical areas after the mid-Miocene, halting the advance of the weathering front in some drier regions. Elsewhere the warmth, humidity and biological productivity have combined to produce younger saprolites with well-defined profiles.

Paradoxically, the most rapid weathering probably takes place in tectonic regions, where a combination of high rainfall, the occurrence of marine sediments (limestones, greywackes) and epithermal igneous rocks, plus stress fracturing of nearly all formations, promote weathering penetration and contribute to high erosion rates (Stallard 1995). However, the steep slopes erode rapidly and become weathering limited, deep weathering is therefore rare. In the humid tropics, steep terrain is subject to frequent landsliding, and the regolith becomes unstable when depths of 5–6 m are reached. As relief and slope are reduced, weathering profiles deepen and there is a need to research the thresholds governing this balance. Observations suggest that, where slopes are reduced below *c*.20°, weathering rates under forest can keep pace with the rate of regolith loss by erosion.

The phenomenon of deep weathering is, therefore, an expression of the formation and survival of materials in equilibrium with near-surface Earth environments. It involves the decay of minerals contained in rocks formed under pressure and in the absence of atmospheric gases, organic acids and micro-organisms, all of which are agents of chemical change. It is also an expression of the fluctuating rates of denudation in time and space. The great stores of saprolite that occur on the continents are, in some areas at least, relicts of the remote past, but weathering processes are continuous and deepening of the weathering mantle occurs where weathering is favoured and rates of denudation low.

References

Bird, M.I., and Chivas, A.R. (1988) Oxygen isotope dating of the Australian regolith, *Nature* 331, 513–516.

Eggleton, R.A. (ed.) (2001) *The Regolith Glossary*, Cooperative Research Centre for Landscape Evolution and Mineral Exploration (CRCLEME), CSIRO, Australia.

Fairbridge, R.W. and Finkl, C.W. Jr. (1980) Cratonic erosional unconformities and peneplains, *Journal of Geology* 88, 69–86.

Fookes, P. (ed.) (1997) *Tropical Residual Soils*, Geological Society Professional Handbook, Bristol: The Geological Society.

Goldberry, R. (1979) Sedimentology of the Lower Jurassic flint clay bearing Mishor Formation, Makhtesh Ramon, Israel, *Sedimentology* 19, 229–251.

Hall, A.M. (1985) Cenozoic weathering covers in Buchan, Scotland and their significance, *Nature* 315, 392–395.

Lidmar Bergström, K. (1989) Exhumed Cretaceous landforms in south Sweden, *Zeitschrift für Geomorphologie NF Supplementband* 72, 21–40.

Löffler, E. (1977) *Geomorphology of Papua New Guinea*, Canberra CSIRO/Australian National University Press.

Mabbutt, J.A. (1965) The weathered land surface of central Australia, *Zeitschrift für Geomorphologie NF* 9, 82–114.

Migoń, P. and Lidmar-Bergström, K. (2001) Weathering mantles and their significance for geomorphological evolution of central and northern Europe since the Mesozoic, *Earth Science Reviews* 56, 285–324.

Millot, G. (1970) *Geology of Clays* (trans. W.R. Farrand and H. Paquet), London: Chapman and Hall.

Ollier, C.D. (1959) A two cycle theory of tropical pedology, *Journal of Soil Science* 10, 137–148.

——(1983) Weathering or hydrothermal alteration, *Catena* 10, 57–59.

Ruxton, B.P. and Berry, L. (1957) Weathering of granite and associated erosional features in Hong Kong, *Geological Society of America Bulletin* 68, 1,263–1,292.

Sheppard, S.M.F. (1977) The Cornubian batholith, SW England: D/H and $^{18}O/^{16}$ studies of kaolinite and other alteration minerals, *Journal of the Geological Society of London* 133, 573–591.

Smith, B.J. and McAlister, J.J. (1995) Mineralogy, chemistry and palaeoenvironmental significance of an

early Tertiary Terra Rossa from Northern Ireland: a preliminary review, *Geomorphology* 12, 63–73.
Stallard, R.F. (1995) Tectonic, environmental and human aspects of weathering and erosion, *Annual Review of Earth and Planetary Sciences* 23, 11–39.
Taylor, G.R. and Eggleton, A. (2001) *Regolith Geology and Geomorphology*, Chichester: Wiley.
Taylor, G.R., Eggleton, R.A., Holzhauer, C.C., Maconachie, L.A., Gordon, M., Brown, M.C. and McQueen, K.G. (1992) Cool climate lateritic and bauxitic weathering, *Journal of Geology* 100, 669–677.
Thomas, M.F. (1994) *Geomorphology in the Tropics*, Chichester: Wiley.

MICHAEL F. THOMAS

DEFLATION

The process by which the wind removes fine material from the surface of a beach or desert. Large particles are left behind as a deflation lag and can cause ARMOURING and STONE PAVEMENT formation. Deflation from largely vegetation-free surfaces can create DUST STORMs and contribute to the formation of various features of wind erosion, including PANs and YARDANGs.

A.S. GOUDIE

DEGLACIATION

Deglaciation means the time period of uncovering of land or water by ice due to glacier retreat, normally forced by a climate change, in contrast to the term glaciation meaning the period of covering of land by ice. Deglaciation is related both to the major retreat of continental or regional-scaled glacier ice masses (large-scale deglaciation), especially during the glaciation phases in the Pleistocene, and for glacier shrinkage during the Holocene neoglaciation, e.g. since the Little Ice Age (small-scale deglaciation).

Deglaciation is triggered through climate changes (long-term) or climate variations (short-term) which impact the GLACIER mass balance due to changes in snow precipitation (accumulation) during winter and energy balance (e.g. temperature, radiation, latent heat release) during summer (ice melting). Increased summer energy input will increase ablation and cause an immediate response as retreat of the glacier front. The dynamic response of the glacier due to positive or negative mass balance of the glacier causes changes in the ice flux (mass transport) from the accumulation area down to the lower ablation area and will result in an advancing or retreating glacier front. The glacier front position will react on this forcing after a certain time period, known as the reaction time and the response time. The reaction time is given as the time lag between when the changes in mass balance occur and the first visible dynamic response of the front, and the longer response time defines the period until the glacier has stabilized to the new mass balance. These timescales are related to the dynamics of the glacier, and the glacier geometry, and can vary from a few years on a small valley glacier to several hundred or even thousands of years on large outlets from an inland ice sheet. The geometry and hypsometry (area-altitude distribution) of the glacier is important for the response. For example, if the Equilibrium Line Altitude (ELA) is raised by 100 m due to a warmer climate the increased area affected by more melting will be larger on a flat, wide glacier, and smaller on a steep, narrow glacier. Higher summer energy input, giving an immediate glacier front retreat and gradually lower ice transport over time, almost always causes deglaciation. During the last decades several energy balance models coupled to glacier-dynamical models have been developed, allowing the spatial and temporal simulation of glacier retreat due to different types and magnitudes of climatic forcing (e.g. Oerlemans 2001) (Figure 36).

The rate of deglaciation thus depends on climatic and topographic factors. Glaciers ending on land will usually get thinner and flatter during deglaciation. Calving glaciers will keep their steep, calving front, but get thinner and retreat much faster than glaciers ending on land. If the glacier is grounding in water, the reduction of mass flux may trigger buoyancy forces to lift up parts of the glacier front, which in turn leads to a massive up-calving of the glacier front. Such a rapid ice retreat occurred in deep fjord areas in western Scandinavia during the deglaciation of the Weichselian ice sheet (e.g. Sollid and Reite 1983).

The high temperature variability during the Pleistocene has caused numerous deglaciation phases in the northern hemisphere (Figure 37). According to present knowledge, there have been more than forty phases of glaciation and deglaciation during the Pleistocene. The last deglaciation was forced by a rapid increase of temperature. The mean Holocene temperature is about 10–13 °C

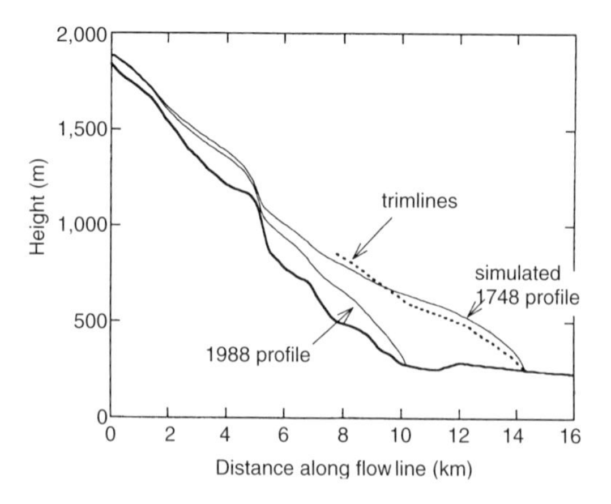

Figure 36 Glacier front positions at the outlet glacier Nigardsbreen, southern Norway (from Oerlemans 2001). The positions are obtained by applying a combined mass balance and dynamic glacier model

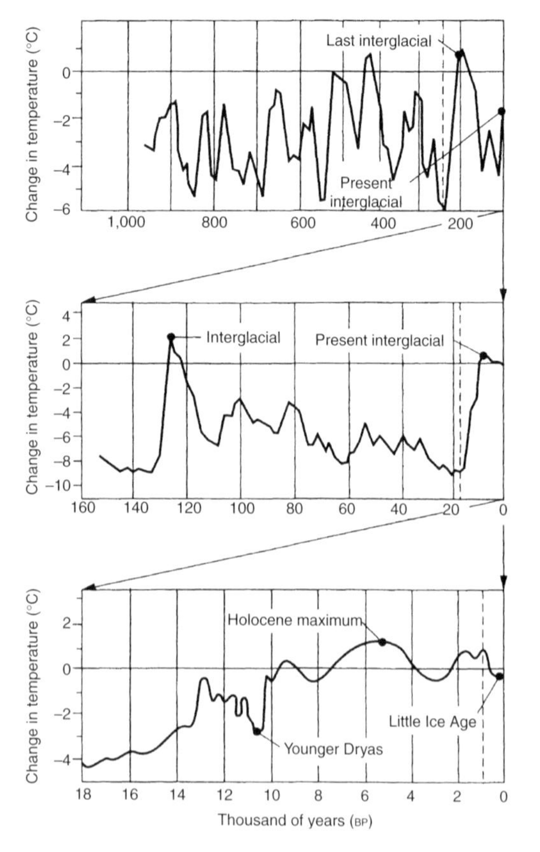

Figure 37 Temperature variation during the Pleistocene and Holocene, indicating glaciation and deglaciation phases (changed from Siegert 2001)

warmer than the mean temperature during full glacial conditions (Figures 37, 38). From the Greenland ice cores several rapid changes, called Dansgaard–Oeschger events, have been observed during the last glaciation (Dansgaard 1993). They all show an extremely rapid temperature increase of about 10° over only a hundred years followed by a slow cooling over several hundred years. At about 10,000 BP the temperature increased quickly again and stabilized at the Holocene temperature level, causing a rapid deglaciation. The forcing mechanisms for these large temperature changes over short time periods are discussed but not yet known. During Holocene the warmest period was in early to mid-Holocene and in many mountain regions the glaciers were probably melted away in the period 8,000–6,000 BP. The climate became colder from about 3,000 BP, starting the increase of glaciers in high mountain areas of the world (NEOGLACIATION), with a culmination in the period between the thirteenth century and about 1750 in Europe (e.g. Nesje *et al.* 2000), or about a hundred years later in some regions of the world. This period is known as the Little Ice Age. Since then deglaciation has prevailed until recent times in most glacial environments of the world. In some mountain regions with valley and cirque glaciers the mass loss has been massive, as for example in the Alps where the glacier retreat has resulted in a nearly 50 per cent reduction of the ice volume since the mid-1800s (e.g. IAHS(ICSI)/UNEP/UNESCO/ WMO 2001). Since the 1990s an accelerated deglaciation is observed in many alpine and in some arctic areas (Arendt *et al.* 2002; Meier and Dyurgerov 2002) and has been attributed to global warming.

Large-scale deglaciation led to a reduced weight on the landmasses, forcing a land heave. Simultaneously, the melting of glacier ice forced sea level to rise. Thus large-scale deglaciation is always related to land emergence and sea-level rise (see ISOSTASY; EUSTASY). During the maximum Weichselian ice extension global sea level was approximately 120 m lower than today. During the last deglaciation of the Weichselian ice sheet, the deglaciated area showed a net land heave, e.g. in Scandinavia. The central Bothnian area has emerged by more than 800 m since the deglaciation. Land heave produces continuously new coastlines with corresponding landforms, such as BEACH RIDGES and coastal ABRASION platforms, indicating the marine limits during a period of time. The spatial relationship between land heave rates and marine limits has been used for relative dating (see DATING METHODS) of ice-recessional landforms. Furthermore, land heave results in subaerial exposure of the former sea bottom, covered by mainly marine clay-rich sediments. Areas covered by marine clays are abundant in areas of deglaciation, such as in Scandinavia and Canada. Having a high nutrient content and being easily arable these areas were of interest for early settlement and agriculture. However, these sediments are highly unconsolidated, and since deglaciation subject to severe GULLYing and prone to landsliding (see LANDSLIDE; QUICKCLAY).

The period of deglaciation is also the period of sediment accumulation by glacier ice and meltwater. Deglaciation is not a continuous process. Especially during the early phase of deglaciation the recession of ice margins frequently halted ('stagnation') or smaller re-advances occurred due to short-term climatic deterioration. The geological time periods of the Younger Dryas and the Preboreal are examples of such short-term climatic variations, which morphologically can be followed by glacial landforms almost throughout southern Finland, central Sweden and coastal Norway (Figure 38).

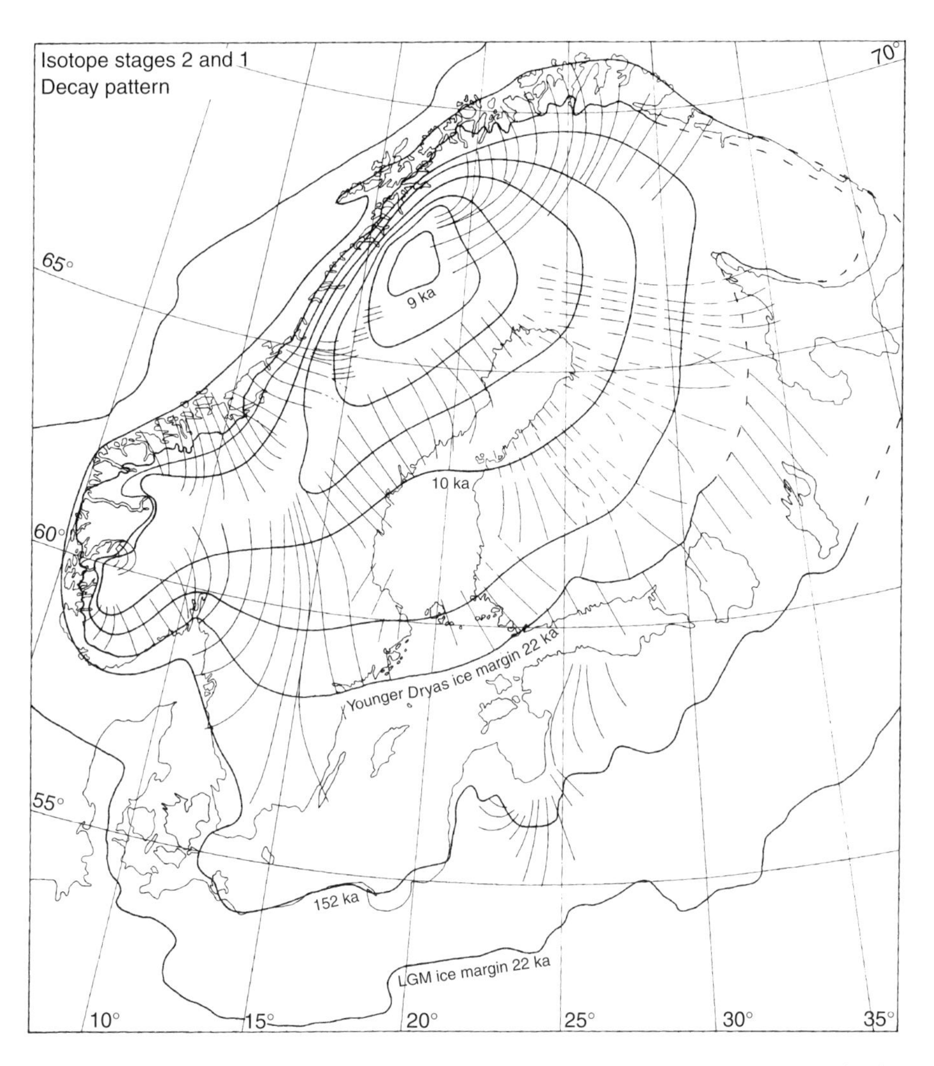

Figure 38 Glacial decay pattern of the Weichselian glaciation in Fennoscandia since the last glacial maximum (LGM, 22 ka) and the onset of the Holocene (*c.*10 ka) (adapted from Kleman *et al.* 1997)

The type of glacial land system (see Plate 33) during deglaciation phases is dependant upon whether the glacier front ends in water (marine/lacustrine) or on land (terrestrial), how fast the land is uncovered (deglaciation rate) and what glacier temperature regime prevailed during deglaciation (glacier thermal regime) (see also Benn and Evans 1998). A fast retreat of glacier tongues or lobes results in an uncovering of subglacial landforms such as fluted surface or drumlinoide forms. These subglacial landforms normally show the very last glacier movement direction. If the deglaciation happens in a permafrost environment, parts of the glacier marginal areas may be cold based. In such environments, glaciers may preserve sediments and landforms derived from earlier glaciation and deglaciation periods. Slow retreats and/or temporary stagnation of the glacier front produce landforms of accumulated glacigenic material, terminal moraines. During stagnation phases the glaciers advance some metres during winter and retreat during summer due to variations of ablation. The winter advances produce small annual moraines. Produced under water they are called DeGeer-moraines in marine environments and cross-valley moraines in mountainous lacustrine environments. Such moraines often build a sequence of landforms, used in deglaciation reconstruction. Glacier margins in permafrost environments are cold based. In this setting, a net-freezing condition along the glacier base prevails (e.g. Boulton 1972), forcing glacigenic material to accumulate in the glacier front area. Ice motion

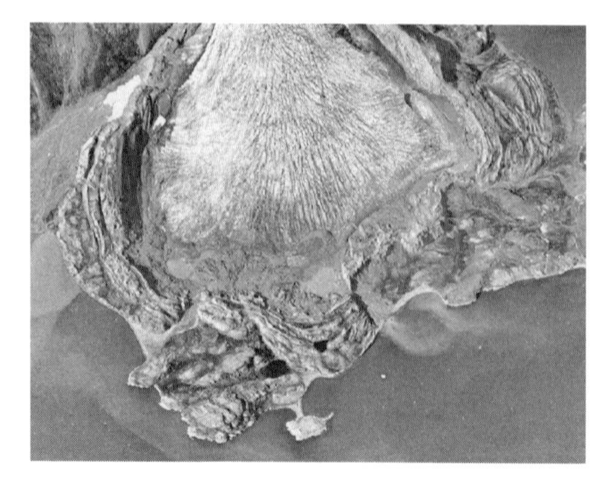

Plate 33 Oblique air photo of recently deglaciated terrain, Erikbreen, northern Spitsbergen (from Etzelmüller *et al.* 1996)

and surface melting leads then to an accumulation of glacigenic sediments on the glacier surface. If the thickness of this layer is larger than the ACTIVE LAYER thickness of the environment (see PERMAFROST), the ice below the layer is preserved, preventing further ablation. These ice-cored moraines can persist over long periods during deglaciation. If the active layer thickness increases or is eroded by e.g. fluvial action, the ice core melts and produces a hummocky moraine terrain proximal to the glacier front area and a clear distinctive border distal from the glacier. Deposited on steep terrain, the ice-cored moraines may continue creeping, producing ROCK GLACIER-like landforms. If the glacier front ends in water, most sediment transported in glacial meltwater is deposited in the vicinity of the glacier front, building up ice-contact deltas. When built up to the water surface, they give an indication of sea level during the deglaciation phase. Especially during the last stages of deglaciation, parts of the ice streams may become decoupled from the active part of the glaciers. This results in the loss of glacier flux and thus dead-ice wastage occurs. In this situation ice can be buried by e.g. glaciflu-vial sediment. Melting of these ice bodies results in typical dead-ice landforms, consisting of hummocky irregular terrain and KETTLES AND KETTLE HOLES. On cold ice caps, meltwater is often routed along the glacier margins, forming channels that mark the ice surface during phases of deglaciation. Swarms of subsequently lower channels along mountain slopes indicate the lowering of the ice surface and the glacier surface slope during different phases of deglaciation.

On continental ice sheets the ice divide did not necessarily correspond with the position of the topographic water divide of the underlying relief. During deglaciation topographic water divides often became ice free before all ice disappeared. This ice could occasionally block the drainage and thus formed ice-dammed lakes that drained over local or regional water divides. Like terminal moraines, lacustrine sediments and landforms such as shorelines bear witness to periods of deglaciation. Sudden outbursts of glacial lakes are called by the Icelandic word *jokullhaup*. In many high mountain areas, e.g. in central Asia, deglaciation of valley glaciers leads to damming of lakes between the glacier front and terminal moraines, which often are ice-cored. These lakes are unstable, and outbursts, often called GLOFs – glacier lake outburst floods – are a potential hazard for lower lying valleys and human settlements and infrastructures. The same risk applies to the situation where valley glaciers block the drainage from minor side valleys.

Research on deglaciation is concentrated on (1) determining the start of deglaciation, (2) the deglaciation rate, and (3) the change of the spatial distribution of the ice body during different phases of deglaciation. Traditionally, scientists concentrated on determining the start and deglaciation rate, by dating of landforms associated to terrestrial or marine/lacustrine glacier margins (see DATING METHODS) and analysing sediment succession building up these landforms. Coring of marine sediments on continental margins and deep ocean basins revealed continuous information on glaciation and deglaciation phases during the Pleistocene (e.g. Elverhøi *et al.* 1995). The Holocene NEOGLACIATION chronology is depicted by core analysis of local sediment sinks such as lakes fed by meltwater from glaciers (e.g. Karlén 1976). The spatial distribution of ice bodies during deglaciation phases is often determined through GEOMORPHOLOGICAL MAPPING of glacial landforms combined with dating methods. The past vertical extent of an ice sheet in its accumulation area is difficult to obtain because of the lack of marked landforms due to low glacier velocities and often cold ice in culmination zones of ice sheets. Recently, exposure dating using cosmogenic isotopes has proved to be a helpful tool in this respect.

Deglaciation has become an important problem for human settlement and sustainable development in many high mountain environments. Especially in many semi-arid areas, such as the eastern slopes of the Andes Mountains and in central Asian mountain ranges, glaciers act as freshwater reservoirs, since meltwater from the glaciers is important for irrigation and water supply. Deglaciation leads to a periodically enhanced runoff. However, due to shrinkage of the water reserve (the glaciers), water availability will be reduced over the long term.

References

Arendt, A.A., Echelmeyer, K., Harrison, W.D, Lingle, C.S. and Valentine, V.B. (2002) Rapid wastage of Alaska glaciers and their contribution to rising sea level, *Science* 297, 382–386.

Benn, D.I. and Evans, D.J.A. (1998) *Glaciers and Glaciations*, London: Arnold.

Boulton, G.S. (1972) The role of thermal regime in glacial sedimentation, *Institute of British Geographers. Special Publication* 4, 1–19.

Dansgaard, W.E.A. (1993) Evidence for general instability of past climate from a 250-kyr ice-core record, *Nature* 364, 218–220.

Elverhøi, A., Anderson, E.S., Trond, D., Hebbeln, D., Spielhagen, R.F., Srendsen, J.I. *et al.* (1995) The growth and decay of late Weichselian ice sheet in western Svalbard and adjacent areas based on provenance studies of marine sediments, *Quaternary Research* 44, 303–316.

Etzelmüller, B., Hagen, J.O., Vatne, G., Ødegård, R.S. and Sollid, J.L. (1996) Glacier debris accumulation and sediment deformation influenced by permafrost, examples from Svalbard, *Annals of Glaciology* 22, 53–62.

IAHS(ICSI)/UNEP/UNESCO/WMO (2001) *Glacier Mass Balance, Bulletin No.* 6, Haeberli, W., Frauenfelder, R. and Hoelzle, M. (eds) World Glacier Monitoring Service, University of Zurich and ETH Zurich.

Karlén, W. (1976) Lacustrine sediments and tree-limit variations as indicators of Holocene climatic fluctuations in Lappland, northern Sweden, *Geografiska Annaler* 58A, 1–34.

Kleman, J., Hättestrand, C., Borgström, I. and Stroeven, A.P. (1997) Fennoscandian palaeoglaciology reconstructed using a glacial geological inversion model, *Journal of Glaciology* 43(144), 283–299.

Meier, M. and Dyurgerov, M. (2002) How Alaska affects the world, *Science* 297, 350–351.

Nesje, A., Dahl, S.O., Andersson, C. and Matthews, J.A. (2000) The lacustrine sedimentary sequence in Sygneskardvatnet, western Norway: a continuous, high-resolution record of the Jostedalsbreen ice cap during the Holocene, *Quaternary Science Reviews* 19, 1,047–1,065.

Oerlemans, J. (2001) *Glaciers and Climate Change*, Lisse: A.A. Balkema.

Siegert, M.J. (2001) *Ice Sheets and Late Quaternary Environmental Change*, Chichester: Wiley.

Sollid, J.L. and Reite, A. (1983) Central Norway, glaciation and deglaciation, in J. Ehlers (ed.) *Glacial Deposits in North-west Europe*, 41–59, Rotterdam: Balkema.

SEE ALSO: dating methods; eustasy; glacier; ice dam, glacier dam; isostasy; moraine; neoglaciation

BERND ETZELMÜLLER AND JON OVE HAGEN

DELL

Small headwater valleys which are characteristically sediment-choked and swampy. Dells frequently occur at the head of deep gorges on plateaux surfaces and may be analogous to DAMBOS. Notable dells have developed on sandstone on the Woronora Plateau of New South Wales, Australia (Young 1986).

They are also known from Eocene beds in the New Forest of southern England, where they may have a periglacial origin and form tributaries to small dry valleys (Tuckfield 1986).

References

Tuckfield, C.G. (1986) A study of dells in the New Forest, Hampshire, England, *Earth Surface Processes and Landforms* 11, 23–40.

Young, A.R.M. (1986) The geomorphic development of dells (upland swamps) on the Woronora Plateau, N.S.W., Australia, *Zeitschrift für Geomorphologie NF* 30, 317–327.

A.S. GOUDIE

DEMOISELLE

A French term used to describe a needle-shaped Earth pillar composed of eroded rock, and capped by a large boulder. The overlying block is typically more resistant than the underlying material, and so tends to protect it from erosion (predominantly by water), as well as helping to maintain its vertical integrity. Demoiselles are commonly found in Alpine areas of highly weathered volcanic breccia or of glacial till. The term is derived from French for 'young lady', and is commonly employed throughout the French Alps. The term *cheminée de fees* (fairies' chimney) is also used frequently in place of demoiselle, while the American synonym of demoiselle is HOODOO.

STEVE WARD

DENDROCHRONOLOGY

Dendrochronology is the study of annual tree rings, with studies based on the measurement of variations in ring widths caused by variations in climate and environment at the time of ring formation. Ring counts and ring-width measurements provide precise calendar dating for ring formation years and a basis for numerous research applications.

The main disciplines included in these applications are: dendroclimatology, dendroarchaeology, art historical research, history, dendroecology and the related disciplines of DENDROGEOMORPHOLOGY, dendroglaciology, dendrohydrology, dendroniveology (snow and ice research), dendropyrology (fire events) and dendroseismology. Tree-ring chronologies have also been used to calibrate the radiocarbon timescale.

Douglass explored the potential of tree rings for climate analysis in 1919, but it was not until 1953 with the discovery of the 4,000-year old Bristlecone pines that the technique drew widespread attention. By the end of the century, chronologies covering nearly 10,000 years had been constructed from matching ring-width patterns of living and dead trees, and dendrochronology was being applied worldwide in research on global change, although as yet work in the tropics has been limited.

Procedures entail either cutting discs from a stem or, more usually and less destructively, the collection of wood cores with a 5 mm increment borer. The core is usually glued to a wood support, with its grain perpendicular to the support, and polished to reveal the ring structures ready for ring-width measurement.

Coring causes mechanical injury. Thus, it is important to obtain permission before taking samples and never to core anything that could be valuable as timber. However, trees compartmentalize wounds and often produce anti-fungicidal substances that generally limit damage; injuries stimulate local growth and holes are callused over in a few years. Studies have indicated that the core hole should be left open and untreated since this could introduce foreign organisms and impede the healing process.

An annual tree ring usually has two growth phases. At the beginning of the growing season conifers produce large, pale, thin-walled earlywood cells; towards the end of the season increasingly small diameter, dark, thick-walled, latewood cells develop. Hardwood trees have a variety of ring forms. Healthy, unstressed trees will produce concentric rings with approximately equal ring widths while stressed trees will form eccentric ring patterns and show narrow, variable ring growth. Trees on slopes are frequently bent, with deciduous species having their central pith displaced towards the downslope side of the stem and conifers to the upslope side (a point to be remembered when coring bent trees).

Problems for the technique, apart from those caused by growth eccentricities, are introduced by non-uniform cell growth due to adverse conditions, with normal cell formation either halted or present over only part of a stem resulting in missing rings. Alternatively, false rings can be produced by late frosts, droughts or other growth-inhibiting events resulting in darkened cells followed by resumption of normal growth before true dark latewood growth marks the end of the growing season. These complications can be mitigated by crossdating.

Crossdating is achieved by matching sample ring-widths using visual and statistical tests. At least two cores are usually collected from each tree, so that they can be crossdated to check for missing or false rings and, where there are none the radii are averaged to show mean annual ring growth. Graph plots of the means of individual trees are then compared and crossdated and a site masterplot created.

Sample depth (number of trees sampled) will change through time affecting the quality of a chronology. Consequently, chronologies may be truncated where there are less than three trees to support a mean curve and, for valid climatic results, curves should contain an absolute minimum of ten trees per site, but thirty or more are desirable. Where sample depth is important, this information should be included on ring-width graphs.

The technique was revolutionized by the advent of computer processing enabling the digitizing of ring-width measurements; rapid plotting and comparison of graphs; rapid application of multi-variant statistics, and radio-densitometric determination of wood density. This latter approach is based on X-ray analysis of changes in cell densities; it is used particularly in climate studies to highlight sensitive reactions of cell densities to temperature variations. It is also used for analysis of tropical species' growth, since these species, rather than always forming annual rings, may produce growth zones reflecting aperiodic precipitation or drought.

Prior to computerization, measurements were made by hand and one approach to crossdating was the use of 'skeleton plotting' based on ring counting, visual assessment of relative per cent ring widths, and identification of event years. Skeleton plots provide dating for, and a visual summary of, the effects of environmental events without the need to make precise ring-width measurements. It is a useful procedure showing major events and growth trends where rapid assessment of limited sample numbers is required.

Apart from the effect of sudden events on tree growth, slow growth changes may occur due to gradual variations in climate or natural reduction in ring width as a tree ages. This latter effect is routinely removed by standardizing (detrending) ring widths using various techniques. Standardization, apart from eliminating age trends, emphasizes event years while removing climatic trends shown by moving averages of the mean ring-width data.

Further reading

Cook, E.R. and Kairiukstis, L.A. (eds) (1990) *Methods of Dendrochronology: Applications in the Environmental Sciences*, Dordrecht: Kluwer Academic Publishers.

Esper, J. and Gärtner, H. (2001) Interpretation of tree-ring chronologies, *Erdkunde* 55, 277–288.

Schweingruber, F.H. (1989) *Tree Rings: Basics and Applications of Dendrochronology*, Dordrecht: Kluwer Academic Publishers.

Schweingruber, F.H., Eckstein, D., Serre-Bachet, F. and Bräker, O.U. (1990) Identification, presentation and interpretation of event years and pointer years in dendrochronology, *Dendrochronologia* 8, 9–38.

VANESSA WINCHESTER

DENDROGEOMORPHOLOGY

Dendrogeomorphology is based on analysis of the annual growth rings of trees or woody plants and their growth forms. It is used to investigate spatial and temporal aspects of Earth surface processes operating during the Holocene at annual to centennial timescales.

The technique is closely associated with dendroclimatology and employs largely the same methods as DENDROCHRONOLOGY. Applications include dating and establishing rates of change and frequencies of storms, FLOODS, LAKE outbursts, river channel changes, frost events and ICE surges, GLACIER movements, snow avalanches (see AVALANCHE, SNOW) MASS MOVEMENTS and FIRES, and show the relationships of events to climate. In addition, tree rings can provide records of events unrelated to climate: volcanic eruptions; earthquakes and TSUNAMIS; environmental management; COMPACTION OF SOIL; water table variations, changes in pollution and saltwater ingression.

Methods, other than those used in dendrochronology, include studies of the age, anatomy, morphology, and structures of tree roots, stems and crowns. Root ring patterns can be used to date sediment aggradation or degradation. Trees respond to increases in soil depth by producing adventitious roots; soil movements cause root structures to bend while degradation leaves roots exposed. Ring counts supply dates for root structures, bends and stem age at ground level while age and distances between features show the scale of events. Eccentric ring patterns develop where roots are part-exposed or when denudation brings them close to the surface. Changes in patterns supported by changes in cell structures can be dated. Before sampling buried or exposed roots, records should be made of all relevant features: positions and orientations of main and adventitious root systems; distances to the ground surface; vegetation cover and soil type.

Stems deformed by site changes produce eccentric ring patterns. Stem discs or cores taken both in the direction of stress and at right angles, show when eccentricity begins and the orientation of patterns provides information on the direction of changes. Injuries to stems or roots produce scarring with local growth being stimulated. A core from an undamaged area near the wound (but avoiding re-growth tissue) will show the number of years elapsed since the event.

Crown development provides information on competition, wind and storm events, snow cover and tree health.

The main problems for dendrogeomorphology where surface age is the focus of interest are to establish the total age of a tree and the length of time taken for a tree to colonize a freshly exposed surface. Core ring counts only show the age of a tree above the coring point; thus to find total age an estimate of the number of years growth below this is required. One method is to cut stem discs near ground level of a number of small trees growing in a range of local microenvironments, correlate the height of the trees with their age and calculate the mean growth to height ratio for the location. Verification of ecesis (colonization) times requires an alternative dating source.

Further reading

Gärtner, H., Schweingruber, F.H. and Dikau, R. (2001) Determination of erosion rates by analysing structural changes in the growth pattern of exposed roots, *Dendrochronologia* 19, 81–91.

Schweingruber, F.H. (1996) *Tree Rings and Environment Dendroecology*, Berne: Paul Haupt.

Strunk, H. (1997) Dating of geomorphological processes using dendrogeomorphological methods, *Catena* 31, 137–151.

Winchester, V. and Harrison S. (2000) Dendrochronology and lichenometry: an investigation into colonization, growth rates and dating on the east side of the North Patagonian Icefield, Chile, *Geomorphology* 34, 181–194.

VANESSA WINCHESTER

DENUDATION

On Earth, two forces counterbalance: uplift (i.e. creation of relief) and denudation. Denudation includes all processes that remove the relief at the surface of the Earth. Denudation acts chemically or physically. Chemical denudation, also termed chemical weathering or chemical erosion, is the slow complete or partial dissolution of rock minerals. Physical denudation or mechanical weathering processes correspond to the removal of solids from the land surface. Quantification is generally expressed by the mean of chemical or physical fluxes (or rates) of denudation, expressed most often in $t\,km^{-2}\,a^{-1}$.

The adaptation of rock minerals to the conditions at the surface of the Earth (Ahnert 1996) releases the most soluble elements and leads generally to the formation of residual minerals, usually clays and hydrous iron or aluminium oxides, that accumulate at the interface between the atmosphere and the lithosphere (the REGOLITH). Examples of chemical denudation reactions are the weathering of calcite, which does not leave any residue

$$CaCO_3 + H_2O + H^+ = HCO_3^- + Ca^{2+}$$

and the weathering of albite, that produces a secondary phase, for example kaolinite:

$$2NaAlSi_3O_8 + 9H_2O + 2H^+ = \\ Al_2Si_2O_5(OH)_4 + 2Na^2 + 4H_4SiO_4$$

These equations show that protons are necessary to attack rock minerals. At the surface of the Earth, these protons are mostly derived from the dissolution of the atmospheric or soil CO_2 in water

$$CO_2 + H_2O = H_2CO_3 = H^+ + HCO_3^-$$

Weathering reactions therefore pump CO_2 from the atmosphere and convert it into bicarbonate ions. Ultimately, these ions, combined with the Ca and Mg ions liberated by rock weathering, will lead to the precipitation of calcite in the ocean by living organisms, allowing the sequestration of atmospheric-derived carbon. Other possible origins for protons include production by organic molecules derived from the degradation of organic matter in soils (humic and fulvic acids) and the oxidation of sulphide minerals producing sulphuric acid.

Like minerals, rocks do not weather at the same rate. A good way of estimating the rate of chemical weathering is by analysing rivers draining a single type of rock, provided that the river dissolved load is corrected from the inputs that do not derive from rock weathering (atmosphere, pollution, biomass). Another approach is based on soil mass budgets. Rates of chemical denudation are extremely variable, ranging from less than $t\,km^{-2}\,a^{-1}$ in high latitude granitic catchments or in the low-lying regions of central Africa, to more than $100\,t\,km^{-2}\,a^{-1}$ for rivers draining basalts at Réunion or Java Island (Louvat and Allègre 1997). Basaltic lithologies thus weather 10 to 100 times faster than granites. At a global scale, it has been shown by Dessert *et al.* (2002), that, even if the outcrops of basalts represent 5 per cent of the emerged surface of the Earth, the flux of CO_2 uptake by basalt weathering is as high as 35 per cent of the total consumption flux by rock weathering. Basalt weathering therefore appears as a major mechanism of atmospheric CO_2 regulation. Carbonate rocks also have high denudation rates, ranging from 10 to $200\,t\,km^{-2}\,a^{-1}$. Saline rocks have the highest chemical denudation rates because they are highly soluble in water. From large river systems, chemical denudation rates (Figure 39) ranging from a few $t\,km^{-2}\,a^{-1}$ for the Zaire, Nile and Siberian rivers to about $50\,t\,km^{-2}\,a^{-1}$ for rivers such as the Mekong, Mackenzie or Brahmaputra have been determined (Summerfield and Hulton 1994). These rates are strongly correlated to the abundance of carbonates within the drainage basin, simply because carbonates and evaporites weather at a faster rate than silicates.

Although lithology is the first controlling factor on chemical weathering rates, other parameters exert a control on chemical denudation. At both small and large scales, chemical weathering rates increase strongly with runoff and temperature. This is especially true for basalt weathering rates which respond at a global scale to an Arrhenius-type law (Dessert *et al.* 2002). At a continental scale, however, several authors have pointed out that the highest chemical denudation rates of silicate rocks are not found in the regions of highest rainfall and temperatures (Edmond *et al.* 1994). The Zaire river has the same chemical denudation rates as the Yenisey (Gaillardet *et al.* 1999). The low chemical denudation rates found in the flat and humid tropical areas contrast with the highly weathered nature of soil material (laterites) that characterize these regions. At a global scale, intensity and flux of chemical denudation of silicates are inversely correlated. This paradox will be explained later.

Physical denudation rates can be estimated by different means: by using rivers, sediment accumulation in reservoirs or sedimentary basins and

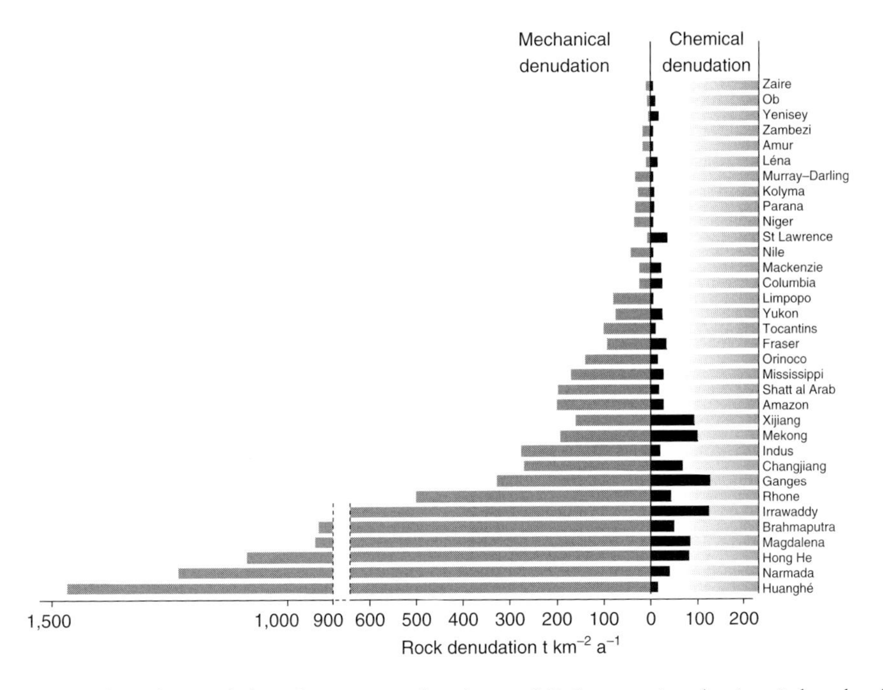

Figure 39 Physical vs chemical denudation rates for the world's largest river basins. Solute load is corrected from atmospheric inputs. Based on Summerfield and Hulton (1994) and Gaillardet *et al.* (1999)

cosmogenic isotopes. Rivers are probably the easiest way of calculating mechanical denudation rates, but this approach suffers from various limitations. Unlike dissolved load, which often fluctuates by a factor of two or so between high water and low water stages, the quantity of solids transported by rivers can vary drastically. A single daily flood event can transport as much sediment as several years of regular sediment flux. In addition, the amount of material transported as bottom sand is generally unknown. Rivers, especially large rivers, store sediments in their floodplains or alluvial fans. For example, it has been shown that two-thirds of the sediments removed from the Andes by the headwaters of the Rio Madeira in South America are stored in the foreland floodplains and never reach either the Amazon, or the sea (Guyot 1993). Finally, the influence of anthropogenic activities has usually resulted in a strong increase of river sediment yield. The best example is the Huanghe river, that transports, today, up to $20 \, g \, l^{-1}$ of sediments to the ocean, while the long-term pre-anthropogenic estimate of Holocene sediment volumes deposited by the river indicates an average suspended load concentration one order of magnitude lower (Milliman and Syvitski 1992).

Mechanical denudation rates (Figure 39) have been computed for the largest river systems by a number of authors (Pinet and Souriau 1988; Milliman and Syvitski 1992; Summerfield and Hulton 1994). They range from numbers lower than $10 \, t \, km^{-2} a^{-1}$ for rivers such as the Siberian (Ob, Yenisey, Kolyma) and tropical (like the Zaire), to numbers higher than $700 \, t \, km^{-2} a^{-1}$ for the largest rivers of Asia (Brahmaputra, Ganges). The world average value is estimated to be $200 \, t \, km^{-2} a^{-1}$, corresponding to $20.10^9 \, t \, a^{-1}$ (Milliman and Syvitski 1992).

The dominant factors that influence mechanical denudation are the erodibility of rocks and relief. High relief areas tend to have high mechanical denudation rates. This is mainly due to slope instabilities and to glacial abrasion. There does not seem to exist any clear relationship between climate (runoff or temperature) and physical erosion, at least on a global scale (Pinet and Souriau 1988; Summerfield and Hulton 1994). Regions of high precipitation and high seasonality of rainfall seem however to exhibit higher mechanical denudation rates (see Goudie 1995). An inverse correlation between suspended yields and basin area is reported by an extensive study of Milliman and Syvitski (1992), possibly showing the importance of sediment storage. The same authors showed that humans are also a major controlling parameter, as fluvial denudation rates during the Holocene are estimated to be less than half the present-day rates. However, for the largest rivers of Asia, there is a remarkably good agreement between present-day fluvial physical denudation rates and long-term rates based on sediment volume (Métivier and Gaudemer 1999) accumulated in the sea. Based on the denudation rates computed on large rivers, Gaillardet *et al.* (1999) have shown that chemical denudation rates of silicate rocks are positively correlated to physical denudation rates. Such a relation is confirmed by cosmogenic nuclides measurements (Riebe *et al.* 2001). This global coupling between chemical and physical fluxes of silicate denudation is explained as follows. With a low mechanical denudation regime, soil development is favoured, leading to a shielding effect of the soil and a negative feedback on the interaction between water and mineral surfaces. This is the transport-limited regime. Conversely, in regions of high mechanical denudation mineral surfaces are continuously exposed, and even if chemical weathering is not intense, the fluxes of released solutes are increased. This is the weathering-limited regime. Overall, the present-day Earth is under the weathering-limited regime.

Total denudation rates are calculated by adding the chemical and physical denudation rates. Using a mean crustal density of $2,700 \, kg \, m^{-3}$, these rates are usually translated in $mm \, a^{-1}$ (Figure 40). For large rivers, landscape downwearing ranges from about $10 \, mm \, a^{-1}$ in regions such as the shield low-lying areas of the Congo craton, Niger basin, Brazilian shield and Australian shield to 100–200 mm per 1,000 years for the Ganges, Indus, Changjiang and Amazon rivers. The Brahmaputra river has the highest rate of total denudation, close to 700 mm per 1,000 years. The mean value for the Earth's surface is 61 mm per 1,000 years (areas of internal drainage excluded). Total denudation rates calculated by mass budgets of riverine products are in good agreement with cosmogenic isotope measurements. Cosmogenic isotopes are produced during the bombardment of cosmic rays and their abundance is a function of the production rate and total erosion rate. This technique has been applied by a number of authors and gives denudation rates, integrated over tens of millions of years in the case of ^{10}Be, which are in general agreement with other techniques. For example, the ^{10}Be derived denudation rates of the Loire, Meuse, Neckar and Regen basins are in the order of magnitude of those found by conventional techniques (Schaller *et al.* 2001).

As a global average, 10–15 per cent of the total denudation is chemical denudation, 75–80 per cent is physical denudation (Figure 39). The ratio of mechanical over chemical fluxes fluctuates widely from about 1 for the Siberian and tropical rivers to 10–20 for mountainous rivers. The present world is therefore dominated by physical denudation, but it may not have been the case for geological periods of low relief.

Atmospheric dust transport from the continent to the ocean is also responsible for the denudation of continents. Aeolian denudation is extremely difficult to quantify and estimates vary between 0.1 to $5.10^9\,t\,a^{-1}$. In deserts, aeolian denudation may be the only process of relief denudation.

References

Ahnert, F. (1996) *Introduction to Geomorphology*, London: Arnold.

Dessert, C., Dupré, B., Gaillardet, J., Francois, L. and Allègre, C. (2003) Basalt weathering laws and global geochemical cycles, *Chemical Geology*. In press.

Edmond, J.M. *et al.* (1994) Fluvial geochemistry and denudation rate of the Guyana shield, *Geochimica et Cosmochimica. Acta* 59, 3,301–3,325.

Gaillardet, J., Dupré, B., Louvat, P. and Allègre, C. (1999) Global silicate weathering of silicates estimated from large river geochemistry, *Chemical Geology* 159, 3–30.

Goudie, A. (1995) *The Changing Earth. Rates of Geomorphological Processes*, Oxford: Blackwell.

Guyot, J.L. (1993) Hydrogéochimie des fleuves de l'Amazonie bolivienne, Ph.D. thesis, Université Bordeaux I.

Louvat, P. and Allègre, C.J. (1997) Present denudation rates at Réunion island determined by river geochemistry: basalt weathering and mass budget between chemical and mechanical erosions, *Geochimica et Cosmochimica. Acta* 61(17), 3,645–3,669.

Métivier, F. and Gaudemer, Y. (1999) Stability of output fluxes of large rivers in South and East Asia during the last two million years. Implications on floodplain processes, *Basin Research* 11, 293–304.

Milliman, J.D. and Syvitski, J.P.M. (1992) Geomorphic/tectonic control of sediment discharge to the ocean: the importance of small mountainous rivers, *Journal of Geology* 100, 525–544.

Pinet, P. and Souriau, M. (1988) Continental erosion and large-scale relief, *Tectonics* 7, 563–582.

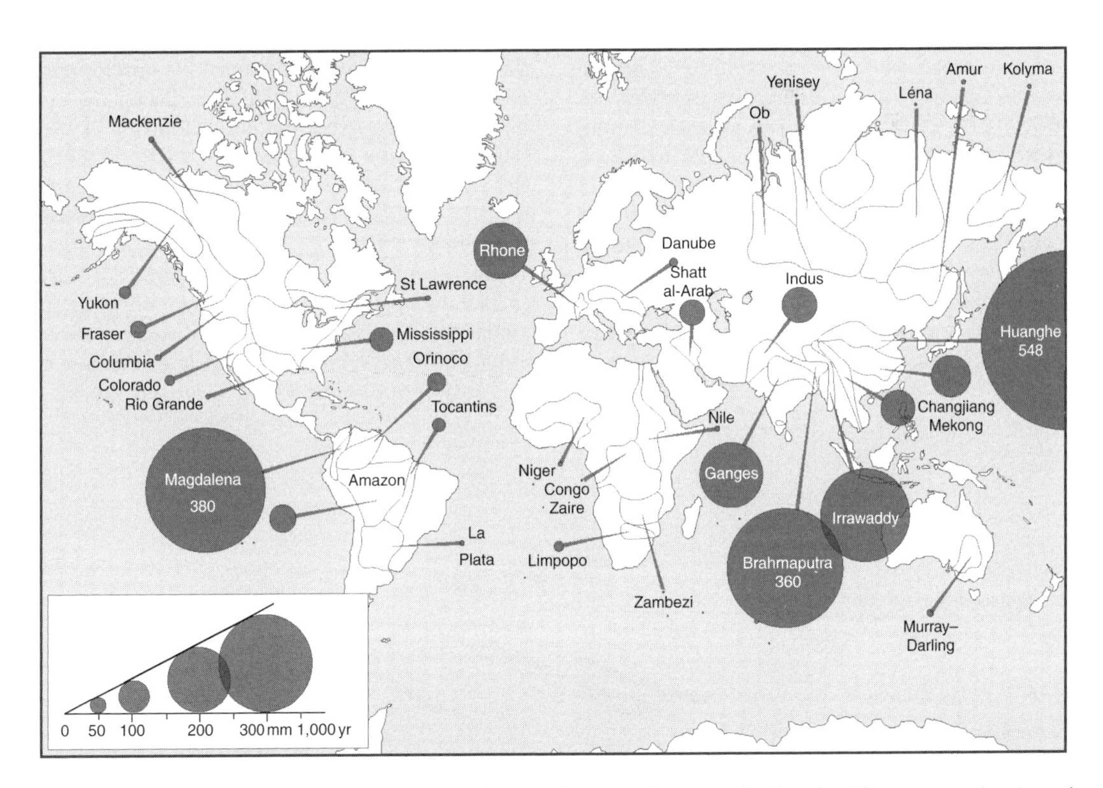

Figure 40 Total denudation rates for the largest drainage basins calculated with a mean density of $2.7\,g\,cm^{-3}$. Based on Summerfield and Hulton (1994) and Gaillardet *et al.* (1999)

Riebe, C.S., Kirchner, J.W., Granger, D.E. and Finkel, R.C. (2001) Strong tectonic and weak climatic control of long-term chemical weathering rates, *Geology* 29, 511–514.
Schaller, M., von Blanckenbourg, F., Hovius, N. and Kubik, P.W. (2001) Large-scale erosion rates from in situ-produced cosmogenic nuclides in European rivers, *Earth and Planetary Science Letters* 188, 441–458.
Summerfield, M.A. and Hulton, N.J. (1994) Natural control of fluvial denudation rates in major world drainage basins, *Journal of Geophysical Research* 99, 13,871–13,883.

Further reading

Summerfield, M.A. (1991) *Global Geomorphology*, London: Longman; NewYork: Wiley.

SEE ALSO: cosmogenic dating

JÉRÔME GAILLARDET

DENUDATION CHRONOLOGY

The explanation of how topographic landscapes came to attain their present form has always been a prime objective of geomorphology. Prior to the 1960s, most workers adopted an essentially historical approach to landscape evolution. The aim was to identify the sequence of episodes, or stages, of erosional development that demonstrated how contemporary landscapes had been sculptured from hypothetical initial topographies that were usually considered fairly uniform and featureless. This sequential approach to topographic evolution, with its focus on DENUDATION, came to be known as denudation chronology. A significant proportion of such studies were based on the Davisian model of landscape evolution and can be termed classical denudation chronology, an historic element of a broad field of study currently known as long-term landscape development or evolutionary geomorphology.

Classical denudation chronology sought to identify evidence of past PLANATION SURFACES and erosional levels in a landscape and to place them in a chronological sequence. Two key concepts were employed. First, that topographic 'flats', bevels and benches, together with accordant ridge and summit levels, represented the remnants of marine platforms, SUBAERIAL low-relief surfaces or terraces produced during past periods of relatively stable BASE LEVEL or 'stillstand'. Second, that there had been a progressive but episodic fall in base level with time, so that the highest features (in terms of elevation) were the oldest. The resulting 'geomorphological staircases' often rose via terraces and benches to the more fragmentary remains of 'summit surfaces' preserved on ridges and CUESTAS (e.g. the 'Schooley Peneplain' of W.M. Davis (Figure 41), the 'Mio-Pliocene Peneplain' of Wooldridge and Linton (Figure 42)), or to even higher surfaces whose former existence was postulated on the basis of the summits of residual hills or 'monadnocks' (Figure 42), or by the projection of the planes of outcropping unconformities (e.g. the 'Fall Zone Peneplane' rising above the Appalachians (Johnson 1931) (Figure 41), the 'Sub-Eocene Surface' of south-east England (Wooldridge and Linton 1955) or the 'Sub-Cretaceous Surface' rising over Wales (e.g. Brown 1960)).

The identification and delimitation of such surfaces was usually based on visual observation, augmented by field mapping, profiling and various kinds of cartographic analysis, including the use of

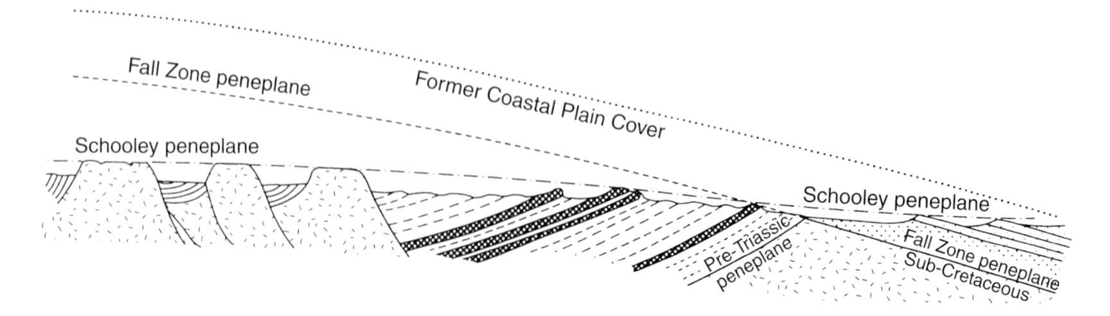

Figure 41 Scheme of landscape development for the eastern USA first advanced by D.W. Johnson in 1928

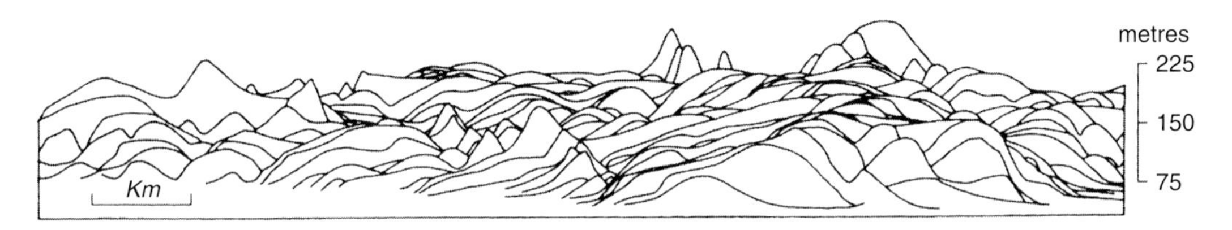

Figure 42 Projected profiles on the Chalk, western margins of Salisbury Plain, southern England, published by C.P. Green in 1974. The residual hills indicate the altitude of the former Sub-Eocene (now Sub-Palaeogene) surface and rise above the Mio-Pliocene Peneplain of Wooldridge and Linton, which Green subdivided into Summit Peneplain: Higher Surface and Summit Peneplain: Lower Surface, dated as Miocene and Pliocene respectively (see Jones 1981)

superimposed and projected profiles (Figure 42), and the resulting sequences could be extended back into the geological past by use of the basic rules of stratigraphy, such as the laws of superposition and of original horizontality and the nature of unconformable relationships. Hence many denudation chronologies were extended through the Neogene into the Palaeogene and even into the Mesozoic (Figure 41). Although this was often claimed to be a strength, because such studies provided a bridge between Quaternary landscape development and stratigraphy, in reality the resultant 'histories' had much closer affinities with historical geology than with geomorphology, especially once contemporary process studies had developed to dominate geomorphology.

Relatively little emphasis was placed on the study of surficial deposits, largely because of the limited availability of analytical techniques and the complexity of such deposits. Often the studies that were undertaken focused on whether or not the identified erosional remnants were of subaerial or marine origin. By contrast, drainage patterns and drainage evolution figured prominently. Drainage-structure relationships, together with the existence of cols, wind-gaps and UNDERFIT STREAMs, were all used to recreate former drainage patterns associated with particular stages of landscape evolution (see Brown 1960). Often the aim of such studies was to recreate the original pattern of sub-parallel consequent rivers that developed on an uplifted and tilted marine plane or were superimposed from overlying cover strata across a fundamental unconformity. The identification and interpretation of discordant drainage therefore became a highly contentious issue and led to great debates about the extent and significance of former marine planation surfaces (see Jones 1981).

It is often assumed that classical denudation chronology evolved from W.M. Davis's exposition of the concept of the CYCLE OF EROSION, but this is incorrect. Long before the Davisian model was first outlined in 1889 others had begun to develop simple, embryonic denudation histories. For example, in Britain there was the work of Ramsay in 1846 and 1864, Jukes in 1862 and Topley in 1875, while in America McGee in 1888 developed an erosional chronology for that very same part of the Appalachians that was to be made classic by W.M. Davis's own detailing of the cycle of erosion concept (Davis 1889).

The Davisian model, as subsequently refined and elaborated (see Davis 1909), with its notions of peneplanation and cyclic change due to variations in base level, clearly fitted in so well with notions of the sequential development of landscapes, that it is no surprise that the two approaches merged. Many high-level marine surfaces were reinterpreted as PENEPLAINs and every attempt was made to place identified 'flats' into discrete groupings on the basis of elevation (Figure 43) in order to recreate cycles and partial cycles which could then be correlated with evidence from adjacent regions on the basis of elevation alone. Thus the cycle of erosion concept invigorated denudation chronology and, in turn, the concept was to survive as a basic element of classical denudation chronology long after it had been discarded by the remainder of geomorphology, following advances in process-based understanding of landform development.

During the first half of the twentieth century, denudation chronology became a major preoccupation of geomorphological studies: in America, under the influence of D.W. Johnson (1931), in Britain, where S.W. Wooldridge was the dominant figure (see Wooldridge and Linton 1955) and in France, where the pioneering study of the Massif Central by H. Baulig (1928) established the blueprint for later studies in Europe. However, there were differences across the Atlantic regarding the emphasis placed on eustatic versus diastrophic mechanisms as the main cause of base level changes, with European studies following the lead of Baulig in favouring eustatic change. This is classically displayed in B.W. Sparks's (1949) interpretation of the morphology of the South Downs cuesta backslopes in southern England as consisting of eight marine levels between the two proven raised beaches below 40 m and the postulated Plio-Pleistocene marine plain at 170–200 m (Figure 43) – a geomorphological staircase encompassing the Pleistocene in no fewer than eleven treads, the majority of which lacked supporting sedimentological evidence.

Denudation chronologies were developed for other areas using modified models of landscape evolution, most dramatically in the case of South Africa where Lester King (1972) produced a classic sequential interpretation of the area between the Drakensberg Escarpment and the Natal Coast (Figure 44), including artistic representations of the landscapes developed at each stage. King's model of landscape evolution represents an amalgam of the views of Davis and Penck; episodic uplift resulting in both downwearing and backwearing, with the parallel retreat of slopes leading to the formation of PEDIMENTS which coalesced to form pediplains through the process of pediplanation. King himself went so far as to state that the morphological and sedimentological sequences identified in Natal provided the basis for a chronological scheme that had global application (King 1967), a grand design published at a time when workers elsewhere were experiencing increasing difficulty in making compatible the numerous 'local histories' that had been identified, as was well shown in David Linton's (1964) valiant attempt to provide a synthesis of Tertiary landscape evolution in Britain.

Since the 1960s there has been increasingly widespread and severe criticism of classical denudation chronology. Some of these criticisms focused on the inadequacies of the Davisian

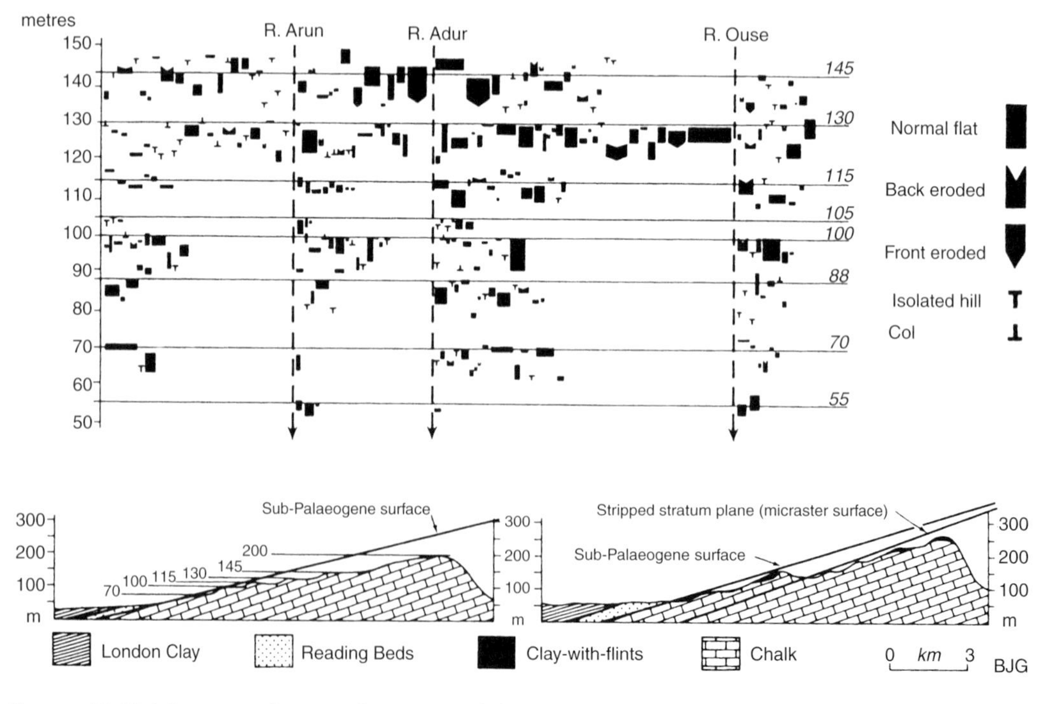

Figure 43 Height-range diagram for erosional levels identified on the South Downs backslopes by Sparks (1949), together with his interpretation of several horizontal marine benches. The two sections compare Sparks's interpretation with that of later workers (see Jones 1981)

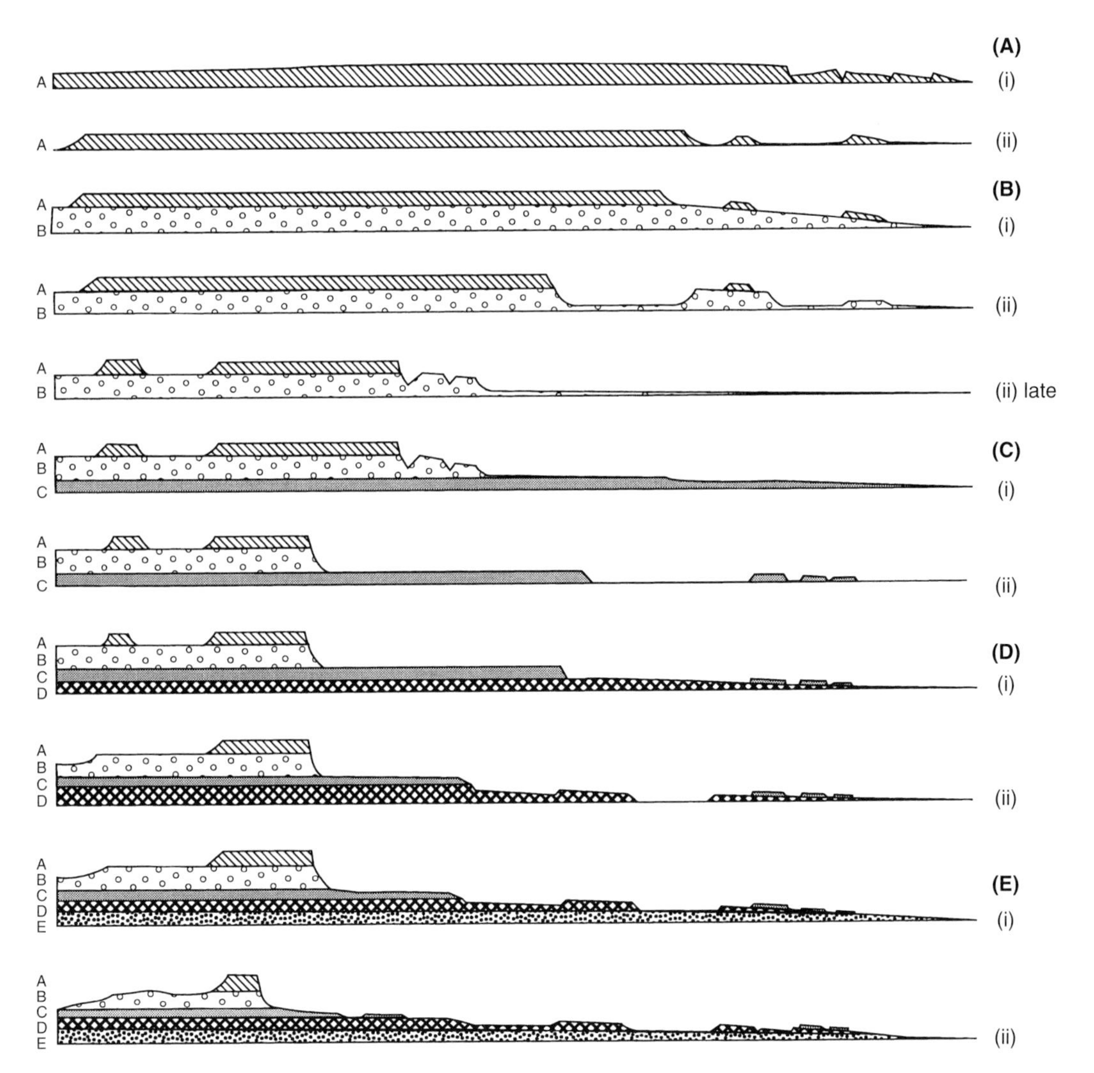

Figure 44 The topographic evolution of the area between the Drakensberg Escarpment and the Natal Coast as detailed by Lester King in 1972. Continental rupture in Jurassic–Cretaceous (A) was followed by pulses of differential uplift in the mid-Cretaceous (up to 1,250 m – B), Miocene (C), end Miocene (up to 800 m – D) and Pliocene (up to 625 m – E), each of which generated incision, backwearing and pediplanation. The result is a stepped and warped landscape in which remnants of the Mesozoic Gondwana Surface, originally at 600 m (A), survive at *c*.3,500 m (E(ii))

model of landscape evolution (Chorley 1965) compared with the approaches of Gilbert, Hack and Penck. Others pointed to the oversimplified theoretical concepts employed and the dangerous over reliance on morphological evidence compared with the far more rigorous, scientific approach adopted by the relatively new but rapidly expanding field of Quaternary Science, with its focus on the analysis of sediments. Yet others pointed to the difficulties of separating base level controls from structural controls and the fundamental problems of correlating surfaces over significant distances simply on the basis of elevation, especially when the origin and age of such 'surfaces' were often largely based on speculation. It came to be recognized that 'flats' or erosional levels could be produced by a wide variety of processes including pediplanation, etchplanation (see ETCHING, ETCHPLAIN AND ETCHPLANATION) and CRYOPLANATION and that contemporary landscapes could contain EXHUMED LANDFORMS, including planes of unconformity that may have been warped.

But, most importantly, the 1960s witnessed the onset of radical changes to prevailing views of the past arising from growing knowledge about global tectonics and Quaternary climate change. The new paradigms indicated 'ceaseless motion' as a characteristic of the lithosphere, together with oscillating climatic conditions and sea levels, and thereby seriously undermined notions of both the formation and survival of surfaces from the distant past, except under special conditions or where LANDSCAPE SENSITIVITY is very low.

As a result of these criticisms, denudation chronology fell into disrepute and almost became a term of abuse. Many of the detailed chronologies came to be dismissed as pure speculation or were demolished after detailed reinvestigation (as in the case of Sparks's (1949) sequence – see Figure 43). However, alternative explanations of landscape development as due to waves of aggression acting on rock sequences offering variable resistance to erosion have proved to be neither edifying nor satisfying. Most landscapes contain conspicuous morphological features (low relief surfaces, bevels, benches) that require explanation and the evolution of topographic landscapes remains a fascination. As a consequence, studies of landscape development have continued, albeit in different form and under new names. Long-term landform/landscape development or evolutionary geomorphology has many of the same aims as denudation chronology but within a much more complex conceptual framework, utilizing analysis of surficial deposits and the ever-growing range of absolute dating techniques. One of the most significant developments has been the attempt to correlate offshore sedimentary sequences with onshore denudation episodes, as pioneered by L.C. King. Some idea of the range of approaches adopted can be gained from Smith *et al.* (1999).

References

Baulig, H. (1928) *Le Plateau central de la France et sa bordure Méditerranéenne, étude morphologique*, Paris: Armand Colin.

Brown, E.H. (1960) *The Relief and Drainage of Wales*, Cardiff: University of Wales Press.

Chorley, R.J. (1965) A re-evaluation of the Geomorphic System of W.M. Davis, in R.J. Chorley and P. Haggett (eds) *Frontiers in Geographical Teaching*, 21–38, London: Methuen.

Davis, W.M. (1889) The rivers and valleys of Pennsylvania, *National Geographic Magazine* 1, 183–253.

——(1909) *Geographical Essays*, Boston: Ginn and Company.

Johnson, D.W. (1931) *Stream Sculpture on the Atlantic Slope*, New York: Columbia University Press.

Jones, D.K.C. (1981) *The Geomorphology of the British Isles: Southeast and Southern England*, London: Methuen.

King, L.C. (1967) *The Morphology of the Earth*, Edinburgh: Oliver and Boyd.

——(1972) *The Natal Monocline*, Durban: University of Natal.

Linton, D.L. (1964) Tertiary landscape evolution, in J.W. Watson and J.B. Sissons (eds) *The British Isles*, 110–130, London: Nelson.

Smith, B.J., Whalley W.B. and Warke, P.A. (eds) (1999) *Uplift, Erosion and Stability: Perspectives on Long-term Landscape Development*, London: Geological Society Special Publication 162.

Sparks, B.W. (1949) The denudation chronology of the dip-slope of the South Downs, *Proceedings of the Geologists Association* 60, 165–215.

Wooldridge, S.W. and Linton, D.L. (1955) *Structure, Surface and Drainage in South-east England*, London: George Philip.

Further reading

Summerfield, M.A. (1991) *Global Geomorphology*, Ch.18, Harlow: Longman.

SEE ALSO: clay-with-flint; dating methods; drainage pattern; duricrust; dynamic equilibrium; erosion; geomorphic evolution; global geomorphology; inselberg; sea level; slope, evolution; tectonic geomorphology

DAVID K.C. JONES

DESERT GEOMORPHOLOGY

The scientific study of deserts and desert landforms had its origins in the latter half of the nineteenth century driven by many external and internal forces including imperialism, colonialism, military adventurism, romance as well as the desire to explore and exploit mineral resources and claim land for agriculture, ranching and grazing.

Some of the earliest descriptive observations on deserts can be found in the works of Greek and Roman geographers (e.g. Herodotus, Aristotle, Seneca, Strabo, Pliny, Ptolemy, among others). Arab and Muslim geographers also wrote about deserts during their various journeys within the Dar-el-Islam which stretched from Morocco to Indonesia. Deserts have also been mentioned in the writings of European travellers that went along the great Silk Road to China (via the deserts of the Middle East and Central Asia) beginning in the thirteenth century (e.g. Marco Polo and others).

The imperial ventures of Spain and Portugal in the New World during the sixteenth century led to some of the earliest descriptive observations of deserts. The Spanish, in the shape of the *conquistadores* and missionaries, were probably the first to venture in the Sonoran and Chihuahuan deserts of North America (e.g. De Vaca, Coronado, De Soto, Father Kino, among others) and provided detailed descriptions of the landforms.

Historical framework I: 1850 to 1950

With colonial aspirations and military adventurism in full swing during the middle part of the nineteenth century, England, France, Germany and the United States sent forth the first wave of soldiers, surveyors and scientists to investigate deserts which lay within their purview (Tchakerian 1995: 2).

With British administration in Egypt, some of the earliest scientific works on desert landforms were conducted from the late 1890s to the late 1920s (see the bibliographies in Goudie 1999) by Cornish, Beadnell, King and Ball, with particular emphasis on general dune forms and processes. The book *Waves of Sand and Snow* by Vaughan Cornish (1914) is one of the earliest detailed studies on bedforms and wind regime. It was Ralph Alger Bagnold (1896–1990), who, during his travels to the Western Desert in Egypt in the late 1920s, became fascinated with dune processes and dynamics and, once discharged from the British Army in 1934, built a wind tunnel in Imperial College, London to study the mechanics of blown sand and dunes. His pioneering research culminated in the now classic book *The Physics of Blown Sand and Desert Dunes* (1941), one of the most significant canons in the geomorphic literature (Goudie 1999: 6). He laid the theoretical foundations of AEOLIAN GEOMORPHOLOGY with his detailed analysis of fluid flow, particle motion, sediment transport and dune forms (particularly barchans and siefs). Bagnold's autobiography, *Wind, War and Sand* (1990), written shortly before his death, is a wonderful journey into the mind and soul of this great scientist, and, along with his earlier monograph of desert travels, *Libyan Sands* (1935), should be read by all desert scholars.

With France gradually extending its colonial empire from Morocco eastwards to Algeria and Tunisia and ultimately to all of the Sahelian countries (Mauritania, Niger, Mali and Chad), the western and central Sahara Desert became the focus of study by French scientists and their African colleagues. The work of the French geoscientists has gone relatively unnoticed outside of western Europe (most have not been translated into English) and some very notable works remain to be read and cited. An excellent synopsis is provided by Goudie (1999). A significant body of works is devoted understandably to aeolian processes and landforms (particularly to dune processes and formation, sand sea dynamics and dune orientations and wind regimes) owing to the fact that some of the world's most extensive sand seas (ergs) (see SAND SEA AND DUNEFIELD) are found in the Western Sahara. Some of the most notable contributors include Rolland, Chudeau, Aufrère (the most prolific), Capot-Rey, Dubief and Clos-Arceduc. Many of their papers were published in the *Travaux Institute de Recherches Sahariennes*, in Algiers. Another significant contributor was Emile Gautier, whose seminal monograph *Le Sahara* (1935) is one of the classic works on the general geomorphology of the Sahara Desert (including the human impact on the environment). This rich tradition of French research in the Sahara has continued in the second half of the twentieth century including works by Birot, Cailleux, Dresch and Tricart.

German colonial interests in south-west Africa (including the Namib and the Kalahari deserts) resulted in a number of expeditions to evaluate the economic potentials for the deserts of southern Africa. The works of the German geographer Passarge are especially significant including his monograph *Die Kalahari* published in 1904. In the deserts of interior Asia, German scientists such as Richthofen were one of the first westerners to recognize the significance of desert dust and LOESS. The Swedish geographer and military opportunist Sven Hedin wrote about the Gobi Desert and its landforms. The Australian arid zone was investigated by a number of explorers and scientists from the mid-nineteenth into the early twentieth centuries, driven largely by ranching and mineral resource evaluation (Cooke *et al.* 1993: 13).

The scientific study of the North American deserts begins with the United States federal exploration of lands west of the Mississippi

River (largely as a result of the acquisition of large swaths of territory beginning with the Louisiana Purchase of 1803 and ending with lands gained from Mexico as a result of the US–Mexican war of 1848). After the end of the US Civil War in 1865, the western surveys (primarily to evaluate the mineral, settlement and railroad prospects) led by King, Hayden, Powell and Wheeler produced some of the first detailed descriptions of the American deserts. John Wesley Powell's 1878 classic monograph *Report on the Lands of the Arid Region of the United States* has extensive descriptions of deserts and desert landforms and was one of the first studies to evaluate the natural resources of the region as well as its suitability for extensive human occupation (something that Powell did not recommend). G.K. Gilbert's influential *Report on the Geology of Henry Mountains* (1877) for the first time showed how process geomorphology (with its foundation in detailed fieldwork and data gathering based on principles from physics, mathematics, statistics) can contribute to the understanding of deserts and desert landforms. Gilbert's contributions to geomorphology are too numerous to cite here (he is considered the 'founder of process geomorphology') but his works on desert fluvial processes, Quaternary lakes and WIND EROSION are particularly noteworthy. In the 1920s and 1930s, Kirk Bryan published numerous influential papers on wind erosion, differential weathering and erosion, pedestal rocks, pediments, and arroyos of the southwestern USA. During the early part of the twentieth century, AEOLIAN PROCESSES were seen as the dominant sculptor of desert landforms, as promulgated by Keyes in 1912 in his 'aeoliana-tion' cycle (Tchakerian 1995: 2). William Morris Davis and his colleagues in a series of papers in the 1930s put to rest the dominating role of wind in the evolution of desert landscapes and accurately pointed out the more substantial role played by weathering, mass movement and fluvial processes. Also in the 1930s, the severe environmental consequences of the 'Dust Bowl' years in the Great Plains of the United States, led many scientists to focus their studies on wind and soil erosion, including those by W.S. Chepil and colleagues, resulting in many publications dealing with the mechanics of aeolian entrainment and transport under different land use activities (Tchakerian 1995: 3).

Historical framework II: 1950 to 2000

Global studies of deserts and desert landforms were rejuvenated in the second half of the twentieth century as a result of a number of technological and theoretical advances. In the following section, some of the major themes are briefly highlighted but detailed consideration and bibliographies in specific desert processes or landforms is left to other contributors to this volume.

Major developments were led by advances, refinement and availability of air photos (after the Second World War) and satellite imagery (during the 1970s), enabling for the first time a continental and global perspective in desert research, including the first comprehensive global dune classification using Landsat imagery (McKee 1979). The emergence of planetary geomorphology has led to more focused attention on desert landforms as terrestrial analogs for arid Mars, particularly in aeolian processes and landforms, canyon formation, sapping and other mass movement processes (e.g. Malin and Edgett 2000).

Increased studies in desert geomorphology in the latter half of the twentieth century have been driven for several reasons: (1) intrinsic fascination as distinct landforms, such as the study of desert PEDIMENTS or STONE PAVEMENTS, ALLUVIAL FANS, sand dunes. Studies in fluvial processes and landforms have been at the forefront of much desert geomorphology research during this period (e.g. Graf 1988); (2) mineral resource potential of desert landforms, such as evaporites (sulphates, nitrates, etc.) from playas, aggregates and groundwater resources in alluvial fans, uranium in DURICRUSTS; (3) desert landforms as indicators for past climatic and ecological change, such as Quaternary lakes and their sediments; (4) the archaeological value of certain desert landforms, such as the study of petroglyphs on rock coatings; and (5) the increased human occupancy and flood hazard risk on or near alluvial fans, ARROYOS, BADLANDS, as well as the environmental hazards associated with increased desert aeolian dust and mineral aerosols and their effects on human health and atmospheric visibility.

The refinement and extension of Bagnold's theoretical foundations have been the primary focus for most aeolian-related research during the past fifty years. This can be seen in the plethora of scientific papers, monographs and symposia proceedings devoted exclusively to aeolian processes and landforms during the past two decades alone

(Tchakerian 1995: 4). Associated with the above refinements were developments in mathematical modelling, computer simulations, and the use of complex, non-linear dynamical systems theory for understanding wind flow and bedforms (e.g. Werner and Kocurek 1999). The resurgence of aeolian geomorphology has also been stimulated by the fact that aeolian sedimentary environments are good analogues for studying hydrocarbon reservoirs in the geologic record. Advances in the understanding of single dune formation and dynamics has now led researchers to tackle the more daunting task of analysing draas (megadunes) and ergs (e.g. Pease and Tchakerian 2002). Other significant developments include the application of luminescence techniques for dating aeolian sands and Quaternary dune systems, and the wide availability of very sophisticated instruments (anemometers, electron microscope, sediment traps) and data loggers for the gathering and analysis of wind data and sediments. Technological advances were also instrumental in heralding numerous studies in aeolian desert dust and mineral aerosols culminating in a series of papers assessing the global distribution of major atmospheric dust sources (e.g. Prospero *et al.* 2002). Research in aeolian processes and landforms has also been driven by concerns for land degradation (DESERTIFICATION) and global environmental change.

The growing scientific interest in desert geomorphology of the past three decades has led to the establishment of many centres of desert research, new journals, international associations, and to UN-sponsored research and publications. This has led to vigorous exchanges of ideas and co-operation among Earth scientists unconstrained by traditional disciplinary boundaries.

References

Cooke, R.U., Warren, A. and Goudie, A.S. (1993) *Desert Geomorphology*, London: UCL Press.
Goudie, A.S. (1999) The history of desert dune studies over the last 100 years', in A.S. Goudie, I. Livingstone and S. Stokes (eds) *Aeolian Environments, Sediments and Landforms*, 1–13, London: Wiley.
Graf, W. (1988) *Fluvial Processes in Dryland Rivers*, Berlin: Springer-Verlag.
McKee, E.D. (ed.) (1979) *A Study of Global Sand Seas*, United States Geological Survey Professional Paper No. 1,052.
Malin, M.C and Edgett, K.S. (2000) Evidence for recent groundwater seepage and surface runoff on Mars, *Science* 288, 2,330–2,335.
Pease, P.P. and Tchakerian, V.P. (2002) Composition and sources of sand in the Wahiba Sand Sea, Sultanate of Oman, *Annals of the Association of American Geographers* 92, 416–434.
Prospero, J.M., Ginoux, P., Torres, O., Nicholson, S.E. and Gill, T.E. (2002) Environmental characterization of global sources of atmospheric soil dust identified with the Nimbus-7 Total Ozone Mapping Spectrometer (TOMS) absorbing aerosol product, *Reviews of Geophysics* 40, 1–29.
Tchakerian, V.P. (1995) The resurgence of aeolian geomorphology, in V.P. Tchakerian (ed.) *Desert Aeolian Processes*, London: Chapman and Hall.
Werner, B.T. and Kocurek, G. (1999) Bedform spacing from defect dynamics, *Geology* 27, 727–730.

Further reading

Thomas, D.S.G. (ed.) (1997) *Arid Zone Geomorphology*, London: Belhaven.

VATCHE P. TCHAKERIAN

DESERT VARNISH

Desert varnish, a paper-thin deposit, drastically darkens the appearance of desert rocks. Any rock type can host desert varnish, so long as its surface remains stable for the thousands of years it usually takes varnish to accrete. Rock varnish is the preferred term where this ROCK COATING occurs in non-desert settings, for example, alpine, Antarctic, Arctic, periglacial, stream, temperate and tropical environments. The term desert varnish is most often used in arid regions.

Like other rock coatings, desert varnish is deposited on rock surfaces and does not derive from the host rock itself. Arrows in the middle image in the middle row of Plate 34 exemplify this discrete contact. Like CASE HARDENING and other rock coatings, many varnishes seen at the surface today actually start in the subsurface in fissures. Varnishes are usually less than 100 μm thick, and even where micro-basins host deposits of a few hundred micrometres, median thicknesses are usually less than 10 μm thick.

Wind does not cause shiny varnish; wind abrades away this relatively soft coating. In fact, the presence or absence of desert varnish is an important clue that a particular desert pavement was not or was made by aeolian deflation. Usually dull in lustre, its occasional sheen comes from a smooth surface micromorphology in combination with manganese enrichment at the very surface of the varnish.

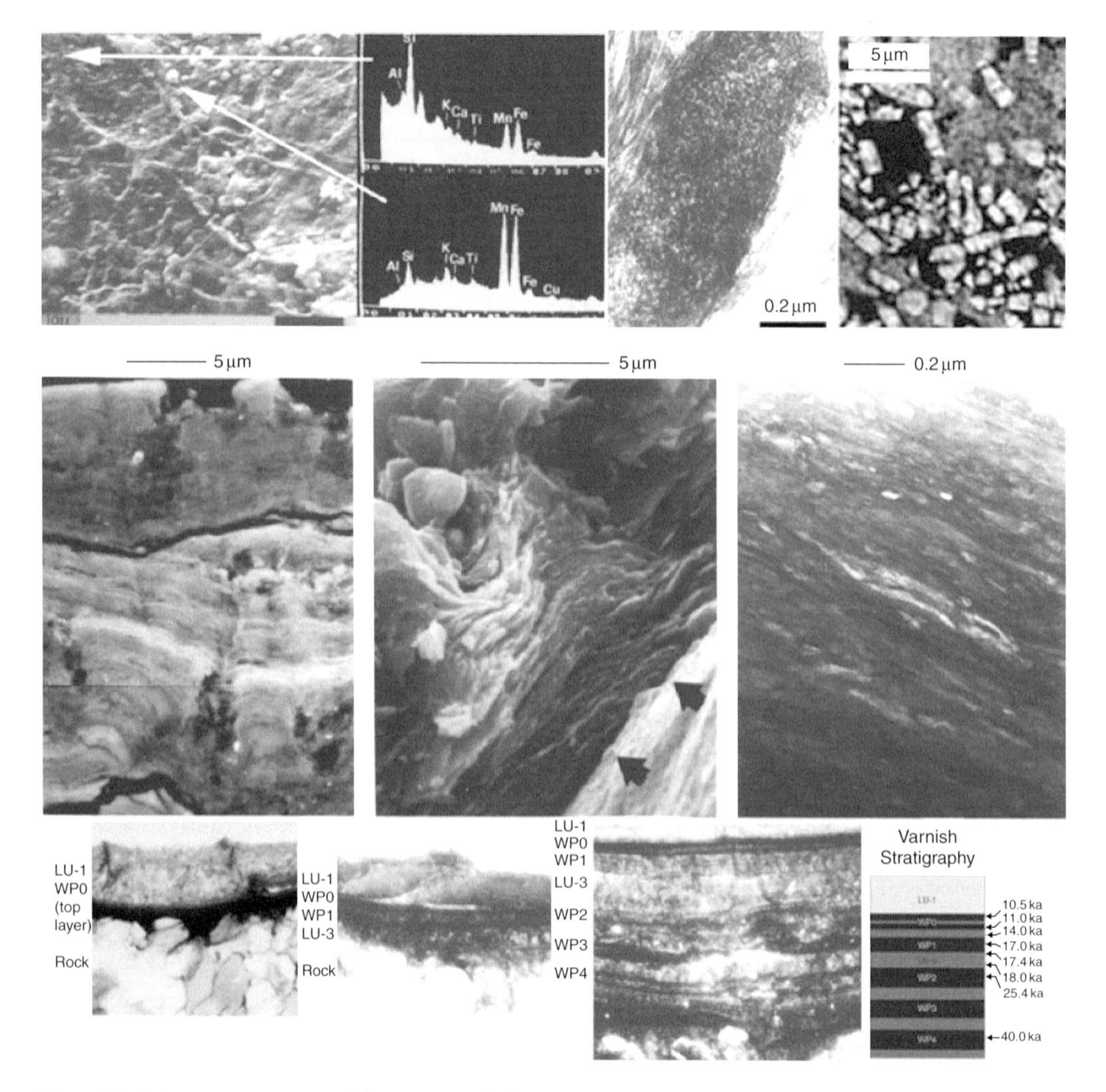

Plate 34 Microscopic views of desert varnish from arid environments. Top row: microscopic evidence for bacterial origin of rock varnish from left to right: secondary electron image of Negev Desert budding bacteria where the bacteria greatly enhanced manganese and iron; transmission electron image of manganese encrusting a bacterial form; and backscattered electron image of bacteria revealed by acid etching. Middle row: layering of desert varnish shown in backscattered (left), secondary (middle), and high resolution transmission electron microscope (right) images. Bottom row: calibration of microlaminations seen in optical microscopic views of ultra-thin sections, where black layers in the varnish represent wet periods that have been calibrated by Tanzhuo Liu's research (Liu *et al.* 2000). The thin sections from left to right show progressively older varnishes with progressively more complex layers from Death Valley, California

Clay minerals are the major ingredient of desert varnish, typically comprising more than half and sometimes as much as 90 per cent. The clay minerals impose the layered structure seen in the middle row of Plate 34. Clays are deposited as dust on rock surfaces, and are then cemented to the host rock by hydroxides and oxides (Potter and Rossman 1979) of manganese (birnessite) and iron (goethite and hematite). Manganese and iron make up about a third of varnish, with typically less than 5 per cent of varnish composed of other components.

The mystery of desert varnish surrounds how to explain the great abundance of manganese, the element that gives varnish its dark brown to black colour. The elemental abundance of manganese in varnish is 50 to 300 times the concentrations found in dust that falls on rock surfaces. Put another way, ratios of manganese to iron are about 1:40 to 1:60 in surrounding soils and dust, but are about 1:1 in varnish.

In the past century, there have been two competing models to explain manganese enrichment. The first was a chemical process favoured by geochemists, whereby naturally acidic rain dissolves manganese in the rock or dust (but not the iron). Then, manganese oxidizes upon exposure to a slightly higher pH. The competing model was a microbial process, whereby bacteria precipitate manganese (Drake *et al.* 1993).

Although bacteria have been cultured from varnish and have made 'artificial varnish' in the laboratory (Dorn 1998), the typically slow rate of varnish growth (on average, about a micrometre per thousand years) makes it very difficult to have confidence that bacteria cultured today in the laboratory make varnish. In fact, the type of gram-positive bacteria most easily cultured from desert varnish today have not yet been identified within varnish layers. To make the matter more difficult, 'biomolecular fossils', such as amino acids generated by these bacteria, exist in both desert varnish and unvarnished weathering materials. Thus, most convincing evidence for a bacterial mechanism is actually seeing manganese enhancement *in situ*. In the upper row of Plate 34, budding bacteria can be seen concentrating manganese and iron.

New high resolution transmission electron microscope evidence (Krinsley 1998) reveals that these chemical and biological models are not truly in competition, but work in tandem. Varnish formation can be explained by a four-step process. Step 1 is the enhancement of varnish (and to a lesser extent iron) by bacteria; the top row in Plate 34 shows manganese-rich sheaths of bacteria. Step 2 is the chemical dissolution of the bacterial sheaths, whereby manganese and iron are broken down into nanometre-sized granules. Step 3 is chemical transport of manganese and iron into clay minerals. Step 4 is the precipitation of unit cells of manganese and iron inside clay minerals. Potter and Rossman (1979) noted that the hexagonal arrangement of oxygens in clay mineral layers form a template for crystallization of the manganese mineral birnessite seen in desert varnish.

Krinsley (1998) shows high resolution imagery revealing all steps in this polygenetic process whereby clay minerals and oxides are co-dependent in varnish formation. Clay provides the overall structure and template for oxide precipitation, while bacteria simply provide a ready source of manganese (and iron) cement. Varnish formation all takes place within a few micrometres of the bacterial source, where the manganese and iron are redistributed with hygroscopic water – all inside layers like those seen in the middle row of Plate 34.

Environmental changes play an important role in the development of desert varnish. Where lichens start to grow, for example, biological acids destroy desert varnish by dissolving the manganese and iron oxides. Where rocks come to exist in a desert pavement, a ground-line band of very thin and shiny varnish forms a circle around a desert pavement clast. But where varnish grows on the tops of boulders less influenced by local environmental changes, regional climatic change plays an important role in varnish formation.

In these boulder-top varnishes wetter climates favour bacterial enhancement, yielding layers that are particularly rich in manganese. Drier climates with more alkaline dust produce layers that are not as rich in manganese. The bottom row in Plate 34 shows desert varnishes from Death Valley, California, where growth of these layers has been calibrated using a combination of numerical dating methods (Liu *et al.* 2000; Zhou *et al.* 2000). Progressively older varnishes show progressively more complex layers. Varnish microlaminations provide archaeologists and geomorphologists with a powerful tool, because they reveal both climatic change and a time signal.

Some of the most interesting aspects of desert varnish surround its minor and trace constituents. Lead, for example, is greatly enhanced in the uppermost micron of the varnish from twentieth-century air pollution. The carbon that is trapped within and underneath varnish shows some potential as a means of radiocarbon dating varnish, but the history of the carbon is usually too complicated to make this technique useful. Mobile trace elements decline progressively over time, as they are leached by hygroscopic and capillary water (Krinsley 1998). Varnish also traps foreign material crushed into rock engravings as a part of religious ceremonies (Whitley *et al.* 1999). New experimental studies of trace isotopes such as ^{7}Be, δ^{13}C and δ^{17}O show potential to reveal new insights into this ubiquitous weathering phenomenon. Some planetary scientists believe that desert varnish exists on Mars, and that Martian varnish might preserve active organisms or at least biological fossils such as those seen in the top row of Plate 34.

References

Dorn, R.I. (1998) *Rock Coatings*, Amsterdam: Elsevier.

Drake, N.A., Heydeman, M.T. and White, K.H. (1993) Distribution and formation of rock varnish in southern Tunisia, *Earth Surface Processes and Landforms* 18, 31–41.

Krinsley, D. (1998) Models of rock varnish formation constrained by high resolution transmission electron microscopy, *Sedimentology* 45, 711–725.

Liu, T., Broecker, W.S., Bell, J.W. and Mandeville, C.W. (2000) Terminal Pleistocene wet event recorded in rock varnish from the Las Vegas Valley, southern Nevada, *Palaeogeography, Palaeoclimatology, Palaeoecology* 161, 423–433.

Potter, R.M. and Rossman, G.R. (1979) The manganese- and iron-oxide mineralogy of desert varnish, *Chemical Geology* 25, 79–94.

Whitley, D.S., Dorn, R.I., Simon, J.M., Rechtman, R. and Whitley, T.K. (1999) Sally's Rockshelter and the archaeology of the vision quest, *Cambridge Archaeological Journal* 9, 221–247.

Zhou, B.G., Liu, T. and Zhang, Y.M. (2000) Rock varnish microlaminations from northern Tianshan, Xinjiang and their paleoclimatic implications, *Chinese Science Bulletin* 45, 372–376.

SEE ALSO: rock coating

RONALD I. DORN

DESERTIFICATION

Desertification is a term for land degradation caused by adverse human impacts in arid, semi-arid and dry-subhumid areas, together called the 'susceptible drylands' (Middleton and Thomas 1997). Hyper arid areas are in general terms not regarded as sites of desertification, since these areas are extremely desert-like due to natural conditions. The term desertification has also been used more widely, for example to refer to land degradation in non-dryland contexts, and there are over a hundred published definitions. There is widespread consensus that the term should be restricted to susceptible drylands, a view embodied in the 1994 UN Convention to Combat Desertification (UNCCD), which by March 2002 had been ratified by the governments of 179 countries. Desertification occurs not just in the developing world, but also under the impact of the inappropriate or excessive use of agricultural technologies in the dryland areas of the developed world. Overall, desertification relates to the unsustainable use of the land in drylands.

The term desertification is widely regarded to have been coined by Aubreville (1949), who used it to describe the environmental impact of forest clearance in West Africa, leading in his opinion to the creation of an ecological desert. It gained renewed use in the 1960s and 1970s (and was sometimes used interchangeably with desertization) when the major Sahel drought at the time led to a decline in biomass, famine and livestock and human deaths in the Sahel zone along the southern margin of the Sahara Desert.

Desertification has both social and environmental components. Social dimensions relate to the pressures and processes that cause people to carry out activities that degrade the land. Today, international efforts to reduce desertification, such as the UNCCD, lay great emphasis, particularly in the developing world, on addressing the social and economic issues that lead people to carry out land use practices that lead to land degradation. Environmental dimensions of desertification are important too, however, and relate to the actual physical processes of land degradation and to distinguishing impacts of natural dryland environmental variability from anthropogenically aggravated changes to land systems. The latter is important since it is possible to confuse the impacts of dryland rainfall variability with changes in soils and vegetation caused by human actions (Thomas and Middleton 1994). These confusions have led, in the past, to overestimates of the physical dimensions and scale of desertification, including for example, inappropriate statements being made about the advance of deserts over productive land (see Hellden 1991; Thomas 1993). The causes and use of the term desertification are complex, however, since natural drought may itself place pressures on social and agricultural systems that in turn lead to land degradation.

Physical processes of desertification

The physical processes of desertification comprise depletion of and damage to the soil, ground water, and to some extent vegetation systems, that reduces their productive capacity or biological potential. The soil provides the geomorphological context of desertification, with loss of productive potential occurring either through the physical loss of soil by water or wind erosion, or internal physical and chemical changes such as compaction, salinization, alkalinization and nutrient depletion. Dryland soils may be thin or skeletal, with generally slow rates of formation due to the limited availability of water for the weathering of underlying bedrock and also to slow build up of organic material. Exceptions may occur in specific geomorphic locations such as valley floors where seasonal water supply may be greater. Natural recovery from soil erosion or internal changes is therefore likely to be slow.

There has been considerable debate regarding whether changes in dryland vegetation systems that are unaccompanied by soil system changes constitute desertification. These debates have arisen for a number of reasons: many savanna systems are 'non-equilibrium' ecosystems that do not achieve a spatial or temporal steady state because of the natural dynamism of dryland climates (Mistry 2000); dryland vegetation systems can be both resilient and adapted to disturbance and can exhibit rapid recoveries; the impacts of droughts and land degradation on vegetation systems can be hard to distinguish; observed vegetation changes caused by human pressures may not be accompanied by soil system changes (Dougill *et al.* 1999), facilitating recovery if land use pressures are reduced or removed. It should be noted however that vegetation depletion can set the scene for desertification processes to take effect via the processes of erosion.

Geomorphological dimensions of desertification

Desertification via erosion processes may be relatively easy to recognize and in susceptible drylands both wind and water erosion can be important. WIND EROSION potential is greatest in areas of low relief and unconsolidated sediments, for example the Canadian Prairies and the midwest of the USA, as witnessed by the occurrence of severe DUST STORMS during the 1930s, while areas of steeper topography are more susceptible to water erosion, for example in the Highlands of Ethiopia and Kenya and around the Mediterranean basin in Europe and North Africa.

Changes within the soil due to human activities are, with the possible exception of salinization, less visible and more insidious than those caused by erosion. Salinization and alkalinization associated with irrigation schemes are widely cited causes of productivity decline in drylands. Other internal changes relate to waterlogging, also associated with irrigation, and the crusting and compaction of soils, increasingly caused by the mechanized cultivation procedures used in agriculture. Nutrient depletion is a notably widespread but often underestimated form of desertification (Thomas and Middleton 1994). Nutrient loss can be caused by the actual physical removal of soil by erosion but is often a function of the clearance of natural vegetation to create fields for cultivation, the subsequent intensity of cultivation or the lack of application of fertilizers, especially in developing world dryland areas.

Assessing the extent and nature of desertification

There are no readily available means of gaining absolute data on the occurrence of desertification. Earth observation via remote sensing can be useful for detecting dimensions of vegetation change, and for distinguishing natural fluctuations due to rainfall variability from changes caused by human actions (e.g. Tucker *et al.* 1991), but even the highest resolution imagery can be too coarse to identify many elements of soil erosion and unsuitable for determining soil internal changes. Thus estimates of the global extent of desertification are likely to be crude, and sometimes highly erroneous (Thomas 1993). Field studies and modelling approaches are important for local and regional investigations of degradation (Mairota *et al.* 1998). Increasingly, as the social dimensions of land degradation are recognized as vital elements of any efforts to ameliorate the problem, land user understanding and knowledge, often untapped, especially in the developing world, are viewed as essential for an effective understanding of where, why and how desertification occurs (Reed and Dougill 2002).

The UN has been the instigator of the few attempts to assess the global scale of desertification. Estimates produced in the 1970s and 1980s have subsequently been heavily criticized in the scientific community for their lack of methodological rigour and for confusing natural cyclic changes in vegetation systems due to drought impacts with desertification caused by human actions.

The most recent global assessment of land degradation caused by soil degradation (GLASOD) was carried out by the International Soil Reference Centre in the Netherlands on behalf of UNEP in the late 1980s and early 1990s. A GIS was used to analyse data collected through a clearly defined, but somewhat qualitative, methodology (Middleton and Thomas 1997). Despite its flaws, GLASOD has provided a database through which assessments of susceptible dryland soil degradation can be analysed in terms of geographical coverage (see Table 10), contributory degradation processes, and relationships to land use.

GLASOD estimates that in the late 1980s and early 1990s approximately 1,030 million hectares, equivalent to 20 per cent of the susceptible drylands, had experienced soil degradation processes caused by human activities. Water erosion was identified as the major physical process of degradation in 48 per cent of this area and wind erosion in 39 per cent. Chemical degradation (mainly salinization, alkalinization and nutrient depletion) was dominant in only 10 per cent of the area, and physical changes such as compaction and crusting in just 4 per cent. These latter figures may well be underestimated given the problems of identification of these less visible processes. The severity of degradation was described by GLASOD as strong or extreme in 4 per cent of the susceptible drylands – meaning that land where the original biotic functions of the soil have been destroyed, and which are irreclaimable without major restorative measures does not occur widely.

What human actions lead to desertification?

Almost any land use in the susceptible drylands has the potential to lead to desertification, if it is conducted to excess or in locations to which it is not well suited. The literature on desertification widely cites four forms of land use as major contributors to the problem: cultivation, irrigation schemes, livestock production and deforestation.

Overcultivation has sometimes been viewed as the main cause of desertification, especially in areas of the developing world where increasing populations have necessitated attempts to increase yields without the resources available for additional fertilizers. Nutrient depletion is therefore a potentially serious issue, with the World Bank attributing declining yields of staple crops in Sahel nations and in parts of South America to this problem. Attempts to increase crop yields through mechanization can lead to soil compaction, increasing runoff and erosion under intensive dryland rainfall events. Aeolian deflation can also be exacerbated by the removal of shelter belts to allow large machinery to be used.

Irrigation systems, whether by canals from storage dams or through centre pivot systems, can cause desertification through waterlogging, salinization and alkalinization, the excessive accumulation of sodium in the soil. High evapotranspiration rates in drylands mean that excessive irrigation can lead readily to the accumulation of salts in the soil. Waterlogging arises when irrigation rates are so high that the water table is raised excessively.

Table 10 Dryland soil degradation (million ha) by continent according to GLASOD

	Susceptible dryland area	Light and moderate desertification	Strong and extreme desertification	Total desertified
Africa	1,286.0	245.3	74.0	319.3
Asia	1,671.8	326.7	43.7	370.4
Australasia	663.3	86.0	1.6	87.6
Europe	299.7	94.6	4.9	99.5
North America	732.4	72.2	7.1	79.3
South America	516.0	72.8	6.3	79.1
Total	5,169.2	897.6	133.7	1,035.2

Source: Derived from data in Middleton and Thomas (1997)

Many dryland regions may be better suited to extensive livestock production than to cultivation. Pastoralism is seen as a contributor to desertification when it is over intensive, which may occur in developed world drylands and in the developing world when traditional practices are replaced by commercial systems. What may result is excessive grazing that can alter plant community composition and reduce overall plant cover, thereby increasing the susceptibility of the land to erosion processes. The role of pastoralism in desertification is however complex and somewhat controversial: as noted earlier vegetation changes may not always be accompanied by soil system changes and therefore may be reversible.

Deforestation in drylands is associated both with the clearance of lands for cultivation and, in developing world contexts, with the collection of wood for use as the dominant domestic fuel. The greatest threat deforestation offers for desertification is in areas of steep slopes. For example, during the twentieth century a tenfold reduction in woodland cover occurred in the Ethiopian Highlands through the effects of expanding cultivation and fuelwood collection: it is not surprising that soil erosion is a major problem in this area.

References

Aubreville, A. (1949) *Climats, forêts et desertification de l'Afrique tropicale*, Paris: Editions géographiques maritimes et colonials.

Dougill, A.J., Thomas, D.S.G. and Heathwaite, A.L. (1999) Environmental change in the Kalahari: integrated land degradation studies for non equilibrium dryland environments, *Annals, Association of American Geographers* 89, 420–422.

Hellden, U. (1991) Desertification: time for a reassessment? *Ambio* 20, 372–383.

Mairota, P., Thornes, J.B. and Geeson, N. (1998) *Atlas of Mediterranean Environments in Europe: The Desertification Context*, Chichester: Wiley.

Middleton, N.J. and Thomas, D.S.G. (1997) *World Atlas of Desertification*, London: UNEP/Edward Arnold.

Mistry, J. (2000) *World Savannas: Ecology and Human Use*, Harlow: Prentice Hall.

Reed, M. and Dougill, A.J. (2002) *Geographical Journal*, 163, 195–210.

Thomas, D.S.G. (1993) Sandstorm in a teacup? Understanding desertification in the 1990s, *Geographical Journal* 159, 318–331.

Thomas, D.S.G. and Middleton, N.J. (1994) *Desertification: Exploding the Myth*, Chichester: Wiley.

Tucker, C.J., Dregne, H.E. and Newcomb, W.W. (1991) Expansion and contraction of the Sahara Desert, 1980 to 1990, *Science* 253, 299–301.

DAVID S.G. THOMAS

DESICCATION CRACKS AND POLYGONS

As evaporation of water from a saturated, fine-grained, cohesive sediment occurs, volume reduction may be accompanied by sufficient tensional stress for rupture to take place. Cracks are thereby formed and may display polygonal patterns. The morphology of the rupture patterns depends both on material properties (structure, degree of packing, moisture content, etc.) and on environmental conditions (temperature, humidity, rate of drying, etc.) (Corte and Higashi 1964). Cracks and polygons of this type are common geomorphic phenomena of both periglacial and arid environments (Maizels 1981), of vertisols and of drained proglacial lakes (Huddart and Bennett 2000).

Cracks tend to be fairly straight or smoothly curved in plan. However, their patterns, lengths, depths, widths and number show great variation. The plan form of blocks between cracks, which is determined by the crack pattern, can be flat, convex, concave or irregular, but the size of blocks tends to increase with the thickness of the material of which they are composed. The number of cracks is generally inversely proportional to sediment size, being greatest in materials rich in silt and clay. Mean crack length is proportional to sediment moisture content. It tends to decrease with time, because new, short cracks continue to be formed during each new cycle of drying. The spacing of cracks may increase with the rate of desiccation and with the proportion of clay present and according to the nature of the clay type. Sediment that is montmorillonite rich, for example, is prone to greater contraction than one with a comparable proportion of kaolinite. The presence of stones in and on fine-grained sediments may also affect the nature of the cracking.

Lachenbruch (1962) identified two common crack patterns. One is an orthogonal pattern in which cracks meet at right angles. The other is a non-orthogonal pattern characterized by triradial intersections that form obtuse angles of around 120°. The former are probably characteristic of inhomogeneous or plastic media in which stress accumulates gradually. Cracks form first at loci of low strength or high stress concentration. Because the cracks do not form simultaneously, a new crack has a tendency to join a pre-existing one orthogonally. The latter systems form in more homogeneous, relatively non-plastic media which are dried uniformly.

Particularly large 'giant desiccation cracks' are especially common on salty, playa surfaces (Neal *et al.* 1968) and may result not only from desiccation, but also because of such factors as salt mobilization, subsidence caused by groundwater withdrawal and seismic activity. The sudden creation of giant fissures can damage engineering structures (Al-Harthi and Bankher 1999; Corwin *et al.* 1991).

References

Al-Harthi, A.A. and Bankher, K.A. (1999) Collapsing loess-like soil in western Saudi Arabia, *Journal of Arid Environments* 41, 381–399.

Corte, A.E. and Higashi, A. (1964) *Experimental Research on Desiccation Cracks in Soil.* US Army Material Command Cold Regions Research and Engineering Laboratory Research Report 66.

Corwin, E.J. Alhadeff, S.C., Oggel, S.P. and Shlemon, R.J. (1991) Earth fissures, urbanisation and litigation: a case study from the Temecula area, southwestern Riverside County, California, *IAHS Publication* 200, 291.

Huddart, D. and Bennett, M.R. (2000) Subsidence structures associated with subaerial desiccation-crack piping and their role in drainage evolution on a drained proglacial lake bed: Hagavant, Iceland, *Journal of Sedimentary Research Section A: Sedimentary Petrology and Processes* 70, 985–993.

Lachenbruch, A.H. (1962) *Mechanisms of Thermal Contraction Cracks and Ice-wedge Polygons in Permafrost*, Geological Society of America Special Paper 70.

Maizels, J.K. (1981) *Freeze–thaw Experiments in the Simulation of Sediment Cracking Patterns*, Bedford College, London, Papers in Geography 13.

Neal, J.T., Langer, A.M. and Kerr, P.G. (1968) Giant desiccation polygons of Great Basin playas, *Geological Society of America Bulletin* 79, 69–90.

A.S. GOUDIE

DEW POND

Dew ponds are closed depressions, often filled with water, usually associated with chalk downland in the southern UK. The constant supply of water to these depressions on permeable substrates and frequently close to the top of slopes has caused considerable discussion over the past two centuries. As the name indicates, some writers have suggested dew (or mist or fog) as the source of replenishment but others believe that rainfall and surface runoff provide sufficient supply. Many dew ponds were dug on agricultural land, occasionally by people employed specifically for this purpose who moved from farm to farm, and the ponds were lined with clay, straw and more recently cement. Today some are dry because of lack of maintenance of the lining. There is considerable confusion in assigning origins to closed depressions in calcareous areas (see KARST) because some are natural DOLINEs or sink holes while others have anthropogenic origins as dew ponds or as pits dug for chalk, marl or clay. Some even have complex origins, forming initially as a doline but subsequently being lined and used as a dew pond.

Further reading

Prince, H.C. (1964) The origin of pits and depressions in Norfolk, *Geography* 49, 15–32.

Pugsley, A.J. (1939) *Dewponds in Fable and Fact: A Collection and Criticism of Exciting Knowledge on these Curiosities*, London: Country Life.

Snow, M. (2002) *Dewponds.* http://www.dewponds.info/(Accessed 18th September 2002).

IAN LIVINGSTONE

DIAGENESIS

All the changes (physical, chemical and biological) undergone by a sediment after its initial deposition, exclusive of weathering and metamorphism (and incorporates processes such as reworking, authigenesis, replacement, leaching, hydration, bacterial action, and the formation of concretions). The term was coined by Gümbel in 1868 (*diagenese*), though no universal definition exists.

Reference

Gümbel, C.W. von (1868) *Geognostische Beschreibung des ostbayerischen Grenzgebirges oder des bayerischen und oberpfalzer Waldgebirges*, Gotha: Justus Perthes; Bavaria: Bayerisches schreibung des Koenigreichs Bayern, v. 2.

Further reading

Larson, G. and Chilingar, G.V. (eds) (1979) *Diagenesis in Sediments and Sedimentary Rocks. Developments in Sedimentology* 25A, Oxford: Elsevier Scientific.

SEE ALSO: lithification

STEVE WARD

DIAMICTITE

Diamictite (paraconglomerate, mixtite, diamixtite) is a non-genetic term for sedimentary rock consisting of sand and/or larger particles resting in a muddy matrix (Flint *et al.* 1960). Its unlithified counterpart is known as diamicton, and diamict is a general term comprising both consolidated and non-consolidated deposits.

Diamictites consist of a wide range of particles with matrix dominating, giving an overall appearance of clasts chaotically dispersed through structureless or laminated mud. Clasts are dropstones sporadically deposited on the soft substrate due to ice rafting or volcanic explosions. They are angular to subrounded, may be slightly imbricated (see IMBRICATION) and often come from remote sources hundreds of kilometres away. If lamination is present, drapes (see DRAPE, SILT AND MUD) around clasts caused by post-depositional compaction (see COMPACTION OF SOIL) occur frequently.

The origin may be by GLACIERs, DEBRIS FLOWs or turbidity currents. Most diamictites are interpreted as lithified tills (tillites) or GLACIMARINE deposits of pre-Quaternary glaciations (Schermerhorn 1974; Hambrey and Harland 1981) because of a range of features resembling tills known from modern glacial environments, such as lack of sorting, grain sizes from clays to blocks, clasts bearing GLACIAL EROSION features (e.g. STRIATIONs and polishing), glacidynamic structures (e.g. shear planes (see SHEAR AND SHEAR SURFACE) and folds), and mixed lithological components corresponding to ice flow paths. Glacial origin of most diamictites is also supported by intimate association with striated bedrock surfaces, varved clays and lithified GLACIFLUVIAL deposits, arctic fauna, periglacial (see PERIGLACIAL GEOMORPHOLOGY) structures and the correspondence to polar regions of the past.

Diamictites of glacial and glacimarine origin are known from many regions of the Earth (Miller 1996: 469–483). The Late Palaeozoic Dwyka Formation occupies extensive areas in southern Africa and consists of facies deposited mainly under disintegrating ice shelves, near the grounding line and in FJORDs by tidewater glaciers (Visser 1991). Also the Lower Proterozoic Gowganda Formation, which covers more than 12,000 km^2 on the Canadian Shield is possibly a glacimarine diamictite (Eyles *et al.* 1985). Terrestrial tillites are known from the Upper Ordovician of north-west Africa (6–8 million km^2), Upper Proterozoic of western Mauritania and north Norway, Permo-Carboniferous of the Transantarctic Mountains, and Upper Palaeozoic of Oman and Brazil.

References

Eyles, C.H., Eyles, N. and Miall, A.D. (1985) Models of glaciomarine sedimentation and their application to the interpretation of ancient glacial sequences, *Palaeogeography, Palaeoclimatology, Palaeoecology* 51, 15–84.

Flint, R.F., Sanders, J.E. and Rodgers, J. (1960) Diamictite, a substitute term for symmictite, *Geological Society of America Bulletin* 71, 1,809.

Hambrey, M.J. and Harland, W.B. (eds) (1981) *Earth's Pre-Pleistocene Glacial Record*, Cambridge: Cambridge University Press.

Miller, J.M.G. (1996) Glacial Sediments, in H.G. Reading (ed.) *Sedimentary Environments: Processes, Facies and Stratigraphy*, 454–484, Oxford: Blackwell Science.

Schermerhorn, L.J.G. (1974) Late Precambrian mixtites: Glacial and/or nonglacial? *American Journal of Science* 274, 673–824.

Visser, J.N.J. (1991) The paleoclimatic setting of the late Paleozoic marine ice sheet in the Karoo Basin of Southern Africa, in J.B. Anderson and G.M. Ashley (eds) *Glacial* Marine Sedimentation: Paleoclimatic Significance, Geological Society of America Special Paper 261, 181–189.

Further reading

Prothero, D.R. and Schwab, F. (1996) *Sedimentary Geology*, New York: W.H. Freeman.

JAN A. PIOTROWSKI

DIAPIR

Vertical intrusions, bulbous or cylindrical in shape, resulting from the upward movement of mobile materials, such as salt (halite), magma, mud and ice, which lie beneath more competent strata (see MUD VOLCANO).

Further reading

Vendeville, B.C., Hart, Y. and Vigneresse, J.L. (eds) (2000) *Salt, Shale and Igneous Diapirs in and around Europe*, Special Publication of the Geological Society of London 174.

A.S. GOUDIE

DIASTROPHISM

A general term for the various types of tectonic processes that change the level, position and altitude of the Earth's surface. It is derived from the Greek word *diastrophos*, which means 'turned', 'twisted' or 'distorted'. There are five classes of diastrophic movement (Chorley *et al.* 1984: 98): (1) orogenesis; (2) epeirogeny; (3) isostasy; (4) igneous (including volcanic); and (5) eustasy.

Reference

Chorley, R.J., Schumm, S.A. and Sugden, D.E. (1984) *Geomorphology*, London: Methuen.

A.S. GOUDIE

DIATREME

Vents and pipes which have been injected through sedimentary strata by the explosive release of magmatic gases and filled with the products of the eruption, and fragments torn from the side of the pipes. Kimberlite pipes are examples of diatremes. Some maars, such as those in Germany, are lakes that are the surface expression of diatremes. They can be rich in environmental information (Narcisi 1996).

Reference

Narcisi, B. (1996) Tephrochronology of a late Quaternary lacustrine record from the Monticchio Maar (Vulture Volcano, southern Italy), *Quaternary Science Reviews* 15, 155–165.

A.S. GOUDIE

DIGITAL ELEVATION MODEL

A numerical description of the ground surface is helpful in addressing many geomorphological problems. Continuous topography may be quantified by a digital elevation model (DEM), any spatial array of terrain heights but most commonly a square mesh, or grid. DEM point spacing, or horizontal resolution, varies from fine (1 m) to coarse (⩾5 km), depending on the application, required level of detail, and the limitations of computer storage and processing (Table 11).

DEM-related nomenclature can be inconsistent. Digital terrain model (DTM), a frequent synonym for DEM, also is applied loosely to any calculated result, such as a map of slope gradient, rather than to the input heights themselves. Digital terrain *modelling*, confusingly also DTM, widely denotes DEM processing or GEOMORPHOMETRY.

For efficient computation and display, terrain heights are arranged in a defined structure (Figure 45), usually a regular grid, a triangulated irregular network (TIN), or digitized height contours or slope lines normal to contours. DEM structures are compromises, each with advantages and drawbacks.

Square-grid, actually rectangular, DEMs (hexagons or equilateral triangles are rare) store height Z as an array of implicit longitude X and latitude Y co-ordinates (Figure 45A). While this regular and discretized (discontinuous) structure is ill matched to the varied intricacies of surface features, a grid optimizes algorithm development, data processing and registration with spacecraft images. Grid DEMs can adapt somewhat to complex terrain by recursively subdividing squares, but computational efficiency declines.

The irregularly distributed heights of a TIN (Figure 45B) are vertices of triangles that vary in shape and size. A TIN is interpolated directly from surveyed points or discrete features that are extracted manually from maps or by computer from a grid or contour DEM. The TIN is adaptive, or surface specific: it aligns with ridges and channels and has many heights in complex areas but few redundant heights in planar terrain. Offsetting these advantages is the storage and processing burden imposed by the explicit X,Y co-ordinates required for each value of Z.

Ground-surface form is neither rectilinear nor triangular. Although not all topography is fluvial, the DEM structure best reflecting processes of erosion and deposition mimics paths of steepest gradient (Moore 1991). Terrain heights at intersecting contours and interpolated slope lines (Figure 45C) comprise an adaptive DEM that defines quadrilateral units of varied size and shape – most significantly the hillside concavities followed by surface flow. Explicit X,Y co-ordinates are necessary.

Early DEMs were created by field survey, manual interpolation of topographical maps, or semi-automated tracing of contours coupled with computer interpolation. Photogrammetric profiling and later optical scanning and automated interpolation of contours replaced these techniques. Grid DEMs now are available for the Earth, its seafloor, and the planet Mars. GTOPO30, compiled from many contour maps from several sources (Gesch *et al.* 1999), covers Earth at 30′ (nominally 1 km) resolution; two older global DEMs, ETOPO5 and TerrainBase,

Table 11 Sources and applications of Digital Elevation Models (DEMs)

DEM spacing	Sources of height measurements	Some geomorphological applications
1–50 m	Contours and stream lines from airphotos and topographic maps at scales 1:5,000 to 1:50,000 Surface-specific heights and stream lines from ground survey by GPS Remotely sensed heights from airborne and spacebome photogrammetry, radar and laser altimetry	Detailed terrain parameterization and visualization Estimates of flood inundation, soil moisture and other distributed-parameter hydrological modelling Spatial analysis of terrain and soil properties Terrain-aspect corrections to remotely sensed data Effects of terrain aspect on patterns of solar radiation, evaporation and vegetation
50–200 m	Contours and stream lines from airphotos and topographic maps at scales 1:50,000 to 1:200,000 Surface-specific heights and stream lines digitized from topographic maps at 1:100,000 scale	Broader scale distributed hydrological modelling Geomorphometric regionalization and analysis Sub-catchment analysis for lumped-parameter modelling and assessment of biodiversity
0.2–5 km	Surface-specific heights and stream lines digitized from topographic maps at scales 1:100,000 to 1:250,000 N.B.: Coarse-scale DEMs often are compiled by local averaging of fine-scale data	Height-dependent representations of surface temperature and precipitation Effects of terrain aspect on precipitation and surface roughness on wind Mapping continental drainage divisions
5–500 km	Surface-specific heights digitized from topographic maps at scales 1:250,000 to 1:1,000,000 (see also, N.B., above) National archives of trigonometric points, bench marks and other ground-surveyed terrain heights	Modelling relation of erosion and sediment distribution to tectonism Orographic barriers for general circulation models Broad-scale height and shaded-relief base maps for non-topographic information

Source: Modified after Hutchinson and Gallant (2000)

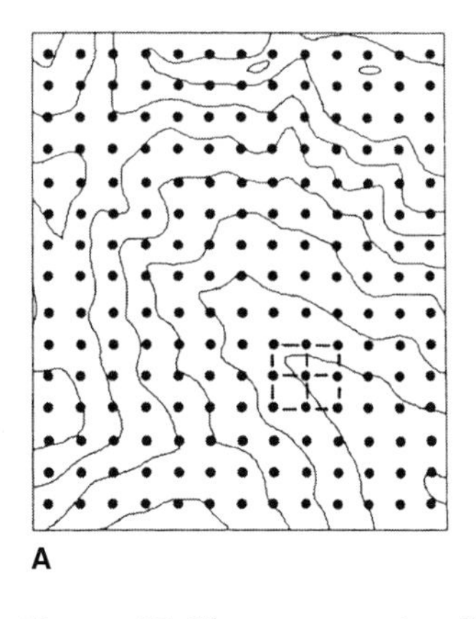

A

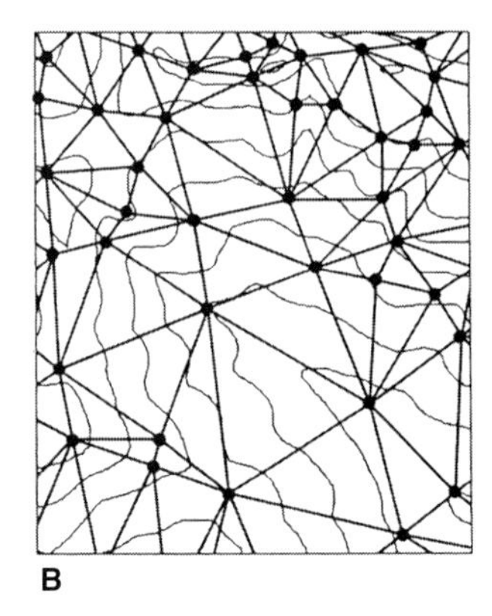

B

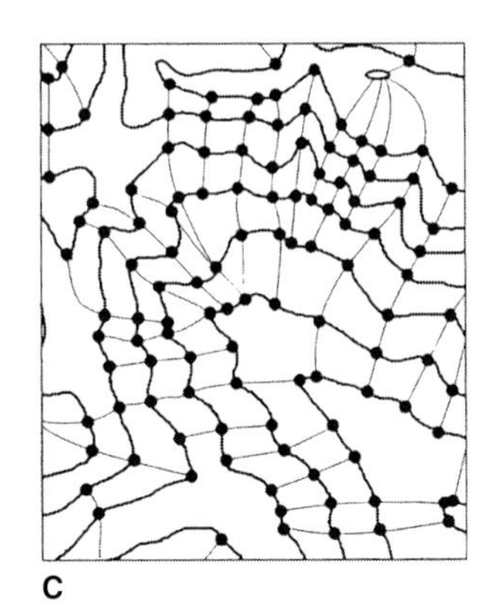

C

Figure 45 Three contrasting DEM structures for part of a small watershed in California. Dots are height locations. A: square grid, showing 3 × 3 subgrid used for neighbourhood operations; B: triangulated irregular network, TIN; C: intersections of 20-m contours (heavy lines) with slope lines (light). Each panel is 780 m across

are spaced at 10 km. The US National Elevation Dataset (NED) is a seamless 1″ (30 m) DEM (Alaska at 2″) assembled from all 55,000 1:24,000- and 1:63,360-scale topographical maps. Japan and the UK are gridded at 50 m, Italy nominally at 230 m, Australia 9″ (250 m), and Germany and other countries at various resolutions. Distribution of most military DEMs, e.g. DTED for the USA is restricted.

Bypassing contour maps, remote sensing (see REMOTE SENSING IN GEOMORPHOLOGY) can generate DEMs from direct measurements of terrain height (Maune 2001). Technologies include digital photogrammetry, the Global Positioning System (GPS), laser-ranging altimetry (LiDAR), synthetic-aperture radar interferometry (InSAR or IfSAR), thermal emission and reflection radiometry (ASTER), and, for bathymetry, deep-towed SONAR. The 3″ (90 m) DEM compiled from the 2000 Shuttle Radar Topography Mission (SRTM) includes 80 per cent of Earth's land surface (www.jpl.nasa.gov/srtm/). A variably spaced (1–12 km) depth grid, devised from radar altimetry inverted to sea-surface gravity and thence to bathymetry, covers Earth's entire seafloor.

Random and systematic flaws degrade DEM quality, defined by horizontal and vertical accuracy and precision of the constituent heights. Much of the error in DEMs derived from contours originates in the maps themselves, which never were intended to supply data of the high quality desirable for geomorphometry. Because contour maps only approximate terrain, just as DEMs approximate the maps, declared levels of DEM quality are merely statistical; locally, accuracy can be low. Heights expressed in integers can have insufficient vertical precision, or resolution, especially in level terrain. Contour-to-grid processing, always a compromise, is a second source of error. Some interpolation algorithms overrepresent contours; others add spurious terracing, closed depressions, and linear artefacts. Nor do advanced methods assure DEM quality. InSAR, LiDAR, and other remotely sensed data all contain errors, some of them severe, that are unique to their technologies. Where 1″ SRTM data reflect the dense tree canopy, they reproduce the ground surface no more faithfully than the 1″ NED.

Most DEMs must be refined for subsequent analysis (Figure 46). Computer processing can create a TIN or grid DEM from scattered heights and convert from one data structure or map projection to another. A DEM may be interpolated to finer – within limits, to avoid creating artefacts – or coarser resolutions for compatibility in merging DEMs and registering terrain height to other data. Bulk processing or point-by-point editing corrects errors or replaces parts of a DEM with data of higher quality. Removing long- or short-wave variation from a DEM by digital filtering can enhance detail or subdue erroneous artefacts of production. In creating a grid DEM intended for hydrological application, a drainage-enforcement algorithm reduces digitizing errors and preserves continuity of streams by incorporating channels, ridge lines, and other slope discontinuities (Hutchinson and Gallant 2000).

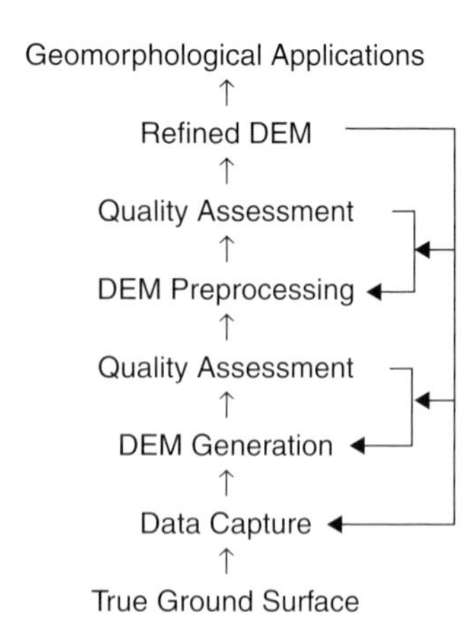

Figure 46 Sequence of activities in preparing a DEM to address geomorphological objectives; modified after Hutchinson and Gallant (2000)

Suitably preprocessed, grid DEMs are used to describe continuous topography through a spatial calculation adopted from digital image-processing, the neighbourhood operation, wherein a result – for example, relief shading – is obtained from adjacent input values. The input from a grid DEM is a compact matrix of heights, usually 3×3, moved through the data in regular increments; calculations for TINs or contour DEMs are on triangles or quadrilateral facets. Neighbourhood operations characterize terrain in three overlapping domains – RELIEF (Z), spatial (X,Y) and three-dimensional (X,Y,Z).

Most MORPHOMETRIC PROPERTIES derived from DEMs are moment statistics of height Z and its first two derivatives, slope gradient and profile curvature. Spatial parameters of terrain pattern and texture, unreferenced to an absolute datum, are more abstract; common X,Y measures are aspect, the compass direction faced by a slope, and contour curvature. Processing DEMs in the X,Y,Z domain captures the most complex properties – roughness, intervisibility, and variance of relief with azimuth.

Calculations on DEMs both visualize and parameterize the ground surface (Table 11). Digital maps in colour or monochrome portray topography, often in oblique perspective, by shaded relief, slope gradient, or aspect. Multispectral data, symbols for GEOMORPHOLOGICAL MAPPING, and other types of information are commonly displayed as overlays on a base of contoured height.

Among important parameters are the eight DEM derivatives calculated across each continent from the GTOPO30 data (Verdin and Greenlee 1998): a hydrologically integrated DEM, slope gradient and aspect, streamflow direction and accumulation, the topographical wetness index, stream networks, and drainage basins. DEM parameters are used to quantify hillside form (see HILLSLOPE, FORM), map landslide susceptibility, conduct TERRAIN EVALUATION, devise LAND SYSTEMS, assess cross-country trafficability, plan military operations, describe remote submarine and extraterrestrial surfaces, model slope evolution, simulate WATERSHED hydrographs, forecast the extent of flooding, and estimate sediment delivery.

References

Gesch, D.B., Verdin, K.L. and Greenlee, S.K. (1999) New land surface digital elevation model covers the Earth, *Eos, Transactions, American Geophysical Union* 80, 69–70; http://edcdaac.usgs.gov/gtopo30/gtopo30.html

Hutchinson, M.F. and Gallant, J.C. (2000) Digital elevation models and representation of terrain shape, in J.P. Wilson and J.C. Gallant (eds) *Terrain Analysis: Principles and Applications*, 29–50, Chichester: Wiley.

Maune, D. (ed.) (2001) *Digital Elevation Model Technologies and Applications, the DEM Users Manual*, American Society for Photogrammetry and Remote Sensing: Bethesda.

Moore, I.D. (ed.) (1991) Digital terrain modelling in hydrology, *Hydrological Processes* 5.

Verdin, K.L. and Greenlee, S.K. (1998) *HYDRO1k Elevation Derivative Database*, http://edcdaac.usgs.gov/ gtopo30/hydro/index.html

Further reading

Discoe, B. (2002) *The Virtual Terrain Project*, http://www.vterrain.org

Gesch, D.B., Oimoen, M.J., Greenlee, S.K., Nelson, Charles, Steuck, Michael and Tyler, Dean (2002) The National Elevation Dataset, *Photogrammetric Engineering and Remote Sensing* 68, 5–11.

Guth, P.L. (2002) *MicroDEM+ for Windows 95 and NT*, http://www.nadn.navy.mil/Users/oceano /pguth/website/microdem.htm

Kumler, M.P. (1994) An intensive comparison of triangulated irregular networks (TINs) and digital elevation models (DEMs), *Cartographica* 31, 2 (Monograph 45).

Pike, R.J. and Dikau, R. (eds) (1995) Advances in geomorphometry, *Zeitschrift für Geomorphologie, Supplementband* 101.

SEE ALSO: allometry; applied geomorphology; complexity in geomorphology; cross profile, valley; engineering geomorphology; equilibrium slope

RICHARD J. PIKE

DILUVIALISM

A form of CATASTROPHISM in which it is believed that the landscape was shaped by Noah's Flood, as reported in the book of Genesis. Before the true origin of glacial deposits was recognized, such materials, called 'drift' were ascribed by workers such as Buckland (Davies 1969) to a great deluge, when 'waves of translation' covered the Earth. By the 1830s the recognition of the complex stratigraphy of the drift and the discovery of the importance of the Ice Age greatly weakened the diluvial viewpoint.

The term diluvial is still sometimes used in the context, for example, of supposedly water-lain loess deposits.

Reference

Davies, G.L. (1969) *The Earth in Decay*, London: Macdonald.

A.S. GOUDIE

DISSOLUTION

In the geomorphological context, dissolution is the process whereby a rock, or parts of a rock, combine with water to form a solution. As the rock dissolves the different minerals disintegrate into individual ions or molecules and these diffuse into the solution. Hence, study of dissolution must focus on specific minerals as opposed to the rocks that are made up from them. Dissolution of a mineral is congruent when all components dissolve together (i.e. no solid remains) and incongruent where only a part of the components dissolve (for example the alumino-silicate minerals where ions are released in reaction with water but retain most of their elements in re-ordered solids such as kaolinite). There is a very wide range of mineral solubility in water, from gibbsite which is virtually insoluble (0.001 $mg\,l^{-1}$ at pH 7) through to halite (360 000 $mg\,l^{-1}$ at pH 7). Rocks made up of minerals with a very low solubility are highly resistant to CHEMICAL WEATHERING, while rocks containing highly soluble minerals such as rock salt are only found at outcrop in the driest places. Between these two extremes are a group of rocks in which dissolution along groundwater flow paths leads to the development of concentrated underground drainage and a landform assemblage known as KARST. Karst develops on silicate and evaporite rocks but is most common on the carbonate rocks, limestone and dolomite. Hence the remainder of this entry discusses the carbonate dissolution process.

The solution chemistry of carbonates is relatively simple as only two major minerals, calcite ($CaCO_3$) and dolomite ($CaMg(CO_3)_2$), are involved. Both are only slightly soluble in pure water (*c.*14 $mg\,l^{-1}$) and the solvent action of natural waters depends on their acid content. Organic and mineral acids may be important in some localities, particularly during the earliest (inception) phase of karstification (Lowe *et al.* 2000) but dissolution of calcite and dolomite is generally dominated by carbonic acid produced by hydration of dissolved carbon dioxide. It is frequently stated that the reaction between carbonic acid and 'insoluble' limestone produces calcium bicarbonate which is soluble. However, this is incorrect as there is no evidence for the existence of calcium bicarbonate molecules in solution. In fact there are three elementary chemical reactions in the dissolution of calcite which proceed in parallel:

$$CaCO_3 + H^+ \leftrightarrow Ca^{2+} + HCO_3^- \quad (1)$$
$$2CaCO_3 + H_2CO_3 \leftrightarrow 2Ca^{2+} + 2HCO_3^- \quad (2)$$
$$CaCO_3 + H_2O \leftrightarrow Ca^{2+} + HCO_3^- + OH^- \quad (3)$$

These can be summarized into:

$$CaCO_3 + H_2O + CO_2 \leftrightarrow Ca^{2+} + 2HCO_3^- \quad (4)$$

Similar processes take place in the dissolution of dolomite and are summarized as:

$$CaMg(CO_3)_2 + 2H_2O + 2CO_2 \leftrightarrow Ca^{2+} + Mg^{2+} + Mg^{2+} + 4HCO_3^- \quad (5)$$

The reactions continue until the forward and reverse rates become equal at which point the system is in equilibrium and the solution is said to be saturated with calcite. Addition of any acid to the system will increase the concentration of hydrogen ions and displace the equilibria in a forward direction. This reduces the concentration of $CO_3{}^{2-}$ and permits more $CaCO_3$ to dissolve so that when equilibrium is re-established the saturated solution has a higher calcium concentration.

In contrast to mechanical erosion processes, these reactions may occur in static water as well as through the range of water velocities. The speed of the reactions, and the amount of mineral dissolved, are controlled by the detailed solution kinetics (discussion of which is beyond the scope of this entry). However, the role of four important factors: carbon dioxide concentrations, temperature, equilibrium conditions and mixing corrosion will be considered briefly.

1 *Carbon dioxide concentrations* The atmospheric concentration of carbon dioxide is close to 0.035 per cent which would yield a saturation value of 70 $mg\,l^{-1}$ at 10 °C under open system conditions. Observed concentrations are frequently higher and it is generally assumed that this is due to the biogenic carbon dioxide in the soil atmosphere. However, the fluctuations in soil carbon dioxide concentrations are frequently more pronounced than those of calcium concentrations at springs and it is possible that ground air carbon dioxide in the subcutaneous zone may provide a relatively stable source.
2 *Temperature* For any fixed carbon dioxide concentration in a gas mixture in contact with water and rock the calcite solubility decreases with increasing temperature at a rate of approximately 1.3 per cent per degree Celsius. However, this effect is usually less significant than carbon dioxide concentrations in the gas phase and reaction rates, both

of which broadly increase with temperature. In addition, regional runoff variations account for a greater proportion of the observed variability in solutional erosion rates than do solute concentration variations.

3 *Equilibrium conditions* The two principal equilibrium conditions under which limestone may be dissolved are the 'open' system in which gas, water and rock are all in contact together such that carbon dioxide is available to replace that used up in the reaction of limestone and carbonic acid, and the 'closed' system in which gas and water come into equilibrium but the gas supply is cut off before contact with rock. Since there is no replacement of carbon dioxide under closed system conditions, the amount of limestone which can be dissolved is less than under open system conditions.
4 *Mixing corrosion* The mixing of two saturated waters produces an unsaturated (aggressive) solution and the mixing of a saturated and an aggressive solution, or of two aggressive solutions, may result in increased aggressivity. In extreme cases the new solution may be capable of dissolving 20 per cent more calcite but 1–2 per cent is more usual in natural waters. Hence, the mixing effect is generally less effective than 'normal' solution and its importance lies in its ability to operate in conditions under which normal solution is impossible, such as narrow fissures and in the phreatic zone.

While the key role of carbonic acid in the dissolution of carbonates was understood by the end of the eighteenth century it was not until towards the end of the twentieth century that details of the equilibrium chemistry of carbonate waters and their importance for speleogenesis and landscape evolution were elucidated, most notably by Dreybrodt (1988, 2000), Palmer (1991) and White (1984).

References

Dreybrodt, W. (1988) *Processes in Karst Systems: Physics, Chemistry and Geology*, Berlin and New York: Springer.

——(2000) Equilibrium chemistry of karst water in limestone terranes, in A. Klimchouk, D.C. Ford, A.N. Palmer and W. Dreybrodt (eds) *Speleogenesis: Evolution of Karst Aquifers*, 126–135, Huntsville, AL: National Speleological Society.

Lowe, D.J., Bottrell, S.J. and Gunn, J. (2000) Some case studies of speleogenesis by sulphuric acid, in A. Klimchouk, D.C. Ford, A.N. Palmer and W. Dreybrodt (eds) *Speleogenesis: Evolution of Karst Aquifers*, 304–308, Huntsville, AL: National Speleological Society.

Palmer, A.N. (1991) The origin and morphology of limestone caves, *Geological Society of America Bulletin* 103, 1–21.

White, W.B. (1984) Rate processes: chemical kinetics and karst landform development, in R. G. LaFleur (ed.) *Groundwater as a Geomorphic Agent*, 227–248, Boston: Allen and Unwin.

SEE ALSO: corrosion

JOHN GUNN

DIVERGENT EROSION

The term was derived from the evolution of INSELBERGS. When a rock outcrop is laid open in the tropics the rainwater runs off very fast. Thus weathering lacks the moisture for further decomposition. In the surrounding zone the water can percolate into the soil and guarantees continued weathering of the rock there. During the contemporaneous lowering of the surface, the rock outcrop is resistant. Often its sides are further exposed, whilst the neighbouring areas are eroded. Mostly this occurs at a similar rate to that at which the rock outcrop at the base of the regolith is decomposed. Thus the rock grows out relatively slowly from the weathering mantle. Therefore diverging erosion follows diverging weathering. Inselbergs rise up to 300 m and even more. They are developed, however, in all sizes, so that sequences are observed in the tropics. The above derivation developed from these observations. Isolated mountains outside of the tropics are as a rule palaeoforms, sometimes exhumed.

The initial stage of the exposure of the rock outcrops has different causes: thinning of the soil cover occurs in special geomorphic positions, quite often in an area where an ESCARPMENT is originating, i.e. where planational processes produced a slight increase in gradient. The first rock outcrops then evolve into inselbergs which lie in front of or on top of escarpments. Eventually rock outcrops are so numerous that an escarpment develops. Sometimes inselbergs occur near rivers or sea coasts. Rapid downwearing produces an initial small rock outcrop by chance, e.g. by tree fall. Other positions for inselbergs are watersheds, large and small. Examples are the prominent inselbergs in the south of Central Australia: Ayers Rock, the Olgas, and Mt. Connor. All these are developed in sedimentary rocks, which shows that divergent weathering is independent from rock hardness as the lithology is more or less the same in these rocks and the neighbouring plain, at least in areas larger than the inselbergs. Quite often a different spacing of fractures is postulated as a reason for special resistance, but tectonic lines should show repetitive patterns and thus a regular spacing of inselbergs, which is not the case. Divergent erosion is sometimes used for different processes controlled by rock hardness, too, especially in the case of inselbergs. But it is always hard to prove this as rock samples for comparison from the deeply weathered plain are difficult to retrieve.

Rock outcrops are generally resistant in the tropics due to the minor importance of physical weathering and the overwhelming power of chemical processes. The first needs water, if at all, only for a short time, while the second can only work with long wetting, preferably with water containing organic acids. Both are missing on bare slopes. Therefore inselbergs and escarpments in the tropics are often very old. After exposure these rock outcrops are only slightly weathered. Even if special forms like runnels, exfoliation sheets, or small caverns developed, the overall form is not changed. At Ayers Rock these weathering forms are nested. Thus they are of different ages, which proves the stability of the slope during a very long time even under changing climates.

Very steep slopes in Sri Lanka are surprisingly stable. This is not only due to rock outcrops but to the rapid movement of the subterraneous water. For this, soil analysis gives an explanation: in thin sections a very high volume of large pores is seen, as is the relatively stable soil texture due to iron and silica minerals in the matrix or even as cutans on pore walls. Thus the internal water movement has good pathways. Soil stability is maintained due to the low swelling capacity of the kaolinite minerals. The rapid water movement is similar to that on rock outcrops. Thus these processes were called internal divergence. Once the soil fabric is disturbed, e.g. by building a street on a steep slope, water movement is blocked and severe slides may occur.

Divergent erosion as a principle in tropical geomorphology shows a discontinuity of erosion in space and time. It is nearly independent of the forces of gravity but dependent on differences in friction. One can consider divergent erosion as a positive feedback mechanism. A threshold of resistance to weathering and erosion is not surpassed in the case of the exposed rock facets. Thus they possess an extremely low sensitivity to change. The ergodic principle is applicable in the humid tropics where planational lowering is still active to different heights on the periphery of inselbergs due to divergent erosion.

References

Bremer, H. (1972) Flußarbeit, Flächen- und Stufenbildung in den feuchten Tropen, *Zeitscrhift für Geomorphologie Supplementband* 14, 21–38.

Bremer, H. and Sander, H. (2002) Inselbergs: geomorphology and geoecology, in S. Porembski and W. Barthlott (eds) *Inselbergs, Biodiversity of Isolated Rock Outcrops in Tropical and Temperate Regions*, Ecological Studies 146, 7–35, Berlin, Heidelberg: Springer.

H. BREMER

DOLINE

Dolines are natural enclosed depressions found in karst landscapes (Ford and Williams 1989). They are subcircular in plan, tens to hundreds of metres in diameter, and can range from a few metres to about a kilometre in width. They are typically a few metres to tens of metres in depth, but some are hundreds of metres deep. Their sides range from gently sloping to vertical, and their overall form can vary from saucer shaped to conical or even cylindrical. Dolines are especially common in terrains underlain by carbonate rocks, and are widespread on evaporites. Some are also found in siliceous rocks such as quartzite. Dolines have long been considered a diagnostic landform of KARST, but this is only partly true. Where there are dolines there is certainly karst, but karst can also be developed subsurface in the hydrogeological network even when no dolines are found on the surface. Dolines have a similar function in karst landscapes to the drainage basin in non-karstic lithologies, in that they drain rainwater from the surface, but in the case of the doline it is discharged underground via an outlet at the lowest point in the doline basin.

The term sinkhole is sometimes used (especially in North America) to refer both to dolines and to depressions where streams sink underground, which in Europe are described by separate terms (including ponor, swallow hole, and stream-sink). Thus the terms doline and sinkhole are not strictly synonymous. Table 12 lists the terms employed by different authors and Figure 47 illustrates six main doline types (both are from and are discussed in more detail in Williams 2003).

Enclosed depressions in karst can be formed by four main mechanisms: DISSOLUTION, collapse, SUFFOSION and regional SUBSIDENCE. In practice the complexity of natural processes often results in more than one mechanism being involved, in which case the doline is polygenetic in origin. A typical case is a depression formed initially by dissolution that later in its development is subject to collapse of its floor into an underlying cave. In such a case, the gentler upper slopes of the doline were formed by dissolution and the steeper lower slopes by collapse.

Solution dolines

The bowl-shaped form of a typical doline indicates that more material has been removed from its centre than from around its margins. Where the principal process responsible for this is dissolution of the bedrock, it follows that there is a mechanism that focuses chemical attack. The amount of limestone that can be removed in solution depends upon two variables: first, the concentration of the solute and, second, the volume of the solvent (in this case the amount of water draining through the doline). Variations in either or both of these variables could be responsible for the focusing of dissolution near the centre of the depression, but if local variation in solute concentration alone were sufficient to explain the occurrence of solution dolines, then they would be found on every type of limestone in a given climatic zone. This is not the case, as illustrated by comparison of landscapes formed on Devonian, Carboniferous, Jurassic and Cretaceous limestones in England, where dolines are most frequently found on Carboniferous limestones and tend to be less prevalent on Cretaceous and Jurassic limestones. It follows, therefore, that local spatial variations in water flow must be responsible for focusing corrosional attack.

The development of dolines of all kinds depends on the ability of water to sink into and flow through karst rocks to outlet springs. The exposure of limestones by erosion provides an input boundary for infiltration of water and a valley incised into the limestone provides an output boundary. Infiltrating rainwater is acidified in the atmosphere and further acidified in the soil. On percolating downwards this water accomplishes most of its dissolutional work within 10 m of the surface. Joints (see JOINTING), faults and bedding-planes vary spatially within the rock because of tectonic history and variations in lithology. Consequently the frequency and interconnectedness of fissures available to transmit flow also varies. Some fissures are more favourable for percolation than others, for example where several joints intersect, and as a result these develop as principal drainage paths. Water flows towards them and as a result they are subjected to still more dissolution by a positive feedback mechanism and so vertical permeability is enhanced. The local surface of water saturation in the EPIKARST is drawn down over the preferred leakage paths similar to cones of depression in the water table over pumped wells; streamlines adjust and resulting flow lines are centripetal and convergent on the preferred drainage zones. By this means solvent flow is focused and, as the surface lowers, the more intensely corroded zones begin to obtain topographic expression as solution dolines. Particularly large solution depressions often occur in the humid tropics where corrosion processes were uninterrupted by Pleistocene glaciations. In these places the term cockpit is sometimes applied to them after a particular style of landscape in Jamaica, where depressions are incised between intervening conical hills.

Although small solution dolines have formed in 15,000 years or so in some mid to high latitude areas that were glaciated in the late Pleistocene, several tens to hundreds of thousands of years are required to develop large solution dolines in limestone. Once formed they may persist in the landscape for several million years provided there is sufficient thickness of limestone for their continued incision. Individual dolines may merge to form compound closed depressions (known as uvalas) and large dolines may subdivide internally into smaller second generation basins. Where all the available space is occupied by depressions, rather like an egg box, the landscape is termed polygonal karst, because the

Table 12 Doline/sinkhole English language nomenclature as used by various authors

<table>
<tr><th colspan="2">Doline-forming processes</th><th>Ford and Williams 1989</th><th>White 1988</th><th>Jennings 1985</th><th>Bogli 1980</th><th>Sweeting 1972</th><th>Culshaw and Waltham 1987</th><th>Beck and Sinclair 1986</th><th>Other terms in use</th></tr>
<tr><td colspan="2">Dissolution</td><td>solution</td><td>solution</td><td>solution</td><td>solution</td><td>solution</td><td>solution</td><td>solution</td><td></td></tr>
<tr><td colspan="2">Collapse</td><td rowspan="3">collapse</td><td>collapse</td><td>collapse</td><td rowspan="2">collapse (fast) or subsidence (slow)</td><td>collapse</td><td>collapse</td><td rowspan="2">collapse</td><td></td></tr>
<tr><td colspan="2">Caprock collapse</td><td>—</td><td>subjacent collapse</td><td>solution subsidence</td><td>—</td><td>interstratal collapse</td></tr>
<tr><td>Dropout</td><td rowspan="2">subsidence</td><td>cover collapse</td><td rowspan="2">subsidence</td><td rowspan="2">alluvial</td><td rowspan="2">alluvial</td><td>subsidence</td><td>cover collapse</td><td></td></tr>
<tr><td>Suffosion</td><td>suffosion</td><td>cover subsidence</td><td>—</td><td>cover subsidence</td><td>ravelled, shakehole</td></tr>
<tr><td colspan="2">Burial</td><td>—</td><td>—</td><td>—</td><td>—</td><td>—</td><td></td><td>—</td><td>filled, palaeo-</td></tr>
</table>

Source: from Williams 2003 modified from Waltham and Fookes 2003

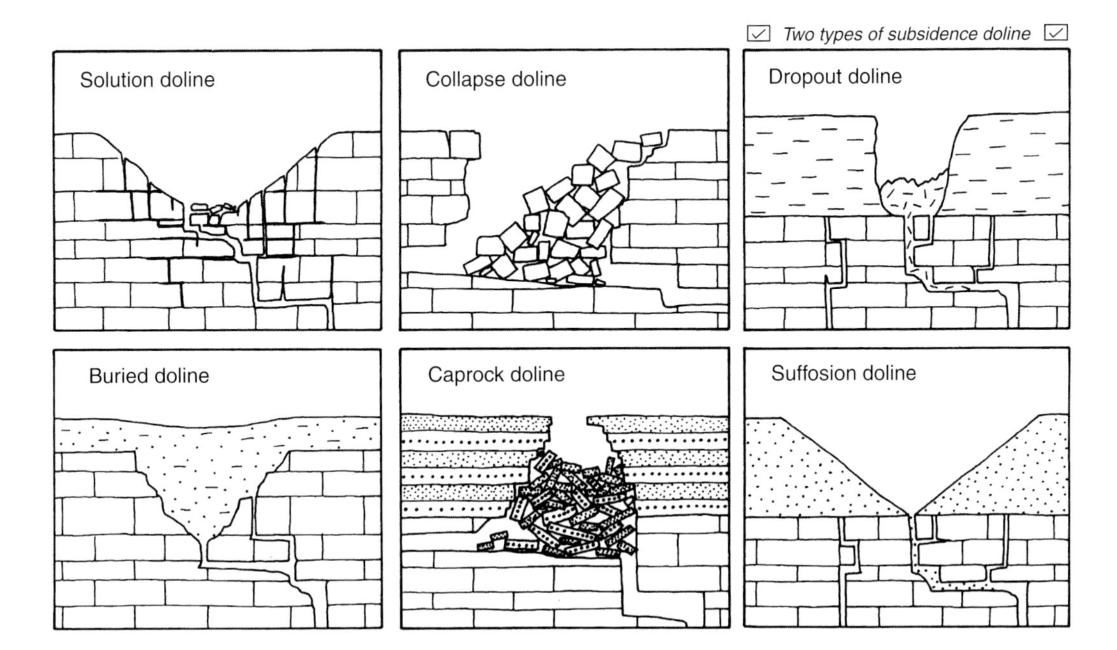

Figure 47 Classification of dolines (after Williams 2003)

topographic divides of the adjoining solution depressions have a polygonal pattern when viewed in plan.

Collapse dolines

Collapse dolines are formed mainly by mechanical processes. There is considerable variation in nomenclature concerning depressions formed mainly by mechanical processes (Table 12), largely because of the variety of materials and processes involved (Waltham 1989). Collapse refers to rapid downward movement of the ground, whereas subsidence refers to gradual movement sometimes without even ripping the surface. These processes can occur in karst bedrock, in caprock that may stratigraphically overly it, and in veneers of unconsolidated sediments. In all cases the collapse has to be preceded by dissolution of the karst rock to form a void into which material can fall. The kind of landforms produced depends upon which of the various materials and processes were involved.

Where collapse dolines form in karst bedrock then the void is commonly part of a cave system. Collapse may occur following undermining from below as the roof of a cavity stopes upwards, ultimately causing the surface above to collapse, or following dissolution from above that weakens the span of a cave roof, causing it to collapse. For example, solutional attack by drainage water near the bottom of a solution doline may combine with upwards stoping of an underlying cave roof to weaken a span from above and below, thereby causing the doline floor to collapse into a cave. Collapse dolines are on average smaller in diameter than solution dolines, although particularly large examples 700 m along their largest axis and up to 400 m deep are known in the Nakanai Mountains of New Britain, Papua New Guinea.

Sometimes a collapse extends from a cave below the modern water-table level, in which case the collapse doline will contain a lake. Such features are known as cenotes after the type-site in the Yucatan Peninsula of Mexico, although similar landforms are found elsewhere, such as in south-east Australia. The deepest known case of a collapse doline containing a lake is the Crveno Jezero (Red Lake) in Croatia, which is 528 m deep from its lowest rim, the bottom of the collapse extending 281 m below the modern level of the nearby Adriatic Sea.

Another process that increases the effective stress on rock arches and subsurface domes is removal of buoyant support by water-table lowering. This increases the effective weight on the span of the roof, resulting in its strength being exceeded and so in its failure and collapse. This occurs because in a fully saturated medium the buoyant force of water is 1 t m^{-3}, and if the water table is lowered by 30 m, the increase in the effective stress on the rocks is 30 t m^{-3}. A gradual lowering of the water table occurs with valley incision, because springs are lowered too, and with them the level of the saturated zone that feeds them. More rapid still is the lowering caused by sea-level fall, a process that occurred frequently in the Pleistocene because of repeated glacio-eustatic (vertical movement of sea level caused by glaciation and deglaciation) fluctuations. This particularly affected karsts well connected to the coast such as in Florida, southeastern Australia and Yucatan, where it probably was a significant influence in the development of cenotes.

If unconsolidated coverbeds are drained by water-table lowering, then consolidation and compression occurs, leading to subsiding of the surface and collapse where clastic sediments span de-watered unsupported arches. This is a common process in Florida where porous sandy formations overlie karstified limestones, and has been exacerbated by groundwater pumping for water supplies, which has still further reduced buoyant support. This process and the resulting incidence of collapse attains dangerous hazardous proportions in karstified areas extensively de-watered by mining activities (Beck and Pearson 1995). These dolines in unconsolidated coverbeds are sometimes referred to as cover collapse sinkholes (Table 12).

Subsidence (suffosion/dropout) dolines

When unconsolidated deposits such as alluvium, glacial moraine, loess or sand mantle karstified rock, the sediments are sometimes evacuated downwards through corrosionally enlarged pipes in the underlying karst, resulting in gradual or rapid SUBSIDENCE of the surface. Hence, the term subsidence doline is sometimes used for any closed depression in unconsolidated deposits, although the term is also used for depressions formed by much larger scale regional subsidence. Often a combination of processes is involved in the development of subsidence dolines including corrosion and collapse of the underlying bedrock, as well as suffosion, mudflow and void collapse in the mantling materials. However, the main process by which the sediment moves is known as suffosion and involves the gradual downwashing of fines by a combination of physical and chemical processes. The topographic consequence of this activity depends on whether the material is cohesive or non-cohesive. In cohesive sediments evacuation of material may proceed for some time without any surface expression. However, a void is formed that enlarges and stopes upwards resulting in a sudden, and sometimes catastrophic, failure of the ground surface. The depression thus formed is called a dropout doline or cover collapse doline. In Britain suffosion dolines formed in glacial boulder clay overlying limestone are widely referred to as shakeholes. Similar features but in more uniform finer grained materials are referred to as cover subsidence sinkholes in the USA.

References

Beck, B.F. and Pearson, F.M. (eds) (1995) *Karst Geohazards: Engineering and Environmental Problems in Karst Terrane*, Rotterdam: Balkema.

Ford, D.C. and Williams, W. (1989) *Karst Geomorphology and Hydrology*, London: Chapman and Hall.

Waltham, A.C. (1989) *Ground Subsidence*, Glasgow: Blackie.

Waltham, A.C. and Fookes, P.G. (2003) Engineering classification of Karst ground conditions, *Quarterly Journal of Engineering Geology and Hydrogeology* 36(2), 101–118.

Williams, P.W. (2003) Dolines, in J. Gunn (ed.) *Encyclopedia of Cave and Karst Science*, London: Routledge.

PAUL W. WILLIAMS

DONGA

Derived from the Nguni word *Udonga*, meaning a wall, it is a term used in southern Africa to describe a gully or BADLAND area caused by severe erosion (Plate 35). Widespread in Lesotho, Zimbabwe, the middleveld of Swaziland, in the Karoo, and Kwazulu-Natal, they are especially prevalent in COLLUVIUM and in deeply weathered bedrock in areas where the mean annual precipitation lies between *c.*600 and 800 mm. Where the materials in which they are developed have high ESP (Exchangeable Sodium Percentage) contents, they may have highly fluted 'organ pipe' sides (Watson *et al.* 1984). Repeated oscillations have taken place in colluvium deposition and palaeosol formation on the one hand, and incision on the other (Botha and Federoff 1995). Causes of incision may include climatic change, and land cover changes brought about by human activities, the latter including the spread of pastoralism and deforestation for iron smelting. Debates about their origin are similar to those that have been raised in connection with the formation of ARROYOs in the American west. Piping

Plate 35 A deep donga developed in highly erodible colluvial material in a valley bottom near St Michael's Mission in central Zimbabwe

(see PIPE AND PIPING) is probably an important process in their development (Rienks *et al.* 2000).

References

Botha, G.A. and Federoff, N. (1995) Palaeosols in Late Quaternary colluvium, northern Kwazulu-Natal, South Africa, *Journal of African Earth Sciences* 21, 291–311.

Rienks, S.M., Botha, G.A. and Hughes, J.C. (2000) Some physical and chemical properties of sediments exposed in a gully (donga) in northern Kwazulu-Natal, South Africa, and their relationship to the erodibility of the colluvial layers, *Catena* 39, 11–31.

Watson, A., Goudie, A.S. and Price-Williams, D. (1984) The palaeoenvironmental interpretation of colluvial sediments and palaeosols of the Late Pleistocene Hypothermal in southern Africa, *Palaeogeography, Palaeoclimatology, Palaeoecology* 45, 225–249.

A.S. GOUDIE

DOWNSTREAM FINING

The characteristic decline in the average size of riverbed material with distance downstream is downstream fining. This may include a full sequence from boulder-sized material close to the river source, through gravel-, sand- and silt-sizes to clay-sized material where the river enters the sea. Many rivers do not have all these changes, and may have gravel- or sand-beds at termination. The downstream decline in grain size was first recognized to follow a negative exponential trend by Sternberg (1875) who explained this by the process of ABRASION. Abrasion is significant in many cases, but laboratory measurement of abrasion rates suggests that abrasion alone is insufficient to account for observed rates of fining. It has long been recognized that smaller sediment particles should move more frequently and further than larger ones during BEDLOAD transport. This selective transport mechanism was questioned during the 1980s when it was found that, in GRAVEL-BED RIVERS, there is only slight size selectivity in bedload transport. Further investigation has shown that even a small degree of size selectivity can cause significant downstream fining over long time periods. Downstream fining is thus best explained as a consequence of size sorting during bedload transport, with abrasion and particle breakdown generally acting as secondary effects that may accelerate the rate of fining.

Downstream fining is also one of the downstream adjustments that takes place in graded river systems (see GRADE, CONCEPT OF), along with changes in bed slope, channel width and depth, and flow velocity. The rate of downstream fining (the degree of concavity of a graph of particle size versus distance downstream) is inversely proportional to the length of the river, such that fining is rapid in short rivers and slow in long ones. Close inspection of bed material size data shows that downstream fining is rarely a smooth, continuous process. Abrupt changes in grain size occur where tributaries enter the river, or close to sediment sources (see TERRACE, RIVER). These perturbations are smoothed out at the scale of the whole river, but demonstrate how river networks route both water and sediment downstream, and cause changes in stream ecology. Particularly notable is the abrupt transition from gravel- (>2 mm) to sand-sized (<2 mm) bed material that occurs in many rivers. This transition can result from the supply of large amounts of sand-sized material to the river, or from complex interactions between sediment movement and flow hydraulics that occur as the percentage of sand in the river bed exceeds about 20 per cent.

References

Sternberg, H. (1875) Untersuchungen über Längen-und Querprofil geschiebeführender Flüsse, *Zeitschrift für Bauwesen* 25, 483–506.

Further reading

Hoey, T.B. and Bluck, B.J. (1999) Identifying the controls over downstream fining of river gravels, *Journal of Sedimentary Research* 69, 40–50.

Sambrook Smith, G.H. and Ferguson, R.I. (1995) The gravel-sand transition along river channels, *Journal of Sedimentary Research* A65, 423–430.

SEE ALSO: channels, alluvial; hydraulic geometry

TREVOR B. HOEY

DRAA (MEGADUNE)

Draa, the Arabic word for 'arm', may be used to denote the largest members of the aeolian bedform hierarchy. The term was first used in English by Wilson (1972).

Draas are also known as compound and complex dunes (Breed and Grow 1979), or megadunes (Warren and Allison 1998). They are typically large bedforms with a spacing exceeding 500 m and a height reaching 200 or 300 m and may occur as linear, crescentic, or star forms. Examples of linear draa occur in the Namib Sand Sea, Rub al Khali of Arabia and the Akchar erg of Mauritania; crescentic draa can be found in the Liwa area of the United Arab Emirates and Saudia Arabia, the Namib Sand Sea, and the Algodones dunefield of California. Draa of star form occur in the Grand Erg Occidental and Oriental of northern Africa, the Namib Sand Sea, and the Gran Desierto of Mexico.

Draas are characterized by superimposed bedforms of dune size, with heights up to 10 m and a spacing of up to 300 m. In some places, e.g. the northern Namib Sand Sea, the superimposed dunes appear to be features contemporary with the main bedform (Bristow *et al.* 2000); elsewhere, e.g. in Wahiba Sands of Oman and in Mauritania, the superimposed dunes represent different generations of dunes, in some cases formed in a wind regime different from that which formed the main draa (Warren and Allison 1998; Lancaster *et al.* in press). Thus crescentic dunes may be superimposed on linear draa, and two or more smaller sets of linear dunes are superimposed on older linear draa.

The large size of draa has been thought to be the product of strong winds (e.g. Wilson 1972), but others have suggested that their large size is a product of long continued development in a wind regime that promotes deposition on the dune (e.g. Lancaster 1988). Their large size indicates persistence over long periods of time and reconstitution times in the order of 1 to 100 ka.

Recent stratigraphic and dating studies suggest that some draa (especially linear draa, which tend to conserve their form over long periods) may be composite landforms constructed by multiple generations of aeolian deposition, stability and reworking. In several areas (e.g. UAE, Oman, Mauritania), the cores of large linear draa are at least 15–22 ka old (Glennie and Singhvi 2002; Lancaster *et al.* in press).

References

Breed, C.S. and Grow, T. (1979) Morphology and distribution of dunes in sand seas observed by remote sensing, in E.D. McKee (ed.) *A Study of Global Sand Seas*, United States Geological Survey Professional Paper 1,052, 253–304.

Bristow, C.S., Bailey, S.D. and Lancaster, N. (2000) Sedimentary structure of linear sand dunes, *Nature* 406, 56–59.

Glennie, K.W. and Singhvi, A.K. (2002) Event stratigraphy, palaeoenvironment and chronology of SE Arabian deserts, *Quaternary Science Reviews* 21, 853–869.

Lancaster, N. (1988) Controls of eolian dune size and spacing, *Geology* 16, 972–975.

Lancaster, N., Kocurek, G., Singhvi, A.K., Pandey, V., Deynoux, M., Ghienne, J.-P. and Lo, K. 2003. Late Pleistocene and Holocene dune activity and wind regimes in the western Sahara of Mauritania, *Geology* 30, 991–994.

Warren, A. and Allison, D. (1998) The palaeoenvironmental significance of dune size hierarchies, *Palaeogeography, Palaeoclimatology, Palaeocology* 137, 289–303.

Wilson, I.G. (1972) Aeolian bedforms – their development and origins, *Sedimentology* 19, 173–210.

SEE ALSO: dune, aeolian

NICK LANCASTER

DRAINAGE BASIN

A drainage basin is an area of land that contributes water and sediment to a specific outlet point on a stream. It is separated from other drainage basins by its drainage divide, a boundary that encircles a basin along its highest, outermost ridge tops. The drainage basin is recognized as a fundamental geomorphological unit (Horton 1932: 350) and is frequently used as the primary landscape unit for hydrological, water supply and ecological investigations and for land management activities.

Drainage basins are EROSION created landforms sculpted predominantly by the actions of flowing water. They may be conceptualized as consisting of two geomorphological components: a set of hillslopes dominated by unconfined OVERLAND

FLOW and a branching network of stream channels conveying concentrated flows. The transition from hillslope to channel has been characterized as both indistinct and distinct. Davis (1899: 495) wrote: 'Although the river and hillside waste-sheet do not resemble each other at first sight, they are only the extreme members of a continuous series; and when this generalization is appreciated, one may fairly extend the "river" all over its basin and up to its very divides.' From this perspective, every point within a drainage basin is located along a flow pathway, and a basin is composed of a branching, space-filling drainage network of flow pathways extending from outlet to basin divide. The alternative viewpoint is that the transition from hillslope to channel is determined by a geomorphological threshold (see THRESHOLD, GEOMORPHIC) of channelization that sets a finite scale for dividing a landscape into valleys and hillslopes. Because a drainage basin may be defined upstream of any point on the land's surface, the delineation of a landscape into specific drainage basins is done for some designated purpose.

Several other terms are used synonymously with drainage basin. In Great Britain, catchment is commonly used, whereas in the United States, watershed is a preferred term. Unfortunately, watershed is an ambiguous term that has historically been used as a synonym for drainage divide, and this usage is retained in Great Britain. For large basins drained by a major river, the term river basin is often used (e.g. Amazon River basin). Drainage basin, catchment and watershed do not inherently imply a particular size of drainage area. However, some government agencies in the United States and others are using these terms in size-based classification systems, such as catchment being smaller than a watershed and watershed being smaller than a basin.

The form and structure characteristics of drainage basins and their associated drainage networks are described by their MORPHOMETRIC PROPERTIES, which can be classified into the categories of size, surface, shape, relief and texture. Drainage area, a variable specifying the amount of land area contained within a drainage divide, is an important basin descriptor and is frequently used as a surrogate for the amount of water and sediment yielded by a drainage basin. Because a basin's drainage network is its most prominent feature, network morphometric properties are also used for drainage basin description. A qualitative indication of drainage basin size is indicated by the stream order of its outlet stream (see STREAM ORDERING). The delineation of drainage basins and determination of their morphometric properties has traditionally been done using topographic maps and manual methods. With DIGITAL ELEVATION MODELS (DEM) and geographic information systems (GIS), watershed delineation, drainage network extraction and the automatic calculation of morphometric properties is possible.

Although drainage basins are fundamental geomorphological units, they may not always prove the best choice for organizing landscapes for research or land management purposes. In landscapes dominated by non-fluvial features such as kettle holes (see KETTLE AND KETTLE HOLE) or aeolian dunes (see DUNE, AEOLIAN) drainage networks and drainage basins are often poorly defined and may be difficult to delineate using either manual or GIS methods.

Drainage basin organization

A drainage basin may be organized into two subsidiary landform units: a set of hillslopes and a drainage network. Although hillslopes may occupy 95 per cent or greater of a basin's area, it is the drainage network that noticeably provides the organization of the hillslopes within a basin. The drainage network is the tree-like structure of flow pathways along which water and sediment are concentrated and delivered to the basin outlet. Drainage networks are comprised of exterior and interior links between successive nodes, where nodes are sources, junctions or the basin outlet. Exterior, or first-order links, connect an upstream source node to a downstream junction. Interior links connect two junctions or a junction to the basin outlet.

Using GIS and DEMs, a space-filling network of flow paths can be delineated within a drainage basin, with external links terminating at the basin's exterior divide or internal divides. There are several subsidiary networks contained within the space-filling drainage network. Many investigators discuss the drainage network in terms of streams and stream system, referring to the blue-line streams on topographical maps. Some have considered the stream network to be synonymous with the channel network, but others view channels as geomorphological features identifiable only from field investigation. A drainage basin will also contain a network of VALLEYs, which may or may not contain streams or channels.

Much of the geomorphological analysis devoted to drainage basins has been with respect to the organization and development of the branching link drainage network structure. In a seminal geomorphology paper, Horton (1945) provided many of the concepts supporting modern geomorphological analysis of drainage basins. He provided the basis for the hierarchical method of stream ordering and laws of drainage composition that with later modifications due to Strahler (1957) and others provide a means to organize the understanding about the topologic and geometric properties of drainage networks. In the Horton/Strahler ordering system, source streams (exterior links) are designated as first order. When two first-order streams join, the stream that continues is designated as second order and, in general, at the junction of two streams of equal order, the order of the downstream segment is increased by one. Low-order tributaries may flow into high-order streams without the order being incremented, and the entire section of stream of same order is referred to as one stream segment for the purposes of quantifying the number of streams, stream length, stream slope and contributing area. HORTON'S LAWS of drainage composition refer to the empirical straight-line relationships between these quantities and stream order on semi-log plots.

Horton (1945: 283) also devised the concept of DRAINAGE DENSITY, which indicates the degree of dissection of a drainage basin into subsidiary hillslopes by its channel network. Horton's concepts of network analysis can be applied to any of the drainage networks including channel, valley and GIS-derived networks.

Probabilistic-topologic approaches to network analysis have been devised that examine both the regularity and randomness of drainage networks (Shreve 1966; Smart 1968). The random topology models can readily explain many of Horton's laws. Horton's laws are not actually laws in the strictest sense, but merely expressions of the most probable states of network composition.

Horton's laws characterize the self-similarity in the organization and structure of river networks. This self-similarity has stimulated the use of fractals to characterize river networks (see FRACTAL). Fractals are objects with self-similar geometry, retaining similar organization and complexity over a range of scales. The planform river network when characterized as a fractal has a fractal dimension between one (linear features) and two (filling a two-dimensional space), that can be related to Horton's bifurcation and length ratios (Tarboton *et al.* 1988; La Barbera and Rosso 1989). Hack (1957) first noted an apparent dimensional inconsistency between the lengths of the mainstream and drainage area of a river basin. This can imply elongation with increasing basin size, an idea inconsistent with self-similarity but that has been advanced by some, or that individual streams are themselves fractal with dimension between 1.1 and 1.2. There have been suggestions that space filling is a constraint on the organization of river networks, because in general they should drain an entire two-dimensional area. This leads to a constraint that implies relationships between Horton's length, bifurcation and area ratios and the fractal dimension of individual streams.

The uniting of drainage network configuration and flow characteristics has been proposed in the theoretical framework of the optimal channel network. An optimal channel network is one in which there is energy minimization in the whole and parts of a drainage network. Three principles of optimal channel networks are that energy expenditure in every link is minimized for the transportation of a given discharge, equal energy is expended per unit area everywhere within the network, and energy expenditure is minimized for the network as a whole. A combination of these principles is sufficient to explain the tree-like structure of drainage networks and the empirical relationships describing network organization.

FIRST-ORDER STREAMS and drainage basins are substantial components of river basins. At the upper end of a first-order channel is an unchannelled HILLSLOPE HOLLOW or zero-order drainage basin. Nearly one-half the length of a river basin's drainage network may be contained in its first-order links, and first-order basins can contain 50 per cent of a river basin's area. It is within such small watersheds that runoff produced on hillslopes concentrates into streamflows that initiate channel formation. Low-order basins exhibit the tightest HILLSLOPE-CHANNEL COUPLING and competition between hillslope processes (see HILLSLOPE, PROCESS) and channel processes.

Basin development and evolution

The expression of drainage basin organization is a spatial characterization of drainage basin condition at a point in time. Drainage basins and networks, however, are not static and change over time due to external influences and internal COMPLEX RESPONSES. To explore drainage basin development, geomorphologists have used three different methods: space-for-time substitution, experimental studies and computer simulation modelling. Early studies of drainage basin evolution were based upon space–time substitution, i.e. the ERGODIC HYPOTHESIS. Maps of different drainage basins in progressive stages of development were ordered to depict a temporally evolving basin undergoing advancing stages of evolution. EXPERIMENTAL GEOMORPHOLOGY has been employed through the use of rainfall simulators raining over 'sandboxes' with drainage system development documented through detailed mapping and time-lapse photography. With advances in computer technology, empirical and theoretical concepts have been implemented into computer models that simulate long-term drainage basin evolution.

Although modes of basin evolution depend upon specific boundary conditions and driving factors, several major steps in basin development can be specified. With the assumption that a basin originates on a smooth surface, its channel network grows through the processes of initiation, elongation and elaboration. After initial development of a skeletal network, a few streams elongate to extend in parallel fashion across much of the length of the surface to form a low drainage density network. Over time, downcutting of these elongated stream channels causes the elaboration of the network through the addition of tributary streams, with the concomitant increase in drainage density. During these initial phases of basin development, drainage density increases rapidly and SEDIMENT LOAD AND YIELD from the basin are high.

Eventually, the channel network reaches a period of maximum extension in which stream elongation through HEADWARD EROSION and infilling by hillslope processes reach an equilibrium condition. Smaller drainage basins become integrated into larger ones through the mechanism of RIVER CAPTURE. As erosion continues and the entire basin drops closer in elevation to BASE LEVEL, the process of network abstraction occurs. Lowered stream slopes reduces STREAM POWER and stream EROSIVITY thereby allowing hillslope processes (see SOIL CREEP and SLOPEWASH) to infill low-order streams and abstract them from the drainage network.

Except in experimental models, rarely will a drainage basin originate on a flat, sloping surface of uniform material, so evolution of real drainage basins will be much more complex than the above model depicts. Also, there is no timescale associated with the evolutionary model described above, but longevity of a drainage basin and its drainage system can be correlated with increasing basin size. Large river basins may persist for tens of millions of years. During such long time frames, changing climatic conditions and tectonic events can so alter conditions that completed evolutionary stages are never fully realized. Channel networks can expand and contract through upstream and downstream migration of channel heads as climate changes over periods of decades to thousands of years. Tectonic events can drop or raise the base level for a drainage basin and initiate a new cycle of headward erosion or halt an existing erosional stage (see TECTONIC GEOMORPHOLOGY).

Also, the channel network evolution model presented above does not account for the causal processes of network development and evolution. Different processes are responsible for channel initiation and evolution in different terrain and environments. In steep terrain, LANDSLIDES may result in the initiation of channels, but in low-gradient basins, headward erosion of the channel head because of changed CONTRIBUTING AREA hydrological conditions may be the primary method of network growth. Though simple descriptive models can specify the overall pattern of evolution, detailed circumstances of basin evolution will be quite variable from one basin to another and may require complex computer simulation models to fully understand.

Geologic and climatic influences

Seeking to explain the regularity exhibited by drainage networks has guided much of the geomorphological interest in these features. Although such explanations provide theoretical bases for network pattern regularity when boundary conditions are uniform in space and invariant over time, actual drainage networks evolve under spatially and temporally varying conditions. Therefore, variability in DRAINAGE PATTERNS can arise because processes defined by geomorphological laws (see LAWS, GEOMORPHOLOGICAL) are operating within non-uniform environmental conditions. In particular, geology and climate have profound effects upon the processes and characteristics of drainage basins and drainage networks. Some have suggested that it would be more beneficial to seek relationships between geology and drainage network characteristics than refine sophisticated theories that disregard such an obvious control (Blöschl and Sivapalan 1995: 282).

The effect of geology upon drainage basin characteristics is difficult to quantify, but geology is nonetheless a controlling factor on drainage basin form and development at multiple scales. Empirical studies have identified relationships between drainage density and bedrock lithology, and rock type can be responsible for many drainage pattern details. Drainage basins underlain by shale or siltstone tend to have higher drainage densities than other lithologies due to low infiltration rates and high production of overland flow. Areas dominated by lithologies with high infiltration rates, such as dune sands, often have poorly defined drainage systems and very low drainage densities. Dendritic stream patterns are common on shales and siltstones, as they are weak rocks that provide limited lithologic resistance to erosion. In geologic formations comprised of stronger rock types, such as sandstones and granites, joint patterns frequently control network pattern because fractured rock along joints has greater ERODIBILITY.

Geologic structures also influence drainage basin shape and drainage network form. At a large scale, river basins may be coincident with geologic or structural basins, with the basin's drainage divide corresponding to the ridgetops of surrounding, uplifted mountain chains. For smaller drainage basins, folded geologic strata can control drainage basin shape and drainage pattern where weaker strata are more readily eroded. Trellis and annular drainage patterns are common in these circumstances. Multiple, low-order streams flow from divides to strike valleys, medium-order streams occupy longitudinal or strike valleys, and master high-order streams run across the strike of more resistant folds in superimposed traversal valleys. Drainage basins may be irregularly shaped with drainage divides following the ridgelines of HOGBACKs or CUESTAs formed by more resistant rock strata. Faults, similar to joints, are areas of weaker, fractured rock and are frequently occupied by streams in superimposed valleys.

Climatic effects upon drainage basin development are promulgated through their controls upon erosion processes. The most critical effect of climate upon drainage basin form and development is through the influence of precipitation and temperature upon vegetation cover. Density of vegetation cover is a predominant control upon erosion and sediment delivery to the drainage network by decreasing soil erodibility. Channel head location, and thereby drainage density, may be dependent upon vegetation because of increased shear stress required for channel formation where vegetation cover is dense. Drainage densities typically are low in arid regions where there is little runoff, are highest in semi-arid regions where sparse vegetation cover does little to prevent channel initiation, are low in moderate precipitation environments where vegetation cover restricts channel development, and can be high even with heavy vegetation cover in areas with high annual rainfall and runoff.

References

Blöschl, G. and Sivapalan, M. (1995) Scale issues in hydrological modeling: a review, *Hydrological Processes* 9, 251–290.

Davis, W.M. (1899) The Geographical Cycle, *Geographical Journal* 14, 481–504.

Hack, J.T. (1957) *Studies of Longitudinal Stream Profiles in Virginia and Maryland*, Washington: US Geological Survey Professional Paper 294B.

Horton, R.E. (1932) Drainage basin characteristics, *Transactions of the American Geophysical Union* 13, 350–361.

——(1945) Erosional development of streams and their drainage basins; hydrophysical approach to quantitative morphology, *Geological Society of America Bulletin* 56, 275–370.

La Barbera, P. and Rosso, R. (1989) On the fractal dimension of stream networks, *Water Resources Research* 25, 735–741.

Shreve, R.L. (1966) Statistical law of stream numbers, *Journal of Geology* 74, 17–37.

Smart, J.S. (1968) Statistical properties of stream lengths, *Water Resources Research* 4, 1,001–1,014.

Strahler, A.N. (1957) Quantitative analysis of watershed geomorphology, *Transactions of the American Geophysical Union* 38, 913–920.

Tarboton, D.G., Bras, R.L. and Rodriguez-Iturbe, I. (1988) The fractal nature of river networks, *Water Resources Research* 24, 1,317–1,322.

Further reading

Gregory, K.J. and Walling, D.E. (1973) *Drainage Basin Form and Process*, New York: Wiley.

Knighton, D. (1998) *Fluvial Forms and Processes*, London: Arnold.

Leopold, L.B., Wolman, M.G. and Miller, J.P. (1964) *Fluvial Processes in Geomorphology*, San Francisco: W.H. Freeman.

Rodiguez-Iturbe, I. and Rinaldo, A. (1997) *Fractal River Basins*, Cambridge: Cambridge University Press.
Schumm, S.A. (1977) *The Fluvial System*, New York: Wiley.

SEE ALSO: GIS

CRAIG N. GOODWIN AND DAVID G. TARBOTON

DRAINAGE DENSITY

Drainage density is defined as the cumulative length of all stream channels in a drainage basin, divided by the drainage basin area. The dimension of drainage density is the inverse of length. In principle, drainage density is a fundamental measure of landscape dissection and half its reciprocal is an average measure of length of overland flow (or distance from divide to the nearest channel). Drainage density has been shown to vary as a function of climate, past climate conditions, biomass, parent material, lithology, relief, time and land use. Unfortunately, no consistent relations have been demonstrated from region to region and Schumm (1997) comes to some discouraging conclusions from a number of detailed basin studies. Nevertheless, drainage density is an important geometric parameter for channel networks as it determines the spacing of channels, the length of hill slopes, the maximum length scale of slope failures and reflects the processes governing landscape dissection. The hydrological response of a channel network is strongly influenced by drainage density and sediment erosion rates have been linked to channel spacing. There are relatively few studies of how drainage density varies with time. Flume table experiments (Schumm *et al.* 1987), studies of land fills (Schumm 1956), glacial tills (Ruhe 1952), coastal terraces (Kashiwaya 1987) and drainage networks on an anticlinal fold which has been progressively uplifted during the past 250,000 years (Talling and Sowter 1999) are representative examples of such studies.

It should be pointed out that the definition of drainage density begs at least two questions:

1 *How is a stream channel defined?* This is a deceptively simple question which does not have a simple answer. Montgomery and Dietrich (1992) suggest that there is an empirically defined topographic threshold for channel head locations which defines the boundary between essentially smooth and undissected slopes and the valley bottoms to which they drain. They derived an experimental relation of drainage area versus local slope for channel heads, unchannelled valleys and low-order channel networks from different study areas. Local slope was measured in the field and drainage area was determined from topographic maps. High slopes generate channels from smaller basin areas and lower slopes require larger basin areas to produce a channel. At the same time, spatial heterogeneity, reflecting the controlling factors listed above, introduces variability into these relations. An empirical, field-based definition of a channel uses the presence of fluvial incision and one or two stream banks, but finger tip tributaries are often indeterminate in the field.
2 *What is the relation between stream channels drawn on maps or stream channels detectable on air photographs and actual stream channels on the ground?* There is a basic stream channel network, which is composed of perennial streams and which expands and contracts with runoff, or there is the active channel network, which is composed of ephemeral, intermittent and perennial streams. In addition, the use of contour crenulations as evidence of the presence of channels will result in the inclusion of parts of valleys that do not contain active channels. Clearly, the largest problems concern the uppermost finger tip tributaries of a drainage basin. When air photos are used, there are further problems of visibility below tree canopies and as always, the scale of the photograph or the map will be a constraint on the resolution achievable. In sum, what is measured by one investigator may not be the same phenomenon as that which is measured by another.

If we assume that the identification and measurement problem can be resolved, a variety of theoretical issues relating to drainage basin characterization and evolution can be broached. Strahler (1956) in developing his view of the drainage basin as an open system tending to achieve a steady state of operation asked how to predict erosional or aggradational response by drainage basins when land use or climate changed. Central to his theoretical discussion was the role of drainage density. He argued that because drainage density is the most valuable scale index with respect to degree of dissection of a basin that it should be possible to express drainage density as a function of several variables that control the evolution of the basin. These variables he deduced in part from Horton (1945) as runoff intensity, an erosion proportionality factor, slope gradient, relief, kinematic viscosity of runoff and acceleration of gravity. Through application of the Buckingham Pi Theorem, he reduced the equation to a function containing four dimensionless groups:

1 the product of drainage density and relief (the ruggedness number)
2 the product of runoff intensity, erosion proportionality factor and slope gradient (the Horton number)
3 the product of runoff intensity, kinematic viscosity and relief (a basin Reynolds number)
4 the square of runoff intensity divided by relief times acceleration of gravity (a basin Froude number).

By solving for drainage density, drainage density is shown to be inversely proportional to relief times a function of the Horton, Reynolds and Froude numbers. The challenge of solving this function has still not been met, though the topic of drainage basin transformation has been put onto a more rational basis.

Melton followed up Strahler's analysis with one paper on drainage basin growth models (1958a) and another on the theory of variable systems (1958b), both of which relied heavily on the assumption of the importance of drainage density. Melton (1958a) demonstrated a close relation between F (stream frequency = number of streams per unit area) and D (drainage density) for mature basins with a wide range of orders, valley side slope angles, climates and rock types. The relation, subsequently known as Melton's Law, is of the form $F = 0.694\ D^2$. Shreve (1967) revisited this relation using links instead of streams and found a related term $K = 0.667$ for topologically random networks. The dimensionless ratio F/D^2 varies inversely with valley side slope and basin relief (where area and channel length are held constant) and is interpreted as a measure of the completeness with which a channel system fills a basin outline. For an ideal basin of 1 mi^2, Melton's Law is postulated to be a growth model. This argument is predicated on the assumption that many basins measured at one point in time can be considered equivalent to the behaviour of a single basin over time. The approach taken in Melton (1958b) is different. He arranges fifteen variables of geomorphic, surficial and climatic elements into two related variable systems on the basis of correlation coefficients for every possible pair in the study of 59 drainage basins. 'Melton's ambitious field program of data collection, coupled with his analysis of the interrelations of the components of a drainage basin and the variables that influence morphology, is a model for future geomorphic studies' (Schumm 1977: 180). The high correlation of drainage density with per cent bare area and a precipitation effectiveness index fits well with the Horton theory of drainage density as a function of the resistivity of the surface to erosive forces, determined in part by vegetation which in turn determines the mean length of overland flow. Keylock (personal communication) has shown that the most frequently cited of Melton's contributions are Melton's Law (1958a) and his correlation structure approach to geomorphology (Melton 1958b), both of which emphasize the importance of drainage density.

References

Horton, R.E. (1945) Erosional development of streams and their drainage basins: hydrophysical approach to quantitative morphology, *Geological Society of America Bulletin* 50, 275–370.

Kashiwaya, K. (1987) Theoretical investigation of the time variation of drainage density, *Earth Surface Processes and Landforms* 12, 39–46.

Melton, M.A. (1958a) Geometric properties of mature drainage systems and their representation in an E4 phase space, *Journal of Geology* 66, 35–56.

——(1958b) Correlation structure of morphometric properties of drainage systems and their controlling agents, *Journal of Geology* 66, 442–460.

Montgomery, D.R. and Dietrich, W.E. (1992) Channel initiation and the problem of landscape scale, *Science* 255, 826–830.

Ruhe, R.V. (1952) Topographic discontinuities of the Des Moines lobe, *American Journal of Science* 250, 46–56.

Schumm, S.A. (1956) Evolution of drainage systems and slopes at Perth Amboy, New Jersey, *Geological Society of America Bulletin* 67, 597–646.

——(1977) *Drainage Basin Morphology*, Pennsylvania: Dowden, Hutchinson and Ross.

——(1997) Drainage density: problems of prediction and application, in D.R. Stoddart (ed.) *Process and Form in Geomorphology*, 15–45, London: Routledge.

Schumm, S.A., Mosley, M.P. and Weaver, W.E. (1987) *Experimental Fluvial Geomorphology*, New York: Wiley.

Shreve, R.L. (1967) Infinite topologically random channel networks, *Journal of Geology* 75, 178–186.
Strahler, A.N. (1956) The nature of induced erosion and aggradation, in W.L. Thomas (ed.) *Man's Role in Changing the Face of the Earth*, 621–638, Chicago: University of Chicago Press.
Talling, P.J. and Sowter, M.J. (1999) Drainage density on progressively tilted surfaces with different gradients, Wheeler Ridge, California, *Earth Surface Processes and Landforms* 24, 809–824.

SEE ALSO: dynamic geomorphology; stream ordering

OLAV SLAYMAKER

DRAINAGE PATTERN

Because river channels concentrate surface flow and erode into landscape more efficiently than other processes, new channels tend to persist from the pattern initially developed and are subsequently hard to alter. A collection of river channels joined together is called a drainage network, how it is laid out on the ground in plan view is called the drainage pattern, and the channels together with all the land surface that drains to the channel is called the DRAINAGE BASIN. Channels when they join normally do so in an accordant manner, the channels join without a sudden break in elevation (sometimes called Playfair's Law), unless they occupy unmodified glacial terrain, in which case a discordant junction is called a HANGING VALLEY. Subsequent adjustments to networks and patterns may occur when rivers are close together, and the divides between them may be broached by erosion or overflow, or underground drainage may divert water from one system to another prior to there being a surface connection of the rivers. Exploitation of geological weakness by surface erosion eventually causes the overall drainage pattern to reflect the patterns of weakness in the underlying rocks. Major joints and fracture zones may influence subsurface as well as surface drainage and tend to localize major channels. Adjustments by divide erosion and breaching (river piracy, RIVER CAPTURE, diversion) will be most common early in the history of a landscape when relative RELIEF is least. Adjustments by underground diversion may take longer to become active features because large subterranean networks, usually developed in soluble bedrock such as carbonates (KARST terrains), are needed to divert substantial drainage (abstraction). Subterranean diversion is favoured by increasing local relief in the drainage which may permit steeper hydraulic gradients between adjacent channels. Drainage patterns which derive their water entirely from regions external to the locality in question – such as the Nile River in Egypt – are called exoreic, and systems which drain to a central closed depression such as the Jordan River to the Dead Sea, and the basin draining to the Great Salt Lake in Utah – are endoreic.

The nineteenth and early twentieth-century geomorphologist W.M. Davis (1889, 1899) developed an elaborate scheme to describe the components of a river drainage network as they related to stages in its physiographic development. Of those terms, those which remain in common use are *consequent* and *subsequent*. *Consequent* streams are those that develop on the initial land surface in response to regional slope and any random surface declivities. Because they must eventually follow regional slope they usually reflect the tectonic framework of uplift, rather than details of the underlying geology. The term has normally been applied to large streams, but can also describe initial drainage on any new surface – such as recently glaciated terrain. *Subsequent* streams describe streams which, through geologic time, have been able to exploit differences in the relative erodibility of the underlying geology as the drainage system incises slowly into the uplifted block of land. Typically they develop along the geological strike exploiting, for example, weak shales or clays exposed between stronger formations (e.g. sandstones or limestones) in a sequence of sedimentary rocks so that long continued weathering and subaerial erosion over CYCLIC TIME etches out a skeleton of the underlying geology – thus the ridges and valleys of the Appalachian Ranges along the eastern side of North America reveal the folded structures; less dramatically the valleys and escarpments of southern England and northern France also reveal the geological structures. The effect is even more dramatic in dry climates with no masking vegetation. In igneous and metamorphic terrain master joints and shear zones may provide weakness to exploit (Figure 48c). Faults and fault zones, with heavily fractured rocks allowing access for weathering agents, are often weak zones in any geological terrain.

Because consequent drainage flows down the regional slope regardless of local variations in geology, such streams are often used to reconstruct the initial stages of a landscape. However,

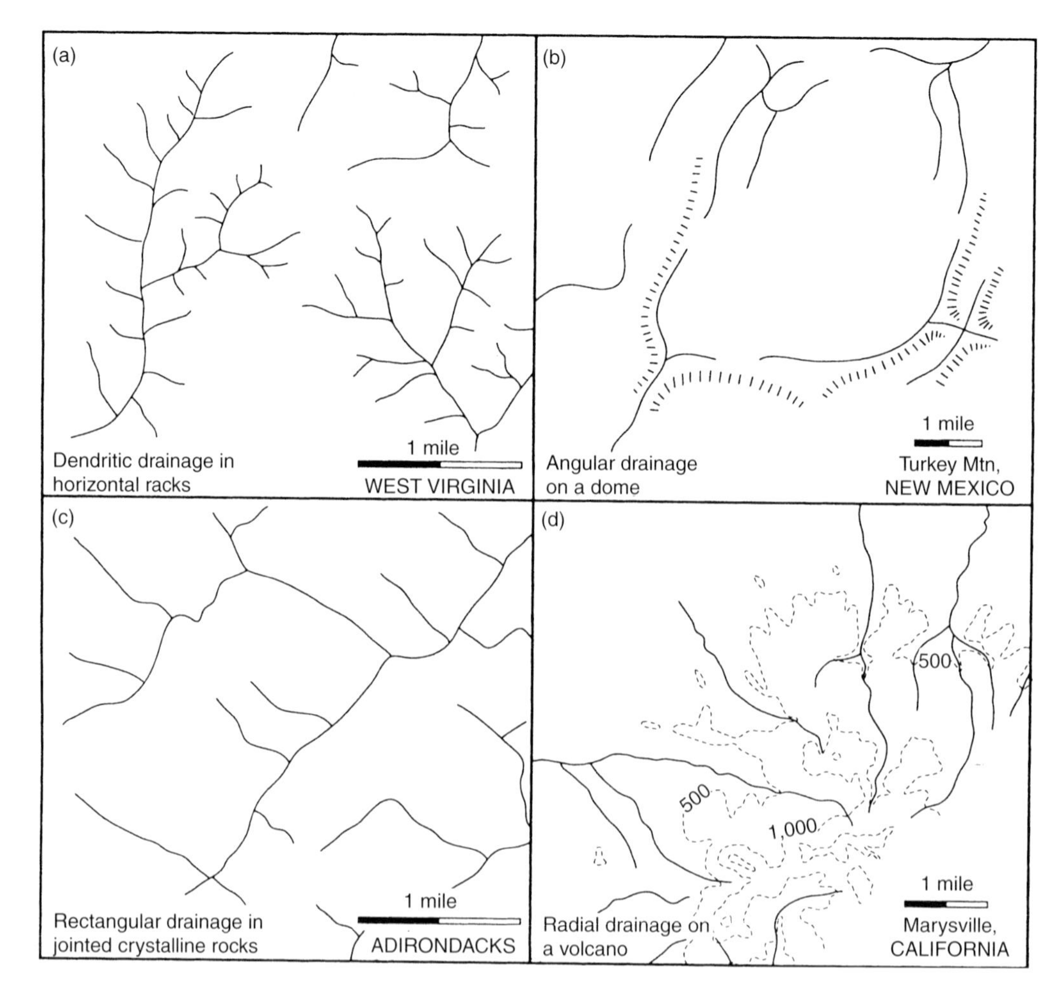

Figure 48 Drainage patterns in relation to topography and geological structures

even consequent streams can be disrupted by continued uplift, with geological structures growing upwards into the overlying streams. If the river channel can erode its bed fast enough to maintain a continuous downslope against the rising land, the river is called antecedent. As a result a river may be seen to have cut a channel, often seen as a deep gorge (see GORGE AND RAVINE), through a prominent topographic ridge around which it might have otherwise been forced to flow. The north to south segments of the Ganges and the Brahmaputra in the Himalaya have been cut in response to, and across the rising folds of, the mountain system.

On occasion though, the rising structure blocks the channel and causes upstream ponding, whose new outlet may provide an entirely new pattern. Complete reversal is possible too. The Amazon originally drained to the Pacific, but its course was reversed by the rising Andean ranges. A related condition, however, is when a regional river system, developed for example on gently tilted sedimentary strata, slowly erodes away that sediment and then erodes into a very different geological underlay. If the sedimentary cover rocks are lying unconformably upon the rocks below, the drainage pattern is said to be superimposed or superposed (Tarr 1890). It is doubtful in practice that either antecedence or superimposition are ever pure conditions because rarely can the full tectonic history of the region be known (Smith *et al.* 1999).

Also, large-scale topographic patterns characterizing the initial topography of the area may be reflected, as for example with radial drainage, such as in the English Lake District where original drainage lines have been greatly accentuated by glacial deepening. Part of a miniature example of radial drainage developed on a volcano is shown in Figure 48d.

Davis developed many terms for other parts of the drainage system as they related to a supposed sequence of drainage and landscape development, and with respect to the original regional slope. These other terms are: insequent, resquent, obsequent; but they have fallen into disuse. Full definitions are available in Lobeck (1939: 171). Of these, insequent streams describe the myriad of streams for which no discernible control can be detected, and which give rise to dendritic patterns (Figure 48a).

Despite the variations in apparent patterns (Figure 48) the patterns that matter most to the operation of the system are the internal structure of connections, and the plan of the DRAINAGE BASIN on the ground. Circular basins concentrate flow more rapidly, and generate larger peak flows than elongate basins. The structural arrangement of channels tends to reflect that of the ground plan – dendritic or vein-like structures being found usually in oval and round basins with homogeneous bedrock (Figure 48a). The Kentucky region with nearly level sedimentary rocks, and lying beyond the glacial limit, has often been used as a basis for comparison with random or randomly generated drainage networks (Mark 1983).

References

Davis, W.M. (1889) The rivers and valleys of Pennsylvania, *National Geographic Magazine* 1, 183–253.

——(1899) The Geographical Cycle, *Geographical Journal* 14, 481–504.

Lobeck, A.K. (1939) *Geomorphology: An Introduction to the Study of Landscapes*, New York: McGraw-Hill.

Mark, D.M. (1983) Relations between field-surveyed channel networks and map-based geomorphometric measures, Inez, Kentucky, *Annals of the Association of American Geographers* 73(3), 358–371.

Smith, B.J., Whalley, W.B. and Warke, P.A. (1999) *Uplift, Erosion and Stability: Perspectives on Long Term Landscape Development*, Geological Society Special Publications, 162, London: Geological Society of London.

Tarr, R.S. (1890) Superimposition of drainage in central Texas, *American Journal of Science* 3rd Series 40, 359–361.

Further reading

Brown, E.H. (1960) *The Relief and Drainage of Wales*, Cardiff: University of Wales.

Johnson, D.W. (1927) *Stream Sculpture on the Atlantic Slope*, New York: Columbia University Press.

Woolridge, S.W. and Linton, D.L. (1955) *Structure, Surface and Drainage in South-East England*, Institute of British Geographers, No. 11, London: George Philip.

KEITH J. TINKLER

DRAPE, SILT AND MUD

A thin deposit of waterlain silt or mud coating a pre-existing morphological feature. Drapes generally grade upwards, sometimes exhibiting internal laminations. They are typically several centimetres thick, though their form and composition vary spatially and temporally, in response to factors including sediment supply, and the hydrological (fluvial or current) regime.

Drapes are indicative of tidal/subtidal settings and are considered one of the most distinctive features of such an environment. Typical tidal regimes exhibit ebb-flood cycles in which one current is more dominant than the other. During periods of tidal dominance various bedforms are produced (e.g. sand bars, ripples and dunes) that are characteristic of the tidal regime. However, there is a period of time during high tide and low tide where no dominant direction of flow exists (termed the slackwater period). During this short period, the suspended sediment of the water may fall and settle on the pre-existing features formed during the dominant tidal period. The subsequent current stage may partly rework the mud or clay drape producing an erosive reactivation surface, though the cohesive nature of the fine clay-rich drapes commonly protects against tidal erosion and preserves the drape. Over time, continued preservation of alternating tidal (sand deposited) and slackwater deposits (mud/clay drapes) produces sand/mud couplets, also called tidal rhythmites. This systematic deposition of tidal rhythmites has allowed detailed reconstruction of past tidal regimes (e.g. Visser 1980), and are particularly distinctive of inshore tidal environments.

Drapes may also form in fluvial environments, particularly within rivers that exhibit seasonal flow and flooding. As flooding wanes, the clay/mud settles on river levees, etc. thus signifying slackwater periods and forming drapes.

Reference

Visser, M.J. (1980) Neap-spring cycles reflected in Holocene subtidal large-scale bedform deposits: a preliminary note, *Geology* 8, 543–546.

STEVE WARD

DRUMLIN

Drumlins are roughly ovoid-shaped hills dominantly composed of glacial debris. Typically, they occur within groups or fields of several thousands, exhibit strong, *en echelon*, long-axis preferred orientation paralleling the main direction of ice flow. The classical shaped drumlin usually has a steeper stoss end and a tapered lee-side, however variants on this shape are perhaps more common than the classical shape itself. Drumlins may vary from 5 to 200 m in height, 10 to 100 m in width, and overall from 100 m to several kilometres in length (see e.g. Mills 1987). Interestingly, few modern drumlins appear from beneath modern-day ice masses other than on James Ross Island, Antarctica, in the proglacial areas of some Icelandic outlet glaciers, and at the Bifertensgletscher, Switzerland. Vast drumlin fields – numbering in thousands – exist, for example, in Canada, Estonia, Finland, Ireland, Germany, Poland, Russia and the USA. The topographic locations within which drumlins are found are many and varied. Drumlins occur in both lowland and highland terrains, beneath ice sheets and valley glaciers, close to terminal MORAINES and may, in places, appear contiguous with these moraines, while elsewhere they occur on the edge of ice sheet centres. Occasionally, a radiating pattern can be observed within a drumlin field that has been interpreted as evidence of basal crevasse infilling owing to divergent ice flow close to an ice margin. It has also been suggested that drumlins, in association with Rogen and fluted moraines, may be related to deformable beds beneath ice sheets, linked to fast basal ice (>500 m a^{-1} mm) and a preferential location within ice streams. Limited relationships appear to occur between drumlins and topography.

Drumlins are composed of a vast range of sediment types of varied provenance, containing an array of sediment structures and forms (Figure 49). In the past, drumlins were mistakenly perceived as being composed almost exclusively of subglacial tills; although many drumlins contain stratified sediment. Drumlins composed of stratified sediments are known, for example, near

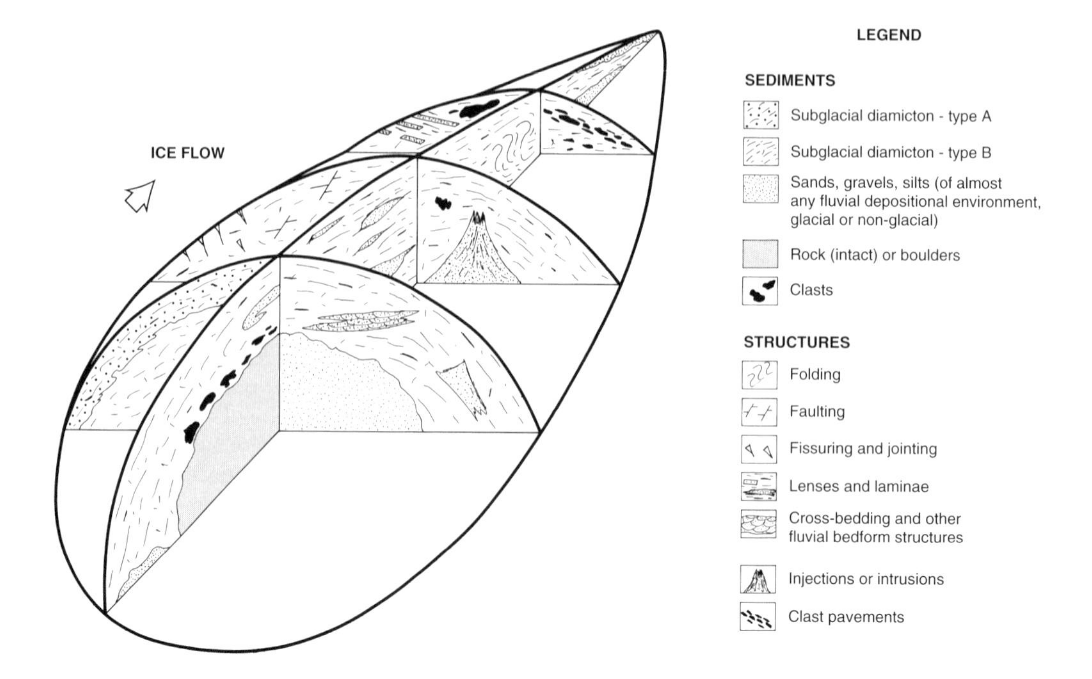

Figure 49 A general model of internal sediments and structures found within drumlins

Velva, North Dakota and Livingstone Lake, Saskatchewan. Also, stratified sediments can occur in individual drumlins that often 'sit' adjacent to till drumlins as in Peterborough, Ontario. Many drumlins have observable cores of bedrock, boulder dykes and other non-glacial nuclei around which subglacial debris has accreted or been emplaced. In some cases, drumlin or drumlinoidal forms can be observed 'carved' from bedrock in the form of roc-drumlins. However, most drumlins do not appear to have obvious cores around which they have been 'built' and these forms remain enigmatic in origin.

Clast fabrics within drumlins appear, in some cases, to follow the outer morphology of the drumlin, while others exhibit a transverse, 'herringbone' style pattern. In many cases the complexity of internal sedimentological structures provides a random fabric orientation. Drumlins exhibit such a wide complexity of form and internal composition that it is impossible to characterize an 'ideal' drumlin. Many drumlins are found lying on top or obliquely across other larger drumlin forms (megadrumlins). Drumlin shapes may vary enormously and may reflect formative processes or simply post-depositional subaerial mass movement. Many drumlin fields progressively change as part of a continuum of bedforms, thus drumlin genesis would appear tied to subglacial environments conducive to Rogen and fluted moraine formation. The question of drumlin formation has attracted an array of research. In terms of drumlin formation, it is germane to consider the 'conditions' that must be met by any formative hypotheses, assuming that a single explanation does exist for such a diverse landform/bedform type.

Any explanation of drumlin formation must address the following issues: (1) the diverse location of drumlins and their propensity to occur in 'fields'; (2) the differing shape and morphology of drumlins; (3) the range of sediment types and structures within drumlins; (4) the existence of rock-cored and non-rock-cored drumlins, often in proximity to each other; (5) the presence of drumlins in bedform continua in some, but not all, cases; (6) the relationship of drumlins to subglacial glacidynamics and hydraulics; (7) the chronology of drumlin formation whether drumlins form simultaneously as a single field or develop into a field by repeated 'overprinting' in a single glacial phase or repetition over several glacial phases; (8) stages of drumlin development whether formed *en masse* or by gradual accretion in a single continuous event or interrupted accretionary events; and, finally, (9) a 'trigger' mechanism(s) that is operative in certain specific conditions yet not under others.

At present, three main groups of drumlin-forming hypotheses can be identified:

1. By moulding of previously deposited material within a subglacial environment in which a limited amount of subglacial meltwater activity occurs (possibly where a frozen bed transforms to a melted bed; Menzies 2002). Meltwater may influence moulding and deformational processes by acting either as a lubricating basal film at the upper ice–bed interface, or as porewater reducing subglacial sediment effective stresses. Debris is moulded by direct deformation of previously deposited sediment (both glacial and non-glacial) into drumlinoidal shapes by basal ice contact following smearing-on or sculpting process(es).
2. By anisotropic differences in the subglacial debris under melted-bed conditions owing to: (a) dilatancy (Smalley and Unwin 1968); (b) porewater dissipation; (c) localized freezing; (d) helicoidal basal ice flow patterns (Aario 1977); or (e) localized subglacial debris deformation (Boulton 1987; Menzies 1989). Within this specific group, meltwater activity is of limited impact, whereas porewater is considered critical in local bed debris deformation. Changing stress field and/or stress/strain histories owing to transient basal glacidynamics locally affecting subglacial debris rheology are the important parameters in determining whether drumlins begin to form or not.
3. By the influence of active basal meltwater (under frozen bed conditions) carving cavities beneath an ice mass and later infilling with assorted but predominantly stratified sediment or by the subglacial meltwater erosion of already deposited sediment at the upper ice–bed interface (Dardis and McCabe 1987; Sharpe 1988), or through the entire drumlin, or the sculpting by fluvial processes of previously deposited sediment (Shaw *et al.* 2000). This hypothesis demands meltwater flow of catastrophic discharges from beneath certain areas of an ice mass across the upper ice–bed interface yet permitting the overall ice mass to remain glacidynamically stable. This form of

drumlin development, as with the hypothesis in (1), requires a two-stage process of initiation, beginning with either a pre-formed cavity or pre-existing sediment at the upper ice–bed interface. The latter stage need not be linked directly to the former stage, therefore in some cases although conditions may be suitable for initiation for the first stage, the second stage may not continue toward the critical point (trigger) of drumlin development.

In all these hypotheses the conditions at the subglacial interface(s) are the key to subsequent drumlin formation and, in the long-term, to drumlin 'survival'. A complex relationship must exist between basal glacidynamics, subglacial sediment rheology, and hydraulics for any particular area of ice bed. Fluctuations in state or stress levels or meltwater production and pathways will affect all other parameters to some extent. Certain fluctuations may cross critical thresholds that cannot be reversed, while others may exhibit varying degrees of hysteresis. The likelihood or otherwise of subglacial conditions occurring in any or all these hypotheses remains a fundamental, ongoing, research problem.

References

Aario, R. (1977) Associations of flutings, drumlins, hummocks and transverse ridges, *Geojournal* 1, 65–72.
Boulton, G.S. (1987) A theory of drumlin formation by subglacial sediment deformation, in J. Menzies and J. Rose (eds) *Drumlin Symposium*, 25–80, Rotterdam: A.A. Balkema.
Dardis, G.F. and McCabe, A.M. (1987) Subglacial sheetwash and debris flow deposits in Late Pleistocene drumlins, Northern Ireland, in J. Menzies and J. Rose (eds) *Drumlin Symposium*, 225–240, Rotterdam: A.A. Balkema.
Glückert, G. (1973) Two large drumlin fields in central Finland, *Fennia* 120.
Menzies, J. (1989) Subglacial hydraulic conditions and their possible impact upon subglacial bed formation, *Sedimentary Geology* 62, 125–150.
——(2002) Ice flow and hydrology, in J. Menzies, (ed.) *Modern and Past Glacial Environments*, 79–130, Oxford: Butterworth-Heinemann.
Mills, H.H. (1987) Morphometry of drumlins in the northeastern and north-central USA, in J. Menzies and J. Rose (eds) *Drumlin Symposium*, 131–147, Rotterdam: A.A. Balkema.
Sharpe, D.R. (1988) Late glacial landforms of Wollaston Peninsula, Victoria Island, Northwest Territories: product of ice-marginal retreat, surge, and mass stagnation, *Canadian Journal of Earth Sciences* 25, 262–279.
Shaw, J., Faragini, D.M., Kvill, D.R. and Rains, R.B. (2000) The Saythabasca fluting field, Alberta, Canada: implications for the formation of large-scale fluting (erosional lineations), *Quaternary Science Reviews* 19, 959–980.
Smalley, I.J. and Unwin, D.J. (1968) Formation and shape of drumlins and their orientation and distribution, *Journal of Glaciology* 7, 377–480.

JOHN MENZIES

DRY VALLEY

A valley which is seldom, if ever at the present time, occupied by an active stream channel. Such valleys occur in a wide range of climatic and lithological environments, including extensive areas of Britain and Europe, where they have often been regarded as a product of intense incision under former periglacial conditions (Büdel 1982). There have been many different hypotheses put forward to explain why such valleys are dry (see Table 13; Goudie 1993).

The uniformitarian hypotheses require no major changes of climate or base level, merely the operation of normal processes through time; the marine hypotheses are related to base-level changes; and the palaeoclimatic hypotheses are associated primarily with the major climatic changes of the Pleistocene. British dry valleys (Plate 36) show a considerable range of shapes and sizes, from mere indentations in escarpments, to great winding chasms like Cheddar Gorge in the Mendips. Many, but not all, are formed in carbonate rocks.

Other dry valleys include those that occur in the world's warm deserts and which are relicts of former pluvial conditions and of extensive groundwater sapping (e.g. the MEKGACHA of the Kalahari (Nash 1996). At the other extreme, there are the famous dry valleys of Antarctica, which were cut by former outlet glaciers draining from the Polar Plateau (Summerfield *et al.* 1999).

References

Büdel, J. (1982) *Climatic Geomorphology*, Princeton, NJ: Princeton University Press.
Goudie, A.S. (1993) *The Nature of the Environment*, Oxford: Blackwell.
Nash, D.J. (1996) On the dry valleys of the Kalahari: documentary evidence for environmental change in central southern Africa, *Journal of African Earth Sciences* 27, 11–25.

Table 13 Hypotheses of dry valley formation

Uniformitarian	
1	Superimposition from a cover of impermeable rocks or sediments
2	Joint enlargement by solution through time
3	Cutting down of major through-flowing streams
4	Reduction in catchment area and groundwater lowering through scarp retreat
5	Cavern collapse
6	River capture
7	Rare events of extreme magnitude
Marine	
1	Non-adjustment of streams to a falling Pleistocene sea level and associated fall of groundwater levels
2	Tidal scour in association with former estuarine conditions
Palaeoclimatic	
1	Overflow from proglacial lakes
2	Glacial scour
3	Erosion by glacial meltwater
4	Reduced evaporation caused by lower temperatures
5	Spring snowmelt under periglacial conditions
6	Runoff from impermeable permafrost

Plate 36 A dry valley, the Manger, developed in the Vale of the White Horse near Wantage, southern England. The valley is developed in Cretaceous chalk and on the left side shows a series of parallel flutes or dells, which some investigators have proposed to be avalanche chutes

Summerfield, M.A., Stuart, F.M., Cockburn, H.A.P., Sugden, D.E., Denton, G.H., Dunai, T. and Marchant, D.R. (1999) Long-term rates of denudation in the Dry Valleys, Transantarctic Mountains, Southern Victoria Land, Antarctica based on in-situ-produced cosmogenic nuclides, *Geomorphology* 27, 113–129.

A.S. GOUDIE

DUNE, AEOLIAN

Aeolian dunes form part of a hierarchical system of bedforms developed in wind-transported sand which comprises: (1) wind ripples (spacing 0.1–1 m); (2) individual simple dunes or superimposed dunes on draa or compound and complex dunes (spacing 50–500 m); and (3) draa or compound and complex dunes (spacing >500 m). Most dunes occur in contiguous areas of aeolian deposits called ergs or sand seas (with an area of >100 km^2). Smaller areas of dunes are called dunefields (see SAND SEA AND DUNEFIELD). The majority of dunes are composed of quartz and feldspar grains of sand size, although dunes composed of gypsum, carbonate and volcaniclastic sand as well as clay pellets also occur.

The formation of areas of dunes is determined by the production of sediment of a range of suitable particle sizes, the availability of this sediment for transport by wind and the transport capacity of the wind (Kocurek and Lancaster 1999). Most dunes are derived from material that has been transported by fluvial or littoral processes. Important sources include marine and lacustrine beaches, dry lake basins, river floodplains and deltas. The availability of sediment for transport by wind is determined by its moisture content, vegetation cover, crusting and cohesion. The transport capacity of the wind is a cubic function of wind speed or surface shear stress above the transport threshold (see AEOLIAN PROCESSES). These conditions are satisfied in two main environments: (1) coastal areas with sandy beaches and onshore winds (e.g. the Atlantic coasts of north-west Europe, the Pacific north-west of North America, south-east and northeastern Australia and southern South Africa); and (2) subtropical and temperate desert areas.

Dune types

Aeolian dunes develop as a result of interactions between a granular bed (sand) and turbulent shearing flow (the atmospheric boundary layer). The resulting landforms are bedforms that are dynamically similar to those developed in subaqueous shearing flows (e.g. rivers, tidal currents). The morphology of aeolian dunes therefore reflects the characteristics of the sediment (primarily its grain size) and the wind (both the local shear stress, which determines local sand transport rates, and the long-term directional variability of the wind regime). Vegetation is a significant factor influencing the morphology of dunes in coastal dunefields as well as those in semi-arid and subhumid regions. Interactions with topographic obstacles may also result in dune formation.

Dunes occur in self-organized patterns that develop over time as the response of sand surfaces to the wind regime (especially its directional variability) and the supply of sand (Werner 1995). Development of these patterns is modulated by the effects of changes in climate and sea level on sediment supply, dune mobility and wind regime characteristics, often resulting in the formation of a series of different generations of dunes. The dune types described below represent the steady-state attractors of the aeolian transport system and can evolve from a wide range of initial conditions. The orientation of dunes with respect to the wind regime is another aspect of the self-organizing nature of the system, in which dunes are oriented to maximize the gross sand transport normal to the crest. Characteristic features of dune patterns include close correlations between the height and spacing of dunes and systematic spatial variations in dune type, orientation and sediment volume.

Despite the variety of different dune types and the multiplicity of local names that have been used to refer to them, satellite images show that dunes of essentially similar form occur in widely separated areas, and occur in five main morphologic types (Figures 50 and 51). The only dune form restricted to coastal areas is the foredune because it is an integral part of the complex of near shore processes forming the beach–dune system (Bauer and Sherman 1999). Three varieties of each dune type can occur: simple (the basic form), compound (superimposition of small dunes of the same type on larger dunes), and complex (superimposition of different dune types on the primary form, (e.g. crescentic dunes on linear dunes).

Relations between the occurrence of different dune morphological types and their wind regime environment indicate that the main control of dune type is the directional variability of the wind regime (Figure 52), which can be characterized by the ratio between the resultant (vector sum) of potential sand transport (RDP or resultant drift potential) and total potential sand transport (DP or drift potential). Sand grain size, vegetation cover, topography and sediment supply play subordinate roles in the majority of cases.

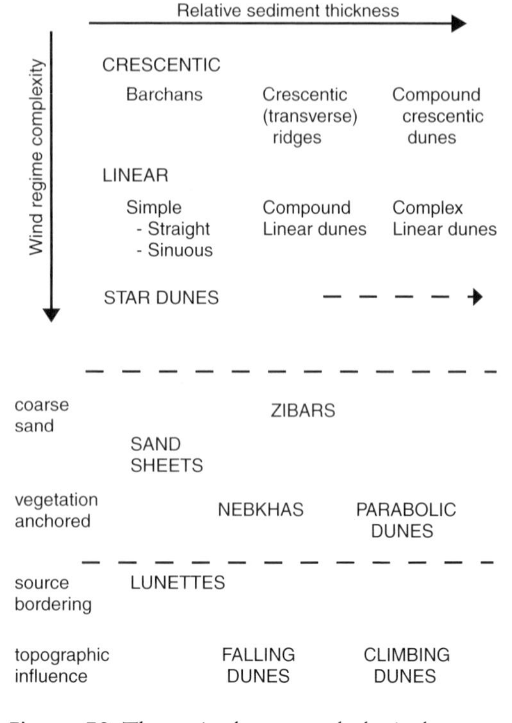

Figure 50 The main dune morphological types

The simplest dune types and patterns form in areas characterized by a narrow range of wind directions (unidirectional wind regime, RDP/DP > 0.8). In the absence of vegetation, the dominant form will be crescentic or transverse dunes with crest lines aligned approximately perpendicular to the dominant wind. Good examples are to be found in coastal areas of Namibia and the United Arab Emirates. Isolated crescentic dunes or barchans occur in areas of limited sand supply, and coalesce laterally to form crescentic or barchanoid ridges that consist of a series of connected crescents in plan view as sand availability increases. Larger forms with superimposed dunes are termed compound crescent dunes (e.g. Algodones Dunes, California; coastal areas of the Namib Sand Sea). In areas of partial vegetation cover and similar wind regimes, parabolic dunes will occur. These dunes are characterized by a U or V shape with a 'nose' of active sand and two partly vegetated arms that trail up wind. They are common in many coastal dunefields and semi-arid inland areas and often develop from localized blowouts in vegetated sand surfaces (Wolfe and David 1997). Both crescentic and parabolic dunes tend to migrate downwind, at a rate that is inversely proportional to their height.

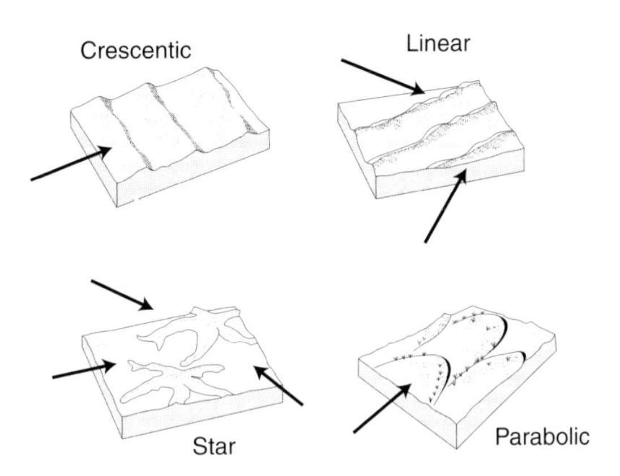

Figure 51 Schematic morphology of major dune morphological types and wind regime environments (modified from Lancaster 1995)

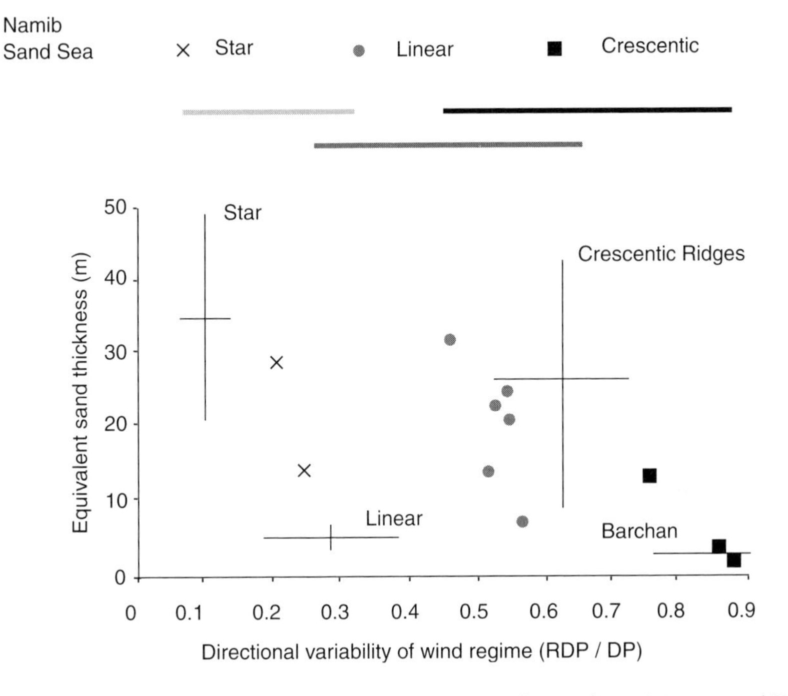

Figure 52 Relations between dune types and wind regimes. Redrawn from Wasson and Hyde (1983) with Namib Sand Sea data superimposed (symbols) and range of wind regimes for three major dune types from Fryberger (1979). Equivalent sand thickness is a measure of the sand available for dune building and represents the thickness of sand if the dunes were levelled. The directional variability of the wind regime is characterized by the ratio between resultant (RDP) and total sand drift potential (DP)

Linear dunes are characterized by their straightness, length (often more than 20 km), sinuous crestline, parallelism, and regular spacing, and high ratio of dune to interdune areas. Many linear dunes consist of a lower gently sloping plinth, often partly vegetated, and an upper crestal area where sand movement is more active. Slip faces develop on the upper part of the dune, their orientation depending on the winds of the season. The average form of the dune may be symmetrical with an approximately triangular profile, but in each season its profile tends to an asymmetric form with a concave stoss slope and a well-developed lee face. Linear dunes occur in areas of bimodal or wide unimodal wind regimes (RDP/DP > 0.4 < 0.8), and appear to be the most widespread dune type worldwide. Simple linear dunes occur in two forms: the straight partially vegetated dunes of the southwestern Kalahari and Simpson–Strezlecki deserts and the more sinuous 'seif'-type dunes of the Sinai and eastern Sahara. Complex linear dunes are best represented by the large (50–200-m high), widely spaced (1–2 km) linear dunes of the Namib Sand Sea.

The origins of linear dunes and their relationship to formative wind directions have been the subject of considerable controversy. A widely held view was that linear dunes form parallel to the prevailing or dominant wind direction. Their parallelism and straightness was believed to result from the existence of boundary layer roller vortices in which helicoidal flow sweeps sand from interdune areas to dunes. However, there is little empirical evidence to support such a model. Field studies of airflow and sediment transport over linear dunes (Bristow *et al.* 2000; Tsoar 1983) suggest that the fundamental mechanism for linear dune formation is the deflection of winds that approach at an oblique angle to the crest to flow parallel to the dune along its lee side and transport sand along the dune. Thus any winds from a 180° sector centred on the dune will be diverted in this manner. Linear dunes tend to extend downwind, as sinuosities in the crest migrate along its length. Evidence for lateral migration, is not conclusive.

Star dunes have a pyramidal shape, with three or four sinuous sharp-crested arms radiating from a central peak and multiple avalanche faces and are the largest dunes in many sand seas, reaching heights of more than 300 m in the eastern Namib Sand Sea and the Grand Erg Oriental in Algeria. The upper parts of many star dunes are very steep with slopes at angles of 15–30°; the lower parts consist of a broad, gently sloping (5–10°) plinth or apron. Small crescentic or reversing dunes may be superimposed on the lower flank and upper plinth areas of star dunes. Comparisons between the distribution of star dunes and wind regimes suggest that they form in multidirectional or complex wind regimes (RDP/DP < 0.3). A strong association between the occurrence of star dunes and topographic barriers has also been noted. Topography may modify regional wind regimes to increase their directional variability, as in the Erg Fachi Bilma or create traps for sand transport, as at Kelso Dunes and Great Sand Dunes.

The development of star dunes is strongly influenced by the high degree of form–flow interaction that occurs as a result of seasonal changes in wind direction, and the existence of a major lee-side secondary circulation. Most of the erosion and deposition involves the reworking of deposits deposited in the previous wind season. Sand, once transported to the dune, tends to stay there and add to its bulk, resulting in dunes that do not change position over time (Lancaster 1989).

Other important dune types include nebkhas or hummock dunes (common in many coastal dunefields) anchored by vegetation, lunettes (often comprised of sand-sized clay pellets) that form downwind of small playas; and a variety of topographically controlled dunes (climbing and falling dunes, echo dunes). Low relief sand surfaces such as sand sheets are common in many ergs and occupy from as little as 5 per cent of the area of the Namib Sand Sea to as much as 70 per cent of the area of Gran Desierto. Sand sheets occur where sediment availability is limited as a result of coarse sand, high water table or vegetation cover. Zibar, or low rolling dunes without slip faces composed of coarse sand, are transitional between sand sheets and crescentic dunes in some dunefields (e.g. Algodones, Skeleton Coast, Namibia).

Dune processes and dynamics

The initiation, development and equilibrium morphology of all aeolian dunes are determined by a complex series of interactions between dune morphology, airflow, vegetation cover and sediment transport rates. In turn, the developing bedforms exert a strong control on local transport rates through form–flow interactions and secondary flow circulations, leading to a dynamic equilibrium between dune morphology and local airflow. In multidirectional wind regimes, the nature of interactions between dune form and airflow change as winds vary direction seasonally, and lee-side secondary flow patterns become important in determining dune morphology and dynamics.

As dunes grow, they project into the atmospheric boundary layer so that they affect the airflow around and over them in a manner similar to isolated hills. Winds approaching the upwind toe of a dune stagnate slightly and are reduced in velocity, but likely not turbulence intensity. On the stoss, or windward slope of the dune, streamlines are compressed and winds accelerate up the slope. The degree of flow acceleration (the speed-up factor) is determined by the aspect ratio and the height of the dune. Wind speed at the crest of the dune is typically 1.1 to 2.5 times that measured immediately upwind of the dune (Figure 53a). Flow acceleration, coupled with effects of stream line curvature, on the windward slopes of dunes give rise to an exponential increase in sediment transport rates (Figure 53b) towards the dune crest (Lancaster *et al.* 1996; McKenna Neuman *et al.* 1997), resulting in erosion of the stoss slope, and a high level of erosion and deposition in crestal areas of linear and star dunes (e.g. Lancaster 1989; Livingstone 1989). Numerical models suggest that the non-linear increase in sediment transport with height on a dune limits dune size and results in an equilibrium dune configuration.

In the lee of the crest of dunes, wind velocities and transport rates decrease rapidly as a result of flow expansion between the crest and brink of the lee or avalanche face and flow separation on the avalanche face itself. There is a complex pattern of flow separation, diversion and re-attachment on the lee slopes of dunes, which is determined by the angle between the wind and the dune crest (angle of attack) and the dune aspect ratio (Walker and Nickling 2002). Secondary flows, including lee-side flow diversion, are especially important where winds approach the dune obliquely, and are an important process on linear and many star dunes.

High angles of attack on high aspect ratio (steep) dunes result in flow separation in the lee, while lower angles of attack result in flow diversion along the lee slope, whereas low aspect ratio dunes are characterized by flow expansion. Flow separation results in the development of an eddy in the lee of the dune, which may have the form of a roller vortex if flow is truly transverse, with the separation cell extending downwind for 4 to 10 times the height of crescentic dunes. When flow is oblique to the dune crest a helical vortex develops. The oblique flow is deflected along the lee slope parallel to the dune crest, with the degree of deflection being inversely proportional to the incidence angle between the crestline and the primary wind. Field studies indicate that the lee-side helical eddy affects the whole of the lee side on simple (5–10-m high) linear dunes, but extends for only 10–20 per cent of the height of large (50–150-m high) complex linear dunes and 40 m high star dunes. Changes in the local incidence angle between primary winds and a sinuous dune crest result in a spatially varying pattern of deposition and along-dune transport on the lee face. Deposition dominates where winds cross the crest line at angles approaching 90°, and erosion or along-dune transport occurs where incidence angles are $<40°$. Studies on flow-transverse dunes indicate that, downwind of and above the flow separation cell, there is a series of wakes that gradually expand, diffuse and mix downwind for a distance of as much as 25 to 30 times the height of the dune (Walker and Nickling 2002).

Flow separation also causes fallout of previously saltating sand grains. Field experiments show that 95 per cent of the sand transported over the crest is deposited within a metre of the crest, with the rate of deposition decreasing exponentially downwind (Nicking *et al.* 2002). High rates of deposition in the immediate lee of the crest result in oversteepening of the slope and avalanching of grains to form slip faces. All sand transported over the crest of crescentic dunes is deposited, so that they are typically 'sand trapping' bedforms. As a result, their movement can be described by: $c = Q/yh$ where c is the migration rate, Q is the bulk volumetric sand transport rate, y is the bulk density of sand and h is dune height.

Challenges and opportunities in dune studies

The past two decades have seen a dramatic change in the level of understanding of dune dynamics and morphology through intensive field studies of processes and synoptic views of sand seas provided by satellite images. As a result, the fundamentals of dune dynamics and the formative factors of major dune types are known in

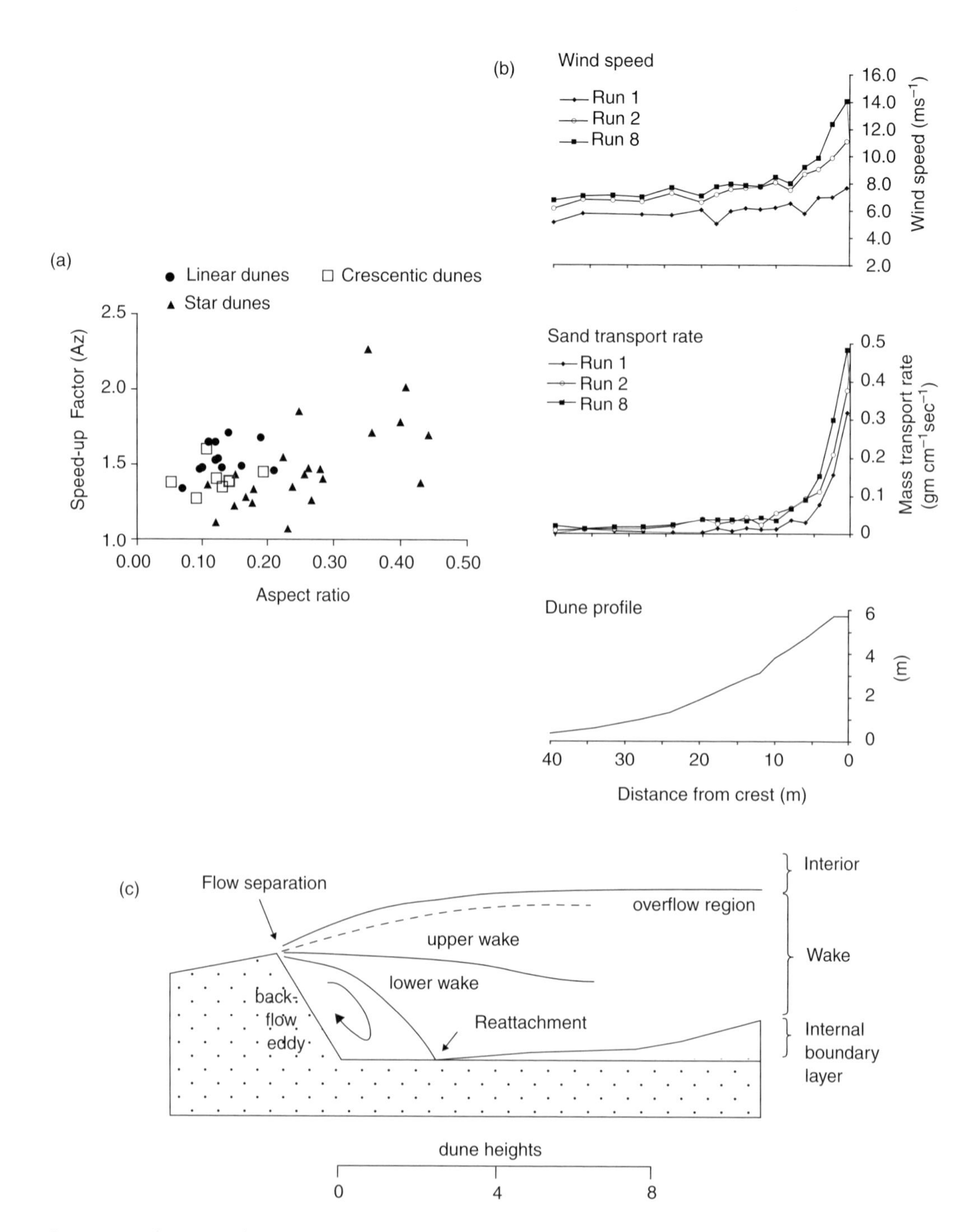

Figure 53 Elements of dune dynamics: (a): velocity speed-up (from Lancaster 1995); (b) winds and sediment transport rates on the stoss slope (from McKenna Neuman *et al*. (1997) (c) lee-side flow separation and wake mixing (from Walker and Nickling 2002)

some detail. Not well known are processes leading to dune initiation, the dynamics of lee-side processes (including avalanching), and the controls of dune size and spacing. It has also proved very difficult to extrapolate the results of short-term studies of dune processes to understanding of long-term or even annual dune dynamics. One promising approach is to develop numerical models of dune and dunefield evolution (Werner 1995). The other is to use ground-penetrating radar to image dune sedimentary structures, which provide a record of the results of dune-forming processes on a variety of timescales and allow empirical models of dune evolution to be developed. A good example of this approach is the five-stage model of linear dune development put forward by Bristow *et al.* (2000).

References

Bauer, B.O. and Sherman, D.J. (1999) Coastal dune dynamics: problems and prospects, in A.S. Goudie, I. Livingstone and S. Stokes (eds) *Aeolian Environments, Sediments and Landforms*, 71–104, Chichester, New York: Wiley.

Bristow, C.S., Bailey, S.D. and Lancaster, N. (2000) Sedimentary structure of linear sand dunes, *Nature* 406, 56–59.

Fryberger, S.G. (1979) Dune forms and wind regimes, in E.D. McKee (ed.) *A Study of Global Sand Seas*, 137–140, United States Geological Survey, Professional Paper 1,052.

Kocurek, G. and Lancaster, N. (1999) Aeolian sediment states: theory and Mojave Desert Kelso Dunefield example, *Sedimentology* 46, 505–516.

Lancaster, N. (1989) The dynamics of star dunes: an example from the Gran Desierto, Mexico, *Sedimentology* 36, 273–289.

——(1995) *Geomorphology of Desert Dunes*, London: Routledge.

Lancaster, N., Nickling, W.G., McKenna Neuman, C.K. and Wyatt, V.E. (1996) Sediment flux and airflow on the stoss slope of a barchan dune, *Geomorphology* 17, 55–62.

Livingstone, I. (1989) Monitoring surface change on a Namib linear dune, *Earth Surface Processes and Landforms* 14, 317–332.

McKenna Neuman, C., Lancaster, N. and Nickling, W.G. (1997) Relations between dune morphology, airflow, and sediment flux on reversing dunes, Silver Peak, Nevada, *Sedimentology* 44, 1,103–1,114.

Nickling, W.G., McKenna Neuman, C. and Lancaster, N. (2002) Grainfall processes in the lee of transverse dunes, Silver Peak, Nevada, *Sedimentology* 49, 191–211.

Tsoar, H. (1983) Dynamic processes acting on a longitudinal (seif) dune, *Sedimentology* 30, 567–578.

Walker, I.J. and Nickling, W.G. (2002) Dynamics of secondary airflow and sediment transport over and in the lee of transverse dunes, *Progress in Physical Geography* 26, 47–75.

Wasson, R.J. and Hyde, R. (1983) Factors determining desert dune type, *Nature* 304, 337–339.

Werner, B.T. (1995) Eolian dunes: computer simulations and attractor interpretation, *Geology* 23, 1,107–1,110.

Wolfe, S.A. and David, P.P. (1997) Parabolic dunes: examples from the Great Sand Hills, southwestern Saskatchewan, *Canadian Geographer* 41, 207–213.

Further reading

Livingstone, I. and Warren, A. (1996) *Aeolian Geomorphology: An Introduction*, Harlow: Addison-Wesley Longman.

Nickling, W.G. and McKenna Neuman, C. (1999) Recent investigations of airflow and sediment transport over desert dunes, in A.S. Goudie, I. Livingstone and S. Stokes (eds), *Aeolian Environments, Sediments and Landforms*, Chichester: Wiley.

Pye, K. and Tsoar, H. (1990) *Aeolian Sand and Sand Dunes*, London: Unwin Hyman.

SEE ALSO: aeolian processes; barchan; draa (megadune); dune mobility

NICK LANCASTER

DUNE, COASTAL

Foredune

Foredunes are shore-parallel dune ridges formed on the top of the backshore by aeolian sand deposition within vegetation. They may range from scattered hummocks or nebkha, relatively flat terraces, to markedly convex ridges. Actively forming foredunes occupy a foremost seaward position, but not all foremost dunes are foredunes. Other dune types may occupy a foremost position on eroding coasts or coasts where foredunes are unable to form. Foredunes generally fall into two main types, incipient and established foredunes.

INCIPIENT FOREDUNE

Incipient foredunes are new, or developing foredunes forming within pioneer plant communities. They may be formed by sand deposition within discrete or relatively discrete clumps of vegetation or individual plants forming shadow dunes, hummocks or nebkha.

Incipient foredunes may also form on the backshore by relatively laterally continuous alongshore growth of pioneer plant seedlings and/or rhizomes in the wrack line or spring high tide region (Hesp 1989). Morphological development principally depends on plant density, height and cover, wind velocity and rates of sand transport (Davies 1980).

Shadow dunes, hummocks, embyro dunes and nebkha all form due to high localized drag within and behind individual plants and clumps of plants. Wind velocities experience rapid deceleration on reaching the plants, local acceleration around the plants, and flow separation behind the plants (Hesp 1999).

Relatively continuous plant canopies variously impact the wind/sand flow depending on plant density, distribution and height. High, dense canopies act to reduce flow velocities very rapidly, and sand transport (saltation and traction) is markedly reduced from the leading edge. In canopies which vary alongshore in density or distribution, foredune morphology also varies (Nickling and Davidson-Arnott 1990). Plant density is increased as wind velocities increase as the vegetation bends and streamlines to the wind.

Incipient foredunes generally display one of three morphological types: ramps, terraces and ridges. Swales (lee dune depressions) are generally created by seaward accretion of a foredune. They develop as low to limited aeolian deposition zones (Hesp 2002).

ESTABLISHED FOREDUNE

Established foredunes develop from incipient foredunes and are commonly distinguished by the growth of intermediate or 'secondary' plant species, and/or by their greater morphological complexity, height, width, age and geographical position.

Foredunes range from very low, and commonly scattered, dunes less than a metre or so in height on some barrier islands dominated by overwash and in areas of limited sediment supply to over 30 m in height in some instances. The morphological development and evolution of established foredunes depends on a number of factors including: sand supply, beach width and fetch, surfzone-beach type, the degree of vegetation cover, plant species present (a function of climate and biogeographical region), the rate of aeolian sand accretion and erosion, the frequency and magnitude of wave and wind forces, the occurrence and magnitude of storm erosion, dune scarping and overwash processes, the medium to long-term beach or barrier state (stable, accreting or eroding), and increasingly, the extent of human interference and use (Davidson-Arnott and Law 1996; Short and Hesp 1982; Hesp 1999).

The wind flow is topographically accelerated over foredunes, particularly up stoss slopes and over crests. However, the variable vegetation cover of foredunes and their topographic variability leads to local decelerations and variations in roughness length (Arens 1997), and these become more pronounced as foredune morphological complexity and vegetation cover increases.

Foredune Plain

Foredunes may gradually, or rapidly, become isolated from accretion and erosion processes by the seaward development of a new incipient foredune which itself may evolve into an established foredune. The original foredune then becomes a relict foredune as it is largely or wholly removed from a foremost beach position. Systematic beach progradation over time frames of tens to thousands of years has led to the development of wide foredune plains.

Blowout

A blowout is a saucer-, cup-, bowl or trough-shaped depression or hollow formed by wind erosion on a pre-existing sand deposit. The adjoining accumulation of sand, the depositional lobe, derived from the depression and possibly other sources, is normally considered part of the blowout (Nordstrom *et al.* 1990).

Blowout morphology may be highly variable, ranging from cigar-shaped, V-shaped, scooped hollow, and cauldron and corridor types, from pits to elongated notches, troughs or broad basins, and saucer and trough blowouts (Cooper 1967). Saucer blowouts are semicircular or saucer-shaped and often appear as shallow dishes. Deeper cup- or bowl-shaped blowouts may evolve from these. Trough blowouts are generally more elongate, with deeper deflation floors and basins, and with steeper, longer erosional lateral walls or slopes.

INITIATION

Blowouts may be initiated in a variety of ways including:

1. wave erosion of dunes followed by wind erosion of dune scarps;
2. die-back of vegetation following dune wave erosion and subsequent wind erosion;
3. wind erosion of overwash hollows and fans;
4. topographic acceleration of airflow over (or through) dunes, dune cols, scarps and cliffs;

5 where the vegetation cover is naturally low, or is weakened, reduced or dies due to a prolonged dry or arid period;
6 vegetation die-back due to soil nutrient depletion;
7 localized aridity (e.g. on dune crests) reducing plant cover;
8 the activities of animals and humans;
9 water erosion;
10 high velocity wind erosion leading to either erosion, or sand inundation and burial.

Once initiated, the subsequent morphologic development may depend on the size of the initial constriction, the height and width of the dune in which the blowout is developing, the degree and type of vegetation cover, the magnitude of regional winds, and the degree of exposure to winds from various directions (Hesp 2002; Jennings 1957).

FLOW DYNAMICS

Flow in saucer blowouts is complex with flow separation occurring around much of the erosion walls. Sand erosion and deposition are also complex as a result of varying wind speeds and directions, although, in general, deflation basins deepen in most blowouts studied. Saucer blowouts commonly grow in length upwind against the prevailing wind.

The flow up trough blowouts is commonly topographically accelerated, and displays marked single and double jets up the deflation basin, corkscrew vortices over the lateral erosional wall crests, and rapid flow deceleration, lateral expansion and flow separation over the depositional lobe. Topographic steering can be significant (Hesp and Hyde 1996).

Parabolic dune

Parabolic dunes (also termed U-dunes, upsiloidal dunes, hairpin dunes) are typically U-and V-shaped dunes characterized by short to elongate, trailing ridges which terminate downwind in U- or V-shaped depositional lobes. The depositional lobes may be simple, relatively featureless sandsheets, or textured with a variety of dune forms (e.g. transverse dunes, barchanoidal dunes, etc). Deflation basins and plains, slacks, seasonal wetlands, ponds, lagoons and gegenwalle ridges occupy the area between the trailing ridges.

INITIATION

Parabolic dunes typically evolve in a number of ways including:

1 from blowouts (Pye 1983). In many cases, the blowout depositional lobe continues to advance downwind forming trailing ridges;
2 evolution from the landward and downwind margins of transgressive sandsheets and dunefield.

Blowouts and parabolic dunes may be formed on both stable (sediment supply balanced) and accreting/prograding (positive sediment supply) coasts which experience occasional or regular high energy wind events (Hesp 2002). They are commonly formed on eroding coasts where the foredune stability is reduced by wave erosion, and subsequent wind erosion (e.g. Ruz and Allard 1994).

MORPHOLOGY

Two principle sub-types of parabolic dune are common: long-walled types and squat, elliptical types. The multiple development of these leads to there being two principle sub-types of parabolic dunefields: long-walled types and imbricate types (Trenhaile 1997).

Long-walled parabolic dunes display long trailing ridges and extensive deflation basins. They are particularly well developed on relatively flat terrain, in regions of low heath or shrubland, high sand supply and strong, more unidirectional winds. Some parabolic dunes display a squat, shorter form, often with more semicircular or elliptical deflation basins. Multiple development results in the dunes overlapping each other in an imbricate fashion. They commonly develop in wetter areas, on flat terrain where deflation depths are limited and/or wind speeds are relatively low, on steep terrain, in less unidirectional or multidirectional wind regimes and/or in dense, tall vegetation where the rate of advance is low and/or migration is impeded.

EVOLUTION

Deflation basins tend to continue to erode until a base level is reached such as the seasonally lowest water-table level, a calcrete (or other cemented/indurated) layer, an armoured surface such as a pebble, shell, pumice or artifact surface. Trailing ridges develop due to trapping of the outside, marginal lateral edge of the depositional lobe as it migrates downwind. This sediment is trapped while the inside (deflation plain) portion of the ridge is eroded. Depositional lobes are arcuate, hairpin, V-shaped, radial or parabolic-shaped depending on wind direction, lobe height and volume, vegetation cover and species type, and speed of migration.

RATES OF MIGRATION

Rates of parabolic dune advance or migration vary considerably depending on the morphology, slope and type (e.g. sandy vs rocky) of terrain the dunes are moving across, the vegetation cover and type (e.g. woodland vs grassland), wind velocities and directional variability of the wind. Dune migration rates range from 0.05 to 25 $m\,yr^{-1}$.

Transgressive dunefield and sheet

Transgressive dunefields and sheets are aeolian sand deposits formed by the downwind or alongshore movement of sand over vegetated to semi-vegetated terrain. Such sheets and dunefields may range from quite small (hundreds of metres in alongshore and landward extent) to draa or megadune size fields. They may be completely unvegetated, partially vegetated or fully vegetated (post-formation) (Nordstrom *et al.* 1990). Sheets display little or no surface dunes; dunefields are covered with a variety of superimposed dune forms. They have also been termed mobile dunes, migratory dunes, mendano and machair.

Transgressive dunefields are particularly well developed on high wind and wave energy (west and south) coasts with significant sediment supply, and in virtually all climatic regions (tropics to the arctic).

TRANSGRESSIVE DUNEFIELD TYPES

At the gross scale, transgressive dunefields may describe tabular forms (including headland bypass dunefields), buttress forms, or climbing, clifftop and falling dunefields.

INITIATION AND DEVELOPMENT

Transgressive dunefields develop for a variety of reasons. They may form, or have formed:

1. as a response to rising sea level and/or climatic change, particularly in the period, 10,000 to 7,000 years BP.
2. in regions of high alongshore and onshore sediment supply, often in high wind and wave energy environments;
3. on coasts experiencing erosion;
4. as continental shelves were exposed and/or climate changed during the Last Glacial;
5. as a response to periods of regional sea-level fall; and
6. on coasts experiencing climatic extremes such as in arid and arctic and subarctic environments, and where vegetation growth may also be limited.

TRANSGRESSIVE DUNEFIELD LANDFORMS

Active transgressive dunefields may extend (and/or migrate) directly alongshore, obliquely onshore or directly onshore. Dunefields migrating alongshore are typically characterized by transverse (and other) dunes extending inland from the backshore. Interdunes are dominated by nabkha, deflation flats, sandsheets and overwash plains and fans. Dunefields migrating obliquely or normally onshore are usually characterized by a small to extensive deflation basin (or plain) or a series of slacks on the upwind or seaward side, a small to extensive, mobile to partially vegetated sandsheet or dunefield, and a long-walled, commonly sinuous 'main' slipface or precipitation ridge on the landward side.

The surfaces of active transgressive dunefields are commonly covered with a variety of dune types (including 'desert' dune types) ranging from simple domes, transverse dunes and BARCHANS, to barchanoidal and sinuous transverse and oblique dunes, to complex aklé or network and star dune forms (e.g. Hunter *et al.* 1983).

Deflation plains and basins typically lie parallel to the shore and are eroded down to a base level such as a CALCRETE pavement, the seasonally lowest water table, older dune surfaces and PALAEOSOLS, carbonate or bedrock.

Active transgressive dunefields may display a variety of generally smaller scale dune forms and environments, including remnant knobs, hummocks, bush pockets, nabkha, shadow dunes and 'rim dune' (around the margins of washover fans) (Nordstrom *et al.* 1990).

Precipitation ridges (long-walled or main slipfaces) commonly occur along the downwind and surrounding margins of transgressive dunefields. Where the dunefields are migrating in one primary direction, they generally have one precipitation ridge. Where they are expanding/migrating landwards and alongshore, they may have two to many precipitation and trailing ridges.

References

Arens, S.M. (1997) Transport rates and volume changes in a coastal foredune on a Dutch Wadden island, *Journal of Coastal Conservation* 3, 49–56.
Cooper, W.S. (1967) *Coastal Sand Dunes of California*, Geological Society of American Memoir 101.
Davidson-Arnott, R.G.D. and Law, M.N. (1996) Measurement and prediction of long-term sediment supply to Coastal Foredunes, *Journal of Coastal Research* 12, 654–663.
Davies, J.L. (1980) *Geographical Variation in Coastal Development*, London: Longman.
Hesp, P.A. (1989) A review of biological and geomorphological processes involved in the initiation and development of incipient foredunes, in C.H. Gimmingham, W. Ritchie, B.B. Willetts and A.J. Willis (eds) *Coastal Sand Dunes*, Proceedings of the Royal Society of Edinburgh, 96B, 181–202.
——(1999) The beach backshore and beyond, in A.D. Short (ed.) *Handbook of Beach and Shoreface Morphodynamics*, 145–170, Chichester: Wiley.
——(2002) Foredunes and blowouts: initiation, geomorphology and dynamics, in P. Gares (ed.) 29th Binghamton Geomorphology Symposium: 'Coastal Geomorphology' *Geomorphology* 48 (1/3): 245–268.
Hesp, P.A. and Hyde, R. (1996) Flow dynamics and geomorphology of a trough blowout, *Sedimentology* 43, 505–525.
Hunter, R., Richmond, B.M. and Alpha, T.R (1983) Storm-controlled oblique dunes of the Oregon Coast, *Geological Society of America Bulletin* 94, 1,450–1,465.
Jennings, J.N. (1957) On the orientation of parabolic or U-dunes, *Geographical Journal* 123, 474–480.
Nickling, W.G. and Davidson-Arnott, R.G.D. (1990) Aeolian sediment transport on beaches and coastal dunes, in R.G.D. Davidson-Arnott (ed.) *Proceedings Symposium on Coastal Sand Dunes*, 1–35, Ottawa: NRC.
Nordstrom, K.F., Psuty, N.P. and Carter, R.W.G. (eds) (1990) *Coastal Dunes: Form and Process*, London: Wiley.
Pye, K. (1983) Coastal dunes, *Progress in Physical Geography* 7(4), 531–557.
Pye, K. and Tsoar, H. (1990) *Aeolian Sand and Sand Dunes*, London: Unwin Hyman.
Ruz, M.-H. and Allard, M. (1994) Coastal dune development in cold climate environments, *Physical Geography* 15(4), 372–380.
Short, A.D. and Hesp, P.A. (1982) Wave, beach and dune interactions in southeast Australia, *Marine Geology* 48, 259–284.
Trenhaille, A.S. (1997) *Coastal Dynamics and Landforms*, Oxford: Oxford University Press.

PATRICK HESP

DUNE, FLUVIAL

Dunes are the commonest sedimentary feature in sand- or silt-bedded streams. They are roughly triangular in profile with a gentle upstream slope and a steeper downstream slope (see DUNE, AEOLIAN). Dune height and wavelength are directly related to water depth. Reaching heights of up to one-third of flow depth they are commonly 0.1 m to 10 m high with a wavelength 4 to 8 times flow depth (Knighton 1998). They frequently form in streams with higher intensity flows than those with RIPPLE bedforms but, like ripples, they migrate downstream through the processes of erosion on the upstream slope followed by deposition on the downstream slope. Separation of flow from the crest of dunes generates large-scale turbulence in rivers and the downstream migration and change of dune form is an important mechanism of bedform adjustment to changing river discharge. Cross-bedding structures resulting from dune migration are often preserved in alluvial deposits and can be used to interpret palaeohydraulic processes.

Reference

Knighton, D. (1998) *Fluvial Forms and Processes*, London: Arnold.

GILES F.S. WIGGS

DUNE MOBILITY

Approximately 20 per cent of the world's drylands are covered by aeolian sands, within which desert dunes (see DUNE, AEOLIAN) occur, while coastal dunes occur in a range of climatic settings. In dryland and coastal settings dunes may be mobile, but many dunes are not mobile in the sense of the dune body migrating across the landscape. Dune mobility therefore needs to be considered in terms of dune setting, dune type, and the nature of aeolian activity upon the dune surface.

The mobility of sand dunes is in broad terms a function of the relationship between the forces of erodibility, affecting the potential of the dune surface to be eroded by the wind, and erosivity, which is the potential of the agencies effecting AEOLIAN PROCESSES to move sediment. Dune mobility may be assessed from climatological data using a dune mobility index (e.g. that of Lancaster 1988) that integrates the forces that affect erodibility, such as P/PET, and those that affect the erosivity, which relate to the wind field. Mobility can also be assessed in the field, by monitoring dune surface change and airflow (e.g. Wiggs *et al.* 1995) or dune movement in the landscape (e.g. Hastenrath 1987)

Since dunes, except those that have become lithified, are mainly comprised of unconsolidated sand-sized sediment, their potential to be moved by mobile air via processes of sediment entrainment may appear to be considerable. Reality is more complex, however, since winds have to exceed a threshold velocity for entrainment to take place, and dune bodies can, once formed, store moisture and act as a host to plants, which can markedly reduce erodibility. Dune mobility can be a discontinuous process too. Winds capable of entraining and moving sediment do not occur continuously, but vary seasonally and daily. For example, sand transport on dunes in the Namib Sand Sea generally occurs in response to fairly persistent but moderate south westerly winds during summer months, while in winter short-lived but high magnitude wind events transport sediment from an easterly direction (Livingstone 1989).

Vegetation and dune mobility

Isolated or widely spaced plants on a dune can lead to localized zones of higher wind velocities as airflow is streamlined around the obstacle (Thomas and Tsoar 1991). But in general, dunes that possess some form of surface vegetation or crusting have, all other things being equal, a lower erodibility than those that do not. Crusts and plants can play several roles in affecting the potential mobility of surface sediments (Wolfe and Nickling 1993): protection of the sediment immediately below the plant or crust, increasing surface roughness and thereby reducing wind velocity, and trapping any moving sediment grains. A partial or discontinuous vegetation cover does not totally exclude sand movement but it may anchor a dune plinth and inhibit dune migration or lateral movement. On partially vegetated dunes, different studies have identified various threshold vegetation covers, ranging from *c.*6 per cent (Marshall 1970) to 30 per cent (Ash and Wasson 1983) above which any aeolian activity ceases. However, the impact of a given cover will vary not only according to plant shape and porosity but to both ambient wind velocities and position on the dune body (Wiggs *et al.* 1995).

Dune size and mobility

All other things being equal, smaller dunes move or experience surface change more quickly than large dunes. This is because, for a given sediment transport event, the volume of sand that can be moved represents a smaller component of the total volume of sand of a large dune than of a small dune. The ability of a dune to retain its form and position as environmental conditions change has been called 'dune memory' by Warren and Kay (1987), with small dunes of low volume having little memory, and therefore adjusting relatively rapidly to wind events, while large dunes with large volumes have 'mega memories' that may record histories spanning millennia.

Mobility of different dune types

Different basic dune forms develop in different wind directional regimes (Fryberger 1979; Thomas 1997). Generally, barchans and transverse dunes form in unimodal sand transporting wind regimes, linear dunes in bimodal or wide unimodal regimes, and star dunes in multimodal regimes, where regime refers to the overall annual directional pattern of sand transporting winds.

These different regimes determine the general types of mobility or, more appropriately activity, of these dune types. Transverse dunes are mobile in the true sense of the word, since with transport for a single direction the dunes are able to migrate. Migration rates differ between and within dunefields according to the available transport energy, but given the principle of dune memory (see above), in any location larger dunes will move more slowly than small dunes, as expressed by

$$c_r = (q_c - q_t)/h\gamma_p$$

where c_r is the migration rate, q_c is the mass transport rate at the dune crest, q_t is the mass transport rate at the dune base, h is dune height and γ_p is the bulk density of the sediment. A number of studies of migration rates have been conducted in different deserts, and are summarized in Thomas (1992) with examples of rates given in Table 14.

Linear dunes are extending forms. Net sediment transport is along the dune in the resultant direction of transport generated by the combined effect of bimodal winds. This can lead to elongation of the dune at the downwind end, but also to some lateral movement if one direction has greater transport potential than the other and if the dune plinth is not anchored by vegetation. Lateral migration can be extremely slow, for example at a rate of 50–100 m over the past 10,000 years as suggested by Rubin (1990) from evidence in the Strezlecki Desert in Australia. Other studies from Namibia and the Sinai Desert

Table 14 Examples of barchan and transverse dune migration rates

Location	Dune height (m)	Migration rate (m yr^{-1})
Barchan dunes, southern Peru	1	32
	7	9
Barchan dunes, Salton Sand Sea, California	3.1	27
	8.2	14
Transverse dunes, Erg Oriental, Saudi Arabia	35	0.3
	240	0.16

Source: Data from various authors

suggest elongation rates may range from less than 2 m to over 14 m per annum.

Star dunes can be regarded as sand accumulating forms that, under the interactive effect of sand transport from at least three directions, gain in volume and height over time. The individual arms of the dune may, on a seasonal basis, behave as if they are transverse or linear forms and display displacements of up to 20 m (Lancaster 1989). If any of the contributory transport directions has a net advantage over the others, some migration of the dune body may occur over time.

References

Ash, J.E. and Wasson, R.J. (1983) Vegetation and sand mobility in the Australian desert dunefield, *Zeitschrift für Geomorphologie Supplementband* 45, 7–25.

Fryberger, S. (1979) Dune form and wind regime, *US Geological Survey Professional Paper* 1,052, 137–169.

Hastenrath, S. (1987) The barchan sand dunes of southern Peru revisited, *Zeitschrift für Geomorphologie NF* 31, 167–178.

Lancaster, N. (1988) Development of linear dunes in the southwestern Kalahari, southern Africa, *Journal of Arid Environments* 14, 233–244.

——(1989) Star dunes, *Progress in Physical Geography* 13, 67–91.

Livingstone, I. (1989) Monitoring surface change on a Namib linear dune, *Earth Surface Processes and Landforms* 14, 317–332.

Marshall, J.K. (1970) Assessing the protective role of shrub-dominated rangeland vegetation against soil erosion by wind, *Proceedings XI International Grassland Congress*, 19–23.

Rubin, D.M. (1990) Lateral migration of linear dunes in the Strezlecki desert, Australia, *Earth Surface Processes and Landforms* 15, 1–14.

Thomas, D.S.G. (1992) Desert dune activity: concepts and significance, *Journal of Arid Environments* 22, 31–38.

——(1997) Sand seas and aeolian bedforms, in D.S.G. Thomas (ed.) *Arid Zone Geomorphology*, 373–412, Chichester: Wiley.

Thomas, D.S.G. and Tsoar, H. (1991) The geomorphological role of vegetation in desert dune systems, in J.B. Thornes (ed.) *Vegetation and Erosion*, 471–489, Chichester: Wiley.

Warren, A. and Kay, S. (1987) The dynamics of dune network, in L. Frostick and I. Reid (eds) *Desert Dunes, Ancient and Modern, Geological Society of London Special publication* 35, 205–212, Oxford: Blackwell.

Wiggs, G.F.S., Thomas, D.S.G., Bullard, J.E. and Livingstone, I. (1995) Dune mobility and vegetation cover in the southwest Kalahari Desert, *Earth Surface Processes and Landforms* 20, 515–529.

Wolfe, S.A. and Nickling, W.G (1993) The protective role of sparse vegetation in wind erosion, *Progress in Physical Geography* 17, 50–68.

DAVID S.G. THOMAS

DUNE, SNOW

Aeolian bedform are common in snow, and include ripples, drifts, barchans, and the like (Cornish 1914).

In recent years the size and importance of various megadunes have become appreciated, particularly in eastern Antarctica. These are transverse features that are oriented perpendicular to the regional katabatic wind direction. Their amplitudes are small (*c*.4 m), but their wavelengths range from 2 to over 4 km, and megadune crests are nearly parallel and 10–100 km in length (Frezzotti *et al.* 2002).

References

Cornish, V. (1914) *Waves of Sand and Snow*, London: Fisher Unwin.

Frezzotti, M., Gandolfi, S. and Urbini, S. (2002) Snow megadunes in Antarctica: sedimentary structure and genesis, *Journal of Geophysical Research* 107, D18, 1-1–1-12.

A.S. GOUDIE

DURICRUST

The word was introduced by Woolnough (1927) who subsequently defined the term thus (Woolnough 1930: 124–125): 'The widespread chemically formed capping in Australia, resting on a thoroughly leached sub-stratum . . . The nature of the deposit varies from a mere infiltration of pre-existing surface rock, to a thick mass of relatively pure chemical precipitate'.

The mineral matter deposited from solution falls into three main groups: (1) aluminous and ferruginous; (2) siliceous; and (3) calcareous and magnesian. Woolnough believed that bedrock was an important influence on the distribution of these three types, which in effect are broadly equivalent to (1) laterites, bauxites, FERRICRETES (see Tardy 1997; Bardossy and Aleva 1990); (2) SILCRETES; (3) CALCRETES, dolocretes.

Because of subsequent work on the individual duricrust types, the *crete*-based terminology of which had been laid down by Lamplugh (1907), Goudie (1973: 5) proposed a modified definition which resulted from a synthesis of various definitions that had already been developed for the individual types, and stressed their essentially subaerial and near-surface origin and nature:

> A product of terrestrial processes within the zone of weathering in which either iron and aluminium sesquioxides (in the case of ferricretes and alcretes) or silica (in the case of silcrete) or calcium carbonate (in the case of calcrete) or other compounds in the case of magnesicrete and the like have dominantly accumulated in and/or replaced a pre-existing soil, rock, or weathered material, to give a substance which may ultimately develop into an indurated mass.

Sometimes duricrusts may incorporate characteristics of more than one type, as with the widespread calsilcretes of the Kalahari.

To understand the origin and development of these geomorphologically important materials some general considerations need to be borne in mind. First, there is the question of the sources of the materials which contribute to the make-up of duricrusts. The primary elements can be derived from at least four main sources: the weathering of bedrock and sediment, inputs from dust and precipitation, plant residues and the dissolved solids in ground water. Then these sources have to be translocated and concentrated either by lateral transfers, or by vertical movements, whether upwards (*per ascensum*) or downward (*per descensum*). Third, the transferred materials need to be precipitated, and here a very wide range of processes come into play. Among the most important of these are changes in chemical equilibria caused by evaporation, by temperature changes, by pressure changes in the soil, air and water systems, by the action of organisms and by miscellaneous changes caused by interactions of different solution types.

Models for the origin of duricrusts normally fall into one of two categories: those involving relative accumulation and those involving absolute accumulation. *Relative accumulations* owe their concentrations to the removal of more mobile components, while *absolute accumulations* owe their concentration to the addition of materials to a profile. However, as McFarlane (1983: 20) has pointed out, the utility of this subdivision depends on important scale considerations. At one extreme the accumulation is entirely relative since laterites would not exist at all were not Fe and Al less readily mobilized during rigorous chemical weathering. At the other extreme, in hand specimens even the residual laterites on interfluves show much addition of Fe, since samples are enriched absolutely in materials which originated above them in the formerly existing column of rock, consumed to provide the residuum.

Furthermore, laterites and silcretes differ in that, while ferricretes can result from either relative or absolute accumulation of iron, silcrete can only form by absolute accumulation. Weathering provides the silica and in some cases the material (a weathering profile, for example) in which the silica is deposited.

Many of the early models of duricrust formation involved vertical processes, and especially the role of capillary rise of solutions from ground water. Vertical process models of this *per ascensum* type were complemented by *per descensum* models, in which it was believed that material leached from the upper part of a profile would accumulate lower down. Some of the material to be leached downward might be added to the top of the profile in the form of inputs of dust, etc.

However, more recently appreciation of the importance of CATENAs and toposequences, and of lateral soil-water movements, has resulted in an increasing concern with lateral transfer models. For example, Stephens (1971) argued that the silcretes of inland Australia formed from silica that was leached during lateritization in the humid upland areas of the east and then transported by rivers to low relief areas lying to the west. Similarly the detrital model of calcrete formation (Goudie 1983: 115) involves the lateral transportation and redeposition of weathered fragments of calcrete, moving from plateaux surfaces to footslopes.

One slightly unusual explanation for duricrust formation is that proposed for the silcretes of parts of Australia, where, it has been suggested, overlying or adjacent basalt sheets have played a role. Even amongst those who have proposed this association there is little agreement as to whether the supposed basaltic effect has been hydrothermal alteration, contact metamorphism, a release of silica from weathered basalt, or a reduction in the migration of pore waters caused by the presence of a basaltic caprock. Some doubt, however, whether such a special mechanism is justified (e.g. Ollier 1991) for what is such a widespread phenomenon.

Another general feature of models of duricrust formation has been the appreciation of the importance of organic processes. For example, in the case of calcrete, laboratory simulations with micro-flora (Krumbein 1968), and studies of petrography which have revealed calcified organic filaments of soil fungi, algae, actinomycetes and root hairs of vascular land plants, have caused the role of organisms to be given the attention they deserve (see Goudie 1996, for a review). In the case of laterite, various organic agencies have also been mooted. Micro-organisms could contribute to both mobilization and precipitation of materials. The transition from geothite to haematite in laterite profiles could be the result of iron bacteria activity, and desilicifying bacteria could be used to remove combined silica (kaolin) from bauxite.

Several factors contribute to the geomorphological importance of duricrusts: the thickness of the profiles, the properties of the different components of the profiles (e.g. their occasional ability to harden on exposure) and the topographic situation in which duricrusts develop. Ferricrete profiles may be as much as 60 m thick. Calcrete profiles in parts of southern Africa, western Australia and the Texan High Plains may exceed 40 m, while in Zaire and Namibia maximum depths of silicification may also be of the order of 50 m.

The hard upper crusts of duricrust profiles form only a limited proportion of the total profile thickness. Typical values for alcrete and ferricrete hardpans are 1–10 m, for calcrete 0.1–10 m (with around 0.3–0.5 m being the most common), while for silcrete values of between 1 and 5 m appear normal.

Beneath the hardpan layer duricrusts display a variety of material types. Ferricretes, for example, often have rather erodible pallid and mottled horizons grading down into more or less coherent bedrock, while calcretes may be underlain by friable nodule horizons, and silcretes by kaolinitic clays. Related to the important geomorphological role of the differences between the properties of hardpans, sub-hardpan zones, weathered bedrock and bedrock in relatively simple profiles, is the role of alternations of different layers in complex profiles.

Another general aspect of duricrusts, which is relevant to their geomorphological impact, is the speed at which they form, and the rapidity with which they may harden on exposure. Rapid formation helps to preserve otherwise relatively ephemeral landforms (e.g. dunes or alluvial terraces). Quick rates of formation tend to be associated with duricrusts that originate through absolute accumulation rather than relative accumulation.

In spite of examples of rapid formation it is nonetheless apparent that for some of the great thicknesses of profiles to have developed, a considerable span of time (10^5–10^7 years) is required, together with a degree of land-surface stability. The Pleistocene was too short and too variable in climate for many of the great duricrust surfaces to have formed, and it may be for these reasons that so many of the world's duricrusts are of Tertiary age, or even earlier.

It is also important to realize that the geomorphological influence of duricrusts will depend to a considerable degree on the stage of evolution which the feature has reached. This affects both the overall thickness, the nature of the constituents, and the degree of induration.

Duricrusts may play a role in relief inversion (Plate 37). In the case of laterite, laterite-covered valleys may become ridges or strings of mesas flanking lower, younger valleys, and pediments may become mesas (McFarlane 1976). The relief of laterite surfaces may be modified by pseudokarstic processes so that the central areas of laterite-capped mesas may become gradually lowered. Thus the periphery stands relatively higher, giving a soup-plate form. Likewise, the tendency for some calcretes and dolocretes to form preferentially in valleys and depressions sometimes leads to inversion of relief in times of greater erosion, whether by water or wind. Examples of such inverted calcrete relief are provided by McLeod (1966).

Summerfield (1978) has also indicated that silcrete can cause relief inversion. In stage 1 of his model silcrete forms in areas subject to inundation and possibly reaches its thickest development in proximity to rivers. In stage 2 rejuvenation of drainage occurs leading to erosion and drainage inversion. Subsequent back-wearing (stage 3) creates silcrete-capped residuals. These may be highly resistant to further destruction by weathering since on a world basis silcretes have a mean silica content of around 96 per cent and may on occasion exceed 99 per cent. Silcrete residuals of Tertiary age are widespread in Europe and Britain, where they are known as sarsen stones.

The presence of duricrust profiles with marked differences in properties between hardpans and some of the more friable and fine-grained materials beneath, creates conditions that favour the formation of PSEUDOKARST produced by subsurface flushing, and in the case of calcrete, solutional effects. Cave formation and roof collapse produce karst-like forms in laterites. Calcretes, because of their high carbonate content and relative solubility, frequently show sinkhole development and pipe formation.

Many workers have used duricrusts as indicators of palaeoclimates, and in broad terms this may be acceptable. Calcretes, for example, are for the most part, though not exclusively, currently forming in semi-arid areas where annual rainfall is around 200–500 mm, so that their presence in various Tertiary sediments in western and central Europe may be used with a fair degree of certainty to infer formerly more arid conditions with an annual water deficit.

Plate 37 A laterite-capped plateau at Panchgani in the Deccan Plateau, India. The laterite acts as a caprock and has resulted from severe tropical weathering acting on Tertiary basalts

Much more controversy surrounds silcrete, however, as indicated by Summerfield (1983), with a range of inferred climatic conditions ranging from extreme arid to humid tropical. Summerfield maintains that silcrete may form under two distinct climatic regimes. He draws a distinction between 'the non-weathering profile' silcretes, which results from localized silica mobility and concentration in high pH environments under a predominantly arid and semi-arid climate, and 'the weathering profile' silcretes whose geochemical and petrographic characteristics are indicative of silicification under a much more humid climate in highly acidic, poorly drained weathering environments.

It is normally accepted that ferricretes and alcretes form under relatively humid conditions. Alternating wet and dry seasons were widely considered to be favourable if not essential to laterite genesis. In particular it was believed that seasonally alternating conditions were necessary for sesquioxide precipitation. However, as McFarlane (1976: 45) has pointed out, there is some evidence for its formation under permanently moist atmospheric conditions.

Duricrusts are widespread features, especially in low latitudes, though relict forms occur in more temperate ones. They have many geomorphological effects. Controversial is the question of their palaeoclimatic significance. They result from a complex interplay of different source materials, transfer processes and precipitation mechanisms in the surface and near-surface environment. In the past the roles of lateral translocations and organic processes have tended to be neglected.

References

Bardossy, G. and Aleva, G.J.J. (1990) *Lateritic bauxites*, Amsterdam: Elsevier.

Goudie, A.S. (1973) *Duricrusts in Tropical and Subtropical Landscapes*, Oxford: Clarendon Press.

——(1983) Calcrete, in A.S. Goudie and K. Pye (eds) *Chemical Sediments and Geomorphology*, 93–131, London: Academic Press.

——(1996) Organic agency in calcrete development, *Journal of Arid Environments* 32, 103–110.

Krumbein, W.E. (1968) Geomicrobiology and geochemistry of the 'Narilime-crust' (Israel), in G. Miller and

G.M. Friedman (eds) *Recent Developments in Carbonate Sedimentology in Central Europe*, 138–147, Heidelberg: Springer Verlag.
Lamplugh, G.W. (1907) Geology of the Zambezi Basin around Batoka Gorge, *Quarterly Journal of the Geological Society of London* 63, 162–216.
McFarlane, M.J. (1976) *Laterite and Landscape*, London: Academic Press.
——(1983) Laterites, in A.S. Goudie and K. Pye (eds) *Chemical Sediments and Geomorphology*, 7–58, London: Academic Press.
McLeod, W.N. (1966) The geology and iron deposits of the Hammersley Range area, Western Australia, *Bulletin Geological Survey of Western Australia* 117.
Ollier, C.D. (1991) Aspects of silcrete formation in Australia, *Zeitschrift für Geomorphologie* 35, 151–163.
Stephens, C.G. (1971) Laterite and silcrete in Australia: a study of the genetic relationships of laterite and silcrete and their companion materials, and their collective significance in the weathered mantle, soils, relief and drainage of the Australian continent, *Geoderma* 5(1), 5–52.
Summerfield, M.A. (1978) The nature and origin of silcrete with particular reference to southern Africa. Unpublished D.Phil. Thesis, University of Oxford.
——(1983) Silcrete, in A.S. Goudie and K. Pye (eds) *Chemical Sediments and Geomorphology*, 59–91, London: Academic Press.
Tardy, Y. (1997) *Petrology of Laterites and Tropical Soils*, Rotterdam: Balkema.
Woolnough, W.G. (1927) The duricrust of Australia, *Journal and Proceedings of the Royal Society of New South Wales* 61, 24–53.
——(1930) Influence of climate and topography in the formation and distribution of products of weathering, *Geological Magazine* 67, 123–132.

A.S. GOUDIE

DUST STORM

A large volume of predominantly silt-sized sediment blown into the atmosphere by a strong wind. The definition most widely used for this type of WIND EROSION OF SOIL event is that devised by meteorologists: a dust-raising event that reduces horizontal visibility to 1,000 m or less.

The entrainment of dust from the ground surface is controlled by the nature of the soil or sediment itself, the nature of the wind, and the presence of any surface obstacles to wind flow. Dust storms can occur in any environment given appropriate conditions of bare, unconsolidated sediment and a strong turbulent wind, but they are most common in deserts and on their margins. Most geomorphologists define dust particles according to the silt/sand boundary (i.e. less than 62.5 μm). The particles that make up desert dust storms are dominated by SiO_2, probably reflecting the importance of quartz in source areas. Grain size, mineralogy and chemical composition can be used to distinguish soil dust from other types of particles in the atmosphere such as those derived from sea salt, volcanoes and smoke particles from fire.

Different terrain types vary greatly in their susceptibility to dust storm occurrence. Important determining factors include the ratio of clay-, silt- and sand-sized particles, the soil moisture content, the compaction of sediments, the presence of particle cements such as salts or organic breakdown products, and the presence of crusts or armoured surfaces (Middleton 1990). The most favourable dust-producing surfaces are areas of bare, loose and mobile sediments containing substantial amounts of sand and silt but little clay. Terrains that satisfy these conditions are most commonly found in geomorphologically active landscapes where tectonic movements, climatic changes and/or human disturbance are responsible for rapid exposure, incision and reworking of sediment formations containing dust. Important sources of dust storms are generally located in specific, relatively small desert environments (Coudé-Gaussen 1984) such as floodplains, alluvial fans, salt pans and devegetated fossil dunes.

The most important meteorological systems capable of generating dust storms are synoptic in scale, dominated by the passage of low pressure fronts with intense baroclinal gradients that are accompanied by very high-velocity winds. Such frontal passage is the dominant dust-generating mechanism in many of the world's dusty regions, including Australia, northern China and Mongolia, Central Asia, the Levant, the Mediterranean coast of north Africa, the Sahelian latitudes of west Africa, the High Plains of the USA and the plains of the Argentine Pampas. Surface cyclones themselves may sweep out gyres of dust when circulation around the low pressure becomes very intense. Other synoptic-scale dust-raising systems include winds generated in areas with steep pressure gradients, such as in the Thar Desert of India and Pakistan and the northwesterly Shamal wind that blows down the Arabian Gulf from Iraq and Kuwait. More localized dust storms occur when katabatic winds deflate mountain foot sediments, as on the northern slopes of Kopet Dag on the Iran-Turkmenistan border, or in California (the Santa Ana wind). The high Andean Altiplano of Chile, north-west Argentina and southern Bolivia experiences strong dust-raising from the upper westerlies and similar upper airflow deflates sediments from the arid Tibetan Plateau. The cold downburst wind of a dry thunderstorm, the classic Haboob of southern Sudan, is perhaps the most common meso-scale dust-raising system, which raises dust at the gust front some kilometres in advance of the towering convective clouds.

Dust transport and deposition

Globally, the amount of material mobilized in dust storms is thought to be around a billion tons a year and up to half of this comes from the Sahara, indicating the geomorphological importance of AEOLIAN PROCESSES in moulding parts of its landscape. The world's two most active dust storm source areas are both in the Sahara: the Bodélé Depression to the south of the Tibesti Mountains and an area covering eastern Mauritania, western Mali and southern Algeria (Goudie and Middleton 2001). Their importance relative to other major global dust sources is indicated in Table 15 which shows maximum mean values of an Aerosol Index (AI) that indicates the intensity of atmospheric dust content. The AI is derived from the satellite-borne Total Ozone Mapping Spectrometer (TOMS) that detects UV-absorbing aerosols in the atmosphere.

Sediments from these and other world dust storm areas are regularly transported over great distances. Saharan dust is transported along three main trajectories: westward over the North Atlantic to North and South America; northward across the Mediterranean to southern Europe and sometimes as far north as Scandinavia, and along easterly trajectories across the eastern Mediterranean to the Middle East. Dust storm material from other major deserts also follows common trajectories, many of which are highly seasonal (Plate 38). They include flows from north-east Asia across the Pacific Ocean and from Mesopotamia down the Arabian Gulf. Dense dust loadings following such trajectories can be discerned on imagery from remote sensing platforms and many techniques have been applied to deposited material in order to detect its source. These include dust mineralogy and elemental composition, scanning electron microscopy of individual grain features and the presence of pollen and foraminifera.

While in the troposphere, dust can have effects on climate through a range of possible influences. Dust outbreaks may affect air temperatures through the absorption and scattering of solar radiation and may cause ocean cooling. Dust-induced changes in atmospheric temperatures and changes in concentrations of potential condensation nuclei may also affect convectional activity and cloud formation, thereby altering rainfall amounts.

Dust aerosols influence the nutrient dynamics and biogeochemical cycling of both marine and terrestrial ecosystems. Much of the material transported over long distances is deposited over the oceans (Prospero 1996) where dust storm sediments provide a major nutrient input. Where deposited on land, dust may affect soil formation. Dust that has a high carbonate content may be a factor in the formation of calcretes and dust

Table 15 Maximum mean AI values for major global dust sources determined from TOMS

Location	
Bodélé Depression of central Sahara	>30
West Sahara in Mali and Mauritania	>24
Arabia (southern Oman/Saudi border)	>21
Eastern Sahara (Libya)	>15
South-west Asia (Makran coast)	>12
Taklamakan/Tarim basin	>11
Etosha Pan (Namibia)	>11
Lake Eyre basin	>11
Mkgadikgadi basin (Botswana)	>8
Salar de Uyuni (Bolivia)	>7
Great Basin of the USA	>5

Source: Goudie and Middleton (2001)

Plate 38 A dust raising event at Disi, south-east Jordan, with dust being raised from a dry playa surface

contributes to the formation of other desert surface coverings such as desert varnish and case hardening of rocks. Salts carried in wind-blown dusts can act as weathering agents and increase the salinity of soils and water bodies.

LOESS is by definition a wind-deposited dust with a median grain size range of 20–30 μm (Tsoar and Pye 1987) and has been estimated to cover up to 10 per cent of the world's land area. Interestingly, however, the occurrence of loess in Africa is very limited, a fact that is surprising given the Sahara's prominence as the world's largest area of contemporary dust storm activity and evidence that suggests it produced more dust during cold phases of the Pleistocene (see below).

Changing frequencies of dust storms

There is considerable evidence that dust storm frequencies can change substantially in response to climatic changes both in the long term (e.g. during the Last Glacial Maximum) and in the short term (e.g. in response to the North Atlantic Oscillation and to drought phases). Analysis of dust in cores taken from deep-sea sediments, ice caps and loess deposits has enabled the reconstruction of long-term changes in dust storm activity. Dust in North Atlantic sediments has been dated back to the early Cretaceous, although aeolian activity in the Sahara appears to have become more active in the late Tertiary and high dust loadings were a particular feature of the Pleistocene in many parts of the world.

Intensification of dust storm activity during glacial periods, such as the Last Glacial Maximum, was probably due in part to lower precipitation, although changes in wind regimes may also have contributed. It has also been suggested that increased atmospheric dust during the Last Glacial Maximum was not only a response to climate change but also a contributory factor to the change, and regional dust loadings are being built into models of climate (Mahowald *et al.* 1999).

Drought is commonly associated with an increase in dust-raising activity as vegetation cover dies off and soils dry out (Brooks and Legrand 2000). In the more recent past, the effects of drought on dust storm activity have sometimes been exacerbated by human influences on the wind erosion system. Human activity has been shown to affect dust storm activity by destabilizing soil surfaces and altering vegetation cover. The most common human impact in this respect is agriculture. Type examples of large areas in semi-arid climates converted from grasslands to cultivation subsequently becoming enhanced dust-producing regions include the Great Plains of the USA in the 1930s, the so-called Dust Bowl, and the Virgin Lands scheme of the former Soviet Union in the 1950s. Other human activities that affect changes in dust storm frequencies through the breakup of wind-resistant surfaces and/or the removal of protective vegetation cover from soils include drainage, construction, vehicle use and military movements. The relationship between dust-raising, environmental change and human impacts has meant that changes in dust storm frequency have been studied as an indicator of DESERTIFICATION.

Dust storm hazards

Airborne dust presents a variety of problems to inhabitants of desert areas. In areas where dust is raised from agricultural fields it represents a serious form of wind erosion while blowing dust and sand can cause considerable damage to crops and natural vegetation by abrasion, which is particularly critical for young shoots when fields are poorly protected by vegetation cover. The reduction in visibility caused by dust storms is a serious hazard to aviation and road transport.

Dust storms are a form of atmospheric pollution and may transmit diseases that affect plants, animals and humans. Fungus carried in Saharan dust has been implicated in disease outbreaks in coral reefs throughout the Caribbean (Smith *et al.* 1996). Micro-organisms blown in dust may settle on the skin, be swallowed or inhaled into respiratory passages. In Arizona, Valley Fever is caused by *Coccidioides immitis*, a common airborne fungus blown by dust storms. Inhalation of fine particles can also aggravate diseases such as asthma, bronchitis and emphysema. These risks to human and ecosystem health have been noted both in dust source areas and in areas of deposition after long-range transport (Griffin *et al.* 2001).

Applied geomorphologists have played a useful role in identifying dust sources and methods for preventing dust entrainment in arid zones. Jones *et al.* (1986) have suggested a general procedure for the assessment of dust hazards in urban areas after their work in the Middle East.

References

Brooks, N. and Legrand, M. (2000) Dust variability over northern Africa and rainfall in the Sahel, in S.J. McLaren and D. Kniverton (eds) *Linking Land*

Surface Change to Climate Change, 1–25, Dordrecht: Kluwer.
Coudé-Gaussen, G. (1984) Le cycle des poussières éoliennes désertiques actuelles et la sédimentation des loess peridésertiques quaternaires, *Bulletin Centre Recherche et Exploration-Production Elf-Aquitaine* 8, 167–182.
Goudie, A.S. and Middleton, N.J. (2001) Saharan dust storms: nature and consequences, *Earth-Science Reviews* 56, 179–204.
Griffin, D.W., Garrison, V.H., Herman, J.R. and Shinn, E.A. (2001) African desert dust in the Caribbean atmosphere: microbiology and public health, *Aerobiologia* 17, 203–213.
Jones, D.K.C., Cooke, R.U. and Warren, A. (1986) Geomorphological investigation, for engineering purposes, of blowing sand and dust hazard, *Quarterly Journal of Engineering Geology* 19, 251–270.
Mahowald, N., Kohfield, K., Hansson, M., Balkanski, Y., Harrison, S.P., Prentice, I.C., Schulz, M. and Rodhe, H. (1999) Dust sources and deposition during the last glacial maximum and current climate: a comparison of model results with paleodata from ice cores and marine sediments, *Journal of Geophysical Research* 104(D13), 15,895–15,916.
Middleton, N.J. (1990) Wind erosion and dust storm prevention, in A.S. Goudie (ed.) *Desert Reclamation*, 87–108, Chichester: Wiley.
Prospero, J.M. (1996) The atmospheric transport of particles to the ocean, in V. Ittekkot, P. Schafer, S. Honjo and P.J. Depetris (eds) *Particle Flux in the Ocean*, 19–52, Chichester: Wiley.
Smith, G.T., Ives, L.D., Nagelkerken, I.A. and Ritchie, K.B. (1996) Caribbean sea fan mortalities, *Nature* 383, 487.
Tsoar, H. and Pye, K. (1987) Dust transport and the question of desert loess formation, *Sedimentology* 34, 139–153.

Further reading

Morales, C. (ed.) (1979) *Saharan Dust*, Chichester: Wiley.
Péwé, T.L. (ed.) (1981) *Desert Dust*, Geological Society of America Special Paper 186.
Pye, K. (1987) *Aeolian Dust and Dust Deposits*, London: Academic Press.

NICHOLAS MIDDLETON

DYE TRACING

A large variety of dyes and related compounds have been used in water tracing and these are generally classified into non-fluorescent and fluorescent dyes, although strictly some of the substances commonly included in the latter category are not dyes but dye intermediaries or optical brightening agents (OBAs). The earliest scientific water-tracing experiments used simple colouring agents which were detected visually. Later it was discovered that some dyes, such as Rhodamine, may be detected on treated cotton hanks. These visual dyes have no advantages and several disadvantages. Visual detection requires high concentrations and observers and of the two dyes which can be absorbed onto cotton hanks, Malachite Green and Rhodamine B, the latter has been shown to be toxic. Hence, they cannot be recommended and are no longer used.

Fluorescent dyes differ from simple colouring agents in that when irradiated at a particular wavelength they emit light at a different, longer, wavelength. Hence, they can be detected in water samples in concentrations invisible to the naked eye; theoretically down to $ng\,l^{-1}$. They have become the most widely used water-tracing substances in limestone areas and have also been used successfully to trace water flow through other fractured rocks, through soils and through peat pipes. There are many fluorescent dyes and they are generally divided into three groups on the basis of their fluorescence spectra: Blue (e.g. Amino G Acid and Optical Brighteners such as Leucophor BS and Tinopal CBS-X and ABP); Green (e.g. Fluorescein (and its disodium salt Uranine), Pyranine and Lissamine) and Orange (e.g. Rhodamine WT, Sulpho-Rhodamine B and Eosine).

Field determination of the green and orange dyes is possible using a portable fluorimeter and in the laboratory a modern scanning spectrofluorimeter can distinguish between different dyes allowing multiple tracing experiments to be undertaken. A further advantage is that several of the green and orange dyes are absorbed by charcoal grains and may be released in the laboratory by an alkaline-alcohol elutant. This is particularly useful if it is necessary to monitor several sites as charcoal bags (variously known as fluocapteurs or receptors) can be deployed at each site and left for up to a week to scavenge dye. Unfortunately the charcoal also scavenges other organic substances which can make interpretation difficult. Treated cotton detectors can be used as fluocapteurs for optical brightening agents.

There are also disadvantages associated with each individual dye. Blue dyes, and especially OBAs suffer from high and very variable background and break down in sunlight; the green dyes fluoresce in the same area as certain organisms and organic substances; certain reds are potentially carcinogenic; and green, and to a lesser extent red, dyes may be lost on sediment.

Further reading

http://www.dyetracing.com. This website provides useful information on dye tracing.

Ward, R.S., Williams, A.T., Barker, J.A., Brewerton, L.J. and Gale, I.N. (1998) Groundwater tracer tests: a review and guidelines for their use in British aquifers, British Geological Survey Report WD/98/19.

JOHN GUNN

DYKE (DIKE) SWARM

A dyke (dike) is a tabular igneous body that was sub-vertical at the time of emplacement.[1] A dyke swarm is a set of coeval dykes which typically display a linear, radiating or arcuate geometry. Dyke swarm compositions range from ultramafic to felsic. The largest swarms are of basaltic composition (with diabasic texture) and are most prominent in basement terranes. Individual diabase dykes can range in width from centimetres to hundreds of metres and in length from metres to 2,000 km or more.

Radiating swarms with radii of about 100 km are associated with individual volcanic centres. Radiating swarms with radii >300 km (referred to as giant radiating dyke swarms) form the feeder systems for large igneous provinces, and are thought to focus above the centres of mantle plumes (Ernst and Buchan 2001: chapters 12 and 19). Magma can be transported both vertically and laterally in dykes. In particular, giant radiating swarms can transport magma laterally more than 2,000 km from the plume centre and can feed sills and volcanic rocks at any distance along their extent. A classic example of a giant radiating swarm is the 1267 Ma Mackenzie swarm of the northern Canadian Shield which fans over an arc of 90° and extends 2,300 km from the focal point (Fahrig 1987). There are also numerous giant radiating swarms on Venus and Mars (Grosfils and Head 1994; Ernst *et al.* 2001).

Many giant linear swarms on Earth may be fragments of giant radiating swarms, which have been dismembered during episodes of continental breakup. However, linear swarms can also be associated with spreading ridges, ophiolite complexes and rift zones. Arcuate portions of otherwise linear or radiating swarms may reflect primary geometry (i.e. changes in the regional stress field) or later deformation. In addition, some arcuate swams occur as a set of ring dykes generated above an intrusion.

In addition to their magmatic significance, dykes can also act as a barrier to groundwater flow, and may localize hydrothermal fluids along their margins. They often weather positively as linear ridges or negatively as troughs due to differential erosion.

Note

1 This differs from the traditional definition of dyke which would allow an originally horizontal sheet to be termed a dyke if it is discordant to the bedding or foliation of its host rocks. Because vertical and horizontal sheets imply fundamentally different stress conditions, originally sub-vertical sheets should be termed dykes and originally sub-horizontal sheets should be referred to as sills, regardless of the degree of discordance. Tabular bodies for which the original orientation cannot be determined or where the dip is intermediate would be termed 'sheets'.

References

Ernst, R.E. and Buchan, K.L. (eds) (2001) *Mantle Plumes: Their Identification Through Time*, Geological Society of America Special Paper 352.

Ernst, R.E., Grosfils, E.B. and Mège, D. (2001) Giant dike swarms: Earth, Venus and Mars, *Annual Review of Earth and Planetary Sciences* 29, 489–534.

Fahrig, W.F. (1987) The tectonic settings of continental mafic dyke swarms: failed arm and early passive margin, in H.C. Halls and W.F. Fahrig (eds) *Mafic Dyke Swarms*, Geological Association of Canada Special Paper 34, 331–348.

Grosfils, E.B. and Head, J.W. (1994) The global distribution of giant radiating dike swarms on Venus: implications for the global stress state, *Geophysical Research Letters* 21, 701–704.

RICHARD E. ERNST AND KENNETH L. BUCHAN

DYNAMIC EQUILIBRIUM

This is a concept which describes a situation of relatively restricted fluctuations about a mean value, together with non-stationarity of that mean. It has been propounded, especially by the American geomorphologist S.A. Schumm (1977, 1991) as a device which permits the short-term variability of 'GRADED TIME' to be viewed alongside the longer trajectory of 'CYCLIC TIME' (most especially for changes in the relief of valleys). In this sense, Schumm, in particular, has sought to bring together the apparently conflicting timescales and explanatory modes which late twentieth-century geomorphologists have often considered to be favoured by G.K. Gilbert and W.M. Davis respectively. The need to reconcile such apparent conflict may be linked to the extremely influential paper by R.J. Chorley (1960) that advocated the shift from an historical emphasis on the long-term and largely predictable progressive change in landforms represented by the Davisian cycle, towards a focus upon the dynamic and process-oriented approach associated with Gilbert and considered to be strongly encouraged by the adoption of systems thinking (see SYSTEMS IN GEOMORPHOLOGY).

At the root of the use of the concept of dynamic equilibrium in late twentieth-century geomorphology is, undoubtedly, Gilbert's 1877 development of the concept of grade (see GRADE, CONCEPT OF), which was then – interestingly – taken up by Davis who incorporated it within his cyclic models (see Chorley *et al.* 1964; Chorley *et al.* 1973). Grade itself has proved one of the thorniest and most elusive geomorphological concepts (cf. Kesseli 1941) and one which, by the year 2000, had more or less vanished from the literature.

Although ideas of equilibria at a variety of timescales and with a variety of theoretical underpinnings characterized much of Anglo-American geomorphology in the latter half of the twentieth century, it became increasingly clear that the concepts work best in very closely defined circumstances, where the physics or chemistry of the situation is relatively unobscured by historically contingent variability which is poorly susceptible to mathematical treatment (cf. Selby's 'strength EQUILIBRIUM SLOPES'). The whole notion of geomorphological equilibria was extensively reviewed by Thorn and Welford (1994; and discussion) and it must remain a matter of opinion whether one accepts their conclusion that the concepts are of central, significant and continuing explanatory value to geomorphology as a whole.

Certainly Schumm's belief in the utility of the notion of dynamic equilibrium has been persistent and pervasive. However, in the light of his equal emphasis on the role of geomorphic thresholds (see THRESHOLD, GEOMORPHIC), it might be argued that a better concept to link the emphases on process and on evolution is his 'Model 2' (Schumm 1977: 12) in which the dynamic equilibrium of cyclic time is replaced, in the same time frame, by dynamic *metastable* equilibrium. Whereas Schumm's dynamic equilibrium (*sensu stricto*) couples a general reduction in relief (that is, a non-stationary mean elevation) with oscillating episodes of cut-and-fill; his dynamic metastable equilibrium assumes long periods of effective stationarity of mean elevation, with variable erosion and aggradation interrupted by abrupt, episodic erosion. This condition would certainly better describe the kind of situation observed in Piceance Creek, Colorado (Schumm 1977: 78–81) where gullying was shown to be episodic. It would also accord more neatly with developing ideas of complex, non-linear models.

It remains the case, however, that both forms of dynamic equilibrium are difficult to identify unambiguously and, further, that neither may be said to add true clarity to our understanding of geomorphic process and form. One central problem is the uncertainty of the temporal and spatial scales at which the 'graded' gives way to the 'cyclic'. Such basic questions of the identification of the crucial timescales and spatial scales of landscape development and the inherent problems of piecing together events on different scales (cf. Schumm and Lichty 1965) remain central (see Church 1996). But whether the concepts of dynamic or dynamic metastable equilibrium are what is needed as the framework to reconcile the process study with the evolutionary one, is far from certain.

References

Chorley, R.J. (1960) Geomorphology and General Systems Theory, *US Geological Survey Professional Paper* 500-B.
Chorley, R.J., Beckinsale, R.P. and Dunn, A.J. (1973) *The History of the Study of Landforms*, vol. 2, London: Methuen.
Chorley, R.J., Dunn, A.J. and Beckinsale, R.P. (1964) *The History of the Study of Landforms*, vol. 1, London: Methuen.
Church, M.A. (1996) Space, time and the mountain – how do we order what we see? in B.L Rhoads and C.E. Thorn (eds) *The Scientific Nature of Geomorphology*, 147–170, Chichester: Wiley.
Kesseli, J.E. (1941) The concept of the graded river, *Journal of Geology* 49, 561–588.
Schumm, S.A. (1977) *The Fluvial System*, Chichester: Wiley.
——(1991) *To Interpret the Earth: Ten Ways to be Wrong*, Cambridge: Cambridge University Press.
Schumm, S.A. and Lichty, R.W. (1965) Time, space and causality in geomorphology, *American Journal of Science* 263, 110–119.
Thorn, C.E. and Welford, M.R. (1994) The equilibrium concept in geomorphology, *Annals of the Association of American Geographers* 84, 666–696 and discussion, 697–709.

Further reading

Stoddart, D.R. (ed.) (1997) *Process and Form in Geomorphology*, London: Routledge.
Summerfield, M.A. (1991) *Global Geomorphology*, Harlow: Longman.

SEE ALSO: complex response; complexity in geomorphology; cycle of erosion; denudation chronology; equilibrium shoreline; geomorphic evolution; punctuated aggradation

BARBARA A. KENNEDY

DYNAMIC GEOMORPHOLOGY

Dynamic geomorphology is defined as an emphasis in geomorphology which treats geomorphic processes as 'gravitational or molecular shear stresses, acting on elastic, plastic or fluid earth materials to produce the characteristic varieties of strain or failure which we recognize as the processes of weathering, erosion, transportation and deposition' (Strahler 1952). This emphasis was first thoroughly exemplified in the work of G.K. Gilbert (1877, 1909, 1914, 1917) and emulated by Brigadier Bagnold (1941) but was largely overlooked by geomorphologists until Horton (1945), Strahler (1952) and Tricart's initiation of the *Revue de Géomorphologie Dynamique* (1950). It is no exaggeration to say that this emphasis is today dominant in Anglo-American and Japanese geomorphology and is often equated with process geomorphology.

The work of G.K. Gilbert is the first seminal antecedent to the study of geomorphic process or dynamic geomorphology. His report on the geology of the Henry Mountains (1877) described the physical erosional processes and derived a system of laws governing progress from initial to adjusted forms; his discussion of the convexity of hill-tops introduced the role of soil creep as a dynamic process; his discussion of transportation of debris by running water was based on flume experimental data; and his paper on hydraulic mining debris in the Sierra Nevada was path-breaking in its recognition of the effects of the passage of a slug of sediment passing through the Sacramento River system over a period of sixty years, from the time of the commencement of gold mining to the time of his publication in 1917. Gilbert, apparently single-handedly, established the paradigm of dynamic or process geomorphology.

Brigadier Bagnold published his monumental *Physics of Blown Sand and Desert Dunes* in 1941. This book remains a sourcebook for students of AEOLIAN PROCESSES. But Bagnold also contributed to the fundamental understanding of beach formation processes and fluvial processes, much of this understanding deriving from his home made 2-m flume which he maintained in his home grounds. The third antecedent of contemporary dynamic geomorphology was the Uppsala school of physical geography in Sweden, where Hjulstrom (1935), Sundborg (1956) and Rapp (1960) transformed our understanding of fluvial process and drainage basin geomorphology. A fourth antecedent is surely the French collaboration of Tricart and Cailleux and the significant influence of Tricart in maintaining the momentum of a journal dedicated to dynamic geomorphology. Tricart's department of applied geomorphology at the University of Strasbourg was unique in continental Europe in its pioneering of this emphasis.

The essential vision of what constituted dynamic geomorphology was developed by Strahler (1952). Shear stresses affecting Earth materials were divided into two major categories: gravitational and molecular. Gravitational stresses activate all downslope movements of matter, hence include all mass movements, all fluvial and glacial processes. Indirect gravitational stresses activate tide- and wave-induced currents and winds. Phenomena of gravitational shear stresses were subdivided according to behaviour of rock, soil, ice, water and air as elastic or plastic solids and viscous fluids.

Molecular stresses are those induced by temperature changes, crystallization and melting, absorption and desiccation, or osmosis. These stresses act in random or unrelated directions with respect to gravity. Surficial creep results from a combination of gravitational and molecular stresses on a slope. Chemical processes of solution and acid reaction were considered separately. Strahler went on to say that a fully dynamic approach requires analysis of geomorphic processes in terms of open systems which tend to achieve steady states of operation and are self-regulatory to a large degree. Finally, he specified that formulation of mathematical models, both by rational deduction and empirical analysis of observational data, to relate energy, mass and time was the ultimate goal of the dynamic approach.

Strahler's motivation was to counteract the heavy emphasis on descriptive, deductive studies of landform development and regional geomorphologies that had come to dominate the subject in the early part of the twentieth century. This dynamic emphasis in geomorphology has also been characterized as functional geomorphology (Chorley 1978) and is contrasted with historical geomorphology.

Strahler (1992: 72–73) describes his encounter with open systems theory (Von Bertalanffy 1950) in the following way: 'It was as if a closed door had opened before me, revealing an entirely new and powerful epistemology of science – a paradigm capable of unifying all dynamic processes and forms that can be observed in the universe.' Strahler (1980) describes the five levels of systems organization which compose his mature reflections on dynamic geomorphology. Level 1, which corresponds closely with his 1952 discussion, concerns the collection of data which are considered potentially useful in understanding the geomorphic system. The data must be quantitative and must be in the fundamental dimensions of mass, length, time, temperature and their products. The system variables are grouped into (a) dynamic, (b) mass-flow, (c) geometry and (d) material property variables. The dynamic variables relate to energy, force and stress; the mass-flow variables express rates of flow of matter; the geometry variables describe size and form, and material property variables include environmental constants and regulators. The second level of analysis relates to morphological elements; the third level examines flow systems of interconnected pathways of transport of energy and matter; the fourth level describes process-form systems, characterized by self-regulation through physical feedback loops; and the fifth level, systems regulated by cybernetic feedback, links natural systems with those regulated and/or disturbed by human intervention. The agenda described is reminiscent of the Chorley and Kennedy (1971) agenda and underlines the close relation between dynamic geomorphology and the general systems framework.

Following Strahler's most important impetus to the development of dynamic geomorphology, the contributions of Schumm, Leopold, Wolman, Gregory and Walling can be seen as setting the seal on the paradigm, especially in the context of fluvial and watershed geomorphology. John Miller, the third of the triumvirate of Leopold and Miller, would surely have had a major influence, perhaps even the greatest influence, had he not died tragically at the age of 39. The reason for such a bold suggestion is that John Miller alone among this group of leaders was a geochemist as well as a geomorphologist and he was attempting to link weathering processes as well as mechanical erosional processes to drainage basin evolution. The themes of dynamic equilibrium, magnitude and frequency (see MAGNITUDE–FREQUENCY CONCEPT) of operation of geomorphic processes, and a strong bias towards fluvial process were to become the hallmark of dynamic geomorphology in the Anglo-American literature. The paradigm became hugely popular partly because of its quantitative rigour, partly because it seemed to provide specific answers in a field where thoughtful arm-waving had become a tradition and partly because it recognized the value of the combination of theory, experiment and practice.

Dynamic geomorphology has now become equated with 'process geomorphology'. A current textbook on process geomorphology notes that valid interpretations of geomorphic history must be based on a thorough understanding of the processes involved in landform development. Geomorphologists therefore must be cognizant of process mechanics prior to analysing how landform history manifests past climatic or tectonic phenomena (Ritter *et al.* 1995). Five basic principles of process geomorphology according to Ritter *et al.* are: (1) a delicate balance or equilibrium exists between landforms and processes. The character of this balance is revealed by considering both landforms and processes as systems or parts of systems; (2) the perceived balance between process and form is created by the interaction of energy, force and resistance; (3) changes in driving force and/or resistance may stress the system beyond the defined limits of stability. When these limits of equilibrium or thresholds are exceeded, the system is temporarily in disequilibrium and a major response may occur. The system will develop a different equilibrium condition adjusted to the new force or resistance controls, but it may establish the new balance in a complex manner; (4) various processes are linked in such a way that the effect of one process may initiate the action of another; and (5) geomorphic analyses can be made over a variety of time intervals. In process studies the time framework utilized has a direct bearing on what conclusions can be made regarding the relation between process and form. Therefore the time framework should be determined by what type of geomorphic analysis is desired.

These principles are clearly related to the earlier dynamic geomorphology of Gilbert, Bagnold, Hjulstrom, Sundborg, Rapp, Cailleux, Tricart and Strahler through the ideas of 'a balanced condition', the centrality of energy, force and resistance, the language of systems theory, complex response and the importance of timescale of study. One implication of the reductionist functionalism of dynamic geomorphology was the emergence of semi-independent geomorphic process schools, such that coastal, slope, glacial, periglacial, karst, aeolian and fluvial geomorphology became more formally differentiated. The demands of learning the mechanics and dynamics of process led to an isolation of the new dynamic geomorphology from those who were more interested in the evolution of landscapes over geological time. The central conundrum was articulated by Church (1980) when he commented that contemporary records of geomorphological processes were not likely to represent long-term behaviour sufficiently well to provide any firm basis for understanding landscape evolution.

Schumm and Lichty (1965) made a significant contribution to the linking of short-term and long-term studies by explicitly recognizing the different status of process variables over cyclic, graded and steady-state timescales. Steady-state timescales were appropriate for process studies; cyclic timescales would be appropriate to geological evolutionary studies and the graded timescale would be appropriate to, perhaps, the Holocene timescale. They suggested a reconciliation between timeless and time-bound aspects of geomorphology by noting that the distinction between cause and effect among geomorphic variables varies with the size of the landform/landscape under consideration as well as with time. Theirs is an interesting and valuable insight, but the problem would seem to arise from an insistence on the idea of balance or equilibrium at the core of dynamic geomorphology. In spite of the power of the dynamic geomorphology paradigm, there remains a tension between the way in which time and space scales of variability are treated and the central assumption of balance and equilibrium. The question at issue is whether the landscape is fundamentally in equilibrium or whether it is fundamentally in a transient state between equilibria which are rarely if ever achieved.

Dynamic geomorphology has revived geomorphology from its pre-Second World War slumber, has connected geomorphology with the other natural sciences of physics, chemistry and biology, and has opened up opportunities for professional accreditation alongside engineers. At the same time, it can be suggested that process geomorphology has, at least for a few decades, lost touch with both traditional geology and geography. With geology in that the diastrophic framework supplied by global plate tectonics has been difficult to marry with local-scale process studies; with geography in that the challenges of global environmental change and the role of human society have been equally difficult to marry with site-scale process studies.

References

Bagnold, R.A. (1941) *The Physics of Blown Sand and Desert Dunes*, London: Methuen.

Chorley, R.J. (1978) Bases for theory in geomorphology, in C. Embleton, D. Brunsden and D.K.C. Jones (eds) *Geomorphology: Present Problems and Future Prospects*, Oxford: Oxford University Press.

Chorley, R.J. and Kennedy, B.A. (1971) *Physical Geography: A Systems Approach*, London: Prentice Hall.

Church, M. (1980) Records of recent geomorphological events, in R.A. Cullingford, D.A. Davidson and J. Lewin (eds) *Time Scales in Geomorphology*, 13–30, New York: Wiley.

Gilbert, G.K. (1877) *Report on the Geology of the Henry Mountains*, Washington, DC: US Geological and Geographical Survey.

——(1909) The convexity of hill tops, *Journal of Geology* 17, 344–350.

——(1914) The transportation of debris by running water, *US Geological Survey Professional Paper* 86.

——(1917) Hydraulic mining debris in the Sierra Nevada, *US Geological Survey Professional Paper* 105.

Horton, R.E. (1945) Erosional development of streams and their drainage basins: hydrophysical approach to quantitative morphology, *Geological Society of America Bulletin* 56, 275–370.

Ritter, D.F., Miller, J.R., Grizel, Y. and Wells, S.G. (1995) Reconciling the roles of tectonics and climate in Quaternary alluvial fan evolution, *Geology* 23, 245–248.

Schumm, S.A. and Lichty, R.W. (1965) Time, space and causality in geomorphology, *American Journal of Science* 263, 110–119.

Strahler, A.N. (1952) Dynamic basis of geomorphology, *Geological Society of America Bulletin* 63, 923–938.
——(1980) Systems theory in physical geography, *Physical Geography* 1, 1–27.
——(1992) Quantitative/dynamic geomorphology at Columbia (1945–1960): a retrospective, *Progress in Physical Geography* 16, 65–84.
Von Bertalanffy, L. (1950) The theory of open systems in physics and biology, *Science* 111, 23–28.

SEE ALSO: cyclic time; dynamic equilibrium; force and resistance concept; graded time

OLAV SLAYMAKER

E

EFFECTIVE STRESS

The difference between total stress (σ) and pore pressure in a material (u), responsible for mobilizing internal friction. Effective stress (σ') is one of the two components of internal stress within a material, alongside pore pressure, and measures the distribution of load carried by the soil over a specific area. The principle of effective stress was developed by Karl Terzaghi between 1923–1936, and is a fundamental theory in soil mechanics. Changes in stress, such as distortion, compression and shearing resistance changes, are due to variations in effective stress. As effective stress values increase, the soil or rock becomes more consolidated, exhibiting a maximum value at complete consolidation and before shear failure. Thus, it is the effective stress that causes important changes in material strength, volume and shape. Long-term SLOPE STABILITY analysis often incorporates effective stress analysis (inclusive of internal stresses), rather than total stress analysis (short-term slope instability).

Further reading

Lade, P.V. and De Boer, R. (1997) The concept of effective stress for soil, concrete, and rock, *Géotechnique* 47(1), 61–78.

Clayton, C.R.I., Muller-Steinhagen, H. and Powrie, W. (1995) Terzaghi's theory of consolidation, and the discovery of effective stress, *Proceedings International Conference on Engineering (Geotechnical Engineering)* 113(4), 191–205.

STEVE WARD

EL NIÑO EFFECTS

Climate oscillations occur at many timescales. For example, in the tropics a sub-annual, intra-seasonal 40–60-day period Madden–Julian oscillation has been identified. At a slightly longer timescale there is a 2 to 2.5 year oscillation in the equatorial jet in the lower stratosphere, and this is called the Quasi-Biennial Oscillation (QBO). Every three to seven years a two or so year-long event occurs which is called the El Niño Southern Oscillation (ENSO). Decadal and interdecadal variability is evident in the North Atlantic Oscillation (NAO). At even longer timescales there are such important phenomena as the Dansgaard–Oeschger Cycles, Bond Cycles and Heinrich events, which occur at century to millennial scales. El Niño is the term used to describe an extensive warming of the upper ocean in the tropical eastern Pacific lasting up to a year or even more. The negative or cooling phase of El Niño is called *La Niña*. El Niño events are linked with a change in atmospheric pressure known as the Southern Oscillation (SO). Because the SO and El Niño are so closely linked, they are often known collectively as the El Niño/ Southern Oscillation or ENSO. The system oscillates between warm to neutral (or cold) conditions every three to four years.

Precipitation and temperature anomalies appear to characterize all El Niño warm episodes. These include:

- The eastward shift of thunderstorm activity from Indonesia to the central Pacific usually results in abnormally dry conditions over north Australia, Indonesia and the Philippines.
- Drier-than-normal conditions are also usually observed over south-eastern Africa and north Brazil.
- During the northern summer season, the Indian monsoon rainfall tends to be less than normal, especially in the north-west of the subcontinent.

- Wetter-than-normal conditions are usual along the west coast of tropical south America, and at subtropical latitudes of North America (the Gulf Coast) and South America (south Brazil to central Argentina).
- El Niño conditions are thought to suppress the development of tropical storms and hurricanes in the Atlantic but to increase the numbers of tropical storms over the eastern and central Pacific Ocean.

In the twentieth century there were around twenty-five warm events of differing strengths, with that of 1997/8 being seen as especially strong. ENSO was relatively quiescent from the 1920s to 1940s.

Severe El Niños, like that of 1997/8, can have a dramatic effect on rainfall amounts. This was shown with particular clarity in the context of Peru (Bendix *et al.* 2000), where normally dry locations suffered huge storms. At Paita (mean annual rainfall 15 mm) there were 1,845 mm of rainfall while at Chulucanas (mean annual rainfall 310 mm) there were 3,803 mm. Major floods resulted (Magilligan and Goldstein 2001).

The Holocene history of El Niño has been a matter of some controversy (Wells and Noller 1999) but Grosjean *et al.* (1997) have discovered more than thirty debris flow events caused by heavy rainfall events between 6.1 and 3.10 Kyr BP in the northern Atacama Desert. The stratigraphy of debris flows has also been examined by Rodbell *et al.* (1999), who have been able to reconstruct their activity over the last 15 Kyr. Between 15 and 7 Kyr BP, the periodicity of deposition was equal to or greater than 15 years and then progressively increased to a periodicity of 2 to 8.5 years. The modern periodicity of El Niño may have been established about 5 Kyr BP, possibly in response to orbitally driven changes in solar radiation (Liu *et al.* 2000). Going back still further, studies of the geochemistry of dated *Porites* corals from the last interglacial of Indonesia have shown that at that time there was an ENSO signal with frequencies nearly identical to the instrumental record from 1856–1976 (Hughen *et al.* 1999).

El Niño events have considerable geomorphological significance. For example, the changes in temperatures of sea water between El Niño and La Niña years have a clear significance for coral reefs (Spencer *et al.* 2000). In 1998, sea surface temperatures in the tropical Indian Ocean were as much as 3–5 °C above normal, and this led to up to 90 per cent coral mortality in shallow areas (Reaser *et al.* 2000). Although other factors may be implicated in coral mortality (e.g. eutrophication, disease, heavy fishing, etc.) large changes in the health of reefs have been noted from remote islands and reefs with low levels of human influence. It would seem that warm conditions between 25 and 29 °C are good for coral growth, but that temperatures above 30 °C are deleterious (McClanahan 2000) and lead to such phenomena as coral bleaching. Bleaching tends to be greatest at shallow depths but a feature of the 1998 event was that it not only caused bleaching of rapidly growing species, but also affected massive species. It also reached to depths as great as 50 m in the Maldives. If global sea temperatures rise as a result of global warming and become closer to the thermal tolerance level of 30 °C, El Niño events of smaller magnitude will be sufficient to cause bleaching. Moreover, the closer the mean sea temperature is to this thermal limit, the longer will be the period for which the tolerances of corals will be exceeded during any El Niño, thereby increasing the likelihood of coral mortality (Souter *et al.* 2000).

Lake levels also respond to El Niño events. El Niño warming in 1997 led to increased rainfall over East Africa that caused Lake Victoria to rise by 1.7 m and Lake Turkana by *c.*2 m (Birkett *et al.* 1999). The abrupt rise in the level of the Caspian Sea (2.5 m between 1978 and 1995) has also been attributed to ENSO phenomena (Arpe *et al.* 2000). Similarly, the 3.7-m rise in the level of the Great Salt Lake (Utah, USA) between 1982 and 1986 was at least partly related to the record rainfall and snowfall in its catchment during the 1982/3 El Niño (Arnow and Stephens 1990). The enormous changes that occur in the areal extent of Lake Eyre in Australia result from ENSO-related changes in inflow, with the greatest flooding occurring during La Niña phases (Kotwicki and Allan 1998).

Some glacier fluctuations are controlled in part by El Niño. Glacier retreat in the tropical Andes can be attributed to increased ablation during the warm phases of ENSO (Francou *et al.* 2000). Conversely, further south, in the southern Andean Patagonia of Argentina, El Niño events have led to increased snow accumulation, causing glaciers to advance so that they create barriers across drainage, creating glacier-dammed lakes (Depetris and Pasquini 2000).

El Niño impacts upon tropical cyclone activity, and the differences in cyclone frequency between El Niño and La Niña years is considerable (Bove *et al.* 1998). For example, the probabilities of at least two hurricanes striking the US is 28 per cent during El Niño years, 48 per cent during neutral years and 66 per cent during La Niña years. There can be very large differences in hurricane landfalls from decade to decade. In Florida, over the period 1851–1996, the number of hurricane landfalls ranged from 3 per decade (1860s, and 1980s) to 17 per decade (1940s) (Elsner and Kara 1999). Given the importance of hurricanes for slope, channel and coastal processes, changes of this type of magnitude have considerable geomorphological significance. Mangroves, for example, are highly susceptible to hurricanes, being damaged by high winds and surges (Doyle and Girod 1997).

Streamflows and sediment yields may also be affected by El Niño events. One area where there have been many investigations of the links between ENSO and streamflow is in the western United States. There is a tendency for the south-west to be wet and the north-east to be dry during the El Niño warm phases (Negative Southern Oscillation Index), and vice versa for La Niña (Cayan *et al.* 1999). There is some evidence that the effect on streamflow is amplified over that on precipitation. A study of sediment yields in southern California showed that during strong El Niño years severe storms and extensive runoff occurred, producing sediment fluxes that exceeded those of dry years by a factor of about five. The abrupt transition from a dry climate to a wet climate in 1969 brought a suspended sediment flux in the rivers of the Transverse Range of 100 million tons, an amount greater than their total flux during the preceding 25-year period (Inman and Jenkins 1999). The wet period from 1978–1983 caused a significant response on alluvial fans and in channels in desert piedmont areas (Kochel *et al.* 1997).

Phases of high sediment yield may themselves have geomorphological consequences. It has been argued, for example, that Holocene beach ridge sequences along the north coast of Peru may record El Niño events that have occurred over the last few thousands of years. The argument (Ortlieb and Machare 1993) is that heavy rainfall causes exceptional runoff and sediment supply to coastal rivers. This, combined with rough sea conditions and elevated sea levels, is potentially favourable for the formation of beach ridge sequences. The high sea levels caused by El Niño, often amounting to 20–30 cm, can contribute to washover of coastal barriers (Morton *et al.* 2000).

Heavy rainfall events associated with ENSO phenomena can cause slope instability. Some of the most distinctive landslides in the south-west of the USA have occurred during El Niño events, and they can be especially serious if the heavy rainfall events occur on slopes that have been subjected to fires associated with previous drought episodes (Swetnam and Betancourt 1990).

On the other hand, exceedingly wet years can in due course cause a great increase in vegetation cover on slopes that may persist for some years and so lead to more stable conditions. In the arid islands of the Gulf of California, for example, plant cover ranges from 0–5 per cent during 'normal' years, but during rainy El Niño periods it rises to 54–89 per cent of the surface available for growth (Holmgren *et al.* 2001). Wet ENSO events can provide rare windows of opportunity for the recruitment of trees and shrubs. Such woodland can be resilient and, once established, can persist.

ENSO can be associated with intensified drought conditions and so can influence the activity of aeolian processes, particularly in areas which are at a threshold for dust entrainment or dune activation. Such areas will be those where in wet years there is just enough vegetation to stabilize ground surfaces. In the USA, dust emissions were greatly reduced in the period 1983–84 following the heavy rainfall of the 1982 El Niño (Lancaster 1997). Likewise, Forman *et al.* (2001) have reconstructed the history of dune movements in the Holocene in the USA Great Plains. They have found that phases of dune activity have been associated with a La Niña-dominated climate state and weakened cyclogenesis over central North America.

References

Arnow, T. and Stephens, D. (1990) Hydrologic characteristics of the Great Salt Lake, Utah, 1847–1986, *US Geological Survey Water-Supply Paper* 2,332, 32.

Arpe, K., Bengtsson, L., Golitsyn, G.S., Mokhov, I.I., Semenov, A. and Sporyshev, P.V. (2000) Connection between Caspian sea level variability and ENSO, *Geophysical Research Letters* 27, 2,693–2,696.

Bendix, J., Bendix, A. and Richter, M. (2000) El Niño 1997/1998 in Nordperu: Anzeichen eines Ökosystem – Wandels? *Petermanns Geographische Mitteilungen* 144, 20–31.

Birkett, C., Murtugudde, R. and Allan, J.A. (1999) Indian ocean climate event brings floods to East Africa's lakes and the Sudd Marsh, *Geophysical Research Letters* 26, 1,031–1,034.

Bove, M.C., Elsner, J.B., Landsea, C.W., Niu, X. and O'Brien, J.J. (1998) Effect of El Niño on US landfalling hurricanes, revisited, *Bulletin American Meteorological Society* 79, 2,477–2,482.

Cayan, D.R., Redmond, K.T. and Riddle, L.G. (1999) ENSO and hydrological extremes in the Western United States, *Journal of Climate* 12, 2,881–2,893.

Depetris, P.J. and Pasquini, A.I. (2000) The hydrological signal of the Perito Moreno Glacier damming of Lake Argentino (Southern Andean Patagonia): the connection to climate anomalies, *Global and Planetary Change* 26, 367–374.

Doyle, T.W. and Girod, G.F. (1997) The frequency and intensity of Atlantic hurricanes and their influence on the structure of the South Florida mangrove communities, in H.F. Diaz and R.S. Pulwarty (eds) *Hurricanes*, 109–120, Berlin: Springer.

Elsner, J.B. and Kara, A.B. (1999) *Hurricanes of the North Atlantic*, New York: Oxford University Press.

Forman, S.L., Oglesby, R. and Webb, R.S. (2001) Temporal and spatial patterns of Holocene dune activity on the Great Plains of North America: megadroughts and climate links, *Global and Planetary Change* 29, 1–29.

Francou, B., Ramirez, E., Cáceres, B. and Mendoza, J. (2000) Glacier evolution in the tropical Andes during the last decades of the 20th century: Chalcaltaya, Bolivia and Antizana, Ecuador, *Ambio* 29, 416–422.

Grosjean, M., Núñez, L., Castajena, I. and Messerli, B. (1997) Mid-Holocene climate and culture changes in the Atacama Desert, Northern Chile, *Quaternary Research* 48, 239–246.

Holmgren, M., Scheffer, M., Ezcurra, E., Gutierrez, J.R. and Mohren, G.M.J. (2001) El Niño effects on the dynamics of terrestrial ecosystems, *Trends in Ecology and Evolution* 16, 89–94.

Hughen, K.A., Schrag, D.P., Jacobsen, S.B. and Hantoro, W. (1999) El Niño during the last interglacial period recorded by a fossil coral from Indonesia, *Geophysical Research Letters* 26, 3,129–3,132.

Inman, D.L. and Jenkins, S.A. (1999) Climate change and the episodicity of sediment flux of small California rivers, *Journal of Geology* 107, 251–270.

Kochel, R.C., Miller, J.R and Ritter, D.F. (1997) Geomorphic response to minor cyclic climate changes, San Diego County, California, *Geomorphology* 19, 277–302.

Kotwicki, V. and Allen, R. (1998) La Niña de Australia – contemporary and palaeo-hydrology of Lake Eyre, *Palaeogeography, Palaeoclimatology, Palaeoecology* 84, 87–98.

Lancaster, N. (1997) Response of eolian geomorphic systems to minor climate change: examples from the southern Californian deserts, *Geomorphology* 19, 333–347.

Liu, Z., Kutzbach, J. and Wu, L. (2000) Modelling climate shift of El Niño variability in the Holocene, *Geophysical Research Letters* 27, 2,265–2,268.

McClanahan, T.R. (2000) Bleaching damage and recovery potential of Maldivian Coral Reefs, *Marine Pollution Bulletin* 40, 587–597.

Magilligan, F.J. and Goldstein, P.S. (2001) El Niño floods and culture change: a late Holocene flood history for the Rio Moquegua, Southern Peru, *Geology* 29, 431–434.

Morton, R.A., Gonzalez, J.L, Lopez, G.I. and Correa, I.D. (2000) Frequent non-storm washover of barrier islands, Pacific coast of Colombia, *Journal of Coastal Research* 16, 82–87.

Ortlieb, L. and Machare, J. (1993) Former El Niño events: records from western South America, *Global and Planetary Change* 7, 181–202.

Reaser, J.K., Pomerance, R. and Thomas, P.P. (2000) Coral bleaching and global climate change: scientific findings and policy recommendations, *Conservation Biology* 14, 1,500–1,511.

Rodbell, D.T., Seltzer, G.O., Anderson, D.M., Abbott, M.B., Enfield, D.B. and Newman, J.H. (1999) An ~15,000-year record of El Niño-driven alluviation in southwestern Ecuador, *Science* 283, 516–520.

Souter, D.W. and Linden, O. (2000) The health and future of coral reef systems, *Ocean and Coastal Management* 43, 657–688.

Spencer, T., Teleki, K.A., Bradshaw, C. and Spalding, M.D. (2000) Coral bleaching in the Southern Seychelles during the 1997–1998 Indian Ocean warm event, *Marine Pollution Bulletin* 40, 569–586.

Swetnam, T.W. and Betancourt, J.L. (1990) Fire-Southern Oscillation Relations in the Southwestern United States, *Science* 249, 1,017–1,020.

Wells, L.E. and Noller, J.S. (1999) Holocene evolution of the physical landscape and human settlement in northern coastal Peru, *Geoarchaeology* 14, 755–789.

A.S. GOUDIE

ELUVIUM AND ELUVIATION

For soils which exist in areas where the water balance is such that rainfall is greater than evaporation, the excess water drains downwards under the influence of gravity. This percolating water can carry material in solution, a process known as leaching. Additionally, material in the form of very fine particles can be moved down in suspension and this is referred to as eluviation – or a 'washing' of particles out of an upper soil horizon, the material being referred to as eluvium. Eluviation can be referred to as mechanical eluviation to distinguish it from losses occuring in solution. When the material becomes redeposited further down the soil profile it is referred to as illuvium, or that material which is washed in to a new lower location by illuviation. The process overall results in the upper soil horizons having a coarser texture, and therefore a greater porosity and permeability, and can result in a finer textured, and sometimes compacted, layer below.

In the lower soil profile, the redeposition of the eluvial material takes place within voids or on the walls of channels, forming a coating of clay round coarser particles in a skin of material where the clay particles are often oriented parallel to each other round the large particle. Such a clay skin is referred to as a cutan, derived from the Latin *cutis*, meaning a skin, coating or rind (cf. cuticle) (Brewer 1964). Such a deposition contributes to the decease of soil pore size and can thus impede further drainage.

Alternatively the eluvial material can be washed out of the soil profile in downslope moving waters such as throughflow or return overland flow. Here the eluvium may be redeposited in or on the soil at the slope foot or washed out of the hillslope system, contributing to the suspended sediment load of rivers and thus forming a constituent of the denudation system of the hillslope in a drainage basin. Whether the eluvial material is redeposited within the soil or reaches the river is largely a matter of the porosity and permeability of the soil and the overall water balance, thus eluviation and hillslope loss to a river is more characteristic of permeable soils in climates with a moderate or high rainfall whereas in less permeable soils and/or with climates with lower rainfall the eluvium is more likely to stay in the soils as redeposited illuvium. Thus, eluviation can form a significant denudation process where there are permeable soils and regoliths. Ruxton (1958) calculated that denudation of hillslopes in the Sudan by eluviation was almost as significant as that by the removal of material in solution, with around 25 per cent of removal by the former and around 35 per cent by the latter.

At the intermediate stage of deposition between the lower soil profile and loss to a river, the formation of clay plains at the foot of slopes along seepage lines can be quite significant. Ruxton (1958) reported such lateral sediment transport and deposition in the form of surface deposits of eluviated clay near the edges of weathered granite domes. The deposits were up to 500 m long, curved round the base of the slope, and up to 150 m wide, though with some longer tongues of deposition where there was evidence of greater water flow and some channelization. Steep (20°) slopes give way sharply to low-angle slopes of deposited fine material below. Here, there was evidently enough rainfall to wash the clay from the higher areas through the bedrock but insufficient runoff to transport the clay further than the slope foot.

Where there are landforms constituted from loose, unstable material, such as sand dunes or even under periglacial conditions with repeated frost heave and downslope movement, eluviation is not a dominant feature as the material is frequently in motion. However, if the material stabilizes – by vegetation growth on dunes or amelioration of climate – then eluviation can occur and this can be referred to as an eluvial phase of development for the soils and associated landform.

References

Brewer, R. (1964) *Fabric and Mineral Analysis of Soils*, Chichester: Wiley.

Ruxton, B.P. (1958) Weathering and sub-surface erosion in granite at the piedmont angle, Balos, Sudan, *Geological Magazine* 95, 353–377.

STEVE TRUDGILL

ENDOKARST

Endokarst consists of the main part of karstic relief and contains carbonate rocks and cave systems (shafts, cavern, etc.). It involves all the underground features of the input karst and of the output karst. It is situated below the EPIKARSTic zone and is fully developed when the karstification is mature (Ford and Williams 1989). It can develop when the acid water from the surface (humic acid from biological activity and carbonic acid from CO_2 exchanges between atmosphere and rainwater) can reach the deepest part of the carbonate (or other karstifiable rocks) layers. It means that the openings of the stratification joints and fractures are large enough to allow the flow of water and suspended material. The development of endokarst is ruled by the competition between dissolution, carbonate precipitation and clogging with non-dissolved particulate material (Rodet *et al.* 1995). Infillings of endokarst conduits are frequently used to date the genesis and the evolution of cave systems (Maire 1990).

The thickness and the lateral extension of the karstifiable layers mark the boundaries of each endokarst. The world's largest cave systems explored by humans are known to exceed 10,000 km^2 and to reach some depths of more than 2,000 m.

References

Ford, D. and Williams, P. (1989) *Karst Geomorphology and Hydrology*, London: Unwin Hyman.
Maire, R. (1990) La haute montagne calcaire, *Karstologia* 3, 731.
Rodet, J., Meyer, R., Dupont J.P., Sayaret, D., Tomat, A. and Viard, J.P. (1995) Relations entre la dissolution des carbonates et le remplissage terrigène dans le karst de la craie en Normandie (France), *Comptes Rendus Academie de Science de Paris*, IIa, 321, 1,155–1,162.

SEE ALSO: chemical weathering; epikarst; ground water; karst; palaeokarst and relict karst

MICHEL LACROIX

ENGINEERING GEOMORPHOLOGY

Engineering geomorphology deals with the geomorphic features of the Earth's surface, with special reference to their engineering properties. These properties include topography, rock units (lithology, rock mass strength, joint spacing, point load range, plasticity characteristics, compaction bearing strength, texture, etc.) soil units, water retention capacity, weathering, mass movement, erosion, etc. The results are very significant in sustainable land management (SLM) through combating processes of land degradation and DESERTIFICATION, and for obtaining a better planning system (see ENVIRONMENTAL GEOMORPHOLOGY).

Engineering geomorphological maps are prepared in a composite form from which collected, processed and stored data, required for a particular project, are extracted and analysed. Derivative maps are also obtained for specific purposes. An engineering geomorphological map (EGM) depicts the morphological and engineering properties of the terrain. It is very useful during the phase of policy or project formulation and implementation, the phase of development and construction, and the continuing management of a development, particularly in the context of civil engineering (see GEOMORPHOLOGICAL MAPPING; TERRAIN EVALUATION).

Various studies have been made in the field of APPLIED GEOMORPHOLOGY that particularly emphasize engineering applications (Cooke and Doornkamp 1974; Hails 1977; Verstappen 1982; Jha and Mandal 1997).

The methodology for the preparation of an engineering map (Table 16) involves three phases: (1) pre-fieldwork; (2) fieldwork; and (3) post-fieldwork. The pre-field phase includes the preparation of the base map from topographical sheets and cadastral maps. Delineation of the area is also done with high-resolution satellite imagery and aerial photographs at a 1:5,000 to 1:25,000 scale.

Table 16 Phases in engineering geomorphological mapping

Pre-fieldwork
Objectives
Framing of delineation rule
Delineation of the study area (topographical and cadastral maps, aerial photographs, and satellite imagery)
Classification criteria
Selection of engineering geomorphological properties
Selection of sites for collection of rock samples
Fieldwork
Field checks, scanning of engineering geomorphological properties
Collection of rock samples, soils, weathering, mass movement and erosional characteristics of rills and gullies
Rock fall, landslides, slope failures and scree zones
Post-fieldwork
Discussion
(Addition/alteration, etc.)
Field mapping corroboration

In the fieldwork phase all the rock units, soil units, tectonic elements, and active geomorphic processes are investigated. Also noted is the water retention capacity of the mass. Ultimately, the areal coverage and locations of occurrences are marked on the base map. The topographic features marked from the toposheets are also updated during field investigation. Rock samples are collected from different litho-stratigraphic units for laboratory analysis to obtain their geo-engineering properties.

The post-fieldwork phase involves laboratory testing of rock samples, transfer of field data relating to mass movement, weathering and erosion, and preparation of the final map and its interpretation. The final map can be prepared by transferring and plotting data obtained from different sources in a synthesized manner, and by using different symbols and colours. During this phase of work, aerial photographs and satellite images are consulted, in addition to the fieldwork, to identify and demarcate the exact location and boundaries of different rock mass strength units, soil units, zones of weathering and areas of active mass movements and erosion, etc. Finally, the engineering geomorphological map for the study area can be prepared.

Engineering geomorphological map preparation (see Figure 54) includes the following physical parameters: topography; rock units; tectonics; soil units; weathering; mass movement; and erosion. The engineering geomorphological map gives special emphasis to the following aspects of the rock mass: strength of intact rock; joint spacing; width of joints; bedding planes; gauge or infilling; the materials and water movements within the rock; point load range; water retention capacity; and compaction bearing strength.

Considering the above attributes of the rock units, the study area can be divided into four categories by superimposing the layering of information on the base map using a Geographic Information System (GIS).

1 Low rock mass strength unit (Rlo)
2 Medium rock mass strength unit (Rme)
3 High rock mass strength unit (Rhi)
4 Very high rock mass strength unit (Rvhi).

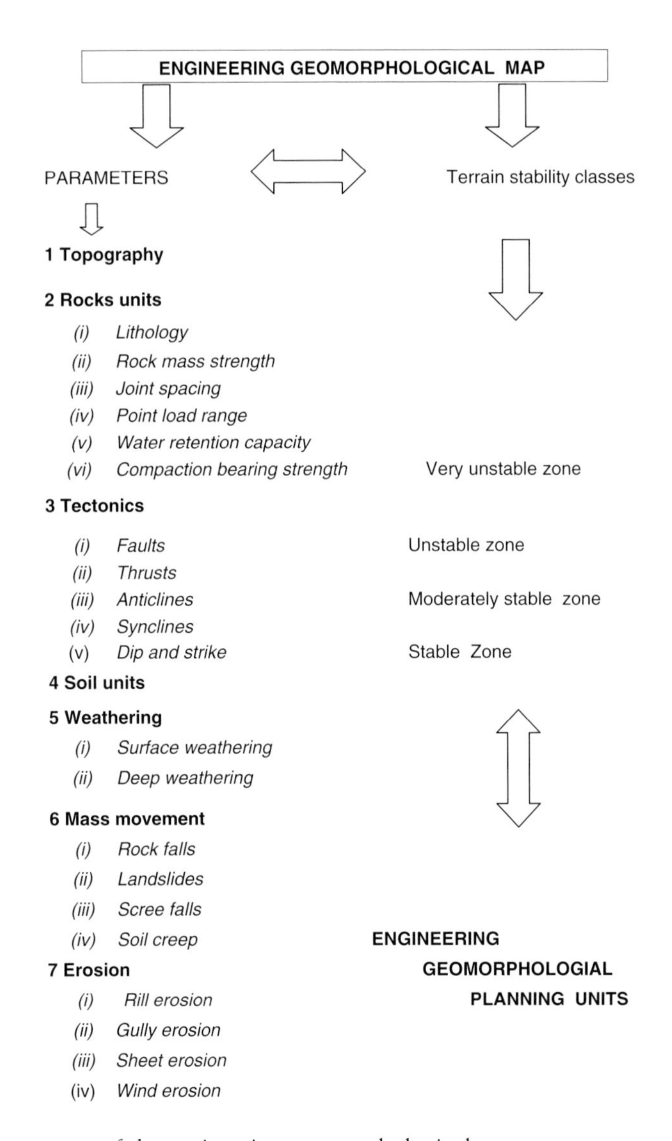

Figure 54 Parameters of the engineering geomorphological map

The engineering geomorphological map shows that all the parameters are functionally interrelated in the sample area. The result can be presented in four terrain stability classes: very unstable zone; unstable zone; moderately stable zone; and stable zone.

An engineering geomorphological map is an important tool in planning and development. It indicates the stability of terrain which is very significant in civil constructions, agriculture, industry, transportation networks and settlement establishment. As this type of map records all sorts of topographical, morphological, and geo-engineering data, so the development and planning of an area should be based on this type of mapping in which the land use and other planning aspects can be regulated according to the stability/suitability of the terrain.

References

Cooke, R.U. and Doornkamp, J. (1974) *Geomorphology in Environmental Management: An Introduction*, Oxford: Clarendon Press.

Hails, J.R. (1977) *Applied Geomorphology*, Amsterdam: Elsevier.

Jha, V.C. and Mandal, U.K. (1997) Drainage Basin in Environmental Management: A Case Study of Eastern Nayer Basin, U.P. Himalayas, in P. Nag, V.K. Kumra and J. Singh (eds) *Geography and Environment*, Vol. 2, 204–225, New Delhi: Concept.

Verstappen, H.Th. (1982) *Applied Geomorphology*, Amsterdam: Elsevier.

VIBHASH C. JHA

ENVIRONMENTAL GEOMORPHOLOGY

Environmental geomorphology is the practical use of geomorphology for the solution of problems where humans wish to transform or to use and change surface processes. According to Coates (1971, 1972–1975), this discipline involves the following issues and themes:

1 the study of geomorphic processes and terrain that affect man, including hazard phenomena such as floods and landslides;
2 the analysis of problems where man plans to disturb or has already degraded the land–water ecosystem;
3 human utilization of geomorphic agents or products as resources, such as water or sand and gravel;
4 how the science of geomorphology can be used in environmental planning and management.

Many other researchers have dealt with environmental geomorphology, in the discussion of both specific topics and the various applications of geomorphology in the forms of APPLIED GEOMORPHOLOGY, ENGINEERING GEOMORPHOLOGY and also engineering geology. It is not necessary to review the available literature here; it is sufficient to mention Tricart (1962, 1973, 1978), Verstappen (1968, 1983), Craig and Craft (1982) and Cendrero *et al.* (1992), among others.

More recently, Panizza (1996) defined *environmental geomorphology* as that area of Earth sciences which examines the *relationships between man and environment*, the latter being considered from the *geomorphological* point of view. It should be further specified that *environment*, in general, is defined (Panizza 1988) as the 'range of physical and biological components that have an effect on life and on the development and activities of living organisms'.

The geomorphological components of the environment may be schematically subdivided into: *geomorphological resources*; and *geomorphological hazards*. Geomorphological resources include both raw materials (related to geomorphological processes) and landforms: both of which are useful to man or may become useful depending on economic, social and technological circumstances. For instance, littoral deposits can become important, economically valuable and considered as geomorphological resources when used for sand quarrying. Similarly, a sea beach can acquire value and be considered as a geomorphological resource when utilized as a seaside resort. A landform can be considered a resource also from the scientific and cultural viewpoint: for example, a marine cliff can be seen as a model of geomorphological evolution.

GEOMORPHOLOGICAL HAZARDS can be defined (Coates 1972–1975) as the 'probability that a certain phenomenon of geomorphological instability and of a given magnitude may occur in a certain territory in a given period of time'. For example, in any one area, the possibility of a certain landslide occurring over a 50-year time span may be assessed. Hazard is therefore a function of the intensity/magnitude and of the frequency/probability of the phenomenon (Varnes 1984). The term 'susceptibility' as used in many mapping procedures (e.g. Brabb *et al.* 1972) corresponds to hazard by equating spatial probabilities to temporal probabilities.

In the context of the relationships with the environment, man represents: *human activity*; and *area vulnerability*. Human activity is the specific action of man which may be summarized under the headings of hunting, grazing, farming, deforestation, utilization of natural resources, engineering works, etc. Man's interventions take place essentially on that thin layer of the Earth's surface which makes up the interface between atmosphere and lithosphere where most energy exchanges and complex phenomena take place (Piacente 1996). Hardly ever are these phenomena confinable within preconstituted and rigid schemes, but nevertheless they can be summarized as follows (Castiglioni 1979): artificial forms, directly modelled by man's activities; works aiming to divert, correct or upgrade natural processes; modifications of natural phenomena, indirectly resulting from man's activities.

Area vulnerability is the complex of the inhabitants and all things that exist as a result of the work of man in a given area and which may be directly or indirectly sensitive to material damage. Included in this complex, we find the population, buildings and structures, infrastructures, economic activity, social organization and any expansion and development programmes planned for an area. In short, it corresponds to an 'exposed element'. Vulnerability can also be defined (Varnes 1984; Einstein 1988) as the level of potential damage (ranging from 0 to 1) to a given exposed element, which is subject to a possible or real phenomenon of a given intensity: we prefer the first definition, which does not imply elements already included in the definition of hazard.

Considering the relationships between geomorphological environment and man, two main possibilities can be examined (Panizza 1992) (Figure 55):

1. Geomorphological resources in relation to human activity, where geomorphological environment is regarded as mainly passive in relation to man (active); in other words, a resource may be altered or destroyed by human activity (e.g. a mountain landscape that has been modified by a bulldozer). We define as *impact* these consequences of human activity on geomorphological resources. It consists of the physical, biological and social changes that human intervention brings about in the environment, the latter term being intended in its geomorphological elements. Therefore, this impact equals the 'product' of human activity and geomorphological resources.
2. Geomorphological hazard in relation to area vulnerability, where geomorphological environment is regarded as mainly active in relation to man (passive): in other words, a hazard may alter or destroy some buildings or infrastructures (e.g. a landslide or river

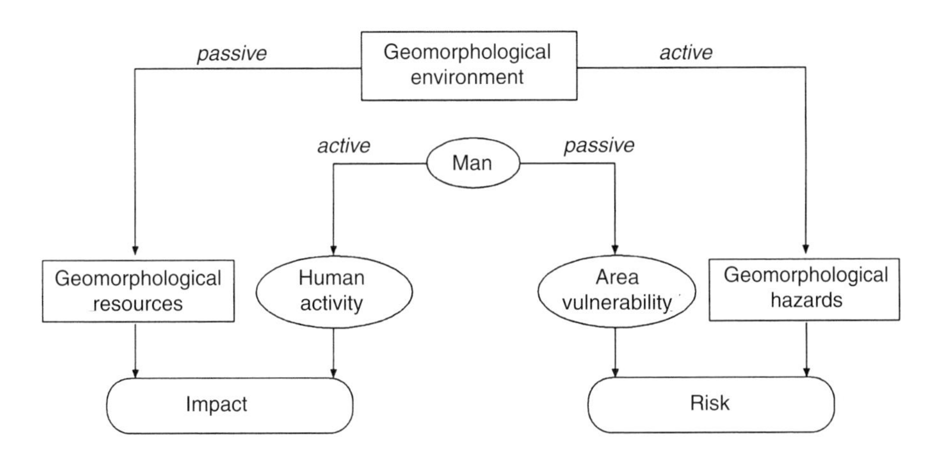

Figure 55 Relationships between geomorphological environment and man

erosion that cause road collapse). We define as *risk* these consequences of geomorphological hazard on a situation of area vulnerability. It is a natural risk connected to a geomorphological hazard: the term refers to the probability that the economic and social consequences of a particular phenomenon reflecting geomorphological instability will exceed a certain threshold. Therefore, this risk is equal to the 'product' of geomorphological hazard and an area's social and economic vulnerability. It corresponds to the term 'specific risk' by Varnes (1984), which expresses the loss due to a particular natural process.

References

Brabb, E.E., Pampeyan, E.H. and Bonilla, M.G. (1972) *Landslide susceptibility in San Mateo county, California*, US Geological Survey Miscellaneous Field Studies Map MF-360, scale 1:62,500.

Castiglioni, G.B. (1979) *Geomorfologia*, Torino: UTET.

Cendrero, A., Luttig, G. and Wolff, F.C. (eds) (1992) *Planning the Use of Earth's Surface*, Berlin: Springer Verlag.

Coates, D. (ed.) (1971) *Environmental Geomorphology*, Proceedings Symposium State University of New York, Binghamton.

——(ed.) (1972–1975) *Environmental Geomorphology and Landscape Conservation*, 3 vols, Stroudsburg, PA: Hutchinson and Ross.

Craig, R.G. and Craft, J.L. (eds) (1982) *Applied Geomorphology*, London: Allen and Unwin.

Einstein, H.H. (1988) *Landslide Risk Assessment Procedure*, International Symposium on Landslides, Lausanne 2, 1,075–1,090.

Panizza, M. (1988) *Geomorfologia applicata. Metodi di applicazione alla Pianificazione territoriale e alla Valutazione d'Impatto Ambientale*, La Nuova Roma: Italia Scientifica.

——(1992) *Geomorfologia*, Bologna: Pitagora.

——(1996) *Environmental Geomorphology*, Amsterdam: Elsevier.

Piacente, S. (1996) Man as geomorphological agent, in M. Panizza (ed.) *Environmental Geomorphology*, Amsterdam: Elsevier.

Tricart, J. (1962) *L'Epiderme de la Terre. Esquisse d'une géomorphologie appliquée*, Paris: Masson.

——(1973) La géomorphologie dans les études integrées d'aménagement du milieu naturel, *Annales de Géographie* 82, 421–453.

——(1978) *Géomorphologie applicab1e*, Paris: Masson.

Varnes, D.J. (1984) *Landslide Hazard Zonation: A Review of Principles and Practice*, Paris: Unesco.

Verstappen, H.Th. (1968) Geomorphology and Environment, inaugural address, Delft: Waltman, 1–23.

——(1983) *Applied Geomorphology*, Amsterdam: Elsevier.

MARIO PANIZZA

EPEIROGENY

In his monograph on Lake Bonneville, Gilbert (1890: 340) formalized the definition of certain tectonic terms: 'Displacements of the Earth's crust which produce mountain ridges are called orogenic...the process of mountain formation is orogeny, the process of continent formation is epeirogeny, and the two collectively are diastrophism.' Bloom (1998: 43) concurred that there was need to describe non-orogenic tectonism (i.e. tectonic movements not associated with mountain belts) and so redefined epeirogeny as: 'continental vertical tectonic movement of low amplitude relative to its wavelength, not within an orogenic belt, that does not deform rocks or the land surface to an extent that is measurable within a single exposure.'

Such broad movements can be either positive (uplift) or negative (subsidence). Epeirogenic uplift can be attributed to MANTLE PLUMES beneath broad areas of continental crust and to such processes as glacio-isostasy. Rates of epeirogeny have generally been thought to be one or two orders of magnitude lower over similar time intervals to rates of orogeny, but studies of present rates of neotectonism suggest this is not invariably the case. In areas of active epeirogeny, river incision may occur (Wisniewski and Pazzaglia 2002).

References

Bloom, A.L. (1998) *Geomorphology*, 3rd edition, New Jersey: Prentice Hall.

Gilbert, G.K. (1890) Lake Bonneville, *United States Geological Survey Monograph* No. 1.

Wisniewski, P.A. and Pazzaglia, F.J. (2002) Epeirogenic controls on Canadian River incision and landscape evolution, Great Plains of Northeastern New Mexico, *Journal of Geology* 110, 437–456.

A.S. GOUDIE

EPIKARST

The epikarst is the uppermost highly weathered layer of karst bedrock beneath the soil (Klimchouk 2000). It is also known as the subcutaneous zone (Williams 1983). Where there is no soil there is still an epikarst, for example beneath limestone pavements and alpine karrenfeld. The epikarst develops because rainwater is acidified by dissolving carbon dioxide in the atmosphere and especially in the soil, thereby producing weak carbonic acid. On percolating downwards from the surface into the bedrock, this water accomplishes most of its dissolutional work within 10 m of the surface, i.e. close to its main source of carbon dioxide. The result is that fissures in the limestone are especially enlarged by corrosion near the surface but taper with depth. Consequently, infiltration of rainwater into the karst is initially rapid, but vertical water flow encounters increasing resistance with depth as fissures become narrower and less frequent. This produces a bottleneck effect after particularly heavy rain, resulting in temporary storage of percolation water in a perched epikarstic aquifer.

Joints, faults and bedding-planes vary spatially within the rock because of tectonic history and variations in lithology. As a result the frequency and interconnectedness of fissures available to transmit flow also varies. Nevertheless, near the surface there is considerable interconnectedness in the horizontal plane; so recharging rainwater tends to be homogenized by lateral mixing. However, in the vertical plane some fissures are more favourable for vertical percolation than others, for example master joints that penetrate numerous beds and especially where several joints intersect. As a result these fissures develop as principal drainage paths. Water in the epikarstic aquifer flows laterally towards them and, as a result, they are subjected to still more dissolution by a positive feedback mechanism and so vertical permeability is enhanced. Water captured within the zone of influence of particular drainage routes becomes increasingly isolated as it percolates downwards from water elsewhere in the epikarst, and so despite the early homogenization it gradually acquires a water quality that reflects the residence time in the epikarst.

The saturated zone in the epikarst is especially well developed after heavy rain, when the epikarstic saturated zone is suspended like a perched aquifer above the main phreatic zone in the karst. The piezometric surface (water table) of the epikarst draws down over a preferred leakage path similar to the cone of depression in the water table over a pumped well. Streamlines adjust and resulting flow within the epikarst is centripetal and convergent on the drainage zone. The diameter of any solution doline that ultimately develops as a consequence of the focused dissolution is determined by the radius of the epikarstic drawdown cone.

References

Klimchouk, A. (2000) The formation of epikarst and its role in vadose speleogenesis, in A.B. Klimchouk, D.C. Ford, A.N. Palmer and W. Dreybrodt (eds) *Speleogenesis: Evolution of Karst Aquifers*, Huntsville: National Speleological Society, 91–99.

Williams, P.W. (1983) The role of the subcutaneous zone in karst hydrology, *Journal of Hydrology* (Netherlands) 61, 45–67.

PAUL W. WILLIAMS

EQUIFINALITY

Equifinality is the principle which states that morphology alone cannot be used to reconstruct the mode of origin of a landform on the grounds that identical landforms can be produced by a number of alternative processes, process assemblages or process histories. Different processes may lead to an apparent similarity in the forms produced. For example, sea-level change, tectonic uplift, climatic change, change in source of sediment or water or change in storage may all lead to river incision and a convergence of form. The usage of the term in this way stems from Chorley (1962) but the related concept of converging landforms was developed earlier by Mortensen (1948), who pointed out that there are many convergences in the landforms of arid and polar regions even though their climates (and therefore by implication their geomorphic process assemblages) are so different.

Perhaps one of the better illustrations of the problem of equifinality concerns the origin of TORS. Four principal theories are held to explain identical landforms: (1) subaerial weathering causes spheroidal modification to the morphology of outcrops produced by differential erosion of the bedrock; (2) exposure of tors is due to a two-stage process: a period of prolonged subsurface groundwater weathering leading to decay of closely jointed rock and spheroidal modification of larger blocks, followed by a period of erosional stripping leading to exposure of the tor at the surface; (3) reduction in area of larger inselbergs by scarp retreat and the formation of pediments; and (4) tors are isolated as a result of freeze–thaw weathering, followed by solifluction over permafrost in a periglacial climate. In the last analysis, it is probable that all four processes are reasonable alternative explanations of the origin of this landform. It is generally accepted that no final descriptive definition is without ambiguity. Hence it is not possible to argue from the presence of this landform alone that a certain sequence of genetic events has occurred. This is the classical concept of equifinality.

Brunsden (1990) in discussing his Proposition 10 – the ability of a landscape to resist impulses of change tends to increase with time – notes that in spite of the existence of complex causes and complex responses in geomorphology 'there is within any tectono-climatic domain a tendency toward an all pervading unity and a repetitive but characteristic geometrical order and regularity.' He explains this tendency as resulting from preferential selection of stable forms; exponential decrease in rate of change; increasing effectiveness of barriers to change; constancy of process; persistence; convergence; over-relaxed systems; self-propagation; preferential fabric relief patterns; and process smoothing and extreme event accumulation.

Haines-Young and Petch (1983) suggest that the concept has been misused in that geomorphologists have invoked equifinality in order to avoid the hard question of specific mode of origin of the landforms in question and, they claim, too rapid an acceptance of equifinality may inhibit the development of general laws or may lead to detailed differences of form being overlooked. They suggest a redefinition of the concept as follows: 'a single landform type is said to exhibit equifinality when it can be shown to arise from a range of initial conditions through the operation of the same causal processes' (1983: 465). In this context, they commend Culling (1957) for his use of the graded stream as an example of equifinality in the sense that whatever the initial conditions, a graded stream will display a similar long profile.

Culling (1987) suggests that the word 'equifinality' is no longer useful because advances in our understanding of dynamic systems have opened up a new and richer world with its own more flexible vocabulary. The idea that a system will strive to arrive at similar positions in phase or state space despite differing initial conditions has become familiar to students of general systems theory. The existence of strange attractors in non-linear dissipative dynamic systems is also reminiscent of the older idea of equifinality, but there are equal evidences of chaotic motion in systems that are fully determined and predictable with accuracy in the immediate future. Because of the ubiquity of noise, all stable systems are transient. It is the recognition of complicated periodic behaviour at points far from equilibrium and its interaction with strange attractors that upsets a simple definition of equifinality. Culling proposes a complex topology of degrees of equifinality, which depart from the definition of equifinality *sensu stricto*. That definition remains 'that upon perturbation a system will eventually return to its initial position'. It is important to realize that such a condition is itself transient. Therein lies the essence of the flexibility of the new approach to the concept of equifinality.

Phillips (1995) discusses the value of viewing landscape evolution as an example of self-organization. Self-organization or self-regulation depend upon the dominance of negative feedback in the system. He points out that geomorphic systems give evidence of both self-organization (as in the case of at-a-station hydraulic geometry) and non-self-organization (as in soil landscape evolution). The challenge for geomorphology is how to distinguish between such systems. Self-similarity and equifinality are incompatible with non-self-organizing systems.

Culling (1987) argues for a variety of looser definitions of equifinality than that of the strict definition above. He suggests that a quasi-equifinality can be defined to include approximate return to initial conditions, by placing a restriction on the allowable magnitude of perturbation, by defining a system domain whose manifold has several local minima and in accepting the retention of certain ergodic and topologic properties as adequate criteria for equifinality.

The debate is reminiscent of the dynamic equilibrium debate of thirty years ago, but with two developments: the level of analytical sophistication has increased and, perhaps more importantly, the debate is open ended and does not yield a unique conclusion. 'In looking once again at the concept of equifinality, it is as if we had opened some magic casement to find, between chance and necessity, one dimension and the next, a whole new world of chaotic motions, strange attractors and periodic windows. With a wild surmise we gaze upon an ocean of discovery between two continents previously thought contiguous' (Culling 1987: 69).

References

Brunsden, D. (1990) Tablets of stone: toward the ten commandments of geomorphology' *Zeitschrift für Geomorphologie Supplementband* 79, 1–37.

Chorley, R.J. (1962) Geomorphology and general systems theory, *US Geological Survey Professional Paper* 500-B.
Culling, W.E.H. (1957) Multicyclic streams and the equilibrium theory of grade, *Journal of Geology* 65, 259–274.
——(1987) Equifinality: modern approaches to dynamical systems and their potential for geographical thought, *Institute of British Geographers Transactions* N.S. 12, 57–72.
Haines-Young, R.H. and Petch, J.R. (1983) Multiple working hypotheses: equifinality and the study of landforms, *Institute of British Geographers Transactions* N.S. 8, 458–466.
Mortensen, H. (1948) Das morphologische Harteverhaltnis Hornfels-Granit, in *Harz. Nachrichten Akademische Wissenschaften Göttingen, Mathematische-Physische Klass*, 8–20.
Phillips, J.D. (1995) Self-organization and landscape evolution, *Progress in Physical Geography* 19, 309–321.

SEE ALSO: non-linear dynamics; tor

OLAV SLAYMAKER

EQUILIBRIUM LINE OF GLACIERS

The position on a glacier where seasonal accumulation equals seasonal ablation is termed the equilibrium line (refer to Figure 56). At this area on a glacier the net mass balance is at zero (no ice mass is lost or gained at this point). A glacier gains mass during winter as snow falls, causing a positive annual mass balance above the equilibrium line. Below the equilibrium line, the annual mass balance of glaciers is negative due to ablation during the summer melt season (Sugden 1982).

The equilibrium line altitude (ELA) for a particular glacier budget year is considered synonymous with the end of summer snowline (EOSS). The altitude of the annual EOSS averaged over many years, defines the steady-state ELA. The annual snowline position with respect to the long-term or steady-state ELA can be used as a surrogate or index of the annual mass balance changes of a glacier. Changes in glacier mass balance are a direct, undelayed response to changes in atmospheric conditions (Fitzharris *et al.* 1997) and hence can be a useful indicator of larger scale changes in global climate.

A commonly used method to work out the ELA calculates the area of the glacier, comparing this to the accumulation area. This is termed the accumulation area ratio (AAR). Glacier studies worldwide have demonstrated that the AAR for glaciers with stable ELA has a value of around 0.6 (Lowe and Walker 1997). Where it is possible to estimate the extent of late Pleistocene glaciers, this method can describe the change in ELA compared with the present, and consequently the changes in climate over time.

References

Fitzharris, B.B., Chin, T.J. and Lamont, G.N. (1997) Glacier balance fluctuations and atmospheric circulation patterns over the Southern Alps, New Zealand, *International Journal of Climatology*, 17, 745–763.
Lowe, J.J. and Walker, M.J.C. (1997) *Reconstructing Quaternary Environments*, 2nd edition, 43–44, Harlow: Addison-Wesley Longman.
Sugden, D. (1982) *Arctic and Antarctic, A Modern Geographical Synthesis*, New Jersey: Barnes and Noble.

BLAIR FITZHARRIS

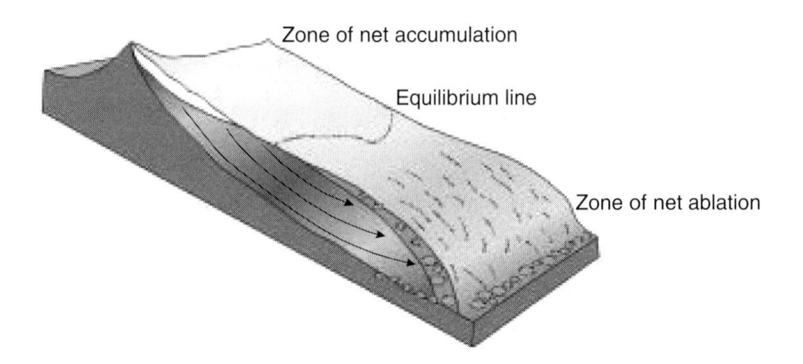

Figure 56 The longitudinal profile of a valley glacier, showing the area of seasonal accumulation and ablation, and the equilibrium line

EQUILIBRIUM SHORELINE

A beach face is in a state of DYNAMIC EQUILIBRIUM when the same amount of sediment is moved landwards by the stronger uprush as is moved seawards by the weaker backrush. This is accomplished by adjustments to the gradient and shape of the BEACH. The gradient is largely determined by the amount of water that percolates into the beach and is lost to the downrush, which is primarily a function of grain size. Pebble or shingle beaches (see SHINGLE COAST) are particularly steep because of rapid percolation and the consequently very weak backrush. Much less water percolates into fine-grained sandy beaches, however, and the weak gravitational effect of a gently sloping beach face is therefore able to compensate for small differences in the onshore and offshore transport rates. There is also a tendency for beach gradient to decrease as wave steepness increases, presumably because the greater velocity of the uprush makes it easier to carry sediment up the slope. The gradient of beaches, or portions of beaches, with the same grain size can therefore vary according to differences in exposure and wave steepness, while temporal variations in wave steepness explain why beach gradient changes during and following storms. There is also evidence to suggest that beach slope is partly determined by the height or energy of the WAVE, which would explain why, for the same wave steepness, slopes are generally greater in low than in high energy environments. The proportion of heavy minerals in a beach, which increases the resistance to removal by backrush, may also be significant in determining the equilibrium gradient.

It is difficult to know if beaches are in quasi-equilibrium, and it can also be difficult to define the slope of beaches that consist of more than one slope element, although it is usually measured in the swash zone where it is essentially linear. A variety of descriptive, empirical and mathematical models are concerned with the relationship between equilibrium beach slope, grain size and wave parameters. The first attempts to model beach gradient were largely statistical and concerned with correlations with wave and sedimentological parameters. Analytical models solve equations for equilibrium gradient on the assumption that there is no net sediment transport, whereas iterative models simulate beach development until an equilibrium slope has developed.

Bruun (see BRUUN RULE) suggested that the shape of equilibrium beach profiles can be represented by the power law:

$$h_x = A\, x^{2/3}$$

where h_x is the depth at a distance x offshore of the mean water line and A is a scale parameter that is largely determined by grain size or fall velocity. If there has been significant sorting, however, coarser sediment can make the shoreward portions of beaches steeper than model predictions, and finer sediment can make the seaward portions more gentle than predicted. Although other workers have provided alternate expressions for the geometry of the equilibrium profile, the use of a single equilibrium equation to represent all beach profiles has been criticized, and the concept of a profile of equilibrium has been questioned.

Beaches adjust their equilibrium morphology and sediments with variations in waves, tides and other influences (Short 1999). Two profiles represent the extremes of a fairly predictable range of forms that can be assumed according to the size or power of the waves. The distinction has been made between reflective profiles with wide berms or swash bars and steep foreshore slopes, sometimes with steps at the breaker line, and dissipative profiles with gentler foreshore slopes and longshore submarine bars (see BAR, COASTAL). Reflective profiles change into dissipative profiles during storms, when large waves move sediment seawards, whereas the reverse occurs when smaller swell waves move sediment back onshore. Frequent changes in wave power are responsible for cycles that are frequently much shorter than that between the two extremes, and wave environments therefore tend to generate globally distinctive beach state characteristics. The mid-latitudes, for example, have persistently high wave power and the beaches are generally kept in a highly dissipative state, whereas beaches in low swell or sheltered environments are normally in reflective states. Beach states also change with tidal level, and the microtidal model has to be modified for areas with a high tidal range. Equilibrium beach profiles and beach states have been modelled as a function of the relative tidal range (RTR) – the ratio of tidal range to breaker height and the dimensionless fall velocity of the sediment, with high RTR values representing tide-dominance and low values wave-dominance.

Coastlines trend towards an equilibrium state in the longshore as well as in the cross-shore direction. The distinction can be made between swash- and drift-aligned equilibrium beach forms (Plates 39 and 40). Swash-aligned beaches are parallel to the incoming wave crests and net LONGSHORE (LITTORAL) DRIFT is at a minimum. Drift-aligned beaches are parallel to the line of maximum drift and sediments can be carried great distances in one direction. Swash-aligned beaches are associated with irregular coasts where longshore transport is impeded, and the important wave trains reach the shoreline almost normally. Drift-aligned beaches develop where the initial coastal outline is fairly regular, or where important sediment-moving waves approach the coast at an angle.

Drift alignment and dynamic beach equilibrium require a constant supply of sediment, as for example when a coastal cell is coupled to the mouth of an ESTUARY. Static beach equilibrium can be attained in several ways (Carter 1988):

1 through swash-alignment, when sediment movement is restricted to cross-shore transport;
2 by strong wave height gradients or the interaction of two wave trains causing the longshore current velocity to become zero; and
3 by the alongshore grading of beach sediments in such a way that the strength of the current at each place is too low to entrain it.

All three situations are common, and in some cases equilibrium is attained through a combination of options.

Plate 39 Swash-aligned beach at San Martinho do Porto, Portugal

Plate 40 Drift-aligned beach, St Petersburg, Florida

References

Carter, R.W.G. (1988) *Coastal Environments*, London: Academic Press.
Short, A.D. (ed.) (1999) *Handbook of Beach and Shoreface Morphodynamics*, Chichester: Wiley.

Further reading

Trenhaile, A.S. (1997) *Coastal Dynamics and Landforms*, Oxford: Oxford University Press.

ALAN TRENHAILE

EQUILIBRIUM SLOPE

Equilibrium slopes are hillslopes which are characterized by an equilibrium of forces that compensate each other. An equilibrium slope exists 'if the amount of material that is removed from an areal unit of the surface per time unit is equal to the amount of material that is supplied to this areal unit during the same time' (Ahnert 1994). This definition follows the conception that a geomorphic system, e.g. a hillslope, is under equilibrium conditions if the 'mass budget' of that system does not change. Hillslopes are in equilibrium conditions if the processes acting upon the hillslope are in equilibrium. Each change of this equilibrium will result in adjustments of the acting processes towards a new equilibrium.

This definition goes back to the fundamental work of Grove Karl Gilbert on the Henry Mountains (Gilbert 1877). In this publication Gilbert introduced the concept of equality of action: 'Erosion is most rapid where the resistance is least, and hence as the soft rocks are worn away the hard are left prominent. The differentiation continues until an equilibrium is reached through the law of declivities (slope gradient). When the ratio of erosive action as dependent on declivities becomes equal to the ratio of resistances as dependent on rock character, there is equality of action.' This situation is called 'dynamic equilibrium' because the system equilibrium is attained by mechanisms of self-regulation, where a change of process components caused by a change of input will result in a compensation between these process components by negative feedbacks (see DYNAMIC EQUILIBRIUM). The negative feedback between processes governing the system causes adjustments to the changes of inputs (e.g. a climatic change or a BASE LEVEL lowering) towards a new equilibrium. In this way the two central aspects (1) of mass transport rates and (2) negative feedback mechanisms were established in geomorphology.

Gilbert (1877) distinguished two types of transport laws of hillslopes: weathering-limited and transport-limited regolith removal. Weathering-limited transport occurs where the weathering rate is lower than the transport capacity of the hillslope forming processes, so that the regolith is removed and slope development is related to the weathering rate of rocks. In this case the slope system is in non-equilibrium, the inverse of equilibrium. The material supply by different slope processes (weathering, slope wash, soil creep, etc. from upslope) is smaller than the potential rate of removal. These form elements can be found in arid and semi-arid environments, in mountain areas and on free faces of cliffs and all stream channels in bedrock. Transport-limited transport occurs where the weathering rate is higher than the transport capacity of the hillslope-forming processes, so that regolith accumulates and the transport processes operate at their full capacities.

The concept of equilibrium slopes has been applied to numerous investigations in geomorphology related to slope evolution by river incision and undercutting followed by mass movements if internal frictional threshold angles of the regolith are crossed. Examples are given, for instance, by the pioneering research of Strahler (1950), who related statistically maximum valley-side slope gradients with the frictional threshold angles of up to 1-m thick regolith cover. Similar equilibrium approaches were published by Young (1972) and Carson (1975).

Further approaches are related to finding characteristic form slope profiles (equilibrium profiles) for a range of transport processes by empirical modelling transport capacity relationships. Kirkby (1971) used this relationship to derive characteristic equilibrium slope profiles for soil creep, soil wash without and with gullying and rivers. Ahnert (1976) developed a more complex computer model which generates five different equilibrium slope profiles related to splash (convex), suspended load-wash (convex–straight–concave), point-to-point wash (rolling) (convex–straight), plastic flow and viscous flow (convex–straight) processes. These models are based on equilibrium assumptions concerning mass transport rates of different processes and feedback mechanisms between them.

There are extreme events that destroy the equilibrium on hillslopes, e.g. by removing the regolith by extreme rainfall, landsliding, gully erosion or vegetation change. In this situation the process rates change significantly in time, which means that the system is in a state of disequilibrium. If the entire slope system has the tendency towards a dynamic equilibrium, negative feedbacks adjust the slope after these external impacts. The period of recovery from this event depends on the constitution of the system itself and on the magnitude of the external impact. If the regolith coverage has been removed and the bare rocks are exposed, system response will result in an adjustment by an increase of the rate of the weathering processes.

In a recent debate Ahnert (1994) and Thorn and Welford (1994) reviewed different equilibrium definitions and concepts in geomorphology. The authors were especially concerned with a high degree of confusion generated by different types of equilibrium concepts used. Based on the very clearly defined concept of Ahnert (1994), Thorn and Welford (1994) suggested the use in the future of a mass-based equilibria concept which is based on field data, namely mass volume and mass flux. They suggested one should abandon the term 'dynamic equilibrium' and use the term 'mass flux equilibrium' to avoid associations with former definitions and concepts and to reach a clearer coupling with other disciplines.

References

Ahnert, F. (1976) A brief description of a comprehensive three-dimensional process-response model of landform development, *Zeitschrift für Geomorphologie, N.F. Supplementband* 25, 29–49.

——(1994) Equilibrium, scale and inheritance in geomorphology, *Geomorphology* 11, 125–140.

Carson, M.A. (1975) Threshold and characteristic angles of straight slopes, *Proceedings of the 4th*

Guelph Symposium on Geomorphology, 19–34, Norwich: Geo Books.
Gilbert, G.K. (1877) *Report on the Geology of the Henry Mountains*, Washington, DC: US Geological and Geographical Survey.
Kirkby, M.J. (1971) Hillslope process-response models based on the continuity equation, *Institute of British Geographers Special Publication* 3, 15–30.
Strahler, A.N. (1950) Equilibrium theory of erosional slopes approached by frequency distribution analysis, Part I and Part II, *American Journal of Science* 248, 673–696, 800–814.
Thorn, C.E. and Welford, M.R. (1994) The equilibrium concept in geomorphology, *Annals of the Association of American Geographers* 84, 666–696.
Young, A. (1972) *Slopes*, London: Longman.

SEE ALSO: dynamic equilibrium; equilibrium shoreline; hillslope, form; hillslope, process

RICHARD DIKAU

ERGODIC HYPOTHESIS

Ergodicity is an idea developed in physics. In studying the movement of molecules in a macroscopic system (such as a room full of air), physicists faced a difficult problem: innumerable molecules move very fast compared with the time taken to observe them. To overcome this problem, they devised the ergodic theorem and the ergodic hypothesis, the word ergodic coming from the Greek *ergon* ('work' or 'energy') and *hodos* ('road') and meaning a 'path of constant energy'.

To appreciate the ideas behind ergodicity, take the case of people in a maze. Now, a maze is a network consisting of a number of links joined at nodes. Imagine one person entering the maze, which has no exit, and wandering around long enough to have entered all possible links at least once. By keeping a record of the path taken by the person, it is possible to calculate two probabilities. First, is the probability of the person's being in a particular link after a given time. Second, is the probability that the person has spent so many minutes in a particular link. Alternatively, in a new experiment, imagine that a large number of people enter the maze (again the maze has no escape route). After sufficient wandering (sufficient for an equilibrium to obtain), the probability of a person's being in a particular link may be given as the ratio of the number of people in that link to the total number of people in the maze. Therefore, by taking an aerial picture of the maze, it is possible to say how much time a person would spend in each link had he or she wandered around the maze for a long time. The first case specifies the relative amount of time spent by one person in each link; the second case specifies the relative number of people in different parts of the maze in an instant of time after equilibrium prevails. The system is ergodic when these two probabilities are the same. In formal language, the statistical properties of a time series of a phenomenon (the individual maze-wanderer) are essentially the same as a set of observations made on a spatial ensemble (the spatial distribution of collective maze-wanderers) at a single time. In other words, ensemble averages can replace time averages in large-scale statistical statements. The individual maze-wanderer exemplifies the ergodic hypothesis, the collective maze-wanderers the ergodic theorem, which states that sampling across an ensemble (the people in the maze) is equivalent to sampling through time for a single system (the lone maze person).

How does this reasoning apply to geomorphology? One might discover that of all slopes in a region, 9 per cent stand at 6 degrees. If ergodic conditions apply, then the ergodic hypothesis would predict that the region would have 6-degree slopes for 9 per cent of its lifetime. In practice, few geomorphic applications of ergodicity make such quantitative statements about time using spatial data. This dearth of applications results largely from the strict conditions required for ergodic arguments to hold, including the difficulty of finding equilibrium landforms in an environment that is constantly changing. The few geomorphological applications that do meet the stringent statistical demands of the ergodic assumptions include studies of geomorphic magnitudes and frequencies, 'threshold' hillslopes, and the growth of drainage basins (see Paine 1985); river channel evolution (Zhang *et al.* 1999); and a general analogy between statistical thermodynamics and the transfer of mass within a landscape (Scheidegger 1991: 254).

A far commoner practice in geomorphology is to study change through time by identifying similar landforms of differing age at different locations, and then arranging them chronologically to create a time sequence or topographic chronosequence. Such space–time, or – more strictly – location–time, substitution has proved salutary in understanding landform development. Two broad types of location–time substitution are used. The first looks at equilibrium ('characteristic') landforms and the second looks at non-equilibrium ('relaxation') landforms.

In the first category of location–time substitution, the assumption is that the geomorphic processes and forms being considered are in equilibrium with landforms and environmental factors. For instance, modern rivers on the Great Plains display relationships between their width–depth ratio, sinuosity and suspended load, which aid the understanding of channel change through time (Schumm 1963). Allometric models are a special case of this kind of location–time substitution (see Church and Mark 1980).

Studies in the second category of location–time substitution, which look at developing or 'relaxation' landforms, bear little relationship to the ergodicity of physics. The argument runs that similar landforms of different ages occur in different places. A developmental sequence emerges by arranging the landforms in chronological order. The reliability of such location–time substitution depends upon the accuracy of the landform chronology. Least reliable are studies that simply assume a time sequence. Charles Darwin, investigating coral-reef formation, thought that barrier reefs, fringing reefs and atolls occurring at different places represented different evolutionary stages of island development applicable to any subsiding volcanic peak in tropical waters. William Morris Davis applied this evolutionary schema to landforms in different places and derived what he deemed was a time sequence of landform development – the Geographical Cycle – running from youth, through maturity, to senility. This seductively simple approach is open to misuse. The temptation is to fit the landforms into some preconceived view of landscape change, even though other sequences might be constructed.

More useful are situations where, although an absolute chronology is unavailable, field observations enable geomorphologists to place the landforms in the correct order. This occasionally happens when, for instance, adjacent hillslopes become progressively removed from the action of fluvial or marine processes at their bases. This has happened along a segment of the South Wales coast, in the British Isles, where the Old Red Sandstone cliffs between Gilman Point and the Taf estuary have been affected by a sand spit growing from west to east (Savigear 1952). In consequence, the westernmost cliffs have been subject to subaerial denudation without waves cutting their bases the longest, while the cliffs to the east are progressively younger.

Relative-age chronosequences depend upon some temporal index that, though not fixing an absolute age of landforms, enables the establishment of an interval scale. For example, the basin hypsometric integral and stream order both measure the degree of fluvial landscape development and are surrogates of time (e.g. Schumm 1956).

The most informative examples of location–time substitution arise where absolute landform chronologies exist. Historical evidence of slope profiles along Port Hudson bluff, on the Mississippi River in Louisiana, southern USA, revealed a dated chronosequence (Brunsden and Kesel 1973). The Mississippi River was undercutting the entire bluff segment in 1722. Since then, the channel has shifted about 3 km downstream with a concomitant stopping of undercutting. The changing conditions at the slope bases have reduced the mean slope angle from 40° to 22°.

Location–time substitution does have pitfalls. First, not all spatial differences are temporal differences because factors other than time exert an influence on landforms. Second, landforms of the same age might differ through historical accidents. Third, equifinality, the idea that different sets of processes may produce the same landform, may cloud interpretation. Fourth, process rates and their controls may have changed in the past, with human impacts presenting particular problems. Fifth, equilibrium conditions are unlikely to have endured for the timescales over which the locational data substitute for time, especially in areas subject to Pleistocene glaciations. Sixth, some ancient landforms are relicts of past environmental conditions and are in disequilibrium with present conditions.

Many geomorphologists substitute space for time to infer the nature of landform change. Only a handful of these adhere to the statistical assumptions of ergodicity. Nonetheless, the loose application of the ergodic reasoning, as seen in location–time substitution, is a productive line of geomorphological enquiry.

References

Brunsden, D. and Kesel, R.H. (1973) The evolution of the Mississippi River bluff in historic time, *Journal of Geology* 81, 576–597.

Church, M. and Mark, D.M. (1980) On size and scale in geomorphology, *Progress in Physical Geography* 4, 342–390.

Paine, A.D.M. (1985) 'Ergodic' reasoning in geomorphology: time for a review of the term? *Progress in Physical Geography* 9, 1–15.
Savigear, R.A.G. (1952) Some observations on slope development in South Wales, *Transactions of the Institute of British Geographers* 18, 31–51.
Scheidegger, A.E. (1991) *Theoretical Geomorphology*, 3rd completely revised edn, Berlin: Springer.
Schumm, S.A. (1956) Evolution of drainage systems and slopes in badlands at Perth Amboy, New Jersey, *Bulletin of the Geological Society of America* 67, 597–646.
——(1963) Sinuosity of alluvial rivers on the Great Plains, *Geoglogical Society of America Bulletin* 74, 1,089–1,099.
Zhang, O., Jin, D. and Chen, H. (1999) An experimental study on temporal and spatial processes of wandering–braided river channel evolution, *International Journal of Sediment Research* 14, 31–38.

Further reading

Burt, T. and Goudie, A. (1994) Timing shape and shaping time, *Geography Review* 8, 25–29.
Thorn, C.E. (1988) *An Introduction to Theoretical Geomorphology*, 46–51, Boston: Unwin Hyman.
Thornes, J.B. and Brunsden, D. (1977) *Geomorphology and Time*, 19–27, London: Methuen.

RICHARD HUGGETT

ERODIBILITY

Erodibility is the resistance of surface material to erosion. It is usually restricted to soils or REGOLITH, and to water or WIND EROSION OF SOIL. Both water and wind erosion are complex processes, but when other factors are constant, erosion rates still vary due to differences in soil resistance. Erodibility is influenced by climate and is a complex, dynamic characteristic, which changes significantly over annual, seasonal or irregular time intervals, or even during a single storm. Nevertheless, the concept is useful for small-scale field or hillslope processes of rainsplash, sheetwash (see SHEET EROSION, SHEET FLOW, SHEET WASH) and rill erosion. It is more difficult to use with processes such as GULLY erosion, which involve very different spatial and temporal controls.

The erosion sub-processes affected by soil erodibility, are entrainment (by which soil particles are picked up), and transport. The relevant properties vary with the erosion process, and affect the erosive force available and resistance. In rainsplash erosion the entraining force is the kinetic energy of raindrop impact (see RAINDROP IMPACT, SPLASH AND WASH), converted to an upward force, while sheet wash and rill erosion involve runoff for both entrainment and transport. Raindrop impact can also disrupt particles and change their resistance to movement. The effectiveness of the upward force depends on soil particle size and mass. Poesen and Savat's (1981) experiments showed an entrainment:particle size relationship resembling the Hjulstrom curve for flowing water, particles between 0.1 and 0.25 mm diameter requiring least energy for movement. These results apply to non-cohesive particles with uniform density, such as quartz sands, where there is a direct relationship between particle size and mass. On such soils, erodibility can be assessed by standard particle size analysis techniques.

The relationship between particle size and erodibility is more complex when the surface is largely composed of aggregates. Aggregates are mixtures of mineral and organic matter, joined by electrostatic charges, microbial muscilages, hydrous oxides and carbonates. These materials and the volume of pore space are quite variable, so aggregate density is also variable. Microaggregate (<0.25 mm diameter) density is usually much higher than for macroaggregates, which may exceed 10 mm diameter (Oades 1993). Some large aggregates have densities below $1\,g.cm^{-1}$, can float on water, and are more easily eroded than small aggregates or mineral particles. The relationship between aggregate size and mass is not linear or direct, and size:entrainment relationships can be very complex.

These relationships are further complicated as aggregates can disintegrate during rainfall, if subjected to stresses exceeding the strength of 'stabilizing' agents. Under rainsplash, three dominant stresses occur. Raindrop energy, which can disrupt aggregates, is most significant, but SLAKING and differential hydration swelling may also cause breakdown. Aggregate strength is derived mainly from *cohesion*, due to electrostatic forces that bond clay and humus particles. Bond strength depends on total clay and humus content, on cation adsorption capacity, and on the cations adsorbed. Bivalent cations, such as calcium, cause flocculation and strong bonding, yielding small, highly resistant aggregates. Monovalent cations, such as sodium, cause particle repulsion, yielding weak, easily dispersed aggregates.

These interactions ensure that aggregates vary greatly in size, shape, stability and density, even in a single soil. As many soils consist largely of aggregates, it is their properties, rather than those of mineral particles, which most strongly influence erodibility. Aggregation is also affected by extrinsic factors such as wetting–drying and freeze–thaw, and by soil organic matter, changing with inputs of plant litter and decomposition by microbial activity. It is normal for regular or irregular seasonal changes in aggregation to be superimposed on short-term aggregate dynamics during or between storm events (Bryan 2000).

This discussion has emphasized the role of particles or aggregates as individual units. In fact, they only behave this way in recently disturbed or dispersive soils. The most common cause of soil disturbance is tillage. After tillage, disturbed soils gradually regain *coherence* due to weathering, compaction and crusting (see CRUSTING OF SOIL) by raindrop impact. This may occur in one rainstorm, or may take many months, but strongly affects erodibility. Erodibility usually declines as soils regain coherence, as sufficient force is required to overcome soil coherence and to entrain loosened particles.

Erodibility of coherent soils is determined by *soil shear strength*, the resistance to *interparticle failure*, defined by the Coulomb equation as:

$$r = c + zy \tan o$$

The active components are internal friction (o) which integrates surface and interlocking friction of particles, and (c) cohesion. Cohesion is explained above, while internal friction depends on the strength, heterogeneity of mineral particles and aggregates, and on overburden pressure (zy) at the point of potential failure. In surface soils acted on by rainsplash, sheetwash and rill erosion, overburden pressure is negligible and shear strength is dominated by cohesion. Both cohesion and internal friction are strongly influenced by soil water content. Cohesion is modified, either positively or negatively, by water molecules between particles. In completely dry soils, these are absent and soils are usually non-coherent and highly erodible. As soil water content increases, thin water films form (often <1 micron in thickness) which are viscous and hold mineral particles together. As these become thicker, viscosity declines and cohesion is reduced. Internal friction also declines in wet soils, as positive pressure in water-filled pores counteracts overburden pressure. The strength of both coherent soils and soil aggregates is thus greatly reduced when soil water content approaches saturation, and erodibility increases.

The role of soil water means that erodibility also depends on soil hydrological properties. The overall control is climatic, but soil hydrological properties determine the proportion of water reaching the surface that infiltrates the soil, and how long the soil remains wet after a storm. Soil *infiltration capacity* depends on surface porosity, but on bare soils under intense rainfall, soil crusting or sealing often produces a thin almost impermeable surface layer. Once water infiltrates, its distribution and ultimate drainage depend on soil permeability, which is controlled by the *soil structure*, the arrangement of soil material and pore space, which determines such features as pore space continuity.

The properties described affect erodibility for all the processes discussed, but the relationship is somewhat different in each case, because of the role of surface water layers. Rainsplash often occurs without any surface water layer, and a significant water layer can reduce or eliminate erosive energy. In sheet wash and rill erosion, it is the *shear stress* exerted by runoff which causes entrainment and transport. As the existence and depth of the surface water layer is determined by the ratio of rainfall:infiltration capacity, this means that for sheetwash and rill erosion, shear stress is partially controlled by soil properties. The magnitude of shear stress exerted by runoff is also affected by water distribution, increasing significantly when runoff is concentrated on a small surface area. Flow concentration depends on surface roughness, which is controlled by vegetation and on bare surfaces by soil particle and aggregate heterogeneity. These properties determine whether, under raindrop impact, the surface becomes progressively rougher or smoother, and will therefore affect the distribution and magnitude of runoff shear stress.

The complexity of the interacting processes and properties involved means that erodibility is not a single, simple, measurable soil property (Lal 1990), but reflects the collective interaction of many soil properties. Nevertheless, it is often necessary to attempt to assess erodibility by one or several simple measures. Many attempts have been made to identify or combine soil properties as *indices of soil erodibility* (Bryan 1968). No single measure is successful in all cases, but several measures can be effective if the precise nature of the dominant erosion process is clearly identified.

Measures of aggregate water stability, such as wet-sieving, are useful for rainsplash erosion, particularly on disturbed soils, while soil shear strength measured with a *vane shear* apparatus can be useful on coherent soils. *Soil consistency*, assessed by Atterberg limits, can indicate crusting potential. As sheetwash and rill erosion are more complex processes, it is more difficult to isolate a reliable index, but measures based on range of aggregate or particle size may be promising. In all cases, however, the high temporal and spatial variability of erodibility must be recognized.

References

Bryan, R.B. (1968) The development, use and efficiency of indices of soil erodibility, *Geoderma* 2, 5–26.

——(2000) Soil erodibility and processes of water erosion on hillslopes, *Geomorphology* 32, 385–415.

Lal, R. (1990) *Soil Erosion in the Tropics: Principles and Management*, New York: McGraw-Hill.

Morgan, R.P.C. (1995) *Soil Erosion and Conservation*, London: Longman.

Oades, J.M. (1993) The role of biology in the formation, stabilization and degradation of soil structure, *Geoderma* 56, 377–400.

Poesen, J. and Savat, J. (1981) Detachment and transportation of loose sand by raindrop splash, Part II: Detachability and transportability measurements, *Catena* 8, 19–41.

RORKE BRYAN

EROSION

Commonly speaking the term erosion (from Latin *erodere* = to gnaw away) is often used to indicate the overall exogenic process or group of processes that are directed at levelling off Earth relief, in contrast with the antagonist endogenic processes (crustal movements and volcanism) that build it up. In this very wide meaning erosion includes: acquiring materials from the higher elevations, moving them from one place to another (*transport*) and leaving them in lowlands (*deposition*).

Actually, it is the opinion of all scientists that erosion cannot include deposition. In fact, in a more technical language, the term erosion – although variously defined – usually excludes the processes whereby transported materials are set down. In the most broad and common of the technical meanings, erosion includes all exogenic processes, excluding WEATHERING and MASS MOVEMENTs, that involve the entrainment of loose weathered materials by a mobile agent, the removal of bedrock particles by the impact of transported materials, the mutual wear of rock fragments in transit and the transportation of acquired materials (Thornbury 1954).

Sometimes the term is restricted by excluding transportation; in this case DENUDATION is the more general term. More rarely, erosion indicates exclusively the entrainment of loose materials by a mobile agent.

Erosional agents and processes

Erosional processes are performed by mobile agents that draw their energy from solar radiation and act in one or more ways, constantly driven by the force of gravity. The principal erosional agents are running water, glaciers, wind and sea waves (Figure 57). In some cases they complete the same process, in some others a given process is accomplished by a distinctive agency that operates according to its physical peculiarities. Beside the cited 'natural' agents, humans can be rightly considered an important erosional agent too. Nowadays anthropogenic activities are so widespread and marked that they deeply modify the Earth's surface, often in an irreversible manner (see ANTHROPOGEOMORPHOLOGY).

The most common processes performed by natural erosional agents are shown in Table 17. The entrainment of rocky particles by erosional agents can be both chemical and mechanical. The first action (CORROSION) implies the work of a solvent and therefore it is typically accomplished by running water or waves; it is less important than mechanical action.

Mechanical removal takes place with different modalities, depending on the erosional agent. *Hydraulic action* comes from the pressure and hydraulic force of flowing water or sea waves that allow the acquisition of rocky particles. CAVITATION is a particular process operated by running water; it is still poorly documented and would represent a mechanism through which hydraulic action has a direct role in bedrock breakage. This process occurs when an increase in flow velocity and the following decrease in pressure cause the formation of bubbles that implode emitting jets of water capable of fracturing solid rocks. Moving ice acquires materials by *plucking* (or *quarrying*); through this process glaciers that move forward can remove large rocky fragments already detached from the bedrock by the freezing of water circulating inside cracks. *Overdeepening* is the process whereby glaciers erode to levels below regional BASE LEVEL related to fluvial systems;

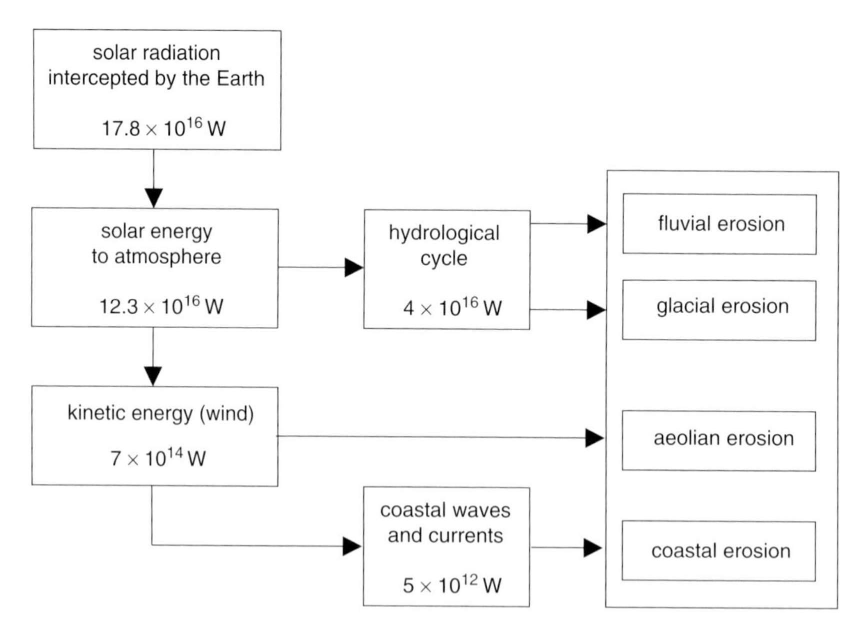

Figure 57 Source and flow of energy available for the different kinds of erosional processes. (From Summerfield 1991: 21, simplified and modified.)

Table 17 Erosional agents and their relevant erosional processes

Erosional Agent	Erosional processes			
	Entrainment of rocky materials	Erosion by transported materials	Wear of transported materials	Methods of transportation
Running water	Hydraulic action (Corrosion)	Abrasion	Attrition	Traction Suspension Solution
Glacier	Plucking or quarrying	Abrasion	Attrition	Traction
Wind	Deflation	Corrosion or abrasion	Attrition	Traction Suspension
Waves and currents	Hydraulic action (Corrosion)	Abrasion	Attrition	Traction Suspension (Solution)

Source: From Thornbury (1954: 47, modified)
Note: The less effective processes are indicated in brackets

however, it would be more correct to consider overdeepening as one of the effects of glacial erosion rather than an erosional process (Castiglioni 1979). The turbulent eddy action of wind is responsible for *deflation* and produces effects that are similar to those deriving from hydraulic action of waters.

Erosion accomplished by transported materials (named ABRASION) is due to the continuous collisions and friction by particles in transit on bedrock. All erosional mobile agents carry out this process. Abrasion due to running water is operated by solid materials of any size (up to large boulders, depending on the flow velocity) that can be transported as BEDLOAD by the water flow. Breaking waves that throw solid particles against the shore perform the same abrasive effect. EVORSION is a particular kind of abrasion due to the action of running water. It is caused by the erosional action of vortices and eddies on stream rocky beds. When a stationary eddy rotates a pebble, a small hollow is produced; this process leads to the formation of POT-HOLES (evorsion hollows), which contribute to valley deepening. In glacial environments abrasion is the friction produced on the bedrock by the debris carried along in the basal parts of glaciers; in its broader meaning it can include STRIATION, i.e. the bedrock *scoring* and *polishing* that reduces the rock surface roughness (Benn and Evans 1998). Aeolian abrasion derives from the repeated impacts of sand grains, silt particles and dust on rock surfaces; it is more properly named *corrasion*.

The wear of transported solid particles (*attrition*) takes place through repeated reciprocal knocking and collisions among the materials in transit: the result is a progressive decrease of particle sizes. At the same time the rocky particles tend to assume particular shapes, depending on the different ways in which each agent accomplishes transportation.

Mobile agents carry out transportation in three different ways. *Traction* consists of the rolling, sliding, pushing or jumping (in which case SALTATION is the specific term) of transported particles that are swept along on or immediately above a bottom surface. *Suspension* is a mode of transportation by water and wind; it implies the holding up of transported particles by the upward currents that develop in turbulent flows like those of running water and moving air (see SUSPENDED LOAD). *Solution* is a kind of 'chemical' transportation that is restricted to the action of running water and waves.

Erosional processes produce distinctive erosional landforms; furthermore, each erosional agent develops its own typical assemblage of landforms, depending on its mode of shaping Earth's relief. Erosional landforms are particularly striking features of the landscape; for this reason, perhaps, they have been considered with more attention than depositional landforms that, although interesting from a morphogenetic point of view, are usually less attractive.

The close links between erosional agency, accomplished process and produced landforms were recognized also by the ancients. Leonardo da Vinci at the end of the fifteenth century wrote: 'Every valley is created by its river and the proportion between one valley and another is the same as that between one river and another.' Once the concept that distinguishing features of landforms depend on the geomorphic process responsible for their development was fully understood, the genetic classification of landforms became possible. This scientific advancement made it possible to study the Earth's physical landscape not only from the descriptive point of view, but also considering the possible interpretation as to its geomorphological history. The genetic interpretation of erosional landforms, however, to be satisfactory must take into account the possible homologous or converging landforms, i.e. those landforms that although generated by different processes show similar features. Moreover it must be kept in mind that the genesis of most erosional landforms cannot be attributed to a single process, although it is rather simple to identify the dominant one.

The work of erosional agents produces peculiar assemblages of landforms that take on distinctive aspects depending on the stage of their development. The recognition that landforms change in time sequentially is the basis of the concept of the *Geographical Cycle* by Davis (1899). Once this concept had imposed itself, geomorphological interpretations made a new step forward. In fact, if properly applied, the geographical cycle affords a useful reference scheme to predict the possible future evolution of Earth's physical landscape (see CYCLE OF EROSION).

Erosion factors

Erosional landform features strictly depend not only on the way the exogenic agents operate, but also on a series of factors that control both the nature and the rate of erosion. The most important factors of erosion are lithology, tectonics, climate, vegetation and humans.

LITHOLOGY

Lithology strongly controls erosional processes as rock ERODIBILITY relies on it; as a consequence it influences the speed of erosional processes. In this perspective rocks are often referred to as 'hard' or 'resistant' or 'weak' and 'non-resistant' to erosional processes. The same erosional process can operate in a differentiated way where resistant rocks crop out next to no resistant rocks: as the erosional process proceeds, an uneven surface originates where more resistant rocks, slowly and hardly eroded, stand higher above less resistant rocks, which are more quickly and easily eroded. To some extent differential erosion can produce INVERTED RELIEF. The effects of differential erosion are particularly evident on stratified and differently erodible rocks. In this case the result of erosion is the formation of steep, abrupt faces of rock that mark the outcrop of the more resistant layers; the steep faces of a CUESTA, the rock terraces of a step-like slope or the scarp of a MESA are typical products of differential erosion. The concept of more or less erodible rocks is a relative one; in fact a rock can be resistant to one process and non-resistant to another. Therefore lithology has an influence also on the typology of the erosional processes.

TECTONICS

Tectonics influences erosional processes in different ways. Faults (see FAULT AND FAULT SCARP) and FOLDS can bring into contact rocks with different erodibility and then favour differential erosion. Furthermore they can directly influence the response of rocks to erosion, thus conditioning erosion rate. In fact rock erodibility depends not only on lithological characteristics but also on rock attitude (dip-slopes are less resistant than scarp-slopes) and on the degree of tectonic deformation (the higher the deformation of rocks, the higher their erodibility). Tectonic joints and faults can influence both the intensity of erosion and the location of the resulting landforms (see JOINTING). For example, fluvial erosion acts more powerfully where joints and faults create zones of weakness in the rocks than in other directions. As a consequence the orientation of stream valleys often follows closely the directions of these discontinuities. The sensitivity of tectonic discontinuities towards erosional processes may be so great that the morphological effects of differential erosion can help in the identification of discontinuities of small entity or affecting plastic lithologies (Belisario *et al.* 1999).

Tectonic uplift has also an important role in controlling the effectiveness of erosional processes. Uplift and erosion, together with relief, are the fundamental components of geomorphodynamic systems (see SYSTEMS IN GEOMORPHOLOGY) and are functionally related to one another in a negative feedback process (Ahnert 1998). When uplift prevails relief increases, and as a consequence erosion rate is faster. Increased erosion rate can eventually balance the building processes; in this case mountains do not change in elevation. When erosion overcomes the effects of uplift, elevations begin to lower and consequently the erosion rate slows down until the whole process comes to an end (Figure 58).

CLIMATE

Climate controls erosional processes both directly and indirectly. The direct control is exerted by the climate elements – temperature, rainfall and wind – that show a wide variability not only from one part to another of our planet, but also within very restricted areas, as for example from one slope to another on the same mountain. This wide variability of climatic conditions affects WEATHERING processes that weaken the rocks, predisposing them to subsequent erosional processes. Furthermore it favours the action of one erosional agent with respect to others: fluvial erosional processes become dominant in shaping the Earth's surface where rainfall amount is enough to guarantee the perennial channelled flow of waters, wind erosion is particularly effective where humidity is low, and glaciers can operate only where temperatures are such as to allow the fall and accumulation of snow. Besides this quite obvious influence, climatic conditions also control the way the different erosional processes operate with each other; these considerations are the basis of CLIMATIC GEOMORPHOLOGY which examines the systems of morphodynamic processes and their reciprocal interactions, in relation to the different climatic conditions.

Climate not only affects the typology of erosional processes but also the different behaviour of rocks. Under different climatic conditions the same rock may exhibit a different degree of resistance towards erosional processes and therefore it can be shaped into a variety of landforms. Granitic rocks are a good example. Depending on the climatic conditions and therefore on the dominant erosional processes they can be eroded into: the sharp peaks of the Monte Bianco massif, the low relief of INSELBERGs that dominate the savannah and steppe regions of South Africa, the ellipsoidal hollows (TAFONI) which originate from chemical corrosion and from the sweeping action of wind, the large round-topped mountains (piton) of tropical regions, etc. The close relationships among climatic conditions, erosional processes and landforms help to reconstruct the climatic variations that occurred in the past by

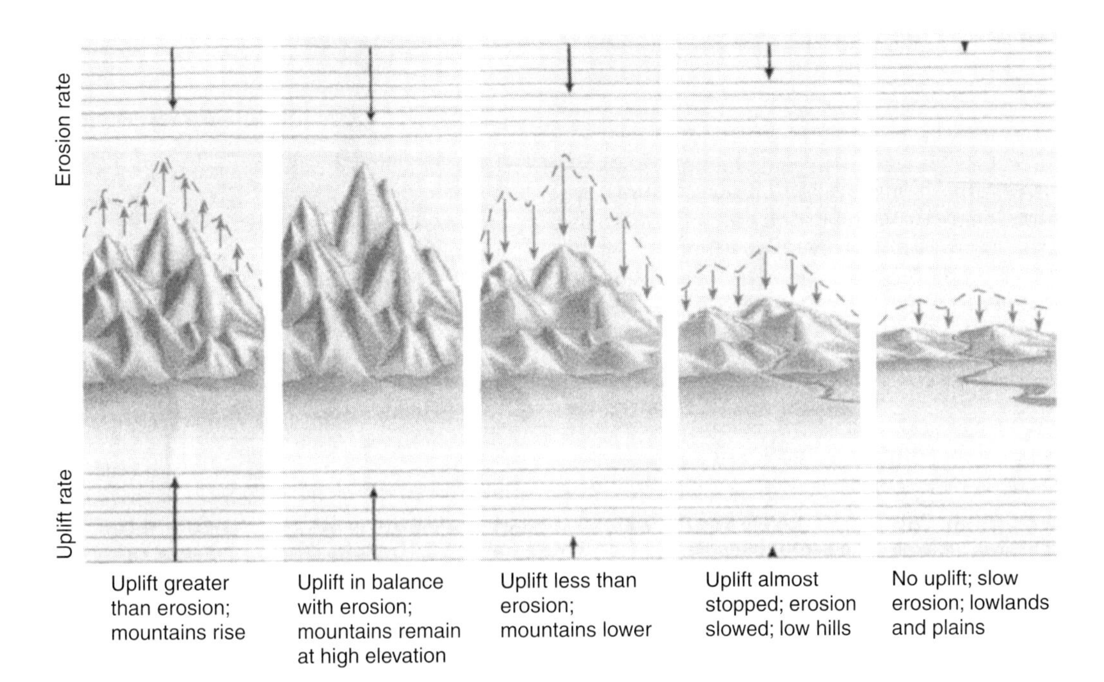

Figure 58 Negative feedback process relating uplift, erosion and mountain elevation. (From Press and Siever 1994: 364, modified)

examining the imprints left on the Earth's landscape by the prevailing erosional processes.

VEGETATION

The indirect influence of climate on erosion is largely related to the way it affects the amount and type of vegetation that, in its turn, has an important control on the EROSIVITY of some erosional agents. A thick vegetation cover inhibits surface runoff, thus restraining the action of running water; furthermore it obstructs the free flow of winds and therefore reduces the effectiveness of aeolian erosional processes. Root structures have a double influence: they can enhance the resistance of loose materials towards erosion or they can cause the breakage of solid bedrock, thus making erosion easier. As a whole, vegetation constrains erosional processes more frequently than it favours them. The limiting effect of vegetation on erosional processes varies as a function of the kind and density of the vegetation cover, and reveals itself because it enhances the stability of surface materials. Vegetation is also an important factor of pedogenesis, as it supplies the organic matter necessary for humus formation; therefore it has a role both in forming soil and in protecting it from erosion. This protective action is of primary importance to inhibit SOIL EROSION. When climatic conditions are such as to assure a dense and persisting vegetation cover erosional processes are slowed down: in these conditions (referred to as biorhexistasy) soils can develop and stay in place; on the contrary when climate is unfavourable to the development of vegetation, erosional processes are largely widespread and soils are easily removed (rhexistasy).

HUMAN IMPACT

If it is true that *humans* are nowadays powerful erosional agents, they are also an important factor of erosion. Anthropogenic activities are sometimes directed to undo or reduce erosional processes accomplished by 'natural' agents, as, for example, in the case of coastal defences built to inhibit sea erosion. More frequently, however, they produce the opposite effect and make erosion rate faster: in very densely inhabited areas, for instance, the extensive use of asphalt and concrete favour surface runoff and then erosion due to running water.

All the erosion-controlling factors play their role together; therefore their effects can interfere with each other in many possible ways so that the overall control on erosional processes is widely differentiated both in space and time. An example that clarifies the response of erosional processes to the complex constraints imposed by erosional factors is afforded by some more careful considerations about soil erosion. Once pedogenetic processes have led to formation of soils, they are exposed to the action of exogenetic erosional agents that start consuming them. When soil erosion proceeds normally, equilibrium conditions are attained: the rate at which soil is eroded equals the rate of soil formation. If this equilibrium is broken, erosional processes can become faster than pedogenetic processes; as a result, soil erosion is accelerated. The conditions more favourable to start *accelerated erosion* occur where weak rocks (such as clay or marls) crop out on areas affected by abundant and irregular precipitation that favours erosion by running water; under these conditions, for instance, BADLANDS originate. Wherever these natural predisposing conditions are added to deforestation and faulty land use connected to anthropogenic activities, accelerated erosion attains its maximum intensity. Under these conditions the soil erosion rate exceeds the soil formation rate. As a result soils get thinner and can completely disappear. In some cases erosion becomes so severe that it can be compared to the process of DESERTIFICATION: the irreversible process whereby soils lose their fertility because of the destructive effects of some anthropogenic activities.

References

Ahnert, F. (1998) *Introduction to Geomorphology*, London: Arnold.
Belisario, F., Del Monte, M., Fredi, P., Funiciello, R., Lupia-Palmieri, E. and Salvini, F. (1999) Azimuthal analysis of stream orientations to define regional tectonic lines, *Zeitschrift für Geomorphologie, Supplementband, NF* 118, 41–63.
Benn, D.I. and Evans, D.J.A. (1998) *Glaciers and Glaciation*, London: Arnold.
Castiglioni, G.B. (1979) *Geomorfologia*, Torino: UTET.
Davis, W.M. (1899) The Geographical Cycle, *Geographical Journal* 14, 481–504.
Press, F. and Siever, R. (1994) *Understanding Earth*, New York: Freeman and Company.
Summerfield, M.A. (1991) *Global Geomorphology*, Harlow: Longman.
Thornbury, W.D. (1954) *Principles of Geomorphology*, New York: Wiley.

Further reading

Fournier, F. (1960) *Climat et érosion la relation entre l'érosion du sol par l'eau et les précipitations atmosphériques*, Paris: Presses Univ. de France.
Haggett, P. (1983) *Geography. A modern Synthesis* (Chapter 10), New York: Harper & Row.
Howard, J.A. and Mitchell, C.W. (1985) *Phytogeomorphology*, New York: Wiley.
Keller, E.A. and Pinter, N. (1996) *Active Tectonics. Earthquakes, Uplift and Landscape*, New Jersey: Prentice Hall.
Scheidegger, A.E. (1979) The principle of antagonism in the Earth's evolution, *Tectonophysics* 55, T7–T10.

SEE ALSO: fluvial geomorphology; glacial erosion; granite geomorphology; tectonic geomorphology; wind erosion of soil

ELVIDIO LUPIA-PALMIERI

EROSIVITY

A measure of the potential ability of a soil to be eroded by particular geomorphological processes. Erosion is a function of erosivity on the one hand and of erodibiltiy (the vulnerability of a soil to erosion) on the other.

Water erosion susceptibility is related to various rainfall erosivity indices. Rainfall intensity, rainfall amount and antecedent conditions are all important controls of erosivity, but as Morgan (1995: 27) has remarked: 'the most suitable expression of the erosivity of rainfall is an index based on the kinetic energy of the rain. Thus the erosivity of a rain storm is a function of its intensity and duration, and of the mass, diameter and velocity of the raindrops'. In recent years RAINFALL SIMULATION has been used to assess the response of soils to storms with different characteristics.

Wind erosion susceptibility has also often been determined using indices based on wind velocities and durations above certain threshold velocities (e.g. Skidmore and Woodruff 1968) and portable wind tunnels have been employed to assess the response of different ground surfaces to different wind velocities.

References

Morgan, R.P.C. (1995) *Soil Erosion and Conservation*, 2nd edition, Harlow: Longman.
Skidmore, E.L. and Woodruff, N.P. (1968) Wind erosion forces in the United States and their use in predicting soil loss, *USDA Agricultural Research Service Handbook* 346.

A.S. GOUDIE

ERRATIC

Erratics are rock fragments carried by a glacier, or in some cases by floating ice, and subsequently deposited at some distance from the outcrop from which they were derived. For this reason their lithology differs from the surrounding rocks and sediments – hence the term 'erratic'. Some erratics are large blocks, that lie free on the surface and form interesting landscape features. Glaciologists, however, mainly use the term for the exotic components embedded in tills (see GLACIAL DEPOSITION), encompassing both large clasts and fine-grained rock fragments.

Scientific investigation of erratics started during the first half of the nineteenth century, when most geologists thought that they were swept into the northlands by the universal flood. At the same time their transportation and distribution by former widespread glaciers was first suggested, and later in the nineteenth century it was universally agreed.

Some erratics form landmarks because of their spectacular dimensions. One of the largest erratics measures 45 m by 20 m by 10 m and is estimated to weigh 16,500 tons. It is part of the Foothills erratic train, a series of boulders stretching over 400 km along the eastern foothills of the Rocky Mountains.

Erratics not only give evidence of the existence of former glaciers; especially erratics in tills provide a powerful tool for many glacial investigations. Such studies are based on the identification of 'indicator erratics'. Indicator erratics are those for which a definite source area is known. They form 'indicator trains' or, in cases of shifting ice divides and ice flow directions within an ice sheet, 'indicator fans' trailing downglacier from the source rock. Indicator trains and fans are enriched in the distinctive component relative to the till underlying or enclosing it. The concentrations of indicator erratics vary systematically along former ice flowlines. Within indicator outcrops, concentrations increase rapidly downglacier, reflecting the addition of new material from the glacier bed, but concentrations drop off rapidly down-ice of the outcrop margin. The up-ice and down-ice limits of an indicator plume are known as the 'head' and the 'tail', respectively.

Erratics can be used to reconstruct the pattern and history of ice flow in studies of ice sheet dynamics as long as erratic transport histories are not blurred by repeated glaciations involving total redistribution of previously deposited materials (Benn and Evans 1998). The study of erratic dispersal patterns furthermore can provide important clues to the location of mineral outcrops or ore bodies, because the erratic plumes are much larger than their bedrock sources, making them easier targets to find (Kujansuu and Saarnisto 1990). In Denmark, Germany and the Netherlands till units of different age show differently composed indicator assemblages and counts of the erratics derived from various western, central or eastern Scandinavian source areas are here successfully employed in stratigraphical investigations (Ehlers 1996).

References

Benn, D.I. and Evans, D.J.A. (1998) *Glaciers and Glaciation*, London: Arnold.

Ehlers, J. (1996) *Quaternary and Glacial Geology*, Chichester: Wiley.

Kujansuu, R. and Saarnisto, M. (eds) (1990) *Handbook of Glacial Indicator Tracing*, Rotterdam: Balkema.

CHRISTINE EMBLETON-HAMANN

ESCARPMENT

The term escarpment, or scarp, has been applied traditionally to a steep, often single slope, of considerable length, that dominates a section of landscape. An escarpment thus can be distinguished from the two flanking walls of canyons. For example, south of Sydney, Australia, the coastal escarpment forms a long, virtually continuous wall, but it is outflanked by canyons which extend more than 100 km further inland. Another notable instance of a valley cut well back from an escarpment is the Sognefjord, which extends about 200 km inland from the coastal edge of the Norwegian highlands. The lengths of escarpments vary from a few kilometres to the subcontinental scale of mega-escarpments, or Great Escarpments, such as the Drakensberg of South Africa, while their heights vary from a few tens of metres to several thousand metres. A distinction is generally drawn between denudational escarpments and fault scarps (see FAULT AND FAULT SCARP), although this may be no simple exercise in areas of essentially homogeneous crystalline rocks.

The majority of denudational escarpments have formed as a result of differential rock resistance to erosion. Probably the most outstanding examples form the sequence of the Vermillion, White, Grey and Pink Cliffs, known as Great Staircase, which rises from the rim of the Grand Canyon of the Colorado River to the high plateaux of southern Utah. The treads of this staircase are cut mainly in softer strata between the sandstones which form the cliffs. Major sandstone escarpments also occur in the Adrar and Borkou areas of the Sahara. But such features are not limited to arid lands, for the great ramparts of the Roraima massif have developed in the humid tropics of Venezuela. Neither are they limited to sequences of sedimentary rocks. Major escarpments and benches also have developed in response to the differential resistance of volcanic strata in the Bushveld of Transvaal, and of sheeted granites intruded through metamorphic rocks in Madagascar. Some escarpments in granitic rocks were initiated at the boundary with less resistant rocks, and have retreated to their present positions. Others, such as the multiple steps in the Sierra Nevada of California, were initiated by differential weathering within a granitic mass.

While many escarpments can be attributed to the great resistance to erosion of particular types of rocks, others show no systematic relationship to lithology. For example, the coastal escarpment south of Sydney extends from sedimentary to metamorphic rocks, and thence to crystalline rocks. Escarpments of this type have developed in response to regional uplift. The most extensive of them, which are sometimes called Great Escarpments, occur on many tectonically PASSIVE MARGINs of continents. Notable examples are the Drakensberg, the Western Ghats of India, the escarpments of east Australia, the Serra da Mantiqueira of Brazil, the coastal escarpment of Norway, the escarpment of east Greenland and the Torngat Mountains of the uplifted margin of Labrador.

Great Escarpments are not limited to passive margins, for they also occur on tectonically active continental margins. Collision of the Australian and Pacific Plates has resulted in some 20 km of uplift in about the last 10 million years along the west coast of the South Island of New Zealand. And, although uplift has been virtually matched by erosion, the flank of the Southern Alps rises in a steep, heavily dissected wall from the narrow coastal plain. Likewise, collision of the Pacific, North American and Cocos Plates has resulted in major uplift that, together with rapid erosion, has produced the great escarpments of the Sierra Madre Occidental and Sierra Madre Del Sur on the western flank of the Mexican highlands.

Less extensive, though nonetheless impressive, escarpments have developed as a result of regional uplift in continental interiors. The classic examples are along the margins of the *Mittelgebirge* of central Europe, such as the Massif Centrale of France and the Erzgebirge of Germany. However, even more impressive escarpments occur along the margins of uplands in central Asia, as for example on the northern side of the Bogda Shan in north-west China. Erosion in response to the regional uplift in north-east Africa has resulted in high escarpments cut largely in volcanic rocks along the west flank of the Ethiopian plateau. The Kaibab Limestone escarpment of the Mogollon Rim on the southern flank of the Colorado Plateau is a notable North American example.

Much of the initial research on escarpments was carried out in the folded terrain of the Appalachians and north-west Europe, where they occur in association with homoclinal CUESTAs and HOGBACKs. As early as 1895 W.M. Davis pointed out escarpments in this type of terrain were second-stage features that did not develop until streams began to extend headwards along the strike of the folds. He attributed the retreat of escarpments not only to erosion on their steep faces, but also to lateral erosion of streams at the foot of escarpments. Although Davis's ideas, and especially his terminology, have been subject to much criticism (see SLOPE, EVOLUTION), the scheme that he proposed a century ago still provides the basis for the interpretation of many scarp and cuesta landscapes. However, major challenges to it have come from German geomorphologists.

Schmitthenner (1920) argued that the most important form of denudation in scarplands (*Schichtstufenlandschaft*) is the lowering of the backslope surface and the breaching of escarpments from the rear. He attributed this primarily to seepage down the dip, and the consequent development of swampy hollows, or *dellen*, by solutional processes. Strong support for the role of seepage leading to the lowering of escarpment crests and to the breaching of them from the rear has come from recent research on the Colorado Plateau and Australia.

Many stairways of multiple escarpments have been attributed to repeated uplift and erosion, but, as independent evidence of repeated uplift is often lacking, alternate hypotheses need consideration. W. Penck (1924) claimed that multiple scarps and benches (*Piedmonttreppen*) could form on a continuously rising and expanding dome. Penck's hypothesis provides a valuable warning that the relationship between tectonics and slope form may be complex (see SLOPE, EVOLUTION).

In recent decades prominent German researchers have expounded climatic explanations of escarpments (see MORPHOGENETIC REGION). Büdel (1982) argued that escarpments in the tropics are essentially the result of deep weathering and the subsequent stripping of regolith, and referred to them as 'etchplain stairways'. He also extended this climatic interpretation to temperate lands by claiming that most of the escarpments of south-west Germany were formed in a similar fashion to etchplain stairways under a past 'tropicoid climate'. The structural influences so clearly expressed in many of these escarpments were dismissed as only an 'arabesque' in the general two-stage development by deep weathering and subsequent stripping. According to this theory, scarplands are sculpted predominantly by areal downwearing, and scarp retreat is minimal.

In striking contrast, conclusions drawn from studies in southern Africa, especially those of L.C. King, emphasize the dominant role of the retreat of escarpments over long distances. According to King (1953), the most important processes are sheet wash on pediments at the foot of scarps, and mass failure and gully erosion on the steep slopes. He argued that 'scarps retreat virtually as fast as nick-point advance up rivers, so that the distribution of successive erosion cycles bears no relationship to the drainage pattern whatsoever'. He argued also that retreating scarps resulted in isostatic (see ISOSTASY) uplift, generally in the form of large-scale warping, initiating a new cycle of scarp retreat.

Although many escarpments may have retreated over considerable distances, some have apparently not done so. The Blue Ridge Escarpment on the east flank of the Appalachian Mountains is a major feature, which lies about 350 km inland from the coast. Rather than having retreated from the coast, however, this escarpment may have been maintained in approximately its present position over many millions of years by uplift along an ancient continental margin preserved deep in the crust. It thus seems to be in a long-term state of dynamic equilibrium in which denudation has been largely balanced by the slow rise of the underlying rocks. On the other hand, some escarpments are essentially fossil features that have been exhumed by erosion (Plate 41). Sedimentary evidence indicates that the Arnhemland Escarpment, cut in Proterozoic rocks in the Northern Territory of Australia, is a coastal cliff-line that was buried during Cretaceous times and subsequently exhumed. Since being exhumed, parts of this escarpment have been almost stationary, and the most active parts have retreated only a few kilometres.

Assessing which explanation is best suited to a particular escarpment may be no simple task, and indeed none of them may be entirely satisfactory. The final appeal must be to field evidence rather than to conceptual constructs.

The rates at which escarpments retreat vary greatly, and seem to depend on lithological resistance, tectonic activity and the intensity of erosion which itself is largely controlled by climate. Many major escarpments are thought to retreat about 1 km per million years. However, the distance of retreat from dated basalts show that escarpments in temperate east Australia have retreated at rates of only about 25 to 170 m per million years, and the rates for some of them have been even slower. Low rates of only about 70 m to 150 m per million years have also been recorded in subarctic Spitzbergen. Average rates of scarp retreat on the Colorado Plateau since Miocene times have been only about 160 m per million years, but earlier than that the rates were about 1.5 km to 4 km per million years. The great reduction has been attributed to the incision of streams and thus the cessation of very active lateral planation at the foot of scarps. Denudation

Plate 41 Proterozoic sandstone capping schists on the Arnhemland Escarpment, Northern Territory, Australia

on the Arnhemland Escarpment increased by an order of magnitude during the late Quaternary, but in this case the change seems to have been climatically controlled. Moreover, the evidence from Arnhemland prompts caution in extrapolating long-term rates of retreat from relatively short-term records.

Although most research has been carried out on escarpments above sea level, submarine escarpments are of far greater magnitude. The fall below the continental shelf off east Australia is more than five times greater than the height of the escarpment onshore, and contrasts of similar, or greater, magnitude occur along most continental margins.

References

Büdel, J. (1982) *Climatic Geomorphology*, Princeton: Princeton University Press.

Davis, W.M. (1895) The development of certain English rivers, *Geographical Journal* 5, 127–146.

King, L.C. (1953) Canons of landscape evolution, *Geological Society of America Bulletin* 64, 721–752.

Penck, W. (1924) *Die Morphologiche Analyse; ein Kapital der Physikalischen Geologie*, Stuttgart: Engelhorn. Trans H. Czech and K.C. Boswell (1954) *Morphological Analysis of Landforms*, London: Macmillan.

Schmitthenner, H. (1920) Die Enstehung der Stufenlandschaft, *Geographische Zeitschrift* 26, 207–229.

Further reading

Battiau-Queney, Y. (1988) Long term landform development of the Appalachian piedmont (USA), *Geografiska Annaler* 70A, 369–374.

Blume, H. (1958) Das morphologische Werk Heinrich Schmitthenners, *Zeitschrift für Geomorphologie* 2, 149–164.

Nott, J.F. (1995) The antiquity of landscapes on the Northern Australian craton; implications for models of longterm landscape evolution, *Journal of Geology* 102, 19–32.

Ollier, C.D. (1985) Morphotectonics of passive continental margins: introduction, *Zeitschrift für Geomorphologie Supplementband* 54, 1–9.

Young, R.A. (1985) Geomorphic evolution of the Colorado Plateau margin in west-central Arizona: a tectonic model to distinguish between the causes of rapid symmetrical scarp retreat and scarp dissection, in J.T. Hack and M. Morisawa (eds) *Tectonic Geomorphology*, 261–278, London: Allen and Unwin.

Young, R.W. and Wray, R.A.L. (2000) Contribution to the theory of scarpland development from observations in central Queensland, Australia, *Journal of Geology* 108, 705–719.

R.W. YOUNG

ESKER

Derived from the Irish word *eiscir* (ridge), an esker is an elongate sinuous ridge composed of glacifluvial sediments and marking the former position of a subglacial, englacial or supraglacial stream. The routing of former meltwater channels in glaciers, and their association with ice-marginal landforms and sediments is indicated by the overall form of eskers. There are four major types of esker (Warren and Ashley 1994: (1) continuous ridges (single or multiple) that document tunnel fills; (2) ice-channel fills produced by the infilling of supraglacial channels; (3) segmented ridges deposited in tunnels during pulsed glacier recession; and (4) beaded eskers consisting of successive subaqueous fans deposited in ice-contact lakes during pulsed glacier recession. In planform, eskers also come in a wide variety of types. These include, single ridges of uniform cross section or of variable volume, single ridges linking numerous mounds (beads) and complex braided or anabranched systems. Although individual eskers or esker networks may stretch over hundreds of kilometres of former ice sheet beds it is unlikely that they formed in tunnel systems of that length. Rather, they were probably deposited in segments in the ice sheet marginal zone of ablation and each segment was added as the ablation zone migrated towards the ice sheet centre. In some locations eskers lie on the bottoms of bedrock meltwater channels (TUNNEL VALLEYs or Nye channels) indicating that erosion by subglacial meltwater was followed by deposition, possibly due to waning discharges, and that subglacial conduits were remarkably stable features.

Most eskers are aligned sub-parallel to glacier flow, reflecting the flow of meltwater towards the glacier margin. Eskers that were deposited as subglacial tunnel fills may be the result of flow in pressurized conduits and therefore may possess up-and-down long profiles where they climb over topographic obstacles. This is due to the fact that flow in the conduits was driven by the ice surface slope rather than by the glacier bed topography (Shreve 1972). If conduits or tunnels switch to atmospheric pressure, as occurs beneath the thinner ice nearer to the glacier snout, then the water will follow the local bed slope and so any resulting eskers will be deposited transverse to glacier flow (*valley eskers*).

The former englacial position of some eskers is indicated by the occurrence of buried glacier ice or almost complete disturbance of the stratified core. Englacial or supraglacial construction of eskers will result in the draping of the features over former subglacial topography after glacier melting. These apparent up-and-down long profiles must not be confused with true subglacial eskers deposited under pressure and therefore also crossing topographic high points. Largely intact internal stratigraphies are typical of subglacial eskers.

Eskers are often well stratified but contain a variety of sediment facies. Particles are usually not far-travelled, most being no further than 15 km from their source outcrop. The wide range of BEDFORMS observed in esker sediments reflect and document the large variations in meltwater discharge on both diurnal and seasonal timescales. Rhythmicity or cyclicity in the sediments is often represented by fining-upwards sequences separated by erosional contacts. Each fining-upward unit records an individual discharge event of maybe only hours in duration. The occurrence of large cross-bedded sequences may document deposition in deltas in subglacial ponds. Where tunnels collapse and/or change shape or streams change size or position, one depositional sequence may be truncated or partially infilled by another. Apparent anticlinal bedding in some eskers is thought to be the result of sediment slumping down the esker flanks as the supporting ice walls melt back.

The segments or beads on eskers are interpreted as the products of ice marginal deposition at the mouths of subglacial tunnels. At each tunnel mouth the glacifluvial sediment being carried through the tunnel or conduit is deposited in SUBAERIAL or subaqueous fans/deltas due to the sudden drop in meltwater velocity as it leaves the confines of its ice walls. Beads may also accumulate in subglacial cavities that develop as offshoots to the main tunnel. Where well-integrated esker networks carry large volumes of debris to the glacier margin they may link up with extensive ice-marginal depositional systems that include ice-contact deltas, subaqueous fans and MORAINES.

References

Shreve, R.L. (1972) Movement of water in glaciers, *Journal of Glaciology* 11, 205–214.

Warren, W.P. and Ashley, G.M. (1994) Origins of the ice-contact stratified ridges (eskers) of Ireland, *Journal of Sedimentary Research* A64, 433–449.

Further reading

Auton, C.A. (1992) Scottish Landform Examples – 6: The Flemington eskers, *Scottish Geographical Magazine* 108, 190–196.

Bannerjee, I. and McDonald, B.C. (1975) Nature of esker sedimentation, in A.V. Jopling and B.C. McDonald (eds) *Glaciofluvial and Glaciolacustrine Sedimentation*, 132–154, SEPM Special Publication 23.

Brennand, T.A. (1994) Macroforms, large bedforms and rhythmic sedimentary sequences in subglacial eskers, south-central Ontario: implications for esker genesis and meltwater regime, *Sedimentary Geology* 91, 9–55.

Evans, D.J.A. and Twigg, D.R. (2002) The active temperate glacial landsystem: a model based on Breiðamerkurjökull and Fjallsjökull, Iceland, *Quaternary Science Reviews* 21(20–22), 2,143–2,177.

Gorrell, G. and Shaw, J. (1991) Deposition in an esker, bead and fan complex, Lanark, Ontario, Canada, *Sedimentary Geology* 72, 285–314.

Price, R.J. (1969) Moraines, sandar, kames and eskers near Breiðamerkurjökull, Iceland, *Transactions of the Institute of British Geographers* 46, 17–43.

——(1973) *Glacial and Fluvioglacial Landforms*, Edinburgh: Oliver and Boyd.

Terwindt, J.H.J. and Augustinus, P.G.E.F. (1985) Lateral and longitudinal successions in sedimentary structures in the Middle Mause esker, Scotland, *Sedimentary Geology* 45, 161–188.

Thomas, G.S.P. and Montague, E. (1997) The morphology, stratigraphy and sedimentology of the Carstairs esker, Scotland, UK, *Quaternary Science Reviews* 16, 661–674.

DAVID J.A. EVANS

ESTUARY

Estuaries are unique ecosystems that provide spawning grounds for many organisms, feeding stops for migratory birds and natural filters to maintain water quality. Estuaries have value to humans for shipping and boating, settlements, erosion protection, recreation, mineral extraction and release of waste materials. Estuaries are generally considered areas where salt water from the ocean mixes with freshwater from land drainage but there are many definitions for the term (see Perillo 1995) reflecting the complex physical and biological processes present. Estuaries may be classified or described on the basis of numerous criteria, including entrance conditions (Cooper 2001), stage of development and degree of infilling (Roy 1984), morphology (Pritchard 1967; Fairbridge 1980), tidal range (Hayes 1975), vertical stratification and salinity structure (Cameron and Pritchard 1963). All these criteria affect the evolution of estuaries and the nature of transport of sediment and biota. The most common definition is Cameron and Pritchard (1963: 306) who define an estuary as 'a semi-enclosed coastal body of water having a free connection with the open sea and within which sea-water is measurably diluted with fresh water derived from land drainage'. The boundaries of an estuary can be defined by salinity (ranging from 0.1‰ at the head of the estuary and 30–35‰ at the mouth) or sedimentary facies and the processes that shape them. For example, Dalrymple *et al.* (1992) defined the upper boundary as the landward limit of tidal facies and the lower boundary as the seaward limit of marine facies.

From a geologic perspective, today's estuaries are recent features. Estuaries are a product of inherited factors (i.e. lithology) that influence the configuration of the estuarine basin and sediment type and availability; broad-scale controls such as climate and sea-level rise that influence rates of discharge and inundation; and, contemporary processes (wave, tide and river) that influence hydrodynamics and sediment transport. The position of estuaries is a result of fluctuations in sea-level rise, with sea-level elevation at or above current levels during interglacial periods and up to 150 m below present levels during glacial periods. Present-day estuaries are the result of sea-level rise and inundation of coastal lowlands following the last glacial period and the sea-level stillstand that began approximately 6,000 years ago. More recent regional sea-level histories have revealed both lowering and rising of sea level from stillstand levels.

Classification

Estuaries can be broadly classified as drowned river valleys, fjords, bar-built, and those formed by faulting or local subsidence (Pritchard 1967; Fairbridge 1980). Drowned river valley estuaries are found along the east coast of the United States (i.e. Delaware Bay and Chesapeake Bay) and in England (i.e. Thames and Mersey estuaries), France (i.e. Seine) and in Australia (i.e. Batesman Bay). Rivers eroded deep V-shaped valleys during the last glacial period that were subsequently inundated when melting ice sheets caused a rise in sea level. The planform and cross section of these estuaries are often triangular or funnel-shaped. In systems where sedimentation rates are less than rates of sea-level rise the river valley topography is maintained. Bar-built estuaries have a geologic history similar to drowned river valleys (the result of glacial incision and subsequent inundation by sea-level rise) but recent marine sediment transport (alongshore or cross shore) results in the creation of a barrier or spit across the mouth. The inlet at the mouth is small relative to the size of the shallow estuary created behind the barrier. In some cases the barrier may restrict exchange of water between the ocean and estuary except during high tides. Examples of this type of estuary can be found in the United States (i.e. Mobile Bay and Pamlico Sound) and in Australia (i.e. Clarence and Narooma Estuaries). Fjords are glacially-scoured U-shaped valleys that were subsequently inundated by a rising sea level. Most fjords possess a shallow rock sill near the mouth that forms an estuarine basin. Fjord-type estuaries are found in upper latitudes (i.e Oslo Fjord, Norway and Puget Sound, USA). Some estuaries are formed in valleys that were created by processes such as faulting (i.e. San Francisco Bay, USA) or subsidence.

Estuaries are located in micro-, meso- and macro-tidal environments. Planform morphology is an important control on the variation of tidal range and the magnitude of the tidal current within an estuary (Nichols and Biggs 1985). Convergence of the estuarine sides causes the tidal wave to compress laterally. In the absence of bed friction, the tidal range will increase. In the presence of friction, the tidal range will decrease. The relationship between convergence and friction control the amplitude of the tide within the estuary. In cases where convergence is greater than friction the tidal range and strength of the tide will increase toward the head of the estuary (hypersynchronous estuaries). In cases where convergence is less than friction the tidal range will decrease throughout the estuary (hyposynchronous estuaries).

Morphology

Wave, tide and river processes control the location of marine and river sediments in the estuary and the morphology of the sedimentary deposits. Conceptual models of estuarine morphology classify estuaries according to the relative contribution of these processes (see Dalrymple *et al.* 1992; Cooper, 1993) and are based, in part, on regional studies of estuarine sedimentation and morphology. These studies include tide-dominated, macro-tidal estuaries (Dalrymple *et al.* 1990) and micro-tidal estuaries in wave-dominated (Roy 1984) and river-dominated (Cooper 1993) environments.

Tide-dominated estuaries are found in macro-tidal environments (tidal range >4 m). They are generally funnel-shaped with wide mouths and high current velocities. Dalrymple *et al.* (1990) characterized the sedimentary characteristics of the macro-tidal Cobequid Bay–Salmon River Estuary, Canada. The axial sands are characterized by the presence of elongate tidal sand bars in the lower sector of the estuary that trend parallel to the dominant current direction. Sand flats and braided channels are located in the middle sector of the basin and a single channel is located in the river-dominated head of the estuary. Tidal currents are at a maximum in the inner part of the estuary. Sediments decrease in size from the mouth to the head. Dominant direction of sediment transport is landward with accumulation in the upper sector at the head of the estuary.

Wave-dominated estuaries are generally found in micro-tidal (tidal range <2 m) environments (Roy 1984). In general, these estuaries have an upper sector near the head, where river processes, sediments and bedforms dominate, a lower sector near the mouth, where wave and tidal processes and marine sediments dominate, and a middle sector, where tidal currents dominate and both river and marine sediments are present. High wave and tidal energies at the mouth of an estuary can deposit sediment and restrict or completely prohibit exchange of water between the ocean and the estuary.

Mixed wave-tide estuaries (such as those in meso-tidal environments with a tidal range of 2–4 m) can be found behind barrier islands (Hayes 1975). The dominant sand bodies in meso-tidal estuaries are the deltas (ebb and flood) formed by tidal inlet processes. Within the estuary are meandering tidal channels and point bars and marsh deposits.

River-dominated estuaries do not display the characteristic downstream facies changes observed in wave- and tide-dominated estuaries, and energy levels may remain similar along the axis of the river valley (Cooper 1993). River-dominated estuaries can range from those completely dominated by river processes (river channels) to those that experience some marine inputs at the mouth.

Shoreline environments

Estuarine shoreline environments often occur in small isolated reaches with different orientations and with great variability in morphology, vegetation and rate of erosion. This variability results from regional differences in fetch characteristics, exposure to dominant and prevailing winds, variations in subsurface stratigraphy, irregular topography inherited from drainage systems, differential erosion of vegetation or clay, peat and marsh outcrops on the surface of the subtidal and intertidal zones, small-scale variations in submergence rates, effects of varying amounts of sediment in eroding formations and effects of obstacles to longshore sediment transport, such as headlands and coves, that define drift compartments (Nordstrom 1992). Differences in the gradient of wave energy between the low-energy (upper) and high-energy (lower) shorelines in an estuary and between the high-energy (windward) and low-energy (leeward) sides of an estuary also contribute to differences in the types of estuarine environments and their dimensions. Saltmarsh is likely to form on alluvium in the upper reaches of the estuary, on the upwind side of the estuary or on the downwind side of the estuary in the lee of headlands that provide protection from breaking waves. Beaches are likely to form on the downwind side of estuaries because there is sufficient energy in the locally generated waves to erode coastal formations or prevent plants from growing in the intertidal zone.

BEACHES may be unvegetated or partially vegetated and are composed of sand, gravel or shell. The dominant processes of sediment reworking on beaches in estuaries is usually locally generated waves, although refracted and diffracted ocean waves may be present. The best development of beaches occurs where relatively high wave energies have exposed abundant unconsolidated sand or gravel in the eroding coastal formations. Adequate source materials occur where these formations are moraine deposits, submerged glacial streams, coarse-grained fluvial deposits, and sand delivered by ocean waves and winds, such as the estuarine shorelines of spits and barrier islands. Beach formation is favoured where high ground protrudes into relatively deep water, where wave refraction and wave energy loss through dissipation on the bay bottom is minimal. Ocean waves that enter the estuary usually create beaches close to the inlets. Sediment transported into the estuary by ocean waves may form spits in the lee of headlands in the estuary. Beaches created by waves generated within estuaries are most common in shoreline re-entrants, where sediments can accumulate over time. Other beaches occur where sand is plentiful on the bayside of barriers enclosing the estuary, particularly on former recurves, subaerial overwash platforms and former oceanside dunes (Nordstrom 1992).

Beaches may form on the bay side of eroding marshes from coarse-grained sediment removed from the eroding substrate. Beaches may precede and favour marsh growth by creating spits that form low-energy environments landward of them. Both processes create a beach ridge shoreline that combines features characteristic of beach shorelines and marsh shorelines. Peat, representing the substrate of former marsh, is often exposed in outcrops on eroding beaches transgressing marshes. These outcrops are resistant because of the presence of fine-grained materials trapped by upward growth of the marsh and the binding effect of vegetation.

Dunes (see DUNE, COASTAL) form within estuaries only where beaches are sufficiently wide to provide a viable sediment source or where the shoreline is stable enough to allow ample time for slow accretion or to prevent wave erosion. Wave energy must be sufficient to prevent colonization of intertidal vegetation, but erosion cannot be too great for aeolian forms to survive. Onshore aeolian transport occurring between moderate-intensity storms may create only a thin aeolian cap on top of the backbeach or overwash platform.

Marshes are components of the intertidal profile affected by waves and tidal flows, and they bear many similarities with beaches, including the potential for cyclic exchange of sediment between the upper and lower parts of the profile and a tendency to buffer energy in a way that resists long-term morphological change (Pethick 1992). Marsh shorelines differ from beaches in that they are characterized by fine sediment sizes, low gradients and dissipative slopes. The occurrence of marshes, like beaches, depends on their environmental setting and mode of origin, defined by factors such as bedrock geology, availability of sediments and recent sea-level rise history (Wood *et al.* 1989). Examples of distinctive morphological marsh units determined by macro-scale differences within estuaries include fluvial marshes, occurring on the upper estuarine margins of rivers; bluff-toe marshes that form at the base of coastal bluffs; backbarrier marshes found behind barrier islands and spits; and transitional marshes where freshwater peatlands are colonized by saltmarsh (Wood *et al.* 1989).

Competition for human resource values of the estuarine shoreline has led to elimination of many natural environments. The conversion of some of the environments (especially bay bottoms and marshes) is now severely restricted by land use controls in many countries, but many estuaries are still threatened by human activities.

References

Cameron, W.M. and Pritchard, D.W. (1963) Estuaries, in M.N. Hill (ed.) *The Sea*, 306–324, New York: Wiley Interscience.

Cooper, J.A.G. (1993) Sedimentation in a river dominated estuary, *Sedimentology* 40, 979–1,017.

——(2001) Geomorphological variability among microtidal estuaries from the wave-dominated South African coast, *Geomorphology* 40, 99–122.

Dalrymple, R.W., Knight, R.J., Zaitlin, B.A. and Middleton, G.V. (1990) Dynamics and facies model of a macrotidal sand-bar complex, Cobequid Bay–Salmon River estuary (Bay of Fundy), *Sedimentology* 37, 577–612.

Dalrymple, R.W., Zaitlin, B.R. and Boyd, R. (1992) Estuarine facies models: conceptual basis and stratigraphic implications, *Journal of Sedimentary Petrology* 62, 1,130–1,146.

Fairbridge, R.W. (1980) The estuary: its definition and geodynamic cycle, in E. Olausson and I. Cato (eds) *Chemistry and Biogeochemistry of Estuaries*, 1–35, New York: Wiley.

Hayes, M.O. (1975) Morphology of sand accumulation in estuaries: an introduction to the symposium, in L.E. Cronin (ed.) *Estuarine Research, Vol. II*, 3–22, New York: Academic.

Nichols, M.M. and Biggs, R.B. (1985) Estuaries, in R.A. Davis (ed.) *Coastal Sedimentary Environments*, 77–186, New York: Springer-Verlag.

Nordstrom, K.F. (1992) *Estuarine Beaches*, London: Elsevier.

Perillo, G.M.E. (ed.) (1995) *Geomorphology and Sedimentology of Estuaries*, New York: Elsevier.

Pethick, J.S. (1992) Saltmarsh geomorphology, in J.R.L. Allen and K. Pye (eds) *Saltmarshes: Morphodynamics, Conservation and Engineering Significance*, 41–62, Cambridge: Cambridge University Press.

Pritchard, D.W. (1967) What is an estuary: physical viewpoint, in G.H. Lauff (ed.) *Estuaries*, 3–5, Washington, DC: American Association for the Advancement of Science.

Roy, P.S. (1984) New South Wales estuaries: their origin and evolution, in B.G. Thom (ed.) *Coastal Geomorphology in Australia*, 99–121, New York: Academic.

Wood, M.E., Kelley, J.T. and Belknap, D.F. (1989) Patterns of sediment accumulation in the tidal marshes of Maine, *Estuaries* 12, 237–246.

N.L. JACKSON

ETCHING, ETCHPLAIN AND ETCHPLANATION

The word 'etching' generally means corroding a surface by aggressive reagents and is used in geomorphology to describe progressive rock decomposition which occurs within deep weathering profiles. In particular, it is applied to situations where rocks differ in their resistance to chemical decay and consequently thickness of a weathering mantle is highly variable over short distances. Removal of products of deep weathering will expose bedrock surface, the topography of which is the direct result of etching, thus it is an 'etched surface'. At an early stage of development of geomorphology, when focus on planation surfaces and peneplains was pre-eminent, etched surfaces were visualized as surfaces of rather low relief and thought of as a special category of a PENEPLAIN, produced by deep-reaching rock decay followed by stripping of weathering mantle. For surfaces of this origin, the term 'etchplain' was proposed by B. Willis and E.J. Wayland, working in East Africa in the 1930s. Accordingly, the process of producing an etchplain through weathering and stripping has later become known as 'etchplanation'.

The impact of the concept of etching and etchplanation on general geomorphology was initially limited, mainly due to the association of etchplains with peneplains, remoteness of original study areas, and restricted access to early publications. Furthermore, no applications for extratropical areas were offered and etchplains were considered as of local importance and specific for low latitudes. Proposals to restrict the usage of the term 'etchplain' to areas of exposed rock, i.e. completely stripped of weathering products, added to the slow progress and general downplaying of the significance of etchplains in geomorphology.

The situation began to change with the arrival of the paper by Büdel (1957) which is noteworthy for a number of reasons, although the very term 'etchplain' was not used. First, Büdel made clear that he was applying the weathering/stripping concept to entire landscapes rather than to limited areas within them or to individual landforms. Second, he suggested that many upland surfaces in middle and high latitudes are inherited Tertiary etchplains, and hence extended the applicability of the concept outside the tropics. Third, he pointed out that transition from the phase of dominant weathering to the phase of dominant stripping might be associated with major environmental changes, whose profound impact for landform development was realized only later. Fourth, it contributed to the appreciation of the concept by the Central European geomorphological community, which was reflected in numerous detailed studies shortly after.

Realization of the crucial role of deep weathering and SAPROLITE development in shaping most tropical landscapes, achieved in the 1960s, led to the expansion of the original ideas of Wayland and Willis, so different types of landscapes could be described. The proposed classification, in the form subsequently modified by its author himself (Thomas 1989), include:

- *Mantled etchplain*: weathering mantle is ubiquitous and virtually no bedrock is exposed. Weathering progressively attacks solid rock at the base of the mantle, moulding the etched surface which is to be exposed later, but the mantle can also be relict.
- *Partly stripped etchplain*: develops from mantled etchplain through selective removal of the weathering mantle and exposure of bedrock surface, but part of the original saprolite remains. The proportion of areas still underlain by saprolite may vary from 10 to almost 100 per cent.
- *Stripped etchplain*: most of the bedrock is exposed from beneath a weathering mantle and only isolated patches of saprolite are left (< 10 per cent of the area). These characteristics conform with the original definition by Wayland.
- *Complex etchplain*: includes a few variants, in which deeply incised valleys may be present (incised etchplain), or removal of saprolite is accomplished by pedimentation (pedimented etchplain), or a new generation of weathering mantles begins to form (re-weathered etchplain).
- *Buried etchplain*: one which has been covered by younger sediments or lava flows.
- *Exhumed etchplain*: one which has been re-exposed after burial.

One important terminological problem has been noted, that a stripped surface is rarely a plain but tends to show some relief, which reflects differential progress of etching (see Figure 59). This happens in particular, if bedrock is lithologically varied or various structures, e.g. fractures, are differentially exploited by weathering. In many granite areas, stripped surfaces are typified by domes, TORS, basins and boulder piles, and to call them 'etchplains' would be inappropriate and misleading. Therefore the term 'etchsurface' is recommended for use wherever evacuation of weathering mantles reveals a varied topography.

Etchplanation, and in particular the transition from the weathering phase to the stripping phase, is commonly linked with major external changes experienced by a landscape, related either to a change in tectonic regime or to environmental changes. It is envisaged that mantled etchplains form and exist during long periods (up to 10^9 yr) of stability, whereas stripping is initiated by uplift, or climatic change towards drier conditions, and is accomplished over much shorter timescales (10^5–10^7 yr). In this view, major external disturbances are essential for the formation of etchplains and static nature of planate landsurfaces might be implied. This position is contradicted by field evidence of geomorphic activity, hence an idea of 'dynamic etchplanation' has been introduced to emphasize ongoing landscape development through etching and stripping (Thomas and Thorp 1985). Key points made are simultaneous weathering and removal of its products, lowering of both interfluves and valley floors, continuous sediment transfer, redistribution and temporal storage of weathering products, and the importance of minor environmental disturbances.

From the 1980s onwards, following the progress in weathering studies, the etching concept has been extended away from tropical plains to the much wider range of settings. Emphasis on the process of deep weathering rather than on the ultimate form of a plain has made it possible to see geomorphic development of many low-latitude mountain ranges of moderate relief as being accomplished by differential etching. Deep weathering is facilitated by strong groundwater movement, steep hydraulic gradient, tensional fracture patterns and numerous lines of weakness within bedrock, whereas landslides play a major part in removal of the saprolite. Realization that

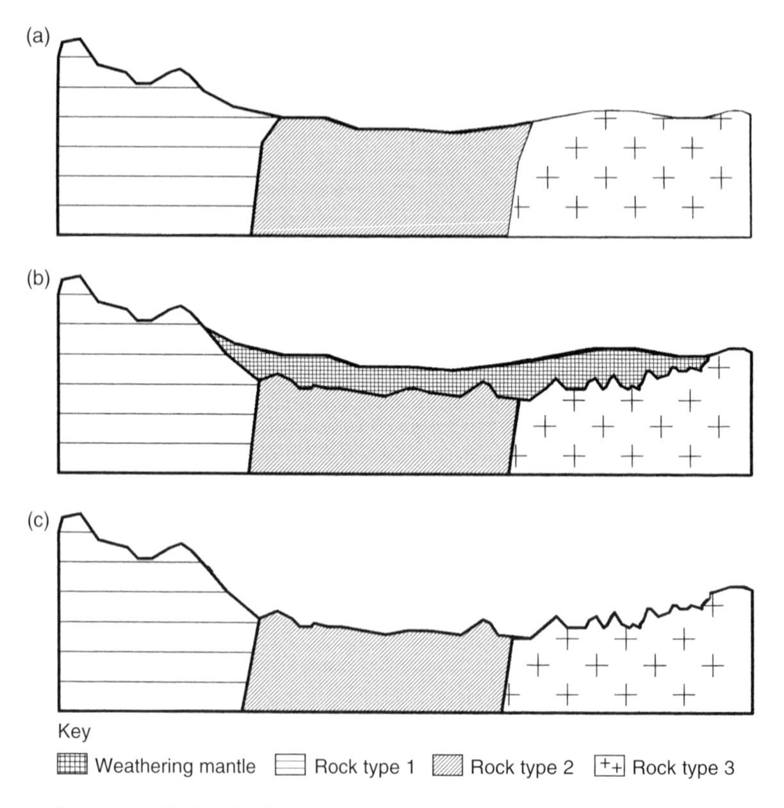

Figure 59 Depending on (a) bedrock characteristics and their susceptibility to (b) selective deep weathering, etching may produce surfaces of (c) various types, for instance inselberg-dotted plains (middle) or multi-convex, hilly areas (right)

formation of thick sandy weathering mantles (see GRUS) can effectively take place outside the tropics has opened the way to interpret several middle to high latitude terrains, with no or extremely remote history of tropical conditions, as etchsurfaces or etchplains, even if the very term has not always been used (Pavich 1989; Lidmar-Bergström 1995; Migoń and Lidmar-Bergström 2001).

Over the years, the idea of etching and etchplanation has evolved from being a mere specific, 'tropical' variant of peneplanation to the status of an autonomic concept, capable of accounting for both planate and topographically complex landsurfaces, integrating tectonic and climatic controls, linking historical and process geomorphology. Despite what the name and early history suggest, it must not be regarded as focusing on explaining the origin of planation surfaces. Nor does it compete with other planation theories, as for instance pedimentation may be a means of stripping. To the contrary, long-term etching and stripping may, and in many places do, lead to the differentiation and increase of relief. Depending on local lithology, tectonic setting and environmental history, long-term etching may transform an initial landscape into a range of topographies, from plains to even mountainous. Therefore, it is the evidence of former or present deep weathering which is a prerequisite for a terrain to be identified as an etchsurface, and not any particular assemblage of landforms.

References

Büdel, J. (1957) Die 'doppelten Einebnungsflächen' in den feuchten Tropen, *Zeitschrift für Geomorphologie N.F.* 1, 201–1, 228.

Lidmar-Bergström, K. (1995) Relief and saprolites through time on the Baltic Shield, *Geomorphology* 12, 45–61.

Migoń, P. and Lidmar-Bergström, K. (2001) Weathering mantles and their significance for geomorphological evolution of central and northern Europe since the Mesozoic, *Earth Science Reviews* 56, 285–324.

Pavich, M.J. (1989) Regolith residence time and the concept of surface age of the Piedmont 'peneplain', *Geomorphology* 2, 181–196.

Thomas, M.F. (1989) The role of etch processes in landform development, *Zeitschrift für Geomorphologie N.F.* 33, 129–142 and 257–274.

Thomas, M.F. and Thorp, M. (1985) Environmental change and episodic etchplanation in the humid tropics: the Koidu etchplain, in I. Douglas and T. Spencer (eds) *Environmental Change and Tropical Geomorphology*, 239–267, London: George Allen and Unwin.

Further reading

Adams, G. (ed.) (1975) *Planation Surfaces* Benchmark Papers in Geology, 22.

Bremer, H. (1993) Etchplanation, Review and Comments of Büdel's model, *Zeitschrift für Geomorphologie N.F. Supplementband* 92, 189–200.

Thomas, M.F. (1994) *Geomorphology in the Tropics*, Chichester: Wiley.

Twidale, C.R. (2002) The two-stage concept of landform and landscape development involving etching: origin, development and implications of an idea, *Earth Science Reviews* 57, 37–74.

SEE ALSO: granite geomorphology; inselberg; planation surface; tropical geomorphology

PIOTR MIGOŃ

EUSTASY

The concept of 'eustatic' changes in sea level, implying vertical displacements of the ocean surface occurring uniformly throughout the world, was introduced by Suess (1885–1909). Global changes in sea level depend in fact on a combination of factors (changes in the quantity of oceanic water, deformation of the shape of the oceanic basin, variations in water density, and dynamic changes affecting water masses) which operate globally, regionally or locally on different timescales.

The quantity of ocean water is controlled mainly by climate, that may cause the development or melting of huge continental ice sheets. According to IPCC (2001), the present volume of continental ice can be estimated at about 29 million cubic kilometres, equivalent to *c.*70 metres water depth in the oceans. At the time of the last glacial maximum, some 20 ka ago, when the global sea level was estimated about 120 metres lower than now, the continental-ice volume must have been more than double the present.

Daly (1934) stressed the importance of changes in sea level and of glaci-isostatic effects (see ISOSTASY) accompanying the last deglaciation phase, with uplift in areas of ice melting and subsidence in a wide peripheral belt. During the last decades, improved global isostatic models based on ice volumes and water depths have been developed (e.g. Lambeck 1993; Peltier 1994), demonstrating that ice-volume changes imply vertical deformation of Earth crust which is highly variable regionally.

Mörner (1976) supported anew the old notion of geoidal changes, suggesting that displacements of the bumps and depressions revealed by satellites on the ocean surface topography could cause differences between coastal areas in the relative sea-level history.

Recently, the analysis of satellite observations, especially by Topex/Poseidon, have shown that the level of the ocean surface can be closely correlated with sea-surface temperature. The resulting sea-surface topography is highly variable, with sea-level rise in certain areas, and sea-level fall in other areas. Steric effects, which depend on the temperature (and density) of the whole water column are also highly variable. Analysis of the dynamic behaviour of water masses and their displacements would bring similar results.

SEA-LEVEL changes are therefore not uniform, but variable over several temporal and spatial scales. Sea level may therefore change from place to place in the ocean, and even more in coastal areas, where hydro-isostatic movements are controlled by the water depth on the continental shelf. In short, there is now general agreement that no coastal area exists where the local sea-level history could be representative of the global eustatic situation. Worldwide or simultaneous sea-level changes, therefore, do not exist. They are an abstraction.

In spite of this field evidence, the concept of eustasy is not an obsolete one, because the estimation of global sea-level changes, even if obtained with some approximation, may have many useful applications in geosciences, in relation to climate, tectonics, paleo-environmental, and also near-future environmental changes. If eustatic variations cannot be specified from coastal field data, the estimation of the changes in the quantity of ocean water is possible using geochemical analysis of marine sediments. The $\delta^{18}O$ content in fossil foraminifera shells cored from the deep ocean floor depends on the salinity and temperature of sea water at the time they lived. If benthic species are selected, temperature changes will be minimized and $\delta^{18}O$ will depend mainly on salinity, i.e. on the quantity of fresh water held up in continental ice sheets. Such calculation makes the estimation of approximate eustatic changes possible, with assumptions, and with an accuracy that depends on the resolution of geochemical measurements, i.e. with an uncertainty range, for sea level, of the order of $\pm$ 10 m. Such precision may seem relatively poor, if compared to what can be obtained at a local scale from the study of former shorelines data. In addition, tectonic, isostatic, steric and hydrodynamic factors are neglected. Nevertheless, continuous oceanic cores have the great advantage that they can cover long time sequences, making approximate estimations of eustatic oscillations possible for periods even longer than the whole Quaternary. According to Milankovitch's astronomical theory, major climatic changes have an astronomical origin, with cycles of near 100 ka for orbital eccentricity, 41 ka for orbital obliquity, and 23 ka and 19 ka for precession phenomena. The age of the climate oscillations deduced from oceanic cores is generally estimated with good accuracy, through calibration with selected astronomical (e.g. insolation at 65°N) curves.

Even with some approximation, eustatic oscillations can be very useful to coastal geomorphologists, e.g. to those who study sequences of datable raised marine terraces in uplifting areas. Each terrace, especially if made of coral reefs, can be considered to have developed when the rising sea level overtook the rising land, and therefore corresponds to a sea-level transgression peak. In this manner, the sea level relative to a stable oceanic floor can be extracted from each dated section if the uplift rate for that section is known (Chappell and Shackleton 1986).

Estimates of eustatic changes during the last century have been attempted by many authors, mainly using tide-gauge records. Discrepancies arising from different analysis methods (for a critical review, see Pirazzoli 1993) remain high enough, however, for IPCC (2001) not to choose between a recent sea-level rise of 1.0 mm yr^{-1}, or an upper bound of 2.0 mm yr^{-1}, or a central value of 0.7 mm yr^{-1} estimated independently from observations and models of sea-level rise components.

Satellite data are more reliable than tide gauges for global calculations. According to Topex/Poseidon, a global sea-level rise of 2.5 $\pm$ 0.2 mm yr^{-1} can be inferred between January 1993 and December 2000 (Cabanes *et al.* 2001). However, El Niño events produce significant oscillations in the global sea level trend and a few decades of additional records would be necessary before a reliable assessment of the present-day eustatic trend can be made with some confidence.

For the next century, eustatic predictions are based on climatic models and scenarios of greenhouse gas emissions. The most recent estimate (IPCC 2001) is of a global sea-level rise of 0.09 to 0.88 m for the period from 1990 to 2100, with a central value of 0.48 m. This estimate includes the variation in the quantity of oceanic water and steric effects, but is exclusive of vertical land motion and hydrodynamic effects. Therefore, it may have very little to do with the relative sea level experienced on a regional basis or at single sites.

References

Cabanes, C., Cazenave, A. and Le Provost, C. (2001) Sea level rise during past 40 years determined from satellite and in situ observations, *Science* 294, 840–842.
Chappell, J. and Shackleton, N.J. (1986) Oxygen isotopes and sea level, *Nature* 324, 137–140.
Daly, A. (1934) *The Changing World of the Ice Age*, New Haven: Yale University Press.
IPCC (2001) *Climate Change 2001: The Scientific Basis*, Cambridge (UK) and New York: Cambridge University Press.
Lambeck, K. (1993) Glacial rebound and sea-level change: an example of a relationship between mantle and surface processes, *Tectonophysics* 223, 15–37.
Mörner, N.A. (1976) Eustasy and geoid changes, *Journal of Geology* 84(2), 123–151.
Peltier, W.R. (1994) Ice age paleotopography, *Science* 265, 195–201.
Pirazzoli, P.A. (1993) Global sea-level changes and their measurement, *Global and Planetary Change* 8, 135–148.
Suess, E. (1885–1909) *Das Antlitz der Erde, 3 vols*, Wien.

Further reading

Pirazzoli, P.A. (1996) *Sea-Level Changes – The Last 20,000 Years*, Chichester: Wiley.

P.A. PIRAZZOLI

EVORSION

The erosion of rock or sediments in a river or streambed, by the impact of clear water carrying no suspended load. The process of evorsion often results in the formation of pot-holes (evorsion pot-holes) within the streambed, due to the action of eddies and vortices. The predominant processes involved in evorsion are hydraulic action and fluid stressing.

Further reading

Aengeby, O. (1952) Recent, subglacial and laterglacial pothole erosion (evorsion), *Lund Studies in Geography, Series A, Physical Geography* 3, 14–24.

STEVE WARD

EXFOLIATION

The shedding of material in scales or layers, it is often used interchangeably with sheeting or onion-skin weathering. Exfoliation of rock has been attributed to various causes including UNLOADING, insolation and HYDRATION (see INSOLATION WEATHERING). It is a process that has some applied significance and is, for example, a consideration in road, tunnel and dam construction where excavation can cause pressure release and fracturing to occur (Bahat *et al.* 1999). Exfoliation occurs at a variety of spatial scales from thin (<cm) scaling from boulders to mega-form some metres in size (Bradley 1963).

Plate 42 As a consequence of pressure release resulting from the erosion of overlying material, this granite near Kyle in Zimbabwe is being broken up into a series of curved sheets which parallel the land surface

References

Bahat, D., Grossenbacher, K. and Karasaki, K. (1999) Mechanism of exfoliation joint formation in granitic rocks, Yosemite National Park, *Journal of Structural Geology* 21, 85–96.
Bradley, W.C. (1963) Large-scale exfoliation in massive sandstones of the Colorado Plateau, *Geological Society of America Bulletin* 74, 519–528.

A.S. GOUDIE

EXHUMED LANDFORM

Landforms that have been covered by sedimentary strata or volcanic rocks and then re-exposed are called exhumed. Exhumed landforms of different age are common within Precambrian shields. Oldest exhumed landforms in Australia are encountered below Proterozoic covers. Flat surfaces are often exhumed from below Lower Palaeozoic rocks on the Laurentian and Baltic shields, while etched (deeply weathered) more or less hilly surfaces extend from below Jurassic or Cretaceous rocks in Minnesota, USA, along parts of the Greenland west coast and in southern Sweden. Hilly relief is exhumed from Neogene sediments in south Poland. Glacially polished surfaces extend from below Upper Precambrian strata in northern Norway, Ordovician strata in the Sahara, and Permian strata on the Gondwana continents. Palaeokarstic features are exhumed from below Carboniferous strata in eastern Canada and in south Germany they make up the Kuppenalb, exhumed from a Cretaceous cover. Exhumed landforms give important information on past processes and their recognition is necessary for correct interpretation of present landscapes. Exhumed denudation surfaces are also important geomorphic markers for studies of Cainozoic uplift and erosion.

Further reading

Ambrose, J.W. (1964) Exhumed palaeoplains of the Precambrian shield of North America, *American Journal of Science* 262, 817–857.

Fairbridge, R.W. and Finkl, C.W. (1980) Cratonic erosional unconformities and peneplains, *Journal of Geology* 88, 69–86.

Lidmar-Bergström, K. (1996) Long term morphotectonic evolution in Sweden, *Geomorphology* 16(1), 33–59.

Migon, P. (1999) Inherited landscapes of the Sudetic Foreland (SW Poland) and implications for reconstructing uplift and erosional histories of upland terrains in Central Europe, in B.J. Smith, W.B. Whalley and P.A. Warke (eds) *Uplift, Erosion and Stability: Perspective on Longterm Landscape Development*, Geological Society London Special Publications, 162, 93–107.

Peulvast, J.-P., Bouchard, M., Jolicoeur, G. and Schroeder, J. (1996) Palaeolandforms and morphotectonic evolution around the Baie de Chaleurs (eastern Canada), *Geomorphology* 16(1), 5–32.

KARNA LIDMAR-BERGSTRÖM

EXPANSIVE SOIL

Most clay soils experience a volume change on wetting and drying. The soils with the relatively inactive clay minerals, such as kaolinite, only produce a modest volume change; but soils containing montmorillonite and other smectite minerals, can have considerable changes of volume; expanding on wetting and shrinking on drying. This causes widespread construction problems because of damage to buildings and other structures, but also accounts for certain geomorphological features such as GILGAI.

The clay particles in soils carry electrical charges, and this accounts for their interesting relationship with water. Clay mineral particles tend to be negatively charged and to attract the cations in the soil water. These cations are hydrated because the negative end of the polarized water molecule is attracted to the charged ion. So, via the activity of the cations, the clay minerals attract water, and this confers on the clay systems the property of plasticity. This is measured via the plasticity index PI, which is low, perhaps around 20, for the inactive kaolinites, illites and chlorites, but high, perhaps up to 200, for the montmorillonites. It is the high PI systems which dominate in expansive soils. The structure of montmorillonite is such that water enters between the clay layers and generates considerable swelling force. The uplift pressure in undisturbed montmorillonite clays can be up to 0.1–0.6 $MN\,m^{-2}$. These expansive pressures easily exceed the loads imposed by small structures such as single-family houses and single-storey schools. It is damage to these small structures which generates the vast costs caused by expansive soil effects. In the USA costs of over $2 billion a year are cited. This is about twice the cost of flood or landslide damage, and more than twenty times the cost of earthquakes.

In the USDA Soil Taxonomy system of soil classification the expansive soils fall into the order Vertisols, and they are defined as cracking soils; mineral soils that have been strongly affected by argillipedoturbation, i.e. mixing by the shrinking and swelling of clays. This normally requires alternate wetting and drying in the presence of >30 per cent clay, much of which, typically, is montmorillonite or some other smectite mineral. If not irrigated, these soils have cracks at least 1 cm wide at a 50 cm depth at some season in most years. Vertisols form the Black Cotton soils of north-west India; they form from the basalts of the Deccan plateau, under the influence of tropical weathering. These soils are classified as Usterts, i.e. ustic (dryish) vertisols; and so are the soils in east Australia which comprise the other large occurrence.

Regions underlain extensively by expansive clays can often be recognized by a distinctive microtopography called gilgai. Where undisturbed by humans, gilgai can be easily distinguished on aerial photography either as an irregular network of microridges, or, where the slope is greater than 1 per cent, as a pattern of downslope linear ridges and troughs. In Australia gilgai relief has been observed to reach over 3 metres. Gilgai microrelief can be used as a rapid means of mapping regions where a significant potential hazard from clay soil expansion can be expected.

The potential volume change of soils is controlled by a number of factors: (1) the type of clay, amount of clay, cations present and clay particle size, (2) density; dense or consolidated soils swell the most, (3) moisture content; dry soils swell more than moist soils, (4) soil structure; remoulded soils swell more than undisturbed soils, (5) loading; schools and houses with lightly loaded foundations are the most susceptible.

There are a variety of tests for expansive soils but one of the most reliable is the oedometer (consolidometer) test. In this test compacted soils are loaded and then wetted and the expansive pressures produced are measured. A simple classification can be produced:

$0–0.15\,MN\,m^{-2}$	= non-critical
$0.15–0.17\,MN\,m^{-2}$	= marginal
$0.17–0.25\,MN\,m^{-2}$	= critical
$>0.25\,MN\,m^{-2}$	= very critical

It is possible to recognize expansive soils in the field; some factors to look out for are:

Under dry soil conditions

- Soil hard and rocklike; difficult/impossible to crush by hand
- Glazed, almost shiny surface where previously cut by scrapers, digger teeth or shovels
- Very difficult to penetrate with auger or shovel
- Ground surface displays cracks occurring in a more or less regular pattern
- Surface irregularities cannot be obliterated by foot pressure

Under wet soil conditions

- Soil very sticky; exposed soil will build up on shoe soles
- Can be easily moulded into a ball by hand; hand moulding will leave a nearly invisible residue on the hands after they dry
- A shovel will penetrate soil quite easily and the cut surface will be very smooth and shiny
- Freshly machine scraped or cut areas will tend to be very smooth and shiny
- Heavy construction equipment will develop a thick soil coating that may impair their function.

These high PI, high montmorillonite soils can be stabilized by lime addition. This causes cation replacement and the soils become more rigid.

Further reading

Chen, F.H. (1988) *Foundations on Expansive Soils*, Amsterdam: Elsevier.

Fanning, D.S. and Fanning, M.C.B. (1989) *Soil: Morphology, Genesis, and Classification*, New York: Wiley.

Proceedings of the 7th International Conference on Expansive Soils, Dallas, Texas, (1992) 2 vols.

Yanagisawa, E., Moroto, N. and Mitachi, T. (eds) (1998) *Problematic Soils*, section on expansive and collapsible soils, 253–384, Rotterdam: Balkema.

IAN SMALLEY

EXPERIMENTAL GEOMORPHOLOGY

Experimental geomorphology is the study, under experimental conditions, of a representation of a selected geomorphological feature or landscape. The term 'representation' is intended to cover full-scale features, scale models (hardware models), and numerical constructs. This definition raises the question 'what constitutes a geomorphological experiment?' Writing in a geomorphological context, Church (1984: 563) defined a scientific experiment as 'an operation designed to discover some principle or natural effect, or to establish or controvert it once discovered'. This activity differs from casual observation in that the phenomena observed are, to a critical degree, controlled by human agency, and from systematically structured observations in that the results must bear on the verity of some conceptual generalization about the phenomena. That definition leads to specific criteria for an experiment:

1. There must be a conceptual model of the processes or relations of interest that will ultimately be supported or refuted by the experiment, giving rise to:
2. Specific hypotheses about landforms or landforming processes that will be established or falsified by the experiment. (If the conceptual

model is a well-developed theory, the hypotheses will constitute exact predictions.)

To test the hypotheses, three further conditions are required:

3 Definitions must be given of explicit geomorphological properties of interest and operational statements of the measurements that will be made on them (sufficiently completely that replicate measurements might be made elsewhere);
4 A formalized schedule must be established of measurements to be made in conditions that are controlled insofar as possible to ensure that the remaining variability be predictable under the research hypotheses;
5 A scheme must be specified for analysis of the measurements that will discriminate the possibilities in (2).

The critical condition is the fourth one. Under this rubric, Church recognized two types of geomorphological experiment:

(a) Intentional, controlled interference with the natural conditions of the landscape in order to obtain unequivocal results about a limited set of processes that change the landscape;
(b) Statistically structured replication of observations in the landscape so that extraneous variability is effectively controlled or averaged over the experimental units.

Experiments of the second type entrain the capacity of statistical experimental design to discriminate and classify information in contexts where variability cannot actively be controlled. Ecologists face problems similar to those posed by geomorphology and have developed a sophisticated understanding of statistical experimentation (e.g. Hairston 1989) in order to deal with them.

The space and time scales associated with the development of most landforms effectively exclude them from experimental study. For this reason a definition of experimental geomorphology given by Schumm *et al.* (1987: 3) includes a careful exemption from strict experimental accountability. They wrote that experimental geomorphology is 'study, under closely monitored or controlled experimental conditions', accepting close observation as sufficient to establish a geomorphological experiment. Slaymaker (1991) similarly accepted 'formally structured', though not actively controlled, field studies as satisfactorily experimental exercises; exercises characterized by Church rather as formal case studies.

Schumm *et al.* (1987) also strongly implied that the objects of experimental manipulation would normally be small, or deliberately reduced scale, examples of field landforms. Experimental control is much more easily arranged in such cases, and their major summary book is entirely preoccupied with model studies. Models represent satisfactory experimental tests of hypotheses designed to explain features of the full-scale landscape provided that formal scaling criteria establish that the results can faithfully be extrapolated. Schumm *et al.* indicated no such requirement, even though there is a long history of scale model investigations in Earth science. Instead, they proposed two other perspectives. They suggested that reduced scale landforms be regarded simply as small prototypes. This does not release the investigator from formal scale constraints for extrapolation. They also explored the concept of model studies as analogues of full-scale systems or as studies in which there remains 'similarity-of-process' (after Hooke 1968, who elided the two perspectives). They argued that the model results might be extrapolated at least qualitatively to increase understanding of the full-scale landscape. Yet they recognized that changes in physical processes that undermine the supposed similarity may occur over large changes in scale. The approach, whilst it may be fruitful of ideas, suffers from inability to achieve exact predictions, or even to confirm unequivocal similarity of process, which disqualifies it as an experimental approach under the criteria given above.

There is, in fact, a tolerably well-developed body of conceptual and empirical studies of scale effects in geomorphology and hydrology (Church and Mark 1980), whilst formal scaling criteria have been investigated for hydrological and sedimentation processes at hillslope, channel and catchment scale. Formal scaling of generic model results (a generic model is one that captures the essential elements of a prototype whilst not conforming in inessential details with any particular prototypical example) ought to be possible in many problems.

In the field, one immediately faces the critical question whether experimental control can be adequately established. Geomorphological processes are driven by weather, which cannot feasibly be controlled at scales beyond that of a plot of order $10^2\,m^2$ (which might be enclosed). But variable forcing by weather is a fundamental feature of geomorphological systems, so one that is driven by artificially controlled weather is in some sense an unrealistic environmental system. Active manipulation should perhaps rather be focused on characteristics of the landform or landscape. Then it will usually be essential to establish a parallel reference or control case in which no manipulation is undertaken, in order that the effects of manipulation of the experimental system may be separated from the effects of variable weather.

It is helpful to differentiate landform and landscape studies. The former present more tractable space and time scales, and are more apt to represent environmental systems sufficiently simple to be amenable to control. At relatively small scales, successful experiments include plot studies of soil erosion, ground surface manipulations to investigate periglacial processes, and local applications of artificial precipitation or drainage adjustments to study effects on erosion or slope stability.

The centre of interest in geomorphology lies in transformations of landscape, which may be addressed through catchment experiments. There is far more experience with them than with any other full-scale experimental arrangement in Earth science (see Rodda 1976, for a historical perspective and critique). Much of the difficulty associated with the use of experimental catchments lies in establishing similarity between a treated catchment and its control, and in deciding how far observed results may be extended to the rest of the landscape. At the base of both issues lies the problem of establishing or measuring similarity between landscapes (Church 2003). Despite the known difficulties, the recognition of small catchments as the fundamental unit for most geomorphological process investigations ensures continuing effort to establish experimentally rigorous investigations at this scale.

Geomorphological experiments may be established inadvertantly. It is important not to overlook the potential value of landform or landscape manipulations undertaken for other purposes that nevertheless can be interpreted satisfactorily in terms of the requirements for an experiment. By this means, far larger systems than might ever be available for deliberate experimental manipulation may be studied. Examples include river rectification, certain water regulation projects, and certain changes in land surface condition. In such cases, it is important to establish a satisfactory reference comparison. Sometimes, this might be a before/after comparison in the same system; otherwise, a parallel reference case must be identified.

Numerical experimentation – the construction and operation of numerical models of geomorphological processes – represents a means by which complete control can indeed be gained over the conditions that drive landscape development. The penalty, of course, is uncertainty whether the numerical model faithfully represents all of the significant processes at work in the world. There also remain significant questions surrounding the means by which model outcomes may be compared with real landscapes, similar to those encountered in comparisons between real landscapes. Nevertheless, numerical modelling holds the promise to be an effective means to establish experimental control in geomorphological studies, especially ones concerning the development of entire landscapes over geomorphologically significant periods of time.

References

Church, M. (1984) On experimental method in geomorphology, in T.P. Burt and D.E. Walling (eds) *Catchment Experiments in Fluvial Geomorphology, 563–580*, Norwich: Geo Books.

——(2003) What is a geomorphological prediction?, in P.R. Wilcock and R.L. Iverson (eds) *Prediction in Geomorphology*, American Geophysical Union, Geophysical Monographs series.

Church, M. and Mark, D.M. (1980) On size and scale in geomorphology, *Progress in Physical Geography* 4, 342–390.

Hairston, N.G. Sr. (1989) *Ecological Experiments*, Cambridge: Cambridge University Press.

Hooke, R.L. (1968) Model geology: prototype and laboratory streams. Discussion, *Geological Society of America Bulletin* 79, 391–394.

Rodda, J.C. (1976) Basin studies, in J.C. Rodda (ed.) *Facets of Hydrology*, 257–297, London: Wiley-Interscience.

Schumm, S.A., Mosley, M.P. and Weaver, W.E. (1987) *Experimental Fluvial Geomorphology*, New York: Wiley-Interscience.

Slaymaker, O. (1991) The nature of geomorphic field experiments, in O. Slaymaker (ed.) *Field Experiments and Measurement Programs in Geomorphology*, 7–16, Rotterdam: Balkema.

MICHAEL CHURCH

EXTRATERRESTRIAL GEOMORPHOLOGY

The term 'extraterrestrial geomorphology' was not included in the 1968 *Encyclopedia of Geomorphology*. Indeed, at first reading, this term might seem to be an oxymoron. Should not a science of Earth forms (geomorphology) exclude those forms that are beyond the Earth (extraterrestrial)? The answer to this question depends on how one views the nature of science. Is a science more about methods and attitudes of study, or is it more about the organized accumulation of facts associated with specific subject matter? While organized fact accumulation might require precise definition as to the location of its subject matter, the methods and attitudes of geomorphology are readily conceived as extending to landforms on Earth-like planets (Baker 1993), if only better to understand Earth's landforms. To the extent that geomorphology emphasizes methods and attitudes for the study of landforms and landscapes, then it is no more restricted to studying Earth's landforms and landscapes than geometry is restricted to measuring the Earth's mathematical form.

Despite the immense excitement of planetary exploration during the 1960s, 1970s, and 1980s, there was a conspicuous lack of attention to planetary surfaces by mainstream geomorphologists. Nearly all the study of newly discovered landscapes was performed by scientists with very little background in geomorphology. More recently, increased attention to extraterrestrial geomorphology is indicated by Dorn's (2002) analysis of citations to late twentieth-century research works in geomorphology. Two of the 'top ten' cited geomorphology papers in recent years were directly concerned with topics in extraterrestrial geomorphology.

Historical and philosophical perspectives

It was not long after Galileo Galilei (1564–1642) first used a telescope to observe curious circular depressions on the Moon that Robert Hooke performed the first known geomorphological experiments to explain the origin of those depressions. Hooke was an intellectual adversary of Sir Isaac Newton, and, unlike Newton, he had a great interest and considerable talent for geology and geomorphology. In 1665 he compared the newly discovered lunar craters to (1) the cooled surface crust of melted gypsum, which was disrupted by bursting bubbles, and (2) the impact of musket balls and mud pellets into a clay-water target material. Using analogy as his mode of reasoning, Hooke hypothesized that the lunar craters formed either by (1) internal heat that melted and disrupted its surface crust (today we would describe this process as volcanism), or (2) impacts by particles from space (today these would be described as meteor impact craters). The controversy over these two origins for lunar craters actually continued until the 1970s, when it was finally resolved in favour of the impact hypothesis on the evidence of the lunar rocks returned by space missions.

Analogical reasoning was extensively employed by the geomorphologist Grove Karl Gilbert in his studies of (1) lunar craters, to which he correctly ascribed an impact origin (Gilbert, 1893), and (2) a crater in northern Arizona (now known as Meteor Crater; Plate 43), to which he incorrectly ascribed a volcanic origin (Gilbert 1896). The limitations of analogic reasoning in extraterrestrial geomorphology continue to the present day, as ably summarized by Mutch (1979):

1 Many landforms cannot be assigned a unique cause. Rather, the same landforms may be generated by different combinations of processes that converge to the same result.
2 The photointerpreter is artificially constrained in his analysis by his range of familiarity with natural landscapes. Because of these limitations, the student of extraterrestrial landforms must know as much as possible about the origin of landforms in general.

Planets, moons, and other objects

The term *planetary geomorphology* (Baker 1984) is also used for many of the topics covered by this

Plate 43 Meteor crater in northern Arizona. With a diametre of 1.2 kilometres, the crater formed about 25,000 years ago when an iron meteor struck the Earth at a velocity of about 11 kilometres per second

article, and it is true that planetary surfaces provide the sites of many landforms and landscapes (Greeley 1994). However, not all planets have rocky surfaces on which there are landforms. Moreover, there are many objects beyond Earth that are not planets, and many of these do indeed have landforms and landscapes. If we hold to the idea that one seeks to compare Earth's geomorphology to that of objects beyond Earth, then extraterrestrial geomorphology seems to be the appropriate term.

While future extraterrestrial geomorphology will surely extend to planetary objects in other solar systems, over a hundred of which have been discovered at the time of this writing, discussion here will be limited to the rocky objects of our own solar system. The inner planets, Mercury, Venus, Earth and Mars, all have rocky surfaces on which the effects of volcanism, tectonics and impact craters are in abundant evidence. Earth has a relatively large moon with a surface dominated by impact craters, the study of which has been directly facilitated by human visitation. Mars has two moons, but these are really captured asteroids, and are similar to many thousands of objects that occur throughout the inner solar system, mainly in the so-called 'asteroid belt' between Mars and Jupiter. The planets of the outer solar system, Jupiter, Saturn, Uranus and Neptune, are all giant gas balls, lacking any surface with landforms. Their satellites, however, are phenomenally rich in landscape complexity. Jupiter has four very large moons, Io, Europa, Ganymede, and Callisto, which form a kind of mini solar system, first discovered by Galileo's telescopic investigations. Ganymede and Callisto have heavily cratered surfaces on ice that is so cold it behaves as rock. Io's surface is dominated by volcanism that is much more active than any volcanism on Earth. Europa has a very young, nearly uncratered surface that is locally deformed because the icy crust of this moon overlies an immense ocean of liquid water. Other satellites of the outer planets are similar to asteroids in their surface character. Miranda, a moon of Uranus, looks to have been totally shattered by impact, and then accreted once more from the shattered remnants. The icy satellites of Uranus and Neptune are so cold that ices of ammonia and other volatiles comprise their rocky surfaces. A type of volcanism, known as 'cryovolcanism', was generated when these ices melted.

Titan, a moon of Saturn, has a diameter equal to about one-half that of Earth. It has an atmosphere that is slightly thicker than that of Earth, and, like Earth, the atmosphere is composed mainly of nitrogen. However, Titan is also extremely cold. The other main gas in its atmosphere is methane, and the great cold would lead to the condensation of that gas as a liquid on the surface of the satellite. Thus, Titan could have an ocean of methane and other hydrocarbons, or the liquid might just occupy lakes in the impact craters of a rocky surface, on which the 'rock' might be water ice. In any case, there is a very complex spacecraft, Cassini, which is scheduled to arrive in the Saturn system in July of 2004. The radar instrument on Cassini will permit penetration of the hazy atmosphere of Titan to reveal, for the first time to human observers, the landforms and landscapes of this haze-shrouded world.

Cratering and volcanism

Impact craters are the most ubiquitous landforms on the rocky and icy planets and satellites. Relatively low densities of impact craters indicate surfaces that are relatively young and unmodified relative to the 4.5 billion-year age of the solar system. Such surfaces comprise the dominant portions only of the large icy satellites Europa and Triton (a satellite of Neptune), the volcanically active satellite Io, and the planets Venus and Earth. In contrast, Mercury, Mars, the moon, and most planetary satellites have much of their surfaces covered with densely cratered terrains that formed over several billion years (Plate 44, p. 357). These surfaces are preserved because of minimal modification by active surficial processes related to atmospheric effects (exogenetic processes) and relatively localized volcanism and tectonic effects (endogenetic processes).

Volcanism is also very common in the solar system, though it does not dominate the landscapes of objects other than Io, Venus and Earth. All the rocky planets do have extensive volcanic plains, however. On Earth, these are hidden beneath ocean waters, and they were emplaced by seafloor spreading volcanism, mostly within the last 100 million years. Mercury has extensive intercrater plains, and both Mars and the moon have lowland plains that show evidence of being covered by lava flows. The lavas that formed these plains all seem to have been highly fluid, probably with basaltic compositions. Venus has some of the most extensive volcanic plains, and some of these are crossed by remarkable lava channels. The longest of these extends over 6,800 kilometres, making it longer than Earth's longest river (Baker *et al.* 1992).

Volcanic constructs, including large cones, shield volcanoes and calderas occur on Venus, Earth and Mars. Olympus Mons, a shield volcano on Mars, measures over 700 kilometres in diameter, and it rises to a height of 25 kilometres. It is only one example of extraterrestrial landforms that are much larger than their counterparts on Earth (Baker 1985). Though most extraterrestrial volcanic landforms are relict, the active volcanism of Io is spectacular. Eruptive plumes from the surface of Io were observed by the Voyager spacecraft to propel debris up to 300 kilometres above the surfaces and to deposit material up to 600 kilometres from the active vents. There are also active eruptive plumes on Triton, but the responsible process is probably more similar to that of a geyser than that of a volcano.

Tectonic landforms

Most of the rocky planet and satellite surfaces show evidence of structural deformation, with various fractures, graben and faults being the most common features. Mercury was deformed very early in its history by immense compressional forces that produced thrust fault landscapes. Mars has immense fracture zones and graben. However, only Earth exhibits the distinctive landforms associated with plate tectonics, including mid-oceanic ridges, transform faults and convergent continental margins with fold and thrust belt mountain ranges. Despite its density, radius and other geophysical similarities to Earth, the planet Venus does not show plate-tectonic landforms. This raises interesting questions about what makes plate tectonics unique to Earth.

Hillslopes and mass movement

Slopes occur on all the rocky planets. On airless bodies, only gravity and impact processes generate slope processes, but the atmospheres of Mars, Venus and Titan invite comparisons to other processes on Earth. A particularly interesting problem is the movement of extremely large (millions of cubic metres) slides or avalanches of rock and debris. Such masses on Earth have very high mobility over flat terrain. The cause of the very high mobility has been ascribed to the cushioning effect of air or water, reducing the effective pressure of the slide mass that would resist broad lateral spreading. However, these types of MASS MOVEMENT occur on the moon, which lacks both air and water. Many examples also occur on Mars, where air and water may have exerted influences.

Aeolian landscapes

While most extraterrestrial surfaces are airless, the atmospheres of Earth, Mars, Venus and Titan invite the interplanetary comparison of aeolian processes and landforms. Mars has the greatest variety of aeolian landforms. Crescent-shaped and transverse DUNES (see DUNE, AEOLIAN), wind ripples, YARDANGS, pitted and fluted rocks, and various dust streaks are all well displayed. There are also remarkable tracks produced by Martian dust devils. Aeolian bedforms also occur on Venus, which has an atmospheric pressure on the land surface that is ninety times that of Earth.

Channels, valleys and fluvial action

Besides Earth, fluvial action seems to have occurred only on Mars, and the most extensive fluvial processes were active in the remote past. The two main varieties of fluvial landforms on Mars are valley networks and outflow channels, morphological attributes of which are reviewed by Baker (1982, 2001). A great many of the valley networks occur in the old cratered highlands of Mars, leading to the view that nearly all of them formed during the heavy bombardment phase of planetary history, prior to 3 or 4 billion years ago. The outflow channels, in contrast, involve the immense upwelling of cataclysmic flood flows from subsurface sources, mostly during later episodes of Martian history. Much of the Martian surface is underlain by a thick ice-rich permafrost zone, a 'cryolithosphere', and the water feeding the outflow channels emerged from beneath this permafrost, possibly associated with volcanic processes (Baker 2001).

One of the most striking recent discoveries is that some water-related landforms on Mars are exceptionally young in age. This fact was prominently demonstrated by images from the Mars Global Surveyor orbiter showing numerous small gullies generated by surface runoff on hillslopes. The gullies were most likely formed by the melting of near-surface ground ice and the resulting debris-flow processes. The gullies are uncratered, and their associated debris-flow fan deposits are superimposed both on aeolian bedforms (dunes or wind ripples) and on polygonally patterned ground, all of which cover extensive areas that are also uncratered. Exceptionally young outflow channels and associated volcanism also occur on Mars. Data from Mars Global Surveyor show that localized water releases, interspersed with lava flows, occurred approximately within the last 10 million years. The huge discharges associated with these floods and the temporally related volcanism should have introduced considerable water into active hydrological circulation on Mars. It is tempting to hypothesize that the young outflow processes and volcanism are genetically related to other very young water-related landforms. The genetic connection for all these phenomena might well be climate change, induced by the water vapour and gases introduced to the atmosphere by both flooding and volcanism (Baker 2001)

Lakes, seas and 'oceans'

On Earth bodies of standing water include (1) lakes, in which the water is surrounded by extensive land areas, (2) seas, in which saline waters cover the greater part of the planetary surface, and (3) the ocean, which is the vast, interconnected body of water that covers about 70 per cent of Earth's surface. For Mars there is no direct geomorphological evidence that the majority of its surface was ever covered by standing water, though the term 'ocean' has been applied to temporary ancient inundations of the planet's northern plains. Although initially inferred from sedimentary landforms on the northern plains, inundation of the northern plains has been tied most controversially to identifications of 'shorelines'. New data indicate the presence of a regionally mantling layer of sediment, which seems to be contemporaneous with the huge ancient flood discharges of the outflow channels. Though the debate over the Martian 'ocean' has received much attention, even more compelling evidence supports the existence of numerous lakes, which were temporarily extant on the surface of Mars at various times in the planet's history.

Glacial and periglacial landforms

Evidence for past glacial activity on Mars is both abundant and controversial. The glacial features are also associated with periglacial landforms, which include debris flows, polygonally patterned ground, thermokarst, frost mounds, pingos and rock glaciers. On Earth most of these landforms develop under climate conditions that are both warmer and wetter than the conditions for cold-based glacial landforms (Baker 2001). The implications for past climatic change on Mars are profound because glaciers require substantial transport of atmospheric water vapour to sustain the snow accumulation that generates the positive mass balance needed for glacial growth.

The glacial landforms of Mars are erosional (grooves, streamlined/sculpted hills, drumlins, horns, cirques and tunnel valleys), depositional (eskers, moraines and kames), and ice-marginal (outwash plains, kettles and glacilacustrine plains). Of course, the landform names are all genetic designations, and ad hoc alternatives have been suggested for many. What is not ad hoc, however, is that all the glacial landforms occur in spatial associations, proximal-to-distal in regard to past ice margins, that would be obvious in a terrestrial setting. Areas of past glaciation on Mars (Kargel and Strom 1992) include the summits of very large volcanoes, uplands surrounding major impact basins (Plate 44), and the polar

Plate 44 Oblique view of the Martian impact basin Argyre, surrounded by mountainous uplands (centre) many of which contain glacial features (Kargel and Strom 1992). Note the high clouds in the Martian atmosphere on the horizon

regions, where the ice caps were much more extensive during portions of post-Noachian time.

The future of geomorphology

It has long been apparent that the modern frontier of geomorphology, both as a matter of physical discovery and as an intellectual challenge, lies in the comparative study of planetary surfaces. This was summarized rather distinctly by Sharp (1980), as follows:

> Planetary exploration has proved to be a two-way street. It not only created interest in Earth-surface processes and features as analogues, it also caused terrestrial geologists to look at Earth for features and relationships better displayed on other planetary surfaces. Impact cratering, so extensive on Moon, Mercury, and Mars, is a well-known example. Another is the huge size of features, such as great landslides and widespread evidence of large-scale subsidence and collapse on Mars, which suggests that our thinking about features on Earth may have been too small-scaled. One of the lessons from space is to 'think big'.

References

Baker, V.R. (1982) *The Channels of Mars*, Austin, TX: University of Texas Press.

Baker, V.R. (1984) Planetary geomorphology, *Journal of Geological Education* 32, 236–246.

——(1985) Relief forms on planets, in A. Pitty (ed.) *Themes in Geomorphology*, 245–259, London: Croom Helm.

——(1993) Extraterrestrial geomorphology: science and philosophy of Earthlike planetary landscapes, *Geomorphology* 7, 9–35.

——(2001) Water and the Martian landscape, *Nature* 412, 228–236.

Baker, V.R., Komatsu, G., Parker, T.J., Kargel, J.S. and Lewis, J.S. (1992) Channels and valleys on Venus: preliminary analysis of Magellan data, *Journal of Geophysical Research* 97, 13,421–13,444.

Dorn, R.I. (2002) Analysis of geomorphology citations in the last quarter of the 20th century, *Earth Surface Processes and Landforms* 27, 667–672.

Gilbert, G.K. (1893) The Moon's face: a study of the origin of its features, *Philosophical Society of Washington Bulletin* 12, 241–292.

——(1896) The origins of hypotheses illustrated by the discussion of a topographic problem *Science*, n.s. 3, 1–13.

Greeley, R. (1994) *Planetary Landscapes*, Dordrecht, The Netherlands: Kluwer Academic.

Kargel, J.S. and Strom, R.G. (1992) Ancient glaciation on Mars, *Geology* 20, 3–7.

Mutch, T.A. (1979) Planetary surfaces, *Reviews of Geophysics and Space Physics* 17, 1,694–1,722.

Sharp, R.P. (1980) Geomorphological processes on terrestrial planetary surfaces, *Annual Review of Earth and Planetary Surfaces* 8, 231–261.

SEE ALSO: astrobleme; crater; geomorphology

VICTOR R. BAKER

F

FABRIC ANALYSIS

Measures one or more parameter of the three-dimensional disposition of elongated rock fragments in sediments. Such fragments have length, breadth and thickness, defined as a, b and c axes. The fragments contain three projection planes, maximum, intermediate and minimum. The maximum plane contains a and b axes, the intermediate a and c axes and the minimum b and c axes. Measurements of the orientation and dip values for axes and planes combined with statistical analysis can identify processes and environments of deposition for many sediment types (Andrews 1971; Dowdeswell and Sharp 1986). For example, in an undisturbed lodgement till the a-axes of pebbles are strongly oriented in the direction of local ice flow and dip slightly up-glacier. On the bed of a river cobbles and boulders frequently exhibit imbricate structure in which the a-axes are normal to the water current and the maximum plane dips upstream. In a storm beach deposit the a-axes of cobbles or shingles (flat cobbles) are usually deposited normal to the direction of wave advance and the maximum plane dips seaward.

References

Andrews, J.T. (1971) Methods in the analysis of till fabrics, in R.P. Goldthwait (ed.) *Till, A Symposium*, 321–327, Columbus: Ohio University Press.

Dowdeswell, J.A. and Sharp, M. (1986) Characterization of pebble fabrics in modern terrestrial glacigenic sediments, *Sedimentology* 33, 699–710.

ERIC A. COLHOUN

FACTOR OF SAFETY

The factor of safety, F, is defined as the ratio of the sum of resisting forces (shear strength) divided by the sum of driving forces (shear stress) of a slope:

$$F = \frac{\text{sum of resisting forces}}{\text{sum of driving forces}}$$

If at a location within a soil mass the shear stress becomes equal to the shear strength of the soil, failure will occur at that point. In this case $F = 1$. Where $F < 1$ the slope is in a condition for failure, where $F > 1$ the slope is likely to be stable.

Shear strength and shear stress were originally expressed by Coulomb in 1776. Shear strength of a soil is its maximum resistance to shear. Its value determines the stability of a slope. The knowledge of the shear strength is an essential prerequisite to any analysis of slope stability and the factor of safety. Coulomb postulated that:

$$\tau_f = c + \sigma \tan \varphi$$

where τ_f = maximum resistance to shear, c = cohesion of the soil, σ = total stress normal to the failure surface, and $\tan \varphi$ = angle of internal friction of the soil.

In 1925 Terzaghi published the fundamental concept of effective stress, $\sigma' = \sigma - u$ (with u = pore-water pressure), that water cannot sustain shear stress and that shear stress in a soil can be resisted only by the skeleton of solid particles at the particle contact points. Shear strength is expressed as a function of effective normal stress as:

$$\tau_f = c' + \sigma' \tan \varphi'$$

in which the parameters c′ (effective cohesion) and φ′ (effective angle of friction) are properties of the soil skeleton.

The factor of safety can then be expressed as

$$F = \frac{c' + \sigma' \tan \varphi'}{\tau_f}$$

This equation can be used for limit equilibrium methods in slope stability analysis (Duncan 1996). The calculation of F requires the description of a potential slip surface which is defined as a mechanical idealization of the failure surface. The critical slip surface is the one with the minimum value of F of all possible slip surfaces included in the limit equilibrium calculation.

Most natural hillslopes prone to landsliding have F values between about 1 and 1.3, 'but such estimates depend upon an accurate knowledge of all the forces involved and for practical purposes design engineers always adopt very conservative estimates of stability' (Selby 1993). In practice the highest uncertainties are related to soil water, especially with the spatial variability of PORE-WATER PRESSURE and seepage.

In geomorphology the factor of safety concept is essential to understand landscape stability. F is considered as ratio of landform strength resistence and the magnitude of impacting forces.

References

Duncan, J.M. (1996) Soil Slope Analysis, in A.K. Turner and R.L. Schuster (ed.) *Landslides. Investigation and Mitigation*, Transportation Research Board, Special Report 247, 36–75, Washington, DC: National Academy Press.

Selby, M.J. (1993) *Hillslope Materials and Processes*, Oxford: Oxford University Press.

Further reading

Craig, R.F. (1994) *Soil Mechanics*, London: Chapman and Hall.

SEE ALSO: shear and shear surface

RICHARD DIKAU

FAILURE

Within geotechnical geomorphology, the term failure implies the occurrence of a dislocation within a material, usually accompanied by detachment of a soil or rock mass. The most common case involves shear (see SHEAR AND SHEAR SURFACE) failure along a well-defined plane of rupture as a LANDSLIDE. In hard sedimentary strata, igneous and metamorphic rocks, detachments usually occur along planes of weakness defined by bedding, joints (see JOINTING), foliation and faults. The potential for movement is greatest where layers dip downslope. The resultant translational landslides are called rockslides. For dry layers, the kinematic criterion for failure is:

$$\phi' < \delta < \beta$$

where ϕ' is the friction angle along the layers, δ is the dip angle, and β the slope angle. The inequality shows that the weak layer must crop out on the slope (i.e. $\delta < \beta$), and the dip angle must exceed the friction angle. In most cases, the temporal variation of all three parameters is typically small. However, frictional resistance along joints can be abruptly reduced when water pressure increases. The steepest stable angle, θ_c, of a rock layer is then:

$$\theta_c = [(\gamma_{sat} - m\gamma_w) / \gamma_{sat}] \tan \phi'$$

where γ_{sat} is the saturated unit weight of the material, γ_w is the unit weight of water, and parameter m is the ratio of the saturated depth to the total slab depth. Similar principles apply to failures in colluvial materials where planar detachments called debris slides often occur at the interface between COLLUVIUM and harder materials below, such as bedrock. Debris slides also occur in glacial materials, for example at the interface between loose unconsolidated ablation till and denser basal till below. In softer rocks, such as clays, shales and mudstones, a lesser degree of structural control exists, and shear surfaces often run oblique to the direction of bedding as rotational failures. In these cases, the above assumption of a plane translational slide is no longer valid.

A more general way to assess the stability or proximity to failure of a soil or rock mass, of any geometrical shape, is the Mohr–Coulomb equation:

$$s = c' + (\sigma - u) \tan \phi'$$

where s is shear strength, c′ is COHESION, σ is total normal stress, u is PORE-WATER PRESSURE and ϕ' the angle of shearing resistance. Parameters c′ and ϕ' are material properties that control shear strength at the ambient EFFECTIVE STRESS. Pore pressure, u, is independent of these parameters, and is a function of moisture recharge from antecedent and ambient climatic events. The overall stability of a mass is assessed from its FACTOR OF SAFETY, $F = s/\tau$, where τ is shear stress. By definition, failure occurs when $F = 1.0$. Failure is most commonly caused by saturation, which causes an increase in shear stress concomitant with a reduction in frictional strength. This explains why so many landslides are associated with major rainstorms or snowmelt.

The above version of the Mohr–Coulomb equation applies to drained failures, which involve no excess pore pressures. In the case of rapid movements in low density, fine-grained, saturated soils (for example, QUICKCLAYS), collapse of the soil structure under shear loads causes significant excess pore pressures to develop. Such failures involve LIQUEFACTION, and must be analysed with reference to undrained (see UNDRAINED LOADING) strength parameters.

Failure may also occur by toppling and buckling of layers, especially in thinly bedded rocks. Toppling involves forward and downslope rotation of layers and is common where strata dip steeply into a slope. For single blocks the toppling criterion is:

$$b/h < \tan \delta$$

where b and h are the breadth and height of the block and δ is the inclination of the block's base. For flexural toppling, which involves downslope rotation and interlayer slip, the criterion is:

$$\alpha < \beta - \phi'$$

where pole angle $\alpha = (90° - \delta)$ is the angle of the normal to the plane, and δ is the dip angle. The inequality shows that toppling is most likely to occur in steeply dipping strata, but may be enhanced where slopes are undercut and steepened. Buckling tends to occur where ductile, thinly bedded rocks, such as argillite and phyllite, dip downslope slightly steeper than the slope angle. When the downslope compressive stress exceeds the bending resistance of the layers, buckling may occur.

Most slope failures involve more than one type of movement. For example, a landslide dominated by plane failure at its base may also involve buckling or forward toppling of material by compression at the toe area, and tensional failure at the headscarp. Transitions from one type of movement to another are also common, for example detachment of a saturated mass as a debris slide, followed by disintegration and fluidization as a DEBRIS FLOW further downslope.

Although the Mohr–Coulomb equation implies abrupt attainment of failure, many landslides probably involve slow creep movements prior to detachment. Deep-seated gravitational movements of the SACKUNG type probably involve prolonged, slow movements at depth. Such mountain scale masses total tens to hundreds of millions of cubic metres of material moving at millimetres to centimetres per year. The surface expression of such movements is typically tension cracks, uphill facing scarps and grabens, or is less clearly defined as masses of broken, dilated rock. Such movements may occur over centuries to millennia without the development of a landslide rupture surface. However, other cases are known to have terminated in large rock avalanches (STURZSTROMS). This suggests that a continuum of slope movement rates and types may occur over time at an individual site, a circumstance which is not easily encompassed by existing methods used to classify and analyse slope movements.

Further reading

Barnes, G.E. (2000) *Soil Mechanics*, 2nd edition, London: Macmillan.

Bovis, M.J. and Evans, S.G. (1996) Extensive deformations of rock slopes in the southern Coast Mountains, British Columbia, Canada, *Engineering Geology* 44, 163–182.

Goodman, R.E. and Bray, J.W. (1976) Toppling of rock slopes, in *Proceedings of a Specialty Conference on Rock Engineering for Foundations and Slopes*, Boulder, CO; New York: American Association of Civil Engineers.

Hoek, E. and Bray, J.W. (1981) *Rock Slope Engineering*, 3rd edition, London: Institution of Mining and Metallurgy.

Turner, A.K. and Schuster, R.L. (eds) (1996) *Landslides: Investigation and Mitigation*, Washington, DC: National Academy Press.

MICHAEL J. BOVIS

FALL LINE

The topographical and geological boundary between an upland region of relatively high resistance crystalline rock and a lower region of weaker rock. Rivers transcending this boundary often develop waterfalls and rapids in parallel. A less frequent and appropriate use of the term is for the point where a river ceases to be tidal. The fall line is thus a geological and geomorphological boundary. Generally, streams and rivers upstream from the fall line have small floodplains and are of low sinuosity, whereas downstream from the fall line rivers and streams tend to possess larger floodplains and display high sinuosity.

The type example of a fall line is the eastern United States region, where the upland Piedmont Plateau (crystalline rock) meets the Atlantic coastal plain (weaker sedimentary rock). The junction is marked by rapids and waterfalls on each of the major rivers that transcend the zone (i.e. the Delaware, Potomac, James, Savannah, etc.).

The steep gradient of the American fall line has been accounted for in three main ways, as reviewed by Renner (1927) and Lobeck (1930: 454). First, the feature can be interpreted as a zone of monoclinal flexing or faulting (though faulting occurs on the fall line in few localities). Second, as an area where the rivers have eroded away the softer rocks of the coastal plain, forming knickpoints at the boundary with the resistant crystalline piedmont rocks. Third, as the intersection of two ancient peneplains, in which the older mid-Mesozoic erosion surface plunges beneath the coastal plain deposits that overlie the younger peneplain. The fall line represents a stripped part of the older peneplain and accounts for its steeper slope.

References

Lobeck, A.K. (1930) *Geomorphology: An Introduction to the Study of Landscapes*, London and New York: McGraw-Hill.

Renner, G.T. Jr (1927) The physiographic interpretation of the Fall Line, *Geographical Review* 17(2), 278–286.

STEVE WARD

FANGLOMERATE

A sedimentary rock consisting of heterogeneous fragments of assorted size deposited in an alluvial fan and subsequently cemented into a solid mass. The term was introduced by Lawson (1913) to describe the coarse upslope parts of ALLUVIAL FAN formations, though the term is also used more generally for conglomerates and breccias deposited on alluvial fans. They are composed of two main facies: water laid deposits and mass flow deposits. Fanglomerates are characterized by their parallel bedding and decreasing particle size downslope, alongside rapid fan thinning.

References

Lawson, A.C. (1913) *The Petrographic Designation of Alluvial Fan Formations*, University of California Publications Department G.7, 325–334.

STEVE WARD

FAULT AND FAULT SCARP

A *fault* is a surface or zone along which one side has moved relative to the other in a direction parallel to the surface or zone. The term is applied to features extending over distances of metres or larger, whereas those at the scale of centimetres are called shear fractures, and those at the scale of millimetres are microfaults. Most of them are brittle structures, although some may represent ductile deformation. For an inclined fault, the fault block above the fault is the hanging wall, and the block below the fault is the footwall. Faults are subdivided into dip-slip (showing slip parallel to the dip of the fault surface), strike-slip (of slip parallel to the strike of the fault surface), and oblique-slip faults (where slip is inclined obliquely on the fault surface). The dip-slip faults include normal faults, on which the hanging wall block moves down relative to the footwall block, and thrust (dipping $<45°$) or reverse faults (dipping $>45°$), on which the hanging wall moves up relative to the footwall block. Strike-slip faults are either right-lateral or left-lateral, depending on the sense of motion of the fault block across the fault from the observer. These faults are commonly planar and vertical. Both dip-slip and strike-slip faults frequently form a linked fault system which, in cross section, consists of flats and ramps, which cut through the footwall and detach a slice of hanging wall rocks.

Some normal faults are concave-upward faults of dip decreasing with increasing depth; they are called listric faults. These can join or turn into a low-angle detachment fault at depth. Small-scale faults parallel to the major fault and showing the same sense of shear are called synthetic faults; those of the conjugate orientation are antithetic faults. A downthrown block bounded on either side by conjugate normal faults is a graben, whereas a relatively uplifted block bounded by two conjugate faults is a horst. A half-graben is a lowered tilted block bounded on one side by a normal fault. Step faults are parallel faults on which the downthrown side is on the same side of each fault. Rotational movements between the two fault blocks result in varying throws along the fault strike, producing either hinge faults, where displacement increases from zero to a maximum along the strike, or pivot (scissor) faults, where one block appears to have rotated about a point on the fault plane. The traces of normal faults are either straight or slightly sinuous, depending on the fault dip, whereas the traces of thrusts are usually highly sinuous due to low-angle intersection with the ground surface.

Large-scale strike-slip faults are called transform faults when building segments of lithospheric plate boundaries, or trancurrent faults when they occur in continental crust and are not parts of plate margins. Strike-slip faults frequently form bends (curved parts of the fault trace) and stepovers, i.e. places where one fault ends and another, *en echelon* fault begins. A left bend or stepover in a right-lateral fault system (restraining bend) induces local compression (uplift; transpression), whereas a right bend or stepover in a right-lateral fault (releasing bend) produces local extension (subsidence; transtension). Displacement at extensional bends and stepovers forms rhomboidal, fault-bounded depressions, called pull-apart basins.

A *fault scarp* is a tectonic landform coincident with a fault plane that has displaced the ground surface. A residual fault scarp is a mature scarp, upon which the original tectonic surface has been obliterated by geomorphic processes. A fault-line scarp, in turn, results from differential weathering and erosion of the rocks on either side of the fault.

Scarps produced by normal faulting are usually located at the contact between bedrock in the footwall and Quaternary sediments in the hanging wall. Scarps associated with reverse faulting in solid bedrock are commonly overhanging and tend to collapse and/or be eroded; they are also more deeply embayed than their normal counterparts. Scarps associated with strike-slip faults are less prominent and are best developed in areas of uneven topography. In loose sediments, however, fold-limb (monoclinal or fold) scarps are formed, and usually are surface expression of blind thrusts.

Active normal fault scarps include: piedmont (simple) scarps, formed in unconsolidated deposits; multiple (complex) scarps, related to formation of a fault splay during a single faulting event; composite (multi-event, compound) scarps, up to a few tens of metres high, formed due to renewed slip on a fault; and splintered scarps, produced due to fault displacement distributed across overlapping *en echelon* segments. A piedmont scarp (Wallace 1977) includes a steep ($>50°$) free face, a moderately inclined (30–40°) debris slope, and a gently inclined (5–10°) wash slope (see SEISMOTECTONIC GEOMORPHOLOGY). Fault scarps in semi-arid climate degrade from gravity-controlled (10^2 yrs), through debris-controlled (10^3 yrs), to wash-controlled (10^5 yrs) slope due to either: decline, replacement, retreat or rounding (Mayer 1986).

Depending on the climatically controlled rate of removal of debris shed from the scarp, the Oregon or Basin and Range-type scarps, typical of semi-arid climate, and the Awatere or New Zealand-type scarps, formed in a more humid climate, have been distinguished. Faulting in bedrock is accompanied by fracturing and brecciation which can seriously modify the bedrock susceptibility to erosion. Fault rocks of contrasting resistance to erosion are typical of the Aegean-type fault scarps (Stewart and Hancock 1988), where normal faults in carbonate bedrock are underlain by different types of alternating compact and incohesive breccias. Degradation of such scarps proceeds differently as compared to the Nevada-type model of piedmont scarp.

Colluvial wedges shed from fault scarps can be dated by: ^{14}C, luminescence, dendrochronological, palynological, tephrochronological and weathering rates techniques. Scarps formed in loose sediments can be modelled mathematically by: linear regression, diffusion modelling and statistical analysis of scarp parameters.

Due to repeated episodes of faulting, bedrock fault escarpments, several hundred metres high, and fault-generated range fronts, several hundreds of kilometres long and up to 1 km high, can form. The range front morphology is determined mainly by the ratio of uplift to erosion. Range fronts in a humid climate may appear more degraded than range fronts with the same uplift rate in an arid climate. Active normal fault-generated mountain fronts frequently display triangular or trapezoidal facets (faceted spurs, flat irons) that form due to uplift and dissection of a normal scarp by gullies and whose bases are parallel to the fault trace. Flights of faceted spurs have been interpreted as a result of either episodic uplift, distributed faulting within the range-bounding fault, or even active landsliding.

References

Mayer, L. (1986) Tectonic geomorphology of escarpments and mountain fronts, in R.E. Wallace (ed.) *Active Tectonics*, 125–35, Washington, DC: National Academy Press.
Stewart, I.S. and Hancock, P.L. (1988) Fault zone evolution and fault scarp degradation in the Aegean region, *Basin Research* 1, 139–152.
Wallace, R.E. (1977) Profiles and ages of young fault scarps, north-central Nevada, *Geological Society of America Bulletin* 88, 1,267–1,281.

Further reading

Bloom, A.L. (1978) *Geomorphology*, Englewood Cliffs, NJ: Prentice Hall.
Burbank, D.W. and Anderson, R.S. (2001) *Tectonic Geomorphology*, Malden: Blackwell.
Cotton, C.A. (1958) *Geomorphology*, Christchurch: Whitcombe and Tombs.
Keller, E.A. and Pinter, N. (1996) *Active Tectonics*, Upper Saddle River, NJ: Prentice Hall.
McCalpin, J.P. (ed.) (1996) *Paleoseismology*, San Diego: Academic Press.
Morisawa, M. and Hack, J.T. (eds) (1985) *Tectonic Geomorphology*, Boston: Allen and Unwin.
Stewart, I.S. and Hancock, P.L. (1990) What is a fault scarp?, *Episodes* 13, 256–263.
Stewart, I.S. and Hancock, P.L. (1994) Neotectonics, in P.L. Hancock (ed.) *Continental Deformation*, 370–409, London: Pergamon Press.
Turner, J.P. (2000) Faults and faulting, in P.L. Hancock and B.J. Skinner (eds) *The Oxford Companion to the Earth*, 342–345, Oxford: Oxford University Press.
Twiss, R.J. and Moores, E.M. (1992) *Structural Geology*, New York: W.H. Freeman.

SEE ALSO: seismotectonic geomorphology

WITOLD ZUCHIEWICZ

FECH-FECH

A term applied in the Sahara Desert to fine silt with a powdery consistency and to fine superficial deposits with a low density that often contain evaporates. Progress across fech-fech is made difficult by the absence of cohesion of particles (where feet or wheels penetrate). Fech-fech can be classified from a genetic point of view into two main types:

- Fech-fech developed on Holocene lacustrine muds or fluvio-lacustrine sediments: soft zones within the lacustrine limestones, with 40 per cent of fine particles ($<20\,\mu$) and a higher content of soluble salts, differentiate these sediments in an environment which is almost exclusively sandy.
- Fech-fech developed on clayey shales: the present-day weathering of shales leads to their superficial expansion, which is accentuated by the incorporation of aeolian detrital particles between the disconnected layers.

In addition to these two types, one also finds fech-fech on Quaternary regs with a denser structure ($1.5\,g\,cm^{-3}$) due to the formation of aggregates of silty and salty sand. Tracks can be preserved for a long time on these soft regs where they are underlain by a sandy layer.

Further reading

Conrad, G. (1969) *L'évolution continentale post-hercynienne du Sahara algérien (Saoura, Erg Chech, Tanezrouft, Ahnet-Mouydir)*, série: Géologie No. 10, Paris: Centre National de la Recherche Scientifique (CNRS).

MOHAMED TAHAR BENAZZOUZ

FERRALLITIZATION

A process characterized by the aggressive leaching of a substrate as a consequence of intense tropical weathering, and in which the net effect is a relative accumulation of iron-rich (and commonly also aluminium-rich) compounds; in particular iron and alumina sesquioxides. Ferrallitization is an *in situ* process during which prolonged or intense weathering causes the breakdown of the primary constituents of a pre-existing soil or rock substrate.

Ferrallitization progresses upon rock substrates by the hydrolysis of their primary minerals. This leads to the individualization of the chemical elements of these minerals, the complete leaching of constituent alkali and alkali Earth elements, and the partial or total leaching of silica. Once breakdown commences and elements released, they become available for removal from the system, whilst less mobile constituents, such as Fe, Al, and Ti, remain behind as residual materials and form sesquioxides. Since these less mobile constituents are present in significant proportions within many common substrate lithologies (e.g. 12–18 per cent Fe_2O_3, 12–18 per cent Al_2O_3 in continental basalt, and 0.5–5 per cent Fe_2O_3, 12–15 per cent Al_2O_3 in granitic materials), the residuum readily becomes enriched in Fe and Al. Any silica remaining in the residuum is present either as corroded primary quartz crystals or grains, or else becomes combined into alteration products (e.g. kaolinite and gibbsite). Since, by definition, laterites (see FERRICRETE) form by *in situ* mineral breakdown, ferrallitization represents a key process in the development of lateritic weathering profiles.

MIKE WIDDOWSON

FERRICRETE

A horizon, at the land surface, made up of the cementation of near-surface materials by iron oxides, and often forming a resistant DURICRUST. Typically between 1–20 m in thickness, it can form laterally extensive sheets which may extend over a few, to hundreds, or even thousands of km^2. Consequently, it is perhaps the most widespread of all the duricrust materials. At outcrop it comprises a massive, interlocking fretwork of iron, and often aluminium compounds (i.e. sesquioxides) that bind together other lithological and pedogenic components.

Ferricrete has a long history of study by geologists, geomorphologists, pedologists and agronomists. Considerable effort has been directed toward determining the conditions under which it forms, and this has proved crucial in advancing many aspects of TROPICAL GEOMORPHOLOGY (Thomas 1994; Widdowson 1997). Moreover, chemical and physical durability of ferricrete has meant that it has often played a prominent role in evolution of tropical, and subtropical landscapes (e.g. McFarlane 1971; Bowden 1987; Widdowson and Cox 1996).

In its broadest sense, the term ferricrete can be used to describe any duricrust material in which the dominant bulk components are iron-rich compounds. However, whilst this may seem a straightforward definition, difficulties arise because the term has been employed to describe a wide range of terrigenous weathering and alteration products resulting from differing processes of formation (Ollier and Galloway 1990). Therefore, it becomes important to understand the differences and, where possible, make distinctions between genetically different types of iron-rich duricrust.

Since the nineteenth and early twentieth centuries, the terms ferricrete ('an iron-rich crust'; Lamplugh 1907) and laterite ('a highly weathered material rich in secondary forms of iron and/or aluminium'; (Buchanan 1807; Babbington 1821; Sivarajasingham *et al.* 1962; Plate 45) have been used interchangeably to describe iron-rich duricrusts of various genetic origins. This has led to considerable confusion. However, the problems of co-ordinating laterite and ferricrete description stem not only from investigation by a variety of different scientific disciplines, but also from the development of extensive anglophone and francophone descriptive terminologies. Nevertheless, it is evident from field studies that the majority of iron-rich duricrusts can be adequately described in terms of two genetically distinct types. Aleva (1994) distinguishes between those duricrusts in which an absolute iron enrichment occurs (i.e. those which receive a net input of iron), and those which attain their elevated iron contents through residual enrichment within the profile (i.e. no net input of iron).

Ferricretes are those duricrusts which incorporate materials non-indigenous to the immediate locality in which the duricrust formed. In many instances the transported materials can be readily

Plate 45 Laterite quarry near Bidar, south-east Deccan, India (with 1 m scale in lower right). Material beneath the indurated duricrust of a laterite profile is excavated, cut into large bricks, and allowed to harden in the sun. This is similar to the material first named 'laterite' by Buchanan (1807)

identified as pebbles or clasts derived from adjacent lithological terranes, or as fragments from indurated layers of earlier generations of laterite or ferricrete (Plate 46). Importantly, the term ferricrete should also be extended to those materials whose constituents have been substantially augmented by the precipitation or capture of elements and compounds from allochthonous fluids (i.e. those derived during the breakdown and mobilization of materials outside the immediate locality of ferricrete formation). Although it is the allochthony of the constituent materials of the ferricrete which justify its appellation, determining whether the introduction of such fluids has taken place, and confirming their allochthony, is often problematic. However, since ferricretes may develop as ferruginous foot slope accumulations or within topographic depressions, they can often be distinguished by the fact that they display an obvious discordance with the underlying substrate lithologies. In effect, they do not display the progressive weathering profile characteristic of many laterite profiles, and instead the ferricrete horizon sits upon relatively unaltered bedrock.

Laterites are iron-rich duricrusts which have formed directly from the breakdown of materials in their immediate vicinity, and so do not contain any readily identifiable allochthonous component. Lateritic duricrusts are typically manifest as the uppermost layers of *in situ* weathering profiles. Where these profiles are fully exposed, such as the widespread examples developed on basalt in western India (Widdowson and Gunnell 1999), they consist of an uninterrupted progression from unaltered bedrock, through the WEATHERING FRONT into SAPROLITE (in which structure and crystal pseudomorphs of the parent rock may still be recognized), and then upward through increasingly altered and iron-enriched zones that culminate as a highly indurated 'tubular' laterite at the top of the profile (Plate 47).

To summarize, ferricrete and laterite are not synonymous terms and should, wherever possible, be used to distinguish between fundamentally different types of iron-rich duricrust. This distinction is particularly important since it places constraints upon the type of processes operating during evolution of a duricrust, and the palaeoclimatic and morphological conditions existing at the time of its development. However, although emphasis is put upon establishing whether the iron component is allochthonous or autochthonous, distinguishing these two types of duricrust, both in the field and in hand specimen, can prove problematic. Problems arise because, once formed, ferricretes can begin to alter and evolve in response to prevailing climatic and groundwater conditions (Bowden 1997) and, over time, begin to exhibit some of the structural and textural features typical of lateritic weathering profiles. In effect these

Plate 46 Granular ferricrete surface comprising allochthonous materials derived from earlier generations of laterite and ferricrete, near Bunbury, Western Australia

Plate 47 Indurated 'tubular' laterite sample from the top of an *in situ* weathering profile near Bunbury, Western Australia

'evolved' ferricretes become modified by a post-depositional weathering and ferrallitization overprint. Conversely, the role of allochthonous groundwater fluids, and associated lateral or downslope transport of elements and compounds, cannot always be excluded in the development of otherwise autochthonous laterite weathering profiles.

More recently, ferricrete and laterite duricrusts, together with Fe-rich palaeosols, have begun to acquire renewed importance as palaeoenvironmental indicators (e.g. Bardossy 1981; Thomas 1994; Tsekhovskii *et al.* 1995). The investigation of such materials within the geological record, together with appropriate mineralogical, geochemical and isotopic studies, can now reveal detailed information regarding past climatic and atmospheric conditions. For instance, geochemical and isotopic analyses of Proterozoic laterites from South Africa (Gutzmer and Beukes 1998), suggest not only an ancient oxidizing atmosphere, but also a hot and humid climate at *c.*2 Ga. Moreover, carbon isotope signatures preserved within these laterites may indicate the presence of an early terrestrial vegetation.

References

Aleva, G.J.J. (compiler) (1994) *Laterites. Concepts, Geology, Morphology and Chemistry*, Wageningen: ISRIC.

Babbington, B. (1821) Remarks on the geology of the country between Tellicherry and Madras, *Transactions of the Geological Society of London* 5, 328–329.

Bardossy, G. (1981) Palaeoenvironments of laterites and lateritic bauxites – effect of global tectonism on bauxite formation, *Proceedings of the International Seminar on Lateritisation Processes* (Trivandrum, India, 11–14 December 1979), 287–294, Rotterdam: Balkema.

Bowden, D.J. (1987) On the composition and fabric of the footslope laterites (duricrust) of Sierra Leone, West Africa, and their geomorpholoical significance, *Zeitschrift für Geomorphologie NF, Supplementband* 64, 39–53.

—— (1997) The geochemistry and development of lateritized footslope benches: the Kasewe Hills, Sierra Leone, in M. Widdowson (ed.) *Palaeosurfaces: Recognition, Reconstruction, and Paleoenvironmental Interpretation*, Geological Society of London Special Publication 120, 295–306.

Buchanan, F. (1807) A journey from Madras through the countries of Mysore, Kanara, and Malabar Vol. 2, 436–461; Vol. 3, 66, 89, 251, 258, 378, London: East India Co.

Gutzmer, J. and Beukes, N.J. (1998) Earliest laterites and possible evidence for terrestrial vegetation in the Early Proterozoic, *Geology* 26, 263–266.

Lamplugh, G.W. (1907) Geology of the Zambezi basin around Batoka Gorge, *Quarterly Journal of the Geological Society of London* 63, 162–216.

McFarlane, M.J. (1971) Lateritization and landscape development in Kyagwe, Uganda, *Quarterly Journal of the Geological Society of London* 126, 501–539.

Ollier, C.D. and Galloway, R.W. (1990) The laterite profile, ferricrete and unconformity, *Catena* 17, 97–109.

Sivarajasingham, S., Alexander, L.T., Cady, J.G. and Cline, M.G. (1962) Laterite, *Advances in Agronomy* 14, 1–60.

Thomas, M.F. (1994) *Geomorphology in the Tropics*, Chichester: Wiley.

Tsekhovskii, Yu G., Shchipakina, I.G. and Khramtsov, I.N. (1995) Lateritic eluvium and its redeposition products as indicators of Aptian–Turonian climate. *Stratigraphy and Geological Correlation* 3(3), 285–294.

Widdowson, M. (ed.) (1997) *Palaeosurfaces: Recognition, Reconstruction, and Paleoenvironmental Interpretation*, Geological Society of London Special Publication 120.

Widdowson, M. and Cox, K.G. (1996) Uplift and erosional history of the Deccan traps, India: evidence from laterites and drainage patterns of the Western Ghats and Konkan Coast, *Earth and Planetary Science Letters* 137, 57–69.

Widdowson, M. and Gunnell, Y. (1999) Lateritization, geomorphology and geodynamics of a passive continental margin: the Konkan and Kanara lowlands of western peninsular India. Special Publication of the the International Association of Sedimentologists 27, 245–274.

SEE ALSO: duricrust

MIKE WIDDOWSON

FIRE

Wildfire is one of the most potent agents of geomorphic change, modifying processes and greatly increasing erosion and deposition rates in virtually all BIOGEOMORPHOLOGY environments. This entry covers the geomorphological role of fire in WEATHERING, soils, HILLSLOPE PROCESSes and river systems.

Although most texts and reviews list fire as an important weathering agent (Blackwelder 1927), most field studies are anecdotal or supported by little data with fewer experimental controls. Thus, the most important insights on fire weathering result from laboratory studies (Goudie *et al.* 1992) where experimentalists have learned that fire weathering depends heavily on rock physical properties, varies with different rock types, is faster in smaller rocks, and fire weathering rates increase with increasing water content.

An example of the impact of fire on rock weathering comes from the April–May 2000 'Coon Creek' wildfire that burned around 37.5 km^2 of the Sierra Ancha Mountains, 32.3 km north of Globe, Arizona – including 25 sandstone and 19 diorite boulders surveyed in 1989 and resurveyed (a) after the burn, (b) after the summer 2000 precipitation season, and again (c) after the winter 2001 snow season (Dorn 2003, Plate 48). When stretched over cumulative boulder areas, erosion immediately after this single fire averaged > 26 mm for sandstone and > 42 mm for diorite. But averages are misleading, because sandstone and diorite boulders expressed bimodal patterns of erosion, where fire-induced weathering generated either (a) no

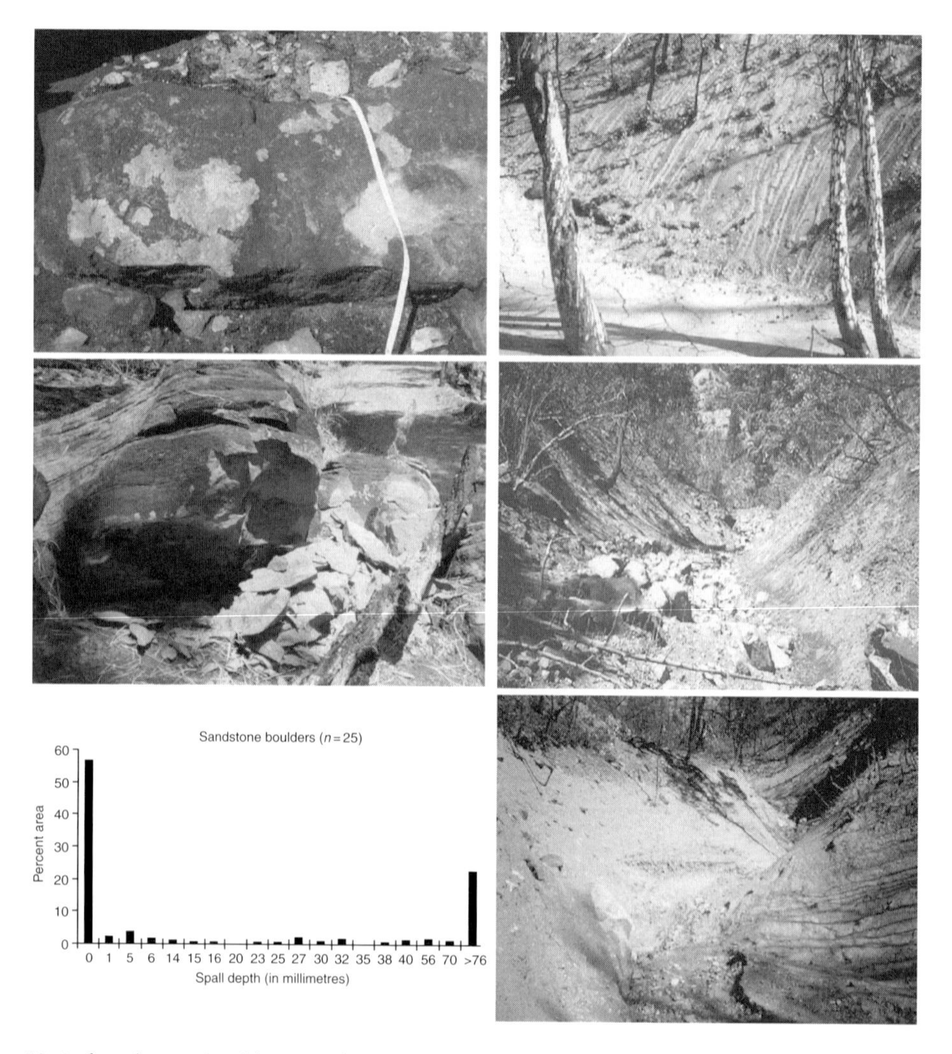

Plate 48 Left column: boulder weathering from the Coon Creek Spring 2000 fire, Sierra Ancha Mountains, Arizona. The top left image shows flaking of millimetre-scale spalls. The centre-left image shows where half of a boulder fragmented as a result of the fire. The graph on the lower left shows the overall bimodal pattern, whereby fire weathering produces erosion of small flakes or extensive slabs. Right column: fire-generated erosion from the 1995 Storm King fire, Colorado, courtesy of the US Geological Survey (Canon *et al.* 2001; see also http://landslides.usgs.gov/html_files/ofr95-508/index.html). The upper right image shows post-fire rill erosion. Other photos show in-channel conditions before (middle right) and after (lower right) passage of a debris flow

erosion to thin, millimetre-scale spalling or (b) massive spalls thicker than 7.6 cm. This field study confirms an earlier experimental finding that fire increases a rock's susceptibility to post-fire weathering and erosion processes (Goudie *et al.* 1992), since summer-time convective storms and subsequent winter snows continued to promote boulder erosion on the order of 1–5 millimetres. In addition to erosion of boulder surfaces, 85-metre- diameter boulders appear to have been fragmented into cm-scale clasts – suggesting that fire can modify hillslope evolution in locations where boulders are important controls on the evolution of slopes.

Wildfire generates extensive changes to soil systems (Morris and Moses 1987), perhaps the most important being the development of HYDROPHOBIC SOILS. Wildfires produce volatile hydrocarbons that penetrate soil up to 15 centimetres and make a water-repellent layer. In addition, fire ash decreases the ability of soils to adsorb water. Field checks involve digging a progressively deeper trench and applying water. Water that does not infiltrate immediately (within 10 seconds) indicates the soil is hydrophobic. Extreme hydrophobicity results in water ponding for more than 30 seconds. On unburned slopes, normal biogeomorphology processes decrease soil erosion, for example, by intercepting raindrop impacts, increasing infiltration and providing structural support. Hydrophobicity from burning decreases infiltration capacity, and increases OVERLAND FLOW and SOIL EROSION.

Even before it rains, burning enhances erosion by dust DEFLATION and dry ravel. Dry ravel is a type of granular MASS MOVEMENT where frictional and collisional particle interactions dominate flow behaviour, all not requiring rainfall. Dry ravel provides sediment to channels from particularly steep slopes, and this process is well documented after southern California fires.

Burning greatly increases surface runoff from precipitation, which increases the volume and velocity of the surface runoff. Higher discharge of surface water flows then result in the formation of RILLS and gullies (see GULLY) on hillsides. Fire-enhanced gullies and rills transport surface runoff and sediment to stream channels. Peak flows in the channel tend to occur with less of a lag time than those observed in unburned watersheds. Flood peaks tend to be much higher and more capable of eroding sediment stored in channels, leading to channel incision.

The sediment load of the fluvial system also changes after a fire. Sediment from a number of different sources may be incorporated into flows progressing down a hillside or channel. Sediment-water flows on burned slopes change the concentration, size distribution and/or composition of the entrained sediment to the point where a change in measurable yield strength takes place; this change is called HYPERCONCENTRATED FLOW. In hyperconcentrated flows, particles are deposited as individual grains from suspension, and the remaining fluid continues to move.

Fires also greatly increase DEBRIS FLOWS (Cannon *et al.* 2001; Swanson 1981). In contrast to streamflow or hyperconcentrated flow, debris flows host a sediment-water mixture that moves as a single phase. Deposition does not separate out particles, so debris-flow deposits have sharp, well-defined flow boundaries. The most recognizable deposits are levees lining flow paths and lobes of material at a flow terminus. Many terms have been used for the processes and deposits of debris flows, including slurry flow, mudflow and debris torrent.

Fire-enhanced debris flows start by landsliding or sediment bulking of surface water flows. Landsliding after burning tends to be more common in colluvial-filled hollows on slopes, where unconsolidated thick deposits of colluvium fail after rainfall. This landslide then mobilizes into a debris flow, where the debris-flow path can then be traced up to a landslide-scar source.

Sediment bulking tends to occur in the surface layer of hydrophobic soils. Hydrophobic soils create a condition where excess water that cannot penetrate deeply saturates the upper few centimetres of soil. This surface material then fails as small-scale debris flows. In addition, water runoff can incorporate so much loose material that sediment concentrations get high enough for the flow to behave as a debris flow. Sediment bulking is probably the most important debris-flow producing process after a fire.

GEOMORPHOLOGICAL HAZARDS are not limited to the first few rainstorms after a fire. Research by Ramon Arrowsmith in the Phoenix, Arizona, region indicates enhanced flash flooding potential decades after a brush fire. Even in forested regions, the supply of loosened material continues to deliver dry ravel sediment, hyperconcentrated flows and debris flows to stream systems for years after a fire.

The link between wildfire and increased erosion leading to large sedimentation events was made as early as 1949 by P.B. Rowe and colleagues working in southern California. They developed the concept of a 'fire–flood sequence' that has been studied extensively in a wide variety of river settings including alpine forests such as Yellowstone (Minshall *et al.* 1998), Mediterranean scrub (Shakesby *et al.* 1993) and even desert ranges (Germanoski and Miller 1995). In Yellowstone, for example, Minshall *et al.* (1998) found extensive RILL development, GULLY formation and MASS MOVEMENTS in burned watersheds during the summer of 1989, when post-fire heavy rains and snowmelt generated widespread 'black water' conditions and increased BEDLOAD and SUSPENDED LOAD. After monitoring Yellowstone streams for a decade after its massive wildfire, Minshall *et al.* (1998) stress that post-fire stream studies can yield misleading insights after only a few years since massive stream reorganization can take place seven to nine years after the fire event.

The study of fire remains associated with soils and sediment, called pedoanthrocology, provides important insight into prehistoric geomorphic changes associated with fires. Studies of fire-induced ALLUVIAL FANS, of fire remains within uneroded soils, and diagenesis of organic remains into such forms as vitrinite and inertinite provide geomorphologists with insights into palaeoecological conditions that may have influenced the geomorphic landscape seen today (Siffedine *et al.* 1994).

References

Blackwelder, E. (1927) Fire as an agent in rock weathering, *Journal of Geology* 35, 134–140.

Cannon, S.H., Kirkham, R.M. and Parise, N. (2001) Wildfire-related debris-flow initiation processes, Storm King Mountain, Colorado, *Geomorphology* 39, 171–188.

Dorn, R.I. (2003) Boulder weathering and erosion associated with a wildfire, Sierra Ancha Mountains, Arizona, *Geomorphology*, 55, 155–171.

Germanoski, D. and Miller, J.R. (1995) Geomorphic response to wildfire in an arid watershed, Crow Canyon, Nevada, *Physical Geography* 16, 243–256.

Goudie, A.S., Allison, R.J. and McClaren, S.J. (1992) The relations between modulus of elasticity and temperature in the context of the experimental simulation of rock weathering by fire, *Earth Surface Processes and Landforms* 17, 605–615.

Minshall, G.W., Brock J.T. and Royer T.V. (1998) Stream ecosystem responses to the 1988 wildfires, *Yellowstone Science* 6(3), 15–22.

Morris, S.E. and Moses, T. (1987) Forest-fire and the natural soil-erosion regime in the Colorado Front Range, *Annals of the Association of American Geographers* 77, 245–254.

Shakesby, R., Coelho, C., Ferreira, A., Terry, J. and Walsh, R. (1993) Wildfire impacts on soil-erosion and hydrology in wet Mediterranean forest, Portugal, *International Journal of Wildland Fire* 3, 95–110.

Siffedine, A. *et al.* (1994) The lacustrine organic sedimentation in tropical humid environment (Carajas, eastern Amazonia, Brazil) – relationship with climatic changes during the last 60,000 years BP, *Bulletin de la Société Géologique de France* 165, 613–621.

Swanson, F.J. (1981) Fire and geomorphic processes, in M.A. Mooney *et al.* (eds) *Fire Regimes and Ecosystem Properties*, 401–420, Washington, DC: US Department of Agriculture General Technical Report WO-26.

RONALD I. DORN

FIRST-ORDER STREAM

STREAM ORDERING is based on the premise that stream size is related to the area contributing to runoff. This provides a method of ranking the relative size of streams within a catchment. The term first-order stream originates from ideas initially proposed by R.E. Horton in the 1930s (Horton 1932, 1945). Horton devised a method of classifying links in a stream network using a system of ordering. Under such a scheme the smallest unbranched streams in a catchment are designated first order. The combination of two first-order streams results in a second-order stream and so forth through successively larger links as additional streams join the network (Figure 60a). This original idea soon led to a proliferation of ordering schemes each providing a development

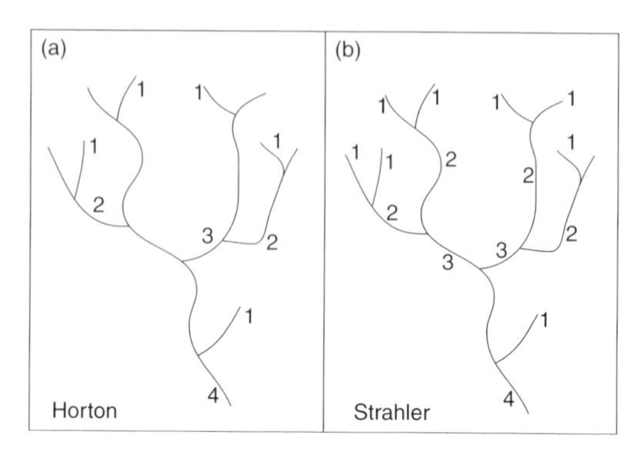

Figure 60 Comparison of stream and segment ordering methods: (a) Horton, (b) Strahler

or refinement of previous ones. Of particular note is the Strahler scheme (Strahler 1952) that begins like the Horton scheme, with the smallest channels being classified as first-order links; but higher order links are only generated when two links of equivalent order are joined (Figure 60b). The highest order generated by this mechanism is often used to classify drainage basins, e.g. a third-order drainage basin. Stream orders vary from the smallest first-order streams to the world's largest rivers that approach twelfth order (Mississippi, Amazon).

The hydrologic response of a stream channel is in part a function of its stream order. Stream order can be used to quantify other aspects of a watershed. These include the Bifurcation Ratio, R_b. The bifurcation ratio (R_b) is defined as the ratio of the number of streams of any order (N_i) to the number of streams of the next highest order. Horton (1945) found that this ratio is relatively constant from one order to another. Values of R_b typically range from the theoretical minimum of 2 to around 6. Typically, the values range from 3 to 5. The bifurcation ratio is calculated as

$$R_b = N_i/N_{i+1}$$

These are important geomorphic parameters that describe the structure and functioning of drainage basins. In the past, calculating these measures was extremely time consuming as catchment boundaries need to be carefully defined. However, these analyses are now routinely undertaken using GIS, which has the potential to provide rapid, accurate and automatic recognition of stream network links (Morris and Heerdegen 1988). This is often based on the topographic definition of streams based on contour crenulation and headwater divide delimitation. In this respect, an advantage of the Strahler scheme is that it retains the same common nomenclature for all similar sized channel links. Thus first-order streams are consistently identified as the smallest channels in a catchment. This is useful because streams with similar attributes, and a similar relative position in the network, are grouped in the same order. Hence, first-order streams tend to have common characteristics. These common characteristics are, however, dependent on the scale at which the channel links are defined, e.g. whether they are mapped from published maps or surveyed in the field. This raises important issues about consistency in definition of network properties (Blyth and Rodda 1973; Mark 1983) and highlights the property that most stream networks are dynamic, so the extent of the network varies in time from storm to storm and across seasons, years and decades.

Topography is not the only criterion used to distinguish first-order channels. First-order streams may also be defined on the basis of flow duration sufficient to sustain aquatic biota year round. In this respect, a first order channel must be by definition permanent, connected to the main stream network and convey runoff from a defined CONTRIBUTING AREA.

The greatest frequency of first-order streams tend to be found in the headwaters of catchments where channels tend to be small, confined, have steeper slopes and individually contribute only small amounts of stream discharge (Wohl 2000). In terms of the overall network, first-order channels defined by a Strahler ordering scheme commonly represent 50–60 per cent of the total stream length in a third-order drainage basin (Strahler 1964). During storms or prolonged wet periods the size of the network will expand and first-order channels may extend up hillslopes as ephemeral water flows are maintained for short periods. The extension of the permanent first-order network beyond the channel head represents a dynamic link. The coupling between the channel head and the network of hillslope hollows upslope usually defines a diffuse topographic network of zero-order basins (Dietrich *et al.* 1987). These zero-order basins are small unchannelled valleys. These form HILLSLOPE HOLLOW networks on slopes which focus runoff and sediment transport via saturated overland flow and gully and debris flows. In general terms, as stream order increases sediment yield per unit area tends to decline as HILLSLOPE-CHANNEL COUPLING becomes less effective.

References

Blyth, K. and Rodda, J.C. (1973) A stream length study, *Water Resources Research* 9, 1,451–1,461.

Dietrich, W.E., Reneau, S.L. and Wilson, C.J. (1987) Overview: zero-order basins and problems of drainage density, sediment transport and hillslope morphology, in *Erosion and Sedimentation in the Pacific Rim*. IAHS Publication 165, 27–37.

Horton, R.E. (1932) Drainage basin characteristics, *Transactions of the American Geophysical Union* 13, 350–361.

——(1945) Erosional development of streams and their drainage basins; hydrophysical approach to quantitative morphology, *Geological Society of America Bulletin* 56, 275–370.

Mark, D.M. (1983) Relations between field-surveyed channel networks and map-based geomorphometric measures, Inez, Kentucky, *Annals of the Association of American Geographers* 73, 358–372.
Morris, D.G. and Heerdegen, R.G. (1988) Automatically derived catchment boundaries and channel networks and their hydrological applications, *Geomorphology* 1, 131–141.
Strahler, A.N. (1952) Hypsometric (area-altitude) analysis of erosional topography, *Geological Society of America Bulletin* 63, 1,117–1,142.
——(1964) Quantitative geomorphology of drainage basins and channel networks, section 4-II, in V.T. Chow (ed.) *Handbook of Applied Hydrology*, 4–39, New York: McGraw-Hill.
Wohl, E. (2000) *Mountain Rivers*, Water Resources Monograph 14, Washington, DC: American Geophysical Union.

SEE ALSO: drainage basin; GIS; runoff generation

JEFF WARBURTON

FISSION TRACK ANALYSIS

Fission track analysis (FTA) is a thermochronometer that provides detailed information on the thermal history of rocks, most usually for temperatures below 350 °C (using zircon) and below 110 °C (using apatite). When a rock has cooled rapidly from its temperature of formation (e.g. a rapidly cooled lava), the technique may provide the age of formation of that rock (hence 'fission track dating') but the technique can be applied in any situation in which low-temperature thermal history is required and the appropriate minerals are present. In geomorphological applications, the technique exploits the increase in temperature with depth in the Earth's crust (the geothermal gradient). This temperature increase means that the low-temperature thermal history of an apatite or a zircon now at the Earth's surface (or in a drill hole) is a record of that mineral's passage through the crust to the sampling point (surface or drill hole). The principal application of FTA in geomorphology is therefore to elucidate the long-term DENUDATION that brings the target mineral(s) to the Earth's surface. For a surface temperature of 20 °C and a geothermal gradient of 25 °C km^{-1}, FTA in apatite provides a denudational history for the upper *c*.4 km of the crust (i.e. below about 110 °C). (URANIUM-THORIUM)/HELIUM ANALYSIS ((U-Th)/He analysis) in apatite provides a shallower denudational history from a lower temperature of *c*.75 °C. In geomorphological studies, in which the final stages of crustal denudation leading to the present topography are of interest, the thermochronometers most often used are the lowest temperature (i.e. shallowest), namely, apatite FTA and (U-Th)/He analysis. If all three low-temperature thermochronometers (i.e. the two in apatite plus zircon FTA) all yield essentially the same ages, then it is clear that denudation (and the associated cooling of the crust through the three thermochronometers' temperature ranges) have occurred very rapidly.

FTA relies on counting the number, and measuring the lengths, of minute damage paths (defects or 'tracks') produced when the heavy daughter products of ^{238}U fission in a mineral's crystal lattice travel away from each other at high speed through the lattice. The tracks are only *c*.5 nm in diameter and are widened slightly by etching in a weak acid during sample preparation, so as to make them visible under microscope. Etched tracks are about 1–2 μm in diameter and up to about 16 μm long. The tracks are produced continuously at a known rate, dependent on the U-content.

In order to reconstruct, from the sample's cooling history, the denudation necessary to bring the sample to the Earth's surface, a knowledge of the geothermal gradient at the time the sample was exhumed is necessary. This geothermal gradient provides the crustal depths associated with the temperatures from which the sample was exhumed. The 'ancient' geothermal gradient is usually unknown and an 'appropriate' geothermal gradient is often assumed based on likely modern analogues of the tectonic and thermal setting of the sample locality at the time of exhumation. If a vertical profile of FT samples is available (for example, from a drill hole or from a mountain side), the gradient of the elevation–age profile provides the geothermal gradient.

In simple terms, the number of tracks is a function of the time since the sample cooled sufficiently for the tracks to be retained (i.e. cooled below about 110 °C in apatite), and the U-content of the mineral in the areas of the grain in which the tracks have been counted and their lengths measured. The fission track age is derived by the application of the standard radiometric dating formula but with the amount of decay product ('daughter') of the dating technique's radioactive decay system replaced by the number of tracks.

Lower and/or more variable rates of denudation through time result in more complex cooling histories, which can be elucidated using frequency distributions of track lengths (the 'track length distribution'). Tracks form continuously as a result of ^{238}U fission but in apatite, for example, they are annealed (repaired) geologically instantaneously above a temperature of about 110 °C. Below 110 °C, apatite fission tracks are only partially annealed and are increasingly retained at temperatures down to surface temperature. This temperature range in which tracks are partially annealed (repaired) is the partial annealing zone (PAZ), and there is a range of views as to the effective lower limit of the PAZ. Strictly, fission tracks may be annealed even at room temperature but some authors set the effective lower boundary of the PAZ at *c*.60 °C. Track annealing is by repair at the ends, resulting in shorter tracks. The duration of the sample's residence in the PAZ is therefore reflected in the frequency distribution of track lengths, shorter track lengths reflecting longer residence time in the PAZ.

Statistical temperature–time paths can be calculated to match the measured fission track age and track length distribution, giving a complete description of thermal history of the apatite below 110 °C, and hence of the sample's trajectory to the surface as a result of denudation. Figure 61 shows the ways in which different fission track ages and track length distributions reflect different cooling histories. In A, the sample cooled very rapidly at 100 Ma ago, and the fission track age (99.8 Ma; the upper number of the three within the plot) is essentially the same as the age of the cooling event. The rapid cooling is reflected in the long mean track length (15 μm; the middle number in the plot) and the very low standard deviation of the track length distribution (1.07 μm; the third number). The track length data have a high, unimodal, narrow distribution in the histogram of the track length distribution.

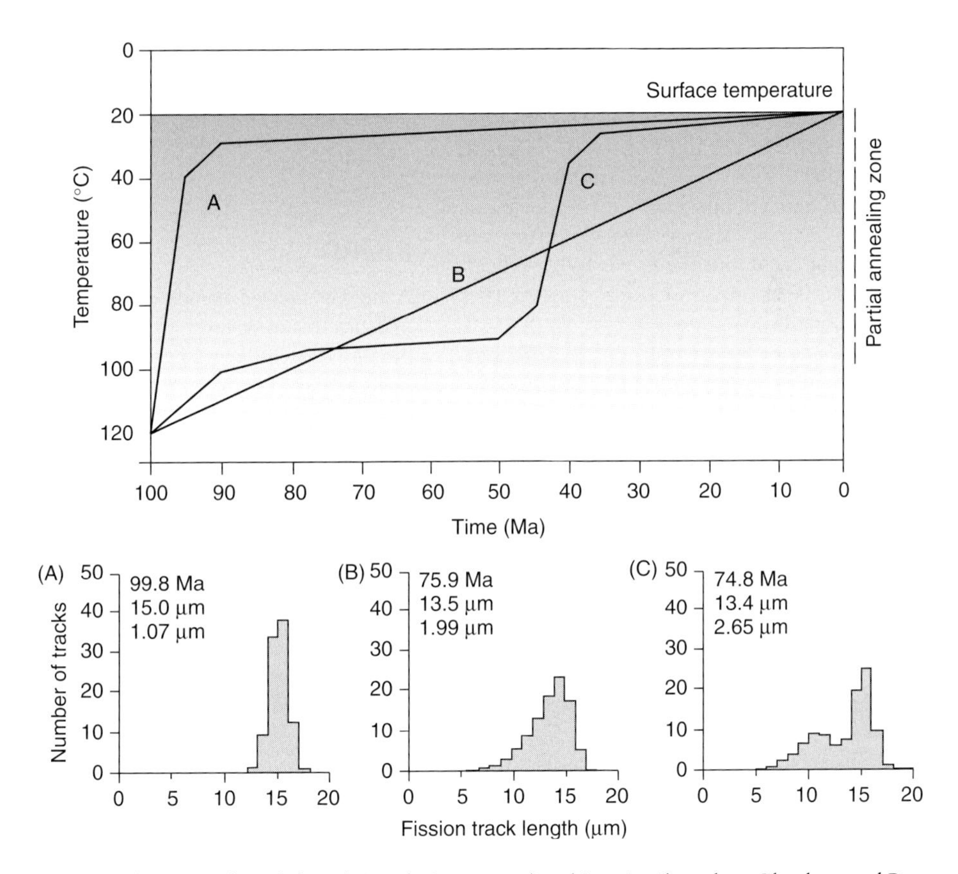

Figure 61 Fission track ages and track length in relation to cooling histories (based on Gleadow and Brown 2000: figure 4.3)

Sample A's very rapid cooling could be the result of very high rates of subaerial denudation (probably combined with ongoing rapid uplift to drive the denudational exhumation) or it could be associated with tectonic denudation in which very high rates of uplift lead to detachment of slabs of crust which slide away by gravity along decollements, thereby cooling the underlying crust. Sample B has experienced steady cooling from 100 Ma ago to the present. The average track length is shorter (13.5 μm) and the distribution broader ($s =$ 1.99 μm), both measures being reflected in the broader histogram with a longer 'tail' into the short track lengths (reflecting the greater time that tracks have spent in the PAZ after formation). Note how B's fission track age (75.9 Ma) bears no obvious relation to the cooling event or to any particular depth in the crust for determining a rate of denudation. The determination of the rate of denudation requires modelling of the cooling history of the sample (in effect determining the cooling history, as in the upper diagram, from the age and track length distribution in the lower diagram). In the more complex cooling history of C (two discrete cooling events: one between 100 and 90 Ma and the second at about 45 Ma), the fission track age (74.8 Ma) relates to neither cooling event. The track length distribution is broad and bimodal, with the upper mode (long track lengths of c.15 μm) reflecting the 45 Ma cooling event and the lower mode reflecting annealing (shortening) of tracks formed after the first cooling event.

There are several inferential and logical steps involved in converting FTA data to a geomorphological history and an amount of denudation (e.g. Gleadow and Brown, 2000). Notwithstanding the uncertainties and assumptions associated with these steps, FTA has been successfully applied to elucidate long-term landscape development in a range of settings. Application in active orogenic settings of FTA in conjunction with higher temperature thermochronometers, such as the $^{40}Ar/^{39}Ar$ system, and lower temperature systems, such as *(uranium-thorium)/helium analysis* in apatite, has very convincingly demonstrated that denudation of these settings is very rapid. When various thermochronometric systems yield the same rates of denudation, it is argued that there is a dynamic equilibrium between denudation and the ongoing tectonic uplift necessary to drive the flux of crust through the Earth's surface where it is removed by denudation at the same rate as uplift. The processes and sequences of events associated with lithospheric extension and subsequent PASSIVE MARGIN development have also been widely elucidated using FTA. FTA data along many passive continental margins, especially the data from closest to the new margin, exhibit rapid cooling events (long, unimodal track length distributions) at about the time of break-up. These FTA data are interpreted in terms of rapid denudation of the nascent or new continental margin at about the time of breakup, in response to one or more of the following: thermally driven active or passive uplift and denudation of the pre-breakup rift shoulders; rapid denudation of the new margin in response to the new BASE LEVEL for denudation that is provided by the formation of a new ocean basin adjacent to the margin; and ongoing flexural isostatic uplift of the new margin in response to this accelerated denudation.

Reference

Gleadow, A.J.W. and Brown, R.W. (2000) Fission-track thermochronology and the long-term denudational response to tectonics, in M.A. Summerfield (ed.) *Geomorphology and Global Tectonics*, 57–75, Chichester: Wiley.

PAUL BISHOP

FJORD

A deeply incised trench or trough excavated in bedrock by long-term glacial erosion and occupied by the sea during periods of glacier recession. Spectacular fjordic scenery occurs along the coasts of British Columbia in Canada, Alaska, southern Chile, Greenland, northern and eastern Iceland, Spitsbergen, Fiordland in New Zealand, the Canadian arctic islands and western Scotland. The longest fjords are Nordvestfjord/Scoresby Sund in Greenland (300 km), Sognefjord in Norway (220 km) and Greely Fjord/Nansen Sound in the Canadian arctic (400 km).

Troughs and fjords have distinctive cross and long profiles, referred to as U-shaped but best approximated by the formula for a parabola:

$$V_d = aw^b$$

where w is the valley half width, V_d is valley depth and a and b are constants. However, true cross profiles deviate from this mathematical parabola largely due to the production of breaks in slope by pulsed erosion through time. These effects have been modelled by Harbor *et al.* (1988) by imparting a valley glacier on a fluvial, V-shaped valley. Basal velocities below a glacier are highest part way up the valley sides and lowest below the glacier margins and centre line. By assuming that the erosion rates are proportional to the sliding velocity, the greatest erosion occurs on the valley sides, thereby causing broadening and steepening of the valley. The development of the steep sides of troughs and fjords is aided by PRESSURE RELEASE or dilatation in the bedrock. This is the development of fractures parallel to the ground surface. Such fractures weaken rock masses, thereby facilitating subsequent subglacial erosion. Dilatation is most likely to take place immediately after deglaciation when the glacier overburden has been removed and freshly eroded rock surfaces are exposed.

Overdeepenings along fjord and trough long profiles separated by sills or thresholds appear to represent areas of increased glacier discharge such as at the junctions of tributary valleys or where fjord narrowing occurs. The area of deepest erosion in a fjord marks the location of the long-term average position of maximum glacier discharge. Fjord mouths are often characterized by STRANDFLATs, likely due to the fact that the erosion capacity of the outlet glaciers is severely reduced due to glacier buoyancy and eventual ice flotation in the sea in addition to the flow divergence induced by the more open topography.

The planform of many fjords clearly reveals fluvial or structural origins. For example, the sinuous forms and dendritic patterns of some fjords suggest that they are glacially overdeepened preglacial fluvial valleys and rectilinear fjord networks have been linked to large-scale structural features such as faults and grabens. Moreover, the close association between linear fjord alignments and intersecting lines of regional fracture have led to purely tectonic theories for fjord initiation. The survivial of preglacial landforms and sediments on upland areas between fjords demonstrates that the deep glacial incision is selective, hence the use of the term *selective linear erosion* to describe the development of fjord and trough systems. It is most likely that pre-existing valley systems, whether fluvial and/or tectonic in origin, will contain thicker ice during glaciations and therefore act as major conduits for glacier flow from the centres of ice dispersal, especially if they are oriented parallel to regional glacier flow. Greater ice thicknesses and concomitant preferential ice flow down such valleys will result in greater frictional heat, increased pressure melting and widespread basal sliding. Conversely, on the plateaux between fjords, cold-based ice will dominate and protect underlying preglacial features from glacial erosion. The occurrence of a preglacial land surface on the plateaux surrounding Sognefjord has allowed the calculation of a fjord erosion rate (Nesje *et al.* 1992). Approximately 7,610 km^3 of material has been removed from the fjord by glacial erosion, yielding erosion rates ranging from 102 to 330 $cm\,kyr^{-1}$ depending upon the amount of time that glaciations have dominated the region.

The dimensions of fjords appear to be scaled to the amount of ice that was discharged through them, several researchers having demonstrated that relationships exist between fjord size and glacier contributing area. The strength of these relationships also varies between regions. For example, Augustinus (1992) demonstrated that fjords in British Columbia are 2.5 times deeper and 2.4 times longer than New Zealand fjords even though the contributing areas are comparable in size. This suggests that glacial erosion is more intense in British Columbia probably due to the fact that water depths are shallower offshore than in New Zealand and are therefore less capable of floating the fjord glaciers. In addition, the lengths of the former British Columbia fjord glaciers were much greater than those of the New Zealand palaeo-glaciers, the former having been nourished by an inland ice sheet.

References

Augustinus, P.C. (1992) Outlet glacier trough size–drainage area relationships, Fjordland, New Zealand, *Geomorphology* 4, 347–361.

Harbor, J.M., Hallet, B. and Raymond, C.F. (1988) A numerical model of landform development by glacial erosion, *Nature* 333, 347–349.

Nesje, A., Dahl, S.O., Valen, V. and Ovstedal, J. (1992) Quaternary erosion in the Sognefjord drainage basin, western Norway, *Geomorphology* 5, 511–520.

Further reading

Benn, D.I. and Evans, D.J.A. (1998) *Glaciers and Glaciation*, 350–356, 362–366, London: Arnold.

England, J. (1987) Glaciation and the evolution of the Canadian high arctic landscape, *Geology* 15, 419–424.

Harbor, J.M. (1992) Numerical modelling of the development of U-shaped valleys by glacial erosion, *Geological Society of America Bulletin* 104, 1,364–1,375.

Holtedahl, H. (1967) Notes on the formation of fjords and fjord valleys, *Geografiska Annaler* 49A, 188–203.

Løken, O.H. and Hodgson, D.A. (1971) On the submarine geomorphology along the east coast of Baffin Island, *Canadian Journal of Earth Sciences* 8, 185–195.
Nesje, A. and Whillans, I.M. (1994) Erosion of Sognefjord, Norway, *Geomorphology* 9, 33–45.
Roberts, M.C. and Rood, R.M. (1984) The role of ice contributing area in the morphology of transverse fjords, British Columbia, *Geografiska Annaler* 66A, 381–393.
Shoemaker, E.M. (1986) The formation of fjord thresholds, *Journal of Glaciology* 32, 65–71.

DAVID J.A. EVANS

FLASH FLOOD

Flash flood denotes an abrupt rise in the discharge of a river or stream, providing an event of short duration. The term has conventionally been associated with ephemeral flow regimes in which the majority of events are rain-fed. The flood is discrete. It impinges on a channel bed that is initially dry and is exhausted within a short interval – a few hours in the case of small drainage basins or a few days where basin size involves longer travel time. Because of this, flash floods are commonly associated with deserts and semi-deserts of low to middle latitudes, flow in high-latitude deserts resulting rather from the slow release of water in the form of snowmelt – a seasonal freshet lasting continuously for weeks or months. However, the term has been used more widely to describe a sudden significant increase in discharge where the annual flow regime is intermittent or even perennial. In environments such as those with a Mediterranean-type climate, flow dwindles seasonally so that flood runoff in summer may add dramatically to a pre-existing trickle. In these circumstances, the perception of an observer will be that the rising limb of the flood hydrograph is steep, occupying tens of minutes rather than hours, a pattern which contrasts with that more typical of runoff during the wet season. To warrant the descriptor, the magnitude of the flood peak will also have been remarkable in causing nuisance or damage, or, indeed, loss of human life. It is debatable whether, here, the flood hydrograph should be described as 'flashy' or 'flashier' in relation to the norm rather than given the epithet 'flash flood'.

In desert or semi-desert environments, where dry channel conditions are the norm, (see WADI) flash floods are usually remarkable regardless of magnitude. Here, in contrast with Mediterranean or temperate environments where events of this type are a feature of the 'dry season', floods are more often than not associated with incursions of monsoonal airmass (as in the Sonoran Desert and the Saudi Arabian peninsula) or with the regular latitudinal shift of the Inter-Tropical Convergence Zone (as in East Africa). In such areas, these are events of the rain season(s), but rainfall is extremely uncertain, making events even more memorable if, by chance, they are witnessed. Although widespread, low intensity rainfall can generate runoff in these areas, flash floods are more likely to be the product of wandering cellular convective storms. The wetted 'trail' or footprint of these is usually only a few kilometres across. Atmospheric dynamics dictate that such storm systems that are capable of releasing rain of sufficient magnitude and intensity will be separated by several tens to several hundreds of kilometres. The likelihood that a drainage basin will receive sufficient rainfall to generate runoff (see RUNOFF GENERATION) depends upon its position in relation to the trajectory of each storm. Small basins (in the order of 10^1 km^2) may experience an event only infrequently, perhaps staying dry even though floods occur in the vicinity. In this case, it may be that a flood series for one basin is developed from events in years that do not contribute to the series of a neighbouring stream. In basins of moderate size (several 10^2 km^2), the storm cell is frequently smaller than the basin. Indeed, in this case, the flood may move down-channel into parts of the basin that have not experienced rain. This is a circumstance that provides the greatest danger for the unwary and is not uncommonly the cause of human mortality, especially where the dry river bed has provided an apparently convenient location for overnight encampment.

The significance of the variable spatial coincidence of storm and drainage basin is that the flood hydrograph can take on a variety of shapes. This is, in part, because different sub-basins may contribute to each event and storms may move up or down catchment, depending on local atmospheric dynamics, so affecting the gathering time of contributions from each tributary. This means that it is more difficult to define a typical flood hydrograph (as in, e.g. unit hydrograph analysis) in a desert or semi-desert setting, not only because the frequency of events is low but also because each runoff event may possess unique characteristics. There is, however, evidence from one drainage basin in the Asir Escarpment of Saudi Arabia that flood volume can be approximated from a parameter such as flood peak discharge, and finite element models of rainfall runoff have been developed with some success for predicting flood waves in small basins in Oman and Arizona. Despite these, the relations between runoff and storm characteristics such as rainfall amount and intensity are often chaotic so that predictability of event frequency and magnitude is low even if rainfall is being monitored by spatially inclusive means such as radar. In all but rare instances where research catchments have been established, rain gauge density and disposition will be either inadequate or, more often, non-existent.

Flash floods are undoubtedly dramatic, if only because of the stark contrast between the event itself and the much longer intervening period when the channel is dry. In southern Israel, at the eastern edge of the Sahara's hyper-arid core, long-term monitoring has revealed that the frequency of events is, on average, much less than one a year, but there can be periods of several years when no runoff occurs. In the semi-arid northern Negev, with a rainfall of *c*.200–300 mm per year, the number of events that occurs in moderate-sized basins ranges from zero to seven. Here, on average, an ephemeral channel is occupied by flow for about 2 per cent of the year, or about seven days.

Perhaps the most dramatic aspect of flash floods is the arrival of a bore. This may be the first that the observer is aware of rainfall, which may have occurred well up-catchment. Field monitoring in semi-arid areas, where vegetation may be sparse at the start of the rain season, has shown that time to ponding is short – typically in the order of a few minutes, depending on the infiltration capacity of the local soils – even under modest rainfall intensities. An observer caught out in the rain undergoes a curious sensation that the ground is moving as a glistening sheet of OVERLAND FLOW slips towards the channel system. Here, high drainage densities, developed in response to the easy and quick shedding of water, ensure rapid concentration of flow and the birth of a flash flood.

A flash flood bore takes on a number of forms. The rapidly advancing 'wall of water' is almost certainly a figment of imagination encouraged by the panic of moving to a place of safety. Indeed, the type of bore most commonly caught on camera is comparatively shallow, with low trailing water-surface slope. However, a few examples have been photographed where the bore reaches a height of about half a metre (Plate 49). Of those few measurements that have been made of bore advance, velocities range from 0.5 to 2 m s^{-1}, the rate depending directly on bore height. This is equivalent to a stiff walking pace for a human being and one might wonder, therefore, what reasons there are for the number of fatalities that are reported. The problem for those unfortunate to be caught napping is that, following the passage of the bore, water levels rapidly increase. One fully documented example has indicated an average rise of a quarter of a metre per minute, so that the water surface was at waist height within two and head height within ten minutes of the start of hydrograph rise. By this time, average flow velocity is in excess of 3 m s^{-1} and increasing to values greater than 5 m s^{-1} (Figure 62).

Plate 49 Flash flood bore in Nahal Eshtemoa, northern Negev, advancing over dry bed at about 2 m s^{-1}. Note that the immediate area has had no rain

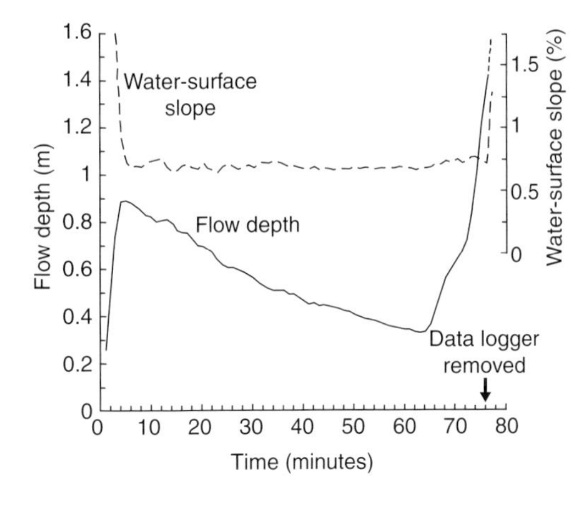

Figure 62 Hydrograph and water surface slope of the flash flood on Nahal Eshtemoa shown in Plate 49

However, a flash flood bore is not stealthy. If it advances over a gravel bed, the cacophony of grains being thrown against each other can be heard several hundred metres away. Flotsam is also characteristic of flash flood bores, the infrequent flow sweeping up LARGE WOODY DEBRIS and other organic matter that has fallen into the channel between events and adding to the general confusion that is already inherent. Indeed, some have reported hearing the clash of tree trunks, etc. several kilometres ahead of the bore's arrival.

Although the bore of a flash flood is its most spectacular feature, another unique but hidden characteristic is the loss of a significant fraction of flow to the dry bed. These are dubbed transmission losses. They are determined in part by the magnitude of the flow and hence the wetted perimeter. For a small heavily gauged ephemeral channel in Arizona, examples show that transmission losses to the bed in each kilometre of channel can account for as much as 6 per cent of the flow. This points to another important characteristic of flash floods in desert and semi-desert settings – many fail to reach the terminal ALLUVIAL FAN.

Further reading

Bull, L.J. and Kirkby, M.J. (eds) (2002) *Dryland Rivers: Processes and Management in Mediterranean Climates*, Chichester: Wiley.

Reid, I., Laronne, J.B., Powell, D.M. and Garcia, C. (1994) Flash floods in desert rivers: studying the unexpected, *EOS, Transactions American Geophysical Union* 75, 452.

Reid, I., Laronne, J.B. and Powell, D.M. (1998) Flashflood and bedload dynamics of desert gravel-bed streams, *Hydrological Processes* 12, 543–557.

IAN REID

FLAT IRON

Term used to designate relic slopes whose morphology resembles a reversed iron. They are also known as talus flat irons or triangular slope facets and develop at the foot of scarps in mesas and cuestas. The slope deposits, which locally contain datable charcoal, ashes or pottery remains, may grade in the distal sector to cover pediments or fluvial or lacustrine terraces.

The most widely accepted genetic model relates the development of talus flat irons to climatic changes. The accumulation processes in the slopes prevail during humid periods whereas the reduction in the vegetation cover during dry periods favours rilling and gullying processes. Successive climate changes give place to different generations of relict slopes whose relative chronology can be inferred from their spatial distribution. Up to five generations of flat irons have been identified in the three main Tertiary basins of Spain. The slope deposits of the flat irons dated with ^{14}C correspond to cold periods. The youngest facet generation fits with Upper Holocene Neoglaciation episodes and the two previous generations correlate to Heinrich events (H_3 and H_4) that indicate cold periods (Gutierrez *et al.* 1998).

Reference

Gutierrez, M., Sancho, C., Arauzo, T. and Peña, J.L. (1998) Evolution and paleoclimatic meaning of the talus flat irons in the Ebro Basin, northeast Spain, in A.S. Alsharham, K.W. Glennie, G.L. Whittle and C.G.St.C. Kendall (eds) *Quaternary Deserts and Climatic Change*, 593–599, Rotterdam: Balkema.

SEE ALSO: slope, evolution

M. GUTIERREZ-ELORZA

FLOOD

A flood is a flow of water greater than the average flow along a river. A flood may be described in terms of its magnitude. On a given river, for example, any discharge exceeding 1,000 m^3/s might be designated a flood. A flood may also be described by its recurrence interval; a 100-yr flood occurs on average once every 100 years. Or a flood may be described as any flow that overtops the banks or LEVEEs along a channel and spreads across the FLOODPLAIN.

Floods along inland rivers result from precipitation or from a damburst. When water rapidly flows downslope from snowmelt, rain-on-snow, or various types of rainfall, the baseflow from subsurface water that keeps some stream channels flowing during dry periods is augmented by runoff (see RUNOFF GENERATION). As discharge increases in the channel during the rising limb of the flood, the channel boundaries may be eroded, and both SUSPENDED LOAD and BEDLOAD sediment transport are likely to increase. Once the input of runoff to the channel declines, sediment transport is likely to decrease and sediment may be deposited along the channel during the falling limb of the flood. Floodwaters in a channel commonly rise more rapidly than they fall during all types of floods, but this difference is most pronounced for damburst floods. Dams built by humans and naturally occurring LANDSLIDES, ice jams, glacial moraines, glacial ice dams, or beaver DAMS may fail suddenly, prompting catastrophic drainage of the water ponded behind the dam. Along with FLASH FLOODS, damburst floods are often the most unexpected and damaging floods. Damburst floods may have a peak discharge more than an order of magnitude larger than the peak discharges created by meteorological floods along a river. This large discharge may generate high values of STREAM POWER that cause substantial erosion and deposition along the flood path. OUTBURST FLOODS generated by the failure of natural dams ponding meltwater from the great continental ice sheets during the Pleistocene shaped such dramatic landscapes as the Channeled SCABLAND of the northwestern United States.

Floods along coastal rivers may also result from STORM SURGES, TSUNAMIS, or other anomalously large waves or tides backflooding upstream from the ocean. Low-relief coastal areas such as those found in Bangladesh may be particularly susceptible to such floods.

The largest measured historical floods generated by precipitation have occurred primarily between 40°N and 40°S latitude, usually near coastal areas where the onshore movement of warm, moist airmasses into the continental interior produces intense and widespread precipitation (Costa 1987). The envelope curve of maximum rainfall-runoff floods is mathematically described by $Q = 90A$ for drainage areas less than 100 square kilometres, and $Q = 850A^{0.357}$ for larger drainage areas, where Q is peak discharge in cubic metres per second and A is drainage area in square kilometres (Herschy 1998).

The importance of a flood relative to smaller flows in shaping channel and valley morphology will depend on the magnitude and duration of the hydraulic forces generated during the flood in comparison to the erosional resistance of the channel boundaries, and on the recurrence interval of the flood. A channel formed on bedrock or very coarse alluvium may have such high boundary resistance that only a flood generates sufficient force to erode the channel boundaries. This effect may be enhanced where a deep, narrow channel and valley geometry concentrate floodwaters such that flow depth increases rapidly with discharge, giving rise to high stream power. In contrast, a channel bordered by a broad floodplain will have much less increase in flow depth with increasing discharge, and the flood may not have a substantially greater capacity for erosion and sediment transport than do smaller flows along the channel. Channels in which geometry and sediment transport reflect primarily floods are likely to have a flashy hydrograph, abundant coarse sediment load, a high channel gradient, highly turbulent flow, and shifting, erodible banks (Kochel 1988). Geomorphic change during floods is likely above a minimum threshold (see THRESHOLD, GEOMORPHIC) of approximately $300\,\mathrm{W\,m^{-2}}$ of unit stream power for alluvial channels (Magilligan 1992). The threshold for bedrock channels may be expressed as $y = 21x^{0.36}$, where y is stream power per unit area and x is drainage area (Wohl *et al.* 2001). In steep channels with abundant sediment, flows may alternate downstream among water-floods, DEBRIS FLOWS and HYPERCONCENTRATED FLOWS.

Measures to reduce hazards to humans associated with floods date back several millennia. Such measures include impoundments to regulate water flow; channelization to increase the flood conveyance of channels; levees to confine floodwaters; warning systems to help alert and evacuate humans at risk; and engineering designs which reduce flood damage to structures. Despite this long history of river engineering and flood mitigation, property damage from floods continues to increase worldwide as population density and building in flood-prone areas increase, and as land uses across drainage basins alter runoff generation. Along rivers where alteration of the natural flow regime has reduced or eliminated floods, aquatic and riparian species adapted to flooding have declined in extent and diversity. River rehabilitation and restoration measures are now being applied to some of these rivers in an attempt to mitigate damages caused by the absence of floods.

References

Costa, J.E. (1987) A comparison of the largest rainfall-runoff floods in the United States with those of the Peoples Republic of China and the world, *Journal of Hydrology* 96, 101–115.

Herschy, R.W. (1998) Floods: largest in the USA, China and the world, in R.W. Herschy and R.W. Fairbridge (eds) *Encyclopedia of Hydrology and Water Resources*, 298–300, Dordrecht: Kluwer Academic.

Kochel, R.C. (1988) Geomorphic impact of large floods: review and new perspectives on magnitude and frequency, in V.R. Baker, R.C. Kochel and P.C. Patton (eds) *Flood Geomorphology*, 169–187, New York: Wiley.

Magilligan, F.J. (1992) Thresholds and the spatial variability of flood power during extreme floods, *Geomorphology* 5, 373–390.
Wohl, E., Cenderelli, D. and Mejia-Navarro, M. (2001) Channel change from extreme floods in canyon rivers, in D.J. Anthony, M.D. Harvey, J.B. Laronne and M.P. Mosley (eds) *Applying Geomorphology to Environmental Management*, 149–174, Highlands Ranch, CO: Water Resources Publications.

SEE ALSO: bankfull discharge; floodout; palaeoflood; sediment rating curve

ELLEN E. WOHL

FLOODOUT

A floodout is a site at the downstream end of a river where channelized flow ceases and floodwaters spill across adjacent, unchannelled, alluvial surfaces. The term has been most widely used in connection with ephemeral channels in arid central Australia (Tooth 1999a) but it has also been applied to discontinuous gullies (see GULLY), intermittent channels and perennial channels in semi-arid, subhumid and humid regions of eastern Australia and southern Africa (Fryirs and Brierley 1998; Tooth *et al.* 2002).

Floodouts form as a result of various factors including downstream decreases in discharge, downstream decreases in gradient, and aeolian or bedrock barriers to flow (Tooth 1999a). These factors commonly act in combination. For example, along many arid or semi-arid rivers, discharge decreases downstream owing to factors such as infiltration into normally dry channel beds, evaporation, hydrograph attenuation and a lack of tributary inflows. Gradient also commonly decreases owing to channel-bed AGGRADATION or lithological/structural factors, such as a change from a harder to a weaker lithology underlying the channel bed. In combination, these discharge and gradient decreases mean that unit STREAM POWER and sediment transport capacity also decrease, which in turn leads to a downstream reduction in the size of the channel and diversion of an increasing proportion of floodwaters overbank. This is often exacerbated by the presence of aeolian or bedrock barriers, such as longitudinal dunes (see DUNE, AEOLIAN) that have formed across the river course. Eventually, the channel loses definition and disappears entirely, and the remaining floodwaters spill across the floodout as a sheet flow (see SHEET EROSION, SHEET FLOW, SHEET WASH). This process is often referred to as 'flooding out' but strictly speaking the term 'floodout' and its derivatives should be used for the fluvial form only.

Floodouts can form in river catchments of widely different scale and thus the areas of floodouts vary considerably (*c.*1–1,000 km^2). The location and shape of floodouts, however, are often strongly influenced by local physiography. In central Australia, for instance, floodouts in the northern Simpson Desert are narrow (< 500 m) features where rivers terminate between longitudinal dunes and occasional bedrock outcrops but on the relatively unconfined Northern Plains they can reach up to several kilometres wide (Tooth 1999a,b).

'Floodout zone' is a related but broader term that encompasses both the lower reaches of the channel and the floodout itself. Geomorphological and sedimentary features commonly associated with floodout zones include distributary channels, splays, waterholes, PANS, PALAEOCHANNELS and various fluvial-aeolian interactions (Tooth 1999a,b). In addition, two basic types of floodout can be distinguished (Tooth 1999a): (1) terminal floodouts, where floodwaters spill across the unchannelled surfaces and eventually dissipate through infiltration or evaporation; and (2) intermediate floodouts, where floodwaters persist across the unchannelled surfaces and ultimately concentrate into small 'reforming channels'. Reforming channels commonly develop where the unchannelled floodwaters become constricted by aeolian deposits or bedrock outcrops, or where small tributaries provide additional inflow, and they either join a larger river or decrease in size downstream before disappearing in another floodout (Tooth 1999a; Tooth *et al.* 2002).

Formation of a floodout is just one possible end result of the broader processes of channel 'breakdown', 'failure' or 'termination' that can also occur where channels disappear in playas, in permanent wetlands, or on the surfaces of ALLUVIAL FANS. Floodouts, however, are predominantly alluvial features which are normally dry except after flood events or heavy local rains, and thus they differ from saline playas or organic-rich, saturated wetlands. Furthermore, the relatively low gradients (< 0.002) and fine-grained deposits typical of floodout zones distinguish them from alluvial fans. Floodout zones have many geomorphological and sedimentological similiarities with 'terminal fans', a term that has been applied to the distal reaches of some inland arid and semi-arid river systems where numerous distributary channels decrease in size downstream and grade into unchannelled, fan-shaped deposits (Mukerji 1976; Kelly and Olsen 1993). Downstream of intermediate floodouts, however, channels can reform, thus showing that floodouts are not necessarily terminal and, as floodouts are often confined laterally by aeolian deposits or bedrock outcrop, neither does alluvial deposition necessarily adopt a fan-shaped form (Tooth 1999a). As such, application of the term 'terminal fan' is inappropriate for many floodouts or floodout zones. In floodout zones, the disappearance of channelized flow means that FLOODPLAINs grade downstream into floodouts. Although the absence of channels makes it difficult to include floodouts within conventional definitions of 'floodplain' or existing floodplain classifications, nevertheless they can be regarded as part of a continuum of floodplain types (Tooth 1999b).

References

Fryirs, K. and Brierley, G.J. (1998) The character and age structure of valley fills in upper Wolumla Creek catchment, South Coast, New South Wales, Australia, *Earth Surface Processes and Landforms* 23, 271–287.

Kelly, S.B. and Olsen, H. (1993) Terminal fans – a review with reference to Devonian examples, in C.R. Fielding (ed.) Current Research in Fluvial Sedimentology, *Sedimentary Geology* 85, 339–374.

Mukerji, A.B. (1976) Terminal fans of inland streams in Sutlej-Yamuna Plain, India, *Zeitschrift für Geomorphologie NF* 20, 190–204.

Tooth, S. (1999a) Floodouts in central Australia, in A. Miller and A. Gupta (eds) *Varieties of Fluvial Form*, 219–247, Chichester: Wiley.

——(1999b) Downstream changes in floodplain character on the Northern Plains of arid central Australia, in N.D. Smith and J. Rogers (eds) *Fluvial Sedimentology VI*, International Association of Sedimentologists, Special Publication 28, 93–112, Oxford: Blackwell Scientific Publications.

Tooth, S., McCarthy, T.S., Hancox, P.J., Brandt, D., Buckley, K., Nortje, E. and McQuade, S. (2002) The geomorphology of the Nyl River and floodplain in the semi-arid Northern Province, South Africa, *South African Geographical Journal* 84(2), 226–237.

Further reading

Bourke, M.C. and Pickup, G. (1999) Fluvial form variability in arid central Australia, in A. Miller and A. Gupta (eds) *Varieties of Fluvial Form*, 249–271, Chichester: Wiley.

Gore, D.B., Brierley, G.J., Pickard, J. and Jansen, J.D. (2000) Anatomy of a floodout in semi-arid eastern Australia, *Zeitschrift für Geomorphologie Supplementband* 122, 113–139.

Mabbutt, J.A. (1977) *Desert Landforms*, Canberra: ANU Press.

SEE ALSO: alluvium; bankfull discharge; flood; hydraulic geometry

STEPHEN TOOTH

FLOODPLAIN

The floodplain is generally considered to be the relatively flat area of land that stretches from the banks of the parent stream to the base of the valley walls and over which water from the parent stream flows at times of high discharge. The sediment that comprises the floodplain is mainly ALLUVIUM derived from the parent stream with minor contributions from aeolian sediment or colluvium from the valley walls. During floods the channel width is increased to include some or all of the floodplain in order to accommodate the increased discharge with relatively smaller increases in velocity and depth than would be the case if the flood discharge were artificially confined within the channel. However, defining the extent of a floodplain at a locality in terms of the area inundated in floods of particular return periods poses problems, since flooding frequency may be a restricting factor. This can be especially problematic in arid and semi-arid areas.

It has been suggested (Wolman and Leopold 1957) that the active floodplain is the area subject to the annual flood (i.e. the highest discharge each year), though this can really only apply to rivers in humid regions. In reality, the active floodplain only forms part of the topographic floodplain, which encompasses the whole valley floor and includes parts of relict floodplains in the form of river terraces (see TERRACE, RIVER) (Plate 50). If the floodplain is defined in terms of the processes (including superfloods) that give rise to it, then the term polygenetic floodplain would apply to most since they result from changes in flow-regime and sediment supply over at least the recent geological past. Nanson and Croke (1992) have proposed the term genetic floodplain, which applies to a generally horizontally bedded landform built from alluvium derived from the present flow-regime of the adjacent stream. This does not take into account the geomorphic history of a floodplain and the processes that have influenced its construction over time, however. A floodplain is a functional part of the whole stream system and forms as

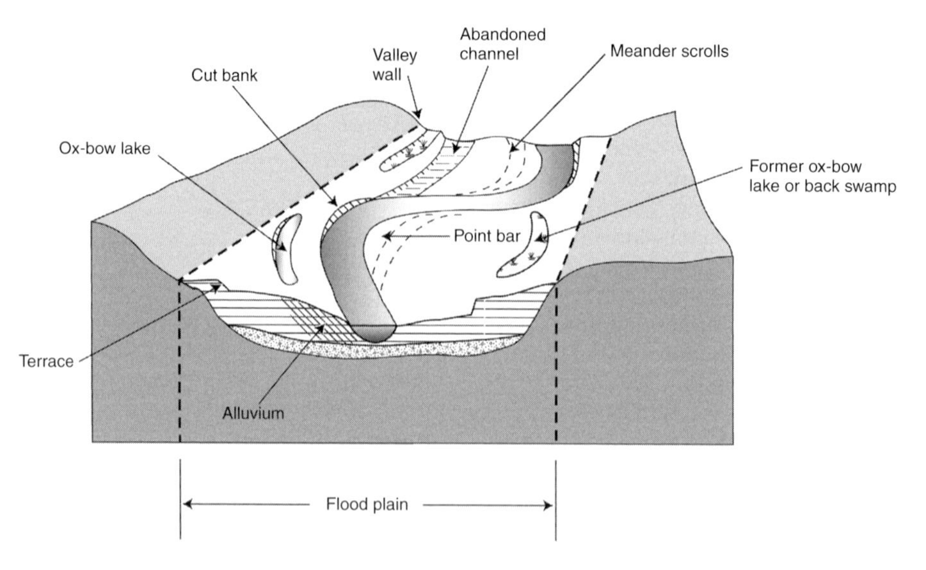

Plate 50 Features of floodplain topography

a byproduct of interrelated processes that, over time, give rise to variable flows and sediment loads derived from the drainage basin.

Floodplain formation

Floodplains are formed by processes that are active both within the channel of the parent stream and during overbank flow. The processes involved are lateral accretion, which takes place within the channel from the formation of bars by movement of relatively coarse bedload; and vertical accretion, which occurs on the floodplain surface due to deposition of finer material from the suspended load during overbank flow. The relative importance of vertical accretion in the formation of floodplains was considered negligible in comparison with within-channel processes (Wolman and Leopold 1957) though it has now been shown that overbank sedimentation can contribute significantly. For example, Nanson and Young (1981) have described the floodplain deposits of streams in New South Wales that have parts of their floodplains dominated by extremely cohesive overbank deposits that prevent the stream from migrating. In lowland rivers in the United Kingdom similar deposits have been described for the Severn (Brown 1987) and the Thames (Lewin 1984) where thick muddy deposits overlie sandy gravels.

Lateral accretion deposits are built up in the channel either as marginal bars that may form in an alternating sequence along relatively straight channels or as point bars that develop on the inside bends of meanders. If the channel migrates laterally by BANK EROSION on one side, the channel dimensions are maintained by compensating deposition on the other bank. The marginal or point bars grow laterally towards the direction of migration and increase in height, with sediment being deposited on them in low-angled, sloping layers overlain by finer material deposited at bankfull and flood stage flow. The sediments within the bars tend to fine upwards as they initially form from the stream's bedload sediment carried by secondary currents from the outer bank region towards the inner bank (Markham and Thorne 1992). Over time, in this way, a stream may rework the entire floodplain sediment as it migrates from one valley side to the other, leaving behind cutoff meanders as oxbow lakes or swamps and traces of the old meander paths as meander scrolls (Plate 50).

Vertically accreted sediment is added to the floodplain surface during overbank flow. As the water in the channel overflows the banks onto the floodplain surface, the flow velocity is reduced as the width of the channel is effectively increased by inclusion of the floodplain. This reduction, in conjunction with an increase in surface roughness if the floodplain is vegetated, causes suspended sediment to be deposited. As the flow of water on the floodplain is slower and shallower than that in the channel, there is a zone of turbulence near to the channel bank which results in a net transfer of momentum and sediment from the channel to the floodplain. The width of the turbulent zone depends on the relative difference in depths between the channel and floodplain flows but influences the nature of sediment deposition near to the channel. The sediment deposited near to the channel bank tends to be coarser and thicker than that deposited further onto the floodplain as transport competence declines with distance from the channel and away from the influence of the turbulent zone (Marriott 1996). However, the amounts and grain sizes deposited depend on sediment supply and duration and depth of overbank flow.

The channel banks can gradually become the highest points on the floodplain as the thicker, coarser overbank deposits build up to form natural levees. The stream may then deposit sediment on its bed during high flows that do not exceed BANKFULL DISCHARGE, resulting in the normal river surface being above the level of the floodplain. Both natural levees and artificial embankments set away from the channel as macrochannel banks to a two-stage channel, afford some protection from flooding. In extreme cases, floodwater may break through the levee, forming a crevasse channel and washing sediment from the channel and reworked from the levee onto the floodplain. These sediments form a crevasse splay of coarser material than the underlying floodplain alluvium. Coarse material can also be transferred from the channel during overbank flow due to the action of convection currents set up by turbulence at bends in the channel. This is because the flow of water on the floodplain tends to travel directly down valley and at meanders the direction of flow within the channel is at an angle to that on the floodplain (Knight and Shiono 1996). Bedload sediment can then be picked up and spread in a lobe downstream from the outer bank of the bend.

In arid and semi-arid areas floodplains are formed mainly during major flood events (superfloods) with recurrence intervals in the region of 10,000 years. These floods bring in material from highland areas of the catchment. Between these events the sediment is reworked by the more frequent relatively minor flash floods that occur in the ephemeral channels rather than new material being added. Studies of the streams of central Australia (Pickup 1991) show different landforms depending on the scale of flooding that gave rise to them. The superfloods result in large sandsheets and sand threads and the contemporary macrochannel system which has levees, FLOODOUTs and floodbasins wherein sediment is reworked during flooding or by aeolian processes in dry periods.

Floodplain classification

Floodplains can be classified according to their morphology rather than the manner in which they were formed or the processes active at present. The floodplain as a whole, together with its parent stream, can be considered as a macroform, whereas the mesoforms are the component parts of the channel, e.g. bar forms, and the floodplain (levees, crevasse channels and splays, backswamps and oxbow lakes). Mesoforms can influence flow and deposition patterns on the macroform. In this classification the microforms are the small-scale structures superimposed on the mesoforms, for example, ripples, dunes, shrinkage cracks (Lewin 1978).

A further classification that takes into account the genesis of the floodplain was suggested by Nanson and Croke (1992) and is based on the relationship between the ability of streams to entrain and transport sediment and the resistance to erosion of the bank sediment. Three classes are recognized: high energy streams with non-cohesive banks, medium energy streams with non-cohesive banks and low energy streams with cohesive banks. Within these classes further levels of classification can be set up based on primary geomorphic factors such as channel cutting and filling and lateral accretion on point bars, and secondary factors such as scroll bar formation and organic (peat) accumulation. The primary geomorphic factors depend on stream power and sediment load and can therefore identify different environments for floodplain formation.

Floodplains respond to changes in channel processes that result from alterations in flow regime and/or sediment supply (Schumm 1977) though the response may be at a slower rate. The sedimentary record of these changes is often incomplete as it is often complicated by episodes of erosion, so although floodplain formation is polygenetic, much of the evidence is destroyed. It could be suggested that, over geological timescales, all floodplains are polygenetic as external influences such as climate and relative base-level change are not constant.

Floodplain sedimentation rates

In humid areas rivers flood every 1–2 years, though it is mainly extraordinary events that are studied and documented due to their catastrophic effects on human activity and property. Studies of flood frequency and sedimentation patterns have, therefore, been carried out with the aim of attempting to predict, and thus avoid, the destructive effects of extreme events. As explained above, the floodplains of many rivers are composed of sediments accumulated from channel and flood activity and many studies record the thickness of deposits in various parts of the floodplain by extraordinary events or they give the nature of the sediment deposited. Generally figures are highly variable and are only useful as a general guide to likely sedimentation rates in the various floodplain environments indicated.

Flooding is an essentially random occurrence and it is difficult to sample satisfactorily from the floodplain surface during a flood. However, borehole data, sediment cores, ^{14}C dating, pollen analysis and radioactive nuclides (e.g. ^{137}Cs and ^{210}Pb) can all be used to estimate sedimentation rates over a period of time. As an example, Brown (1987) used some of these techniques to estimate that sedimentation rates on the River Severn floodplain in the UK over the past 10,000 years have been around 1.4 mm per year.

Floodplains act as storage space or sediment sinks for alluvial sediment. While they are being stored the sediment may be reworked by fluvial, aeolian, biological and/or pedological agents, often over considerable periods of time. The stored sediment may subsequently be eroded and re-incorporated into the sediment budget of the drainage basin. The residence time of sediment in storage will vary according to factors such as surface topography, climate and vegetation and the relative return frequency of major flood events. Originally, sediment storage on floodplains was studied so that SEDIMENT BUDGETs of drainage basins could be calculated; now, however, interest in the storage of sediment that was originally part of the suspended load of the parent stream has increased with an awareness of the ability of contaminants such as heavy metal ions and radionuclides to adhere to and be transported with this fine material.

Summary

Understanding the processes involved in floodplain formation is important because of the interaction between human activity and floodplain environments. These processes include both within-channel and overbank processes which rely on the interaction between channel and floodplain flows during flooding, and which account for the distribution of different sediment grain sizes across the floodplain. Some recent work has investigated the rate of sedimentation and sediment storage on floodplains using a multiproxy approach. As floodplains act as sinks for alluvial sediments, this work is particularly useful for studies of contamination.

References

Brown, A.G. (1987) Holocene floodplain sedimentation and channel response of the lower River Severn, *Zeitschrift für Geomorphologie NF* 31, 293–310.
Knight, D.W. and Shiono, K. (1996) River channel and floodplain hydraulics, in M.A. Anderson, D.E Walling and P.D. Bates (eds) *Floodplain Processes*, 139–181, Chichester: Wiley.
Lewin, J. (1978) Floodplain Geomorphology, *Progress in Physical Geography* 2, 408–437.
——(1984) British meandering rivers; the human impact, in C.M. Elliott (ed.) *River Meandering*, 362–369, New York: American Society of Civil Engineers.
Markham, A.J. and Thorne, C.R. (1992) Geomorphology of gravel-bed river bends, in P. Billi, R.D. Hey, C.R. Thorne and P. Tacconi (eds) *Dynamics of Gravel Bed Rivers*, 433–450, Chichester: Wiley.
Marriott, S.B. (1996) Analysis and modelling of overbank deposits, in M.A. Anderson, D.E. Walling and P.D. Bates (eds) *Floodplain Processes*, 63–93, Chichester: Wiley.
Nanson, G.C. and Croke, J.C. (1992) A genetic classification of floodplains, *Geomorphology* 4, 459–486.
Nanson, G.C. and Young, R.W. (1981) Overbank deposition and floodplain formation on small coastal streams of New South Wales, *Zeitschrift für Geomorphologie NF* 25, 332–347.
Pickup, G. (1991) Event frequency and landscape stability on the floodplain systems of arid Central Australia, *Quaternary Science Reviews* 10, 463–473.
Schumm, S.A. (1977) *The Fluvial System*, New York: Wiley.
Wolman, M.G. and Leopold, L.B. (1957) River floodplains: some observations on their formation, *US Geological Survey Professional Paper* 282C, 87–107.

Further reading

Anderson, M.A., Walling, D.E. and Bates, P.D. (eds) (1996) *Floodplain Processes*, Chichester: Wiley.
Knighton, D. (1998) *Fluvial Forms and Processes: A New Perspective*, London: Edward Arnold.
Richards, K. (1982) *Rivers: Form and Process in Alluvial Channels*, London: Methuen.

SUSAN B. MARRIOTT

FLOW REGULATION SYSTEMS

Flow regulation systems is a general expression including all works constructed along a watercourse in order to regulate the flow in the channels. Flow regulation systems can be planned for maintaining steady-state conditions along a river or for avoiding uncontrolled erosion and/or sedimentation processes. On the other hand, they can be planned for obtaining a more constant distribution of the discharge in the channel by reducing both highwater and minimum flow peaks.

The devices utilized for avoiding accelerated erosion or deposition in the watercourse consist of hydraulic works capable of maintaining flow velocities within suitable ranges of values, which should be adequately chosen for each river. Besides this, a whole set of other measures are planned and constructed for opportunely directing the flow stream in order to avoid the occurrence of maximum velocities in proximity of the banks (see BANK EROSION). The most commonly used remedial measures consist in the direct protection of the banks by means of walls, sheet piles, prefabricated structures, gabions, anchored geotextiles, loose debris and biofixing plants (e.g. planting cuttings, sowing herbaceous species, etc.).

Groynes are typical works which direct the water flow; they are made up of structures stretching both downflow and with a certain angle with respect to the mean flow and can be either linear or composite (such as sledge hammers or bayonet joints). Furthermore, various arrangements of boulders in the river bed are works which increase friction by reducing velocity and increasing turbulence.

Finally, check dams are works which reduce excessive riverbed slope profiles. They break the river's profile into several stretches with lower inclinations, fixing the level of non-erodable weir crests and introducing artificial steps in the longitudinal profile of the watercourse. Besides being constructed with very heterogeneous materials, check dams can be of various patterns and dimensions. When they produce a real impoundment upstream they should rather be considered proper DAMS. To this regard, it should be mentioned that in the Alps many dams constructed for hydroelectric purposes work also as reservoirs for the regulation of water discharge during considerable highwater or minimum flow events. Apart from the regulation of flow, dams across great rivers usually have multipurpose functions, such as production of electric energy, navigation control, irrigation supply and flood control (Jansen *et al.* 1979).

Check dams are used also in watercourses capable of transporting and depositing large amounts of sediments (see BEDLOAD). In this case, they retain a certain amount of sediment in the upper part of the basins and reduce solid transport during highwater and minimum flow phases. Since the early 1950s various kinds of open weirs have been planned and constructed; among them, the most used ones are fissure-weirs and filtering weirs. The latter can be comb-like, with windows, network patterned, etc. (Figure 63). These particular check dams accomplish a two-fold purpose: (1) to retain most of the sediment during highwater phases and release it later during a low-flow phase; (2) to retain the coarse debris (including floating tree-trunks etc.) in order to avoid damage to the hydraulic structures downstream. Since the late 1980s check dams with low-angle filtering intake accompanied by a drainage gallery have been constructed along watercourses subject to overconcentrated flows ascribable to DEBRIS FLOWS.

In order to regulate the flow rate distribution during the year, other kinds of works are utilized. Among these, flood attenuation basins should be mentioned. They consist in hydraulic works which connect the watercourse to sufficiently large artificial basins capable of working as retaining reservoirs during river spates. These flood protection structures, also defined as flow-control weirs with energy dissipators, consist of a downstream regulation dam which allows the passage of a flow not superior to a prefixed value,

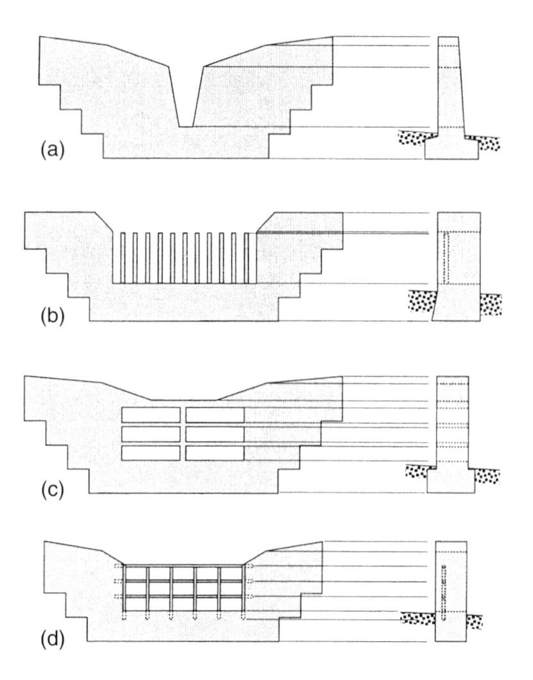

Figure 63 Examples of (a) fissure-weirs; and filtering weirs; (b) comb-like; (c) with windows; (d) network patterned

even in the occurrence of a greater flow. The exceeding volume of water is stored in the basin and returned to the river when the highwater phase is dwindling (Bell and Manson 1998). Flood protection structures can also be equipped with an upstream check dam in order to avoid discharge water from entering the flood attenuation basin when flow rate is below a certain average value. In this way the sedimentation of the bedload usually transported is avoided and, consequently, so is the progressive reduction of the basin's retention capacity. For example, along the central course of the Yangtze River, once this watercourse started to wander in its alluvial plain, a flood control weir was constructed which allows highwater flow to be diverted into a large dissipator basin provided along one of its tributaries (Jingjiang River and Dongting Lake). These hydraulic facilities, built in 1953 and extended in 1990, are made up of a 54-hole barrage, 1,054 m long, and can store up to 8,000 $m^3 s^{-1}$ of water in order to prevent inundations in the stretch located immediately downstream of the catchment works, thus protecting the town of Wuhan and the plain of Janghan from flooding.

Fillways and diversion canals are natural or, more often, artificial watercourses which receive a portion of a river's highwater in order to divert it into another basin or give it back to the same river, downstream of a critical stretch. Diversion canals are characterized by constant flow whereas fillways are utilized only occasionally. In the case of a diversion canal which subtracts water from a river and gives it back downstream of a critical point, usually an inhabited centre, the bypass canal should be long enough to avoid impoundment problems (raised hydrostatic levels due to a return effect starting from the confluence).

Other works which effect a reduction in flow rate levels, consist in modifications of the geometrical features of the river bed (see HYDRAULIC GEOMETRY). These solutions should be implemented with care as they are the result of engineering viewpoints and seldom do they take into sufficient account the geomorphological features of the canal. The usual procedure involves deepening, widening and redressing the canal. In the case of reshaping and widening of a canal's section, a decrease of the stream velocity takes place with consequent sedimentation and rising of the river bed which, therefore, requires periodical dredging. Other cases of course modification consist in canal straightening. For example, the best known cases of straightening are those in the lower course of the Mississippi River, which were carried out during the 1930s and 1940s by means of cutoffs of the meandering course for a total length of 210 km. The consequence of these modifications was an average increase of the riverbed gradient (Winkley 1982) which contributed to an increase of bedload, a rise of the canal's width/depth ratio and a tendency to change from a MEANDERING course to a braided (see BRAIDED RIVER) one. Thus, navigation problems arose due to the decreased depth of the river bed and, as a consequence, expensive dredging and bank protection works were necessary in order to mitigate the problem which still persists in this stretch of the river.

Among the works which regulate flow, artificial embankments should also be mentioned; they are the first fluvial works of a certain entity ever realized by humans. Indeed, it is well known that the construction of considerable embankments in the Po Plain, in northern Italy, started during the Roman age, although traces of partial embankments date back even to the previous Etruscan epoch (Marchetti 2002).

In the alluvial plains of economically advanced regions, flood protection works are connected to dense canal networks with draining and irrigation purposes which allow intensive farming activities in the plain areas and, at the same time, reduce the risk of flooding in urban areas.

Extreme cases of flow regulation systems consist in the implementation of real changes of the hydrographic network of a region by means of artificial diversions of important watercourses. In the ex-Soviet Union, in a vast territory characterized by north- and east-bound large rivers which flow through largely infertile cold regions and extensive drought-stricken areas, impressive diversions were planned and partially implemented. For example, the southward diversions of the rivers Peciora and Irtys, carried out in order to avoid the drying up of Lake Aral following the heavy water exploitation of Syrdarja and Amudarja, has produced, on the one hand, the drying up of vast northern territories covered by taiga and, on the other hand, the swamping of large areas to the south.

References

Bell, F. and Manson, T.R. (1998) The problems of flooding in Ladysmith, Natal, South Africa, in M. Eddleston and J.G. Manud (eds) *Geohazards and*

Engineering Geology, Engineering Geology Special Publication 14.
Jansen, P., van Bendegom, L., Berg., J., de Vries, M. and Zanen, A. (1979) *Principles of River Engineering*, London: Pitman.
Marchetti, M. (2002) Environmental changes in the central Po plain (Northern Italy) due to fluvial modifications and man's activities, *Geomorphology* 44(3–4), 361–373.
Winkley, B.R. (1982) Response of the Lower Mississippi to river training and realignment, in R.D. Hey, J.C. Bathurst and C.R. Thorne (eds) *Gravel-bed Rivers*, 659–680, Chichester: Wiley.

Further reading

Brookes, A. (1985) River channellization: traditional engineering methods, physical consequences and alternative practices, *Progress in Physical Geography* 9, 44–73.
Church, M. (1995) Geomorphic response to river flow regulation: case studies and time-scales, *Regulated Rivers, Research and Management* 11, 3–22.
Gregory, K.J. (1995) Human activity and palaeohydrology, in K.J. Gregory, L. Starkel and V.R. Baker (eds) *Global Continental Palaeohydrology*, 151–172, Chichester: Wiley.
Petts, G.E. (1984) *Impounded Rivers: Perspectives for Ecological Management*, Chichester: Wiley.
Sear, D.A., Darby, S.E., Thorne, C.R. and Brookes, A.B. (1994) Geomorphological approach to stream stabilization and restoration: case study of the Mimmshall Brook, Hertfordshire, *Regulated Rivers, Research and Management* 9, 205–223.

SEE ALSO: flood; fluvial erosion quantification; river restoration

MAURO MARCHETTI

FLOW VISUALIZATION

The human eyes have some difficulty in perceiving the displacement of air, water and of most fluids. However, looking at fluid interfaces (surface boils in river) or particles suspended in a fluid (such as snow flakes or air bubbles) aids in the recognition of structured bodies within the moving fluid. Flow visualization provides a set of tools that have led to significant advances in our understanding of fluid dynamics. Flow visualization tools allow us to track and follow individual turbulent flow structures as they develop (Lagrangian reference frame) or to define flow parameters at one specific location within the flow, such as flow recirculation boundaries (Eulerian reference frame). As a result, these tools have become indispensable in the study of the structured motion of fluids on and within the globe's surface.

Sketches of turbulent flows made by Leonardo da Vinci (1452–1519) demonstrate that flow visualization has long fascinated the imagination of scientists. However, in order to complement quantitative flow monitoring, flow visualization techniques were developed primarily in laboratory studies of fluid mechanics during the last half-century. In the later 1960s, the use of flow visualization led to a major breakthrough in the understanding of turbulent boundary layer structure. Kline *et al.* (1967) (using air bubbles) and Corino and Brodkey (1969) (using neutrally buoyant particles) showed that the flow near a boundary, albeit turbulent in nature, exhibit structured patterns and mechanisms. Today, a wide range of visualization techniques is routinely used in laboratory studies (see the *Atlas of Flow Visualization* and the proceedings from several *International Symposiums on Flow Visualisation* for in-depth reviews of techniques and results from wind tunnels and water flumes experiments).

Most flow visualization techniques rely on the presence of a foreign tracer in the flow. These are often suspended particles or dye/smoke injected at specific locations. The use of tracers relies on the fact that the fluid motion can be inferred from the tracer movement matching that of the fluid. This implies a clear understanding of the mode of introduction or generation of the tracer in the flow, of the relationship between tracer and fluid motion, and of the physical significance of the observed tracer motion. In highly turbulent flows, cameras are used to aid the interpretation of flow patterns from moving particles. The use of dye/smoke present the advantage of providing a quick expression of the flow structure, but can diffuse rapidly away from the source of injection and, thus, reduce the tracing distance.

As well as in laboratory studies, flow visualization techniques can be used in aeolian, fluvial and costal environments where tracers often naturally occur. Matthes (1947), for example, provided an extensive classification of flow structures found in a river based on visual observations of surface boils, waves, turbidity differences and other visual indices. Roy *et al.* (1999) described two flow visualization techniques used in the natural environment. The first uses the turbidity difference to describe the development of flow structures at the shear layer between two merging streams. The other involves the injection of a milky white fluid to visualize shedding motions from the recirculating flow region in the lee of the obstacle.

Two different approaches of *quantitative flow visualization* exist. The first involves sampling velocity over a dense grid using single point velocity meters, such as electromagnetic current meters or acoustic Doppler velocimeters. This grid is then used to create maps of the turbulent parameters of interest (Bennett and Best 1995). As the measurements are not taken simultaneously, this approach provides a frozen picture of the general flow patterns. The second approach it to take velocity measurements using several single probes simultaneously. Such a set-up allows space–time velocity matrices to be created from which space–time velocity coherence and footprints of flow structures can be observed and described within the region covered by the velocity metres (Buffin-Bélanger *et al.* 2000).

The increasing use of multi-point velocity measurement techniques, such as acoustic Doppler profiling (Wewetzer *et al.* 1999) and particle image velocimetry (Bennett *et al.* 2002), is bound to create new breakthroughs in our understanding of flow structure. These techniques rely on the measurement of velocity from embedded particles in the moving fluid and allow the temporal variability of the flow to be described quantitatively at one point as well as the spatial variability of flow patterns in time. Hence, these techniques combine the qualitative realism of flow visualization with quantitative velocity measurements.

Our ability to use computational fluid dynamics (CFD) to simulate complex flows is increasing dramatically. This improvement gives rise to more and more sophisticated *numerical flow visualization* (Lane *et al.* 2002). Traditional flow visualization also complements CFD in allowing us to compare numerical results to natural flow behaviour.

References

Bennett, S.J. and Best, J.L. (1995) Mean flow and turbulence structure over fixed, two-dimensional dunes: implications for sediment transport and dune stability, *Sedimentology* 42, 491–514.

Bennett, S.J., Pirim, T. and Barkdoll, B.D. (2002) Using simulated emergent vegetation to alter stream flow direction within a straight experimental channel, *Geomorphology* 44, 115–126.

Buffin-Bélanger, T., Roy, A.G. and Kirkbride, A.D. (2000) On large-scale flow structures in a gravel-bed river, *Geomorphology* 32, 417–435.

Corino, E.R. and Brodkey, R.S. (1969) A visual investigation of the wall region in turbulent flow, *Journal of Fluid Mechanics* 37, 1–30.

Kline, S.J., Reynolds, W.C., Schraub, F.A. and Runstadler, P.W. (1967) The structure of turbulent boundary layers, *Journal of Fluid Mechanics* 30, 741–773.

Lane, S.N., Hardy, R.J., Elliot, L. and Ingham, D.B. (2002) High-resolution numerical modelling of three-dimensional flows over complex river bed topography, *Hydrological Processes* 16, 2,261–2,272.

Matthes, G.H. (1947) Macroturbulence in natural stream flow, *Transactions of American Geophysical Union* 28, 255–265.

Roy, A.G., Biron, P.M., Buffin-Bélanger, T. and Levasseur, M. (1999) Combined visual and quantitative techniques in the study of natural turbulent flows, *Water Resources Research* 35, 871–877.

Wewetzer, S.F.K., Duck, R.W. and Anderson, J.M. (1999) Acoustic Doppler current profiler measurements in coastal and estuarine environments: examples from the Tay Estuary, Scotland, *Geomorphology* 29, 21–30.

Further reading

Atlas of Flow Visualization, Vols 1, 2 and 3. Visualization Society of Japan (eds), Tokyo.

Van Dyke, M. (1988) *An Album of Fluid Motion*, Standford, CA: The Parabolic Press.

Tavoularis, S. (1986) Techniques for turbulence measurement, in *Encyclopedia of Fluid Mechanics*, 1,207–1,255, Texas: Gulf Publishing Company.

SEE ALSO: boundary layer

THOMAS BUFFIN-BÉLANGER AND
ALISTAIR D. KIRKBRIDE

FLUIDIZATION

A geomaterial becoming a fluid, behaving like a fluid, usually associated with high-speed debris flows such as very mobile rockslides and NUÉE ARDENTE. It may be that fluidization within a rock avalanche mass can account for the properties of catastrophic landslides. The fluidization arises from the interaction of energetic particles and trapped air under pressure. In a fluidized system granular material is supported by air (or any other gas, but usually air). The geomorphological relevance of fluidization is to ground failure in which debris flows occur, usually as long run systems apparently supported by air. Some classic landslides, e.g. Blackhawk, Elm, Frank, Saidmarrah, etc., fall into this category. One of the most striking properties of debris is its relatively high fluidity; debris flows with 80–90 per cent granular solids by weight can move in sheets about 1 m thick over surfaces with slopes of 5–10 degrees. The high fluidity suggests that the debris is 'fluidized' in the sense that this term is used by the chemical engineers. In fluidization, the interstitial fluid moves so rapidly upwards through the granular solids that these solids are suspended.

There seem to be three possibilities for debris flow mechanisms: either the flow moves essentially as a mass, supported by an air cushion, which allows long travel; or the system is fluidized in the classical sense and air and particles are interacting to keep the system mobile; or the mobility is ensured by particles interacting with each other in the high energy system. There are factors which support all three views. The Blackhawk landslide fell about 1,000 m from the mountain of the same name in south California in some prehistoric period. Possibly the slide moved almost as a single unit, gaining a nearly frictionless ride on a cushion of compressed air. This theory appears to be supported by the marginal ridges of debris formed where material was dropped as air leaked from the edges of the slide mass, and by debris cones on the landslip formed by air leakages blasting up through holes in the main mass.

There were no witnesses to the Blackhawk event; the Elm rockslide in Switzerland was closely observed. In September 1881 a large part of the Plattenberg mountain fell about 400 m and landed near the village of Elm. A vast amount of rubble crashed to the valley floor, bounced 100 m up the opposite wall, turned and – in less than a minute – careered down the valley for over 1 km before coming to a sudden stop. It is feasible that the Elm rockslide could have travelled down the valley on a cushion of air but eyewitnesses suggest that some fluidization mechanism was more likely; the surface of the slide was observed 'boiling' in great turbulence, and parts of the slide ran into houses, suggesting a great overall mobility.

Further reading

Brunsden, D. and Prior, D.B. (eds) (1984) *Slope Instability*, Chichester: Wiley.

Shreve, R.L. (1968) *The Blackhawk Landslide*, Geological Society of America Special Paper 108.

Voight, B. (ed.) (1978) *Rockslides and Avalanches, I: Natural Phenomena*, Amsterdam: Elsevier.

Waltham, T. (1978) *Catastrophe: The Violent Earth*, London: Macmillan.

SEE ALSO: liquefaction

IAN SMALLEY

FLUVIAL ARMOUR

'Armour' is one of several terms applied to clastic deposits in which the surface layer is coarser than the substrate (see BOULDER PAVEMENT). The phenomenon is widespread in gravel-bed rivers, where maximum surface and subsurface grain sizes may be similar but the respective median diameters differ by a factor of 2 to 4 because fine material is largely absent from the surface. It can also occur in sand-bed rivers that contain a little gravel, which becomes concentrated on the surface as a stable lag deposit. The traditional explanation is preferential winnowing of finer sediment from the surface during degradation, for example below dams which cut off the gravel flux from upstream so that the river erodes its bed to regain a capacity BEDLOAD. This degradation is self-limiting because as the surface coarsens, the transport capacity of the flow declines and the bed becomes immobile. This 'static' armouring has been investigated in flume experiments with no sediment feed and has been modelled mathematically; see Sutherland (1987) for a good review.

Coarse surface layers also exist in unregulated rivers with an ongoing sediment supply and peak flows which can transport all sizes of bed material. This 'mobile armour' allows the channel to be in equilibrium (neither degrading nor aggrading, neither coarsening nor fining) despite the size-selective nature of bedload transport, since intrinsically less mobile coarse fractions are preferentially available for transport whereas potentially mobile fine fractions are mainly hidden in the subsurface (Parker and Klingeman 1982). The armour forms by vertical winnowing during active bedload transport: entrainment of coarse clasts during floods creates gaps which are filled mainly by finer grains. Extreme floods may wash out the armour, but it re-forms during intermediate flows in most environments. In ephemeral streams there may be no such flows and armouring is generally absent (Laronne *et al.* 1994).

Mobile armour helps reduce bedload flux to match a restricted supply, with static armour as the limiting case when supply is cut off completely (Dietrich *et al.* 1989; Parker and Sutherland 1990). Changes in grain packing, as well as size distribution, are involved in this self-regulation; they include imbrication of coarser clasts and the development of pebble clusters, stone cells and transverse steps. The river bed is thus a degree of freedom in the adjustment of alluvial channels (see CHANNEL, ALLUVIAL) towards grade (see GRADE, CONCEPT OF).

References

Dietrich, W.E., Kirchner, J.F., Ikeda, H. and Iseya, F. (1989) Sediment supply and the development of the coarse surface layer in gravel-bedded rivers, *Nature* 340, 215–217.

Laronne, J.B., Reid, I., Frostick, L.C. and Yitshak, Y. (1994) The non-layering of gravel streambeds under ephemeral flood regimes, *Journal of Hydrology* 159, 353–363.

Parker, G. and Klingeman, P.C. (1982) On why gravel bed streams are paved, *Water Resources Research* 18, 1,409–1,423.

Parker, G. and Sutherland, A.J. (1990) Fluvial armor, *Journal of Hydraulics Research* 28, 529–544.

Sutherland, A.J. (1987) Static armour layers by selective erosion, in C.R. Thorne, J.C. Bathurst and R.D. Hey (eds) *Sediment Transport in Gravel-bed Rivers*, 141–169, Chichester: Wiley.

ROB FERGUSON

FLUVIAL EROSION QUANTIFICATION

The quantification of fluvial erosion, as for any other geomorphic process, is a way to make more precise and objective the assessment of the morphological changes that affect the Earth's relief. Beside this purpose, fluvial erosion quantification has a critical importance in the field of APPLIED GEOMORPHOLOGY because it can help the elaboration of erosion rate prediction models that are useful in the evaluation of GEOMORPHOLOGICAL HAZARDS.

Fluvial erosion consists of the entrainment and transport (by solution, suspension and traction) of particles that make up the stream bed or the stream banks (see EROSION). To quantify fluvial erosion, therefore, means to assess the amount of materials that a stream is capable of wearing away from bedrock or alluvial channels. These materials, however, become part of the total solid load and it is almost impossible to differentiate them from those delivered to the stream and deriving from the denudation processes acting within the whole basin. In other words, although streams are powerful erosional agents, they operate in conjunction with other exogenetic agents and with unconcentrated surface waters in particular; therefore drainage basins are acknowledged as the fundamental geomorphic units and rivers as their main elements along which energy is available. Consequently the quantification of fluvial erosion must be understood as quantification of the overall DENUDATION affecting both the slopes and the channels of the drainage basins.

The estimates of denudation affecting the slopes are often based on the direct determination of the amount of sediments removed from small sample areas or erosion field plots, or on the measurement of ground surface lowering and calculation of the volume of sediment dislodged. The evaluations of SOIL EROSION obtained in this way, however, are strictly dependent on the peculiarities of the studied slopes; therefore they have only local significance and can lead to misleading conclusions when they are extended to larger areas. In spite of this limit, a large number of field data on soil loss are useful in developing erosion prediction equations when they are plotted against several erosion controlling factors: the UNIVERSAL SOIL LOSS EQUATION is the most famous of them.

Channel erosion is usually evaluated by surveying the modifications of channel form and calculating the volume of material removed. The methods used include measurements to reference pegs and periodical controls of both channel cross section and long profile. Long-term variations can also be estimated by comparing aerial photographs of different periods.

As the material removed from slopes and channels of a drainage basin is the source for fluvial transport, the total amount of sediment load (see SEDIMENT LOAD AND YIELD) at the main river mouth can measure the intensity of denudation affecting the whole basin. Actually, most of the attempts directed towards the determination of erosion rate are based on the assessment of stream load quantity. Such assessment can be approached in different ways. One approach consists in the measurement of all the transported materials at the recording stations, that is to say the direct measurement of dissolved load, SUSPENDED LOAD and BEDLOAD. The indirect approaches, instead, lead to the stream load prediction through theoretical formulae or multiple regressions.

The field determination of dissolved load is obtained by portable instruments that measure certain water quality parameters, such as conductivity and pH. More often the dissolved solid content is determined in laboratories by evaporation of known volumes of water and weighing the residue. Suspended load concentration is obtained by measuring the turbidity of water samples collected by specially developed devices, ranging in complexity from simple dip-bottles to sophisticated apparatus; the total suspended load is then obtained multiplying the suspended load concentration by the discharge. The assessment of bedload is extremely difficult; many measuring apparatuses have been developed, like slot traps, collecting basin, basket samplers, etc., but none has been universally accepted as adequate for the determination of bedload.

Direct measurements of total solid load by rivers have encountered many problems; among them there are the high costs of instruments, the running expenses, and the alteration of pattern of flow and transport by the presence of the sampler, which can distort especially bedload data. One more problem is sampling both in time and space. Observations made at given time intervals could miss the extreme events; furthermore it may require many years of record before data are enough to be significant. The choice of sampling site must consider the accessibility of instruments, the lack of interference and the planning of a dense instrumentation network.

The theoretical estimation of solid load implies the derivation of specific formulae based on the characteristic of channel flow and of the transported materials. This procedure is unsuitable to predict suspended load, as it is essentially a non-capacity load, but it has been tentatively followed to predict bedload; however none of the derived formulae would seem to offer a completely satisfactory prediction.

An indirect method largely used to evaluate fluvial erosion takes into account the data available on suspended load (the most systematically measured at recording stations) and leads to significant regressions that relate suspended load to several parameters which express the principal factors influencing the spatial pattern of sediment production (Table 18). Once obtained these equations are used to predict suspended load of rivers lacking a recording station. Although suspended load values are a partial measure of erosion processes in drainage basins, they have been used also to obtain world maps of denudation rate (Fournier 1960).

Table 18 Some examples of multivariate regressions between suspended load and controlling variables

Author	Region	Equation
Fournier (1960)	Temperate alpine areas	$\log E = 2.65 \log (p^2 P^{-1}) + 0.46 H_m \tan \phi - 1.56$ E = suspended sediment yield (tonnes km^{-2} year^{-1}); P = precipitation in month of maximum precipitation (mm); P = mean annual precipitation (mm); H_m = mean elevation of basin (m); ϕ = mean basin slope (°)
Jansen and Painter (1974)	Humid microthermal climatic areas	$\log \mathbf{S} = -5.073 + 0.514 \log \mathbf{H} + 2.195 \log \mathbf{P} - 3.706 \log \mathbf{V} + 1.449 \log \mathbf{G}$ **S** = suspended sediment yield (tonnes km^{-2} year^{-1}); **H** = altitude (m.a.s.l.); **P** = mean annual precipitation (mm); **V** = measure of vegetation cover; **G** = estimate of proneness to erosion
Jansen and Painter (1974)	Temperate climates	$\log \mathbf{S} = 12.133 - 0.340 \log \mathbf{Q} + 1.590 \log \mathbf{H} + 3.704 \log \mathbf{P} + 0.936 \log \mathbf{T} - 3.495 \log \mathbf{C}$ **S** = suspended sediment yield (tonnes km^{-2} year^{-1}); *Q* = annual discharge (10^3 m^3 km^{-2}); **H** = altitude (m a.s.l.); **P** = mean annual precipitation (mm); **T** = average annual temperature (°C); **C** = natural vegetation index
Ciccacci *et al.* (1986)	Italy	$\log \mathbf{TU} = 2.79687 \log \mathbf{D} + 0.13985\ \Delta a + 1.05954$ $r^2 = 0.96128$ **Tu** = suspended sediment yield (tonnes km^{-2} year^{-1}); **D** = drainage density (km km^{-2}); **Δa** = hierarchical anomaly index

References

Ciccacci, S., Fredi, P., Lupia-Palmieri, E. and Pugliese, F. (1986) Indirect evaluation of erosion entity in drainage basins through geomorphic, climatic and hydrological parameters, *International Geomorphology*, 2, 33–48, Chichester: Wiley.

Fournier, F. (1960) *Climat et érosion: la relation entre l'érosion du sol par l'eau et les précipitations atmosphériques*, Paris: Presses Univ. de France.

Jansen, J.M.L. and Painter R.B. (1974) Predicting sediment yield from climate and topography, *Journal of Hydrology* 21, 371–380.

Further reading

Cooke, R.U. and Doornkamp, J.C. (1990) *Geomorphology in Environmental Management*, Oxford: Clarendon Press.

Gregory, K.J. and Walling D.E. (1973) *Drainage Basin Form and Process*, London: Edward Arnold.

ELVIDIO LUPIA-PALMIERI

FLUVIAL GEOMORPHOLOGY

Fluvial geomorphology is strictly the geomorphology of rivers. As rivers have always held a prominent role in the study of landforms, it is not surprising that debates about fluvialism, as to whether rivers could produce their valleys, continued to rage early in the nineteenth century until uniformitarianism prevailed, whence temperate areas were seen as the result of rain and rivers. At the end of the nineteenth century rivers were seen as central to the Davisian normal cycle of erosion which came to exert a dominant influence upon geomorphology for the first half of the twentieth century (Gregory 2000). It took time to appreciate that there was insufficient understanding of fluvial processes and, with hindsight, it has been suggested that the way in which G.K.Gilbert approached rivers, including significant contributions on the transport of debris by flowing water (Gilbert 1914), could have provided at least an additional, if not an alternative, approach to that developed by Davis.

Until the mid-twentieth century fluvial geomorphology was dominated by attempts to interpret landscapes in terms of phases of river evolution with emphasis placed upon terraces as indicating sequences of valley development, and erosion or planation surfaces employed to reconstruct stages of landscape development. Questioning the basis for such reconstructions, and realizing a need for a greater focus upon fluvial processes, was answered by *Fluvial Processes in Geomorphology* (Leopold *et al.* 1964) which came to have a dramatic influence upon the way in which fluvial geomorphology was subsequently pursued. In addition to increasing interest in hydrological processes (see HYDROLOGICAL GEOMORPHOLOGY) and leading to recognition that the drainage basin could be regarded as the fundamental geomorphic unit (Chorley 1969), it provided the basis for an expansion of research on the contemporary fluvial system. Emphasis was effectively placed upon seven different themes (Gregory 1976) which were: drainage network morphometry; drainage basin characteristics particularly in relation to statistical models of water and sediment yield; links between morphology and process in the hydraulic geometry of river channels; and the controls upon river channel patterns; together with theoretical approaches to the fluvial system; investigations of the significance of dynamic contributing areas in runoff generation; and finally, ways of analysing changes in fluvial systems including the PALAEOHYDROLOGY-river metamorphosis approach. At this stage in the rapid development of fluvial geomorphology, textbooks were produced including emphases on dynamics and morphology (Morisawa 1968), form and process (Richards 1982; Morisawa 1985; Knighton (1984), 1998), rivers and landscape (Petts and Foster 1985), and the fluvial system (Schumm 1977a). There was thus a progressive development of fluvial geomorphology; in his conclusion to *The Fluvial System*, Schumm (1977a) contended that landscape, like science itself, proceeds by episodic development. In the second part of the twentieth century the development of fluvial geomorphology proceeded episodically, and the result of research progress made has been to demonstrate, in turn, exactly how episodic fluvial systems can be.

Comparatively few explicit definitions of fluvial geomorphology have been given in papers and books but five (Table 19) indicate core themes and to some extent intimate how the subject has evolved, expanded and progressed since the 1960s.

In the course of the development of fluvial geomorphology since the middle of the twentieth century, three sequential phases can be visualized: one of import and expansion, one of consolidation, and one of innovation. The first saw import and utilization of understanding and techniques from other disciplines including hydraulics, hydrology, sedimentary geology and engineering. Clarification and refinement of field

Table 19 Some definitions of fluvial geomorphology

Definition	Source
Fluvial geomorphology has as its object of study not only individual channels but also the entire drainage system	Kruska and Lamarra (1973) cited by Schumm (1977b)
A primary objective of fluvial geomorphology must be to contribute to *explanation* of relationships among the physical properties of flow in mobile-bed channels, the mechanics of sediment transport driven by the flow, and the alluvial channel forms created by spatially differentiated sediment transport	Richards (1987)
Geomorphology is the study of Earth surface forms and processes; fluvial phenomena – those related to running water	Graf (1988)
The study of changing river channels is the domain of fluvial geomorphology. ... Fluvial geomorphology is a field science; classification and description are at the heart of this science	Petts (1995)
The science that seeks to investigate the complexity of behaviour of river channels at a range of scales from cross sections to catchments; it also seeks to investigate the range of processes and responses over a very long timescale but usually within the most recent climatic cycle	Newson and Sear (1998), cited by Dollar (2000)

approaches together with modelling methods including stochastic, deterministic (Werritty 1997) and experimental (Schumm *et al.* 1985) approaches enabled the developing foundation to focus upon equilibrium concepts throughout several distinct sub-branches of fluvial geomorphology established as the outcome of this phase. The independence or dependence of variables involved in research investigations was anticipated to vary according to the steady, graded, or cyclic timescale (Schumm and Lichty 1965) of the investigation being considered. Once established with a significant number of practitioners, fluvial geomorphology experienced a second phase, one of internal consolidation, characterized by investigations of changes over time as referred to different timescales, embracing palaeohydrology, and river channel adjustments as instigated by the effects of land use changes impacting upon the channel. These investigations meant that controls upon change of the fluvial system were explored, including thresholds, complex response and sensitivity. The third phase of innovation, aided by new technology of remote sensing and GPS, has seen the development of exciting links between investigations undertaken at several spatial scales, together with equally innovative developments linking studies of process with landscape development. This phase has been one of export whereby results from fluvial geomorphology are contributing to multidisciplinary projects and making a distinctive input to management problems (e.g. Thorne *et al.* 1997).

Against this background of development of the subject, a choice for fluvial geomorphology was suggested to exist by Smith (1993) because he perceived the discipline to be at a crossroads, requiring major changes in ways of thinking and operating, so that he proposed it needed to move forward and to adopt the ways of the more competitive sectors of the Earth and biosciences. However this need may have been overstated (Rhoads 1994) in view of the vitality shown by recent publications in fluvial geomorphology, and by the way in which collaboration and multidisciplinary activity have increased, complemented by attention being devoted to the scientific foundation of the subject. Rhoads (1994: 588) sees the most critical challenge facing fluvial geomorphologists as that of devising effective strategies for integrating a diverse assortment of research, spanning a broad range of spatial and temporal scales.

From the ideas that prompted Smith's challenge and Rhoads's response, it is possible to seek a general definition; the broad range of research approaches that exists; and ways in which collaboration is now possible. Embracing earlier proposals (Table 19) a general definition for fluvial geomorphology could be that it investigates the fluvial system at a range of spatial scales from the basin to specific within channel locations; at timescales ranging from processes during a single flow event to long-term Quaternary change; undertaking studies which involve explanation of the relations among physical flow properties, sediment transport and channel forms; of the changes that occur both within and between rivers; and that it can provide results which contribute in the sustainable solution of river channel management problems.

Although developed at different times and progressed significantly since the seven themes suggested in 1976 (Gregory 1976), there is now a range of research approaches which are the branches of fluvial geomorphology occupying most practitioners at any one time. These can now be envisaged as focused on components of the fluvial systems, process mechanics, temporal change and management applications.

Components of the fluvial system

Studies of components of the fluvial system have been concerned particularly with morphology of elements of that system across the range of spatial scales from in-channel locations to the complete drainage basin. Particularly significant investigations focused on relations between form and process in fluvial systems and upon the controls upon morphology. This has required definitions which can be applied in different basins including those for channel capacity, channel planform, floodplain extent, and drainage density of the channel network; and at each of these levels there have been attempts to establish equilibrium relations between indices of process and measurements of fluvial system form. Some of the earliest developments in fluvial geomorphology were concerned with analysis of *drainage networks* using techniques of drainage basin morphometry and with the relationship between *channel capacity* and the frequency of the bankfull discharge which was thought to exercise a major control upon channel morphology. In these and other components of the fluvial system it has now been appreciated that the links between form and process and the associated explanation is more complex than at first thought. Thus drainage networks could not easily be related to discharge and channel processes, and the relationship between channel capacities and controlling discharge has been the subject of considerable research, particularly the way in which networks generate the Geomorphic unit hydrograph (Rodriguez-Iturbe and Rinaldo 1998). Relations between dimensions of cross-sectional area and width can be used to provide a basis for discharge estimation at ungauged sites (Wharton *et al.* 1989). In addition research has focused upon *river channel patterns*, upon the controls on single thread and multi-thread patterns and what determines the thresholds between them. The floodplain is also controlled by the interaction between recent hydrological and sediment history together with the characteristics of the local area; and the variability of floodplain characteristics has been reflected in the definition of the river corridor as well as of the floodplain itself. Three major floodplain classes, based on stream power and sediment characteristics, have been recognized (Nanson and Croke 1992), further subdivided into a combination of thirteen floodplain orders and suborders, namely:

1 High energy non-cohesive floodplains: disequilibrium landforms which erode either completely or partially as a result of infrequent extreme events.
2 Medium energy non-cohesive floodplains: in dynamic equilibrium with the annual to decadal flow regime of the channel and not usually affected by extreme events. Preferred mechanism of floodplain construction is by lateral point bar accretion or braid channel accretion.
3 Low energy cohesive floodplains: usually associated with laterally stable single-thread or anastomosing channels. Formed primarily by vertical accretion of fine-grained deposits and by infrequent channel avulsion.

As more is known about each of the several spatial scales of investigation of the fluvial system it is appreciated that the question of explanation relies upon the controls that apply to each particular spatial level. Thus it is necessary to see how the flows and sediment transport are significant in relation to each of the spatial levels of the fluvial system, and how they interrelate. Furthermore, when focusing on the integrity of the fluvial system, it is necessary to consider how a hierarchy of interrelated components makes up the river basin channel structure. Any such structure needs to take account of the progress made by biologists and aquatic ecologists in this regard and an original framework (Downs and Gregory 2003: Chapter 3) involves seven nested scales which are drainage *basin*, basin *zones*, valley *segments*, stream *reaches*, *channel unit*, *within channel*, and channel *environment at a point*. In addition there are a number of environmental flows that have been defined (Dollar 2000) including those that maintain a channel morphology and which have been specified for the purposes of practical application.

Process mechanics

Process mechanics began with hydrological analyses whereby fluvial processes were examined from the standpoint of analysis of stream hydrographs and their generation, so that investigations of dynamic contributing areas became a major reason for field experiments based in small experimental catchments. On the basis of the considerable progress made, attention then moved to the sediment budget, and to suspended sediment, bedload and solute loads. Analysis previously dominated by simple rating curves, which assumed a linear relationship between suspended sediment concentration and discharge, was refined once it was shown that because of hysteretic effects and sediment supply problems, the relationships were much more complex so that earlier estimates of rates of denudation had to be revised. Bedload transport had been very difficult to measure so that transport equations often tended to assume a capacity load, whereas along many rivers the transport of material was supply limited. Advances in instrumentation enabled continuous recording devices to be used providing the basis for more complete explanatory analysis of transport rates; and continuous measurements of channel bank erosion could be the basis for more precise relationships between erosion rates and the controlling variables. Such studies facilitated more detailed investigations within the channel and these have been concerned with the entrainment of bed material, the patterns of flow and sediment movement at confluences, and the controls upon in-channel change. A particularly fruitful theme has derived from the interrelationships with ecology because aquatic ecologists have investigated river channels in relation to instream habitat conditions and aquatic plant distributions; combination of such results with geomorphological data has promoted biogeomorphological investigations of river channels. Along rivers bordered by riparian trees, or flowing through forested areas, the investigation of CWD (coarse woody debris) has become important because such CWD exerts an influential control upon the channel processes, the morphology and ecological characteristics. Numerous investigations have been undertaken considering the impact of CWD upon channel morphology, demonstrating the extent and significance of wood often as debris dams, together with the impact on channel processes, the reasons for spatial variations, as well as the stability of dams and their persistence together with the management implications.

Temporal change

Study of temporal changes had been a long tradition in fluvial geomorphology through the interpretation of past stages of development based on river terrace sequences, but it was not until 1954 that the idea of palaeohydrology was proposed (Leopold and Miller 1954). This contrasted with earlier approaches because it was retrodictive in approach, utilizing understanding that had been gained of contemporary processes, and it was exemplified by Quaternary palaeohydrology suggested by Schumm (1965) and augmented by ideas of river metamorphosis (Schumm 1977a). Palaeohydrology evolved (Gregory 1983) to utilize knowledge of contemporary processes applied to the past, whereas river metamorphosis similarly employed contemporary relationships between channel form and process as a basis for interpreting river channel resulting from a range of causes including dam and reservoir construction, land use change including urbanization and, particularly, as a consequence of channelization. Such human-induced channel changes were found to be extensive (e.g. Brookes and Gregory 1988) and were superimposed upon the impact of shifts in sequences of climate which in some parts of the world, such as Australia, led to the alternation of periods of drought-dominated and flood-dominated regimes. Analysis of river channel changes was initially confined to particular reaches affected, often downstream from the influencing factor, but they have subsequently been analysed in the context of the entire basin with attention accorded to the spatial distribution of adjusting channels and emphasis given to the extent to which they are potentially able to recover to their former condition (e.g. Fryirs and Brierley 2000). The considerable progress achieved as a result of studies of channel change includes ways in which thresholds can be identified, reaction and relaxation times (Graf 1977), patterns of palaeohydrological change in different parts of the world (Benito and Gregory 2003) all culminating in further, more informed understanding of the ways in which fluvial processes relate to environmental change (Brown and Quine 1999) and how fluvial systems and sedimentary sequences reflect shifts of climate, both short or longer term during the Quaternary (Maddy *et al.* 2001). A particularly effective way in which studies of the past have been successful has been the analysis of palaeofloods (see HYDROLOGICAL GEOMORPHOLOGY), overcoming not only inaccuracies in estimating the ages of floods and in reconstructing flood discharges, but also allowing incorporation of palaeoflood data into flood frequency analysis, in order to analyse the effects of climatic shifts and non-stationarity. The results of palaeoflood hydrology have been of practical application by bringing very specific benefits in the design or retrofitting of dams or other floodplain structures.

Management applications

Palaeoflood analysis is just one example of ways in which fluvial geomorphology provides applications to management. Many other applications have been developed and initially were very problem and reach specific, including estimation of sediment yield and the possibility of gullying and channel change (Schumm 1977a), progressing through consequences of particular impacts such as channelization (Brookes 1988), leading to improved procedures for management (Brookes and Shields 1996) and then to comprehensive statements of ways in which fluvial geomorphology can be applied to river engineering and management (Thorne *et al.* 1997). Particular emphasis has been placed upon river restoration and how fluvial geomorphology is able to contribute significantly to restoration projects (Brookes 1995). One aspect of restoration to be considered is what is natural (Graf 1996) and therefore what should be the objective for a particular restoration project. It is being appreciated that in all cases where human impact affects fluvial systems, current knowledge of the way in which the system has evolved can illuminate the way in which management is undertaken. It is also the case that some aspects of human activity are now being substantially reversed and the implications of dam removal (Heinz III Center 2002) is one topic of current interest.

It may seem from the foregoing outline of approaches in fluvial geomorpology that they have become increasingly reductionist and diverse, but there are a number of ways in which there has been integration and collaborative activity tending to unify fluvial geomorphology – although remembering that fluvial geomorphology is not as independent of other disciplines as it once was. Thus analysis of sediment slugs showing how waves of sediment are transmitted through a fluvial system has implications for management, and for interpretation of temporal change as well as for the understanding of contemporary channels' forms and processes. In addition, the use of fallout radionuclides including Cs-137 and Pb-210 has enabled precise dating of specific fluvial changes including floodplain sedimentation (e.g. Walling and He 1999). Based upon knowledge of a number of specific cases, the limits of explanation and prediction have been emphasized, including ten ways to be wrong (Schumm 1991). This introduces the idea of risk so that the fluvial system can be seen in terms of the incidence of twenty-eight geomorphic hazards which may occur, associated with drainage networks, hillslopes, main channels, piedmont and plain areas (Schumm 1988). From each of the above themes, clear signs are emerging of more integrated investigations, for example linking process and morphology in bank stability and modes of channel adjustment which involve bank erosion. There is great awareness of, and interaction between, the range of spatial scales investigated. Such linking analysis is now being facilitated by enhanced remote sensing techniques, and GPS which enhances the detail of data capture and the speed of analysis. Progress is now being made towards enhanced conceptual models which seek to model aspects of the fluvial system in ways not previously possible (Coulthard *et al.* 1999). The outstanding challenge is for further understanding to be achieved of the way in which information can be linked from one timescale to another.

It is inevitable that the investigation of rivers should have expanded greatly, as one of the most researched fields of geomorphology; it is now becoming more integrated but not necessarily strictly confined to geomorphology, as links with ecology, engineering and hydraulics and sedimentary geology prove to be very worthwhile, and multidisciplinary approaches are increasingly common. The concerns expressed by Smith (1993) are being met, and fluvial geomorphology is now sufficiently well founded to address major questions including what has been suggested to be the greatest challenge: 'to understand the way in which short timescale and small space scale processes operate to result in long timescale and large space scale behaviour' (Lane and Richards 1997: 258).

References

Benito, G. and Gregory, K.J. (2003) *Palaeohydrology: Understanding Global Change*, Chichester: Wiley.

Brookes, A. (1988) *Channelized Rivers. Perspectives for Environmental Management*, Chichester: Wiley.

——(1995) River channel restoration: theory and practice, in A. Gurnell and G. Petts (eds) *Changing River Channels*, 369–388, Chichester: Wiley.

Brookes, A. and Gregory, K.J. (1988) Channelization, river engineering and geomorphology, in J.M. Hooke (ed.) *Geomorphology in Environmental Planning*, 145–168, Chichester: Wiley.

Brookes, A. and Shields, F.D. (eds) (1996) *River Channel Restoration. Guiding Principles for Sustainable Projects*, Chichester: Wiley.

Brown, A.G. and Quine, T. (eds) (1999) *Fluvial Processes and Environmental Change*, Chichester: Wiley.

Chorley, R.J. (1969) The drainage basin as the fundamental geomorphic unit, in R.J. Chorley (ed.) *Water, Earth and Man*, 77–100, London: Methuen.

Coulthard, T.J., Kirkby, M.J. and Macklin, M.G. (1999) Modelling the impacts of Holocene environmental change in an upland river catchment, using a cellular automaton approach, in A.G. Brown and T. Quine (eds) *Fluvial Processes and Environmental Change*, 31–46, Chichester: Wiley.

Dollar, E.J. (2000) Fluvial Geomorphology, *Progress in Physical Geography* 24, 385–406.

Downs, P.W. and Gregory, K.J. (2003) *River Channel Management*, London: Arnold.

Fryirs, K. and Brierley, G. (2000) A geomorphic approach to the identification of river recovery potential, *Physical Geography* 21, 244–277.

Gilbert, G.K. (1914) The transportation of debris by running water, *US Geological Survey Professional Paper* 86.

Graf, W.L. (1977) The rate law in fluvial geomorphology, *American Journal of Science* 277, 178–191.

——(1988) *Fluvial Processes in Dryland Rivers*, Berlin: Springer-Verlag.

——(1996) Geomorphology and policy for restoration of impounded American rivers: what is 'natural'? in B.L. Rhoads and C.E. Thorn (eds) *The Scientific Nature of Geomorphology*, 443–473, Chichester: Wiley.

Gregory, K.J. (1976) Changing drainage basins, *Geographical Journal* 142, 237–247.

——(1983) (ed.) *Background to Palaeohydrology*, Chichester: Wiley.

——(2000) *The Changing Nature of Physical Geography*, London: Arnold.

Heinz III Center (2002) *Dam Removal. Science and Decision Making*, Washington: The Heinz Center.

Knighton, D. (1998) *Fluvial Forms and Processes*, 2nd edition, London: Arnold.

Kruska, J. and Lamarra, V.A. (1973) Use of drainage patterns and densities to evaluate large scale land areas for resource management, *Journal of Environmental Systems* 3, 85–100.

Lane, S.N. and Richards, K.S. (1997) Linking river channel form and process: time, space and causality revisited, *Earth Surface Processes and Landforms* 22, 249–260.

Leopold, L.B. and Miller, J.P. (1954) Postglacial chronology for alluvial valleys in Wyoming, *United States Geological Survey Water Supply Paper* 1,261, 61–85.

Leopold, L.B., Wolman, M.G. and Miller, J.P. (1964) *Fluvial Processes in Geomorphology*, San Francisco: Freeman.

Maddy, D., Macklin, M.G. and Woodward, J.C. (eds) (2001) *River Basin Sediment Systems: Fluvial Archives of Environmental Change*, Amsterdam: Balkema.

Morisawa, M.E. (1968) *Streams: Their Dynamics and Morphology*, New York: McGraw-Hill.

——(1985) *Rivers: Form and Process*, London: Longman.

Nanson, G.C. and Croke, J.C. (1992) A genetic classification of floodplains, *Geomorphology* 4, 459–486.

Newson, M.G. and Sear, D.A. (1998) The role of geomorphology in monitoring and managing river sediment systems, *Journal of the Chartered Institution of Water and Environmental Management* 12, 18–24.

Petts, G.E. (1995) Changing river channels: the geographical tradition, in A. Gurnell and G. Petts (eds) *Changing River Channels*, 1–23, Chichester: Wiley.

Petts, G.E. and Foster, I.D. (1985) *Rivers and Landscape*, London: Arnold.

Rhoads, B.W. (1994) Fluvial geomorphology, *Progress in Physical Geography* 18, 588–608.

Richards, K.S. (1982) *Rivers: Form and Process in Alluvial Channels*, London: Methuen.

——(1987) Fluvial geomorphology, *Progress in Physical Geography* 11 432–457.

Rodriguez-Iturbe, I. and Rinaldo, A. (1998) *Fractal River Basins*, Cambridge: Cambridge University Press.

Schumm, S.A. (1965) Quaternary Palaeohydrology, in H.E. Wright and D.G. Frey (eds) *The Quaternary of the United States*, 783–794, Princeton: Princeton University Press.

——(1977a) *The Fluvial System*, New York: Wiley.

——(1977b) Applied fluvial geomorphology, in J.R. Hails (ed.) *Applied Geomorphology*, 119–156, Amsterdam: Elsevier.

——(1988) Geomorphic hazards – problems of prediction, *Zeitschrift für Geomorphologie Supplementband* 67, 17–24.

Schumm, S.A. (1991) *To Interpret the Earth: Ten Ways to be Wrong*, Cambridge: Cambridge University Press.
Schumm, S.A. and Lichty, R.W. (1965) Time, space and causality in geomorphology, *American Journal of Science* 263, 110–119.
Schumm, S.A., Mosley, M.P. and Weaver, W.E. (1985) *Experimental Fluvial Geomorphology*, Chichester: Wiley.
Smith, D.G. (1993) Fluvial geomorphology: where do we go from here? *Geomorphology* 7, 251–262.
Thorne, C.R., Hey, R.D. and Newson, M.D. (eds) (1997) *Applied Fluvial Geomorphology for River Engineering and Management*, Chichester: Wiley.
Walling, D.E. and He, Q. (1999) Changing rates of overbank sedimentation on the floodplains of British rivers during the past 100 years, in A.G. Brown and T. Quine (eds) *Fluvial Processes and Environmental Change*, 207–222, Chichester: Wiley.
Werritty, A. (1997) Chance and necessity in geomorphology, in D.R. Stoddart (ed.) *Process and Form in Geomorphology*, 312–327, London: Routledge.
Wharton, G., Arnell, N.W., Gregory, K.J. and Gurnell, A.M. (1989) River discharge estimated from channel dimensions, *Journal of Hydrology* 106, 365–376.

KENNETH J. GREGORY

FOLD

Structures that originally were planar, like a sedimentary bed, but which have been bent by horizontal or vertical forces in the Earth's crust. Folds can occur at scales that range from mountain ranges to small crumples only a few centimetres long. They can be gentle or severe, depending on the nature and magnitude of the applied forces and the ability of the beds to be deformed. When beds are upfolded or arched (Figure 64a) they are called *anticlines*, whereas downfolds or troughs are termed *synclines*. A *monocline* is a step-like bend in otherwise horizontal or gently dipping beds. Folds can be asymmetrical in shape and, when the deformation is particularly intense, they can be overturned. They also tend to occur in groups rather than in isolation and some mountains consist of a folded belt. Folds can result from various processes: compression in the crust, uplift of a block beneath a cover of sedimentary rock so that the cover becomes draped over the rising block, and from gravitational sliding and folding where layered rocks slide down the flanks of a rising block and crumple. The process of folding may involve either small-scale shearing along many small fractures or flowage by plastic deformation of the rock.

Large folds can have a substantial influence upon landform development. Folding can also occur rapidly, creating vertical increases in elevation of up to 10 m 1,000 y^{-1}. However, when folds cease to grow, the influence of erosion becomes increasingly more important than the original shape of the fold. Thus, in the case of an anticline, initially it will form an area of upstanding, often donal relief. However, as it is eroded, the uppermost strata may be cut through. If the older rock that forms the core is of low resistance the result is a *breached anticline* in which a series of inward-facing escarpments rise above a central lowland. A classic example of this is the Weald of southern England (Jones 1999) (Figure 64b). Conversely, as in the Paris and London Basins, the rim of a syncline will possess outward facing scarps. Folding has major impacts on river systems. If a fold develops across a stream course the river, if it can cut down quickly enough, can maintain its course transverse to the developing structure. Such a channel is said to be *antecedent*.

Antecedence is, however, only one explanation for drainage that cuts across anticlinal structures. An alternative model (Alvarez 1999) is that fold ridges emerge from the sea in sequence, with the erosional debris from each ridge piling up against the next incipient ridge to emerge, gradually extending the coastal plain seaward. The new coastal plain, adjacent to each incipient anticline, provides a level surface on which a newly elongated river could cross the fold, positioning it to cut a gorge as the fold grew. This mechanism is in effect a combination of antecedence and superimposition. The model has been applied both to the Appalachians of the USA and the Apennines of Italy.

On actively developing anticlinal folds, drainage density varies according to the gradient of the evolving slopes. However, the form of the relationship between gradient and drainage density is process-dependent. Talling and Sowter (1999) suggest that a positive correlation occurs when erosion results from overland flow, while a negative correlation occurs when erosion is dominated by shallow mass-wasting. A traditional description of drainage in folded areas such as the Jura is provided by Tricart (1974).

Where sedimentary rocks are tilted by folding there may be a succession of lithologies exposed that have differing degrees of resistance. River channel incision will tend to be more effective on

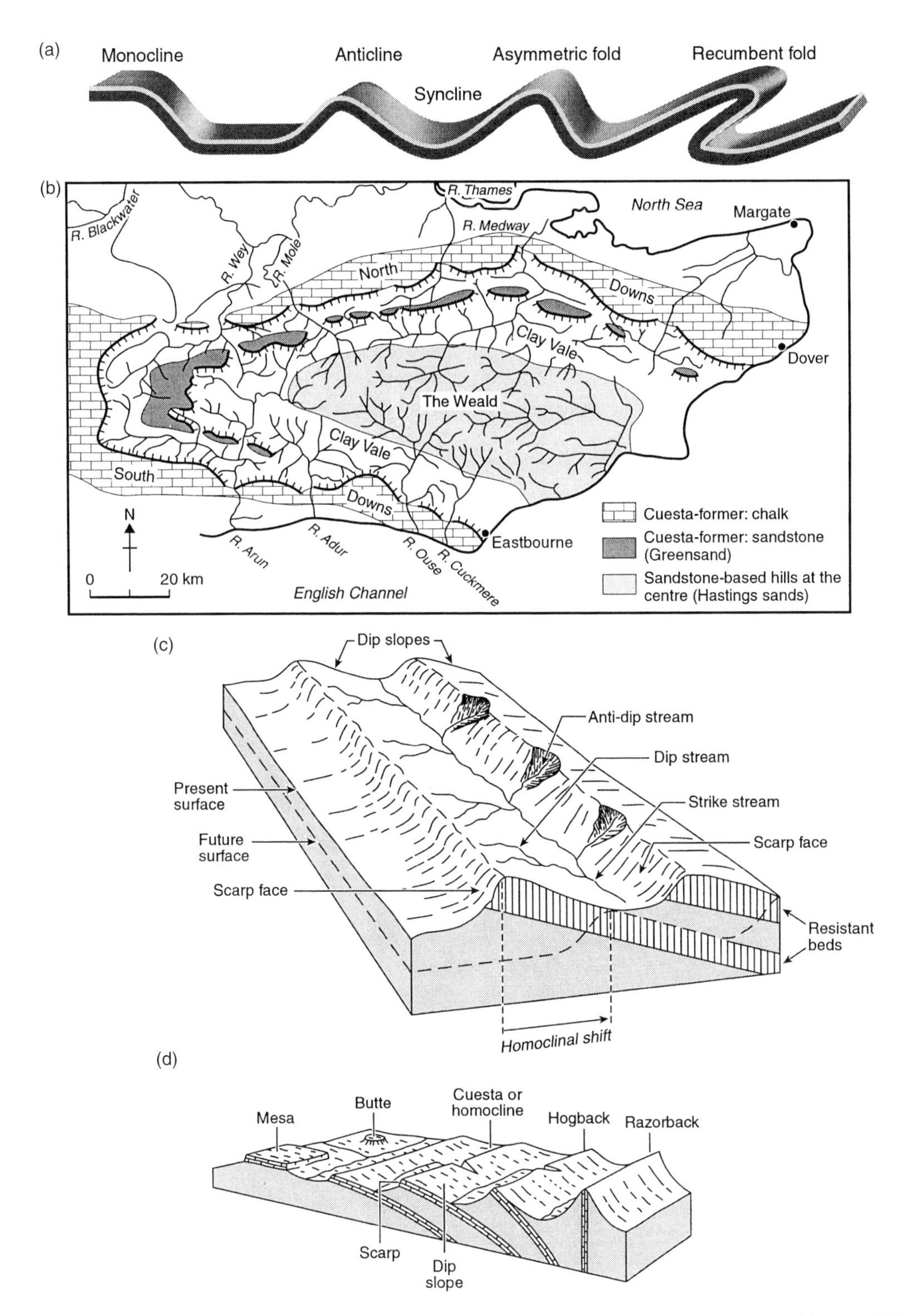

Figure 64 Folds and their relationship to relief: (a) some major types of fold; (b) the Wealden Anticline of south-east England; (c) drainage and slope forms associated with dipping strata; (d) drainage and slope forms associated with strata of progressively steepening dip

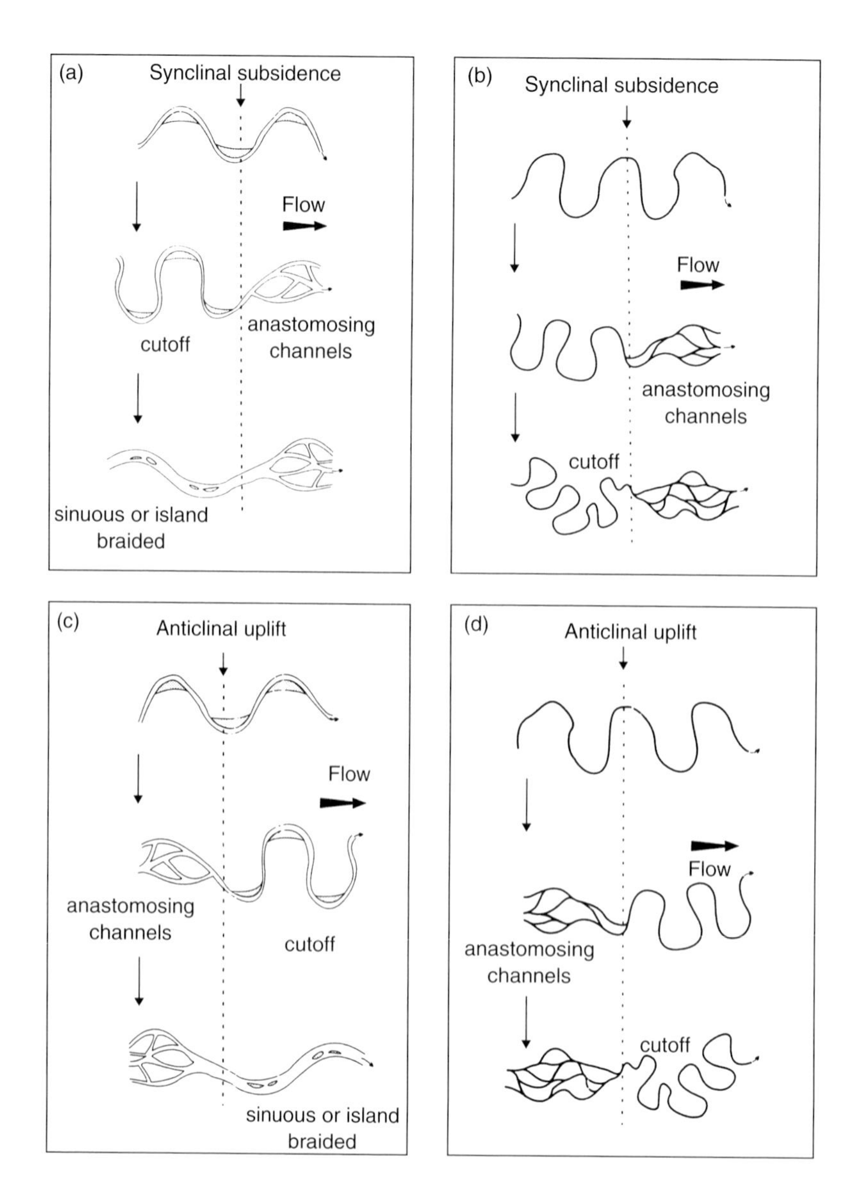

Figure 65 River channel patterns associated with synclinal and anticlinal activity: (a) and (c) = mixed-load meandering channels; (b) and (d) = suspended-load meandering channels (modified from Ouchi in Schumn *et al.* 2000, figures 3.10, 3.11, 3.12 and 3.13)

less resistant beds leading to the development of a *strike valley* (Figure 64c), flanked on the up-dip side by a dip slope and on the down-dip side by an escarpment. The roughly parallel strike streams will be joined at high angles by short *dip streams* and *anti-dip streams*. As downward incision occurs, the rivers will migrate laterally by a process known as *homoclinal shifting*.

The angle of dip also influences topographic form (Figure 64d). Resistant beds in very gently dipping or horizontal beds, form flat-topped plateaux (MESAS). Modestly dipping beds create a CUESTA, whereas steeply dipping strata produce a ridge known as a HOGBACK.

In recent years tectonic geomorphologists (see, for example, Burbank and Anderson 2001) have taken a great interest in how the style and rate of folding affects landform evolution. In particular, Schumm *et al.* (2000) have described how terrace formation, channel form, the locations of degradation and aggradation, valley long profiles, and the spatial distribution of flooding, may be related to folding activity. Figure 65 shows the response of stream channel form to anticlinal uplift and synclinal subsidence for mixed-load and suspended-load streams. Streams flowing across zones of uplift (live anticlines) may show deformed terraces and convex sections of long profile.

References

Alvarez, W. (1999) Drainage on evolving fold-thrust belts: a study of transverse canyons in the Apennines, *Basin Research* 11, 267–284.
Burbank, D.W. and Anderson, R.S. (2001) *Tectonic Geomorphology*, Oxford: Blackwell Science.
Jones, D.K.C. (1999) On the uplift and denudation of the Weald, *Geological Society of London Special Publication* 162, 25–43.
Schumm, S.A., Dumon, J.F. and Holbrook, J.M. (2000) *Active Tectonics and Alluvial Rivers*, Cambridge: Cambridge University Press.
Talling, P.J. and Sowter, M.J. (1999) Drainage density on progressively tilted surfaces with different gradients, Wheeler Ridge, California, *Earth Surface Processes and Landforms* 24, 809–824.
Tricart, J. (1974) *Structural Geomorphology*, London: Longman.

A.S. GOUDIE

FORCE AND RESISTANCE CONCEPT

Materials at the Earth's surface are subjected to a whole range of applied forces (inputs of energy) both physical and chemical. Whereas in Newtonian mechanics applied force (or stress) and reaction (or strain) are thought of as equal and opposite and as essentially simultaneous, the outcome of an application of unit force to Earth materials often cannot be readily predicted. The asymmetry between energy input and response which is almost universally experienced in geomorphology is due partly to the multifaceted nature of resistance and partly to the significance of the specific sequence of energy inputs.

B.W. Sparks was especially concerned to stress that the resistance of rocks to geomorphic processes was always contingent upon the precise manner in which energy was applied. An obvious example is the very different resistance of limestones under chemical or mechanical attack (cf. Sparks 1971). This difficulty is well exemplified by M.J. Selby's elaborate quantification of the ROCK MASS STRENGTH of hillslopes, which nevertheless can only be invoked when the applied force environment is closely defined (Selby 1993: 104). Similarly, the difficulty in writing rational physical equations to describe fluvial flow and sediment transport may be directly traced to the non-uniform expression of channel boundary resistances.

Coupled with this first asymmetry is the widespread observation that energy inputs of equal magnitude do not result in equal amounts of geomorphic work. The non-linearity (see NON-LINEAR DYNAMICS) of process and response is of profound significance in all historical sciences. The most common manifestation of this asymmetry in geomorphology is in hysteresis loops of discharge and sediment plots. The applied force of a given discharge will generally neither entrain nor carry the same volume or calibre of sediment on the rising and falling limbs of a flood. The asymmetry is produced by the variable quantity and quality of sediment available to be moved: that is, on the temporally specific state of the channel boundary. The study of slope failures also (cf. Schumm and Chorley 1964) provides examples of small inputs of energy which propel a system across a resistance threshold (see THRESHOLD, GEOMORPHIC).

References

Schumm, S.A. and Chorley, R.J. (1964) The fall of Threatening Rock, *American Journal of Science* 262, 1,041–1,054.
Selby, M.J. (1993) *Hillslope Materials and Processes*, 2nd edition, Oxford: Oxford University Press.
Sparks, B.W. (1971) *Rocks and Relief*, London: Longman.

SEE ALSO: Goldich weathering series; magnitude–frequency concept; non-linear dynamics; rock mass strength; threshold, geomorphic

BARBARA A. KENNEDY

FOREST GEOMORPHOLOGY

Forest geomorphology is a specialized area of research focused both on the interaction between forest ecosystem dynamics and landform processes, and the effects of forest management activities on the rates and thresholds of geomorphic process events. As a specialty within the field of geomorphology, the study of forest associated process dynamics has broad application to the study of ecosystem dynamics, endangered species and conservation studies, and paleo-landscapes.

Several features distinguish geomorphic processes in forest environments from those in other vegetation types. Forest ecosystems play a prominent role in many aspects of sediment production, transport and storage. Entrainment of soil by windthrow of trees, the binding effect on soils and regolith by tree root strength, and soil displacement by burrowing animals, are among many examples of how sediment movement is influenced by forests. Standing trees and fallen logs on slopes in forested terrain act as temporary sediment storage 'traps', while forest management activities, such as road construction and the removal of trees, transport and disturb sediment production in other ways. Forest composition plays a major role in the interception, evapotranspiration, infiltration, and runoff of precipitated water, and, as in the sediment examples cited, profoundly influences the type, frequency and mechanisms controlling mass wasting and slope stability. Ultimately, such influences of forest vegetation profoundly affect the development and operation of drainage basins, adding a significant organic component to sediment load, hydrologically influencing the discharge response of the channel network, even influencing channel development through bank resistance or flow deflection. Added to these considerations are natural cycles within the forest ecosystem such as wildfire, floods, disease or insect impacts, and human disturbance factors, which can affect the linkage of geomorphic processes, as well as their magnitude and frequency.

The diversity of objectives, approaches and professional disciplines of those who conduct forest-related process studies has resulted in poor communication among scientists and land managers who share many common interests. In recognition of the need to promote interdisciplinary dialogue, and approach complex environmental research within a spatial and temporal scale appropriate to the changes in forest succession affecting process-response, the International Council of Scientific Unions created the International Geosphere–Biosphere Program (IGBP) which recognizes several prominent forest biomes. Geomorphologists are placed prominently among the scientific teams that continue to illuminate the linkages between landforms and ecosystems.

History and development of forest geomorphology

While late nineteenth and early twentieth-century landscape theory focused on denudation and landform development, early geomorphologists were aware of variation in landscape appearance under varying climate and vegetation conditions. Cotton (1942) explains variations in landscape form as climatic 'accidents', including the resultant vegetation influences, as complications of the 'Geographical Cycle' theorized by William Morris Davis (1899). In a similar vein, Birot (1968) further expands upon Davis, and adds vegetation and soil factors to illustrate the influences of climatic variation. Peltier (1950) also refers to Davis's cycle, but places emphasis on the variations in geomorphic process activity under a variety of forest biomes (selva, savanna, boreal). Peltier credits Professor Kirk Bryan with many of these concepts of 'vegetation modified process'.

Hack and Goodlett (1960) illustrated the relationships between process and forest structure to identify relict features within a forested landscape, and heralded the concept of coexisting ecological and geomorphic equilibriums contributing to landform genesis in humid temperate forests, while Douglas (1968) described the effects in humid tropical forests, and added human disturbance factors. Chorley used examples such as these to illustrate the benefits of multidisciplinary approaches in geomorphology using a 'systems' approach to integrate information from widely divergent sources at different temporal and spatial scales. Chorley's 1962 work encouraged scientists to make contributions across disciplinary lines, and US Geological Survey ecologist Sigafoos (Sigafoos and Hendrick, 1961; Sigafoos 1964; Hupp and Sigafoos 1982) dated trees to determine the temporal and spatial activity of glaciers, floods and blockfields (Alestalo 1971). Successive works were absorbed by the resource management community, which funded research concentrated on the effects of timber harvest practices, road and bridge construction and land use change.

Hydrology, sediment budgets and channel stability in forested watersheds

The application of concepts and techniques from geomorphology to ecosystem studies and terrain analysis represents a great opportunity for the discipline, given the need for an interdisciplinary approach to complex environmental problems.

Large-scale forest management, such as practised by government agencies in the United States, has resulted in numerous controversies between the economic, recreation and conservation interests. In the northwestern Unites States, the practice of large-scale total tree removal, called 'clearcutting', is controversial because of ecological, hydrological and erosion concerns. The US Department of Agriculture, accustomed to years of soil erosion studies at its experiment stations, created a network of experimental forests in the 1960s. Paired watershed studies were conducted to evaluate resulting sediment and water yields resulting from various management treatments. Fredriksen (1970) described the effects of traditional 'clearcutting' using roads, tree removal using a unique cable system to completely suspend the trees as they were cut, and a control basin that remained fully forested. This brief research report sparked considerable interest both in the USA and elsewhere, and a number of research investigations have followed the recovery progress of these small watersheds in the H.J. Andrews experimental forest in western Oregon over the past fifty years (Swanson and Jones 2001).

The natural and management effects of large woody debris on channel morphology and sediment transport in forested streams has been of considerable research interest. Natural log debris in stream channels often produces 'log steps' or 'organic knickpoints' that produce a 'step-pool' profile in mountainous forest streams. These pools act as sediment and nutrient traps, provide fish and invertebrate habitats, and may persist for a long time (Swanson and Lienkaemper 1978). The study of channel stability in forested streams has taken on special significance due to the effects that disturbance within such channels has upon the spawning cycle of anadromous fish. Several species of salmon, steelhead and char have been listed as 'endangered' because of loss of spawning gravel habitat, or loss of access to headwaters. The origin of gravel spawning beds, and their preservation, has occupied substantial geomorphic research including the delivery of gravel from regolith by mass wasting, winnowing of fines by floods, the effects of woody debris on in-channel storage, and the effects of surface 'armour layer' development on entrainment and transport. Stream ecologists, working with fluvial geomorphologists, can provide insights into physical habitat processes and sensitivity to disturbance. Several outstanding classification and inventory schemes have been produced regionally to predict habitat sensitivity and stability (Brussock *et al.* 1985).

Forested slopes have been shown to exhibit lower runoff, increased interflow, and greater stability arising from the root strength of the trees. In general, soils are subsequently deeper than on unforested slopes under similar conditions, resulting in conditions that 'trigger' mass wasting events of both natural and man-induced origins (O'Loughlin 1974). Moss and Rosenfeld (1978) have shown that mass wasting events have the potential to alter the composition and character of the forest community structure in predictable ways, thus leading to a model of interrelationships between landform features and vegetation community characteristics.

On a larger scale, Caine and his co-workers (Swanson *et al.* 1988) demonstrated that ecosystem behaviour can be predicted by a better understanding of how landforms affect those processes. They illustrate that ecosystem–terrain interactions often take multiple forms, with patterns imposed by one set of interactions often coexisting in time and space with other sets. The linkage between ecosystem development and landform stability incorporates both geomorphic and biological events of varying magnitude and frequencies, such as wildfire, floods and landslides. Rosenfeld (1998) illustrates that threshold events, such as exceptional storms, can have predictable 'triggering' effects based on morphology, forest composition and management history. Thus, anticipating the effects of global change on forest biomes are realistic objectives.

Recognition of the complex interdisciplinary nature of landform–forest interactions, and the significance of these linkages in the assessment of human impacts and global change, has been included in the principal themes of the Earth Systems Science Committee, established by the National (US) Aeronautics and Space Administration in 1986. These themes have been incorporated in the international Earth Observation System, and in global research designs for the International Space Station. Geomorphologists will continue to be integral members of ground-based research teams quantifying the linkages between terrestrial processes and forest ecosystems. The US National Academy of Sciences has established 'Long Term Ecological Reserves', with a minimum research planning term of two hundred years. Other nations have expressed similar plans, and a global network of sites, focused on major forest biomes, is a major scientific objective.

As forest geomorphology becomes established as a significant sub-field within the discipline, the need for interdisciplinary education has become apparent. Several sessions dedicated to forest geomorphology have been held by the International Association of Geomorphologists (IAG), and at least one formal graduate programme has been established.

References

Alestalo, J. (1971) Dendrochronological interpretation of geomorphic processes, *Fennia* 105, 1–140.

Birot, P. (1968) *The Cycle of Erosion in Different Climates* (English trans.), London: B.T. Batsford.

Brussock, P.P., Brown, A., and Dixon, J.C. (1985) Channel form and ecosystem models, *Water Resources Bulletin* 21, 859–866.

Chorley, R.J. (1962) Geomorphology and general systems theory, *US Geological Survey Professional Paper* 500B, 1–10.

Cotton, C.A. (1942) *Climatic Accidents in Landscape Making*, 2nd edition, New York: Wiley.

Davis, W. M. (1899) The Geographical Cycle, *Geographical Journal* 14, 481–504.

Douglas, I. (1968) Natural and man made erosion in the humid tropics of Australia, Malaysia, and Singapore, Publication of Staff Members, Centre for SE Asian Studies, University of Hull, 2nd Series, No. 2, 17–29.

Fredricksen, R.L. (1970) Erosion and sedimentation following road construction and timber harvest on unstable soils in three small western Oregon watersheds, *USDA Forest Service Research Paper PNW-104*.

Hack, J. and Goodlett, J.C. (1960) Geomorphology and forest ecology of a mountain region in the Central Applachians, *US Geological Survey Professional Paper* 347.

Hupp, C.R. and Sigafoos, R.S. (1982) Plant growth and block-field movement in Virginia, *US Forest Service, General Technical Report* PNW-141, 78–85.

Moss, M.R. and Rosenfeld, C.L. (1978) Morphology, mass wasting and forest ecology of a post-glacial reentrant valley in the Niagara escarpment, *Geografiska Annaler* 60A, 161–174.

O'Loughlin, C.L. (1974) A study of tree root strength deterioration following clearfelling, *Canadian Journal of Forest Research* 4, 107–114.

Peltier, L.C. (1950) The geographical cycle in periglacial regions as it is related to climatic geography, *Annals of the Association of American Geographers* 40, 219–236.

Rosenfeld, C.L. (1998) Storm induced mass-wasting in the Oregon Coast Range, USA, in J. Kalvoda, and C. Rosenfeld (eds) *Geomorphological Hazards in High Mountain Areas*, 167–176, Dordrecht: Kluwer.

Sigafoos, R.S. (1964) Botanical evidence of floods and flood-plain deposition, *US Geological Survey Professional Paper* 485-A.

Sigafoos, R.S. and Hendrick, E.L. (1961) Botanical evidence of the modern history of Nisqually Glacier, Washington, *US Geological Survey Professional Paper* 387-A.

Swanson, F.J. and Lienkaemper, G.W. (1978) Physical consequences of large organic debris in Pacific Northwest streams, *USDA Forest Service General Technical Report* PNW-69.

Swanson, F.J. and Jones, J.A. (2001) Geomorphology and Hydrology of the H.J. Andrews Experimental Forest, Blue River, Oregon. PNW Forest Range Experimental Station, Portland. Authors describe numerous studies focused on sediment yield and routing, water yield, nutrient flux and re-vegetation studies conducted in these experimental watersheds from 1953 to present. This information is updated at the following internet address: http://www.fsl.orst.edu/lter

Swanson, F.J., Krantz, T.K., Caine, N. and Woodhouse, R.G. (1988) Landform effects on ecosystem patterns and processes, *Bioscience* 38, 92–98.

Further reading

Deithich, W.E., Dunne, T., Humphrey, N.F. and Reid, L.M. (1982) Construction of sediment budgets for drainage basins, *USDA Forest Service General Technical Report* PNW-141, 5–24.

Froehlich, H.A. (1973) Natural and man-caused slash in headwater streams, *Logger's Handbook*, vol. 33, Portland: Pacific Logging Congress.

Harden, D., Ugolini, F. and Janda, R. (1982) Weathering and soil profile development as tools in sediment budget and routing studies, *USDA Forest Service General Technical Report* PNW-141, 150–154.

Kelsey, H.M. (1982) Hillslope evolution and sediment movement in a forested headwater basin, *USDA Forest Service General Technical Report* PNW-141, 86–96.

Marston, R.A. (1982) The geomorphic significance of log steps in forest streams, *Annals of the Association of American Geographers* 72, 99–108.

SEE ALSO: applied geomorphology; bar, river; channel, alluvial; climatic geomorphology; landslide; mass movement; sediment routing; step-pool system; threshold, geomorphic

CHARLES L. ROSENFELD

FORMATIVE EVENT

An important idea in geomorphological science is that morphological stability or stasis can be interrupted by brief, instantaneous, episodes of erosion or deposition when significant morphological change takes place (Erhart 1955; Butler 1959; Ager 1976; Gould 1982; Reading 1982; Dott 1983).

A second important idea is that all geomorphological processes are made up of discrete events of varying frequency, magnitude, duration and sequencing characteristics. If we are to understand landform change it can only be with respect to the characteristics of the events which cause change (Brunsden 1996).

An event is a period of activity of a process, at any place. Events may be classified according to their role in landform evolution. An effective event is one that exceeds the resistance or tolerance of a system and does work. Following Wolman and Gerson (1978) this is measured by the ratio of the event to the mean annual condition of erosion, denudation rate or deposition. Small but frequent events cause morphological change in a cumulative way. All that is required is time. A crucial component is the sequence in which events of different potential effectiveness occur. A very effective event may have considerable feedback effect on succeeding events. If all the available work has been done, later events may perform below their energy potential. If the effective event unlocks potential energy (e.g. by creating steep slopes) it may build in to the system further progressive and diffusive change. If the event prepares a threshold (see THRESHOLD, GEOMORPHIC) condition and is followed by another effective event there may follow unusual or rare forms of change.

It is therefore helpful to use the term formative event. A formative event is an event, of a certain frequency and magnitude, which controls the form of the land. If it does more work than the cumulative everyday event the landform it produces will persist (perhaps for long periods) despite the modifying effects of the more frequent events. It may require another formative event to obliterate the landforms produced or such an event may reinforce the effect. Multiple glaciation of a valley is an example, the 'U' form, once produced by a glacial 'event', may remain for millions of years, surviving all changes in the environmental controls. The word 'persistence' describes the length of time a landform survives as a diagnostic element of the landform assemblage.

References

Ager, D.V. (1976) *The Nature of the Stratigraphic Record*, London: Macmillan.

Brunsden, D. (1996) Geomorphological events and landform change. The centenary lecture to the Department of Geography, University of Heidelberg, *Zeitshrift für Geomorphologie NF* 40, 273–288.

Butler, B.E. (1959) Periodic phenomena in landscapes as a basis for soil studies, Melbourne, CSIRO, *Soil Publication* 14.

Dott, R.H. Jr. (1983) Episodic sedimentation: how normal is average? How rare is rare? Does it matter? *Journal of Sedimentary Petrology* 53 (1), 5–23.

Erhart, H. (1955) 'Biostasie' et 'rhexistasie' ésquisse d'une théorie sur le rôle de la pédogenèse en tant que phénomène géologique, *Comptes Rendus Academie de Sciences* 241, 1,218–1,220.

Gould, S.J. (1982) Darwinism and the expansion of evolutionary theory, *Science* 216, 385–387.

Reading, H.G. (1982) Sedimentary basins and global tectonics, *Proceedings Geologists' Association* 93, 321–350.

Wolman, M.G. and Gerson, R. (1978) Relative scales of time and effectiveness in watershed geomorphology, *Earth Surface Processes and Landforms* 3, 189–208.

DENYS BRUNSDEN

FRACTAL

Sciences such as geomorphology are concerned with inherently variable phenomena – things that are not exactly predictable or repeatable because of the sensitive dependence on initial conditions that many geomorphological systems exhibit (i.e. CHAOS THEORY and the 'butterfly effect'), and because of the contingencies among the elements that make up the system (see SELF-ORGANIZED CRITICALITY). While detailed long-term predictions of such systems are impossible, we should be able to provide some statistical bounds within which the future (or past) should lie. Moreover, while actual events may be unpredictable, they are obviously not unexplainable. Fractals are fundamental components of the methods that are required when analysing or modelling such systems, methods that are amenable to complex, non-linear dynamical systems (see COMPLEXITY IN GEOMORPHOLOGY).

Consider that our primary evidence of the past lies in the patterns we observe today – be it on hillslopes or in river valleys. If the means by which we attempt to characterize those patterns are not able to capture the true complexities of the systems, then how are we to turn back the hands of time and develop an understanding of the Earth's past (Werner 1995)? Fractal patterns, chaos and self-organization can provide the null hypotheses against which process-based interpretations can be tested.

The field of fractals emerged primarily from the writings of one person – Benoit Mandelbrot (1967, 1982) – to become a mainstream field of research. Along with chaos theory and self-organization, it has dramatically altered our view of nature, and of geomorphology (Turcotte 1992).

Fractals are the unique patterns left behind by the unpredictable movements of the world at work. The branching patterns of rivers and trees, the coastline of Britain, the pebbles in a stream, the spatial distribution of earthquake epicentres – all these can exhibit fractal patterns. Fractal objects show similar details on many different scales. Imagine, for example, the rough bark of a tree viewed through successively more powerful magnifications. Each magnification reveals more details of the bark's rugosity. In many geomorphological phenomena, such as river networks and coastlines, this fractal self-similarity has long been observed (e.g. Burrough 1981). This means that, as we peer deeper into a fractal image, the shapes seen at one scale are similar to the shapes seen in the detail at another scale. Fractals are formally defined as objects that are self-similar (Baas 2002).

The measure which most people use to quantify fractal scaling and self-similarity is the fractal dimension or D. The fractal dimension is a number that reflects the way in which the phenomenon fills the surrounding space. The fractal dimension of an object is a measure of its degree of irregularity considered at (theoretically) all scales, and it can be a fractional amount greater than the classical geometrical dimension of the object. The fractal dimension is related to how fast the estimated measurement of the object increases as the measurement device becomes smaller. A higher fractal dimension means the object is more irregular, and the estimated measurement increases more rapidly. For objects of classical geometry, such as lines or curves, the geometric or topological dimension and the fractal dimension are the same. The quantification of fractal patterns led to the discovery that many phenomena, when plotted using appropriate transformations, can be described using a power law (1/*f* systems).

An important concept tied to the fractal dimension is that of spatial autocorrelation. If nearby conditions on a surface are very similar to each other, then we call that positive spatial autocorrelation. If nearby conditions on the surface are the opposite, then we call that negative spatial autocorrelation. Spatial autocorrelation is zero when there is no apparent relation between nearby conditions. A low fractal dimension for a surface (e.g. 2.1) indicates that the self-similarity exhibits high positive spatial autocorrelation. A high fractal dimension (e.g. 2.9) indicates that the self-similarity exhibits high negative spatial autocorrelation. A fractal dimension in the middle of the range (2.5) indicates that no spatial autocorrelation exists. Brownian motion is a classic example of a fractal at the middle range – it is a process with zero memory of where it came from and no knowledge of where it will go next.

Although, theoretically, labelling something a fractal implies that it exhibits self-similarity across all scales, in fact most natural objects possess a limited form of self-similarity – between certain limits or resolutions, the object behaves in a fractal-like manner. These are often called fractal elements, and it is possible that an object may possess multiple fractal elements. Many scientists now consider the boundaries at which fractal behaviour is observed to be important, for those boundaries clearly distinguish process limits. However, does knowledge about the limits to the form of a phenomena necessarily allow us to make statements about the limits of the process which is responsible for creating that form? The answer to that question remains unanswered, although it is at the heart of most fractal research.

One of the main reasons for the increased interest in the fractal dimension (D) is the awareness that dissipative dynamical systems and fractal spaces (and time) are linked – that we now have a theoretical basis with which to link form (e.g. D) and process (e.g. self-organized criticality). The lack of such a link has long been one of the criticisms levelled at fractal studies (e.g. Mark and Aronson 1984), so the discovery that a link can be made is an important step forward in fractal research. However, while self-organized critical models developed in a computer have been very successful at mimicking many varied systems, the unequivocal existence of self-organized criticality in real systems has yet to be confirmed. Geomorphic concepts, such as negative feedback, static equilibrium, and the concept of the graded stream, are all similar to the concept of self-organized criticality. These existing concepts provide an explanation for many geomorphic phenomena without the need to invoke a mechanism such as self-organized criticality. Many geomorphometric measures also are not statistically related to the fractal dimension – that alone indicates that fractals are not capturing all the aspects of a landscape that geomorphologists consider important (Klinkenberg 1992).

Earthquakes and avalanches are two of the more visible manifestations of self-organized critical systems. Their statistical properties, such as size distributions, generally obey power-scaling laws – they follow a fractal distribution (Bak 1996). If a form is found to be a fractal form then certain statistical properties follow. A fractal form has no one scale dominant – it is scale invariant – and its second moment is theoretically infinite. Conversely, a form which is not scale-invariant, a 'Gaussian' form, can be completely described by a few statistical moments. A fractal form will be characterized by rare intermittent events that, from a process point of view, dominate the statistical record. Thus, one of the challenges that fractal studies are attempting to meet is the characterization of such statistically intractable events or forms (e.g. Xu *et al.* 1993). Furthermore, such statistical properties mean that obtaining enough data with which to compare the predictions of models against reality is a not an easy process (Baas 2002).

Fractal concepts have been applied extensively in fluvial geomorphology (e.g. Rodriguez-Iturbe and Rinaldo 1997); there are several different aspects which can be studied. The most obvious is: what is the true length of a river? One could also, while considering the entire basin, examine the form of the river network within the basin. At a higher resolution, the actual planform of the river can be considered (i.e. quantifying sinuosity). At these scales one must not only consider the river itself, but also the river valley form and its effects on the geometry of the river. Going even further down the scaling hierarchy, studies of the fractal characteristics of river bedforms can also be made.

It has been found that many allometric relations observed in nature are not dimensionally consistent (Church and Mark 1980). For example, dimensional analysis would conclude that the length of the mainstream channel of a river should be proportional to the square root of the area of the basin. Most studies have observed that, in fact, the mainstream channel length is proportional to the 0.6th root of the area. Mandelbrot interpreted that as a fractal finding: if a river meanders such that it has a fractal dimension of 1.2, then the length–area relation (known in the literature as Hack's relation) should be to the 0.6th power (1.2 divided by 2).

Power laws, which are the signature of fractals, have been experimentally observed over a wide range of scales in probability distributions describing river basin morphology. Some of the observed fractal distributions have been:

- Horton's power laws of bifurcation and length.
- Stream lengths follow a power-law distribution.
- The cumulative total drainage area contributing to any link follows a power-law distribution.
- The mean of the local slope of the links of a drainage network scales in a fractal manner as a function of the cumulative area.

The fact that 'fractal' rivers exist in so many regions implies that fractal growth processes occur in every environment.

If we accept that river networks and topography can sometimes be characterized as fractals, then we must question why that occurs and what processes are responsible. The simplistic explanation is that scale-invariant form is the result of scale-invariant processes (e.g. Burrough 1981). Does this necessarily mean that a scale-invariant process operates over all scales, as the scale-invariant spatial form appears similar over all scales? We know that this can't be the case – consider the processes such as chemical weathering, frost action and soil creep that operate only at the microscale level. Obviously, the assumption of a one-to-one correspondence between the scale of the form and the scale of the process cannot hold. Self-organization provides the means of getting around this assumption. Large- and small-scale spatial structures emerge through the operation of small-scale processes. Simple rules at one level can lead to complex behaviour at a higher level, behaviour which is referred to as emergent behaviour. We do not have to program in the complex behaviour; it just appears as a consequence of the actions of the agents at the smaller scale.

Fractals provide an out from the constraints of Euclidean geometry, and capture the patterns of nature in an intuitive way. Experiments have shown how our perceptions of roughness agree very well with the measured fractal dimension of the object. Fractal geometry has shifted research agendas: while strict quantitative measurement, measurement that values quantifiable features like distance and degrees of angles, is still important, it is now recognized that measures also need to embrace the qualities of things – their texture complexity and holistic patterning. Chaos, self-organization and fractals have allowed us to step away from simple linear deterministic models and step towards models which capture the essence of predictably unpredictable natural systems.

References

Baas, A.C.W. (2002) Chaos, fractals and self-organization in coastal geomorphology: simulating dune landscapes in vegetated environments, *Geomorphology* 48, 309–328.
Bak, P. (1996) *How Nature Works*, New York: Copernicus.
Burrough, P.A. (1981) Fractal dimensions of landscapes and other environmental data, *Nature* 294, 240–242.
Church, M. and Mark, D.M. (1980) On size and scale in geomorphology, *Progress in Physical Geography* 4, 342–390.
Klinkenberg, B. (1992) Fractals and morphometric measures: is there a relationship? *Geomorphology* 5, 5–20.
Mandelbrot, B.B. (1967) How long is the coast of Britain? Statistical self-similarity and fractal dimensions, *Science* 156, 636–638.
——(1982) *The Fractal Geometry of Nature*, San Francisco: Freeman.
Mark, D.M. and Aronson, P.B. (1984) Scale-dependent fractal dimensions of topographic surfaces: an empirical investigation, with applications in geomorphology, *Mathematical Geology* 16, 671–683.
Rodriguez-Iturbe, I. and Rinaldo, A. (1997) *Fractal River Basins (Chance and Self-Organization)*, Cambridge: Cambridge University Press.
Turcotte, D.L. (1992) *Fractals and Chaos in Geology and Geophysics*, Cambridge: Cambridge University Press.
Werner, B.T. (1995) Eolian dunes: computer simulation and attractor interpretation, *Geology* 23, 1,107–1,110.
Xu, T., Moore, I.D. and Gallant, J.C. (1993) Fractals, fractal dimensions and landscapes – a review, *Geomorphology* 8, 245–262.

BRIAN KLINKENBERG

FRAGIPAN

A natural subsurface horizon found deep in the soil profile, which has been altered by pedogenic processes responsible for restricting the entry of water and roots into the soil matrix. Fragipans possess a higher bulk density than the horizons above, contain very little organic matter, are brittle when moist and exhibit slaking properties when immersed in water. Thickness ranges from 15–200 cm, enough to allow sufficient impact upon plant growth so that roots and water are unable to penetrate 60 per cent of the horizon. Fragipans develop mostly in mid-latitude, medium texture, acid materials overlying albic or argillic soil horizons, and with udic or aquic moisture regimes. Fragipans occur mainly beneath forest vegetation, in cultivated or virgin soils within various parent materials including glacial drift, loess, colluvium, lacustrine deposits and alluvium, though they are not found in calcareous deposits. Fragipans consistently possess an abrupt upper boundary at a depth of 30–100 cm beneath the ground surface (Witty and Knox 1989), and often exhibit evidence of soil formation.

The origins of fragipans are poorly understood, though three main formation mechanisms exist. These are: physical ripening during desiccation of initially slurried material; clay bridging; and bonding by an amorphous component (including Si, Al, and Fe). Unfortunately, fragipan is a generally poorly defined term, with many examples of fragipans worldwide unidentified due to their vague definition in the field.

Reference

Witty, J.E. and Knox, E.G. (1989) Identification, role in soil taxonomy, and worldwide distribution of fragipans, in N.E. Smeck and E.J. Ciolkosz (eds) *Fragipans: Their Occurrence, Classification, and Genesis*, SSSA Special Publication Number 24, 1–10.

Further reading

Smeck, N. E. and Ciolkosz, E. J. (eds) (1989) *Fragipans: Their Occurrence, Classification, and Genesis*, SSSA Special Publication Number 24.

STEVE WARD

FREEZE–THAW CYCLE

A freeze–thaw cycle is a cycle in which temperature fluctuates both above and below 0 °C. Field measurements indicate that most freeze–thaw cycles per year occur in the climates of low annual temperature range, which are dominated by diurnal or cyclonic fluctuations (French 1996: 26). These conditions are met in subpolar oceanic locations (e.g. Jan Mayen in the northern hemisphere, Kerguelen Islands or South Georgia in the southern hemisphere) and in intertropical high mountain environments (e.g. Andes, East Africa mountains). Among all cold environments, the least number of freezing and thawing days occur at high latitude and in continental climates, which are dominated by seasonal temperature regimes. In all areas, most cycles occur in the upper 0–5 cm of the ground and only the annual cycle occurs at depths in excess of 20 cm.

Freeze–thaw cycles have important effects on soils, like FROST HEAVE, frost sorting or frost creep. It is important to distinguish between seasonally frozen ground and PERMAFROST (Washburn 1973: 15). In non-permafrost environments, the depth of seasonal freezing increases with increasing latitude, the range being from a few millimetres to more or less 3 metres. In permafrost regions, the active layer is the upper part of the ground that undergoes seasonal freezing and thawing.

With respect to rock frost weathering (see FROST AND FROST WEATHERING), alternate freezing and thawing is much more damaging than continued cold, and the effectiveness of frost is dependent on the frequency of temperature fluctuations about the freezing point in the presence of water (Ollier 1984: 125). Nevertheless, the number of freeze–thaw cycles undergone by materials unfortunately cannot be used as a direct measure of frost action effectiveness for several reasons. First, the use of air temperatures to define cycles is not satisfactory at all, since significant differences exist between air and ground temperature. This can be caused for example by the insulating effect of snow or by insolation on dark rock surfaces (Washburn 1973: 58). Second, the exact freezing temperature across which the oscillations should be measured is difficult to define, as all the water contained in soils and rocks does not freeze instantaneously, nor always at 0 °C but at negative temperatures, because, for example, of the capillary forces existing in the porous media or the supercooling phenomenon. Freezing temperature can also be lowered in presence of salts or clay. Freezing has been reported to begin at temperatures lower than −10 °C in the case of rocks characterized by very small pores.

Finally, what constitutes a freeze–thaw cycle is debatable, as some authors define specific minimum negative temperatures that have to be reached for most of the rock-absorbed water to freeze, or minimum durations for the periods at negative and positive temperatures between successive cycles. For example, according to different studies, one cycle is completed when the hourly rock temperature changes from ≥ + 1 °C to ≤ − 1 °C and then back to ≥ + 1 °C (Lewkowicz 2001: 359), or when a fall below −2 °C is followed by a rise above +2 °C (Matsuoka 1991: 276). Although these thresholds have been defined in order to take into account the actual stresses undergone by the rock as accurately as possible, they make any comparison of cycle frequencies reported in different studies very difficult.

Other important components of freeze–thaw cycles with respect to frost weathering are the duration of freezing (the time period during which negative temperatures persist), the intensity of freezing (the extent of temperature decrease below 0 °C) and the rate of freezing (the rapidity or slowness with which temperature decreases below 0 °C) (McGreevy and Whalley 1982: 158). The influence of these three parameters is quite controversial.

As far as the intensity of freezing is concerned, since the greatest part of pore water freezes between 0 °C and −5 °C, volume expansion causing frost weathering of rock occurs mostly in this temperature range (McGreevy and Whalley 1982: 159; Matsuoka 1991: 272). This explains why frost decay rates do not change significantly between freeze–thaw cycles reaching minimum temperatures of −8 or −30 °C.

The impact of freezing duration has to be viewed in relation to the intensity of freezing. It is the pore sizes that determine the freezing point of water within rocks. Thus freezing occurs over a range of gradually decreasing temperatures and rocks undergo some stress only if the required critical temperatures have been reached, and for a period long enough so that the temperature change propagates from the rock surface into the centre of a block or into a rockwall. There must indeed be time for the transfer of the necessary latent heat to cause the freezing or thawing of the water in the rock (Ollier 1984: 125). The duration of the period at minimum temperature has been considered by laboratory work as completely insignificant (under constant temperature and if the freezing front stopped progressing, no breaking strain can be built up) or quite important (in an open system with a constant unfrozen water supply, segregation ice lenses may keep growing by unfrozen water migrations under constant temperature conditions). On the other hand, in field studies carried out in alpine environments where wedging (see FROST AND FROST WEATHERING) of a massive rock mass is the predominant decay process, freezing intensity and duration have been considered as fundamental parameters as they are responsible for the depth reached by the freezing front. Only long freezing periods, with stagnation of the freezing front at depths between 10 and 50 cm, are able to furnish large slabs in addition to small blocks (Coutard and Francou 1989: 415).

Various rates of freezing can favour various weathering mechanisms and lead to different degrees and types of rock decay in the same rock type. Quick cooling favours bursting and wedging effects, as more pressure is built up in pores and cracks when no time is left for water migration to occur and to relieve some of this pressure. On the contrary, slow cooling offers optimal conditions for the formation of segregation ice lenses and for scaling effects. Numerous works report higher degrees of decay after quicker frosts although some studies argue that freezing rate is not a particularly critical parameter (McGreevy and Whalley 1982: 158), or stress on the quite complex impact of freezing rates, making the evaluation of its importance difficult (Matsuoka 1991: 272). According to Matsuoka, slow freezing in an open system results in prolonged water migration toward the freezing front and, hence, in rising ice force. In contrast, in a closed system, rapid freezing favours a large ice growth strain, because pore ice contracts with time.

Rates of freezing measured in natural environments generally range between 0.2 and 4 °C per hour. However, laboratory simulation usually favours quick cooling rates (in order to accelerate COMMINUTION and the achievement of decay results) and values higher than 10 °C per hour are not uncommon. Results obtained by such experimentation may not reflect natural environmental processes.

Freeze–thaw cycles have been the subject of data collection in the field and of laboratory investigations, testing the impact of different temperature regimes on frost susceptibility. A large variety of cycle characteristics have been used, but the two main types reflect a daily moderate freezing regime (down to −8 °C) characteristic of polar maritime regions and a more intense and prolonged freezing regime (down to −30 °C) characteristic of polar continental areas.

References

Coutard, J.P. and Francou, B. (1989) Rock temperature measurements in two alpine environments, implications for frost shattering, *Arctic and Alpine Research* 21(4), 399–416.

French, H.M. (1996) *The Periglacial Environment*, Harlow: Longman.

Lewkowicz, A.G. (2001) Temperature regime of a small sandstone tor, Latitude 80 °N, Ellesmere Island, Nunavut, Canada, *Permafrost and Periglacial Processes* 12, 351–366.

McGreevy, J.P. and Whalley, W.B. (1982) The geomorphic significance of rock temperature variations in cold environments: a discussion, *Arctic and Alpine Research* 14(2), 157–162.

Matsuoka, N. (1991) A model of the rate of frost shattering: application to field data from Japan, Svalbard and Antarctica, *Permafrost and Periglacial Processes* 2, 271–281.

Ollier, C. (1984) *Weathering*, Harlow: Longman.

Washburn, A.L. (1973) *Periglacial Processes and Environments*, London: Arnold.

Further reading

Lautridou, J.P. and Ozouf, J.C. (1982) Experimental frost shattering: 15 years of research at the Centre de Géomorphologie du CNRS, *Progress in Physical Geography* 6(2), 215–232.

Prick, A. (1995) Dilatometric behaviour of porous calcareous rock samples undergoing freeze–thaw cycles. Some new results, *Catena* 25(1–4), 7–20.

SEE ALSO: experimental geomorphology; frost and frost weathering; mechanical weathering; periglacial geomorphology; weathering

ANGÉLIQUE PRICK

FRINGING REEF

The morphology and genesis of CORAL REEFS varies significantly. They may be divided into ATOLL reefs, barrier reefs and fringing reefs (Nunn 1994). The youngest and most ephemeral of the three forms are fringing reefs, which also often lack the breadth, continuity and species diversity of atoll and barrier reefs. In addition, because they are located nearest the land – and indeed cannot exist distant from it – fringing reefs are those which are usually most affected by humans.

Development of fringing reefs

Unlike atoll reefs and barrier reefs, most fringing reefs formed as discrete units only during the most recent period of postglacial sea-level rise. Most began growing from shallow depths on the flanks of a tropical coastline when ocean-water temperatures (and other factors) at the end of the glacial period became suitable for reef growth. Encouraged by sea-level rise, the nascent fringing reefs began growing upwards and exist as living entities today only if they were able to 'keep up' or ‘catch up’ with sea level during the transgression (Neumann and MacIntyre 1985).

Once sea level reached its maximum level during the Holocene (about 5,000 cal. yr BP), ‘keep-up’ fringing reefs would have stopped growing mainly upwards and would have begun growing laterally, an ecological transformation involving a change in coral species distribution. Branching corals in particular would slowly have been replaced by other species adapted to outward rather than upward growth. A classic study is that of Hanauma Reef which began growing about 7,000 years ago on the inner flanks of an ancient volcanic crater on the Hawaiian island Oahu (Easton and Olson 1976).

On the other hand, ‘catch-up’ fringing reefs would not by definition have been able to keep pace with rising sea level and may have ‘caught up’ only when sea level was falling during the late Holocene. In such cases the change from upward to outward growth may have occurred more recently.

The outward growth of a fringing reef is constrained by the slope angle of the coastline from which it rises. On steeply sloping coasts, like those of the central Pacific island Niue, it is no surprise that fringing reefs are barely noticeable (and have little role in shoreline protection), often no more than a few metres in width. On coasts which slope more gently, fringing reefs may reach several hundred metres in breadth and have well-defined morphological zones (see below).

Some writers like Davis (1928) believed that a fringing reef was part of a genetic continuity and would eventually become a barrier reef and finally, when the land from the flanks of which the reef rose was submerged, an atoll reef. This is valid in only a general sense but did not take into account the effects of sea-level changes and the fact that, at the end of each Quaternary glacial period, fringing reefs re-grew. Such writers often equated the presence of fringing reefs with a coastline that had just begun sinking and, where a barrier reef was found farther offshore, would often cite a complex series of tectonic (rather than sea-level) movements to explain the association.

Morphology of fringing reefs

Fringing reefs have morphological characteristics that are shared with atoll and barrier reefs and others which are not. Along their outer, submarine slopes, fringing reefs have slopes of talus derived from the mechanical erosion of the reef edifice. Owing to the youth of fringing reefs and the comparative shallowness of the adjoining seafloor (usually a lagoon floor), these talus slopes are generally less voluminous than the equivalent features off barrier or atoll reefs. Similarly, owing to the wave energy being generally less along the fronts of fringing reefs (because waves hitting fringing reefs are commonly generated within a lagoon or are residual waves reduced in amplitude from crossing a barrier reef), reef growth and coral diversity on the outer reef crest is generally less than on barrier or atoll reefs. Yet, where a fringing reef faces directly into the ocean, these features and others are of the same size as on barrier or atoll reefs. A good example is the south coast of Tongatapu Island in the South Pacific where the south-east trade winds drive swells straight onto the narrow fringing reef which has well-developed spur-and-groove morphology along its front and an impressively high algal (*Porolithon*) ridge (Nunn and Finau 1995).

Behind the outer reef crest of fringing reefs is generally found a reef flat several tens of metres broad in which there are comparatively few living corals but an abundance of fossil reef, often planed down from a higher level. A good example is from New Caledonia (Cabioch *et al.* 1995). Particularly if the fringing reef has been significantly affected by humans (see below) the back reef area may be covered with seagrass beds or the alga *Halimeda*, sometimes terrigenous sand, all of which inhibit reef growth and may in consequence reduce the supply of calcareous sand to adjacent shorelines.

At the back of many fringing reefs is a ‘boat channel’ eroded in the reef surface at the point where freshwater comes out of the adjacent land. Freshwater springs are common in such places.

Emerged fringing reefs

Along those coasts where coral-reef upgrowth was able to keep pace with postglacial sea-level rise, and the sea level exceeded its present level during the middle Holocene, it is expected that fringing reefs would have grown above their present levels and that remnants of such ‘emerged’ fringing reefs would now be visible to testify to this. The morphology of emerged fringing reefs is often comparable to that of their modern counterparts although many are much reduced by erosion.

In the Hawaiian Islands, for example, many years of searching for emerged fringing reefs bore fruit only quite recently (Grigg and Jones 1997).

Human impact on fringing reefs

Fringing reefs are those most vulnerable to deleterious human impact. Many bear the brunt of indirect impacts like pollution and sedimentation from adjacent land areas. Direct impacts, particularly along coasts where fringing reefs are central to subsistence economies or to recreational activities, include overexploitation of edible reef organisms, trampling by humans, physical damage from boat anchors, and even poisoning or dynamiting for easy kills of large numbers of reef fish.

References

Cabioch, G., Montaggioni, L.F. and Faure, G. (1995) Holocene initiation and development of New Caledonian fringing reefs, SW Pacific, *Coral Reefs* 14, 131–140.

Davis, W.M. (1928) *The Coral Reef Problem*, Special Publication 9, Washington, DC: American Geographical Society.

Easton, W.H. and Olson, E.A. (1976) Radiocarbon profile of Hanauma Reef, Oahu, Hawaii, *Geological Society of America, Bulletin* 87, 711–719.

Grigg, R.W. and Jones, A.T. (1997) Uplift caused by lithospheric flexure in the Hawaiian Archipelago as revealed by elevated coral deposits, *Marine Geology* 141, 11–25.

Neumann, A.C. and MacIntyre, I. (1985) Reef response to sea-level rise: keep-up, catch-up or give-up, in *Proceedings of the 5th International Coral Reef Congress* 3, 105–110.

Nunn, P.D. (1994) *Oceanic Islands*, Oxford: Blackwell.

——and Finau, F.T. (1995) Late Holocene emergence history of Tongatapu island, South Pacific, *Zeitschrift für Geomorphologie* 39, 69–95.

SEE ALSO: coral reef

PATRICK D. NUNN

FROST AND FROST WEATHERING

Frost action is a collective term describing a number of distinct processes which result mainly from alternate freezing and thawing of water in pores and cracks of soil, rock and other material, usually at the ground surface. It is widely believed that frost action is the fundamental characteristic of present-day periglacial environments. Frost-action processes probably achieve their greatest intensity and importance in such areas.

In soils, FROST HEAVE, NEEDLE-ICE formation, frost creep and thermal contraction cracking are very common frost-related processes. The term *cryoturbation* refers to all soil movements due to frost action (French 1996).

Frost weathering (also called *frost shattering, congelifraction, gelifraction* or *gelivation*) contributes to the *in situ* mechanical breakdown of rocks by various processes. The conventional view is that rock decay is due to the fact that when water freezes it expands by about 9 per cent. This creates pressures, calculated to be around 2,100 kg cm^{-2} at $-22\,°C$, that are higher than the tensile strength of rock (generally less than 250 kg cm^{-2}). However, this process rarely induces critical pressures, reached only when freezing occurs in a closed system with a very high rock moisture content (about 90 per cent). Such conditions are not common in natural environments, but when occurring, the volume expansion effect may cause rock *bursting*.

A more realistic model, also applicable to soils, is the segregation ice model (Hallet *et al.* 1991), which treats freezing in rock as closely analogous to slow freezing in fine-grained soils. When water freezes in rock or soil, the ice nuclei attract unfrozen water from the adjoining pores and capillaries. Tensions are primarily the result of these water migrations to growing ice lenses (Prick 1997). Frost weathering is induced by the progressive growth of microcracks and relatively large pores wedged open by ice growth. In the segregation ice model, low saturation in hydraulically connected pores (open system) does not preclude water migration and crack growth. The detachment of thin rock pieces by the growth of ice lenses is called *scaling*.

Frost wedging refers to rock fracturing associated with the freezing of water in existing planes of weakness, i.e. cracks and joints. Wedging can be caused by the volume expansion of water turning into ice in cracks, or by hydraulic pressure. According to this second process, the freezing front penetration in a rockwall induces a freezing of the most external part of the crack first, creating a solid plug of ice. In depth, where the saturated crack is thinner (and the freezing point thus lower), some water can be trapped under pressure by the ice growing further in from the rock surface and so contribute to crack growth outwards and downwards. In both cases, the thinner the crack, the quicker and the more severe the frost has to be in order to cause a wedging effect.

The rate at which frost shattering occurs depends on climatic factors and rock characteristics. Among climatic factors, the most important ones are the number of FREEZE–THAW CYCLES and the availability of moisture. Some thermal characteristics of the freeze–thaw cycle can also have some importance, like the freezing rate or the duration of the freezing period.

The water availability in the environment and the rock moisture content are certainly the most critical elements for defining the susceptibility of this environment to frost action (Matsuoka 1990). Laboratory experiments have shown that the amount of disintegration in rocks supplied with abundant moisture is greater than that in similar rocks containing less moisture. For this reason, dry tundra areas and cold deserts may undergo less extreme frost weathering than moister environments.

If some particular locations are characterized by a continuous and abundant water supply (for example a block sitting next to a lake shore or to a melting snow patch), a large majority of blocks exposed in cold-climate environments experience neither close to saturation conditions (because of insufficient water supply), nor a dry state (intense drying is rare).

A critical degree of saturation can be defined as a threshold moisture level for each rock type (Prick 1997): only when moisture exceeds this level will the material be damaged by frost. This parameter reflects the influence of rock characteristics on frost susceptibility and defines the part of the porous medium that has to be free of water in order not to build up a breaking strain.

The nature and characteristics of the rock are indeed a crucial factor for frost susceptibility. Rocks such as tough quartzites and igneous rocks tend to be most resistant, while porous and well-bedded sedimentary rocks, such as shales, sandstones and chalk, tend to be least resistant. Among the rock characteristics influencing frost weathering, the most determinant ones are: the rock specific surface area, permeability, porosity, pore size distribution, and mechanical strength.

A large specific surface area (i.e. internal surface of the porous media) induces a larger contact area between rock and water and therefore enhances a higher susceptibility to frost decay. A high rock permeability, by allowing easy and quick water migration, prevents critical pressures to build up (Lautridou and Ozouf 1982). Rocks with a very poor porosity are not frost susceptible: experimentation showed that rocks with a porosity of less than 6 per cent are little damaged after several hundreds of freeze–thaw cycles (Lautridou and Ozouf 1982); further research showed that this threshold value can be considered as a valuable, but rough estimate.

Pore size distribution (also called *porosimetry*) can influence frost susceptibility in various ways. Rock porous media characterized mainly by large pores (macroporosity) will tend to be frost resistant, as macroporosity favours a good permeability. Unimodal porous media (i.e. characterized by one predominant pore size) offer ideal conditions for segregation ice formation; rocks with such a pore size distribution tend to be susceptible to any type of freezing (even with a moisture content far below saturation and with a slow cooling rate) and will undergo an increased decay as freezing/thawing goes on. Multimodal porous media (i.e. characterized by pores of various sizes) are not favourable to the set up of large-scale water migrations; rocks with such a pore size distribution tend to be frost susceptible only with high moisture content, preferably in the case of a quick freezing.

Among ROCK MASS STRENGTH parameters, tensile strength has a considerable influence on rock frost decay (Matsuoka 1990). Crack density and width often influence water penetration in the bedrock and allow wedging to take place.

Frost action is one component of *cryogenic weathering*, i.e. the combination of weathering processes, both physical and chemical that operate in cold environments either independently or in combination. Many aspects of cryogenic weathering are not fully understood, but the processes other than frost that may be efficient decay agents are: HYDRATION (see WETTING AND DRYING WEATHERING), thermal fatigue (see INSOLATION WEATHERING), SALT WEATHERING, CHEMICAL WEATHERING, ORGANIC WEATHERING and PRESSURE RELEASE (particularly in recently deglaciated areas). Solutional effects are present in limestone and KARST terrain exists in PERMAFROST regions. The dominance of frost action among these processes is considered as doubtful, but the definition of the exact role of each of these processes in the different cold environments and in the different periods of the year is problematic.

Frost weathering characteristically produces angular fragments of various sizes. In periglacial areas, cryogenic weathering determines the formation of some extensive features like blockfields (see BLOCKFIELD AND BLOCKSTREAM), GRÈZE

LITÉES, SCREES, TALUS slopes or ROCK GLACIERS. Its action is also often crucial for MASS MOVEMENT processes like rock avalanches and rock falls.

The predominant size to which rocks can be ultimately reduced by frost action is generally thought to be silty particles with grain sizes between 0.01 and 0.05 mm in diameter. Experimentation on mineral particles indicated that frost weathering occurs within the layer of unfrozen water adsorbed on the surface of these particles. The minerals' susceptibility to weathering depends not so much on their mechanical strength as on the thickness and properties of this unfrozen water film. Decay occurs when this water film becomes thinner than the dimensions of the microcracks and defects that characterize the surface of mineral particles. The protective role of the stable film of unfrozen water is highest with silicates, such as biotite and muscovite, and lowest with quartz. Experimentation results indicate that under cold conditions the ultimate size reduction of quartz (0.01–0.05 mm) is smaller than for feldspar (0.1–0.5 mm), a reversal of what is assumed for temperate or warm environments (Konishev and Rogov 1993).

Frost weathering is studied both in the field and in the laboratory. The most commonly used techniques are: visual observation and photographic documentation of the decay evolution, weight loss, frost-shattered debris characterization, assessment of mechanical properties (like tensile or compressive strength) or elasticity properties (Young's modulus), ultrasonic testing, evolution of porosity and pore size distribution, dilatometry, and crack opening assessment.

Some field studies have been undertaken with the aim of increasing the availability of data upon rock temperature and moisture content. The lack of such data has up to now been a considerable impediment to a definition of the exact role of frost action in cryogenic weathering and to the realization of laboratory simulation using thermal and moisture regimes likely to occur in natural environments. Other studies focus on the rate of bedrock weathering by frost action (Lautridou and Ozouf 1982) and on the definition of predictive models (Matsuoka 1990). Laboratory simulation and modelling identify the climatic conditions and rock characteristics that emphasize frost action efficiency and so define the exact role of the various weathering mechanisms (e.g. Hallet *et al.* 1991; Prick 1997).

A major gap remains between field and laboratory research (Matsuoka 2001). This is due to a difference in the size of the study object (small blocks in the laboratory, but rockwalls sometimes in the field), in the type of rock material (intact soft rocks with medium or high porosity are overrepresented in laboratory simulations, but jointed massive rocks with low porosity are very common in cold environments) and thus in the type of frost weathering process taken into account (mostly bursting or scaling in the laboratory simulations; mostly wedging in the field). Wedging may sometimes be the only frost weathering process acting on fractured rock characterized by a low porosity and a high mechanical strength for the individual blocks. This may lead to *macrogelivation*, i.e. frost weathering at a large scale, acting mainly through the crack system, as opposed to *microgelivation*, which refers to frost decay acting in the porous media of individual small-sized blocks. This further illustrates the inadequacy of a simplistic view of frost weathering.

References

French, H.M. (1996) *The Periglacial Environment*, Harlow: Longman.

Hallet, B., Walder, J.S. and Stubbs, C.W. (1991) Weathering by segregation ice growth in microcracks at sustained subzero temperatures: verification from an experimental study using acoustic emissions, *Permafrost and Periglacial Processes* 2, 283–300.

Konishev, N.V. and Rogov, V.V. (1993) Investigations of cryogenic weathering in Europe and Northern Asia, *Permafrost and Periglacial Processes* 4, 49–64.

Lautridou, J.P. and Ozouf, J.C. (1982) Experimental frost shattering: 15 years of research at the Centre de Géomorphologie du CNRS, *Progress in Physical Geography* 6(2), 215–232.

Matsuoka, N. (1990) The rate of bedrock weathering by frost action: field measurements and a predictive model, *Earth Surface Processes and Landforms* 15, 73–90.

——(2001) Microgelivation versus macrogelivation: towards bridging the gap between laboratory and field frost weathering, *Permafrost and Periglacial Processes* 12, 299–313.

Prick, A. (1997) Critical degree of saturation as a threshold moisture level in frost weathering of limestones, *Permafrost and Periglacial Processes* 8, 91–99.

Further reading

White, S.E. (1976) Is frost action really only hydration shattering? A review, *Arctic and Alpine Research* 8(1), 1–6.

Yatsu, E. (1988) *The Nature of Weathering: An Introduction*, Tokyo: Sozosha.

SEE ALSO: experimental geomorphology; mechanical weathering; periglacial geomorphology; weathering

ANGÉLIQUE PRICK

FROST HEAVE

Frost heave is best known from the wintertime uplift of the ground surface, familiar to dwellers in cold climates, which is evidenced by jammed gateways, uneven roads, cracked foundations and the breaking-up of road surfaces in the spring thaw. These effects are not ascribable to the expansion of water that occurs on its freezing (9 per cent). They are due to the movement of water into the soil that is freezing, with the formation of accumulations of ice – increasing the soil volume, giving displacement (the 'heave'). These ice structures are called 'lenses', 'schlieren' or ice 'masses' and known collectively as segregation ice (because each is larger than pore size and has been segregated from the soil pore structure). Segregation ice is not ice from entrapped snow, buried glacial ice or buried lake or marine ice, although it may reach cubic metres in size. Its nature and extent depends on the nature of the granular soil material and a variety of local factors (drainage conditions, temperature regime, depth in the ground, etc.). Thus frost heave is commonly uneven, giving rise to irregularities of the ground surface (bumpiness) usually recurring year after year or, in PERMAFROST, persisting for many years. The forces generated by the heaving material can be very large.

Segregation ice and thus frost heave is an expression of the fundamental thermodynamic behaviour of a porous medium on freezing; this thermodynamic behaviour is ultimately responsible for the main properties and characteristics associated with soils in cold climates. As a consequence frost heave has enormous economic (geotechnical) significance; overcoming its effects is the essential problem for construction of buildings, roads, airports and pipelines in the cold regions.

The processes associated with frost heave largely explain the origin of most terrain forms occurring naturally and characteristically in cold climates – so-called 'periglacial' (see PERIGLACIAL GEOMORPHOLOGY) features. Boulders ('growing stones') are heaved to the ground surface by annual cycles of freezing and rearrangement of soil particles at thawing. Incremental frost heave is an important process in formation of PATTERNED GROUND, such as stone circles and stone polygons, where stones and boulders are heaved in particular directions (as a function of temperature and other factors) to give rise to conspicuously ordered surface arrangements. PINGOs, features occurring locally in regions with permafrost, look like volcanic cones and are elevated by the large, hidden central core of ice. They are the product of a particular thermal regime, commonly involving the gradual freezing of previously unfrozen ground below a receding water body. The frost heave process is largely responsible for lifting the above-surface material in pingos to elevations of tens of metres, so the forces developed must be large.

The instability of slopes, and the development of certain forms of SOLIFLUCTION, mudflows and landslides are ascribable to the excess water released on thawing of frost-heaved soils with their ice segregations, and the associated high PORE-WATER PRESSURES. Not infrequently the volume of segregated ice exceeds the volume of water the soil can hold in the thawed state by a factor of two or more. This accounts for a greatly weakened state of the newly thawed soil.

Fundamentally, the water moves toward a zone of freezing in the soil because of thermodynamic potentials arising with the growth of ice crystals in small spaces. Although the thermodynamic principles have been recognized for more than a century (and also describe, for example, crystallization phenomena in solutions, the formation of ice crystals or of water droplets in the atmosphere, or the nucleation of bubbles in liquids) the significance for soils has been realized fully only in recent decades. The thermodynamic potential may be regarded, with some simplification, as a pressure, and is referred to by different terms in different branches of science and technology. The pressure of the water falls with temperature in freezing soil, so that there is a gradient from warm (unfrozen) to cold (frozen). However, the pressure of the ice in the ice segregations rises as the temperature falls, and it is demonstrably this pressure which causes frost heave. Furthermore, the thermodynamic relations require ice to form in larger spaces and pores first. As a consequence, there is an unfrozen water content of frozen soils, decreasing with temperature as progressively smaller pores are filled by ice, and which is, therefore, a function of pore size distribution.

The significance of the soil water accumulating (the process of frost heave) and then freezing in this way over a range of temperatures down to several degrees below 0 °C, is that the thermal and mechanical properties of the frozen soils are highly temperature dependent. Frozen, heaved soils are prone to creep in a manner rather similar to glaciers but with lower rates of deformation; this is probably the cause of certain large vegetation-covered solifluction terraces on slight slopes. The grain-size and pore size distribution of a soil are crucial to its behaviour when frozen because they control the (unfrozen) water content of the specific soil. The release of latent heat of freezing of the water effectively controls the heat capacity of the soil; thermal conductivity is also modified (though less so) because of the difference in thermal conductivity between ice and water. The thermal diffusivity, which is the ratio of thermal conductivity to heat capacity, is consequently highly temperature dependent and controlled by the pore size distribution – that is, by the nature of the soil (clay, silt or sand, or combinations of particle sizes), and the amount of frost heave.

The thermal diffusivity controls such phenomena as the depth to which winter freezing occurs, and the depth of summer thawing (the ACTIVE LAYER) above permafrost. Indeed the distribution of permafrost itself (ground remaining frozen year in, year out), in depth and in time (and in response to climate or microclimate change), depends substantially on its thermal diffusivity. Terrain features, ascribable to frost heave and associated with the comings and goings of permafrost, include ALASes, palsa and THERMOKARST.

Counteracting effects of frost heave and subsequent thawing added billions of dollars to the cost of the transAlaska oil pipeline. The forces generated (CRYOSTATIC PRESSURES) by frost heave around gas pipelines in permafrost regions threaten their stability and thus their financial viability. Avoidance of frost heave is the main reason for added costs of infrastructure in general in the cold regions; these added costs are greatest in the 25 per cent of the Earth's land surface underlain by permafrost but are also a major factor in construction (especially of highways and airports) in the further 20 per cent or so which has significant winter frost penetration, and consists largely of highly populated temperate lands.

When Taber (1918) first clearly demonstrated that the geotechnical problem of frost heave was due to the migration of water with accumulation of excess ice in the frozen ground, he paved the way for Beskow's classic work (1935) on frost heave and its significance in relation to the local environment (soil type, groundwater conditions, confining pressures, etc.). In 1943 the remarkable study by Edlefsen and Andersen (resulting from the wartime collaboration of two scientists in different fields) established the thermodynamic interpretation, which substantiates the largely empirical approach that has been used by geotechnical specialists concerned with engineering (Andersland and Ladanyi 1995) for cold regions development in the broadest sense. Agronomists too, have an important involvement. Today, geocryology, the study of the ground surface regions in freezing climates (Williams and Smith 1989) notably developed in Russia (Yershov 1998), is concerned mainly with the effects of frost heave, a phenomenon first recognized some two hundred and fifty years ago (see Beskow 1935).

References

Andersland, O.B. and Ladanyi, B. (1995) *An Introduction to Frozen Ground Engineering*, London: Chapman and Hall.

Beskow, G. (1935) Tjällyftningen och tjällyftningen med særskild hensyn til vägar och järnvägar. *Sveriges Geologiska Undersökning. Avh. och Uppsats., Ser. C, 375, årsbok 26, 3.* (Available in translation: Soil freezing and frost heave with special application to roads and railroads, in P.B. Black and M.J. Hardenberg (eds) (1991) Historical perspectives in frost heave research, *US Army, Cold Reg. Res. and Engg. Labs*, Special Report 91–23, Hanover, NH.

Taber, S. (1918) Surface heaving caused by segregation of water forming ice crystals, *Engineering News Record* 81, 683–684.

Williams, P.J. and Smith, M.W. (1989) *The Frozen Earth. Fundamentals of Geocryology*, Cambridge: Cambridge University Press.

Yershov, E.D. (ed. P.J. Williams) (1998) *General Geocryology* (translated from: Obschaya geokriologiya, Nedra 1990), Cambridge: Cambridge University Press.

PETER J. WILLIAMS

G

GENDARME

A needle-shaped rock pinnacle located along a mountain ridge or arête. The term gendarme is universal, yet employed predominantly in alpine geomorphology and mountaineering. Gendarme shares its name with a French policeman, as both may block one's passage and hinder progress. They are commonly found in the Alps, such as Pic de Roc gendarme in Chamonix, French Alps. However, similar forms exist in other mountainous regions, such as Bryce Canyon, USA. Gendarmes are also referred to as rock pinnacles and aiguilles, yet are generally more pointed and larger than an aiguille.

SEE ALSO: rock and earth pinnacle and pillar

STEVE WARD

GEOCRYOLOGY

The study of Earth materials having a temperature below 0 °C. Washburn (1979) recognized that it was sometimes taken to include glaciers, but argued that it was more specifically used as a term for PERIGLACIAL GEOMORPHOLOGY and PERMAFROST phenomena. Indeed, the subtitle of his magisterial volume *Geocryology* was 'A survey of periglacial processes and environments'. In that volume he studied such phenomena as frozen ground, FROST AND FROST WEATHERING, PATTERNED GROUND, avalanches (see AVALANCHE, SNOW), SOLIFLUCTION, NIVATION and THERMOKARST.

Reference

Washburn, A.L. (1979) *Geocryology: A Survey of Periglacial Processes and Environments*, London: Edward Arnold.

A.S. GOUDIE

GEODIVERSITY

The geodiversity concept first appeared in Australia (especially Tasmania), and received wider recognition, even if always not proper understanding, in the mid-1990s. This robust geodiversity concept has been poorly developed yet in methodological terms. The most popular definition of geodiversity was put forward by the Australian Natural Heritage Charter (AHC 2002):

> Geodiversity means the natural range (diversity) of geological (bedrock), geomorphological (landform) and soil features, assemblages, systems and processes. Geodiversity includes evidence of the past life, ecosystems and environments in the history of the earth as well as a range of atmospheric, hydrological and biological processes currently acting on rocks, landforms and soils.

Geodiversity is now being used in a very holistic way to emphasize the links between geosciences, wildlife and people in one environment or system. The above definition can be supplemented with the statement that geodiversity also embraces quantitative and qualitative topics or indicators at any timescale which make it possible to distinguish marked peculiarities of a georegion, a spatial unit of an unspecified taxonomic rank. This means that bedrock, landforms and soils can be classified by at least two important categories: uniqueness and representativeness. From the geomorphological diversity perspective, an outstanding landform is a feature which is rare, unique, an exceptionally well-expressed example of its kind, or otherwise of special importance within a georegion. A representative landform, in turn, may be either rare or common, but is considered significant as a well-developed or well-exposed example of its kind. A landform type or system can be characterized by an isotropic entity in terms of topographic shape, physical contents, morphogenetic controls and processes, as well as time of formation.

The term geodiversity is commonly used in two meanings, simpler and broader. The first refers to the total range, or diversity, of geological, geomorphic and soil phenomena, and treats geodiversity as an objective, value-neutral property of a real geosystem. In this case a statement of the diversity is made, but the geosystem is not assessed in terms of what kind of geodiversity it is: low or high? The other usage conveys the idea that geodiversity refers specifically to particular geosystems that are in themselves diverse or complex, and thus does not apply to systems which are uniform or have low internal diversity. An example can be the valley of a river flowing through mountains, uplands and lowlands, filled with a wealth of valley, channel and bedforms, and therefore showing high geodiversity, whereas an area of lowland without any streams, basins and/or hummocks has low geodiversity. Questions about the measure of geodiversity are troublesome. Which area displays higher geodiversity: one in which there are 15 mogotes, or another featuring 5 volcanoes, 5 glaciers and 5 river valleys? Or another: has geodiversity increased or diminished in an area transformed by numerous and extensive man-made changes? Landforms are defined by their surface contours and that is why some people claim that the disturbance of significant landform contours (e.g. by excavation) will by definition degrade their geodiversity values, while others see this morphological disturbance as enrichment of geodiversity. Obviously, this situation calls for some clear-cut criteria of geodiversity. One of the possible solutions is a hierarchical classification of landforms: morphoclimatic zone (polar), morphogenetic zone (mountain), morphosystem (glacial system), type of relief (depositional relief), set of landforms (morainic landforms), and single form (terminal moraine). This classification is a function of complexity (see COMPLEXITY IN GEOMORPHOLOGY) reduction. One might argue that an increase in complexity entails an increase in geodiversity, and variations in this relationship are a matter of two functions: asymptotic and exponential.

Because geodiversity is valuable from a variety of perspectives (intrinsic, ecological, geoheritage, as well as scientific, educational, social, cultural, tourist, etc.) it should undergo geoconservation as a result of which it is possible to create GEOSITES for present and future generations.

It should be added that the term geodiversity is analogous to the term biodiversity, which is used to denote species, genetic and ecosystem diversity. It is important to note that the only analogy is that both involve a diversity of phenomena and beyond this self-evident similarity, no further analogies between the nature of ecological and geomorphic processes are expressed or implied. For example, both processes contrast strongly in their timescales. Ecosystems with plant or animal life cycles of tens to hundreds of years do not closely parallel the much longer term active or relict geosystems with weathering, erosion and sedimentation, or Earth internal processes such as seismic or volcanic activity and plate tectonics controlled by processes acting over many thousands or millions of years.

Reference

AHC (2002) *Australian Natural Heritage Charter for the Conservation of Places of Natural Heritage Significance*, Australian Heritage Commission in association with Australian Committee for IUCN, Sydney.

ZBIGNIEW ZWOLINSKI

GEOINDICATOR

The concept of geoindicators was put forward by the International Union of Geological Sciences in 1992. The task of the IUGS working group was to draw up an inventory of indicators to be measured and evaluated under any programme of abiotic environment monitoring. The inventory is not supposed to be a universal standard, but rather to provide a list for the selection by environment monitoring teams of those indicators that can be usefully employed with reference to their study area and time period. Thus, while the list of twenty-seven geoindicators is finite, their choice for the description of environmental change is free. Each geoindicator was evaluated relative to a set of checklist parameters: name, description, significance, human/natural cause, applicable environment, types of monitoring sites, spatial scale, method of measurement, frequency of measurement, limitations of data, applications to past and future, possible thresholds, key references, other information sources, related issues and overall assessment (see Table 20).

Table 20 Geoindicators: natural* vs. human influences**, and utility for reconstructing past environments***

Geoindicator	N*	H**	P***
Coral chemistry and growth patterns	High	High	High
Desert surface crusts and fissures	High	Moderate	Low
Dune formation and reactivation	High	Moderate	Moderate
Dust storm magnitude, duration and frequency	High	Moderate	Moderate
Frozen ground activity	High	Moderate	High
Glacier fluctuations	High	Low	High
Groundwater quality	Moderate	High	Low
Groundwater chemistry in the unsaturated zone	High	High	High
Groundwater level	Moderate	High	Low
Karst activity	High	Moderate	High
Lake levels and salinity	High	High	Moderate
Relative sea level	High	Moderate	High
Sediment sequence and composition	High	High	High
Seismicity	High	Moderate	Low
Shoreline position	High	High	High
Slope failure (landslides)	High	High	Moderate
Soil and sediment erosion	High	High	Moderate
Soil quality	Moderate	High	High
Streamflow	High	High	Low
Stream channel morphology	High	High	Low
Stream sediment storage and load	High	High	Moderate
Subsurface temperature regime	High	Moderate	High
Surface displacement	High	Moderate	Moderate
Surface water quality	High	High	Low
Volcanic unrest	High	Low	High
Wetlands extent, structure and hydrology	High	High	High
Wind erosion	High	Moderate	Moderate

Source: After ITC (1995)

From the point of view of geomorphology, especially dynamic geomorphology, the geoindicator concept seems to be particularly well-suited to determine changes in morphogenetic and sedimentary environments or, broadly speaking, in geosystems. Just like systems theory or allometric analysis, the geoindicator concept has also been adapted from biological sciences. Geoindicators are measures of surface and near-surface geological processes and phenomena that tend to change significantly in less than a hundred years, and which supply crucial information for estimating the state of the environment. This definition specifies the time interval concerned as under a hundred years, which means that geoindicators embrace those processes and phenomena that are highly variable at a short timescale. Hence geoindication will not cover processes involving slow change, like metamorphism or large-scale sedimentation. Geoindicators should answer such questions as, e.g.:

- How often does a process occur?
- What is the rate of river load transport?
- How stable is an individual landform?
- Is the given landform still active, or is it a remnant of an earlier developmental stage?

This way of question formulation determines the specific character of geoindicators: they can express the magnitude, frequency, and rate and/or behaviour trend of an event, process or phenomenon. This means that geoindicators can have widespread application in present-day geomorphological research and, when backed up by paleoenvironmental research, they can provide an excellent basis for forecasting studies. It is especially important when one considers the last decades with their climate change and the consequences it has for the operation of most geosystems throughout the globe. This characterization of geoindicators can be extended to include interactions between the abiotic and biotic environments as well as the fact that it is possible to use geoindicators for different-sized areas to measure extreme, secular and predominant events and to observe natural and man-made processes. Altogether, geomorphologists will find that they have acquired a research tool which is bound to bring about methodological changes in their field.

Reference

ITC (1995) *Tools for assessing rapid environmental changes. The 1995 geoindicator checklist*, International Institute for Aerospace Survey and Earth Sciences, Enschede, Publication Number 46.

Further reading

Berger, A.R. and Iams, W.J. (eds) (1996) *Geoindicators. Assessing Rapid Environmental Changes in Earth Sciences*, Rotterdam: A.A. Balkema.

ZBIGNIEW ZWOLINSKI

GEOMORPHIC EVOLUTION

Geomorphic evolution at its simplest means the mode of change of landform or geomorphic system over time. Qualitative theories continue to dominate geomorphology but a quantitative theory of landform evolution is becoming a central challenge. Traditional qualitative models of landform evolution include the geographical cycle (Davis 1899), Penckian morphological analysis (Penck 1924), the semi-arid erosion cycle (King 1962) and climatogenetic geomorphology (Büdel 1977). These four models represent the framework and options within which landscape evolution were considered from about 1890 until the 1960s. Each of them (except for Penckian morphological analysis) is still in vogue among those who are interested in landscape evolution at the regional scale. The geographical cycle (Davis 1899) is still widely celebrated as a uniquely effective pedagogic device. The orderly evolution of landscape through the stages of youth, maturity and old age, and its interruption at widely separate points in time by massive tectonic uplift, is intuitively appealing. Davis claimed that his model embraced the five factors of structure, process, stage, relief and texture of dissection, but much of the literature says that he considered only the first three. The single major problem with this model was the complete absence of field measurements to confirm or reject assumptions in the model. Nevertheless, there are few better models available to interpret, in a qualitative way, the massive erosional unconformities of, for example, the Grand Canyon of the Colorado River. The geographical cycle has never been proven wrong, but it has been bypassed rather than replaced.

Lester King's subaerial cycle of erosion (King 1962) is perhaps the only serious competitor with the geographical cycle as an interpreter of large-scale, low gradient erosion surfaces. King, strongly influenced by his observations on African escarpments and plateau surfaces, framed his model around the notion of the parallel retreat of scarps. He also attempted, with debatable success, to link his model with the global plate tectonics framework that was evolving during his productive career. His concept of CYMATOGENY (arching of extensive land surfaces with little rock deformation) was a necessary addition to the traditional concepts of orogeny and epeirogeny, and flies in the face of conventional plate tectonics, where massive horizontal movements are favoured. His attempts (King 1962) to correlate pre-Tertiary erosion surfaces globally have met with little debate (except for the critique of cymatogeny) because so few geomorphologists are working at this scale.

A third interesting model of geomorphic evolution is provided by Julius Büdel (1977) and known as the climatogenetic model. The major elements of this model have been interpreted for English readers by Hanna Bremer. The underlying premise is that landscapes are composed of several RELIEF GENERATIONS and the challenge is to recognize, order and distinguish these relief generations It is unfortunate that the major references in the English literature have been sceptical of the model and have failed to give a balanced review (Bremer 1984). Twidale (1976) provides a refreshing summary of the relevance of etchplains (see ETCHING, ETCHPLAIN AND ETCHPLANATION) in Australian landscape evolution.

The Penckian model (Penck 1924) was called morphological analysis. The underlying premise of his analysis is that the rate of uplift, and variations in that rate of uplift over time, dictate landform evolution. His ideas were not taken seriously in

Germany, but they were widely promulgated in Anglo-America because of Davis's interest and opposition to the model. Details of the slope processes discussed are hard to verify and understand because of the lack of field data. But in its championing of endogenic processes and its time independent emphasis, this model was strongly differentiated from the first three. Time-independent models (in which the idea of evolution sits uncomfortably) have been promoted by G.K. Gilbert (1877) and J.T. Hack (1960).

The dichotomy between historical evolutionary studies and functional geomorphology implies that these two approaches do not fit easily together. Indeed, Bremer has said that geomorphology is developing along two lines: the origin of landforms is primarily being studied in continental Europe with climatogenic or tectogenic causes in the foreground. In the English-speaking world the study of geomorphic processes prevails.

Discussions by Schumm (1973), Twidale (1976), Brunsden (1980; 1993) and Ollier (1991) have attempted to reconcile these apparently contradictory positions within a largely qualitative dialogue. The essential contributions to this more recent discussion are the concept of geomorphic thresholds and complex geomorphic response (Schumm), formative events, relaxation time and landform persistence (Brunsden), the understanding that pre-Tertiary landscapes are still decipherable (Twidale), the importance of reconciling plate tectonic theory and morphological evidence (Ollier) and the disequilibrium of all landscapes influenced by Quaternary glaciation (Church and Slaymaker 1989).

A quantitative theory of landform evolution, by contrast with the theories discussed above, requires that the storage and flux rates of water, its flow paths and pressure fields be quantitatively related to their controls and that the boundary conditions of climate, rock properties, topography and stratigraphy be known. But by far the bulk of research on geomorphic evolution has taken place at meso- and micro-scales. And this is where the basic disjuncture in geomorphic thinking has been most evident. Systems modelling and mathematical modelling has tended to drive geomorphic discussion towards the smaller scale landforms, and geomorphic evolution has become, for example, slope evolution, or channel evolution or shoreline evolution.

The work of Ahnert (1967 et seq.) and Kirkby (1971 et seq.) is instructive in that they have been able to satisfy the requirements of quantitative theory by limiting the scale of their models and establishing precise boundary conditions to simulate real world slopes and basins. From 1967 to 1977, Frank Ahnert developed a series of models that used empirical equations to deal with possible ways of relating waste production, delivery and removal at a point on a slope. His final model was a three-dimensional process-response model of landform development. From 1971 to the present, Kirkby has developed increasingly integrated models of slope and drainage basin development, many of them using differential equations that constrain mass balance and thereby maintain continuity These models have had difficulty in dealing realistically with such phenomena as landslides (too rapid) and storage accumulation (too slow), but they represent the cutting edge of modelling in geomorphology from a slope process perspective.

Hydrogeomorphologists, such as Dunne, Dietrich, Montgomery and Church, have led the movement from micro-scale modelling of fluvial process towards a meso-scale modelling of drainage basins, in which they couple slope and channel processes and exploit the drainage network properties to produce more realistic dynamic drainage basin models. Howard (1994) poses a series of critical questions around the landscape modelling project. What is the simplest mathematical model that will simulate morphologically realistic landscapes? What are the effects of initial conditions and inheritance on basin form and evolution? What are the relative roles of deterministic and random processes in basin evolution? Do processes and forms in the drainage basin embody principles of optimization and, if so, why? Is there some basin characteristic form that is invariant in time even under a change in the relative role of the chief land-forming processes. The development of drainage basins requires at least two superimposed processes. He called them soil creep and water flow; in the language of the modellers, one must be a diffusional creep-like mass wasting process capable of eroding the land surface even for vanishingly small contributing areas. Such a process requires an increase in gradient downslope because of its loss of efficiency with increasing area.

The other is an advective fluvial process that increases in efficiency with increasing contributing area. The interplay of these processes produces a combination of convex and concave landforms. By enforcing continuity of flow and continuity of sediment through a coupled system of partial differential equations, the rate of change of elevation can be made dependent on the net flux of sediments as forced by linear increase in discharge. This fundamental step in the understanding of the self-organization of landscape depends on the coupling of the developing landscape with flow rate.

Willgoose *et al.* (1992) presented a catchment evolution model that was essentially a process-response model sensitive to the erosional development of river basins and their channel networks. The model describes the long-term changes in elevation with time that occur in a drainage basin as a result of large-scale mass transport processes. The mass transport processes modelled are tectonic uplift, fluvial erosion, creep, rain splash and landslides. Individual landslides are not modelled but the aggregate effect of many landslides is. The model explicitly differentiates between the part of the basin that is channel and the part that is hillslope. A channel initiation function provided by Dietrich (Dietrich *et al.* 1992) defines a threshold beyond which a channel is formed.

Both dynamic equilibrium and transient states can be modelled in this way. Howard (1994) has noted that the erosion, transport and depositional processes, especially in the river channels, have been greatly oversimplified in the Willgoose *et al.* model and he has generated both alluvial and non-alluvial channel versions of his own model. A more fundamental criticism is that the model does not clarify the linkages of fundamental aspects of the dynamics and the existence of general scaling relations in the network and the landscape itself. Hence the search for improved understanding through analysis of the fractal characteristics of river basins, particularly scale invariance, self-similarity and self-affinity. Multifractality has become a valuable property to identify changing domains of specific process sets (Montgomery and Dietrich 1994).

Understanding of the variety of modes of geomorphic evolution at a variety of spatial and temporal scales is the best evidence of progress in the field. For a number of years at the beginning of this century, researchers were expected to adopt a single model and to stick with it. As a result, the field stagnated under the influence of a single paradigm. In the contemporary state of geomorphology, one of the large issues within models of evolution at the site and basin scale relates to the relation between deterministic and probabilistic modelling.

References

Ahnert, F. (1967) The role of the equilibrium concept in the interpretation of landforms of fluvial erosion and deposition, in P. Macar (ed.) *L'Evolution des Versants*, 23–41, Liège: University of Liège.

Bremer, H. (1984) Twenty one years of German geomorphology, *Earth Surface Processes and Landforms* 9, 281–287.

Brunsden, D. (1980) Applicable models of long term landform evolution, *Zeitschrift für Geomorphologie Supplementband* 36, 16–26.

——(1993) Barriers to geomorphological change, in D.S.G. Thomas and R.J. Allison (eds) *Landscape Sensitivity*, 7–12, Chichester: Wiley.

Büdel, J. (1977) *Klima-Geomorphologie*, Berlin, Stuttgart: Borntraeger.

Church, M. and Slaymaker, O. (1989) Disequilibrium of Holocence sediment yield in glaciated British Columbia, *Nature* 337, 452–454.

Davis, W.M. (1899) The Geographical Cycle, *Geographical Journal* 14, 481–504.

Dietrich, W., Wilson, C.J., Montgomery, D.R., McKean, J. and Bauer, R. (1992) Erosion thresholds and land surface morphology, *Geology* 20, 675–679.

Gilbert, G.K. (1877) *Report on the Geology of the Henry Mountains*. US Geographical and Geological Survey of the Rocky Mountain Region. Washington, DC: US Government Printing Office.

Hack, J.T. (1960) Interpretation of erosional topography in humid temperate regions, *American Journal of Science* 258A, 80–97.

Howard, A.D. (1994) A detachment limited model of drainage basin evolution, *Water Resources Research* 30, 2,261–2,285.

King, L.C. (1962) *The Morphology of the Earth*, Edinburgh: Oliver and Boyd.

Kirkby, M.J. (1971) Hill slope process response models based on the continuity equation, in D. Brunsden (ed.) *Slopes: Form and Process*, 15–30, Institute of British Geographers Special Publication 3.

Montgomery, D.R. and Dietrich, W.E. (1994) Landscape dissection and drainage slope thresholds, in M.J. Kirkby (ed.) *Process Models and Theoretical Geomorphology*, 221–246, New York: Wiley.

Ollier, C.D. (1991) *Ancient Landforms*, London: Belhaven.

Penck, W.D. (1924) *Die Morphologische Analyse: Ein Kapital der Physikalischen Geologie*, Geographische Abhandlungen, 2 Reihe, Heft 2. Stuttgart: Engelhorn.

Schumm, S.A. (1973) Geomorphic thresholds and complex responses of drainage systems, in M. Morisawa (ed.) *Fluvial Geomorphology*, 299–310, Binghamton: Publications in Geomorphology.

Twidale, C.R. (1976) On the survival of paleoforms, *American Journal of Science* 276, 77–95.

Willgoose, G.R., Bras, R.L. and Rodriguez-Iturbe, I. (1992) The relationship between catchment and hill slope properties: implications of a catchment evolution model, *Geomorphology* 5, 21–38.

SEE ALSO: dynamic geomorphology; fractal; geomorphology

OLAV SLAYMAKER

GEOMORPHOLOGICAL HAZARD

A significant practical contribution of geomorphology is the identification of stable landforms and sites with a low probability of catastrophic or progressive involvement with natural or man-induced processes adverse to human occupance or use. Hazards exist when landscape developing processes conflict with human activity, often with catastrophic results. People are killed and property is destroyed or damaged by extreme geomorphic events, and the toll has become greater as human activity has stretched to areas that were avoided in the past. As the population of the Earth has more than doubled from the three billion of 1960, annual losses due to disasters have grown more than ten fold (Bruce 1993).

Tragic examples abound: in 1970 a cyclonic storm surge pushed three to five metres of water into the low deltas entering the Bay of Bengal. The surge, and riverine flooding caused by discharge blocking, resulted in the deaths of an estimated 300,000 to 500,000 people in Bangladesh; an earthquake-induced debris flow descending the flanks of Mt. Huarascan that same year buried over 25,000 people in Peru; a 1991 storm-induced mudflow overwhelmed a concrete drainage channel, killing an estimated 7,000 people in the Philippines; and despite more than fifty years of comprehensive flood control, the 1993 floods along the Mississippi River were the costliest in American history.

Geomorphologists are increasingly engaged in the mapping and modelling of geophysical, hydrological and surficial material characteristics which expose areas to rupture, failure, fire, inundation, drought, erosion or submergence. Coupled with land use and human infrastructure analyses they examine the location, value, exposure and vulnerability of the human environment to hazard damage (see ENVIRONMENTAL GEOMORPHOLOGY). When the population density and demographics are added, potential casualties and emergency services needs may be forecast. This requires the integration of social scientists who focus upon social, technical, administrative, political, legal and economic forces which structure a society's strategies and policies for risk management (i.e. prevention, mitigation, preparedness, prediction and warning, and recovery), public awareness, emergency training, regulation and social insurance. Such a comprehensive approach would have been nearly inconceivable in the past, but with the advent of computerized geographic information systems (GIS) such mapping, modelling, and decision support systems are becoming more commonplace (Carrara and Guzzetti 1995).

Perplexing questions centre upon the apparent increase in frequency of catastrophic geomorphic hazards. Accurate statistical analyses of such infrequent occurrences require an observation period of well in excess of a century (Berz 1993), while consistent reporting of most types of disasters have a much shorter history. Monitoring techniques and measurement scales (e.g. Richter, Beaufort), remote sensing, and communications have only recently allowed the global reporting of events in comparable terms. A study of volcanic eruptions by Simpkin *et al.* (1981) concludes that the reported increase in volcanic activity over the past 120 years is almost certainly due to improved reporting and communications technology; they even report a reduction in 'apparent' activity during the two world wars. Despite the growing influence of global databases, scientific consortia, and the news media, additional factors may be influencing the growing number of reported disasters.

While considerable scientific debate lingers around the issue of global climatic change, geomorphologists are well aware of other indirect effects of human activity (Rosenfeld 1994b). Certainly the deforestation of large areas has caused landslides and increased both the frequency and peak flows of flood events in many areas, while overgrazing has accelerated drought effects and erosion. Groundwater withdrawal and irrigation diversions have affected natural vegetation and micro-climates of some regions, and have even induced earthquakes in some cases.

As climate models become increasingly realistic, their mathematical results consistently point to a more hazardous world in the future. That increasing concentrations of greenhouse gases in the atmosphere, primarily resulting from the burning of fossil fuels, is changing the radiation balance and perhaps the climate is consistent with recent disaster experience. There is general agreement among atmospheric scientists that a 'warmer' world would be a 'wetter' world, with no increase in the number of days with rain, but with more intense rainfall. Combined with the hydrologic effects of land use changes, the frequency and severity of floods would surely increase, especially in the monsoon climates of south Asia where flooding already reaches catastrophic proportions. Drought effects in sub-Saharan Africa, South America and Australasia could occur more frequently and be more severe as a result of intensified El Niño–Southern Oscillation events. Resultant sea-level rise could pose additional storm surge or tsunami risk to heavily populated low-lying coastal regions such as lower Egypt, Bangladesh and many Pacific islands, along with the loss of most freshwater resources in the latter case. At higher latitudes, global warming may induce profound effects upon the human use and occupancy of land underlain by permafrost. Regardless of the causes, the impacts of anticipated changes in extreme weather hazards as a result of global climatic change, and their implications for human activity, demand the attention of geomorphologists.

Observational framework: natural hazards paradigms

Early academic research into natural hazards was characterized by an emphasis on human response to natural events. American geographer Gilbert White (1974) proposed the following research paradigm:

1 estimate the extent of human occupancy in areas subject to natural hazards;
2 determine the range of possible adjustment by social groups to those extreme events;
3 examine how people perceive the extreme events and resultant hazards;
4 examine the process of choosing damage-reducing adjustments;
5 estimate the effects of varying public policy upon that choice process.

This view emphasizes human response to specific catastrophic events, focusing on only extreme events, and implying that rational decisions are made based on cultural perceptions. Subsequent studies have increased the importance of risk assessment and the vulnerability of a population based upon the probability of an event. Although these views appear to reduce the role of the geomorphologist, the evaluation of a site with respect to specific risk lies at the very heart of hazard research.

Burton *et al.* (1978) suggest ranking the significance of potential hazards by evaluating the physical parameters of an event in terms that are obvious to geomorphologists:

1 magnitude: high to low
2 frequency: often to rare
3 duration: long to short
4 areal extent: widespread to limited
5 speed of onset: rapid to slow
6 spatial dispersion: diffuse to concentrated
7 temporal interval: regular to random.

Although qualitative, this view recognizes that events can range from intensive (such as a storm surge briefly affecting a stretch of coastline) to pervasive (such as the erosional effects of global sea-level rise).

Causal linkages are inherent in the notion of *geomorphic* hazards, where an extreme event may initiate other exceptional events of another type. Thus geomorphic hazards that are associated with landform response may be 'triggered' by climatic, hydrological, geophysical or man-induced events. Landslides may be causally linked with earthquakes, volcanic eruptions, heavy precipitation or construction activity. Pervasive linkages can result from land use change within a watershed affecting the magnitude and frequency of discharge within the stream.

Planners and developers often focus on a particular site or region, where mapping of hazard areas evaluates the potential risks for all potential hazards in such locations. Most physical scientists shun this 'hazardousness of place' concept, not wanting to venture beyond their own areas of expertise. Many social scientists characterize the actual hazard event only by its immediate physical effects, concentrating only on the societal response.

Geomorphologists recognize that the high-magnitude, low-frequency catastrophic events (large earthquakes, hurricanes) capture the attention due to the immediacy of large casualty and financial losses, but that events of moderate frequency (landslides, floods) often do as much or more collective damage. Geomorphic hazards tend to be more at the pervasive end of the hazards continuum, have slower speed of onset, longer duration, more widespread areal extent, more diffuse spatial dispersion, and more regular temporal interval. Exceptions such as slope failure exist, but in general landform change occurs over the long term at slow rates. Nevertheless geomorphologists should adopt a hazard paradigm in an effort to promote compatibility within this area of complex, and essential, interdisciplinary research.

Geomorphic hazards research

Gares *et al.* (1994) use the paradigm suggested by Burton *et al.* (1978) to discuss the role of geomorphic hazards research with respect to specific hazards. They illustrate the great variety of processes involved and suggest geomorphic evaluation in terms of the following aspects:

1 the dynamics of the physical process;
2 the prediction of the rate or occurrence;
3 the determination of the spatial and temporal characteristics;
4 an understanding of people's perception of the impact of the occurrence;
5 knowledge of how the physical aspects can be used to formulate adjustments to the event.

Geomorphologists vary widely in their definition of geomorphic hazards. Gares *et al.* (1994) limit their inclusion only to those process actions that gradually shape landforms, not agents of catastrophic change that arise from the consequences of geophysical, hydrological or atmospheric hazards, although many of these hazards result in geomorphic events. Wolman and Miller (1960) recognized that low-frequency, high-magnitude events often produce spectacular damage and geomorphic change, but events of moderate magnitude often do as much work (damage, change) cumulatively over the long run. As an encyclopedic entry, our definition will be necessarily inclusive of all agents of surficial change, pervasive to episodic.

Pervasive processes, such as soil erosion, are minimal in natural environments, but accelerate greatly with human disturbance such as forest clearing or agricultural tilling. Perception of soil erosion as a hazard involves farmers who lose crops to sheet wash and gullies, water managers and engineers who suffer siltation in reservoirs or canals, fishermen whose catch is reduced by silt and turbidity, and all who suffer from reduced crop and water yields. Soil erosion rarely results in direct loss of life, but it has a widespread distribution, high remediation costs, and long-term effects on water and food production. Despite more than seventy years of soil erosion mitigation research, countless thousands suffer malnutrition due to lost soil productivity.

Numerous geomorphic processes have causal links to volcanic eruptions. Geophysicists and volcanologists monitor eruptive precursors, prior eruptive history and distrubution of past eruptive products to assist disaster managers with warnings about the type and magnitude of imminent risk. Geomorphologists contribute to post-eruption mitigation efforts, as impacts such as pyroclastic flows and ash fall frequently result in unstable slopes, lahars clog stream channels, and overall sediment yield is greatly increased. Siltation and debris loading of streams radiating out from affected areas result in reduced channel capacity, increasing the frequency of overbank flooding, causing flow deflections and bank erosion. Applications of geomorphology include erosion control and engineering impact analysis, along with mitigation and recovery planning. Since the 1980s geomorphologists have made significant contributions following the eruptions of Mt. St Helens, USA, Mt. Pinatubo, Philippines and Nevada del Ruiz, Colombia.

Heavy rainfall can saturate soils causing rapid debris flows and mudflows. In 1938, such events in Japan were triggered by typhoon rains, resulting in the loss of more than 130,000 homes and over 2,000 lives. The magnitude of this loss prompted government attention focused on landslide control, and similar rains in 1976 affected less than 2,000 homes and cost 125 lives. Similar reductions have occurred with other catastrophic events. The horrific death toll experienced in Bangladesh due to storm surge and flooding in 1970 was reduced by thirty to fifty times during a cyclonic storm surge of similar magnitude in 1985 because a satellite-based early warning system prompted the evacuation of island and coastal dwellers. These two examples come from opposite ends of Asia's economic spectrum. Rosenfeld (1994a) points out that economically developed countries often suffer the greatest economic losses, while their lesser developed counterparts endure the highest loss of life. In developing nations, there is often a conscious decision to allocate resources toward economic development, at the risk of underfunding disaster mitigation, often with the effect of greater loss of both infrastructure and human lives.

In some instances, disaster mitigation strategies and international relief efforts may actually be partially responsible for rising losses. In many developed countries, state-backed hazard insurance programmes are designed to encourage the use of hazard zoning and the implementation of damage-resistant building codes to reduce the demand for structural control measures. However this may have actually encouraged the development of hazard-prone areas through the combined effect of lower land costs and cheap indemnity. Thus, the 'insured' transfers the risk to the 'insurer' and may dismiss the concern for loss prevention measures.

Geomorphologists have the opportunity to demonstrate the nature of geomorphic hazards (Figure 66), map landform or surface material conditions that have hazard potential, and recognize the effects of human modification of natural conditions which could result in increased hazard potential. As scientists, we are reluctant to translate this knowledge into arguments for the adoption of specific mitigation or management strategies, and thus are less than proficient at applying our information base. Most land use managers, planners, developers and government decision-makers rely on 'on the job' training to develop expertise in the interpretation of technical information for risk assessment and disaster reduction. Often this is prompted only in response to significant losses.

Automated monitoring networks and advanced computer modelling techniques are giving us new tools to test alternative hazard mitigation strategies. Geomorphologists must be willing to embrace new technologies which will permit them to exercise their specialized talents globally, interface more readily with professionals outside the Earth science community (for example social scientists and engineers), and associate more closely with monitoring networks and scientific unions to ensure that major events are anticipated by identifying their physical precursors. Natural hazards research is obviously an interdisciplinary field involving a range of physical scientists with social scientists assessing the human dimension of the problems. Given the limited observed record of hazard events in most regions, a geomorphological approach, where the areal extent, and perhaps the frequency, of events can be determined from the landscape, is essential. The geomorphological approach may also encourage the 'nature knows best' path to designing hazard mitigation strategies in balance with the dynamics of processes within the region. In the final analysis, occupance of hazard-prone areas is both physically and economically self-regulating. It is the function of science, as a servant of society, to identify those limitations and point the way toward minimizing the disastrous consequences of 'learning our lessons' in nature's way.

References

Berz, G. (1993) The insurance industry and IDNDR: common interests and tasks, *IDNDR Newsletter 15*, Observatorio Vesuviano, 8–11.

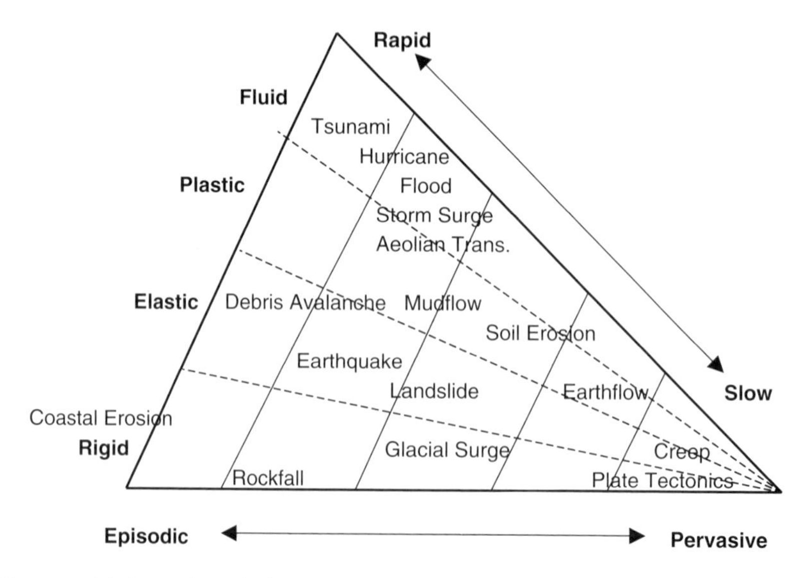

Figure 66 Geomorphic hazards include pervasive processes, which may be imperceptible to individuals, to episodic events which may have frequencies of occurrence below thresholds deemed significant by planners, but which may have significant magnitude. These processes involve the full spectrum of stress/strain modulus, and involve virtually every speciality within the discipline

Bruce, J.P. (1993) Natural disasters and global change, *IDNDR Newsletter 15*, Observatorio Vesuviano, 3–8.
Burton, I., Kates, R.W. and White, G.F. (1978) *The Environment as Hazard*, Oxford: Oxford University Press.
Carrara, A. and Guzzetti, F. (eds) (1995) *Geographical Information Systems in Assessing Natural Hazards*, Dordrecht: Kluwer.
Gares, P.A., Sherman, D.J. and Nordstrom, K.F. (1994) Geomorphology and Natural Hazards, *Geomorphology* 10, 1–18.
Rosenfeld, C.L. (1994a) Flood hazard reduction: GIS maps survival strategies in Bangladesh, *Geographical Information Systems* 2(3), 29–39.
——(1994b) The geomorphological dimensions of natural disasters, *Geomorphology* 10, 27–36.
Simpkin, T., Seibert, T., McClelland, L., Bridge, D., Newhall, C. and Latter, J. (1981) *Volcanoes of the World*, Stroudsburg, PA: Hutchinson Ross.
White, G.F. (1974) Natural hazards research: concepts, methods and policy implications, in G.F White (ed.) *Natural Hazards: Local, National, Global*, 3–16, Oxford: Oxford University Press.
Wolman, M.G. and Miller, J.P (1960) Magnitude and frequency of forces in geomorphic processes, *Journal of Geology* 68, 54–74.

CHARLES L. ROSENFELD

GEOMORPHOLOGICAL MAPPING

Geomorphological mapping encompasses one of a group of techniques under the general category of TERRAIN EVALUATION employed to record systematically the shape (or morphology), landforms, landscape-forming processes and materials that constitute the surface of the Earth. Lee (2001) identifies three forms of geomorphological map:

1 Regional surveys of terrain conditions, either for land use planning or in baseline studies for environmental impact assessment (e.g. the 1:25,000 scale maps of Torbay, Doornkamp (1988)).
2 General assessments of resources or geohazards at scales between 1:50,000 and 1:10,000 (e.g. Bahrain Surface Materials Resources Survey, Doornkamp *et al.* (1980); ground problems in the Suez City area, Egypt, Jones (2001)).
3 Specific-purpose large-scale surveys to delineate and characterize particular landforms (e.g. the 1:500 scale investigations around the Channel Tunnel portal, Folkestone, Griffiths *et al.* (1995)).

The initial stage of geomorphological mapping involves factually recording ground shape through a process of morphological mapping. This requires the production of a map on which the land surface is subdivided into planar facets separated by gradual changes or sharp breaks in slope. On the map the changes and breaks in slope are identified as either concave or convex in nature and recorded using decorated lines, a system first established by Savigear (1965). Arrows with a numeric value in degrees indicate the slope angle and downslope direction of the planar facets. Once the morphology has been recorded a geomorphological interpretation is undertaken whereby details of the contemporary and relict landforms and geomorphological processes are added to the map. In addition, data on the nature of materials and hydrology of the area are noted. Geomorphological interpretation can allow a suite of derivative maps to be produced, e.g. resource maps and landscape genesis maps. Standard symbols to be used on all these maps are contained in Cooke and Doornkamp (1990), although, Demek and Embleton (1978) provide a more comprehensive collection of symbols that allow subtle differences in the landscape to be highlighted. However, in many situations the geomorphological maps are produced as unique products with a bespoke legend.

The techniques used to compile the data involve both field survey and, where possible, examination of remote sensing information. The main form of remote sensing analysis has traditionally been through the interpretation of vertical pairs of aerial photographs viewed stereoscopically. An initial preliminary morphological map and geomorphological interpretation is produced using aerial photographs but this should normally be subject to 'ground-truth' mapping in the field. With the advent of higher resolution satellite images this preliminary mapping stage increasingly is being carried out using data from the array of new satellite-based scanners.

A two-person team normally undertakes the field mapping. The main requirement for the production of effective geomorphological maps is an accurate base map at a suitable scale. The base map may be a standard survey map depicting man-made and natural features including ground topography, or a spatially corrected ortho-photo. The field data should be compiled directly on the base map. Spatial data and slope information can be obtained through a simple tape, compass and clinometer survey, using more sophisticated land survey techniques, use of global positioning systems, or a suitable combination of these methods. The geomorphological, materials and hydrological data are noted on maps and recorded in field notebooks where appropriate.

Whilst geomorphological mapping has been used for general landscape investigations, it has been employed most successfully by applied geomorphologists, particularly for engineering studies. Brunsden *et al.* 1975, articulated the aims of geomorphological mapping for highway engineering:

1 Identification of the general terrain characteristics of the route corridor, including suggestion of alternative routes and location of hazards.
2 Defining the 'situation' of the route corridor, for example identifying influences from beyond the boundary of the corridor.
3 Provision of a synopsis of geomorphological development of the site, including location of materials for use in construction and location of processes affecting safety during and after construction.
4 Definition of specific hazards, e.g. landsliding, flooding, etc.
5 Description of drainage characteristics, location and pattern of surface and subsurface drainage, nature of drainage measures required.
6 Slope classification, according to steepness, genesis and stability.
7 Characterization of nature and extent of weathering, also susceptibility to mining subsidence and erosion.
8 Definition of geomorphological units, to act as a framework for a borehole sampling plan and to extend the derived data away from the sample points.

Although these aims were developed specifically for highway projects they represent an appropriate checklist for all geomorphological mapping programmes undertaken for civil engineering projects.

References

Brunsden, D., Doornkamp, J.C., Fookes, P.G., Jones, D.K.C. and Kelly, J.M.N. (1975) Large scale geomorphological mapping and highway engineering design, *Quarterly Journal of Engineering Geology* 8, 227–253.

Cooke, R.U. and Doornkamp, J.C. (1990) *Geomorphology in Environmental Management*, 2nd edition, Oxford: Oxford University Press.

Demek, J. and Embleton, C. (eds) (1978) *Guide to Medium-scale Geomorphological Mapping*, Stuttgart: International Geographical Union.

Doornkamp, J.C. (ed.) (1988) *Planning and Development: Applied Earth Science Background, Torbay*, Nottingham: MI Press.

Doornkamp, J.C., Brunsden, D., Jones, D.K.C. and Cooke, R.U. (1980) *Geology, Geomorphology and Pedology of Bahrain*, Norwich: GeoBooks.

Griffiths, J.S., Brunsden, D., Lee, E.M. and Jones, D.K.C. (1995) Geomorphological investigation for the Channel Tunnel and Portal, *Geography Journal*, 161, 257–284.

Jones, D.K.C. (2001) Ground conditions and hazards: Suez City development, Egypt, in J.S. Griffiths (ed.) *Land Surface Evaluation for Engineering Practice*, Geological Society Engineering Geology Special Publication 18, 159–170.

Lee, E.M. (2001) Geomorphological mapping, in J.S. Griffiths (ed.) *Land Surface Evaluation for Engineering Practice*, Geological Society Engineering Geology Special Publication 18, 53–56.

Savigear, R.A.G. (1965) A technique of morphological mapping, *Annals of the Association of American Geographers* 53, 514–538.

Further reading

Fookes, P.G. (1997) Geology for engineers: the geological model, prediction and performance, *Quarterly Journal of Engineering Geology* 30, 290–424.

JAMES S. GRIFFITHS

GEOMORPHOLOGY

Definition and scope

Geomorphology is the area of study leading to an understanding of and appreciation for landforms and landscapes, including those on continents and islands, those beneath oceans, lakes, rivers, glaciers and other water bodies, as well as those on the terrestrial planets and moons of our Solar System. Contemporary geomorphologic investigations are most commonly conducted within a scientific framework (see Rhoads and Thorn 1996) although academic, applied or engineering interests may motivate them. A broad range of alternative research methodologies have been employed by geomorphologists, and past attempts to impose a systematic structure on the discipline have yielded stifling tendencies and overt resistance. Geomorphologists frequently profess to innate aesthetic appreciation for the complex diversity of Earth-surface forms, and, in this regard, a fitting definition of geomorphology is simply 'the science of scenery' (Fairbridge 1968).

Past and present concerns have focused on the description and classification of landforms (including their geometric shape, topologic attributes and internal structure), on the dynamical processes characterizing their evolution and existence, and on their relationship to and association with other forms and processes (geomorphic, hydro-climatic, tectonic, biotic, anthropogenic, extraterrestrial, or otherwise). Geomorphology is an empirical science that attempts to formulate answers to the following fundamental questions. What makes one landform distinct from another? How are different landforms associated? How did a particular landform or complex landscape evolve? How might it evolve in the future? What are the ramifications for humans and human society?

Modern geomorphology is currently subdivided and practised along the lines of specialized domains. *Fluvial geomorphology*, for example, is concerned with flowing water (primarily in the form of rivers, streams and channels) and the work it accomplishes during its journey through the terrestrial phase of the hydrologic cycle. A very broad spectrum of interests are subsumed within fluvial geomorphology, ranging from the influence of turbulence on the entrainment, transport and deposition of sediment particles at the finest scale, to the mechanics of MEANDERING, POINT BAR formation and FLOODPLAIN development at middle scales, to the nature and character of DRAINAGE BASIN evolution at the coarsest scales. Within the other substantive areas of geomorphology are: *hillslope geomorphologists*, who boast expertise on the geotechnical properties of soil and rock, the mechanics of LANDSLIDES, and the movement of water within the ground; *tectonic geomorphologists*, who study neotectonic (see NEOTECTONICS) stress fields, continental-scale sedimentary basins and active/passive margin landscapes; *glacial and periglacial geomorphologists*, who are interested in alpine and continental glaciers, PERMAFROST and other cold-climate forms or processes that involve ice, snow and frost; *karst geomorphologists*, who deal with soluble rocks (e.g. limestone) and chemical processes of DISSOLUTION that lead to landforms such as gorges, caverns and underground streams; *coastal geomorphologists*, who study nearshore, lacustrine and marine systems where oscillatory, rather than unidirectional, flow processes dominate; and *aeolian geomorphologists*, who study the transport of sand and dust by wind, mostly in desert or semi-arid environments, but also along beaches, over agricultural fields and on the moon and Mars. Other subspecialities include: soils geomorphology, biogeomorphology (zoogeomorphology), climatic geomorphology, tropical geomorphology, desert geomorphology, mountain geomorphology, extraterrestrial (planetary) geomorphology, remote-sensing geomorphology, experimental geomorphology, environmental geomorphology, forest geomorphology, applied geomorphology, engineering geomorphology and anthropogeomorphology.

Major themes and concepts

Landforms are dynamic entities that evolve through time as a consequence of characteristic suites of processes acting upon Earth-surface materials. Geomorphologists are concerned with documenting and unravelling the mysteries of this process-form interaction. Relevant knowledge includes not only the manner and direction of landform evolution (progressive or cyclic, slow or rapid), but also the processes that dominate or direct the evolution (type, intensity) as well as the mutual adjustments and feedbacks that occur between the forms and processes as energy and matter are cycled through the landscape. To better understand these complex interrelationships, geomorphologists have proposed various conceptual themes or templates to aid in organizing their thinking. Among these are:

1. *Endogenic–exogenic forces* Geomorphic systems are governed by dynamic controls that may be internally produced (endogenic) or externally imposed (exogenic) upon the system. Tectonic, volcanic and isostatic activities are manifestations of endogenic forces within Earth, whereas rainfall and meteorite showers are exogenic forces. The spatial and temporal scales of the geomorphic system influence the types of endogenic–exogenic forces that are relevant. The control–volume, force–balance approach in fluid mechanics is an analogue to this concept.
2. *Destructive–constructive action* Some geomorphic processes create landforms (e.g. volcanic cones, meteorite impact craters, termite mounds) whereas other processes (e.g. chemical weathering, rainwash, human activity) destroy landforms or cause widespread denudation. More typically, most geomorphic

processes both create and destroy landforms simultaneously. For example, flowing water in a river will erode the outer bank of a meander bend while depositing sediment on the inner bank in the form of a point bar. Similarly, a glacier can erode, excavate and sculpt the surface upon which it moves while also depositing sediment in the form of eskers and moraines.

3 *Erosional–depositional forms* Some landforms are sculpted by erosion of pre-existing materials (e.g. bedrock canyons, roches moutonnées) whereas others are built via deposition of new material on existing substrate (e.g. deltas, lava flows). Yet others are hybrid features formed through both erosion and deposition locally (e.g. impact craters) or maintained by an intricate balance between erosion and deposition at different positions on the same form (e.g. migrating sand dunes).

4 *Stress–strength relationships* Most geomorphic processes induce landscape change by stressing the system, as with flowing fluids, chemical reactions, tectonic motion or the prolonged action of gravity. The materials upon which these processes act have the ability to resist change because of inherent properties that provide strength (e.g. mineralogy, cohesion, structure, relative placement). Geomorphologists have generally devoted more effort to measuring processes than to investigating how material strength is a complementary and counterbalancing factor (exceptions are many, and they include the efforts of the Japanese school to elucidate systematically the nature of rock control on geomorphic evolution, of the many coastal geomorphologists interested in rocky coasts, of several geomorphological geologists concerned with relict landforms and ancient landscapes, and various engineering geomorphologists who study slope failures).

5 *Polygenesis and inheritance* Landscapes consist of landform assemblages that are rarely simple. Complex suites of polygenetic forms may coexist in the same location if, for example, multiple processes are active contemporaneously or when particularly resistant relict forms are inherited from prior eras. The latter are progressively modified by contemporary processes that also create new forms to produce palimpsest landscapes.

The integrated sum of exogenic–endogenic forces and destructive–constructive actions working in concert to create erosional–depositional forms according to dominant stress–strength relationships will dictate whether landscape RELIEF will be enhanced or reduced over a given time interval. At one end of the spectrum are the steep mountain and valley systems of the globe (e.g. Himalayas), and at the other are the extensive abyssal plains of the deep ocean as well as the PENEPLAIN of William Morris Davis. Geomorphologists have identified several additional themes and concepts that serve to strengthen the theoretical foundation of their science. These include scale, causality, equilibrium, equifinality, thresholds, magnitude–frequency, landscape memory and relaxation. Readers should consult the references and other entries in this encyclopedia for detailed discussions.

Early historical development

The subject matter of geomorphology has occupied human thinking for thousands of years, and early writings on landforms can be traced to the time of the ancient Greek, Roman, Arab and Chinese philosophers. Aristotle (384–322 BC) and Strabo (54 BC–AD 25), for example, had keen insights into the origin of springs, the work of rivers and the importance of earthquakes and volcanoes. Nevertheless, the history of geomorphology (see Chorley *et al.* 1964; Tinkler 1985, 1989) is typically traced back only as far as the European Renaissance because few written documents about geomorphic knowledge remain from the period prior to the sixteenth century. During the Renaissance, most studies of Earth were conducted from a naturalist, philosophical perspective because specialized academic disciplines had not evolved and scientific methods were not widely known. Leonardo da Vinci, Bernard Palissy, Nathanael Carpenter, Bernhard Varenius, Thomas Burnet and Nicolaus Steno are among the key figures from this period, and unwittingly they began to lay the foundation for the science of geomorphology. Unfortunately, this was also a time when the Church exerted powerful control over academic thinking, and the predominant objective of learned men was to reconcile their day-to-day observations of natural processes with strict religious orthodoxy and bibliolatry. The biblical scholar, James Ussher, Archbishop of Armagh, decreed that Earth was created on Sunday, 23 October 4004 BC and that the Flood of the Old Testament began in 2349 BC, and in so doing, he may well have imposed the most stifling proclamation on the developmental history of the Earth sciences. All evolutionary processes, by definition, had now to be contemplated within the constraints of a 6,000-year Earth history, and to think otherwise was heresy. Unsurprisingly, the dominant interpretation of Earth-surface processes invariably involved catastrophes, cataclysms and disasters such as global deluges and seismic convulsions.

The period following the Renaissance and into the early nineteenth century was one of scepticism, controversy and debate. It was also one that witnessed several changes that bear directly on the development of geomorphology as an academic discipline. The first was the evolution of specialized areas of study such as biology, physics, astronomy, mathematics, hydraulics and geology, and this set the stage for various subdisciplines, such as petrology, mineralogy, paleontology, stratigraphy and geomorphology, to be spawned. Second was a slow transformation in academic discourse away from the unassailable validity of belief systems and authoritarianism toward a standard of proof based on empiricism and observable evidence. Third was the development of increasingly sophisticated instrumentation and measurement technologies and protocols. Fourth was the enhanced mobility of people and information, thereby facilitating greater exposure to new and interesting environments and ideas. And fifth was the gradual acceptance of gradualism (see UNIFORMITARIANISM) in favour of CATASTROPHISM. Two dominant factions emerged during this period. The Neptunists (or Wernerians) followed the ideas of a German mineralogist, Abraham Gottleb Werner, who contended that rocks on Earth originated from mechanical and chemical processes in a universal ocean. The Plutonists (or Vulcanists) stressed the importance of intrusive and extrusive volcanic processes in rock formation. Key figures during this period include members of the 'French School', such as Jean Étienne Guettard, Nicolas Desmarest and Jean-Baptiste Lamarck, as well as the Swiss geologist, Horace Benedict de Saussure.

James Hutton, credited by some as the founder of modern geomorphology, was a Plutonist who argued vehemently for the importance of gradual subaerial denudation across millennia. His uniformitarian ideas, expressed in well-known phrases such as 'the present is the key to the past' and 'no vestige of a beginning, no prospect of an end', were revolutionary because they shifted the focus of attention away from catastrophic events of 'creation' toward continuous, everyday agents of erosion. Unfortunately, Hutton's teachings were not warmly received by the conservative cognoscenti of that time. After Hutton's death, his friend and colleague, John Playfair, published a book that explained and expanded Hutton's writings, and by the beginning of the nineteenth century, a slow conversion to gradualism was taking hold. Cyclic and timeless theories of landscape evolution were coming into vogue. Three schools of thought regarding landform evolution emerged. DILUVIALISM represented a transformed extension of the catastrophist lineage, and diluvialists such as Reverend William Buckland and Reverend Adam Sedgwick believed that huge floods carved many surface features. Structuralists, such as Henry Thomas de la Beche and John Phillips contended that structural controls were paramount to understanding landscape genesis (see STRUCTURAL LANDFORM), while also acknowledging that both catastrophic and gradual processes could yield substantial erosion. Fluvialists, in contrast, argued for the dominance of rivers and streams in wearing away the landscape through slow, but continuous action.

A chief proponent of fluvialism and UNIFORMITARIANISM was Sir Charles Lyell, whose *Principles of Geology* went into twelve editions after original publication in 1830. Lyell based his arguments on careful observations and measurements, and effectively attacked the notions of theological reconciliation, catastrophism and diluvialism. His writings on uniformitarianism incorporated four distinct notions: (1) uniformity of law (the laws of nature are immutable); (2) uniformity of process (processes operative today were also operative in the past, and exotic causes need only be invoked unusually); (3) uniformity of rate (gradualism); and (4) uniformity of state (change is endlessly cyclical and directionless). The publication of Lyell's *Principles* engendered considerable debate, and the period through to the middle of the nineteenth century witnessed both conflict and compromise regarding the importance of fluvial action, pluvial denudation, marine dissection, iceberg drift and glaciation (see GLACIAL THEORY) as agents of erosion. Indeed, even Lyell began to expound the virtues of marine dissection above fluvial degradation. In part, this was due to the existence of various unexplainable observations such as major unconformities in the stratigraphic record and huge ERRATICs in unexpected places. The powerful action of the sea presented an expedient solution because submarine processes could not be observed or measured directly, and theorization could proceed unbridled. Nevertheless, most of these seemingly contradictory theories incorporated at least some common elements and themes and, invariably, they were cast within a framework of uniformitarianism rather than catastrophism.

By the mid-1870s, some consensus was beginning to emerge about the multifaceted and complex nature of landscape evolution. The marine planation theory of Sir Andrew Crombie Ramsay, for example, proposed that the action of waves and currents in the ocean was not to dissect the sea bottom, but to level off bathymetric protuberances thereby producing marine plains. Upon emergence through tectonic activity, subaerial forces become active and fluvial erosion proceeds to carve out valleys and denude landscapes. Support for this theory came from the many accordant summit heights in the highlands of Wales and England, as well as from the marine abrasion studies of Baron Ferdinand von Richthofen in China. Concurrently, the glacial theories (see GLACIAL THEORY) of Ignace Venetz, Jean de Charpentier and Louis Agassiz were receiving widespread acceptance decades after their introduction, albeit with climatic and glaciofluvial amendments. This was a significant development in geomorphology because environmental dynamism (see DYNAMIC GEOMORPHOLOGY) was implicit to these theories. Gradualist and neo-catastrophist perspectives could both be accommodated under this new framework because uniformity of process (the nature of past and present processes are the same) did not necessarily imply that the intensities and rates of process action could not vary.

At the conclusion of the nineteenth century, geomorphology was poised to begin its emergence as a modern scientific discipline. The word 'geomorphology' had already been coined in the mid-1800s (Tinkler 1985: 4), and several textbooks on exclusively geomorphic matters had been written. As an area of academic study, geomorphology was experiencing legitimate interest under the guise of 'physiography' or 'physiographical geology'. Centres of expertise were arising in many different countries within and outside Europe, all with subtly different identities and separate agendas. British geomorphologists, for example, spent considerable effort on compiling complex denudation chronologies linked to marine processes and periods of tectonic stability/instability and sea-level fluctuation. German geomorphologists (e.g. Hettner, A. Penck, Walther) became interested in the influence of climate as a consequence of conducting research in the Alps as well as in the subhumid tropics. The North American school, in contrast, was dominated by fluvialism buoyed by indisputable evidence derived from the great explorations of the largely unvegetated, semi-arid West. John Wesley Powell's trips into the Grand Canyon and his reports on the Colorado Plateau and Uinta Mountains provided powerful testimony to the efficacy of rivers to erode landscapes. Grove Karl Gilbert's studies on the mechanics of fluvial erosion, sediment transport and turbulence are exemplars of the elegant application of the scientific method. He also investigated the origin of pediments and lateral planation, and in recognition of his many contributions, Gilbert is often identified as the first truly process-oriented American geomorphologist. Indeed, it is largely due to the efforts of Powell, Gilbert, Dana and Dutton and various other United States Geological Survey employees that the North America school became the dominant force in the development of geomorphology at the turn of the century.

Twentieth-century developments

Geomorphology in the twentieth century experienced rapid evolution and growth, and six overlapping phases of development can be identified. These are little more than crude caricatures, and the reader is referred to Chorley *et al.* (1973) and Beckinsale and Chorley (1991) for detailed discussions of the key figures and their substantive contributions. The *historical* phase, roughly from 1890–1930, was dominated by William Morris Davis and his many disciples. Davis's deductively derived model, 'The Geographical Cycle', envisioned serial evolution of landscapes beginning with rapid tectonic uplift followed by progressive denudation in characteristically distinct stages of 'youth', 'maturity' and 'old age'. It served as the genetic template upon which reconstructive narratives of landscape evolution were hung, with relatively little concern for the mechanical and chemical processes responsible for erosion and deposition. Nevertheless, these denudation chronologies spawned keen interest in tectonic geomorphology as well as an appreciation for the importance of unravelling the historical sequence of steps that ultimately manifest themselves as a contemporary landscape.

The *regionalist* phase (1920–1950) was characterized by detailed and thorough investigations of regional landscapes, both in the conventional mid-latitudes of North America and Europe as well as in globally remote areas (e.g. tropics, deserts, high latitudes). Increasingly, these regionally based studies yielded data about landforms and landform assemblages that could not be easily explained within the framework of Davis's geographical cycle, especially his contentions about the 'normalcy' of the humid, mid-latitudes. Although Davis found support within Britain and France, many European schools remained unconvinced by Davis's teachings. Walther Penck, for example, proposed an alternative model of landscape evolution that highlighted the importance of the relative rates of uplift and denudation in controlling landform geometry. Another German geomorphologist, J. Büdel, stressed the dominance of climatic controls and proposed the concept of etchplanation (see ETCHING, ETCHPLAIN AND ETCHPLANATION) and MORPHOGENETIC REGIONS. Climatic geomorphology was also practised by Louis Peltier in North America and it was later championed by J. Tricart and A. Caillieux in France. In this way, Davis's unifying ideas gradually fell into disfavour, and geomorphology became an empirically driven scientific confederacy of polyglot regionalist schools.

The *quantitative* phase (1940–1970) reflected a broader trend within many of the Earth sciences toward enhanced use of sophisticated technologies (often derived from the war effort) to measure, describe and analyse the surface features of Earth. R.E. Horton's publications on stream networks and drainage basin processes are classically identified as the precursor to this quantitative movement, but the foundational works of Bagnold, Gilbert, Hjulstrom, Leighly, Rubey and Shields, among many others, are rightfully acknowledged. These early 'quantifiers' were concerned to understand landforms and geomorphic processes in deterministic or probabilistic, but testable, ways rather than on the basis of deductively derived heuristic models that ultimately yielded little predictive power. Logical positivism was the dominant philosophy and reductionism was the overriding methodological approach. As a consequence, geomorphology became increasingly fragmented and specialized, with fewer and fewer connections between the sub-specializations as well as pronounced distancing from its mother disciplines of geography and geology. Fortunately, connections to other allied disciplines such as fluid mechanics, engineering hydrology, statistics, thermodynamics, meteorology, pedology and agricultural physics were being cultivated, and these provided a theoretical and conceptual richness upon which geomorphologists could draw, if so inclined.

The *systems* phase (1960–1980) in the development of geomorphology was inaugurated by the introduction of general systems theory into the conceptual toolkit of geomorphology by Richard J. Chorley, which was a logical outgrowth of the quantitative phase. The quantification of prior decades was basically of two genres: (a) statistical 'black-box' description (e.g. Horton's Law of Stream Numbers); and (b) detailed measurement and interpretation of dynamical processes (e.g. Strahler 1952). The former proved unrewarding in terms of providing insight into geomorphic behaviour, whereas the latter were typically conducted at a scale that was too small to be relevant to landscape evolution. The systems approach alleviated the 'black-box' quandary by describing geomorphic behaviour in terms of energy and mass flows, equilibrium tendencies, relaxation times and thresholds (see THRESHOLD, GEOMORPHIC) of response. A large number of concepts, such as ALLOMETRY, entropy and ergodicity (see ERGODIC HYPOTHESIS), were borrowed from other disciplines as theoretical templates. These were applied to a broad range of geomorphic systems with varying degrees of success, but the large number of journal articles and textbooks containing box-and-arrow plots attests to the popularity of this approach during the systems phase. Unfortunately, there was an irresistible tendency to equate system behaviour with geomorphic process, much to the detriment of dynamical process investigations.

Since about the 1980s, geomorphology has entered a phase of increasing *reconciliation* and *unification* that signals its arrival as a mature modern science. Introspective debates about catastrophic versus uniformitarian ideas, quantitative-deterministic/stochastic versus qualitative- historical methodologies, and geographical versus geological disciplinary roots are taking place not for purposes of disciplinary leadership or hegemonic posturing, but rather in consequence of the pragmatic need for geomorphology to assert an identity distinct from that of other Earth sciences (e.g. geology, geography, sedimentology, stratigraphy, paleontology) as well as to understand the complex spectrum of conceptual ideas upon which geomorphology is founded (e.g. Rhoads and Thorn 1996). Many extremist ideas of prior eras have been reintroduced into the literature as softened compromises (e.g. NEOCATASTROPHISM, neo-historicism, neo-regionalism) to provide balance to the uniformitarian-style fluvialism that dominated the quantitative and system phases. Invariably, these conceptual ideas were discussed in the context of factual evidence and with a view toward generating insight into unusual geomorphic features or terrain that belie conventional explanation (e.g. Baker 1981). The modern-day geomorphologist has a deep appreciation for the importance of slowly acting processes in concert with large-magnitude, low-frequency events in leaving imprints on the landscape, for the utility of detailed process-mechanical studies as well as historical reconstructions of landform assemblages in unravelling the complexities of the present-day surface, for the interconnectivity between the various subspecializations of geomorphology and allied Earth and engineering sciences, and for the complementarities among twenty-first century technological capacities when combined with a field geomorphologist's keen sense of the lie of the land.

Future directions

Geomorphology in the twenty-first century will continue to mature as a science and assert its importance among the Earth-science disciplines. The issue of scale will remain a dominant topic of investigation and discourse, and it will be richly informed by expanding concerns about tectonic and structural controls on geomorphic systems over long time periods (i.e. megageomorphology), the evolution of lunar and Martian surfaces (i.e. planetary geomorphology), the intricate linkages between geomorphic and biogeochemical systems, and the hierarchically nested versus scale-invariant nature of geomorphic systems. The term 'neogeomorphology' was recently coined (Haff 2002) to suggest that a new or modern form of geomorphology may be evolving – one which, of necessity, takes into account the sobering fact that humans now displace more soil and rock per year than rivers, glaciers and wind combined (Hooke 2000). The pace of anthropogenically driven landscape alteration, whether direct or indirect (e.g. via global warming), is likely only to increase in the future. And, because there are no analogues for such pronounced surface modification in the stratigraphic record, the relevance and utility of geomorphology (with its traditional focus on process–form interaction) to the planning and environmental management communities seems assured. Geomorphologists already play central roles in mandated environmental impact assessments involving construction, mining and forestry, and increasingly their expertise (in conjunction with biologists and botanists) is utilized in landscape reclamation, rehabilitation and restoration efforts involving streams, wetlands and coastal dunes.

In the quest for a deeper understanding of the past (retrodictive) and future (predictive) evolution of Earth's surface, geomorphologists are becoming increasingly reliant on sophisticated technologies. These include: new dating methods (e.g. cosmogenic radionuclides, optical- and thermo-luminescence, rock varnish, lichenometry) that yield the relative ages of landform elements and thereby unfold the historical sequence of events that produced the landscape; novel remote-sensing techniques (e.g. interferometric synthetic-aperature radar, lidar, ground-penetrating radar, time-domain reflectrometry) to measure and monitor a broad range of surface and subsurface attributes; advanced computational methods involving more powerful hardware, more efficient software codes, and more easily integrated and interoperable data platforms (e.g. Geographical Information Systems, Digital Elevation Models); and enhanced satellite coverage to provide synoptic information about inaccessible regions and across large distances with ever-increasing accuracy regarding absolute location and relative movement via the Global Positioning Systems. In addition, the means to communicate information and ideas virtually instantaneously to the entire community of geomorphologists has been greatly facilitated by the World Wide Web and by various national and international organizations such as the International Association of Geomorphologists (IAG) that maintain electronic bulletin boards and membership/address lists. For the first time in its long developmental history, geomorphology has the potential to become a truly global enterprise in terms of both coverage and participation.

References

Baker, V.R. (ed.) (1981) *Catastrophic Flooding: The Origin of the Channelled Scablands*, Stroudsburg, PA: Dowden, Hutchinson and Ross.
Beckinsale, R.P. and Chorley, R.J. (1991) *The History of the Study of Landforms or the Development of Geomorphology: Volume 3, Historical and Regional Geomorphology 1890–1950*, New York: Routledge.
Chorley, R.J., Beckinsale, R.P. and Dunn, A.J. (1973) *The History of the Study of Landforms or the Development of Geomorphology: Volume 2, The Life and Work of William Morris Davis*, London: Methuen.
——Dunn, A.J. and Beckinsale, R.P. (1964) *The History of the Study of Landforms or the Development of Geomorphology: Volume 1, Geomorphology Before Davis*, London: Methuen.
Fairbridge, R.W. (1968) *The Encyclopedia of Geomorphology*, New York: Reinhold.
Haff, P.K. (2002) Neogeomorphology, *EOS, Transactions of the American Geophysical Union* 83(29), 310.
Hooke, R. LeB. (2000) On the history of humans as geomorphic agents, *Geology* 28, 843–846.
Rhoads, B.L. and Thorn, C.E. (eds) (1996) *The Scientific Nature of Geomorphology*, Chichester: Wiley.
Strahler, A.N. (1952) Dynamic basis for geomorphology, *Geological Society of America Bulletin* 63, 923–938.
Tinkler, K.J. (1985) *A Short History of Geomorphology*, London: Croom Helm.
——(ed.) (1989) *History of Geomorphology, from Hutton to Hack*, London: Unwin Hyman.

Further reading

Leopold, L.B., Wolman, M.G. and Miller, J.P. (1964) *Fluvial Processes in Geomorphology*, San Francisco: Freeman.
Ritter, D.F., Kochel, R.C. and Miller, J.R. (1995) *Process Geomorphology*, 3rd edition, Dubuque, IA: William C. Brown.
Scheidegger, A.E. (1970) *Theoretical Geomorphology*, Berlin: Springer-Verlag.
Schumm, S. (1991) *To Interpret the Earth: Ten Ways to be Wrong*, Cambridge: Cambridge University Press.
Yatsu, E. (2002) *Fantasia in Geomorphology*, Tokyo: Sozosha.

BERNARD O. BAUER

GEOMORPHOMETRY

Dealing with quantitative analysis of the land surface, geomorphometry is a central theme in both theoretical and applied geomorphology (Pike and Dikau 1995). It is also diverse, dealing both with landforms and with the land surface as a one-sided rough surface, with vertical position a unique function of horizontal location. It could also be referred to as the combination of 'landform morphometry' and 'land surface morphometry'. It does not include surveying, photogrammetry and profiling, which provide the raw data for geomorphometry, but some knowledge of these is essential in considering error margins (Richards 1990: 36–41). Morphometry itself is a broader field, important not only in various aspects of Earth science, but also in engineering, biology and medicine. Each field of application has things to teach the others (Pike 2000), so long as the specifics of the original application are remembered. Geomorphometry inspired the idea of statistical FRACTALs, following difficulties in specifying 'how long is a coastline?'

Where individual landforms can be defined and distinguished from their surroundings, a series of MORPHOMETRIC PROPERTIES can be measured to provide a multivariate characterization of the landform. Such analysis is labelled 'specific geomorphometry'. This is distinct from 'general geomorphometry' of the land surface as in spectral or fractal analysis, or the study of surface derivatives and their interrelations (Evans 1980). General geomorphometry was extremely difficult before the introduction of computers: today it may involve the processing of very large DIGITAL ELEVATION MODELS (DEMs), and has many applications in digital terrain modelling (Pike 2000). Specific geomorphometry has a much longer history, starting with measurements of lunar craters and of coastal sinuosity in the nineteenth century.

The two aspects of geomorphology are not completely distinct, first because some specific landforms such as slopes or hillslopes and drainage networks are so widespread on Earth that their specific geomorphometry acquires a general importance: and second, because some techniques of general geomorphometry can be applied to specific landforms (Evans 1987). This permits analysis of variation within a landform (distributional analysis), not just generalization of its overall characteristics, and is more useful in the context of modelling.

General geomorphometry; surface derivatives

General geomorphometry starts with the altitude (elevation) of the surface – its height above sea level. This has major effects on climate and thus on surface processes. The frequency distribution of altitude (hypsometry) tells us quite a lot about the land surface. In the pre-computer era, this was summarized by its range (relief) and by the hypsometric integral – the relation of mean altitude above minimum, to this range. Relief varies with the size of area considered, and reaches several km (ridge to valley) in high mountain areas: the total range for the Earth is 8,852 + 11,033 m, i.e. 19.9 km. Hypsometric integral is around 0.5 for topography with sharp ridges and valleys, approaching but not reaching 1.0 for a plateau with few deep valleys, or 0.0 for a lowland with a few high hills. Evans (1972) suggested that instead of ranges and extremes, the use of standard statistical concepts – standard deviation and skewness – was both more economic and provided more stable statistics, influenced by the whole body of the distribution rather than by the extremes.

Ohmori (1993) found that hypsometric curves of mountainous areas such as Japan tend to be S-shaped or concave, giving integrals between 0.15 and 0.50. They can be simulated from empirically based relations between uplift, altitude, altitude dispersion and denudation rates. Hypsometric curves vary considerably with extent of area considered, and whether headwaters, large erosional basins or areas including depositional plains are analysed. Fuller understanding of landscape development is obtained by considering dimensional indices (mean and standard deviation of altitude) and not just dimensionless indices.

Gradient (slope angle) is the second local value of a surface that is very important in geomorphology and hydrology. It provides the stress to generate mass movements, and gives energy to surface flows. Engineers prefer percentages, i.e. 100 × (tangent of angle), but geomorphologists prefer to measure gradient in degrees. Mean gradient is of primary interest, but standard deviation and skewness of the point-by-point distribution of gradients tell us much about the regional topography. Fluvially dissected hill areas with slopes near some threshold tend to have low standard deviations of gradient, while glaciated mountains with cliffs, valley floors, terraced areas and often plateau remnants and benches have high standard deviations.

Lowland areas tend to have a few steep slopes and many gentle ones; their gradients are positively skewed. In mountain areas, the opposite applies as slopes approach gradients limited by slope stability and ROCK MASS STRENGTH. For example, in the Japanese mountains, on igneous and sedimentary rocks, the mode of gradients becomes sharper as altitude increases. The mode is at 33 to 37 degrees in all three ranges of the Japan Alps (central Honshu) above 1,000 m, up to 2,800 m (Figure 67; Katsube and Oguchi 1999). Mean gradients increase with altitude, to maxima of 32 to 35 degrees above 2,000 m. In the high relief of the north-west Himalaya, on crystalline rocks, gradients range from 0 to 60 degrees, with modes of 33–37 and means of 30–34 degrees (Burbank *et al.* 1996) despite varying uplift and denudation rates. These distributions may reflect a dynamic equilibrium with landsliding removing fractured rock and river gradients increasing to transport this. The similarity to Japan may be deceptive in that averaging over several hundred metres reduces the Himalayan measurements.

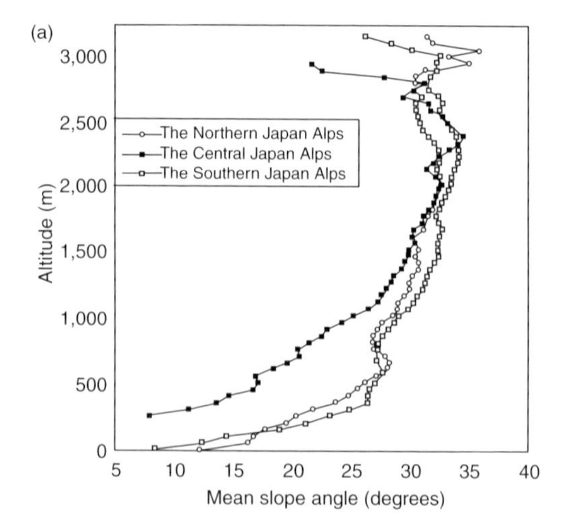

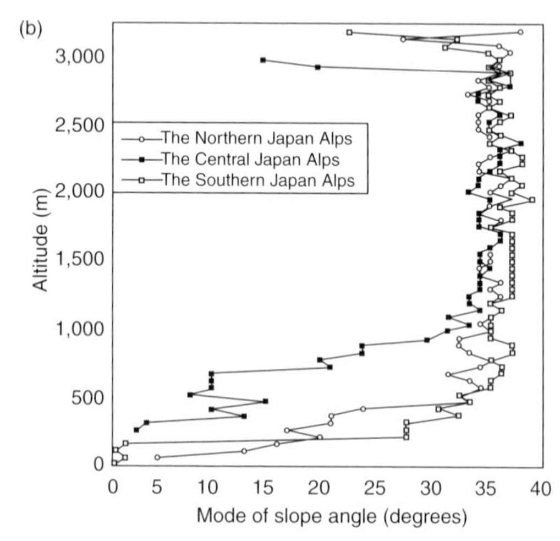

Figure 67 Altitudinal change in (a) mean and (b) modal slope angle (gradient) in three divisions of the Japan Alps

Source: Reproduced from Katsube and Oguchi (1999) with permission from the Association of Japanese Geographers

Gradient is defined as the rate of change of altitude in the direction where that rate is maximized (this is 'true gradient' as opposed to 'apparent gradient' in an arbitrary direction along a profile). This immediately implies a related variable, aspect – the direction or azimuth of this true gradient. Gradient and aspect form a closely related pair, the vector defining surface slope. Aspect, modulated by gradient, has considerable influence on slope climate (mesoclimate), especially solar radiation and exposure to wind. Although slope vectors can be analysed as poles to planes tangential to the surface, that approach ignores the different ways in which gradient and aspect affect surface processes. Aspect is a circular variable (0 ≡ 360 degrees) and it is easy to produce misleading results by applying ordinary linear statistics; it should be summarized by vector, directional or circular statistics, and related to other variables through its sine and cosine, in Fourier Series Analysis.

Rates of change of gradient and aspect in turn define components of curvature, the second derivative of the surface. Evans (1980) defined profile convexity as rate of change of gradient (with negative values representing concavity) and followed earlier geomorphologists such as Young (1972) in expressing this in degrees per 100 m. Plan (contour) convexity is thus the rate of change of aspect. Both variables are encountered in models of surface runoff (see RUNOFF GENERATION). Tangential curvature and other definitions have also been used: mathematically, curvature has three independent components. Of the many ways in which surface curvature can be defined, here we are concerned with those related to the gravity field, which is of central importance in geomorphology. As standard deviation of plan convexity measures the intricacy of contours, it expresses DRAINAGE DENSITY: this relationship requires further investigation.

Surface roughness or ruggedness is a broad concept, covering mean and variability of gradient, and variability of curvature in both profile and plan.

Altitude and its first and second derivatives provide local variables, conceptually related to points although in practice small neighbourhoods are used in their measurement. Context or position on the surface is also important, especially in relation to runoff. Contributing area upslope (per unit width of contour) controls the potential runoff that can be generated, and is used in models and applications (Lane *et al.* 1998; Wilson and Gallant 2000).

Other point aspects of the surface are those which are topologically special, in terms of position: these are summits, saddles and pits, and can be further subdivided in relation to the pattern of higher and lower land in the vicinity. Ridges, valleys and breaks of slope provide linear features at which slope either reverses or changes abruptly. Topological and other linear aspects of the surface are considered under DRAINAGE BASINS. At special points or lines, some derivatives may be indeterminate as gradient passes through zero: notably, aspect and plan convexity/curvature. Plains are areas of zero gradient and again aspect is indeterminate: their extent, however, varies with the vertical resolution of the data (e.g. altitude in metre units, or in tenth-metres, etc.).

General geomorphometry gives an appearance of objectivity, but it involves choice of data source, of horizontal and vertical resolution, and of algorithms for interpolation, smoothing and derivative calculation. Most important of all is definition of areas for which statistical summaries are to be provided. Map sheets or tiles of data are easiest to use, but natural regions may be more appropriate. Islands are the most obvious, but there are two complementary ways in which the land surface may be subdivided into exhaustive, non-overlapping areas. These are drainage basins, and 'mountains' bounded by valleys and low passes.

Spatial series, and complexity

Altitude is a positively autocorrelated variable, that is it defines a generally smooth surface. The rate of decline of autocorrelation with separation is thus an important property, and forms a basis for spectral analysis (Pike and Rozema 1975). This relates to the use of geostatistics and FRACTALS. They provide highly simplified models poorly suited to subaerial topography.

The land surface is complex and its morphometry varies from area to area with rock type and structure, climatic variables and their history, and tectonic history. Attempts to compress its variability into two or three statistical dimensions meet with difficulties. Multivariate studies show that at least nine dimensions (Table 21) are largely independent of each other.

Specific geomorphometry

Taking measurements of landforms requires their precise definition (what is/is not. . .?) and

Table 21 Statistical dimensions of (a) the Wessex land surface, England for 53 areas, and (b) the French land surface, for 72 areas

Property	Statistical descriptor (key variable)	Dimension
	(a) Wessex (Evans)	*(b) France (Depraetere)*
Gradient	Mean gradient	1. Relief
Massiveness	Skewness of altitude	4. Skewness of altitude (and 5.)
Level	Mean altitude	* in 1.
Profile convexity	Skewness of profile convexity	2. Convexity, cols and depressions
Orientation	Weighted vector strength (modulo 180°)	–
Plan convexity	Standard deviation of plan convexity	* in 1.
Altitude-convexity	Correlation of altitude with profile convexity	3. Convexity, crests and slopes
(Profile) variability	Standard deviation of gradient	*in 1.
Directedness	Weighted vector strength (modulo 360°)	–
		5. Skewness of gradient

Notes: All areas 10 × 10 km and analysed from 50 m grids. Numbers in (b) give the rank order of factors
Source: From Evans, in Hergarten and Neugebauer 1999

complete delimitation by a closed outline; here it may be difficult to achieve consistency between researchers. Although specific geomorphometry has been more subjective than general geomorphometry, work has now started on recognition and delimitation of landforms on DEMs by objective criteria. In specific geomorphometry, variables are defined specifically for each landform type. Commonly these include size (length, width, height, area, volume), gradient and shape (often ratios between size variables). The number of possible indices is increased where landforms are subdivided into several parts, e.g. volcano (or impact feature) outer slopes, craters and central peaks. Position (often a surrogate for climate) and geology are sometimes included as potential controlling variables. The more definable landforms include those listed at the end. Each of these has a body of geomorphometric literature. Landform shape and spatial pattern (position relative to others of the same type) were discussed by Jarvis and Clifford (in Richards 1990).

Evans (1987) distinguished eight stages in a specific morphometric study: conceptualization; definition; delimitation; measurement; calculation of indices; analysis of statistical frequency distributions; mapping and spatial analysis; interrelation of attributes; and assessing meaning. Analysis can be in terms of distributions of point variables (altitude, slope and curvature, as discussed above), sets of indices or measurements characterizing each landform (the most common approach), or fitting equations to the whole form or a selected part, outline or profile. In that residuals from such equations usually exhibit spatial pattern, simple equations are rarely good models for landforms.

General concepts in specific geomorphometry include symmetry (radial or axial), scale and the relation of size to shape. The latter can be isometric (shape does not vary with size, expected values of all ratios remain the same) or allometric (see ALLOMETRY; shape changes systematically, often as a power function of size). Scale is fundamental to both general and specific geomorphometry (Dietrich and Montgomery 1998; Wood 1996). Most landforms are defined with specific scales in mind, usually with something like a tenfold range in linear size. Within a particular landform type, different attributes (size, gradient) scale smoothly with each other. Sometimes scale breaks are discovered, and these reveal process thresholds – as for the central features of impact craters (Figure 68; Pike 1980).

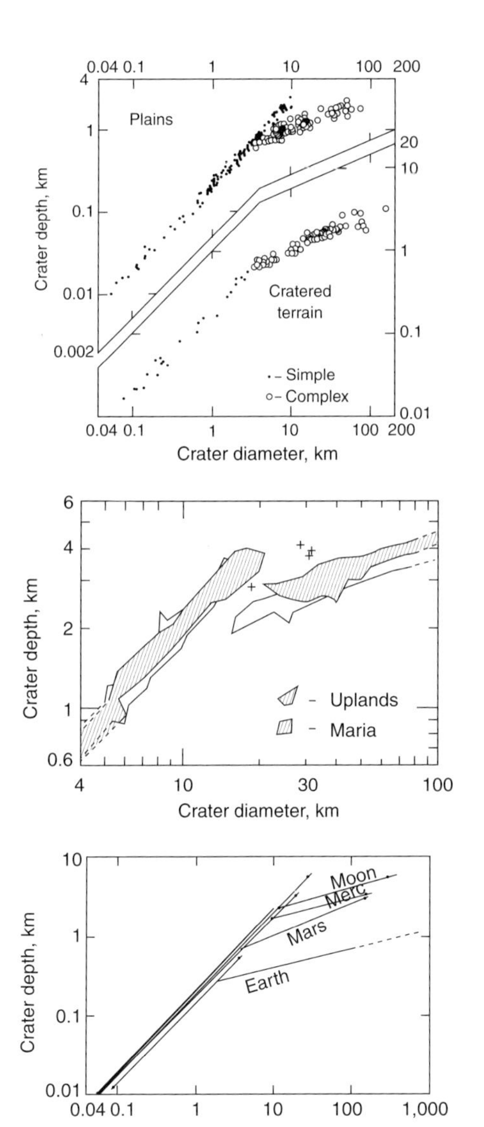

Figure 68 Breaks in the crater depth:diameter scaling relation, illustrating the morphologic transition from simple to complex craters (a) 230 craters on Mars, showing larger simple craters on plains than on 'cratered terrain'; (b) based on 203 mare craters and 136 upland craters on the moon. Simple craters follow a similar relation for maria and for uplands (as for the two divisions of Mars), but complex craters average 12 per cent deeper in uplands; (c) summary of the relationships on three planets and the moon. The transition size increases as gravity decreases

Source: Reproduced from Pike (1980: figures 6, 9 and 2) with permission

References

Burbank, D.W., Leland, J., Fielding, E., Anderson, R.S., Brozovic, N., Reid, M.R. and Duncan, C. (1996) Bedrock incision, rock uplift and threshold hillslopes in the northwestern Himalayas, *Nature* 379, 505–510.

Dietrich, W.E. and Montgomery, D.R. (1998) Hillslopes, channels and landscape scale, in G. Sposito (ed.) *Scale Dependence and Scale Invariance in Hydrology*, 30–60, Cambridge: Cambridge University Press.

Evans, I.S. (1972) General geomorphometry, derivatives of altitude, and descriptive statistics, in R.J. Chorley (ed.) *Spatial Analysis in Geomorphology*, 17–90, London: Methuen.

——(1980) An integrated system of terrain analysis and slope mapping, *Zeitschrift für Geomorphologie N.F. Supplementband* 36, 274–295.

——(1987) The morphometry of specific landforms, in V. Gardiner (ed.) *International Geomorphology 1986* Part II, 105–124, Chichester: Wiley.

Hergarten, S. and Neugebauer, H.J. (eds) (1999) *Process Modelling and Landform Evolution*, Lecture Notes in Earth Sciences, 78, Berlin: Springer.

Katsube, K. and Oguchi, T. (1999) Altitudinal changes in slope angle and profile curvature in the Japan Alps: a hypothesis regarding a characteristic slope angle, *Geographical Review of Japan B* 72, 63–72.

Lane, S.N., Richards, K.S. and Chandler, J.H. (eds) (1998) *Landform Monitoring, Modelling and Analysis*, Chichester: Wiley.

Ohmori, H. (1993) Changes in the hypsometric curve through mountain building and denudation, *Geomorphology* 8, 263–277.

Pike, R.J. (1980) Control of crater morphology by gravity and target type: Mars, Earth, Moon, *Proceedings, Lunar and Planetary Science Conference* 11, 2,159–2,189.

Pike, R.J. (2000) Geomorphometry – diversity in quantitative surface analysis, *Progress in Physical Geography* 24, 1–20.

Pike, R.J. and Dikau, R. (eds) (1995) Geomorphometry, *Zeitschrift für Geomorphologie N.F. Supplementband* 101.

Pike, R.J. and Rozema, W.J. (1975) Spectral analysis of landforms, *Annals of the Association of American Geographers* 64, 499–514.

Richards, K.S. (ed.) (1990) Form, in A. Goudie (ed.) *Geomorphological Techniques*, 31–108, London: Unwin Hyman.

Wilson, J.P. and Gallant, J.C. (eds) (2000) *Terrain Analysis: Principles and Applications*, New York: Wiley.

Wood, J. (1996) Scale-based characterization of digital elevation models, in D. Parker (ed.) *Innovations in GIS 3*, 163–175, London: Taylor and Francis.

Young, A. (1972) *Slopes*, Edinburgh: Oliver and Boyd.

SEE ALSO: hillslope, form; hillslope, process; slope, evolution; and the landforms: alluvial fan; atoll; cave; channel, alluvial (hydraulic geometry); cirque, glacial; crater; doline; drumlin; dune, aeolian; fjord; inselberg; karren; lake; landslide; palsa; pingo; river delta; tafoni; tor; volcano; yardang

IAN S. EVANS

GEOSITE

Geosites (synonyms: geotopes, Earth science sites, geoscience sites) are portions of the geosphere that present a particular importance for the comprehension of Earth history. They are spatially delimited and from a scientific point of view clearly distinguishable from their surroundings. More precisely, geosites are defined as geological or geomorphological objects that have acquired a scientific (e.g. sedimentological stratotype, relict moraine representative of a glacier extension), cultural/historical (e.g. religious or mystical value), aesthetic (e.g. some mountainous or coastal landscapes) and/or social/economic (e.g. aesthetic landscapes as tourist destinations) value due to human perception or exploitation. Various groups of geosites are generally specified in the reference literature: structural, petrological, geochemical, mineralogical, palaeontological, hydrogeological, sedimentological, pedological and geomorphological geosites. In the last case, they are also called geomorphological sites or geomorphosites. Some anthropic objects (e.g. mines) are also considered as geohistorical sites. Geosites can be single objects (e.g. springs, lava streams) and larger systems (e.g. river systems, glacier forefields, coastal landscapes). Active geosites allow the visualization of geo(morpho)logical processes in action (e.g. river systems, active volcanoes), whereas passive geosites testify to past processes; in this case, they have a particular patrimonial value as Earth memory (landscape evolution, life history and climate variations).

Geosites may be modified, damaged, and even destroyed, by natural processes and anthropogenic actions. In order to avoid damage and destruction, geosites need conservation. Conservation strategies are generally based on inventories of geosites requiring the development of assessment methods. Assessment is based on criteria such as integrity (whether the object is complete), exemplarity (to what extent the geosite is representative of the geology or geomorphology of a region or country), rarity (in the space of reference or in scientific terms), legibility (whether it is easily visible scientifically), accessibility (for pedagogic activities), vulnerability, paleogeographical value (its contribution to the history of the Earth), aesthetic value, and cultural/historical value. Geodiversity is a criterion used for assessing groups of geosites: geodiversity is higher where there is a concentration of different objects in a given space facilitating visits and protection. Several quantitative or qualitative procedures for evaluation exist in the literature.

Some countries have adopted specific legislation for geosite conservation (e.g. Great Britain has individuated Regionally Important Geological/Geomorphological Sites – RIGS). Generally, geosite conservation is relatively high in developed countries but low in developing countries.

Further reading

Actes du premier symposium international sur la protection du patrimoine géologique, Digne-les-Bains, 11–16 juin 1991, Mém. Soc. Géol. France, N.S., 165, 1994.

Barettino, D., Vallejo, M. and Gallego, E. (eds) (1999) *Towards the Balanced Management and Conservation of the Geological Heritage in the New Millenium*, III International Symposium ProGEO on the Conservation of the Geological Heritage, Madrid: Ed. Sociedad Geológica de España.

O'Halloran, D., Green, C., Harley, M., Stanley, M. and Knill, J. (eds) (1994) *Geological and Landscape Conservation*, Proceedings of the Malvern International Conference 1993, London: The Geological Society.

Wilson, R.C.L. (ed.) (1994) *Earth Heritage Conservation*, London: The Geological Society and The Open University.

SEE ALSO: geodiversity; landscape sensitivity

EMMANUEL REYNARD

GILGAI

A form of micro-relief consisting of mounds and depressions arranged in random to ordered patterns (Verger 1964). There is a great variety of forms and they occur on a range of swelling clay and texture-contrast soils that have thick subsoil clay horizons. They tend to occur on level or gently sloping plains in areas subject to cycles of intense wetting and drying. Gilgai is an Australian aboriginal word meaning 'small waterhole' (Hubble *et al.* 1983) and some seasonal ponding of water does occur in some of the closed depressions of the larger forms.

The mechanisms of gilgai development involves swelling and shrinking of clay subsoils under a severe seasonal climate. A widely adopted hypothesis for their formation is as follows (Hubble *et al.* 1983: 31):

> when the soil is dry, material from the surface and the sides of the upper part of major cracks

> falls into or is washed into the deeper cracks, so reducing the volume available for expansion on rewetting of the subsoil. This creates pressures which are revealed by heaving of the soil between the major cracks which, once established, tend to be maintained on subsequent drying. This process is repeated, with the result that the subsoil is progressively displaced, a mound develops between the cracks, and the soil surface adjacent to the cracks is lowered to form depressions.

However, some gilgai are linear forms, known colloquially as 'Adams furrows', 'black-men's furrows', 'stripy country' and 'wavy country' (Hallsworth *et al.* 1955). Beckmann *et al.* (1973: 365) see surface runoff and soil heaving as working together to produce such features, particularly on pediment slopes.

In the Kimberley there are individual linear gilgai up to 2 km long and it is possible that in their case aeolian processes have contributed to their development (Goudie *et al.* 1992).

References

Beckmann, G.G., Thompson, C.H. and Hubble, G.D. (1973) Australian landform example no. 22: linear gilgai, *Australian Geographer* 12, 363–366.

Goudie, A.S., Sands, M.J.S. and Livingstone, I. (1992) Aligned linear gilgai in the west Kimberley District, Western Australia, *Journal of Arid Environments* 23, 157–167.

Hallsworth, E.G., Robertson, G.K. and Gibbons, F.R. (1955) Studies in pedogenesis in New South Wales VIII. The 'Gilgai' soils, *Journal of Soil Science* 6, 1–34.

Hubble, G.D., Isbell, R.F. and Nortcote, K-H. (1983) Features of Australian soils, in Division of Soils, CSIRO, *Soils an Australian Viewpoint*, 17–47, Melbourne: Academic Press.

Verger, F. (1964) Mottureaux et gilgais, *Annales de Géographie* 73, 413–430.

A.S. GOUDIE

GIS

A Geographic Information System (GIS) can be defined as a system of hardware and software used for the capture, storage, management, retrieval, display and analysis of geographic data. GIS's have been around since the 1960s, when, independently, initiatives in Canada (the Canadian GIS (CGIS), developed under the direction of Roger Tomlinson within the Canadian Federal Department of Agriculture) and in the United States (the Laboratory for Computer Graphics at Harvard University, established under the direction of Howard Fisher) resulted in the development of computer-based geographic information systems as we now know them (Foresman 1998).

A GIS uses two fundamental data types: spatial data, consisting of points (e.g. a sample location, a spot height), lines (e.g. the bank of a river, a break in a slope), areas (e.g. a drumlin, a drainage basin) and cells or rasters (e.g. a pixel from a satellite image), and attribute or descriptor data, consisting of characteristics associated with the spatial data (e.g. the elevation of the spot height, the area of the drainage basin).

While initial efforts during the 1970s and early 1980s saw the development of many one-off systems, the emergence in the 1980s of dominant players such as Environmental Systems Research Institute (ESRI) and Intergraph produced a shift from building systems to application development. While geomorphologists were among the first to appreciate the power that computers could bring to scientific analyses (e.g. Chorley 1972), a lack of spatial data and appropriate analytical procedures meant that, initially, such systems were not widely used within the scientific community. As well, the widespread acceptance of GIS required time for scientists to learn, apply and review the new technology in light of contemporary research problems. However, the rise in the computing power of personal computers, coupled with an increased availability of spatial data, meant that by the late 1980s GIS had become a tool used by many geomorphologists.

The earliest links between GIS and geomorphological research can be traced back to applications involving digital elevation models. DIGITAL ELEVATION MODELS (DEMs) are the digital representation of elevation, and while contours have traditionally been used to graphically represent topography on printed maps, in a GIS either a regular tessellation such as gridded cells (i.e. a raster representation, typically referred to as a DEM) or an irregular tessellation such as a Triangulated Irregular Network (TIN) (i.e. a vector representation consisting of elevation points and connecting lines forming triangular planar regions) are the preferred means for representing topography (Weibel and Heller 1991). The main concerns surrounding any digital representation of topography relate to issues around fidelity, accuracy and resolution (Moore *et al.* 1991). While TINs, with their variable sampling structure and their ability to represent important landform features such as peaks, ridges, cliffs and valleys, are capable of more accurately capturing the complexity of topography than are DEMs, DEMs remain the favoured means of representing topography for geomorphologists because of the ease with which analytical procedures can be applied to them (e.g. moving windows of 3 by 3 cells within which attributes such as slope and aspect can be quickly calculated).

The resolution of the data available to the analyst determines the areal extent of the study area that is appropriately analysed, and the type of features that can be identified. Initially, much digital data was of coarse resolution that limited the nature of the analysis (e.g. to regional or macro analyses). As higher resolution data become more commonplace, geomorphologists are better able to study processes at finer spatial and temporal scales, and to examine the role that scale plays in physical processes (Walsh *et al.* 1998).

Extraction of drainage features from any digital representation of topography remains a complex process, however. While with TINs the extraction of the drainage network can be an easy process, determination of the direction and accumulation of surface waters remains a geometrically complex task. For DEMs, the presence of artifactual sinks – either through errors in the elevation values or as a result of the discretization of the elevation values – complicates the process, and much effort has been directed at automatically recognizing and removing such features from DEMs (e.g. Maidment 1993).

Geomorphology is concerned with the form, the materials and the processes from which landforms are created. Some geomorphologists claim that a full understanding of materials and process can be obtained by focusing studies on the form of the landscape (Speight 1974) – a view shared by others in fields such as fractals. GIS are the ideal tools with which to study form, and it is not surprising to find that computer-based morphometric analysis of Digital Elevation Models (DEMs) has long been a very active area of research (e.g. Dikau 1989; Pike 1988; see GEOMORPHOMETRY). Early work focused on the derivation of topographic properties of watersheds, including the derivation of slope, curvature, channel links and drainage areas, all properties that can be mathematically derived from a DEM. This early work focused on examining morphologic patterns, rather than the physical processes that control them. While the link between form and process remains tentative, even at present, we are seeing an increasing sophistication in the application of quantitative approaches to the study of landform, along with the integration of more complex process models within Geographic Information Systems.

The use of GIS in geomorphology can be conceptually classified into four general types of analyses (Vitek *et al.* 1996; see also Walsh *et al.* 1998). These include: (1) landform measurement, (2) landform mapping, (3) process monitoring, and (4) landscape and process modelling. Often an application will involve several levels of GIS analysis. Process monitoring and landscape modelling, for example, will routinely require landform measurement and mapping in order to define initial parameters and establish the spatial distribution of controlling factors of interest. Although the utilization of GIS is not always required for carrying out these types of geomorphological analyses, given suitable digital data and processing capabilities, a GIS can provide a platform for automating these functions. This automation can greatly enhance the spatial/temporal scope and resolution of many geomorphological investigations. A general description and research example for each of the four different types of GIS utilization in geomorphology is provided below.

At the most basic level, a GIS may be used to perform fundamental measurements of landform features. This includes the enumeration of landform features, and making measurements of landform length, area and volume. Often it is some relation between fundamental measures that is of interest. Some common geomorphological examples include counting the number of landslides per region area (event frequency), measuring the length of channels per unit area (drainage density), taking the ratio of horizontal to vertical hillslope or channel lengths (slope), and calculating the areal extent of glaciers in an alpine catchment (glacial coverage). Carrying out these types of landform measurements is usually easily and efficiently accomplished using most standard GIS applications. By automating such measurements in a GIS environment, the spatial scope and resolution of the analysis can be expanded beyond what could be accomplished using traditional manual techniques. For example, Fontana and Marchi (1998) used this type of GIS analysis in order to evaluate the intensity of localized erosional processes in two alpine drainage basins of the Dolomite Mountains in northeastern Italy. In their work the combination of two landscape measurements of erosion potential was considered – the contributing drainage area (indication of flow concentration occurrence) and local slope (index of flow erosivity). By using the automated measurement capabilities of the GIS, these hydrologic parameters were calculated over entire drainage basin areas of many square kilometres using a high-resolution ten by ten square metre grid base. This type of analysis provided highly localized information on sediment erosion potential within the alpine catchments, information that would have been impossible to obtain using manual methods.

The second way in which GIS is utilized in geomorphology is in the development and analysis of landform maps (see TERRAIN EVALUATION; GEOMORPHOLOGICAL MAPPING). Landform mapping is commonly used in geomorphology in order to characterize landscapes and relate landform distribution to spatial patterns of physical, chemical and biological geomorphic processes. Through the use of basic GIS-based mapping tools, users can rapidly produce and modify landform maps. The typical map manipulation and analysis functions that are included in most GIS applications further enhance landform map production and interpretation. Such functions include map generalization and simplification, map overlay, spatial query and browsing, and various algorithms for the analysis of spatial patterns and relationships. Computer automation is becoming increasingly necessary in landform mapping because of the large amount of digital geographic data that is available and the rapidly increasing rate of digital geographic data acquisition. Bishop *et al.* (1998) used GIS-based mapping for studying large-scale geomorphic processes acting on the Nanga Parbat Himalaya massif of northern Pakistan. A digital elevation model and multispectral remote sensing data were used to study the structural geology and surface geomorphology of this extensive and remote mountain environment. Landform maps, a major component of the investigation, were developed using GIS software and the integration of some massive digital spatial data sets comprised of both surface and subsurface remote sensing data. Assessments of denudation rates and sediment storage were quantified and glacial, fluvial and mass movement processes were reconstructed for the massif by analysing the three-dimensional form and spatial surface characteristics of the mapped landscape in this study. Walsh *et al.* (1998) and Bishop *et al.* (1998) both stress the growing importance of the integration of remote sensing and GIS spatial analysis in order to solve complex, large-scale geomorphic problems.

The analytic capabilities of GIS are well suited to studies of river channel dynamics. At a fundamental level, GIS is an efficient tool for mapping channel features including, for example, sand and gravel bars, vegetated islands, channel banks, woody debris jams and historic (abandoned) channels. Commonly, channel features are digitized from existing maps, orthophotos, aerial photographs or satellite images and coded in the GIS as line or polygon features. Data collected in the field may also be included. If a scale or co-ordinate system has been defined, GIS is an efficient tool for quickly examining spatial relations between, and for measuring the length, width and area of digitized features. If maps or imagery are available for different dates, more advanced spatial and temporal overlay analysis can be performed such as lateral migration, loss or gain of riparian surfaces, aggradation or degradation of sediment, and changes in channel planform. Changes can be examined visually, or summarized and tabulated in a database, while rates of change may be additionally determined if exact dates are known. Similarly, if sufficient historical data are available, GIS may further be used to show trends in morphologic development over time, and even predict landform evolution. The capability of most GIS to collate data sources with different scales and co-ordinate systems is key to these types of analysis. In recent years, GIS has been used increasingly to study the relation between (morphologic) form and (hydraulic) process in river channels using fully distributed topographic models of the channel bed and banks for different dates (cf. Lane *et al.* 1994). Topographic information may be derived from conventional cross-section surveys, tacheometry, photogrammetry or depth soundings. The data are imported to the GIS in order to produce either a TIN or DEM, and then are overlaid in order to produce maps and volumetric summaries of channel scour (erosion) and fill (deposition). The net difference between channel scour and fill may then be used to infer rates of sediment transport within the framework of a sediment budget.

The seamless integration of environmental models with GIS is the ultimate goal of many researchers (Raper and Livingstone 1996). Since landscape and process modelling requires the storing, retrieving and analysing of spatio-temporal data sets, it is not surprising that GIS can play a significant role in this area. GIS enhances the process as it assists in the derivation, manipulation, processing and visualization of such geo-referenced data. Boggs *et al.* (2000) used an integrated approach to look at landform evolution in a catchment that could be subject to impacts from mining activities. Landform evolution models typically require extensive parameterization, often involving both hydrology and sediment transport models, and using an integrated environment allows for the rapid production of modified input scenarios. Therefore, a much wider array of impact scenarios can be made, and the environmental implications of decisions made today can be modelled over the long term, which should lead to better management decisions.

References

Bishop, M.P., Shroder Jr, J.F., Sloan, V.F., Copland, L. and Colby, J.D. (1998) Remote sensing and GIS technology for studying lithospheric processes in a mountain environment, *Geocarto International* 13(4), 75–87.

Boggs, G.S., Evans, K.G., Devonport, C.C., Moliere, D.R. and Saynor, M.J. (2000) Assessing catchment-wide mining-related impacts on sediment movement in the Swift Creek catchment, Northern Territory, Australia, using GIS and landform-evolution modelling techniques, *Journal of Environmental Management* 59, 321–334.

Chorley, R.J. (ed.) (1972) *Spatial Analysis in Geomorphology*, New York: Harper and Row.

Dikau, R. (1989) The application of a digital relief model to landform analysis in geomorphology, in J. Raper (ed.) *Three Dimensional Applications in Geographic Informations Systems*, 51–77, London: Taylor and Francis.

Fontana, G.D. and Marchi, L. (1998) GIS indicators for sediment sources study in alpine basins, in K. Kovar, U. Tappeiner, N. Peters and R. Craig (eds) *Hydrology, Water Resources and Ecology in Headwaters*, 553–560, IAHS Publication no. 248.

Foresman, T.W. (ed.) (1998) *The History of Geographic Information Systems, Perspectives from the Pioneers*, Upper Saddle River, NJ: Prentice Hall.

Lane, S.N., Chandler, J.H. and Richards, K.S. (1994) Developments in monitoring and modelling small-scale river bed topography, *Earth Surface Processes and Landforms* 19, 349–368.

Maidment, D.R. (1993) GIS and hydrological modelling, in M.F. Goodchild, B.O. Parks and L.T. Steyaert (eds) *Environmental Modeling with GIS*, Chapter 14, New York: Oxford University Press.

Moore, I.D. Grayson, R.B. and Ladson, A.R. (1991) Digital terrain modelling: a review of hydrological, geomorphological and biological applications, *Hydrological Processes* 5, 3–30.

Pike, R. (1988) The geometric signature: quantifying landslide-terrain types from digital elevation models, *Mathematical Geology* 20, 491–511.

Raper, J. and Livingstone, D. (1996) High-level coupling of GIS and environmental process modeling, in M. Goodchild, L.T. Steyart, B.O. Parks, C. Johnston, D. Maidment, M. Crane and S. Glendinning (eds) *GIS and Environmental Modeling: Progress and Research Issues*, Fort Collins, CO: GIS World, Inc.

Speight, J.G. (1974) A parametric approach to landform regions, *Institute of British Geography Special Publication*, No. 7, 213–230.

Vitek, J.D., Giardino, J.R. and Fitzgerald, J.W. (1996) Mapping geomorphology: a journey from paper maps, through computer mapping to GIS and virtual reality, *Geomorphology* 16, 233–249.

Walsh, S.J., Butler, D.R. and Malanson, G.P. (1998) An overview of scale, pattern, process relationships in geomorphology: a remote sensing and GIS perspective, *Geomorphology* 21, 183–205.

Weibel, R. and Heller, M. (1991) Digital terrain modelling, in D.J. Maguire, M.F. Goodchild and D.W. Rhind (eds) *Geographic Information Systems: Principles and Applications* Vol. 1, 269–297, Harlow: Longman Scientific.

BRIAN KLINKENBERG, ERIK SCHIEFER AND DARREN HAM

GLACIAEOLIAN (GLACIOAEOLIAN)

The association between glaciation (past and present) and aeolian processes and forms. Glaciation comminutes rock fragments by grinding and so produces material, including silt and sand, that can then be transported by wind to create dunes and LOESS deposits. Such materials are also available to cause wind abrasion in glacial and near-glacial environments, thereby producing wind-moulded pebbles and cobbles (VENTIFACTS) and streamlined ridges (YARDANGS) and grooves. Deflation from glacial deposits can create STONE PAVEMENTS (Derbyshire and Owen 1996).

The fine materials blown across proglacial plains and beyond create dunes and sandsheet deposits (coversands). These are widespread in Canada and Central Europe. The facies progression from coversands through to sandy loess and then loess is well documented from the proglacial forelands around the northern hemisphere. Above all, however, the thick loess deposits of Central Europe, Central Asia, China, New Zealand, the Argentinian Pamas and the mid-USA, may at least in part be indirect products of glacial sediment supply.

Ice sheets and glaciers may themselves affect air pressure conditions and generate high velocity winds that contribute to the power of aeolian processes in their vicinity.

Reference

Derbyshire, E. and Owen, L.A. (1996) Glacioaeolian processes, sediments and landforms, in J. Menzies (ed.) *Past Glacial Environments: Sediments, Forms and techniques, Vol. 2, Glacial Environments*, 213–237, Oxford: Butterworth-Heinemann.

A.S. GOUDIE

GLACIAL DEPOSITION

Glacial deposition occurs when debris is released from glacial transport at the margin or the base of a GLACIER. Narrow definitions include only primary sedimentation directly from the ice into the position of rest, but broader definitions include secondary processes such as deposition through water and resedimentation of glacigenic materials by flowage. In addition, release of material onto the glacier surface is sometimes referred to as supraglacial deposition, but if the glacier is still moving this is only a temporary stage in the sediment transport process.

Glacial deposition produces characteristic sediments and landforms, and landscapes of glacial deposition exist across large areas of the mid-latitudes formerly covered by ice sheets. The term till is generally used to describe sediments deposited by glaciers, and replaces the term boulder clay, which was commonly used in the past. Landforms created by glacial deposition are called MORAINES. Glacial sediments are often unstable in non-glacial environments and are subject to reactivation under PARAGLACIAL conditions. Glacially deposited materials are therefore an important debris source in geomorphic processes in postglacial and proglacial environments.

The characteristics of glacial sediments reflect both the processes of their deposition and also the processes of GLACIAL EROSION and entrainment by which the material was originally produced. A substantial literature exists on the classification of glacial sediments and the processes by which they form (e.g. Schlüchter 1979; van der Meer 1987; Goldthwait and Matsch 1988) and several convenient summaries of depositional processes have been produced (e.g. Whiteman 1995). Following a broad definition, the main mechanisms of deposition include:

1. release of debris by melting or sublimation of the surrounding ice
2. lodgement of debris by friction against a substrate
3. deposition of material from meltwater (glacifluvial deposition)
4. chemical precipitation
5. flow and resedimentation of deposited material
6. glacitectonic processes (see GLACITECTONICS).

Different processes of sedimentation are dominant in different parts of a glacier. In supraglacial locations, material can be released by ablation of the ice surface. This occurs most commonly by melting, and is referred to as melt-out, but sublimation can make a significant contribution in cold arid environments (Shaw 1988). Englacial debris, and debris in basal ice penetrating to the surface, is then exposed on the ice surface and can contribute to a supraglacial sediment layer. This supraglacial sediment is liable to redistribution by flow, wash and mass movements on the ice surface as ablation continues. Resedimented material derived from flow of supraglacial debris, sometimes referred to as 'flow-till', can make up a substantial proportion of glacial deposits in environments where supraglacial sedimentation occurs (e.g. Boulton 1968). The extent of supraglacial sedimentation depends on the debris content of the ice, the ablation rate and the rate of removal of sediment from the surface by processes such as wash, deflation and mass movement. Supraglacial sediment can be deposited at the ground surface when the glacier retreats or disintegrates.

At glacier margins material can be released by ablation and dumped directly into the proglacial environment. Sediment can be released directly from within the ice, carried to the margin supraglacially and dropped over the margin as if over the end of a conveyor belt, or brought to the margin by water flowing from the interior or surface of the glacier. Sediment can also be transferred to the margin by deformation of subglacial material (e.g. Boulton *et al.* 1995). The amount of sediment that is transported to the margin is primarily a function of the size and speed of the glacier, the glacier's erosive capability, the erodibility of the substrate and the input of sediment from extra-glacial sources such as rockfalls or tephra. The characteristics of the sediments and landforms produced by deposition at the margin depend on whether the margin occurs on land or in water, and the processes and environments of their formation are reflected in their morphology and sedimentological structure. Reactivation and resedimentation of material at the margin by movement of the ice or by fluvial and mass movement processes is common.

In subglacial environments, material can be released by ablation either in cavities or in contact with the bed, or can be lodged against the bed by moving ice. Release of basal sediment by melt-out or sublimation beneath moving ice can produce conditions conducive to lodgement, to the development of thick basal till, and to subglacial deformation of the released material. Lodgement of sediment against a rigid bed occurs when the friction of the clast against the bed outweighs the tractive power of the ice and material is released from the ice by either pressure melting or plastic deformation of ice around clasts. Lodgement can also occur against the upper surface of a deforming bed if the overlying ice is moving faster than the deforming layer. Deposition within a deforming subglacial layer occurs when deformation cannot remove all of the material that is being supplied to the layer and deformation ceases either throughout the layer or for a certain thickness at its lower margin.

Subglacial deposits can also include chemical precipitates, although these are not always considered in the context of glacial sediments. Chemical precipitates such as calcite can be deposited when solute-rich waters freeze. In carbonate environments, such as where limestone bedrock is present, comminuted carbonate rocks contribute to a highly reactive rock flour. Meltwater produced at the bed can take carbonate into solution, and if the water subsequently refreezes the carbonate can be released as a precipitate onto bedrock or basal debris (e.g. Souchez and Lemmens 1985).

Many glaciers release sediment into water, and GLACIMARINE and GLACILACUSTRINE sediments form an important part of the glacial sediment record. Release of sediment into water can produce different effects from terrestrial sedimentation. Sediment can be released directly from the ice, by discharge of sediment-rich meltwater, and by the melting or breakup of icebergs. The characteristics of subaqueous sediments and landforms reflect aqueous as well as glacial processes and conditions. Where a glacier is grounded on its bed beneath the water, glacial deposition can occur beneath the margin by lodgement and melt-out. However, material that is dumped from the front of the glacier into water, or from the base of the glacier where the glacier is floating, forms a deposit that is not strictly a glacial deposit, as its character upon settling will be controlled largely by aqueous sedimentation processes. The principal features of ice margins in water include subaqueous moraines caused both by pushing of proglacial sediments and release of sediment from the glacier; subaqueous grounding line fans formed from material emerging from beneath the glacier into the water at the grounding line; ice contact fan deltas that form when grounding line fans grow and emerge at the water surface; and a distal proglacial zone in which sediment settles out from suspension in the water and rains out from icebergs drifting away from the ice margin. Useful reviews of sedimentation at marine and lacustrine glacier margins include those provided by Dowdeswell and Scourse (1990) and Powell and Molnia (1989).

The mechanisms of glacial deposition impart specific characteristics to the deposited material. Fabric, particle size and shape characteristics, and consolidation have all been used to infer depositional processes and glacier characteristics from glacial sediments. The distribution of structures in deformed sediments can be used to reconstruct former ice sheets, and Boulton and Dobbie (1993) suggested that consolidation characteristics of formerly subglacial sediments can be used to infer basal melting rates, subglacial groundwater flow patterns, ice overburden, basal shear stress, ice-surface profiles and the amount of sediment removed by erosion. Glacial deposits can thus provide valuable information about glacial and climatic history.

References

Boulton, G.S. (1968) Flowtills and related deposits on some Vestspitsbergen Glaciers, *Journal of Glaciology* 7, 391–412.

Boulton, G.S.,and Dobbie, K.E. (1993) Consolidation of sediments by glaciers: relations between sediment geotechnics, soft-bed glacier dynamics and subglacial ground-water flow, *Journal of Glaciology* 39, 26–44.

Boulton, G.S., Caban, P.E. and van Gijssel, K. (1995) Groundwater flow beneath ice sheets: part I – large scale patterns, *Quaternary Science Reviews* 14, 545–562.

Dowdeswell, J.A. and Scourse, J.D. (1990) *Glacimarine Environments: Processes and Sediments*, Geological Society Special Publication 53, Bath: Geological Society.

Goldthwait, R.P. and Matsch, C. (1988) *Genetic Classification of Glaciogenic Deposits*, Rotterdam: Balkema.

Powell, R.D. and Molnia, B.F. (1989) Glacimarine sedimentary processes, facies and morphology of the south-southeast Alaska shelf and fjords, *Marine Geology* 85, 359–390.

Schlüchter, C. (ed.) (1979) *Moraines and Varves*, Rotterdam: A.A. Balkema.
Shaw, J. (1988) Sublimation till, in R.P. Goldthwait and C. Matsch (eds) *Genetic Classification of Glaciogenic Deposits*, 141–142, Rotterdam: A.A. Balkema.
Souchez, R.A. and Lemmens, M. (1985) Subglacial carbonate deposition: an isotopic study of a present-day case, *Palaeogeography, Palaeoclimatology, Palaeoecology* 51, 357–364.
van der Meer, J.J.M. (ed.) (1987) *Tills and Glaciotectonics*, Rotterdam: A.A. Balkema.
Whiteman, C.A. (1995) Processes of terrestrial deposition, in J. Menzies (ed) *Modern Glacial Environments*, 293–308, Oxford: Butterworth-Heinemann.

Further reading

Benn, D.I. and Evans, D.J.A. (1998) *Glaciers and Glaciation*, London: Arnold.
Bennett, M.R. and Glasser, N.F. (1996) *Glacial Geology*, London: Wiley.
Hambrey, M.J. (1994) *Glacial Environments*, London: UCL Press.
Knight, P.G. (1999) *Glaciers*, Cheltenham: Nelson Thornes.

SEE ALSO: glacier; moraine

PETER G. KNIGHT

GLACIAL EROSION

Glaciers cause erosion in a variety of ways (Bennett and Glasser 1996). First of all, glaciers can be likened to conveyor belts. If a rockfall puts coarse debris on to a glacier surface, for example, or if frost-shattering sends down a mass of angular rock fragments on to the glacier surface, it can then be transported, almost whatever its size, down valley. Second, beneath glaciers there is often a very considerable flow of meltwater. This may flow under pressure through tunnels in the ice at great speed, and may be charged with coarse debris. Such subglacial streams are highly effective at eroding the bedrock beneath a glacier. This can contribute to the excavation of TUNNEL VALLEYs. Meltwater may also cause chemical erosion. Third, although glacier ice itself might not cause marked erosion of a rock surface by *abrasion*, when it carries coarse debris at its base some abrasion can occur. This grinding process has been observed directly by digging tunnels into glaciers, but there is other evidence for it: rock beneath glaciers may be *striated* or scratched, and much of the debris in glaciers is ground down to a fine mixture of silt and clay called *rock flour*.

Glaciers also cause erosion by means of *plucking*. If the bedrock beneath the glacier has been weathered in preglacial times, or if the rock is full of joints, the glacier can detach large particles of rock. As this process goes on, moreover, some of the underlying joints in the rock may open up still more as the overburden of dense rock above them is removed by the glacier. This is a process called *pressure release*.

As debris-laden ice grinds and plucks away the surface over which it moves, characteristic landforms are produced which give a distinctive character to glacial landscapes. Of the features resulting from glacial quarrying, one of the most impressive is the CIRQUE. This is a horseshoe-shaped, steep-walled, glaciated valley head. As cirques evolve they eat back in the hill mass in which they have developed. When several cirques lie close to one another, the divide separating them may become progressively narrowed until it is reduced to a thin, precipitous ridge called an *arête*. Should the glaciers continue to whittle away at the mountain from all sides, the result is the formation of a pyramidal *horn*.

With valley glaciation the lower ends of spurs and ridges are blunted or truncated; the valleys assume a U-shaped configuration; they become more linear; and hollows or troughs are excavated in their floors. Many high-latitude coasts, such as those of Norway, New Zealand and western Scotland, are flanked by narrow troughs, called FJORDs, which differ from land-based glacial valleys in that they are submerged by the sea.

Fjards are related to fjords. They are coastal inlets associated with the glaciation of a lowland coast, and therefore lacking the steep walls characteristic of glacial troughs. A good example of a fjard coast is that of Maine, USA.

A further erosional effect of valley glacier is the breaching of watersheds, for when ice cannot get away down a valley fast enough – perhaps because its valley is blocked lower down by other ice or because there is a constriction – it will overflow at the lowest available point, a process known as *glacial diffluence*. The result of this erosion is the creation of a *col*, or a gap in the watershed.

Tributary valleys to a main glacial trough have their lower ends cut clean away as the spurs between them are ground back and truncated. Furthermore, the floor of a trunk glacier is deepened more effectively than those of feeders from the side or at the head, so that after a period of prolonged glaciation such valleys are left hanging above the main trough. Such *hanging valleys* have often become the sites of waterfalls.

The development of an ice sheet tends to scour the landscape. In Canada there are vast expanses of territory where the Pleistocene glaciers scoured and cleaned the land surface, removing almost all the soil and superficial deposits and exposing the joint and fracture patterns of the ancient crystalline rocks beneath. Streamlined and moulded rock ridges develop, including *roches moutonnées*. In parts of Scandinavia and New England these may be several kilometres long and have steepened faces of more than 100 m in height. They are interspersed with scoured hollows which may be occupied by small lakes when the ice sheet retreats. In western Scotland relief that is dominated by this mixture of rock ridges and small basins is called *knock and lochan* topography. Areas of calcareous rocks that have been scoured by ice often display LIMESTONE PAVEMENTS.

Considerable debate has been attached to the question of rates of glacial erosion (Summerfield and Kirkbride 1992) and some have argued (see GLACIAL PROTECTIONISM) that glaciers can protect the underlying surface from erosion. Much depends on the context in which a glacier or ice sheet occurs, but even for an area like the Laurentide Shield there is controversy (see Braun 1989).

A range of methods has been used to measure amounts of glacial erosion. These have included:

1 The use of artificial marks on rock surfaces later scraped by advancing ice.
2 The installation of platens to measure abrasional loss.
3 Measurements of the suspended, solutional and bedload content of glacial meltwater streams and of the area of the respective glacial basins.
4 The use of sediment cores from lake basins of known age which are fed by glacial meltwater.
5 Reconstructions of preglacial or interglacial land surfaces.
6 Estimates of the volume of glacial drift in a given region and its comparison with the area of the source region of that drift.
7 Cosmogenic nuclides (Colgan *et al.* 2002).

Many published rates of glacial erosion are high, typically 1,000–5,000 $m^3 km^{-2} a^{-1}$, but they vary greatly according to the measuring techniques that are used (Warburton and Beecroft 1993) and to the nature of the environment (Embleton and King 1968). Clearly, geographical location is an important control of the rate of glacial erosion. Some areas have characteristics that limit the power of glacial erosion (e.g. resistant lithologies, low relief, frozen beds), but other areas suffer severe erosion (e.g. non-resistant lithologies, proximity to fast ice streams, thawed beds, etc.).

References

Bennett, M.R. and Glasser, N.F. (1996) *Glacial Geology: Ice Sheets and Landforms*, Chichester: Wiley.
Braun, D.D. (1989) Glacial and periglacial erosion of the Appalachians, *Geomorphology* 2, 233–256.
Colgan, P.M., Bierman, P.R., Mickelson, D.M. and Caffree, M. (2002) Variation in glacial erosion near the southern margin of the Laurentide ice sheet, south-central Wisconsin, USA; implications for cosmogenic dating of glacial terrains, *Geological Society of America Bulletin* 1,114, 1,581–1,591.
Embleton, C. and King, C.A.M. (1968) *Glacial and Periglacial Geomorphology*, London: Arnold.
Summerfield, M.A. and Kirkbride, M.P. (1992) Climate and landscape response, *Nature* 355, 306.
Warburton, J. and Beecroft, I. (1993) Use of meltwater stream material loads in the estimation of glacial erosion rates, *Zeitschrift für Geomorphologie*, 37, 19–28.

A.S. GOUDIE

GLACIAL ISOSTASY

'ISOSTASY' (= equal standing) refers to an equal distribution of pressure or mass within a column of rock that extends from the surface of the Earth to its interior. This concept of equilibrium is, however, perturbed by forces at the surface and within the Earth which create unequal mass distributions. With time, such inequalities are compensated for by adjustments within the lithosphere and aesthenosphere, and the rate and relaxation time for this recovery to equilibrium is a function of the Earth's rheology. The term 'glacial isostasy' is thus used to define the adjustments of the Earth's surface to the growth and disappearance of ice sheets (Andrews 1974). For example, 22,000 years ago the surface of the Earth differed fundamentally from today's geography, as large ice sheets, several kilometres thick, covered much of Canada and most of the area centred on the Baltic (the Fennoscandian ice sheet), large areas of Great Britain, and many

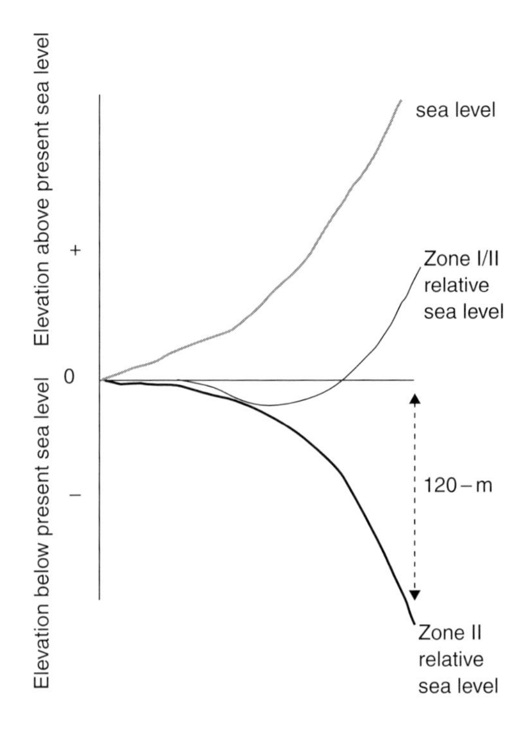

Figure 69 Schematic depiction of changes of relative sea level at sites within the borders of former ice sheets (Zone I), at sites distant (>1,000 km) from the maximum ice extent (Zone II), and sites in the transition between Zones I and II

other parts of the world (Denton and Hughes 1981). Together these ice sheets and glaciers extracted about 120 m of water from the global ocean, thus reducing the water load on the seafloor. During the last 15,000 years or so the ice sheets melted, resulting in a reduction in the load of ice sheets on continents, but the resulting meltwater has added to the load over ocean basins (Peltier 1980). Thus 'glacial isostasy' in its most complete definition includes the global-wide changes in the differential elevation of land due to ice sheet removal, and changes in relative sea level around all the world's coastlines caused by the unique combination of ice and water loading and/or unloading at each location (Figure 69, Zones I, I/II, and II) (Clark *et al.* 1978).

Observations on changes in sea level within formerly glaciated areas have been made for over 150 years, but the link between the removal of the ice load, crustal recovery and isostasy were made in north-west Europe and North America by the mid to late nineteenth century (Andrews 1974). Although significant research was carried out in the early part of the twentieth century studies of glacial isostasy bloomed in the period after ~1960 due in no small part to the development of radiocarbon dating, and to increased levels of field research in Arctic Canada, Greenland and Svalbard. In these areas materials to date the changes of sea level through time (molluscs, whalebone and driftwood) were abundant. The combination of a technique (^{14}C dating) plus exploration of vast tracts of formerly glaciated areas resulted in an explosion of data on changes in sea level and on the delimitation of former ice sheet margins through time (Andrews 1970; Blake 1975; Dyke 1998; Dyke and Peltier 2000; Forman *et al.* 1995). In the mid- to late 1970s these data-rich field observations attracted the interests of the geophysicists who now had both the mathematical tools and the computer power to tackle what is a global Earth-science problem with many ramifications (Peltier and Andrews 1976).

Studies of 'glacial isostasy' represent a significant interplay between workers in several different fields (Figure 70), including (1) the glacial geologist who maps the time-dependant changes in ice sheet extent (and more problematically, thickness); (2) the Quaternary scientist working on changes in sea level from sites within the margins of present ice sheets (near-field sites), to those at and just beyond the former ice sheet limits, and finally to workers reconstructing changes in sea level at sites far-distant (the far-field) from ice margins (say the central Pacific Islands); (3) glaciologists who combine data from (1) above with knowledge of the physics of ice to model changes in ice sheet extent and volume; (4) the geophysicists who 'tune' the behaviour of the Earth's rheology to match the data from inputs (1) and (2). The schematic interplay between these disciplines (Figure 70) has resulted in a series of successive approximations of each of the key components. This process started in 1976 (Peltier and Andrews 1976) and is still continuing.

The rheology of the Earth is most frequently modelled as a self-gravitating viscoelastic (Maxwell) solid (Cathles 1975), although a case can be made for a more complex rheology where the response is a nonlinear (power > 1) function, not unlike the behaviour of glacial ice. A simple 2-D model consists of a lithosphere of some thickness which overlies a fluid aesthenosphere. The lithosphere is rigid with the application of a small

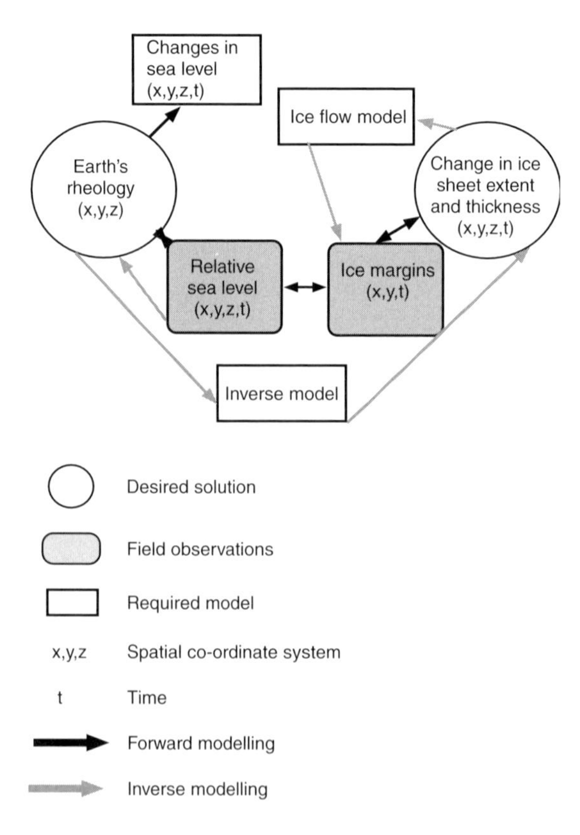

Figure 70 Schematic diagram on interactions between data and models

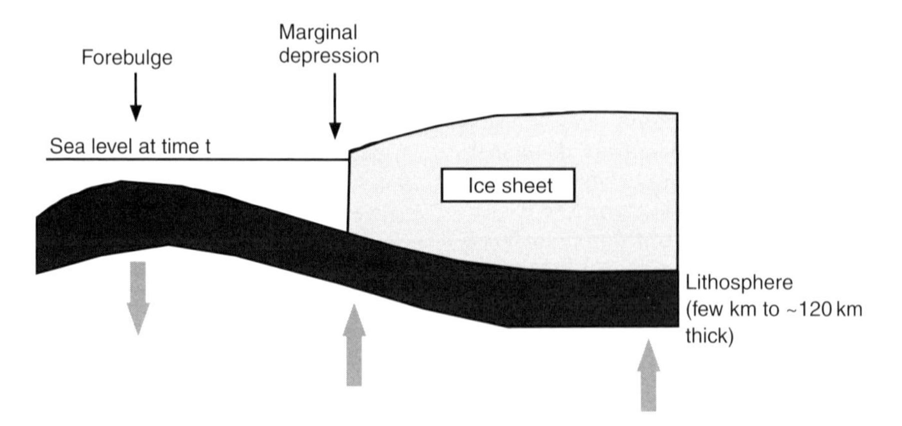

Figure 71 Two-dimensional model of an Earth with a lithosphere and aesthenosphere. The distance between the ice margin and the forebulge is a function of the thickness and elastic properties of the lithosphere. The senses of motion are those during the retreat phase of an ice sheet

load ('small' being a function of the lithosphere thickness, but on old continental shields the load may have to exceed a diameter of ~300 km, whereas on Iceland, with the mantle virtually at the surface, then the load diameter is probably a few kilometres to a few tens of kilometres at most for isostatic compensation by flow to be induced. As an ice sheet grows on land at some point the lithosphere will bend and material will be displaced by flow. It is generally assumed that the flow can be approximated by a layered Newtonian viscous fluid with a viscosity in the range of 10^{22} poises. The 2-D model of the Earth's response to a glacial load (Figure 71) shows that at the margins of the ice sheet the load is partly supported by the lithosphere so that there is a depression at the ice margin but at some distance from the ice sheet there is a zone of uplift in the forebulge. Upon retreat of the ice sheet the forebulge will collapse, hence there will be a rise in relative sea level.

Figure 70 shows the elements of the problem. Glacial isostasy has been examined in terms of 'forward models' and 'inverse models' in a full 3-D global model. In reality there is an ongoing iteration between the field scientists and the modellers so that both approaches are required (Lambeck 1995; Lambeck *et al.* 1998; Peltier 1994; Peltier 1996). In the 'forward' case, the explicit data are the positions of the ice sheet margins through time (x, y, t), and changes in relative sea level at a suite of sites from Zones I, I/II, and II. What then has to be approximated is the changes in thickness within an ice sheet (x, y, z, t). The application of this time-dependent load to a model of the Earth's rheology will result in changes in sea level (required model) where the predicted changes in sea level include not only the obvious fall of sea level in Zone I, but also account for the transfer of mass (meltwater) from the melting ice sheets to the oceans. Disagreements between the observed relative sea levels (Figure 70) and the predicted sea levels could be related to either an incorrect Earth rheology or an incorrect estimate of changes in the ice sheets in all four dimensions. In contrast, inverse modelling is an attempt to develop a model of the global ice sheet changes by taking the observed relative sea-level data, assuming a rheology, and then using these data to reconstruct the changes in the ice sheets. Appropriate ice flow models can then be applied to see if the reconstructions are glaciologically feasible and whether the reconstructed ice sheets match the data, in this case the mapped and dated ice sheet margins.

References

Andrews, J.T. (1970) A geomorphological study of post-glacial uplift with particular reference to Arctic Canada, Institute of British Geographers Special Publication No. 1.

Andrews, J.T. (1974) *Glacial Isostasy*, Stroudburg, PA: Dowden, Hutchinson and Ross.

Blake, J.W. (1975) Radiocarbon age determination and postglacial emergence at Cape Storm, Southern Ellesmere Island, Arctic Canada, *Geografiska Annaler* 57, 1–71.

Cathles, L.M. III (1975) *The Viscosity of the Earth's Mantle*, Princeton: Princeton University Press.

Clark, J.A., Farrell, W.E. and Peltier, W.R. (1978) Global changes in postglacial sea level: a numerical calculation, *Quaternary Research* 9, 265–287.

Denton, G.H. and Hughes, T.J. (1981) *The Last Great Ice Sheets*, New York: Wiley.

Dyke, A.S. (1998) Holocene delevelling of Devon Island, Arctic Canada: implications for ice sheet geometry and crustal response, *Canadian Journal of Earth Science* 35, 885–904.

Dyke, A.S. and Peltier, W.R. (2000) Forms, response times and variability of relative sea-level curves, glaciated North America, *Geomorphology* 32, 315–333.

Forman, S.L., Lubinski, D., Miller, G.H., Snyder, J., Matishov, G., Korsun, S. and Myslivets, V. (1995) Postglacial emergence and distribution of late Weichselian ice-sheet loads in the northern Barents and Kara seas, Russia, *Geology* 23, 113–116.

Lambeck, K. (1995) Constraints on the Late Weichselian Ice Sheet over the Barents Sea from observation of raised shorelines, *Quaternary Science Reviews* 14, 1–16.

Lambeck, K., Smither, C. and Johnston, P. (1998) Sea-level change, glacial rebound and mantle viscosity for northern Europe, *Geophysical Journal International* 143, 102–144.

Peltier, W.R. (1980) Models of glacial isostasy and relative sea level, *Dynamics of Plate Interiors* 1, 111–128.

Peltier, W.R. (1994) Ice age paleotopography, *Science* 265, 195–201.

——(1996) Mantle viscosity and ice-age ice sheet topography, *Science* 273, 1,359–1,364.

Peltier, W.R. and Andrews, J.T. (1976) Glacial-isostatic adjustment: I The forward problem, *Geophysical Journal Royal Astronomical Society* 46, 605–646.

JOHN T. ANDREWS

GLACIAL PROTECTIONISM

The belief that the erosive power of rain and rivers far exceeds that of glacier ice, and that the presence of glaciers in a region protects the landscape from much more effective fluvial attack (Davies 1969). Glaciers were thought, following Ruskin, to rest in depressions like custard in a pie dish, rather than to erode the basins. Proponents of this theory included the British geologists J.W. Judd, T.G. Bonney, E.J. Garwood and S.W. Wooldridge.

In some areas, where ice stream velocities are low and relief is limited, the erosive role of glaciers may well be passive rather than active. Some degree of protection may be afforded.

Reference

Davies, G.L. (1969) *The Earth in Decay*, London: Macdonald.

A.S. GOUDIE

GLACIAL THEORY

The belief that in former times glaciers had been more extensive than they are today. Some suggestions as to this had originally been made at the end of the eighteenth century. In 1787 de Saussure recognized erratic boulders of palpably Alpine rocks on the slopes of the Jura Ranges, and Hutton reasoned that such far-travelled boulders must have been glacier-borne to their anomalous positions. Playfair extended these ideas in 1802, but it was in the 1820s that the *Glacial Theory*, as it came to be known, really became widely postulated. Venetz, a Swiss engineer, proposed the former expansion of the Swiss glaciers in 1821, and his ideas were supported and strengthened by Charpentier in 1834. The poet Goethe expressed the idea of 'an epoch of great cold' in 1830. However, the ideas of both Venetz and Charpentier were extended and widely publicized by their fellow countryman, Louis Agassiz, who was one of the originators of the term *Eiszeit* or *Ice Age*. In Norway Esmark put forward similar ideas in 1824, and in 1832 Bernhardi went so far as to suggest that the great German Plain had once been affected by glacier ice advancing from the North Polar region.

In spite of this convergence of opinion from numerous sources, these ideas were not easily accepted or assimilated into prevailing dogma, and for many years it was still believed that glacial till, called drift, and isolated boulders, called erratics, were the result of marine submergence, much of the debris, it was thought, having been carried on floating icebergs. Sir Charles Lyell noted debris-laden icebergs on a sea-crossing to America, and found that such a source of the drift was more in line with his belief in the power of current processes – UNIFORMITARIANISM – than a direct glacial origin.

Even towards the end of the nineteenth century some opposition still remained. In 1892, for instance, H.H. Howorth produced his massive neocatastrophist *The Glacial Nightmare and the Flood – a second appeal to common sense from the extravagance of some recent geology*, and tried to return to a fundamentalist-catastrophic interpretation of the evidence.

Conversely, others were overenthusiastic about the glacial theory and Agassiz himself postulated that glaciers reached the humid tropics in South America.

Further reading

Chorley, R.J., Dunn, A.J. and Beckinsale, R.P. (1964) *History of the Study of Landforms*, Vol. 1, London: Methuen.
Davies, G.L. (1969) *The Earth in Decay*, London: Macdonald.

A.S. GOUDIE

GLACIDELTAIC (GLACIODELTAIC)

The discharge of sediment-laden streams from melting glaciers into lakes or fjords usually results in accumulation of a delta. The morphology and structure of the delta depends on several factors. These include nature of the ice margin; a vertical glacier front calving into a fjord contrasts strongly with a gently sloping ice surface descending into a shallow lake overlying dead ice. The sources of meltwater, flowing directly from the ice or entering the water body at the surface, deeper within or at the base of the ice as overflows, interflows or underflows, are also important. The quantity and particle size of sediment, the channel gradient, and the location of deposition close to or distant from the ice margin all influence the form, structure and sedimentary characteristics of the delta.

Deltas formed by discharge of glacial meltwater are located either in the proglacial environment or in ice-contact situations. In the proglacial environment they are deposited in lakes and fjords, and distinct types referred to as Hjulstrom, Gilbert and Salisbury deltas have been identified. In ice-contact situations deltas may form in supraglacial lakes, subglacial water bodies and lakes of the terminoglacial environment.

Hjulstrom deltas occur either in lakes or fjords where outwash sheets (the Icelandic sandur) of gravel and sand enter shallow water with gently sloping fronts. They may also form fans or small deltas in lakes on the ice surface. The delta gravels and sands are the flood traction loads of braided sandur channels or of fan channels that on deposition in standing water become smaller in grain size away from the point of discharge. Beyond the delta front, which advances into the water as a series of lobes due to changes in channel discharge points, the suspended silts and clays are deposited as prodelta mud.

The Salisbury type delta is intermediate in type and size between the Hjulstrom and Gilbert-type deltas. It forms where high-energy flow from a subglacial tunnel mouth supplies material rapidly via sheet- and streamfloods, and topset beds accrete very rapidly.

The classic glacigenic delta is the Gilbert type that may be formed on ice, in proglacial lakes and in fjords where the water is deep enough to allow development of a distinct structure that includes bottomset, foreset and topset beds. The bottomset beds occur beneath and beyond the foreset beds and extend as prodelta clays to merge with the lake – or fjord floor clays. The bottomset beds result when the suspended sediment carried in turbid flows beyond the delta front settles. The beds are generally laminated. The laminae reflect grain-size variations between fine sand, silt and clay layers due to deposition that may be controlled by seasons, short periods or single events. The sediments become finer away from the point of discharge. They may contain occasional dropstones derived from floating ice, and convoluted laminations due to disturbance of the delta face by slumping, sediment flowage and turbidity currents. Sediment loading and dewatering will also produce convoluted laminations.

The foreset beds dip steeply (*c*.25–30°) away from the source of the glacial sands and gravels and prograde into the lake by the intermittent avalanching down the delta front of cohesionless debris flows. The foresets decrease slightly in slope and sediment size towards the delta face, and individual foresets tend to decrease in grain size upwards. Channels that result from erosion of the delta face by turbidity flows may be cut into the foreset beds. As the delta builds up to water level the glacifluvial sandur deposits extend onto the surface as topset beds. The topset gravels and sands are coarser than those of the foresets, are gently inclined (2–5°), relatively thin (*c*.1–2 m) and exhibit cut-and-fill structures related to the shifting courses of the braided channels of the sandur.

Small ponds and lakes develop on the decaying marginal and terminal parts of glaciers. Meltwater from surface streams or shallow depth within the ice may form small deltas in the shallow lakes. The delta sediments consist of inclined beds of gravel and sand interbedded with unsorted debris flow sediments from the ice surface. Subsequent melting of the underlying and laterally supporting ice causes slumping and flowage of the sediments with the development of fault and fold structures. Where preserved, the sediments form delta-kames.

Subglacial and englacial streams may enter bodies of standing water at or near the base of the ice. Reduction in stream velocity will cause sedimentation of the sands and gravels to form a delta. Decay of the ice may result in formation of a delta-kame.

More extensive lakes are frequently formed in the ice contact terminoglacial environment of glaciers and ice sheets. Where large meltwater streams enter lakes or fjords, deltas of the Gilbert type are formed. Bottomset beds may be formed where the water body is relatively large but most of the delta sediments consist of foreset beds of sand and gravel. Topset beds only develop where a sandur forms between the meltwater outlet and water body. Interstratification of debris flow deposits and foreset sands and gravels is common. Removal of ice support during decay causes collapse and pitting of the ice proximal delta margin. Such deltas have been described as kamiform and where subglacial stream outlets are numerous along an ice edge many may be developed and may coalesce laterally as a kame moraine. The Salpausselka Moraines formed in southern Finland during the final readvance stage, the Younger Dryas, of the last glaciation are over 600 km in length and largely form delta moraines.

In southern Finland the three major Salpausselka delta moraines – formed in the Baltic Ice Lake before the postglacial rise in sea level drowned the Baltic Sea – record retreat stages of the ice sheet margin. Where deltas are formed sequentially inland along fjord margins the stages of glacier retreat in the valley and relative sea-level rise in the fjord can be detected. When ice barriers impounded glacial lakes in upland valleys and formed deltas and shorelines at a number of water levels, as in Glenroy,

Scotland, the delta surfaces and shorelines record the stages in draining of the glacial lake.

Further reading

Benn, D.I. and Evans, D.J.A. (1998) *Glaciers and Glaciation*, London: Arnold.
Bennett, M.R. and Glasser, N.F. (1996) *Glacial Geology: Ice Sheets and Landforms*, Chichester: Wiley.
Brodzikowski, K. and van Loon, A.J. (1991) *Glacigenic Sediments*, Amsterdam: Elsevier.
Drewry, D. (1986) *Glacial Geologic Processes*, London: Arnold.

ERIC A. COLHOUN

GLACIER

Glaciers are accumulations of snow and ice on the Earth's surface. They form predominantly at high latitude (polar regions) and at high elevation (on mountains). Here, two meteorological variables, temperature and precipitation, combine to yield conditions where the annual amount of snowfall (predominantly during the cold or wet season) outweighs the annual amount of snow melt (predominantly during the warm or dry season). Under these conditions, consecutive annual snow layers develop, one on top of the other, the pressures of which force snow at depth to change structure and density (recrystallization). These conditions first produce firn/névé (density of 0.400–0.830 $kg\,m^{-3}$) and, when pores are sealed off to create air bubbles, ice (density of 0.830–0.917 $kg\,m^{-3}$). These changes can occur within one year and at shallow depths in maritime regions or take hundreds to thousands of years and considerable depths in continental locations. Glaciers are said to have formed when glacier flow occurs, that is when ice is thick enough to deform plastically under its own weight (Paterson 1994).

Glacier systematics

Glaciers are commonly ordered using their geomorphological or thermal characteristics, both of which yield a tripartite system. The common characteristic of ICE SHEETs and ice caps (smaller versions of the former, i.e. <50,000 km^2) is that they are so thick relative to the landscape relief on top of which they rest, that their surface topography and flow direction are unconstrained by the underlying topography (except near the ice margin). Their surface morphology dominantly shows the presence of ice domes, the dome-shaped central regions where ice forms and flows outward towards its perimeter, and outlet glaciers and ice streams, the narrow flow-parallel bands of faster flowing ice that are the primary routes by which the ice is evacuated in marginal areas (Plate 51A). Two ice sheets and numerous ice caps occur in the polar regions, the latter typically on upland plateaux.

The margins of ice sheets and ice caps, especially in Antarctica, frequently become afloat in the ocean that surrounds them. These floating sections, or ice shelves, normally cover shallow marine embayments or continental shelves with many islands. Eventually, slabs of ice break off from the ice cliffs that border ice shelves, thus producing (tabular) ICEBERGS.

The surface morphology and flow of glaciers (alpine glaciers, mountain glaciers) follow the morphology of the subglacial landscape. The morphology of ice fields, thinner than ice caps and lacking the characteristic ice dome, characteristically shows mountain uplands covered by ice, except for the highest ridges and peaks (see NUNATAK). Valley glaciers, elongated features confined by valley walls, occur as outlets of the ice-covered uplands and as glaciers on their own (Plate 51B). Their heads (highest section) often start in bedrock depressions (see CIRQUE) and their snouts (lowest section) reside within a valley. However, when the glacier emanates and terminates immediately beyond the valley mouth (i.e. beyond its lateral constraint), the ice tongue spreads outward to form a wide piedmont lobe, forming a piedmont glacier (Plate 51C). If outlet glaciers and valley glaciers extend offshore they form tidewater glaciers (Plate 51A). Cirque glaciers or glacierets are located within, or just barely extend beyond, the amphitheatre-like basins in which they form, and they tend to be wide rather than long (Plate 51D). Glaciers on steep mountain slopes above bedrock depressions, hanging glaciers or ice aprons, primarily loose mass as blocks of ice become detached and avalanche downslope. Sometimes, these and other avalanches supply the mass necessary for rejuvenated or regenerated glaciers to exist in locations where precipitation alone would be insufficient to support a glacier (Hambrey 1994; Sharp 1988; Sugden and John 1976).

Plate 51 (a) outlet glacier, Antarctica; (b) valley glacier, Sweden; (c) piedmont glacier, Antarctica; and (d) cirque glacier, Sweden

Ordering glaciers by their thermal characteristics yields the (high) polar and temperate glacier-type end members, and the subpolar glacier type for those that combine both former characteristics. The ice in polar glaciers is below freezing throughout and meltwater is absent, except, maybe, for surface melting during short periods in summer. The ice in temperate glaciers is close to melting throughout, except, usually, for surface freezing during winter. Hence, the glacier has a temperature close to its melting point, which is 0 °C for pure ice at atmospheric pressure, but occurs at lower temperatures (pressure melting) when pressurized at depth. However, many glaciers experience both freezing and melting conditions; they are subpolar or polythermal glaciers. In polar regions they experience abundant surface melting and heating throughout the summer and in temperate regions they have a thick cold surface layer that is maintained throughout the summer (Ahlmann 1935).

Mass balance of glaciers

Most glaciers are nourished by snowfall and depleted by snow and ice melt. Because the conditions favourable for snow to fall and snow and ice to melt are related to the temperature of the atmosphere over the glacier surface, and because the mean annual temperature decreases with increasing altitude (lapse rate), the surface of a glacier normally shows two or more predictable zones

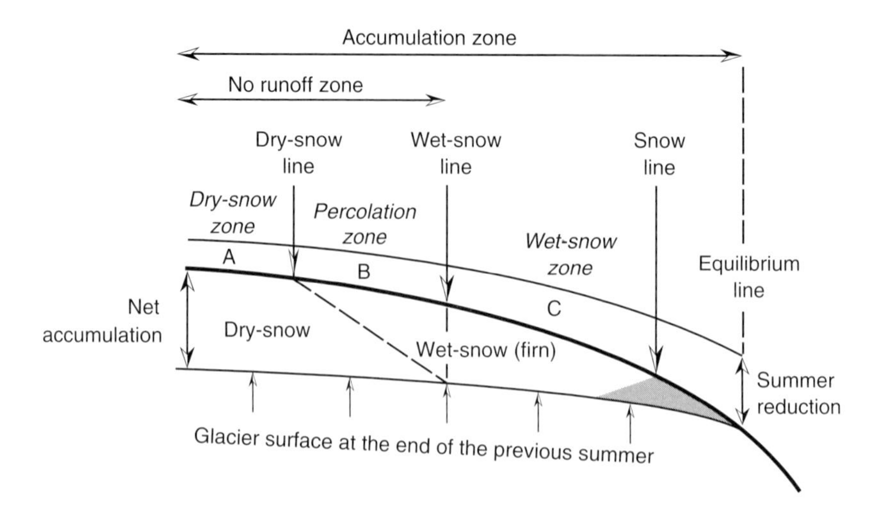

Figure 72 The surface structure of the accumulation area. The heavy line is the glacier surface at the end of the summer, the light line is the previous winter surface. The grey-shaded region is superimposed ice. Volume reduction by (A) recrystallization, (B) melt (and refreezing at depth) and (C) melt and runoff. Modified from Paterson (1994: 10)

(Figure 72). In the highest elevation zone, the accumulation zone, temperatures are at a minimum yielding conditions where the snowpack that accumulates during the winter or wet season is not melted entirely during the summer or dry season, thus creating an annual mass gain (accumulation). In the lowest elevation zone, the ablation zone, temperatures are higher and last winter's snowpack plus an additional amount of the underlying solid ice melts, creating an annual mass loss (ablation). The MASS BALANCE OF GLACIERS denotes the balance between the amount of snow remaining after the summer integrated across the accumulation area (and converted to the amount of water it represents when melted, m w.e. or metre water equivalent) and the amount of solid ice lost underneath the snowpack integrated across the ablation area (in m w.e.). When mass balance is positive, and more and more mass is added each year, the glacier will grow and expand. Conversely, when mass balance is negative over many years, the glacier will shrink and contract (retreat). When ice reaches considerable thickness a positive feedback mechanism occurs – because of the higher elevations the conditions for ice accumulation improve, especially close to the snout. The boundary between the accumulation and ablation zones is a relatively narrow zone, or line, where the annual mass balance is zero, the EQUILIBRIUM LINE OF GLACIERS (which exists on all glaciers that are not strongly out of equilibrium with contemporary climate, i.e. where the whole glacier surface becomes an accumulation area or ablation area). Some glaciers have additional boundaries, lines, in their accumulation areas (Figure 72). The dry-snow line borders the dry-snow zone, a region of extreme climate (in polar areas and at extremely high elevations), where no surface melt occurs, even in summer. The wet-snow line is the lower limit of the percolation zone, where snow melts at the surface and refreezes at depth during the summer. The refreezing of meltwater occurs at depth because the snowpack is initially cold, but the refreezing process releases heat and warms the snowpack until it reaches melting temperatures throughout by the end of the summer at the wet-snow line. Strictly taken, above the wet-snow line there is no mass loss (no runoff zone). The snow line is the lower boundary of the wet-snow zone, where all remaining snow at the end of the ablation season is at the melting temperature. The firn line, not shown, is the boundary between ice and firn at the end of the summer, and may coincide with the snow line (only in temperate regions where snow may transform to firn in one summer can this latter situation occur). Although there is mass loss throughout the wet-snow zone during the summer, it is characterized by a positive annual mass balance. Meltwater that has refrozen at depth, forming ice lenses, may become particularly extensive and form superimposed ice layers underneath the firn in the wet-snow zone. These can visually crop out at the surface of a glacier at the end of the summer season between the snow line and the equilibrium line (Oerlemans 2001; Paterson 1994).

Alternative important components in the annual mass balance of glaciers are mass gain by avalanching and rime, and mass loss through avalanching, calving and sublimation.

Glacier flow

Structural glaciology, the geomorphology of glacier surfaces, shows that bodies of ice experience differential movements between the glacier bed and the surface, between the lateral margins and the glacier centre, and between different locations along its longitudinal profile. The most conspicuous features are crevasses, foliation structures, and band- or wave ogives (Forbes bands) (Hambrey 1994: 61–69; Paterson 1994: 173–190; Sugden and John 1976: 71–78). For example, a visible sign of glacier flow, except for these ice surface structures, is the creation of a BERGSCHRUND, an opening between (stagnant ice on) the valley wall and the glacier that pulls away from the wall as it moves downslope (Figure 73).

Crevasses, which are fissures in the ice surface with a typical depth of 25–30 m, occur abundantly on glacier surfaces, are often consistent in their direction over limited distances but vary considerably along the length of a glacier, and

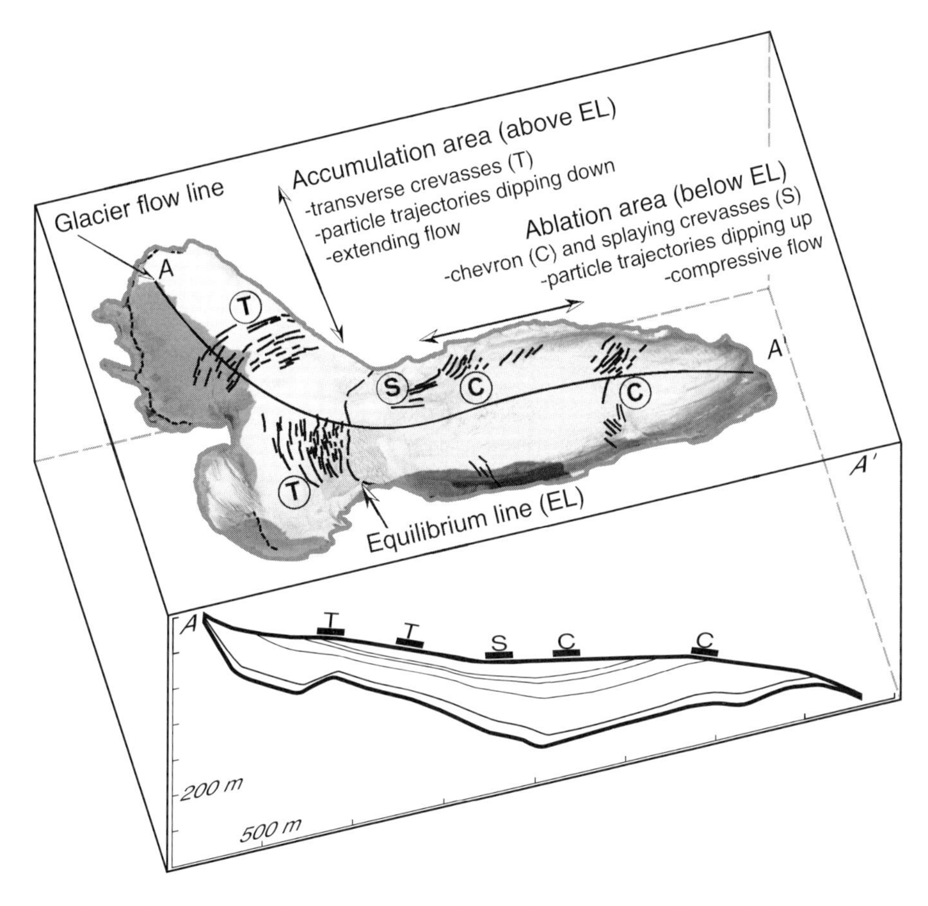

Figure 73 Glacier flow from surface structure and particle trajectories. Example is from Storglaciären, northern Sweden. Particle trajectories modified from Pohjola (1996). Note the location of the bergschrund (dashed) in the upper part of the accumulation area

will form where at least one principle stress is tensile and exceeds the tensile strength of the ice. The direction of crevasses can be predicted from a straightforward geometric analysis of stress distributions. Hence, the pattern of crevasses found in many accumulation areas of glaciers, transverse crevasses, can be related to the existence of extending flow (when ice flow accelerates downslope; i.e. where the glacier bed steepens, where the glacier is joined by another branch, on the outside of a glacier bend, and at the grounding line, where glaciers become afloat). Conversely, most glacier ablation areas reveal a pattern of crevasses, splaying crevasses, which are expected when the stress situation is dominantly compressive (when ice flow decelerates downslope; i.e. where the glacier bed flattens or rises, where the valley widens, on the inside of a bend, and at the snout; Hambrey 1994: 56). Finally, where the stress situation is neither dominantly extending, nor compressive, the lateral drag of the valley walls results in a third predictable stress and crevasse pattern, chevron crevasses, a feature common to middle sections of alpine glaciers. The width and depth of the crevasses depend primarily on the temperature of the ice, i.e. they are widest and deepest in polar glaciers.

Foliation structures result from previous structures that have been compressed (pure shear) and stretched (simple shear) to attain a direction that dominantly is parallel to ice flow. The original structures include the near-horizontal layering during accumulation (and the formation of ice lenses) and vertical or inclined elongated structures, such as crevasses, and point structures (moulins). Foliation structures, therefore, are consistent with differential motion through the ice body.

Forbes bands, or ogives, can form as glacier ice flows through an ice fall. Ice falls are causally related to the occurrence of very steep glacier beds. Here, the tensional stresses result in transverse crevassing and thinning while the ice moves through the fall. However, summer and winter conditions result in a modification of the ice as it moves through. Ice ablation during the summer results in thinner ice (and dirty because of dust collection) arriving at the foot of the fall. Precipitation during the winter, on the other hand, fills the crevasses with snow, resulting in thicker and cleaner ice arriving at the foot of the fall. For ice falls where the ice moves through within a year's time span, the resulting situation at the foot of the fall, where flow is compressive and crevasses are closed, are alternating ridges of blue ice (winter) and depressions of white ice (summer). These structures are convex in the direction of ice flow (even though the crevasses were slightly concave), indicating the differential motion in the transverse direction and, because of the crevassing, in the longitudinal direction.

The geomorphology of glacier surfaces, therefore, was the initial guidance in an understanding of some of the basic characteristics of glacier motion by the middle of the century (e.g. Tyndall 1860). Modern advancements in understanding the flow of glaciers, dating from the past fifty years, were mainly theoretical in nature (Glen 1952; Nye 1952; Weertman 1957), realizing that ice behaves like a crystalline solid, leading to theorems that have subsequently been verified experimentally (e.g. Raymond 1971).

From basic mass balance principles it must follow that a glacier of constant shape and volume (steady-state situation) must transfer all the annual mass gain in the accumulation area to the ablation area to compensate for the annual mass loss. For each transverse cross section through the glacier it is pertinent that the discharge of ice through that section per year balances the integrated annual mass balance across the up-glacier surface. Hence, this requires the discharge to be at a maximum at the equilibrium line, and normally this equates to having the thickest ice and highest ice flow velocities there, and implies the existence of extending flow in the accumulation area and compressive flow in the ablation area. Snow that falls in the accumulation area is successively compressed by subsequent layers of snow (until glacier ice is formed, which is incompressible), yielding a small velocity component towards the glacier bed. A cube of solid ice, when subjected to the range of pressures by the burden of overlying snow, firn and ice (typically less than 1 bar), will deform (stretching), thus departing on a trajectory that parallels, but is dipping slightly away from, the glacier surface (Figure 73). To satisfy mass continuity, this yields a convergence of flow in the accumulation area, and, typically, a thickening of the glacier downslope. Conversely, a cube of ice in the ablation area follows a trajectory which parallels, but is slightly directed towards, the slope of the ablating ice surface (Figure 73). This results in a divergence of flow in the ablation area, and, typically, a thinning of the glacier towards its lateral margin and snout.

As mentioned, the movement of ice is a gravity-driven internal deformation of the ice body. Because the deformation is at a maximum where the pressures are at a maximum, the deformation occurs primarily close to the glacier bed ('dragging' the overlying ice along). Because the amount of deformation higher up in the ice body, although dramatically less than at its bed, occurs in addition to the deformation for deeper ice layers, the velocity vector of ice flow close to the surface is measurably higher than at its bed.

Like other solids, ice deforms more readily when it is close to its PRESSURE MELTING POINT, a trait of temperate glaciers. Typical for pressure melting conditions is the presence of water (fluid state) and ice (solid), side by side, in the glacier body. The presence of water at the ice–bedrock interface is of importance in the motion of glaciers because its lubricating effect (especially when pressurized) facilitates the sliding of ice over its substrate (basal sliding). When ice covers sediment, basal sliding will normally be insignificant, but enhanced deformation of the water-saturated sediment may occur. Basal sliding would be very effective over a smooth bedrock surface, but, often, there are bedrock protrusions which hamper the effectiveness of the sliding process. When the obstructions are relatively small ($<10^{-2}$–10^{-1} m-size), ice melts on the upstream pressurized side, flows around the obstacle as water, and refreezes in its wake as the pressure drops, thus producing heat that can be used to help melting the ice on the upstream side. This experimental verification of the presence of regelation ice in the wake of obstacles is one of the modern findings of glacier flow. When the obstructions are relatively large (10^{-1}–10^{1} m-size), they create pressures so large that ice will deform more readily around it (by enhanced plastic deformation), a process that is less effective than regelation and thus more hampering to ice flow.

As noted, the effectiveness of glacier motion is in large part dependent on the temperature of the ice. Because most of the motion occurs in the basal ice as deformation and between the ice and the bedrock by sliding, it is specifically the subglacial temperature which is of interest. Temperate glaciers are warm-based (wet-), and their surface velocities integrate basal sliding and effective ice deformation. Polar glaciers are cold-based (dry-), which inhibits basal sliding and their surface velocities only reflect the integrated effect of ice deformation (at sub-optimal temperatures). Subpolar glaciers may have a bed at the pressure melting point and behave like temperate glaciers, or otherwise are cold-based and behave like polar glaciers.

References

Ahlmann, H.W. (1935) Contribution to the physics of glaciers, *Geographical Journal* 86, 97–113.

Glen, J.W. (1952) Experiments on the deformation of ice, *Journal of Glaciology* 2, 111–114.

Hambrey, M.J. (1994) *Glacial Environments*, London: UCL Press.

Nye, J.F. (1952) The mechanics of glacier flow, *Journal of Glaciology* 2, 82–93.

Oerlemans, J. (2001) *Glaciers and Climate Change*, Lisse: A.A. Balkema.

Paterson, W.S.B. (1994) *The Physics of Glaciers*, Oxford: Pergamon Press.

Pohjola, V.A. (1996) Simulation of particle paths and deformation of ice structures along a flow-line on Storglaciären, Sweden, *Geografiska Annaler* 78A, 181–192.

Raymond, C.F. (1971) Flow in a transverse section of Athabasca glacier, Alberta, Canada, *Journal of Glaciology* 10(58), 55–84.

Sharp, R.P. (1988) *Living Ice: Understanding Glaciers and Glaciation*, Cambridge: Cambridge University Press.

Sugden, D.E. and John, B.S. (1976) *Glaciers and Landscape: A Geomorphological Approach*, London: Edward Arnold.

Tyndall, J. (1860) *Glaciers of the Alps*, London: John Murray.

Weertman, J. (1957) On the sliding of glaciers, *Journal of Glaciology* 3, 33–38.

Further reading

Lliboutry, L. (1964–1965) *Traité de Glaciologie* (2 vols), Paris: Masson.

SEE ALSO: glacial deposition; glacial erosion; moraine

ARJEN P. STROEVEN

GLACIFLUVIAL (GLACIOFLUVIAL)

Glacifluvial is an adjective that applies to the processes, sediments and landforms produced by water flowing on, in and/or under glaciers and away from glacier snouts. Consequently, glacifluvial environments may occur in supraglacial, englacial, subglacial, ice-marginal and proglacial locations of alpine and continental glaciers both present and past. Proglacial, glacifluvial environments are often transitional to fluvial environments when the processes, sediments and landforms become dominated by non-glacial tributary inflows rather than the annual rhythm of melting ice (Lundqvist 1985).

Meltwater supply and pathways

Water enters a glacier from melting ice, snow melt, rainfall, hillslope runoff and the release of stored water. Consequently, most glacifluvial environments are dominated by strong seasonal, diurnal and episodic discharge variations. The release of stored water from subglacial, englacial, supraglacial or ice-marginal lakes and reservoirs produces floods several orders of magnitude greater than 'normal' melt-related flows. Such floods and megafloods (peak discharges estimated at 10^6–10^7 m^3.s^{-1} compared to 10^5 m^3.s^{-1} for the Amazon today) are known as OUTBURST FLOODS or jökulhlaups.

The path of meltwater through a glacier is determined by water supply (amount and location), ice temperature and dynamics and basal substrate. Pathways include: (1) supraglacial meandering channels; (2) englacial passages; (3) subglacial channels cut up into the ice (ice tunnels) or down into the bed, broad flows (sheets), linked-cavities, films, canals or aquifers; and (4) proglacial channels, broad flows and jets (into standing water – lakes, oceans).

Glacifluvial processes

Glacifluvial processes include erosion, transport and deposition. The type, rate and effectiveness of meltwater erosion are influenced by the nature of the basal substrate (sediment, bedrock), meltwater supply and pathway, and sediment supply. Mechanical erosion is effective because rapidly flowing meltwater in channels and broad flows is turbulent and carries a high sediment load (see SEDIMENT LOAD AND YIELD). Mechanical erosion includes: (1) hydraulic action – the force of water against its bed lifts or drags loose debris into the flow; (2) CAVITATION – shock waves and microjets resulting from the formation and implosion of bubbles of vapour (cavities) cause rock pitting or loosen mineral grains; and (3) abrasion – the impact and grinding of rock particles carried by the flow on themselves (attrition) and on flow boundary materials causes wear. Chemical erosion, or DISSOLUTION, occurs when meltwater removes soluble minerals in bedrock and debris. Subglacial dissolution is particularly effective because freshly abraded bedrock and debris present a high surface area for chemical reactions and solutes are continually flushed from the system. Together, meltwater erosion processes act to fracture, round and wear down bedrock and rock fragments and result in a variety of erosional landforms.

Landforms resulting from meltwater erosion

Meltwater erosion produces meltwater channels, bedrock erosion marks (s-forms) and may form some DRUMLINS, flutings, ribbed and hummocky terrain (Shaw 1996).

Meltwater channels may form (1) along the ice margin, (2) proglacially, or (3) subglacially. These channels exist in a variety of sizes (cm to km wide, cm to hundreds of m deep, decimetres to tens of km long) and substrates (bedrock, sediment); channels associated with past ice sheets are typically larger than those associated with alpine glaciers today. They may be differentiated by their slope and relationship to other glacial landforms. Ice-marginal (lateral) channels typically form parallel to the ice margin of cold-based glaciers, are often left perched and nested on valley sides during glacier retreat and their slope approximates that of the glacier surface at the time of formation. Advancing glaciers may divert rivers parallel to their ice front forming URSTROMTÄLER. Proglacial channels can form braided patterns (multiple shallow channels separated by bars) across outwash plains and always follow downslope paths (Plate 52a). Catastrophic drainage of ice-dammed or proglacial lakes has produced (1) trench-like valleys or spillways, and (2) networks of dry channels and waterfalls (see SCABLAND; Plate 52b). Subglacial channels may follow upslope paths, cross-cut drumlins and contain ESKERS (Plate 52c). TUNNEL VALLEYS or channels are long (can be >100 km), wide (<4 km), flat-bottomed, overdeepened (<100 m), often radial or anabranched subglacial channel systems that formed under ancient ice sheets. They can be buried by thick fills, muting their topographic expression. They may have formed catastrophically by meltwater erosion during the waning, channelized stages of subglacial megafloods (Shaw 1996).

Bedrock erosion marks, or s-forms, come in a variety of shapes and sizes (cm to km scale) and are found on the beds of bedrock channels and on broad bedrock plains. They are classified according to their shape and to the direction of formative

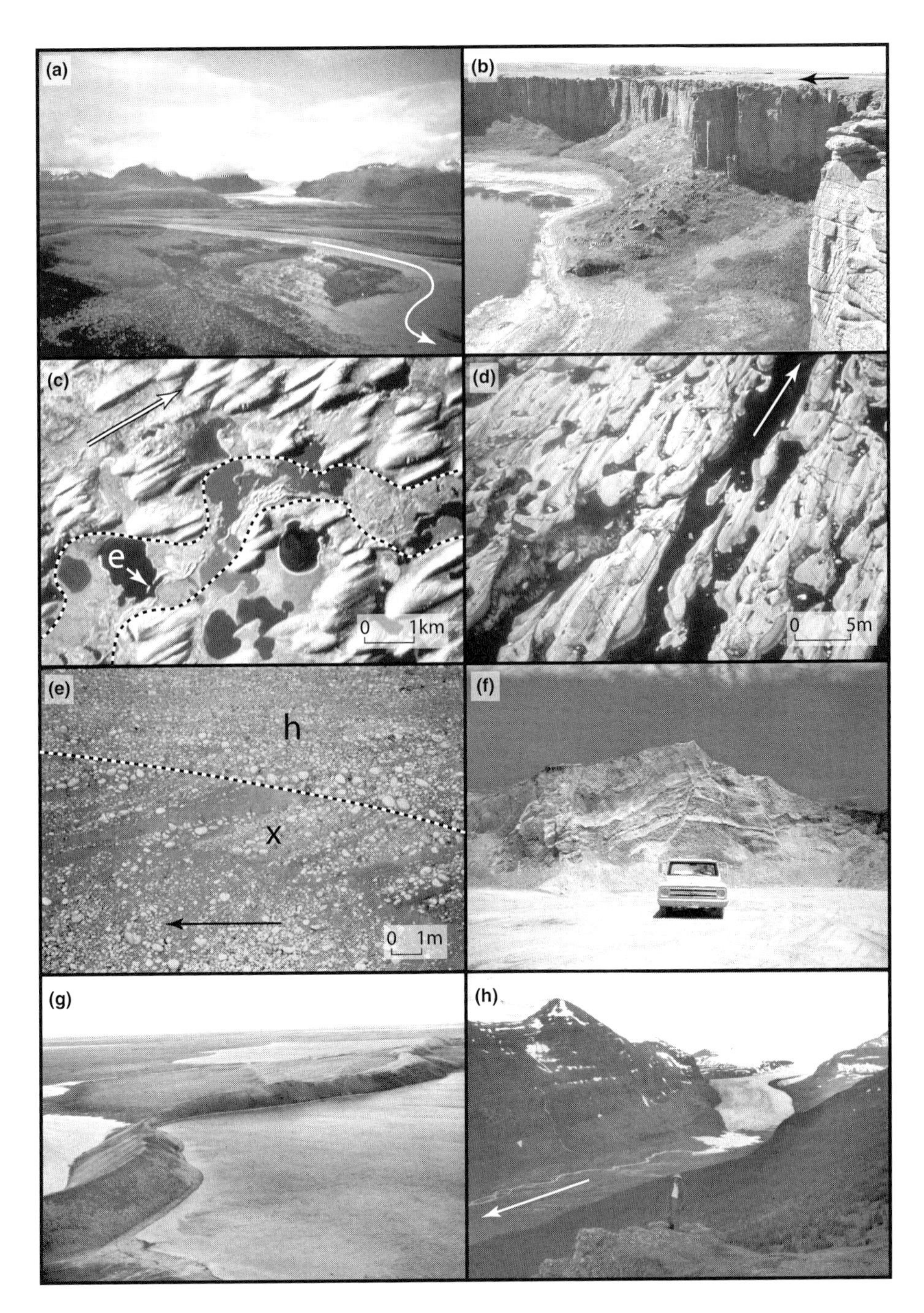

Plate 52 (a) braided outwash on Icelandic sandur (Skeiđaràrsandur, courtesy C. Simpson); (b) dry falls along Shonkin Sag spillway (USA, backwall ~55 m high); (c) cavity-fill drumlins truncated by tunnel channel (dashes) containing esker [e] (Livingstone Lake, Canada; aerial photograph A14509–77 © Her Majesty the Queen in Right of Canada, Centre for Topographic Information (Ottawa), Natural Resources Canada); (d) bedrock erosion marks (French River, Canada; Shaw 1996, © Wiley and Sons Ltd. Reproduced with permission); (e) cross-stratified [x] and horizontally-bedded [h] cobbles (Harricana esker, Canada); (f) faulted sand and gravel lithofacies (Campbellford esker, Canada); (g) sinuous esker (Victoria Island, Canada); (h) valley train (Saskatchewan glacier, Canada). Arrows indicate direction of formative water flow

flow. Mussel shell-shaped scours (muschelbrüche), sickle-shaped scours (sichelwannen; Plate 52d), comma-shaped scours and transverse troughs form mainly transverse to flow direction. Rock drumlins, rat-tails (tapering rock ridges extending from resistant rock knobs), flutes (spindle-shaped scours), furrows (Plate 52d) and cavettos (channels on vertical walls) form parallel to flow. POT-HOLES record vertical scour. Together s-forms often exhibit a directional consistency, and a hierachical and systematic arrangement consistent with erosion by turbulent subglacial megafloods (Shaw 1996; Figure 74).

Drumlins are elongated hills of various sizes (up to ~2 km long, tens of m high, hundreds of m wide) and shapes (inverted spoon, parabolic, transverse asymmetrical, spindle-shaped; Plate 52c). They occur in en-echelon patterns and in fields spanning hundreds of kilometres. Some flutings are long (tens of km), narrow (hundreds of m) remnant ridges that occur downflow from escarpments. Ribbed terrain is composed of fields of coalescent and subparallel, convex-upflow and crenulate-downflow ridges (up to 30 m high) that are formed transverse to flow. Hummocky terrain is identified as a field of mounds (<10 m high and ~100 m diameter) and hollows. These subglacial bedforms are often transitional to one another and the material within them may be truncated at the landsurface. Consequently, some drumlins (inverted spoon-shaped drumlins), flutings, ribbed and hummocky terrain are attributed to erosion by turbulent subglacial megafloods (Shaw 1996; Figure 74).

Glacifluvial sediment

Glacifluvial sediment is mainly derived from (1) material supplied to the glacier surface from valley side rock falls, debris flows and avalanches, or (2) meltwater erosion of bedrock, sediment or debris-rich ice along its flow path. Typically, sediment transport varies with glacial environment and discharge, being greatest in subglacial channel and broad flows and during OUTBURST FLOODS. Deposition rates are generally greatest at the ice margin where meltwater issues from ice tunnels onto open outwash plains or into standing-water bodies.

Glacifluvial sediment is deposited in BEDFORMS which vary in scale from centimetres (e.g. ripples) to hundreds of metres (e.g. bars or macroforms). Bedforms are preserved in the sedimentary record as lithofacies (Plate 52e). Lithofacies are sedimentary units distinguished by their physical and/or chemical characteristics such as colour, texture, structure and mineralogy. By applying experimental relationships derived from pipe and flume studies, lithofacies are used to reconstruct former bedforms, flow conditions (see PALAEOHYDROLOGY) and glacifluvial environments. For example, cross-laminated sand records ripples deposited during low flow, whereas horizontally bedded gravel may record the movement of gravel sheets across the tops of macroforms during flood flows (Plate 52c).

Glacifluvial sediment may be found in ice-contact environments (supraglacial, englacial, subglacial and ice-marginal) or proglacial environments beyond an apron of stagnant ice. It shows many of the same characteristics as sediment of non-glacial rivers – both are typically characterized by sorted and stratified sand and gravel lithofacies. However, glacifluvial deposits differ from non-glacial fluvial deposits in several ways. (1) Glacifluvial sediment is typically coarse grained (boulder through sand sizes) as the flow velocity is generally too high for settling of the finest particles (silt and clay); such particles become trapped in lateral and distal standing-water bodies. (2) Course-grained lithofacies may exhibit relatively poor sorting (large grain-size range) and rudimentary bedding when rapidly deposited during the waning stages of an outburst flood. (3) Glacifluvial sediments typically exhibit abrupt changes in lithofacies (Plate 52e) due to pronounced seasonal and episodic changes in flow regime. (4) Ice-contact glacifluvial deposits frequently include flow deposits (diamicton) and till balls (eroded ice deposits) and exhibit structures indicative of shearing, faulting (Plate 52f), slumping and subsidence. These characteristics develop due to the proximity to a moving glacier and to the melting of buried or supporting ice.

Landforms resulting from meltwater deposition

Meltwater deposition forms ice-contact and proglacial landforms. Ice-contact landforms may include drumlins, ribbed and hummocky terrain, some crevasse-fill ridges and De Geer MORAINES, eskers, KAMES, kame terraces, outwash fans and some end, grounding-line and interlobate moraines. Proglacial landforms include outwash plains, valley trains and sandurs. The main differences between these two categories are (1) ice-contact landforms may contain structures

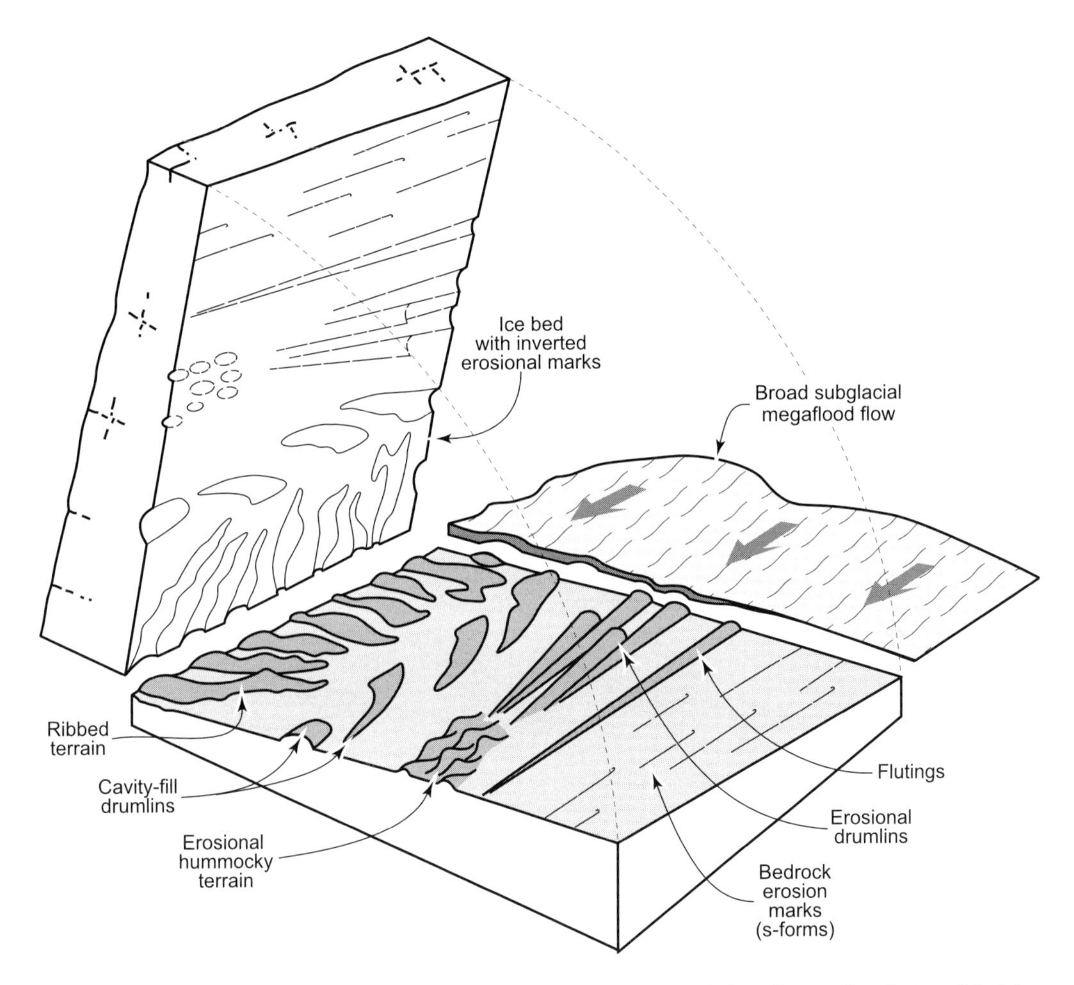

Figure 74 A model for subglacial landforms produced by broad subglacial megafloods (modified from Shaw 1996)

indicative of proximity to a moving glacier (e.g. thrust faults and shears) and/or the melting or removal of buried or supporting ice (faulting, slumping, pitted surfaces; see KETTLE AND KETTLE HOLES), and (2) proglacial landforms may contain sediment indicative of both glacifluvial and non-glacial processes.

Some drumlins (spindle, transverse asymmetrical and parabolic forms; Plate 52c), ribbed and hummocky terrain may have formed by deposition into cavities incised into the ice base during subglacial megafloods (Shaw 1996). As the broad megafloods waned, sediment carried in the flow was rapidly deposited into flow parallel, transverse and non-directional cavities forming cavity-fill drumlins, ribbed and hummocky terrain respectively (Figure 74). These landforms are composed of glacifluvial sediment with a sedimentary architecture that conforms to the landform shape.

Crevasse-fill ridges are linear, low ridges (up to ~10 m high) that are arranged in a pattern that mimics the radial and transverse crevasse patterns of a glacier. They formed by the infilling of crevasses within or at the base of glaciers (Figure 75). De Geer moraines are linear, low (<10 m) ridges that occur in fields subparallel to one another and in locations where the glacier was in contact with a standing-water body. They may form at the glacier margin during punctuated glacier retreat or in subglacial crevasses. Both ridge types may contain glacifluvial sediment.

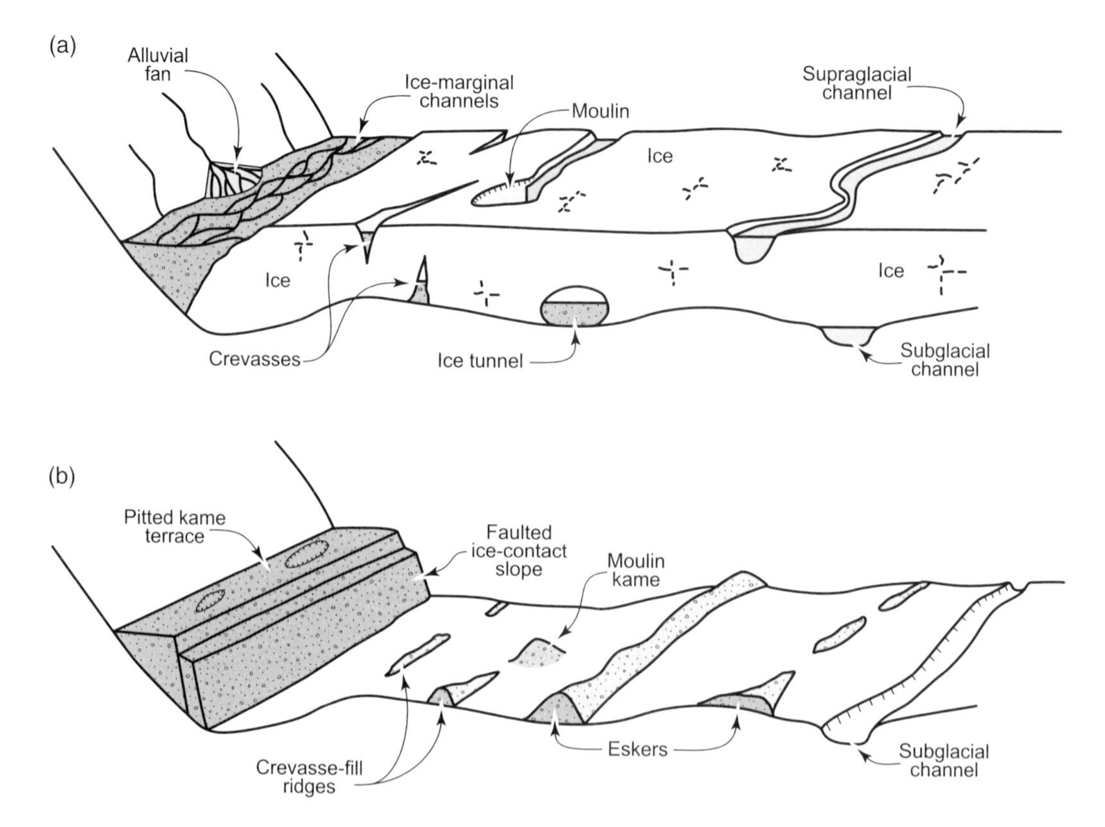

Figure 75 Development of glacifluvial ice-contact landforms (a) before and (b) after stagnant ice melt

Eskers form narrow, sinuous ridges (Plate 52g) or a series of ridges separated by broader beads. They occur in a range of sizes (m to tens of m high, m to hundreds of m wide, tens of m to hundreds of km long). They can be located in valleys or follow upslope paths. They may occur in isolation or in groups forming subparallel, deranged (not aligned with regional ice flow) or dendritic (tree-like) patterns. Eskers record the location of past ice-walled streams – mainly subglacial streams as the ridge-form is often lost when supraglacial or englacial channel deposits are lowered to the ground during glacier melt (Figure 74). Narrow ridges are tunnel deposits, whereas beads contain macroform, fan or delta sediments that formed where tunnel flow expanded into a subglacial cavity or at the ice margin. Strings of beads may indicate the punctuated retreat of the glacier front. Deranged eskers likely formed under stagnant ice. Long (hundreds of km), dendritic esker systems may have (1) developed in long and persistent (perhaps operating for hundreds of years) subglacial tunnels in stagnant ice or (2) formed almost instantaneously (over perhaps weeks or months) during the drainage of large subglacial reservoirs or supraglacial lakes.

Kames are steep-sided, variously shaped mounds of sand and gravel that were originally deposited with two or more ice-contact margins (Figure 75). Examples include moulin deposits (moulin kames) and small deltas or fans deposited at or on the ice margin (delta kames). Kame terraces are linear, often pitted benches of sand and gravel deposited by braided rivers which flowed between the valley side and the ice margin. They need not be altitudinally matched on both sides of a valley glacier.

Outwash fans are fan-shaped bodies of downstream-fining sediment with their apex at a meltwater portal. Deposition on land results in subaerial outwash fans and deposition in water (at a grounding line) results in subaqueous outwash fans. Adjacent outwash fans may coalesce forming a ridge along the ice front – an end moraine (on land) or grounding-line moraine (in water), or between ice lobes – an interlobate moraine. Past ice sheets have left many such glacifluvial moraines. Subaerial outwash fans often grade downflow into proglacial stream deposits.

Outwash plains are planar landforms containing proglacial stream deposits. Proglacial streams are often braided (Plate 52a, h) as high sediment loads, fluctuating discharge and a lack of vegetative anchoring results in a high degree of channel instability. Channels vary in width (m to hundreds of m) and bars vary in size (m to hundreds of m long, m relief). Braided streams continually evolve with each successive flood by channel scour and fill, bar development and overbank deposition. Where numerous proglacial streams issue from the ice front onto an open lowland an extensive outwash plain known as a sandur is formed (Plate 52a). When proglacial streams are hemmed in by valley sides in mountainous terrain, deposition is focused along the valley axis resulting in thick valley fills and a linear outwash plain called a valley train (Plate 52h). Where proglacial rivers enter standing-water bodies deltas (see GLACIDELTAIC) may form.

Outwash plains typically exhibit downflow changes in their morphology and composition reflecting a lessening of glacier influence and a decrease in energy away from the glacier snout. Close to the glacier (proximal) coarse gravel devoid of vegetation is arranged into longitudinal bars separated by a few large channels. The surface is often pitted. Grain size is highly variable reflecting strong melt and flood cycles. With increasing distance from the ice front transverse bars separated by a complex network of braided channels become prevalent, grain size decreases (sand and gravel are present) and grain roundness increases due to selective sorting and abrasion during transport. Lithofacies variation within and between beds is diminished as inflowing non-glacial tributaries dampen discharge fluctuations. Distally (furthest from source), flow in shallow braided channels, sheets or meandering streams deposits sand.

Megafloods and climate change

A broad suite of glacial landforms are now attributed to megafloods from past ice sheets (Figure 74; other explanations are also debated, see DRUMLINS; SUBGLACIAL GEOMORPHOLOGY). The discharge of such enormous quantities of freshwater across continents and into the oceans caused sea level to rise and may have modified ocean and atmospheric circulation, heralding climate change. As meltwater drainage, ice temperature and dynamics (movement) are linked, glacifluvial sediments and landforms contain a record of ice and water behaviour that is essential in our quest to understand climate change.

References

Lundqvist, J. (1985) What should be called glaciofluvium? *Striae* 22, 5–8.

Shaw, J. (1996) A meltwater model of Laurentide subglacial landscapes, in S.B. McCann and D.C. Ford (eds) *Geomorphology Sans Frontières*, 183–226, Chichester: Wiley.

Further reading

Benn, D.I. and Evans, D.J.A. (1998) *Glaciers and Glaciation*, London: Arnold.

Drewry, D. (1986) *Glacial Geologic Processes*, London: Arnold.

SEE ALSO: esker; drumlin; glacier; kame; meltwater and meltwater channel; outburst flood; scabland; subglacial geomorphology; tunnel valley

TRACY A. BRENNAND

GLACILACUSTRINE (GLACIOLACUSTRINE)

Modern and ancient glacilacustrine deposits tend to be variable in grain size, mineralogy, bedding thickness and sedimentary structures, reflecting the broad range of settings in which they accumulate. Glacial lakes may originate from ice erosion of bedrock, in depressions of glacial deposits, or be impounded behind drainage barriers composed of moraine, outwash or ice (Hutchinson 1957). Today, lakes of glacial origin outnumber all other lake types combined. However, most present-day glacial lakes owe their origin to Pleistocene glacial activities and are now under no direct influence of glaciers. Thus it is useful to distinguish glacial lakes and their deposits from those of glacier-fed lakes, and to divide the latter into those bordered by an actively calving glacier (ice-contact or proglacial lakes) and those located downstream (non-contact or distal lake) (Ashley *et al.* 1985).

Most of the material deposited in glacial lakes comes from sediment in suspension and bedload in glacial meltwater streams. Additional contributions may be derived from slope processes delivering sediment directly into the lake (slope wash, avalanching, debris torrents, for example), atmospheric precipitation (including volcanic events), hydrochemical precipitation, biogenic activity, upwelling of material from groundwater flow, and resuspension from bottom current activity.

Deltas form where a meltwater stream or the glacier itself enters a lake. Sudden flow expansion causes an abrupt decrease in stream velocity and competence, which in turn results in rapid deposition of coarser material (see GLACIDELTAIC (GLACIODELTAIC)). At ice margins, other glacilacustrine sediments are also deposited, including subaquatic flow tills, formed by gravity deposits from debris-rich glacier ice standing in a lake. Icebergs can release particles either individually, dropstones, or in conical debris mounds on the lake floor.

The bulk of sediment discharge into a glacial lake comes from glacial streams during the spring and summer-melt period. Concentrations of suspended sediments are highly variable, with values ranging from a few $mg\,l^{-1}$ to $g\,l^{-1}$ in extreme cases. Density differences between inflowing stream waters and glacial lakes result largely from differences in suspended sediment concentrations and temperature. With strong density contrasts, the incoming stream water will maintain its integrity and flow into the lake as a discrete density current, either as an overflow (if its density is less than the lake water), an interflow (strong thermal stratification may result in flow along the thermocline), or underflow (if the inflowing water is more dense). The highly seasonal and weather-dependent nature of glacial-river discharge, temperature and suspended sediment concentration, together with the normal seasonal evolution of lake thermal structure, result in changing and often complex mixing and sedimentation patterns at different stages of the year. The resulting rhythmic deposition of sediments is a signature of many ice-contact and distal glacier-fed lakes.

Turbid underflows, high-density currents generated by underflowing sediment laden river water which produce quasi-continuous currents, and episodic surge-type currents formed by subaqueous slumping (velocities may range up $1\,ms^{-1}$) both transfer suspended sediment and a large quantity of bedload directly to deeper parts of the lake floor. A distinctive suite of graded deposits often characterized by ripple-drift and cross-laminations result. In lakes where underflows dominate, the descent of turbidity currents down the basin sides may inhibit deposition and in places may cause active erosion.

When and where underflow activity is not evident, such as during winter months or due to fluctuations in discharge, settling of particles takes place from sediment suspended in the water column. The resulting deposits, normally only a few millimetres to centimetres thick, grade from silty-clay at the base to fine clay at the top. They often terminate abruptly with a sharp contact, due to a new underflow influx of coarse material. In the most distal areas of glacial lakes, variations in sediment inflow may be sufficiently damped to give rise to homogeneous clays.

A signature of many glacial lake floor deposits are 'rhythmites'. These are pairs (couplets), composed of light-coloured, silt layers, representing spring flood or storm deposits, and dark, clay layers, with higher organic content, representing quiet deposition under winter ice. The contact between the two layers may be gradational, but more often it is sharp. Multiple laminations may occur within the more proximal silt layers, reflecting short-term fluctuations (hours and days) in sediment influx and dispersal. Local factors, load and volume of the meltwater stream, the depth of the lake and relief of its floor, the strength of the currents and the distance from the point of entry into the lake, affect the thickness of the couplets (Menounos 2002). A recurring theme in discussions of rhythmites is their periodicity. De Geer (1912) introduced the term 'varves' to describe annual couplets. Non-annual glacilacustrine rhythmites can be formed from sudden fluctuations of discharge and sediment load, sometimes from OUTBURST FLOODS, cold and warm spells of a non-annual nature, episodic slope activity, or periodic action of storms stirring up lake waters (Sturm 1979). Great care must be taken to establish a reliable, independent chronology for rhythmites, especially if they are to be used as a geochronological tool (Brauer and Negendank 2002). Varved glacilacustrine deposits have been used to interpret high-resolution records of paleoenvironmental conditions; notably, climate, glacial activity, mineralogy of drainage areas, and changes in water level, temperature and trophic state (see, for example, Karlen 1976; Leonard 1986).

Shoreline processes in glacial lakes are similar to those in lakes in other environments. Lake waters standing at particular levels create strandlines with wave-eroded scarps, beaches, small deltas and terraces. Coarse-washed gravel, cobble and boulder deposits may accumulate where waves erodc older glacigenic (e.g. till) deposits. In glacial lakes, wave activity may be inhibited for part of the year by the presence of ice cover. The effects of movement of ice cover against the shore, due either to thermal expansion or wind coupling, produce small ice-push features, which may reach heights up to a few metres. The inclination of glacial strandlines (commonly 1 or $2\,\mathrm{m\,km^{-1}}$) gives important insight into the rebound and tilting since ice unloaded certain areas.

Water levels in many ice-contact lakes fluctuate widely, a consequence of meltwater filling and subsequent ice-dam collapse and drainage. This has important effects on lake-bottom sediments, through scouring and slumping, as well as ancillary effects due to changing wave base, iceberg grounding and adjustments of distribution patterns of suspended sediments.

References

Ashley, G.M., Shaw, J. and Smith, N.D. (1985) Glacial sedimentary environments, Society of Paleontologists and Mineralogists, Short Course 15, Tulsa, OK.

Brauer, A. and Negendank, J.F.W. (2002) The value of annually laminated lake sediments in paleoenvironmental reconstruction, *Quaternary International* 88, 1–3.

De Geer, G. (1912) A geochronology of the last 12,000 years, 11th International Geological Congress (Stockholm, 1910) 1, 241–1, 258.

Hutchinson, G.E. (1957) *A Treatise on Limnology. Geography, Physics and Chemistry*, New York: Wiley.

Karlen, W. (1976) Lacustrine sediments and tree-limit variations as indicators of Holocene climatic fluctuations in Lappland, Northern Sweden, *Geografiska Annaler* 58A, 1–34.

Leonard, E. (1986) Varve studies at Hector Lakes, Alberta, Canada, and their relationship to glacial activity and sedimentation, *Quaternary Research* 25, 199–214.

Menounos, B. (2002) Climate, fine-sediment transport linkages, Coast Mountains, British Columbia, Ph.D. Thesis, Department of Geography, The University of British Columbia, Vancouver, Canada.

Sturm, M. (1979) Origin and composition of clastic varves, in C. Schlüchter (ed.) *Moraines and Varves*, 281–285, Rotterdam: Balkema.

SEE ALSO: glacier; glacideltaic; glacifluvial

CATHERINE SOUCH

GLACIMARINE (GLACIOMARINE)

Here, the term glacimarine is preferred to other alternatives (glacial marine, glacial-marine and glaciomarine) because etymologically, words with Latin roots are joined with an 'i'. An inclusive definition is also preferred here, where the term is taken to encompass the environment, processes and deposits including landforms, sedimentary systems, stratigraphy and life forms. Glacimarine systems include a combination of glacial and marine processes that produce a penecontemporaneous mixture of primarily siliciclastic and biogenic sedimentary deposits. Terrestrial sediment is introduced by ice rafting and rainout of debris (IRD); by fluvial transport feeding turbid overflow plumes with eventual suspension settling of particles; by mass flows and rockfalls from ice contact and shoreline subaerial systems; by aeolian transport with eventual settling through water (perhaps via sea ice); and by shoreline and shelf processes such as longshore transport. Glacimarine settings occur within a range of climatic (and glaciological) regimes from polar, to subpolar, to cool temperate, and encompass fjords and nearshore areas, continental shelves and the deep sea.

Grounding-line depositional systems are formed at the contact of a glacier with the seafloor. These deposits take the form of a bank (morainal bank (less-favoured alternatives: moraine, submarine moraine and moraine bank)), a fan (grounding-line fan (less-favoured alternatives: subwash fan, glacimarine fan and submarine ice-contact fan)) and a wedge (grounding-line or grounding-zone wedge and trough-mouth fan (less-favoured alternatives: till tongue, till delta, subglacial delta and diamict apron)). Grounding-line systems include a mixture of facies: till, glacimarine diamicton (stratified or massive), gravelly mud (laminated, e.g. cyclopels and cyclopsams, or massive), poorly or well-sorted sand and gravel (stratified or massive), and interlaminated sand and mud (e.g. turbidites) (see Further reading for details).

Till with various modifiers (e.g. waterlain till and paratill), has been used as the genetic term for glacimarine diamicts; however, till is best reserved for deposition directly from glaciers without modification such as by flowage or by currents during rainout. Thus glacimarine diamict is preferred, and if genetic interpretations are possible, then such terms as debris flow deposit (or debrite), or rainout diamict (produced from ice rafting), or ice-keel turbate (produced by keels of icebergs or sea ice) may be used. Specific environmental terms for the rainout diamicts may be shelfstone diamict or bergstone diamict, depending on whether their debris source is an ice shelf or icebergs, respectively.

Beyond these ice-contact systems that extend two to several kilometres from a grounding line, are ice-proximal (to ~10 km from a grounding line) and ice-distal glacimarine systems (to thousands of km from a grounding line, e.g. Heinrich layers). These distances are relative to grounding lines and may be within an ice shelf or an iceberg zone. The main glacial components are from IRD, suspension settling and, more rarely, wind transport. Deposits are either gravelly mud or diamict depending on relative accumulation rates of IRD and matrix sediment, which often is from meltwater streams. The matrix is stratified under higher current strengths and sedimentation rates or under continuous ice cover (which control the degree of bioturbation), and is otherwise massive. However, extremely high sedimentation rates with few bottom currents can produce massive deposits.

Ice rafting occurs via three forms of ice and, if possible, recognizing the distinction is useful, such as by: ice shelves and floating glacier-tongues – ISRD, icebergs – IBRD, and sea ice – SIRD. Sea-ice rafting perhaps should be excluded from glacimarine systems because it is not strictly glacial and may occur under non-glacial conditions. However, often distinguishing SIRD from other IRD is impossible and thus it is commonly included in the glacimarine system. Ensuring that particle rafting is not by tree roots or kelp holdfasts is also important. A French term, *glaciel* has been suggested for sediment containing IBRD and SIRD, but is not commonly used.

Biogenic components in glacimarine deposits become more common with lower terrestrial sediment flux and meltwater; that is, either with distance from a glacier terminus or in colder climates. The geologically significant components include various macrofossils and microfossils, but diatoms commonly dominate and often form diatomaceous mudstone and diatomaceous ooze (diatomite). Marine productivity and diversity may depend on sea-ice extent, thickness and seasonal longevity, on sea water temperature and salinity changes, and on current up-welling (including polynya); thus these records contain high-resolution climate signals.

Morphologically significant forms produced in glacimarine settings include: fjords, cross-shelf troughs (or submarine troughs or sea valleys), inter-ice-stream ridges, mega-lineations (large-scale forms like flutes), flutes, grounding-line systems, iceberg and sea-ice scours, ploughs, or wallows, and striated boulder pavements.

The glacimarine environment includes sedimentary systems and processes that are typical of lower latitude settings, such as deltas, fan deltas, estuaries, tidal flats, linear sandy shorelines, shelves and deep water systems that commonly may include indicators of ice action described above. It includes lags, erosional surfaces, hiatuses and condensed sections produced from reworking by marine currents, from sediment starvation under large ice shelves or in ice distal areas during glacial retreat, and from isostatic rebound. By analogy with terrestrial glacial outwash and lacustrine systems, paraglacial marine settings occur where glaciers terminate on land, but their products of glacial rock flour accumulate as marine mud, perhaps including SIRD.

Further reading

Anderson, J.B. (1999) *Antarctic Marine Geology*, Cambridge: Cambridge University Press.

Anderson, J.B. and Ashley, G.M. (eds) (1991) *Glacial Marine Sedimentation: Paleoclimatic Significance*, Special Paper 261, Boulder, CO: Geological Society of America.

Davies, T.A., Bell., T., Cooper, A.K., Josenhans, H., Polyak, L., Solheim, A., Stoker, M.S. and Stravers, J.A. (eds) (1997) *Glaciated Continental Margins: An Atlas of Acoustic Images*, New York: Chapman and Hall.

Dowdeswell, J.A. and Ó Cofaigh, C. (eds) (2002) *Glacier Influenced Sedimentation on High Latitude Continental Margins*, Special Publication No. 203, London: Geological Society.

Dowdeswell, J.A. and Scourse, J.D. (eds) (1990) *Glacimarine Environments: Processes and Sediments*, Special Publication No. 53, London: Geological Society.

Molnia, B.F. (ed.) (1983) *Glacial-Marine Sedimentation*, New York: Plenum Press.

Powell, R.D. and Elverhøi, A. (eds) (1989) Modern Glacimarine Environments: Glacial and Marine Controls of Modern Lithofacies and Biofacies, *Marine Geology* 85, III-416.

Syvitski, J.P.M., Burrell, D.C. and Skei, J.M. (1987) *Fjords: Processes and Products*, Berlin: Springer-Verlag.

ROSS D. POWELL

GLACIPRESSURE (GLACIOPRESSURE)

The term, 'Glaciopressure' was introduced by Panizza (1973) to indicate the pressure of ice on the narrow part of a valley, which is particularly intense at the confluence of glacial tongues in the areas affected by Pleistocene glacier advances. It caused rock deformations in correspondence with surfaces of structural discontinuity, like strata, fissures, etc., favouring the formation of sliding surfaces. In fact, some landslides which took place in the late Glacial and Post Glacial were observed in the Alps, and particularly in the Dolomite region: they were triggered by a tensional discharge following the loss of pressure previously exerted on the rocky slopes by two or more glaciers merging in a valley narrow. Even if the fall of large slope portions can directly affect human settlements or obstruct a whole valley, with the negative resulting consequences, the extremely high risk degree assumed by this type of phenomenon is purely theoretical. Indeed, the long time span from the withdrawal of the glacial network to the present has practically produced the total exhaustion of these events.

References

Panizza, M. (1973) Glaciopressure implications in the production of landslides in the Dolomitic area, *Geologia Applicata à Idrogeologia* 8(1), 28–298.

MARIO PANIZZA

GLACIS D'ÉROSION

Glacis d'érosion are a form of PEDIMENT, a gently inclined slope of transportation and/or erosion that truncates rock and connects eroding slopes or scarps to areas of sediment deposition at lower levels (Oberlander 1989). Two fundamental types of pediment are recognized by Oberlander (1989): glacis d'érosion, which truncate softer rocks adjacent to a more resistant upland; and 'true' pediments, where there is no change in lithology between upland and pediment.

The name glacis d'érosion is derived from the work of French geomorphologists who studied examples of these landforms on the northern margin of the Sahara Desert, where they are particularly well developed on the flanks of the Atlas Mountains (Coque 1960). These landforms truncate weak materials such as poorly indurated Tertiary sediments, and tend to be veneered by alluvial gravels, indicating the importance of fluvial processes in their creation (Dresch 1957). The glacis piedmonts of the Atlas Mountains have a distinctive morphology, consisting of a series of coalescing flattened cones whose apices occur where stream channels debouch from upland drainage basins. The glacis long profiles range from nearly rectilinear to concave; the latter form having a slope of about 10 degrees at the top, dropping to about 3 degrees or less at the base.

Glacis d'érosion often exhibit multiple levels, or terraces, which can be traced back into the upland drainage basin where they form river terraces (see TERRACE, RIVER) (Plakht *et al.* 2000). These forms, known as stepped or nested glacis (Coque and Jauzein 1967), are formed as older glacis are incised by stream channels, which then form a younger glacis at a lower level, the new glacis being inset within the older glacis. The resulting landform appears similar in form to a telescopically segmented ALLUVIAL FAN, leading some workers to suggest that both landforms result from similar responses to environmental change (White 1991). The slope profiles of stepped glacis tend to converge downslope, the gradients decreasing from oldest to youngest. The oldest glacis are often only present as narrow residual ridges or outlying mesas, as planation of lower glacis have progressively removed upper glacis. Coque and Jauzein (1967) suggest that the number of glacis in Tunisia decreases systematically towards the south (Plate 53). Five glacis are present around Tunis and the High Steppe, south of Gafsa there are only four, the highest being present only as a few outliers.

Glacis d'érosion are thought to be erosional surfaces formed by fluvial action, cutting sequences of rocks that are easily eroded relative to the rocks of the adjacent upland. Supporting evidence for this fluvial model comes from the fact that stepped glacis are often paired on either side of contemporary channels rather like paired river terraces, and the fact that glacis are almost always covered with a layer of alluvium. This alluvial cover can be up to 15 m thick, though it rarely exceeds 10 m. Lower (younger) glacis tend to have thinner alluvial cover, and the alluvium tends to decrease in thickness towards the distal edge of the piedmont. The alluvium tends to be poorly sorted at the top of the glacis, becoming

Plate 53 A series of stepped glacis d'érosion developed on the southern flank of Djebel Sehib, southern Tunisia

better sorted downslope. In more arid areas, the alluvial cover of the glacis is frequently cemented by calcium carbonate or gypsum, forming an indurated DURICRUST. The role of duricrust in the formation of glacis is uncertain, although it may play an important part in preservation of older glacis.

Coque (1962) ascribes the formation of glacis in North Africa to slope retreat resulting from climate change; specifically a succession of humid and arid phases known to have affected the Sahara Desert during the Quaternary. He envisages a sequence of lateral planation during a humid phase, when moisture was sufficient to produce enough debris to balance the carrying capacity of streams, allowing them to erode laterally. This was followed by incision during an arid phase, when downcutting was promoted by lower sediment load in the streams. A return to more humid conditions resulted in renewed lateral planation at a lower level, forming a new glacis inset within the one above. This model is a gross oversimplification of the COMPLEX RESPONSE which river channels are now known to exhibit in response to environmental changes, but it is still generally believed that changes in the fluvial system resulting from climate changes are the basic trigger for formation of stepped glacis. The fact that the stepped glacis converge downslope indicates that changes in base level are unlikely to be involved in their formation. The widespread distribution of glacis across areas of different structural setting also rules out NEOTECTONICS as a major factor in their formation.

References

Coque, R. (1960) L'evolution des versants en Tunisie présaharienne, *Zeitschrift für Geomorphologie Supplementband* 1, 172–177.

Coque, R. (1962) *La Tunisie Présaharienne. Etude Géomorphologique*, Paris: Armand Colin.

Coque, R. and Jauzein, A. (1967) The geomorphology and Quaternary geology of Tunisia, in L. Martin (ed.) *Guidebook to the Geology and History of Tunisia*, 227–257, Tripoli: Petroleum Exploration Society of Libya.

Dresch, J. (1957) Pediments et glacis d'érosion, pédiplains et inselbergs, *Information Géographique* 22, 183–196.

Oberlander, T.M. (1989) Slope and pediment systems, in D.S.G. Thomas (ed.) *Arid Zone Geomorphology*, 56–84, London: Belhaven.

Plakht, J., Patyk-Kara, N. and Gorelikova, N. (2000) Terrace pediments in Makhtesh Ramon, central Negev, Israel, *Earth Surface Processes and Landforms* 25, 29–39.

White, K. (1991) Geomorphological analysis of piedmont landforms in the Tunisian Southern Atlas using ground data and satellite imagery, *Geographical Journal* 157, 279–294.

SEE ALSO: alluvial fan; desert geomorphology; pediment

KEVIN WHITE

GLACITECTONIC CAVITY

Glacitectonic cavities are narrow and subhorizontal openings generated in bedrock by traction under a flowing GLACIER (Schroeder *et al.* 1986). The parallel walls take the form of irregular chevrons that follow the vertical joint pattern, while the roofs follow stratification planes. In some cases, well-compacted lodgement till forms the roof. The irregular floor of galleries is usually covered by debris issued from localized roof or walls failures.

Located less than 20 m below the surface, glacitectonic cavities can be hundreds of metres long, but typically less than 3 m wide and less than 10 m high. As their presence is only revealed by chance, from excavation work or local roof failures, they constitute hazardous constraints in

URBAN GEOMORPHOLOGY, especially in eastern Canada (Schroeder 1991).

Glacitectonic cavities are found below planar topographies, within sub-horizontal limestone or thinly bedded shale. Movement and weight of a flowing inlandsis, possibly aided by dissolution along the stratification planes, allows rock sheets to slide one on the other, leading to the spreading apart of vertical joints and to the creation of glacitectonic voids.

References

Schroeder, J. (1991) Les cavernes à Montréal, du glaciotectonisme à l'aménagement urbain, *Canadian Geographer* 35(1) 9–23.

Schroeder, J., Beaupré, M. and Cloutier, M. (1986) Ice-push caves in platform limestones of the Montreal area, *Canadian Journal of Earth Sciences* 23, 1,842–1,851.

JACQUES SCHROEDER

GLACITECTONICS (GLACIOTECTONICS)

Glacitectonic deformation may be defined as 'the structural deformation as a direct result of glacier movement or loading' (INQUA Work Group on Glacier Tectonics 1988). This term was first introduced by Slater (1926), and re-examined by Banham (1975). This topic has been studied a great deal in recent years, and a number of collections of papers on the subject have been published (Aber 1993; Warren and Croot 1994) and an online bibliography (http://www.emporia.edu/s/www/earthsci/biblio/biblio.htm). Additionally, recent textbooks on glacial geology include detailed sections on glacitectonic deformation (Benn and Evans 1997; van der Wateren 1995).

However, prior to the 1980s, glacitectonic deformation was thought to be a rare phenomenon, and was studied as a distinct field in glacial sedimentology. This view was first challenged when Boulton and Jones (1979) suggested that a significant proportion of glacier motion occurred not in the ice, but in a saturated weak deforming layer beneath the ice. These results showed how glacitectonic deformation was an integral part of the glacial environment, and not an unusual occurrence.

There are two types of glacitectonic deformation formed by the action of a moving glacier (Figure 76):

(a) *Proglacial deformation* which takes place at the glacier margin and is characterized by pure shear and compressional tectonics, i.e. open folds, thrusts and nappes. This results in the formation of push moraines;

(b) *Subglacial deformation* which takes place beneath the glacier and is characterized by simple shear and extensional tectonics, i.e. attenuated folds, boudins and augens, and results in the formation of deformation till and/or flutes and drumlins.

Similar styles of deformation can also occur within the ice itself (Hart 1998) as well as within permafrost (Astakhov *et al.* 1996).

Proglacial glacial tectonic structures have been relatively well studied because of their accessibility. In fact, the large number of studies of these features led many workers in the past to consider only proglacial deformation in discussions of glacitectonics. In contrast, subglacial deformation has had the least study because of the logistical

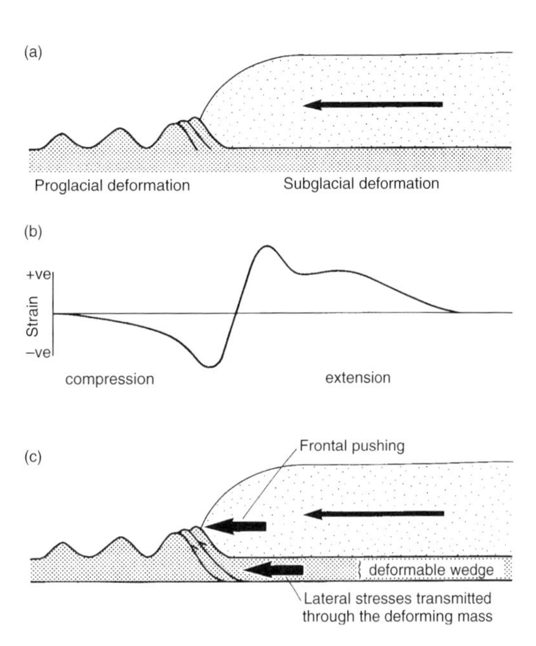

Figure 76 (a) schematic diagram showing the positions of subglacial and proglacial glacitectonic deformation; (b) the theoretical pattern of longitudinal strain; (c) schematic diagram of the forces producing proglacial deformation: frontal pushing and compressive stresses transmitted through a subglacial deformable wedge (after Hart 1998)

problems involved in subglacial process studies, but the number of studies has dramatically risen in the past ten years.

Additionally, deformation also occurs within the glacial environment as a result of gravitation instabilities associated with stagnant ice and is known as dead-ice tectonics. Typical features include ice-collapse structures in an outwash plain, debris-flow mobilizations of till, and instability of subglacial sediments to produce 'crevasse infill' structures. These features are not glacitectonic structures *sensu stricto*, but may reflect the presence of saturated till in the subglacial environment, and so may be associated with subglacial deformation.

Proglacial glacitectonic deformation

Proglacial glacitectonic deformation is generally characterized by large-scale compressional folds and thrusts. The usual result of the proglacial deformation processes is to produce a topographic ridge transverse to the ice margin called a push moraine. There is often a basin up-glacier from which the material of the ridge has been removed. However, they do not always have a topographic expression. Many proglacial structures have been subsequently overridden by ice and so have become incorporated into drift deposits and have little or no topographic expression.

Push moraines are very common and can occur on scales ranging in height from 0.5 to 50 m, and in length from 1 m to several kilometres. Push moraines are associated with both contemporary glaciers and Quaternary glaciations (as well as pre-Quaternary glaciations) (see reference list). A recent review of push moraines is by Bennett (2001).

It has been argued by numerous workers that proglacial deformation can be modelled as thin-skin thrust tectonics, and the processes involved in the formation of push moraines are similar to mountain building in hard rock tectonic terrains. Using the work of Hubbert and Rubey (1959), many researchers have argued that glacitectonic nappes move along incompetent rock units or planes of weakness due to high pore-water pressures.

Although there are many processes associated with proglacial glacitectonic deformation, they can be generally divided into two types:

1 *'Foreland only' deformation* Where there is no deforming bed present, deformation may only take place in the foreland by the deformation of pre-existing sediments. This may typically include sandur sediments in terrestrial environments and shallow marine or fjord sediments in marine environments.
2 *'Deformable wedge' deformation* Where there is a deforming bed present, the subglacial and proglacial environment can be modelled as a deformable wedge which deforms by gravity spreading driven by the ice (Figure 76c). Deformation occurs due to both down-ice thrusting of the glacier into the foreland, as well as the horizontal component of the glacier's effective pressure (normal pressure minus pore-water pressure) transmitted through the subglacial layer into the foreland.

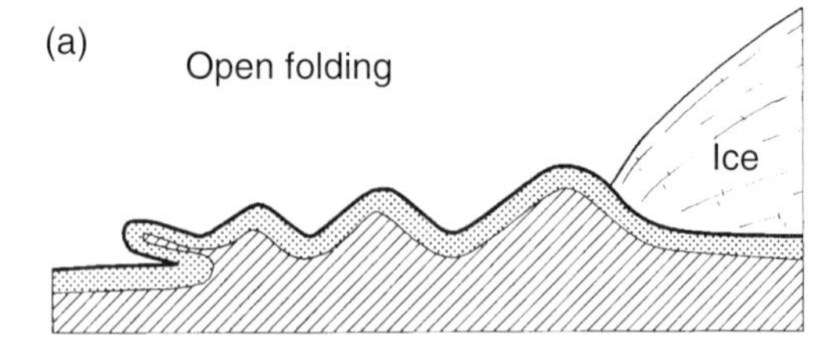

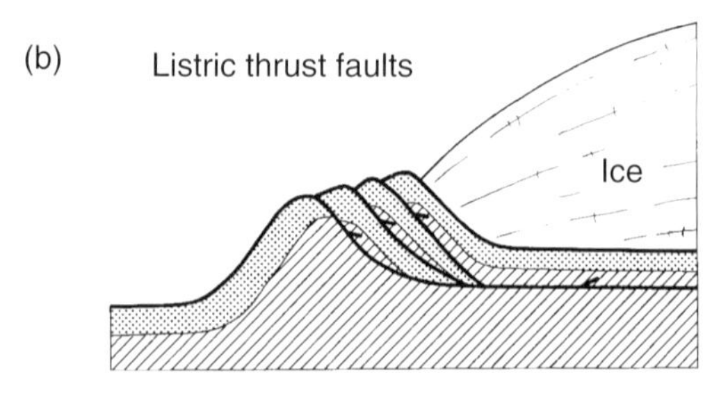

Figure 77 Schematic diagrams of different proglacial push moraines: (a) open folding; (b) listric thrust folding (after Hart and Boulton 1991)

Deformation of sediments (in both styles of push moraine) range from ductile (open folding) to brittle (thrust faulting and thrust nappes) (Figure 77). These styles of deformation reflect both the competancy of the material and increasing longitudinal compression from the simple folding to more complex nappe structures. Deformation structures are also found at the base of thrust faults and nappes, with tectonic breccias associated with brittle deformation and shear zones formed associated with ductile deformation.

Subglacial glacitectonic deformation

Although there are fewer studies of the subglacial environment due to its inaccessibility, there is still a considerable body of literature on subglacial deformation. Early descriptions of deformation structures in till included folds, laminations and blocks, boudins or rafts of soft sediments; such till was called 'deformation till'.

Subglacial deformation can occur beneath warm-based glaciers, when meltwater released from the glacier bed cannot easily escape from the system, so that pore-water pressures in the subglacial sediments build up and sediment strength is reduced:

$$\tau = (P_i - P_w) \tan \varphi$$

where τ is basal friction, P_i is ice overburden pressure, P_w is pore-water pressure and tan φ is the angle of internal friction (Coulomb's Law).

STUDY METHODS

In recent years subglacial deformation has been studied by three ways: (1) *in situ* process studies; (2) geophysical techniques; and (3) sedimentology. These methods have been discussed in detail in Hart and Rose (2001).

In situ subglacial process studies consist of the monitoring of the subglacial environment by inserting instruments into the subglacial bed via hot water drilled boreholes. This is a relatively simple technique and has been used on about ten modern glaciers including Breiðamerkurjökull (Iceland), Ice Stream B (Antarctica), Trapridge Glacier (Canada), Black Rapids Glacier (Alaska), Storglaciären (Sweden) and Bakaninbreen (Svalbard). These studies reveal the average thickness of the deforming layer is 0.5 m and indicate that deformation does occur beneath most of these glaciers. However, the importance of the effects on basal motion due to subglacial deformation ranges between 100 per cent at Black Rapids Glacier, Alaska to 13 per cent at Ice Stream B, Antarctica. Although the reason for the difference is not yet known, it has been suggested that the granulometry of the subglacial sediment may account for the difference in behaviour within the deforming layer. The glaciers with coarse-grained till appear more likely to have a higher percentage of basal motion due to sediment deformation, whilst those with more clay-rich lithologies may have only very thin deforming layers.

In addition, the presence of a deforming bed over large areas has been identified by seismic investigations in Antarctica, in particular beneath Ice Stream B and the Rutford Ice Stream.

However, most studies of subglacial deformation have been based on sedimentological studies from both modern and Quaternary glacial sequences. Most researchers have argued that the subglacial deforming layer behaves as a shear zone, which is a narrow band of high overall ductile shear located between sub-parallel walls. This deformation results in three features that will be discussed briefly below: deformation till, deformation structures and subglacial bedforms.

Deformation till

Hart and Boulton (1991) have argued that the resultant till from subglacial deformation is deformation (or deforming bed) till, which is a primary till formed from a combination of both deformation and deposition. It forms by direct melt-out of debris from the ice above, advection from till up-glacier, and changes in the thickness of the deforming layer. Where layers of deformation till are accreted on one another this is known as constructional deformation. In contrast, where the deforming layer thickens (due to changes in effective pressure, or large rafts of bedrock being thrusted into the shear zone), this is known as excavational deformation.

Deformation structures

Features typical of shear zones include: folds, boudins, augens, rotated clasts and tectonic laminations (Plate 54). The latter form from the attenuation of perturbations of the base of the deforming layer, producing ungraded laminations. However, these features will only be visible if they are formed from the mixing of sediments with different lithologies or competencies, under relatively low to medium simple shear. At very high shear strains these tectonic features can become homogenized and so macro-scale structures may not be visible. Instead criteria such as a specific till fabric (low strength associated with a thick deforming layer, high strength associated with a constrained deforming layer), or specific micromorphology structures (evidence of rotation or shears) may be used as a distinguishing criterion.

Subglacial bedforms

It has also been argued that subglacial streamlined bedforms (lineations, flutes and drumlins) are a product of subglacial deformation (Boulton

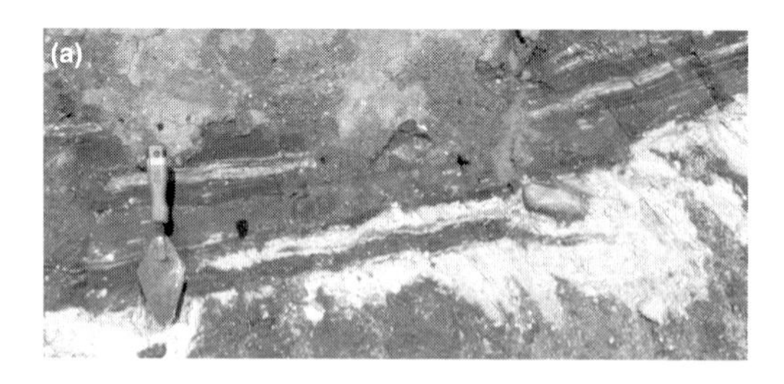

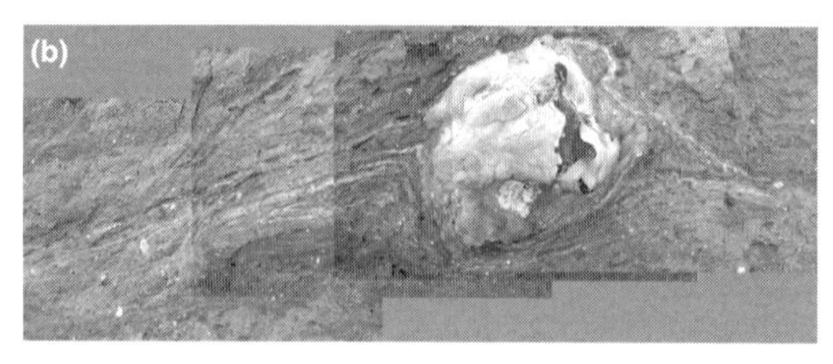

Plate 54 Examples of subglacial deformation: (a) chalk being attenuated to form tectonic laminations, West Runton, Norfolk; (b) chalk laminations flowing around on obstacle (flint clast), Weybourne, Norfolk (photographs by Kirk Martinez)

1987). These features form due to the presence of more competent masses (or cores) within the deforming layer which act as obstacles to flow. Where the core of the drumlin is weak then deformation structures will be seen, but these are relatively rare, and instead most drumlins have a competent core and a carapace composed of deformation till.

Using this sedimentological data, a number of authors have argued for wide spread subglacial deformation beneath the Pleistocene glaciers where the ice moved over the unconsolidated rocks of the European (and British) and Laurentide ice sheets.

Conclusions

Glacitectonic deformation is a fundamental process in glacier behaviour and a key component in the proglacial and subglacial sediment/ice deposition, erosion and transport system. There are very few modern glaciers that do not show evidence for proglacial deformation at their margins, and subglacial *in situ* process studies have revealed that subglacial deformation is also a common process. In addition, studies of Quaternary sediments demonstrate that such processes were also widespread in the past.

As a result, any study of the glacial environment needs to take glacitectonics into consideration, and future research needs to focus on the geotechnical properties of till to further understand the links between sediment behaviour and ice dynamics.

References

Aber, J.S. (ed.) (1993) *Glaciotectonics and Mapping Glacial Deposits*, Canadian Plains Research Centre, University of Regina.

Astakhov, V.I., Kaplyanskaya, F.A. and Tarnogradsky, V.D. (1996) Pleistocene permafrost of West Siberia as a deformable glacier bed, *Permafrost and Periglacial Processes* 7, 165–191.

Banham, P.H. (1975) Glaciotectonic structures: a general discusion with particular reference to the Contorted drift of Norfolk, in A.E. Wright and F. Moseley (eds) *Ice Ages, Ancient and Modern*, 69–94, Liverpool: Seel House Press.

Benn, D.I. and Evans, D.J.A. (1997) *Glaciers and Glaciation*, London: Arnold.

Bennett, M.R. (2001) The morphology, structural evolution and significance of push moraines, *Earth-Science Reviews* 53, 197–236.

Boulton, G.S. (1987) A theory of drumlin formation by subglacial deformation, in J. Menzies and J. Rose (eds) *Drumlin Symposium*, 25–80, Rotterdam: Balkema.

Boulton, G.S. and Jones, A.S. (1979) Stability of temperate ice caps and ice sheets resting on beds of deformable sediment, *Journal of Glaciology* 24, 29–44.

Hart, J.K. (1998) The deforming bed/debris-rich basal ice continuum and its implications for the formation of glacial landforms (flutes) and sediments (melt-out till), *Quaternary Science Reviews* 17, 737–754.

Hart, J.K. and Boulton, G.S. (1991) The interrelationship between glaciotectonic deformation and glaciodeposition, *Quaternary Science Reviews* 10, 335–350.

Hart, J.K. and Rose, J. (2001) Approaches to the study of glacier bed deformation, *Quaternary International* 86, 45–58.

Hubbert, M.K. and Rubey, W.W. (1959) Role of fluid pressure in mechanics of overthrust faulting, *Geological Society of America Bulletin* 70, 115–166.

Slater, G. (1926) Glacial tectonics as reflected in disturbed drift deposits, *Geologists' Association Proceedings* 37, 392–400.

Warren, W.P. and Croot, D.G. (eds) (1994) *Formation and Deformation of Glacial Deposits*, Rotterdam: Balkema.

Wateren, van der F.M. (1995) Processes of glaciotectonism, in J. Menzies (ed.) *Modern Glacial Environments: Processes, Dynamics and Sediments*, 309–333, Oxford: Butterworth-Heinemann.

Further reading

Croot, D.G. (ed.) (1988) *Glaciotectonics: Forms and Processes*, Rotterdam: Balkema.

JANE K. HART

GLINT

A marked gemorphological line dividing (the Canadian and the Baltic) SHIELDs from neighbouring stable platforms (the Great Plains and the Russian Plains, respectively) in the northern hemisphere. It is manifest in an ESCARPMENT which extends hundreds of kilometres and rises 20–100 m above the shield. The front of this escarpment is called a glint line. Pre-Pleistocene DENUDATION and, more significantly, differential scouring of expanding ICE SHEETs during the Pleistocene is responsible for glint formation. The Palaeozoic (Ordovician, Silurian) limestones, dolomites and sandstones of the tableland are more resistant to glacial erosion than the weathered Precambrian igneous and metamorphic rocks of the shield. Forward pushing ice was temporarily halted by the escarpment and thus allowed deep scouring at its base. After ice retreat meltwater accumulated in the depressions and glint lakes originated.

Glint is an Estonian term of Germanic origin. Once it denoted cliffs along coasts. The Baltic–Ladoga Glint extends from the islands of Estonia along the southern coast of the Gulf of Finland. Lake Ladoga and Lake Oniega occupy the depressions. Although the term is not in use in Canada and the United States, the glint line is also present there. Some major lakes, including the Great Bear, the Great Slave, Lake Winnipeg and the Great Lakes are all glint lakes, a subclass of ice-scoured lakes. Niagara Falls is the best known example of waterfalls along the glint line.

DÉNES LÓCZY

GLOBAL GEOMORPHOLOGY

The term global geomorphology embodies the notions of studying landform development at large spatial and temporal scales, of emphasizing global variations in landforms and geomorphic processes, of investigating the interactions between the land surface and other components of the Earth system, and of appreciating the particular combination of conditions for landform genesis on Earth compared with the other solid planetary bodies of the Solar System.

In focusing attention on large-scale phenomena and change over long periods of time, global geomorphology is concerned primarily either with the development of very large individual landform features, such as an entire mountain range, or with the assemblage of smaller individual landforms making up whole landscapes. At these large spatial and temporal scales the internal geomorphic processes of volcanism and tectonics generally become more significant in relation to surface geomorphic processes. A further consequence of looking at the macroscale is that short-term measurements of geomorphic processes on their own provide relatively little insight into the nature and rates of processes responsible for landscape genesis. Although numerical models (see MODELS) have been developed to investigate large-scale landscape change, data relevant to testing such models generally have to pertain to periods of thousands to millions of years, rather than the timescale of years or decades of modern process measurements.

A methodological consequence of these long timescales is that the approach to global geomorphology is predominantly historical where the emphasis is on explaining the conditions and processes responsible for development over time of a single major landform, or a regional or larger scale landscape. This contrasts with the dominance of the functional approach in small-scale surface process geomorphology where the main interest is understanding the adjustment of form to process over short periods of time.

Another distinction between global geomorphology and smaller scale approaches to landform analysis is the frame of reference that must often be employed. At the small scale it is usually sufficient to know local slope gradients and height differences, such as between interfluves and river channels, rather than absolute changes in elevation with respect to sea level. In global geomorphology, however, constraining changes in absolute elevation of the land surface above sea level is required in order to relate changes in regional topography through time to rates of crustal uplift and rates of denudation.

Historical context

A global approach to geomorphology is not new. An important theme in the study of the landforms up to the nineteenth century was the attempt to understand the origin and history of the surface of the Earth as a whole. For instance, one of the elements of Charles Lyell's concept of UNIFORMITARIANISM, was the notion of a steady-state Earth in which uplift in some areas was 'balanced' by subsidence in others, and where changes in the location of uplift and subsidence occurred over time, but the overall form of the Earth's surface did not change substantially.

Lyell's idea of regions of crustal uplift and subsidence was taken up by Charles Darwin who, during the earlier part of his career dominated by writings on geological subjects, sought to develop a global synthesis relating uplift of the continents to processes such as volcanism and mountain building. More specifically, Darwin adopted Lyell's notion of regions of subsidence in developing his own theory of coral atoll formation in which he envisaged coral reefs growing upwards from the substrate of volcanic islands grouped in broad regions of what he inferred was subsiding ocean crust. In developing his coral reef theory, Darwin provided an object lesson in historical methodology applied to understanding landform development by suggesting how spatial patterns of a range of related landforms – in this case, volcanic islands, barrier reefs, fringing reefs and atolls – could, with careful observation and reasoning, be viewed as representing stages in the development of a single landform through time.

This strategy was taken up and extended by William Morris Davis who, in his CYCLE OF EROSION, sought to develop a general evolutionary scheme for landscape development, where the form of a landscape was seen as a product of the rock structures present and the surface geomorphic processes operating, but predominantly as a function of stage of development. Although acknowledging complications arising from variations in climatic conditions, and particularly from intermittent crustal uplift, Davis's evolutionary model depended heavily on the reality of distinctive landscapes being created at different stages of the cycle of erosion. In its simplest form this involved rapid uplift from close to sea level of a low relief land surface, its progressive incision by river systems creating maximum local relief, and then an extended period of interfluve lowering with respect to valley bottoms until a low relief surface close to sea level, or PENEPLAIN, was restored.

Although often heavily dependent on particular interpretations of landscape features, and thus far less secure than Darwin's earlier exemplary treatment of coral atoll formation, the evolutionary approach advocated by Davis became the dominant strategy of global geomorphology through the first half of the twentieth century, at least amongst geomorphologists in Britain and North America. In Germany a different model of landscape change was developed by Walther Penck who emphasized the importance of the interplay between external erosional processes and internal tectonic processes causing uplift. Although more in sympathy with modern approaches to understanding large-scale landscape development, Penck's more complex approach to landform analysis never achieved the influence of the simple version of Davis's evolutionary model.

The idea of the development over millions of years of extensive, low relief erosion surfaces graded to sea level led to the development of DENUDATION CHRONOLOGY, a method of landscape analysis in which low relief erosion surfaces at different elevations were interpreted in terms of falls in base level resulting either from eustatic sea-level change (global sea-level fall), or from tectonic uplift of the land surface. Throughout the mid-twentieth century much emphasis was placed in correlating supposed remnants of particular surfaces considered to have resulted from specific uplift events. Taken to its ultimate extent, the correlation of such erosion surface remnants across the continents was seen as potentially being a chronological replacement for stratigraphy where the sedimentary record was absent or incomplete. The fullest development of denudation chronology as a methodological basis for global geomorphology is perhaps to be found in the work of Lester King who interpreted flights of low relief surfaces across different continents as representing synchronous episodes of continental uplift of global extent.

The decline in interest in global geomorphology and the corresponding move towards studies of small-scale surface process geomorphology from the 1950s occurred for two main reasons. One was the wholly inadequate dating control that was usually available for the denudation chronologies presented, the other a lack of understanding of the tectonic and surface geomorphic processes that could create erosion surfaces graded to sea level, and then subsequently preserve remnants of them when base level fell.

Renewed interest in global geomorphology

Although the revolution in Earth sciences arising from PLATE TECTONICS might have been expected to reinstate interest in global geomorphology, little attention was paid by most geomorphologists to this integrative global-scale model when it was formulated in the late 1960s and early 1970s, presumably because the focus by that time was on quantitative approaches to small-scale surface geomorphic processes. A renewed concern with global geomorphology is really only evident from the 1980s, and it occurred for a number of reasons. One was the growing availability of satellite remote sensing imagery which made evident the large-scale components of the Earth's landforms. Although initially used primarily to explore regional and subcontinental-scale landform associations, by the 1990s satellite data was being used to create digital elevation models (DEM) of the land surface at horizontal resolutions down to a few metres. This added to the growing number of digital topographic data sets being created from national archives of topographic data. By 2001 the Shuttle Radar Topography Mission had collected high resolution radar-based elevation data covering the Earth's surface between latitudes ~60°S to 60°N.

At the same time that Earth-orbiting satellites were providing images of terrestrial landscapes, there was a flood of remote sensing imagery from planetary missions, such as the Viking missions to Mars and the Voyager missions to the outer planets in the 1970s, the Magellan mission to Venus in the 1980s and the Mars Global Surveyor which provided high resolution images. Understanding of the tectonics, volcanism, surface processes and climatic history of these planetary bodies relied heavily on the interpretation of their landforms, where possible by comparisons with supposed terrestrial analogues (for instance, the outflow channels on Mars which were seen to have many similarities to the landscapes of catastrophic flooding in the Channeled Scabland of eastern Washington, USA). At the same time, the scale of landforms seen in planetary imagery pointed to the insights to be gained by studying terrestrial forms with a similar global perspective (see EXTRATERRESTRIAL GEOMORPHOLOGY).

Another important reason for increased interest in global geomorphology was the development of the computing capability necessary to numerically model regional-scale landscapes over geological time spans. Although small catchment/channel slope-scale surface process models have been developed by geomorphologists and hydrologists since the 1960s, numerical models of regional-scale landscape evolution incorporating tectonic deformation and isostasy, as well as surface processes, have been under active development only since the late 1980s.

Constraining such models requires data on denudation rates for time spans of millions of years relevant to long-term landscape development. The increasing availability of such information from the 1980s as a result of new geochronological methods and data sources is another reason for the revived interest in global geomorphology. Hydrocarbon exploration along continental margins has provided a wealth of data on rates of sediment deposition from which denudation rates on the adjacent hinterland can be estimated, at least where the sediment source area and its changes over time can be constrained. More important, however, has been the application of thermochronological techniques to infer denudational histories and denudation rates. A range of low-temperature techniques such as $^{39}Ar/^{40}Ar$ dating, fission-track thermochronology (see FISSION TRACK ANALYSIS) and helium thermochronology can now provide information on the cooling history of rocks in the upper few kilometres of the Earth's crust. This information on the timing and rate of cooling can be converted into estimates of denudation rates averaged over periods of millions of years since it is the progressive stripping of crust by denudation that is largely responsible for shallow crustal cooling. These data provide information on broad regional patterns of denudation, but they can now be related to more local denudation rates by coupling with data from cosmogenic isotope analysis (see COSMOGENIC DATING) which provides denudation rates over timescales of thousands to hundreds of thousands of years.

Key issues in global geomorphology

The most obvious issue in global geomorphology is to understand the gross variations in the Earth's continental topography and how this topography has changed over time. Why, for instance, is 82 per cent of the world's land surface over 4,000 m above sea level concentrated in the Tibetan Plateau? And what is the origin of the large area of anomalously high topography extending across southern Africa and into the adjacent Atlantic Ocean. Answers to these questions require an understanding of the interaction of internal and external processes over periods of millions of years. Crucial to answering such questions are data on changes on the elevation of the land surface over time, since the timing of the uplift of the Tibetan Plateau, for instance, is key to understanding the cause of such uplift.

Unfortunately, constraining such surface uplift has proved very difficult, not least because denudation in uplifted terrain tends to remove evidence that would be indicative of prior elevations. However, various techniques have been developed to infer past elevations in addition to the obvious strategy of using shoreline or shallow marine deposits where present. These include inferring temperature (and therefore indirectly elevation) change from the characteristics of fossil leaves on the basis that specific plant types have particular temperature tolerances and that surface uplift will elevate fossils into cooler climatic zones. This approach has been used to infer surface uplift in mountain ranges such as the Himalayas, but it requires detailed information on global and regional climatic changes which would also produce vertical shifts in climatic zones. Another approach is to use the elevation-dependent fractionation of oxygen in precipitation across mountain ranges which can be incorporated into carbonate sediments, but of considerable potential is basalt vesicle ratio analysis. This technique uses the effect of atmospheric pressure of the relative size of gas bubbles at the top and bottom of individual lava flows to infer the atmospheric pressure, and hence elevation, at the time of eruption. Notwithstanding these and other techniques, constraining changes over time in the absolute elevation of the land surface remains a problematic but fundamental issue in global geomorphology.

Another important issue is the coupling of onshore and offshore records of denudation and deposition. The growth in offshore hydrocarbon exploration along continental margins since the 1970s has greatly expanded our knowledge of their depositional history, but it has also raised the question of what controls the supply of sediment from the adjacent continental hinterland. Answering this question requires information not just on the mobilization and transport of sediment from onshore to offshore but also on tectonic mechanisms and the isostatic response to changes in crustal loading as mass is transferred offshore. Although largely irrelevant to small-scale surface process geomorphology, ISOSTASY assumes a critical role in global geomorphology since, at these larger spatial and temporal scales, flexure of the lithosphere in response to denudational unloading can have important effects on the mode of landscape development.

A further key theme in global geomorphology is the coupling between internal and external processes. Although the influence of tectonic mechanisms on surface processes through the construction of relief has been long understood, the way in which spatial variations in denudation rates can affect patterns of tectonic deformation was only fully appreciated in the 1990s. This is evident in the commonly found strike-parallel pairing of metamorphic facies in mountain ranges as a result of higher rates of denudation (and therefore greater depths of exposure) on the wetter, windward side compared with the drier, leeward side. Modelling of patterns of crustal deformation as a result of spatial variations in denudation rates has further emphasized the two-way interaction between surface and internal geomorphic processes.

The role of the land surface in interactions between tectonics and climate has also received attention in attempts to understand the long-term geological controls over the concentration of atmospheric carbon dioxide and hence, through the greenhouse effect, global climate. The key process here is the weathering of silicate minerals, a reaction which draws down CO^2 from the atmosphere. As global topography and relief changes as a result of interactions between tectonics, climate and landscape development, the global rate of CO^2 drawdown would be expected to vary, although the operation of these interactions are far from fully understood.

Finally, comparative planetary geomorphology provides the key perspective for global geomorphology. Looking at landscape development on other planetary bodies shifts our perspective from viewing terrestrial landforms as 'normal', and emphasizes that the Earth's landforms have arisen from a particular combination of its size, its composition, its distance from the sun, the composition and density of its atmosphere, and its age. The great majority of planetary bodies have surfaces dominated by impact craters, making impact cratering the dominant geomorphic process in the Solar System (although most occurred in the first 500 to 600 Ma from the birth of the Solar System around 4.5 Ga ago). The critical factor for Earth is the surface temperatures that it experiences which encompass the range over which water can exist as a solid, a liquid and a gas. This enables Earth to have an active hydrological cycle which is key to many geomorphological processes. Also Earth's size and composition means that it has a high enough internal temperature to melt rock and therefore permit volcanism and the convection that helps power plate tectonics. Earlier in its history, Mars also probably experienced short-lived episodes when there was a fairly active hydrological cycle including oceans; this is the main period of the channel formation on Mars. By contrast, the high surface temperatures on Venus, largely resulting from an intense greenhouse effect associated with a dense CO^2–rich atmosphere, has prevented the existence of liquid water; thus its surface is dominated by the effects of volcanism and impact cratering.

Further reading

Burbank, D.W. and Anderson, R.S. (2001) *Tectonic Geomorphology*, Malden, MA: Blackwell Science.

Ellis, M. and Merritts, D. (1994) *Tectonics and Topography*, Washington, DC: American Geophysical Union.

Greeley, R. (1994) *Planetary Landscapes*, 2nd edition, London: Chapman and Hall.

Stüwe, K. (2002) *Geodynamics of the Lithosphere*, Berlin: Springer-Verlag.

Summerfield, M.A. (1991) *Global Geomorphology*, London: Longman.

——(ed.) (2000) *Geomorphology and Global Tectonics*, Chichester: Wiley.

MIKE SUMMERFIELD

GLOBAL WARMING

There is now a widespread appreciation that the build-up of greenhouse gases in the atmosphere (carbon dioxide, methane, nitrous oxide, CFCs, etc.) will create an enhanced greenhouse effect that will cause global warming. Details of the degree of warming that will occur and of the associated changes in other climatic variables are provided in the reports of the Intergovernmental Panel on Climate Change (2001). If such changes occur over coming decades certain landscapes and geomorphological processes will be modified (Table 22).

Some landscapes, 'geomorphological hot spots', will be especially sensitive because they are located in zones where it is forecast that climate will change to an above average degree. In the high latitudes of Canada or Russia the degree of warming may be three or four times greater than the global average. It may also be the case with respect to some critical areas where particularly substantial changes in precipitation may result from global warming. For example, various scenarios suggest the High Plains of the United States of America will become markedly drier. Other landscapes will be highly sensitive because certain landscape-forming processes are so closely controlled by climatic conditions. If such landscapes are close to particular climatic thresholds then quite modest amounts of climatic change can flip them from one state to another. In this entry attention will be paid to some of these hot spots.

Tundra and permafrost terrains

High latitude tundra and PERMAFROST terrains may be regarded as one of these sensitive zones. They are likely to undergo especially substantial temperature change. In addition, the condition of permafrost is particularly closely controlled by temperature conditions. By definition it cannot occur where mean annual temperatures are positive, and the latitudinal limits of different types of permafrost can be related to varying degrees of negative temperatures. Thus the equatorward limit of continuous permafrost may approximate to the −5 °C isotherm and the equatorward limit of discontinuous or sporadic permafrost to the −2 °C isotherm. It is likely that the latitudinal limits of permafrost will be displaced polewards by 100 to 250 km for every 1 °C rise in mean annual temperature. The quickest loss of permafrost would occur in terrains underlain by surface material with low ice contents. The slowest response would be in ice-rich materials, which require more heat to thaw. Snow or the presence of thick, insulating organic layers (i.e. peat) might also buffer the effects of increased surface temperatures in some areas.

There is historical evidence that permafrost can degrade speedily. For instance, during the warm 'optimum' of the Holocene (*c*.6,000 years ago) the southern limit of discontinuous permafrost in the Russian Arctic was up to 600 km north of its present position (Koster 1994). Similarly, researchers have demonstrated that along the Mackenzie Highway (Canada), between 1962 and 1988, the southern fringe of the discontinuous zone had moved north by about 120 km in response to an increase over the same period of 1 °C mean annual temperature (Kwong and Tau 1994).

Table 22 Some geomorphologic consequences of global warming

Hydrologic
Increased evapotranspiration loss
Increased percentage of precipitation as rainfall at expense of winter snowfall
Increased precipitation as snowfall in very high latitudes
Possible increased risk of cyclones (greater spread, frequency and intensity)
Changes in state of peatbogs and wetlands
Less vegetational use of water because of increased CO_2 effect on stomatal closure

Vegetational controls
Major changes in latitudinal extent of biomes
Reduction in boreal forest, increase in grassland, etc.
Major changes in altitudinal distribution of vegetation types (*c.*500 m for 3 °C)
Growth enhancement by CO_2 fertilization

Cryospheric
Permafrost, decay, thermokarst, increased thickness of active layer, instability of slopes, river banks, and shorelines
Changes in glacier and ice-sheet rates of ablation and accumulation
Sea-ice melting

Coastal
Inundation of low-lying areas (including wetlands, deltas, reefs, lagoons, etc.)
Accelerated coast recession (particularly of sandy beaches)
Changes in rate of reef growth
Spread of mangrove swamp

Aeolian
Increased dust storm activity and dune movement in areas of moisture deficit

Soil erosion
Changes in response to changes in land use, fires, natural vegetation cover, rainfall erosivity, etc.
Changes resulting from soil erodibility modification (e.g. sodium and organic contents)

Subsidence
Desiccation of clays under conditions of summer drought

Woo *et al.* (1992) made certain predictions based on the assumption that a greenhouse warming of 4–5 °C causes a spatially uniform increase in surface temperature of the same magnitude over northern Canada. They suggested that permafrost in over half of what is now the discontinuous zone could be eliminated, that the boundary between continuous and discontinuous permafrost might shift northwards by hundreds of kilometres and that a warmer climate could ultimately eliminate continuous permafrost from the whole of the mainland of North America, restricting its presence only to the Arctic Archipelago.

In areas where rapid permafrost melting occurs, the consequences will be legion. They include ground subsidence (THERMOKARST), increased erosion of shorelines and riverbanks, and an increase in debris flow activity and other forms of slope instability.

High latitude areas may also be particularly susceptible to changes in precipitation and runoff. Areas which are currently very dry, because the air is so cold, may become moister as warmer winters cause more snow to fall, thereby creating a likelihood of increased summer runoff. In somewhat warmer environments, where substantial winter snowfall occurs, there might be a tendency in a warmer world for a decrease in the proportion of winter precipitation that falls as snow. There would thus be greater winter rainfall and runoff, but less overall precipitation to enter snowpacks to be held over until spring snowmelt. This in turn would have adverse consequences both for late spring and summer runoff levels in rivers and for soil moisture levels. Other factors may also modify runoff. For example, as permafrost thaws, groundwater recharge may increase and surface runoff decrease.

Glaciers and ice sheets

Glaciers and ice sheets will be highly susceptible to a rise in temperature. Although there has been considerable debate as to whether or not polar ice caps might respond catastrophically to global warming because of an increase in ablation, accelerated melting of tidewater snouts, the cliffing of termini by a rising sea level, or the removal of the buttressing effects of ice shelves as they melt (Huybrechts *et al.* 1990), it is probably valley glaciers in alpine situations which will respond most quickly and markedly to climatic warming. Such glaciers are highly responsive, as is made evident by their frequent and rapid fluctuations during the Neoglacials of the Holocene. Although topographic controls and changes in precipitation and cloudiness are significant controls of glacier state, it is highly likely that most alpine glaciers will show increasing rates of retreat in a warmer world. Indeed, given the rates of retreat (20–70 cm year) experienced in many mountainous areas in response to the warming episode since the 1880s, it is probable that many glaciers will disappear altogether, from areas as diverse as the Highlands of East Africa or the Southern Alps of New Zealand.

Desert margins

The history of desert margins indicates that in the past they too have been sensitive to environmental change. This in turn suggests that they are likely to be susceptible to future environmental changes. Thus closed depressions have fluctuated repeatedly from being dry and saline to being full and fresh. Valley bottoms and hillsides have alternated between cut-and-fill, and dunefields have at some times been mobile and at other times stable (Forman *et al.* 2001). Many dry regions will suffer large diminutions in runoff (Arnell 1999), with annual totals likely to be reduced by over 60 per cent. Indeed Shiklomanov (1999) has suggested that in arid and semi-arid areas an increase in mean annual temperature by 1 to 2 °C and a 10 per cent decrease in precipitation could reduce annual river runoff by up to 40–70 per cent.

In the case of closed depressions, the dating of high water levels in lakes in the tropics and subtropics shows that many of them have had a complex history during the Holocene and that their water levels have varied considerably. High levels were a feature of the Saharan region around 8,000 years ago, a time when global temperatures were probably slightly greater than today. Very large numbers of freshwater deposits date from this time, even in the dry heart of the Sahara. Some stream courses (e.g. the Wadi Howar) were active.

In the case of river and slope systems, they too have fluctuated between phases of stability or alluviation and phases of erosion and incision. Even over the last century or so the valley systems of the American south-west, called ARROYOs, have experienced trenching and filling episodes in response to climatic and other stimuli (e.g. land use change). Of particular importance have been changes in the amount and intensity of precipitation. Crucial in this respect is the response of vegetation cover to rainfall events, for in semi-arid areas it is not only highly dependent on moisture availability but also controls the erodibility of the ground surface (Elliot *et al.* 1999).

Changes in precipitation and evapotranspiration rates also have a marked impact on aeolian environments and processes. Rates of deflation, sand and dust entrainment and dune formation are closely related to soil moisture conditions and vegetation cover. Areas that are at present marginal with respect to aeolian processes will be particularly susceptible, and this has been made evident, for example, through recent studies of the semi-arid portions of the United States (e.g. the High Plains). Repeatedly throughout the Holocene they have flipped from a state of vegetated stability to states of drought-induced surface instability. Thermoluminescent and optical dates have made evident their sensitivity to quite minor perturbations. Geomorphologists, using the output from General Circulation Models (GCMs), combined with a dune mobility index which incorporates wind strength and the ratio of mean annual precipitation to potential evapotranspiration, have shown that with global warming, sand dunes and sandsheets on the Great Plains are likely to become reactivated over a significant part of the region, particularly if the frequencies of wind speeds above the threshold velocity for sand movement were to increase by even a moderate amount (Muhs and Maat 1993). The same applies to dust storm generation in the Great Plains and the Canadian Prairies, where the application of GCMs shows that conditions comparable to the devastating dust-bowl years of the 1930s are likely to be experienced.

Tropical coastlines

Tropical coastlines are a further very sensitive environment with respect to future climatic change. This is for three main reasons: the relationship between tropical cyclone activity and the sea-surface temperature (SST), the temperature tolerances of coral reefs, and the effects that temperature change and sea-level rise have upon mangrove swamps.

Tropical cyclones are important agents of geomorphological change. They scour out river channels, deposit debris fans, cause slope failures, build up or break down coastal barriers, transform the nature of some coral islands (either building them up or erasing them), and change the turbidity and salinity of lagoons. Were their frequency, intensity and geographical spread to change it would have significant implications. It is not, however, entirely clear just how much these important characteristics will change. Intuitively one would expect cyclone activity to become more frequent, intense and extensive if sea-surface temperatures were to rise, because SST is a clear control of where they develop. Indeed, there is a threshold at about 26.5–27.0 °C. However, the Intergovernmental Panel and some individual scientists are far from convinced that global warming will invariably stimulate cyclone activity.

Coral reefs may be sensitive to warming, partly because of the role that cyclones play in their evolution, partly because their growth can be retarded or accelerated because of changes in SSTs, and partly because their existence is so closely related to sea level.

In the 1980s there were widespread fears that if rates of sea-level rise were high (perhaps 2 to 3 m or more by 2100) then coral reefs would be unable to keep up and submergence of whole atolls might occur. Particular concern was expressed about the potential fate of Pacific Island groups, and of the Maldives in the Indian Ocean. However, with the reduced expectations for the degree of sea-level rise that may occur, there has arisen a belief that coral reefs may survive and even prosper with moderate rates of sea-level rise. As is the case with marshes and other wetlands, reefs are dynamic features that may be able to respond adequately to sea-level rise. It is also important to realize that their condition depends on factors other than the rate of submergence.

Increased sea-surface temperatures could have deleterious consequences for corals which are near their thermal maximum. Most coral species cannot tolerate temperatures greater than about 30 °C and even a rise in seawater temperature of 1–2 °C could adversely affect many shallow-water coral species. Increased temperatures in recent years have been identified as a cause of widespread coral bleaching (loss of symbiotic zooxanthellae). Those corals stressed by temperature or pollution might well find it more difficult to cope with rapidly rising sea levels than would healthy coral. Moreover, it is possible that increased ultraviolet radiation because of ozone layer depletion could aggravate bleaching and mortality caused by global warming. Various studies suggest that coral bleaching was a widespread feature in the warm years of the 1980s and 1990s (Goreau and Hayes 1994).

However, Kinsey and Hopley (1991) believe that few of the reefs in the world are so close to the limits of temperature tolerance that they are likely to fail to adapt satisfactorily to an increase in ocean temperature of 1–2 °C, provided that there are not very many more short-term temperature deviations. Indeed, in general they believe that reef growth will be stimulated by the rising sea levels of a warmer world, and they predict that reef productivity could double in the next hundred years from around 900 to 1,800 million tonnes per year. They do, however, point to a range of subsidiary factors that could serve to diminish the increase in productivity: increased cloud cover in a warmer world could reduce calcification because of reduced rates of photosynthesis; increased rainfall levels and hurricane activity could cause storm damage and freshwater kills; and a drop in seawater pH might adversely affect calcification.

However, reef accretion is not the sole response of reefs to sea-level rise, for reef tops are frequently surmounted by small islands (cays and motus) composed of clastic debris. Such islands might be very susceptible to sea-level rise. On the other hand, were warmer seas to produce more storms, then the deposition of large amounts of very coarse debris could in some circumstances lead to their enhanced development. However, the situation is complex, and in some cases potential vertical reef accretion could be reduced by storm attack. One also needs to consider changes in tropical storm frequency as well as changes in tropical storm magnitude, for high storm frequencies might change the relative importance of corals and calcareous algae (Spencer 1994).

Other coastlines

There are other coastlines that will also be substantially modified by sea-level rise resulting from global warming. These include sandy beaches, miscellaneous types of saltmarsh and areas of land subsidence.

Sandy beaches are held to be sensitive because of the so-called BRUUN RULE (Bruun 1962, see Plate 55). This predicts future rates of coastal erosion in response to rising sea level. Bruun envisaged a profile of equilibrium in which the volume of material removed during shoreline retreat is transferred onto the adjacent shoreface/inner shelf, thus maintaining the original bottom profile and nearshore shallow conditions. With a rise in sea level additional sediment has to be added to the below-water portion of the beach profile. One source of such material is beach erosion, and estimates of beach erosion of *c.*100 m for every 1 m rise in sea level have been postulated. However, although the concept is intuitively appealing, it is also difficult to confirm or quantify without precise bathymetric surveys and integration of complex nearshore profiles over a long period of time. Moreover, an appreciable time-lag may occur in shoreline response which is highly dependent upon local storm frequency. Furthermore, the model is essentially a two-dimensional one in which the role of longshore sediment movement is not considered. It is also assumed that no substantial offshore leakage of sediment occurs. Accurate determination of sediment budgets in three dimensions is still replete with problems. Whatever the problems of modelling, however, sandy beaches will tend to disappear from locations where they are already narrow and backed by high ground or swamp and marsh, but will probably tend to persist where they can retreat across wide beach ridge plains.

Saltmarshes, including MANGROVE SWAMPS, are potentially highly vulnerable to sea-level rise, particularly where sea defences and other barriers prevent the landward migration of marshes as sea-level rises. However, saltmarshes are dynamic features and in some situations may well be able to cope, even with quite rapid rises of sea level. Indeed, some important sediment trapping plants may extend their range in response to warming. Such plants include mangroves (e.g. in New Zealand) and also *Spartina anglica* (e.g. in northern Europe). They would tend to lead to an acceleration in marsh accretion.

One way of attempting to predict the effects of increasing rates of sea-level rise is to study those areas where the rates of sea-level rise are currently high because of subsidence. On the coast of south-east England, where rise occurs at a rate of 5 mm per year, saltmarshes appear to cope. Sediments eroded from the outer edge appear to contribute to the sediments which are accreted on the inner marsh surface. Moreover, UK saltmarshes have current rates of accretion that are the same order of magnitude as, or greater than, the predicted rates of sea-level rise.

Plate 55 The main railway line between France and Spain near Barcelona. Note the severe erosion of the coastline, and the abandoned track in the foreground. Sandy beaches of this type will be especially sensitive to the effects of accelerated sea-level rise associated with global warming

Reed (1990) suggests that saltmarshes in riverine settings may receive sufficient inputs of sediment that they are able to accrete sufficiently rapidly to keep pace with projected rises of sea level. Likewise, some vegetation associations, e.g. *Spartina* swards, may be relatively more effective than others at encouraging accretion, and organic matter accumulation may itself be significant in promoting vertical build-up of some marsh surfaces. For marshes that are dependent upon inorganic sediment accretion, increased storm activity and beach erosion which might be associated with the greenhouse effect could conceivably mobilize sufficient sediments in coastal areas to increase their sediment supply.

One particular type of marsh that may be affected by anthropogenically accelerated sea-level rise is the mangrove swamp. Mangroves may respond rather differently to other marshes because their main plants are relatively long-lived trees and shrubs. This means that the speed of zonation change will be less. The degree of disruption is likely to be greatest in microtidal areas, where any rise in sea level represents a larger proportion of the total tidal range than in macrotidal areas. However, the setting of mangrove swamps will be very important in determining how they respond. River-dominated systems with large allochthonous sediment supply will have faster rates of shoreline progradation and deltaic plain accretion and so may be able to keep pace with relatively rapid rates of sea-level rise. By contrast, in reef settings in which sedimentation is primarily autochthonous, mangrove surfaces are less likely to be able to keep up with sea-level rises (Ellison and Stoddart 1990).

The ability of mangrove propagules to take root and become established in intertidal areas subjected to a higher mean sea level is in part dependent on species. In general the larger propagule species (e.g. *Rhizophora* spp) can become established in rather deeper water than can the smaller (e.g. *Avicennia* spp). The latter has aerial roots which project only vertically above tidal muds for short distances.

Mangrove colonization and migration would also be influenced by salinity conditions so that any speculations about mangrove response to sea-level rise must also incorporate allowance for change in rainfall and freshwater runoff.

In arid areas, such as the Middle East, great lengths of coastline are fringed by low level salt-plains (SABKHA). These features are generally regarded as equilibrium forms that are produced by depositional processes (e.g. alluvial siltation, aeolian inputs, evaporite formation, faecal pellet deposition) and planation processes (e.g. wind erosion and storm surge effects). They tend to occur at or about high tide level. Because of the range of depositional processes involved in their development they might be able to adjust to a rising sea level but quantitative data on present and past rates of accretion are sparse.

A crucial issue with all types of wetlands is the nature of the hinterland. Under natural conditions many marshes and swamps are backed by low-lying estuarine and alluvial land which could be displaced if a rising sea level were to drive the marshes landward. However, in many parts of the world sea defences, bunds and other structures have been built at the inner margins and these will prevent colonization of the hinterland. Experiments are now being conducted to see whether saltmarsh development can be promoted by the deliberate breaching of sea defences.

One final type of sensitive coastal environment is that where coastal submergence is taking place. The combination of local submergence with global sea-level rise will make these coasts especially prone to inundation. Some areas are subject to natural subsidence as a result of sediment loading on the crust (e.g. deltas) or because of tectonic processes, but some key areas are subjected to accelerated (i.e. anthropogenic) subsidence. This is brought about primarily by mining of ground water or hydrocarbons and can be especially serious in the case of coastal mega-cities, portions of which are either close to or beneath current sea level (e.g. Bangkok and Tokyo).

Although there may still be uncertainties about whether global warming will occur and about the various impacts of such warming should it occur, and although the degree of climatic and sea-level change that is being postulated might at first sight appear relatively modest, it would be wrong to be complacent about the potential geomorphological impacts brought about by global warming. Our knowledge of how geomorphological systems have reacted to the climatic fluctuations of the Holocene, and our knowledge of the intimate relationships between some geomorphological processes and climatic conditions, both lead us to the conclusion that some environments will respond in a manner that will be substantial in degree and which will have numerous consequences for human occupation of these environments.

References

Arnell, N. (1999) The impacts of climate change on water resources, in *Climate Change and its Impacts*, 14–17, Bracknell: UK Meteorological Office.

Bruun, P. (1962) Sea-level rise as a cause of shore erosion, *American Society of Civil Engineers Proceedings: Journal of Waterways and Harbors Division* 88, 117–130.

Elliot, J.G., Gellis, A.C. and Aby, S.C. (1999) Evolution of arroyos: incised channels of the southwestern United States, in S.E. Darby and A. Simon (eds) *Incised River Channels*, 153–185, Chichester: Wiley.

Ellison, J.C. and Stoddart, D.R. (1990) Mangrove ecosystem collapse during predicted sea level rise: Holocene analogues and implications, *Journal of Coastal Research* 7, 151–165.

Forman, S., Oglesby, R. and Webb, R.S. (2001) Temporal and spatial patterns of Holocene dune activity on the Great Plains of North America: megadroughts and climate links, *Global and Planetary Change* 29, 1–29.

Goreau, T.L. and Hayes, R.L. (1994) Coral bleaching and ocean 'hot spots', *Ambio* 23, 176–180.

Huybrechts, P., Litreguilly, A. and Reels, N. (1990) The Greenland ice sheets and greenhouse warming, *Palaeolgeography, Palaeoclimatology, Palaeoecology* 89, 399–412.

Intergovernmental Panel on Climate Change (2001) *Climate Change 2001: The Scientific Basis*, Cambridge: Cambridge University Press.

Kinsey, D.W. and Hopley, D. (1991) The significance of coral reefs as global carbon sinks – response to greenhouse, *Palaeogeography, Palaeoclimatology, Palaeoecology* 89, 363–377.

Koster, E.A. (1994) Global warming and periglacial landscapes, in N. Roberts (ed.) *The Changing Global Environment*, 150–172, Oxford: Blackwell.

Kwong, Y.T.J. and Tau, T.Y. (1994) Northward migration of permafrost along the Mackenzie Highway and climatic warming, *Climatic Change* 26, 399–419.

Muhs, D.R. and Maat, P.B. (1993) The potential response of aeolian sands to greenhouse warming and precipitation reduction on the Great Plains of the United States, *Journal of Arid Environments* 25, 351–361.

Reed, D.J. (1990) The impact of sea level rise on coastal saltmarshes, *Progress in Physical Geography* 14, 465–481.

Shiklomanov, I.A. (1999) Climate change, hydrology and water resources: the work of the IPCC, 1988–1994, in J.C. van Dam (ed.) *Impacts of Climate Change and Climate Variability on Hydrological Regimes*, 8–20, Cambridge: Cambridge University Press.

Spencer, T. (1994) Tropical coral islands – an uncertain future, in N. Roberts (ed.) *The Changing Global Environment*, 190–209, Oxford: Blackwell.

Woo, M.K., Lewkowicz, A.G. and Rouse, W.R. (1992) Response of the Canadian permafrost environment to climate change, *Physical Geography* 13, 287–317.

A.S. GOUDIE

GOLDICH WEATHERING SERIES

The types and proportions of various minerals in a weathering profile are usually quite different from the original bedrock. Some minerals seem to survive more or less unaltered even after being subjected to prolonged WEATHERING, while others decompose more rapidly. In many weathering studies, the silicate minerals, the primary constituents of igneous and metamorphic rocks, are arranged into an order of susceptibility to chemical weathering. The most commonly cited order was first proposed by S.S. Goldich (1938), based on a detailed study of the mineralogic changes of granitoid rocks during weathering (Figure 78).

Goldich (1938) concluded that minerals which form at high temperatures and pressures (olivine, amphiboles, pyroxenes, calcium plagioclase), and hence are the first to precipitate, are markedly less stable and weather much more quickly than minerals which crystallize at lower temperatures and pressures (sodium plagioclase, potassium feldspar, micas and quartz) (Figure 78). This sequence is the reverse of BOWEN'S REACTION SERIES, which ranks minerals in their order of crystallization from a melt.

CHEMICAL WEATHERING reactions are with the cations that bind the silica structural units together. Thus the relative strength between the oxygen and cations in each mineral and the structure of the bonding are both significant. The isolated Si-O tetrahedra in olivine are the least stable in weathering; while quartz, which is completely formed of interlocking silica tetrahedra with no intervening cations, is the most stable. If muscovite and the plagioclases are disregarded, the order of the Goldich weathering series coincides with the classification of silicate structures, based on increasing Si–O–Si bonds from zero in olivine to four in quartz.

Although Goldich's series has widespread applications and usually works well, local exceptions have been documented.

Reference

Goldich, S.S. (1938) A study of rock weathering, *Journal of Geology* 46, 17–58.

SEE ALSO: Bowen's reaction series; chemical weathering

CATHERINE SOUCH

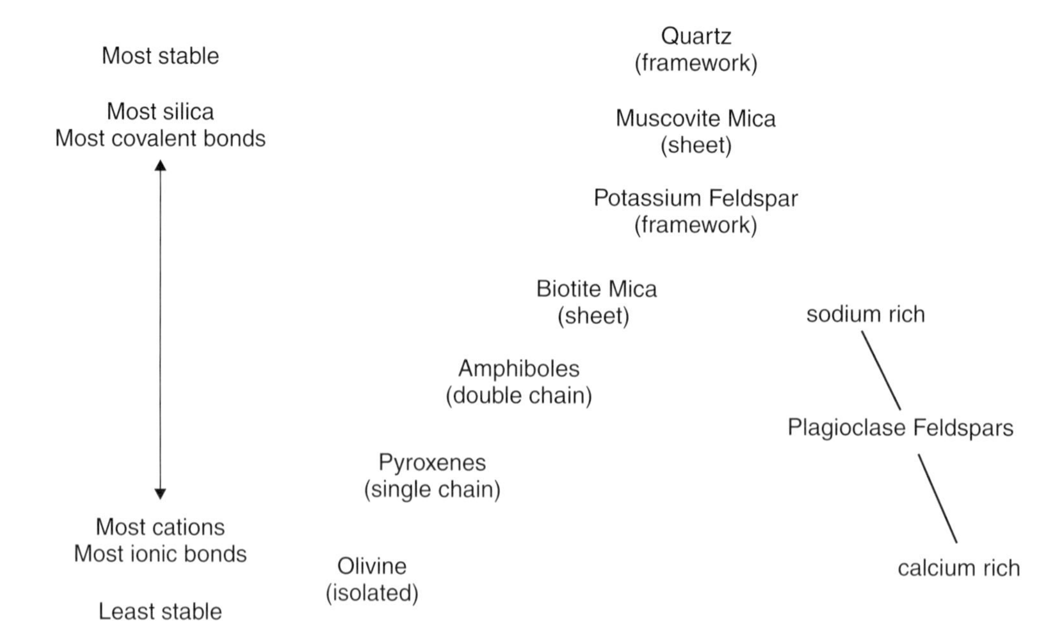

Figure 78 Goldich weathering series

GORGE AND RAVINE

Gorges, which may be hundreds of metres deep, are caused either by incision of a river against an uplifting landmass, the superimposition of a channel across resistant rock, the outburst of floodwaters across a landscape, or by the headward retreat of a KNICKPOINT or WATERFALL (Rashleigh 1935; Derricourt 1976; Tinkler *et al.* 1994; van der Beek *et al.* 2001). Ravines are much smaller gashes (the order of metres to tens of metres wide and deep) cut into the weak bedrock, or frequently into superficial sediments such as glacial deposits or deeply weathered horizons. The term ravine is frequently used in the context of soil erosion and land degradation, and a ravine, or ravine network, will have steep, weakly consolidated side slopes, flat channel bottoms characterized by a heavy sediment load, and a clear break of slope with the surface above. Present academic literature seems to find the technical use of the word limited to south and east Asia (Raj *et al.* 1999), otherwise it is used as a synonym for gully. It may occur as the generic part of a place name: Yamuna Ravine in India, Elk Ravine, New Hampshire, USA.

Neither of the terms *gorge* nor *ravine* is well defined in the literature, although a Gorge (e.g. the Three Rivers Gorge in western China) is generally understood to be typical of rivers of larger sizes. The word 'ravine' tends to imply a small deeply incised channel in a low-order drainage basin. Both terms imply a river deeply incised below the surrounding landscape, local slope processes being unable to reduce the side slopes at the same rate that the river is incising into the terrain. Thus there is often little sensitivity to the local topography. Large, deep gorges require mechanically strong country rock, although the typically steep valley slopes may still be susceptible to failure by rock fall and rockslides. Gorges are frequently found in areas where drainage is antecedent upon actively growing fold systems such as the Himalayan ranges, or where it is superimposed (superposed) upon more resistant rocks from weaker cover rocks.

In exceptionally large river systems the term gorge usually refers to the deeply incised, and often scarcely accessible inner gorge (Kelsey 1988). Bedrock channels often contain an inner channel, where bedload transport rates are highest, and erosion is enhanced. Because of the large variations in discharge which have occurred in many high latitude drainage basins during the Quaternary, it is unclear to what extent the inner channel of a river system carrying large glacial outflow discharges becomes the inner gorge of its non-glacial successor. Excavations for the Boulder Dam on the Black Canyon of the Colorado revealed an unsuspected inner gorge up to 25 m below flanking bedrock edges to the channel (Legget 1939: 322–323).

Catastrophic scale outburst floods of glacial stored waters are another mechanism, unsuspected until recent decades, for the formation of gorges (Baker 1978; O'Connor 1993; Rathburn 1993; Knudsen *et al.* 2001). Such floods may have been repeated many times during the Quaternary, their cumulative sum affect being what we now see.

Scheidegger *et al.* (1994) argue for strong structural control by large-scale regional joints and fault systems, in the geographical layout of large gorges. However overall trends in gorge orientation usually owe their origin to regional scale topographic trends (Baker 1978; Rathburn *et al.* 1993), to which structural control merely adds local detail. Subsequent river erosion may generate entrenched or incised meandering patterns unrelated to local or regional structure.

Buried gorges are not uncommon in glaciated terrains, many being found in the Great Lakes region of eastern Canada (Davis 1884; Karrow and Terasmae 1970; Greenhouse and Karrow 1994). The infill of permeable glacial sediments within a bedrock gorge often produces localities favourable for groundwater exploration (Farvolden 1969).

References

Baker, V.R. (1978) Paleohydraulics and hydrodynamics of Scabland Floods, in V.R. Baker and D. Nummedal (eds) *The Channeled Scabland*, 59–80, Washington, DC: NASA.

Davis, W.M. (1884) Gorges and waterfalls, *American Journal of Science* 28, 123–132.

Derricourt, R.M. (1976) Retrogression rate of the Victoria Falls and the Batoka Gorge, *Nature* 264, 23–25.

Farvolden, R.N. (1969) Bedrock channels of southern Alberta, in J.G. Nelson and M.J. Chambers (eds) *Geomorphology: Process and Methods in Canadian Geography*, 243–255, Toronto: Methuen.

Greenhouse, J.P. and Karrow, P.F. (1994) Geological and geophysical studies of buried valleys and their fills near Elora and Rockwood, Ontario, *Canadian Journal of Earth Sciences* 31, 1,838–1,848.

Karrow, P.F. and Terasmae, J. (1970) Pollen-bearing sediments of the St. Davids buried valley at the Whirlpool, Niagara Gorge, Ontario, *Canadian Journal of Earth Sciences* 7, 539–542.

Kelsey, H.M. (1988) The formation of inner gorges, *Catena* 15, 433–458.

Knudsen, K.L., Sowers, J.M., Ostenaa, D.A. and Levish, D.R. (2001) Evaluation of glacial outburst flood hypothesis for the Big Lost River, Idaho, *Ancient Floods, Modern Hazards*, Washington, DC: American Geophysical Union.

Legget, R.E. (1939) *Geology and Engineering*, New York: McGraw-Hill.

O'Connor, J.E. (1993) Hydrology, hydraulics, and geomorphology of the Bonneville Flood, *Geological Society of America Special Paper* 274.

Raj, R., Maurya, D.M. and Chamyal, L.S. (1999) Tectonic control on distribution and evolution of ravines in the lower Mahi Valley, Gujarat, *Journal of the Geological Society of India* 53(6), 669–674.

Rashleigh, E.C. (1935) *Among the Waterfalls of the World*, London: Jarrolds.

Rathburn, S.L. (1993) Pleistocene cataclysmic flooding along the Big Lost River, east central Idaho, *Geomorphology* 8, 305–319.

Scheidegger, A.E. (1994) On the genesis of river gorges, *Transactions, Japanese Geomorphological Union* 15(2), 91–110.

Tinkler, K.J., Pengelly, J.W., Parkins, W.G. and Asselin, G. (1994) Postglacial recession of Niagara Falls in relation to the Great Lakes, *Quaternary Research* 42, 20–29.

Van der Beek, P., Pulford, A. and Braun, J. (2001) Cenozoic landscape development in the Blue Mountains (SE Australia): lithological and tectonic controls on Rifted Margin Morphology, *Journal of Geology* 109, 35–56.

KEITH J. TINKLER

GPS

The Global Positioning System (GPS) is a constellation of satellites developed by the US Department of Defense to provide precise positioning and navigation information. GPS receivers determine position through repeated measurements of digitally tagged radio signals from the satellites. Conceived for military purposes, the commercial applications for positioning information have blossomed. Analysts have suggested that the global GPS market is worth over US$16 billion (in 2002). Among the varied users of GPS are geomorphologists requiring geo-referenced positioning information for field terrain. However, the wide variety of systems available and the enormous range in cost and capability requires caution on the part of the user and an ability to assimilate a multitude of jargon and proprietary software.

Initially, the Department of Defense used a procedure termed Selective Availability to dither the precise time code and degrade the accuracy of the signal. This has been set to zero since May 2000, improving reliability and consistency though applications requiring sub-metre accuracy (and hence all fieldwork requiring elevation data) continue to require differential GPS (DGPS). DGPS relies on a static reference receiver at a known control point which logs bias errors over the same time period that another receiver (the 'rover') is occupying the points of interest. The measured errors are used to correct the rover position either by downloading and 'post-processing' the data, or by receiving corrections via radio telemetry (known as 'real time kinematic'). The control point can be operated by the user or be a commercial ground station broadcasting corrections. The reference frame for GPS output is the World Geodetic System 1984 (WGS-84), a geocentric system returning ellipsoid co-ordinates in latitude and longitude. Altitude is derived as elevation above the ellipsoid and some knowledge is needed to integrate GPS-derived height data with existing levelling data or to translate positions into a local datum. Fortunately there are many textbooks providing technical details (e.g. Hofmann-Wellenhof *et al.* 2001). Relatively few papers consider explicitly GPS applications in geomorphology (Cornelius *et al.* 1994; Fix and Burt 1995; Higgitt and Warburton 1999) but an increasing number make routine use of GPS as part of the data-gathering procedure. Four broad areas of application can be identified:

Rectification

Global referencing is essential in most geomorphological research. GPS can assist observations where detailed maps are lacking or it can be used for registering ground markers to analyse aerial photographs or remotely sensed imagery. This is useful for assessing change in sequential imagery such as the dynamics of land degradation (Gillieson *et al.* 1994). In terrain remote from conventional benchmarks, GPS can save much time in establishing the elevation of sample points.

Detailed topographic survey

The speed of GPS data capture offers scope for producing accurate digital elevation models (DEMs) of moderately sized field areas. The abundance of points in a GPS survey generates topographic attributes which can be used as input in hydrological models. A related commercial development is 'precision agriculture' where GPS receivers mounted to farm vehicles produce detailed information about spatial variations in crop yields or soil conditions. One consequence of precision, as highlighted by Wilson *et al.* (1998), is the recognition that calculated topographic attributes are sensitive to the resolution and distribution of survey points. By implication, estimation of topographic variables from a limited number of survey points may be prone to large errors. A dense network of GPS survey points around a catchment can provide a more enlightened summary about the statistical distribution of slope characteristics.

Measuring change in landforms

GPS is ideal for measuring sequential change in landform characteristics. Geologists have made extensive use of networks of high precision GPS for identifying ground movements associated with earthquakes and volcanic eruptions. Geomorphological applications are apparent in neotectonics and landslide research. Where budgets are more restrictive, repeat surveys provide similar information. This has been used to construct detailed maps of river channel change (Brasington *et al.* 2000). As the object of geomorphological study is usually inanimate, there are no parallels to ecological applications that examine animal behaviour. The methodology to determine grazing patterns by fitting ungulates with GPS collars might be adapted to keep track of students during field trips!

Geomorphological mapping

Where acquisition of elevation data are not critical, GPS can be an effective mapping tool. The outline of geomorphological features (e.g. the edge of river terraces or landslides) or point patterns (e.g. glacial erratics) can be obtained speedily in terrain where conventional surveying is impractical and the features cannot be determined sufficiently from aerial photography. GPS software has a facility to tag attribute information to the data and can be integrated into GIS. It should be remembered that the GPS receiver requires an unobstructed path to the satellites and hence mapping in mountainous terrain, urban environments or under forest cover can be problematic.

In each of the categories above, the speed and frequency of positioning is the essential difference enabling GPS to provide data that would be difficult or impossible to derive from conventional surveying methods. As such, GPS is not a technique producing completely new data but rather an application that improves accuracy and/or frequency of measurement coupled with efficient data-processing capability. The cost of GPS receivers spans at least two orders of magnitude. High precision GPS is not only expensive but requires a thorough understanding of surveying principles and the equipment can be bulky. Mapping grade GPS is highly portable and can be operated by a single user where safety considerations allow. The required accuracy should dictate the specification of GPS but its subsequent application in geomorphology is wide-ranging.

References

Brasington, J., Rumsby, B.T. and McVey, R.A. (2000) Monitoring and modelling morphological change in a braided gravel-bed river using high resolution GPS-based survey, *Earth Surface Processes and Landforms* 25, 973–990.

Cornelius, S.C., Sear, D.A. and Craver, S.J. (1994) GPS, GIS and geomorphological field work, *Earth Surface Processes and Landforms* 19, 777–787.

Fix, R.E. and Burt, T.P. (1995) Global Positioning System: an effective way to map a small area or catchment, *Earth Surface Processes and Landforms* 20, 817–828.

Gillieson, D.S., Cochrane, J.A. and Murray, A. (1994) Surface hydrology and soil movement in an arid karst – the Nullabor Plain, Australia, *Environmental Geology* 23, 125–133.

Higgitt, D.L. and Warburton, J. (1999) Applications of differential GPS in upland fluvial geomorphology, *Geomorphology* 31, 411–439.

Hofmann-Wellenhof, B., Lichtenegger, H. and Collins, J. (2001) *GPS: Theory and Practice*, 5th edition, Heidelberg: Springer-Verlag.

Wilson, J.P., Spangrud, D.J., Nielsen, G.A., Jacobsen, J.S. and Tyler, D.A. (1998) Global positioning system sampling intensity and pattern effects on computed topographic attributes, *Soil Science Society of America Journal* 62, 1,410–1,417.

DAVID HIGGITT

GRADE, CONCEPT OF

Since the end of the seventeenth century (Dury 1966; Chorley 2000), various engineers have been concerned both with the regulation of natural rivers and with the operation and construction of artificial channels. This required an interest in geometric stability or equilibrium (grade) which tended to be roughly constant or subjected to limited oscillation over a recognized period of time. Such stability could arise from some sort of balance between, for example, fluid shear stress and material resistance, or some equalization between those sedimentary processes (e.g. cut-and-fill) which control channel morphology. The concept entered mainstream geomorphology through G.K. Gilbert (1877), for whom the major geometrical evidence of the graded state was a smooth, concave-up river long profile. Grade was also accommodated within the Davisian cycle, and for Davis (1902) the elimination of breaks of slope was the hallmark of the graded condition. A major contribution to understanding the concept of grade was made by Mackin (1948) who defined a graded river as: 'one in which, over a period of years, slope and channel characteristics are delicately adjusted to provide, with available discharge . . . just the velocity required for the transportation of the load supplied from the drainage basin'.

In 1965 Schumm and Lichty introduced the concept of a time-span intermediate between the longer interval of 'cyclic time' and the shorter period of 'steady time'. They defined graded time as 'a short span of cyclic time during which a graded condition or dynamic equilibrium exists'. Later, Schumm (1977) saw a graded stream as 'a process-response system in steady-state equilibrium, and the equilibrium is maintained by self-regulation or negative feedback, which operates to counteract or reduce the effects of external change on the system so that it returns to an equilibrium condition'.

References

Chorley, R.J. (2000) Classics in physical geography revisited, *Progress in Physical Geography* 24, 563–578.

Davis, W.M. (1902) Base level, grade and peneplain, *Journal of Geology* 10, 77–111.

Dury, G.H. (1966) The concept of grade, in G.H. Dury (ed.) *Essays in Geomorphology*, 211–233, London: Heinemann.

Gilbert, G.K. (1877) *Report on the Geology of the Henry Mountains*, Washington, DC: US Geological Survey.
Mackin, J.H. (1948) Concept of the graded river, *Geological Society of America Bulletin* 59, 463–512.
Schumm, S.A. (1977) *The Fluvial System*, New York: Wiley.
Schumm, S.A. and Lichty, R.W. (1965) Time, space and causality in geomorphology, *American Journal of Science* 263, 110–119.

A.S. GOUDIE

GRADED TIME

The most concise description of graded time is derived from Mackin (1948):

> A graded river is one in which, over a period of years, slope and channel characteristics are delicately adjusted to provide, with available discharge, just the velocity required for the transportation of the load supplied from the drainage basin. The graded stream is a system in equilibrium; its diagnostic characteristic is that any change in any of the controlling factors will cause a displacement of the equilibrium in a direction that will tend to absorb the effect of the change.

It is clear that 'graded time' is the time over which a stream is in balance in this way.

A second, less useful, term is the 'time to grade' (see GRADE, CONCEPT OF) or the time that it takes for a river to attain a graded condition. This is not a simple idea because the graded condition is not reached throughout all parts of a system at the same time. In rivers, for example, W.M. Davis (1902) stated that grade would be attained first in the lower reaches and then extend upstream. It would also be attained first in the most adjustable materials.

The term, graded time, therefore has a spatial dimension. Davis (1899) stated that when the trunk streams were graded the stage of early maturity had been reached, when the smaller headwaters were graded maturity was well advanced and when even the wet river rills and the waste mantle were graded the stage of old age had been attained.

The idea that once grade had been achieved the balance of forces, sediment loads and forms would remain adjusted even though the landscape was still being slowly lowered has always been an uncomfortable element of the geographical cycle. The timeless aspect of the concept of grade does not sit happily within the timebound cyclical framework.

References

Davis, W.M. (1899) The Geographical Cycle, *Geographical Journal* 14, 481–504.
——(1902) Base level, grade and peneplain, *Journal of Geology* 10, 77–111.
Mackin, J.H. (1948) Concept of the graded river, *Geological Society of America Bulletin* 59, 463–512.

DENYS BRUNSDEN

GRANITE GEOMORPHOLOGY

Granite terrains of the world, whether in lowland, upland or mountain settings, often have distinctive morphology, different from one typical for the surrounding country rock. Although it would probably be impossible to find a landform endemic for granite, many are most prominent if bedrock is granitic. Examples include boulders, TORS, INSELBERGS, BORNHARDTS, INTERMONTANE BASINS, and a range of microforms such as WEATHERING PITS or TAFONI (Twidale 1982). They usually form through selective bedrock weathering, either in subsurface (see DEEP WEATHERING) or at the topographic surface, followed by evacuation of the loose products of rock disintegration. However, there is no 'standard' granite landscape, as these can be significantly different, even if located adjacent to each other. Granite is known to support extensive plains of extreme flatness and, by contast, high-mountain, highly dissected terrains. In spite of widespread presence of a weathering mantle, bedrock frequently crops out at the topographic surface and tors and boulder fields are characteristic landmarks. Granite is typically, but by no means universally, more resistant to weathering and erosion than surrounding country rock, and therefore tends to form upland terrains and to support topographic steps.

Lithological and structural properties of granite, such as mineral composition, texture and joint density, which are often highly variable within a single granite intrusion, are the keys to understanding the selectivity of weathering and the prominence of many small- and medium-scale granite landforms.

Granites are usually fairly regularly jointed according to an orthogonal pattern, i.e. they are cut by three subsets of joints perpendicular to themselves, which delimit cuboid block compartments. As fractures guide movement of ground water through the rock mass, weathering acts most efficiently along joints and preferentially attacks the sides and edges of joint-bound cubes, which results in their progressive rounding and the typical multi-convex appearance of many granite landscapes. In the subsurface, weathering attack along joints transforms sharp-edged blocks into rounded core stones surrounded by a thoroughly disintegrated mass. Furthermore, because of variable joint density over short distances (< 10 m) significant differences in the intensity of rock disintegration may occur. Less fractured parts are left standing as rock pillars or castellated tors, whereas adjacent more closely jointed compartments are disintegrated into block rubble or GRUS. Evacuation of weathered material reveals a range of topographically negative features, common for granite areas. These include rock basins developed either at joint intersections or between master joints, and linear joint-guided valleys.

Many post-orogenic granites are typically very massive, with large-scale SHEETING joints being dominant. Joint spacing in such granites can be extremely wide, more than 10 m apart. In these areas topography usually follows the curvature of sheeting planes, bornhardts are common, and minor weathering features on rock surfaces often grow to gigantic dimensions.

Rock texture is equally important. Coarse variants of granite with abundant large phenocrysts of K-feldspar usually support a varied, rough relief, with big boulders, inselbergs, and intervening basins. The majority of domed inselbergs and bornhardts seems to be built of massive, coarse-grained granite. Likewise, minor features on rock surfaces are best developed within coarse granite. Finer variants tend to give rise to a more subdued topography, often with frequent angular tors.

Another factor important for the development of granite topography is mineralogical and chemical composition of the rock, including proportions between quartz and feldspar, between different types of feldspar, silica content, and proportions between potassium, sodium and calcium. Potassium-rich granites tend to be more resistant and therefore often form higher ground and give rise to spectacular inselberg landscapes, whereas granites with high plagioclase content typically underlie gently rolling terrains and low ground (Brook 1978; Pye *et al.* 1986).

In areas, where high precipitation and humidity levels favour deep weathering, subsurface decomposition of granite becomes crucial in the evolution of topography (Twidale 1982). Granite terrains, except for those in arid areas or in high mountains, usually carry a spatially extensive, thick mantle of weathering residuals. A very wide range of thicknesses has been reported, from only a few to as much as 200–300 m (Ollier 1984). There are different types of weathering mantles developing on granite, but rather shallow grus and more advanced geochemically, kaolinite-rich covers are most typical. This division likely reflects environmental conditions during weathering, including climatic conditions, their change through time and geomorphic stability of the surface. What both categories of granite weathering mantles have in common though, is the rough topography of the weathering mantle/bedrock interface (i.e. WEATHERING FRONT), attributable to the selectivity of deep weathering, and the usually sharp nature of this boundary. Therefore, stripping of the pre-weathered material often reveals complicated bedrock topography, with numerous low domes, isolated boulders and tor-like bedrock projections separated by basins and linear hollows. Indeed, many granite landscapes are interpreted to be the product of two-stage development, with the phase, or phases, of deep selective weathering followed by stripping and exposure of weathering front topography (Plate 56). The presence of bornhardts, tors and rounded boulders is occasionally used to infer the two-stage evolution, even if no remnants of any weathering mantle are left and no independent evidence exists that such ever existed. However, examples from areas with a long history of aridity such as the Namib Desert demonstrate that deep weathering is not a necessary precursor to the development of multi-convex granite topography, which primarily reflects structural control (Selby 1982).

The majority of detailed studies has concentrated on prominent medium-scale landforms such as boulder fields, tors, bornhardts and inselbergs, and pediments, or distinctive minor features of rock surfaces. Analyses of entire landform assemblages and their evolution through time are fewer. One of the attempts has been made by Thomas (1974) who distinguished multi-concave, multi-convex and stepped or

(a)

(b)

Plate 56 Granite landscapes share many of their characteristics regardless of the climatic zone in which they occur. Both the granite landscape of (a) the Erongo massif in arid Namibia and (b) the humid Estrela Mountains in central Portugal are dominated by massive domes, big rounded boulders scattered around and basins formed through selective joint-guided weathering

multistorey landscapes. In multi-concave terrains, topographic basins of various sizes are dominant features. Their occurrence may be related to either inliers of less resistant granite or to the occurrence of initially more jointed rock compartments (Thorp 1967; Johansson *et al.* 2001). Multi-convex terrains are those typified by closely spaced domes, or similar upstanding rock masses, so there is little space left for basins to develop. They are common in homogeneous, poorly jointed intrusions, where lines of structural weakness available for exploitation by weathering are few. Another type of multi-convex landscape is one dominated by low hills weathered throughout, possibly with a solid rock core.

Stepped landscapes are characterized by the presence of topographic scarps separating successive levels or 'storeys'. They typically occur in areas subjected to recent, but moderate uplift which was proceeding concurrently with weathering and stripping. Since the scarps are apparently not tectonically controlled, it is proposed that they form due to reduced rates of advance of the weathering front at progressively higher topographic levels, whereas their exact location reflects the occurrence of a more massive granite (Wahrhaftig 1965; Bremer 1993). In each of these landscape types, spatial patterns of individual landforms are largely controlled by lithology and structure.

Specific landform assemblages typify ring complexes, made of concentrically arranged intrusions of granite and other rocks, differing in mineralogy and texture, and intersected by dykes. Depending on the susceptibility of particular complex-forming rock units to weathering and erosion, a concentric pattern of uplands alternating with basins develops. Most resistant dykes form linear ridges, sculpted into jagged rock crests.

In addition, granite landscapes may take the form of a plain, either rock-cut or deeply weathered, as it is common in Australia, parts of Africa, or Scandinavia. Within strongly uplifted and highly dissected areas, a mountainous all-slope topography evolves (Twidale 1982). In both cases, structural control is less obvious and its influence surpassed by the high efficacy of planation or dissection.

Granite geomorphology has played an important part in CLIMATIC GEOMORPHOLOGY, and especially in the attempts to use specific landforms as indicators of specific climatic conditions. For instance, claims have been made that granite domes evolve in the humid tropics, boulder heaps are more typical for seasonally dry areas, whereas small-scale flutings indicate hot and humid conditions (Wilhelmy 1958). Moreover, the apparent durability of granite and its ability to withstand high compressive and tensile stresses have been used to support the claim that granite landforms, once formed under distinctive environmental conditions, may survive many subsequent environmental changes. In some Central European studies, minor granite landforms have been used to establish the chronology of denudation and environmental change since the mid-Tertiary. Increasing recognition of pervasive structural and lithological control on the evolution of granite landforms, as well as of the crucial role of subsurface weathering, have seriously undermined the basis of the climatic approach to granite geomorphology. At present, a consensus appears to have been reached that the evolution and appearance of granite landscapes are primarily controlled by structure, and similarities of structures explain why granite landform assemblages in contrasting geographical settings often look very much the same.

Many of the geomorphologically classic landforms and landscapes are underlain by granite. Examples include the tors of Dartmoor in southwest England, domes and U-shaped glacial valleys of the Yosemite National Park in Sierra Nevada, USA, sugar-loaf hills in Rio de Janeiro, African inselberg landscapes of Nigeria, Kenya and Namibia, fluted coastal outcrops in the Seychelles, and the Wave Rock in Western Australia.

References

Bremer, H. (1993) Etchplanation, review and comments of Büdel's model, *Zeitschrift für Geomorphologie N.F., Supplementband* 92, 189–200.

Brook, G.A. (1978) A new approach to the study of inselberg landscapes, *Zeitschrift für Geomorphologie N.F., Supplementband* 31, 138–160.

Johansson, M., Migoń, P. and Olvmo, M. (2001) Joint-controlled basin development in Bohus granite, SW Sweden, *Geomorphology* 40, 145–161.

Ollier, C.D. (1984) *Weathering*, London: Longman.

Pye, K., Goudie, A.S. and Watson, A. (1986) Petrological influence on differential weathering and inselberg development in the Kora area of Central Kenya, *Earth Surface Processes and Landforms* 11, 41–52.

Selby, M.J. (1982) Form and origin of some bornhardts of the Namib Desert, *Zeitschrift für Geomorphologie N.F., Supplementband* 26, 1–15.

Thomas, M.F. (1974) Granite landforms: a review of some recurrent problems of interpretation, in *Institute of British Geographers, Special Publication* 7, 13–37.

Thorp, M. (1967) Closed basins in Younger Granite Massifs, northern Nigeria, *Zeitschrift für Geomorphologie N.F., Supplementband* 11, 459–480.

Twidale, C.R. (1982) *Granite Landforms*, Amsterdam: Elsevier.

Wahrhaftig, C. (1965) Stepped topography of the southern Sierra Nevada, *Geological Society of America Bulletin* 76, 1,165–1,190.

Wilhelmy, H. (1958) *Klimamorphologie der Massengesteine*, Braunschweig: Westermann.

Further reading

Gerrard, J. (1986) *Rocks and Landforms*, London: Unwin Hyman.

Godard A., Lagasquie, J.-J. and Lageat, Y. (2001) *Basement Regions*, Berlin: Springer.

Lageat, Y. and Robb, L.J. (1984) The relationships between structural landforms, erosion surfaces and the geology of the Archaean granite basement in the Barberton region, Eastern Transvaal, *Transactions Geological Society of South Africa* 87, 141–159.

Twidale, C.R. (1993) The research frontier and beyond: granitic terrains, *Geomorphology* 7, 187–223.

PIOTR MIGOŃ

GRANULAR DISINTEGRATION

Granular disintegration is the physical disintegration of rock into individual grains and rock crystals. The product of granular disintegration is usually coarse-grained, loose debris, which can be easily removed by erosive agents such as wind, water and gravity. This form of rock breakdown occurs commonly in coarse-grained rocks such as sandstone, dolerite and granite. Clay-rich rocks are thought to be particularly susceptible (Smith *et al.* 1994).

The surface grains and rock crystals which become detached may be unweathered and unaltered. They may also remain *in situ* but could be easily removed by light brushing with the hand. Where the product of granular disintegration remains *in situ* and accumulates, a gritty SAPROLITE is produced, known as GRUS. When loose material is removed, the fresh surface beneath may be pitted and uneven. The loose material may accumulate as a sandy deposit.

There are a number of mechanical and chemical mechanisms of granular disintegration and it is likely that the process can be attributed to several or all of these. It is equally likely that more than one of the mechanisms operates simultaneously in many cases:

1 *Solution of soluble cement* Sandstones cemented by soluble calcareous material are particularly susceptible to granular disintegration due to this mechanism.
2 *Stress induced by volumetric expansion* The growth of salt and ice crystals leads to a volumetric expansion. Under certain conditions, this can produce sufficient force to rupture the rock and this is most likely to occur at locations of weakness such as grain boundaries. There is ample evidence that rocks readily disintegrate in salt-rich environments due to salt crystallization (Evans 1970). Experimental work has also shown salt to be extremely effective in the

physical breakdown of rock (e.g. Goudie *et al.* 1970). Chemical weathering processes such as hydrolysis may involve expansion of minerals sufficient to produce crystal fracture.

3 *Release of residual stress* This is stress in rock due to primary crystallization or lithification. These stresses exist in a balanced state in unweathered rock. However, residual stresses can become unbalanced, and therefore released, by erosion, weathering and mass movement. The stresses generated can be large enough to cause crack propagation (e.g. Bock 1979).

4 *Water adsorption* Repeated wetting and drying may be responsible for the disintegration of fine-grained rocks such as mudstone (see SLAKING). Water molecules are absorbed onto mineral surfaces and may produce force sufficient to prise particles apart.

References

Bock, H. (1979) A simple failure criterion for rough joints and compound shear surfaces, *Engineering Geology* 14, 241–254.

Evans, I.S. (1970) Salt crystallisation and rock weathering: a review, *Revue de Géomorphologie Dynamique* 19, 153–177.

Goudie, A.S., Cooke, R.U. and Evans, I.S. (1970) Experimental investigation of rock weathering by salts, *Area* 2, 42–48.

Smith, B.J., Magee, R.W. and Whalley, W.B. (1994) Breakdown patterns of quartz sandstone in a polluted urban environment, Belfast, Northern Ireland, in D.A. Robinson and R.B.G. Williams (eds) *Rock Weathering and Landform Evolution*, 131–150, Chichester: Wiley.

Further reading

Cooke, R.U. (1981) Salt weathering in deserts, *Proceedings of the Geologists' Association* 92, 1–16.

Yatsu, E. (1988) *The Nature of Weathering: An Introduction*, Tokyo: Sozosha.

DAWN T. NICHOLSON

GRAVEL-BED RIVER

An alluvial river in which the average diameter of bed materials exceeds 2 mm. An upper grain-size limit is seldom identified, but channels with beds predominantly composed of very large, essentially immobile boulders (>256 mm) may be regarded as a distinct type or subcategory, especially if they exhibit step-pool morphology (see STEP-POOL SYSTEM). The primary distinction is with SAND-BED RIVERS (bed material 0.063–2 mm). Gravel-bed rivers dominate in upland and piedmont settings where the sediment supplied to the channel is coarse and poorly sorted. With distance downstream, bed materials become smaller (see DOWNSTREAM FINING) and an abrupt gravel–sand transition often terminates the gravel reach.

Although gravel-bed rivers transport significant quantities of sand, much of it in suspension, the proportion of BEDLOAD transport is apt to be higher than in sand-bed channels. Parker (in press) usefully defines a limit case wherein the median bed-particle size is greater than 25 mm and bedload transport dominates. Examination of such channels reveals that the boundary shear stresses generated by modest flows (for example, bankfull) are barely capable of moving median grain-sizes. In sharp contrast to sand-beds, gravel-bed channels are therefore characterized by hydraulic stresses that rarely exceed the entrainment thresholds of particles exposed at the bed surface, and large floods are required to generate significant sediment transport. Sediment yield is limited by the competence of flows to move the coarse load, rather than the availability of mobile sediments per se. This is a definitive characteristic of gravel-bed rivers, though in many environments vertical sorting of the bed material (ARMOURING) does significantly limit the availability of potentially mobile, subsurface sediments.

Close to the threshold for motion, the coarse armour layer remains intact and transport involves individual grain movements across its largely unbroken surface. Once rotated out of bed pockets by lift and drag forces, particles roll and bounce across the bed, intermittently stopping in stable positions from where they may be entrained again by instantaneous turbulent stresses. Particles cover relatively short distances, potentially falling into stable interstices or pockets in the armour layer. This marginal transport regime dominates during most floods, with the armour layer moderating sediment supply and grain velocities. However, as flow intensity increases, larger areas of the armour layer are breached, the number of particles in motion rises and, during exceptional floods, most of the bed may be mobile. Even during mass transport, flows are seldom sufficiently deep to form mobile bedforms of the geometry found in sand-bed channels (steep ripples and dunes), but low-amplitude forms known as gravel sheets are common, and their passage generates bedload pulses.

A number of small-scale bedforms are recognized in gravel-bed rivers and are important because they influence near-bed hydraulics and, like bed armour, moderate sediment supply and entrainment. Pebble clusters that form when large obstacle clasts distort the flow and impede the passage of other clasts, are repeating, streamlined features. Transverse ribs are regularly spaced, linear ridges of coarse clasts that form perpendicular to the flow under supercritical conditions. Stone cells are reticulate structures that may form where transverse ribs and pebble clusters intersect.

These micro-bedforms protrude above the bed surface and therefore contribute to overall flow resistance, as do channel-scale grain accumulations (bars and riffles) that retard the passage of water. However, in contrast to sand-bed channels where bedforms dominate boundary resistance, grain roughness is regarded as the dominant component in gravel-bed rivers (see ROUGHNESS).

The longitudinal profiles of gravel-bed rivers typically exhibit significant concavity that reflects adjustment to downstream fining and the associated reduction in competence required to transport a given load. Channel gradients therefore vary significantly from as much as 0.1 to as little as 0.001.

Cross-sectional form is determined by numerous variables in addition to bed-material size. Nevertheless, for a given discharge gravel-bed rivers do tend to be shallower than sand-bed channels and have higher width–depth ratios. This reflects the dominance of bedload transport and a lack of fine-grained, floodplain deposition, that together promote lateral instability and channel widening. Local variability of width and depth is exacerbated in gravel-bed rivers by well-developed riffle-pool sequences. Analogous bed topography is evident in some sand-bed channels, bedrock channels, and as step-pools in boulder-bed channels, but riffle-pools are best developed in gravelly channels with heterogeneous bed materials.

Gravel-bed rivers may be straight, meandering, anabranching (see ANABRANCHING AND ANASTOMOSING RIVER) or braided (see BRAIDED RIVER). To the extent that bedload transport dominates, and stabilizing, cohesive, floodplain sediments are lacking, gravel-bed channels tend to exhibit larger meander wavelengths and a propensity to wander or braid. Wandering is a type of anabranching that represents a transitional stage between meandering and braiding, with some sinuosity, low-level braiding and stable mid-channel islands. Wandering channels tend to have lower slopes and less abundant bedload than fully braided channels. Relative to sand-beds, gravel-bed channels require steeper slopes to generate full braiding.

Bed material size is a fundamental control of river form and function, and characteristic morphological and process attributes do justify the general distinction that is made by geomorphologists between gravel- and sand-bed rivers. The presence of a gravel–sand transition in many rivers and the widely reported deficiency of fluvial sediments in the range 1 to 4 mm – the so-called 'grain-size gap' – reinforce this binary categorization. However, all gravel-bed rivers contain sand, and many gravel-bed rivers transport larger volumes of sand than gravel. The sand is apparent to varying degrees as a patchy surface veneer (for example in pools) and in the subsurface matrix. This suggests that the two-fold, sand versus gravel, classification is rather simplistic and potentially limiting. It may obscure important attributes that are peculiar to channels containing particular mixtures of sand and gravel. Indeed, there is increasing evidence that understanding channel hydraulics, sediment transport and the formation of fluvial deposits in 'gravel-bed' rivers depends upon explicit recognition that bimodal gravel and sand mixtures often dominate the bed materials (e.g. Sambrook-Smith 1996).

References

Parker, G. (in press) Transport of gravel and sediment mixtures, in *Sedimentation Engineering*, American Society of Civil Engineers, Manual 54.

Sambrook-Smith, G.H. (1996) Bimodal fluvial bed sediments: origin, spatial extent and process, *Progress in Physical Geography* 20, 402–417.

Further reading

Simons, D.B. and Simons, R.K. (1987) Differences between gravel- and sand-bed rivers, in C.R. Thorne,

J.C. Bathurst and R.D. Hey (eds) *Sediment Transport in Gravel-bed Rivers*, 3–15, Chichester: Wiley.
Kleinhans, M.G. (2002) *Sorting Out Sand and Gravel: Sediment Transport and Deposition in Sand-gravel Bed Rivers*, Netherlands Geographical Studies 293, Utrecht: Royal Dutch Geographical Society.

SEE ALSO: armouring; bedload; downstream fining; roughness; sand-bed river

STEPHEN RICE

GRÈZE LITÉE

Stratified TALUS deposits displaying well-developed cm-thick beds and composed of small angular clasts (Guillien 1951). They have also been referred to as éboulis ordonnés and stratified screes, though some authors find slight differences between these terms (mostly related to the slope gradient and the mean clast size). The most diagnostic features of grèzes litées are (1) the internal structure of the deposit, organized in parallel beds of around 10 to 25 cm thick, and (2) the small size of the clasts as a result of very frequent freezing and thawing cycles on a gelivable (frost susceptible) rock substratum.

The sedimentary structure shows alternating matrix-rich (matrix-supported) and openwork (clast-supported) beds. In many cases openwork beds show fining upward textures. In longitudinal sections the base of the matrix-rich beds is affected by festoons, with increasing size downslope, and even with the development of lobate fronts (Bertran *et al.* 1992). Undulations are relatively frequent in frontal sections. The presence of blocks within the clast-supported beds defines another more heterogeneous type of talus deposit called groizes litées. In carbonate-rich deposits the presence of carbonate cemented crusts is relatively frequent, as a result of percolation and water circulation.

Most authors consider that grèzes litées are better developed in limestone areas, at the foot of large vertical or sub-vertical cliffs. This is the case of Charentes (France), where the best examples have been studied, and many other localities in the Alps and the Pyrenees. However, they have also been described in crystalline, volcanic and metamorphic rock areas (for instance, in the Chilean Andes, Vosges, the French Central Massif and the Atlas in Morocco). These deposits can be up to 40 m thick. In all cases the grèzes litées are most frequent in middle latitudes, with a periglacial climate. In these regions the annual number of freezing and thawing cycles is high (even more than 200 days per year), providing the best conditions to break down the rocks and to accumulate large volumes of debris. Under these conditions the cliffs erode backward rapidly and they are partially fossilized by the grèzes litées.

Grèzes litées have been described in a wide range of slope gradients (between 5 and 35°), though, in general, gentler than in ordinary talus with non-stratified screes, thus excluding an origin based only on gravity. Most active grèzes litées are located on sunny aspects or in snow-free hillslopes. Pleistocene deposits, related to former cold-climate phases, are located in almost any aspect, depending not only on the altitude but also on local topography and wind direction (García-Ruiz *et al.* 2001).

Several hypotheses have been used to explain the development of grèzes litées. Tricart and Cailleux (1967) stressed the importance of in-mass transport (especially solifluction) accompanied by pipkrake (needle-ice) activity. Bertran *et al.* (1992) and Francou (1988) confirm the decisive role of continual burial of stone-banked sheets. This implies the existence of large solifluction sheets in which pipkrakes cause a vertical sorting of the material, displacing the clasts towards the surface. The movement of the front of the stone-banked sheets produces the accumulation of continuous layers of clast-supported and matrix-supported beds. The slow mass movement is responsible for the occurrence of frontal and lateral festoons and undulations. The presence of debris flows also contributes to the characteristic alternating structure (Van Steijn *et al.* 1995), though the limits and continuity of the beds can be poorly developed. Slopewash processes are almost completely excluded as the main mechanism, since most of the rock fragments are oriented parallel to the slope gradient. Furthermore, the absence of rills, longitudinal sorting and cross-bedding suggests the inability of overland flow to redistribute the debris along the talus.

References

Bertran, P., Coutard, J.P., Francou, B., Ozouf, J.C. and Texier, J.P. (1992) Données nouvelles sur l'origine du litage des grèzes: implications paleoclimatiques, *Géographie Physique et Quaternaire* 46, 97–112.

Francou, B. (1988) Éboulis stratifiés dans les Hautes Andes Centrales du Pérou, *Zeitschrift für Geomorphologie* 32, 47–76.
García-Ruiz, J.M., Valero, B., González-Sampériz, P., Lorente, A., Martí-Bono, C., Beguería, S. and Edwards, L. (2001) Stratified scree in the Central Spanish Pyrenees: palaeoenvironmental implications, *Permafrost and Periglacial Processes* 12, 233–242.
Guillien, Y. (1951) Les grèzes litées de Charente, *Revue Géographique des Pyrénées et du Sud-Ouest* 22, 153–162.
Tricart, J. and Cailleux, A. (1967) *Le modelé des régions périglaciaires*, Paris: SEDES.
Van Steijn, H., Bertran, P., Francou, B., Hétu, B. and Texier, J.P. (1995) Models for the genetic and environmental interpretations of stratified slope deposits: review, *Permafrost and Periglacial Processes* 6, 125–146.

JOSÉ M. GARCÍA-RUIZ

GROUND WATER

Groundwater processes

Ground water is an important source of water for domestic use, irrigation and industrial uses and concern about the quantity and quality of ground water withdrawals is global in nature. It is a critical link in the hydrologic cycle, as it is a major source of water in rivers and lakes. Ground water is water under positive (greater than atmospheric) pressure in the saturated zone. The fluctuating water table marks the upper boundary of saturation in unconfined aquifers. Recharge can occur by infiltration of rainwater or snowmelt and by horizontal or vertical seepage from surface-water bodies. Ground water leaves the system by discharge into rivers, lakes or the ocean, by transpiration from deep-rooted plants, or by evaporation when the water table is close to the surface. Ground water is in continual motion, with velocities that are typically less than 1 m day^{-1}.

The most important geologic factors controlling the movement of ground water are lithology, stratigraphy and structure, and combinations of these conditions produce a great variety of groundwater flow patterns. The term aquifer is used to define a geologic unit that can store and transmit enough water to be hydrologically or economically significant. Layers of rock which are impermeable are termed aquicludes and semi-permeable rocks, which retard the flow, are termed aquitards. An unconfined aquifer is open to the atmosphere and its hydrostatic level is the water table. In a confined aquifer water is held between confining layers (aquitards or aquicludes) and is not vertically connected to the atmosphere.

In humid regions, ground water may be an important contributor to streamflow, with water entering the channel by effluent seepage to form the baseflow discharge. If ground water input is significant, the streams will be characterized by relatively low temporal flow variability. By contrast, in arid regions streamflow often percolates into permeable beds to contribute to the water table. Such streams are referred to as influent.

Ground water as a geomorphological agent

Ground water is a significant geomorphological agent in many environments, both arid and humid, hot and cold. It influences cave formation in karst terrains; water chemistry and surface morphology of playas or PANS; the erosion of rock faces and formation of alcoves and caves; cliff retreat and mass movement; and canyon growth by basal sapping processes. Ground water can impede wind erosion in arid areas where the water table lies close to the surface and can affect dune type. Over time, the role of ground water in geomorphic development is strongly affected by fluctuations in the height of the water table which result from climate change and human pumpage. Excessive groundwater withdrawal and falling water tables can lead to surface subsidence, vegetation death and dune mobilization.

The term KARST is given to limestone terrains that include such distinctive landforms as caves, springs, blind valleys and dolines. The dominant erosional process is dissolution and the region is typified by lack of surface water and the development of stream sinks or dolines. A unique pattern of drainage results from karst processes. Solution creates and enlarges voids, which then integrate to allow the transmission of large amounts of water underground, thereby promoting further solution. In karst areas underground drainage is developed at the expense of surface flow networks. Solution and weakening of silicic rocks to form karst-like topography has also been noted in arid environments, such as the Bungle Bungle of north-western Australia, but it is uncertain whether such landforms are wholly or partly inherited from more humid climatic periods.

Ground water plays a role in mass movement and channel formation by the process known as sapping. Concentrated seepage caused by ground water convergence is capable of slowly eroding materials at valley head or cliff bases, undermining overlying structures, and causing failure and headward retreat. The term spring sapping is often used when a point-source spring is involved, whereas seepage erosion may be employed where the groundwater discharge is less concentrated. Computer modelling suggests that scalloped escarpments develop where groundwater flow is diffuse, whereas elongation into channels or canyons results from higher and more concentrated seepage discharges, often associated with growth updip along fracture systems characterized by higher hydraulic conductivities. In rocks which are susceptible, chemical weathering renders the rocks even more permeable. Although many sapping networks, for instance those of coastal Italy and the Colorado Plateau, have developed in highly jointed bedrock, field research by Schumm *et al.* (1995) illustrates that similar networks can develop in highly permeable sands without significant structural controls.

Common morphological characteristics of valleys in which sapping plays a dominant role include amphitheatre-shaped headwalls, relatively constant valley width from source to outlet,

Plate 57 Long Canyon and Cow Canyon are tributaries to the Colorado River, developed in the Navajo Sandstone. The morphology of these valleys, with theatre-shaped heads and relatively constant valley width from source to outlet, is consistent with their formation by groundwater sapping

high and often steep valley sidewalls, a degree of structural control, short and stubby tributaries, and a longitudinal profile which is relatively straight (Plate 57). Similarly, simulation models (Howard 1995) of groundwater sapping produce canyons which are weakly branched, nearly constant in width and terminate in rounded headwalls.

Interest in groundwater outflow processes as an important factor in valley network development was stimulated by imagery of Mars which revealed features that were broadly similar in morphology to those on Earth (Laity and Malin 1985). It is now widely believed that many valleys on Mars were probably the result of erosion by groundwater sapping (Gulick 2001), although the actual mechanism is still subject to conjecture and debate. On Earth, an ever-increasing number of research papers illustrate that groundwater sapping is a global process, which can occur in a number of diverse lithologic and hydrologic settings. Valleys formed by sapping processes have been identified in Libya, Egypt, England, the Netherlands, the United States (Vermont; the Colorado plateau; Hawaii; Florida), New Zealand, Japan and Botswana. Valleys and escarpments which maintain the characteristic forms outlined above, but which lack modern seepage, may be relict from previously wetter climates (for instance, in Egypt). Other systems may include both active and relict components.

In addition to forming valleys by headward erosion, sapping at zones of groundwater discharge also contributes to the backwasting of scarps. Slopes are undermined and collapse owing to the removal of basal support by fluid flow which weakens rock at sites of concentrated seepage or diffuse discharge. These processes have received particular attention when considering scarp retreat in sandstone-shale sequences of the American south-west. Additionally, slopes of sandstone, granite, tuff or other massive rock form may be modified to include alveolar weathering or tafoni. The term 'dry sapping' has been applied to the formation of these features, for although the rock surfaces may be encrusted by salts, they do not appear to be damp. By contrast, larger alcoves formed by 'wet sapping' show wet surfaces on at least a seasonal basis (Howard and Selby 1994).

Boulders and inselberg landscapes in arid regions, when exposed by mantle stripping, are collectively referred to as etch forms. Such landforms may have developed over periods of 100 My or more and had an origin beneath deep mantles (tens or hundreds of metres in depth) which weathered as a response to ground water, under the control of geothermal heat. It has been proposed that the residual rock masses formed at the basal weathering front and were later exposed as the regolith was stripped.

Playa surfaces in deserts vary considerably owing to the range of unique hydrologic environments. Where ground water discharges seasonally or perennially onto the playa surfaces salt crystallization is characteristic and salt crusts of varying thickness form. The surface expression of groundwater discharge and salt crystallization includes extremely irregular micro-topography, polygonal forms and salt ridges. Solution phenomena such as pits and sinkholes may occur. Beneath the surface, the sediments are usually wet, soft and sticky. Spring mounds, elevated forms which may have a central pool, form where the water table is higher than the playa surface.

Landscape changes associated with groundwater overdraft

When more water is withdrawn from an aquifer by pumping than can be returned by natural recharge, the system is considered to be in overdraft. Such conditions have geomorphic impacts. Overdraft conditions have led to measurable SUBSIDENCE of the ground (one to ten metres) in such areas as Mexico City, Tokyo, Hanoi, the Central Valley of California, the Houston–Galveston area of Texas, and Las Vegas, Nevada. Surface geomorphic expression includes the development of fissure systems, such as those at Yucca dry lake, Nevada, where parallel fissures are as much as 2 km long and perhaps 500 m deep. In Bangkok, Thailand, subsidence averages 1.5–2.2 cm/yr, but has occurred at rates as high as 10 cm/yr, causing damage to buildings and infrastructure. As the city is almost at sea level, the most serious impact has been flooding at the end of the rainy season.

Ground water plays a significant role in aeolian and fluvial systems of deserts. In channel systems, ground water forced to the surface by faulting or bedrock may flow at the surface for short distances in zones marked by dense phreatophytic vegetation – plants whose root systems draw water directly from ground water, and which dominate the riparian habitat. Phreatophytes affect hydraulic roughness and depositional processes, and their loss owing to water table drawdown often precedes episodes of stream widening.

Dune systems also change in response to a decline in water table elevation. In a 'wet aeolian system' the water table lies at or close to the surface and various stabilizing agents, such as vegetation, deflation lags, or cements allow accumulation while the system remains active. Drawdown of the water table may lead to a change to a 'dry aeolian system', where neither the water table nor vegetation exerts any significant influence, and surface behaviour is largely controlled by aerodynamic configuration (Kocurek 1998). In the Mojave Desert of California, sand released from degrading nebkhas (vegetation-anchored dunes) has reaccumulated downwind in migrating sand streaks and barchan dunes. Problems to nearby settlement include dune encroachment and blowing dust episodes.

References

Gulick, V.C. (2001) Origin of the valley networks on Mars; a hydrological perspective, *Geomorphology* 37, 241–268.
Howard, A.D. (1995) Simulation modeling and statistical classification of escarpment planforms, *Geomorphology* 12, 187–214.
Howard, A.D. and Selby, M.J. (1994) Rock Slopes, in A.D. Abrahams and A.J. Parsons (eds) *Geomorphology of Desert Environments*, 123–172, London: Chapman and Hall.
Kocurek, G. (1998) Aeolian system response to external forcing factors – a sequence stratigraphic view of the Sahara Region, in A.S. Alsharhan, K.W. Glennie, G.L. Whittle and C.G. St C. Kendall (eds) *Quaternary Deserts and Climatic Change*, 327–337, Rotterdam: A.A. Balkema.
Laity, J.E. and Malin, M.C. (1985) Sapping processes and the development of theater-headed valley networks in the Colorado Plateau, *Geological Society of America Bulletin* 96, 203–217.
Schumm, S.A., Boyd, K.F., Wolff, C.G. and Spitz, W.J. (1995) A ground-water sapping landscape in the Florida Panhandle, *Geomorphology* 12, 281–297.

Further reading

Kochel, R.C. and Piper, J.F. (1986) Morphology of large valleys on Hawaii; evidence for ground water sapping and comparisons with Martian valleys, *Journal of Geophysical Research* 91, E175–192.
Laity, J.E. (in press) Ground water drawdown and destabilization of the aeolian environment in the Mojave Desert, California, *Physical Geography*.
Luo, W., Arvidson, R.E., Sultan, M., Becker, R., Crombie, M.K., Sturchio, N. and Zeinhom, E.A. (1997) Groundwater sapping processes, Western Desert, Egypt, *Geological Society of America Bulletin* 109, 43–62.
Péwé, Troy L. (1990) Land subsidence and earth-fissure formation caused by ground water withdrawal in Arizona, in C.G. Higgins and D.R. Coates (eds) *Ground Water Geomorphology; The Role of Subsurface Water in Earth-surface Processes and Landforms*, Geological Society of America Special Paper 252, 218–233.
Young, R.W. (1987) Sandstone landforms of the tropical East Kimberley region, northwestern Australia, *Journal of Geology* 95, 205–218.

SEE ALSO: canyon; etching, etchplain and etchplanation; karst; pan; sandstone geomorphology; spalling

JULIE E. LAITY

GROYNE

Groynes are shore-perpendicular structures that are emplaced to control sand movement along a beach by altering processes in the swash and surf zones and providing a physical barrier to sediment moved as littoral drift.

Groynes change patterns of wave-refraction, wave-breaking and surf-zone circulation, generate rip currents, trap sediments on the updrift beach, reduce sediment inputs to the downdrift beach and redirect sediment offshore. Geomorphic effects include creation of wider beaches with steeper foreshores on the updrift sides, narrower beaches with flatter foreshores on the downdrift sides, lobate deposition zones downdrift of the tips of the structures and pronounced breaks in shoreline orientation (Everts 1979). The locally wider updrift beaches can enhance aeolian transport, and the subaerial portions of the groynes can form effective traps for blown sand, increasing the potential for creation of dunes on their updrift side (Nersesian *et al.* 1992; Nordstrom 2000). However, shoreline recession rates may be greatly increased on the downdrift side of groynes (Everts 1979; Nersesian *et al.* 1992), leading to truncation of beaches and dunes and loss of habitat.

Shortening, lowering or notching of existing groynes or construction of permeable pile groynes or submerged groynes have been suggested to allow for some sediment to bypass the structures in order to reduce downdrift erosion rates. Permeable pile groynes can reduce the longshore current while eliminating the effect of the structurally induced rip current, creating a more linear shoreline than occurs with an impermeable groyne and creating an underwater terrace that can reduce the erosion potential of waves crossing it (Trampenau *et al.* 1996). Submerged groynes retain the original aesthetics of the landscape, and allow beach traffic to proceed unimpeded, but their effects have been poorly studied (Aminti *et al.* 2003). T-groynes, built with a short, shore-parallel seaward end, are favoured in some areas to reduce scour and redirect rip currents, thereby reducing unwanted sedimentation offshore, but they can leave the beach in the centre badly depleted (McDowell *et al.* 1993).

Groynes can be used to best advantage when they are located where (1) sediment transport diverges from a nodal region; (2) there is no source of sand, such as downdrift of a breakwater or jetty; (3) transport of sand downdrift is undesirable; (4) the longevity of BEACH NOURISHMENT must be increased; (5) an entire reach will be stabilized; and (6) currents are especially strong at inlets (Kraus *et al.* 1994). Groynes also have considerable recreational value for fishing because they create new habitat and provide access to deep water. Combined pier/groyne structures have been built to enhance this value.

New groynes or alterations to existing groynes are now often included in beach nourishment plans, but groynes have been banned or strongly discouraged in some management policies (Truitt *et al.* 1993; Kraus *et al.* 1994). Instances of removal of groynes have been reported (McDowell *et al.* 1993), but there is little documentation of the results on beach change. Alterations to groynes to allow for some bypass of sediment are more common than removal and are better documented (Rankin and Kraus 2003).

References

Aminti, P., Cammelli, C., Cappietti, L., Jackson, N.L., Nordstrom, K.F. and Pranzini, E. (2003) Evaluation of beach response to submerged groin construction at Marina di Ronchi, Italy using field data and a numerical simulation model, *Journal of Coastal Research*, Special Issue, in press.

Everts, C.H. (1979) Beach behaviour in the vicinity of groins – two New Jersey field examples, *Coastal Structures* 79, 853–867, New York: American Society of Civil Engineers.

Kraus, N.C., Hanson, H. and Blomgren, S.H. (1994) Modern functional design of groin systems, *Coastal Engineering: Proceedings of the Twenty-fourth Coastal Engineering Conference*, 1,327–1,342, New York: American Society of Civil Engineers.

McDowell, A.J., Carter, R.W.G. and Pollard, H.J. (1993) The impact of man on the shoreline environment of the Costa del Sol, southern Spain, in P.P. Wong (ed.) *Tourism vs Environment: The Case for Coastal Areas*, 189–209, Dordrecht: Kluwer Academic Publishers.

Nersesian, G.K., Kraus, N.C. and Carson, F.C. (1992) Functioning of groins at Westhampton Beach, Long Island, New York, *Coastal Engineering: Proceedings of the Twenty-third Coastal Engineering Conference*, 3,357–3,370, New York: American Society of Civil Engineers.

Nordstrom (2000) *Beaches and Dunes of Developed Coasts*, Cambridge: Cambridge University Press.

Rankin, K.L. and Kraus, N.C. (eds) (2003) Functioning and design of coastal groins: the interaction of groins and the beach: processes and planning, *Journal of Coastal Research*, Special Issue, in press.

Trampenau, T., Göricke, F. and Raudkivi, A.J. (1996) Permeable pile groins, *Coastal Engineering 1996: Proceedings of the Twenty-fifth International Conference*, 2,142–2,151, New York: American Society of Civil Engineers.

Truitt, C.L., Kraus, N.C. and Hayward, D. (1993) Beach fill performance at the Lido Beach, Florida groin, in D.K. Stauble and N.C. Kraus (eds) *Beach Nourishment: Engineering and Management Considerations*, 31–42, New York: American Society of Civil Engineers.

KARL F. NORDSTROM

GRUS

A product of *in situ* GRANULAR DISINTEGRATION of coarse-grained rocks characterized by its specific grain-size distribution, where the sand (0.1–2.0 mm) and gravel (> 2.0 mm) fraction predominate and may constitute up to 100 per cent of the total. The percentage of finer particles liberated by weathering is often negligible (Migoń and Thomas 2002). Thus, grus is not associated with any particular bedrock, although some rocks, e.g. mudstones, are unlikely to produce grus because of their grain-size composition. Granitic rocks, gneiss and migmatites are parent rocks that typically break down into grus.

The term is also used by sedimentologists to describe a product of accumulation of weathering-derived, poorly sorted, angular quartz and feldspar grains that have been subjected to very limited transport, usually towards the base of an outcrop. Such a sedimentary veneer of grus is particularly widespread in arid and semi-arid areas, where slope wash redistributes products of rock disintegration across PEDIMENTs.

Grus as the product of current superficial weathering of rock outcrops should not be confused with grus weathering mantles, which may be many metres thick and can be found in geological records. Grus saprolites may be defined as *in situ* weathering profiles, consisting almost entirely, or predominantly, of grus throughout, that grades into unweathered parent rock. Grus may occur at the base of a deep weathering profile and would represent a transitional stage in alteration of solid rock into a clayey weathering mantle, although there is evidence that many tropical deep weathering profiles do not have a basal zone of grusification and the transition zone is very thin. 'Arenaceous mantles' and 'sandy saprolites' are usually used as synonyms of grus mantles.

Grus saprolites are diversified in terms of their internal structure, depth and lithology. Many of them are homogeneous throughout, yet some contain frequent core stones, zones of more advanced breakdown along fractures or show sharp lateral or vertical contacts between weathered and unweathered parent rock. Core stones within grus profiles may be as large as 3–4 m across and be either closely spaced, separated by weathered fractures, or in isolation in an otherwise strongly disintegrated rock mass. Various depths of grus saprolites are reported and profiles more than 10 m deep are not uncommon. Mineralogical changes associated with grusification are usually slight and the content of secondary clay minerals in grus profiles is often insignificant (<2 per cent). Among clays, interstratified minerals, kaolinite, halloisyte and vermiculite are the most common. The occurrence of gibbsite is reported, but its percentage is usually low and is likely to represent a transitional stage in the formation of kaolinite.

The origin of deep grus saprolites is still unclear and several mechanisms have been suggested to be responsible for opening of microfractures within and between the grains in the near-surface zones (Pye 1985; Irfan 1996). Microfracturing results from de-stressing of quartz and feldspars during weathering and may be enhanced by expansion of biotite after its HYDRATION. Development of intergranular porosity in response to partial solution along grain boundaries and transgranular microcracks may be an important contributing agent. However, advanced chemical processes play rather a subordinate, if any, role as indicated by minor amounts of secondary clay, preservation of easily weatherable minerals such as biotite and plagioclase, and limited degree of corrosion of quartz and feldspar grains.

Grus mantles are particularly widespread in areas of temperate climate, but they in fact occur in a variety of climatic zones and may be found in every climatic regime, both humid and semi-arid (Migoń and Thomas 2002). In low latitudes they occur alongside products of more advanced alteration, such as ferrallitic saprolites, as for instance in south-east Brazil. This distribution contradicts the claim, often made in the past, that production of grus is primarily controlled by climatic conditions, and that it is specific for a humid temperate climate. Generalization is further inhibited by the possibility that many grus mantles are not the result of weathering under contemporary climatic conditions, but are inherited from a geological past and different climatic regimes, and by the fact that many grus profiles are evidently truncated.

From a geomorphological point of view, grus mantles occur in three major settings. First, they are common within elevated plateaux and uplands, beneath gentle upper slopes and along valley sides. Second, they occur in hilly and inselberg landscapes, but hills may either be weathered throughout into grus or have only their lower slopes underlain by a grus mantle. Third, they are associated with highly dissected mountain areas. In some subtropical mountains, watershed ridges, spurs and isolated hills are very often deeply weathered and only the cores of unweathered bedrock protrude as massive domes from the widespread saprolitic mantle (Thomas 1994).

The common association of thick grus mantles with areas of moderate to high relief, often with a recent history of uplift, across the world's morphoclimatic belts, implies that dissected terrains of moderate relief are particularly suitable for thick grus to develop. This is because of free drainage, strong hydraulic gradient, tensional stress and rock dilatation. On the other hand, surface instability prevents grus profiles from attaining geochemical and mineralogical maturity. It has therefore been proposed that the deep grus phenomenon is a response of weathering systems to rapid relief differentiation, whether by tectonics, erosion, or both, and associated enhancement of groundwater circulation, although it is not exclusive to such settings (Migoń and Thomas 2002).

References

Irfan, Y.T. (1996) Mineralogy, fabric properties and classification of weathered granites in Hong Kong, *Quarterly Journal of Engineering Geology* 29, 5–35.

Migoń, P. and Thomas, M.F. (2002) Grus weathering mantles – problems of interpretation', *Catena* 49, 5–24.
Pye, K. (1985) Granular disintegration of gneiss and migmatites, *Catena* 12, 191–199.
Thomas, M.F. (1994) *Geomorphology in the Tropics*, Chichester: Wiley.

Further reading

Dixon, J.C. and Young, R.W. (1981) Character and origin of deep arenaceous weathering mantles on the Bega batholith, Southeastern Australia, *Catena* 8, 97–109.
Lidmar-Bergström, K., Olsson, S. and Olvmo, M. (1997) Palaeosurfaces and associated saprolites in southern Sweden, in M. Widdowson (ed.) *Palaeosurfaces: Recognition, Reconstruction and Palaeoenvironmental Interpretation*, Geological Society Special Publication 120, 95–124.

SEE ALSO: granite geomorphology; weathering

PIOTR MIGOŃ

GULLY

'Gully' can refer correctly, if uncommonly, to clefts down cliffs and to several sorts of seafloor channels. Those uses stem logically from the root sense of a narrow passageway, following derivation from the Latin *gula,* meaning throat, the French *goulet,* meaning a narrow entry or passage (including into a harbour or bay), and the Middle English *golet,* or gullet.

Minor uses aside, however, *gully* predominantly denotes a small and narrow but relatively deeply incised stream course, difficult to cross or to ascend, for which words like valley and gorge are too grandiose. It ideally connotes a young cut, with steep sides and a steep headwall, that has been carved out of unconsolidated regolith, typically by ephemeral flow from rainstorms or meltwater. However, these are not required attributes and exceptions abound.

Gullies are very variable in terms of processes of initiation and growth, as well as conditions of substrate, vegetation and climate, so they vary greatly in appearance and can show distinct regional differences. Thus the literature contains diverse usages and many local synonyms, including DONGA, vocaroca, ramp and lavaka. Among the variations, gullies may be slit-like to lobate (expanded at the head end), and continuous or discontinuous (depending on whether or not the gully has become connected at grade to the main drainage system). They can grow downward from mid-hillslope positions or from the hill-toe up, or they can develop along valley floors. Most gullies have a relatively simple, single thalweg, but gully heads can split during headward retreat, thereby creating a dendritic shape, and once in a while two branches rejoin head-to-head, creating a ring valley around a central pedestal or hillock.

Gully has distinct but poorly defined connotations of size, and even vaguer implications of cross-sectional shape. A gully is bigger than a RILL, which is a small entrenched rivulet, small enough to be crossed by a wheeled vehicle or to be eliminated by ploughing. An ARROYO (or wadi or barranca) represents an entrenched stream, not necessarily very deep, that has a somewhat wider floor than a gully – one might hope to drive up an arroyo, but probably not up a gully. A gully has a greater width to depth ratio than a slot canyon, and is neither as deep nor as wide as a box canyon or a gorge (see GORGE AND RAVINE), all of which would also likely have rock walls, unlike typical gullies. A gully is ideally narrower and shallower than a ravine, although no size limits have been specified. Gullies and gorges can share equally steep and enclosing walls, but ravines can be more V-shaped in cross section. Floors of ravines are ideally less enclosed than floors of gullies, but need not offer easier access. Ravines can be cut in regolith or rock. Overall, gully and ravine overlap considerably (the French *ravine* explicitly includes both gullies and larger valleys). In popular usage, gullies, ravines and gorges are perhaps best separated by their implications of lethality: a fall into a gorge could easily be fatal, whereas only the terminally unlucky would die by falling into a gully, and falls into ravines are unpredictable. Gullies might therefore be considered to range approximately from 5 or 10 m long, 1 or 1.5 m wide, and about as deep, arguably up to the order of several hundred metres long, many tens of metres wide at the original ground level, and perhaps twenty or thirty metres deep. Use of 'gully' at the larger end of that spectrum seems most supportable when there is a gradation in size from similar but smaller gullies nearby.

Probably the most dramatic and common causes of gullies involve human misuse of the land, where sites are made vulnerable to erosion by deforestation or by a more general devegetation, via logging, burning, overgrazing, or establishment of fields (especially when on hillsides and when unterraced or ploughed down the slope rather than across it). Additional proximal causes related to humans include runoff along paths or tracks that run straight downhill, regrading of hillsides and diversion or concentration of runoff from roads or building sites uphill. Gullying can also be initiated when soil compaction tips the balance from infiltration to runoff, or following mass movement after a hillside is undercut or overloaded.

However, there are also many natural causes for gullying. Exceptionally intense or prolonged storms are a very common culprit in erosion, but critical increases in rainfall can also come about from such climate changes as increased annual rainfall, or increased storminess with no net increase in rainfall. Critical levels of devegetation can be reached by aridification or by natural fires.

Plate 58 Examples of gully erosion in hills of weak saprolite in Madagascar. (a) Intense gully erosion. Concave-up runoff profiles are replacing smoothly convex hill profiles that formed by infiltration and chemical weathering. (b) Gullies at various stages of evolution. The biggest gully has expanded headward up dip, through the ridge crest. Ridge-crest cattle trails attest to endemic overgrazing in Malagasy hills. Many of these gullies receive no runoff from upslope. (c) These long and narrow gullies apparently represent entrenchment of a pre-existing dendritic drainage system. (d) Erosion dominated by runoff generated within the gully

Gully erosion may also be initiated by natural slope collapse, varying in style from deep slumps to soil slips, which can be triggered by rainstorms, earthquakes or undercutting by springs or rivers. Another cause is the collapse of pipes (see PIPE AND PIPING), which are natural subsurface passageways through soils, enlarged by rapid drainage along burrows, root voids, interconnected soil pores and the like. Critical increases in runoff relative to infiltration may happen naturally due to the plugging of a soil with fines or (especially in laterites) hardening due to exposure and drying following devegetation. Gullying can also be caused by incision or headward extension of first-order streams, as a result of uplift of headwater regions, subsidence downstream, a fall in base level, an increase in runoff and discharge, or breaching of a dam or a sill.

Note that the causes that make a site vulnerable to gullying can be very different from the specific initiators of erosion, such as when fires remove vegetation, but erosion starts only when sufficiently intense or prolonged rains attack the site. Thus, gullies may easily have different proximal and ultimate causes, and causes may operate at the gully, or upstream or downstream. Thus also, many gullies show threshold-related behaviour rather than linear responses to processes. Overall, gullying tends to be diagnostic of recently disrupted or non-equilibrium landscapes, whether perturbed by natural or cultural agents.

Once erosion is initiated, it can continue by a variety of processes, and the processes may change during growth. Rain attack and slope wash on the walls can be surprisingly effective, as can spalling or collapse of the walls from multiple wetting/drying cycles. Stream incision along the gully floor and concomitant undercutting of sidewalls are major growth processes. Another very effective process is headward retreat of a waterfall at the head end, which is notably associated with gullies that formed due to a lowering of base level. However, some gullies form entirely from rain that falls within the gully itself and have little or no exterior catchment area and no streams or rivulets flowing into them. In some situations, such as in thick saprolite in Madagascar, once the gullies have cut deeply enough to intersect wet zones near bedrock, groundwater seepage out of the bases of the headwall and adjacent sidewalls keeps the bases of those parts of the walls moist and vulnerable to further erosion and spalling, hence causing a dramatic increase in growth around the head end (Plate 58). This in turn creates very distinctive lobate or teardrop shapes, with broad arcuate heads and tiny exits. In these instances, the ultimate size of a gully is determined by the limits of the supply of subsurface water rather than the limits of the watershed on the surface. Gullies that grow by seepage and sapping can in some situations cut far back into high ground, effectively becoming amphitheatre-headed valleys. On occasion, they can even grow backward through ridge crests, for example when layering in the regolith dips through a ridge and delivers GROUND WATER from one side of a hill to the other.

In most cases, the shape and position of a gully reflect its causes and growth processes. For example, collapse of pipes, downcutting along stream beds, and headward retreat by springs or waterfalls create very long gullies, whereas growth by seepage and sapping can lead to a lobate shape with a broad and arcuate headscarp. Gullies that have grown along tracks and paths may lie along the crests of hill spurs, if those provided the easiest routes up the hill.

Most valley-side gullies represent new or future extensions of drainage networks into hillsides. Thus gullies commonly cut deeply into high ground, but they typically start within it and rarely pass through it at a low level, unlike most gorges. However, arroyo-like valley-floor gullies (which could form in the floor of a gorge) typically represent renewed incision of pre-existing drainages. Because they form most easily in unconsolidated material, gullies are common not only in colluvium on hillsides and alluviated valley floors but also in loess, till, outwash, loose fine volcaniclastics, laterite, saprolite and anthropogenic fill. The low resistance to erosion of these materials means that long-lived gullies are most typically associated with infrequent flow, which gives gullies a common but non-causal and non-exclusive association with relatively dry climates. (In contrast, ravines are more likely to have resistant walls and permanent flow.) It also means that gullies can grow extraordinarily rapidly. Thick regolith is generally a prerequisite for impressively deep gullies.

Left to themselves, gullies will eventually stop expanding, albeit possibly only when they have consumed their entire upslope watershed or have run out of erosible material. At that time, it will either have become a box canyon or ravine, or, if still cut into regolith, the upper walls will crumble back and the floor and lower walls will fill in and be buried, and the whole will become overgrown. Such a gully will become less angular and more rounded, with smoothly concave longitudinal and transverse profiles. Most will ultimately evolve into minor hillside hollows or reentrants.

Possibilities for remediating a gully include diverting drainage away from its head end; revegetating its walls and floors and/or the surrounding hillside; contour ploughing the surrounding hillside to promote infiltration rather than runoff; establishing small sediment-trapping dams along the floor, filling it in, and regrading the entire hillside. Nevertheless, the remedy should be matched to the specific causes and growth processes dominating each gully: disturbance of the surface (e.g. contour ploughing) will not be helpful if removal of vegetation is the critical initiator of erosion, and diversion of surface drainage may merely create a bigger problem somewhere else. If the gully is growing by sapping at the base of the walls, then work on the upper walls and the surrounding hillside may be at best a waste of energy, and effort should instead be concentrated on burying the wet zone, for example by promoting deposition along the gully floor. Overall, gullies are easier to prevent than to cure.

Further reading

Harvey, M.D., Watson, C.C. and Schumm, S.A. (1985) Gully erosion, *US Department of the Interior, Bureau of Land Management, Technical Note* 366, 1–181.

Higgins, C.G. (1990) Gully development, in C.G. Higgins and D.R. Coates (eds) *Groundwater Geomorphology*, Geological Society of America Special Paper 252, 18–58.

Ireland, H.A., Sharpe, C.F.S. and Eargle, D.H. (1939) Principles of gully erosion in the Piedmont of South Carolina, *US Department of Agriculture Technical Bulletin* 633.

NWSCA (National Water and Soil Conservation Authority) (1985) Soil erosion in New Zealand, *Soil and Water* 21(4), supplement.

Wells, N.A, and Andriamihaja, B. (1993) The initiation and growth of gullies in Madagascar: are humans to blame?, *Geomorphology*, 8, 1–46.

NEIL A. WELLS

GUYOT

Although occasionally used to refer to any sizeable underwater ocean-floor edifice, the term 'guyot' should be confined to those that are flat-topped and were once above the ocean surface. This is to distinguish them from seamounts which are underwater volcanoes that have never been above the ocean surface.

The flat tops of guyots were thought for many years to be erosional – an expression of wave truncation of their summits during an island's slow submergence. The first to be studied in detail were in the Hawaiian chain, and submergence and summit truncation were thought to be natural and unavoidable consequences for an island moving along the chain (Hess 1946).

As ideas of Earth-surface mobility became fashionable in the 1960s, so it was realized that the distribution of guyots about mid-ocean ridges (seafloor spreading centres) was significant. Most such guyots had evidently originated at the mid-ocean ridge and then, following a period as a subaerial volcanic island (or atoll), they were submerged as they moved down the ridge's steep flanks and became guyots.

Another important step in the understanding of the significance of guyots came when they were found to be mixed in with atolls in various Pacific island groups like the Marshall Islands, Austral, Tuamotu and several in Kiribati (and some in the Indian and Atlantic Oceans). It has become clear that these guyots were once ATOLLS. That they are no longer is due to various reasons, including the morphology of the ocean floor in these regions (particularly the presence of intraplate swells) and oceanographic factors (principally temperature) which inhibit coral growth (Menard 1984).

Guyot morphology and location

The existence of low-relief surfaces on the summits of guyots is clear evidence for most authors of wave truncation, specifically shoreline erosion (at typical rates of 1 km/Ma) coincident with island subsidence (e.g. Vogt and Smoot 1984). Coral reefs on some guyots demonstrate that they are drowned atolls (see below). Phosphorites on certain Pacific guyots (at depths of 550–1,100 m) also derive from subaerial avian phosphorites (Cullen and Burnett 1986) demonstrating that these islands were once above the ocean surface.

Following observations and insights of Charles Darwin, it was proposed in 1982 that an oceanographic threshold existed in the Hawaiian chain which explained why atolls became converted to guyots at around 29 °N. It was proposed that at the 'Darwin Point', the gross carbonate production by corals was no longer sufficient for atoll reefs to regrow during periods of sea-level rise and so the atoll which had once existed became drowned (Grigg 1982). More recently it has been argued that other factors such as climate and sea-level history, palaeolatitude, seawater temperature and light all contribute to the Darwin Point which has shifted in the Hawaii region between 24 and 30°N within the last 34 Ma (Flood 2001).

Although guyots are commonly located at the older ends of hotspot island chains where these cross the Darwin Point, other guyots are located in equally instructive locations. For example, the morphology of guyots which have been pulled down into the Tonga–Kermadec Trench (southwest Pacific) has given us insights into the nature of tectonic processes across oceanic plate convergent boundaries (Coulbourn *et al.* 1989).

References

Coulbourn, W.T., Hill, P.J. and Bergerson, D.D. (1989) Machias Seamount, Western Samoa: sediment remobilisation, tectonic dismemberment and subduction of a guyot, *Geo-Marine Letters* 9, 119–125.

Cullen, D.J. and Burnett, W.C. (1986) Phosphorite associations on seamounts in the tropical southwest Pacific Ocean, *Marine Geology* 71, 215–236.

Flood, P.G. (2001) The 'Darwin Point' of Pacific Ocean atolls and guyots: a reappraisal, *Palaeogeography, Palaeoclimatology, Palaeoecology* 175, 147–152.

Grigg, R.W. (1982) Darwin Point: a threshold for atoll formation, *Coral Reefs* 1, 29–34.

Hess, H.H. (1946) Drowned ancient islands of the Pacific Basin, *American Journal of Science* 244, 772–791.

Menard, H.W. (1984) Origin of guyots: the *Beagle* to *Seabeam*, *Journal of Geophysical Research* 89(B13), 11,117–11,123.

Vogt, P.R. and Smoot, N.C. (1984) The Geisha Guyots: multibeam bathymetry and morphometric interpretation, *Journal of Geophysical Research* 89(B13), 11,085–11,107.

PATRICK D. NUNN

GYPCRETE

The accumulation of gypsum ($CaSO_4 \cdot 2H_2O$) within a soil or sediment profile leads to the formation of gypsic horizons, which are called gypcrete. Gypcrete is a member of the dryland DURICRUST family, which also includes CALCRETE and SILCRETE. As gypcrete is more soluble than other duricrusts it seldom produces MESA landscapes with a gypsum CAPROCK. The process of gypcrete formation and its global distribution differs from that of other duricrusts. Nevertheless, calcrete and gypcrete may occur in the same profile and can be associated with salts such as halite. Gypcrete may also be host to one of the most commonly recognized forms of gypsum which is the desert rose.

Gypsum, the building material in the formation of gypcrete, is a very common mineral. It can be found throughout the world and forms under present-day evaporitic conditions in inland salt lakes (PANs), coastal salt flats (SABKHAs), springs, CAVEs, organic rich submarine sediments and in dryland soils. It is most commonly associated with massive sedimentary bedrock sequences, in particular those of evaporitic lake or sea basins. Gypsum formation generally requires evaporation of water and due to its solubility forms preferentially in areas of very low rainfall. It may also produce GYPSUM KARST if massive gypsum horizons are subjected to significant precipitation. Surficial gypsum does occur in all arid regions including the polar deserts, but massive gypcrete appears to be restricted to the drier subtropical desert regions of North Africa, the Middle East, southwestern Africa, southwestern America, South America, Central Asia and Western Australia.

Significant primary sources of gypsum formation are pans and sabkhas, which may also host gypcrete. These environments often feature shallow groundwater tables that are subject to substantial evaporation rates, which lead to the formation of evaporites above the water table in close proximity or at the surface. An upward migration of water and formation of gypsum is described as *per ascensum* gypcrete formation. When shallow ground water evaporates into a sandy substrate around a pan or sabkha margin, desert rose gypcrete forms. These crystals can be up to 30 cm in size and can be joined to form a single massive horizon.

Pans may also be subject to pronounced surface salinity gradients between the point of freshwater input and the point of saline brine formation at the centre. Such gradients are often accompanied by distinct evaporitic zones, which in a circular pan are arranged in concentric belts. Under such conditions, sulphate formation often follows the formation of carbonates and precedes the precipitation of chlorides and other salts. Pan surface gypsum may form hard gypsum crusts, which can develop a thrust polygon pattern.

Pan or sabkha environments may also accumulate unconsolidated fine gypsum crystals and powdery gypsum soils on their surface. Such freshly precipitated gypsum may not always form a hardened surface crust, but may be subject to aeolian dispersal (see AEOLIAN PROCESSES) Gypsum deflation may lead to gypsiferous lunette dunes, in particular at pan margins and will lead to the accumulation of gypsum-rich dust in the downwind environment. Aeolian dispersal of gypsum from pans is common and may be relatively rapid as indicated by significant burial of Roman artefacts in Tunisia (Drake 1997). It may also produce regional-scale gypcrete as demonstrated in the Namib Desert region (Eckardt *et al.* 2001).

Pedogenic accumulations form during sporadic rain, which dissolves surface dust and reprecipitates gypsum in stable soils or sediment profiles. Regolith cover in particular traps gypsum dust and gradually incorporates gypsum into the stable subsurface soil or sediments below the STONE PAVEMENT. It has been suggested that pavement surfaces are displaced upward during the process of gypsum dust entrapment (McFadden *et al.* 1987). The external and primary aeolian input of gypsum into a soil results in the relative downward migration of gypsum into the profile and is described as the *per descensum* mode of gypcrete formation. This process generally takes place in the absence of ground water. The resulting gypcrete can be massive in character, and may partly consolidate the surface regolith of a stone pavement. Stone pavement surfaces that cover significant gypcrete accumulations sometimes develop polygonal surface patterns.

The various crusts outlined above may differ considerably in terms of thickness, strength, composition and purity. The structure of gypcrete may range from powdery, nodular to massive horizons that vary in thickness from a few centimetres to many metres. Gypsum crystals may vary in size from microcrystalline (powdery) to massive (desert rose) and may include alabasterine morphologies as well as transparent lenticular clasts. A single crust may undergo multiple stages of formation with reworking, removal, production and storage of crust occurring simultaneously. This can produce a spatially complex and varied morphology. As a result no typical gypcrete profile exists. Gypsum crusts can however be defined as 'accumulations at or within 10 m of the land surface from 0.10 m to 5.0 m thick containing more than 15 per cent by weight of gypsum and at least 5.0 per cent by weight more gypsum than the underlying bedrock' (Watson 1985).

The formation of gypcrete is not only determined by climate but also by the provision of the elements, which produce gypsum. In particular sulphate is not as common as the elements required to form silcrete or calcrete. Sulphur isotopes have demonstrated that the formation of gypsum is dependent on the supply of dissolved sulphate or pre-existing sulphate accumulations such as bedrock gypsum. Gypcrete formation is thereby directly linked to the regional sulphur cycle (Eckardt 2001). Dissolved sulphate in surface or ground water leading to the formation of gypsum may be derived from the dissolution of sulphates or sulphides in the bedrock, the marine atmospheric contribution of sea spray, marine dimethyl sulphide (CH_3SCH_3) also known as DMS (Eckardt and Spiro 1999), gaseous hydrogen sulphide (H_2S) or the dissolution of aeolian sulphate from terrestrial gypsum dust sources.

We still have little information on exact rates of gypcrete formation and the response of gypcrete to climatic change (Plate 59). Attempts have been made to examine the micropetrography of gypcrete and to infer palaeoenvironmental conditions from such observations (Watson 1988). Dating of gypsum and gypcrete is possible using U-series dating and thermal luminescence techniques.

Plate 59 Example of a desert rose, the most commonly recognized form of gypsum. It forms with the evaporation of shallow ground water into a sandy substrate around a pan or sabkha margin

We have also been able to map gypcrete using remote sensing. Due to the distinct spectral response of gypsum in the mid-infrared region of the spectrum (2.08–2.35 μm) it can be separated from many other dryland features (White and Drake 1993).

Some powdery gypcrete accumulations may attain purities which make them attractive to mining while other deposits, in particular those fed by ground-water in Namibia and Australia, are known to be associated with high concentrations of uranium (Carlisle1978). In some areas pure gypcrete is also being used as a road paving material.

Gypcrete formation needs to be examined in the context of regional dryland processes, which include a combination of fluvial and aeolian processes. An understanding of pan or sabkha chemistry and hydrology is particularly important, as these are significant aeolian point sources of gypsum dispersal.

References

Carlisle, D. (1978) *The distribution of calcretes and gypsum in SW USA and their uranium favorability based on a study of deposits in Western Australia and South West Africa*, US Dept of Energy Subcontract, Open File Report, 76–022–E.

Drake, N.A. (1997) Recent aeolian origin of surficial gypsum crusts in southern Tunisia: geomorphological, archaeological and remote sensing evidence, *Earth Surface Processes and Landforms* 22, 641–656.

Eckardt, F.D. (2001) Sulphur isotopic applications, example: origin of sulphates, *Progress in Physical Geography*, 24(4), 512–519.

Eckardt, F.D. and Spiro, B. (1999) The origin of sulphur in gypsum and dissolved sulphate in the Central Namib Desert, Namibia, *Sedimentary Geology* 123 (3–4), 255–273.

Eckardt, F.D., Drake, N.A, Goudie, A.S. White, K. and Viles, H. (2001) The role of playas in the formation of pedogenic gypsum crusts of the Central Namib Desert, *Earth Surface Processes and Landforms* 26, 1,177–1,193.

McFadden, L.D., Wells, S.G. and Jercinovich, M.J. (1987) Influences of eolian and pedogenic processes on the origin and evolution of desert pavements, *Geology* 5(15), 504–508.

Watson, A. (1985) Structure, chemistry and origins of gypsum crusts in southern Tunisia and the central Namib Desert, *Sedimentology* 32, 855–875.

——(1988) Desert gypsum crusts as palaeoenvironmental indicators: a micropetrographic study of crusts from southern Tunisia and the central Namib Desert, *Journal of Arid Environments* 15(1), 19–42.

White, K.H and Drake, N.A. (1993) Mapping the distribution and abundance of gypsum in south-central Tunisia from Landsat Thematic Mapper data, *Zeitschrift für Geomorphologie* 30(3), 309–325.

Further reading

Cooke, R.U., Warren, A. and Goudie, A.S. (1993) *Desert Geomorphology*, London: UCL Press.

Watson, A. and Nash, D. (1997) Desert crusts and varnishes, in D.S.G. Thomas (ed.) *Arid Zone Geomorphology*, 69–107, Chichester: Wiley.

SEE ALSO: duricrust; gypsum karst; pan; sabkha

FRANK ECKARDT

GYPSUM KARST

Karst associated with gypsum and anhydrite rocks is generally referred to as 'Gypsum Karst' and has received little appreciation by geomorphologists if compared to the normal (limestone) KARST. However, gypsum karst is widely spread in the world where the global gypsum-anhydrite outcrop exceeds 7 million km^2, the largest areas being in the northern hemisphere, particularly in the United States, Russia and the Mediterranean basin.

Due to the high solubility of calcium sulphate, the gypsum karst life cycle is commonly far shorter than that of carbonate karst. The average experimental values for gypsum degradation within the Mediterranean area were 0.91 mm/1,000 mm of rain (Cucchi *et al.* 1998) and therefore no outcrop of such rock may survive more than a few hundred thousand years if exposed to the meteorological agents. The actual evolution depends greatly upon the geological history of the particular region so that intra-Messinian and even older gypsum karst may have been preserved until the present.

Exposed karst forms

Medium- to large-sized gypsum karst landforms (DOLINES, BLIND VALLEYS, polje-like depressions) are very similar in genesis and morphology to those found on carbonate rocks, while meso-, micro- and nano-forms may sometimes be peculiar to a gypsum environment. The differences in meso-, micro- and nano-forms are normally the direct consequence of the fact that the size of the crystals in different gypsum outcrops may range from over a metre to a fraction of a mm, while in the carbonate rocks the crystal size is normally around a mm.

The most peculiar gypsum karst meso-form are 'tumulos', while 'weathering crusts' are amongst the micro-forms. All of them develop in gypsum formations characterized by a crystal size of 1–10 cm, which is normal for the Messinian gypsum in the Mediterranean area. Their evolution is produced by the increase of volume of the superficial gypsum stratum induced by the dis-aggregation of the rock texture, as a consequence of the anisotropic behaviour of the gypsum crystals with respect to temperature changes (Calaforra 1998).

Finally, whereas in carbonate rocks some of the micro- and most of the nano-forms are the result of biological activity, the outcrop of very large gypsum crystals (up to 1 m or more in length) together with their high solubility allows for the evolution of nano-forms, the morphology of which is simply controlled by the structure of the crystal lattice (Forti 1996).

Deep karst forms

In gypsum karst the single active speleogenetic mechanism is simple dissolution. Therefore, here deep forms are not so varied as in carbonate karst, where plenty of different speleogenetic mechanisms are active. Moreover, most of the dissolution–erosion forms (pits, canyons, domes, scallops, large collapse chambers) are quite similar to those present in carbonate ones.

Gypsum CAVES are generally very simple linear or crudely dendritic caves that directly connect sink points and resurgence. They are commonly referred to as 'through caves' and consist of a principal drainage tube running along the water table with few and short, often subvertical, effluents; through caves are common in almost every entrenched and denuded gypsum karst area.

The deepest gypsum caves currently known rarely exceed 200 m in depth, being far shallower than those in carbonate karst: the reason is that always in mountainous regions, where the potential drained depth is greatest, gypsum formations are fragmented and do not favour the development of such vertically extensive sequences as do carbonates.

For the same reason the length of a gypsum cave rarely exceeds 2–5 km even if, in peculiar hydrogeological conditions (basal and/or lateral injection and dispersed inputs), complex dendritic 2- or 3-dimensional (multistorey) maze caves may develop up to several tens of kilometres. Podolia (Ukraine) is the 'type' region in which such caves have been explored and studied.

Chemical deposits

Chemical deposits (Hill and Forti 1997) are rather uncommon if compared with those present in carbonate caves: this depends mainly on the scarce chemical reactivity of gypsum. Normally they consist of calcite and gypsum and the local prevalence of one or the other mineral depends on climate.

Calcite SPELEOTHEMS show no morphologic peculiarities to distinguish them from similar deposits in limestone caves. Although, in most cases, their depositional mechanism is unlike that which dominates in a limestone environment (supersaturation due to CO_2 loss) being the product of the incongruent dissolution of gypsum by water with a high initial carbon dioxide content. Incongruent dissolution also explains the existence of unique forms like 'calcite blades' and 'half calcite bubbles'.

Gypsum speleothems have a more ubiquitous distribution. They present obvious morphological differences compared to calcite ones, due to their distinct genetic mechanism, which involves supersaturation due to evaporation. This genetic mechanism is also responsible for several unique forms such as 'gypsum balls', 'gypsum hollow stalagmites' and 'gypsum powder'.

Climatic influence on the chemical deposits

In gypsum caves climatic factors have a strong influence upon calcite and/or gypsum deposits. The completely different depositional mechanisms (incongruent dissolution for calcite and evaporation for gypsum) are influenced in very different ways by climatic variables: therefore climate strictly controls (far more than in carbonate karst) what chemical deposit can develop in a given gypsum cave. This close relationship with climate gives deposits preserved in the gypsum environment a potentially very great importance on the basis of their application to palaeoclimatology (Figure 79) and as indicators for present-day climatic changes.

References

Calaforra, Chordi, J.M. (1998) *Karstologia de yesos*, Universidad de Almeria, Spain.

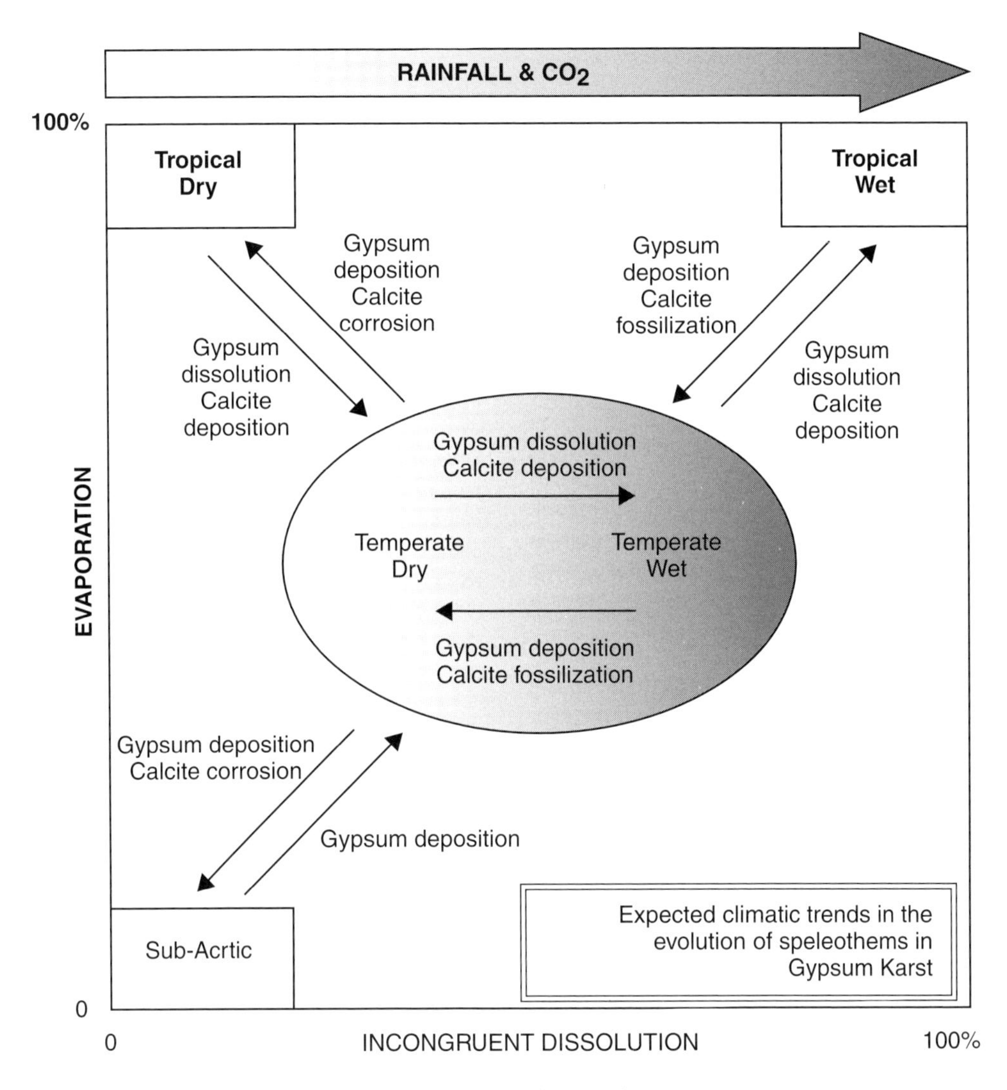

Figure 79 Expected climatic trends in the evolution of speleothems in gypsum caves

Cucchi, F., Forti, P. and Finocchiaro, F. (1998) Gypsum degradation in Italy with respect to climatic, textural and erosional conditions, in J. James. and P. Forti (eds) *Karst Geomorphology*, 41–49, Geografia Fisica e Dinamica Quaternaria supplement III, v.4.

Forti, P. (1996) Erosion rate, crystal size and exokarst microforms, in J.J. Fornos. and A. Gines (eds) *Karren Landforms*, 261–276, Universitat de Illes Balears, Mallorca.

Hill, C. and Forti, P. (1997) *Cave Minerals of the World*, Huntsville, AL: National Speleological Society.

Further reading

Klimchouk, A., Lowe, D., Cooper, A. and Sauro, U. (eds) (1996) Gypsum karst of the world, *International Journal of Speleology* 23(3–4).

PAOLO FORTI

H

HALDENHANG

Haldenhang is a German geomorphic expression introduced by W. Penck (1924, see translation 1953) for a 'basal slope – the less steep slope found at the foot of a rock wall, usually beneath an accumulation of talus'.

Figure 80 illustrates the formation of a haldenhang at the foot of a rock wall: all parts of the rock face but one are subject to erosion through rockfall, namely its base. The material there cannot fail because there is no gradient beneath it. This results in a parallel retreat of the rock face, during which its foot moves gradually upwards. Thus a rock slope of lower inclination appears, the haldenhang. Underneath a rapidly weathering rock face, the haldenhang may be covered by rockfall debris, forming a SCREE or TALUS. Over time the rock face will suffer incremental reduction in its height finally to be replaced by the haldenhang.

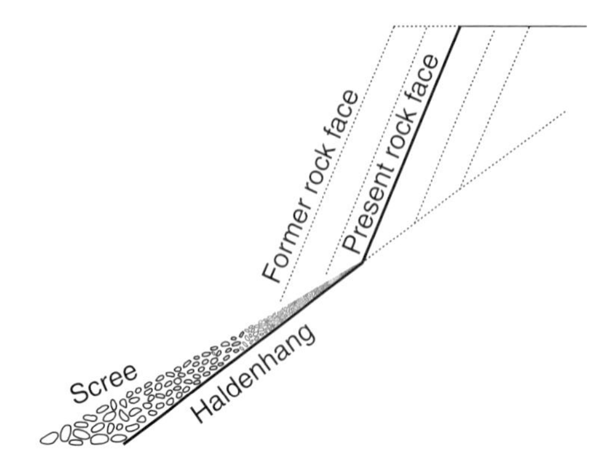

Figure 80 Formation of a haldenhang at the foot of a rock wall

W. Penck built his observations on haldenhang formation into his classic model of landform evolution and used them for explaining the erosional processes responsible for the downwearing of a relief (see SLOPE, EVOLUTION). According to Penck (1953), waning slope development starts with steep valley slopes being replaced by haldenhang of less inclination. Weathering of the surface material of the haldenhang will eventually produce finer material susceptible to creep and rainwash. Thus erosion of the haldenhang starts, resulting in a parallel retreat of the haldenhang and the development of a lower slope segment at its base. The whole process of slope retreat by mutual consumption produces concave and progressively lower slopes.

Reference

Penck, W. (1953) *Morphological Analysis of Land Forms*, trans. H. Czech and K.C. Boswell, London: Macmillian.

CHRISTINE EMBLETON-HAMANN

HAMADA

Large rocky, unvegetated plateaux spread over dozens of kilometres in the Sahara, the Australian deserts and Libya, e.g. Hamadas of Dra, and in the Guir, north-western Sahara (Mabbutt 1977).

The surface of hamadas shows STONE PAVEMENTS which could either be residual and result from the disintegration of rock formations below or consist of boulders transported only short distances, forming a reg. The reg acts as a protective layer to the underlying formations of the hamada which represent forms of stabilized relief. The hamadas of the Sahara rest on erosion surfaces of varied age: Cretaceous, Oligocene, Miocene, Pliocene or Quaternary (Conrad 1969). The hamadas of the northern Sahara are characterized by the extension onto the southern Atlas piedmont of the detrital and lacustrine facies of the *torba*, which is a non-stratified sediment, interrupted by one or several levels of silicified dolomitic limestones, called the *hamadienne carapace*. The geochemical interpretation of the torba and the carapace is that it formed by continental sedimentation in a lacustrine environment as indicated by the abundance of neoformed attapulgite, dolomite and calcite.

References

Conrad, G. (1969) *L'évolution continentale post-hercynienne du Sahara algérien (Saoura, Erg Chech, Tanezrouft, Ahnet-Mouydir)*, série Géologie No. 10, Paris: Centre National de la Recherche Scientifique, CNRS.
Mabbutt, J.A. (1977) *Desert Landforms, An Introduction to Systematic Geomorphology*, Cambridge, MA: MIT Press.

MOHAMED TAHAR BENAZZOUZ

HANGING VALLEY

A tributary valley in which the floor at the lower end is notably higher than the floor of the main valley in the area of junction. Hanging valleys are a hallmark of GLACIAL EROSION in mountains, because the greater bulk of the trunk glacier was able to cut a larger valley cross section than those of smaller tributary glaciers, in consequence of which the floor of the main valley was eroded to a lower level. The relationship between the size of valley glacier troughs and ice discharge was first noticed by A. Penck (1905) who termed it as the 'law of adjusted cross-sections'. Indeed, geomorphometric assessments undertaken in more recent years strongly support his idea that discharge and trough size are mutually adjusted (Benn and Evans 1998: 365).

Hanging valleys have a variety of forms. In high mountain areas the cross profile will exhibit the typical U-shape of glacial erosion. If the tributary valley was not glaciated or only occupied by thin, cold-based ice the preglacial V-shape might prevail. In some cases a waterfall cascades over the lip into the main valley, but in high mountain areas headward erosion of the tributary stream has usually cut a narrow gorge into the lower reaches of the hanging valley floor.

Outside previously glaciated areas hanging valleys sometimes occur along youthful fault scarps or along coasts where the rate of cliff retreat is higher than the adjustment potential of the smaller streams, e.g. in the chalk cliffs in the south of England. Hanging valleys can also develop in karst areas where surface streams flow directly on the groundwater surface. If the main river is downcutting rapidly, the water level will be progressively lowered and the smaller tributaries will eventually turn into DRY VALLEYs hanging above the main valley.

References

Benn, D.I. and Evans, D.J.A. (1998) *Glaciers and Glaciation*, London: Arnold.
Penck, A. (1905) Glacial features in the surface of the Alps, *Journal of Geology* 13, 1–17.

CHRISTINE EMBLETON-HAMANN

HEADWARD EROSION

Headward erosion is the process by which a stream extends upstream towards the catchment divide. Headward erosion occurs at a range of scales, from rills to large rivers, and in all environments. Drylands have been central to research into headward erosion because conditions that favour GULLY and ARROYO development are frequently found in these regions. Headward erosion may also be caused by RIVER CAPTURE.

In any channel network, approximately half of the total length of channels is in un-branched (first-order) fingertip tributaries. Environmental changes that promote channel extension therefore have a large potential impact on the landscape. During discharge events channel heads may advance great distances upslope, or retreat downslope if the hollow refills. In extreme cases, gullies can grow in length by tens of metres per year, and may also incise their channels creating steep ravine banks (Bull and Kirkby 2002). One possible end result of these processes is the creation of BADLANDs, where there is little or no remaining land suitable for agriculture.

Headward erosion occurs at the channel head (see CONTRIBUTING AREA). In terms of landscape dynamics, the channel head is one of the most important elements of the coupled hillslope-channel system. The location of the channel head controls the distance to the catchment divide and therefore influences the drainage density and average hillslope length of a catchment (although bifurcation frequency, confluence angles and tributary spacing are also important) (Bull and Kirkby 2002). The position of the channel head is controlled by the balance of sediment supply and sediment removal (Kirkby 1980; Dietrich and Dunne 1993). A change in any factor that influences this balance, such as fluctuations in climate or land use, alter the surface erodibility, sediment supply and runoff rates and may therefore result in headward erosion.

Channel extension results from a complex array of processes that reflect variations in slope, soil type, soil thickness, vegetation type and vegetation density. These processes include overland flow, pipe initiation and collapse, mass failures and hillslope processes (which have the reverse effect to headward erosion by infilling the channels).

Overland flow occurs in small rills or as sheets of moderate depth over large surfaces. For erosion to occur the rate of rainfall must be sufficient to produce runoff, and the shear stress produced by the moving water must exceed the resistance of the soil surface. Erodibility is a function of the permeability of the surface, the physical and chemical properties that determine the cohesiveness of the soil, and the vegetation.

In some areas there is a close association between piping (see PIPE AND PIPING and headward erosion. The erosive effects of flow through subsurface channels may result in TUNNEL EROSION and subsequent collapse to cause headward erosion. Piping intensity reflects a critical interaction between climate conditions, soil/regolith characteristics and local hydraulic gradients.

Mass failures also occur at channel heads to cause headward erosion. Failure of steep channel heads occurs when the driving forces exceed resisting forces. Channel heads are loaded by three different forces: (1) the weight of the soil, (2) the weight of water added by infiltration or a rise in the water table, and (3) seepage forces of percolating water (Bradford and Piest 1977). The change in water content is important because it has a strong influence on the shearing resistance of the soil. The shear strength is also influenced by freeze–thaw cycles and wetting–drying cycles. Vertical tension cracks tend to decrease overall stability by reducing cohesion, and when these are filled with water the pore-water pressure increases dramatically, often resulting in failure.

Hillslope processes such as rainsplash, wetting and drying cycles and frost action operate to infill channels, and hence reverse headward erosion. For incisions to grow the rate of sediment transport out of an incision must also exceed the rate of sediment input at the same point, otherwise filling will occur. The inter-rill hillslope processes involved are rainsplash and rainflow. Both processes depend on raindrop impact to detach soil material. Mass failures may also act to fill channels if failed material is not removed, but builds up at the base of headcuts.

Traditionally there are two conceptual approaches to understanding processes operating at the channel head, the stability approach (Smith and Bretherton 1972) and the threshold approach (Horton 1945). The stability approach emphasizes that the channel head represents the point where sediment transport increases faster than linearly downslope. This usually requires wash processes to dominate. The threshold approach takes the view that the channel head represents a point at which processes not acting upslope become important. The balance of sediment still determines whether the channel head becomes stable or migrates, but changing process domains drive incision. However, it is not clear whether there is always a change in process at the headcut, or whether a change in the intensity of the process operating, or a variation in the spatial distribution causes incision. The different approaches tend to be better suited to different environments and determine the two extremes of a range of factors that combine to produce channel heads. These models assist our understanding and prediction of headward erosion.

References

Bradford, J.M. and Piest, R.F. (1977) Gully wall stability in Loess derived alluvium, *Journal of the American Soil Science Society* 41, 115–122.

Bull, L.J. and Kirkby, M.J. (eds) (2002) *Dryland Rivers: Hydrology and Geomorphology of Semi-Arid Channels*, Chichester: Wiley.

Dietrich, W.E. and Dunne, T. (1993) The channel head, in K. Beven and M.J. Kirkby (eds) *Channel Network Hydrology*, 175–219, London: Wiley.

Horton, R.E. (1945) Erosional development of streams and their drainage basins; hydrophysical approach to quantitative morphology, *American Geological Society Bulletin* 56, 275–370.

Kirkby, M.J. (1980) The stream head as a significant geomorphic threshold, in D.R. Coates and A.D. Vitek (eds) *Thresholds in Geomorphology*, 53–73, London: Allen and Unwin.
Smith, T.R. and Bretherton, F.P. (1972) Stability and the conservation of mass in drainage basin evolution, *Water Resources Research* 8, 1,506–1,529.

SEE ALSO: arroyo; badland; donga; gully; pipe and piping; tunnel erosion

LOUISE BRACKEN (NÉE BULL)

HIGH-ENERGY WINDOW

Neumann (1972) suggested that in the mid-Holocene on tropical coasts there was a period when wave energy was greater than now. This occurred during the phase when the present sea level was being first approached by the Flandrian (Holocene) transgression and prior to the protective development of coral reefs. The 'window' may have operated on a more local scale on individual reefs with waves breaking not on margins of an extensive reef flat as now, but more extensively over a shallowly submerged reef top prior to the development of the reef flat (Hopley 1984).

References

Hopley, D. (1984) The Holocene 'high energy window' on the central Great Barrier Reef, in B.G. Thom (ed.) *Coastal Geomorphology in Australia*, 135–150, Sydney: Academic Press.
Neumann, A.C. (1972) Quaternary sea level history of Bermuda and the Bahamas, *American Quaternary Association Second National Conference Abstracts*, 41–44.

A.S. GOUDIE

HILLSLOPE-CHANNEL COUPLING

Fluxes from hillslopes to the channel system are controlled by the connectivity of process domains between different elements of the catchment system. Brunsden (1993) defines coupled systems as being ones where there is a free transmission of energy between elements, for example where a river channel directly undercuts a hillslope, whereas decoupled systems are ones where a barrier is present, for example in the case of a FLOODPLAIN buffering the input of the hillslope to the channel. The extent to which a hillslope is coupled to the channel is thus a function of any factor that affects its connectivity, and may relate to spatial variability of properties such as soil texture or vegetation cover (see OVERLAND FLOW). A floodplain may cause a hillslope to be strongly coupled to the channel if it has a low enough infiltration rate, or at times when it is already saturated. The main channel may be decoupled from the hillslope by the presence of minor channels running along the edge of floodplains (YAZOO channels) or human-made drainage channels.

The strength of coupling may affect the type of process that occurs on either side of the boundary. A channel directly undercutting a hillslope may cause the local gradient to be steep enough to initiate RILLS or gullies (see GULLY) on the hillslope, or may lead to failure of the base of the slope (e.g. Harvey 1994). In all cases, the amount of sediment fed into the channel will increase, and may cause it to avulse (see AVULSION) or change its planform (see BRAIDED RIVERS). The rate of removal of sediment from the base of a hillslope relative to its supply by processes on the slope will also affect the form of slope evolution (see SLOPE, EVOLUTION) in the longer term. Coupled slopes will tend to have more convex lower profiles whereas decoupled slopes will encourage deposition at the slope base leading to concave lower profiles. Strongly coupled slopes will also be more sensitive to changes elsewhere in the catchment system.

Consideration of the strength of coupling may also be important in an APPLIED GEOMORPHOLOGY context. SLOPE STABILITY from undercutting is again an important process here, while Burt and Haycock (1993) discuss the impact of floodplain buffers on water quality, for example due to pollutants carried by runoff (see RUNOFF GENERATION) from hillslopes.

References

Brunsden, D. (1993) The persistence of landforms, *Zeitschrift für Geomorphologie, Supplementband* 93, 13–28.
Burt, T.P. and Haycock, N.E. (1993) The sensitivity of rivers to nitrate leaching: the effectiveness of near-stream land as a nutrient retention zone, in D.S.G. Thomas and R.J. Allison (eds) *Landscape Sensitivity*, 261–272, Chichester: Wiley.
Harvey, A.M. (1994) Influence of slope/stream coupling on process interactions on eroding gully slopes, in M.J. Kirkby (ed.) *Process Models and Theoretical Geomorphology*, 247–270, Chichester: Wiley.

Further reading

Harvey, A.M. (2002) Effective timescales of coupling within fluvial systems, *Geomorphology* 44, 175–201.

Michaelides, K. and Wainwright, J. (2002) Modelling the effects of hillslope-channel coupling on catchment hydrological response, *Earth Surface Processes and Landforms* 27, 1,441–1,457.

JOHN WAINWRIGHT AND
KATERINA MICHAELIDES

HILLSLOPE, FORM

What are hillslopes?

Most of the Earth's surface is occupied by hillslopes. Hillslopes therefore constitute a basic element of all landscapes (Finlayson and Statham 1980) and a fundamental component of geomorphologic systems (see SYSTEMS IN GEOMORPHOLOGY). However, there is an 'amazing absence of any precise definition' of hillslopes (Schumm and Mosley 1973; Dehn *et al.* 2001). Hillslopes have a very large variety of sizes and forms; and several more or less synonymous terms are used to describe the phenomenon hillslope, e.g. valley slope, hillside slope, mountain flank. The description of hillslope form is a fundamental problem in geomorphology (see GEOMORPHOMETRY).

Generally, a hillslope is a landform unit, that is, a part of the Earth's surface, with specific characteristics (see LAND SYSTEM). As a basic characterization, a hillslope can be defined as an *inclined landform unit* with a slope angle larger than a lower threshold β_{min} (delimiting hillslopes from plains) and smaller than a higher threshold β_{max} (delimiting hillslopes from vertical walls like cliffs or overhangs), which is *limited by an upper and a lower landform unit* (Dehn *et al.* 2001). A definition of hillslopes additionally has to include position within the landscape as an *external context*. A valley, for example, can only exist with its accompanying hillslopes. Moreover, *size and scale* context are important properties for definitions of hillslopes: a hiker in Grand Canyon might identify the components of the valley side as an individual hillslope itself, whereas a pilot flying over the scene defines the whole canyon side as a hillslope. Hillslopes are formed as the result of hillslope processes (see HILLSLOPE PROCESS) acting over different timescales. Therefore, hillslopes are units, where downslope component of gravitational stress (g sin β) plays a dominant role for acting hydrologic and geomorphologic processes. However, a hillslope is usually the product of a variety of processes interacting in space and time; therefore, hillslopes form sequences of hillslope units with different characteristics (compare Figure 81; see SLOPE, EVOLUTION).

Therefore, fundamental properties for the definition of a hillslope are: (1) local geometry, (2) external landform relationships, (3) scale, and (4) related processes. The utilization of these fundamental properties into a definition of hillslope depends on the perception or the specific application, that is, a specific semantic model for hillslopes (Dehn *et al.* 2001). In geomorphometry, hillslope forms are usually described as arrangements of individual hillslope units. This concept facilitates description and classification of hillslopes, and enables modelling of interaction of hillslope form with forming geomorphologic processes. Hillslope analysis therefore incorporates two major connected aspects: the decomposition of a hillslope profile into units, and the aggregation of a hillslope by arrangements of form units. These analysis steps are carried out in three dimensions for a hillslope or, in a simplified way, two dimensionally for a hillslope profile. The related terminology used here is listed in Table 23.

Hillslope units

Hillslope analysis is carried out by subdivision of a hillslope into different hillslope units. There have been several approaches to standardize hillslope units using qualitative terms (e.g. Speight 1990). Most commonly, a hillslope is described by a series of basic units describing changes in slope, curvature and processes along the hillslope profile.

- Ridge/crest/interfluve: convex/rectilinear unit; most stable unit in landscape, if of considerable width; mainly vertical water transport; more poorly drained soils.
- Shoulder/upper midslope: convex element; unstable unit due to erosion processes; minimum soil thickness.
- Backslope/midslope: usually rectilinear segment; unstable unit; intensive lateral drainage; sediment transport; soils of varying depth.
- Footslope/lower midslope: concave element; sediment deposition; unstable unit; soil thickness tends to increase.
- Toeslope/floodplain: concave/rectilinear unit; sediment input from upstream and hillslope; unstable unit; thicker soils.

Table 23 Basic components and terminology for hillslope analysis

Hillslope component	Definition
Hillslope profile	Flowline connecting drainage divide with thalweg
Hillslope toposequence	Arrangement of hillslope units within the hillslope
(Hill)slope unit	Part of hillslope with specific characteristics: segment or element
Segment	Unit of homogeneous slope angle
Element	Unit of homogeneous curvature
Convex element	Element with a downslope increase in angle
Concave element	Element with a downslope decrease in angle
Maximum segment	Segment, steeper than units above and below
Minimum segment	Segment, gentler than units above and below
Crest segment	Segment bounded by downward slopes in opposite directions
Basal segment	Segment bounded by upward slopes in opposite directions
Irregular unit	Slope unit with frequent changes of both angle and curvature

Source: Young (1972, modified and extended)

In the direction of contours, hillslopes are usually stratified into the elements HILLSLOPE HOLLOWS, spurs (or noses), and rectilinear valley sideslopes using plan curvature. Another method for hillslope unit classification is based on the position within the DRAINAGE BASIN: Young (1972: 4) distinguishes the 'component slopes': valley-head slopes, spur-end slopes and valley side slopes. Speight (1990) provided an exhaustive list of nomenclatures of different land elements, including many hillslope units. Hillslope units therefore generally incorporate different aspects of land surface form: (1) slope angle, (2) curvature, (3) position within drainage basin and (4) position within hillslope. These properties are used to derive quantitative models of hillslope units (see below). Moreover, hillslope units are related to different geomorphic processes and REGOLITH properties (see above, compare Speight 1990). This leads to the utilization of hillslope units for soil-landscape modelling (see SOIL GEOMORPHOLOGY), formalized for example in the concept of the CATENA.

One of the early quantitative approaches in hillslope analysis, based entirely on geomorphometric properties, was established by Savigear (1952) using the components profile intercept (constant slope gradient), slope segment (constant slope gradient consisting of several profile intercepts), and slope element (constant convex or concave curvature). These units are delimited by breaks of slope, which are characterized by a distinct change in slope gradient. This approach has been extended and quantified by Young (1972), who subdivided the slope into convex, concave and rectilinear units (Table 23).

Those early approaches mostly concentrated on quantitative description of hillslope profiles; however, a hillslope is not simply a linear feature, but a two-dimensional landform unit within the three-dimensional space, acting as a boundary layer of a three-dimensional lithological body. Characterization of local hillslope form is therefore based on the land surface derivatives: gradient, which has two components, slope angle and aspect angle; and curvature, which is usually described by two components in profile and contour or tangential directions (see MORPHOMETRIC PROPERTIES). Curvature can be classified into convex, concave and rectilinear surfaces (Young 1972). Hence, the combination of three slope profile curvature characteristics and three plan curvature characteristics leads to nine possible hillslope units, which are defined by Dikau (1989) as basic form elements of the landscape (Figure 81). They deliver a disjunctive description of the hillslope surface into units of homogeneous curvature characteristics.

Slope angle has been used to describe hillslopes by slope segments (Table 23). Young (1972: 173) compared several classifications of slope angle and proposed a system of seven classes: 0°–2° level to very gentle; 2°–5° gentle; 5°–10° moderate; 10°–18° moderately steep; 18°–30° steep; 30°–45° very steep; >45° precipitous to vertical/overhanging. Limiting angles describe the range of slope

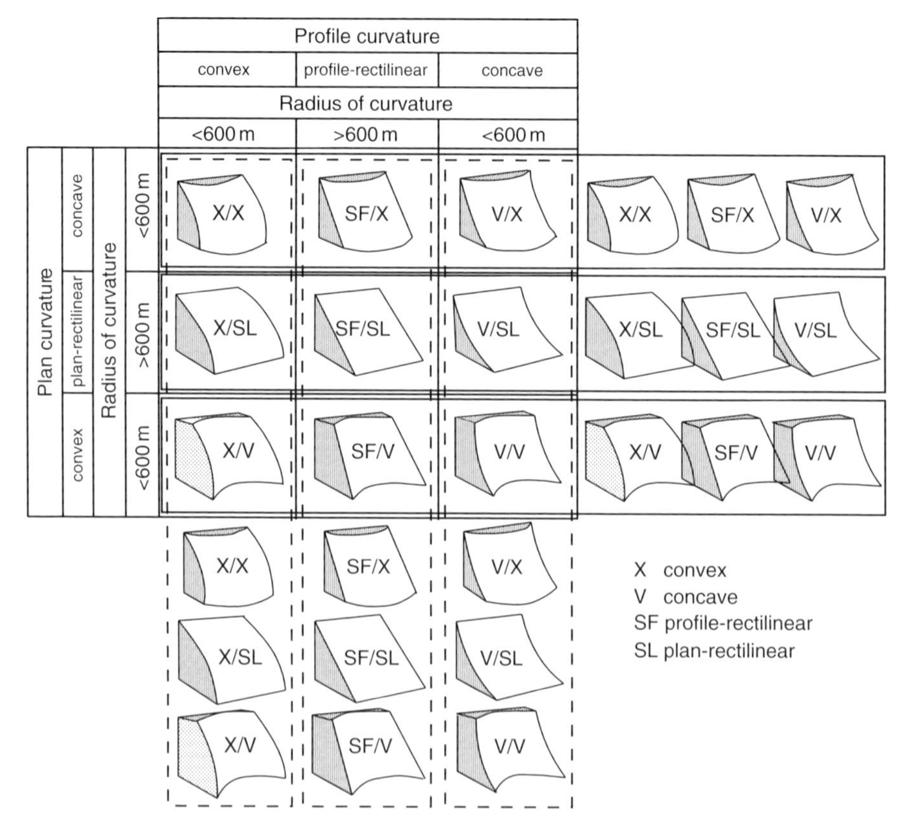

Figure 81 Fundamental hillslope form elements classified by plan and profile curvature (Dikau 1989)

angles within which specific slope forms occur. They include maximum and minimum limiting angles, which are related to the environmental conditions and the corresponding geomorphic processes. The angle of repose (see RESPOSE, ANGLE OF) defines the maximum angle for a given granular material type. Young (1972: 165) gives some figures for limiting angles of hillslope units under different environmental and lithological conditions.

An additional parameter for description of hillslope units is the position in the toposequence (Table 23) of the hillslope. Usually qualitative terms as upper slope, mid-slope, and lower slope are applied. Young (1972) used neighbourhood relationships to upper/lower units to describe hillslope units (see e.g. maximum and minimum segment in Table 23). However, for more complex profiles, these measures fail to describe absolute hillslope position, and quantitative rules, derived from total hillslope length or height, need to be introduced.

Investigations on hillslope forms are centred on process-form relationships, i.e. the explanation of a specific slope form (slope unit) by hillslope evolution, and contemporary processes. Young (1972: 92) gives a series of classical explanations of convex, concave and rectilinear slopes. Convex slope elements generally indicate erosion processes, which increase with slope length (surface wash), additionally soil creep and weathering has been identified as the dominating process regime for convex slopes. Rectilinear slopes generally indicate spatially homogeneous erosion conditions, i.e. the slope retreats parallel, or static transport units. Concave parts of hillslopes are explained by sediment accumulation due to constant base levels and/or by surface wash as an analogue to graded river profiles. However, as hillslopes are complex phenomena with an evolutionary history over long timescales, many interactions of processes and components can occur. Therefore, such simple assumptions generally do not match with the specific hillslope case and can only be used as guidelines.

Hillslope profiles and toposequences

Complete hillslopes are often represented by *hillslope profiles*. According to Young (1972) and Parsons (1988) a hillslope profile can be defined as a line on a land surface, connecting a starting point at the drainage divide with an end point at the thalweg, following the direction of the steepest slope. Hillslope profiles have been used to characterize various types of terrain using typical distribution of slope angles. Differences in the frequency distributions of slope angle are related to lithology (material resistance), climate (stress through rainfall, temperature), and evolutionary state of the slope (see limiting angles above). Hillslope profiles usually cover several process domains. Often the upper section of a slope is characterized by erosion, the middle section by transport and the basal section by deposition. Therefore, *toposequences* are used to describe the arrangement of different units within the hillslope profile. Characterization of these sequences delivers information about the slope system. It can be used to classify hillslopes. A toposequence may include one simple slope (single-sequence, e.g. a rectilinear element connecting ridge and valley), or two or more units (multi-sequence, e.g. convexo-concave slopes) (Speight 1990: 14). For multi-sequences, the order of the slope units (e.g. XMV for convex–rectilinear–concave slope) and the proportional length of the same units can be used to characterize the whole profile (Young 1972: 189).

The description of slope units within the context of the entire two-dimensional slope in three-dimensional space is also part of a toposequential analysis. Dikau (1989) and Schmidt and Dikau (1999) used parameters such as the neighbourhood relationship, distance to drainage divide, or height difference to the drainage channel to classify and aggregate complex hillslope systems.

Different models of hillslope profile form have been developed, which relate a specific hillslope toposequence to evolutionary history and contemporary geomorphic processes. Wood (1942) introduced the term 'waxing slopes' for convex hillslope units on crests developed by weathering processes acting on a cliff top. Likewise, Wood defines 'waning slopes' as depositional concave hillslope units developing at the base of a SCREE by sediment sorting due to aquatic hillslope processes.

King (see Young 1972: 37) developed a classical four-unit toposequential model based on work of Wood (1942). The crest (waxing slope) is a convex element of little erosion by weathering and creep processes. The evolution of the whole hillslope is driven by an active scarp segment of steep slope angle (rill erosion, mass movements). The downslope debris slope segment is formed by sediment provided by the scarp and determined by the angle of repose of the coarser material. The PEDIMENT (waning slope) is a rectilinear-concave, upward erosional element, produced by surface wash that connects to alluvial plains. Dalrymple *et al.* (1969) developed King's toposequence into a nine-unit slope model (Figure 82). The sequence consists of three low erosional upslope units, an intensively erosional unit (4), a transformational midslope (5), the depositional colluvial footslope (6) and three low-angled units associated with fluvial work. Conceptual hillslope models like these contribute to an understanding of the function of hillslope units and hillslope sequences, and can be utilized to classify hillslopes according to dominant process regimes. Numerical hillslope models (see MODELS) are used today to simulate process behaviour and evolution predicted by conceptual hillslope models, and thereby contribute to understanding of hillslope form and evolution based on current knowledge of process physics.

Measurement and analysis of hillslopes

There exist numerous techniques to measure hillslope froms. A selection of direct manual methods based on field observations and indirect measurements from maps and aerial photographs are described in Goudie (1990). Advances in computer technologies and the availability of DIGITAL ELEVATION MODELS (DEMs) have significantly revolutionized hillslope form analysis in the last decades. GIS technologies (see GIS) with algorithms to calculate morphometric properties, including slope gradient, slope curvature and flow paths, are now common tools for geomorphometric hillslope analysis (Schmidt and Dikau 1999). Raster-based DEMs are available at increasingly higher resolutions and, through the development of satellite data, on a global extent. As a result many numerical geomorphometric applications are raster-based. A typical GIS for hillslope analysis includes the following components (Table 24). Local morphometric properties of hillslopes

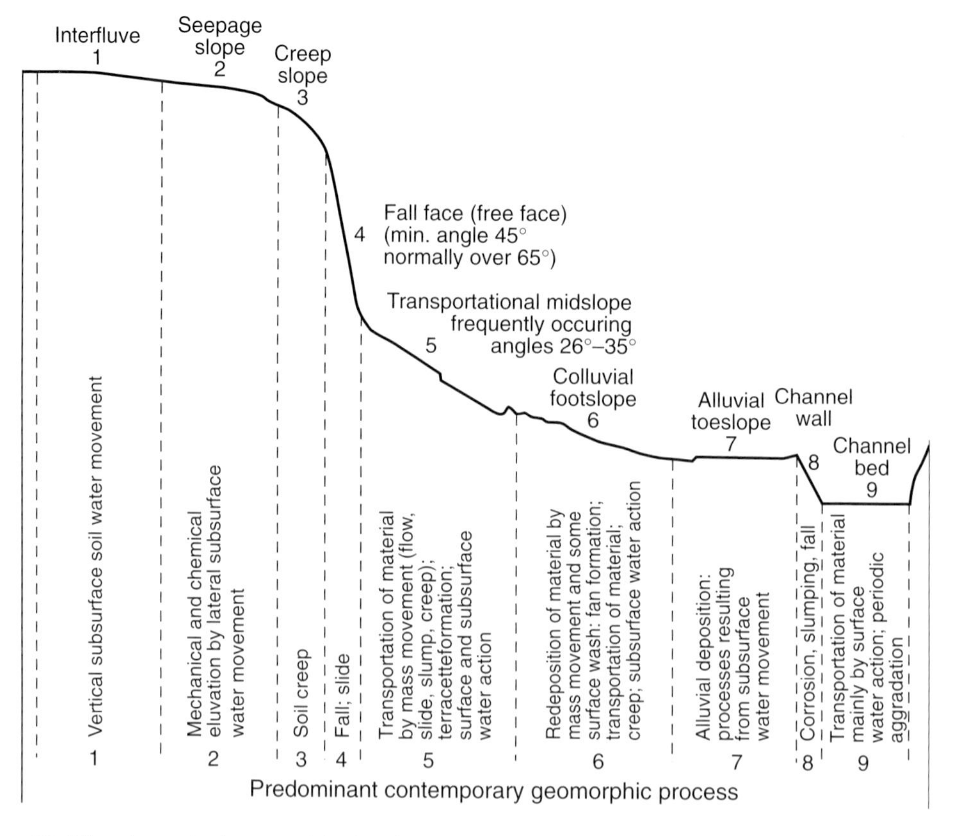

Figure 82 The nine-unit slope model of Dalrymple *et al.* (1969) (modified)

(height, curvature, gradient; see Schmidt and Dikau 1999) are generated from gridded DEMs through local interpolation, whereas complex parameters are based on flow routing algorithms. As slope and curvature are strongly dependent on scale, effects of DEM resolution on these parameters have to be considered, and preferably a specific scale for calculation of derivatives should be chosen. Hillslope units (areal geomorphometric objects after Schmidt and Dikau 1999) can be derived by GIS-based classification of slope, curvature and hillslope position (Dikau 1989). Linear hillslope profiles can be derived directly from a DEM by flow routing algorithms (Rasemann *et al.* 2003). Hillslopes as toposequences (geomorphometric objects of higher level after Schmidt and Dikau 1999) are derived by combining hillslope units according to their properties in gradient, curvature, position, and neighbourhood relationships. Hillslopes are characterized using representative geomorphometric parameters (Schmidt and Dikau 1999): frequency distributions and statistical moments of slope angle have been used to describe different types of slope profiles (Young 1972; Schumm and Mosley 1973). However, modern GIS technologies allow the calculation of many more hillslope parameters, including toposequential characteristics (Schmidt and Dikau 1999).

References

Dalrymple, J.B., Blong, R.J. and Conacher, A.J. (1969) A hypothetical nine-unit landsurface model, *Zeitschrift für Geomorphologie* 12, 60–76.

Dehn, M., Gärtner, H. and Dikau, R. (2001) Principles of semantic modeling of landform structures, *Computers and Geosciences* 27, 1,005–1,010.

Dikau, R. (1989) The application of a digital relief model to landform analysis in geomorphology, in J. Raper (ed.) *Three-dimensional Applications in Geographical Information Systems*, 51–77, London: Taylor and Francis.

Finlayson, B. and Statham, I. (1980) *Hillslope Analysis*, London: Butterworth.

Table 24 Hillslope analysis in a GIS

Hillslope parameters and objects	Algorithm	Input
Local geomorphometric parameters – local geometry		
Slope angle and aspect, profile and plan curvature	Local interpolator	DEM
Complex primary geomorphometric parameters – hillslope position		
Flowdirection	Local classifier	DEM
Upslope contributing area	Flow routing	Flowdirection
Downslope/upslope flowlength	Flow routing	Flowdirection
Hillslope position	Algebraic operation	Flowlength
Geomorphometric objects – hillslope units		
Form elements	Classification	Curvature
Slope segments	Classification	Slope angle
Hillslope units	Classification	Hillslope position
Geomorphometric objects – hillslopes		
Hillslope profile/flowpath	Flow routing	Flowdirection
Hillslope toposequence	Neighbourhood analysis	Form elements, slope segments
Representative parameters – hillslope characteristics		
Frequency distribution	Grid overlay	Hillslope profiles, Hillslope toposequences
Statistical moments of parameters	Grid overlay	Hillslope profiles, Hillslope toposequences

Notes: Local geomorphometric parameters are derived through a local interpolator. Complex geomorphometric parameters, which are related to landscape position, are derived through flow routing. Hillslope units and hillslope profiles are derived as geomorphometric objects. Hillslopes as a toposequential arrangement of units can be analysed by topological relationships of hillslope units

Goudie, A. (ed.) (1990) *Geomorphological Techniques*, London: Unwin Hyman.

Parsons, A.J. (1988) *Hillslope Form*, London: Routledge.

Rasemann, S., Schmidt, J., Schrott, L. and Dikau, R. (2003) Geomorphometry in mountain terrain, in M.P. Bishop and J.F. Schroder (eds) *Geographic Information Science (GIScience) and Mountain Geomorphology*, Berlin: Praxis Scientific Publishing.

Savigear, R.A.G. (1952) Some observations on slope development in South Wales, *Transactions Institute of British Geographers* 18, 31–51.

Schmidt, J. and Dikau, R. (1999) Extracting geomorphometric attributes and objects from digital elevation models – semantics, methods and future needs, in R. Dikau and H. Saurer (eds) *GIS for Earth Surface Systems*, 153–173, Stuttgart: Schweizerbarth.

Schumm, S.A. and Mosley, M.P. (eds) (1973) *Slope Morphology. Benchmark Papers in Geology*, Stroudsburg, PA: Dowden, Hutchinson and Ross.

Speight, J.G. (1990) Landform, in R.C. McDonald, R.F. Isbell, J.G. Speight and J. Walker (eds) *Australian Soil and Land Survey: Field Handbook*, 9–57, Melbourne: Inkata Press.

Young, A. (1972) *Slopes*, London: Longman.

Wood, E.B. (1942) The development of hillside slopes, *Proceedings of the Geological Association* 53, 128–140.

Further reading

Ahnert, F. (1970) An approach towards a descriptive classification of slopes, *Zeitschrift für Geomorphologie, N.F. Supplementband* 9, 71–84.

Carson, M.A. and Kirkby, M.J. (eds) (1972) *Hillslope – Form and Process*, Cambridge: Cambridge University Press.

Pike, R. and Dikau, R. (eds) (1995) Advances in geomorphometry, *Zeitschrift für Geomorphologie N.F. Supplementband* 101, Stuttgart: Schweizerbarth.

SEE ALSO: catena; cliff, coastal; digital elevation model; drainage basin; geomorphometry; GIS; hillslope hollow; hillslope, process; morphometric properties; pediment; repose, angle of; slope, evolution; valley.

RICHARD DIKAU, STEFAN RASEMANN AND JOCHEN SCHMIDT

HILLSLOPE HOLLOW

Hillslope hollows are elongate depressions within the bedrock of regolith mantled hillslopes. They have no obvious stream channel but serve as drainage lines that are integrated with the drainage network by either subsurface or surface topography. They belong to a morphological spectrum ranging from small low-relief upland depressions, a few tens of metres in length, through dells which may extend 200–300 m in length (Ahnert 1998), to valley head depressions, and DRY VALLEYs which have no active channels but are clearly of fluvial origin. Such features are found in many different countries, in a wide range of morphoclimatic and lithological conditions, and have many different modes of origin; thus giving rise to a varied and at times contradictory terminology. For example, United Kingdom usage represents a 'dell' as 'a small well-wooded stream or river valley' while in many other countries a dell is defined as 'a small dry valley with no trace of linear, fluvial erosion' and of periglacial origin (Fairbridge 1968: 250). Other forms of hollow occurring on hillslopes that are not integrated with the drainage network (e.g. individual landslide scars or fault sags) are excluded from this discussion. Reneau and Dietrich (1987) provide a literature review on hillslope hollows.

Hillslope hollows within the substrate are often filled with colluvium and other regolith material (Plate 60) and may or may not exhibit a depression in the landsurface. Because they can lack surface expression, the term 'colluvium-filled bedrock depression' has been suggested as a more accurate descriptive term for these features. In one study of 80 hillslope hollows exposed in road cuts, 37 were found to be associated with concave depressions in the slope surface, 35 occurred beneath planar slopes with no surface depression and 8 occurred on spurs. The cross-sectional form of the depression within the bedrock can be either V-shaped or broad-based (Crozier *et al.* 1990). Hillslope hollows are located headward of first-order channels or join higher order channels in positions similar to that occupied by first-order channels. Tsukamoto (1973) has used the term 'zero-order basin' to describe landsurface hollows, emphasizing their hydrological integration with the drainage network.

Plate 60 V-shaped colluvium-filled bedrock depression, in Pliocene marine sediment, Taranaki, New Zealand

The configuration of hillslope hollows in valley head settings (Figure 83) can be related to contemporary processes (Ahnert 1998; Montgomery and Dietrich 1989). The main criteria used to differentiate the various valley head forms are gradient, number of convergent hollows, and shape. There are four basic types that can be further subdivided on the basis of number of contributing hollows. Shallow gentle hollows (Figure 83a) are wide with a low gradient head and commonly a downslope topographic threshold. During prolonged rainstorms, this form of hollow generally produces saturated overland flow. Steep narrow hollows (Figure 83b), often with a head cut in the colluvial fill, are dominated by seepage and regolith landsliding. Funnel-shape valley heads (Figure 83c) usually result from the convergence of multiple steep hollows. Spring-sapping valley heads (Figure 83d) tend to be circular in shape with a distinctive low angle floor separated from convergent hollows by a marked break in slope at the spring line.

Several processes have been suggested as capable of producing hillslope hollows in different geomorphic settings (see Crozier *et al.* 1990), including landsliding, subaerial fluvial dissection, subsoil percoline erosion, gelifluction and seasonal meltwater and a combination of periglacial and fluvial processes. Cotton and Te Punga, (1955) demonstrate that hillslope hollows are products of alternating morphoclimatic regimes. They conclude that former stream channels, initially incised during the last interglacial became modified by periglacial mass movement processes in the subsequent glacial episode and eventually infilled by periglacial deposits and loess. Under present-day conditions these colluvial infills are being removed by shallow landsliding. Another cyclic infilling and evacuation process, but in this case involving landsliding as the initial hollow generating process (Figure 84),

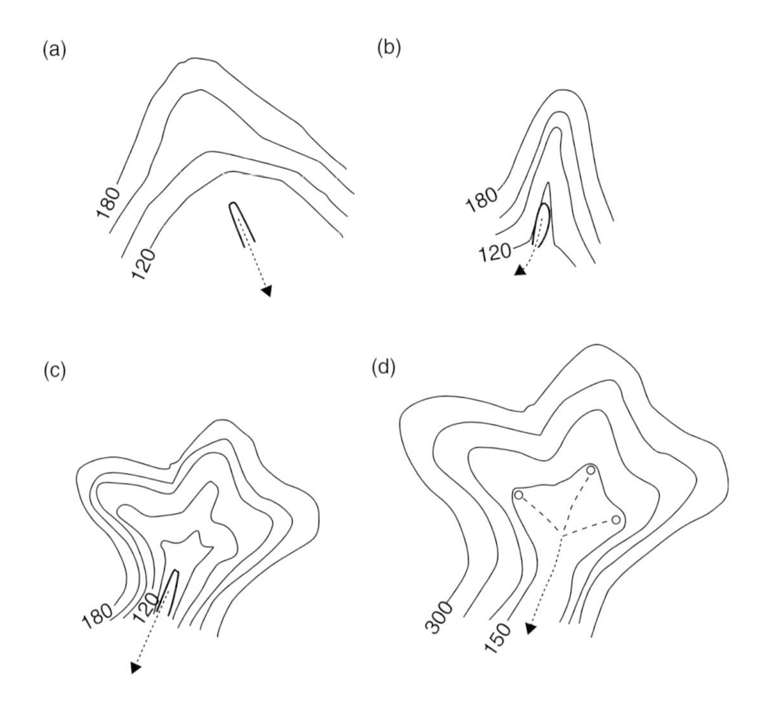

Figure 83 Contour pattern for types of valley head hollows: (a) shallow gentle hollow; (b) steep narrow hollow; (c) funnel-shaped valley head with three convergent hollows; (d) spring-sapping valley head (based on descriptions of Ahnert (1998) and Montgomery and Dietrich (1989))

has been described by Dietrich and Dunne (1978).

During rainstorms, the geometry of hillslope hollows directs both surface and subsurface runoff towards the centre line of the hollow. Accumulation of water within the hollow is a function of the contributing area and the ratio of side slope to thalweg gradients. During prolonged rainstorm events the hollows may become preferentially saturated acting as a source area for saturated overland flow. Regolith-filled hollows are also a preferential location for the generation of debris flows and debris slides. Compared to other hillslope locations, perched water tables are more readily established and accumulation of sediment surpasses the critical thickness required for failure. The scale of landsliding that occurs under these conditions is a function of the number of convergent hillslope hollows and their volume of stored sediment.

References

Ahnert, F. (1998) *Introduction to Geomorphology*, London: Arnold.

Cotton, C.A. and Te Punga, M.T. (1955) Solifluxion and periglacially modified landforms of Wellington, New Zealand, *Transactions Royal Society, New Zealand* 82(5), 1,001–1,031.

Crozier, M.J., Vaughan, E.E. and Tippett, J.M. (1990) Relative instability of colluvium-filled bedrock depressions, *Earth Surface Processes and Landforms* 15, 329–339.

Dietrich, W.E. and Dunne, T. (1978) Sediment budget for a small catchment in mountainous terrain, *Zeitschrift für Geomorphologie Supplementband* 29, 191–206.

Fairbridge, R.W. (1968) Dell, in R.W. Fairbridge (ed.) *The Encyclopedia of Geomorphology*, 250–252, New York: Reinhold.

Montgomery, D.R. and Dietrich, W.E. (1989) Source areas, drainage density and channel initiation, *Water Resources Research* 26, 1,907–1,918.

Reneau, S.L. and Dietrich, W.E. (1987) The importance of hollows in debris flow studies: examples from Marin County, California, in J.E. Costa and G.F. Wieczorek (eds) *Debris flows/Avalanches; Processes, Recognition, and Mitigation*, Geological Society of America, *Reviews in Engineering Geology* 7, 1–26.

Tsukamoto, Y. (1973) Study on the growth of stream channel, (1) Relation between stream channel growth and landslides occurring during heavy rainstorm, *Shin-Sabo* 25, 4–13.

MICHAEL J. CROZIER

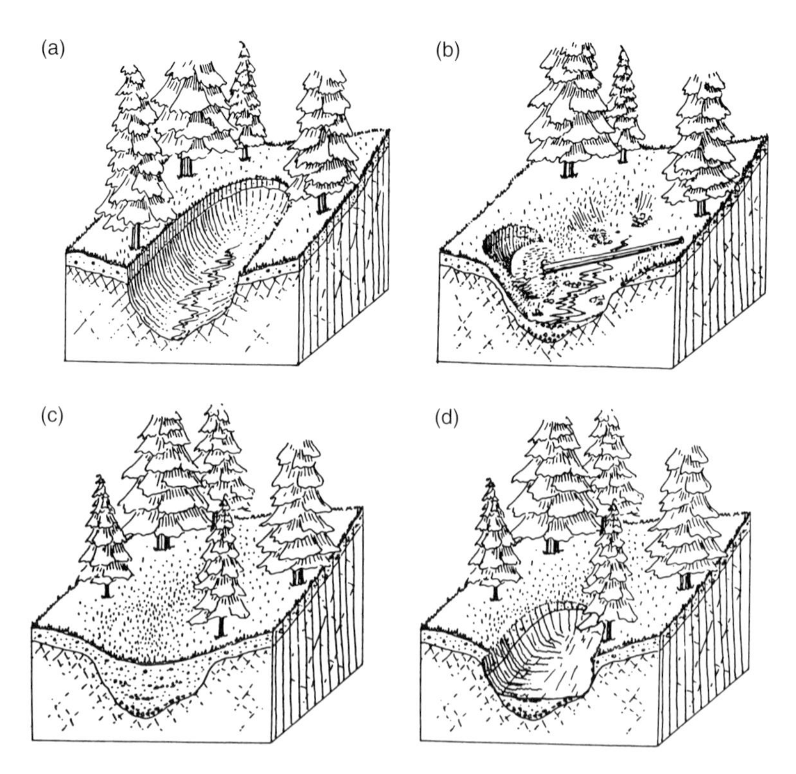

Figure 84 A model for the origin and evolution of hillslope hollows (based on Dietrich and Dunne (1978)): (a) bedrock landslide produces initial hollow; (b) peripheral debris fills hollow and is sorted by fluvial processes; (c) filled hollow becomes a site of concentrated subsurface flow and potential debris slide; (d) evacuation by debris slide

HILLSLOPE, PROCESS

The form that hillslopes take is a product of the materials of which they are made and the forces that act upon them. While the gross form of the landscape is determined by many factors, at the local scale a range of characteristic geomorphic processes shapes hillslopes. At the same time, hillslope morphology acts as a major influence on the occurrence, magnitude and nature of the processes themselves. Thus form and process tend toward a mutual adjustment. Hillslope processes are those geomorphic processes that involve the entrainment, transport and deposition of material from, over and on slopes. Their net effect is the transfer of material to lower parts of the landscape. This occurs either under the influence of gravity alone or, more commonly, with the additional incorporation of varying amounts of water. Flowing ice and wind also contribute to hillslope form, but operate at scales that are greater than the hillslope (see GLACIFLUVIAL; AEOLIAN PROCESSES; WIND EROSION OF SOIL).

A useful distinction can be drawn between two sets of processes that differ with respect to the role that water plays in the entrainment phase. First, MASS MOVEMENTS involve the entrainment of material due to the effect of gravity. The impetus for movement of material is derived from the potential energy inherent in the material by virtue of its elevation above BASE LEVEL, and the magnitude of the potential energy gradient induced by the inclination of both hillslope surface and strata. Water may play a crucial role as a preparatory or triggering factor, but it is the balance of geomechanical stresses within the mass of material that determines whether entrainment occurs (see FACTOR OF SAFETY). Once failure has occurred water is also significant in determining the nature of subsequent transport. In a second set of hillslope processes both entrainment and transport are effected directly by the kinetic energy of moving water.

The term mass movement can be applied to a broad spectrum of processes that are often classified in terms of the type of material involved (e.g. bedrock, debris or earth) and the type of movement (e.g. fall, topple, slide, spread and flow). This range of types of movement reflects an increasing significance of water in the transport phase; falls are almost exclusively gravitational, while, at the other extreme, flows almost always require the presence of water. A similar spectrum exists with respect to the importance of water for entrainment of material, i.e. the initiation of mass movement. In some instances failure may occur simply due to the effects of gravity, e.g. a rockfall. More commonly, water will have some influence on the balance of stresses within the slope material. Water is an important factor in preparing soil and regolith material for mass movement, and can often also be a triggering factor. The effects of water in various forms of physico-chemical WEATHERING produce a slow reduction in shear strength as rock is transformed into an engineering soil and rendered increasingly susceptible to gravitational stresses. More dynamically, the behaviour of water within hillslope materials frequently triggers failure. Variation in the height of groundwater tables can alter the balance of stresses by both increasing the mass of material (increased shear stress) and reducing its shear strength as a result of elevated PORE-WATER PRESSURES or reduced COHESION.

The second set of processes involves the entrainment of material through the energy imparted by flowing or impacting water. While not the direct cause of entrainment, gravity plays an important role in determining the velocity of flow, and hence its energy, and the direction of transport. The energy of a raindrop impact (see RAINDROP IMPACT, SPLASH AND WASH) is capable of detaching soil particles, which may be transported by the resulting splash. Although splash may be in all directions, the net effect is of downslope transport. Greater volumes of material may be entrained and transported by water flowing over the surface of unprotected soils (see SHEET EROSION, SHEET FLOW, SHEET WASH; OVERLAND FLOW). The mass of material that can be entrained will be controlled by the shear stress that is applied to the soil or regolith surface, which will be determined by the relationship between the HYDRAULIC GEOMETRY of the flow and the ROUGHNESS of the surface. The distance over which material of a given size can be transported is proportional to the kinetic energy of the flow. Where microtopographic convergences induce concentration into linear and turbulent flow, a threshold (see THRESHOLD, GEOMORPHIC) is reached and RILLS are formed. Once initiated, rills act to further concentrate flow. This concentrated flow can result in the formation of a GULLY, giving an example of positive feedback. Another important process involving flowing water is TUNNEL EROSION, which occurs when subsurface flow of water through the soil/regolith matrix is of sufficient velocity to initiate particle entrainment (see PIPE AND PIPING).

Although useful to illustrate the differing roles that water may play in hillslope processes, the distinction between mass movement and aquatic processes is often indistinct. The boundary between a highly fluid mass movement (e.g. an earthflow) and a sediment-laden stream may not always be easily identified. Strictly speaking, a rheologic distinction can be made between Newtonian and non-Newtonian flows; the former will involve both vertical and horizontal sorting of clasts during transport, while the latter implies matrix support and an absence of sorting. In reality the removal of material from slopes will occur as a result of a complex interaction between a range of different processes. A more important distinction can be made between those geomorphic processes that are diffuse and those that are not.

Diffusivity

Diffusivity is the property of being spread or dispersed, of not being specifically associated with any one place. Diffusive geomorphic processes, therefore, are those that are widely distributed in space. More importantly, diffusivity refers to the dissipation of energy. Thus, diffusive geomorphic processes can also be defined as those that occur where energy is dissipated over large areas (e.g. SOIL CREEP, sheet wash), while linear processes are characterized by the concentration of energy in a discrete unit of space (e.g. DEBRIS FLOW, RILLS). Consistent with the MAGNITUDE–FREQUENCY CONCEPT, low energy geomorphic processes occur frequently and with a wide spatial distribution. Conversely, high magnitude processes – those that concentrate large amounts of energy – occur more rarely and in a limited number of places. Hence, although diffuse geomorphic processes are often unspectacular, over longer periods they can accomplish large amounts of geomorphic work. Indeed, the predominance of either diffusive or non-diffusive processes has a large bearing on hillslope form (see HILLSLOPE, FORM).

Diffusivity tends to produce convex hillslope profiles. This is because of the adjustment of form to the energy available for entrainment and transport of material. On upper slopes close to drainage divides, with minimal catchment area, energy remains dissipated. Only small amounts of material can be transported. As catchment area increases with increasing distance from the divide, so too does available energy, and thus the amount of material that can be transported. The net effect of increasing entrainment and removal with greater distance from the drainage divide is the development of a convex long profile. At some distance from the drainage divide, available energy will be sufficient for the initiation of concentrated processes. This represents a threshold between diffusive and non-diffusive processes. It also represents a morphological threshold, and below this point long profiles are typically concave.

Characteristic form is therefore a reflection of process domain, and characteristic hillslope forms can be illustrated with reference to the processes that formed them. This may be on a local scale, with diffusivity implying that different processes are predominant on different parts of hillslopes (see, for example, the nine unit hillslope model developed by Dalrymple *et al.* (1969)). Available energy determines whether diffuse or concentrated processes can occur, and therefore how much geomorphic work can be done. There is thus a zone in which diffuse processes, accomplishing smaller amounts of work, predominate. These zones tend to remain constant; diffuse processes produce a characteristic low energy form, which further determines that only these low energy processes occur. Similarly, high energy processes tend to maintain the form that is necessary for their initiation until there is no longer sufficient potential energy available for their initiation (see PENEPLAIN).

However, there is a temporal element in this distinction. Given a sufficiently long period, many hillslope processes – especially mass movements – can be treated as spatially diffuse. For example, within a brief period landslides can be seen as singularities, concentrated in one particular place. Through time, however, as the sites of individual failures shift in space reflecting the availability of susceptible material, every part of the hillslope may be subjected to this process. It is important to note, however, that diffusivity in this case applies only in the sense of spatial dsitribution. The characteristic forms that are associated with the dissipation of energy through diffusive geomorphic processes will not necessarily occur. Indeed, hillslope form will generally be linked to the dominant geomorphic process, whether this is diffusive or concentrated.

At regional scales, the suite of processes that dominate will be determined by climatic and tectonic boundary conditions (see CLIMATIC GEOMORPHOLOGY, TECTONIC GEOMORPHOLOGY). Slope angle has an important control on the manifestation of gravity. Both slope angle and other MORPHOMETRIC PROPERTIES are influenced by tectonic phenomena and lithology, in addition to the action of hillslope processes themselves. Because of the role played by water in many geomorphic processes, climate is especially important, and both the amount and variability of rainfall will influence the type of processes that occur. Soils and vegetation, as products of climatic and geological phenomena, are important, particularly in their effect on slope hydrology. Mass movements dominate on steeper slopes. Aquatic processes tend to dominate in arid to subhumid climates with gentler slopes. Despite the greater availability of water in more humid environments, initiation of aquatic processes is often precluded by protective vegetation and deeper soils with greater capacity for infiltration and subsurface runoff.

Both vegetation and soil are especially susceptible to anthropogenic influence. Although anthropogenic, in many areas tillage is recognized as an important geomorphic process in its own right (see, for example, Govers *et al.* 1994). It is diffuse, with low magnitude, and has a pronounced effect on the form of hillslopes where it occurs. Because it is mechanically initiated, tillage is neither a mass movement as defined here nor an aquatic process. A mass of material is physically displaced with each application of the plough, and in this respect tillage might be considered to be analogous with a diffuse mass movement such as soil creep. Importantly, however, the mechanical disruption and displacement of soil material can also be seen as equivalent to physical weathering, providing easily erodible material for the operation of aquatic processes.

References

Dalrymple, J.B., Blong, R.J. and Conacher, A.J. (1969) An hypothetical nine unit landsurface model, *Zeitschrift für Geomorphologie Supplementband* 12, 61–76.

Govers, G., Vandaele, K., Desmet, P.J.J., Poesen, J.W.A. and Bunte, K. (1994) The role of tillage in soil redistribution on hillslopes, *European Journal of Soil Science* 45, 469–478.

Further reading

Abrahams, A.D. (ed.) (1986) *Hillslope Processes*, Boston: Allen and Unwin.
Carson, M.A. and Kirkby, M.J. (1972) *Hillslope Form and Process*, London: Cambridge University Press.
Crozier, M.J. (1989) *Landslides: Causes, Consequences and Environment*, London: Routledge.
Dunne, T. and Leopold, L.B. (1978) *Water in Environmental Planning*, San Francisco: W.H. Freeman.
Gilbert, G.K. (1909) The convexity of hillslopes, *Journal of Geology* 17, 344–350.
Selby, M.J. (1982) *Hillslope Materials and Processes*, Oxford: Oxford University Press.
Varnes, D.J. (1978) Slope movement types and processes, in R.L. Schuster and R.J. Krizek (eds) *Landslides: Analysis and Control*, Special Report 176, 11–33, Washington, DC: Transportation Research Board, National Research Council.

SEE ALSO: freeze–thaw cycle; landslide; solifluction; threshold, geomorphic; unloading

NICK PRESTON

HOGBACK

A sharp, crested ridge of hard rock, with steeply dipping (>20°) strata and steep near-symmetrical slopes. Hogbacks form as a result of slow differential erosion over time of alternating hard and soft strata. The soft rock is preferentially eroded, leaving steeply angled, slowly eroding resistant rock in place. The term is therefore derived from the resultant feature resembling a hog's back when viewed in planform. Examples of such features include Hogback Ridge, North Dakota, USA, Mount Rundle, Canadian Rockies and Gaishörndl, Austria.

STEVE WARD

HOLOCENE GEOMORPHOLOGY

The Holocene – or 'wholly recent' – epoch is the youngest phase of Earth history, which began with the end of the last large-scale glaciation on northern hemisphere continents other than Greenland. For this reason it is sometimes also known as the post-glacial period. In reality, however, the Holocene is one of many interglacials which have punctuated the late Cenozoic Ice Age. The Holocene is conventionally defined as beginning 10,000 radiocarbon (^{14}C) years ago, which is equivalent to about 11,500 calendar years. The term 'Holocene' was introduced by Gervais in 1869 and was accepted as part of valid geological nomenclature by the International Geological Congress in 1885. The International Union for Quaternary Research (INQUA) has a Commission devoted to the study of the Holocene, and several IGCP projects have been based around environmental changes during the Holocene. Since 1991 there has also existed a journal dedicated exclusively to Holocene research (J. Matthews, ed., *The Holocene*). A potted history of the Holocene can be found in Roberts (1998).

During the Holocene, the Earth's climates and landscapes took on their modern natural form. Geomorphological change was especially rapid during the first few millennia, with DEGLACIATION of the ICE SHEETs remaining over Scandinavia and Canada, and SEA LEVELs rising to within a few metres of their modern elevations in most parts of the world. Because soil formation and vegetation development lagged behind the often rapid shifts in climate, many landscapes – both temperate and tropical – experienced a phase of temporary geomorphic instability during this deglacial climatic transition (Thomas and Thorp 1995; Edwards and Whittington 2001). In river valleys, there were major changes in the discharge of both water and sediment, and many streams and rivers which experienced increased discharges at the end of the last glacial period are now underfit (see UNDERFIT STREAM). As a consequence of these changes, rates of denudation and sediment flux were frequently above the long-term geological norm at the start of the Holocene (Figure 85). A range of different sedimentary 'archives', including ALLUVIAL FAN, FLOODPLAIN and deltaic/estuarine deposits, can be used to establish long-term changes in rates of sediment accumulation and hence upstream soil erosion. On the other hand, river valleys are not closed systems, and quantitative SEDIMENT BUDGET calculations are more easily achieved by exploiting lake sequences. LAKEs act as receptacles for materials eroded from their catchments, and dated cores can be used to calculate volumes of sediment deposited per unit time (Dearing 1994).

In most coastal regions, recognizably modern shoreline configurations were achieved around

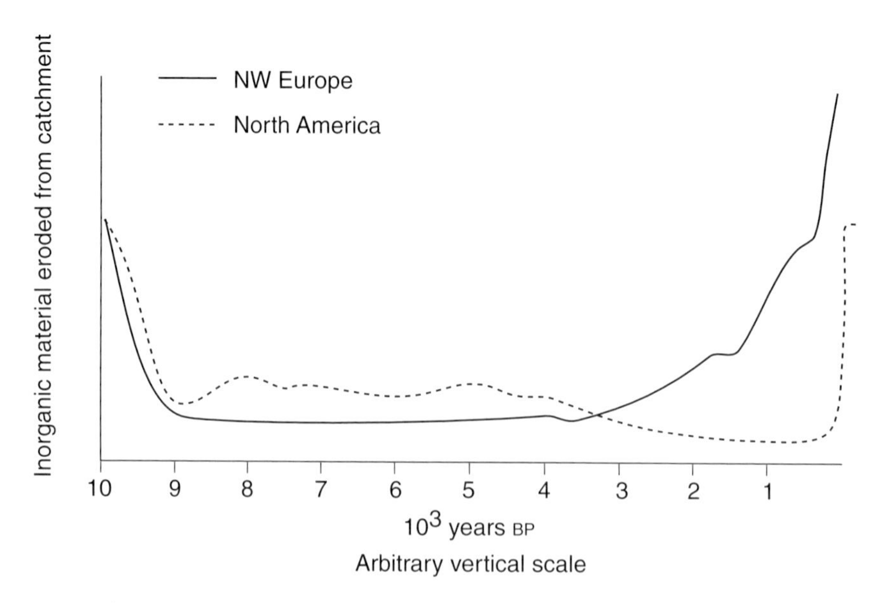

Figure 85 Generalized records of Holocene erosion based on sediment influx into lake basins (various sources, partly based on Dearing 1994)

7,000 cal. yr BP, with the main exceptions being in some high latitude regions such as Hudson Bay, where glacio-isostatic (see GLACIAL ISOSTASY) uplift has led to a continuing fall in sea levels during the Holocene. Elsewhere, rising sea levels during the early Holocene led to river valleys being drowned, with the end of this TRANSGRESSION representing the time of maximum marine incursion inland. Since then, stabilized sea levels and fluvially derived sediment discharge have led to a reversal in this trend, with the land pushing seawards at the mouths of major rivers such as the Rhône. This process has left many ancient harbour cities, such as Ephesus, Miletus and Troy in western Turkey, now stranded several km inland from the coast (Figure 86).

Various attempts have been made to subdivide the Holocene, usually on the basis of inferred climatic changes. Blytt and Sernander, for instance, proposed a scheme of alternating cool-set and warm-dry phases based on shifts in peat stratigraphy in northern Europe. In many temperate regions there is evidence of a 'thermal optimum' during the early-to-mid part of the Holocene. However, the clearest climatically induced environmental changes within the Holocene took place in the tropics and subtropics. One of the most important sources of palaeoclimatic (see PALAEOCLIMATE) and palaeohydrological (see PALAEOHYDROLOGY) information in low- and mid-latitude regions derives from non-outlet lakes, which can act like giant rain gauges. In East Africa, for example, lake levels were markedly higher and their waters markedly less saline between ~10,000 and ~6,000 cal. yr BP (Gasse 2000). On the other side, aeolian activity (see AEOLIAN PROCESSES) in regions such as the Saharan, Arabian and Thar deserts was greatly reduced and many sand dunefields (see SAND SEA AND DUNEFIELD) were inactive at this time. Rainfall in these regions increased by between 150 and 400 mm pa as tropical convectional rains moved further northwards, linked to a general strengthening of the African and Asian monsoonal system during the early Holocene.

By contrast with these climatically induced environmental changes, human impact has become an increasingly important agency in the creation and modification of landscapes during the later part of the Holocene. A critical point came when *Homo sapiens* turned to farming as the basis for human subsistence. The adoption and intensification of agriculture during the Holocene has led to widespread conversion of natural woodland or grassland into farming land, and this in turn has caused land degradation

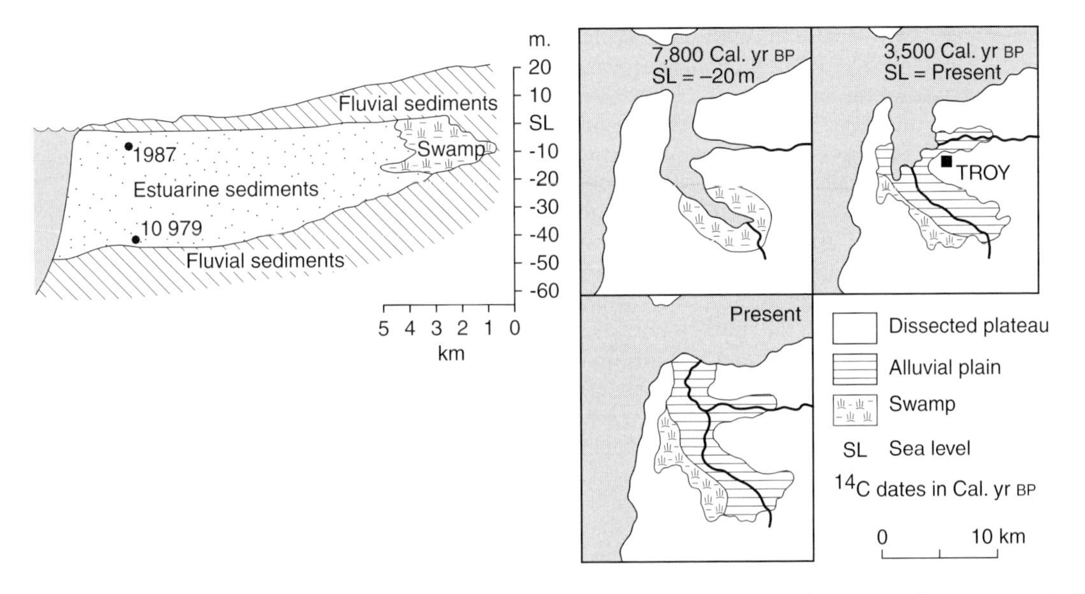

Figure 86 Geomorphological reconstructions in the vicinity of Troy, north-west Turkey, during the Holocene (based on Kraft *et al.* 1980)

through accelerated soil erosion and salinization. As a consequence, a much larger proportion of fluvial suspended sediment (see SUSPENDED LOAD) now originates from topsoil compared to bedrock sources than was previously the case. The same lake-sediment records that show high rates of flux during the major Pleistocene–Holocene climatic transition, typically show increases during the late Holocene associated with increasing human impact and land-use conversion (Figure 85). On the other hand, the dates for the onset of anthropogenically increased erosion rates vary from region to region, occurring earlier in Europe and in South and East Asia, but later in New World continents such as Australia and the Americas. At Frains Lake in the American Midwest, for example, soil erosion rates increased by two orders of magnitude to over 5 t/ha/yr^{-1} during the decade following the arrival of European settlement in 1830, but then stabilized at around 0.5–1 t/ha/yr^{-1} as forest clearance gave way to cropland (Davis 1976).

Fluctuations in climate have been superimposed upon the increasing human impact on geomorphic systems during the late Holocene. Particularly notable among these was the so-called 'Little Ice Age' from AD ~1400 to ~1850, when temperatures in Europe and the North Atlantic fell sufficiently for GLACIERs to advance down-valley in the Alps and other mountains (Plate 61). This climatic deterioration brought a higher risk of geomorphological hazards including LANDSLIDEs, avalanches, glacier outbursts and

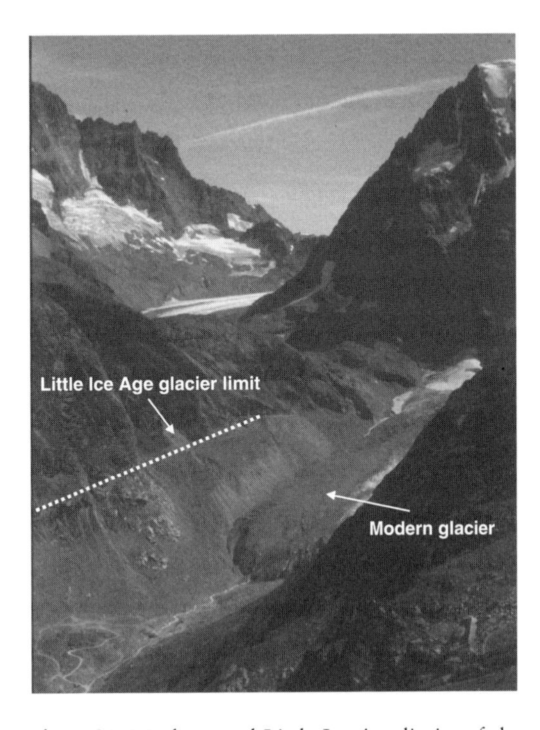

Plate 61 Modern and Little Ice Age limits of the Lower Arolla glacier, Switzerland

other floods (Grove 2003), with comparable periods of extreme floods and droughts occurring in dryland parts of the world.

It is during the Holocene that modern boundary conditions for the Earth system have come into existence. Consequently the Holocene represents a key baseline for assessing human impact on Earth surface and atmospheric processes, in our attempts to tease apart the relative roles of natural and human agencies in rates of landscape change. It also provides the time frame over which long-term magnitude–frequency relationships (see MANGNITUDE–FREQUENCY CONCEPT) can be assessed and the return period for extreme events, such as floods, can be calculated (Benito *et al.* 1998; Knox 2000).

References

Benito, G., Baker, V.R. and Gregory, K.J. (eds) (1998) *Palaeohydrology and Environmental Change*, Chichester: Wiley.

Davis, M.B. (1976) Erosion rates and land use history in southern Michigan, *Environmental Conservation* 3, 139–148.

Dearing, J. (1994) Reconstructing the history of soil erosion, in N. Roberts (ed.) *The Changing Global Environment*, 242–261, Oxford: Blackwell.

Edwards, K.J. and Whittington, G. (2001) Lake sediments, erosion and landscape change during the Holocene in Britain and Ireland, *Catena* 42, 143–173.

Gasse, F. (2000) Hydrological change in the African tropics since the Last Glacial Maximum, *Quaternary Science Reviews* 19, 189–212.

Grove, J.M. (2003) *The Little Ice Ages: Ancient and Modern*, Routledge: London.

Knox, J.C. (2000) Sensitivity of modern and Holocene floods to climate change, *Quaternary Science Reviews* 19, 439–458.

Kraft, J.C., Kayan, I. and Erol, O. (1980) Geomorphic reconstructions in the environs of ancient Troy, *Science* 209, 776–782.

Roberts, N. (1998) *The Holocene. An Environmental History*, 2nd edition, Blackwell: Oxford.

Thomas, M.F. and Thorp, M.B. (1995) Geomorphic response to rapid climatic and hydrologic change during the Late Pleistocene and Early Holocene in the humid and sub-humid-tropics, *Quaternary Science Reviews* 14, 101–124.

NEIL ROBERTS

HONEYCOMB WEATHERING

Honeycomb weathering is a type of CAVERNOUS WEATHERING. The terms honeycomb, stone lattice, stone lace and alveolar weathering have been used as synonyms. Honeycombs are associated in particular with arid and coastal environments, though they are also a feature of some building stones in urban environments. They may also occur on Mars (Rodriquez-Navarro 1998). Many honeycombs are seemingly caused by SALT WEATHERING (Mustoe 1982; Rodriguez-Navarro *et al.* 1999). They are composed of pits, commonly some centimetres deep, that are developed so close together as to be separated by a narrow wall only millimetres thick. They are known from a wide range of rock types, including sandstones, limestone, schists, gneiss, greywacke, arkose and metavolcanics. In favourable environments they can form in a matter of decades (Mottershead 1994).

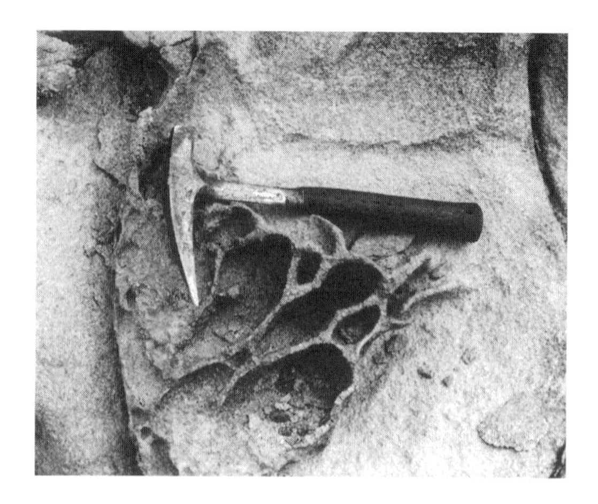

Plate 62 A small cluster of alveoles or honeycomb weathering forms developed on granite gneiss in a salty and foggy coastal environment on the south coast of Namibia, near Luderitz

References

Mottershead, D.N. (1994). Spatial variations in intensity of alveolar weathering of a dated sandstone structure in a coastal environment, Weston-Super-Mare, UK, in D.A. Robinson and R.B.G. Williams (eds) *Rock Weathering and Landforms Evolution*, 151–174, Chichester: Wiley.

Mustoe, G.E. (1982). The origin of honeycomb weathering, *Geological Society of America Bulletin* 93, 108–115.

Rodriguez-Navarro, C. (1998) Evidence of honeycomb weathering on Mars, *Geophysical Research Letters* 25, 3,249–3,252.

Rodriquez-Navarro, C., Doehne, E. and Sebastian, E. (1999) Origins of honeycomb weathering; the role of salts and wind, *Geological Society of America Bulletin* 111, 1,250–1,255.

A.S. GOUDIE

HOODOO

A common North American term for a pillar of eroded rock which is capped by a resistant rock layer, protecting a column of more erodable material beneath. They are effectively remnants of steep slopes as they are being eroded back, forming as the less resistant material is eroded away by water. The overlying hard caprock (generally a boulder or cobble) maintains the form's vertical integrity. Hoodoos are common in BADLAND morphology, and are typically formed in sedimentary rock (e.g. Bryce Canyon, Utah, USA) although examples exist in volcanoclastics (e.g. San Juan Mountains of Colorado, USA) and unconsolidated glacifluvial materials (e.g. Norway).

SEE ALSO: demoiselle

STEVE WARD

HORST

A relatively upraised fault block bounded by sharply defined and sometimes parallel reverse faults, though more commonly conjugate normal faults and opposing dips. The formation of the horst can be due to both the rifting and compressive movement of these marginal normal faults. Horsts are generally elongate ridge-like structures, with a plateau form on the uplifted horst block surface. Horst is a German term, and means retreat. Converse to horsts are graben (singular and plural). These are relatively low-standing fault blocks once again bounded by opposing normal faults, and occurring between zones of extension or compression. Half-grabens are grabens that are bounded by a normal fault on one side only. Areas of alternating uplifted and down-dropped fault blocks are thus referred to as horst and graben structures, and are associated with RIFT VALLEY AND RIFTING. In these regions, horsts are often the predominant sediment source into the down-dropped graben and any basins within. Examples of horst structures include the Black Forest and Harz Mountains in Germany, and the Vosges of eastern France. A famous graben structure is the Rhine Graben in Germany.

Further reading

Jaroszewski, W. and Kirk, W. L. (1984) *Fault and Fold Tectonics*, Chichester: Ellis Horwood.

SEE ALSO: fault and fault scarp

STEVE WARD

HORTON'S LAWS

In 1945, in one of the most significant twentieth-century contributions to geomorphology, the American engineer, Robert E. Horton endeavoured to express both the hierarchical arrangement and density of drainage networks in quantitative terms. In this he was explicitly following what he termed the 'ocular observation' (Horton 1945: 280) of the Scottish mathematician John Playfair (1802: 102).

> Every valley appears to consist of a main trunk, fed from a variety of branches, each running in a valley proportioned to its size, and all of them together forming a system of vallies, communicating with one another, and having such a nice adjustment of their declivities that none of them join the principal valley either on too high or too low a level.

Horton proceeded by introducing, first, the concept of STREAM ORDERING (1945: 281), in the following fashion:

> (U)nbranched fingertip tributaries are always designated as of order 1, tributaries or streams of the 2nd order receive branches or tributaries of the 1st order, but these only; a 3rd order stream must receive one or more tributaries of the 2nd order but may also receive 1st order tributaries. A 4th order stream receives branches of the 3rd and usually of lower orders, and so on. Using this system, the order of the main stream is the highest.

Somewhat unfortunately, Horton then developed an approach whereby the 'parent stream' had to be identified from source to mouth (Horton 1945: figure 7), something which is tricky and undoubtedly subjective (Figure 87a). A.N. Strahler (1957) proposed a modification of Horton's scheme (Figure 87b) which is less subjective and has been almost universally applied since 1957.

From the stream-ordering system, Horton proceeded to calculate two indices which he found to evince such regularity that they have become known as Horton's Laws (they are equally apparent whichever ordering scheme is used). The first observation was that the number of streams of different orders tended to follow an inverse geometric sequence. The ratio between each order being termed r_b, the bifurcation ratio. The second was that the average length of streams tended to increase as order rose. One set of Horton's data (and the value of r_b and r_l) are given in Table 25. If the data for stream numbers and lengths are plotted on semi-logarithmic paper, straight lines can be drawn showing roughly constant ratios throughout any one basin.

One further law – of stream slopes – was described by Horton (1945: 295), showing a geometric decrease in channel slope with increasing stream order; and yet two further laws (of basin area, which increases regularly with order) and the constant of channel maintenance were introduced by Schumm (1956). Horton's three basic 'laws' were set out (1945: 84 and figure 6) as:

Law of stream numbers: $N_o = r_b^{\ (s-o)}$
Law of stream lengths: $l_s = l_1 r_l^{0-1}$
Law of stream slopes: $s_c = s_1 \sqrt{r_s}^{\ (0-1)}$

Where: N_o is the number of streams in the drainage basin
r_b is the bifurcation ratio
s is the order of the main stream
o is the order of a given class of tributaries
l_s is the length of the main stream
l_1 is the average length of first-order streams
r_l is the length ratio
s_c is the slope of the main stream
s_1 is the average slope of first-order streams
r_s is the slope ratio

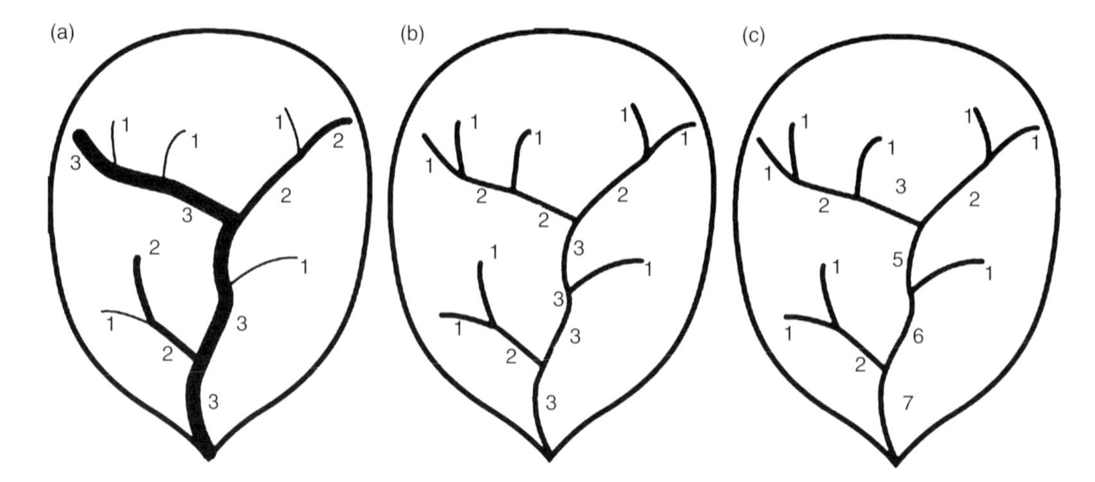

Figure 87 A drainage network with channels ordered: (a) according to Horton (1945); (b) according to Strahler (1957); (c) according to Shreve (1966)

Table 25 Drainage net, upper Hiwassee River

Order	No. of streams	Bifurcation ratio r_b	Average length, miles	Ratio of length, r_l
1	146		0.49	
		4.6		2.6
2	32		1.28	
		3.6		2.85
3	9		3.65	
		3.6		3.37
4	2		12.30	

Source: from Horton (1945: figure 7)

Schumm's additions may be stated in his words as: 'the mean drainage-basin areas of streams of each order tend to approximate closely a direct geometric series in which the first term is the mean area of the first-order basins' (1956: 606); and 'the relationship between mean drainage-basin areas of each order and mean channel lengths of each order is a linear function whose slope. . . is equivalent to the area in square feet necessary. . . for the maintenance of 1 foot. . . of channel' (1956: 607).

As Rodríguez-Iturbe and Rinaldo (1997: 6) make clear, the Hortonian laws, over the years, have been considered variously as demonstrating: that basins show a regular, evolutionary process; that basins demonstrate the development of purely topologically random networks; and that they say nothing whatsoever except that virtually all networks show these relationships. This last point is linked to the intuitively puzzling fact that Horton's Laws emerge whether basins are described by Horton's or by Strahler's methods.

However, there is a more fundamental difficulty with Horton's analysis as described in 1945: the categories of stream channels plotted on the x axis of the semi-logarithmic plots are, of course, *ordinal* numbers. As Horton's definition, quoted above and Figures 87a and b make plain, a 'second-order' basin may contain any number of 'first-order' channels between 2 and (effectively) ∞. Whilst it is quite proper to identify basins of different orders and to express numbers and averages in each class of variables and even ratios of values between classes, it is not proper to conduct the mathematical operations of multiplication or division using ordinal numbers and, in consequence, the 'regressions' shown on Horton's 1945 plots are simply spurious.

However, Horton introduced two other quantitative measures: DRAINAGE DENSITY (D_d) – the length of stream channels per unit area – and stream frequency (F_s), or the number of streams per unit area. Melton (1958) demonstrated that these two terms could be linked by a constant relationship:

$$F_s = 0.694\ D_d^2$$

This is now known as Melton's Law (Rodríguez-Iturbe and Rinaldo 1997: 8). These relationships described the degree of dissection of a landscape (see Kennedy 1978).

But the importance of the emphasis on stream frequency, involving actual counting of entities, was crucially developed by R. Shreve (1966) as the concept of basin magnitude, which is simply the number of first-order streams (or exterior links). Figure 87c shows the nature of this, true, ordering system, where magnitude is a ratio number that is susceptible of full mathematical manipulation.

Despite the difficulties with the Horton/Strahler ordering system, it is far easier to establish a sample of (say) fourth order basins than of 10 magnitude ones and a huge volume of research, especially although not exclusively from the 1950s and 1960s, focused on Hortonian relationships in generally fourth- or fifth-order basins. Church and Mark (1980) showed how this focus had tended to produce an apparent scale-dependence in the relationship between basin area and drainage density, which is actually isometric.

Nevertheless, the thrust of recent work on the FRACTAL nature of drainage basins, notably that summarized by Rodríguez-Iturbe and Rinaldo (1997), has been to substantiate the universality of the two fundamental Hortonian Laws: for example, their discussion of Optimal Channel Networks (OCNS) is shown to support both the Bifurcation Law and the Length ratio (1997: 278–279). Indeed, there is a major discussion of Hortonian networks which 'can be interpreted as signs of the fractal structure of the underlying network' (1997: 498). It is worth stressing, however, that what has endured have been the geometrically regular ratios of stream numbers and lengths between adjacent stream orders, rather than the spurious 'regression' plots which were such a feature of mid-twentieth century investigations of Horton's Laws.

References

Church, M.A. and Mark, D.M. (1980) On size and scale in geomorphology, *Progress in Physical Geography* 4, 342–390.

Horton, R.E. (1945) Erosional development of streams and their drainage basins; hydrophysical approach to quantitative morphology, *Geological Society of America Bulletin* 56, 275–370.

Kennedy, B.A. (1978) After Horton, *Earth Surface Processes and landforms* 3, 219–232.

Melton, M.A. (1958) Correlation structure of morphometric properties of drainage systems and their controlling agents, *Journal of Geology* 66, 442–460.

Playfair, J. (1802) *Illustrations of the Huttonian Theory of the Earth*, London: Cadell and Davies. Reprinted in facsimile, G.W. White (ed.) (1964), New York: Dover.

Rodríquez-Iturbe, I. and Rinaldo, A. (1997) *Fractal River Basins: Chance and Self-organization*, Cambridge: Cambridge University Press.
Schumm, S.A. (1956) Evolution of drainage systems and slopes in badlands at Perth Amboy, New Jersey, *Geological Society of America Bulletin* 67, 597–646.
Shreve, R.L. (1966) Statistical law of stream numbers, *Journal of Geology* 74, 17–37.
Strahler, A.N. (1957) Quantitative analysis of watershed geomorphology, *EOS Transactions American Geophysical Union* 38, 912–920.

Further reading

Knighton, D. (1998) *Fluvial Forms and Processes*, 2nd edition, London: Arnold.
Schumm, S.A. (ed.) (1977) *Drainage Basin Morphology*, Stroudsburg, PA: Dowden, Hutchinson and Ross.

SEE ALSO: drainage density; fractal; laws, geomorphological; stream ordering

BARBARA A. KENNEDY

HUMMOCK

Small mounds of low relief, which cover the ground surface, are common where fine-grained soils overlie PERMAFROST. Most hummocks are circular, and 1 to 2 m in diameter. They are domed, with a vertical relief of up to 25 cm, but usually less than 15 cm. The ACTIVE LAYER is thickest beneath the hummock centres, and thinnest at the circumference. The base of the active layer is bowl-shaped. Segregated ice lenses, subparallel to the base of the active layer, are characteristically abundant directly beneath hummocks, and this ice-rich zone is commonly also rich in organic material. Hummocks are generally stable features, which may persist for thousands of years.

At the surface, organic material accumulates around the edges of hummocks, but the centres may be bare of vegetation (mud hummocks) or covered by peat or vascular plants (earth hummocks). The soil in hummocks is frost-susceptible, and may contain little sand. Where the clay content is low, the soil may liquefy in response to small changes in moisture content of stress, and be extruded at the ground surface. Such mudboils may occur in fields of hummocks, but in general the clay content is of the order of 40 to 50 per cent, and sufficient to prevent liquefaction.

The hummock form is maintained by soil circulation within each feature, driven by moisture redistribution during freezing and thawing (Mackay 1980). The soil circulation proceeds by upwards movement in the middle of hummocks, spreads to the circumference near the surface, and slides downwards at the edges, near the base of the active layer. The upward movement is driven by convection, due to the contrast between mud of relatively low density, and enclosing sediment, where the mud is supersaturated by melting of ice lenses. The movement at the base of the active layer is associated with heave towards the hummock centre during upfreezing at the base of the active layer, and settlement down the bowl-shaped frost table as ice lenses thaw. At the surface, soil is driven outward by heave and subsidence during freezing and thawing over an inclined plane. These three processes are constrained by the requirement for conservation of mass. Evidence for the circulation has been provided by movement of markers at the ground surface, and by involutions in the soil stratigraphy when viewed in cross section. The importance of the bowl-shaped frost table on hummock form is demonstrated by the disappearance of hummocks in the years following forest fires, when the active layer deepens, and their reappearance with subsequent vegetation regeneration as the active layer thins.

In the boreal forest, trees on hummocks are tilted, and are commonly located near the edges of hummocks. Tilting of the trees is associated with development of the ice-rich zone at the base of the active layer and accompanying heave of the hummocks. Hummocks are the complementary feature for fine-grained soil of sorted circles in coarser materials.

Reference

Mackay, J.R. (1980) The origin of hummocks, western Arctic coast, Canada, *Canadian Journal of Earth Sciences* 17, 996–1,006.

C.R. BURN

HYDRATION

Hydration is the uptake of the entire water molecule by a mineral. For example, calcium sulphate (anhydrite $CaSO_4$) is hydrated to gypsum $CaSO_4{\cdot}H_2O$. This results in the mineral swelling. In a confined space, hydration pressures can be up to 100 Mpa, weakening the rock. In cold climates, White (1976) felt that much freeze–thaw weathering could actually be hydration shattering, with the forces of hydration as high as 2,000 kg cm^{-3}.

Widely occurring is the conversion of iron oxide (Haematite Fe_2O_3) into iron hydroxides variously cited as being in a poorly defined crystal form as $Fe(OH)_3$, as goethite 2FeOOH or as limonite ($2Fe_2O_3 \cdot 3H_2O$). The formation of these iron hydroxides involves considerable volume increases.

Alumino-silicate minerals can become subject to hydration through the formation of hydrated aluminium oxide. HYDROLYSIS can be seen as more important than hydration because it is the products of hydrolysis which are hydrated. For the formation of the hydrated aluminium oxide from microcline, a potassium-containing feldspar:

$$KAlSi_3O_8 + H_2O \rightarrow HAlSi_3O_8 + K^+ + OH^- \quad \text{(hydrolysis)}$$

Microcline

$$2HAlSi_3O_8 + 11H_2O \rightarrow Al_2O_3 + 6H_4SiO_4 \quad \text{(hydrolysis)}$$

$$Al_2O_3 + 3H_2O \rightarrow Al_2O_3 \cdot 3H_2O \quad \text{(hydration)}$$

However, it should be stressed that it is the hydration which facilitates the physical disintegration through volume expansion, weakening the mineral structure.

In addition to this formation of new hydrated minerals, more complex layered minerals can take up water between their layers and this can also be referred to as hydration. The plate-like minerals, such as mica, can be subject to expansion and physical disintegration when water penetrates between the plates. Water can be incorporated into a clay crystal lattice and, especially in the open lattice of montmorillonite clays, involving increases in volume of around $0.5\,cm^3\,g^{-1}$.

Reference

White, S.E. (1976) Is frost action really only hydration shattering? *Arctic and Alpine Research* 8, 1–6.

STEVE TRUDGILL

HYDRAULIC GEOMETRY

Hydraulic geometry of a river is the quantitative (mathematical and graphical) description of the channel cross section size and shape, fluid-flow properties and sediment-transport characteristics, in relation to the discharge being conducted by the channel. As such, every river channel, with rigid or deformable boundaries, has a hydraulic geometry. It is a descriptive device, derived from the empirical relations of regime 'theory' developed to aid canal design in India early last century (Lacey 1929). These ideas were first introduced into geomorphology by Leopold and Maddock (1953) who proposed the term 'hydraulic geometry' for this descriptor of the morphodynamics of alluvial channels.

The general equations of hydraulic geometry proposed by Leopold and Maddock (1953) necessarily are selective, reflecting relations among variables that were routinely measured or easily derived from such measurements made at US gauging stations:

$$\begin{aligned} w &= aQ^b \\ d &= cQ^f \\ v &= kQ^m \\ s &= gQ^z \\ n &= tQ^y \\ ff &= hQ^p \\ Q_{susp} &= rQ^j \end{aligned}$$

where w, d, v, s, n, ff and Q_{susp} are respectively width, mean depth, mean velocity, water-surface slope, flow resistance (Manning's n or D'Arcy Weisbach ff) and suspended-sediment load. An important missing element of this seven-variable set is bedload transport but these measured data are rarely available.

Implicit in the specification of these equations of hydraulic geometry are the following notions:

1. Discharge, Q, is the dominant independent variable in the hydraulic geometry;
2. The relations between the independent and dependent variables can be described as simple power functions;
3. As power functions, the logarithm of the dependent variables plot against the logarithm of discharge as a straight-line graph (that is, there is a linear relationship between the order-of-magnitude increases in the pairs of variables);
4. The existence of these orderly hydraulic-geometry relations implies an underlying set of processes reflecting the operation of equilibrium in the morphodynamic system;
5. Because continuity must be satisfied in fluid flow it follows from the rules of algebra that:

   $Q = wdv = (aQ^b)\,(cQ^f)\,(kQ^m)$
   and that $ack = 1$ and $b + f + m = 1$

The adjustment of channel morphology and hydraulics in relation to changes in discharge has been considered in two quite different contexts: at-a-station hydraulic geometry and downstream hydraulic geometry.

At-a-station hydraulic geometry

The at-a-station hydraulic geometry describes how the channel geometry and hydraulics of flow change as discharge increases at an individual channel cross section of a river. Thus it describes the way in which the flow fills the often effectively rigid channel boundary as discharge changes over time. An example of this type of hydraulic geometry is shown in Figure 88 for the Fraser River in western Canada. At-a-station hydraulic geometry is only defined for discharges up to the channel-filling (or bankfull) stage.

Downstream hydraulic geometry

Discharge in a river also increases as tributaries join the main stem in the downstream direction and add flow to the fluvial system. The downstream hydraulic geometry describes how this spatially increasing discharge enlarges and shapes the channel and alters the properties of the streamflow. In order to allow for comparisons between channel sections these changes are referred to as a discharge of constant return period or to a consistent relative stage. The most common reference discharge is bankfull discharge, which is often taken to be the channel-forming discharge. An example of this type of hydraulic geometry is shown in Figure 89 for Oldman River in western Canada.

Theoretical context and interpretation of hydraulic geometry

Conventional hydraulic geometry describes a partial picture of adjustments in the fluvial system but contains little information about the controls on such adjustments. When responding to changes in discharge, an alluvial stream must satisfy at least three sets of physical relations: continuity, flow resistance, and sediment transport. The first relation is definitional but the other two relations are only understood in a qualitative

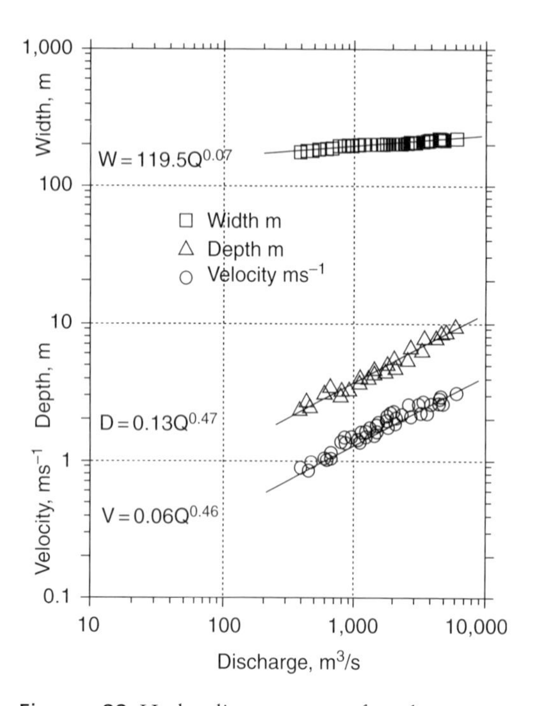

Figure 88 Hydraulic geometry for the Fraser River, Canada

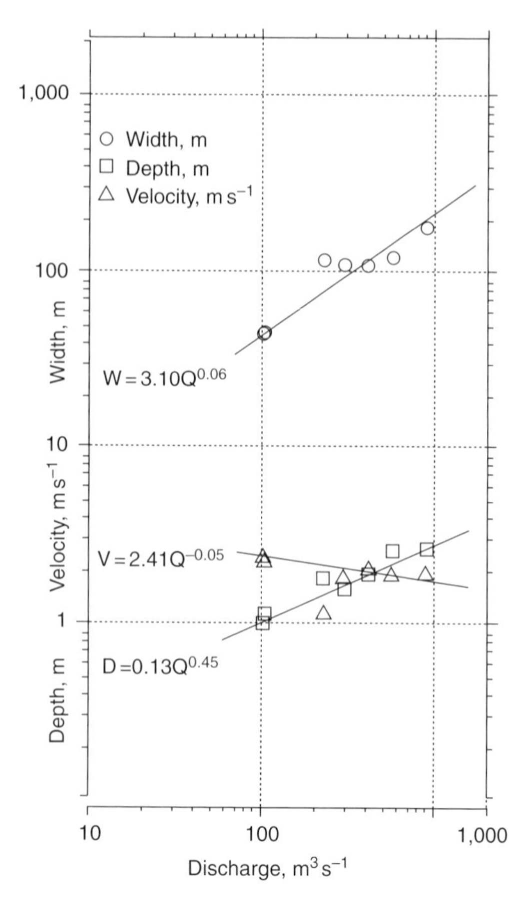

Figure 89 Hydraulic geometry for the Oldman River, Canada

sense. For this reason, and also because channels are free to adjust in ways other than by changes in width, depth and mean velocity (Hey 1988), the hydraulic geometry of alluvial channels generally is regarded as being indeterminate. Nevertheless, these physical relations do inform our interpretation of the hydraulic geometry at a qualitative level of analysis.

The primary difference between the two hydraulic geometries is that, unlike the downstream case, for the at-a-station case the boundary materials and the water-surface slope essentially remain constant as discharge changes. The principal control on the hydraulic geometry is the shape of the channel cross section formed at high discharges. Channel form in turn largely is determined by the strength of the boundary materials. If the boundary materials are cohesive and strong (mud-dominated rivers, for example), banks can develop which are steep and high. In such cases, as discharge increases, channel width will change much more slowly than flow depth and velocity. The rate of increase in flow velocity depends on the changing relative roughness of the channel. Typically, but not always, flow resistance declines as the effects of roughness elements are drowned by the increasing discharge. Thus the exponents on the velocity/discharge relation are relatively high. A different geomorphic response can be expected in the case of channels with weak non-cohesive banks (such as sand-dominated channels). In this case bank height is limited by material strength and the capacity for mean depth changes to accommodate increases in discharge is small. Change in velocity is also highly constrained by the conservative adjustment in flow depth and as a result channel width changes greatly to accommodate the discharge increases in such channels. Table 26 shows typical exponent values in the at-a-station hydraulic geometry equations for a variety of channel types.

In the case of downstream hydraulic geometry, adjustments of river channels to increases in bankfull discharge also reflect spatial changes in water-surface slope and the size of boundary-materials along the channel. A major control in this case is the balance that is struck between the impelling and resisting forces acting in the downstream direction. Although downstream increases in discharge and declining size of boundary material tend to work together to accelerate the flow, the forces for this change are almost equally countered by the decline in boundary shear stress related to declining water-surface slope. As a result, downstream hydraulic geometry is characterized by constant or declining mean velocity downstream and by the necessity for changes in discharge to be almost fully accommodated by adjustments in channel width and mean flow-depth alone. As before, the apportioning of the discharge change between width and velocity depends largely on boundary material strength. In a mud-dominated channel which can support steep and high channel banks the exponent on depth will be high and that on width low. In contrast, a sand-bed channel with weak banks will accommodate most of the increased bankfull

Table 26 Selected values of exponents in the equations of hydraulic geometry of river channels

Channel locality and type	At-a-station values			Downstream values		
	b	f	m	b	f	m
Mid-west USA (Leopold and Maddock 1953)	0.26	0.40	0.34	0.50	0.40	0.10
Mid-west USA (Carlston 1969)				0.46	0.38	0.16
Ephemeral streams, semi-arid USA (Leopold and Miller 1956)	0.25	0.41	0.33	0.50	0.30	0.20
Upper Salmon River, Idaho (Emmett 1975)				0.54	0.34	0.12
R. Bollin Dean, coarse-bed cohesive banks (Knighton 1974)	0.12	0.40	0.48	0.46	0.16	0.38
British gravel-bed rivers (Charleton *et al.* 1978)				0.45	0.40	0.15
Columbia River, Canada, canal-like anastomosed sand channels, cohesive banks (Tabata 2002)	0.10	0.66	0.24			

Notes: $W = aQ^b$, $D = cQ^f$, $V = kQ^m$

discharge downstream by widening the channel and the exponent for depth will be correspondingly low. Table 26 shows typical exponent values in the downstream hydraulic geometry equations for a variety of channel types.

Comprehensive reviews of all aspects of the hydraulic geometry of natural stream channels are available in fluvial geomorphology texts such as those by Knighton (1998), Richards (1982) and Leopold *et al.* (1964).

Limitations of hydraulic geometry

The power of conventional hydraulic geometry is its facility to generalize the process of channel adjustment in terms of simple functions so that the morphology of channels can be compared readily, one with another. This facility is also its primary limitation: it ascribes to the channel adjustment process the simple behaviour of simple functions which in detailed reality, however, may be very complex. Given the typical statistical noise in measurements of hydraulics and morphology of natural channels, power functions provide a simple and robust model for the hydraulic geometry but there is no independent theoretical justification for their use. The power-function model is merely a convenient approximation of reality.

Geomorphologists who recognize the limitations of power functions and have suggested hydraulic geometries based on alternative linearizing models (such as the log-quadratic model) include Richards (1973), Knighton (1975) and Ferguson (1986). Still others have taken multivariate statistical approaches to characterizing hydraulic geometry (Bates 1990; Rhoads 1992).

All these mathematical models of hydraulic geometry imply, however, that the channel adjustment process is continuous when in reality it is often markedly discontinuous. For example, many channels exhibit channel-in-channel morphology reflecting the capacity of low flows to shape the basal boundary (long duration offsetting low flow magnitude). Others exhibit within-channel benches reflecting the unequal effectiveness of higher discharge ranges of the hydrologic regime in shaping the channel (Woodyer 1968). Further, although low discharges essentially flow over rigid boundaries in many channels, increases in discharge at a section eventually lead to the onset of sediment entrainment and channel scour and this fundamentally alters the morphodynamics of the channel. Indeed such scour-related discontinuities in channel adjustments to changes in discharge may be an important part of the adjustment regime but may be completely obscured by the use of hydraulic geometry (Hickin 1995).

In the case of downstream hydraulic geometry the point-source addition of tributary flow and sediment to the fluvial system is a fundamental spatial and process discontinuity that is essentially a step-function process, not the continuous adjustment approximated by power functions.

References

Bates, B.C. (1990) A statistical log piecewise linear model of at-a-station hydraulic geometry, *Water Resources Research* 26, 109–118.

Carlston, C.W. (1969) Downstream variations in the hydraulic geometry of streams: special emphasis on mean velocity, *American Journal of Science* 267, 499–510.

Charleton, F.G., Brown, P.M. and Benson, R.W. (1978) The hydraulic geometry of some gravel rivers in Britain, *Hydraulics Research Station Report*, IT 180.

Emmett, W.W. (1975) The channels and waters of the Upper Salmon River area, Idaho, *US Geological Survey Professional Paper* 870A.

Ferguson, R.I. (1986) Hydraulics and hydraulic geometry, *Progress in Physical Geography* 10, 1–31.

Hey, R.D. (1988) Mathematical models of channel morphology, in M.G. Anderson (ed.) *Modelling Geomorphological Systems*, 99–125, Chichester: Wiley.

Hickin, E.J. (1995) Hydraulic geometry and channel scour: Fraser River, B.C., Canada, in E.J. Hickin (ed.) *River Geomorphology*, 155–167, Chichester: Wiley.

Knighton, A.D. (1974) Variation in width-discharge relation and some implications for hydraulic geometry, *Geological Society of America Bulletin* 85, 1,069–1,076.

——(1975) Variations in at-a-station hydraulic geometry, *American Journal of Science* 275, 186–218.

——(1998) *Fluvial Forms and Processes: A New Perspective*, London: Arnold.

Lacey, C. (1929) Stable channels in alluvium, *Proceedings of the Institution of Civil Engineers* 229, 259–384.

Leopold, L.B. and Maddock, T. (1953) The hydraulic geometry of stream channels and some physiographic implications, *United States Geological Survey Professional Paper* 252.

Leopold, L.B. and Miller, J.P. (1956) Ephemeral streams – hydraulic factors and their relation to the drainage net, *United States Geological Survey Professional Paper* 282A.

Leopold, L.B., Wolman, M.G. and Miller, J.P. (1964) *Fluvial Processes in Geomorphology*, San Francisco: Freeman.

Rhoads, B.L. (1992) Statistical models of fluvial systems, *Geomorphology* 5, 433–455.

Richards, K.S. (1973) Hydraulic geometry and channel roughness – a non-linear system, *American Journal of Science* 273, 877–896.

Richards, K.S. (1982) *Rivers: Form and Process in Alluvial Rivers*, London: Methuen.
Tabata, K.K. (2002) Character and conductivity of anastomosing channels, upper Columbia River, British Columbia, Canada, M.Sc. Thesis, Department of Geography, Simon Fraser University, BC, Canada.
Woodyer, K.D. (1968) Bankfull frequency in rivers, *Journal of Hydrology* 6, 114–142.

Further reading

Kellerhals, R., Neill, C.R. and Bray, D.I. (1972) Hydraulic and geomorphic characteristics of rivers in Alberta, *Research Council of Alberta, River Engineering and Surface Hydrology Report* 72–1, 16–18.

SEE ALSO: channel, alluvial; fluvial geomorphology

EDWARD J. HICKIN

HYDRO-LACCOLITH

A mound of ice formed by frost heaving of frozen underground water, resembling a laccolith in section. The term hydro-laccolith is synonymous to the terms ice laccolith and PINGO. However, they differ from pingos in that they are seasonal forms (whereas pingos are perennial), and differ from ice laccoliths in that they do not form within the active layer of permafrost ground. Hydro-laccoliths range in size between 1 and 10 m diameter, and are usually less than 2 m in height.

Further reading

French, H.M. (1996) *The Periglacial Environment*, Harlow: Longman.

SEE ALSO: periglacial geomorphology

STEVE WARD

HYDROCOMPACTION

The compaction and reduction in volume of soils and sediments that occurs when their moisture content is increased. It is also known as 'collapse compression', 'hydrocompression', 'hydroconsolidation' and 'saturation shrinkage' (Charles 1994). The process causes ground subsidence when unconsolidated sediments of low density are wetted, as for example by the application of irrigation water. It is a feature of arid and semi-arid lands where materials such as wind-blown loess or certain alluvial sediments above the water table are not normally wetted below the root zone and have high void ratios. When dry, such materials may have sufficient strength to support considerable effective stresses without compacting. However, when they are wetted, their intergranular strength is weakened because of the rearrangement of their particles. The associated subsidence may create fissures in the ground and is a process that needs to be considered during the construction of canals, pipelines, dams and irrigation schemes (Al-Harthi and Bankher 1999).

References

Al-Harthi, A.A. and Bankher, K.A. (1999) Collapsing loess-like soil in western Saudi Arabia, *Journal of Arid Environments* 41, 383–399.
Charles, J.A. (1994) Collapse compression of fills on inundation, in K.R. Saxena (ed.) *Geotechnical Engineering: Emerging Trends in Design and Practice*, 353–375, Rotterdam: Balkema.

A.S. GOUDIE

HYDROLOGICAL GEOMORPHOLOGY

Hydrological geomorphology is literally the interface between hydrology, the science of water and geomorphology, the study of landforms and their causative processes. It is particularly surface water hydrology that interacts with geomorphology although recently there has been an increasing convergence between research in geomorphology and in groundwater hydrology and hydrogeology (Brown and Bradley 1995).

When geomorphology emerged, more than a century ago, with a focus upon the morphology of landscape and the study of landforms, the attractive CYCLE OF EROSION promulgated by W.M. Davis provided a focus for many approaches to geomorphology adopted in the first half of the twentieth century. In the second half of the century, investigation of processes prevailed to a much greater extent (Gregory 2000) and to achieve this in relation to research on land areas, involving the study of flowing water, there was a need to take a much greater interest in hydrology. At first a period of familiarization saw the publication of books written by geographers (e.g. Ward 1966) and geomorphologists (e.g. Gregory and Walling 1973) but then progress was made towards the research interface between geomorphology and hydrology where the geomorphologist could make significant contributions. Hydrology was for long the science of water, with comparatively little attention given to water quality, but increasing attention given to landscape-forming processes, and to hydrological influences upon those processes, naturally led to original and innovative contributions being made by geomorphologists at the interface of hydrology with geomorphology.

Many interface investigations have accounted for the growth of hydrological geomorphology and these include at least four types. In addition to relationships between drainage basin characteristics and basin hydrological response, geomorphologists have made particular contributions in the investigation of *runoff producing areas* and the dynamic ways in which such areas contribute to the generation of stream hydrographs, including networks of subsurface pipes as well as headwater drainage systems and the modelling of their role in runoff production (Beven and Kirkby 1993). Such investigations often employed results from small experimental catchments which were also useful in relation to research on *sediment dynamics*. Geomorphological interest in the sediment area arose from the requirement for rates of erosion or denudation to relate to landscape development. In the case of suspended sediment production and transport, whereas simple rating curves relating discharge and suspended sediment concentrations had previously been used for analysis, it was demonstrated how analysis of sediment hydrographs could be employed to advance understanding and explanation of the mechanics of erosion. Later research by geomorphologists could therefore focus on the mechanics of production of such sediment in relation to the range of sediment sources and sediment producing areas. Similar contributions, made by geomorphologists to investigate and refine relationships employed to model bedload and solute transport, benefited from studies of the generation of solutes from catchment areas and of the entrainment of bedload in different channel situations.

Contributions at the level of the drainage basin have arisen first because of interest in the *temporal changes* that have occurred. As such changes cannot be based entirely on continuous hydrological records used in hydrology, other techniques are necessary to reconstruct past hydrological changes and these are used in geomorphology. The approach of palaeohydrology (Schumm 1965), which has been defined as 'the science of the waters of the earth, their composition, distribution and movement on ancient landscapes from the beginning of the first rainfall to the beginning of continuous hydrological records' (Gregory 1996), has been developed significantly so that the broad picture of past hydrological changes has been reconstructed for different parts of the world (Benito and Gregory 2003). Such reconstructions can provide potentially useful background for studies of global change and of basin management. A particularly valuable and successful approach has been the investigation and analysis of PALAEOFLOODS based upon the recognition of remnants of flood deposits, often as slackwater deposits, because this affords information on flood frequency which extends well beyond the period of instrumental record but may significantly affect the way in which flood frequency is analysed and relationships are established (Baker 2003). Partly as a result of the results obtained from investigations of temporal change, geomorphological contributions have become significant in relation to *river basin management*. Approaches are increasingly required to be integrated and should therefore include consideration of the range of human impacts throughout the basin (Downs *et al.* 1991) but in addition there is a need for sustainable approaches both at the basin level (NRC 1999) and in relation to the restoration of specific river reaches (Brookes and Shields 1996).

A number of the salient contributions made by geomorphologists have illuminated understanding of particular components of the hydrological cycle so that aspects of a hydrological geomorphology have emerged. This has, paradoxically, led to the fudging of the original definition of geomorphology. No longer focused primarily upon landforms, geomorphology is involved in contributions to hybrid fields where some of the most innovative research occurs and where multidisciplinary approaches can be optimized. Thus hydromorphology was suggested by Scheidegger (1973) as the geomorphological study of water and its effects, which includes coastal as well as fluvial hydrogeomorphology, in which there is a range of ways in which applications can be made (Gregory 1979). Such multidisciplinary fields also include biogeomorphology and, although there is no precise definition of it, hydrological geomorphology is an area of interaction which continues to offer promising research and applied opportunities.

References

Baker, V.R. (2003) Palaeofloods and extended discharge records, in G. Benito and K.J. Gregory (eds) *Palaeohydrology. Understanding Global Change*, Chichester: Wiley.

Benito, G. and Gregory, K.J. (2003) *Palaeohydrology. Understanding Global Change*, Chichester: Wiley.

Beven, K. and Kirkby, M.J. (eds) (1993) *Channel Network Hydrology*, Chichester: Wiley.

Brookes, A. and Shields, F.D. (eds) (1996) *River Channel Restoration. Guiding Principles for Sustainable Projects*, Chichester: Wiley.

Brown, A.G. and Bradley, C. (1995) Geomorphology and groundwater: convergence and diversification, in A.G. Brown (ed.) *Geomorphology and Groundwater*, 1–20, Chichester: Wiley.

Downs, P.W., Gregory, K.J. and Brookes, A. (1991) How integrated is river basin management? *Environmental Management* 15, 299–309.

Gregory, K.J. (1979) Hydrogeomorphology: how applied should we become? *Progress in Physical Geography* 3, 84–101.

——(1996) Introduction, in J. Branson, A.G. Brown and K.J. Gregory (eds) *Global Continental Changes: The Context of Palaeohydrology*, 1–8, London: Geological Society.

——(2000) *The Changing Nature of Physical Geography*, London: Arnold.

Gregory, K.J. and Walling, D.E. (1973) *Drainage Basin Form and Process*, London: Arnold.

NRC Committee on Watershed Management (National Research Council) (1999) *New Strategies for America's Watersheds*, Washington, DC: National Academy Press.

Scheidegger, A.E. (1973) Hydrogeomorphology, *Journal of Hydrology* 20, 193–215.

Schumm, S.A. (1965) Quaternary palaeohydrology, in H.E. Wright and D.G. Frey (eds) *The Quaternary of the United States*, 783–794, Princeton: Princeton University Press.

Ward, R.C. (1966) *Principles of Hydrology*, London: McGraw-Hill.

KENNETH J.GREGORY

HYDROLYSIS

Hydrolysis is a chemical reaction of a compound with water. As opposed to HYDRATION where the water is absorbed into the compound, in hydrolysis (or 'splitting by water') both the water and the compound split up and recombine. The water is thus a reactant and not merely a solvent. For example, the reaction between a potassium-containing feldspar and water:

$$\underset{\text{feldspar}}{KAlSi_3O_8} + \underset{\text{water}}{H_2O} \rightarrow \underset{\text{silicic acid}}{HAlSi_3O_8} + \underset{\text{potassium}}{K^+} + \underset{\text{hydroxyl}}{OH^-} \quad (1)$$

The mineral releases potassium and the water splits into OH^- and H^+, the H^+ combining with the aluminosilicate from the mineral. The production of hydroxyl (OH^-) ions in solution means that the pH of the water rises. This is illustrated by grinding a mineral to powder in water and measuring the pH or 'abrasion pH'. For the more reactive minerals, this is between 8 and 11, with 8 for calcite and feldspars 8–10.

This reaction can take place in pure water at neutral pH (7). However, if the water is acidified by additional H^+ ions so the pH is below 7, the weathering reaction is accelerated. The prevailing form of acidification is by carbon dioxide:

$$CO_2 + H_2O \rightarrow H_2CO_3 \rightarrow H^+ + HCO_3^- \quad (2)$$

with, for calcite:

$$CaCO_3 \rightarrow Ca^{2+} + CO_3^{2-} \rightarrow Ca^{2+} + 2(HCO_3^-) \quad (3)$$

The mineral has combined with the constituents of water, giving a free mineral ion in the water (Ca^{2+}) with one source of HCO_3^- from OH^- in the water and CO_2 and the other source of HCO_3^- from the H^+ from the water in Equation 2 combined with the CO_3^{2-} from the calcite.

Hydrolysis is thus a fundamental weathering process and it can be readily appreciated that as there are many organic sources of CO_2 through respiration and decomposition, then many such reactions are biologically originated.

STEVE TRUDGILL

HYDROPHOBIC SOIL (WATER REPELLENCY)

A soil that resists wetting by water for periods ranging from a few seconds up to days or even weeks. The reduced affinity for water is caused by a coating of long-chained organic molecules on soil particles and/or by the presence of hydrophobic (water repellent) interstitial matter. Such matter is released from a wide variety of plants through mechanical wear from leaf surfaces, decomposition of litter, release via roots and vaporization followed by condensation onto soil particles during burning, or from soil fungi and micro-organisms. The effect is temporally variable and is usually most pronounced after prolonged dry spells. Although mostly associated with semi-arid and areas with Mediterranean-type climates, it is now known to occur in a wide range of climates, including temperate and arctic-alpine environments. The potential geomorphological impacts include the restriction of soil water movement to preferential pathways; increased OVERLAND FLOW; enhanced streamflow responses to rainstorms; enhanced total streamflow; enhanced splash detachment by raindrop impact (see RAINDROP IMPACT, SPLASH AND WASH); increased SOIL EROSION by both wind and water; and increased erosion by dry creep (movement by loose, dry surface material on steep slopes). In contrast, water-repellent organic material in well-developed soil aggregates can help to stabilize them, thereby reducing soil ERODIBILITY. These impacts, however, have been largely inferred rather than demonstrated under field conditions.

Further reading

Dekker, L.W. and Ritsema, C.J. (1994) How water moves in a water repellent sandy soil. 1. Potential and actual water repellency, *Water Resources Research* 30, 2,507–2,517.

Doerr, S.H., Shakesby, R.A. and Walsh, R.P.D. (2000) Soil hydrophobicity: its causes, characteristics and hydro-geomorphological significance, *Earth-Science Reviews* 51, 33–65.

Shakesby, R.A., Doerr, S.H. and Walsh, R.P.D. (2000) The erosional impact of soil hydrophobicity: current problems and future research directions, *Journal of Hydrology* 231–232, 178–191.

SEE ALSO: fire

RICHARD A. SHAKESBY

HYPERCONCENTRATED FLOW

A flowing mixture of water and sediment transitional between a debris flow and muddy streamflow. The terms hyperconcentrated flow, hyperconcentrated flood flow, and hyperconcentrated streamflow are synonymous. The term was originally used for streamflow with sediment concentrations between 40–80 per cent by weight or 20–60 per cent sediment by volume. Rheologically the fluid appears to be slightly plastic but flows like water (Pierson and Costa 1987). Such flows are gravitationally driven, non-uniform mixtures of debris and water. They possess fluvial characteristics yet are capable of carrying very high sediment loads. Hyperconcentrated flows show clast support from grain-dispersive forces, dampened turbulence and buoyancy (implying yield strength). Sediment deposition appears to be by rapid grain-by-grain settling at the base and margins of the flow. Resultant deposits are usually either massive or display weak, near-horizontal stratification. Hyperconcentrated flows are common in volcanic environments where eruptions release large volumes of water from crater lakes or from melting of ice and snow, and when debris flows evolve downstream into hyperconcentrated flows.

Reference

Pierson, T.C. and Costa, J.E. (1987) A rheologic classification of subaerial sediment-water flows, in J.E. Costa and G.P. Wieczorek (eds) *Debris Flows/Avalanches: Process, Recognition and Mitigation, Geological Society of America Reviews in Engineering Geology* 7, 1–12.

VINCENT E. NEALL

HYPSOMETRIC ANALYSIS

Hypsometric analysis is the study of the distribution of topographic surface area with respect to altitude. The area-altitude relationship is described by the hypsometric curve (hypsographic curve) that is expressed by the function $y = f(x)$. In its absolute formulation, this curve is obtained by plotting on ordinate elevations and depths, from the top of the highest mountain to the maximum depth of abyssal trenches, and on abscissa the values of topographic surface areas. It is a cumulative curve: the abscissa of any point on it expresses the total area lying above the elevation of the corresponding ordinate.

The *absolute hypsometric curve* can be constructed for any area of land, from a small portion to the entire planet. Its use, however, is unsatisfactory when it is necessary to compare areas of different sizes and relief. To overcome this difficulty, percentage hypsometric analysis can be used, as it affords a method for expressing the area-altitude relationships in a dimensionless form (Langbein 1947).

The *percentage hypsometric curve* was used by Strahler (1952) to analyse erosional topography of drainage basins that are the basic geomorphic unit. This curve is represented by the function $y = f(x)$, but x and y are dimensionless parameters: x is the ratio of the area a above a given contour line and the whole basin area (A) and y is the ratio of the height h between the basin mouth and the contour that defines the lower limit of the area a and the total height range in the basin (H). Obviously, x and y vary between 0 and 1. These curves can be compared irrespective of true scale as they express merely the way the landmass is distributed from base to top.

Integrating the function between the limits of $x = 0$ and $x = 1$ (or simply measuring on the graph the area under the curve) the *hypsometric integral* is obtained. It is expressed in percentage units and represents the ratio of the landmass volume of a given drainage basin to the volume of the reference solid with base equal to the basin area and height equal to the total height range in the basin. In other words the hypsometric integral measures the percentage volume of earth material remaining after the erosion of an original landmass having volume equal to the reference solid.

In their classical interpretation, Strahler's hypsometric curves and integrals identify quantitatively the stages of the Davisian geomorphic cycle. Convex curves, with hypsometric integrals higher than 0.60, indicate the inequilibrium stage of 'youth'. Smoothly S-shaped curves that cross approximately the centre of the diagram and have integrals ranging from 0.60 to 0.40 express the equilibrium stage of 'maturity' or the 'old stage'. Strongly concave curves with very low integrals result only where monadnock masses are present.

Further studies delineated another interpretation of the area-altitude analysis: the hypsometric curves express not only the stage of the 'geomorphic cycle' but also the complexity of denudational processes and the rate of the geomorphological changes in drainage basins. Such changes take place through subsequent stages of dynamic equilibrium between tectonic uplift and DENUDATION (Ciccacci *et al.* 1992); therefore each basin is marked by a hypsometric curve which is mostly a function of the denudational process type. Convex curves with a high integral refer to basins in which stream erosion is the most vigorous denudational process. Concave curves with a low integral mark basins mainly affected by intensive slope processes. Finally, hypsometric curves with an integral close to 0.5 are characteristic of basins where stream erosion balances the effectiveness of slope processes.

Actually, the classic interpretation of the hypsometric curves matches the morphodynamic characters of drainage basins in tectonically stable regions; the same interpretation is rather unsuited to explain the plano-altimetric configurations of regions affected by recent or active tectonics (Ohmori 1993; D'Alessandro *et al.* 1999).

References

Ciccacci, S., D'Alessandro, L., Fredi, P. and Lupia-Palmieri, E. (1992) Relations between morphometric characteristics and denudational processes in some drainage basins of Italy, *Zeitschrift für Geomorphology N.F.* 36, 53–67.

D'Alessandro, L., Del Monte, M., Fredi, P., Lupia-Palmieri, E. and Peppoloni, S. (1999) Hypsometric analysis in the study of Italian drainage basin morphoevolution, *Transactions Japanese Geomorphological Union* 20–23, 187–201.

Langbein, W.B. (1947) Topographic characteristics of drainage basins, *US Geological Survey, Water Supply Paper* 968-C, 125–157.

Ohmori, H. (1993) Changes in the hypsometric curve through mountain building resulting from concurrent tectonics and denudation, *Geomorphology* 8, 263–277.

Strahler, A.N. (1952) Hypsometric (area-altitude) analysis of erosional topography, *Geological Society of America Bulletin* 63, 1,117–1,142.

ELVIDIO LUPIA-PALMIERI

I

ICE

Ice is the solid phase of water, i.e. a chemical compound of two positive univalent ions of hydrogen and one negative bivalent oxygen (H_2O). Water and ice are responsible for most of the processes leading to landscape sculpturing. In water an isosceles triangle is formed by the three nuclei of the H_2O and the apex angle is equal to 105° 3′. The oxygen nucleus is at the apex and the hydrogen nuclei (protons) at the other two corners of the molecule. Two of the eight electrons of the oxygen atom rotate close to the nucleus and another two have eccentric orbits which also contain the electrons from one of the hydrogen atoms. The oxygen nucleus and one proton are enclosed by each of these orbits. The other four electrons circle in two other eccentric orbits. For this reason, four eccentric orbits radiate tetrahedrally from the oxygen nucleus, which is completely screened by the electron orbits. Some of the positive charge of the protons is not screened completely. The eccentric electron orbits provide an excess negative charge in the direction of the two orbits without protons. Small negative charges at the oxygen end of the molecule and equally small positive charges at the hydrogen end determine that the H_2O molecules are slightly charged. Thus, a weak electrostatic bond is formed between the hydrogen end of one molecule and the oxygen end of another (the hydrogen bond) and it is for this reason that each molecule of water is surrounded by four others in a regular tetrahedral spatial arrangement. With respect to any particular molecule, the hydrogen bond is more than ten times weaker than the covalent bond. Therefore, mixtures of molecules are easily formed and broken by the reduction or addition of energy. In the liquid phase, the molecules are in motion and can be of variable distances, one to another, but, in ice, stable hydrogen bonds are created between the surrounding molecules, thereby generating the hexagonal crystal structure (Figure 90a, b). This is because in negative temperatures the energy of the system is lower. It gives pure ice a symmetrical lattice arrangement and a lower density ($0.91668\,g\,cm^{-3}$) than pure water ($0.999984\,g\,cm^{-3}$ – at the temperature 0 °C and the normal atmospheric pressure 1013 hPa). The molecules within the ice crystal are organized in layers of hexagonal rings. The atoms in a ring are not in one plane but two (Figure 90b). Spacing between two layers (0.276 nm) is much larger than between two planes of atoms (0.0923 nm). Adjacent layers form mirror images of each other. The optic axis (also called the *c-axis* or *principal axis*) of the ice crystal is perpendicular to the *basal plane*, i.e. the plane of a layer of hexagonal rings of atoms. Within the hexagonal lattice, three crystallographic *a-axes*, separated by 60° from each other, form the basal plane (Figure 90a, c).

A dozen crystallographic kinds of ice are known in nature and laboratory experiments but only two are observed in the Earth's natural conditions. The most common is the hexagonal crystalline ice which exists in temperatures between 0 °C and *c.* −70 °C and pressures up to *c.*210 MPa. Ice in the cubic crystalline form has been found at very low temperatures (below *c.* −70 °C) and very low pressures in the upper parts of the troposphere. Other ice polymorphs can appear in very high pressures (higher than 800 MPa) and at various positive temperatures. Such 'hot ices' can reach density higher than $1.3\,g\,cm^{-3}$ and might be present in the lithosphere as films of solid water, only a few molecules thick, chemically bonded

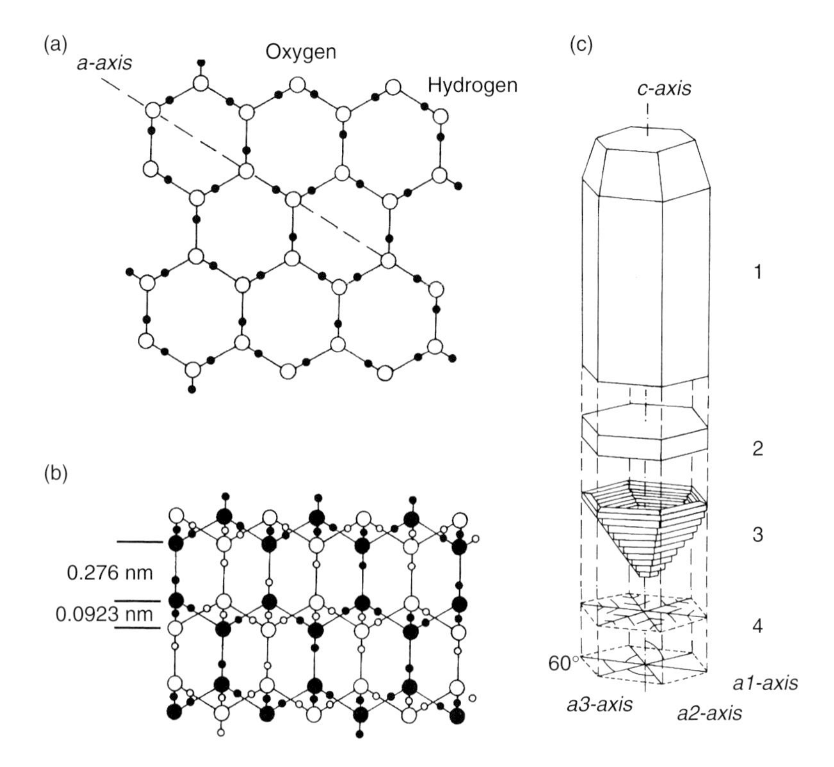

Figure 90 Structure of ice and forms of crystals (modified from Jania 1988). Models of the ice lattice structure: (a) basal plane, (b) view perpendicular to the c-axis. Types of the hexagonal ice crystals (c): 1 – column (prism), 2 – plate, 3 – cup, 4 – snow star

with other minerals. It is thought that these may affect certain properties of rocks, e.g. their susceptibility to weathering after exposure at the land surface.

The increase of volume of freezing water, by *c.*10 per cent, is responsible for important geomorphic processes (e.g. MECHANICAL WEATHERING; SOLIFLUCTION). The higher specific volume of the 'common' hexagonal ice compared with water, especially in respect of the relatively more dense sea water, enables ice to flow. Floating ice can be also a geomorphic agent (e.g. ICEBERGS). *Drift ice deposits* resulting from melting of debris-rich icebergs are recorded not only in deep sea deposits but also in shallow bays and proglacial lakes.

Six isotopes of the oxygen and three isotopes of hydrogen have been noted in natural conditions. Their various combinations result in thirty-six isotopic kinds of water and ice. The most important are: $H_2{}^{16}O$ (common water); $HD^{16}O$ ('heavy water' with deuter – D) and $H_2{}^{18}O$ (water with 'heavy' oxygen – ^{18}O). The last two are minor admixtures (0.03 per cent and 0.2 per cent, respectively) of ocean waters which are the most important source of water vapour in the atmosphere and consequently of snow and ice on land. Molecules of water with the heavier stable isotopes evaporate less rapidly and condense slightly faster from the vapour. The isotopic composition of snow precipitation depends on the temperature of evaporation of the sea water and the temperature of condensation (re-sublimation). The proportion of contents of heavy isotopes is a fraction of their concentration relative to the 'light' isotopes ($^{18}O/^{16}O$ or D/H) in a sample of snow (ice). In respect of its deviation (δ) from the ratio in the 'standard mean ocean water' (SMOW) it is expressed in parts per thousand (‰). The average annual values of $\delta^{18}O$ in snow samples in the polar areas correlates closely (in linear regression) with mean annual air temperatures there. It follows that the stable isotope content in samples from deep ice cores is a good source of data on air temperatures in the past (paleotemperatures).

Despite the fact that the hexagonal crystalline ice (Figure 90c) is the most common on the Earth, perfectly symmetrical hexagonal ice crystals are in the minority in the natural environment. Varied factors, which influence crystallization of water, determine that ice crystals are often incomplete or defective. First, natural waters are very rarely chemically pure. They have dissolved admixtures which lead to the phase change (crystallization) at temperatures lower than 0.16 °C – the figure usually regarded as the freezing temperature. Water in a supercooled state (i.e. being in the liquid form below the freezing point temperature) rarely freezes spontaneously. It is able to freeze only when crystallization centres are available within the liquid (e.g. small particles of ice or other minerals). When water freezes, H_2O molecules arrange themselves in a lattice of crystals. When cooling is slow in still waters, other molecules are removed to outside the walls of the ice crystals. Fast crystallization and turbulent movement of water cause incorporation of molecules of other chemical species into the lattice of ice crystals and their structure becomes defective. In general, optic axes of ice crystals are oriented perpendicularly to the 'cooling front' or the freezing surface. Vertical ice crystals are characteristic of lake ice and *naled ice*, whereas *rime ice* may have c-axes oriented toward the wind direction. *Glacier ice* and the anchor and frazil river ice usually have randomly oriented crystal optic axes.

Defects within the crystal lattice are regarded as factor facilitated recrystallization and ice crystals become reoriented when new directions of shear stress are applied. Under high cryostatic pressure, atmospheric gas molecules (e.g. N_2, O_2, CO_2) from air bubbles, trapped between glacier ice crystals, are incorporated into their lattice in the form of clatrates. When such ice crystals appear on the surface or in front of a glacier, where pressure is lower, these gas molecules are released from the lattice to either meltwater or the atmosphere.

Ice under natural conditions may variously be regarded as a mineral, sediment or a rock (Shumskyi 1955: 15–16). The hardness of ice varies with temperature. At 0 °C, its hardness is 1.5 on the Mohs' scale (as is talc and gypsum), whereas at −40 °C, the ice hardness is as much as 4 (equivalent of the hardness of fluorite). It is clear that glacier erosion would be impossible without the regelation process at the glacier bed, i.e. melting (under pressure) at the proximal side of subglacial obstacles and refreezing in their lee side.

Ice is common on the planet Earth and is also present on Mars, Pluto and satellites of Jupiter (e.g. Europa, Callisto) and Saturn. However, the presence of water in vapour, liquid and solid forms makes the Earth a planet unique to the Solar System. An irregular envelope containing ice encircles the Earth and is called the cryosphere. The term, introduced by Dobrowolski (1923), derives from the Greek (*kryos* – cold). The cryosphere exists only in an intermingled state with the lithosphere, hydrosphere and atmosphere, where water appears in the form of ice (snow cover, glaciers, sea and fresh waters ice, ground ice and ice in the atmosphere). The total mass of all ice on the Earth is estimated at 2.5×10^{16} metric tons. This is present on the surface or in the near-surface layers of the crust of our planet and occupies an area of *c.* 73.4×10^6 km^2. It covers 14.2 per cent of the total area of the globe with annual fluctuations between 10.5 and 17.9 per cent. Ice, therefore, can hardly be other than a vastly important geomorphic agent, especially in polar and temperate climates. Considering all forms of natural ice, GLACIERS are the most important agents of relief changes. Total volume of glaciers and ice sheets constitutes about 97.7 per cent of all natural forms of ice on Earth, while subsurface ice is only 2.1 per cent (Kotlyakov 1984: 347–348).

Atmospheric ice, in the form of stars, plates or columns (Figure 90c), forms a snow cover when deposited on the ground. This mixture of ice crystals and their inclusions of air and water (in warmer environments) normally has a density between 0.05 and 0.5 g cm^{-3}. A seasonal snow cover affects slope processes, including rapid and spectacular mass movement as avalanches (see AVALANCHE, SNOW).

In areas where accumulation of snow exceeds its melting, snow is transformed into glacier ice. The chain of processes leading to the metamorphism of a fresh snow cover into this specific form of polycrystalline 'ice rock' differs in warm (temperate) environments (with presence of water within the snow cover) and in dry cold ones. Infiltration of meltwater into the snowpack accelerates the transformation. In cold conditions sublimation plays an important role in the rounding of crystals and their growth. Despite the environmental differences, four steps of transformation of snow into polycrystalline glacier ice are usually distinguished: the first is from fresh snow (porosity *c.* 95 per cent) to old snow (density 0.3–0.5 g cm^{-3}) due to setting, compaction and rounding of crystals (step 1); further densification and compaction by pressure of overlying layers leads (step 2) from old snow (porosity *c*.50 per cent) into coarse-grained snow called *firn* (density $0.55\,g\,cm^{-3}$) about a year later. Firn has a low porosity, but is still permeable. Transformation from firn to the white (new) glacier ice (step 4) takes a longer time in cold climates, where intergranular movements (due to compaction), combined with sintering (recrystallization by way of sublimation) and plastic deformation, dominate. In warmer climates, owing to liquid water infiltration, metamorphosis is faster; this is due to refreezing and, of course, settling. In extremely cold and dry conditions, such processes takes hundreds or thousands of years (*c*.4,000 years at the Vostok Station, East Antarctica). In temperate and subpolar climates, it lasts decades. New glacier ice has a density higher than $0.83\,g\,cm^{-3}$, and, owing to closure of the pores into air bubbles between ice crystals, very low permeability. At this stage, crystals usually have dimensions of a couple of millimetres and their c-axes are randomly oriented. The last step in the growth of crystals and densification of glacier ice occurs through dynamic metamorphism and requires the deformations caused by *glacier flow* (see GLACIER). The old glacier ice may have densities of as much as $0.9\,g\,cm^{-3}$ and crystal sizes can have diameters larger than 10 cm (Plate 63). The optic axes of dynamically transformed glacier ice crystals reveal a predominant orientation, which is perpendicular to the terminal stress field. However, the *basal ice* of glaciers possesses a different structure. Most glaciers have an irregular bed with resistant rock protuberances. Melting develops when the sliding sole of a glacier passes such an obstacle in its bedrock (in thermal conditions close to the pressure melting point of ice). On the lee side of the obstacle, water refreezes, forming a layer of REGELATION ice and basal debris can be incorporated into this. The formation of debris-rich basal ice is observed in many temperate and polythermal glaciers. The isotopic composition of basal ice layers suggests an accretion of ice crystals from supercooled water from the subglacial drainage (Titus *et al.* 1999: 43). The debris content incorporated into basal ice (a basal moraine) is different, is very variable and depends on a number of factors. The basal debris zone tends to be thicker in the cold-based ice sheets of Greenland and Antarctic (up to16 m) than in temperate or polythermal glaciers, where its thickness varies from 0.4 m to a couple of metres. Debris concentration in the thick dispersed basal debris zones reaches 7–12 per cent (by volume), whereas in the thin basal zones (<1 m) of warm-based glaciers, the concentration can exceed 50 per cent (Menzies 2002: table 6.2). In contrast, the englacial concentration of debris is generally very low (<1 per cent). Glacier sediment discharge depends on the debris concentration in the basal ice zone and the basal sliding velocity.

An ICE SHEET is a shield-like, broad ice mass of continent scale (larger than $50{,}000\,km^2$). It can be thick enough (thousands of metres) to cover any irregularity in topography of its bed completely. Ice flow is generally organized in a quasi-radial pattern from one or more centres: *ice domes*. The Antarctic Ice Sheet is the contemporary example ice mass of such dimensions: *c*.$13 \times 10^6\,km^2$, with volume of $30.11 \times 10^6\,km^3$. The ice cover of Antarctica has five ice domes which reach an elevation of 4,000 m a.s.l. The maximum thickness of ice (4,776 m) has been found in East Antarctica and its mean thickness is 2,160 m (Drewry 1983: 4). The Greenland Ice Sheet covers *c*.$1.75 \times 10^6\,km^2$ and has a mean ice thickness of 1,790 m (volume: $2.74 \times 10^6\,km^3$). Two well-pronounced ice domes are distinguished in Greenland and the northern dome reaches an elevation of 3,236 m a.s.l. (Van der Ween 1999). An ice dome has a

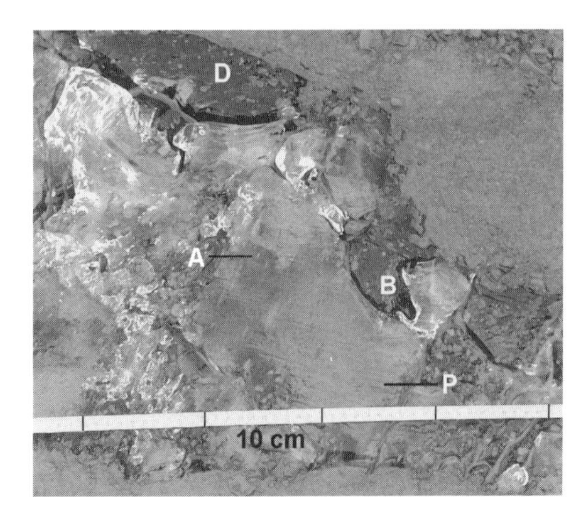

Plate 63 Large crystals of the glacier ice from bottom layers of Horn Glacier, Spitsbergen (intervals of 10 cm are marked). Note inclusions of air bubbles (A), cross section of the crystal basal planes (P), plastically deformed crystal boundaries (B) and presence of debris admixtures (D). Photo by Jacek Jania

convex shape and ice flow is predominantly vertical (downward) within it.

An ICE STREAM is a zone within an ice sheet or ice cap in which the ice flows significantly faster than on its sides. Theoretically, an ice stream has no rock boundaries; however, the course of the majority of ice streams is determined by their bedrock topography. Antarctic ice streams drain up to 90 per cent of mass accumulation in the interior (Paterson 1994: 301). The best known ice streams are on Siple Coast in Antarctica (Ice Streams A, B, C, D and E) and Jakobshavn Glacier, West Greenland. The measured velocities of Ice Stream B exceed 800 m a^{-1} whereas on the neighbouring ice sheet area they are less than 10 m a^{-1} (Van der Ween 1999). The surface velocity of Jakobshavn Glacier reaches almost 8 km per year near its terminus (Fahnestock *et al.* 1993: 1,532). In both cases, basal sliding is responsible for the fast flow of ice. High basal velocities of ice streams are thought to produce intense erosion of glacier beds.

In PERMAFROST areas several kinds of ground ice are present and ice bodies have a different origin, size, shape and internal structure. They might reasonably be classified into two genetic groups: (1) those originated from ground water and (2) those from water migrating from the surface to the ground. Within the first group, the largest ice masses are called hydro-laccolites. Their thickness can be as much as 20–40 m and their diameters can be dozens or even hundreds of metres (e.g. PINGO). Smaller forms are usually lenticular. *Ice lenses* appear in relatively dry ground or in sediments which have low permeability (e.g. tills). Owing to mineralization of water, which migrates within the sediments and to the higher than atmospheric pressure within the ground, crystallization of ice within the ground takes place in temperatures below the freezing point. Ice lenses grow upwards and have a horizontally laminated structure with inclusions of mineral particles. Ice wedges (see ICE WEDGE AND RELATED STRUCTURES) are the most common forms of the various ice bodies which result from water infiltration.

Ice jams on rivers most often take place in the winter and spring season and are a notable feature of large rivers which flow northwards in the northern hemisphere (Siberia, Canada). They are often caused by the earlier melting of snow in the southern parts of drainage basins and blockage of water by the longer preserved river ice cover in the northern reaches. Jamming of ice transport occurs as it lodges beneath the superficial ice cover of the frazil and anchor ice and thereby drainage blockage is typical for winter periods. The impediment may be so severe that wide areas of the adjacent floodplains become inundated. Detached drifting ice blocks may locally cause erosion of the floodplain sediments.

Specific ice masses originate in caves. The *ice cave* usually has a thick layer of laminated infiltration ice and icicles (ice stalactites and stalagmites). Ice caves are common in KARST areas within a permafrost zone. They are also known from locations where ground is not perennially frozen. In such cases, the very specific microclimate of the cave is responsible for the formation of annual infiltration ice layers: there is only one entrance located in the upper part of the cave and during winter, cold air flows down through the cave, while lighter warm air in spring and summer cannot penetrate into the system. Percolated meltwaters from snow or precipitation are frozen on the floor of the ice cave.

Tunnels formed in glaciers by englacial and subglacial drainage systems are called glacier caves. Vertical tunnels which transfer water from the glacier surface to subglacial channel are termed *ice shafts* or MOULINs. Ice shafts develop on planes of discontinuity within the glacier ice as crevasses and shear planes.

References

Dobrowolski, A.B. (1923) *Historja nauturalna lodu* [*Natural history of ice*], Warszawa: Kasa im. Mianowskiego.

Drewry, D. (1983) *Antarctica: Glaciological and Geophysical Folio*, Cambridge: Scott Polar Research Institute.

Fahnestock, M., Bindschadler, R., Kwok, R. and Jezek, K. (1993) Greenland Ice Sheet surface properties and ice dynamics from ERS-1 SAR imagery, *Science* 262, 1,530–1,534.

Jania, J. (1988) *Zrozumiec lodowce* [*Understanding glaciers*], Katowice: Wydawnictwo Slask.

Kotlyakov, V.M. (ed.) (1984) *Glyatsiologicheskiy slovar* [*Glaciological dictionary*], Leningrad: Gidrometeoizdat.

Menzies, J. (ed.) (2002) *Modern and Past Glacial Environments*, Oxford: Butterworth-Heinemann.

Paterson, W.S.B. (1994) *The Physics of Glaciers*, Oxford: Pergamon.

Shumskyi, P.A. (1955) *Osnovy structurnogo ledovedeniya* [*Principles of structural glaciology*], Moskva: Izdatelstvo Akademii Nauk SSSR; Trans. (1964) *Principles of Structrual Glaciology*, New York: Dover.

Titus, D.D., Larson, G.J., Strasser, J.C., Lawson, D.E., Evenson, E.B. and Alley, R.B. (1999) Isotopic

composition of vent discharge from the Matanuska Glacier, Alaska: implication for the origin of basal ice, *Geological Society of America, Special Paper* 337, 37–44.
Van der Ween, C.J. (1999) *Fundamentals of Glacier Dynamics*, Rotterdam: A.A. Balkema.

Further reading

Hooke, R.L. (1998) *Principles of Glacier Mechanics*, Englewood Cliffs, NJ: Prentice Hall.
Martini, I.P., Brookfield, M.E. and Sadura, S. (2001) *Principles of Glacial Geomorphology and Geology*, Upper Saddle River, NJ: Prentice Hall.
Petrenko, V.F. and Whitworth, R.W. (1999) *Physics of Ice*, Oxford: Oxford University Press.

SEE ALSO: glacier; ice sheet; ice stream; ice wedge and related structures; permafrost

JACEK JANIA

ICE AGES (INTERGLACIALS, INTERSTADIALS AND STADIALS)

Ice ages have occurred throughout geological time: as, for example, during the Precambrian, the Ordovician, Permo-Triassic and the Cenozoic (Tertiary), of which the Quaternary is a part. During the Permo-Triassic ice age, the ice sheets were centred on the Antarctic continent, located across the southern polar area. Its principal glacial deposit, the Dwyka tillite, is found in South America, southern Africa, southern India and south Australia, a distribution that shows the fragmentation of the super continent (Gondwanaland) where glaciation occurred by 'continental drift'. The Cenozoic ice ages commenced when Antarctica experienced major glaciation about 34 Ma ago (Ma = millions of years). But, in a geomorphological sense, the term 'ice age', incorporating many ice ages and interglacials, most appropriately refers to the Quaternary ice ages, when major geomorphic changes occurred.

Evolution of the present boundary conditions of the Earth's climate, namely the location of the continents, seaways and mountain ranges, closely match the evolution of the climate system and ice ages. As the southern continents 'drifted' away from Gondwanaland, India collided with Asia to produce the largest landform on Earth, namely the Tibetan Plateau, a feature so high that it modulated the circulation of the high level westerlies that led to waves on the circum polar vortex. These drew cold arctic air to Canada and Europe. Cold continents adjacent to warm oceans were the ideal combination for initiating the major Quaternary ice ages. Once the closure of the Straits of Panama (about 4 Ma) had occurred, a zonal circulation of ocean water between the Atlantic and Pacific ceased, to be replaced by meridional warm water circulation in the Atlantic – the provider of precipitation to grow the ice sheets – and the start of the intensification of North Atlantic Deep Water formation, the driver of oceanic thermohaline circulation.

The standard definition of the Quaternary is that it commenced 1.8 Ma ago, a view tied to a stratotype (type-site) at Vrica, near Crotone, Calabria, Italy. This horizon lies at the top of the Olduvai magnetic reversal Subchron. But, an alternative view would place the start at about 2.5 Ma, because it coincides with the first record in marine sediments in the North Atlantic. This shows that ice sheets had grown large enough to reach tidewater lands where they launched icebergs into the ocean. When these melted, their load of ice rafted debris (IRD) was released onto the ocean floor. In China, it coincides with the contact between the 'Red Clay' and the first, Wucheng Loess; and also coincides with the base of the earliest, pre-Nebraskan, glaciation of North America.

The Quaternary ice ages are characterized by lowered temperatures when the snowline in mid-latitude regions was lowered by up to 1 km. This led to the growth of large mid-latitude ice sheets in Canada and Scandinavia, while the water they abstracted from the oceans lowered sea level by up to 135 m. But how did these large ice sheets grow, because they have no modern analogues? Three main theories have been proposed. (1) *Highland origin and windward growth*: based on the former Laurentide ice sheet in Canada and then applied to the Fenno-Scandian ice sheet. It is suggested that ice grew initially in the highland of Ungava and Quebec, then spread towards its source of precipitation from the Gulf of Mexico, before forming a large ice dome up to 3 km thick or more, centred over Hudson Bay. (2) *Instantaneous glacierization* involved rapid vertical growth of ice from large snow patches distributed over wide areas, including the low ground of Keewatin west of Hudson Bay. It led to a thinner multi-domed ice sheet, capable of rapid response to climate change. (3) *Marine-based ice sheets* are conceptually based on the West

Antarctic ice sheet and involved mid-latitude ice sheets centred on shallow sea areas such as Hudson Bay or the Baltic Sea, where large ice domes grew. These, drained by ice streams, were buttressed by floating ice shelves which, on destabilization, no longer provided support, so that the ice domes collapsed and surged into the ocean. Whatever the mode of ice sheet growth, it is clear that they were large enough to reach the tidewater regions of the North Atlantic Ocean on several occasions when they launched armadas of icebergs. On melting, these released the glacial material they were carrying which was deposited in the ocean as discrete layers of ice rafted debris (IRD). Such episodes are known as Heinrich Events and they occurred on an irregular 5 to 7 ka cycle.

South of the ice sheets vegetation zones were compressed towards the equator. Accompanying this was a migration of the fauna and flora and in geomorphic processes. Mid-latitudes experienced periglacial climates and atmospheric circulation was more vigorous as the consequence of steeper poleward climate gradients. Low latitudes experienced widespread aridity. Ice ages were diversified by brief climatic ameliorations called interstadials separated by colder stadials. Interglacials supported no mid-latitude ice sheets, had relatively high sea levels and a fauna and flora similar to the present (Holocene) interglacial. During the past million years or so, ice ages have commonly lasted about 100,000 years. But the length of interglacials appears to have been variable and is controversial. The interglacial about 400,000 years ago may have lasted for the best part of 60,000 years; but duration of the 'last interglacial' about 125 ka is controversial: according to some, it lasted about 10,000 years, but others believe it was twice as long. The current interglacial has already lasted 11,600 years, during which time several minor fluctuations in climate have occurred, such as the Little Ice Age.

Figure 91 shows that some fifty ice ages and fifty interglacials occurred in the last 2.5 Ma. This contrasts with earlier, classical, views that maintained that only four ice ages occurred in the 'Great Quaternary Ice Age', a view based on the record of Alpine glacial advances shown by four outwash terraces in Bavaria by Penck and Bruckner in 1910. Their view was reinforced in America and elsewhere by four major groups of glacial deposits.

Their paradigm lasted for over sixty years. But evidence of greater complexity came from an unexpected source, namely marine deposits of the deep open ocean, where long sequences recorded the ice ages of the entire Quaternary. These consist of muds composed of microfossils of plankton (planktic) and benthos (benthic) which secreted their calcareous shells in oxygen isotopic equilibrium with the ocean water they inhabited. Initially it was believed that variability in the ratio of ^{18}O and ^{16}O in foraminifera microfossils was primarily an indicator of the ocean temperatures in which the organisms grew, and only to a lesser extent was the isotopic composition of the ocean involved. Subsequently it was shown that the isotopic composition, not the temperature, of the ocean was the primary control on oxygen isotope

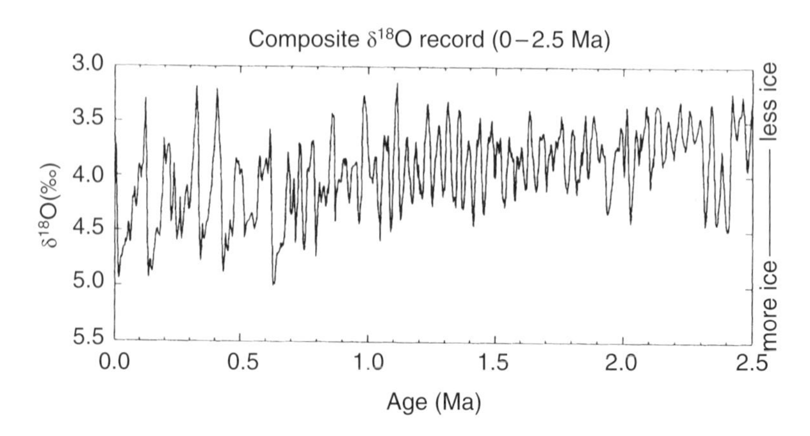

Figure 91 Composite record of the ice ages for the last 2.5 million years, based on oxygen isotope variability (from Crowley and North 1991)

ratio. Furthermore, because bottom water temperatures remained constantly cold, then benthic foraminifera would only record the oxygen isotopic composition of the ocean. Thus, because the primary control on this was the volume of isotopically light ice locked up on the continents during ice ages, then the benthic signal would provide an ice volume indication: that is, a record of the ice ages and interglacials. Moreover, because this was altered as the sea rose and fell as continental ice volume changed, the benthic signal was also one of sea-level change. Recently, it was shown that bottom water temperatures had varied; thus ^{18}O benthic ratios could no longer provide precise ice volume or sea-level information. Fortunately this was resolved by temperature determinations based on Magnesium/Calcium ratios in Ostracoda, that allowed discrete ice volume and temperature parameters to be identified separately.

Oxygen isotope signals are subdivided into *oxygen isotope stages* and *sub-stages*. Interglacials are numbered backwards in time with odd numbers, starting with the present (Holocene) interglacial as stage 1. Similarly ice ages are numbered backwards with time with even numbers, starting with the last ice age or Last Glacial Maximum as stage 2. Considerable confusion has been caused by the designation of an oxygen isotope stage 3 in the middle of the last 100 ka cycle. This arose because it was once thought that the 41 ka tilt frequency was the main ice age pacing (below). So in reality, the present (Holocene) interglacial is stage 1 and the last interglacial centred on ~125 ka is sub-stage 5e, the warmest part of stage 5, when conditions more or less corresponded with the present. Stage 3 is not an interglacial but merely a general climatic amelioration punctuated by millennial scale interstadials (below) during the last glacial cycle.

A means of providing a chronology for the ice ages and interglacials was provided by fixed points in the cores where changes in the Earth's magnetism occurred. Of these, the most important reversal is the Matuyama/Brunhes reversal at 0.78 Ma (780 ka). By assuming constant sedimentation, ages were provided for the ice age and interglacial boundaries. This was reinforced when it was found that the predicted age of the 'last interglacial' of 125 ka was supported by Uranium 234–Thorium 230 ages on uplifted coral reefs that showed a relatively high sea level at that time. Since then, the same has been confirmed for the age of the interglacial at about 300 ka, oxygen isotope stage 9. The last interglacial shoreline is found throughout the world as coral reefs, raised beaches or shore-platforms. Detailed records of sea-level fluctuations have been established in regions with uplifted coral reefs, such as Barbados, Tahiti and the Huon Peninsula of New Guinea.

Inspection of oxygen isotope signal reveals a number of cycles or pacings (Figure 91). These are the well-known pacings caused by the orbital movement of the Earth around the sun, and occur at 100 ka (eccentricity of the orbit), 41 ka (variation of the ecliptic or tilt of the Earth's axis), 23 ka and 19 ka (precession of the equinoxes that varies the distance between the Earth and the sun at the summer solstice). Relating the changing amount of solar energy received at given latitudes to the oxygen isotope (ice volume) record, however, had to await confirmation of these pacings in high sedimentation marine cores from the Indian Ocean. The astronomical and oxygen isotope records are not an exact match, because of a time-lag between orbital forcing and the climatic response as ice sheets grew or decayed to change the oxygen isotopic composition of the ocean. By using phase relationships (leads and lags), the oxygen isotope record has been 'tuned' by the predicted orbital changes. This controversial method is illustrated by the debate over the commencement age and duration of the 'last interglacial' (oxygen isotope sub-stage 5e) which Uranium–Thorium ages on stalagmite shows has been underestimated. It may be resolved because the shorter estimate is based on orbital calculations for 65°N (the 'sensitive latitude' for mid-latitude ice sheets), whereas the longer one is more consistent with orbital calculations for 65°S.

Despite ongoing debate about the precision of ice age and interglacial timing, it would seem that the origin of the ice ages has been discovered. Changes in temperature caused by changes in the orbit of the Earth seem to be the pacemaker of climate change. But how was orbital forcing transformed into actual climate change? Several questions remain before the matter can be settled. The earlier part of the Quaternary record displays a strong 41 ka tilt pacing with no trace of the 100 ka eccentricity pacing that only becomes distinct after 700 ka (Figure 91). This was when the large mid-latitude ice sheets grew and extended south to Ohio, London, Berlin and Kiev for the first time. What was the cause of the later

Quaternary 100 ka cycles? Perhaps further uplift of the Tibetan Plateau intensified the climate forcing? More likely is that continental erosion had progressively removed extensive layers of weathered rock to reveal more resistant bedrock. Low gradient ice sheets occur on soft deformable beds, but more resistant rocks support ice domes with steep marginal profiles. These depressed the Earth's crust, while continuing to receive more snow at their summits, thus prolonging the ice age. This may account for the transition from the 41 ka to the 100 ka world. Not unrelated to this is the large mismatch between the forcing provided by the 100 ka pacing and the response of the climate system (Figure 92). While there is a predictable and strong response of the climate system in the tilt and precessional bands, the disproportionate response in the eccentricity band remains unexplained. Could it be because of the development of large ice domes (above), or perhaps the real role of eccentricity is to modulate the tilt and precession forcings?

Amplification of orbitally forced climate change occurred by enhanced or decreased greenhouse gas concentrations of carbon dioxide and methane in the atmosphere. These are clearly implicated in climate change as is shown by measurements of past atmospheres preserved within bubbles of air trapped in ice layers of the Antarctic and Greenland ice sheets (Figure 93). Higher concentrations of greenhouse gases correspond with interglacials, while they are lowered during the ice ages. Oxygen isotope or Deuterium analyses of ice from cores through the Greenland and Antarctic (Figure 93j) ice sheets, provide a record of temperature changes at the surface of the ice sheet. But the age of the ice layers and the bubbles of air within them is not the same because air bubbles provide a greenhouse gas record at the time when they were sealed by the weight of overlying snow and firn. Differences of up to sixty years occur in high snowfall regions such as Greenland, but up to 1,200 years in the centre of Antarctica where only about 2 cm of snow falls

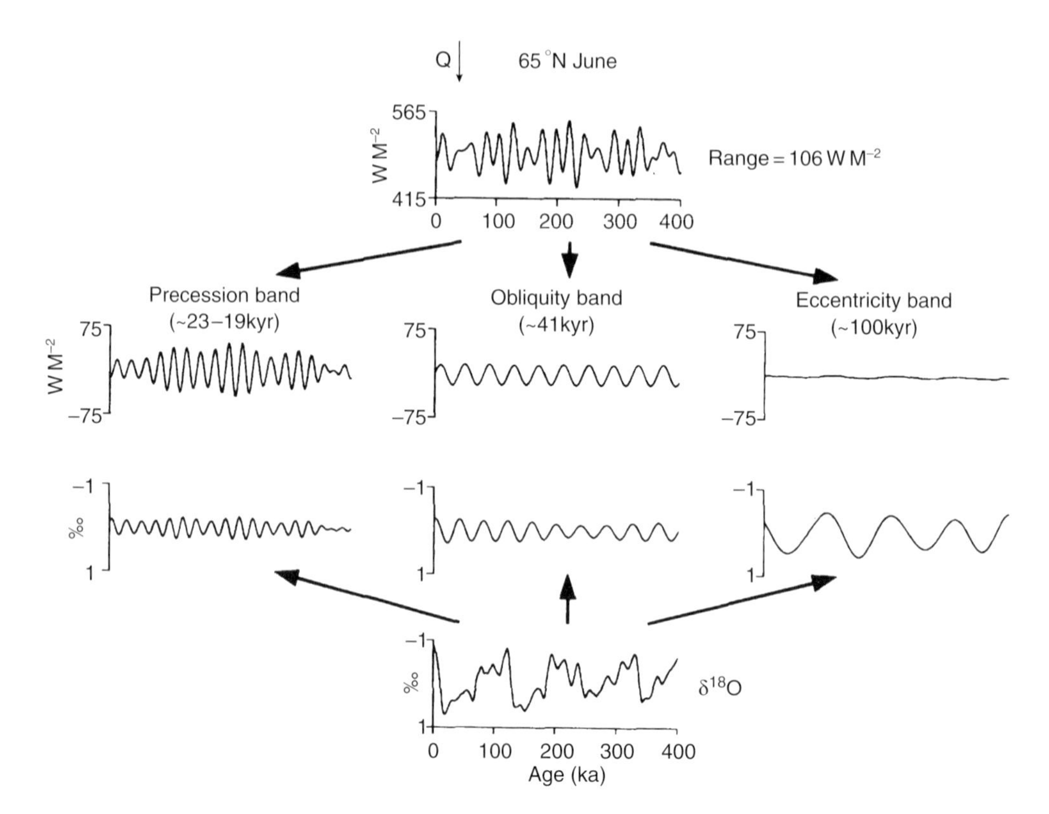

Figure 92 The 100,000 cycle problem shown by partitioning the radiation and climate time series into their dominant periodic components (from Imbrie *et al.* 1993)

annually. This may conceal important phase relationships between temperature and greenhouse gases. Fortunately the sharp methane spikes in the record allow good correlation between interhemispheric ice cores. The record of temperature and greenhouse gases now extends back more than 400,000 years at Vostok Station, Antarctica, through four major ice ages and five interglacials; whereas the record at Greenland Stations only extends back beyond the last interglacial.

Not only do palaeotemperature records from ice sheets reveal the main orbital pacings, they also display strong millennial ones (Figure 93). Moreover, unlike the sinusoidal shape of the

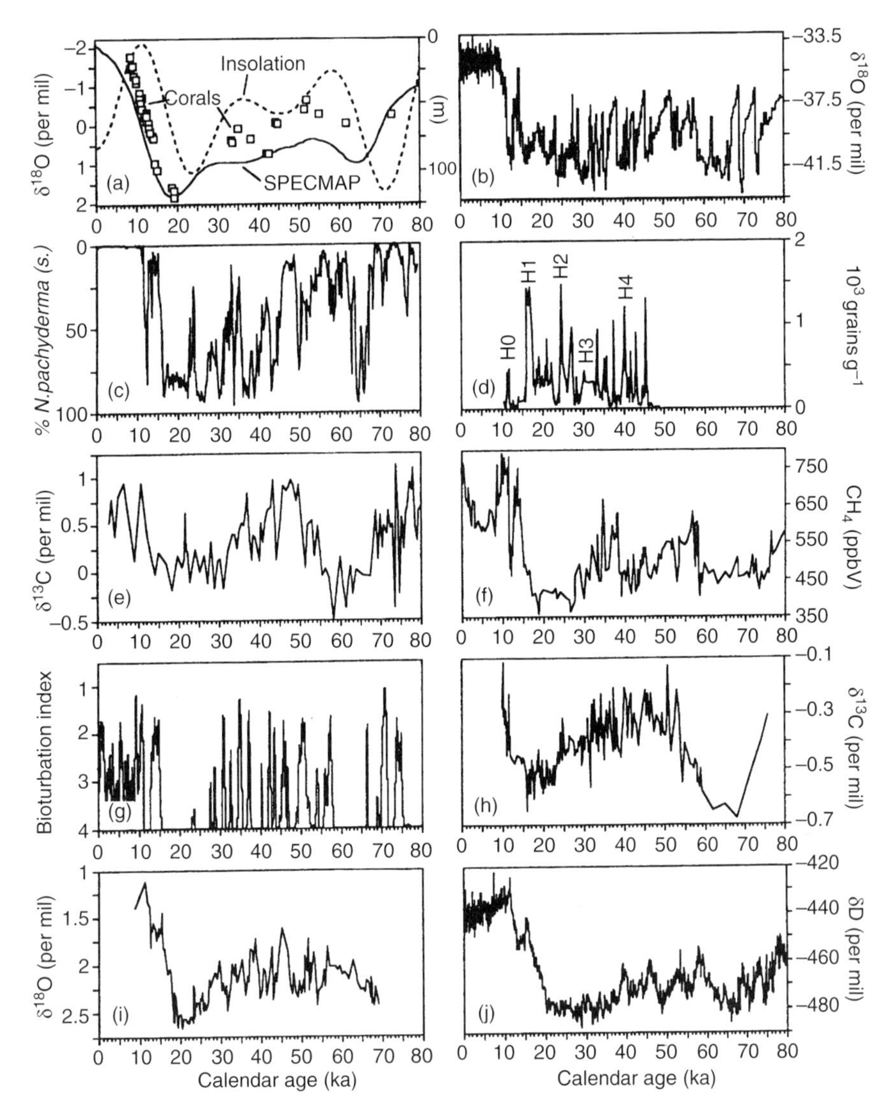

Figure 93 Millennial pacings (Alley and Clark 1999). (a) marine oxygen isotope signal (SPECMAP), insolation (orbital forcing) and sea level from dated corals; (b) Greenland ice sheet project (GRIP) temperatures; (c) sea surface temperatures from *N. pachyderma*, off western Ireland; (d) ice rafted debris and Heinrich Events; (e) tropical Atlantic 13C variability; (f) methane concentrations in the Greenland Ice Sheet Project 2 (GISP2); (g) changes in sediment type, Santa Barbara Basin, California; (h) north-east Pacific 13C variability; (i) southern ocean sea surface temperatures from oxygen isotopes; (j) deuterium (temperature) variability, Vostok Station, South Pole

orbital pacings, the millennial ones are square-waved in shape, and show abrupt changes in temperature. Some of these represent changes in temperature of up to 10 °C in less than a decade. Matching millennial records from high sedimentation marine core records also correspond with sea surface temperature variability (Figure 93c). These are estimated from the palaeoecological requirements of planktic fossils, notably the polar *Neogloboquadrina pachyderma* (with sinistral coiling). These millennial records are joined by other records: for example, continental pollen records in Europe, by the monsoon record from the Arabian Sea, sea surface temperatures from the Sulu and China Seas, changes in biological productivity or ocean ventilation inferred from ^{13}C millennial variability (Figure 93h), and changes in water masses from offshore California (Figure 93g).

Such millennial pacings are global in character and they show that the Earth's climate changed on short and abrupt timescales. The principal pacing, that underpins all the major changes in climate over the past 110,000 years, is that at 1,450 (sometimes stated as 1,500) years. It corresponds to pacings in the production rate of the cosmogenic isotope Berylium 10 (^{10}Be) also recorded in ice sheets. ^{10}Be is produced in the atmosphere by cosmic ray bombardment. Higher quantities are produced when the solar magnetic field is at its weakest and not shielding the Earth's atmosphere from cosmic ray bombardment. Conversely, lower quantities are produced when the solar magnetic field is at its strongest and shielding the Earth from cosmic bombardment. Therefore, lower quantities of ^{10}Be correspond to a stronger and warmer Sun and vice versa. Thus variability in solar output appears to control the 1,450 climate cycle. The 1,450 pacing occurs throughout the last glacial cycle as well as the current (Holocene) interglacial. Its amplitude is greater during the unstable geographical configuration of the ice ages because of large ice sheets, low sea levels and an isostatically depressed crust. But its interglacial mode is manifested in cycles of sea ice formation in the North Atlantic, the record of alpine glacier fluctuation, and in high sedimentation continental records throughout the world.

Cycles of ice core interstadials are known as Daansgard–Oeschger cycles, named after the Danish and Swiss scientists who discovered them. Some of them correspond to the less perfect record of interstadials on the continents, where evidence from pollen and insect faunas show climatic amelioration during ice ages. The Bølling–Allerød interstadial, between about 15 and 13 ka ago, is well known and was followed by the Younger Dryas stadsial. This dramatic reversal to cold conditions just before the Holocene may have been forced by a major meltwater flux to the North Atlantic from glacial Lake Aggasiz in North America, or it may be the result of solar forcing. Cycles of Heinrich Events (Figure 93) are known as Bond cycles, and they contain packages of Dansgaard–Oeschger interstadials. Heinrich Events may have been caused by mechanisms internal to the ice sheets, but their synchronous nature around the North Atlantic suggests they may have been forced by changes in climate.

Comparison of Greenland and Antarctic ice cores shows broad similarities, but also some important differences. One school of thought believes that deep water thermohaline ocean circulation in the Atlantic carries climate signals between the two hemispheres and is a prime control on the global climate system, with the signals originating in the North Atlantic. Others see the climate signal originating in the tropical Pacific Ocean, its global transmission taking place as millennial scale phenomena similar to El Niño events.

D.Q. BOWEN

References

Alley, R.B. and Clark, P.U. (1999) The Deglaciation of the Northern Hemisphere: A Global Perspective, *Annual Reviews of Earth and Planetary Sciences* 27, 149–182.

Crowley, T.J. and North, G.R. (1991) *Paleoclimatology*, New York and Oxford: Oxford University Press.

Imbrie, J. Berger, A., Boyle, E.A., Clemens, S.C., Duffy, A., Howard, W.R. *et al.* (1993) On the structure and origin of major glaciation cycles, 2: the 100,000 cycle, *Paleoceanography* 8, 699–736, American Geophysical Union.

Further reading

Alverson, K.D., Bradley, R.S. and Pederson, T. (eds) (2003) *Paleoclimate, Global Change and the Future*, Berlin: Springer.

Bradley, R.S. (1999) *Paleoclimatology*, London: Academic Press.

Clark, P.U., Webb, R.S. and Keigwin, L.D. (1999) *Mechanisms of Global Change at Millennial Time Scales*, American Geophysical Union Monograph 112, Washington, DC.

Cronin, Thomas M. (1999) *Principles of Paleoclimatology: Perspectives in Paleobiology and Earth History*, New York: Columbia University Press.

ICE DAM, GLACIER DAM

Glacier dams are barriers of ice that act as a seal for water impounding glacial lakes. Water may be impounded within the GLACIER (*englacially* and *subglacially*), on the surface of the glacier (*supraglacially*), or in lakes formed at the glacier's edge (*marginally* or *proglacially*). Subaerial glacial lakes (marginal or proglacial lakes) are confined by ice on one side and by topographical barriers on the other. Subglacial lakes may form a cupola above the glacier bed at subglacial geothermal areas, accompanied by a depression in the glacier's surface, or be situated in a bedrock hollow beneath a relatively flat dome-shaped glacial surface. Supraglacial lakes are isolated in depressions on the glacier surface.

In general, ice-dammed lakes of every type can drain in episodic bursts. The glacier's surface and the water pressure potential slope towards the lake. The ice-dammed lakes receive water inflow and are gradually made to expand. Basal water pressure will increase and the lake level will be raised. Eventually, the hydraulic seal of the ice dam will be ruptured at the glacier base, the hydraulic seal opened, and seepage causes enlargement of the drainage system, initiating a flood under the surrounding ice. After discharge has begun, pressure from the ice constricts the passageway, and water flow at an early stage in the flood correlates primarily with enlargement of the ice tunnel due to heat from friction against the flowing water and to thermal energy stored in the lake. Increasing as an approximate exponent of time over a matter of hours or days, the discharge falls quickly after peaking. The recession stage of the hydrograph sets in when tunnel deformation begins to exceed enlargement by melting. Fluctuations in the thickness of the blocking ice, due to climatic variations or surges, may modify the outburst cycle or even stop bursts completely.

Occasionally, glacial outbursts are triggered by flotation of the ice dam. Rather than initial drainage from the lake being localized in one narrow conduit, the water is suddenly released as a sheet flow, surging downhill and propagating a subglacial pressure wave, which exceeds the ice overburden and lifts the glacier in order to create space for the water. In this instance, discharge increases faster than can be explained by conduits expanding due to melting.

Marginal lakes at subpolar glaciers, where the ice barrier is frozen to the bed, are typically breached as water spills over the top of the dam into supraglacial channel that melts into a bigger breach – commonly at the juxtaposition of the glacier and a rock wall.

Glacial outbursts (*jökulhlaup* in Icelandic) can have pronounced geomorphological impact, since they scour river courses and inundate floodplains. Outbursts result in enormous erosion, for they carry huge loads of sediment and imprint the landscape, with deep canyons, channelled SCABLANDS, ridges standing parallel to the direction of flow, sediment deposited on outwash plains, coarse boulders strewn along riverbanks, kettleholes where massive ice blocks have become stranded and melted, and breached terminal moraines. Some modern outbursts have produced flood waves in coastal waters (TSUNAMIS). In the North Atlantic, outburst sediment dumped onto the continental shelf and slope has been transported far away by turbid currents. Outburst floods wreak havoc along their paths, threatening people and livestock, destroying vegetated lowlands, devastating farms, disrupting infrastructure such as roads, bridges and power lines, and threatening hydroelectric plants on glacially fed rivers.

Further reading

Tweed, F.S. and Russell, A.J. (1999) Controls on the formation and sudden drainage of glacier-impounded lakes: implications for jökulhlaup characteristics, *Progress in Physical Geography* 23, 79–110.

HELGI BJORNSSON

ICE SHEET

An ice sheet is a large dome-shaped ice mass ($>50,000\,km^2$ in area) that exhibits a generally radial flow pattern which is predominantly unconstrained by the underlying topography. If the ice mass is less than $50,000\,km^2$, it is usually referred to as an ice cap. Ice sheets are comprised of a single ice dome or series of coalescent domes that represent the highest parts of the ice sheet surface. In the interior regions of the three present-day ice sheets in East Antarctica, West Antarctica and Greenland, ice thickness often exceeds several thousand metres. These central accumulation areas (or dispersal centres) are marked by ice divides which delineate neighbouring catchment areas (analogous to a drainage divide in fluvial systems).

Towards the margins of an ice sheet, the underlying topography becomes increasingly more important in channelling flow and fast moving outlet GLACIERS and ICE STREAMS may develop in deep subglacial troughs. Other ice streams may arise irrespective of the bedrock topography. Thus, flow within an ice sheet is generally divided into slow 'sheet' flow, typical of ice domes, and fast 'stream' flow, which occurs in outlet glaciers and ice streams. This may be an oversimplification because some ice streams and outlet glaciers have smaller tributaries of intermediate velocity extending well into the ice sheet interior. Ice streams and outlet glaciers are known to be a key control on overall ice sheet stability because of their capacity to rapidly drain large volumes of ice.

Once established, ice sheets respond to climate forcing and also influence the climate system over a global as well as local scale. They represent a major obstacle to atmospheric circulation and produce some of the largest regional anomalies in albedo and radiation balance (Clark *et al.* 1999). Ice sheets also store and release considerable amounts of freshwater and are one of the main regulators of continental water balance and global sea-level change. For example, during the Last Glacial Maximum (18,000–21,000 yr BP), global SEA LEVEL was around 120 m lower as continental ice sheets developed in the high to mid-latitudes of the northern hemisphere, covering large parts of North America (e.g. Laurentide Ice Sheet) and Europe (e.g. British and Scandinavian Ice Sheets). It has been discovered that rapid discharges of ICEBERGS and meltwater (see MELTWATER AND MELTWATER CHANNEL) associated with these former ice sheets exerted a profound and often abrupt impact on ocean circulation and climate (cf. Clark *et al.* 1999). There is also compelling evidence that some former ice sheets (e.g. Laurentide Ice Sheet) were characterized by relative instability during the last glacial cycle, indicating that even the largest ice masses are highly dynamic systems (Boulton and Clark 1990). A major challenge for contemporary ice sheet research lies in predicting the response of the potentially susceptible West Antarctic Ice Sheet to future changes in sea level and climate, particularly global warming.

References

Boulton, G.S. and Clark, C.D. (1990) A highly mobile Laurentide ice sheet revealed by satellite images of glacial lineations, *Nature* 346, 813–817.

Clark, P.U., Alley, R.B. and Pollard, D. (1999) Northern Hemisphere ice-sheet influences on global climate change, *Science* 286, 1,104–1,111.

Further reading

Benn, D.I. and Evans, D.J.A. (1998) *Glaciers and Glaciation*, London: Arnold.

Paterson, W.S.B. (1994) *The Physics of Glaciers*, 3rd edition, Oxford: Elsevier Science.

SEE ALSO: glacier; glacial deposition; glacial erosion; glacial theory; ice

CHRIS R. STOKES

ICE STAGNATION TOPOGRAPHY

Often also referred to as 'dead-ice topography', this is undulatory to hummocky terrain composed of a wide range of glacigenic sediments that accumulate on the surface of a melting glacier by the processes of melt-out, mass movement and glacifluvial reworking (Boulton 1972). Hummocks are often randomly distributed and of different dimensions, giving a chaotic appearance, but some linearity may be imparted by former ice structures. Although the term ice stagnation topography has been traditionally associated with deglaciated terrain, large areas of such supraglacial landform assemblages often still contain buried glacier ice, sometimes many thousands of years in age. In glaciers that carry relatively large debris loads, englacial and supraglacial sediment-landform associations develop over long periods of time and are subject to numerous cycles of reworking, re-mobilization and topographic inversion before they are finally lowered to the glacier substrate. The wide range of processes involved in the production of ice stagnation topography, including mass flowage, meltwater reworking, lacustrine sedimentation folding and faulting, result in the occurrence of a variety of sediment assemblages.

Although the term 'hummocky moraine' has been used to describe a wide range of landforms, it has most recently been restricted to moundy, irregular topography deposited by the melt-out of debris-mantled glaciers and therefore also relates to stagnating ice but is not strictly synonymous with the term ice stagnation topography. For example ice stagnation topography often encompasses large areas of glacifluvially reworked materials or supraglacial KAME and kettle (see

KETTLE AND KETTLE HOLE) topography and even complex ESKER systems and ice-walled lake plains. In some circumstances the melt-out process results in the partial preservation of englacial structures, specifically alternating debris-rich and debris-poor ice layers, and the production of transverse linear elements called 'controlled moraine'. The preservation potential of controlled moraines is poor but they are very striking elements of freshly deglaciated terrain at high latitudes where glacier snouts are melting down very slowly and supraglacial sediment reworking is locally restricted.

The processes of differential ablation, multicyclic debris reworking and topographic inversion and complex meltwater drainage development on a debris-rich glacier will result in the production of a thick supraglacial debris mantle that can decouple glacier response to climate forcing. Ice wastage or stagnation can therefore lag behind climatic inputs by decades or even thousands of years. As meltwater systems gradually open up large subglacial and englacial drainage conduits so the developing ice stagnation topography may become perforated by numerous moulins, ponds and lakes. Referred to as glacier karst (Clayton 1964), these water-filled depressions often coalesce to produce large supraglacial lakes.

References

Boulton, G.S. (1972) Modern arctic glaciers as depositional models for former ice sheets, *Journal of the Geological Society of London* 128, 361–393.

Clayton, L. (1964) Karst topography on stagnant glaciers, *Journal of Glaciology* 5, 107–112.

Further reading

Benn, D.I. (1992) The genesis and significance of 'hummocky moraine': evidence from the Isle of Skye, Scotland, *Quaternary Science Reviews* 11, 781–799.

Evans, D.J.A. and Twigg, D.R. (2002) The active temperate glacial landsystem: a model based on Breiðamerkurjökull and Fjallsjökull, Iceland, *Quaternary Science Reviews* 21(20–22), 2,143–2,177.

Eyles, N. (1979) Facies of supraglacial sedimentation on Icelandic and alpine temperate glaciers, *Canadian Journal of Earth Sciences* 16, 1,341–1,361.

Fleisher, P.J. (1986) Dead-ice sinks and moats: environments of stagnant ice deposition, *Geology* 14, 39–42.

Gravenor, C.P. and Kupsch, W.O. (1959) Ice disintegration features in western Canada, *Journal of Geology* 67, 48–64.

Gustavson, T.C. and Boothroyd, J.C. (1987) A depositional model for outwash, sediment sources, and hydrologic characteristics, Malaspina Glacier, Alaska: a modern analog of the southeastern margin of the Laurentide Ice Sheet, *Geological Society of America Bulletin* 99, 187–200.

Ham, N.R. and Attig, J.W. (1996) Ice wastage and landscape evolution along the southern margin of the Laurentide Ice Sheet, north-central Wisconsin, *Boreas* 25, 171–186.

Huddart, D. (1999) Supraglacial trough fills, southern Scotland: origins and implications for deglacial processes, *Glacial Geology and Geomorphology* http://boris.qub.ac.uk/ggg/papers/full/1999/rp04.html

Johnson, M.D., Mickelson, D.M., Clayton, L. and Attig, J.W. (1995) Composition and genesis of glacial hummocks, western Wisconsin, USA, *Boreas* 24, 97–116.

Nakawo, M., Raymond, C.F. and Fountain, A. (2000) *Debris-Covered Glaciers*, IAHS Publication no. 264.

Paul, M.A. (1983) The supraglacial landsystem, in N. Eyles (ed.) *Glacial Geology*, 71–90, Oxford: Pergamon.

Price, R.J. (1969) Moraines, sandar, kames and eskers near Breiðamerkurjökull, Iceland, *Transactions of the Institute of British Geographers* 46, 17–43.

Thomas, G.S.P., Connaughton, M. and Dackombe, R.V. (1985) Facies variation in a late Pleistocene supraglacial outwash sandur from the Isle of Man, *Geological Journal* 20, 193–213.

DAVID J.A. EVANS

ICE STREAM

An ice stream is a zone of an ICE SHEET that flows much faster than the surrounding ice. Typical ice sheet velocities are of the order of tens of $\mathrm{m\,a^{-1}}$ but ice stream velocities range from hundreds to several thousands of $\mathrm{m\,a^{-1}}$. Ice streams display a broad range of characteristics and behaviour but they are generally large features with widths of tens of km and lengths of hundreds of km. Their large size and profligate ice flux dominate ice sheet discharge and it is for this reason that they are viewed as a critical control on ice sheet mass balance and stability.

A defining characteristic of ice streams is that they are bordered by slower moving ice. This creates heavily crevassed lateral shear margins that aid their identification. If the fast-flowing ice is bordered at the surface by rock walls, it is usually referred to as an outlet glacier. It should be noted, however, that many ice streams show characteristics of both along their length. To add to their complexity, some ice streams appear to be fed by numerous smaller tributaries that penetrate up to 1,000 km into the interior of the ice sheet (Bamber *et al.* 2000).

Ice streams have been investigated in the East Antarctic, West Antarctic and Greenland Ice Sheets and have also been hypothesized in many palaeo-ice sheets. They occur in a variety of settings and their fast flow may be achieved through a variety of flow mechanisms. At one end of the spectrum, some ice streams occupy deep bedrock troughs and they are often characterized by steep surface slopes and high driving stresses. A large component of their fast flow arises from rapid basal sliding but, under exceptionally high driving stresses, thermally enhanced deformation of a thick basal ice layer may be the dominant flow mechanism (cf. Iken *et al.* 1993). In contrast, some ice streams have low surface slopes, low driving stresses and do not appear to be constrained by the underlying topography. Their behaviour is more enigmatic but their fast flow appears to be related to a metres thick layer of soft, saturated sediment that deforms beneath the ice stream and/or provides an effective surface for basal sliding (Engelhardt and Kamb 1998). Other ice streams show characteristics of both of these end-member behaviours and many more ice streams remain to be studied.

References

Bamber, J.L., Vaughan, D.G. and Joughin, I. (2000) Widespread complex flow in the interior of the Antarctic Ice Sheet, *Science* 287, 1,248–1,250.

Engelhardt, H. and Kamb, B. (1998) Basal sliding of Ice Stream B, West Antarctica, *Journal of Glaciology* 44, 223–230.

Iken, A., Echelmeyer, K., Harrison, W. and Funk, M. (1993) Mechanisms of fast flow in Jakobshavns Isbræ, West Greenland: Part I. Measurements of temperature and water level in deep boreholes, *Journal of Glaciology* 39, 15–25.

Further reading

Bentley, C.R. (1987) Antarctic ice streams: a review, *Journal of Geophysical Research* 92, 8,843–8,858.

Stokes, C.R. and Clark, C.D. (2001) Palaeo-ice streams, *Quaternary Science Reviews* 20, 1,437–1,457.

SEE ALSO: glacier; ice; ice sheet

CHRIS R. STOKES

ICE WEDGE AND RELATED STRUCTURES

Ice wedges are among the most common forms of ground ice within continuous PERMAFROST. They are commonly 1.0–1.5 m wide at the top and up to 4 m deep (Plate 64), although extreme examples up to 4 m wide and 10 m deep are reported from Siberia. In plan the linear ice wedge structures link to form polygonal patterns with a mesh size of several tens of metres, termed ice-wedge polygons or tundra polygons. The surface expression is a linear furrow due to differential settlement of the active layer immediately above the ice wedge.

Ice wedges result from thermal contraction cracking of the permafrost. Low winter air temperatures, and lack of insulating snow, lead to rapid cooling of the ground and the development of tensile stresses. Lachenbruch (1962) showed that cracking is most likely when the ground temperature is below approximately −20 °C, though slightly lower temperatures may be necessary in sands and gravels than is the case in silts and clays. Entry of snow and hoarfrost prevents crack closure as the permafrost warms during the following summer, leaving thin (often less than 1 mm; Mackay 1992) veins of ice penetrating into the permafrost. These provide lines of weakness for further cracking in subsequent years, leading to progressive widening into ice-wedges over time. Wedge growth associated with individual cracking events is generally less than 1 mm, and cracking may occur only periodically, so that a 1-m wide ice wedge may take thousands of years to form (Harry and Gozdzik 1988).

Epigenetic ice wedges develop below a stable ground surface and are younger than the adjacent frozen host sediments, while syngenetic ice wedges form contemporaneously with the slow

Plate 64 Upper portion of an ice wedge, Ellesmere Island, Canada. Note the active layer above the wedge is approximately 0.75 m deep at this location

accumulation of host sediments in subaerial permafrost environments. Where aeolian sediment transport is active, sand or silt – rather than snow and hoarfrost – may enter open thermal contraction cracks, resulting in the formation of sand-filled wedge structures termed sand wedges (Murton *et al.* 2000).

A mean annual air temperature value of between −6 °C and −8 °C has been widely used as the warm limit for the development of thermal contraction cracking and the formation of ice wedges (Péwé 1966), although variations in seasonal extreme temperatures may lead to a range of limiting mean annual temperatures. Thawing of permafrost and contained ice wedges is associated with slumping of sediments to fill the resulting voids, thus preserving the wedge forms within the host sediments. Such 'pseudomorphs' or 'casts' are important stratigraphic markers in sedimentary sequences, providing evidence for the former existence of permafrost (Svensson 1988). Host sediments may be upturned against the sides of casts, marking former compression during summer ground warming, or there may be downward slumping of adjacent sediments into the cast.

Ice-wedge casts and sand wedges may be preserved within sedimentary units (intraformational), between sedimentary units (interformational) or they may penetrate downwards from the present ground surface (supraformational). Intraformational casts suggest episodic sediment accumulation in a permafrost environment and are common in fluvial gravel trains (Seddon and Holyoak 1985). Interformational wedge casts indicate a major change in sedimentary environment separated by a phase of permafrost. Supraformational casts may be visible in aerial photographs as polygonal 'crop marks' (Svensson 1988).

References

Harry, D.G. and Gozdzik, J.S. (1988) Ice wedges: growth, thaw transformation and palaeoenvironmental significance, *Journal of Quaternary Science* 3, 39–55.
Lachenbruch, A.H. (1962) Mechanics of thermal contraction cracks and ice-wedge polygons in permafrost, *Geological Society of America Special Paper* 70.
Murton, J.B., Worsley, P. and Gozdzik, J. (2000) Sand veins and wedges in cold Aeolian environments, *Quaternary Science Reviews* 19, 899–922.
Mackay, J.R. (1992) The frequency of ice-wedge cracking (1967–1987) at Garry Island, western Arctic coast, *Canadian Journal of Earth Sciences* 29, 236–248.
Péwé, T.L. (1966) Palaeoclimatic significance of fossil ice-wedges, *Biuletyn Peryglacjalny* 15, 65–73.
Seddon, M.B. and Holyoak, D.T. (1985) Evidence of sustained regional permafrost during deposition of fossiliferous Late Pleistocene sediments at Stanton Harcourt, Oxfordshire, England, *Proceedings of the Geologists' Association* 96, 53–71.
Svensson, H. (1988) Ice-wedge casts and relict polygonal patterns in Scandinavia, *Journal of Quaternary Science* 3, 57–67.

Further reading

French, H.M. (1996) *The Periglacial Environment*, 2nd edition, Harlow: Longman.

CHARLES HARRIS

ICEBERG

Icebergs are floating pieces of glacier ice that are found in marine and lake waters. They are produced when ice breaks off from the margins of glaciers and ice sheets that terminate as ice cliffs in the sea and lakes. Icebergs vary in length from metres to kilometres – those less than 5–10 m in length are referred to as bergy bits. Larger icebergs are usually flat-topped or tabular in shape, whereas smaller bergs, and those in the later stages of melting and breakup, are of irregular form. The ice in icebergs comes originally from snowfall on the glacier surface and is transformed into glacier ice as a result of pressure during burial. This ice may be hundreds or even thousands of years old by the time it calves into the sea. Icebergs are less dense than water because of the presence of air bubbles that are trapped within them during formation. This density difference makes them buoyant, and about 80–90 per cent of their bulk is hidden below the water surface. Iceberg keels can reach hundreds of metres deep. Once icebergs have broken off from their parent glacier, they drift under the influence of ocean currents and, to a lesser extent, wind.

Reference

Wadhams, P. (2000) *Ice in the Ocean*, Amsterdam: Gordon and Breach.

JULIAN A. DOWDESWELL

ICING

Sheets of ice formed on the ground or on lake or river ice by the freezing of successive flows of water discharging from the ground or rising up through fractures in an antecedent ice cover. Icings are not restricted to PERMAFROST areas, but are common in the discontinuous permafrost zone, especially in carbonate terrain. Icings may be a serious hazard for road traffic, occurring persistently at the base of road cuts and at creek crossings. Most icings melt during the first few weeks of summer, but some extensive features may persist for years.

River icings are the most extensive features, and may extend over tens of km^2. The spatial extent of the icing depends on the discharge rate of water to the ice surface, water temperature, air temperature and channel slope (Hu and Pollard 1997). Icings are characteristically layered, with each bed the product of a discharge event. The basal portion of these layers may be discoloured by the solutes expelled and concentrated during freezing. Icings may extend outside the channel used by the river in summer, and may leave a dusting of precipitate on the local vegetation as they melt.

Icing blisters form if hydraulic pressure lifts an overlying ice layer. In continuous permafrost, such blisters have been observed at the edges of pingos and in residual ponds of drained lake beds, where water is expelled during permafrost aggradation (Mackay 1997). Such features are similar to frost blisters, but lack overlying ground.

References

Hu, X. and Pollard, W.H. (1997) The hydrologic analysis and modelling of river icing growth, North Fork Pass, Yukon Territory, Canada, *Permafrost and Periglacial Processes* 8, 279–294.

Mackay, J.R. (1997) A full-scale experiment (1978–1995) on the growth of permafrost by means of lake drainage, western Arctic coast: a discussion of the method and some results, *Canadian Journal of Earth Sciences* 34, 17–33.

C.R. BURN

ILLUVIATION

Illuviation is a process by which material removed from one horizon is deposited in another horizon of a soil (Foth 1984; Ritter 1986; Soil Science Society of America 1987). Usually the direction of transport is from an upper to a lower horizon. The lower horizon, an illuvial horizon, is considered a zone of accumulation and concentration for material. The material can be precipitated from solution or deposited from suspension. Depending on the conditions of soil formation, different illuviated constituents are present. For example, an acid soil formed under forest vegetation may exhibit a dark-coloured illuvial horizon containing quantities of sesquioxides and mineral-organic complexes as well as clay minerals and clay-size material. These materials have been eluviated or transported from an overlying horizon that may be significantly depleted in these constituents. A soil in a prairie region may exhibit only clay increases in the illuvial horizon or zone of accumulation. For contrast, see ELUVIUM AND ELUVIATION.

References

Foth, H.D. (1984) *Fundamentals of Soil Science*, New York: Wiley.

Ritter, D.F. (1986) *Process Geomorphology*, 2nd edition, Dubuque, IA: William Brown.

Soil Science Society of America (1987) *Glossary of Soil Science Terms*, Madison, WI.

CAROLYN G. OLSON

IMBRICATION

Imbrication is the orientation of an assembly of rocks or pebbles in one predominant direction (usually upstream), as a result of flow movement. Imbrication is common on gravel beaches, on bars and outwash fans in braided rivers, and glacial tills. Additionally, imbrication is often a key for interpreting facies. The term imbrication also refers to the near parallel overlapping and orientation of a series of lesser thrust faults, directed towards the source of stress.

Further reading

Byrne, J.V. (1963) Variations in fluvial gravel imbrication, *Journal of Sedimentary Petrology* 33, 467–469.

STEVE WARD

INHERITANCE

Landforms and landscapes have different lifetimes, which generally increase proportionally to the size of a landform. Therefore, many small landforms are formed and destroyed relatively quickly, over timescales of 10^3–10^4 yr, especially if the substratum is soft. By contrast, landforms built of hard, resistant rocks such as certain INSELBERGs could be very durable and may survive for much longer, over timescales of 10^5–10^7 yr. Likewise, regional landscapes such as extensive PLANATION SURFACES may have considerable lifetimes, measurable in millions of years. With such protracted histories, the bearing of present-day environmental conditions on their appearance and development is limited, and inheritance of the past becomes crucial.

Inherited landscapes are those that were formed in the past, under influences of a different external environment and/or within a tectonic regime different from the recent one, and have retained their characteristic features up to now, because later processes have proved to be inadequate to obliterate earlier sets of type landforms. Hence, there is no universal, worldwide applicable boundary between 'inherited' and 'contemporary'. In high latitudes of northern Europe and North America everything that predates Pleistocene continental glaciation is usually considered 'inherited', so that inherited landscapes are older than 1–1.5 Ma. By contrast, in Australia and South America some landscapes are likely to be inherited from the period preceding the breakup of Gondwana in the Mesozoic (Ollier 1991; Twidale 1994). In Central Europe, in turn, major change in the course of geomorphic evolution is often associated with growing tectonic instability from the Palaeogene/Neogene boundary onwards. Therefore, 'inherited' landforms would be those which originated within a general regime of tectonic quiescence and are older than ~20 Ma. However, in the specific context of Holocene morphogenesis, Pleistocene periglacial landscapes will also be regarded as inherited. It has to be emphasized that, strictly speaking, each landscape is recent in the sense that current processes do some action on it; yet the rate of change could be so negligible that the origin of gross features of such a landscape can be traced well back beyond the present day. In other situations, inherited forms may occur very close to relief features of recent origin, forming a palimpsest of landforms of contrasting origins and ages (Starkel 1987; Brunsden 1993a). In the case of the Fennoscandian Shield, where exhumed elements play a significant part, the range of ages of inherited landscapes spans from the Precambrian through the Jurassic, Cretaceous, late Tertiary up to the Pleistocene (Lidmar-Bergström 1995).

The phenomenon of landform and landscape inheritance has been recognized in geomorphology for a long time, though the very term 'inheritance' has not always been used. A German geomorphologist, S. Passarge (1919), made a distinction between *Vorzeitformen* (i.e. inherited) and *Jetztzeitformen* (i.e. moulded at present). Realization that the history of environmental change is very complex itself has led to the concept of 'generations of relief' (Büdel 1977). Accordingly, landscapes observed nowadays comprise units inherited from various epochs of the past, each recognizable through its characteristic set of landforms.

Inherited landscapes should not be equated with persistent landscapes as elaborated by Brunsden (1993b). The latter are landscapes which maintain their principal characteristics over time, and they do this for two main reasons. The first is that they have undergone very little change since their period of origin. The second is that persistent characteristics are maintained either because they are renewed or because they are uniformly changed. Thus while they maintain their original morphology, the currently observed landforms are not necessarily old. Therefore 'steady-state' landscapes are persistent. Zones of high relief may be maintained, for example, by erosion being compensated by isostatic adjustment. Hence, while every inherited landscape is persistent, persistent landscapes are not themselves necessarily inherited from the distant past.

Evidence for inheritance of land surfaces comes from three main sources. The most convincing one is offered by the occurrence of sediments of known age within denudational landscapes; these parts of landscapes are then at least contemporaneous with the sediments. Remnants of weathering mantles, including DURICRUSTs, are essentially of the same significance as the sediments, although the age of weathering residuals is much more difficult to establish. The third line of evidence is provided by the landforms themselves. In many cases there exists a sharp boundary between landscape units which helps to differentiate between older (inherited) and younger landforms. Examples include glacial U-shaped valleys breaching former watershed ridges, antecedent valleys cut into older surfaces and sea cliffs truncating subaerial landforms of long geomorphic history.

The survival of ancient landscapes means little geomorphological change, and these low rates of long-term denudation may be causally linked to one or more factors (Twidale 1999; Migoń and Goudie 2001). Subsiding areas retain their inherited geomorphological features for longer than areas subjected to surface uplift because erosional dissection and water divide lowering within the former is less effective. Moreover, as dissection proceeds and more rock mass is eroded, isostatic recovery (see ISOSTASY) becomes increasingly important, promoting further erosion and destruction of any remnants of old surfaces. However, rapid surface uplift with limited dissection seems capable of elevating a palaeosurface without transforming it into an all-slope topography. An inherited landscape will then be present at a high elevation.

Further reasons for inheritance may include climatic conditions unfavourable to rapid progress of denudation, high bedrock resistance, distance from base level or other lines of active erosion, and the protective role of durable duricrust blankets. In formerly glaciated countries the protective role of cold-based ice or the location within ice divides may play an additional part. A separate category is temporary burial and later re-exposure of an ancient landscape.

References

Brunsden, D. (1993a) Barriers to geomorphological change, in D.S.G. Thomas and R.J. Allison (eds) *Landscape Sensitivity*, 7–12, Chichester: Wiley.

——(1993b) The persistence of landforms, *Zeitschrift für Geomorphologie N.F. Supplementband* 93, 13–28.

Büdel, J. (1977) *Klima-Geomorphologie*, Stuttgart: Borntraeger.

Lidmar-Bergström, K. (1995) Relief and saprolites through time on the Baltic Shield, *Geomorphology* 12, 45–61.

Migoń, P. and Goudie, A.S. (2001) Inherited landscapes of Britain – possible reasons for survival, *Zeitschrift für Geomorphologie* 45(4), 417–441.

Ollier, C.D. (1991) *Ancient Landforms*, London: Belhaven.

Passarge, S. (1919) Die Vorzeitformen der deutschen Mittelgebirgslandschaften, *Petermann's Geographische Mitteillungen* 65, 41–46.

Starkel, L. (1987) The role of inherited forms in the present-day relief of the Polish Carpathians, in V. Gardiner (ed.) *International Geomorphology 1986*, Part II, 1,033–1,045, Chichester: Wiley.

Twidale, C.R. (1994) Gondwanan (Late Jurassic and Cretaceous) palaeosurfaces of the Australian Craton, *Palaeogeography, Palaeoclimatology, Palaeoecology* 112, 157–186.

——(1999) Landforms ancient and recent: the paradox, *Geografiska Annaler* 81A, 431–441.

SEE ALSO: climato-genetic geomorphology

PIOTR MIGOŃ

INITIATION OF MOTION

The initiation of sediment transport (also known as incipient motion or the critical condition) describes when one or more particles are moved from a stationary bed. In river or air flows, the force balance for a particle resting on the surface can be resolved into three components: a drag force acting parallel to the bed, known as the shear stress (τ), an upward lift force produced by turbulence and the downward frictional force caused by the particle's weight. In 1936, Shields showed that the onset of particle motion is defined as $\tau_c^* = \tau_c/(\rho_s - \rho)gD$ where τ_c^* is the dimensionless shear stress at the critical condition, τ_c is the tangential fluid shear stress at the boundary as grain movement begins and ρ_s, ρ, g and D are the sediment density, fluid density, acceleration due to gravity and grain diameter at the bed surface, respectively. Using a series of laboratory flume experiments with sediment of uniform grain size, Shields showed that the value of τ_c^* lies within a narrow range for hydraulically rough and turbulent flow (mean $\tau_c^* \approx 0.046$ when corrected for sidewall effects and form drag). Consequently, the Shields equation stated above simplifies to $\tau_c \alpha D$ which implies that a stronger fluid force will move coarser particles. Some authors refer to this finding as the principle of 'selective entrainment'.

Although intuitively attractive, it has now been shown that there is a frequency distribution of τ_c^* values for any particular grain size. This is because the three-dimensional forces of turbulent shear stress that impact on the bed vary through space and time and the potential for movement of a particular grain is strongly dependent on the physical arrangement of the bed surface (e.g. grain pivoting angle, degree of packing and sorting, mixture of grain shapes). The variation in τ_c^* is most marked over coarse-grained and hydraulically rough boundaries where different particle sizes protrude or are 'hidden' behind and below other particles on the bed surface. Researchers have shown that τ_c^* decreases with an increase in relative particle size (i.e. the ratio between the size of the surface particle and the ambient grains). So long as there is an overrepresentation of coarse sediment on the bed surface, sediment entrainment may achieve what is termed 'equal mobility' whereby all particles in the surface may be moved regardless of their absolute size or weight and the BEDLOAD grain size distribution matches that of the bed subsurface.

Further reading

Buffington, J.M. and Montgomery, D.R. (1997) A systematic analysis of eight decades of incipient motion studies, with special reference to gravel-bedded rivers, *Water Resources Research* 33, 1,993–2,029.

Kennedy, J.F. (1995) The Albert Shields story, *Journal of Hydraulic Engineering*, American Society of Civil Engineers 121, 766–771.

Lavelle, J.W. and Mofjeld, H.O. (1987) Do critical stresses for incipient motion and erosion really exist? *Journal of Hydraulic Engineering*, American Society of Civil Engineers 113, 370–385.

SEE ALSO: downstream fining; fluvial armour; mobile bed

PHILIP J. ASHWORTH

INLAND DELTA

The term was coined by cartographers in the late nineteenth century to describe a morphologically unusual reach of the Niger River in the interior of west Africa (Mali), which they termed the Niger Inland Delta. Here, in the vicinity of the confluence (see CONFLUENCE, CHANNEL AND RIVER JUNCTION) of the Niger and Bani Rivers, the two rivers divide to form a network of anastomosed (see ANABRANCHING AND ANASTOMOSING RIVER) channels spread over a FLOODPLAIN approximately 100 km wide and 200 km long. The channels terminate in Lakes Debo and Korientze, which are ponded against aeolian dunes on the southern edge of the Sahel. A single channel forms beyond the dunefield. The adjective 'inland' was added to distinguish this feature from the delta (see RIVER DELTA) at the mouth of the Niger. A form of the term was also applied to the terminus of the Okavango River in the endoreic Kalahari Basin of central southern Africa (Botswana). After meandering through a 10 to 15 km wide, 100 km long corridor known as the Panhandle, the Okavango River abruptly divides into a number of radial, distributary channels extending over a floodplain some 50,000 km^2 in area, known as the Okavango Delta. The adjective 'inland' was omitted in this case because the Okavango River, unlike the Niger, does not reach the ocean, so there is no risk of confusion. A geomorphologically similar feature is also developed on the White Nile River (Bahr el Jebel) in Sudan, where the river divides into several distributary channels forming a triangle 520 km long and 150 km along its base. The term 'delta' has not been applied in this case, however, and the feature is known by its Arabic name, the Sudd.

The Niger Inland Delta, Okavango Delta and the Sudd have several features in common. They occur in actively subsiding half-grabens, possibly related to incipient rifting. The multiple channels of these inland deltas form in response to the loss of confinement of flow as the rivers enter the grabens. They have relatively low gradients: 3 cm km^{-1} for the Niger, 10 cm km^{-1} for the Sudd and 28 cm km^{-1} for the Okavango. River discharge is seasonal and the flood wave moves slowly down the floodplain, taking three months to cross the Niger Inland Delta, four months to cross the Sudd and five months to cross the Okavango Delta. There is considerable evapotranspirational loss of water – 98 per cent in the case of the Okavango.

Whilst the channel geometry of these inland deltas is superficially similar to those on deltas at river mouths, the sedimentation processes are fundamentally different. In river deltas, sedimentation occurs primarily on the delta front as the distributary channels lose their ability to transport sediment on entering standing-water bodies. River deltas therefore build outwards into water bodies. In contrast, on inland deltas sediment deposition occurs primarily as a result of flood water spilling from channels and spreading laterally as sheet flow. Inland deltas therefore aggrade vertically. Local AGGRADATION around channels results in instability, causing AVULSION which gives rise to new channels. In this way, sediment is spread uniformly across the delta surface. This sedimentary style is more akin to that occurring on ALLUVIAL FANs, and these inland deltas constitute a variety of low gradient alluvial fan.

Further reading

Howell, P., Lock, M. and Cobb, S. (1988) *The Jonglei Canal*, Cambridge: Cambridge University Press.

McCarthy, T.S. (1993) The great inland deltas of Africa, *Journal of African Earth Sciences* 17, 275–291.

Makaske, B. (1998) *Anastomosing Rivers: Forms, Process and Sediments*, Faculteit Ruimtelijke Wetenschappen, Universiteit Utrecht.

Stanistreet, I.G. and McCarthy, T.S. (1993) The Okavango fan and the classification of subaerial fan systems, *Sedimentary Geology* 85, 115–133.

SEE ALSO: alluvial fan; megafan; river delta

T.S. McCARTHY

INSELBERG

Inselberg is a descriptive term, derived from German (*inselberg* literally means 'island hill') and adopted in English, used to describe a hill which stands in isolation and rises sharply above the level of the surrounding plain. It was first coined by W. Bornhardt, a German naturalist travelling in East Africa at the turn of the nineteenth century, to emphasize visual similarities between steep-sided hills dotting an otherwise flat savanna and islands rising from the sea. Inselbergs often tend to occur in groups, to form inselberg landscapes.

It is generally accepted that the term applies to a hill produced by lowering of the surface around it. Therefore volcanoes and small tectonic horsts are normally not called inselbergs (Kesel 1973). It was also noted that the term is rather infrequently used to describe isolated hills built of flat-lying sedimentary rocks such as MESA or BUTTE, hence the impression arises that inselberg landscapes are restricted to basement rock areas. There have been attempts to make the definition more strict by using certain quantitative constraints for inselbergs. These included minimal height of 15 m, length to width ratio not exceeding 4 : 1, or minimal distance to the nearest neighbour of 0.8 km (Faniran 1974). None of these proposals has been universally accepted and there is quite a degree of freedom in deciding which hill is to be called an inselberg, which makes comparative analyses of inselberg landscapes difficult.

Visual differences provide the basis for classification of inselbergs, most readily applicable to granite hills. Three main types are distinguished (Thomas 1978; Twidale 1981). Domed inselbergs have slopes convex-upward and are built of massive, poorly jointed rock, with little REGOLITH at the surface. They are called BORNHARDTS, but it is important to note that the two terms are not equivalents. Hills having all the characteristics of bornhardts listed above can also occur within hilly and mountainous terrains, and hence are not by any means inselbergs. Examples include 'sugarloaf' hills in Rio de Janeiro, or granite domes in the Yosemite National Park in the USA. Castellated inselbergs are built of well-jointed but bedrock-rooted rock and their detailed morphology consists of towers, pillars and walls, separated by joint-aligned avenues. Boulder inselbergs are composed of loose boulders, chaotically lying one upon another. They typically form through advanced degradation of a domed or castellated inselberg. In addition, some inselbergs, especially in metamorphic rocks, may be conical, and there are also hills built entirely of SAPROLITE.

Inselbergs tend to be built of hard, resistant igneous and metamorphic rocks such as granite or gneiss. They are particularly common in coarse-grained and poorly jointed granites, rich in potassium feldspar (Pye *et al.* 1986). They are also known from gabbro, syenite, rlyolite and migmatite. However, there are relatively few instances where the base of the hillslope coincides with a lithological boundary, nor are differences in mineralogical composition between the hill and the plain significant (Brook 1978). The majority of inselbergs show joint- rather than lithological control. Bedrock to build an inselberg is typically massive, with few open fractures and the predominance of tight SHEETING joints. The latter are often arranged concentrically to form a structural dome, the outline of which is followed by the topography. It is generally assumed, although usually not proved, that joint density around a hill is higher than within the hill, and the existence of an inselberg reflects primary variations in the degree of rock mass fracturing. Another manifestation of joint control is seen in plan. Outlines of inselbergs frequently follow master joints (Twidale 1982; Selby 1982).

Notwithstanding the impression that inselbergs are only supported by igneous rocks, spectacular inselberg assemblages can also be built of sedimentary rocks, especially of massive sandstone and conglomerate. Ayers Rock (Uluru) in central Australia, one of the most famous inselbergs

Plate 65 The granite inselberg of Sptizkoppe in the Namib Desert is built of extremely massive – and therefore most resistant – granite, which accounts for its impressive height of more than 600 m

worldwide, is made of steeply dipping arkosic sandstone, whereas the adjacent Olgas are composed of massive conglomerate. Further examples of isolated sandstone towers have been reported from certain arid areas, such as the Sahara Desert in Niger and Mali, or the deserts of south-west Jordan. Interestingly, sandstone inselbergs in these areas are not associated with any actively retreating escarpments but rather have originated due to advanced dissection of a former plateau.

The origin of inselbergs had been a matter of hot debate, but it is now accepted that they may form in various ways, and contrasting evolutionary pathways may produce landforms looking superficially very similar. There are at least three theories present in the literature. Perhaps the most universally accepted one holds that inselbergs are products of two-stage development involving differential deep weathering in the first phase and stripping of the weathering mantle in the second one, which leaves an unweathered rock compartment exposed at the surface (Thomas 1965). Reasons for survival of such compartments include significantly wider spacing of joints, the presence of primary poorly fractured mass, enrichment in quartz and/or potassium feldspar, or petrological differences. Validity of the two-stage hypothesis has been confirmed in deep excavations and quarries in equatorial Africa, where massive hill-like features up to 50 m high have been found within a thick mantle of decomposed bedrock. In reality, the average thickness of weathering mantles seems insufficient to account for the height of many inselbergs, which may exceed 200–300 m. Therefore their exposure is more likely to have been accomplished in many stages of weathering and stripping (Twidale and Bourne 1975). Minor landforms on hillslopes such as flared slopes and platforms are considered as the evidence of multiphase exposure.

Another hypothesis invokes scarp retreat across unweathered, but possibly differentially jointed, bedrock and links the origin of inselbergs with the cyclic development of relief and retreat of big escarpments separating denudational landscapes of various ages (King 1949). Clustering of inselbergs in front of major escarpments might validate this theory, but other people argue for an important role of deep weathering in the development of scarps and inselbergs too (Bremer 1993).

The massive nature of many prominent inselbergs found in the semi-arid and arid zone, and the occurrence of more jointed compartments around their flanks, suggest that exposure and growth in height may not necessarily be associated with deep weathering. Long-term lowering of differentially jointed bedrock probably accounts for the origin of the spectacular inselbergs of the Namib Desert (Selby 1982) and may be applicable elsewhere. In the same way, some minor granite intrusions may have been exposed as inselbergs as the surrounding less resistant schist has been completely eroded away.

Inselbergs, once exposed or isolated, undergo further development and are the scene of competing processes of continuous growth and destruction. Many authors emphasize that inselbergs are very durable, long-lived landforms because their surfaces shed rainwater and remain dry, being therefore immune to chemical weathering. At the same time, because they are built of poorly jointed bedrock, they are resistant against physical weathering too. Fast runoff from the slopes of an inselberg provides additional moisture to its footslopes, so the rate and intensity of weathering are enhanced. Episodic removal of regolith cover from around the inselberg may then result in the increase of its height, providing the lowering of its summit proceeds at a slower rate. In Australia, it is argued that inselbergs have been rising for millions of years and their top surfaces may date back to the Mesozoic (Twidale and Bourne 1975; Twidale 1978).

How inselbergs are reduced in height and extent depends on their jointing patterns. Massive domed inselbergs are subject to mega-exfoliation due to pressure release and opening of sheeting joints. Individual slabs are separated from the underlying rock mass and fall or slide off the slopes, forming big debris cones mantling lower slopes. Ongoing exfoliation will gradually reduce the surface area of an inselberg. Rock fall is also typical for slope development of sandstone inselbergs in deserts. In orthogonal patterns, vertical joints open too, topples occur, and the summit part assumes ruiniform relief. Minor weathering features play an important part in the development of inselbergs (Watson and Pye 1985). Selective weathering along fractures and growth of TAFONI reduces rock mass strength and facilitates mass movement, whereas horizontal surfaces are destroyed by enlargement of WEATHERING PITS. Caves and massive overhangs are frequently reported from granite inselbergs and develop either through mechanical widening of fractures, preferential weathering along sheeting joints, or chaotic accumulation of big boulders on lower slopes.

The discussion about the origin of inselbergs has a direct bearing on the issue of their significance in geomorphology, especially in CLIMATIC GEOMORPHOLOGY and CLIMATO-GENETIC GEOMORPHOLOGY. Two positions emerge from the literature. One holds that inselbergs occur all around the world and cannot be considered as indicators of environments, present or past (Kesel 1973; King 1975). King argued that the process of scarp retreat is climate-independent, hence inselbergs are not dependent on climatic conditions either. By contrast, a German school of geomorphology has maintained that inselbergs are specific products of landscape development in the seasonally humid tropics and develop through deep weathering and stripping, which are the processes acting at their highest efficacy in this zone. Consequently, inselbergs present in middle or high latitudes, or in arid areas, would be relict landforms, inherited from the geological past.

Although it is probably true that humid tropical environments with ubiquitous deep weathering favour the development of inselbergs, a claim that inselbergs are by definition 'tropical' landforms is most likely wrong. Inselbergs are present in many desert areas of the world, including the long-lived ones such as the Namib, and in many evidence for inheritance from previously humid conditions is clearly lacking. Nevertheless, the origin of those present in central and northern Europe, as well as in North America, is usually traced back to the Early Tertiary, when climate was warmer and wetter, and deep weathering was widespread. It is worth noting that even if these inselbergs are indeed ancient Tertiary features, it does not mean that tropical climate was essential for their origin.

Inselberg landscapes have been reported from all around the world, including parts of Antarctica, but they are probably most widespread in Africa, within extensive tracts of crystalline rock terrain. Classic examples are known from Nigerian savannas, East African plains in Kenya and Tanzania, Zimbabwe, South Africa, Namib Desert and Angola. Namibian inselbergs, such as the almost 700-m high Spitzkoppe, belong to the highest in the world. There are also examples from the Sahara and its southern margin, from Sudan, Niger and Libya. Inselbergs are also common in Australia, especially in the central and western part of the continent. Further examples are known from the Indian Peninsula and basement areas of South America. Numerous, purportedly inherited, inselberg landscapes have been described in Europe, including Germany, the Czech Republic, Poland, Hungary and Scandinavia.

References

Bremer, H. (1993) Etchplanation, review and comments of Büdel's model, *Zeitschrift für Geomorphologie Supplementband N.F.* 92, 189–200.

Brook, G.A. (1978) A new approach to the study of inselberg landscapes, *Zeitschrift für Geomorphologie Supplementband N.F.* 31, 138–160.

Faniran, A. (1974) Nearest-neighbour analysis of inter-inselberg distance: a case study of inselbergs of south-western Nigeria, *Zeitschrift für Geomorphologie Supplementband N.F.* 20, 150–167.

Kesel, R.H. (1973) Inselberg landform elements: definition and synthesis, *Revue Géomorphologie Dynamique* 22, 97–108.

King, L.C. (1949) A theory of bornhardts, *Geographical Journal* 113, 83–87.

——(1975) Bornhardt landforms and what they teach, *Zeitschrift für Geomorphologie* 19, 299–318.

Pye, K., Goudie, A.S. and Watson, A. (1986) Petrological influence on differential weathering and inselberg development in the Kora area of Central Kenya, *Earth Surface Processes and Landforms* 11, 41–52.

Selby, M.J. (1982) Form and origin of some bornhardts of the Namib Desert, *Zeitschrift für Geomorphologie Supplementband N.F.* 26, 1–15.

Thomas, M.F. (1965) Some aspects of the geomorphology of domes and tors in Nigeria, *Zeitschrift für Geomorphologie Supplementband N.F.* 9, 63–82.

——(1978) The study of inselbergs, *Zeitschrift für Geomorphologie Supplementband N.F.* 31, 1–41.

Twidale, C.R. (1978) On the origin of Ayers Rock, central Australia, *Zeitschrift für Geomorphologie Supplementband N.F.* 31, 177–206.

——(1981) Granite inselbergs: domed, block-strewn and castellated, *Geographical Journal* 147, 54–71.

——(1982) *Granite Landforms*, Amsterdam: Elsevier.

Twidale, C.R. and Bourne, J.A. (1975) Episodic exposure of inselbergs, *Geological Society of America Bulletin* 86, 1,473–1,481.

Watson, A. and Pye, K. (1985) Pseudokarstic microrelief and other weathering features on the Mswati Granite (Swaziland), *Zeitschrift für Geomorphologie Supplementband N.F.* 29, 285–300.

PIOTR MIGOŃ

INSOLATION WEATHERING

Insolation weathering (thermoclasty or thermal stress fatigue) is the rupturing of rocks and minerals primarily as a result of large daily temperature changes in dry environments which lead to temperature gradients within the rock mass. Fires can operate in a similar way, though the temperature extremes are greater (see FIRE). Areas that are heated expand relative to the cooler portions of the rock and stresses are thereby set up.

In igneous rocks, which contain many different types of mineral (polymineralcy) with different coefficients and directions of expansion, such stresses are enhanced. Moreover, the varying colours of minerals exposed at the surface (polychromacy) will cause differential heating and cooling.

Daily temperature cycles under desert conditions may exceed 50 °C, and during the heat of the day rock surfaces may occasionally exceed a temperature of 80 °C. However, rapid cooling takes place at night, creating, it has been thought, high tensile stresses in the rock. Desert travellers, like David Livingstone, have claimed to hear rocks splitting with sounds like pistol shots in the cool evening air – certainly, split rocks are evident on many desert surfaces. Insolation was a mechanism that found considerable favour with pioneer desert geomorphologists (e.g. Hume 1925; Walther 1997).

At first sight the process of insolation weathering seems a compelling and attractive mechanism of rock disintegration. However, doubt has been cast upon its effectiveness upon a variety of grounds (Schattner 1961). The most persuasive basis for doubting its power was provided by early experimental work in the laboratory by geomorphologists like Blackwelder, Griggs and Tarr. They all found that simulated insolation produced no discernable disintegration of dry rock, but that when water was used in the cooling phase of a weathering cycle disintegration was evident. This highlighted the importance of the presence of water. Likewise, studies of ancient buildings and monuments in dry parts of North Africa and Arabia showed very little sign of decay except in areas, for example close to the Nile, where moisture was present (Barton 1916). Indeed, there are many situations where there is moisture in deserts (e.g. where there is fog, dew and groundwater seepage). When water combines chemically with the more susceptible minerals in a rock they may swell, producing a sufficient increase in volume to cause the outer layers of rock to be lifted off as concentric shells, a process called EXFOLIATION. Thus, some of the weathering that used to be attributed to insolation may now be attributed to chemical changes produced by moisture, including HYDRATION.

However, the importance of insolation cannot be dismissed entirely. The early experimental work had grave limitations: the blocks used were very small and unconfined, the temperature cycles were unrealistic and only a limited range of rock types was used (Rice 1976). Moreover, engineering and ceramic studies have shown that a threshold value for thermal shock approximates to a rate of temperature change of 2 °C min^{-1}. Datalogger studies show that such rates can occur, not least in polar regions (Hall and André 2001). In addition, consideration of fracture patterns observed on rock in cold, dry environments appears to show very similar forms to those produced in thermal shock experiments in the laboratory (Hall 1999). Finally, because of the temperature response of calcite crystals, marble seems to be especially prone to thermal degradation (Royer-Carfagni 1999)

References

Barton, D.C. (1916) Notes on the disintegration of granite in Egypt, *Journal of Geology* 24, 382–393.

Hall, K. (1999) The role of thermal stress fatigue in the breakdown of rock in cold regions, *Geomorphology* 31, 47–63.

Hall, K. and André M.-F. (2001) New insights into rock weathering from high-frequency rock temperature data: an Antarctic study of weathering by thermal stress, *Geomorphology* 41, 23–35.

Hume, W.F. (1925) *Geology in Egypt*, Vol. 1, Cairo: Government Press.

Rice, A. (1976) Insolation warmed over, *Geology* 4, 61–62.

Royer-Carfagni, G.F. (1999) On the thermal degradation of marble, *International Journal of Rock Mechanics and Mining Sciences* 36, 119–126.

Schattner, I. (1961) Weathering phenomena in the crystalline of the Sinai in the light of current notions, *Bulletin Research Council of Israel* 106, 247–266.

Walther, J. (1997) *The Law of Desert Formation – Present and Past* (translated by G. Meyer and edited by E. Gischler and K.W. Glennie), University of Miami, Geological Milestones, Vol. 4.

A.S. GOUDIE

INTEGRATED COASTAL MANAGEMENT

Coastal geomorphology and associated coastal resources are subject to increasing pressure from human impact. In a global context, this is significant for two reasons. First, the majority of the world's population currently lives near the coast and the proportion of coastal dwellers is projected to increase in the future. The world's coastal population has been variously estimated as over half living within 60 km of the coast (UNCED 1992: 17.3); 1.2 billion within 100 km of the coast (Nicholls and Small 2002, based on 1990 data); or 3.2 billion within 200 km and two-thirds within 400 km (Hinrichsen 1998). Second, humans have a high dependence on coastal resources. According to key researchers in resource economics (Costanza *et al.* 1997) the coastal biome currently contributes over 40 per cent of the total global flow value of ecosystem services.

These global population pressures and human dependence on the coast require appropriate management strategies which recognize the dynamic nature of the coast at various geomorphic timescales and across different spatial dimensions. While longer timescales may be less problematic from a management perspective, the identification of rapid geomorphic change may be difficult to separate from human-induced change. For example, coastal wetlands may be subject to human impacts of reclamation, development or subsidence through groundwater withdrawal acting simultaneously with natural processes of wetland loss such as local relative sea-level change or sediment compaction. It is also important to recognize that local coastal impacts may have broader spatial linkages to marine or land-based processes. For example, coastal pollution, erosion or accretion near a river mouth may be attributed to poor catchment management practices rather than localized coastal processes.

For these reasons, there has been a recognition that the coast, along with other types of environment, needs to be managed in a holistic rather than a sectoral manner. This has given rise to an increasing acceptance of integrated resource management in general, and more specifically to integrated coastal management (ICM). Although the concept of coastal management has been around for more than thirty years, particularly with the introduction of coastal legislation to the United States in 1972, it was the United Nations Conference on Environment and Development (UNCED) (also known as the 'Earth Summit') held in Rio de Janeiro in 1992 that created an international push for the adoption of 'integrated' coastal management with an agreement that coastal states should 'commit themselves to integrated management and sustainable development of coastal areas and the marine environment under their national jurisdiction' (UNCED 1992: Agenda 21: 17.5).

Similar objectives are contained within the United Nations Framework Convention on Climate Change (1992) which outlines the need to develop integrated plans for coastal management. In the following year (1993), the Council of the Organisation for Economic Cooperation and Development and the first World Coastal Conference adopted and produced guidelines for the integrated management of coastal resources. These *inter alia* required that nations developing their ICM should take into account the traditional, cultural and historical perspectives and conflicting interests and uses (IPCC 1994). In order to define ICM it is useful to draw on earlier definitions from the IPCC (1994: 40), Cicin-Sain and Knecht (1998: 39) and also from the Joint Group of Experts on the Scientific Aspects of Marine Environmental Protection (GESAMP 1996: 2).

Integrated Coastal Management is a continuous and dynamic process incorporating feedback loops which aims to manage human use of coastal resources in a sustainable manner by adopting a holistic and integrative approach between terrestrial and marine environments; levels and sectors of government; government and community; science and management; and sectors of the economy.

It is important to realize that although there are a number of definitions of ICM, there is still some confusion over the use of other related coastal management terms and definitions. Cicin-Sain and Knecht (1998) and Burbridge (1999) note that there has been a major change in emphasis away from coastal 'zone' or 'area' management towards 'integrated' coastal management. Cicin-Sain and Knecht argue that the terms integrated coastal zone management (ICZM), integrated marine and coastal area management and integrated coastal management (ICM) all refer to the same concept and they adopt ICM for reasons of consistency and simplicity. Others still use the term ICZM (see Salomons *et al.* 1999) although some authors in the same volume (Burbridge 1999; Harvey 1999) prefer the term ICM. In a major review of the coastal management literature Sorensen (1997) distinguishes between ICM as a concept or field of study and ICZM as a programme which has the task of defining the boundaries of the coastal 'zone'. On balance it appears that although the use of the term 'zone' was originally intended to be flexible, it can also be interpreted as prescriptive if the identification of boundary conditions mitigate against the need to integrate across them. For this reason, the use of ICM is becoming more acceptable and common in the literature.

There is now a global trend toward a more integrated approach for coastal management which incorporates linkages between activities in coastal lands and waters. A decade ago, the Earth Summit (UNCED 1992) recognized the need for a new approach to marine and coastal area management developed at the national, subregional, regional and global levels. It also commented that any new approach to coastal management should be integrated in content, and precautionary and anticipatory in its scope. Subsequently, there have been various international attempts to develop guidelines for ICM, stressing the importance of strengthening and harmonizing cross-sectoral management. While there are different approaches for achieving ICM, most agree that horizontal and vertical integration and co-ordination must be part of any attempt to achieve ICM.

Cicin-Sain and Knecht (1998) suggest that the rationale for an integrated approach is first to examine the effects of ocean and coastal use, as well as activities further inland, on ocean and coastal environments; and second to examine the effects that ocean and coastal users can have on one another. The World Coast Conference puts it simply as the need for co-operation between all responsible actors involved in coastal management (IPCC 1994: 25). The key elements of integration in coastal management can be defined as follows:

- Intergovernmental integration (vertical integration) between different levels of government such as national, provincial or state and local governments;
- Intersectoral integration (horizontal integration) between different government sectors: such as industry, conservation, recreation, tourism, beach protection and integration of policies between different sectors of the economy;
- Community integration with government producing effective community participation and involvement in coastal management;
- Spatial integration between management of the land, ocean and coast;
- Integration between science and management particularly between different disciplines; scientists and managers; including economic, technical and legal approaches to coastal management;
- International integration between nations on trans-boundary coastal management issues.

There are a number of requirements for the success of ICM. There is a need for a long-term strategy with clearly defined national objectives and guiding principles for coastal managers. It is also necessary to have a defined authority with responsibility for the strategy along with monitoring of key performance indicators and most importantly there is a need for the political will for implementation of ICM. Various conceptual models (e.g. Cicin-Sain and Knecht 1998: 58) of the different stages in the ICM process all emphasize the cyclical nature of the process and the need for continual re-evaluation. The implication of this is a need to proceed through all stages at least once before the success of ICM can be properly evaluated. This is likely to take a number of years.

Since the Earth Summit and the World Coast Conference, ICM has been adopted by many nations with coastal management programmes and associated legislation. Cicin-Sain and Knecht (1998) conclude from their cross-national ICM survey in 1996 that there were approximately 150 national ICM efforts globally, including the following countries: United States, United Kingdom, Belize, Brazil, Costa Rica, Ecuador, Sri Lanka, Turkey, Australia, Canada, Italy, China, Mexico, Nigeria, Venezuela and Pohnpei State (Federated States of Micronesia). Subsequently, Sorensen has created a global data base (www.uhi.umb.edu) which, in 2000, contained a total of 385 'ICM efforts' comprising 250 in 87 countries, plus 100 in the United States and 35 internationally. However, care is needed in interpreting Sorensen's data on ICM efforts which contain a mixture of programmes, policy statements and feasibility studies.

Notwithstanding the increasing global number of ICM efforts, there is a paucity of data on their success; for example, Cicin-Sain and Knecht's comment that their 1996 survey results provided scanty evidence on the extent of ICM implementation and effectiveness. They concluded that it was very difficult to produce a model of successful ICM because there was a lack of objective evaluation data for any of the ICM examples they described (Cicin-Sain and Knecht 1998: 294). Sorensen (1997) comments that there is uncertainty in our knowledge about the important implementation phase of ICM programmes which comes after the adoption phase of various plans and policies. Burbridge (1997) makes a distinction between developing ICM initiatives and assessing their success in meeting stated goals and he suggests that there are not many good examples of fully developed ICM strategies, plans or practices that extend beyond a local or problem-specific level (Burbridge 1997: 181). It is also important to note that there are significant socio-economic and political differences between coastal nations which need to be considered in assessing ICM achievements.

Thus the concept of ICM is internationally accepted as an appropriate method for sustainable management of coastal resources. The concept has evolved from a realization that a sectoral approach to coastal management is fundamentally flawed. There is now a global proliferation of what has been loosely termed ICM efforts although there is scope for more rigorous survey of these. Finally, it is clear that the international acceptance of ICM as an approach is not matched by definable criteria or models to judge its success or best practice.

References

Burbridge, P.R. (1997) A generic framework for measuring success in Integrated Coastal Management, *Ocean and Coastal Management* 37(2), 175–189.

——(1999) The challenge of demonstrating the socio-economic benefits of Integrated Coastal Management, in W. Salomons, R.K. Turner, L.D. de Lacerda and S. Ramachandran (eds) *Perspectives on Integrated Coastal Zone Management*, 35–53, Berlin: Springer-Verlag.

Cicin-Sain, B. and Knecht, R.W. (1998) *Integrated Coastal and Ocean Management: Concepts and Practices*, Washington, DC: Island Press.

Costanza, R., d'Arge, R., de Groot, R., Farber, S., Grasso, M., Hannon, B., Linburg, K., Naeem, S., O'Neill, R.V., Paruelo, J., Raskin, R.G., Sutton, P. and van den Belt, M. (1997) The value of the world's ecosystem services and natural capital, *Nature* 387, 253–260.

GESAMP (IMO/FAO/UNESCO-IOC/WMO/WHO/IAEA/UN/UNEP Joint Group of Experts on the Scientific Aspects of Marine Environmental Protection) (1996) *The Contributions of Science to Coastal Zone Management*, GESAMP Reports and Studies 61.

Harvey, N. (1999) Australian Integrated Coastal Management: a case study of the Great Barrier Reef, in W. Salomons, R.K. Turner, L.D. de Lacerda and S. Ramachandran (eds) *Perspectives on Integrated Coastal Zone Management*, 279–296, Berlin: Springer-Verlag.

Hinrichsen, D. (1998) *Coastal Waters of the World: Trends, Threats and Strategies*, Washington, DC: Island Press.

Intergovernmental Panel on Climate Change (IPCC) (1994) *Preparing to Meet the Coastal Challenges of the twenty-first Century. Report of the World Coast Conference, 1–5 November 1993*, The Hague: Ministry of Transport, Public Works and Water Management.

Nicholls, R.J. and Small, C. (2002) Improved estimates of coastal population and exposure to hazards released, *EOS, Transactions, American Geophysical Union* 83(28), 301–305.

Salomons, W., Turner, R.K., de Lacerda, L.D. and Ramachandran, S. (eds) (1999) *Perspectives on Integrated Coastal Zone Management*, Berlin: Springer-Verlag.

Sorenson, J. (1997) National and international efforts at Integrated Coastal Zone Management: definitions, achievements, and lessons, *Coastal Management* 25, 3–41.

United Nations Conference on Environment and Development (UNCED) (1992) *Agenda 21 and the UNCED Proceedings*, New York: Oceania Publications.

United Nations Framework Convention on Climate Change (UNFCC) (1992).

NICK HARVEY

INTERDUNE

Interdune areas are the depressions between dunes. Interdune areas occur in a variety of shapes, which reflect the morphology of the dunes with which they are associated. Fields of crescentic dunes typically have ellipsoidal interdune areas that are elongate with the dune crestline. Extremely long interdune corridors occur with linear dunes, whereas more irregular, interconnected interdune areas accompany complicated dune shapes such as star dunes. There is a complete gradation from interdune flats that cover a greater portion of the field than the dunes, to interdune depressions between the lee and the stoss (upwind) slopes of adjacent dunes.

At least for transverse dunes, the formation of an interdune area can be visualized by watching a fixed point at the base of the stoss slope of a migrating dune. The interdune area begins at this erosional point and extends downwind as an interdune BOUNDING SURFACE as the stoss slope migrates. The surface continues to extend downwind until it is buried by the lee deposits of the next dune upwind, which has now migrated to the fixed point.

Although easy to visualize, the explanation for interdune areas remains controversial, and is intimately linked with the explanation of how dunes come to be regularly spaced. In one hypothesis, dune spacing is a function of the fluid dynamics established with airflow over dunes. With transverse dunes, there is flow separation at the dune brink, creating a separation cell in the immediate lee, and flow reattachment at a distance of a few dune heights from the brink (Walker and Nickling 2002). At the reattachment point, a new BOUNDARY LAYER is created. Wind speed and shear stress increase downwind within this new boundary layer, owing to the downward flux of momentum from the overlying, relatively high-speed flow. As long as the surface winds are accelerating, there is at least potential deflation of the interdune area (Kocurek *et al.* 1991). The boundary layer, however, is also expanding vertically, and at some point this expansion overwhelms the effects of the momentum flux, and wind speed and shear stress within the boundary should decrease. At that point, deposition (and initiation of a new dune) can begin. In this way, the spacing of the next dune downwind and the length of the interdune area is defined by the fluid dynamics established by the upwind dune. Field data, however, show that flow speed and shear stress reach a steady-state condition that favours sediment bypass, and not deceleration and deposition (Frank and Kocurek 1996). Adequate data do not yet exist to evaluate the role of turbulence, which may play a more significant role in the spacing of subaqueous dunes (Nelson *et al.* 1995).

In an alternative view, fields of dunes are viewed as self-organizing and the interdune areas exist by default. Computer simulations have modelled the formation of all major dune types as a function of the direction and duration of sand-transporting winds, and independent of the creation of an internal boundary layer (Werner 1995). In this hypothesis, dunes begin as small, randomly spaced collections of sand that merge and grow larger as they migrate, progressing to a steady-state condition where little change is possible because all the dunes are about the same size and, therefore, migrating at about the same speed. Initiation of sand dunes and their development to a steady-state field on Padre Island, Texas, strongly resemble the model (Kocurek *et al.* 1991). The interdune areas in this model are then simply the areas where the dunes are not (Werner and Kocurek 1999).

Regardless of their origin, interdune areas contain some of the most diverse features seen in dunefields. The first division is between those in which accumulation occurs, and those that are deflationary, in which older dune accumulations, reg or bedrock can be exposed. For deflationary interdune areas, or ones in which sediment bypass only occurs, the interdune flats essentially form a continuous surface over which the field of dunes migrate. Characteristic of dunefields where the water table controls the level of deflation (wet aeolian systems), corrugated surfaces develop on the interdune floor that reflect the stratification types of underlying dune accumulations.

Where sediment accumulation occurs on the interdune surface, these sediments can be classified as dry-, damp-, and wet-surface deposits (Kocurek 1981). Dry-surface, sandy accumulations are typically those from wind ripples, grainfall from the upwind dune and satellite dunes. Because the surface is dry and subject to deflation, these interdune accumulations within dry aeolian systems are usually restricted to interdune depressions that exist within the low-wind-speed separation cell. Extensive dry interdune flats, however, are common and these may contain concentrations of less-easily transported coarse grains, typically organized into granule ripples and zibar dunes. Sandy deposits are usually concentrated around sediment-trapping vegetation.

Where the interdune surface is wet or damp, typically because the water table is near the surface, a major division occurs between those that contain evaporate minerals and those that do not. Interdune areas in temperate climates (e.g. coastal fields) typically lack evaporites and the deposits are dominated by ADHESION structures, wrinkle marks, fluid-escape structures and subaqueous ripples, mud drapes, algal mats, channels and other features formed during ponding. In arid climates, wet- or damp-surface interdune areas are nearly always characterized by evaporites. These may occur as interdune SABKHAs, which show a characteristic crinkly sediment texture that results from deposition by salt ridges. Because interdune surfaces in wet aeolian systems have a high capillary water content or cementing evaporites, accumulation can occur on these flats in spite of flow acceleration within the interdune boundary layer, and commonly these interdune areas contain the finest sediment within the field.

References

Frank, A. and Kocurek, G. (1996) Toward a model of airflow on the lee side of aeolian dunes, *Geomorphology* 17, 47–54.
Kocurek, G. (1981) Significance of interdune deposits and bounding surfaces in aeolian dune sands, *Sedimentology* 28, 753–780.
Kocurek, G., Townsley, M., Yeh, E., Havholm, K. and Sweet, M.L. (1991) Dune and dunefield development on Padre Island, Texas, with implications for interdune deposition and water-table-controlled accumulation, *Journal of Sedimentary Petrology* 62, 622–635.
Nelson, J.M., Shreve, R.L., McLean, S.R. and Drake, T.G. (1995) Role of near-bed turbulence structure in bedload transport and bedform mechanics, *Water Resources Research* 31, 2,071–2,086.
Walker, I.J. and Nickling, W.G. (2002) Dynamics of secondary airflow and sediment transport over and in the lee of transverse dunes, *Progress in Physical Geography* 26, 47–75.
Werner, B. (1995) Eolian dunes: computer simulations and attractor interpretation, *Geology* 23, 1,107–1,110.
Werner, B. and Kocurek, G. (1999) Bedform spacing from defect dynamics, *Geology* 27, 727–730.

Further reading

Kocurek, G. (1996) Desert aeolian systems, in H.G. Reading (ed.) *Sedimentary Environments: Processes, Facies and Stratigraphy*, 125–153, Oxford: Blackwell.

SEE ALSO: adhesion; boundary layer; bounding surface; dune, aeolian; sabkha

GARY KOCUREK

INTERFLUVE

Interfluves are areas of relatively high ground that lie between adjacent valleys in a drainage basin. In its most literal sense of land 'between rivers', the term interfluves refers to undissected ridges that lie between streams. However, it is often used in a more general sense to describe generally high terrain (including lower order streams and/or VALLEYS) lying between major river systems. It has also been applied to land between glacial valleys and subglacial streams. In addition, the concept of interfluves also has application in studies of ancient sedimentary sequences, where it is used to distinguish paleo-uplands from incised valleys.

In many parts of the world the spacing between ridges and valleys is remarkably regular. For example, the coastal hills of northern California, USA, contain many examples of ridge–valley topography in which first-order DRAINAGE BASINs flanking a relatively linear main valley are quite evenly spaced (e.g. see figure 1B of Dietrich *et al.* 2003). This is also true, for example, of hollows along quartzite ridges in the ridge-and-valley province of Pennsylvania, USA. On a larger scale, many linear mountain belts show a strikingly regular space of main valleys (Hovius 1996). Neither the origins of this regularity nor the controls on valley spacing are well understood, but some clues emerge from analyses of process dynamics. Many have argued, for example, that the transition from hillslope to valley (in the absence of obvious lithologic or structural controls) reflects a switch in process dominance (Gilbert 1909; Smith and Bretherton 1972; Tarboton *et al.* 1992). Valleys, according to this argument, represent landscape zones dominated by processes that are fed by upslope surface or groundwater discharge, and whose effectiveness therefore grows with increasing drainage area (sometimes abruptly, as with Horton's (1945) concept of an OVERLAND FLOW erosion threshold). Such 'concentrative' processes can include overland flow erosion, groundwater (see GROUNDWATER) sapping, glacial scour and some forms of landsliding (see LANDSLIDE). Closer to drainage divides, processes that do not have this upstream dependence (e.g. mass movement processes such as SOIL CREEP) hold sway.

Theories based on this process-dominance concept (or the related process-threshold concept; see Kirkby 1994) successfully explain observed relationships between gradient and drainage area around valley and channel heads (Kirkby 1987; Montgomery and Dietrich 1989). They do not, however, account explicitly for the spacing between ridges and interfluves, nor do they explain the instabilities that lead to spontaneous formation of channels and valleys in the first place. The problem of spontaneous formation of channel networks was first analysed mathematically by Smith and Bretherton (1972). Using linear stability analysis, they demonstrated the conditions under which spontaneous formation of incipient channels could occur under steady, uniform sheet flow on an undissected slope. One useful result of this analysis was the notion that there will be a tendency toward CHANNELIZATION when sediment transport capacity increases more than linearly with discharge per unit contour width, so that two units of flow together carry more than twice the sediment of one. The simplifications in the Smith–Bretherton analysis, however, did not allow for prediction of the incipient spacing between channels. This shortcoming was overcome in later work (Loewenherz 1991; Izumi and Parker 2000). For example, Izumi and Parker (2000) used linear stability analysis to predict incipient channel spacing in the case of sub-critical sheet flow (see SHEET EROSION, SHEET FLOW, SHEET WASH) on a convex-upward surface of cohesive sediment. Their analysis predicted that incipient channel spacing should depend on flow depth and roughness. Using reasonable values for these parameters they were able to account for incipient channel spacing on the order of tens of metres, a value not too dissimilar from observed channel spacings of the order of 100 metres (Montgomery and Dietrich 1989; Dietrich and Dunne 1993). Numerical models of drainage basin development have also been able to reproduce hillslope–valley topography and have revealed some important controls on hillslope scale and valley density (see, for example, Willgoose *et al.* 1991; Howard 1997; Tucker and Bras 1998).

As noted above, the concept of interfluves often refers not just to individual ridges, but also to much larger areas of land lying between incised valleys. An example is the Colorado Plateau region, USA, where expanses of high-elevation, low-relief terrain are cut by abrupt, steep-sided canyons (the largest being the Grand Canyon in northern Arizona). These less-dissected upland surfaces, though still subject to active fluvial erosion, can be considered interfluves by virtue of their lesser degree of incision.

References

Dietrich, W.E. and Dunne, T. (1993) *The Channel Head, Channel Network Hydrology*, Chichester: Wiley.

Dietrich, W.E., Bellugi, D.G., Sklar, L.S., Stock, J.D., Heinsath, A.M. and Roering, J.J. (2003) Geomorphic transport laws for predicting landscape form and dynamics, in P.R. Wilcock and R.M. Iverson (eds) *Prediction in Geomorphology*, 103–132, Washington, DC. American Geophysical Union.

Gilbert, G.K. (1909) The convexity of hilltops, *Journal of Geology* 17, 344–350.

Horton, R.E. (1945) Erosional development of streams and their drainage basins; hydrophysical approach to quantitative morphology, *Geological Society of America Bulletin* 56, 275–370.

Hovius, N. (1996) Regular spacing of drainage outlets from linear mountain belts, *Basin Research* 8, 29–44.

Howard, A.D. (1997) Badland morphology and evolution: interpretation using a simulation model, *Earth Surface Processes and Landforms* 22, 211–227.

Izumi, N. and Parker, G. (2000) Linear stability analysis of channel inception: downstream-driven theory, *Journal of Fluid Mechanics* 419, 239–262.

Kirkby, M.J. (1987) Modelling some influences of soil erosion, landslides and valley gradient on drainage density and hollow development, *Catena Supplement* 10, 1–14.

——(1994) Thresholds and instability in stream head hollows: a model of magnitude and frequency for wash processes, in M.J. Kirkby (ed.) *Process Models and Theoretical Geomorphology*, 295–314, New York: Wiley.

Loewenherz, D.S. (1991) Stability and the initiation of channelized surface drainage; a reassessment of the short wavelength limit, *Journal of Geophysical Research, B, Solid Earth and Planets* 96, 8,453–8,464.

Montgomery, D.R. and Dietrich, W.E. (1989) Source areas, drainage density, and channel initiation, *Water Resources Research* 25, 1,907–1,918.

Smith, T.R. and Bretherton, F.P. (1972) Stability and the conservation of mass in drainage basin evolution, *Water Resources Research* 8, 1,506–1,529.

Tarboton, D.G., Bras, R.L. and Rodriguez, I.I. (1992) A physical basis for drainage density, *Geomorphology* 5, 59–76.

Tucker, G.E. and Bras, R.L. (1998) Hillslope processes, drainage density, and landscape morphology, *Water Resources Research* 34, 2,751–2,764.

Willgoose, G., Bras, R.L. and Rodriguez, I.I. (1991) A coupled channel network growth and hillslope evolution model; 1, Theory, *Water Resources Research* 27, 1,671–1,684.

SEE ALSO: drainage density; hillslope-channel coupling; hillslope, form; hillslope, process; valley

GREG TUCKER

INTERMONTANE BASIN

Topographic basins of various shapes and sizes are common within many uplands and low to medium-height mountains, not affected by very recent differential tectonics. They are surrounded by higher terrain on all sides and drained by rivers which typically leave the basin floor through a narrow valley. Bounding slopes are often steep and there may exist a sharp junction between the floor and the marginal slope. In the majority of cases active tectonics does not contribute to the development of the basins, hence they are features due to differential denudation. Little accumulation normally takes place within intermontane basins, except for thin alluvium along river courses and localized slope-derived deposits.

The location of intermontane basins is often influenced by lithology and structure. Basin outlines tend to follow structural lines in the bedrock, e.g. ancient faults or regional fractures, or lithological boundaries, or master joints cross each other within the basin floor. The origin of the basins can then be ascribed to selective weathering and erosion, which exploit weaker rock compartments more effectively. For example, Thorp (1967) demonstrated the existence of structure-controlled basins in granite massifs of Nigeria.

However, Bremer (1975) maintains that many basins in tropical areas cannot be explained by unequal resistance and applies the concept of divergent weathering, which is topographically rather than rock-controlled. An initial depression receives and holds more water than its drier surroundings, hence weathering mantle develops, increasing in thickness and maturity. Occasional stripping of the SAPROLITE from the adjacent slope exposes the bedrock surface, which remains dry and sheds more water into the depression. This enhances local contrast in the intensity of weathering and leads to progressive deepening of the basin and steepening of its bounding slopes, as weathering products are episodically eroded away. According to Bremer, lateral enlargement of the basins plays a minor part. Basin floors may be flat and retain a weathering mantle, or they may have some relief moulded at the WEATHERING FRONT. Low hills, groups of boulders and occasional tors may occur. Thus, basins develop through localized deep weathering and are examples of long-term, two-stage landform development.

Basins developing in the way outlined above are not to be confused with basins forming basin-and-range terrains. The latter are typical tectonic features and originate due to downfaulting of crustal blocks in areas subjected to extensional regime. Moreover, they carry a thick fill of sediment washed down from surrounding uplifted terrain.

References

Bremer, H. (1975) Intramontane Ebenen, Prozesse der Flächenbildung, *Zeitschrift für Geomorphologie Supplementband* N.F. 23, 26–48.

Thorp, M. (1967) Closed basins in Younger Granite Massifs, northern Nigeria, *Zeitschrift für Geomorphologie Supplementband* N.F. 11, 459–480.

Further reading

Johansson, M., Migoń, P. and Olvmo, M. (2001) Joint-controlled basin development in Bohus granite, SW Sweden, *Geomorphology* 40, 145–161.

Thomas, M.F. (1974) Granite landforms: a review of some recurrent problems of interpretation, in *Institute of British Geographers, Special Publication* 7, 13–37.

PIOTR MIGOŃ

INVERTED RELIEF

A type of surface morphology within which former valley floors constitute the most elevated parts of the landscape whereas pre-existing valley sides and divides have been lowered to the extent that they now form topographic lows. Relief inversion occurs when materials on the valley floors are, or become, more resistant to erosion than those underlying the adjacent slopes. This may be the case with lava flows filling concave landforms, and with DURICRUSTs developed through the induration of alluvial or lake sediments. Silcretes, calcretes and ferricretes particularly often occur in inverted position.

Typical landforms due to relief inversion are sinuous flat-topped ridges indicating former valley courses. As erosion proceeds they are reduced to isolated elongated plateaux or MESAs capped by a remnant of the more resistant material. Recognition of inverted relief may have important economic implications as former placer deposits would now occur in elevated position.

Large-scale relief inversion is often hypothesized in folded terrains (see FOLD). It is often observed that drainage lines follow axes of anticlines whereas synclines underlie ridges on both sides of a valley which seems to be in contrast to original relief. The explanation holds that once erosion reaches the softer core of an anticline, it accelerates and outpaces any lowering of adjacent synclinal structures, leading to relief inversion. Erosion hollows of the Negev Desert in Israel (*makhteshim*) are examples of this kind of inversion. A more complicated scenario invokes planation of an original mountain range in the first phase and subsequent development of inverted relief at the expense of the planation surface.

Further Reading

Goudie, A.S. (1985) Duricrusts and landforms, in K.S. Richards, R.R. Arnett and S. Ellis (eds) *Geomorphology and Soils*, 37–57, London: George Allen and Unwin.

Pain, C.F. and Ollier, C.D. (1995) Inversion of relief – a component of landscape evolution, *Geomorphology* 12, 151–165.

PIOTR MIGOŃ

ISLAND ARC

Around the western edge of the Pacific, many islands are arranged as festoons of arcs, some distance from the continent, in either single or double rows of islands. Other island arcs are the Indonesian, Caribbean, Middle America and Scotia Arcs. The arc radius ranges from about 4,000 km for Java–Sumatra to 1,500 km for Japan, and only 250 km for the hairpin Banda Arc.

The arcs face different directions: the Aleutian Arc is convex to the south, Sumatra–Java to the west, Mariana to the east, and Papua New Guinea to the north. Some 'straight arcs' such as the Solomon Islands and the Tonga–Kermadec Trench have many of the attributes of arcs without the curvature.

Some arcs are separated from the continent by backarc basins (Japan). The Scotia Arc has no backing continent, only the Pacific seafloor. The Caribbean Arc is backed by the straight Middle America Arc. The Mariana Arc is separated from the Philippine Arcs by a complex of backarc basins.

The standard arrangement of landforms in a simple arc is: continent; backarc basin; arc – forearc basin; accretionary prism; trench; ocean (Figure 94). The simplest arcs are lines of active volcanoes (e.g. Kurile Arc), which may be subdivided on the basis of underlying rocks into volcanic or continental based. The commonest type of double arcs comprises two rows of volcanoes, one older than the other. The older, usually outer line may be wholly or partially covered by limestone. The Sumatra–Java Arc is volcanic, with an outer arc of sedimentary rocks.

Topographically, the front of an island arc is generally rising while the back of the arc is sinking. In New Britain the front of the arc is rising, with flights of uplifted coral terraces, while the other side has drowned coasts, indicating sinking. Many arcs have large, mountainous islands built of sedimentary, metamorphic and granitic rocks (Japan, Papua New Guinea). They were eroded to a planation surface before the modern mountains were uplifted. The Virgin Islands display an erosion surface at about 300 m, and many terraces. Uplift occurred within the past 2 million years.

On the ocean side of arcs lie deep trenches. The greatest known depth in the ocean is the Mariana Deep (11,035 m). Traverses of trenches reveal numerous normal faults and graben. Trenches contain variable amounts of sediment: part of the Chile Trench is virtually empty; the Aleutian Trench contains up to 4 km of undeformed horizontal layers of sediment. An area of relatively shallow water separates the islands of the arc from the trench bottom. This is called the arc–trench gap. It is usually over 100 km wide, and is 570 km in the eastern Aleutian Arc. It is underlain by thick, generally horizontal sediments of the forearc basin.

At the outer edge of the forearc basin there is a marked break in slope, and the steep slope bounding the trench may constitute an accretionary wedge. The accretionary wedge (or prism) is a package of highly deformed sediment, and perhaps oceanic basalt, presumed to be scraped off the downgoing slab and accumulated at the edge of the overriding slab. Alternatively, the deformation structures can be interpreted as gravity-tectonic structures with décollement (also called detachment faulting) and thrusting. The whole area from the volcanic arc to the ocean, including the trench, is called the forearc. An alternative plate tectonic scenario is that the sediment, instead of being scraped off, is subducted beneath the arc.

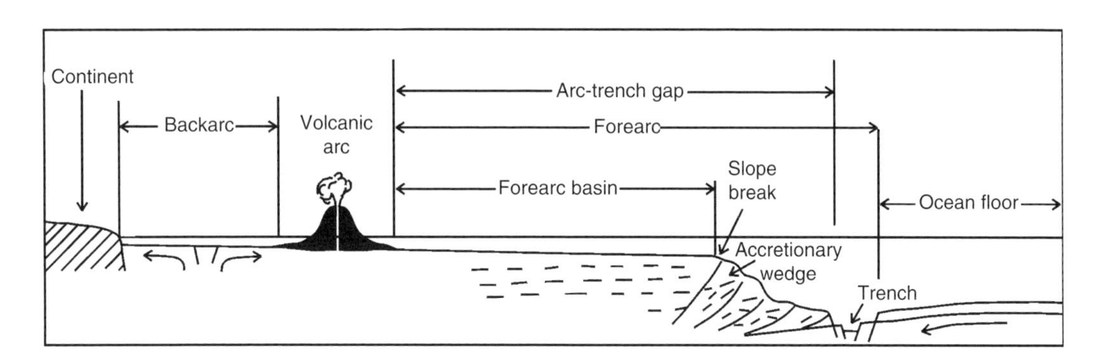

Figure 94 Nomenclature of features in a typical island arc

Behind island arcs are shallow seas known as backarc basins. Backarc basins are spreading sites (Taylor 1995). The Japan backarc basin has up to 2 km of sediment. The curved shape of the arc is not easily related to the angular pattern of multiple spreading sites behind the arc (e.g. Scotia Arc). Nor is the curve related to the approaching Pacific Plate: part of the Aleutian Arc is parallel to the movement, and part is almost perpendicular.

Active volcanoes on arcs frequently erupt with great violence, as at Krakatau (1883), Mont Pelée (1902) and Mount Pinatubo (1991). Stratovolcanoes are the commonest type, with numerous calderas. Andesitic volcanoes predominate in arcs, but some notable volcanoes are basaltic, including Mt. Fuji, Japan. Basalt makes up 70 per cent of the South Sandwich Islands. Arcs with volcanoes were formed in the past, as they are today, sometimes making double arcs. In Papua New Guinea the New Britain Arc has Quaternary volcanoes on the inside and the Palaeogene arc on the outside, but the New Ireland Arc has inner Palaeogene volcanoes and an outer arc of Quaternary volcanoes.

In some arcs (New Hebrides, Middle America) normal faults break the Earth's surface into fault blocks. Normal faults are also common in the trenches and backarc basins. Strike-slip faulting is important in some arcs, with lateral displacements of many kilometres along faults roughly parallel to the arc. These faults are a major feature in the geomorphology of Sumatra and Java. The Middle America Arc is offset in Nicaragua by a strike-slip fault oblique to the arc, which is a continuation of the Clipperton Fracture of the Pacific floor.

The American arcs are somewhat different from the rest. The Caribbean Arc lies east of the Caribbean Plate. This is bounded to the north by what looks like streaked-out bits of North America, consisting of continental rocks and making the islands of the Greater Antilles as far as Puerto Rico. The southern side is similarly made of rocks like those of South America. The true arc is the north–south part of the Leeward Islands, a typical double arc with the Limestone Caribees to the east and Volcanic Caribees to the west. The western side of the Caribbean Plate is bounded by the Middle America Arc, with the Middle America Trench on the Pacific side. The trench has no accretionary prism, and sediments are horizontal with no evidence of compression. The arc has some old continental rocks, and on the western side old volcanic rocks. Topographically the arc is a plateau tilted up to the west, but there is a down-faulted strip on the Pacific side, and this is where the current volcanic arc lies. The Isthmus of Panama is not part of the volcanic arcs, but consists of block-faulted continental rocks.

The Scotia Arc lies east of the Scotia Plate, which has continental rocks from South America streaked out on the northern side, and Antarctic rocks on the southern side. The South Sandwich islands make a true arc, with a spreading site behind that separates the Scotia Plate from a much smaller Sandwich Plate which is less than 8 million years old. The islands are volcanic, rugged and glaciated. The Scotia Plate is separated from the Pacific by a transform fault.

A negative gravity anomaly lies over trenches, and a positive anomaly on the continental side about 115 km distant. Arcs are characterized by earthquakes. The foci sometimes appear to fall in a zone (called the Benioff zone), about 50 km thick and reaching depths of several hundred kilometres. This dips towards the concave side at varied angles, and is vertical under the Mariana Arc. It is presumed to mark a slab being subducted at the trench. In some arcs (e.g. Aleutians) the Benioff zone emerges not at the trench, but on the arc–trench gap. The relation between arc, trench, gravity anomalies, earthquakes and volcanoes is in fact quite variable.

Some arcs run aground onto continents. The Sumatra Arc goes via the Andaman Islands to Burma (Myanmar), and continues as the landbound arc of the Himalayas of about the same size. The Aleutian Arc continues in Alaska; and the southern arm of the Caribbean Arc in Colombia. This is significant in theories that relate island arcs to mountain building, a major concept in PLATE TECTONICS. Since backarc spreading increases the space between continent and arc it might seem difficult to make the two collide, but Hamilton (1988) wrote that island arcs are 'conveyor-belted' towards subduction zones, so that island arcs collide with one another and with continents. Island arc concepts are often used to interpret mainland features including ancient structures and modern mountains such as the Apennines.

The ruling theory of plate tectonics explains island arcs as places where oceanic crust is subducted at trenches. This produces the paradox of having a compressive mechanism, when the normal faults in trenches, islands and backarc basins all indicate tension. Other suggested mechanisms of arc formation include geotumours, mantle diapirism, subsided blocks at the trench, and surge tectonics.

References

Hamilton, W.B. (1988) Plate tectonics and island arcs, *Geological Society of America Bulletin* 100, 1,503–1,527.

Taylor, B. (ed.) (1995) *Backarc Basins, Tectonics and Magmatism*, New York: Plenum.

Further reading

Wezel, F.-C. (1986) *The Origin of Arcs*, Amsterdam: Elsevier.

The Island Arc (Blackwell) is devoted to arcs and related topics and appears four times a year.

SEE ALSO: plate tectonics

CLIFF OLLIER

ISOSTASY

A term introduced in 1882 and derived from the Greek words *iso* and *stasis* that means equal standing. It is used to describe a condition to which the Earth's crust and mantle tend, in the absence of disturbing forces – a condition of rest and quiet in which the lighter crust floats on the dense underlying mantle (Watts 2001). However, isostasy is disturbed by such processes as sedimentation, erosion, volcanism and the waxing and waning of ice sheets (see GLACIAL ISOSTASY). In its simplest form the isostasy concept envisages that rigid blocks of crust, buoyantly supported by the underlying fluid medium of the mantle, are free to move vertically until their weight is balanced by their buoyancy ('isostastic equlibrium'). Differences in the density and thickness of the crust are to a considerable extent responsible for variations in the isostatic adjustment of the lithosphere. If it is in isostatic equilibrium, one portion of the lithosphere will stand higher than another, either because of a lower density (the so-called Pratt model) or because it is of the same density but thicker (the Airy model), or from a combination of both.

Isostatic equilibrium is disturbed by various geomorphological processes. Erosion makes crustal blocks thinner and lighter in weight so that an eroding mountain will tend to rise to maintain equilibrium. Conversely, deposition of sediments, as for example in a delta, represents an added load, so that sinking tends to occur. Extension of the crust by rifting thins it whereas compression in mountain-building thickens it. Loading and unloading by waxing and waning glaciers also affects isostatic equilibrium as do changes in the volume of water in the ocean basins or in lakes – hydro-isostasy (Bloom 1967). Similarly, the extrusion of large amounts of volcanic material can weigh down the crust and cause subsidence to occur.

Isostasy is central to understanding a range of geomorphological phenomena. For example, areas that were once loaded by great ice sheets (e.g. Fennoscandia) are now areas of uplift as they recover from the load of the ice. Conversely areas peripheral to the ice sheets, which bulged up during glacials, are now areas of subsidence (e.g. the southern North Sea area). Likewise, the development of GUYOTS, seamounts and atolls in the Pacific may be related to subsidence caused by volcanic eruptions loading the crust (McNutt and Menard 1978). Load-induced vertical movements have a profound effect on deltas. In particular, subsidence increases the overall water depth and so increases the accommodation space that is available for the prograding sediments. The relief characteristics of rift valleys, features caused by crustal extension, also show the effects of isostasy through their association with broad topographic swells. The rift flanks of passive margins show evidence of uplifted flanks and erosional unloading (as in Namibia) and the presence of rift flank uplifts may explain the deflection of drainage systems towards the continental interiors (e.g. the Kalahari) (Gilchrist and Summerfield 1990). The shorelines of great palaeo-lakes, like Lake Bonneville in the southwest USA, have deformed as the crust has adjusted to the removal of the weight of the lake as it desiccated (Crittenden 1967).

The continental shelves were depressed as the weight of water from rising postglacial sea levels (the Flandrian Transgression) was applied to them. Finally, erosion that deepens and widens river valleys but does not erode the peaks to the same degree, will reduce the mass of an area and so drive isostatic uplift. As a consequence, the altitude of the peaks could increase at the same time that the mean height of a region is decreasing (Molnar and England 1990).

References

Bloom, A.L. (1967) Pleistocene shorelines: a new test of isostasy, *Geological Society of America Bulletin* 78, 1,477–1,494.
Crittenden, M.D. (1967) New data on the isostatic deformation of Lake Bonneville, *US Geological Survey Professional Paper* 454-E.
Gilchrist, A.R. and Summerfield, M.A. (1990) Differential denudation and flexural isostasy in formation of rifted-margin upwarps, *Nature* 346, 739–742.
McNutt, M. and Menard, H.W. (1978) Lithospheric flexure and uplifted atolls, *Journal of Geophysical Research* 83, 1,206–1,212.
Molnar, P. and England, P. (1990) Late Cainozoic uplift of mountain ranges and global climatic change: chicken or egg? *Nature* 346, 29–34.
Watts, A.B. (2001) *Isostasy and Flexure of the Lithosphere*, Cambridge: Cambridge University Press.

A.S. GOUDIE